P9-CSG-589

Linear System Theory and Design

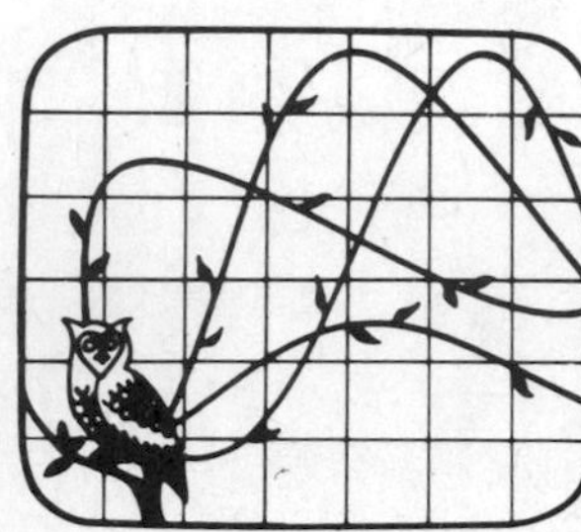

HRW
Series in
Electrical and
Computer Engineering

M. E. Van Valkenburg, Series Editor

L. S. Bobrow ELEMENTARY LINEAR CIRCUIT ANALYSIS
C. T. Chen LINEAR SYSTEM THEORY AND DESIGN
D. J. Comer DIGITAL LOGIC AND STATE MACHINE DESIGN
C. H. Durney, L. D. Harris, C. L. Alley ELECTRIC CIRCUITS: THEORY AND ENGINEERING APPLICATIONS
G. H. Hostetter, C. J. Savant, Jr., R. T. Stefani DESIGN OF FEEDBACK CONTROL SYSTEMS
S. Karni and W. J. Byatt MATHEMATICAL METHODS IN CONTINUOUS AND DISCRETE SYSTEMS
B. C. Kuo DIGITAL CONTROL SYSTEMS
B. P. Lathi MODERN DIGITAL AND ANALOG COMMUNICATION SYSTEMS
C. D. McGillem and G. R. Cooper CONTINUOUS AND DISCRETE SIGNAL AND SYSTEM ANALYSIS Second Edition
A. Papoulis CIRCUITS AND SYSTEMS: A MODERN APPROACH
S. E. Schwarz and W. G. Oldham ELECTRICAL ENGINEERING: AN INTRODUCTION
A. S. Sedra and K. C. Smith MICROELECTRONIC CIRCUITS
M. E. Van Valkenburg ANALOG FILTER DESIGN

Linear System Theory and Design

Chi-Tsong Chen

Professor, Department of Electrical Engineering
State University of New York at Stony Brook

HOLT, RINEHART AND WINSTON

New York Chicago San Francisco Philadelphia
Montreal Toronto London Sydney Tokyo
Mexico City Rio de Janeiro Madrid

This text is a major revision of *Introduction to Linear System Theory* by Chi-Tsong Chen, originally published in 1970. © 1970 by Holt, Rinehart and Winston

Address correspondence to:
383 Madison Avenue, New York NY 10017

Library of Congress Cataloging in Publication Data

Chen, Chi-Tsong.
Linear system theory and design.

Bibliography: p.
Includes index.
1. System analysis. 2. System design. I. Title.
QA402.C442 1984 003 83-12891
ISBN 0-03-060289-0

Printed in the United States of America
4 5 6 7 038 9 8 7 6 5 4 3 2

CBS COLLEGE PUBLISHING
Holt, Rinehart and Winston
The Dryden Press
Saunders College Publishing

To
Beatrice
and
Janet, Pauline, Stanley

Contents

*May be omitted without loss of continuity.

Preface

This text is intended for use at the senior-graduate level in university courses on linear systems and multivariable system design. It may also be used for independent study and reference by engineers and applied mathematicians. The mathematical background assumed for this book is a working knowledge of matrix manipulation and an elementary knowledge of differential equations. The unstarred sections of this book have been used, for over a decade, in the first graduate course on linear system theory at the State University of New York at Stony Brook. The majority of the starred sections were developed during the last three years for a second course on linear systems, mainly on multivariable systems, at Stony Brook and have been classroom tested at a number of universities.

With the advancement of technology, engineers have become interested in designing systems that are not merely workable but also the best possible. Consequently, it is important to study the limitations of a system; otherwise, one might unknowingly try to design an impossible system. Thus, a thorough investigation of all the properties of a system is essential. In fact, many design procedures have evolved from such investigations. This text is devoted to this study and the design procedures developed thereof. This is, however, not a control text per se, because performance criteria, physical constraints, cost, optimization, and sensitivity problems are not considered.

This text is a revised and expanded edition of *Introduction to Linear System Theory* which discussed mostly the state variable approach and was published in 1970. Since then, several important developments have been made in linear system theory. Among them, the geometric approach and the transfer-function matrices in fractional forms, called the matrix-fraction description, are most pertinent to the original text. The geometric approach is well covered in W. M. Wonham's *Linear Multivariable Control: A Geometric Approach*, 2d ed., Springer-Verlag, New York, 1979 and is outside the scope of this text. Hence the new material of this edition is mainly in the transfer-function matrix in fractional form. Because of this addition, we are able to redevelop, probably more simply in concepts and computations, the results of the state variable approach and establish a fairly complete link between the state-variable approach and the transfer-function approach.

We aim to achieve two objectives in the presentation. The first one is to develop major results and design procedures using simple and efficient methods. Thus the presentation is not exhaustive; only those concepts which are essential in the development are introduced. For example, the Smith-McMillan form is not used in the text and is not discussed. The second objective is to enable the reader to employ the results developed in the text. Consequently, most results are developed in a manner suitable for numerical computation and for digital computer programming. We believe that solving one or two problems of each topic by hand will enhance the understanding of the topic and give confidence in the use of digital computers. With the introduction of the row searching algorithm (Appendix A), which has been classroom tested, this is possible even for multivariable systems, as long as their degrees are sufficiently small.

The level of mathematics used in this edition is about the same as that of the original edition. If concepts and results in modern algebra more extensive than those in Chapter 2 are introduced, some results in the text can be developed more elegantly and extended to more general settings. For example, the Jordan form can be established concisely but abstractly by using the concepts of invariance subspaces and direct sum. Its discussion can be found in a large number of mathematical texts and will not be repeated here. In view of our objectives, we discuss the computation of the required basis and then develop the Jordan form. By using some concepts in abstract algebra, such as ring, principal ideal domain, and module, the realization problem (Chapter 6) can be developed more naturally and some results in this text can be extended to delay differential equations, linear distributed systems, and multidimensional systems. These are extensively studied in *Algebraic System Theory*, which was initiated by R. E. Kalman in the late 1960s and has extended in recent years most of the results in this text to linear systems over rings. The concepts used in algebraic system theory are less familiar to engineering students and require more mathematical sophistication and will not be discussed. All the results and design procedures in this text are developed by using only elementary concepts and results in linear algebra.

The results in this text may eventually be implemented on digital computers. Because of the finite word length, the sensitivity of problems and the stability of algorithms become important on computer computations. These problems are complex and extensively discussed in texts on numerical analysis. In our development, we will take note of these problems and remark briefly wherever appropriate.

The arrangement of the topics in this text was not reached without any difficulty. For example, the concepts of poles and zeros seem to be best introduced in Chapter 4. However, their complete treatments require irreducible realizations (Chapter 6) and coprime fractions of transfer-function matrices (Appendix G). Moreover, the concept of zeros is used only in Section 9-6. Hence it was decided to create an appendix for the topic. The coprimeness of polynomials and polynomial matrices might be inserted in the main text. This, however, will digress too much from the state-variable approach; thus the topic was grouped in an appendix.

The logical sequences of various chapters and appendixes are as follows:

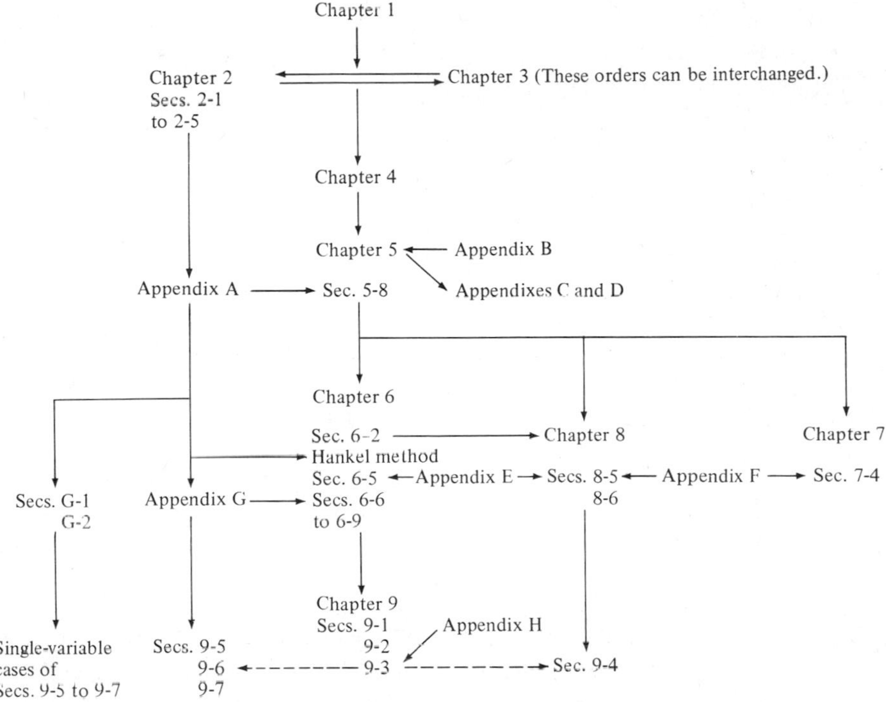

The material connected by a broken line is not essential in the development. The logical dependencies among Chapters 6, 7, 8, and 9 are loose, and their various combinations can be adopted in one- or two-semester courses. When I teach a one-semester course at Stony Brook, the *unstarred sections* of the following chapters are covered:

Chapter 1
Chapter 2
Chapter 3 (Skip Theorem 3-1 and its corollary.)
Chapter 4 (Skip Theorem 4-11.)
Chapter 5 (Emphasize the time-invariant part by skipping Theorems 5-2, 5-5, and 5-11.)
Chapter 6
Chapter 7
Chapter 8

We emphasize the exact meanings of theorems and their implications; hence the proofs of a number of theorems are skipped. For example, we prove only Theorems 2-1 and 2-2 in Chapter 2. We skip the proofs of Theorems 4-1, 4-2, and others. In the second course, we cover the following:

Appendix A
Section 5-8, controllability and observability indices
Hankel method (Section 6-4 and method II of Section 6-5)
Appendix E
Singular value decomposition method (Method I of Section 6-5)
Appendix G
Sections 6-6 to 6-9
Starred sections of Chapter 7
Appendix H
Chapter 9

Those who are interested in quick access to the design methods using the transfer-function matrix in fractional form may proceed from Sections 2-1 to 2-5, Appendixes A and G, and then to Sections 9-5 to 9-7, or only their single-variable cases.

The problem sets form an integral part of the book. They are designed to help the reader understand and utilize the concepts and results covered. In order to retain the continuity of the main text, some important results are stated in the problem sets. A solutions manual is available from the publisher.

The literature on linear system theory is very extensive. The length of this text, however, is limited. Hence the omission of some significant results is inevitable and I would like to apologize for it. I am indebted to many people in writing this book. Kalman's work and Zadeh and Desoer's book *Linear System Theory* form the foundation of the original edition of this book. Rosenbrock's and Wolovich's works are essential in developing the present edition. I have benefited immensely in my learning from Professor C. A. Desoer. Even to this date, I can always go to him whenever I have questions. For this, I can never express enough of my gratitude. To Professors B. J. Leon, E. J. Craig, I. B. Rhodes, P. E. Barry (first edition) and to Professors M. E. Van Valkenburg, W. R. Perkins, D. Z. Zheng (present edition), I wish to express my appreciation for their reviews and valuable suggestions. I would like to thank President F. Zhang and Professor K. W. You of Chengdu University of Science and Technology, Professor S. B. Park of Korea Advanced Institute of Science and Technology, Professor T. S. Kuo of National Taiwan University, and Professor S. K. Chow of National Sun Yat-Sen University, Taiwan, for providing opportunities for me to lecture on an earlier draft of Chapter 9 and Appendix G. I especially appreciate the opportunity at Chengdu University to interact with several faculty members, especially Professor L. S. Zhang, from various universities in China; their suggestions have improved considerably the presentation of the text. I am grateful to many of my graduate students, specially C. Waters, C. H. Hsu (the first edition), I. S. Krishnarao, Y. S. Lai, C. C. Tsui and S. Y. Zhang (the present edition) whose assistance in the form of dissertations and discussions has clarified considerably my understanding of the subject matter. I am grateful to Mrs. V. Donahue, C. LaGuardia, T. Marasco, and F. Trace for typing various drafts of this text, to Mr. P. Becker of Holt, Rinehart and Winston and the staff of Cobb/Dunlop for their assistance in the production and to Professors S. H. Wang, K. W. You, and D. Z. Zheng, visiting scholars from the People's Republic of China, for their proofreading. My special thanks go to my wife Beatrice and my children Janet, Pauline, and Stanley for their support during the writing of this text.

Chi-Tsong Chen

Glossary of Symbols

Q.E.D.	End of the proof of a theorem.
∎	This symbol denotes the end of a statement or an example.
$\mathbf{A}, \mathbf{B}, \mathbf{P}, \ldots$	Capital boldface letters denote matrices.
$\mathbf{u}, \mathbf{y}, \boldsymbol{\alpha}, \ldots$	Lowercase boldface letters denote vectors.
$u, y, \alpha, \ldots$	Lowercase italic and Greek type denote scalar-valued functions or scalars. Capital italic letters are also used in Chapter 9 and Appendix G to denote scalars.
$\mathscr{L}$	Laplace transform.
$\hat{\mathbf{u}}(s), \hat{\mathbf{y}}(s), \hat{\mathbf{G}}(s), C(s)$	If a letter is used in both time and frequency domains, circumflex will be used to denote the Laplace transform such as $\hat{\mathbf{u}}(s) = \mathscr{L}[\mathbf{u}(t)]$ and $\hat{\mathbf{G}}(s) = \mathscr{L}[\mathbf{G}(t)]$. If a letter is used in only one domain, no circumflex will be used, for example, $C(s)$.
$\nu(\mathbf{A}), \ldots$	The nullity of the constant matrix $\mathbf{A}$.
$\delta(\hat{\mathbf{G}}(s))$, deg $\hat{\mathbf{G}}(s)$	The degree of the rational matrix $\hat{\mathbf{G}}(s)$.
$\mathbf{A}', \mathbf{x}', \ldots$	The transpose of the matrix $\mathbf{A}$ and the vector $\mathbf{x}$.
$\mathbf{A}^*, \mathbf{x}^*, \ldots$	The complex-conjugate transpose of the matrix $\mathbf{A}$ and the vector $\mathbf{x}$.
det $\mathbf{A}, \ldots$	The determinant of $\mathbf{A}$.
$\mathbb{C}$	The field of complex numbers.
$\mathbb{R}$	The field of real numbers.
$\mathbb{R}(s)$	The field of rational functions of s with coefficients in $\mathbb{R}$.
$\mathbb{R}[s]$	The set of polynomials of s with coefficients in $\mathbb{R}$.
$\rho(\mathbf{A})$, rank $\mathbf{A}$	The rank of $\mathbf{A}$. If $\mathbf{A}$ is a constant matrix, the rank is

	defined over $\mathbb{C}$ or $\mathbb{R}$. If $\mathbf{A}$ is a rational or polynomial matrix, the rank is defined over $\mathbb{R}(s)$.
deg det $\mathbf{A}(s)$	The degree of the determinant of $\mathbf{A}(s)$.
diag $\{\mathbf{A}, \mathbf{B}, \mathbf{C}\}$	A diagonal matrix with $\mathbf{A}$, $\mathbf{B}$, and $\mathbf{C}$ as block diagonal elements as $\begin{bmatrix} \mathbf{A} & \mathbf{0} & \mathbf{0} \\ \mathbf{0} & \mathbf{B} & \mathbf{0} \\ \mathbf{0} & \mathbf{0} & \mathbf{C} \end{bmatrix}$ where $\mathbf{A}$, $\mathbf{B}$, and $\mathbf{C}$ are matrices, not necessarily square and of the same order.
$\triangleq$	Equals by definition.
$\frac{d}{dt}\mathbf{A} \triangleq \left(\frac{d}{dt}a_{ij}\right)$,	
$\mathscr{L}[\mathbf{A}] \triangleq (\mathscr{L}[a_{ij}]), \ldots$	When an operator is applied to a matrix or a vector, it means that the operator is applied to every entry of the matrix or the vector.
$\dot{\mathbf{x}} \triangleq \frac{d}{dt}\mathbf{x}$	

Linear System Theory and Design

1 Introduction

1-1 The Study of Systems

The study and design of a physical system can be carried out by empirical methods. We apply various signals to the physical system and measure its responses. If the performance is not satisfactory, we adjust some of its parameters or connect to it some compensator to improve its performance. This design is guided by past experience, if any, and proceeds by cut and try. This approach has undoubtedly succeeded in designing many physical systems.

The empirical method may become unsatisfactory if the specifications on the performance become very precise and stringent. It may also become inadequate if physical systems become very complicated or too expensive or too dangerous to be experimented. In these cases, analytical methods become indispensable. Analytical study of physical systems roughly consists of four parts: modeling, development of mathematical-equation description, analysis, and design. The distinction between physical systems and models are basic in engineering. In fact, circuits or control systems studied in any textbook are models of physical systems. A resistor with a constant resistance is a model; the power limitation of the resistor does not appear in the resistance. An inductor with a constant inductance is again a model; in reality, the inductance may vary with the amount of current flowing through it. Modeling is a very important problem, for the success of the design depends upon whether the physical system is properly modeled.

A physical system may have different models depending on the questions asked and the different operational ranges used. For example, an electronic amplifier may be modeled differently at high and low frequencies. A spaceship may be modeled as a particle in the study of its trajectory; however, it must be modeled as a rigid body in the maneuvering. In order to develop a suitable model of a physical system, a thorough understanding of the physical system and its operational range is essential. In this book, we shall refer to models of physical systems as *systems*. Hence a physical system is a device or a collection of devices existing in the real world; a system is a model of a physical system.

Once a system (a model) is found for a physical system, the next step in the study is to develop, by applying various physical laws, mathematical equations to describe the system. For example, we apply Kirchhoff's voltage and current laws to electrical systems and Newton's laws to mechanical systems. The equations that describe systems may assume many forms; they may be linear equations, nonlinear equations, integral equations, difference equations, differential equations, or others. Depending on the question asked, one form of equation may be preferable to another in describing the same system. In conclusion, a system may have many different mathematical-equation descriptions, just as a physical system may have many different models.

Once a mathematical description of a system is obtained, the next step in the study involves analyses—quantitative and/or qualitative. In the quantitative analysis, we are interested in exact responses of systems due to the application of certain input signals. This part of the analysis can be easily carried out by using a digital or an analog computer. In the qualitative analysis, we are interested in the general properties of the system, such as stability, controllability, and observability. This part of analysis is very important, because design techniques may often evolve from this study.

If the response of a system is found to be unsatisfactory, the system has to be improved or optimized. In some cases, the responses of systems can be improved by adjusting certain parameters of the systems; in other cases, compensators have to be introduced. Note that the design is carried out on the model of a physical system. However, if the model is properly chosen, the performance of the physical system should be correspondingly improved by introducing the required adjustments or compensators.

1-2 The Scope of the Book

The study of systems may be divided into four parts: modeling, setting up mathematical equations, analysis, and design. The development of models for physical systems requires knowledge of each particular field and some measuring devices. For example, to develop models for transistors requires the knowledge of quantum physics and some laboratory setup. Developing models for automobile suspension systems requires actual testing and measurements; it cannot be achieved by the use of pencil and paper alone. Thus, the modeling problem should be studied in connection with each specific field and cannot be properly covered in this text. Hence, we shall assume that models of physical systems, or systems, are available to us in this text.

The mathematical equations which will be used in this text to describe systems are mainly limited to

$$\hat{\mathbf{y}}(s) = \hat{\mathbf{G}}(s)\hat{\mathbf{u}}(s) = \mathbf{N}(s)\mathbf{D}^{-1}(s)\hat{\mathbf{u}}(s) = \mathbf{F}^{-1}(s)\mathbf{H}(s)\hat{\mathbf{u}}(s) \tag{1-1}$$

and

$$\dot{\mathbf{x}}(t) = \mathbf{A}\mathbf{x}(t) + \mathbf{B}\mathbf{u}(t) \tag{1-2a}$$

$$\mathbf{y}(t) = \mathbf{C}\mathbf{x}(t) + \mathbf{E}\mathbf{u}(t) \tag{1-2b}$$

and their extensions to the time-varying case. Equation (1-1) describes the relationship between input $\mathbf{u}$ and output $\mathbf{y}$ in the Laplace-transform domain and is called the *input-output description* or the *external description* in the frequency domain. The matrix $\hat{\mathbf{G}}(s)$ is called the *transfer-function matrix*, and its elements are limited to rational functions. In the design, $\hat{\mathbf{G}}(s)$ will be factored as ratios of two polynomial matrices, such as $\mathbf{N}(s)\mathbf{D}^{-1}(s)$ or $\mathbf{F}^{-1}(s)\mathbf{H}(s)$, and is said to be in the polynomial fraction form. The set of two equations in (1-2) is called a *dynamical equation* or *state-variable equation.* Equation (1-2a) is a set of first-order differential equations. If (1-2) is used to describe a system, it is called the *dynamical-equation description*, the *state-variable description*, or the *internal description.* These two types of equations will be developed from the concepts of linearity and time invariance. We shall then show, by examples, how they can be used to describe systems.

The major portion of this text is devoted to the analysis and design centered around Equations (1-1) and (1-2). Analysis can be divided into quantitative and qualitative. The former can now be delegated to analog or digital computers; hence we emphasize the latter. Various properties of these equations will be thoroughly investigated. Their relationships will be established. A number of design procedures will then be developed from the study.

The design problem studied in this book is not exactly the design of feedback control systems. In our design, we are interested in the exact conditions, without any constraints on the complexity of compensators, to achieve certain design objectives, such as stabilization or pole placement: What are the minimum degrees which compensators must possess to achieve such design? We are also interested in developing simple and efficient procedures to carry out the design. In the design of control systems, the ultimate purpose is to design a system to meet some performance criteria, such as the rise time, overshoot, steady-state error, and others, with or without constraints on the degrees of compensators. A design problem with constraints is significantly different from the one without constraints. In this text, the performance of control systems is not considered. Hence our study is not a complete study of the design of control systems, although most of the results in this text are basic and useful in its design. For a discussion of the design of control systems, see References 3, S46, and S96. In this section we give a brief description of the contents of each chapter.

We review in Chapter 2 a number of concepts and results in linear algebra. The objective of this chapter is to enable the reader to carry out similarity transformations, to solve linear algebraic equations, and to compute functions of a matrix. These techniques are very important, if not indispensable, in analysis and design of linear systems.

In Chapter 3 we develop systematically the input-output description and the state-variable description of linear systems. These descriptions are developed from the concepts of linearity, relaxedness, causality, and time invariance. We also show, by examples, how these descriptions can be set up for systems. Mathematical descriptions of composite systems and discrete-time equations are also introduced. We also discuss the well-posedness problem in the feedback systems.

In Chapter 4 we study the solutions of linear dynamical equations. We also show that different analysis often leads to different dynamical-equation descriptions of the same system. The relation between the input-output description and the state-variable description is also established.

We introduce in Chapter 5 the concepts of controllability and observability. The importance of introducing these concepts can be seen from the networks shown in Figure 1-1. Their transfer functions are both equal to 1. There is no doubt about the transfer function of the network shown in Figure 1-1(b); however, we may ask why the capacitor in Figure 1-1(a) does not play any role in the transfer function. In order to answer this question, the concepts of controllability and observability are needed. These two concepts are also essential in optimal control theory, stability studies, and the prediction or filtering of signals. Various necessary and sufficient conditions for a dynamical equation to be controllable and observable are derived. We also discuss the canonical decomposition of a dynamical equation and introduce an efficient and numerically stable method of reducing a dynamical equation to an irreducible one.

In Chapter 6 we study irreducible realizations of rational transfer-function matrices. The problem is to find a controllable and observable linear time-invariant dynamical equation that has a prescribed rational matrix. Its solution is indispensable in analog and digital computer simulations. It also offers a method of synthesizing a rational matrix by using operational-amplifier circuits. This result is also needed in establishing the link between the state-variable approach and the transfer-function approach in the design of linear time-invariant systems.

The practical implications of the concepts of controllability and observability

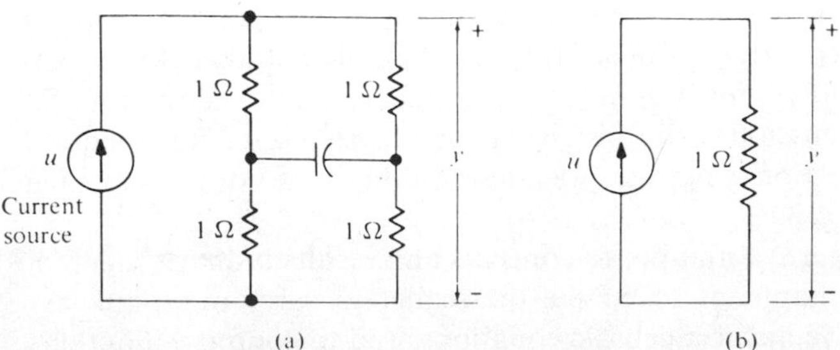

Figure 1-1 Two different networks with the same transfer function, 1.

are studied in Chapter 7. We show that if a dynamical equation is controllable, the eigenvalues of the equation can be arbitrarily assigned by introducing state feedback with a constant gain matrix. If a dynamical equation is observable, its state can be generated by designing a state estimator with arbitrary eigenvalues. Various design procedures are introduced. The separation property is also established.

We study in Chapter 8 a qualitative property of linear systems. This comes under the heading of stability, which is always the first requirement to be met in the design of a system. We introduce the concepts of bounded-input bounded-output stability, stability in the sense of Lyapunov, asymptotic stability, and total stability. Their characterizations and relationships are studied. We also discuss the Lyapunov theorem and then use it to establish the Routh-Hurwitz criterion. Their counterparts in the discrete-time case are also studied.

In the last chapter we study various problems associated with linear, time-invariant composite systems. One of them is to study the implication of pole-zero cancellation of transfer functions. For example, consider two systems with the transfer functions $1/(s-1)$ and $(s-1)/(s+1)$ connected in three different ways, as shown in Figure 1-2. We show why the system in Figure 1-2(b) can be studied from its composite transfer function, but the systems in Figure 1-2(a) and (c) cannot. We also study stabilities of single-variable and multivariable feedback systems. We then study the design of compensators by using transfer-function matrices in the fractional form. We study two feedback configurations: unity feedback and plant input-output feedback connections. We design compensators with minimum degrees to achieve arbitrary pole placement and to achieve asymptotic tracking and disturbance rejection. We reestablish in the transfer-function approach essentially the results developed in the state-variable approach and complete the link between the two approaches.

A total of eight appendixes are introduced. Their relationships with various chapters are covered in the Preface.

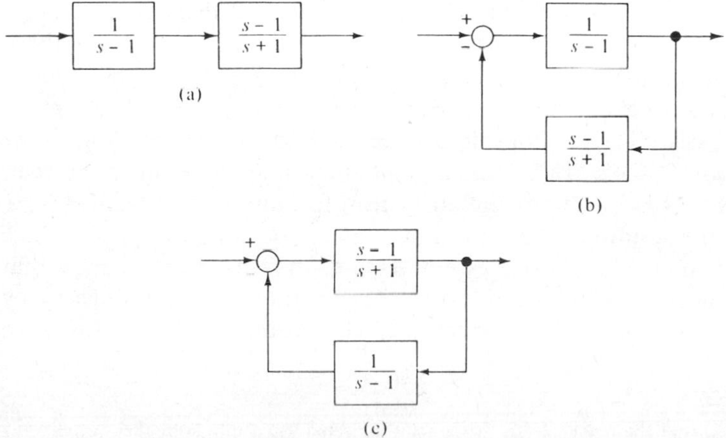

Figure 1-2 Three different connections of $1/(s-1)$ and $(s-1)/(s+1)$.

2

Linear Spaces and Linear Operators

2-1 Introduction

In this chapter we shall review a number of concepts and results in linear algebra that are essential in the study of this text. The topics are carefully selected, and only those which will be subsequently used are introduced. The purpose of this chapter is to enable the reader to understand the mechanism of similarity transformation, to solve linear algebraic equations, to find Jordan-form representations of square matrices, and to compute functions of a matrix, in particular, exponential functions of a matrix (see Section 2-9, Concluding Remarks[1]).

In Section 2-2 we introduce the concepts of field and linear space over a field. The fields we shall encounter in this book are the field of real numbers, the field of complex numbers, and the field of rational functions. In order to have a representation of a vector in a linear space, we introduce, in Section 2-3, the concept of basis. The relationship between different representations of the same vector is established. In Section 2-4, we study linear operators and their representations. The concept of similarity transformation is embedded here. In Section 2-5, the solutions of a set of linear algebraic equations are studied. The concepts of rank and nullity are essential here. In Section 2-6, we show that every square matrix has a Jordan-form representation; this is achieved by introducing eigenvectors and generalized eigenvectors as basis vectors. In

[1] It is recommended that the reader keeps the concluding remarks in mind, for they provide the reader with motivations for studying the mathematical theorems introduced in this chapter.

Section 2-7, we study functions of a square matrix. The minimal polynomial and the Cayley-Hamilton theorem are introduced. In the last section, the concepts of inner product and norm are introduced.

This chapter is intended to be self-contained. The reader is assumed to have some basic knowledge of matrix theory (determinants, matrix addition, multiplication, and inversion). The matrix identities introduced below will also be used. Let $\mathbf{A}$, $\mathbf{B}$, $\mathbf{C}$, $\mathbf{D}$ be $n \times m$, $m \times r$, $l \times n$, and $r \times p$ constant matrices, respectively. Let $\mathbf{a}_i$ be the ith *column* of $\mathbf{A}$, and let $\mathbf{b}_j$ be the jth *row* of $\mathbf{B}$. Then we have

$$\mathbf{AB} = [\mathbf{a}_1 \quad \mathbf{a}_2 \cdots \mathbf{a}_m] \begin{bmatrix} \mathbf{b}_1 \\ \mathbf{b}_2 \\ \vdots \\ \mathbf{b}_m \end{bmatrix} = \mathbf{a}_1\mathbf{b}_1 + \mathbf{a}_2\mathbf{b}_2 + \cdots + \mathbf{a}_m\mathbf{b}_m \tag{2-1}$$

$$\mathbf{CA} = \mathbf{C}[\mathbf{a}_1 \quad \mathbf{a}_2 \cdots \mathbf{a}_m] = [\mathbf{Ca}_1 \quad \mathbf{Ca}_2 \cdots \mathbf{Ca}_m] \tag{2-2}$$

$$\mathbf{BD} = \begin{bmatrix} \mathbf{b}_1 \\ \mathbf{b}_2 \\ \vdots \\ \mathbf{b}_m \end{bmatrix} \mathbf{D} = \begin{bmatrix} \mathbf{b}_1\mathbf{D} \\ \mathbf{b}_2\mathbf{D} \\ \vdots \\ \mathbf{b}_m\mathbf{D} \end{bmatrix} \tag{2-3}$$

These identities can be easily checked. Note that $\mathbf{a}_i\mathbf{b}_i$ is an $n \times r$ matrix, it is the product of an $n \times 1$ matrix $\mathbf{a}_i$ and a $1 \times r$ matrix $\mathbf{b}_i$.

The material presented here is well known and can be found in References 5, 38, 39, 43 to 45, 77, 86, and 116.[2] However our presentation is different. We emphasize the difference between a vector and its representations [see Equation (2-12) and Definition 2-7]. After stressing this distinction, the concepts of matrix representation of an operator and of the similarity transformation follow naturally.

2-2 Linear Spaces over a Field

In the study of mathematics we must first specify a collection of objects that forms the center of study. This collection of objects or elements is called a *set*. For example, in arithmetic, we study the set of real numbers. In boolean algebra, we study the set $\{0, 1\}$, which consists of only two elements. Other examples of sets include the set of complex numbers, the set of positive integers, the set of all polynomials of degree less than 5, the set of all 2×2 real constant matrices. In this section when we discuss a set of objects, the set could be any one of those just mentioned or any other the reader wishes to specify.

Consider the set of real numbers. The operations of addition and multiplication with the commutative and associative properties are defined for the set. The sum and product of any two real numbers are real numbers. The set has elements 0 and 1. Any real number α has an additive inverse $(-\alpha)$ and

[2] Numbers correspond to the References at the end of the book.

a multiplicative inverse ($1/\alpha$, except $\alpha = 0$) in the set. Any set with these properties is called a *field*. We give a formal definition of a field in the following.

Definition 2-1

A field consists of a set, denoted by $\mathscr{F}$, of elements called *scalars* and two operations called addition "+" and multiplication "·"; the two operations are defined over $\mathscr{F}$ such that they satisfy the following conditions:

1. To every pair of elements α and β in $\mathscr{F}$, there correspond an element $\alpha + \beta$ in $\mathscr{F}$ called the *sum* of α and β, and an element $\alpha \cdot \beta$ or $\alpha\beta$ in $\mathscr{F}$, called the *product* of α and β.
2. Addition and multiplication are respectively commutative: For any α, β in $\mathscr{F}$,
$$\alpha + \beta = \beta + \alpha \qquad \alpha \cdot \beta = \beta \cdot \alpha$$
3. Addition and multiplication are respectively associative: For any α, β, γ in $\mathscr{F}$,
$$(\alpha + \beta) + \gamma = \alpha + (\beta + \gamma) \qquad (\alpha \cdot \beta) \cdot \gamma = \alpha \cdot (\beta \cdot \gamma)$$
4. Multiplication is distributive with respect to addition: For any α, β, γ in $\mathscr{F}$,
$$\alpha \cdot (\beta + \gamma) = (\alpha \cdot \beta) + (\alpha \cdot \gamma)$$
5. $\mathscr{F}$ contains an element, denoted by 0, and an element, denoted by 1, such that $\alpha + 0 = \alpha$, $1 \cdot \alpha = \alpha$ for every α in $\mathscr{F}$.
6. To every α in $\mathscr{F}$, there is an element β in $\mathscr{F}$ such that $\alpha + \beta = 0$. The element β is called the *additive inverse*.
7. To every α in $\mathscr{F}$ which is not the element 0, there is an element γ in $\mathscr{F}$ such that $\alpha \cdot \gamma = 1$. The element γ is called the *multiplicative inverse*. ∎

We give some examples to illustrate this concept.

Example 1

Consider the set of numbers that consists of 0 and 1. The set $\{0, 1\}$ does not form a field if we use the usual definition of addition and multiplication, because the element $1 + 1 = 2$ is not in the set $\{0, 1\}$. However, if we *define* $0 + 0 = 1 + 1 = 0$, $1 + 0 = 1$ and $0 \cdot 1 = 0 \cdot 0 = 0$, $1 \cdot 1 = 1$, then it can be verified that $\{0, 1\}$ with the defined addition and multiplication satisfies all the conditions listed for a field. Hence the set $\{0, 1\}$ with the *defined operations* forms a field. It is called the field of *binary numbers*. ∎

Example 2

Consider the set of all 2×2 matrices of the form

$$\begin{bmatrix} x & -y \\ y & x \end{bmatrix}$$

where x and y are arbitrary real numbers. The set with the usual definitions of matrix addition and multiplication forms a field. The elements 0 and 1 of the field are, respectively,

$$\begin{bmatrix} 0 & 0 \\ 0 & 0 \end{bmatrix} \quad \text{and} \quad \begin{bmatrix} 1 & 0 \\ 0 & 1 \end{bmatrix}$$

Note that the set of all 2×2 matrices does not form a field. ∎

From the foregoing examples, we see that the set of objects that forms a field could be anything so long as the two operations can be defined for these objects. The fields we shall encounter in this book are fortunately the most familiar ones: the field of real numbers, the field of complex numbers, and the field of rational functions with real coefficients. The additions and multiplications of these fields are defined in the usual ways. The reader is advised to show that they satisfy all the conditions required for a field. We use $\mathbb{R}$ and $\mathbb{C}$ to denote the field of real numbers and the field of complex numbers, respectively, and use $\mathbb{R}(s)$ to denote the field of rational functions with real coefficients and with indeterminate s. Note that the set of positive real numbers does not form a field because it has no additive inverse. The set of integers and the set of polynomials do not form a field because they have no multiplicative inverse.[3]

Before introducing the concept of vector space, let us consider the ordinary two-dimensional geometric plane. If the origin is chosen, then every point in the plane can be considered as a vector: it has direction as well as magnitude. A vector can be shrunk or extended. Any two vectors can be added, but the product of two points or vectors is *not* defined. Such a plane, in the mathematical terminology, is called a *linear space*, or a *vector space*, or a *linear vector space*.

Definition 2-2

A linear space over a field $\mathscr{F}$, denoted by $(\mathscr{X}, \mathscr{F})$, consists of a set, denoted by $\mathscr{X}$, of elements called *vectors*, a field $\mathscr{F}$, and two operations called *vector addition* and *scalar multiplication*. The two operations are defined over $\mathscr{X}$ and $\mathscr{F}$ such that they satisfy all the following conditions:

1. To every pair of vectors $\mathbf{x}_1$ and $\mathbf{x}_2$ in $\mathscr{X}$, there corresponds a vector $\mathbf{x}_1 + \mathbf{x}_2$ in $\mathscr{X}$, called the sum of $\mathbf{x}_1$ and $\mathbf{x}_2$.
2. Addition is commutative: For any $\mathbf{x}_1, \mathbf{x}_2$ in $\mathscr{X}$, $\mathbf{x}_1 + \mathbf{x}_2 = \mathbf{x}_2 + \mathbf{x}_1$.
3. Addition is associative: For any $\mathbf{x}_1$, $\mathbf{x}_2$, and $\mathbf{x}_3$ in $\mathscr{X}$, $(\mathbf{x}_1 + \mathbf{x}_2) + \mathbf{x}_3 = \mathbf{x}_1 + (\mathbf{x}_2 + \mathbf{x}_3)$.
4. $\mathscr{X}$ contains a vector, denoted by $\mathbf{0}$, such that $\mathbf{0} + \mathbf{x} = \mathbf{x}$ for every $\mathbf{x}$ in $\mathscr{X}$. The vector $\mathbf{0}$ is called the zero vector or the origin.
5. To every $\mathbf{x}$ in $\mathscr{X}$, there is a vector $\bar{\mathbf{x}}$ in $\mathscr{X}$, such that $\mathbf{x} + \bar{\mathbf{x}} = \mathbf{0}$.
6. To every α in $\mathscr{F}$, and every $\mathbf{x}$ in $\mathscr{X}$, there corresponds a vector $\alpha\mathbf{x}$ in $\mathscr{X}$ called the *scalar product* of α and $\mathbf{x}$.
7. Scalar multiplication is associative: For any α, β in $\mathscr{F}$ and any $\mathbf{x}$ in $\mathscr{X}$, $\alpha(\beta\mathbf{x}) = (\alpha\beta)\mathbf{x}$.

---

[3] A set with all properties of a field except property 7 in Definition 2-1 is called a *ring* or, more precisely, a commutative ring with (multiplicative) identity. The set of integers forms a ring, as does the set of polynomials with real coefficients.

8. Scalar multiplication is distributive with respect to vector addition: For any α in $\mathscr{F}$ and any $\mathbf{x}_1, \mathbf{x}_2$ in $\mathscr{X}$, $\alpha(\mathbf{x}_1 + \mathbf{x}_2) = \alpha\mathbf{x}_1 + \alpha\mathbf{x}_2$.
9. Scalar multiplication is distributive with respect to scalar addition: For any α, β in $\mathscr{F}$ and any $\mathbf{x}$ in $\mathscr{X}$, $(\alpha + \beta)\mathbf{x} = \alpha\mathbf{x} + \beta\mathbf{x}$.
10. For any $\mathbf{x}$ in $\mathscr{X}$, $1\mathbf{x} = \mathbf{x}$, where 1 is the element 1 in $\mathscr{F}$. ∎

Example 1

A field forms a vector space over itself with the vector addition and scalar multiplication defined as the corresponding operations in the field. For example, $(\mathbb{R}, \mathbb{R})$ and $(\mathbb{C}, \mathbb{C})$ are vector spaces. Note that $(\mathbb{C}, \mathbb{R})$ is a vector space but not $(\mathbb{R}, \mathbb{C})$. (Why?) We note that $(\mathbb{R}(s), \mathbb{R}(s))$ and $(\mathbb{R}(s), \mathbb{R})$ are also vector spaces, but not $(\mathbb{R}, \mathbb{R}(s))$. ∎

Example 2

The set of all real-valued piecewise continuous functions defined over $(-\infty, \infty)$ forms a linear space over the field of real numbers. The addition and scalar multiplication are defined in the usual way. It is called a *function space.* ∎

Example 3

Given a field $\mathscr{F}$, let $\mathscr{F}^n$ be all n-tuples of scalars written as columns

$$\mathbf{x}_i = \begin{bmatrix} x_{1i} \\ x_{2i} \\ \vdots \\ x_{ni} \end{bmatrix} \tag{2-4}$$

where the first subscript denotes various components of $\mathbf{x}_i$ and the second subscript denotes different vectors in $\mathscr{F}^n$. If the vector addition and the scalar multiplication are defined in the following way:

$$\mathbf{x}_i + \mathbf{x}_j = \begin{bmatrix} x_{1i} + x_{1j} \\ x_{2i} + x_{2j} \\ \vdots \\ x_{ni} + x_{nj} \end{bmatrix} \qquad \alpha\mathbf{x}_i = \begin{bmatrix} \alpha x_{1i} \\ \alpha x_{2i} \\ \vdots \\ \alpha x_{ni} \end{bmatrix} \tag{2-5}$$

then $(\mathscr{F}^n, \mathscr{F})$ is a vector space. If $\mathscr{F} = \mathbb{R}$, $(\mathbb{R}^n, \mathbb{R})$ is called the n-dimensional *real vector space*; if $\mathscr{F} = \mathbb{C}$, $(\mathbb{C}^n, \mathbb{C})$ is called the n-dimensional *complex vector space*; if $\mathscr{F} = \mathbb{R}(s)$, $(\mathbb{R}^n(s), \mathbb{R}(s))$ is called the n-dimensional *rational vector space.* ∎

Example 4

Consider the set $\mathbb{R}_n[s]$ of all polynomials of degree less than n with real coefficients[4]

$$\sum_{i=0}^{n-1} \alpha_i s^i$$

[4] Note that $\mathbb{R}(s)$, with parentheses, denotes the field of rational functions with real coefficients; whereas $\mathbb{R}[s]$, with brackets, denotes the set of polynomials with real coefficients.

Let the vector addition and the scalar multiplication be defined as

$$\sum_{i=0}^{n-1} \alpha_i s^i + \sum_{i=0}^{n-1} \beta_i s^i = \sum_{i=0}^{n-1} (\alpha_i + \beta_i) s^i$$

$$\alpha \left(\sum_{i=0}^{n-1} \alpha_i s^i \right) = \sum_{i=0}^{n-1} (\alpha \alpha_i) s^i$$

It is easy to verify that $(\mathbb{R}_n[s], \mathbb{R})$ is a linear space. Note that $(\mathbb{R}_n[s], \mathbb{R}(s))$ is not a linear space. ∎

Example 5

Let $\mathscr{X}$ denote the set of all solutions of the homogeneous differential equation $\ddot{x} + 2\dot{x} + 3x = 0$. Then $(\mathscr{X}, \mathbb{R})$ is a linear space with the vector addition and the scalar multiplication defined in the usual way. If the differential equation is not homogeneous, then $(\mathscr{X}, \mathbb{R})$ is not a linear space. (Why?) ∎

We introduce one more concept to conclude this section.

Definition 2-3

Let $(\mathscr{X}, \mathscr{F})$ be a linear space and let $\mathscr{Y}$ be a subset of $\mathscr{X}$. Then $(\mathscr{Y}, \mathscr{F})$ is said to be a *subspace* of $(\mathscr{X}, \mathscr{F})$ if under the operations of $(\mathscr{X}, \mathscr{F})$, $\mathscr{Y}$ itself forms a vector space over $\mathscr{F}$. ∎

We remark on the conditions for a subset of $\mathscr{X}$ to form a subspace. Since the vector addition and scalar multiplication have been defined for the linear space $(\mathscr{X}, \mathscr{F})$, they satisfy conditions 2, 3, and 7 through 10 listed in Definition 2-2. Hence we need to check only conditions 1 and 4 through 6 to determine whether a set $\mathscr{Y}$ is a subspace of $(\mathscr{X}, \mathscr{F})$. It is easy to verify that if $\alpha_1 \mathbf{y}_1 + \alpha_2 \mathbf{y}_2$ is in $\mathscr{Y}$ for any $\mathbf{y}_1, \mathbf{y}_2$ in $\mathscr{Y}$ and any α_1, α_2 in $\mathscr{F}$, then conditions 1 and 4 through 6 are satisfied. Hence we conclude that a *set* $\mathscr{Y}$ *is a subspace of* $(\mathscr{X}, \mathscr{F})$ *if* $\alpha_1 \mathbf{y}_1 + \alpha_2 \mathbf{y}_2$ *is in* $\mathscr{Y}$, *for any* $\mathbf{y}_1, \mathbf{y}_2$ *in* $\mathscr{Y}$ *and any* α_1, α_2 *in* $\mathscr{F}$.

Example 6

In the two-dimensional real vector space $(\mathbb{R}^2, \mathbb{R})$, every straight line passing through the origin is a subspace of $(\mathbb{R}^2, \mathbb{R})$. That is, the set

$$\begin{bmatrix} x_1 \\ \alpha x_1 \end{bmatrix}$$

for any fixed real α is a subspace of $(\mathbb{R}^2, \mathbb{R})$. ∎

Example 7

The real vector space $(\mathbb{R}^n, \mathbb{R})$ is a subspace of the vector space $(\mathbb{C}^n, \mathbb{R})$. ∎

2-3 Linear Independence, Bases, and Representations

Every geometric plane has two coordinate axes, which are mutually perpendicular and of the same scale. The reason for having a coordinate system is to have some reference or standard to specify a point or vector in the plane. In this section, we will extend this concept of coordinate to general linear spaces. In linear spaces a coordinate system is called a *basis*. The basis vectors are generally not perpendicular to each other and have different scales. Before proceeding, we need the concept of linear independence of vectors.

Definition 2-4

A set of vectors $\mathbf{x}_1, \mathbf{x}_2, \ldots, \mathbf{x}_n$ in a linear space over a field $\mathscr{F}$, $(\mathscr{X}, \mathscr{F})$, is said to be *linearly dependent* if and only if there exist scalars $\alpha_1, \alpha_2, \ldots, \alpha_n$ in $\mathscr{F}$, not all zero, such that

$$\alpha_1\mathbf{x}_1 + \alpha_2\mathbf{x}_2 + \cdots + \alpha_n\mathbf{x}_n = \mathbf{0} \tag{2-6}$$

If the only set of α_i for which (2-6) holds is $\alpha_1 = 0, \alpha_2 = 0, \ldots, \alpha_n = 0$, then the set of vectors $\mathbf{x}_1, \mathbf{x}_2, \ldots, \mathbf{x}_n$ is said to be *linearly independent*. ∎

Given any set of vectors, Equation (2-6) always holds for $\alpha_1 = 0, \alpha_2 = 0, \ldots, \alpha_n = 0$. Therefore, in order to show the linear independence of the set, we have to show that $\alpha_1 = 0, \alpha_2 = 0, \ldots, \alpha_n = 0$ is the *only* set of α_i for which (2-6) holds; that is, if any one of the α_i's is different from zero, then the right-hand side of (2-6) cannot be a zero vector. If a set of vectors is linearly dependent, there are generally infinitely many sets of α_i, not all zero, that satisfy Equation (2-6). However, it is sufficient to find one set of α_i, not all zero, to conclude the linear dependence of the set of vectors.

Example 1

Consider the set of vectors $\mathbf{x}_1, \mathbf{x}_2, \ldots, \mathbf{x}_n$ in which $\mathbf{x}_1 = \mathbf{0}$. This set of vectors is always linearly dependent, because we may choose $\alpha_1 = 1, \alpha_2 = 0, \alpha_2 = 0, \ldots, \alpha_n = 0$, and Equation (2-6) holds. ∎

Example 2

Consider the set of vector $\mathbf{x}_1$ which consists of only one vector. The set of vector $\mathbf{x}_1$ is linearly independent if and only if $\mathbf{x}_1 \neq \mathbf{0}$. If $\mathbf{x}_1 \neq \mathbf{0}$, the only way to have $\alpha_1\mathbf{x}_1 = \mathbf{0}$ is $\alpha_1 = 0$. If $\mathbf{x}_1 = \mathbf{0}$, we may choose $\alpha_1 = 1$. ∎

If we introduce the notation

$$\alpha_1\mathbf{x}_1 + \alpha_2\mathbf{x}_2 + \cdots + \alpha_n\mathbf{x}_n \triangleq [\mathbf{x}_1 \quad \mathbf{x}_2 \quad \cdots \quad \mathbf{x}_n] \begin{bmatrix} \alpha_1 \\ \alpha_2 \\ \vdots \\ \alpha_n \end{bmatrix}$$
$$\triangleq [\mathbf{x}_1 \quad \mathbf{x}_2 \quad \cdots \quad \mathbf{x}_n]\boldsymbol{\alpha} \tag{2-7}$$

then the linear independence of a set of vectors can also be stated in the following definition.

Definition 2-4′

A set of vectors $\mathbf{x}_1, \mathbf{x}_2, \ldots, \mathbf{x}_n$ in $(\mathscr{X}, \mathscr{F})$ is said to be linearly independent if and only if the equation

$$[\mathbf{x}_1 \quad \mathbf{x}_2 \quad \cdots \quad \mathbf{x}_n]\boldsymbol{\alpha} = 0$$

implies $\boldsymbol{\alpha} = \mathbf{0}$, where every component of $\boldsymbol{\alpha}$ is an element of $\mathscr{F}$ or, correspondingly, $\boldsymbol{\alpha}$ can be considered as a vector in $\mathscr{F}^n$. ∎

Observe that linear dependence depends not only on the set of vectors but also on the field. For example, the set of vectors $\{\mathbf{x}_1, \mathbf{x}_2\}$, where

$$\mathbf{x}_1 \triangleq \begin{bmatrix} \dfrac{1}{s+1} \\ \\ \dfrac{1}{s+2} \end{bmatrix} \qquad \mathbf{x}_2 \triangleq \begin{bmatrix} \dfrac{s+2}{(s+1)(s+3)} \\ \\ \dfrac{1}{s+3} \end{bmatrix}$$

is linearly dependent in the field of rational functions with real coefficients. Indeed, if we choose

$$\alpha_1 = -1 \qquad \text{and} \qquad \alpha_2 = \frac{s+3}{s+2}$$

then $\alpha_1 \mathbf{x}_1 + \alpha_2 \mathbf{x}_2 = \mathbf{0}$. However, this set of vectors is linearly independent in the field of real numbers, for there exist no α_1 and α_2 in $\mathbb{R}$ that are different from zero, such that $\alpha_1 \mathbf{x}_1 + \alpha_2 \mathbf{x}_2 = \mathbf{0}$. In other words, $\mathbf{x}_1$ and $\mathbf{x}_2$ are linearly independent in $(\mathbb{R}^2(s), \mathbb{R})$, but are linearly dependent in $(\mathbb{R}^2(s), \mathbb{R}(s))$. ∎

It is clear from the definition of linear dependence that if the vectors $\mathbf{x}_1, \mathbf{x}_2, \ldots, \mathbf{x}_n$ are linearly dependent, then at least one of them can be written as a linear combination of the others. However, it is not necessarily true that every one of them can be expressed as a linear combination of the others.

Definition 2-5

The maximal number of linearly independent vectors in a linear space $(\mathscr{X}, \mathscr{F})$ is called the *dimension* of the linear space $(\mathscr{X}, \mathscr{F})$. ∎

In the previous section we introduced the n-dimensional real vector space $(\mathbb{R}^n, \mathbb{R})$. The meaning of n-dimensional is now clear. It means that in $(\mathbb{R}^n, \mathbb{R})$ there are, at most, n linearly independent vectors (over the field $\mathbb{R}$). In the two-dimensional real vector space $(\mathbb{R}^2, \mathbb{R})$, one cannot find three linearly independent vectors. (Try!)

Example 3

Consider the function space that consists of all real-valued piecewise continuous functions defined over $(-\infty, \infty)$. The zero vector in this space is the one which is identically zero on $(-\infty, \infty)$. The following functions, with $-\infty < t < \infty$,

$$t, t^2, t^3, \dots$$

are clearly elements of the function space. This set of functions $\{t^n, n=1, 2, \dots\}$ is linearly independent, because there exist no real constants, α_i's, not all zero, such that

$$\sum_{i=1} \alpha_i t^i \equiv \mathbf{0}$$

There are infinitely many of these functions; therefore, the dimension of this space is infinity. ∎

We assume that all the linear spaces we shall encounter are of finite dimensions unless stated otherwise.

Definition 2-6

A set of linearly independent vectors of a linear space $(\mathscr{X}, \mathscr{F})$ is said to be a *basis* of $\mathscr{X}$ if every vector in $\mathscr{X}$ can be expressed as a unique linear combination of these vectors. ∎

Theorem 2-1

In an n-dimensional vector space, *any* set of n linearly independent vectors qualifies as a basis.

Proof

Let $\mathbf{e}_1, \mathbf{e}_2, \dots, \mathbf{e}_n$ be any n linearly independent vectors in $\mathscr{X}$, and let $\mathbf{x}$ be an arbitrary vector in $\mathscr{X}$. Then the set of $n+1$ vectors $\mathbf{x}, \mathbf{e}_1, \mathbf{e}_2, \dots, \mathbf{e}_n$ is linearly dependent (since, by the definition of dimension, n is the maximum number of linearly independent vectors we can have in the space). Consequently, there exist $\alpha_0, \alpha_1, \dots, \alpha_n$ in $\mathscr{F}$, not all zero, such that

$$\alpha_0 \mathbf{x} + \alpha_1 \mathbf{e}_1 + \alpha_2 \mathbf{e}_2 + \cdots + \alpha_n \mathbf{e}_n = \mathbf{0} \tag{2-8}$$

We claim that $\alpha_0 \neq 0$. If $\alpha_0 = 0$, Equation (2-8) reduces to

$$\alpha_1 \mathbf{e}_1 + \alpha_2 \mathbf{e}_2 + \cdots + \alpha_n \mathbf{e}_n = \mathbf{0} \tag{2-9}$$

which, together with the linear independence assumption of $\mathbf{e}_1, \mathbf{e}_2, \dots, \mathbf{e}_n$, implies that $\alpha_1 = 0, \alpha_2 = 0, \dots, \alpha_n = 0$. This contradicts the assumption that not all $\alpha_0, \alpha_1, \dots, \alpha_n$ are zero. If we define $\beta_i \triangleq -\alpha_i/\alpha_0$, for $i = 1, 2, \dots, n$, then (2-8) becomes

$$\mathbf{x} = \beta_1 \mathbf{e}_1 + \beta_2 \mathbf{e}_2 + \cdots + \beta_n \mathbf{e}_n \tag{2-10}$$

This shows that every vector $\mathbf{x}$ in $\mathscr{X}$ can be expressed as a linear combination of $\mathbf{e}_1, \mathbf{e}_2, \dots, \mathbf{e}_n$. Now we show that this combination is unique. Suppose there is another linear combination, say

$$\mathbf{x} = \tilde{\beta}_1 \mathbf{e}_1 + \tilde{\beta}_2 \mathbf{e}_2 + \cdots + \tilde{\beta}_n \mathbf{e}_n \tag{2-11}$$

Then by subtracting (2-11) from (2-10), we obtain

$$\mathbf{0} = (\beta_1 - \tilde{\beta}_1)\mathbf{e}_1 + (\beta_2 - \tilde{\beta}_2)\mathbf{e}_2 + \cdots + (\beta_n - \tilde{\beta}_n)\mathbf{e}_n$$

which, together with the linear independence of $\{\mathbf{e}_i\}$, implies that

$$\beta_i = \tilde{\beta}_i \qquad i = 1, 2, \dots, n$$

This completes the proof of this theorem. Q.E.D.

This theorem has a very important implication. In an n-dimensional vector space $(\mathscr{X}, \mathscr{F})$, if a basis is chosen, then every vector in $\mathscr{X}$ can be uniquely represented by a set of n scalars $\beta_1, \beta_2, \dots, \beta_n$ in $\mathscr{F}$. If we use the notation of (2-7), we may write (2-10) as

$$\mathbf{x} = [\mathbf{e}_1 \quad \mathbf{e}_2 \quad \cdots \quad \mathbf{e}_n]\boldsymbol{\beta} \tag{2-12}$$

where $\boldsymbol{\beta} = [\beta_1, \beta_2, \dots, \beta_n]'$ and the prime denotes the transpose. The $n \times 1$ vector $\boldsymbol{\beta}$ can be considered as a vector in $(\mathscr{F}^n, \mathscr{F})$. Consequently, *there is a one-to-one correspondence between any n-dimensional vector space* $(\mathscr{X}, \mathscr{F})$ *and the same dimensional linear space* $(\mathscr{F}^n, \mathscr{F})$ *if a basis is chosen for* $(\mathscr{X}, \mathscr{F})$.

Definition 2-7

In an n-dimensional vector space $(\mathscr{X}, \mathscr{F})$, if a basis $\{\mathbf{e}_1, \mathbf{e}_2, \dots, \mathbf{e}_n\}$ is chosen, then every vector $\mathbf{x}$ in $\mathscr{X}$ can be uniquely written in the form of (2-12). $\boldsymbol{\beta}$ is called the *representation* of $\mathbf{x}$ with respect to the basis $\{\mathbf{e}_1, \mathbf{e}_2, \dots, \mathbf{e}_n\}$. ∎

Example 4

The geometric plane shown in Figure 2-1 can be considered as a two-dimensional real vector space. Any point in the plane is a vector. Theorem 2-1 states that

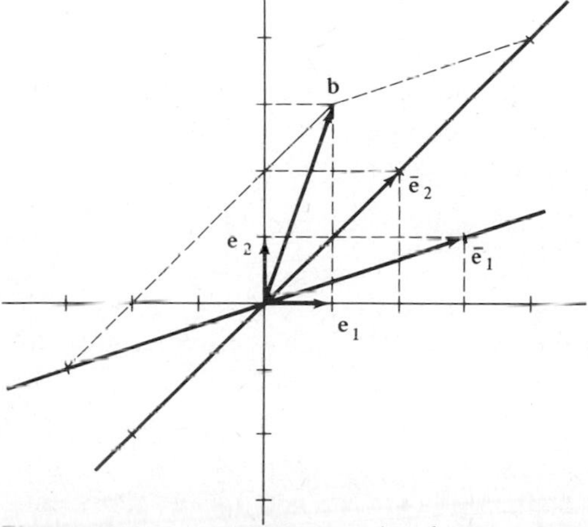

Figure 2-1 A two-dimensional real vector space.

Table 2-1 Different Representations of Vectors

Bases \ Vectors	$\mathbf{b}$	$\bar{\mathbf{e}}_1$	$\bar{\mathbf{e}}_2$	$\mathbf{e}_1$	$\mathbf{e}_2$
$[\mathbf{e}_1 \quad \mathbf{e}_2]$	$\begin{bmatrix}1\\3\end{bmatrix}$	$\begin{bmatrix}3\\1\end{bmatrix}$	$\begin{bmatrix}2\\2\end{bmatrix}$	$\begin{bmatrix}1\\0\end{bmatrix}$	$\begin{bmatrix}0\\1\end{bmatrix}$
$[\bar{\mathbf{e}}_1 \quad \bar{\mathbf{e}}_2]$	$\begin{bmatrix}-1\\2\end{bmatrix}$	$\begin{bmatrix}1\\0\end{bmatrix}$	$\begin{bmatrix}0\\1\end{bmatrix}$	$\begin{bmatrix}\frac{1}{2}\\-\frac{1}{4}\end{bmatrix}$	$\begin{bmatrix}-\frac{1}{2}\\\frac{3}{4}\end{bmatrix}$

any set of two linearly independent vectors forms a basis. Observe that we have not only the freedom in choosing the directions of the basis vectors (as long as they do not lie in the same line) but also the magnitude (scale) of these vectors. Therefore, given a vector in $(\mathbb{R}^2, \mathbb{R})$, for different bases we have different representations of the *same* vector. For example, the representations of the vector $\mathbf{b}$ in Figure 2-1 with respect to the basis $\{\mathbf{e}_1, \mathbf{e}_2\}$ and the basis $\{\bar{\mathbf{e}}_1, \bar{\mathbf{e}}_2\}$ are, respectively, $[1 \quad 3]'$ and $[-1 \quad 2]'$ (where the "prime" symbol denotes the transpose). We summarize the representations of the vectors $\mathbf{b}, \bar{\mathbf{e}}_1, \bar{\mathbf{e}}_2, \mathbf{e}_1$, and $\mathbf{e}_2$ with respect to the bases $\{\mathbf{e}_1, \mathbf{e}_2\}$, $\{\bar{\mathbf{e}}_1, \bar{\mathbf{e}}_2\}$ in Table 2-1.

Example 5

Consider the linear space $(\mathbb{R}_4[s], \mathbb{R})$, where $\mathbb{R}_4[s]$ is the set of all real polynomials of degree less than 4 and with indeterminate s. Let $\mathbf{e}_1 = s^3$, $\mathbf{e}_2 = s^2$, $\mathbf{e}_3 = s$, and $\mathbf{e}_4 = 1$. Clearly, the vectors $\mathbf{e}_i$, $i = 1, 2, 3, 4$, are linearly independent and qualify as basis vectors. With this set of basis vectors, the vector $\mathbf{x} = 3s^3 + 2s^2 - 2s + 10$ can be written as

$$\mathbf{x} = [\mathbf{e}_1 \quad \mathbf{e}_2 \quad \mathbf{e}_3 \quad \mathbf{e}_4] \begin{bmatrix} 3 \\ 2 \\ -2 \\ 10 \end{bmatrix}$$

hence $[3 \quad 2 \quad -2 \quad 10]'$ (where the "prime" denotes the transpose) is the representation of $\mathbf{x}$ with respect to $\{\mathbf{e}_1, \mathbf{e}_2, \mathbf{e}_3, \mathbf{e}_4\}$. If we choose $\bar{\mathbf{e}}_1 = s^3 - s^2$, $\bar{\mathbf{e}}_2 = s^2 - s$, $\bar{\mathbf{e}}_3 = s - 1$, and $\bar{\mathbf{e}}_4 = 1$ as the basis vectors, then

$$\mathbf{x} = 3s^3 + 2s^2 - 2s + 10 = 3(s^3 - s^2) + 5(s^2 - s) + 3(s - 1) + 13 \cdot 1$$
$$= [\bar{\mathbf{e}}_1 \quad \bar{\mathbf{e}}_2 \quad \bar{\mathbf{e}}_3 \quad \bar{\mathbf{e}}_4] \begin{bmatrix} 3 \\ 5 \\ 3 \\ 13 \end{bmatrix}$$

Hence the representation of $\mathbf{x}$ with respect to $\{\bar{\mathbf{e}}_1, \bar{\mathbf{e}}_2, \bar{\mathbf{e}}_3, \bar{\mathbf{e}}_4\}$ is $[3 \quad 5 \quad 3 \quad 13]'$. ∎

In this example, there is a sharp distinction between vectors and representations. However, this is not always the case. Let us consider the n-dimensional real vector space $(\mathbb{R}^n, \mathbb{R})$, complex vector space $(\mathbb{C}^n, \mathbb{C})$, or rational vector space $(\mathbb{R}^n(s), \mathbb{R}(s))$; a vector is an n-tuple of real, complex, or real rational functions, written as

$$\mathbf{x} = \begin{bmatrix} \beta_1 \\ \beta_2 \\ \vdots \\ \beta_n \end{bmatrix}$$

This array of n numbers can be interpreted in two ways: (1) It is defined as such; that is, it is a vector and is independent of basis. (2) It is a representation of a vector with respect to some fixed unknown basis. Given an array of numbers, unless it is tied up with some basis, we shall always consider it as a vector. However we shall also introduce, unless stated otherwise, the following vectors[5]:

$$\mathbf{n}_1 = \begin{bmatrix} 1 \\ 0 \\ 0 \\ \vdots \\ 0 \\ 0 \end{bmatrix}, \quad \mathbf{n}_2 = \begin{bmatrix} 0 \\ 1 \\ 0 \\ \vdots \\ 0 \\ 0 \end{bmatrix}, \quad \ldots, \quad \mathbf{n}_{n-1} = \begin{bmatrix} 0 \\ 0 \\ 0 \\ \vdots \\ 1 \\ 0 \end{bmatrix}, \quad \mathbf{n}_n = \begin{bmatrix} 0 \\ 0 \\ 0 \\ \vdots \\ 0 \\ 1 \end{bmatrix} \qquad \textbf{(2-13)}$$

as the basis of $(\mathbb{R}^n, \mathbb{R})$, $(\mathbb{C}^n, \mathbb{C})$, and $(\mathbb{R}^n(s), \mathbb{R}(s))$. In this case, an array of numbers can be interpreted as a vector or the representation of a vector with respect to the basis $\{\mathbf{n}_1, \mathbf{n}_2, \ldots, \mathbf{n}_n\}$, because with respect to this particular set of bases, the representation and the vector itself are identical; that is,

$$\begin{bmatrix} \beta_1 \\ \beta_2 \\ \vdots \\ \beta_n \end{bmatrix} = [\mathbf{n}_1 \quad \mathbf{n}_2 \quad \cdots \quad \mathbf{n}_n] \begin{bmatrix} \beta_1 \\ \beta_2 \\ \vdots \\ \beta_n \end{bmatrix} \qquad \textbf{(2-14)}$$

Change of basis. We have shown that a vector $\mathbf{x}$ in $(\mathscr{X}, \mathscr{F})$ has different representations with respect to different bases. It is natural to ask what the relationships are between these different representations of the same vector. In this subsection, this problem will be studied.

Let the representations of a vector $\mathbf{x}$ in $(\mathscr{X}, \mathscr{F})$ with respect to $\{\mathbf{e}_1, \mathbf{e}_2, \ldots, \mathbf{e}_n\}$ and $\{\bar{\mathbf{e}}_1, \bar{\mathbf{e}}_2, \ldots, \bar{\mathbf{e}}_n\}$ be $\boldsymbol{\beta}$ and $\bar{\boldsymbol{\beta}}$, respectively; that is,[6]

$$\mathbf{x} = [\mathbf{e}_1 \quad \mathbf{e}_2 \quad \cdots \quad \mathbf{e}_n]\boldsymbol{\beta} = [\bar{\mathbf{e}}_1 \quad \bar{\mathbf{e}}_2 \quad \cdots \quad \bar{\mathbf{e}}_n]\bar{\boldsymbol{\beta}} \qquad \textbf{(2-15)}$$

In order to derive the relationship between $\boldsymbol{\beta}$ and $\bar{\boldsymbol{\beta}}$, we need either the information of the representations of $\bar{\mathbf{e}}_i$, for $i = 1, 2, \ldots, n$, with respect to the

[5] This set of vectors is called an *orthonormal* set.

[6] One might be tempted to write $\boldsymbol{\beta} = [\mathbf{e}_1 \quad \mathbf{e}_2 \quad \cdots \quad \mathbf{e}_n]^{-1}[\bar{\mathbf{e}}_1 \quad \bar{\mathbf{e}}_2 \quad \cdots \quad \bar{\mathbf{e}}_n]\bar{\boldsymbol{\beta}}$. However, $[\mathbf{e}_1 \quad \mathbf{e}_2 \quad \cdots \quad \mathbf{e}_n]^{-1}$ may not be defined as can be seen from Example 5.

basis $\{\mathbf{e}_1, \mathbf{e}_2, \ldots, \mathbf{e}_n\}$, or the information of the representations of $\mathbf{e}_i$, for $i = 1, 2, \ldots, n$, with respect to the basis $\{\bar{\mathbf{e}}_1, \bar{\mathbf{e}}_2, \ldots, \bar{\mathbf{e}}_n\}$. Let the representation of $\mathbf{e}_i$ with respect to $\{\bar{\mathbf{e}}_1, \bar{\mathbf{e}}_2, \ldots, \bar{\mathbf{e}}_n\}$ be $[p_{1i} \quad p_{2i} \quad p_{3i} \quad \cdots \quad p_{ni}]'$; that is,

$$\mathbf{e}_i = [\bar{\mathbf{e}}_1 \quad \bar{\mathbf{e}}_2 \quad \cdots \quad \bar{\mathbf{e}}_n] \begin{bmatrix} p_{1i} \\ p_{2i} \\ \vdots \\ p_{ni} \end{bmatrix} \triangleq \mathbf{E}\mathbf{p}_i \qquad i = 1, 2, \ldots, n \tag{2-16}$$

where $\mathbf{E} \triangleq [\bar{\mathbf{e}}_1 \quad \bar{\mathbf{e}}_2 \quad \cdots \quad \bar{\mathbf{e}}_n]$, $\mathbf{p}_i \triangleq [p_{1i} \quad p_{2i} \quad \cdots \quad p_{ni}]'$. Using matrix notation, we write

$$[\mathbf{e}_1 \quad \mathbf{e}_2 \quad \cdots \quad \mathbf{e}_n] = [\mathbf{E}\mathbf{p}_1 \quad \mathbf{E}\mathbf{p}_2 \quad \cdots \quad \mathbf{E}\mathbf{p}_n] \tag{2-17}$$

which, by using (2-2), can be written as

$$\begin{aligned} [\mathbf{e}_1 \quad \mathbf{e}_2 \quad \cdots \quad \mathbf{e}_n] &= \mathbf{E}[\mathbf{p}_1 \quad \mathbf{p}_2 \quad \cdots \quad \mathbf{p}_n] \\ &= [\bar{\mathbf{e}}_1 \quad \bar{\mathbf{e}}_2 \quad \cdots \quad \bar{\mathbf{e}}_n] \begin{bmatrix} p_{11} & p_{12} & \cdots & p_{1n} \\ p_{21} & p_{22} & \cdots & p_{2n} \\ \vdots & \vdots & & \vdots \\ p_{n1} & p_{n2} & \cdots & p_{nn} \end{bmatrix} \\ &\triangleq [\bar{\mathbf{e}}_1 \quad \bar{\mathbf{e}}_2 \quad \cdots \quad \bar{\mathbf{e}}_n]\mathbf{P} \end{aligned} \tag{2-18}$$

Substituting (2-18) into (2-15), we obtain

$$\mathbf{x} = [\bar{\mathbf{e}}_1 \quad \bar{\mathbf{e}}_2 \quad \cdots \quad \bar{\mathbf{e}}_n]\mathbf{P}\boldsymbol{\beta} = [\bar{\mathbf{e}}_1 \quad \bar{\mathbf{e}}_2 \quad \cdots \quad \bar{\mathbf{e}}_n]\bar{\boldsymbol{\beta}} \tag{2-19}$$

Since the representation of $\mathbf{x}$ with respect to the basis $\{\bar{\mathbf{e}}_1, \bar{\mathbf{e}}_2, \ldots, \bar{\mathbf{e}}_n\}$ is unique, (2-19) implies

$$\bar{\boldsymbol{\beta}} = \mathbf{P}\boldsymbol{\beta} \tag{2-20}$$

where

$$\mathbf{P} = \begin{bmatrix} i\text{th column: the} \\ \text{representation of} \\ \mathbf{e}_i \text{ with respect to} \\ \{\bar{\mathbf{e}}_1, \bar{\mathbf{e}}_2, \ldots, \bar{\mathbf{e}}_n\} \end{bmatrix} \tag{2-21}$$

This establishes the relationship between $\bar{\boldsymbol{\beta}}$ and $\boldsymbol{\beta}$. In (2-16), if the representation of $\bar{\mathbf{e}}_i$ with respect to $\{\mathbf{e}_1, \mathbf{e}_2, \ldots, \mathbf{e}_n\}$ is used, then we shall obtain

$$\boldsymbol{\beta} = \mathbf{Q}\bar{\boldsymbol{\beta}} \tag{2-22}$$

where

$$\mathbf{Q} = \begin{bmatrix} i\text{th column: the} \\ \text{representation of} \\ \bar{\mathbf{e}}_i \text{ with respect to} \\ \{\mathbf{e}_1, \mathbf{e}_2, \ldots, \mathbf{e}_n\} \end{bmatrix} \tag{2-23}$$

Different representations of a vector are related by (2-20) or (2-22). Therefore, given two sets of bases, if the representation of a vector with respect to one set of bases is known, the representation of the same vector with respect to the other set of bases can be computed by using either (2-20) or (2-22). Since $\bar{\boldsymbol{\beta}} = \mathbf{P}\boldsymbol{\beta}$ and $\boldsymbol{\beta} = \mathbf{Q}\bar{\boldsymbol{\beta}}$, we have $\bar{\boldsymbol{\beta}} = \mathbf{P}\mathbf{Q}\bar{\boldsymbol{\beta}}$, for all $\bar{\boldsymbol{\beta}}$; hence we conclude that

$$\mathbf{P}\mathbf{Q} = \mathbf{I} \qquad \text{or} \qquad \mathbf{P} = \mathbf{Q}^{-1} \tag{2-24}$$

Example 6

Consider the two sets of basis vectors of $(\mathbb{R}_4[s], \mathbb{R})$ in Example 5. It can be readily verified that

$$[\mathbf{e}_1 \quad \mathbf{e}_2 \quad \mathbf{e}_3 \quad \mathbf{e}_4] = [\bar{\mathbf{e}}_1 \quad \bar{\mathbf{e}}_2 \quad \bar{\mathbf{e}}_3 \quad \bar{\mathbf{e}}_4] \begin{bmatrix} 1 & 0 & 0 & 0 \\ 1 & 1 & 0 & 0 \\ 1 & 1 & 1 & 0 \\ 1 & 1 & 1 & 1 \end{bmatrix} \triangleq [\bar{\mathbf{e}}_1 \quad \bar{\mathbf{e}}_2 \quad \bar{\mathbf{e}}_3 \quad \bar{\mathbf{e}}_4]\mathbf{P}$$

and

$$[\bar{\mathbf{e}}_1 \quad \bar{\mathbf{e}}_2 \quad \bar{\mathbf{e}}_3 \quad \bar{\mathbf{e}}_4] = [\mathbf{e}_1 \quad \mathbf{e}_2 \quad \mathbf{e}_3 \quad \mathbf{e}_4] \begin{bmatrix} 1 & 0 & 0 & 0 \\ -1 & 1 & 0 & 0 \\ 0 & -1 & 1 & 0 \\ 0 & 0 & -1 & 1 \end{bmatrix} \triangleq [\mathbf{e}_1 \quad \mathbf{e}_2 \quad \mathbf{e}_3 \quad \mathbf{e}_4]\mathbf{Q}$$

Clearly, we have $\mathbf{PQ} = \mathbf{I}$ and $\bar{\boldsymbol{\beta}} = \mathbf{P}\boldsymbol{\beta}$, or

$$\begin{bmatrix} 3 \\ 5 \\ 3 \\ 13 \end{bmatrix} = \mathbf{P} \begin{bmatrix} 3 \\ 2 \\ -2 \\ 10 \end{bmatrix} = \begin{bmatrix} 1 & 0 & 0 & 0 \\ 1 & 1 & 0 & 0 \\ 1 & 1 & 1 & 0 \\ 1 & 1 & 1 & 1 \end{bmatrix} \begin{bmatrix} 3 \\ 2 \\ -2 \\ 10 \end{bmatrix}$$ ∎

2-4 Linear Operators and Their Representations

The concept of a function is basic to all parts of analysis. Given two sets $\mathscr{X}$ and $\mathscr{Y}$, if we assign to each element of $\mathscr{X}$ one and only one element of $\mathscr{Y}$, then the rule of assignments is called a *function*. For example, the rule of assignments in Figure 2-2(a) is a function, but not the one in Figure 2-2(b). A function is usually denoted by the notation $f: \mathscr{X} \to \mathscr{Y}$, and the element of $\mathscr{Y}$ that is assigned to the element x of $\mathscr{X}$ is denoted by $y = f(x)$. The set $\mathscr{X}$ on which a function is defined is called the *domain* of the function. The subset of $\mathscr{Y}$ that is assigned to some element of $\mathscr{X}$ is called the *range* of the function. For example, the

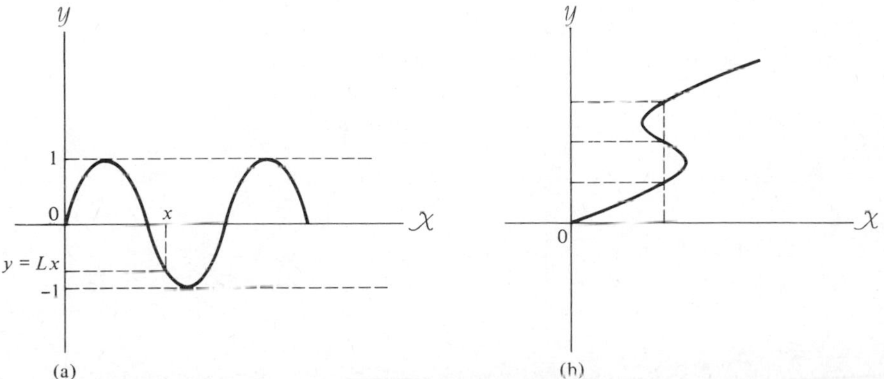

Figure 2-2 Examples in which (a) the curve represents a function and (b) the curve does not represent a function.

domain of the function shown in Figure 2-2(a) is the positive real line, the range of the function is the set $[-1, 1]$, which is a subset of the entire real line $\mathscr{Y}$.

The functions we shall study in this section belong to a restricted class of functions, called *linear functions*, or more often called *linear operators, linear mappings*, or *linear transformations.* The sets associated with linear operators are required to be linear spaces over the same field, say $(\mathscr{X}, \mathscr{F})$ and $(\mathscr{Y}, \mathscr{F})$. A linear operator is denoted by $L:(\mathscr{X}, \mathscr{F}) \rightarrow (\mathscr{Y}, \mathscr{F})$. In words, L maps $(\mathscr{X}, \mathscr{F})$ into $(\mathscr{Y}, \mathscr{F})$.

Definition 2-8

A function L that maps $(\mathscr{X}, \mathscr{F})$ into $(\mathscr{Y}, \mathscr{F})$ is said to be a *linear operator* if and only if

$$L(\alpha_1 \mathbf{x}_1 + \alpha_2 \mathbf{x}_2) = \alpha_1 L\mathbf{x}_1 + \alpha_2 L\mathbf{x}_2$$

for any vectors $\mathbf{x}_1, \mathbf{x}_2$ in $\mathscr{X}$ and any scalars α_1, α_2 in $\mathscr{F}$. ■

Note that the vectors $L\mathbf{x}_1$ and $L\mathbf{x}_2$ are elements of $\mathscr{Y}$. The reason for requiring that $\mathscr{Y}$ be defined over the same field as $\mathscr{X}$ is to ensure that $\alpha_1 L\mathbf{x}_1$ and $\alpha_2 L\mathbf{x}_2$ be defined.

Example 1

Consider the transformation that rotates a point in a geometric plane counterclockwise 90° with respect to the origin as shown in Figure 2-3. Given any two vectors in the plane, it is easy to verify that the vector that is the sum of the two vectors after rotation is equal to the rotation of the vector that is the sum of the two vectors before rotation. Hence the transformation is a linear transformation. The spaces $(\mathscr{X}, \mathscr{F})$ and $(\mathscr{Y}, \mathscr{F})$ of this example are all equal to $(\mathbb{R}^2, \mathbb{R})$. ■

Example 2

Let $\mathscr{U}$ be the set of all real-valued piecewise continuous functions defined over $[0, T]$ for some finite $T > 0$. It is clear that $(\mathscr{U}, \mathbb{R})$ is a linear space whose dimension is infinity (see Example 3, Section 2-3). Let g be a continuous function defined over $[0, T]$. Then the transformation

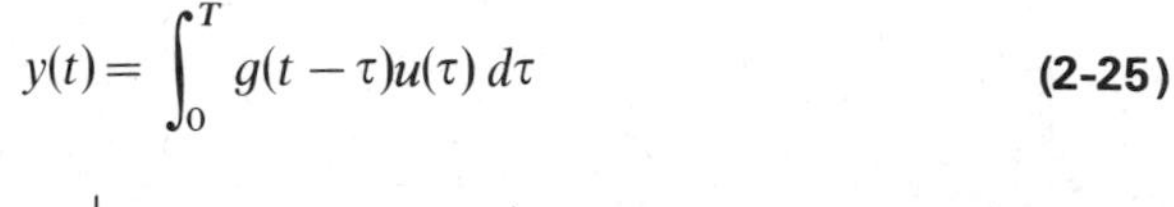

$$y(t) = \int_0^T g(t-\tau)u(\tau)\, d\tau \qquad \textbf{(2-25)}$$

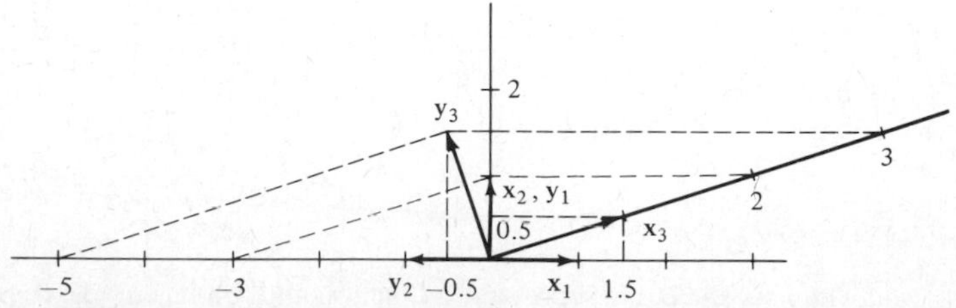

Figure 2-3 The transformation that rotates a vector counterclockwise 90°.

is a linear transformation. The spaces $(\mathscr{X}, \mathscr{F})$ and $(\mathscr{Y}, \mathscr{F})$ of this example are all equal to $(\mathscr{U}, \mathbb{R})$. ■

Matrix representations of a linear operator. We see from the above two examples that the spaces $(\mathscr{X}, \mathscr{F})$ and $(\mathscr{Y}, \mathscr{F})$ on which a linear operator is defined may be of finite or infinite dimension. We show in the following that every linear operator that maps finite-dimensional $(\mathscr{X}, \mathscr{F})$ into finite-dimensional $(\mathscr{Y}, \mathscr{F})$ has matrix representations with coefficients in the field $\mathscr{F}$. If $(\mathscr{X}, \mathscr{F})$ and $(\mathscr{Y}, \mathscr{F})$ are of infinite dimension, a representation of a linear operator can still be found. However, the representation will be a matrix of infinite order or a form similar to (2-25). This is outside the scope of this text and will not be discussed.

Theorem 2-2

Let $(\mathscr{X}, \mathscr{F})$ and $(\mathscr{Y}, \mathscr{F})$ be n- and m-dimensional vector spaces, respectively, over the same field. Let $\mathbf{x}_1, \mathbf{x}_2, \ldots, \mathbf{x}_n$ be a set of linearly independent vectors in $\mathscr{X}$. Then the linear operator $L:(\mathscr{X}, \mathscr{F}) \to (\mathscr{Y}, \mathscr{F})$ is uniquely determined by the n pairs of mappings $\mathbf{y}_i = L\mathbf{x}_i$, for $i = 1, 2, \ldots, n$. Furthermore, with respect to the basis $\{\mathbf{x}_1, \mathbf{x}_2, \ldots, \mathbf{x}_n\}$ of $\mathscr{X}$ and a basis $\{\mathbf{u}_1, \mathbf{u}_2, \ldots, \mathbf{u}_m\}$ of $\mathscr{Y}$, L can be represented by an $m \times n$ matrix $\mathbf{A}$ with coefficients in the field $\mathscr{F}$. The ith column of $\mathbf{A}$ is the representation of $\mathbf{y}_i$ with respect to the basis $\{\mathbf{u}_1, \mathbf{u}_2, \ldots, \mathbf{u}_m\}$.

Proof

Let $\mathbf{x}$ be an arbitrary vector in $\mathscr{X}$. Since $\mathbf{x}_1, \mathbf{x}_2, \ldots, \mathbf{x}_n$ are linearly independent, the set of vectors qualifies as a basis. Consequently, the vector $\mathbf{x}$ can be expressed uniquely as $\alpha_1\mathbf{x}_1 + \alpha_2\mathbf{x}_2 + \cdots + \alpha_n\mathbf{x}_n$ (Theorem 2-1). By the linearity of L, we have

$$\begin{aligned} L\mathbf{x} &= \alpha_1 L\mathbf{x}_1 + \alpha_2 L\mathbf{x}_2 + \cdots + \alpha_n L\mathbf{x}_n \\ &= \alpha_1 \mathbf{y}_1 + \alpha_2 \mathbf{y}_2 + \cdots + \alpha_n \mathbf{y}_n \end{aligned}$$

which implies that for any $\mathbf{x}$ in $\mathscr{X}$, $L\mathbf{x}$ is uniquely determined by $\mathbf{y}_i = L\mathbf{x}_i$, for $i = 1, 2, \ldots, n$. This proves the first part of the theorem.

Let the representation of $\mathbf{y}_i$ with respect to $\{\mathbf{u}_1, \mathbf{u}_2, \ldots, \mathbf{u}_m\}$ be $[a_{1i} \quad a_{2i} \quad \cdots \quad a_{mi}]'$; that is,

$$\mathbf{y}_i = [\mathbf{u}_1 \quad \mathbf{u}_2 \quad \cdots \quad \mathbf{u}_m] \begin{bmatrix} a_{1i} \\ a_{2i} \\ \vdots \\ a_{mi} \end{bmatrix} \qquad i = 1, 2, \ldots, n \tag{2-26}$$

where the a_{ij}'s are elements of $\mathscr{F}$. Let us write, as in (2-17) and (2-18),

$$\begin{aligned} L[\mathbf{x}_1 \quad \mathbf{x}_2 \quad \cdots \quad \mathbf{x}_n] &= [\mathbf{y}_1 \quad \mathbf{y}_2 \quad \cdots \quad \mathbf{y}_n] \\ &= [\mathbf{u}_1 \quad \mathbf{u}_2 \quad \cdots \quad \mathbf{u}_m] \begin{bmatrix} a_{11} & a_{12} & \cdots & a_{1n} \\ a_{21} & a_{22} & \cdots & a_{2n} \\ \vdots & \vdots & & \vdots \\ a_{m1} & a_{m2} & \cdots & a_{mn} \end{bmatrix} \\ &\triangleq [\mathbf{u}_1 \quad \mathbf{u}_2 \quad \cdots \quad \mathbf{u}_m]\mathbf{A} \end{aligned} \tag{2-27}$$

Note that the elements of **A** are in the field $\mathscr{F}$ and the ith column of **A** is the representation of $\mathbf{y}_i$ with respect to the basis of $\mathscr{Y}$. With respect to the basis $\{\mathbf{x}_1, \mathbf{x}_2, \ldots, \mathbf{x}_n\}$ of $(\mathscr{X}, \mathscr{F})$ and the basis $\{\mathbf{u}_1, \mathbf{u}_2, \ldots, \mathbf{u}_m\}$ of $(\mathscr{Y}, \mathscr{F})$, the linear operator $\mathbf{y} = L\mathbf{x}$ can be written as

$$[\mathbf{u}_1 \quad \mathbf{u}_2 \quad \cdots \quad \mathbf{u}_m]\boldsymbol{\beta} = L[\mathbf{x}_1 \quad \mathbf{x}_2 \quad \cdots \quad \mathbf{x}_n]\boldsymbol{\alpha} \tag{2-28}$$

where $\boldsymbol{\beta} \triangleq [\beta_1 \quad \beta_2 \quad \cdots \quad \beta_m]'$ and $\boldsymbol{\alpha} \triangleq [\alpha_1 \quad \alpha_2 \quad \cdots \quad \alpha_n]'$ are the representations of **y** and **x**, respectively. After the bases are chosen, there are no differences between specifying **x**, **y**, and $\boldsymbol{\alpha}$, $\boldsymbol{\beta}$; hence in studying $\mathbf{y} = L\mathbf{x}$, we may just study the relationship between $\boldsymbol{\beta}$ and $\boldsymbol{\alpha}$. By substituting (2-27) into (2-28), we obtain

$$[\mathbf{u}_1 \quad \mathbf{u}_2 \quad \cdots \quad \mathbf{u}_m]\boldsymbol{\beta} = [\mathbf{u}_1 \quad \mathbf{u}_2 \quad \cdots \quad \mathbf{u}_m]\mathbf{A}\boldsymbol{\alpha} \tag{2-29}$$

which, together with the uniqueness of a representation, implies that

$$\boldsymbol{\beta} = \mathbf{A}\boldsymbol{\alpha} \tag{2-30}$$

Hence we conclude that if the bases of $(\mathscr{X}, \mathscr{F})$ and $(\mathscr{Y}, \mathscr{F})$ are chosen, the operator can be represented by a matrix with coefficients in $\mathscr{F}$. Q.E.D.

We see from (2-30) that the matrix **A** gives the relation between the representations $\boldsymbol{\alpha}$ and $\boldsymbol{\beta}$, not the vectors **x** and **y**. We also see that **A** depends on the basis chosen. Hence, for different bases, we have different representations of the same operator.

We study in the following an important subclass of linear operators that maps a linear space $(\mathscr{X}, \mathscr{F})$ into itself; that is, $L:(\mathscr{X}, \mathscr{F}) \to (\mathscr{X}, \mathscr{F})$. In this case, the same basis is always used for these two linear spaces. If a basis of $\mathscr{X}$, say $\{\mathbf{e}_1\ \mathbf{e}_2, \ldots, \mathbf{e}_n\}$, is chosen, then a matrix representation **A** of the linear operator L can be obtained by using Theorem 2-2. For a different basis $\{\bar{\mathbf{e}}_1, \bar{\mathbf{e}}_2, \ldots, \bar{\mathbf{e}}_n\}$, we shall obtain a different representation $\bar{\mathbf{A}}$ of the same operator L. We shall now establish the relationship between **A** and $\bar{\mathbf{A}}$. Consider Figure 2-4; **x** is an arbitrary vector in $\mathscr{X}$; $\boldsymbol{\alpha}$ and $\bar{\boldsymbol{\alpha}}$ are the representations of **x** with respect to the basis $\{\mathbf{e}_1, \mathbf{e}_2, \ldots, \mathbf{e}_n\}$ and the basis $\{\bar{\mathbf{e}}_1, \bar{\mathbf{e}}_2, \ldots, \bar{\mathbf{e}}_n\}$, respectively. Since the vector $\mathbf{y} = L\mathbf{x}$ is in the same space, its representations with respect to the bases chosen, say $\boldsymbol{\beta}$ and $\bar{\boldsymbol{\beta}}$, can also be found. The matrix representations **A** and $\bar{\mathbf{A}}$ can be computed by using Theorem 2-2. The relationships

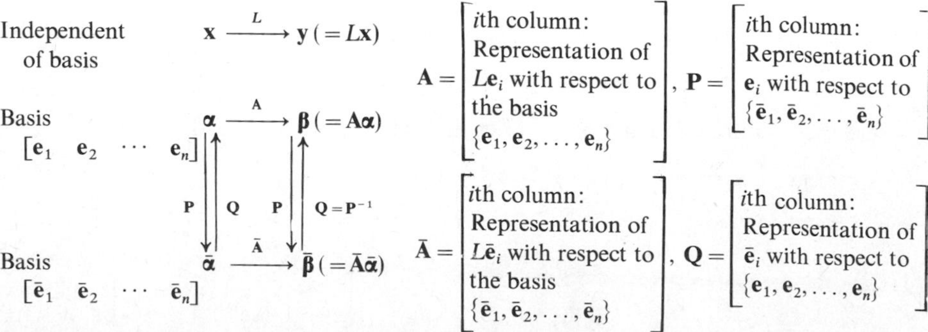

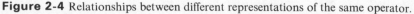

Figure 2-4 Relationships between different representations of the same operator.

between $\boldsymbol{\alpha}$ and $\bar{\boldsymbol{\alpha}}$ and between $\boldsymbol{\beta}$ and $\bar{\boldsymbol{\beta}}$ have been established in (2-20); they are related by $\bar{\boldsymbol{\alpha}} = \mathbf{P}\boldsymbol{\alpha}$ and $\bar{\boldsymbol{\beta}} = \mathbf{P}\boldsymbol{\beta}$, where $\mathbf{P}$ is a nonsingular matrix with coefficients in the field $\mathscr{F}$ and the ith column of $\mathbf{P}$ is the representation of $\mathbf{e}_i$ with respect to the basis $\{\bar{\mathbf{e}}_1, \bar{\mathbf{e}}_2, \ldots, \bar{\mathbf{e}}_n\}$. From Figure 2-4, we have

$$\bar{\boldsymbol{\beta}} = \bar{\mathbf{A}}\bar{\boldsymbol{\alpha}} \qquad \text{and} \qquad \bar{\boldsymbol{\beta}} = \mathbf{P}\boldsymbol{\beta} = \mathbf{P}\mathbf{A}\boldsymbol{\alpha} = \mathbf{P}\mathbf{A}\mathbf{P}^{-1}\bar{\boldsymbol{\alpha}}$$

Hence, by the uniqueness of a representation with respect to a specific basis, we have $\bar{\mathbf{A}}\bar{\boldsymbol{\alpha}} = \mathbf{P}\mathbf{A}\mathbf{P}^{-1}\bar{\boldsymbol{\alpha}}$. Since the relation holds for any $\bar{\boldsymbol{\alpha}}$, we conclude that

$$\bar{\mathbf{A}} = \mathbf{P}\mathbf{A}\mathbf{P}^{-1} = \mathbf{Q}^{-1}\mathbf{A}\mathbf{Q} \tag{2-31a}$$

or

$$\mathbf{A} = \mathbf{P}^{-1}\bar{\mathbf{A}}\mathbf{P} = \mathbf{Q}\bar{\mathbf{A}}\mathbf{Q}^{-1} \tag{2-31b}$$

where $\mathbf{Q} \triangleq \mathbf{P}^{-1}$.

Two matrices $\mathbf{A}$ and $\bar{\mathbf{A}}$ are said to be *similar* if there exists a nonsingular matrix $\mathbf{P}$ satisfying (2-31). The transformation defined in (2-31) is called a *similarity transformation.* Clearly, *all the matrix representations (with respect to different bases) of the same operator are similar.*

Example 3

Consider the linear operator L of Example 1 shown in Figure 2-3. If we choose $\{\mathbf{x}_1, \mathbf{x}_2\}$ as a basis, then

$$\mathbf{y}_1 = L\mathbf{x}_1 = [\mathbf{x}_1 \quad \mathbf{x}_2]\begin{bmatrix} 0 \\ 1 \end{bmatrix} \qquad \text{and} \qquad \mathbf{y}_2 = L\mathbf{x}_2 = [\mathbf{x}_1 \quad \mathbf{x}_2]\begin{bmatrix} -1 \\ 0 \end{bmatrix}$$

Hence the representation of L with respect to the basis $\{\mathbf{x}_1, \mathbf{x}_2\}$ is

$$\begin{bmatrix} 0 & -1 \\ 1 & 0 \end{bmatrix}$$

The representation of $\mathbf{x}_3$ is

$$\begin{bmatrix} 1.5 \\ 0.5 \end{bmatrix}$$

It is easy to verify that the representation of $\mathbf{y}_3$ with respect to $\{\mathbf{x}_1, \mathbf{x}_2\}$ is equal to

$$\begin{bmatrix} 0 & -1 \\ 1 & 0 \end{bmatrix}\begin{bmatrix} 1.5 \\ 0.5 \end{bmatrix} = \begin{bmatrix} -0.5 \\ 1.5 \end{bmatrix}$$

or

$$\mathbf{y}_3 = [\mathbf{x}_1 \quad \mathbf{x}_2]\begin{bmatrix} -0.5 \\ 1.5 \end{bmatrix}$$

If, instead of $\{\mathbf{x}_1, \mathbf{x}_2\}$, we choose $\{\mathbf{x}_1, \mathbf{x}_3\}$ as a basis, then from Figure 2-3,

$$\mathbf{y}_1 = L\mathbf{x}_1 = [\mathbf{x}_1 \quad \mathbf{x}_3]\begin{bmatrix} -3 \\ 2 \end{bmatrix} \qquad \text{and} \qquad \mathbf{y}_3 = L\mathbf{x}_3 = [\mathbf{x}_1 \quad \mathbf{x}_3]\begin{bmatrix} -5 \\ 3 \end{bmatrix}$$

Hence the representation of L with respect to the basis $\{\mathbf{x}_1, \mathbf{x}_3\}$ is

$$\begin{bmatrix} -3 & -5 \\ 2 & 3 \end{bmatrix}$$

The reader is advised to find the **P** matrix for this example and verify $\bar{\mathbf{A}} = \mathbf{PAP}^{-1}$. ∎

In matrix theory, a matrix is introduced as an array of numbers. With the concepts of linear operator and representation, we shall now give a new interpretation of a matrix. Given an $n \times n$ matrix **A** with coefficients in a field $\mathscr{F}$, if it is not specified to be a representation of some operator, we shall consider it as a linear operator that maps $(\mathscr{F}^n, \mathscr{F})$ into itself.[7] The matrix **A** is independent of the basis chosen for $(\mathscr{F}^n, \mathscr{F})$. However, if the set of the vectors $\mathbf{n}_1, \mathbf{n}_2, \ldots, \mathbf{n}_n$ in Equation (2-13) is chosen as a basis of $(\mathscr{F}^n, \mathscr{F})$, then the representation of the linear operator **A** is identical to the linear operator **A** (a matrix) itself. This can be checked by using the fact that the ith column of the representation is equal to the representation of $\mathbf{An}_i$ with respect to the basis $\{\mathbf{n}_1, \mathbf{n}_2, \ldots, \mathbf{n}_n\}$. If $\mathbf{a}_i$ is the ith column of **A**, then $\mathbf{An}_i = \mathbf{a}_i$. Now the representation of $\mathbf{a}_i$ with respect to the basis (2-13) is identical to itself. Therefore we conclude that the representation of a matrix (a linear operator) with respect to the basis (2-13) is identical to itself. For a matrix (an operator), Figure 2-4 can be modified as in Figure 2-5. The equation $\mathbf{Q} = [\mathbf{q}_1 \quad \mathbf{q}_2 \quad \cdots \quad \mathbf{q}_n]$ follows from the fact that the ith column of **Q** is the representation of $\mathbf{q}_i$ with respect to the basis $\{\mathbf{n}_1, \mathbf{n}_2, \ldots, \mathbf{n}_n\}$. If a basis $\{\mathbf{q}_1, \mathbf{q}_2, \ldots, \mathbf{q}_n\}$ is chosen for $(\mathscr{F}^n, \mathscr{F})$, a matrix **A** has a representation $\bar{\mathbf{A}}$. From Figure 2-5, we see that the matrix representation $\bar{\mathbf{A}}$ may be computed either from Theorem 2-2 or from a similarity transformation. In most of the problems encountered in this book, it is always much easier to compute $\bar{\mathbf{A}}$ from Theorem 2-2 than from using a similarity transformation.

Example 4

Consider the following matrix with coefficients in $\mathbb{R}$:

$$L = \mathbf{A} = \begin{bmatrix} 3 & 2 & -1 \\ -2 & 1 & 0 \\ 4 & 3 & 1 \end{bmatrix}$$

Let

$$\mathbf{b} = \begin{bmatrix} 0 \\ 0 \\ 1 \end{bmatrix}$$

[7] This interpretation can be extended to nonsquare matrices.

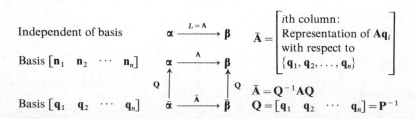

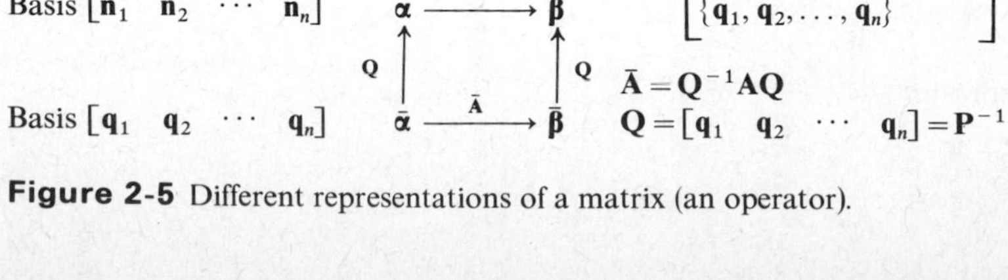

Independent of basis: $\boldsymbol{\alpha} \xrightarrow{L=\mathbf{A}} \boldsymbol{\beta}$

Basis $[\mathbf{n}_1 \quad \mathbf{n}_2 \quad \cdots \quad \mathbf{n}_n]$: $\boldsymbol{\alpha} \xrightarrow{\mathbf{A}} \boldsymbol{\beta}$

$\mathbf{Q} \uparrow \qquad \uparrow \mathbf{Q}$

Basis $[\mathbf{q}_1 \quad \mathbf{q}_2 \quad \cdots \quad \mathbf{q}_n]$: $\bar{\boldsymbol{\alpha}} \xrightarrow{\bar{\mathbf{A}}} \bar{\boldsymbol{\beta}}$

$\bar{\mathbf{A}} = \begin{bmatrix} i\text{th column:} \\ \text{Representation of } \mathbf{Aq}_i \\ \text{with respect to} \\ \{\mathbf{q}_1, \mathbf{q}_2, \ldots, \mathbf{q}_n\} \end{bmatrix}$

$\bar{\mathbf{A}} = \mathbf{Q}^{-1}\mathbf{AQ}$

$\mathbf{Q} = [\mathbf{q}_1 \quad \mathbf{q}_2 \quad \cdots \quad \mathbf{q}_n] = \mathbf{P}^{-1}$

Figure 2-5 Different representations of a matrix (an operator).

Then[8] $\mathbf{Ab}=\begin{bmatrix}-1\\0\\1\end{bmatrix}$ $\mathbf{A}^2\mathbf{b}=\begin{bmatrix}-4\\2\\-3\end{bmatrix}$ $\mathbf{A}^3\mathbf{b}=\begin{bmatrix}-5\\10\\-13\end{bmatrix}$

It can be shown that the following relation holds (check!):

$$\mathbf{A}^3\mathbf{b}=5\mathbf{A}^2\mathbf{b}-15\mathbf{Ab}+17\mathbf{b} \tag{2-32}$$

Since the set of vectors $\mathbf{b}$, $\mathbf{Ab}$, and $\mathbf{A}^2\mathbf{b}$ are linearly independent, it qualifies as a basis. We compute now the representation of $\mathbf{A}$ with respect to this basis. It is clear that

$$\mathbf{A}(\mathbf{b})=[\mathbf{b}\quad \mathbf{Ab}\quad \mathbf{A}^2\mathbf{b}]\begin{bmatrix}0\\1\\0\end{bmatrix}$$

$$\mathbf{A}(\mathbf{Ab})=[\mathbf{b}\quad \mathbf{Ab}\quad \mathbf{A}^2\mathbf{b}]\begin{bmatrix}0\\0\\1\end{bmatrix}$$

and

$$\mathbf{A}(\mathbf{A}^2\mathbf{b})=[\mathbf{b}\quad \mathbf{Ab}\quad \mathbf{A}^2\mathbf{b}]\begin{bmatrix}17\\-15\\5\end{bmatrix}$$

The last equation is obtained from (2-32). Hence the representation of $\mathbf{A}$ with respect to the basis $\{\mathbf{b}, \mathbf{Ab}, \mathbf{A}^2\mathbf{b}\}$ is

$$\bar{\mathbf{A}}=\left[\begin{array}{c:c:c}0 & 0 & 17\\1 & 0 & -15\\0 & 1 & 5\end{array}\right]$$

The matrix $\bar{\mathbf{A}}$ can also be obtained from $\mathbf{Q}^{-1}\mathbf{AQ}$, but it requires an inversion of a matrix and n^3 multiplications. However, we may use $\bar{\mathbf{A}}=\mathbf{Q}^{-1}\mathbf{AQ}$, or more easily, $\mathbf{Q}\bar{\mathbf{A}}=\mathbf{AQ}$, to check our result. The reader is asked to verify

$$\begin{bmatrix}0 & -1 & -4\\0 & 0 & 2\\1 & 1 & -3\end{bmatrix}\begin{bmatrix}0 & 0 & 17\\1 & 0 & -15\\0 & 1 & 5\end{bmatrix}=\begin{bmatrix}3 & 2 & -1\\-2 & 1 & 0\\4 & 3 & 1\end{bmatrix}\begin{bmatrix}0 & -1 & -4\\0 & 0 & 2\\1 & 1 & -3\end{bmatrix}$$

Example 5

We extend Example 4 to the general case. Let $\mathbf{A}$ be an $n\times n$ square matrix with real coefficients. If there exists a real vector $\mathbf{b}$ such that the vectors $\mathbf{b}$, $\mathbf{Ab}, \ldots, \mathbf{A}^{n-1}\mathbf{b}$ are linearly independent and if $\mathbf{A}^n\mathbf{b}=-\alpha_n\mathbf{b}-\alpha_{n-1}\mathbf{Ab}-\cdots-\alpha_1\mathbf{A}^{n-1}\mathbf{b}$ (see Section 2-7), then the representation of $\mathbf{A}$ with respect to the basis $\{\mathbf{b}, \mathbf{Ab}, \ldots, \mathbf{A}^{n-1}\mathbf{b}\}$ is

$$\bar{\mathbf{A}}=\begin{bmatrix}0 & 0 & \cdots & 0 & -\alpha_n\\1 & 0 & \cdots & 0 & -\alpha_{n-1}\\0 & 1 & \cdots & 0 & -\alpha_{n-2}\\\vdots & \vdots & & \vdots & \vdots\\0 & 0 & \cdots & 0 & -\alpha_2\\0 & 0 & \cdots & 1 & -\alpha_1\end{bmatrix} \tag{2-33}$$

∎

[8] $\mathbf{A}^2 \triangleq \mathbf{AA}$, $\mathbf{A}^3 \triangleq \mathbf{AAA}$.

A matrix of the form shown in (2-33) or its transpose is said to be in the *companion form*. See Problem 2-26. This form will constantly arise in this text.

As an aid in memorizing Figure 2-5, we write $\bar{\mathbf{A}} = \mathbf{Q}^{-1}\mathbf{A}\mathbf{Q}$ as

$$\mathbf{Q}\bar{\mathbf{A}} = \mathbf{A}\mathbf{Q}$$

Since $\mathbf{Q} = [\mathbf{q}_1 \quad \mathbf{q}_2 \quad \cdots \quad \mathbf{q}_n]$, it can be further written as

$$[\mathbf{q}_1 \quad \mathbf{q}_2 \quad \cdots \quad \mathbf{q}_n]\bar{\mathbf{A}} = [\mathbf{A}\mathbf{q}_1 \quad \mathbf{A}\mathbf{q}_2 \quad \cdots \quad \mathbf{A}\mathbf{q}_n] \tag{2-34}$$

From (2-34), we see that the ith column of $\bar{\mathbf{A}}$ is indeed the representation of $\mathbf{A}\mathbf{q}_i$ with respect to the basis $\{\mathbf{q}_1, \mathbf{q}_2, \ldots, \mathbf{q}_n\}$.

We pose the following question to conclude this section: Since a linear operator has many representations, is it possible to choose one set of basis vectors such that the representation is nice and simple? The answer is affirmative. In order to give a solution, we must first study linear algebraic equations.

2-5 Systems of Linear Algebraic Equations

Consider the set of linear equations:

$$\begin{array}{c} a_{11}x_1 + a_{12}x_2 + \cdots + a_{1n}x_n = y_1 \\ a_{21}x_1 + a_{22}x_2 + \cdots + a_{2n}x_n = y_2 \\ \cdots\cdots\cdots\cdots\cdots\cdots\cdots\cdots\cdots\cdots \\ a_{m1}x_1 + a_{m2}x_2 + \cdots + a_{mn}x_n = y_m \end{array} \tag{2-35}$$

where the given a_{ij}'s and y_i's are assumed to be elements of a field $\mathscr{F}$ and the unknown x_i's are also required to be in the same field $\mathscr{F}$. This set of equations can be written in matrix form as

$$\mathbf{A}\mathbf{x} = \mathbf{y} \tag{2-36}$$

where
$$\mathbf{A} \triangleq \begin{bmatrix} a_{11} & a_{12} & \cdots & a_{1n} \\ a_{21} & a_{22} & \cdots & a_{2n} \\ \vdots & \vdots & & \vdots \\ a_{m1} & a_{m2} & \cdots & a_{mn} \end{bmatrix} \qquad \mathbf{x} \triangleq \begin{bmatrix} x_1 \\ x_2 \\ \vdots \\ x_n \end{bmatrix} \qquad \mathbf{y} \triangleq \begin{bmatrix} y_1 \\ y_2 \\ \vdots \\ y_m \end{bmatrix}$$

Clearly, $\mathbf{A}$ is an $m \times n$ matrix, $\mathbf{x}$ is an $n \times 1$ vector, and $\mathbf{y}$ is an $m \times 1$ vector. No restriction is made on the integer m; it may be larger than, equal to, or smaller than the integer n. Two questions can be raised in regard to this set of equations: first, the existence of a solution and, second, the number of solutions. More specifically, suppose the matrix $\mathbf{A}$ and the vector $\mathbf{y}$ in Equation (2-36) are given; the first question is concerned with the condition on $\mathbf{A}$ and $\mathbf{y}$ under which at least one vector $\mathbf{x}$ exists such that $\mathbf{A}\mathbf{x} = \mathbf{y}$. If solutions exist, then the second question is concerned with the number of linearly independent vectors $\mathbf{x}$ such that $\mathbf{A}\mathbf{x} = \mathbf{y}$. In order to answer these questions, the rank and the nullity of the matrix $\mathbf{A}$ have to be introduced.

We have agreed in the previous section to consider the matrix $\mathbf{A}$ as a linear operator which maps $(\mathscr{F}^n, \mathscr{F})$ into $(\mathscr{F}^m, \mathscr{F})$. Recall that the linear space $(\mathscr{F}^n, \mathscr{F})$ that undergoes transformation is called the domain of **A**.

Definition 2-9

The *range* of a linear operator **A** is the set $\mathscr{R}(\mathbf{A})$ defined by

$$\mathscr{R}(\mathbf{A}) = \{\text{all the elements } \mathbf{y} \text{ of } (\mathscr{F}^m, \mathscr{F}) \text{ for which there exists at least one vector } \mathbf{x} \text{ in } (\mathscr{F}^n, \mathscr{F}) \text{ such that } \mathbf{y} = \mathbf{A}\mathbf{x}\}$$ ∎

Theorem 2-3

The range of a linear operator **A** is a subspace of $(\mathscr{F}^m, \mathscr{F})$.

Proof

If $\mathbf{y}_1$ and $\mathbf{y}_2$ are elements of $\mathscr{R}(\mathbf{A})$, then by definition there exist vectors $\mathbf{x}_1$ and $\mathbf{x}_2$ in $(\mathscr{F}^n, \mathscr{F})$ such that $\mathbf{y}_1 = \mathbf{A}\mathbf{x}_1$, $\mathbf{y}_2 = \mathbf{A}\mathbf{x}_2$. We claim that for any α_1 and α_2 in $\mathscr{F}$, the vector $\alpha_1\mathbf{y}_1 + \alpha_2\mathbf{y}_2$ is also an element of $\mathscr{R}(\mathbf{A})$. Indeed, by the linearity of **A**, it is easy to show that $\alpha_1\mathbf{y}_1 + \alpha_2\mathbf{y}_2 = \mathbf{A}(\alpha_1\mathbf{x}_1 + \alpha_2\mathbf{x}_2)$, and thus the vector $\alpha_1\mathbf{x}_1 + \alpha_2\mathbf{x}_2$ is an element of $(\mathscr{F}^n, \mathscr{F})$. Hence the range $\mathscr{R}(\mathbf{A})$ is a subspace of $(\mathscr{F}^m, \mathscr{F})$ (see the remark following Definition 2-3). Q.E.D.

Let the ith column of **A** be denoted by $\mathbf{a}_i$; that is, $\mathbf{A} = [\mathbf{a}_1 \quad \mathbf{a}_2 \quad \cdots \quad \mathbf{a}_n]$, then the matrix equation (2-36) can be written as

$$\mathbf{y} = x_1\mathbf{a}_1 + x_2\mathbf{a}_2 + \cdots + x_n\mathbf{a}_n \tag{2-37}$$

where x_i, for $i = 1, 2, \ldots, n$, are components of **x** and are elements of $\mathscr{F}$. The range space $\mathscr{R}(\mathbf{A})$ is, by definition, the set of **y** such that $\mathbf{y} = \mathbf{A}\mathbf{x}$ for some **x** in $(\mathscr{F}^n, \mathscr{F})$. It is the same as saying that $\mathscr{R}(\mathbf{A})$ is the set of **y** with $x_1, x_2, \ldots, x_n$ in (2-37) ranging through all the possible values of $\mathscr{F}$. Therefore we conclude that $\mathscr{R}(\mathbf{A})$ *is the set of all the possible linear combinations of the columns of* **A**. Since $\mathscr{R}(\mathbf{A})$ is a linear space, its dimension is defined and is equal to the maximum number of linearly independent vectors in $\mathscr{R}(\mathbf{A})$. Hence, *the dimension of* $\mathscr{R}(\mathbf{A})$ *is the maximum number of linearly independent columns in* **A**.

Definition 2-10

The *rank* of a matrix **A**, denoted by $\rho(\mathbf{A})$, is the maximum number of linearly independent columns in **A**, or equivalently, the dimension of the range space of **A**. ∎

Example 1

Consider the matrix

$$\mathbf{A} = \begin{bmatrix} 0 & 1 & 1 & 2 \\ 1 & 2 & 3 & 4 \\ 2 & 0 & 2 & 0 \end{bmatrix}$$

The range space of **A** is all the possible linear combinations of all the columns of **A**, or correspondingly, all the possible linear combinations of the first two columns of **A**, because the third and the fourth columns of **A** are linearly dependent on the first two columns. Hence the rank of **A** is 2. ∎

The rank of a matrix can be computed by using a sequence of elementary transformations (see Appendix A). This is based on the property that the rank of a matrix remains unchanged after pre- or postmultiplications of elementary matrices (Theorem 2-7). Once a matrix is transformed into the upper triangular form as shown in (A-6), then the rank is equal to the number of nonzero rows in (A-6). From the form, it is also easy to verify that the number of linear independent columns of a matrix is equal to the number of independent rows. Consequently, if **A** is an $n \times m$ matrix, then

$$\begin{aligned}\text{rank } \mathbf{A} &= \text{no. of linear independent columns} \\ &= \text{no. of linear independent rows} \\ &\leq \min(n, m)\end{aligned} \tag{2-38}$$

The computation of the rank of a matrix on digital computers, however, is not a simple problem. Because of limited accuracies on digital computers, rounding errors always arise on numerical computations. Suppose a matrix, after transformation, becomes

$$\begin{bmatrix} 1 & 2 \\ 0 & \varepsilon \end{bmatrix}$$

where ε is a very small number, say, 10^{-10}. This ε may arise from the given data (assuming no rounding errors) or from rounding errors. If ε arises from rounding errors, we should consider ε as zero, and the matrix has rank 1. If ε is due to the given data, we cannot consider it as a zero, and the matrix has rank 2. To determine what value is small enough to be considered as a zero is a complicated problem in computer computations. For problems encountered on matrix computation, the reader is referred to References S181, S182, S200, and S212.

In matrix theory, the rank of a matrix is defined as the largest order of all nonvanishing minors of **A**. In other words, the matrix **A** has rank k if and only if there is at least one minor of order k in **A** that does not vanish and every minor of order higher than k vanishes. This definition and Definition 2-10 are, in fact, equivalent; the proof can be found, for example, in Reference 43. A consequence is that a square matrix has full rank if and only if the determinant of the matrix is different from zero; or correspondingly, *a matrix is nonsingular if and only if all the rows and columns of the matrix are linearly independent.*

With the concepts of range space and rank, we are ready to study the existence problem of the solutions of $\mathbf{Ax} = \mathbf{y}$.

Theorem 2-4

Consider the matrix equation $\mathbf{Ax} = \mathbf{y}$, where the $m \times n$ matrix **A** maps $(\mathscr{F}^n, \mathscr{F})$ into $(\mathscr{F}^m, \mathscr{F})$.

1. Given $\mathbf{A}$ and given a vector $\mathbf{y}$ in $(\mathscr{F}^m, \mathscr{F})$, there exists a vector $\mathbf{x}$ such that $\mathbf{Ax} = \mathbf{y}$ if and only if the vector $\mathbf{y}$ is an element of $\mathscr{R}(\mathbf{A})$, or equivalently,

$$\rho(\mathbf{A}) = \rho([\mathbf{A} \vdots \mathbf{y}])$$

2. Given $\mathbf{A}$, for every $\mathbf{y}$ in $(\mathscr{F}^m, \mathscr{F})$, there exists a vector $\mathbf{x}$ such that $\mathbf{Ax} = \mathbf{y}$ if and only if $\mathscr{R}(\mathbf{A}) = (\mathscr{F}^m, \mathscr{F})$, or equivalently, $\rho(\mathbf{A}) = m$.

Proof

1. It follows immediately from the definition of the range space of $\mathbf{A}$. If the vector $\mathbf{y}$ is not an element of $\mathscr{R}(\mathbf{A})$, the equation $\mathbf{Ax} = \mathbf{y}$ has no solution and is said to be *inconsistent*.
2. The rank of $\mathbf{A}$, $\rho(\mathbf{A})$, is by definition the dimension of $\mathscr{R}(\mathbf{A})$. Since $\mathscr{R}(\mathbf{A})$ is a subspace of $(\mathscr{F}^m, \mathscr{F})$, if $\rho(\mathbf{A}) = m$, then $\mathscr{R}(\mathbf{A}) = (\mathscr{F}^m, \mathscr{F})$. If $\mathscr{R}(\mathbf{A}) = (\mathscr{F}^m, \mathscr{F})$, then for any $\mathbf{y}$ in $(\mathscr{F}^m, \mathscr{F})$, there exists a vector $\mathbf{x}$ such that $\mathbf{Ax} = \mathbf{y}$. If $\rho(\mathbf{A}) < m$, there exists at least one nonzero vector $\mathbf{y}$ in $(\mathscr{F}^m, \mathscr{F})$, but not in $\mathscr{R}(\mathbf{A})$, for which there exists no $\mathbf{x}$ such that $\mathbf{Ax} = \mathbf{y}$. Q.E.D.

The design of compensators to achieve various design objectives can be reduced to the solution of linear algebraic equations. Hence this theorem is very important in our application. If $\mathbf{A}$ is an $m \times n$ matrix and if $\rho(\mathbf{A}) = m$, then all rows of $\mathbf{A}$ are linearly independent and $\mathbf{A}$ is said to have a *full row rank*. If $\mathbf{A}$ has a full row rank, no matter what column $\mathbf{y}$ is appended to $\mathbf{A}$, the rank of $[\mathbf{A} \vdots \mathbf{y}]$ cannot increase and is equal to $\rho(\mathbf{A})$. In other words, $\mathbf{y}$ lies in the space spanned by the columns of $\mathbf{A}$ and, consequently, can be written as a linear combination of the columns of $\mathbf{A}$. Hence if $\mathbf{A}$ has a full row rank, for any $\mathbf{y}$, there exists a $\mathbf{x}$ such that $\mathbf{Ax} = \mathbf{y}$. Similarly, if $\rho(\mathbf{A}) = n$, the $m \times n$ matrix $\mathbf{A}$ is said to have a *full column rank*. If $\mathbf{A}$ has a full column rank, then every $1 \times n$ vector $\mathbf{a}$ will lie in the space spanned by the rows of $\mathbf{A}$ and, consequently, can be written as a linear combination of the rows of $\mathbf{A}$. In other words, there exists a $1 \times m$ vector $\mathbf{b}$ such that $\mathbf{bA} = \mathbf{a}$.

After we find out that a linear equation has at least one solution, it is natural to ask how many solutions it may have. Instead of studying the general case, we discuss only the homogeneous linear equation $\mathbf{Ax} = \mathbf{0}$.

Definition 2-11

The *null space* of a linear operator $\mathbf{A}$ is the set $\mathscr{N}(\mathbf{A})$ defined by

$$\mathscr{N}(\mathbf{A}) = \{\text{all the elements } \mathbf{x} \text{ of } (\mathscr{F}^n, \mathscr{F}) \text{ for which } \mathbf{Ax} = \mathbf{0}\}$$

The dimension of $\mathscr{N}(\mathbf{A})$ is called the *nullity* of $\mathbf{A}$ and is denoted by $\nu(\mathbf{A})$. ∎

In other words, the null space $\mathscr{N}(\mathbf{A})$ is the set of all solutions of $\mathbf{Ax} = \mathbf{0}$.[9] It is easy to show that $\mathscr{N}(\mathbf{A})$ is indeed a linear space. If the dimension of $\mathscr{N}(\mathbf{A})$, $\nu(\mathbf{A})$, is 0, then $\mathscr{N}(\mathbf{A})$ consists of only the zero vector, and the only

[9] It is also called the *right* null space of $\mathbf{A}$. The set of all $\mathbf{y}$ satisfying $\mathbf{yA} = \mathbf{0}$ will be called the left null space of $\mathbf{A}$. See Problem 2-51.

solution of $\mathbf{Ax}=\mathbf{0}$ is $\mathbf{x}=\mathbf{0}$. If $\nu(\mathbf{A})=k$, then the equation $\mathbf{Ax}=\mathbf{0}$ has k linearly independent vector solutions.

Note that the null space is a subspace of the domain $(\mathscr{F}^n, \mathscr{F})$, whereas the range space is a subspace of $(\mathscr{F}^m, \mathscr{F})$.

Example 2

Consider the matrix

$$\mathbf{A}=\begin{bmatrix} 0 & 1 & 1 & 2 & -1 \\ 1 & 2 & 3 & 4 & -1 \\ 2 & 0 & 2 & 0 & 2 \end{bmatrix}$$

which maps $(\mathbb{R}^5, \mathbb{R})$ into $(\mathbb{R}^3, \mathbb{R})$. It is easy to check that the last three columns of $\mathbf{A}$ are linearly dependent on the first two columns of $\mathbf{A}$. Hence the rank of $\mathbf{A}$, $\rho(\mathbf{A})$, is equal to 2. Let $\mathbf{x}=[x_1 \quad x_2 \quad x_3 \quad x_4 \quad x_5]'$. Then

$$\mathbf{Ax}=x_1\begin{bmatrix}0\\1\\2\end{bmatrix}+x_2\begin{bmatrix}1\\2\\0\end{bmatrix}+x_3\begin{bmatrix}1\\3\\2\end{bmatrix}+x_4\begin{bmatrix}2\\4\\0\end{bmatrix}+x_5\begin{bmatrix}-1\\-1\\2\end{bmatrix}$$

$$=(x_1+x_3+x_5)\begin{bmatrix}0\\1\\2\end{bmatrix}+(x_2+x_3+2x_4-x_5)\begin{bmatrix}1\\2\\0\end{bmatrix} \tag{2-39}$$

Since the vectors $[0 \quad 1 \quad 2]'$ and $[1 \quad 2 \quad 0]'$ are linearly independent, we conclude from (2-39) that a vector $\mathbf{x}$ satisfies $\mathbf{Ax}=\mathbf{0}$ if and only if

$$x_1+x_3+x_5=0$$
$$x_2+x_3+2x_4-x_5=0$$

Note that the number of equations is equal to the rank of $\mathbf{A}$, $\rho(\mathbf{A})$. The solution $\mathbf{x}$ of $\mathbf{Ax}=\mathbf{0}$ has five components but is governed by only two equations; hence three of the five components can be arbitrarily assigned. Let $x_3=1$, $x_4=0$, $x_5=0$; then $x_1=-1$ and $x_2=-1$. Let $x_3=0$, $x_4=1$, $x_5=0$; then $x_1=0$ and $x_2=-2$. Let $x_3=0$, $x_4=0$, $x_5=1$; then $x_1=-1$ and $x_2=1$. It is clear that the three vectors

$$\begin{bmatrix}-1\\-1\\1\\0\\0\end{bmatrix} \quad \begin{bmatrix}0\\-2\\0\\1\\0\end{bmatrix} \quad \begin{bmatrix}-1\\1\\0\\0\\1\end{bmatrix}$$

are linearly independent, and that every solution of $\mathbf{Ax}=\mathbf{0}$ must be a linear combination of these three vectors. Therefore the set of vectors form a basis of $\mathscr{N}(\mathbf{A})$ and $\nu(\mathbf{A})=3$. ∎

We see from this example that the number of equations that the vectors of $\mathscr{N}(\mathbf{A})$ should obey is equal to $\rho(\mathbf{A})$ and that there are n components in every vector of $\mathscr{N}(\mathbf{A})$. Therefore $n-\rho(\mathbf{A})$ components of the vectors of $\mathscr{N}(\mathbf{A})$ can be arbitrarily chosen. Consequently, there are $n-\rho(\mathbf{A})$ linearly independent vectors in $\mathcal{N}(\mathbf{A})$. Hence we conclude that $n \quad \rho(\mathbf{A})-\nu(\mathbf{A})$. We state this as a theorem; its formal proof can be found, for example, in References 43 and 86.

Theorem 2-5

Let $\mathbf{A}$ be an $m \times n$ matrix. Then

$$\rho(\mathbf{A})+\nu(\mathbf{A})=n$$

Corollary 2-5

The number of linearly independent vector solutions of $\mathbf{Ax}=\mathbf{0}$ is equal to $n-\rho(\mathbf{A})$, where n is the number of columns in $\mathbf{A}$, and $\rho(\mathbf{A})$ is the number of linearly independent columns in $\mathbf{A}$. ∎

This corollary follows directly from Theorem 2-5 and the definition of the null space of $\mathbf{A}$. It is clear that if $\rho(\mathbf{A})=n$, then the only solution of $\mathbf{Ax}=\mathbf{0}$ is $\mathbf{x}=\mathbf{0}$, which is called the *trivial solution*. If $\rho(\mathbf{A})<n$, then we can always find a nonzero vector $\mathbf{x}$ such that $\mathbf{Ax}=\mathbf{0}$. In particular, if $\mathbf{A}$ is a square matrix, then $\mathbf{Ax}=\mathbf{0}$ *has a nontrivial solution if and only if* $\rho(\mathbf{A})<n$, *or equivalently,* $\det(\mathbf{A})=0$, where det stands for the determinant.

We introduce three useful theorems to conclude this section.

Theorem 2-6 (Sylvester's inequality)

Let $\mathbf{A}$, $\mathbf{B}$ be $q \times n$ and $n \times p$ matrices with coefficients in the same field. Then

$$\rho(\mathbf{A})+\rho(\mathbf{B})-n \leq \rho(\mathbf{AB}) \leq \min(\rho(\mathbf{A}), \rho(\mathbf{B}))$$

Proof

The composite matrix $\mathbf{AB}$ can be considered as two linear transformations applied successively to $(\mathcal{F}^p, \mathcal{F})$ as shown in Figure 2-6. Since the domain of $\mathbf{AB}$ is $\mathcal{R}(\mathbf{B})$ and the range of $\mathbf{AB}$ is a subspace of $\mathcal{R}(\mathbf{A})$, we have immediately $\rho(\mathbf{AB}) \leq \min(\rho(\mathbf{A}), \rho(\mathbf{B}))$ by using (2-38). From Figure 2-6 we have $\rho(\mathbf{AB})=$

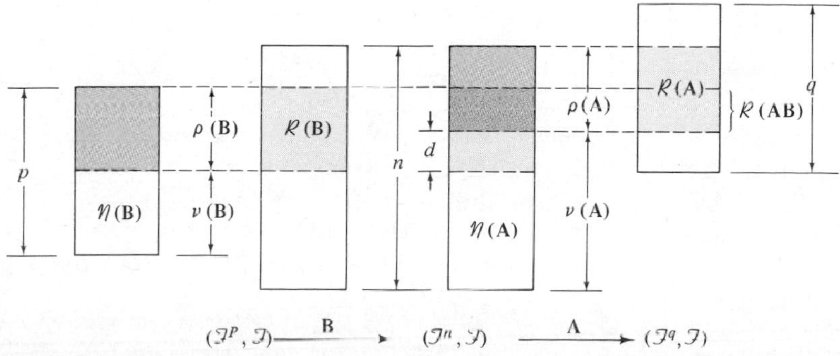

Figure 2-6 A composite transformation.

$\rho(\mathbf{B}) - d$, where d is the dimension of the intersection of $\mathscr{R}(\mathbf{B})$ and $\mathscr{N}(\mathbf{A})$.[10] The dimension of $\mathscr{N}(\mathbf{A})$ is $n - \rho(\mathbf{A})$; hence, $d \leq n - \rho(\mathbf{A})$. Consequently, $\rho(\mathbf{AB}) \geq \rho(\mathbf{B}) - n + \rho(\mathbf{A})$. Q.E.D.

If $\mathbf{B}$ is an $n \times n$ matrix and nonsingular, then

$$\rho(\mathbf{A}) + \rho(\mathbf{B}) - n = \rho(\mathbf{A}) \leq \rho(\mathbf{AB}) \leq \min(\rho(\mathbf{A}), n) = \rho(\mathbf{A})$$

Hence we have the following important theorem:

Theorem 2-7

Let $\mathbf{A}$ be an $m \times n$ matrix. Then

$$\rho(\mathbf{AC}) = \rho(\mathbf{A}) \qquad \text{and} \qquad \rho(\mathbf{DA}) = \rho(\mathbf{A})$$

for any $n \times n$ and $m \times m$ nonsingular matrices $\mathbf{C}$ and $\mathbf{D}$. ∎

In words, the rank of a matrix will not change after the pre- or postmultiplication of a nonsingular matrix. Because of this property, gaussian elimination or the row-searching algorithm discussed in Appendix A can be used to compute the rank of a matrix.

Theorem 2-8

Let $\mathbf{A}$ be an $m \times n$ matrix with coefficients in a field, and let $\mathbf{A}^*$ be the complex conjugate transpose of $\mathbf{A}$. Then

1. $\rho(\mathbf{A}) = n$ if and only if $\rho(\mathbf{A}^*\mathbf{A}) = n$, or equivalently,

$$\det(\mathbf{A}^*\mathbf{A}) \neq 0$$

2. $\rho(\mathbf{A}) = m$ if and only if $\rho(\mathbf{AA}^*) = m$, or equivalently,

$$\det(\mathbf{AA}^*) \neq 0$$ ∎

Note that $\mathbf{A}^*\mathbf{A}$ is an $n \times n$ matrix and $\mathbf{AA}^*$ is an $m \times m$ matrix. In order to have $\rho(\mathbf{A}) = n$, it is necessary to have $n \leq m$. This theorem will be proved by using Definition 2-4′. More specifically, we use the fact that if $\mathbf{A}\boldsymbol{\alpha} = \mathbf{0}$ implies $\boldsymbol{\alpha} = \mathbf{0}$, then all the columns of $\mathbf{A}$ are linearly independent, and $\rho(\mathbf{A}) = n$, where n is the number of the columns of $\mathbf{A}$.

Proof

1. *Sufficiency:* $\rho(\mathbf{A}^*\mathbf{A}) = n$ implies $\rho(\mathbf{A}) = n$. We show that $\mathbf{A}\boldsymbol{\alpha} = \mathbf{0}$ implies $\boldsymbol{\alpha} = \mathbf{0}$ under the assumption of $\rho(\mathbf{A}^*\mathbf{A}) = n$, where $\boldsymbol{\alpha}$ is an $n \times 1$ vector. If $\mathbf{A}\boldsymbol{\alpha} = \mathbf{0}$, then $\mathbf{A}^*\mathbf{A}\boldsymbol{\alpha} = \mathbf{0}$, which, with the assumption $\rho(\mathbf{A}^*\mathbf{A}) = n$, implies $\boldsymbol{\alpha} = \mathbf{0}$. Hence we conclude that $\rho(\mathbf{A}) = n$. *Necessity:* $\rho(\mathbf{A}) = n$ implies $\rho(\mathbf{A}^*\mathbf{A}) = n$. Let $\boldsymbol{\alpha}$ be an $n \times 1$ vector. We show that $\mathbf{A}^*\mathbf{A}\boldsymbol{\alpha} = \mathbf{0}$ implies $\boldsymbol{\alpha} = \mathbf{0}$ under the

[10] The intersection of two linear spaces is a linear space.

assumption of $\rho(\mathbf{A})=n$. The equality $\mathbf{A}^*\mathbf{A}\boldsymbol{\alpha}=\mathbf{0}$ implies $\boldsymbol{\alpha}^*\mathbf{A}^*\mathbf{A}\boldsymbol{\alpha}=\mathbf{0}$. Let $\mathbf{A}\boldsymbol{\alpha}=[\beta_1 \quad \beta_2 \quad \cdots \quad \beta_m]'$. Then

$$\boldsymbol{\alpha}^*\mathbf{A}^*=[\beta_1^* \quad \beta_2^* \quad \cdots \quad \beta_m^*]$$

and
$$\boldsymbol{\alpha}^*\mathbf{A}^*\mathbf{A}\boldsymbol{\alpha}=|\beta_1|^2+|\beta_2|^2+\cdots+|\beta_m|^2$$

Hence $\boldsymbol{\alpha}^*\mathbf{A}^*\mathbf{A}\boldsymbol{\alpha}=\mathbf{0}$ implies $\beta_i=0$, for $i=1, 2, \ldots, m$; or, equivalently, $\mathbf{A}\boldsymbol{\alpha}=\mathbf{0}$, which, in turn, implies $\boldsymbol{\alpha}=\mathbf{0}$ from the assumption of $\rho(\mathbf{A})=n$. Therefore we conclude that $\rho(\mathbf{A}^*\mathbf{A})=n$.

2. This part can be similarly proved or directly deduced from the foregoing by using the fact $\rho(\mathbf{A})=\rho(\mathbf{A}^*)$. Q.E.D.

2-6 Eigenvectors, Generalized Eigenvectors, and Jordan-Form Representations of a Linear Operator

With the background of Section 2-5, we are now ready to study the problem posed at the end of Section 2-4. We discuss in this section only linear operators that map $(\mathbb{C}^n, \mathbb{C})$ into itself with the understanding that the results are applicable to any operator that maps a finite-dimensional linear space over $\mathbb{C}$ into itself. The reason for restricting the field to the field of complex numbers will be seen immediately.

Let $\mathbf{A}$ be an $n \times n$ matrix with coefficients in the field $\mathbb{C}$. We have agreed to consider $\mathbf{A}$ as a linear operator that maps $(\mathbb{C}^n, \mathbb{C})$ into $(\mathbb{C}^n, \mathbb{C})$.

Definition 2-12

Let $\mathbf{A}$ be a linear operator that maps $(\mathbb{C}^n, \mathbb{C})$ into itself. Then a scalar λ in $\mathbb{C}$ is called an *eigenvalue* of $\mathbf{A}$ if there exists a nonzero vector $\mathbf{x}$ in $\mathbb{C}^n$ such that $\mathbf{A}\mathbf{x}=\lambda\mathbf{x}$. Any nonzero vector $\mathbf{x}$ satisfying $\mathbf{A}\mathbf{x}=\lambda\mathbf{x}$ is called an *eigenvector* of $\mathbf{A}$ associated with the eigenvalue λ.[11] ∎

In order to find an eigenvalue of $\mathbf{A}$, we write $\mathbf{A}\mathbf{x}=\lambda\mathbf{x}$ as

$$(\mathbf{A}-\lambda\mathbf{I})\mathbf{x}=0 \tag{2-40}$$

where $\mathbf{I}$ is the unit matrix of order n. We see that for any fixed λ in $\mathbb{C}$, Equation (2-40) is a set of homogeneous linear equations. The matrix $(\mathbf{A}-\lambda\mathbf{I})$ is an $n \times n$ square matrix. From Corollary 2-5, we know that Equation (2-40) has a nontrivial solution if and only if $\det(\mathbf{A}-\lambda\mathbf{I})=0$. It follows that *a scalar λ is an eigenvalue of $\mathbf{A}$ if and only if it is a solution of* $\Delta(\lambda) \triangleq \det(\lambda\mathbf{I}-\mathbf{A})=0$. $\Delta(\lambda)$ is a polynomial of degree n in λ and is called the *characteristic polynomial* of $\mathbf{A}$. Since $\Delta(\lambda)$ is of degree n, the $n \times n$ matrix $\mathbf{A}$ has n eigenvalues (not necessarily all distinct).

[11] It is also called a *right eigenvector of* $\mathbf{A}$. If a $1 \times n$ nonzero vector $\mathbf{y}$ exists such that $\mathbf{y}\mathbf{A}=\lambda\mathbf{y}$, then $\mathbf{y}$ is called a *left eigenvector* of $\mathbf{A}$ associated with λ.

Example 1

Consider the matrix

$$\mathbf{A}=\begin{bmatrix}1 & -1\\ 2 & -1\end{bmatrix} \tag{2-41}$$

which maps $(\mathbb{R}^2, \mathbb{R})$ into itself. We like to check whether Definition 2-12 can be modified and applied to a linear operator that maps $(\mathbb{R}^n, \mathbb{R})$ into $(\mathbb{R}^n, \mathbb{R})$. A modified version of Definition 2-12 reads as a scalar λ in $\mathbb{R}$ is an eigenvalue of $\mathbf{A}$ if there exists a nonzero vector $\mathbf{x}$ such that $\mathbf{Ax}=\lambda\mathbf{x}$. Clearly λ is an eigenvalue of $\mathbf{A}$ if and only if it is a solution of $\det(\lambda\mathbf{I}-\mathbf{A})=0$. Now

$$\det(\lambda\mathbf{I}-\mathbf{A})=\det\begin{bmatrix}\lambda-1 & 1\\ -2 & \lambda+1\end{bmatrix}=\lambda^2+1$$

which has no real-valued solution. Consequently, the matrix $\mathbf{A}$ has no eigenvalue in $\mathbb{R}$.

Since the set of real numbers is a part of the field of complex numbers, there is no reason that we cannot consider the matrix $\mathbf{A}$ in (2-41) as a linear operator that maps $(\mathbb{C}^2, \mathbb{C})$ into itself. In so doing, then the matrix $\mathbf{A}$ has eigenvalues $+i$ and $-i$ where $i^2 \triangleq -1$. ∎

The constant matrices we shall encounter in this book are all real-valued. However in order to ensure the existence of eigenvalues, we shall consider them as linear operators under the field of complex numbers.

With these preliminaries, we are ready to introduce a set of basis vectors such that a linear operator has a diagonal or almost diagonal representation. We study first the case in which all the eigenvalues of $\mathbf{A}$ are distinct; the case where $\mathbf{A}$ has repeated eigenvalues will then be studied.

Case 1: All the eigenvalues of **A** are distinct

Let $\lambda_1, \lambda_2, \ldots, \lambda_n$ be the eigenvalues of $\mathbf{A}$, and let $\mathbf{v}_i$ be an eigenvector of $\mathbf{A}$ associated with λ_i, for $i=1, 2, \ldots, n$; that is, $\mathbf{A}\mathbf{v}_i=\lambda_i\mathbf{v}_i$. We shall use the set of vectors $\{\mathbf{v}_1, \mathbf{v}_2, \ldots, \mathbf{v}_n\}$ as a basis of $(\mathbb{C}^n, \mathbb{C})$. In order to do so, we have to show that the set is linearly independent and qualifies as a basis.

Theorem 2-9

Let $\lambda_1, \lambda_2, \ldots, \lambda_n$ be the distinct eigenvalues of $\mathbf{A}$, and let $\mathbf{v}_i$ be an eigenvector of $\mathbf{A}$ associated with λ_i, for $i=1, 2, \ldots, n$. Then the set $\{\mathbf{v}_1, \mathbf{v}_2, \ldots, \mathbf{v}_n\}$ is linearly independent (over $\mathbb{C}$).

Proof

We prove the theorem by contradiction. Suppose $\mathbf{v}_1, \mathbf{v}_2, \ldots, \mathbf{v}_n$ are linearly dependent; then there exist $\alpha_1, \alpha_2, \ldots, \alpha_n$ (not all zero) in $\mathbb{C}$ such that

$$\alpha_1\mathbf{v}_1+\alpha_2\mathbf{v}_2\cdots+\alpha_n\mathbf{v}_n=\mathbf{0} \tag{2-42}$$

We assume $\alpha_1 \neq 0$. If $\alpha_1=0$, we may reorder λ_i in such a way that $\alpha_1 \neq 0$.

Equation (2-42) implies that

$$(\mathbf{A}-\lambda_2\mathbf{I})(\mathbf{A}-\lambda_3\mathbf{I})\cdots(\mathbf{A}-\lambda_n\mathbf{I})\left(\sum_{i=1}^{n}\alpha_i\mathbf{v}_i\right)=\mathbf{0} \tag{2-43}$$

Since $(\mathbf{A}-\lambda_j\mathbf{I})\mathbf{v}_i=(\lambda_i-\lambda_j)\mathbf{v}_i \quad \text{if } j\neq i$

and $(\mathbf{A}-\lambda_i\mathbf{I})\mathbf{v}_i=\mathbf{0}$

the left-hand side of (2-43) can be reduced to

$$\alpha_1(\lambda_1-\lambda_2)(\lambda_1-\lambda_3)\cdots(\lambda_1-\lambda_n)\mathbf{v}_1=\mathbf{0}$$

By assumption the λ_i's, for $i=1,2,\ldots,n$, are all distinct; hence the equation

$$\alpha_1\prod_{i=2}^{n}(\lambda_1-\lambda_i)\mathbf{v}_1=\mathbf{0}$$

implies $\alpha_1=0$. This is a contradiction. Thus, the set of vectors $\{\mathbf{v}_1,\mathbf{v}_2,\ldots,\mathbf{v}_n\}$ is linearly independent and qualifies as a basis. Q.E.D.

Let $\hat{\mathbf{A}}$ be the representation of $\mathbf{A}$ with respect to the basis $\{\mathbf{v}_1,\mathbf{v}_2,\ldots,\mathbf{v}_n\}$. Recall from Figure 2-5 that the ith column of $\hat{\mathbf{A}}$ is the representation of $\mathbf{A}\mathbf{v}_i=\lambda_i\mathbf{v}_i$ with respect to $\{\mathbf{v}_1,\mathbf{v}_2,\ldots,\mathbf{v}_n\}$—that is, $[0 \ \cdots \ 0 \ \lambda_i \ 0 \ \cdots \ 0]'$, where λ_i is located at the ith entry. Hence the representation of $\mathbf{A}$ with respect to $\{\mathbf{v}_1,\mathbf{v}_2,\ldots,\mathbf{v}_n\}$ is

$$\hat{\mathbf{A}}=\begin{bmatrix}\lambda_1 & 0 & 0 & \cdots & 0\\ 0 & \lambda_2 & 0 & \cdots & 0\\ 0 & 0 & \lambda_3 & \cdots & 0\\ \vdots & \vdots & \vdots & & \vdots\\ 0 & 0 & 0 & \cdots & \lambda_n\end{bmatrix} \tag{2-44}$$

This can also be checked by using a similarity transformation. From Figure 2-5, we have

$$\mathbf{Q}=[\mathbf{v}_1 \quad \mathbf{v}_2 \quad \cdots \quad \mathbf{v}_n]$$

Since

$$\begin{aligned}\mathbf{AQ}=\mathbf{A}[\mathbf{v}_1 \quad \mathbf{v}_2 \ \cdots \ \mathbf{v}_n]&=[\mathbf{A}\mathbf{v}_1 \quad \mathbf{A}\mathbf{v}_2 \quad \cdots \quad \mathbf{A}\mathbf{v}_n]\\ &=[\lambda_1\mathbf{v}_1 \quad \lambda_2\mathbf{v}_2 \quad \cdots \quad \lambda_n\mathbf{v}_n]=\mathbf{Q}\hat{\mathbf{A}}\end{aligned}$$

we have $\hat{\mathbf{A}}=\mathbf{Q}^{-1}\mathbf{AQ}$

We conclude that if the eigenvalues of a linear operator $\mathbf{A}$ that maps $(\mathbb{C}^n, \mathbb{C})$ into itself are all distinct, then by choosing the set of eigenvectors as a basis, the operator $\mathbf{A}$ has a diagonal matrix representation with the eigenvalues on the diagonal.

Example 2

Consider

$$\mathbf{A}=\begin{bmatrix}1 & -1\\ 2 & -1\end{bmatrix}$$

The characteristic polynomial of **A** is λ^2+1. Hence the eigenvalues of **A** are $+i$ and $-i$. The eigenvector associated with $\lambda_1=i$ can be obtained by solving the following homogeneous equation:

$$(\mathbf{A}-\lambda_1\mathbf{I})\mathbf{v}_1=\begin{bmatrix}1-i & -1\\ 2 & -1-i\end{bmatrix}\begin{bmatrix}v_{11}\\ v_{21}\end{bmatrix}=\mathbf{0}$$

Clearly the vector $\mathbf{v}_1=[1 \quad 1-i]'$ is a solution. Similarly, the vector $\mathbf{v}_2=[1 \quad 1+i]'$ can be shown to be an eigenvector of $\lambda_2=-i$. Hence the representation of **A** with respect to $\{\mathbf{v}_1,\mathbf{v}_2\}$ is

$$\hat{\mathbf{A}}=\begin{bmatrix}i & 0\\ 0 & -i\end{bmatrix}$$

The reader is advised to verify this by a similarity transformation. ∎

Case 2: The eigenvalues of A are not all distinct

Unlike the previous case, if an operator **A** has repeated eigenvalues, it is not always possible to find a diagonal matrix representation. We shall use examples to illustrate the difficulty that may arise for matrices with repeated eigenvalues.

Example 3

Consider

$$\mathbf{A}=\begin{bmatrix}1 & 0 & -1\\ 0 & 1 & 0\\ 0 & 0 & 2\end{bmatrix}$$

The eigenvalues of **A** are $\lambda_1=1$, $\lambda_2=1$, and $\lambda_3=2$. The eigenvectors associated with λ_1 can be obtained by solving the following homogeneous equations:

$$(\mathbf{A}-\lambda_1\mathbf{I})\mathbf{v}=\begin{bmatrix}0 & 0 & -1\\ 0 & 0 & 0\\ 0 & 0 & 1\end{bmatrix}\mathbf{v}=\mathbf{0} \tag{2-45}$$

Note that the matrix $(\mathbf{A}-\lambda_1\mathbf{I})$ has rank 1; therefore, two linearly independent vector solutions can be found for (2-45) (see Corollary 2-5). Clearly, $\mathbf{v}_1=[1 \quad 0 \quad 0]'$ and $\mathbf{v}_2=[0 \quad 1 \quad 0]'$ are two linearly independent eigenvectors associated with $\lambda_1=\lambda_2=1$. An eigenvector associated with $\lambda_3=2$ can be found as $\mathbf{v}_3=[-1 \quad 0 \quad 1]'$. Since the set of vectors $\{\mathbf{v}_1,\mathbf{v}_2,\mathbf{v}_3\}$ is linearly independent, it qualifies as a basis. The representation of **A** with respect to $\{\mathbf{v}_1,\mathbf{v}_2,\mathbf{v}_3\}$ is

$$\hat{\mathbf{A}}=\begin{bmatrix}1 & 0 & 0\\ 0 & 1 & 0\\ 0 & 0 & 2\end{bmatrix}$$ ∎

In this example, although **A** has repeated eigenvalues, it can still be diagonalized. However, this is not always the case, as can be seen from the following example.

Example 4

Consider

$$\mathbf{A}=\begin{bmatrix}1 & 1 & 2\\ 0 & 1 & 3\\ 0 & 0 & 2\end{bmatrix} \tag{2-46}$$

The eigenvalues of $\mathbf{A}$ are $\lambda_1=1$, $\lambda_2=1$, and $\lambda_3=2$. The eigenvectors associated with $\lambda_1=1$ can be found by solving

$$(\mathbf{A}-\lambda_1\mathbf{I})\mathbf{v}=\begin{bmatrix}0 & 1 & 2\\ 0 & 0 & 3\\ 0 & 0 & 1\end{bmatrix}\mathbf{v}=\mathbf{0}$$

Since the matrix $(\mathbf{A}-\lambda_1\mathbf{I})$ has rank 2, the null space of $(\mathbf{A}-\lambda_1\mathbf{I})$ has dimension 1. Consequently, we can find only one linearly independent eigenvector, say $\mathbf{v}_1=[1 \quad 0 \quad 0]'$, associated with $\lambda_1=\lambda_2=1$. An eigenvector associated with $\lambda_3=2$ can be found as $\mathbf{v}_3=[5 \quad 3 \quad 1]'$. Clearly the two eigenvectors are not sufficient to form a basis of $(\mathbb{C}^3, \mathbb{C})$. ∎

From this example, we see that if an $n\times n$ matrix $\mathbf{A}$ has repeated eigenvalues, it is not always possible to find n linearly independent eigenvectors. Consequently, the $\mathbf{A}$ cannot be transformed into a diagonal form. However, it is possible to find a special set of basis vectors so that the new representation is *almost* a diagonal form, called a *Jordan canonical form*. The form has the eigenvalues of $\mathbf{A}$ on the diagonal and either 0 or 1 on the superdiagonal. For example, if $\mathbf{A}$ has an eigenvalue λ_1 with multiplicity 4 and an eigenvalue λ_2 with multiplicity 1, then the new representation will assume one of the following forms.

$$\begin{bmatrix}\lambda_1 & 0 & 0 & 0 & 0\\ 0 & \lambda_1 & 0 & 0 & 0\\ 0 & 0 & \lambda_1 & 0 & 0\\ 0 & 0 & 0 & \lambda_1 & 0\\ 0 & 0 & 0 & 0 & \lambda_2\end{bmatrix} \quad \begin{bmatrix}\lambda_1 & 1 & 0 & 0 & 0\\ 0 & \lambda_1 & 0 & 0 & 0\\ 0 & 0 & \lambda_1 & 0 & 0\\ 0 & 0 & 0 & \lambda_1 & 0\\ 0 & 0 & 0 & 0 & \lambda_2\end{bmatrix} \quad \begin{bmatrix}\lambda_1 & 1 & 0 & 0 & 0\\ 0 & \lambda_1 & 0 & 0 & 0\\ 0 & 0 & \lambda_1 & 1 & 0\\ 0 & 0 & 0 & \lambda_1 & 0\\ 0 & 0 & 0 & 0 & \lambda_2\end{bmatrix}$$

$$\begin{bmatrix}\lambda_1 & 1 & 0 & 0 & 0\\ 0 & \lambda_1 & 1 & 0 & 0\\ 0 & 0 & \lambda_1 & 0 & 0\\ 0 & 0 & 0 & \lambda_1 & 0\\ 0 & 0 & 0 & 0 & \lambda_2\end{bmatrix} \quad \begin{bmatrix}\lambda_1 & 1 & 0 & 0 & 0\\ 0 & \lambda_1 & 1 & 0 & 0\\ 0 & 0 & \lambda_1 & 1 & 0\\ 0 & 0 & 0 & \lambda_1 & 0\\ 0 & 0 & 0 & 0 & \lambda_2\end{bmatrix} \tag{2-47}$$

Which form it will assume depends on the characteristics of $\mathbf{A}$ and will be discussed in the next subsection. The matrices in (2-47) are all of block-diagonal form. The blocks on the diagonal are of the form

$$\begin{bmatrix} \lambda & 1 & 0 & \cdots & 0 & 0 \\ 0 & \lambda & 1 & \cdots & 0 & 0 \\ \vdots & \vdots & \vdots & & \vdots & \vdots \\ 0 & 0 & 0 & \cdots & 1 & 0 \\ 0 & 0 & 0 & \cdots & \lambda & 1 \\ 0 & 0 & 0 & \cdots & 0 & \lambda \end{bmatrix} \tag{2-48}$$

with the *same* eigenvalue on the main diagonal and 1's on the diagonal just above the main diagonal. A matrix of this form is called a *Jordan block* associated with λ. A matrix is said to be in the *Jordan canonical form*, or the *Jordan form*, if its principal diagonal consists of Jordan blocks and the remaining elements are zeros. The fourth matrix in (2-47) has two Jordan blocks associated with λ_1 (one with order 3, the other with order 1) and one Jordan block associated with λ_2. A diagonal matrix is clearly a special case of the Jordan form: all of its Jordan blocks are of order 1.

Every matrix which maps $(\mathbb{C}^n, \mathbb{C})$ into itself has a Jordan-form representation. The use of Jordan form is very convenient in developing a number of concepts and results; hence it will be extensively used in the remainder of this chapter.

Derivation of a Jordan-form representation.[12] In this subsection, we discuss how to find a set of basis vectors so that the representation of $\mathbf{A}$ with respect to this set of basis vectors is in a Jordan form. The basis vectors to be used are called the *generalized eigenvectors.*

Definition 2-13

A vector $\mathbf{v}$ is said to be a *generalized eigenvector* of grade k of $\mathbf{A}$ associated with λ if and only if[13]

$$(\mathbf{A} - \lambda\mathbf{I})^k\mathbf{v} = \mathbf{0}$$

and

$$(\mathbf{A} - \lambda\mathbf{I})^{k-1}\mathbf{v} \neq \mathbf{0}$$ ∎

Note that if $k = 1$, Definition 2-13 reduces to $(\mathbf{A} - \lambda\mathbf{I})\mathbf{v} = \mathbf{0}$ and $\mathbf{v} \neq \mathbf{0}$, which is the definition of an eigenvector. Hence the term "generalized eigenvector" is well-justified.

Let $\mathbf{v}$ be a generalized eigenvector of grade k associated with the eigenvalue

[12] This section may be skipped without loss of continuity. However, it is suggested that the reader glances through it to gain a better feeling about the Jordan-form representation.

[13] $(\mathbf{A} - \lambda\mathbf{I})^k \triangleq (\mathbf{A} - \lambda\mathbf{I})(\mathbf{A} - \lambda\mathbf{I}) \cdots (\mathbf{A} - \lambda\mathbf{I})$ (k terms), $(\mathbf{A} - \lambda\mathbf{I})^0 \triangleq \mathbf{I}$.

λ. Define

$$\begin{aligned} \mathbf{v}_k &\triangleq \mathbf{v} \\ \mathbf{v}_{k-1} &\triangleq (\mathbf{A}-\lambda\mathbf{I})\mathbf{v} = (\mathbf{A}-\lambda\mathbf{I})\mathbf{v}_k \\ \mathbf{v}_{k-2} &\triangleq (\mathbf{A}-\lambda\mathbf{I})^2\mathbf{v} = (\mathbf{A}-\lambda\mathbf{I})\mathbf{v}_{k-1} \\ &\cdots\cdots\cdots\cdots\cdots \\ \text{and} \qquad \mathbf{v}_1 &\triangleq (\mathbf{A}-\lambda\mathbf{I})^{k-1}\mathbf{v} = (\mathbf{A}-\lambda\mathbf{I})\mathbf{v}_2 \end{aligned} \tag{2-49}$$

This set of vectors $\{\mathbf{v}_1, \mathbf{v}_2, \ldots, \mathbf{v}_k\}$ is called a *chain of generalized eigenvectors* of length k.

Let $\mathcal{N}_i$ denote the null space of $(\mathbf{A}-\lambda\mathbf{I})^i$, that is, $\mathcal{N}_i$ consists of all $\mathbf{x}$ such that $(\mathbf{A}-\lambda\mathbf{I})^i\mathbf{x} = \mathbf{0}$. It is clear that if $\mathbf{x}$ is in $\mathcal{N}_i$, then it is in $\mathcal{N}_{i+1}$. Hence $\mathcal{N}_i$ is a subspace of $\mathcal{N}_{i+1}$, denoted as $\mathcal{N}_i \subset \mathcal{N}_{i+1}$. Clearly, the $\mathbf{v}$ defined in Definition 2-13 is in $\mathcal{N}_k$ but not in $\mathcal{N}_{k-1}$. In fact, for $i = 1, 2, \ldots, k$, $\mathbf{v}_i = (\mathbf{A}-\lambda\mathbf{I})^{k-i}\mathbf{v}$ defined in (2-49) is in $\mathcal{N}_i$ but not in $\mathcal{N}_{i-1}$. Indeed, we have

$$(\mathbf{A}-\lambda\mathbf{I})^i\mathbf{v}_i = (\mathbf{A}-\lambda\mathbf{I})^i(\mathbf{A}-\lambda\mathbf{I})^{k-i}\mathbf{v} = (\mathbf{A}-\lambda\mathbf{I})^k\mathbf{v} = \mathbf{0}$$

and

$$(\mathbf{A}-\lambda\mathbf{I})^{i-1}\mathbf{v}_i = (\mathbf{A}-\lambda\mathbf{I})^{i-1}(\mathbf{A}-\lambda\mathbf{I})^{k-i}\mathbf{v} = (\mathbf{A}-\lambda\mathbf{I})^{k-1}\mathbf{v} \neq \mathbf{0}$$

hence $\mathbf{v}_i$ is in $\mathcal{N}_i$ but not in $\mathcal{N}_{i-1}$.

Let $\mathbf{A}$ be an $n \times n$ matrix and have eigenvalue λ with multiplicity m. We discuss in the following how to find m linearly independent generalized eigenvectors of $\mathbf{A}$ associated with λ. This is achieved by searching chains of generalized eigenvectors of various lengths. First we compute ranks of $(\mathbf{A}-\lambda\mathbf{I})^i$, $i = 0, 1, 2, \ldots$, until rank $(\mathbf{A}-\lambda\mathbf{I})^k = n - m$. In order not to be overwhelmed by notations, we assume $n = 10$, $m = 8$, $k = 4$, and the ranks of $(\mathbf{A}-\lambda\mathbf{I})^i$, $i = 0, 1, 2, 3, 4$, are as shown in Table 2-2. The nullity ν_i is the dimension of the null space $\mathcal{N}_i$, and is equal to, following Theorem 2-5, $n - \rho(\mathbf{A}-\lambda\mathbf{I})^i$. Because of $\mathcal{N}_0 \subset \mathcal{N}_1 \subset \mathcal{N}_2 \subset \cdots$, we have $0 = \nu_0 \leq \nu_1 \leq \nu_2 \leq \cdots \leq \nu_k = m$.

Now we are ready to find $m = 8$ linearly independent eigenvectors of $\mathbf{A}$ associated with λ. Because of $\mathcal{N}_3 \subset \mathcal{N}_4$ and $\nu_4 - \nu_3 = 1$, we can find one and only one linearly independent vector $\mathbf{u}$ in $\mathcal{N}_4$ but not in $\mathcal{N}_3$ such that

$$\mathbf{B}^4\mathbf{u} = \mathbf{0} \qquad \text{and} \qquad \mathbf{B}^3\mathbf{u} \neq \mathbf{0}$$

where $\mathbf{B} \triangleq \mathbf{A} - \lambda\mathbf{I}$. From this $\mathbf{u}$, we can generate a chain of four generalized

Table 2-2 Chains of Generalized Eigenvectors, where $\mathbf{B} \triangleq \mathbf{A} - \lambda\mathbf{I}$

Rank	Nullity
$\rho(\mathbf{A}-\lambda\mathbf{I})^0 = 10$	$\nu_0 = 0$
$\rho(\mathbf{A}-\lambda\mathbf{I}) = 7$	$\nu_1 = 3$
$\rho(\mathbf{A}-\lambda\mathbf{I})^2 = 4$	$\nu_2 = 6$
$\rho(\mathbf{A}-\lambda\mathbf{I})^3 = 3$	$\nu_3 = 7$
$\rho(\mathbf{A}-\lambda\mathbf{I})^4 = 2$	$\nu_4 = 8$

$\nu_4 - \nu_3 = 1$	$\nu_3 - \nu_2 = 1$	$\nu_2 - \nu_1 = 3$	$\nu_1 - \nu_0 = 3$	No. of independent vectors in $\mathcal{N}_i$ but not in $\mathcal{N}_{i-1}$
		$\mathbf{w}_2 = \mathbf{w}$, $\mathbf{v}_2 = \mathbf{v}$,	$\mathbf{w}_1 = \mathbf{B}\mathbf{w}$, $\mathbf{v}_1 = \mathbf{B}\mathbf{v}$	Two chains with length 2
$\mathbf{u}_4 = \mathbf{u}$	$\mathbf{u}_3 = \mathbf{B}\mathbf{u}$	$\mathbf{u}_2 = \mathbf{B}^2\mathbf{u}$	$\mathbf{u}_1 = \mathbf{B}^3\mathbf{u}$	One chain with length 4
$\mathcal{N}_4$	$\mathcal{N}_3$	$\mathcal{N}_2$	$\mathcal{N}_1$	

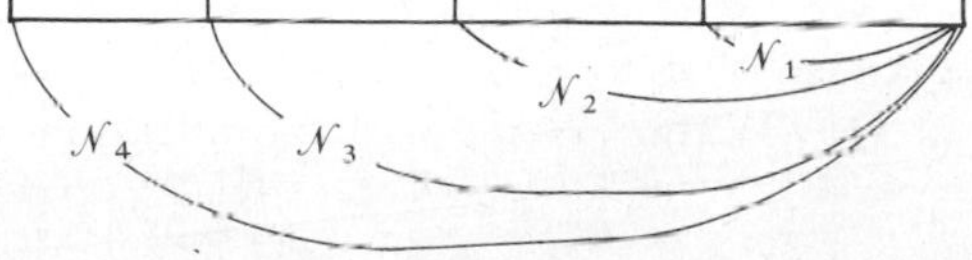

eigenvectors as

$$\mathbf{u}_1 = \mathbf{B}^3\mathbf{u} \qquad \mathbf{u}_2 = \mathbf{B}^2\mathbf{u} \qquad \mathbf{u}_3 = \mathbf{Bu} \qquad \mathbf{u}_4 = \mathbf{u} \tag{2-50}$$

Because of $\mathcal{N}_2 \subset \mathcal{N}_3$ and $v_3 - v_2 = 1$, there is only one linearly independent vector in $\mathcal{N}_3$ but not in $\mathcal{N}_2$. The $\mathbf{u}_3$ in (2-50) is such a vector; therefore we cannot find any other linearly independent vector in $\mathcal{N}_3$ but not in $\mathcal{N}_2$. Consider now the vectors in $\mathcal{N}_2$ but not in $\mathcal{N}_1$. Because $v_2 - v_1 = 3$, there are three such linearly independent vectors. Since $\mathbf{u}_2$ in (2-50) is one of them, we can find two vectors $\mathbf{v}$ and $\mathbf{w}$ such that $\{\mathbf{u}_2, \mathbf{v}, \mathbf{w}\}$ are linearly independent and

$$\mathbf{B}^2\mathbf{v} = \mathbf{0} \qquad \mathbf{Bv} \neq \mathbf{0}$$

and

$$\mathbf{B}^2\mathbf{w} = \mathbf{0} \qquad \mathbf{Bw} \neq \mathbf{0}$$

From $\mathbf{v}$ and $\mathbf{w}$, we can generate two chains of generalized eigenvectors of length 2 as shown in Table 2-2. As can be seen from Table 2-2, the number of vectors in $\mathcal{N}_1$ is equal to $v_1 - v_0 = 3$, hence there is no need to search other vector in $\mathcal{N}_1$. This completes the search of eight generalized eigenvectors of $\mathbf{A}$ associated with λ.

Theorem 2-10

The generalized eigenvectors of $\mathbf{A}$ associated with λ generated as in Table 2-2 are linearly independent.

Proof

First we show that if $\{\mathbf{u}_2, \mathbf{v}, \mathbf{w}\}$ is linearly independent, then $\{\mathbf{u}_1, \mathbf{v}_1, \mathbf{w}_1\}$ is linearly independent. Suppose $\{\mathbf{u}_1, \mathbf{v}_1, \mathbf{w}_1\}$ is not linearly independent, then there exist c_i, $i = 1, 2, 3$, not all zero, such that $c_1\mathbf{u}_1 + c_2\mathbf{v}_1 + c_3\mathbf{w}_1 = \mathbf{0}$. However, we have

$$\mathbf{0} = c_1\mathbf{u}_1 + c_2\mathbf{v}_1 + c_3\mathbf{w}_1 = c_1\mathbf{Bu}_2 + c_2\mathbf{Bv} + c_3\mathbf{Bw} = \mathbf{B}(c_1\mathbf{u}_2 + c_2\mathbf{v} + c_3\mathbf{w}) \triangleq \mathbf{By}$$

Since $\mathbf{y}$ is a vector in $\mathcal{N}_2$, the only way to have $\mathbf{By} = \mathbf{0}$ is that $\mathbf{y} = \mathbf{0}$. Since $\{\mathbf{u}_2, \mathbf{v}, \mathbf{w}\}$ is linearly independent by assumption, $\mathbf{y} = \mathbf{0}$ implies $c_i = 0$, $i = 1, 2, 3$. This is a contradiction. Hence if $\{\mathbf{u}_2, \mathbf{v}, \mathbf{w}\}$ is linearly independent, so is $\{\mathbf{u}_1, \mathbf{v}_1, \mathbf{w}_1\}$.

Now we show that the generalized eigenvectors $\{\mathbf{u}_i, i = 1, 2, 3, 4; \mathbf{v}_j, \mathbf{w}_j, j = 1, 2\}$ are linearly independent. Consider

$$c_1\mathbf{u}_1 + c_2\mathbf{u}_2 + c_3\mathbf{u}_3 + c_4\mathbf{u}_4 + c_5\mathbf{v}_1 + c_6\mathbf{v}_2 + c_7\mathbf{w}_1 + c_8\mathbf{w}_2 = \mathbf{0} \tag{2-51}$$

The application of $\mathbf{B}^3 = (\mathbf{A} - \lambda\mathbf{I})^3$ to (2-51) yields

$$c_4\mathbf{B}^3\mathbf{u}_4 = \mathbf{0}$$

which implies, because of $\mathbf{B}^3\mathbf{u}_4 \neq \mathbf{0}$, $c_4 = 0$. Similarly, we can show $c_3 = 0$ by applying $\mathbf{B}^2$ to (2-51). With $c_3 = c_4 = 0$, the application of $\mathbf{B}$ to (2-51) yields

$$c_2\mathbf{u}_2 + c_6\mathbf{v}_2 + c_8\mathbf{w}_2 = \mathbf{0}$$

which implies, because of the linear independence of $\{\mathbf{u}_2, \mathbf{v}_2, \mathbf{w}_2\}$, $c_2 = c_6 = c_8 = 0$. Finally, we have $c_1 = c_5 = c_7 = 0$ following the linear independence of $\{\mathbf{u}_1, \mathbf{v}_1, \mathbf{w}_1\}$. This completes the proof of this theorem. Q.E.D.

Theorem 2-11

The generalized eigenvectors of $\mathbf{A}$ associated with different eigenvalues are linearly independent. ∎

This theorem can be proved as in Theorem 2-10 by applying repetitively $(\mathbf{A} - \lambda_i\mathbf{I})^k(\mathbf{A} - \lambda_j\mathbf{I})^l$. The proof is left as an exercise.

Now we discuss the representation of $\mathbf{A}$ with respect to $\mathbf{Q} \triangleq [\mathbf{u}_1 \ \ \mathbf{u}_2 \ \ \mathbf{u}_3 \ \ \mathbf{u}_4 \ \ \mathbf{v}_1 \ \ \mathbf{v}_2 \ \ \mathbf{w}_1 \ \ \mathbf{w}_2 \ \ \mathbf{x} \ \ \mathbf{x}]$. The last two vectors are the eigenvectors of $\mathbf{A}$ associated with other eigenvalues. The first four columns of the new representation $\hat{\mathbf{A}}$ are the representations of $\mathbf{A}\mathbf{u}_i$, $i = 1, 2, 3, 4$, with respect to $\{\mathbf{u}_1, \mathbf{u}_2, \mathbf{u}_3, \mathbf{u}_4, \mathbf{v}_1, \mathbf{v}_2, \mathbf{w}_1, \mathbf{w}_2, \mathbf{x}, \mathbf{x}\}$. Because $(\mathbf{A} - \lambda\mathbf{I})\mathbf{u}_1 = \mathbf{0}$, $(\mathbf{A} - \lambda\mathbf{I})\mathbf{u}_2 = \mathbf{u}_1$, $(\mathbf{A} - \lambda\mathbf{I})\mathbf{u}_3 = \mathbf{u}_2$, and $(\mathbf{A} - \lambda\mathbf{I})\mathbf{u}_4 = \mathbf{u}_3$, we have $\mathbf{A}\mathbf{u}_1 = \lambda\mathbf{u}_1 = \mathbf{Q}[\lambda \ \ 0 \ \ 0 \ \ \cdots \ \ 0]'$, $\mathbf{A}\mathbf{u}_2 = \mathbf{u}_1 + \lambda\mathbf{u}_2 = \mathbf{Q}[1 \ \ \lambda \ \ 0 \ \ 0 \ \ \cdots \ \ 0]'$, $\mathbf{A}\mathbf{u}_3 = \mathbf{u}_2 + \lambda\mathbf{u}_3 = \mathbf{Q}[0 \ \ 1 \ \ \lambda \ \ 0 \ \ \cdots \ \ 0]'$, and $\mathbf{A}\mathbf{u}_4 = \mathbf{u}_3 + \lambda\mathbf{u}_4 = \mathbf{Q}[0 \ \ 0 \ \ 1 \ \ \lambda \ \ 0 \ \ \cdots \ \ 0]'$, where the prime denotes the transpose. Proceeding similarly, the new representation $\hat{\mathbf{A}}$ can be obtained as

$$\hat{\mathbf{A}} = \left[\begin{array}{cccc:cc:cc}
\lambda & 1 & 0 & 0 & 0 & 0 & 0 & 0 \\
0 & \lambda & 1 & 0 & 0 & 0 & 0 & 0 \\
0 & 0 & \lambda & 1 & 0 & 0 & 0 & 0 \\
0 & 0 & 0 & \lambda & 0 & 0 & 0 & 0 \\
\hdashline
0 & 0 & 0 & 0 & \lambda & 1 & 0 & 0 \\
0 & 0 & 0 & 0 & 0 & \lambda & 0 & 0 \\
\hdashline
0 & 0 & 0 & 0 & 0 & 0 & \lambda & 1 \\
0 & 0 & 0 & 0 & 0 & 0 & 0 & \lambda \\
 & & & & & & & & \ddots
\end{array}\right] \tag{2-52}$$

This is a Jordan-form matrix. Note that the number of Jordan blocks associated with λ is equal to $\nu_1 = 3$, the dimension of the null space of $(\mathbf{A} - \lambda\mathbf{I})$.[14]

Example 4 (Continued)

Consider

$$\mathbf{A} = \begin{bmatrix} 1 & 1 & 2 \\ 0 & 1 & 3 \\ 0 & 0 & 2 \end{bmatrix}$$

---

[14] This number is called the *geometric multiplicity* of λ in Reference 86. In other words, geometric multiplicity is the number of Jordan blocks, and the (algebraic) multiplicity is the sum of the orders of all Jordan blocks associated with λ.

Its eigenvalues are $\lambda_1 = 1, \lambda_2 = 1, \lambda_3 = 2$. An eigenvector associated with $\lambda_3 = 2$ is $\mathbf{v}_3 = [5 \quad 3 \quad 1]'$. The rank of $(\mathbf{A} - \lambda_1\mathbf{I})$ is 2; hence we can find only one eigenvector associated with λ_1. Consequently, we must use generalized eigenvectors. We compute

$$\mathbf{B} \triangleq (\mathbf{A} - \lambda_1\mathbf{I}) = \begin{bmatrix} 0 & 1 & 2 \\ 0 & 0 & 3 \\ 0 & 0 & 1 \end{bmatrix}$$

and
$$(\mathbf{A} - \lambda_1\mathbf{I})^2 = \begin{bmatrix} 0 & 1 & 2 \\ 0 & 0 & 3 \\ 0 & 0 & 1 \end{bmatrix}\begin{bmatrix} 0 & 1 & 2 \\ 0 & 0 & 3 \\ 0 & 0 & 1 \end{bmatrix} = \begin{bmatrix} 0 & 0 & 5 \\ 0 & 0 & 3 \\ 0 & 0 & 1 \end{bmatrix}$$

Since $\rho\mathbf{B}^2 = 1 = n - m$, we stop here. We search a $\mathbf{v}$ such that $\mathbf{B}^2\mathbf{v} = \mathbf{0}$ and $\mathbf{B}\mathbf{v} \neq \mathbf{0}$. Clearly, $\mathbf{v} = [0 \quad 1 \quad 0]'$ is such a vector. It is a generalized eigenvector of grade 2. Let

$$\mathbf{v}_2 \triangleq \mathbf{v} = \begin{bmatrix} 0 \\ 1 \\ 0 \end{bmatrix} \qquad \mathbf{v}_1 \triangleq (\mathbf{A} - \lambda_1\mathbf{I})\mathbf{v} = \begin{bmatrix} 0 & 1 & 2 \\ 0 & 0 & 3 \\ 0 & 0 & 1 \end{bmatrix}\begin{bmatrix} 0 \\ 1 \\ 0 \end{bmatrix} = \begin{bmatrix} 1 \\ 0 \\ 0 \end{bmatrix}$$

Theorems 2-10 and 2-11 imply that $\mathbf{v}_1, \mathbf{v}_2$, and $\mathbf{v}_3$ are linearly independent. This can also be checked by computing the determinant of $[\mathbf{v}_1 \quad \mathbf{v}_2 \quad \mathbf{v}_3]$. If we use the set of vectors $\{\mathbf{v}_1, \mathbf{v}_2, \mathbf{v}_3\}$ as a basis, then the ith column of the new representation $\hat{\mathbf{A}}$ is the representation of $\mathbf{A}\mathbf{v}_i$ with respect to the basis $\{\mathbf{v}_1, \mathbf{v}_2, \mathbf{v}_3\}$. Since $\mathbf{A}\mathbf{v}_1 = \lambda_1\mathbf{v}_1, \mathbf{A}\mathbf{v}_2 = \mathbf{v}_1 + \lambda_1\mathbf{v}_2$, and $\mathbf{A}\mathbf{v}_3 = \lambda_3\mathbf{v}_3$, the representations of $\mathbf{A}\mathbf{v}_1, \mathbf{A}\mathbf{v}_2$, and $\mathbf{A}\mathbf{v}_3$ with respect to the basis $\{\mathbf{v}_1, \mathbf{v}_2, \mathbf{v}_3\}$ are, respectively,

$$\begin{bmatrix} \lambda_1 \\ 0 \\ 0 \end{bmatrix} \qquad \begin{bmatrix} 1 \\ \lambda_1 \\ 0 \end{bmatrix} \qquad \begin{bmatrix} 0 \\ 0 \\ \lambda_3 \end{bmatrix}$$

where $\lambda_1 = 1, \lambda_3 = 2$. Hence we have

$$\hat{\mathbf{A}} = \left[\begin{array}{cc:c} 1 & 1 & 0 \\ 0 & 1 & 0 \\ \hdashline 0 & 0 & 2 \end{array}\right] \tag{2-53}$$

This can also be obtained by using the similarity transformation

$$\hat{\mathbf{A}} = \mathbf{Q}^{-1}\mathbf{A}\mathbf{Q}$$

where
$$\mathbf{Q} = [\mathbf{v}_1 \quad \mathbf{v}_2 \quad \mathbf{v}_3] = \begin{bmatrix} 1 & 0 & 5 \\ 0 & 1 & 3 \\ 0 & 0 & 1 \end{bmatrix}$$ ∎

Example 5

Transform the following matrix into the Jordan form:

$$\mathbf{A}=\begin{bmatrix}3 & -1 & 1 & 1 & 0 & 0\\ 1 & 1 & -1 & -1 & 0 & 0\\ 0 & 0 & 2 & 0 & 1 & 1\\ 0 & 0 & 0 & 2 & -1 & -1\\ 0 & 0 & 0 & 0 & 1 & 1\\ 0 & 0 & 0 & 0 & 1 & 1\end{bmatrix} \tag{2-54}$$

1. Compute the eigenvalues of **A**.[15]

$$\det(\mathbf{A}-\lambda\mathbf{I})=[(3-\lambda)(1-\lambda)+1](\lambda-2)^2[(1-\lambda)^2-1]=(\lambda-2)^5\lambda$$

Hence **A** has eigenvalue 2 with multiplicity 5 and eigenvalue 0 with multiplicity 1.

2. Compute $(\mathbf{A}-2\mathbf{I})^i$, for $i=1,2,\ldots,$ as follows:

$$\mathbf{B}\triangleq(\mathbf{A}-2\mathbf{I})=\begin{bmatrix}1 & -1 & 1 & 1 & 0 & 0\\ 1 & -1 & -1 & -1 & 0 & 0\\ 0 & 0 & 0 & 0 & 1 & 1\\ 0 & 0 & 0 & 0 & -1 & -1\\ 0 & 0 & 0 & 0 & -1 & 1\\ 0 & 0 & 0 & 0 & 1 & -1\end{bmatrix} \qquad \begin{array}{l}\rho(\mathbf{A}-2\mathbf{I})=4\\ \nu_1=6-4=2\end{array}$$

$$(\mathbf{A}-2\mathbf{I})^2=\begin{bmatrix}0 & 0 & 2 & 2 & 0 & 0\\ 0 & 0 & 2 & 2 & 0 & 0\\ 0 & 0 & 0 & 0 & 0 & 0\\ 0 & 0 & 0 & 0 & 0 & 0\\ 0 & 0 & 0 & 0 & 2 & -2\\ 0 & 0 & 0 & 0 & -2 & 2\end{bmatrix} \qquad \begin{array}{l}\rho(\mathbf{A}-2\mathbf{I})^2=2\\ \nu_2=4\end{array}$$

$$(\mathbf{A}-2\mathbf{I})^3=\begin{bmatrix}0 & 0 & 0 & 0 & 0 & 0\\ 0 & 0 & 0 & 0 & 0 & 0\\ 0 & 0 & 0 & 0 & 0 & 0\\ 0 & 0 & 0 & 0 & 0 & 0\\ 0 & 0 & 0 & 0 & -4 & 4\\ 0 & 0 & 0 & 0 & 4 & -4\end{bmatrix} \qquad \begin{array}{l}\rho(\mathbf{A}-2\mathbf{I})^3=1\\ \nu_3=5\end{array}$$

Since $\rho(\mathbf{A}-2\mathbf{I})^3=n-m=1$, we stop here. Because $\nu_3-\nu_2=1$, we can find a generalized eigenvector **u** of grade 3 such that $\mathbf{B}^3\mathbf{u}=\mathbf{0}$ and $\mathbf{B}^2\mathbf{u}\neq\mathbf{0}$. It is

---

[15] We use the fact that

$$\det\begin{bmatrix}\mathbf{A} & \mathbf{B}\\ \mathbf{0} & \mathbf{C}\end{bmatrix}=\det\mathbf{A}\det\mathbf{C}$$

where **A** and **C** are square matrices, not necessarily of the same order.

easy to verify that $\mathbf{u}=[0\ \ 0\ \ 1\ \ 0\ \ 0\ \ 0]'$ is such a vector. Define

$$\mathbf{u}_1 \triangleq \mathbf{B}^2\mathbf{u} = \begin{bmatrix} 2 \\ 2 \\ 0 \\ 0 \\ 0 \\ 0 \end{bmatrix} \qquad \mathbf{u}_2 \triangleq \mathbf{B}\mathbf{u} = \begin{bmatrix} 1 \\ -1 \\ 0 \\ 0 \\ 0 \\ 0 \end{bmatrix} \qquad \mathbf{u}_3 \triangleq \mathbf{u} = \begin{bmatrix} 0 \\ 0 \\ 1 \\ 0 \\ 0 \\ 0 \end{bmatrix}$$

This is a chain of generalized eigenvectors of length 3. Because of $\nu_2 - \nu_1 = 2$, there are two linearly independent vectors in $\mathcal{N}_2$ but not in $\mathcal{N}_1$. The vector $\mathbf{u}_2$ is one of them. We search a vector $\mathbf{v}$ which is independent of $\mathbf{u}_2$ and has the property $\mathbf{B}^2\mathbf{v}=\mathbf{0}$ and $\mathbf{B}\mathbf{v} \neq \mathbf{0}$. It can be readily verified that $\mathbf{v} = [0\ \ 0\ \ 1\ \ -1\ \ 1\ \ 1]'$ is such a vector. Define

$$\mathbf{v}_1 \triangleq \mathbf{B}\mathbf{v} = \begin{bmatrix} 0 \\ 0 \\ 2 \\ -2 \\ 0 \\ 0 \end{bmatrix} \qquad \mathbf{v}_2 = \mathbf{v} = \begin{bmatrix} 0 \\ 0 \\ 1 \\ -1 \\ 1 \\ 1 \end{bmatrix}$$

Now we have found five generalized eigenvectors of $\mathbf{A}$ associated with $\lambda = 2$.

3. Compute an eigenvector associated with $\lambda_2 = 0$. Let $\mathbf{w}$ be an eigenvector of $\mathbf{A}$ associated with $\lambda_2 = 0$; then

$$(\mathbf{A} - \lambda_2\mathbf{I})\mathbf{w} = \begin{bmatrix} 3 & -1 & 1 & 1 & 0 & 0 \\ 1 & 1 & -1 & -1 & 0 & 0 \\ 0 & 0 & 2 & 0 & 1 & 1 \\ 0 & 0 & 0 & 2 & -1 & -1 \\ 0 & 0 & 0 & 0 & 1 & 1 \\ 0 & 0 & 0 & 0 & 1 & 1 \end{bmatrix} \mathbf{w} = \mathbf{0}$$

Clearly, $\mathbf{w} = [0\ \ 0\ \ 0\ \ 0\ \ 1\ \ -1]'$ is a solution.

4. With respect to the basis $\{\mathbf{u}_1, \mathbf{u}_2, \mathbf{u}_3, \mathbf{v}_1, \mathbf{v}_2, \mathbf{w}\}$, $\mathbf{A}$ has the following Jordan-form representation:

$$\hat{\mathbf{A}} = \left[\begin{array}{ccc|cc|c} 2 & 1 & 0 & 0 & 0 & 0 \\ 0 & 2 & 1 & 0 & 0 & 0 \\ 0 & 0 & 2 & 0 & 0 & 0 \\ \hline 0 & 0 & 0 & 2 & 1 & 0 \\ 0 & 0 & 0 & 0 & 2 & 0 \\ \hline 0 & 0 & 0 & 0 & 0 & 0 \end{array}\right] \tag{2-55}$$

5. This may be checked by using

$$\hat{\mathbf{A}} = \mathbf{Q}^{-1}\mathbf{A}\mathbf{Q} \qquad \text{or} \qquad \mathbf{Q}\hat{\mathbf{A}} = \mathbf{A}\mathbf{Q}$$

where $\mathbf{Q} = [\mathbf{u}_1 \quad \mathbf{u}_2 \quad \mathbf{u}_3 \quad \mathbf{v}_1 \quad \mathbf{v}_2 \quad \mathbf{w}]$

$$= \begin{bmatrix} 2 & 1 & 0 & 0 & 0 & 0 \\ 2 & -1 & 0 & 0 & 0 & 0 \\ 0 & 0 & 1 & 2 & 1 & 0 \\ 0 & 0 & 0 & -2 & -1 & 0 \\ 0 & 0 & 0 & 0 & 1 & 1 \\ 0 & 0 & 0 & 0 & 1 & -1 \end{bmatrix}$$ ∎

In this example, if we reorder the basis $\{\mathbf{u}_1, \mathbf{u}_2, \mathbf{u}_3, \mathbf{v}_1, \mathbf{v}_2, \mathbf{w}\}$ and use $\{\mathbf{w}, \mathbf{v}_2, \mathbf{v}_1, \mathbf{u}_3, \mathbf{u}_2, \mathbf{u}_1\}$ as a new basis, then the representation will be

$$\hat{\mathbf{A}} = \begin{bmatrix} 0 & 0 & 0 & 0 & 0 & 0 \\ 0 & 2 & 0 & 0 & 0 & 0 \\ 0 & 1 & 2 & 0 & 0 & 0 \\ 0 & 0 & 0 & 2 & 0 & 0 \\ 0 & 0 & 0 & 1 & 2 & 0 \\ 0 & 0 & 0 & 0 & 1 & 2 \end{bmatrix} \tag{2-56}$$

This is also called a Jordan-form representation. Comparing it with Equation (2-55), we see that the new Jordan block in (2-56) has 1's on the diagonal just below the main diagonal as a result of the different ordering of the basis vectors. In this book, we use mostly the Jordan block of the form in (2-55). Certainly everything discussed for this form can be modified and be applied to the form given in Equation (2-56).

A Jordan-form representation of any linear operator $\mathbf{A}$ that maps $(\mathbb{C}^n, \mathbb{C})$ into itself is unique up to the ordering of Jordan blocks. That is, the number of Jordan blocks and the order of each Jordan block are uniquely determined by $\mathbf{A}$. However, because of different orderings of basis vectors, we may have different Jordan-form representations of the same matrix.

2-7 Functions of a Square Matrix

In this section we shall study functions of a square matrix or a linear transformation that maps $(\mathbb{C}^n, \mathbb{C})$ into itself. We shall use the Jordan-form representation extensively, because in terms of this representation almost all properties of a function of a matrix can be visualized. We study first polynomials of a square matrix, and then define functions of a matrix in terms of polynomials of the matrix.

Polynomials of a square matrix. Let $\mathbf{A}$ be a square matrix that maps $(\mathbb{C}^n, \mathbb{C})$ into $(\mathbb{C}^n, \mathbb{C})$. If k is a positive integer, we define

$$\mathbf{A}^k \triangleq \mathbf{A}\mathbf{A}\cdots\mathbf{A} \qquad (k \text{ terms}) \tag{2-57a}$$

and

$$\mathbf{A}^0 \triangleq \mathbf{I} \tag{2-57b}$$

where $\mathbf{I}$ is a unit matrix. Let $f(\lambda)$ be a polynomial in λ of finite degree; then $f(\mathbf{A})$ can be defined in terms of (2-57). For example, if $f(\lambda)=\lambda^3+2\lambda^2+6$, then

$$f(\mathbf{A}) \triangleq \mathbf{A}^3+2\mathbf{A}^2+6\mathbf{I}$$

We have shown in the preceding section that every square matrix $\mathbf{A}$ that maps $(\mathbb{C}^n, \mathbb{C})$ into itself has a Jordan-form representation, or equivalently, there exists a nonsingular constant matrix $\mathbf{Q}$ such that $\mathbf{A}=\mathbf{Q}\hat{\mathbf{A}}\mathbf{Q}^{-1}$ with $\hat{\mathbf{A}}$ in a Jordan canonical form. Since

$$\mathbf{A}^k=(\mathbf{Q}\hat{\mathbf{A}}\mathbf{Q}^{-1})(\mathbf{Q}\hat{\mathbf{A}}\mathbf{Q}^{-1})\cdots(\mathbf{Q}\hat{\mathbf{A}}\mathbf{Q}^{-1})=\mathbf{Q}\hat{\mathbf{A}}^k\mathbf{Q}^{-1}$$

we have

$$f(\mathbf{A})=\mathbf{Q}f(\hat{\mathbf{A}})\mathbf{Q}^{-1} \qquad \text{or} \qquad f(\hat{\mathbf{A}})=\mathbf{Q}^{-1}f(\mathbf{A})\mathbf{Q} \tag{2-58}$$

for any polynomial $f(\lambda)$.

One of the reasons to use the Jordan-form matrix is that if

$$\mathbf{A}=\begin{bmatrix} \mathbf{A}_1 & \mathbf{0} \\ \mathbf{0} & \mathbf{A}_2 \end{bmatrix} \tag{2-59}$$

where $\mathbf{A}_1$ and $\mathbf{A}_2$ are square matrices, then

$$f(\mathbf{A})=\begin{bmatrix} f(\mathbf{A}_1) & \mathbf{0} \\ \mathbf{0} & f(\mathbf{A}_2) \end{bmatrix} \tag{2-60}$$

This can be easily verified by observing that

$$\mathbf{A}^k=\begin{bmatrix} \mathbf{A}_1^k & \mathbf{0} \\ \mathbf{0} & \mathbf{A}_2^k \end{bmatrix}$$

Definition 2-14

The minimal polynomial of a matrix $\mathbf{A}$ is the monic polynomial[16] $\psi(\lambda)$ of least degree such that $\psi(\mathbf{A})=\mathbf{0}$. ∎

Note that the $\mathbf{0}$ in $\psi(\mathbf{A})=\mathbf{0}$ is an $n\times n$ square matrix whose entries are all zero. A direct consequence of (2-58) is that $f(\mathbf{A})=\mathbf{0}$ if and only if $f(\hat{\mathbf{A}})=\mathbf{0}$. Consequently, the matrices $\mathbf{A}$ and $\hat{\mathbf{A}}$ have the same minimal polynomial, or more generally, *similar matrices have the same minimal polynomial.* Computing the minimal polynomial of a matrix is generally not a simple job; however, if the Jordan-form representation of the matrix is available, its minimal polynomial can be readily found.

Let $\lambda_1, \lambda_2, \ldots, \lambda_m$ be the distinct eigenvalues of $\mathbf{A}$ with multiplicities $n_1, n_2, \ldots, n_m$, respectively. It is the same as saying that the characteristic

[16] A monic polynomial is a polynomial the coefficient of whose highest power is 1. For example, $3x+1$ and $-x^2+2x+4$ are not monic polynomials, but x^2-4x+7 is.

polynomial of **A** is

$$\Delta(\lambda) \triangleq \det(\lambda\mathbf{I} - \mathbf{A}) = \prod_{i=1}^{m} (\lambda - \lambda_i)^{n_i} \tag{2-61}$$

Assume that a Jordan-form representation of **A** is

$$\hat{\mathbf{A}} = \begin{bmatrix} \hat{\mathbf{A}}_1 & \mathbf{0} & \cdots & \mathbf{0} \\ \mathbf{0} & \hat{\mathbf{A}}_2 & \cdots & \mathbf{0} \\ \vdots & \vdots & & \vdots \\ \mathbf{0} & \mathbf{0} & \cdots & \hat{\mathbf{A}}_m \end{bmatrix} \tag{2-62}$$

where the $n_i \times n_i$ matrix $\hat{\mathbf{A}}_i$ denotes all the Jordan blocks associated with λ_i.

Definition 2-15

The largest order of the Jordan blocks associated with λ_i in **A** is called the *index* of λ_i in **A**. ∎

The multiplicity of λ_i is denoted by n_i, the index of λ_i is denoted by $\bar{n}_i$. For the matrix in (2-52), $n_1 = 8, \bar{n}_1 = 4$; for the matrix in (2-53), $n_1 = \bar{n}_1 = 2, n_2 = \bar{n}_2 = 1$; for the matrix in (2-55), $n_1 = 5, \bar{n}_1 = 3, n_2 = \bar{n}_2 = 1$. It is clear that $\bar{n}_i \leq n_i$.

Theorem 2-12

The minimal polynomial of **A** is

$$\psi(\lambda) = \prod_{i=1}^{m} (\lambda - \lambda_i)^{\bar{n}_i}$$

where $\bar{n}_i$ is the index of λ_i in **A**.

Proof

Since the matrices **A** and $\hat{\mathbf{A}}$ have the same minimal polynomial, it is the same as showing that $\psi(\lambda)$ is the polynomial with least degree such that $\psi(\hat{\mathbf{A}}) = \mathbf{0}$. We first show that the minimal polynomial of $\hat{\mathbf{A}}_i$ is $\psi_i(\lambda) = (\lambda - \lambda_i)^{\bar{n}_i}$. Suppose $\hat{\mathbf{A}}_i$ consists of r Jordan blocks associated with λ_i. Then

$$\hat{\mathbf{A}}_i = \operatorname{diag}(\hat{\mathbf{A}}_{i1}, \hat{\mathbf{A}}_{i2}, \ldots, \hat{\mathbf{A}}_{ir})$$

and

$$\psi_i(\hat{\mathbf{A}}_i) = \begin{bmatrix} \psi_i(\hat{\mathbf{A}}_{i1}) & \mathbf{0} & \cdots & \mathbf{0} \\ \mathbf{0} & \psi_i(\hat{\mathbf{A}}_{i2}) & \cdots & \mathbf{0} \\ \vdots & \vdots & & \vdots \\ \mathbf{0} & \mathbf{0} & \cdots & \psi_i(\hat{\mathbf{A}}_{ir}) \end{bmatrix}$$

$$= \begin{bmatrix} (\hat{\mathbf{A}}_{i1} - \lambda_i\mathbf{I})^{\bar{n}_i} & \mathbf{0} & \cdots & \mathbf{0} \\ \mathbf{0} & (\hat{\mathbf{A}}_{i2} - \lambda_i\mathbf{I})^{\bar{n}_i} & \cdots & \mathbf{0} \\ \vdots & \vdots & & \vdots \\ \mathbf{0} & \mathbf{0} & \cdots & (\hat{\mathbf{A}}_{ir} - \lambda_i\mathbf{I})^{\bar{n}_i} \end{bmatrix} \tag{2-63}$$

If the matrix $(\hat{\mathbf{A}}_{ij}-\lambda_i\mathbf{I})$ has dimension n_{ij}, then we have

$$\underset{(n_{ij}\times n_{ij})}{(\hat{\mathbf{A}}_{ij}-\lambda_i\mathbf{I})}=\begin{bmatrix}0 & 1 & 0 & \cdots & 0\\ 0 & 0 & 1 & \cdots & 0\\ \vdots & \vdots & \vdots & & \vdots\\ 0 & 0 & 0 & \cdots & 1\\ 0 & 0 & 0 & \cdots & 0\end{bmatrix} \tag{2-64a}$$

$$(\hat{\mathbf{A}}_{ij}-\lambda_i\mathbf{I})^2=\begin{bmatrix}0 & 0 & 1 & 0 & \cdots & 0\\ 0 & 0 & 0 & 1 & \cdots & 0\\ \vdots & \vdots & \vdots & \vdots & & \vdots\\ 0 & 0 & 0 & 0 & \cdots & 1\\ 0 & 0 & 0 & 0 & \cdots & 0\\ 0 & 0 & 0 & 0 & & 0\end{bmatrix} \tag{2-64b}$$

$$(\hat{\mathbf{A}}_{ij}-\lambda_i\mathbf{I})^{n_{ij}-1}=\begin{bmatrix}0 & 0 & 0 & \cdots & 0 & 1\\ 0 & 0 & 0 & \cdots & 0 & 0\\ \vdots & \vdots & \vdots & & \vdots & \vdots\\ 0 & 0 & 0 & \cdots & 0 & 0\end{bmatrix} \tag{2-64c}$$

and
$$(\hat{\mathbf{A}}_{ij}-\lambda_i\mathbf{I})^k=\mathbf{0} \qquad \text{for any integer } k\geq n_{ij} \tag{2-64d}$$

By definition, $\bar{n}_i$ is the largest order of the Jordan blocks in $\hat{\mathbf{A}}_i$, or equivalently, $\bar{n}_i=\max(n_{ij}, j=1,2,\ldots,r)$. Hence $(\hat{\mathbf{A}}_{ij}-\lambda_i\mathbf{I})^{\bar{n}_i}=\mathbf{0}$ for $j=1,2,\ldots,r$. Consequently, $\psi_i(\hat{\mathbf{A}}_i)=\mathbf{0}$. It is easy to see from (2-63) and (2-64) that if $\psi_i(\lambda)=(\lambda-\alpha)^k$ with either $\alpha\neq\lambda_i$ or $k<\bar{n}_i$, then $\psi_i(\hat{\mathbf{A}}_i)\neq\mathbf{0}$. Hence we conclude that $\psi_i=(\lambda-\lambda_i)^{\bar{n}_i}$ is the minimal polynomial of $\mathbf{A}_i$. Now we claim that $f(\hat{\mathbf{A}}_i)=\mathbf{0}$ if and only if f is divisible without remainder by ψ_i, denoted as $\psi_i|f$. Indeed, if $\psi_i|f$, then f can be written as $f=\psi_i h$, where h is the quotient polynomial, and $f(\mathbf{A}_i)=\psi_i(\hat{\mathbf{A}}_i)h(\hat{\mathbf{A}}_i)=\mathbf{0}\cdot h(\hat{\mathbf{A}}_i)=\mathbf{0}$. If f is not divisible without remainder by ψ_i, then f can be written as $f=\psi_i h+g$ where g is a polynomial of degree less than $\bar{n}_i$. Now $f(\hat{\mathbf{A}}_i)=\mathbf{0}$ implies $g(\hat{\mathbf{A}}_i)=\mathbf{0}$. This contradicts the assumption that ψ_i is the minimal polynomial of $\hat{\mathbf{A}}_i$, for g is a polynomial of degree less than that of ψ_i and $g(\hat{\mathbf{A}}_i)=\mathbf{0}$. With these preliminaries, the theorem can be readily proved. From (2-62), we have $\psi(\hat{\mathbf{A}})=\operatorname{diag}(\psi(\hat{\mathbf{A}}_1)\quad\psi(\hat{\mathbf{A}}_2)\quad\cdots\quad\psi(\hat{\mathbf{A}}_m))$. Since $\psi(\hat{\mathbf{A}}_i)=\mathbf{0}$ if and only if ψ contains the factor $(\lambda-\lambda_i)^{\bar{n}_i}$, we conclude that the minimal polynomial of $\hat{\mathbf{A}}$ and, correspondingly, of $\mathbf{A}$ is

$$\prod_{i=1}^{m}(\lambda-\lambda_i)^{\bar{n}_i} \qquad \text{Q.E.D.}$$

Example 1

The matrices

$$\begin{bmatrix}3 & 0 & 0 & 0\\ 0 & 3 & 0 & 0\\ 0 & 0 & 3 & 0\\ 0 & 0 & 0 & 1\end{bmatrix}\qquad\begin{bmatrix}3 & 1 & 0 & 0\\ 0 & 3 & 0 & 0\\ 0 & 0 & 3 & 0\\ 0 & 0 & 0 & 1\end{bmatrix}\qquad\begin{bmatrix}3 & 1 & 0 & 0\\ 0 & 3 & 1 & 0\\ 0 & 0 & 3 & 0\\ 0 & 0 & 0 & 1\end{bmatrix}$$

all have the same characteristic polynomial $\Delta(\lambda)=(\lambda-3)^3(\lambda-1)$; however, they have, respectively, $(\lambda-3)(\lambda-1)$, $(\lambda-3)^2(\lambda-1)$, and $(\lambda-3)^3(\lambda-1)$ as minimal polynomials. ∎

Because the characteristic polynomial is always divisible without remainder by the minimal polynomial, we have the following very important corollary of Theorem 2-12.

Corollary 2-12 (Cayley-Hamilton theorem)

Let $\Delta(\lambda) \triangleq \det(\lambda\mathbf{I}-\mathbf{A}) \triangleq \lambda^n + \alpha_1\lambda^{n-1} + \cdots + \alpha_{n-1}\lambda + \alpha_n$ be the characteristic polynomial of $\mathbf{A}$. Then

$$\Delta(\mathbf{A}) = \mathbf{A}^n + \alpha_1\mathbf{A}^{n-1} + \cdots + \alpha_{n-1}\mathbf{A} + \alpha_n\mathbf{I} = \mathbf{0}$$ ∎

The Cayley-Hamilton theorem can also be proved directly without using Theorem 2-12 (see Problems 2-39 and 2-40).

The reason for introducing the concept of minimal polynomial will be seen in the following theorem.

Theorem 2-13

Let $\lambda_1, \lambda_2, \ldots, \lambda_m$ be the distinct eigenvalues of $\mathbf{A}$ with indices $\bar{n}_1, \bar{n}_2, \ldots, \bar{n}_m$. Let f and g be two polynomials. Then the following statements are equivalent.

1. $f(\mathbf{A}) = g(\mathbf{A})$.
2. Either $f = h_1\psi + g$ or $g = h_2\psi + f$, where ψ is the minimal polynomial of $\mathbf{A}$, and h_1 and h_2 are some polynomials.
3. $$f^{(l)}(\lambda_i) = g^{(l)}(\lambda_i) \qquad \text{for } l = 0, 1, 2, \ldots, \bar{n}_i - 1;\ i = 1, 2, \ldots, m \qquad \textbf{(2-65)}$$
   where $f^{(l)}(\lambda_i) \triangleq \left.\dfrac{d^l f(\lambda)}{d\lambda^l}\right|_{\lambda=\lambda_i}$ and $g^{(l)}(\lambda_i)$ is similarly defined.

Proof

The equivalence of statements 1 and 2 follows directly from the fact that $\psi(\mathbf{A}) = \mathbf{0}$. Statements 2 and 3 are equivalent following

$$\psi(\lambda) = \prod_{i=1}^{m} (\lambda - \lambda_i)^{\bar{n}_i}$$ Q.E.D.

In order to apply this theorem, we must know the minimal polynomial of $\mathbf{A}$. The minimal polynomial can be obtained by transforming $\mathbf{A}$ into a Jordan form or by direct computation (Problem 2-42). Both methods are complicated. Therefore it is desirable to modify Theorem 2-13 so that the use of the minimal polynomial can be avoided.

Corollary 2-13

Let the characteristic polynomial of $\mathbf{A}$ be

$$\Delta(\lambda) \triangleq \det(\lambda\mathbf{I}-\mathbf{A}) = \prod_{i=1}^{m} (\lambda - \lambda_i)^{n_i}$$

Let f and g be two arbitrary polynomials. If

$$f^{(l)}(\lambda_i)=g^{(l)}(\lambda_i) \qquad \text{for } l=0,1,2,\ldots,n_i-1 \quad i=1,2,\ldots,m \tag{2-66}$$

then $f(\mathbf{A})=g(\mathbf{A})$. ∎

This follows immediately from Theorem 2-13 by observing that the condition (2-66) implies (2-65). The set of numbers $f^{(l)}(\lambda_i)$, for $i=1,2,\ldots,m$ and $l=0,1,2,\ldots,n_i-1$ (there are totally $n=\sum_{i=1}^{m} n_i$) are called *the values of f on the spectrum of* **A**. Corollary 2-13 implies that any two polynomials that have the same values on the spectrum of **A** define the same matrix function. To state it in a different way: Given n numbers, if we can construct a polynomial which gives these numbers on the spectrum of **A**, then this polynomial defines uniquely a matrix-valued function of **A**. It is well known that given any n numbers, it is possible to find a polynomial $g(\lambda)$ of degree $n-1$ that gives these n numbers at some preassigned λ. Hence if **A** is of order n, for any polynomial $f(\lambda)$, we can construct a polynomial of degree $n-1$,

$$g(\lambda)=\alpha_0+\alpha_1\lambda+\cdots+\alpha_{n-1}\lambda^{n-1} \tag{2-67}$$

such that $g(\lambda)=f(\lambda)$ on the spectrum of **A**. Hence any polynomial of **A** can be expressed as

$$f(\mathbf{A})=g(\mathbf{A})=\alpha_0\mathbf{I}+\alpha_1\mathbf{A}+\cdots+\alpha_{n-1}\mathbf{A}^{n-1}$$

This fact can also be deduced directly from Corollary 2-12 (Problem 2-38).

Corollary 2-13 is useful in computing any polynomial and, as will be discussed, any function of **A**. If **A** is of order n, the polynomial $g(\lambda)$ can be chosen as in (2-67) or as any polynomial of degree $n-1$ with n *independent* parameters. For example, if all eigenvalues, λ_i, $i=1,2,\ldots,n$, of **A** are distinct, then $g(\lambda)$ can be chosen as

$$g(\lambda)=\sum_{i=0}^{n-1}\beta_i\prod_{\substack{j=1\\ j\neq i}}^{n}(\lambda-\lambda_j)$$

or

$$g(\lambda)=\sum_{i=0}^{n-1}\beta_i\prod_{j=1}^{i}(\lambda-\lambda_j)$$

In conclusion, the form of $g(\lambda)$ can be chosen to facilitate the computation.

Example 2

Compute $\mathbf{A}^{100}$, where

$$\mathbf{A}=\begin{bmatrix}1 & 2\\ 0 & 1\end{bmatrix}$$

In other words, given $f(\lambda)=\lambda^{100}$, compute $f(\mathbf{A})$. The characteristic polynomial of **A** is $\Delta(\lambda)=\det(\lambda\mathbf{I}-\mathbf{A})=(\lambda-1)^2$. Let $g(\lambda)$ be a polynomial of degree $n-1=1$, say

$$g(\lambda)=\alpha_0+\alpha_1\lambda$$

Now, from Corollary 2-13, if $f(\lambda)=g(\lambda)$ on the spectrum of $\mathbf{A}$, then $f(\mathbf{A})=g(\mathbf{A})$. On the spectrum of $\mathbf{A}$, we have

$$\begin{aligned} f(1)&=g(1) & (1)^{100}&=\alpha_0+\alpha_1 \\ f'(1)&=g'(1) & 100\cdot(1)^{99}&=\alpha_1 \end{aligned}$$

Solving these two equations, we obtain $\alpha_1=100$ and $\alpha_0=-99$. Hence

$$\mathbf{A}^{100}=g(\mathbf{A})=\alpha_0\mathbf{I}+\alpha_1\mathbf{A}=-99\begin{bmatrix}1 & 0\\ 0 & 1\end{bmatrix}+100\begin{bmatrix}1 & 2\\ 0 & 1\end{bmatrix}=\begin{bmatrix}1 & 200\\ 0 & 1\end{bmatrix}$$

Obviously $\mathbf{A}^{100}$ can also be obtained by multiplying $\mathbf{A}$ 100 times or by using a different $g(\lambda)$ such as $g(\lambda)=\alpha_0+\alpha_1(\lambda-1)$ (Problem 2-33). ∎

Functions of a square matrix

Definition 2-16

Let $f(\lambda)$ be a function (not necessarily a polynomial) that is defined on the spectrum of $\mathbf{A}$. If $g(\lambda)$ is a polynomial that has the same values as $f(\lambda)$ on the spectrum of $\mathbf{A}$, then the matrix-valued function $f(\mathbf{A})$ is defined as $f(\mathbf{A})\triangleq g(\mathbf{A})$. ∎

This definition is an extension of Corollary 2-13 to include functions. To be precise, functions of a matrix should be defined by using the conditions in (2-65). The conditions in (2-66) are used because the characteristic polynomial is easier to obtain than the minimal polynomial. Of course, both conditions will lead to the same result.

If $\mathbf{A}$ is an $n\times n$ matrix, given the n values of $f(\lambda)$ on the spectrum of $\mathbf{A}$, we can find a polynomial of degree $n-1$,

$$g(\lambda)=\alpha_0+\alpha_1\lambda+\cdots+\alpha_{n-1}\lambda^{n-1}$$

which is equal to $f(\lambda)$ on the spectrum of $\mathbf{A}$. Hence from this definition we know that every function of $\mathbf{A}$ can be expressed as

$$f(\mathbf{A})=\alpha_0\mathbf{I}+\alpha_1\mathbf{A}+\cdots+\alpha_{n-1}\mathbf{A}^{n-1}$$

We summarize the procedure of computing a function of a matrix: Given an $n\times n$ matrix $\mathbf{A}$ and a function $f(\lambda)$, we first compute the characteristic polynomial of $\mathbf{A}$, say

$$\Delta(\lambda)=\prod_{i=1}^{m}(\lambda-\lambda_i)^{n_i}$$

Let

$$g(\lambda)=\alpha_0+\alpha_1\lambda+\cdots+\alpha_{n-1}\lambda^{n-1}$$

where $\alpha_0, \alpha_1, \ldots, \alpha_{n-1}$ are n unknowns. Next we use the n equations in (2-66) to compute these α_i's in terms of the values of f on the spectrum of $\mathbf{A}$. Then we have $f(\mathbf{A})=g(\mathbf{A})$. We note that other polynomial $g(\lambda)$ of degree $n-1$ with n independent parameters can also be used.

Example 3

Let

$$\mathbf{A}_1 = \begin{bmatrix} 0 & 0 & -2 \\ 0 & 1 & 0 \\ 1 & 0 & 3 \end{bmatrix}$$

Compute $e^{\mathbf{A}_1 t}$. Or equivalently, if $f(\lambda) = e^{\lambda t}$, what is $f(\mathbf{A}_1)$?

The characteristic polynomial of $\mathbf{A}_1$ is $(\lambda - 1)^2(\lambda - 2)$. Let $g(\lambda) = \alpha_0 + \alpha_1 \lambda + \alpha_2 \lambda^2$. Then

$$f(1) = g(1) \qquad e^t = \alpha_0 + \alpha_1 + \alpha_2$$

$$f'(1) = g'(1) \qquad te^t = \alpha_1 + 2\alpha_2$$ (note that the derivative is with respect to λ, not t)

$$f(2) = g(2) \qquad e^{2t} = \alpha_0 + 2\alpha_1 + 4\alpha_2$$

Solving these equations, we obtain $\alpha_0 = -2te^t + e^{2t}$, $\alpha_1 = 3te^t + 2e^t - 2e^{2t}$, and $\alpha_2 = e^{2t} - e^t - te^t$. Hence, we have

$$e^{\mathbf{A}_1 t} = g(\mathbf{A}_1) = (-2te^t + e^{2t})\mathbf{I} + (3te^t + 2e^t - 2e^{2t})\mathbf{A}_1 + (e^{2t} - e^t - te^t)\mathbf{A}_1^2$$

$$= \begin{bmatrix} 2e^t - e^{2t} & 0 & 2e^t - 2e^{2t} \\ 0 & e^t & 0 \\ -e^t + e^{2t} & 0 & 2e^{2t} - e^t \end{bmatrix}$$ ∎

Example 4

Let

$$\mathbf{A}_2 = \begin{bmatrix} 0 & 2 & -2 \\ 0 & 1 & 0 \\ 1 & -1 & 3 \end{bmatrix}$$

Its characteristic polynomial is $\Delta(\lambda) = (\lambda - 1)^2(\lambda - 2)$, which is the same as the one of $\mathbf{A}_1$ in Example 3. Hence we have the same $g(\lambda)$ as in Example 3. Consequently, we have

$$e^{\mathbf{A}_2 t} = g(\mathbf{A}_2) = \begin{bmatrix} 2e^t - e^{2t} & 2te^t & 2e^t - 2e^{2t} \\ 0 & e^t & 0 \\ e^{2t} - e^t & -te^t & 2e^{2t} - e^t \end{bmatrix}$$ ∎

Example 5

Given

$$\underset{(n \times n)}{\hat{\mathbf{A}}} = \begin{bmatrix} \lambda_1 & 1 & 0 & \cdots & 0 \\ 0 & \lambda_1 & 1 & \cdots & 0 \\ \vdots & \vdots & \vdots & & \vdots \\ 0 & 0 & 0 & \cdots & 1 \\ 0 & 0 & 0 & \cdots & \lambda_1 \end{bmatrix} \tag{2-68}$$

The characteristic polynomial of $\hat{\mathbf{A}}$ is $(\lambda - \lambda_1)^n$. Let the polynomial $g(\lambda)$ be of the form

$$g(\lambda)=\alpha_0+\alpha_1(\lambda-\lambda_1)+\alpha_2(\lambda-\lambda_1)^2+\cdots+\alpha_{n-1}(\lambda-\lambda_1)^{n-1}$$

Then the conditions in (2-66) give immediately

$$\alpha_0=f(\lambda_1),\quad \alpha_1=f'(\lambda_1),\quad \ldots,\quad \alpha_{n-1}=\frac{f^{(n-1)}(\lambda_1)}{(n-1)!}$$

Hence,

$$f(\hat{\mathbf{A}})=g(\hat{\mathbf{A}})=f(\lambda_1)\mathbf{I}+\frac{f'(\lambda_1)}{1!}(\hat{\mathbf{A}}-\lambda_1\mathbf{I})+\cdots+\frac{f^{(n-1)}(\lambda_1)}{(n-1)!}(\hat{\mathbf{A}}-\lambda_1\mathbf{I})^{n-1}$$

$$=\begin{bmatrix} f(\lambda_1) & f'(\lambda_1)/1! & f''(\lambda_1)/2! & \cdots & f^{(n-1)}(\lambda_1)/(n-1)! \\ 0 & f(\lambda_1) & f'(\lambda_1)/1! & \cdots & f^{(n-2)}(\lambda_1)/(n-2)! \\ 0 & 0 & f(\lambda_1) & \cdots & f^{(n-3)}(\lambda_1)/(n-3)! \\ \vdots & \vdots & \vdots & & \vdots \\ 0 & 0 & 0 & \cdots & f(\lambda_1) \end{bmatrix} \tag{2-69}$$

Here in the last step we have used (2-64).

If $f(\lambda)=e^{\lambda t}$, then

$$e^{\hat{\mathbf{A}}t}=\begin{bmatrix} e^{\lambda_1 t} & te^{\lambda_1 t} & t^2e^{\lambda_1 t}/2! & \cdots & t^{n-1}e^{\lambda_1 t}/(n-1)! \\ 0 & e^{\lambda_1 t} & te^{\lambda_1} & \cdots & t^{n-2}e^{\lambda_1 t}/(n-2)! \\ \vdots & \vdots & \vdots & & \vdots \\ 0 & 0 & 0 & \cdots & e^{\lambda_1 t} \end{bmatrix} \tag{2-70}$$

Note that the derivatives in (2-69) are taken with respect to λ_1, not to t. ∎

A function of a matrix is defined through a polynomial of the matrix; therefore, the relations that hold for polynomials can also be applied to functions of a matrix. For example, if $\mathbf{A}=\mathbf{Q}\hat{\mathbf{A}}\mathbf{Q}^{-1}$, then

$$f(\mathbf{A})=\mathbf{Q}f(\hat{\mathbf{A}})\mathbf{Q}^{-1}$$

and if

$$\mathbf{A}=\begin{bmatrix} \mathbf{A}_1 & \mathbf{0} \\ \mathbf{0} & \mathbf{A}_2 \end{bmatrix}$$

then

$$f(\mathbf{A})=\begin{bmatrix} f(\mathbf{A}_1) & \mathbf{0} \\ \mathbf{0} & f(\mathbf{A}_2) \end{bmatrix} \tag{2-71}$$

for any function f that is defined on the spectrum of $\mathbf{A}$. Using (2-69) and (2-71), any function of a Jordan-canonical-form matrix can be obtained immediately.

Example 6

Consider

$$\mathbf{A}=\left[\begin{array}{ccc:cc} \lambda_1 & 1 & 0 & 0 & 0 \\ 0 & \lambda_1 & 1 & 0 & 0 \\ 0 & 0 & \lambda_1 & 0 & 0 \\ \hdashline 0 & 0 & 0 & \lambda_2 & 1 \\ 0 & 0 & 0 & 0 & \lambda_2 \end{array}\right] \tag{2-72}$$

If $f(\lambda)=e^{\lambda t}$, then

$$f(\mathbf{A})=e^{\mathbf{A}t}=\left[\begin{array}{ccc|cc} e^{\lambda_1 t} & te^{\lambda_1 t} & t^2e^{\lambda_1 t}/2! & 0 & 0 \\ 0 & e^{\lambda_1 t} & te^{\lambda_1 t} & 0 & 0 \\ 0 & 0 & e^{\lambda_1 t} & 0 & 0 \\ \hline 0 & 0 & 0 & e^{\lambda_2 t} & te^{\lambda_2 t} \\ 0 & 0 & 0 & 0 & e^{\lambda_2 t} \end{array}\right] \tag{2-73}$$

If $f(\lambda)=(s-\lambda)^{-1}$, where s is a complex variable, then

$$f(\mathbf{A})=(s\mathbf{I}-\mathbf{A})^{-1}$$

$$=\left[\begin{array}{ccc|cc} \frac{1}{s-\lambda_1} & \frac{1}{(s-\lambda_1)^2} & \frac{1}{(s-\lambda_1)^3} & 0 & 0 \\ 0 & \frac{1}{s-\lambda_1} & \frac{1}{(s-\lambda_1)^2} & 0 & 0 \\ 0 & 0 & \frac{1}{s-\lambda_1} & 0 & 0 \\ \hline 0 & 0 & 0 & \frac{1}{s-\lambda_2} & \frac{1}{(s-\lambda_2)^2} \\ 0 & 0 & 0 & 0 & \frac{1}{s-\lambda_2} \end{array}\right] \tag{2-74}$$

■

Functions of a matrix defined by means of power series. We have used a polynomial of finite degree to define a function of a matrix. We shall now use an infinite series to give an alternative expression of a function of a matrix.

Definition 2-17

Let the power series representation of a function f be

$$f(\lambda)=\sum_{i=0}^{\infty}\alpha_i\lambda^i \tag{2-75}$$

with the radius of convergence ρ. Then the function f of a square matrix $\mathbf{A}$ is defined as

$$f(\mathbf{A})\triangleq\sum_{i=0}^{\infty}\alpha_i\mathbf{A}^i \tag{2-76}$$

if the absolute values of all the eigenvalues of $\mathbf{A}$ are smaller than ρ, the radius of convergence; or the matrix $\mathbf{A}$ has the property $\mathbf{A}^k=\mathbf{0}$ for some positive integer k. ■

This definition is meaningful only if the infinite series in (2-76) converges. If $\mathbf{A}^k=\mathbf{0}$ for some positive integer k, then (2-76) reduces to

$$f(\mathbf{A})=\sum_{i=0}^{k-1}\alpha_i\mathbf{A}^i$$

If the absolute values of all the eigenvalues of **A** are smaller than ρ, it can also be shown that the infinite series converges. For a proof, see Reference 77.

Instead of proving that Definitions 2-16 and 2-17 lead to exactly the same matrix function, we shall demonstrate this by using Definition 2-17 to derive (2-69).

Example 7

Consider the Jordan-form matrix $\hat{\mathbf{A}}$ given in (2-68). Let

$$f(\lambda) = f(\lambda_1) + f'(\lambda_1)(\lambda - \lambda_1) + \frac{f''(\lambda_1)}{2!}(\lambda - \lambda_1)^2 + \cdots$$

then

$$f(\hat{\mathbf{A}}) \triangleq f(\lambda_1)\mathbf{I} + f'(\lambda_1)(\hat{\mathbf{A}} - \lambda_1\mathbf{I}) + \cdots + \frac{f^{(n-1)}(\lambda_1)}{(n-1)!}(\hat{\mathbf{A}} - \lambda_1\mathbf{I})^{n-1} + \cdots \tag{2-77}$$

Since $(\hat{\mathbf{A}} - \lambda_1\mathbf{I})^i$ is of the form of (2-64), the matrix function (2-77) reduces immediately to (2-69). ∎

Example 8

The exponential function

$$e^{\lambda t} = 1 + \lambda t + \frac{\lambda^2 t^2}{2!} + \cdots + \frac{\lambda^n t^n}{n!} + \cdots$$

converges for all finite λ and t. Hence for any **A**, we have

$$e^{\mathbf{A}t} = \sum_{k=0}^{\infty} \frac{1}{k!} t^k \mathbf{A}^k \tag{2-78}$$

∎

A remark is in order concerning the computation of $e^{\mathbf{A}t}$. If $e^{\mathbf{A}t}$ is computed by using Definition 2-16, a closed-form matrix can be obtained. However, it requires the computation of the eigenvalues of **A**. This step can be avoided if the infinite series (2-78) is used. Clearly, the disadvantage of using (2-78) is that the resulting matrix may not be in a closed form. However, since the series (2-78) converges very fast, the series is often used to compute $e^{\mathbf{A}t}$ on a digital computer.

We derive some important properties of exponential functions of matrices to close this section. Using (2-78), it can be shown that

$$e^{0} = \mathbf{I}$$

$$e^{\mathbf{A}(t+s)} = e^{\mathbf{A}t}e^{\mathbf{A}s} \tag{2-79}$$

$$e^{(\mathbf{A}+\mathbf{B})t} = e^{\mathbf{A}t}e^{\mathbf{B}t} \quad \text{if and only if } \mathbf{AB} = \mathbf{BA} \tag{2-80}$$

In (2-79), if we choose $s = -t$, then from the fact that $e^0 = \mathbf{I}$, we have

$$[e^{\mathbf{A}t}]^{-1} = e^{-\mathbf{A}t} \tag{2-81}$$

By differentiation, term by term, of (2-78), we have

$$\frac{d}{dt}e^{\mathbf{A}t} = \sum_{k=1}^{\infty} \frac{1}{(k-1)!} t^{k-1}\mathbf{A}^k = \mathbf{A}\left(\sum_{k=0}^{\infty} \frac{1}{k!} t^k \mathbf{A}^k\right)$$
$$= \mathbf{A}e^{\mathbf{A}t} = e^{\mathbf{A}t}\mathbf{A} \tag{2-82}$$

Here we have used the fact that functions of the same matrix commute (see Problem 2-36).

The Laplace transform of a function f defined on $[0, \infty)$ is defined as

$$\hat{f}(s) \triangleq \mathscr{L}[f(t)] \triangleq \int_0^\infty f(t)e^{-st}\,dt \tag{2-83}$$

It is easy to show that

$$\mathscr{L}\left[\frac{t^k}{k!}\right] = s^{-(k+1)}$$

By taking the Laplace transform of (2-78), we have

$$\mathscr{L}(e^{\mathbf{A}t}) = \sum_{k=0}^{\infty} s^{-(k+1)}\mathbf{A}^k = s^{-1}\sum_{k=0}^{\infty}(s^{-1}\mathbf{A})^k \tag{2-84}$$

It is well known that the infinite series

$$f(\lambda) = (1-\lambda)^{-1} = 1 + \lambda + \lambda^2 + \cdots = \sum_{k=0}^{\infty} \lambda^k$$

converges for $|\lambda| < 1$. Now if s is chosen sufficiently large, the absolute values of all the eigenvalues of $s^{-1}\mathbf{A}$ are smaller than 1. Hence from Definition 2-17, we have

$$(\mathbf{I} - s^{-1}\mathbf{A})^{-1} = \sum_{k=0}^{\infty}(s^{-1}\mathbf{A})^k \tag{2-85}$$

Hence from (2-84) we have

$$\mathscr{L}(e^{\mathbf{A}t}) = s^{-1}(\mathbf{I} - s^{-1}\mathbf{A})^{-1} = (s\mathbf{I} - \mathbf{A})^{-1} \tag{2-86}$$

In this derivation, Equation (2-86) holds only for sufficiently large s. However, it can be shown by analytic continuation that Equation (2-86) does hold for all s except at the eigenvalues of $\mathbf{A}$. Equation (2-86) can also be established from (2-82). Because of $\mathscr{L}[dh(t)/dt] = s\mathscr{L}[h(t)] - h(0)$, the application of the Laplace transform to (2-82) yields

$$s\mathscr{L}(e^{\mathbf{A}t}) - e^{\mathbf{0}} = \mathbf{A}\mathscr{L}(e^{\mathbf{A}t})$$

or

$$(s\mathbf{I} - \mathbf{A})\mathscr{L}(e^{\mathbf{A}t}) = \mathbf{I}$$

which yields immediately (2-86). For the matrices in (2-73) and (2-74), we can also readily establish (2-86).

2-8 Norms and Inner Product[17]

All the concepts introduced in this section are applicable to any linear space over the field of complex numbers or over the field of real numbers. However, for convenience in the discussion, we restrict ourself to the complex vector space $(\mathbb{C}^n, \mathbb{C})$.

The concept of the norm of a vector $\mathbf{x}$ in $(\mathbb{C}^n, \mathbb{C})$ is a generalization of the idea of length. Any real-valued function of $\mathbf{x}$, denoted by $\|\mathbf{x}\|$, can be defined as a norm if it has the properties that for any $\mathbf{x}$ in $(\mathbb{C}^n, \mathbb{C})$ and any α in $\mathbb{C}$

1. $\|\mathbf{x}\| \geq 0$ and $\|\mathbf{x}\| = 0$ if and only if $\mathbf{x} = \mathbf{0}$.
2. $\|\alpha\mathbf{x}\| = |\alpha| \, \|\mathbf{x}\|$.
3. $\|\mathbf{x}_1 + \mathbf{x}_2\| \leq \|\mathbf{x}_1\| + \|\mathbf{x}_2\|$.

The last inequality is called the *triangular inequality*.

Let $\mathbf{x} = [x_1 \quad x_2 \quad \cdots \quad x_n]'$. Then the norm of $\mathbf{x}$ can be chosen as

$$\|\mathbf{x}\|_1 \triangleq \sum_{i=1}^{n} |x_i| \tag{2-87}$$

or

$$\|\mathbf{x}\|_2 \triangleq \left(\sum_{i=1}^{n} |x_i|^2 \right)^{1/2} \tag{2-88}$$

or

$$\|\mathbf{x}\|_\infty \triangleq \max_i |x_i| \tag{2-89}$$

It is easy to verify that each of them satisfies all the properties of a norm. The norm $\|\cdot\|_2$ is called the *euclidean norm*. In this book, the concept of norm is used mainly in the stability study; we use the fact that $\|\mathbf{x}\|$ *is finite if and only if all the components of* $\mathbf{x}$ *are finite*.

The concept of norm can be extended to linear operators that map $(\mathbb{C}^n, \mathbb{C})$ into itself, or equivalently, to square matrices with complex coefficients. The norm of a matrix $\mathbf{A}$ is defined as

$$\|\mathbf{A}\| \triangleq \sup_{\mathbf{x} \neq \mathbf{0}} \frac{\|\mathbf{A}\mathbf{x}\|}{\|\mathbf{x}\|} = \sup_{\|\mathbf{x}\|=1} \|\mathbf{A}\mathbf{x}\|$$

where "sup" stands for supremum, the largest possible number of $\|\mathbf{A}\mathbf{x}\|$ or the least upper bound of $\|\mathbf{A}\mathbf{x}\|$. An immediate consequence of the definition of $\|\mathbf{A}\|$ is, for any $\mathbf{x}$ in $(\mathbb{C}^n, \mathbb{C})$,

$$\|\mathbf{A}\mathbf{x}\| \leq \|\mathbf{A}\| \, \|\mathbf{x}\| \tag{2-90}$$

The norm of $\mathbf{A}$ is defined through the norm of $\mathbf{x}$; hence it is called an *induced norm*. For different $\|\mathbf{x}\|$, we have different $\|\mathbf{A}\|$. For example, if $\|\mathbf{x}\|_1$ is used, then

$$\|\mathbf{A}\|_1 = \max_j \left(\sum_{i=1}^{n} |a_{ij}| \right)$$

[17] May be skipped without loss of continuity. The material in this section is used only in Chapter 8, and its study may be coupled with that chapter.

where a_{ij} is the ijth element of **A**. If $||\mathbf{x}||_2$ is used, then

$$||\mathbf{A}||_2 = (\lambda_{\max}(\mathbf{A}^*\mathbf{A}))^{1/2}$$

where $\mathbf{A}^*$ is the complex conjugate transpose of **A** and $\lambda_{\max}(\mathbf{A}^*\mathbf{A})$ denotes the largest eigenvalue of $\mathbf{A}^*\mathbf{A}$ (see Appendix E). If $||\mathbf{x}||_\infty$ is used, then

$$||\mathbf{A}||_\infty = \max_i \left(\sum_{j=1}^{n} |a_{ij}| \right)$$

These norms are all different, as can be seen from Figure 2-7.

The norm of a matrix has the following properties

$$||\mathbf{A}+\mathbf{B}|| \leq ||\mathbf{A}|| + ||\mathbf{B}|| \tag{2-91}$$

$$||\mathbf{AB}|| \leq ||\mathbf{A}|| \, ||\mathbf{B}|| \tag{2-92}$$

These inequalities can be readily verified by observing

$$||(\mathbf{A}+\mathbf{B})\mathbf{x}|| = ||\mathbf{Ax}+\mathbf{Bx}|| \leq ||\mathbf{Ax}|| + ||\mathbf{Bx}|| \leq (||\mathbf{A}|| + ||\mathbf{B}||)||\mathbf{x}||$$

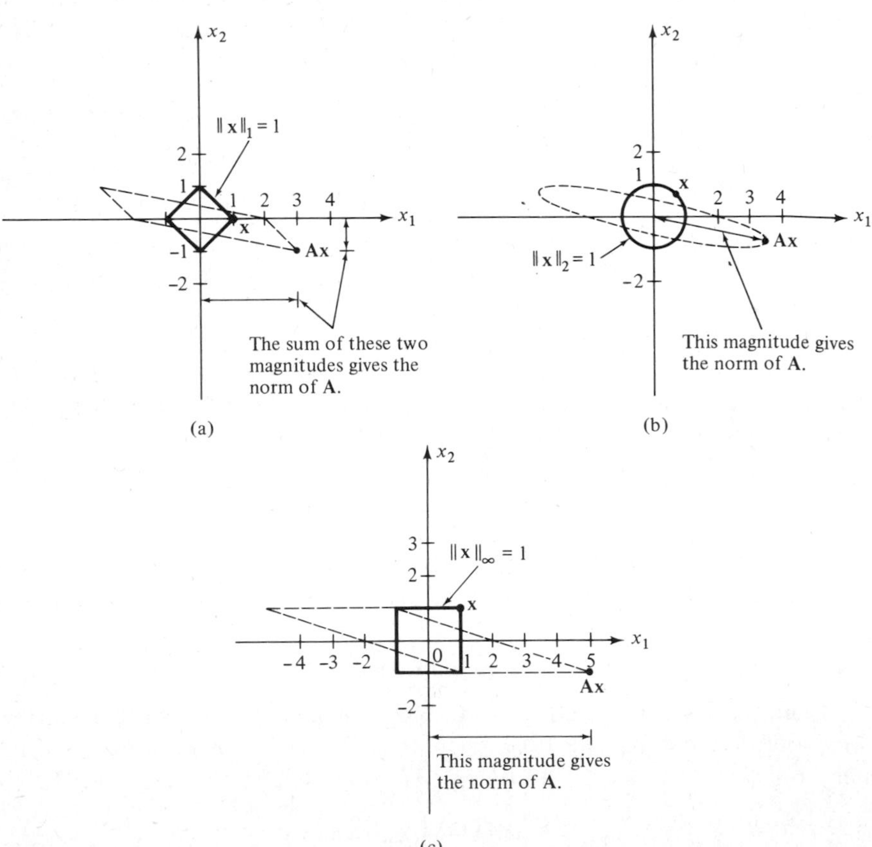

Figure 2-7 Solid lines denote **x**; broken lines denote **Ax**, where $\mathbf{A} = \begin{bmatrix} 3 & 2 \\ -1 & 0 \end{bmatrix}$

(a) $||\mathbf{A}||_1 = 4$. (b) $||\mathbf{A}||_2 = 3.7$. (c) $||\mathbf{A}||_\infty = 5$.

and $$||\mathbf{ABx}|| \le ||\mathbf{A}||\,||\mathbf{Bx}|| \le ||\mathbf{A}||\,||\mathbf{B}||\,||\mathbf{x}||$$

for any $\mathbf{x}$.

The norm is a function of a vector. Now we shall introduce a function of two vectors, called the *scalar product* or *inner product*. The inner product of two vectors $\mathbf{x}$ and $\mathbf{y}$ in $(\mathbb{C}^n, \mathbb{C})$ is a complex number, denoted by $\langle \mathbf{x}, \mathbf{y} \rangle$, having the properties that for any $\mathbf{x}, \mathbf{y}$ in $(\mathbb{C}^n, \mathbb{C})$ and any α_1, α_2 in $\mathbb{C}$,

$$\begin{aligned} &\overline{\langle \mathbf{x}, \mathbf{y} \rangle} = \langle \mathbf{y}, \mathbf{x} \rangle \\ &\langle \alpha_1 \mathbf{x}_1 + \alpha_2 \mathbf{x}_2, \mathbf{y} \rangle = \bar{\alpha}_1 \langle \mathbf{x}_1, \mathbf{y} \rangle + \bar{\alpha}_2 \langle \mathbf{x}_2, \mathbf{y} \rangle \\ &\langle \mathbf{x}, \mathbf{x} \rangle > 0 \qquad \text{for all } \mathbf{x} \neq \mathbf{0} \end{aligned}$$

where the "overbar" denotes the complex conjugate of a number. The first property implies that $\langle \mathbf{x}, \mathbf{x} \rangle$ is a real number. The first two properties imply that $\langle \mathbf{x}, \alpha\mathbf{y} \rangle = \alpha \langle \mathbf{x}, \mathbf{y} \rangle$.

In the complex vector space $(\mathbb{C}^n, \mathbb{C})$, the inner product is always taken to be

$$\langle \mathbf{x}, \mathbf{y} \rangle = \mathbf{x}^*\mathbf{y} = \sum_{i=1}^{n} \bar{x}_i y_i \qquad \textbf{(2-93)}$$

where $\mathbf{x}^*$ is the complex conjugate transpose of $\mathbf{x}$. Hence, for any square matrix $\mathbf{A}$, we have

$$\langle \mathbf{x}, \mathbf{Ay} \rangle = \mathbf{x}^*\mathbf{Ay}$$

and
$$\langle \mathbf{A}^*\mathbf{x}, \mathbf{y} \rangle = (\mathbf{A}^*\mathbf{x})^*\mathbf{y} = \mathbf{x}^*\mathbf{Ay}$$

Consequently we have

$$\langle \mathbf{x}, \mathbf{Ay} \rangle = \langle \mathbf{A}^*\mathbf{x}, \mathbf{y} \rangle \qquad \textbf{(2-94)}$$

The inner product provides a natural norm for a vector $\mathbf{x}$: $||\mathbf{x}|| = (\langle \mathbf{x}, \mathbf{x} \rangle)^{1/2}$ In fact, this is the norm defined in Equation (2-88).

Theorem 2-14 (Schwarz inequality)

If we define $||\mathbf{x}|| = (\langle \mathbf{x}, \mathbf{x} \rangle)^{1/2}$, then

$$|\langle \mathbf{x}, \mathbf{y} \rangle| \le ||\mathbf{x}||\,||\mathbf{y}||$$

Proof

The inequality is obviously true if $\mathbf{y} = \mathbf{0}$. Assume now $\mathbf{y} \neq \mathbf{0}$. Clearly we have

$$0 \le \langle \mathbf{x} + \alpha\mathbf{y}, \mathbf{x} + \alpha\mathbf{y} \rangle = \langle \mathbf{x}, \mathbf{x} \rangle + \bar{\alpha} \langle \mathbf{y}, \mathbf{x} \rangle + \alpha \langle \mathbf{x}, \mathbf{y} \rangle + \alpha\bar{\alpha} \langle \mathbf{y}, \mathbf{y} \rangle \qquad \textbf{(2-95)}$$

for any α. Let $\alpha = -\langle \mathbf{y}, \mathbf{x} \rangle / \langle \mathbf{y}, \mathbf{y} \rangle$; then (2-95) becomes

$$\langle \mathbf{x}, \mathbf{x} \rangle \ge \frac{\langle \mathbf{x}, \mathbf{y} \rangle \langle \mathbf{y}, \mathbf{x} \rangle}{\langle \mathbf{y}, \mathbf{y} \rangle} = \frac{|\langle \mathbf{x}, \mathbf{y} \rangle|^2}{\langle \mathbf{y}, \mathbf{y} \rangle}$$

which gives the Schwarz inequality. Q.E.D.

2-9 Concluding Remarks

In this chapter we have reviewed a number of concepts and results in linear algebra which are useful in this book. The following three main topics were covered:

1. *Similarity transformation.* The basic idea of similarity transformation (change of basis vectors) and the means of carrying out the transformation are summarized in Figures 2-4 and 2-5. Similarity transformations can be carried out (a) by computing $\hat{\mathbf{A}} = \mathbf{PAP}^{-1} = \mathbf{Q}^{-1}\mathbf{AQ}$, or (b) more easily, by using the concept of representation: the ith column of $\hat{\mathbf{A}}$ is the representation of $\mathbf{Aq}_i$ with respect to the basis $\{\mathbf{q}_1, \mathbf{q}_2, \ldots, \mathbf{q}_n\}$. The second method will be constantly employed in the remainder of this book.
2. *Jordan-form representation of a matrix.* For each eigenvalue of $\mathbf{A}$ with multiplicity m, there are m linearly independent generalized eigenvectors. Using these vectors as a basis, the new representation of $\mathbf{A}$ is in a Jordan canonical form. A systematic procedure for searching these generalized eigenvectors can be found in Reference S43. If all eigenvalues are distinct, then a Jordan form reduces to a diagonal form.

   Jordan-form representation is usually developed by introducing invariant subspaces and their direct sum. We bypass these concepts and concentrate on the search of the required basis vectors. The interested reader may correlate our derivation with the concepts of invariant subspaces and their direct sum.
3. *Functions of a square matrix.* Three methods of computing a function of a matrix $f(\mathbf{A})$, where $\mathbf{A}$ is an $n \times n$ constant matrix, were introduced. (a) Use Definition 2-16: First compute the eigenvalues of $\mathbf{A}$, and then find a polynomial $g(\lambda)$ of degree $n-1$ that is equal to $f(\lambda)$ on the spectrum of $\mathbf{A}$, then $f(\mathbf{A}) = g(\mathbf{A})$. (b) Use the Jordan canonical form of $\mathbf{A}$: Let $\mathbf{A} = \mathbf{Q}\hat{\mathbf{A}}\mathbf{Q}^{-1}$. Then $f(\mathbf{A}) = \mathbf{Q}f(\hat{\mathbf{A}})\mathbf{Q}^{-1}$, where $\hat{\mathbf{A}}$ is in the Jordan form and $f(\hat{\mathbf{A}})$ is computed in (2-69) and (2-71). (c) Use Definition 2-17.

Remarks are in order regarding the computer computation of the topics covered in this chapter. A problem is said to be ill-conditioned if small changes in data will lead to large changes in solutions. In engineering terminology, if a problem is very sensitive to the variations of data, then the problem is ill-conditioned; otherwise, it is well-conditioned. A procedure for solving a problem, or an algorithm, is said to be numerically stable if numerical errors inside the procedure will not be amplified; otherwise, the algorithm is said to be numerically unstable. The condition of a problem and the numerical stability of an algorithm are two independent concepts. Clearly, whenever possible, numerically stable methods should be used to solve a problem. There are a large number of texts on the subject of computer computations. The reader is referred to, e.g., References S138, S181, S182, S200, and S212. Several well-tested computer programs such as LINPACK and EISPACK are discussed in References S82, S103, and S182.

The solution of linear algebraic equations is a basic topic in computer computation. The gaussian elimination without any pivoting is numerically unstable and should be avoided. The gaussian elimination with partial pivoting is a numerically stable method. Its numerical stability can be further improved by using complete pivoting or by using the Householder transformation. However, according to Reference S138, from the point of view of overall performance, which includes efficiency, accuracy, reliability, generality, and ease of use, the gaussian elimination with partial pivoting is satisfactory for most general matrices.

The gaussian elimination with partial pivoting can be used to transform a matrix into a triangular form (see Appendix A), and the rank of the matrix can then be determined. However, because of the difficulty in determining how small a number should be considered as a zero, ambiguity always occurs. The numerical property of this process can be improved by using Householder transformations. It can be further improved if Householder transformations are employed together with pivoting (see Appendix A). The most reliable method of computing the rank of a matrix is by using the singular value decomposition (see Appendix E). Although the problem of determining whether a small number is a zero or not remains, the decomposition does provide a "distance" to a matrix of lower rank. A subroutine to achieve the singular value decomposition is available in LINPACK. The singular value decomposition however is quite expensive, and according to Reference S82, its use is often oversimplified; the results depend highly on balancing.[18] Furthermore, comparable results are often obtainable with less cost by using Householder transformations with pivoting.

The eigenvalues of **A** are the roots of the characteristic polynomial of **A**. Once the characteristic polynomial is known, the roots can be solved by using subroutines for solving the roots of polynomials. However a polynomial may be ill-conditioned in the sense that small changes in coefficients may cause large changes in one or more roots. Hence this procedure of computing the eigenvalues may change a well-conditioned problem into an ill-conditioned problem and should be avoided. The most reliable method of computing the eigenvalues is the so-called QR method. The method transforms **A** into the Hessenberg form and then carries out a sequence of QR factorizations (see Problems A-5 and A-6 and References S181 and S200). If it is necessary to compute the characteristic polynomial of a matrix, several methods, including the Leverrier algorithm (Problem 2-39), are discussed in References S82 and S181 (see also Reference S208).

The Jordan canonical form is useful in developing a number of concepts and results (see, e.g., Problems 2-22, 2-23, 2-37, and 2-45). It is also useful in solving Riccati equations (Problem 2-46). Computer computation of a Jordan canonical form, however, is an ill-conditioned problem. For a discussion of this problem, see Reference S108. For a computer program, see Reference S124.

[18] Roughly speaking, *balancing* or *scaling* is to make all entries of a matrix be of comparable size or the norms of some column and row to be equal. After balancing, the results of computation can often be improved. See References S82, S84, S103 and S202.

Problems

2-1 With the usual definition of addition and multiplication, which of the following sets forms a field?

a. The set of integers
b. The set of rational numbers
c. The set of all 2×2 real matrices
d. The set of polynomial of degree less than n with real coefficients

2-2 Is it possible to define rules of addition and multiplication such that the set $\{0, 1, 2\}$ forms a field?

2-3 Given the set $\{a, b\}$ with $a \neq b$. Define rules of addition and multiplication such that $\{a, b\}$ forms a field. What are the 0 and 1 elements in this field?

2-4 Why is $(\mathbb{C}, \mathbb{R})$ a linear space but not $(\mathbb{R}, \mathbb{C})$?

2-5 Let $\mathbb{R}(s)$ denote the set of all rational functions with real coefficients. Show that $(\mathbb{R}(s), \mathbb{R}(s))$ and $(\mathbb{R}(s), \mathbb{R})$ are linear spaces.

2-6 Which of the following sets of vectors are linearly independent?

a. $\begin{bmatrix} 4 \\ -9 \\ 1 \end{bmatrix}, \begin{bmatrix} 2 \\ 13 \\ 10 \end{bmatrix}, \begin{bmatrix} 2 \\ -4 \\ 1 \end{bmatrix}$ in $(\mathbb{R}^3, \mathbb{R})$

b. $\begin{bmatrix} 1+i \\ 2+3i \end{bmatrix}, \begin{bmatrix} 10+2i \\ 4-i \end{bmatrix}, \begin{bmatrix} -i \\ 3 \end{bmatrix}$ in $(\mathbb{C}^2, \mathbb{R})$

c. e^{-t}, te^{-t}, e^{-2t} in $(\mathscr{U}, \mathbb{R})$, where $\mathscr{U}$ denotes the set of all piecewise continuous functions defined on $[0, \infty)$.

d. $3s^2+s-10, \ -2s+3, \ s-5$ in $(\mathbb{R}_3[s], \mathbb{R})$

e. $\dfrac{3s^2-12}{2s^3+4s-1}, \dfrac{4s^5+s^3-2s-1}{1}, \dfrac{1}{s^2+s-1}$ in $(\mathbb{R}(s), \mathbb{R})$

2-7 Is the set in Problem 2-6b linearly independent in $(\mathbb{C}^2, \mathbb{C})$? Are the sets in Problem 2-6d and e linearly independent in $(\mathbb{R}(s), \mathbb{R}(s))$?

2-8 What are the dimensions of the following linear spaces?

a. $(\mathbb{R}, \mathbb{R})$ **b.** $(\mathbb{C}, \mathbb{C})$ **c.** $(\mathbb{C}, \mathbb{R})$ **d.** $(\mathbb{R}(s), \mathbb{R}(s))$ **e.** $(\mathbb{R}(s), \mathbb{R})$

2-9 Show that the vectors $\mathbf{x}_1, \mathbf{x}_2, \ldots, \mathbf{x}_k$ are linearly dependent in $(\mathbb{R}^n(s), \mathbb{R}(s))$ if and only if there exist polynomials $c_i(s)$, $i = 1, 2, \ldots, k$, not all zero, such that

$$c_1(s)\mathbf{x}_1 + c_2(s)\mathbf{x}_2 + \cdots + c_k(s)\mathbf{x}_k = \mathbf{0}$$

See Equation (G-31) in Appendix G.

2-10 Show that the set of all 2×2 matrices with real coefficients forms a linear space over $\mathbb{R}$ with dimension 4.

2-11 In an n-dimensional vector space $(\mathscr{X}, \mathscr{F})$, given the basis $\mathbf{e}_1, \mathbf{e}_2, \ldots, \mathbf{e}_n$, what is the representation of $\mathbf{e}_i$ with respect to the basis?

2-12 Consider Table 2-1. Suppose the representations of $\mathbf{b}, \bar{\mathbf{e}}_1, \bar{\mathbf{e}}_2, \mathbf{e}_1$, and $\mathbf{e}_2$ with respect to the basis $\{\mathbf{e}_1, \mathbf{e}_2\}$ are known, use Equation (2-20) to derive the representations of $\mathbf{b}, \bar{\mathbf{e}}_1, \bar{\mathbf{e}}_2, \mathbf{e}_1$, and $\mathbf{e}_2$ with respect to the basis $\{\bar{\mathbf{e}}_1, \bar{\mathbf{e}}_2\}$.

2-13 Show that similar matrices have the same characteristic polynomial, and consequently, the same set of eigenvalues. [*Hint:* det (**AB**) = det **A** det **B**.]

2-14 Find the **P** matrix in Example 3, Section 2-4, and verify $\bar{\mathbf{A}} = \mathbf{PAP}^{-1}$.

2-15 Given

$$\mathbf{A} = \begin{bmatrix} 2 & 1 & 0 & 0 \\ 0 & 2 & 1 & 0 \\ 0 & 0 & 2 & 0 \\ 0 & 0 & 0 & 1 \end{bmatrix} \qquad \mathbf{b} = \begin{bmatrix} 0 \\ 0 \\ 1 \\ 1 \end{bmatrix} \qquad \bar{\mathbf{b}} = \begin{bmatrix} 1 \\ 1 \\ 1 \\ 1 \end{bmatrix}$$

what are the representations of **A** with respect to the basis $\{\mathbf{b}, \mathbf{Ab}, \mathbf{A}^2\mathbf{b}, \mathbf{A}^3\mathbf{b}\}$ and the basis $\{\bar{\mathbf{b}}, \mathbf{A}\bar{\mathbf{b}}, \mathbf{A}^2\bar{\mathbf{b}}, \mathbf{A}^3\bar{\mathbf{b}}\}$, respectively? (Note that the representations are the same!)

2-16 What are the ranks and nullities of the following matrices?

$$\mathbf{A}_1 = \begin{bmatrix} 4 & 1 & -1 \\ 3 & 2 & -3 \\ 1 & 3 & 0 \end{bmatrix} \qquad \mathbf{A}_2 = \begin{bmatrix} 0 & 1 & 0 \\ 0 & 0 & 0 \\ 0 & 0 & 1 \end{bmatrix} \qquad \mathbf{A}_3 = \begin{bmatrix} 1 & 2 & 3 & 4 & 5 \\ 2 & 3 & 4 & 1 & 2 \\ 3 & 4 & 5 & 0 & 0 \end{bmatrix}$$

2-17 Find the bases of the range spaces and the null spaces of the matrices given in Problem 2-16.

2-18 Are the matrices

$$\begin{bmatrix} s^3 + s^2 & s^2 + 1 \\ s & 1 \end{bmatrix} \qquad \begin{bmatrix} s^2 + 1 & 1 \\ s^2 & 1 \end{bmatrix}$$

nonsingular in the field of rational functions with real coefficients $\mathbb{R}(s)$? For every s in $\mathbb{C}$, the matrices become numerical matrices with elements in $\mathbb{C}$. For every s in $\mathbb{C}$, are the matrices nonsingular in the field of complex numbers $\mathbb{C}$?

2-19 Does there exist a solution for the following linear equations?

$$\begin{bmatrix} 3 & 3 & 0 \\ 2 & 1 & 1 \\ 1 & 2 & -1 \end{bmatrix} \begin{bmatrix} x_1 \\ x_2 \\ x_3 \end{bmatrix} = \begin{bmatrix} 6 \\ 3 \\ 3 \end{bmatrix}$$

If so, find one.

2-20 Consider the set of linear equations

$$\mathbf{x}(n) = \mathbf{A}^n\mathbf{x}(0) + \mathbf{A}^{n-1}\mathbf{b}u(0) + \mathbf{A}^{n-2}\mathbf{b}u(1) + \cdots + \mathbf{Ab}u(n-2) + \mathbf{b}u(n-1)$$

where **A** is an $n \times n$ constant matrix and **b** is an $n \times 1$ column vector. Given any $\mathbf{x}(n)$ and $\mathbf{x}(0)$, under what conditions on **A** and **b** will there exist $u(0), u(1), \ldots, u(n-1)$ satisfying the equation? *Hint:* Write the equation in the form

$$\mathbf{x}(n) - \mathbf{A}^n\mathbf{x}(0) = [\mathbf{b} \quad \mathbf{Ab} \quad \cdots \quad \mathbf{A}^{n-1}\mathbf{b}] \begin{bmatrix} u(n-1) \\ u(n-2) \\ \vdots \\ u(0) \end{bmatrix}$$

2-21 Find the Jordan-canonical-form representations of the following matrices:

$$\mathbf{A}_1 = \begin{bmatrix} 1 & 4 & 10 \\ 0 & 2 & 0 \\ 0 & 0 & 3 \end{bmatrix} \qquad \mathbf{A}_2 = \begin{bmatrix} 0 & 1 & 0 \\ 0 & 0 & 1 \\ -2 & -4 & -3 \end{bmatrix}$$

$$\mathbf{A}_3 = \begin{bmatrix} 0 & 4 & 3 \\ 0 & -150 & -120 \\ 0 & 200 & 160 \end{bmatrix} \qquad \mathbf{A}_4 = \begin{bmatrix} 0 & 4 & 3 \\ 0 & 20 & 16 \\ 0 & -25 & -20 \end{bmatrix}$$

$$\mathbf{A}_5 = \begin{bmatrix} \frac{7}{2} & \frac{21}{2} & 14 \\ -\frac{1}{2} & -\frac{3}{2} & -2 \\ -\frac{1}{2} & -\frac{3}{2} & -2 \end{bmatrix} \qquad \mathbf{A}_6 = \begin{bmatrix} 0 & 1 & 1 & 1 & 1 \\ 0 & 0 & 1 & 1 & 1 \\ 0 & 0 & 0 & 1 & 1 \\ 0 & 0 & 0 & 0 & 1 \\ 0 & 0 & 0 & 0 & 0 \end{bmatrix}$$

$$\mathbf{A}_7 = \begin{bmatrix} 0 & 1 & 0 & 0 \\ 0 & 0 & 1 & 0 \\ 0 & 0 & 0 & 1 \\ 4 & -4 & -3 & 4 \end{bmatrix}$$

2-22 Let λ_i for $i = 1, 2, \ldots, n$ be the eigenvalues of an $n \times n$ matrix $\mathbf{A}$. Show that

$$\det \mathbf{A} = \prod_{i=1}^{n} \lambda_i$$

2-23 Prove that a square matrix is nonsingular if and only if there is no zero eigenvalue.

2-24 Under what condition will $\mathbf{AB} = \mathbf{AC}$ imply $\mathbf{B} = \mathbf{C}$? ($\mathbf{A}$ is assumed to be a square matrix.)

2-25 Show that the Vandermonde determinant

$$\begin{bmatrix} 1 & 1 & \cdots & 1 \\ \lambda_1 & \lambda_2 & \cdots & \lambda_n \\ \lambda_1^2 & \lambda_2^2 & \cdots & \lambda_n^2 \\ \vdots & \vdots & & \vdots \\ \lambda_1^{n-1} & \lambda_2^{n-1} & \cdots & \lambda_n^{n-1} \end{bmatrix}$$

is equal to $\prod_{1 \le i < j \le n} (\lambda_j - \lambda_i)$.

2-26 Consider the matrix

$$\mathbf{A} = \begin{bmatrix} 0 & 1 & 0 & \cdots & 0 \\ 0 & 0 & 1 & \cdots & 0 \\ \vdots & \vdots & \vdots & & \vdots \\ 0 & 0 & 0 & \cdots & 1 \\ -\alpha_n & -\alpha_{n-1} & -\alpha_{n-2} & \cdots & -\alpha_1 \end{bmatrix}$$

Show that the characteristic polynomial of $\mathbf{A}$ is

$$\Delta(\lambda) \triangleq \det(\lambda\mathbf{I} - \mathbf{A}) = \lambda^n + \alpha_1\lambda^{n-1} + \alpha_2\lambda^{n-2} + \cdots + \alpha_{n-1}\lambda + \alpha_n$$

If λ_1 is an eigenvalue of $\mathbf{A}$ [that is, $\Delta(\lambda_1) = 0$], show that $[1 \;\; \lambda_1 \;\; \lambda_1^2 \;\; \cdots \;\; \lambda_1^{n-1}]'$ is an eigenvector associated with λ_1. [The matrix $\mathbf{A}$ is called the *companion matrix* of the polynomial $\Delta(\lambda)$. It is said to be in the *Frobenius* form in the numerical analysis literature.]

2-27[19] Consider the matrix shown in Problem 2-26. Suppose that λ_1 is an eigenvalue of the matrix with multiplicity k; that is, $\Delta(\lambda)$ contains $(\lambda-\lambda_1)^k$ as a factor. Verify that the following k vectors,

$$\begin{bmatrix} 1 \\ \lambda_1 \\ \lambda_1^2 \\ \vdots \\ \lambda_1^{n-1} \end{bmatrix} \begin{bmatrix} 0 \\ 1 \\ 2\lambda_1 \\ \vdots \\ (n-1)\lambda_1^{n-2} \end{bmatrix} \begin{bmatrix} 0 \\ 0 \\ 1 \\ \vdots \\ \binom{n-1}{2}\lambda_1^{n-3} \end{bmatrix} \begin{bmatrix} 0 \\ 0 \\ 0 \\ \vdots \\ \binom{n-1}{3}\lambda_1^{n-4} \end{bmatrix} \cdots \begin{bmatrix} 0 \\ 0 \\ 0 \\ \vdots \\ \binom{n-1}{k-1}\lambda_1^{n-k} \end{bmatrix}$$

where
$$\binom{n-1}{i} \triangleq \frac{(n-1)(n-2)\cdots(n-i)}{1\cdot 2\cdot 3\cdots i} \qquad i\geq 1$$

are generalized eigenvectors of $\mathbf{A}$ associated with λ_1.

2-28 Show that the matrix $\mathbf{A}$ in Problem 2-26 is nonsingular if and only if $\alpha_n \neq 0$. Verify that its inverse is given by

$$\mathbf{A}^{-1} = \begin{bmatrix} -\alpha_{n-1}/\alpha_n & -\alpha_{n-2}/\alpha_n & \cdots & -\alpha_1/\alpha_n & -1/\alpha_n \\ 1 & 0 & \cdots & 0 & 0 \\ 0 & 1 & \cdots & 0 & 0 \\ \vdots & \vdots & & \vdots & \vdots \\ 0 & 0 & \cdots & 1 & 0 \end{bmatrix}$$

2-29 Show that the determinant of the $m \times m$ matrix

$$\begin{bmatrix} s^{k_m} & -1 & 0 & \cdots & 0 & 0 \\ 0 & s^{k_{m-1}} & -1 & \cdots & 0 & 0 \\ 0 & 0 & s^{k_{m-2}} & \cdots & 0 & 0 \\ \vdots & \vdots & \vdots & & \vdots & \vdots \\ 0 & 0 & 0 & \cdots & s^{k_2} & -1 \\ \beta_m(s) & \beta_{m-1}(s) & \beta_{m-2}(s) & \cdots & \beta_2(s) & s^{k_1}+\beta_1(s) \end{bmatrix}$$

is equal to

$$s^n + \beta_1(s)s^{n-k_1} + \beta_2(s)s^{n-k_1-k_2} + \cdots + \beta_m(s)$$

where $n = k_1 + k_2 + \cdots + k_m$ and $\beta_i(s)$ are arbitrary polynomials.

2-30[20] Show that the characteristic polynomial of the matrix

$$\left[\begin{array}{ccccc|ccccc} 0 & 1 & \cdots & 0 & 0 & 0 & 0 & \cdots & 0 & 0 \\ 0 & 0 & \cdots & 0 & 0 & 0 & 0 & \cdots & 0 & 0 \\ \vdots & \vdots & & \vdots & \vdots & \vdots & \vdots & & \vdots & \vdots \\ 0 & 0 & \cdots & 0 & 1 & 0 & 0 & \cdots & 0 & 0 \\ -a_{11n_1} & -a_{11(n_1-1)} & \cdots & -a_{112} & -a_{111} & -a_{12n_2} & -a_{12(n_2-1)} & \cdots & -a_{122} & -a_{121} \\ \hline 0 & 0 & \cdots & 0 & 0 & 0 & 1 & \cdots & 0 & 0 \\ 0 & 0 & \cdots & 0 & 0 & 0 & 0 & \cdots & 0 & 0 \\ \vdots & \vdots & & \vdots & \vdots & \vdots & \vdots & & \vdots & \vdots \\ 0 & 0 & \cdots & 0 & 0 & 0 & 0 & \cdots & 0 & 1 \\ -a_{21n_1} & -a_{21(n_1-1)} & \cdots & -a_{212} & -a_{211} & -a_{22n_2} & -a_{22(n_2-1)} & \cdots & -a_{222} & -a_{221} \end{array}\right]$$

[19] See Reference 6.

[20] See Reference S17.

is given by

$$\det \begin{bmatrix} \Delta_{11}(s) & \Delta_{12}(s) \\ \Delta_{21}(s) & \Delta_{22}(s) \end{bmatrix} = \Delta_{11}(s)\Delta_{22}(s) - \Delta_{12}(s)\Delta_{21}(s)$$

where

$$\Delta_{ii}(s) = s^{n_i} + a_{ii1}s^{(n_i-1)} + \cdots + a_{ii(n_i-1)}s + a_{iin_i}$$
$$\Delta_{ij}(s) = a_{ij1}s^{(n_j-1)} + \cdots + a_{ij(n_j-1)}s + a_{ijn_j}$$

Note that the submatrices on the diagonal are of the companion form (see Problem 2-26); the submatrices not on the diagonal are all zeros except the last row.

2-31 Find the characteristic polynomials and the minimal polynomials of the following matrices:

$$\begin{bmatrix} \lambda_1 & 1 & 0 & 0 \\ 0 & \lambda_1 & 1 & 0 \\ 0 & 0 & \lambda_1 & 0 \\ 0 & 0 & 0 & \lambda_2 \end{bmatrix} \quad \begin{bmatrix} \lambda_1 & 1 & 0 & 0 \\ 0 & \lambda_1 & 1 & 0 \\ 0 & 0 & \lambda_1 & 0 \\ 0 & 0 & 0 & \lambda_1 \end{bmatrix} \quad \begin{bmatrix} \lambda_1 & 1 & 0 & 0 \\ 0 & \lambda_1 & 0 & 0 \\ 0 & 0 & \lambda_1 & 0 \\ 0 & 0 & 0 & \lambda_1 \end{bmatrix} \quad \begin{bmatrix} \lambda_1 & 1 & 0 & 0 \\ 0 & \lambda_1 & 0 & 0 \\ 0 & 0 & \lambda_1 & 1 \\ 0 & 0 & 0 & \lambda_1 \end{bmatrix}$$

What are the multiplicities and indices? What are their geometric multiplicities?

2-32 Show that if λ is an eigenvalue of $\mathbf{A}$ with eigenvector $\mathbf{x}$, then $f(\lambda)$ is an eigenvalue of $f(\mathbf{A})$ with the same eigenvector $\mathbf{x}$.

2-33 Repeat the problems in Examples 2 and 3 of Section 2-7 by choosing, respectively, $g(\lambda) = \alpha_0\lambda + \alpha_1(\lambda - 1)$ and $g(\lambda) = \alpha_0(\lambda - 1) + \alpha_1(\lambda - 1)^2(\lambda - 2) + \alpha_2(\lambda - 2)$.

2-34 Given

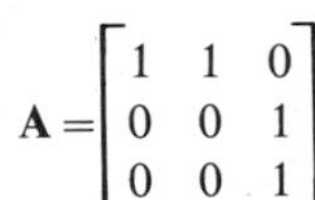

$$\mathbf{A} = \begin{bmatrix} 1 & 1 & 0 \\ 0 & 0 & 1 \\ 0 & 0 & 1 \end{bmatrix}$$

Find $\mathbf{A}^{10}$, $\mathbf{A}^{103}$, and $e^{\mathbf{A}t}$.

2-35 Compute $e^{\mathbf{A}t}$ for the matrices

$$\begin{bmatrix} 1 & 4 & 10 \\ 0 & 2 & 0 \\ 0 & 0 & 3 \end{bmatrix} \quad \begin{bmatrix} 0 & 4 & 3 \\ 0 & -150 & -120 \\ 0 & 200 & 160 \end{bmatrix}$$

by using Definition 2-16 and by using the Jordan-form representation.

2-36 Show that functions of the same matrix commute; that is,

$$f(\mathbf{A})g(\mathbf{A}) = g(\mathbf{A})f(\mathbf{A})$$

Consequently, we have $\mathbf{A}e^{\mathbf{A}t} = e^{\mathbf{A}t}\mathbf{A}$.

2-37 Let

$$\mathbf{C} = \begin{bmatrix} \lambda_1 & 0 & 0 \\ 0 & \lambda_2 & 0 \\ 0 & 0 & \lambda_3 \end{bmatrix}$$

Find a matrix $\mathbf{B}$ such that $e^{\mathbf{B}} = \mathbf{C}$. Show that if $\lambda_i = 0$ for some i then the matrix $\mathbf{B}$ does not exist. Let

$$\mathbf{C}=\begin{bmatrix}\lambda & 1 & 0\\ 0 & \lambda & 0\\ 0 & 0 & \lambda\end{bmatrix}$$

Find a matrix $\mathbf{B}$ such that $e^{\mathbf{B}}=\mathbf{C}$. [*Hint:* Let $f(\lambda)=\log\lambda$ and use (2-69).] Is it true that for any nonsingular matrix $\mathbf{C}$, there exists a matrix $\mathbf{B}$ such that $e^{\mathbf{B}}=\mathbf{C}$?

2-38 Let $\mathbf{A}$ be an $n\times n$ matrix. Show by using the Cayley-Hamilton theorem that any $\mathbf{A}^k$ with $k\geq n$ can be written as a linear combination of $\{\mathbf{I}, \mathbf{A}, \ldots, \mathbf{A}^{n-1}\}$. If the degree of the minimal polynomial of $\mathbf{A}$ is known, what modification can you make?

2-39 Define

$$(s\mathbf{I}-\mathbf{A})^{-1}\triangleq\frac{1}{\Delta(s)}[\mathbf{R}_0s^{n-1}+\mathbf{R}_1s^{n-2}+\cdots+\mathbf{R}_{n-2}s+\mathbf{R}_{n-1}]$$

where $\Delta(s)\triangleq\det(s\mathbf{I}-\mathbf{A})\triangleq s^n+\alpha_1s^{n-1}+\alpha_2s^{n-2}+\cdots+\alpha_n$ and $\mathbf{R}_0, \mathbf{R}_1, \ldots, \mathbf{R}_{n-1}$ are constant matrices. This definition is valid because the degree in s of the adjoint of $(s\mathbf{I}-\mathbf{A})$ is at most $n-1$. Verify that

$$\alpha_1=-\frac{\operatorname{tr}\mathbf{A}\mathbf{R}_0}{1}\qquad \mathbf{R}_0=\mathbf{I}$$

$$\alpha_2=-\frac{\operatorname{tr}\mathbf{A}\mathbf{R}_1}{2}\qquad \mathbf{R}_1=\mathbf{A}\mathbf{R}_0+\alpha_1\mathbf{I}=\mathbf{A}+\alpha_1\mathbf{I}$$

$$\alpha_3=-\frac{\operatorname{tr}\mathbf{A}\mathbf{R}_2}{3}\qquad \mathbf{R}_2=\mathbf{A}\mathbf{R}_1+\alpha_2\mathbf{I}=\mathbf{A}^2+\alpha_1\mathbf{A}+\alpha_2\mathbf{I}$$

$$\cdots\cdots\cdots\cdots\cdots\cdots\cdots\cdots\cdots\cdots\cdots\cdots\cdots\cdots$$

$$\alpha_{n-1}=-\frac{\operatorname{tr}\mathbf{A}\mathbf{R}_{n-2}}{n-1}\qquad \mathbf{R}_{n-1}=\mathbf{A}\mathbf{R}_{n-2}+\alpha_{n-1}\mathbf{I}=\mathbf{A}^{n-1}+\alpha_1\mathbf{A}^{n-2}+\cdots+\alpha_{n-2}\mathbf{A}+\alpha_{n-1}\mathbf{I}$$

$$\alpha_n=-\frac{\operatorname{tr}\mathbf{A}\mathbf{R}_{n-1}}{n}\qquad \mathbf{0}=\mathbf{A}\mathbf{R}_{n-1}+\alpha_n\mathbf{I}$$

where tr stands for the *trace* and is defined as the sum of all the diagonal elements of a matrix. This procedure of computing α_i and $\mathbf{R}_i$ is called the *Leverrier algorithm.* [*Hint:* The right-hand-side equations can be verified from $\Delta(s)\mathbf{I}=(s\mathbf{I}-\mathbf{A})(\mathbf{R}_0s^{n-1}+\mathbf{R}_1s^{n-2}+\cdots+\mathbf{R}_{n-2}s+\mathbf{R}_{n-1})$. For a derivation of the left-hand-side equations, see Reference S185.]

2-40 Prove the Cayley–Hamilton theorem. (*Hint:* Use Problem 2-39 and eliminate $\mathbf{R}_{n-1}, \mathbf{R}_{n-2}, \ldots,$ from $\mathbf{0}=\mathbf{A}\mathbf{R}_{n-1}+\alpha_n\mathbf{I}$.)

2-41 Show, by using Problem 2-39,

$$(s\mathbf{I}-\mathbf{A})^{-1}=\frac{1}{\Delta(s)}[\mathbf{A}^{n-1}+(s+\alpha_1)\mathbf{A}^{n-2}+(s^2+\alpha_1s+\alpha_2)\mathbf{A}^{n-3}+\cdots$$
$$+(s^{n-1}+\alpha_1s^{n-2}+\cdots+\alpha_{n-1})\mathbf{I}]$$

2-42 Let

$$(s\mathbf{I}-\mathbf{A})^{-1}=\frac{1}{\Delta(s)}\text{Adjoint}\,(s\mathbf{I}-\mathbf{A})$$

and let $m(s)$ be the monic greatest common divisor of all elements of Adjoint $(s\mathbf{I}-\mathbf{A})$. Show that the minimal polynomial of $\mathbf{A}$ is equal to $\Delta(s)/m(s)$.

2-43 Let all eigenvalues of $\mathbf{A}$ be distinct and let $\mathbf{q}_i$ be a (right) eigenvector of $\mathbf{A}$ associated with λ_i, that is, $\mathbf{A}\mathbf{q}_i=\lambda_i\mathbf{q}_i$. Define $\mathbf{Q} \triangleq [\mathbf{q}_1 \quad \mathbf{q}_2 \quad \cdots \quad \mathbf{q}_n]$ and define

$$\mathbf{P} \triangleq \mathbf{Q}^{-1} \triangleq \begin{bmatrix} \mathbf{p}_1 \\ \mathbf{p}_2 \\ \vdots \\ \mathbf{p}_n \end{bmatrix}$$

where $\mathbf{p}_i$ is the ith row of $\mathbf{P}$. Show that $\mathbf{p}_i$ is a left eigenvector of $\mathbf{A}$ associated with λ_i, that is, $\mathbf{p}_i\mathbf{A}=\lambda_i\mathbf{p}_i$.

2-44 Show that if all eigenvalues of $\mathbf{A}$ are distinct, then $(s\mathbf{I}-\mathbf{A})^{-1}$ can be expressed as

$$(s\mathbf{I}-\mathbf{A})^{-1}=\sum \frac{1}{s-\lambda_i}\mathbf{q}_i\mathbf{p}_i$$

where $\mathbf{q}_i$ and $\mathbf{p}_i$ are right and left eigenvectors of $\mathbf{A}$ associated with λ_i.

2-45 A matrix $\mathbf{A}$ is defined to be *cyclic* if its characteristic polynomial is equal to its minimal polynomial. Show that $\mathbf{A}$ is cyclic if and only if there is only one Jordan block associated with each distinct eigenvalue.

2-46[21] Consider the matrix equation

$$\mathbf{PEP}+\mathbf{DP}+\mathbf{PF}+\mathbf{G}=\mathbf{0}$$

where all matrices are $n \times n$ constant matrices. It is called an algebraic *Riccati equation.* Define

$$\mathbf{M}=\begin{bmatrix} -\mathbf{F} & -\mathbf{E} \\ \mathbf{G} & \mathbf{D} \end{bmatrix}$$

Let

$$\mathbf{Q}=\begin{bmatrix} \mathbf{Q}_1 & \mathbf{Q}_2 \\ \mathbf{Q}_3 & \mathbf{Q}_4 \end{bmatrix}$$

consist of all generalized eigenvectors of $\mathbf{M}$ so that $\mathbf{Q}^{-1}\mathbf{M}\mathbf{Q}=\mathbf{J}$ is in a Jordan canonical form. We write

$$\begin{bmatrix} -\mathbf{F} & -\mathbf{E} \\ \mathbf{G} & \mathbf{D} \end{bmatrix}\begin{bmatrix} \mathbf{Q}_1 & \mathbf{Q}_2 \\ \mathbf{Q}_3 & \mathbf{Q}_4 \end{bmatrix}=\begin{bmatrix} \mathbf{Q}_1 & \mathbf{Q}_2 \\ \mathbf{Q}_3 & \mathbf{Q}_4 \end{bmatrix}\begin{bmatrix} \mathbf{J}_1 & \mathbf{0} \\ \mathbf{0} & \mathbf{J}_2 \end{bmatrix}$$

Show that if $\mathbf{Q}_1$ is nonsingular, then $\mathbf{P}=\mathbf{Q}_3\mathbf{Q}_1^{-1}$ is a solution of the Riccati equation.

2-47 Give three different norms of the vector $\mathbf{x}=[1 \quad -4 \quad 3]'$.

2-48 Verify the three norms of $\mathbf{A}$ in Figure 2-7.

2-49 Show that the set of all piecewise continuous complex-valued functions defined over $[0, \infty)$ forms a linear space over $\mathbb{C}$. Show that

$$\langle g, h\rangle \triangleq \int_0^\infty g^*(t)h(t)\,dt$$

[21] See Reference S4.

qualifies as an inner product of the space, where g and h are two arbitrary functions of the space. What is the form of the Schwarz inequality in this space?

2-50 Show that an $n \times n$ matrix $\mathbf{A}$ has the property $\mathbf{A}^k = \mathbf{0}$ for $k > m$ if and only if $\mathbf{A}$ has eigenvalue 0 with multiplicity n and index m. A matrix with the property $\mathbf{A}^k = \mathbf{0}$ is called a *nilpotent matrix*. [*Hint:* Use Equation (2-64) and Jordan canonical form.]

2-51 Let $\mathbf{A}$ be an $m \times n$ matrix. Show that the set of all $1 \times m$ vectors $\mathbf{y}$ satisfying $\mathbf{yA} = \mathbf{0}$ forms a linear space, called the left null space of $\mathbf{A}$, of dimension $m - \rho(\mathbf{A})$.

3

Mathematical Descriptions of Systems

3-1 Introduction

The very first step in the analytical study of a system is to set up mathematical equations to describe the system. Because of different analytical methods used, or because of different questions asked, we may often set up different mathematical equations to describe the same system. For example, in network analysis, if we are interested in only the terminal properties, we may use the impedance or transfer function to describe the network; if we want to know the current and voltage of each branch of the network, then loop analysis or node analysis has to be used to find a set of differential equations to describe the network. The transfer function that describes only the terminal property of a system may be called the *external* or *input-output description* of the system. The set of differential equations that describes the internal as well as the terminal behavior of a system may be called the *internal* or *state-variable description* of the system.

In this chapter we shall introduce the input-output description and the state variable description of systems from a very general setting. They will be developed from the concepts of linearity, relaxedness, time invariance, and causality. Therefore they will be applicable to any system, be it an electrical, a mechanical, or a chemical system, provided the system has the aforementioned properties.

The class of systems studied in this book is assumed to have some input terminals and output terminals. The inputs, or the causes, or the excitations $\mathbf{u}$ are applied at the input terminals; the outputs, or the effects, or the responses $\mathbf{y}$ are measurable at the output terminals. In Section 3-2 we show that if the input **u** and the output **y** of a system satisfy the linearity property, then they can be related by an equation of the form

$$\mathbf{y}(t) = \int_{-\infty}^{\infty} \mathbf{G}(t, \tau)\mathbf{u}(\tau)\, d\tau \tag{3-1a}$$

If the input and output have, in addition, the causality property, then (3-1a) can be reduced to

$$\mathbf{y}(t) = \int_{-\infty}^{t} \mathbf{G}(t, \tau)\mathbf{u}(\tau)\, d\tau \tag{3-1b}$$

which can be further reduced to

$$\mathbf{y}(t) = \int_{t_0}^{t} \mathbf{G}(t, \tau)\mathbf{u}(\tau)\, d\tau \tag{3-1c}$$

if the system is relaxed at t_0. Equation (3-1) describes the relation between the input and output of a system and is called the *input-output description* or the *external description* of the system. We also introduce in Section 3-2 the concepts of time invariance and the transfer function. In Section 3-3 the concept of state is introduced. The set of equations

$$\dot{\mathbf{x}}(t) = \mathbf{A}(t)\mathbf{x}(t) + \mathbf{B}(t)\mathbf{u}(t) \tag{3-2a}$$

$$\mathbf{y}(t) = \mathbf{C}(t)\mathbf{x}(t) + \mathbf{E}(t)\mathbf{u}(t) \tag{3-2b}$$

that relates the input **u**, the output **y**, and the state **x** is then introduced. The set of two equations of the form (3-2) is called a *dynamical equation*. If it is used to describe a system, it is called the *dynamical-equation description* or *state-variable description* of the system. We give in Section 3-4 many examples to illustrate the procedure of setting up these two mathematical descriptions. Comparisons between the input-output description and the dynamical-equation description are given in Section 3-5. We study in Section 3-6 the mathematical descriptions of parallel, tandem, and feedback connections of two systems. Finally, the discrete-time versions of Equations (3-1) and (3-2) are introduced in the last section.

We are concerned with descriptions of systems that are models of actual physical systems; hence all the variables and functions in this chapter are assumed to be real-valued. Before proceeding, we classify a system as a single-variable or a multivariable system according to the following definition.

Definition 3-1

A system is said to be a *single-variable system* if and only if it has only one input terminal and only one output terminal. A system is said to be a *multivariable system* if and only if it has more than one input terminal or more than one output terminal. ∎

The references for this chapter are 24, 27, 31, 53, 60, 68, 70, 73, 92, 97, 109, and 116. The main objective of Sections 3-2 and 3-3 is to introduce the concepts of linearity, causality, time invariance, and the state, and to illustrate their importance in developing the linear equations. They are not introduced very rigorously. For a more rigorous exposition, see References 60, 68, 109, and 116.

3-2 The Input-Output Description

The input-output description of a system gives a mathematical relation between the input and output of the system. In developing this description, the knowledge of the internal structure of a system may be assumed to be unavailable to us; the only access to the system is by means of the input terminals and the output terminals. Under this assumption, a system may be considered as a "black box," as shown in Figure 3-1. Clearly what we can do to a black box is to apply all kinds of inputs and measure their corresponding outputs, and then try to abstract key properties of the system from these input-output pairs.

We digress at this point to introduce some notations. The system shown in Figure 3-1 is assumed to have p input terminals and q output terminals. The inputs are denoted by $u_1, u_2, \ldots, u_p$ or by a $p \times 1$ column vector $\mathbf{u} = [u_1 \quad u_2 \quad \cdots \quad u_p]'$. The outputs or responses are denoted by $y_1, y_2, \ldots, y_q$ or by a $q \times 1$ column vector $\mathbf{y} = [y_1 \quad y_2 \quad \cdots \quad y_q]'$. The time interval in which the inputs and outputs will be defined is from $-\infty$ to $+\infty$. We use $\mathbf{u}$ or $\mathbf{u}(\cdot)$ to denote a vector function defined over $(-\infty, \infty)$; $\mathbf{u}(t)$ is used to denote the value of $\mathbf{u}$ at time t. If the function $\mathbf{u}$ is defined only over $[t_0, t_1)$, we write $\mathbf{u}_{[t_0, t_1)}$.

If the output at time t_1 of a system depends only on the input applied at time t_1, the system is called an *instantaneous* or *zero-memory system*. A network that consists of only resistors is such a system. Most systems of interest, however, have memory; that is, the output at time t_1 depends not only on the input applied at t_1, but also on the input applied before and/or after t_1. Hence, if an input $\mathbf{u}_{[t_1, \infty)}$ is applied to a system, unless we know the input applied before t_1, the output $\mathbf{y}_{[t_1, \infty)}$ is generally not uniquely determinable. In fact, for different inputs applied before t_1, we will obtain different output $\mathbf{y}_{[t_1, \infty)}$, although the same input $\mathbf{u}_{[t_1, \infty)}$ is applied. It is clear that such an input-output pair, which lacks a unique relation, is of no use in determining the key properties of the system. Hence, in developing the input-output description, before an input is applied, the system must be assumed to be *relaxed* or *at rest*, and that *the output is excited solely and uniquely by the input applied thereafter*. If the concept of energy is applicable to a system, the system is said to be relaxed at time t_1 if no energy is stored in the system at that instant. As in the engineering literature, we shall assume that every system is relaxed at time $-\infty$. Consequently if an input $\mathbf{u}_{(-\infty, \infty)}$ is applied at $t = -\infty$, the corresponding output

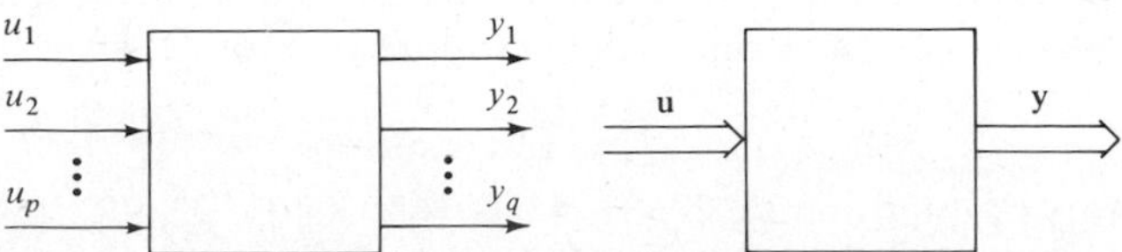

Figure 3-1 A system with p input terminals and q output terminals.

will be excited solely and uniquely by $\mathbf{u}$. Hence, under the relaxedness assumption, it is legitimate to write

$$\mathbf{y} = H\mathbf{u} \tag{3-3}$$

where H is some operator or function that specifies uniquely the output $\mathbf{y}$ in terms of the input $\mathbf{u}$ of the system. In this book, we shall call a system that is initially relaxed at $-\infty$ an *initially relaxed system*—or a *relaxed system*, for short. Note that Equation (3-3) is applicable only to a relaxed system. In this section, whenever we talk about the input-output pairs of a system, we mean only those input-output pairs that can be related by Equation (3-3).

Linearity. We introduce next the concept of linearity. This concept is exactly the same as the linear operators introduced in the preceding chapter.

Definition 3-2

A relaxed system is said to be *linear* if and only if

$$H(\alpha_1 \mathbf{u}_1 + \alpha_2 \mathbf{u}_2) = \alpha_1 H\mathbf{u}_1 + \alpha_2 H\mathbf{u}_2 \tag{3-4}$$

for any inputs $\mathbf{u}_1$ and $\mathbf{u}_2$ and for any real numbers α_1 and α_2. Otherwise the relaxed system is said to be *nonlinear*. ∎

In engineering literature, the condition of Equation (3-4) is often written as

$$H(\mathbf{u}_1 + \mathbf{u}_2) = H\mathbf{u}_1 + H\mathbf{u}_2 \tag{3-5}$$

$$H(\alpha \mathbf{u}_1) = \alpha H\mathbf{u}_1 \tag{3-6}$$

for any $\mathbf{u}_1$, $\mathbf{u}_2$ and any real number α. It is easy to verify that the condition given in (3-4) and the set of conditions in (3-5) and (3-6) are equivalent. The relationship in (3-5) is called the property of *additivity*, and the relationship in (3-6) is called the property of *homogeneity*. If a relaxed system has these two properties, the system is said to satisfy the *principle of superposition*. The reader may wonder whether or not there is redundancy in (3-5) and (3-6). Generally, the property of homogeneity does not imply the property of additivity, as can be seen from the following example.

Example 1

Consider a single-variable system whose input and output are related by

$$y(t) = \begin{cases} \dfrac{u^2(t)}{u(t-1)} & \text{if } u(t-1) \neq 0 \\ 0 & \text{if } u(t-1) = 0 \end{cases}$$

for all t. It is easy to verify that the input-output pair satisfies the property of homogeneity but not the property of additivity. ∎

The property of additivity, however, almost implies the property of homogeneity. To be precise, the condition $H(\mathbf{u}_1 + \mathbf{u}_2) = H\mathbf{u}_1 + H\mathbf{u}_2$ for any $\mathbf{u}_1$ and

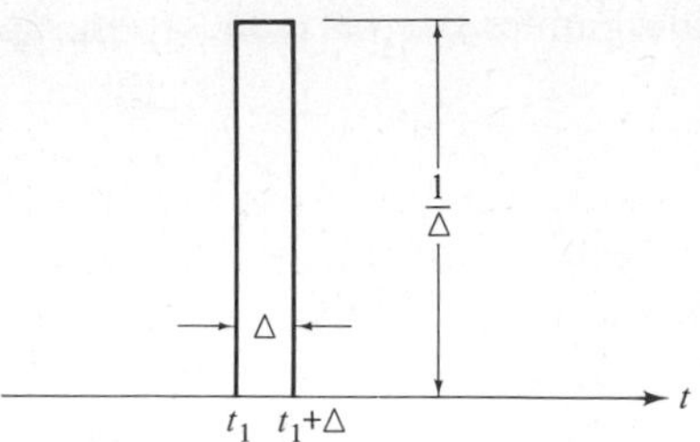

Figure 3-2 A pulse function $\delta_\Delta(t-t_1)$.

$\mathbf{u}_2$ implies that $H(\alpha\mathbf{u}_1)=\alpha H\mathbf{u}_1$, for any rational number α (see Problem 3-9). Since any real number can be approximated as closely as desired by a rational number, if a relaxed system has the continuity property that $\mathbf{u}_n \to \mathbf{u}$ implies $H\mathbf{u}_n \to H\mathbf{u}$, then the property of additivity implies the property of homogeneity.

We shall develop in the following a mathematical description for a linear relaxed system. Before proceeding we need the concept of the delta function or impulse function. We proceed intuitively because a detailed exposition would lead us too far astray.[1] First let $\delta_\Delta(t-t_1)$ be the pulse function defined in Figure 3-2; that is,

$$\delta_\Delta(t-t_1)=\begin{cases} 0 & \text{for } t<t_1 \\ \dfrac{1}{\Delta} & \text{for } t_1 \le t < t_1+\Delta \\ 0 & \text{for } t \ge t_1+\Delta \end{cases}$$

Note that $\delta_\Delta(t-t_1)$ has unit area for all Δ. As Δ approaches zero, the limiting "function"

$$\delta(t-t_1) \triangleq \lim_{\Delta\to 0} \delta_\Delta(t-t_1)$$

is called the *impulse function* or the *Dirac delta function* or simply *δ-function*. Thus the delta function $\delta(t-t_1)$ has the properties that

$$\int_{-\infty}^{\infty} \delta(t-t_1)\,dt = \int_{t_1-\varepsilon}^{t_1+\varepsilon} \delta(t-t_1)\,dt = 1$$

for any positive ε and that

$$\int_{-\infty}^{\infty} f(t)\delta(t-t_1)\,dt = f(t_1) \tag{3-7}$$

for any function f that is continuous at t_1.

With the concept of impulse function, we are ready to develop a mathematical description for relaxed linear systems. We discuss first single-variable systems; the result can then be easily extended to multivariable systems. Con-

[1] For a rigorous development of the subsequent material, the theory of distributions is needed; see Reference 96.

sider a relaxed single-variable system whose input and output are related by

$$y = Hu$$

As shown in Figure 3-3, every piecewise continuous input can be approximated by a series of pulse functions. Since every pulse function can be described by $u(t_i)\delta_\Delta(t-t_i)\Delta$, we can write the input function as

$$u \doteq \sum_i u(t_i)\delta_\Delta(t-t_i)\Delta$$

If the input-output pairs of the relaxed system satisfy the linearity property, then we have[2]

$$y = Hu \doteq \sum_i (H\delta_\Delta(t-t_i))u(t_i)\Delta \tag{3-8}$$

Now as Δ tends to zero, the approximation tends to an exact equality, the summation becomes an integration and the pulse function $\delta_\Delta(t-t_i)$ tends toward a δ-function. Consequently, as $\Delta \to 0$, Equation (3-8) becomes

$$y = \int_{-\infty}^{\infty} (H\delta(t-\tau))u(\tau)\,d\tau \tag{3-9}$$

Now if $H\delta(t-\tau)$ is known for all τ, then for any input, the output can be computed from (3-9). The physical meaning of $H\delta(t-\tau)$ is that it is the output of the *relaxed* system due to an impulse function input applied at time τ. Define

$$H\delta(t-\tau) = g(\cdot, \tau) \tag{3-10}$$

Note that g is a function of two variables—the second variable denoting the time at which the δ-function is applied and the first variable denoting the time at which the output is observed. Since $g(\cdot, \tau)$ is the response of an impulse function, it is called the *impulse response* of the system. Using (3-10), we can write the output at time t as

$$y(t) = \int_{-\infty}^{\infty} g(t, \tau)u(\tau)\,d\tau \tag{3-11}$$

[2] The condition for interchanging the order of H and the summation is disregarded. For a discussion of this problem, see Reference S125, pp. 2–6.

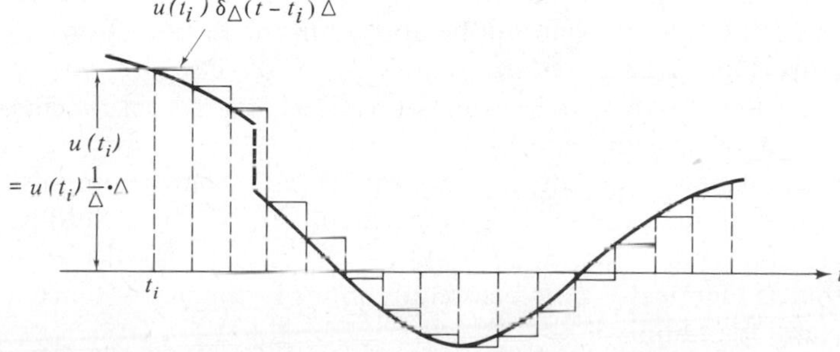

Figure 3-3 Pulse-function approximation of an input function.

Hence if $g(\cdot, \tau)$ for all τ is known then for any input u, the output can be computed from (3-11). In other words, a linear relaxed system is completely described by the superposition integral (3-11), where $g(\cdot, \tau)$ is the impulse response of the system, and theoretically it can be obtained by direct measurements at the input and the output terminals of the system.

If a system has p input terminals and q output terminals, and if the system is initially relaxed at $-\infty$, the input-output description (3-11) can be extended to

$$\mathbf{y}(t) = \int_{-\infty}^{\infty} \mathbf{G}(t, \tau)\mathbf{u}(\tau)\, d\tau \qquad \textbf{(3-12)}$$

where
$$\mathbf{G}(t, \tau) = \begin{bmatrix} g_{11}(t, \tau) & g_{12}(t, \tau) & \cdots & g_{1p}(t, \tau) \\ g_{21}(t, \tau) & g_{22}(t, \tau) & \cdots & g_{2p}(t, \tau) \\ \vdots & \vdots & & \vdots \\ g_{q1}(t, \tau) & g_{q2}(t, \tau) & \cdots & g_{qp}(t, \tau) \end{bmatrix}$$

and $g_{ij}(t, \tau)$ is the response at time t at the ith output terminal due to an impulse function applied at time τ at the jth input terminal, the inputs at other terminals being identically zero. Equivalently, *g_{ij} is the impulse response between the jth input terminal and the ith output terminal.* Hence $\mathbf{G}$ is called the *impulse-response matrix* of the system.

Although the input and output of a relaxed linear system can be related by an equation of the form in Equation (3-12), the equation is not readily applicable because the integration required is from $-\infty$ to ∞ and because there is no way of checking whether or not a system is initially relaxed at $-\infty$. These difficulties will be removed in the subsequent development.

Causality. A system is said to be *causal* or *nonanticipatory* if the output of the system at time t does not depend on the input applied after time t; it depends only on the input applied before and at time t. In short, the past affects the future, but not conversely. Hence, if a relaxed system is causal, its input and output relation can be written as

$$\mathbf{y}(t) = H\mathbf{u}_{(-\infty, t]} \qquad \textbf{(3-13)}$$

for all t in $(-\infty, \infty)$. If a system is not causal, it is said to be *noncausal* or *anticipatory*. The output of a noncausal system depends not only on the past input but also on the future value of the input. This implies that a noncausal system is able to *predict* the input that will be applied in the future. For real physical systems, this is impossible. Hence *causality is an intrinsic property of every physical system.* Since in this book systems are models of physical systems, all the systems we study are assumed to be causal.

If a relaxed system is linear and causal, what can we say about the impulse response matrix, $\mathbf{G}(t, \tau)$, of the system? Every element of $\mathbf{G}(t, \tau)$ is, by definition, the output due to a δ-function input applied at time τ. Now if a relaxed system is causal, the output is identically zero before any input is applied. Hence a linear system is causal if and only if

$$\mathbf{G}(t, \tau) = \mathbf{0} \qquad \text{for all } \tau \text{ and all } t < \tau \qquad \textbf{(3-14)}$$

Consequently, the input-output description of a linear, causal, relaxed system becomes

$$\mathbf{y}(t) = \int_{-\infty}^{t} \mathbf{G}(t, \tau)\mathbf{u}(\tau)\, d\tau \tag{3-15}$$

Relaxedness. Recall that the equation $\mathbf{y} = H\mathbf{u}$ of a system holds only when the system is relaxed at $-\infty$, or equivalently, only when the output $\mathbf{y}$ is excited solely and uniquely by $\mathbf{u}_{(-\infty,\infty)}$. We may apply this concept to an arbitrary t_0.

Definition 3-3

A system is said to be *relaxed* at time t_0 if and only if the output $\mathbf{y}_{[t_0,\infty)}$ is solely and uniquely excited by $\mathbf{u}_{[t_0,\infty)}$. ∎

This concept can be easily understood if we consider an *RLC* network. If all the voltages across capacitors and all the currents through inductors are zero at time t_0, then the network is relaxed at t_0. If a network is not relaxed at t_0 and if an input $\mathbf{u}_{[t_0,\infty)}$ is applied, then part of the response will be excited by the initial conditions; and for different initial conditions, different responses will be excited by the same input $\mathbf{u}_{[t_0,\infty)}$. If a system is known to be relaxed at t_0, then its input-output relation can be written as

$$\mathbf{y}_{[t_0,\infty)} = H\mathbf{u}_{[t_0,\infty)} \tag{3-16}$$

It is clear that if a system is relaxed at $-\infty$, and if $\mathbf{u}_{(-\infty,t_0)} \equiv \mathbf{0}$, then the system is still relaxed at t_0. For the class of systems whose inputs and outputs are linearly related, it is easy to show that *the necessary and sufficient condition for a system to be relaxed at* t_0 *is that* $\mathbf{y}(t) = H\mathbf{u}_{(-\infty,t_0)} = \mathbf{0}$ *for all* $t \geq t_0$. In words, if the net effect of $\mathbf{u}_{(-\infty,t_0)}$ on the output after t_0 is identically zero, the system is relaxed at t_0.

Example 2

A unit-time-delay system is a device whose output is equal to the input delayed one unit of time; that is, $y(t) = u(t-1)$ for all t. The system is relaxed at t_0 if $u_{[t_0-1,t_0)} \equiv 0$, although $u_{(-\infty,t_0-1)} \neq 0$. ∎

Now if a system whose inputs and outputs are linearly related is known to be relaxed at t_0, then the input-output description reduces to

$$\mathbf{y}(t) = \int_{t_0}^{\infty} \mathbf{G}(t, \tau)\mathbf{u}(\tau)\, d\tau$$

Hence, if a system is described by

$$\mathbf{y}(t) = \int_{t_0}^{t} \mathbf{G}(t, \tau)\mathbf{u}(\tau)\, d\tau \tag{3-17}$$

we know that the input-output pairs of the system satisfy the linearity property and that the system is causal and is relaxed at t_0.

A legitimate question may be raised at this point: Given a system at time t_0, how do we know that the system is relaxed? For a system whose inputs and outputs are linearly related, this can be determined without knowing the previous history of the system, by the use of the following theorem.

Theorem 3-1

A system that is describable by

$$\mathbf{y}(t) = \int_{-\infty}^{\infty} \mathbf{G}(t, \tau)\mathbf{u}(\tau)\, d\tau$$

is relaxed at t_0 if and only if $\mathbf{u}_{[t_0, \infty)} \equiv \mathbf{0}$ implies $\mathbf{y}_{[t_0, \infty)} \equiv \mathbf{0}$. ∎

Proof

Necessity: If a system is relaxed at t_0, the output $\mathbf{y}(t)$ for $t \geq t_0$ is given by

$$\int_{t_0}^{\infty} \mathbf{G}(t, \tau)\mathbf{u}(\tau)\, d\tau$$

Hence, if $\mathbf{u}_{[t_0, \infty)} \equiv \mathbf{0}$, then $\mathbf{y}_{[t_0, \infty)} \equiv \mathbf{0}$. *Sufficiency:* We show that if $\mathbf{u}_{[t_0, \infty)} \equiv \mathbf{0}$ implies $\mathbf{y}_{[t_0, \infty)} \equiv \mathbf{0}$, then the system is relaxed at t_0. Since

$$\mathbf{y}(t) = \int_{-\infty}^{\infty} \mathbf{G}(t, \tau)\mathbf{u}(\tau)\, d\tau = \int_{-\infty}^{t_0} \mathbf{G}(t, \tau)\mathbf{u}(\tau)\, d\tau + \int_{t_0}^{\infty} \mathbf{G}(t, \tau)\mathbf{u}(\tau)\, d\tau$$

the assumptions $\mathbf{u}_{[t_0, \infty)} \equiv \mathbf{0}$, $\mathbf{y}_{[t_0, \infty)} \equiv \mathbf{0}$ imply that

$$\int_{-\infty}^{t_0} \mathbf{G}(t, \tau)\mathbf{u}(\tau)\, d\tau = \mathbf{0} \qquad \text{for all } t \geq t_0$$

In words, the net effect of $\mathbf{u}_{(-\infty, t_0)}$ on the output $\mathbf{y}(t)$ for $t \geq t_0$ is zero, and hence the system is relaxed at t_0. Q.E.D.

An implication of Theorem 3-1 is that given a system at time t_0, if the system is known to be describable by

$$\int_{-\infty}^{\infty} \mathbf{G}(t, \tau)\mathbf{u}(\tau)\, d\tau$$

the relaxedness of the system can be determined from the behavior of the system after t_0 without knowing the previous history of the system. Certainly it is impractical or impossible to observe the output from time t_0 to infinity; fortunately, for a large class of systems, it is not necessary to do so.

Corollary 3-1

If the impulse-response matrix $\mathbf{G}(t, \tau)$ of a system can be decomposed into $\mathbf{G}(t, \tau) = \mathbf{M}(t)\mathbf{N}(\tau)$, and if every element of $\mathbf{M}$ is analytic (see Appendix B) on $(-\infty, \infty)$, then the system is relaxed at t_0 if and only if for some fixed positive ε, $\mathbf{u}_{[t_0, t_0+\varepsilon)} \equiv \mathbf{0}$ implies $\mathbf{y}_{[t_0, t_0+\varepsilon)} \equiv \mathbf{0}$.

Proof

If $\mathbf{u}_{[t_0, \infty)} \equiv \mathbf{0}$, the output $\mathbf{y}(t)$ of the system is given by

$$\mathbf{y}(t) = \int_{-\infty}^{t_0} \mathbf{G}(t, \tau)\mathbf{u}(\tau)\, d\tau = \mathbf{M}(t) \int_{-\infty}^{t_0} \mathbf{N}(\tau)\mathbf{u}(\tau)\, d\tau \qquad \text{for } t \geq t_0$$

Since

$$\int_{-\infty}^{t_0} \mathbf{N}(\tau)\mathbf{u}(\tau)\, d\tau$$

is a constant vector, the analyticity assumption of $\mathbf{M}$ implies that $\mathbf{y}(t)$ is analytic on $[t_0, \infty)$. Consequently, if $\mathbf{y}_{[t_0, t_0+\varepsilon)} \equiv \mathbf{0}$ and if $\mathbf{u}_{[t_0, \infty)} \equiv \mathbf{0}$, then $\mathbf{y}_{[t_0, \infty)} \equiv \mathbf{0}$ (see Appendix B), and the corollary follows from Theorem 3-1. Q.E.D.

This is an important result. For any system that satisfies the conditions of Corollary 3-1, its relaxedness can be easily determined by observing the output over any nonzero interval of time, say 10 seconds. If the output is zero in this interval, then the system is relaxed at that moment. It will be shown in the next chapter that the class of systems that are describable by rational transfer-function matrices or linear time-invariant ordinary differential equations satisfies the conditions of Corollary 3-1. Hence Corollary 3-1 is widely applicable.

We give an example to illustrate that Theorem 3-1 does not hold for systems whose inputs and outputs are not linearly related.

Example 3

Consider the system shown in Figure 3-4; the input is a voltage source, and the output is the voltage across the nonlinear capacitor. If the electric charge stored in the capacitor at time t_0 is either 0, q_1, or q_2, the output will be identically zero if no voltage is applied at the input. However, the system is not necessarily relaxed at t_0, because if an input is applied, we may obtain different outputs depending on which initial charge the capacitor has. ∎

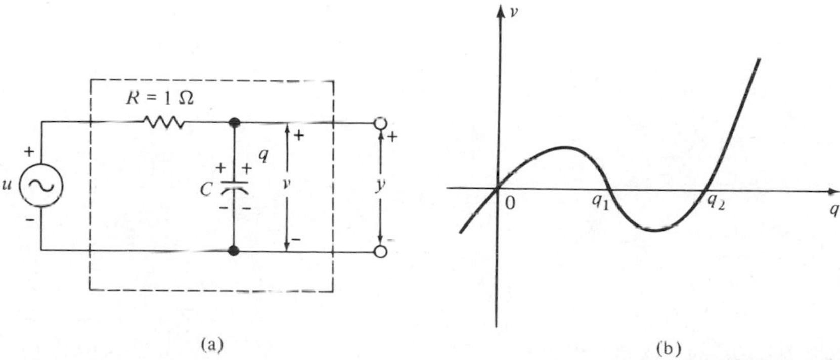

Figure 3-4 A nonlinear network. (a) The network. (b) The characteristic of the nonlinear capacitor C.

Time invariance. If the characteristics of a system do not change with time, then the system is said to be *time invariant, fixed,* or *stationary*. In order to define it precisely, we need the concept of a shifting operator Q_α. The effect of the shifting operator Q_α is illustrated in Figure 3-5. The output of Q_α is equal to the input delayed by α seconds. Mathematically it is defined as $\bar{u} \triangleq Q_\alpha u$ if and only if $\bar{u}(t) = u(t-\alpha)$ or $\bar{u}(t+\alpha) = u(t)$ for all t.

Definition 3-4

A relaxed system is said to be *time invariant* (or *stationary* or *fixed*) if and only if

$$HQ_\alpha \mathbf{u} = Q_\alpha H\mathbf{u} \tag{3-18}$$

for any input $\mathbf{u}$ and any real number α. Otherwise the relaxed system is said to be *time varying*. ∎

The relation $HQ_\alpha \mathbf{u} = Q_\alpha H\mathbf{u}$ can also be written as $HQ_\alpha \mathbf{u} = Q_\alpha \mathbf{y}$, which implies that if an input is shifted by α seconds, the waveform of the output remains the same except for a shift by α seconds. In other words, no matter at what time an input is applied to a relaxed time-invariant system, the waveform of the output is always the same.

If a relaxed linear system is known to be time invariant, what condition does that impose on its impulse response? The impulse response $g(\cdot, \tau)$ is the output due to a δ-function input applied at time τ; that is, $g(\cdot, \tau) = H\delta(t-\tau)$. If the system is time invariant, then we have

$$\begin{aligned} Q_\alpha g(\cdot, \tau) &= Q_\alpha H\delta(t-\tau) = HQ_\alpha \delta(t-\tau) \\ &= H\delta(t-(\tau+\alpha)) = g(\cdot, \tau+\alpha) \end{aligned}$$

Now by the definition of Q_α, the equation $Q_\alpha g(\cdot, \tau) = g(\cdot, \tau+\alpha)$ implies $g(t, \tau) = g(t+\alpha, \tau+\alpha)$, which holds for any t, τ, and α. By choosing $\alpha = -\tau$, we have $g(t, \tau) = g(t-\tau, 0)$ for all t, τ. Hence *the impulse response $g(t, \tau)$ of a relaxed, linear, time-invariant system depends only on the difference of t and τ.* Extending

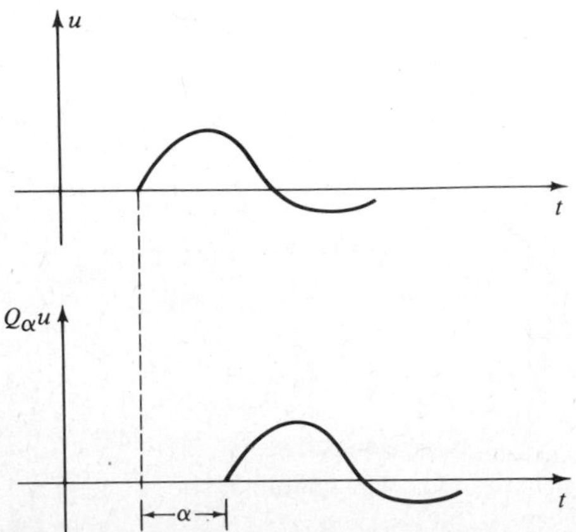

Figure 3-5 The effect of a shifting operator on a signal.

this to the multivariable case, we obtain[3]

$$\mathbf{G}(t, \tau) = \mathbf{G}(t-\tau, 0) = \mathbf{G}(t-\tau)$$

for all t and τ. Hence, if a system is described by

$$\mathbf{y}(t) = \int_{t_0}^{t} \mathbf{G}(t-\tau)\mathbf{u}(\tau)\, d\tau \tag{3-19}$$

we know that its input-output pairs satisfy the linearity, causality, and time-invariance properties; furthermore, the system is relaxed at time t_0. In the time-invariant case, the initial time t_0 is always chosen, without loss of generality, to be 0 and the time interval of interest is $[0, \infty)$. Note that $t_0 = 0$ is the instant we start to consider the system or to apply the input $\mathbf{u}$. If $t_0 = 0$, Equation (3-19) becomes

$$\mathbf{y}(t) = \int_0^t \mathbf{G}(t-\tau)\mathbf{u}(\tau)\, d\tau = \int_0^t \mathbf{G}(\tau)\mathbf{u}(t-\tau)\, d\tau \tag{3-20}$$

The second equality of (3-20) can be easily verified by changing the variables. The integration in (3-20) is called the *convolution integral*. Since $\mathbf{G}(t-\tau) = \mathbf{G}(t, \tau)$ represents the responses at time t due to δ-function inputs applied at time τ, $\mathbf{G}(t)$ represents the responses at time t due to δ-function inputs applied at $\tau = 0$. Following (3-14), a linear time-invariant system is causal if and only if $\mathbf{G}(t) = \mathbf{0}$ for all $t < 0$.

Transfer-function matrix. In the study of the class of systems that are describable by convolution integrals, it is of great advantage to use the Laplace transform, because it will change a convolution integral in the time domain into an algebraic equation in the frequency domain. Let $\hat{\mathbf{y}}(s)$ be *the Laplace transform* of $\mathbf{y}$; that is,[4]

$$\hat{\mathbf{y}}(s) \triangleq \mathscr{L}(\mathbf{y}) = \int_0^\infty \mathbf{y}(t)e^{-st}\, dt$$

Since $\mathbf{G}(t-\tau) = \mathbf{0}$ for $\tau > t$, the upper limit of the integration in (3-20) can be set at ∞; hence, from (3-20), we have

$$\begin{aligned}
\hat{\mathbf{y}}(s) &= \int_0^\infty \left(\int_0^\infty \mathbf{G}(t-\tau)\mathbf{u}(\tau)\, d\tau \right) e^{-st}\, dt \\
&= \int_0^\infty \left(\int_0^\infty \mathbf{G}(t-\tau)e^{-s(t-\tau)}\, dt \right) \mathbf{u}(\tau)e^{-s\tau}\, d\tau \\
&= \int_0^\infty \mathbf{G}(v)e^{-sv}\, dv \int_0^\infty \mathbf{u}(\tau)e^{-s\tau}\, d\tau \\
&\triangleq \hat{\mathbf{G}}(s)\hat{\mathbf{u}}(s)
\end{aligned} \tag{3-21}$$

[3] Note that $\mathbf{G}(t, \tau)$ and $\mathbf{G}(t-\tau)$ are two different functions. However, for convenience, the same symbol $\mathbf{G}$ is used.

[4] If $\mathbf{y}$ contains delta functions at $t = 0$, the lower limit of the integration should start from $0-$ to include the delta functions in the transform.

Here we have changed the order of integration, changed the variables, and used the fact that $\mathbf{G}(t)=\mathbf{0}$ for $t<0$. As defined in (3-21), $\hat{\mathbf{G}}(s)$ is the Laplace transform of the impulse-response matrix; that is,

$$\hat{\mathbf{G}}(s)=\int_0^\infty \mathbf{G}(t)e^{-st}\,dt$$

and is called the *transfer-function matrix* of the system. For single-variable systems, $\hat{\mathbf{G}}(s)$ reduces to a scalar and is called the *transfer function.* Hence the *transfer function is the Laplace transform of the impulse response*; it can also be defined, following (3-21), as

$$\hat{g}(s)=\left.\frac{\mathscr{L}[y(t)]}{\mathscr{L}[u(t)]}\right|_{\substack{\text{the system is}\\ \text{relaxed at } t=0}}=\left.\frac{\hat{y}(s)}{\hat{u}(s)}\right|_{\substack{\text{relaxed}\\ \text{at } t=0}} \tag{3-22}$$

where the circumflex (^) over a variable denotes the Laplace transform of the same variable; for example,

$$\hat{\mathbf{x}}(s)\triangleq\mathscr{L}[\mathbf{x}]\triangleq\int_0^\infty \mathbf{x}(t)e^{-st}\,dt$$

We see that the familiar transfer functions are the input-output descriptions of systems. It is important to note that this input-output description is obtained under the relaxedness assumption of a system; hence, if the system is not relaxed at $t=0$, the transfer function cannot be directly applied. Thus *whenever a transfer function is used, the system is always implicitly assumed to be relaxed at* $t=0$.

A transfer function is not necessarily a rational function of s. For example, the impulse response $g(t)$ of the unit-time-delay system introduced in Example 2 is $\delta(t-1)$, and its transfer function is e^{-s}, which is not a rational function of s. However, the transfer functions we shall study in this book are exclusively rational functions of s. In fact, we study only a special class of rational functions.

Definition 3-5

A rational function $\hat{g}(s)$ is said to be *proper* if $\hat{g}(\infty)$ is a finite (zero or nonzero) constant. $\hat{g}(s)$ is said to be *strictly proper* if $\hat{g}(\infty)=0$. A rational matrix $\hat{\mathbf{G}}(s)$ is said to be proper if $\hat{\mathbf{G}}(\infty)$ is a finite (zero or nonzero) constant matrix. $\hat{\mathbf{G}}(s)$ is said to be strictly proper if $\hat{\mathbf{G}}(\infty)=\mathbf{0}$. ∎

For example, $\hat{g}(s)=s^2/(s-1)$ is not proper; $\hat{g}(s)=s^2/(s^2-s+2)$ is proper and $\hat{g}(s)=s^2/(s^3-s)$ is strictly proper. It is clear that if $\hat{g}(s)=N(s)/D(s)$, $\hat{g}(s)$ is proper if and only if $\deg N(s)\leq \deg D(s)$; $\hat{g}(s)$ is strictly proper if and only if $\deg N(s)<\deg D(s)$, where deg stands for the degree of a polynomial. A rational matrix is proper if and only if all of its elements are proper. A rational matrix is strictly proper if and only if all of its elements are strictly proper.

If a transfer function is not proper, high-frequency noises will be greatly amplified and will overwhelm information-bearing signals. Hence improper rational functions are hardly used in practice.[5] In this book, we study only proper rational functions and proper rational matrices.

3-3 The State-Variable Description

The concept of state. The input-output description of a system is applicable only when the system is initially relaxed. If a system is not initially relaxed, say at time t_0, then the equation $\mathbf{y}_{[t_0,\infty)} = H\mathbf{u}_{[t_0,\infty)}$ does not hold. In this case the output $\mathbf{y}_{[t_0,\infty)}$ depends not only on the input $\mathbf{u}_{[t_0,\infty)}$ but also on the initial conditions at t_0. Hence, in order to determine the output $\mathbf{y}_{[t_0,\infty)}$ uniquely, in addition to the input $\mathbf{u}_{[t_0,\infty)}$ we need a set of initial conditions at t_0. This set of initial conditions is called the *state*. Hence the state at t_0 is the information that, together with the input $u_{[t_0,\infty)}$, determines *uniquely* the output $y_{[t_0,\infty)}$. For example, in newtonian mechanics, if an external force (input) is applied to a particle (system) at time t_0, the motion (output) of the particle for $t \geq t_0$ is not uniquely determinable unless the position and velocity at time t_0 are also known. How the particle actually attained the position and velocity at t_0 is immaterial in determining the motion of the particle after t_0. Hence the set of two numbers, the position and velocity at time t_0, is qualified to be called the state of the system at time t_0. Note that the set of two numbers, the position and the momentum at t_0, is also qualified as the state.

Definition 3-6

The *state* of a system at time t_0 is the amount of information at t_0 that, together with $\mathbf{u}_{[t_0,\infty)}$, determines uniquely the behavior of the system for all $t \geq t_0$. ∎

By the behavior of a system, we mean all responses, including the state, of the system. If the system is a network, we mean the voltage and current of every branch of the network. Hence from the state at t_0, the state at all $t > t_0$ can be computed. If the state at $t_1 > t_0$ is known, we need $\mathbf{u}_{[t_1,\infty)}$, rather than $\mathbf{u}_{[t_0,\infty)}$, in computing the behavior of the system for $t \geq t_1$. Hence the state at t_1 summarizes the essential information about the past input $\mathbf{u}_{(-\infty,t_1)}$ which is needed in determining the future behavior of the system. We note that different input $\mathbf{u}^i_{(-\infty,t_1)}$, $i = 1, 2, \ldots$, may yield the same state at t_1. In this case, even though $\mathbf{u}^i_{(-\infty,t_1)}$, $i = 1, 2, \ldots$, are different, their effects on the future behavior of the system are identical.

We give some examples to illustrate the concept of state.

Example 1

Consider the network shown in Figure 3-6. It is well known that if the initial current through the inductor and the initial voltage across the capacitor are

[5] The exceptions are transfer functions of some transducers such as tachometers and accelerometers. See Reference S46.

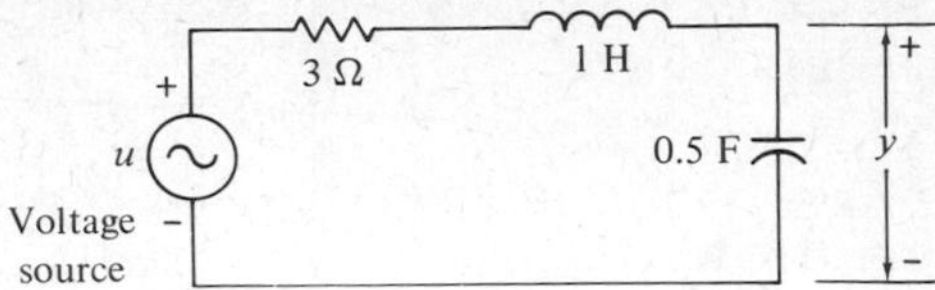

Figure 3-6 A network.

known, then for any driving voltage the behavior of the network can be determined uniquely. Hence the inductor current and the capacitor voltage qualify as the state of the network. ∎

Example 2

We consider again the network in Example 1. The transfer function from u to y of the network can be easily found as

$$\hat{g}(s) = \frac{2}{(s+1)(s+2)} = \frac{2}{s+1} - \frac{2}{s+2}$$

Hence the impulse response of the network is

$$g(t) = 2e^{-t} - 2e^{-2t} \tag{3-23}$$

which is the inverse Laplace transform of the transfer function $\hat{g}(s)$. Now we apply an input $\mathbf{u}_{[t_0, \infty)}$ to the network. If the network is relaxed at t_0, the output is given by

$$y(t) = \int_{t_0}^{t} g(t-\tau)u(\tau)\,d\tau \qquad \text{for } t \geq t_0$$

If the network is not relaxed at t_0, the output must be computed from

$$y(t) = \int_{-\infty}^{t} g(t-\tau)u(\tau)\,d\tau = \int_{-\infty}^{t_0} g(t-\tau)u(\tau)\,d\tau + \int_{t_0}^{t} g(t-\tau)u(\tau)\,d\tau \qquad \text{for } t \geq t_0 \tag{3-24}$$

because the input that had been applied before t_0 may still have some effect on the output after t_0 through the energy stored in the capacitor and inductor. We consider now the effect on $y_{[t_0, \infty)}$ due to the unknown input $u_{(-\infty, t_0)}$. From (3-23) we have

$$\int_{-\infty}^{t_0} g(t-\tau)u(\tau)\,d\tau = 2e^{-t}\int_{-\infty}^{t_0} e^{\tau}u(\tau)\,d\tau - 2e^{-2t}\int_{-\infty}^{t_0} e^{2\tau}u(\tau)\,d\tau$$
$$\triangleq 2e^{-t}c_1 - 2e^{-2t}c_2 \tag{3-25}$$

for $t \geq t_0$, where

$$c_1 \triangleq \int_{-\infty}^{t_0} e^{\tau}u(\tau)\,d\tau \qquad \text{and} \qquad c_2 \triangleq \int_{-\infty}^{t_0} e^{2\tau}u(\tau)\,d\tau$$

Note that c_1 and c_2 are independent of t. Hence if c_1 and c_2 are known, the output after $t \geq t_0$ excited by the unknown input $u_{(-\infty, t_0)}$ is completely determinable. From (3-24) and (3-25) we have

$$y(t_0) = 2e^{-t_0}c_1 - 2e^{-2t_0}c_2 \tag{3-26}$$

Taking the derivative of (3-24) with respect to t yields[6]

$$\dot{y}(t) = -2e^{-t}c_1 + 4e^{-2t}c_2 + g(0)u(t) + \int_{t_0}^{t} \frac{\partial}{\partial t} g(t-\tau)u(\tau)\, d\tau$$

which, together with $g(0) = 0$, implies that

$$\dot{y}(t_0) = -2e^{-t_0}c_1 + 4e^{-2t_0}c_2 \tag{3-27}$$

Solving for c_1 and c_2 from (3-26) and (3-27), we obtain

$$c_1 = 0.5e^{t_0}(2y(t_0) + \dot{y}(t_0))$$
$$c_2 = 0.5e^{2t_0}(y(t_0) + \dot{y}(t_0))$$

Hence if the network is not relaxed at t_0, the output $y(t)$ is given by

$$y(t) = (2y(t_0) + \dot{y}(t_0))e^{-(t-t_0)} - (y(t_0) + \dot{y}(t_0))e^{-2(t-t_0)}$$
$$+ \int_{t_0}^{t} g(t-\tau)u(\tau)\, d\tau \qquad \text{for } t \geq t_0$$

We see that if $y(t_0)$ and $\dot{y}(t_0)$ are known, the output after $t \geq t_0$ can be uniquely determined even if the network is not relaxed at t_0. Hence the set of numbers $y(t_0)$ and $\dot{y}(t_0)$ qualifies as the state of the network at t_0. Clearly, the set $\{c_1, c_2\}$ also qualifies as the state of the network. ∎

In this example, we see that the effect of the input over the infinite interval $(-\infty, t_0)$ is summarized into two numbers $\{y(t_0), \dot{y}(t_0)\}$ or $\{c_1, c_2\}$; hence the concept of state is very efficient and powerful.

Example 3

Consider the network shown in Figure 3-7. It is clear that if all the capacitor voltages are known, then the behavior of the network is uniquely determinable for any applied input. Hence the set of capacitor voltages x_1, x_2, and x_3 qualifies as the state of the network. Let us examine the network more carefully. If we

6 $\frac{d}{dt} \int_{t_0}^{t} g(t-\tau)u(\tau)\, d\tau = g(t-\tau)u(\tau)\Big|_{\tau=t} + \int_{t_0}^{t} \frac{\partial}{\partial t} g(t-\tau)u(\tau)\, d\tau$

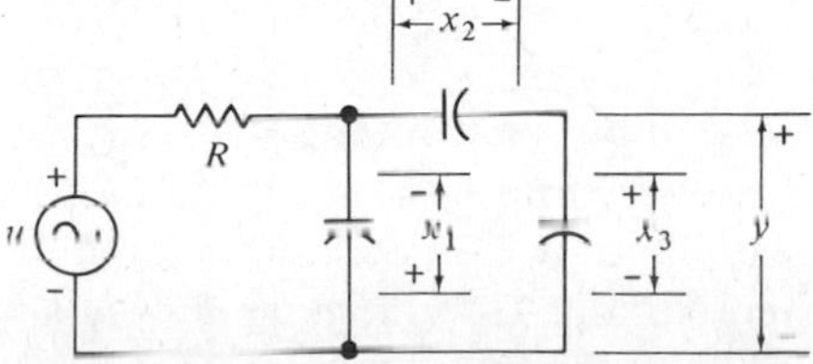

Figure 3-7 A network with a loop that consists of capacitors only.

apply the Kirchhoff voltage law to the loop that consists of three capacitors, we have $x_1(t)+x_2(t)+x_3(t)=0$ for all t. It implies that if any two of x_1, x_2, and x_3 are known, the third one is also known. Consequently, if any two of the capacitor voltages are known, the behavior of the network is uniquely determinable for any input applied thereafter. In other words, any two of the three capacitor voltages qualify as the state. If all the three capacitor voltages are chosen as the state, then there is a redundancy. In choosing the state of a system, it is desirable to choose a state that consists of the least number of variables. How to pick the state with the least number of variables for general *RLC* networks will be studied in the next section. ∎

Example 4

A unit-time-delay system is a device whose output $y(t)$ is equal to $u(t-1)$ for all t. For this system, in order to determine $y_{[t_0,\infty)}$ uniquely from $u_{[t_0,\infty)}$, we need the information $u_{[t_0-1,t_0)}$. Hence the information $u_{[t_0-1,t_0)}$ is qualified to be called the state of the system at time t_0. ∎

From these examples, we may have the following observations concerning the state of a system. First, the choice of the state is not unique. For the network shown in Figure 3-6, the state may be chosen as the inductor current and the capacitor voltage, or chosen as $y(t_0)$ and $\dot{y}(t_0)$ or c_1 and c_2. For the network shown in Figure 3-7, any two of the three capacitor voltages can be chosen as the state. Different analyses often lead to different choices of state. Second, the state chosen in Example 1 is associated with physical quantities, whereas in Example 2 the state is introduced for mathematical necessity. Hence, the state of a system is an auxiliary quantity that may or may not be easily interpretable in physical terms. Finally the state at each instant may consist of only a finite set of numbers, as in Examples 1, 2, and 3, or consist of an infinite set of numbers, as in Example 4. Note that there is an infinite number of points between $[t_0-1, t_0)$; hence the state of this example consists of an infinite set of numbers.

In this book we study only the class of systems whose states may be chosen to consist of a finite number of variables. The state of a system can then be represented by a finite-dimensional column vector $\mathbf{x}$, called the *state vector*. The components of $\mathbf{x}$ are called *state variables*. The linear space in which the state vector ranges is denoted by Σ. Since state variables are usually real-valued, and since we study only systems with a finite number of state variables, the state spaces we encounter in this book are the familiar finite-dimensional real vector space$(\mathbb{R}^n, \mathbb{R})$.

Dynamical equations. In addition to the input and output of a system we have now the state of the system. The state at time t_0 is, by definition, the required information at t_0 that, together with input $\mathbf{u}_{[t_0,\infty)}$, determines uniquely the behavior (output and state) of the system for all $t \geq t_0$. *The set of equations that describes the unique relations between the input, output, and state is called a* *dynamical equation.* In this book, we study only the dynamical equations of the form

$$\dot{\mathbf{x}}(t)=\mathbf{h}(\mathbf{x}(t), \mathbf{u}(t), t) \qquad \text{(state equation)} \tag{3-28a}$$

$$\mathbf{y}(t)=\mathbf{g}(\mathbf{x}(t), \mathbf{u}(t), t) \qquad \text{(output equation)} \tag{3-28b}$$

or, more explicitly,

$$\begin{aligned}
\dot{x}_1(t)&=h_1(x_1(t), x_2(t), \ldots, x_n(t), u_1(t), u_2(t), \ldots, u_p(t), t)\\
\dot{x}_2(t)&=h_2(x_1(t), x_2(t), \ldots, x_n(t), u_1(t), u_2(t), \ldots, u_p(t), t)\\
&\cdots\cdots\cdots\cdots\cdots\cdots\cdots\cdots\cdots\cdots\cdots\cdots\\
\dot{x}_n(t)&=h_n(x_1(t), x_2(t), \ldots, x_n(t), u_1(t), u_2(t), \ldots, u_p(t), t)
\end{aligned} \tag{3-29a}$$

$$\begin{aligned}
y_1(t)&=g_1(x_1(t), x_2(t), \ldots, x_n(t), u_1(t), u_2(t), \ldots, u_p(t), t)\\
y_2(t)&=g_2(x_1(t), x_2(t), \ldots, x_n(t), u_1(t), u_2(t), \ldots, u_p(t), t)\\
&\cdots\cdots\cdots\cdots\cdots\cdots\cdots\cdots\cdots\cdots\cdots\cdots\\
y_q(t)&=g_q(x_1(t), x_2(t), \ldots, x_n(t), u_1(t), u_2(t), \ldots, u_p(t), t)
\end{aligned} \tag{3-29b}$$

where $\mathbf{x}=[x_1 \quad x_2 \quad \cdots \quad x_n]'$ is the state, $\mathbf{y}=[y_1 \quad y_2 \quad \cdots \quad y_q]'$ is the output, and $\mathbf{u}=[u_1 \quad u_2 \quad \cdots \quad u_p]'$ is the input. The input $\mathbf{u}$, the output $\mathbf{y}$, and the state $\mathbf{x}$ are real-valued vector functions of t defined over $(-\infty, \infty)$. In order for (3-28) to qualify as a dynamical equation, we must assume that for any initial state $\mathbf{x}(t_0)$ and any given $\mathbf{u}$, Equation (3-28) has a unique solution. A sufficient condition for (3-28) to have a unique solution for a given $\mathbf{u}$ and a given initial state is that h_i and $\partial h_i/\partial x_j$ are continuous functions of t for $i, j=1, 2, \ldots, n$; see References 24, 77, and 92. If a unique solution exists in (3-28), it can be shown that the solution can be solved in terms of $\mathbf{x}(t_0)$ and $\mathbf{u}_{[t_0, t)}$. Hence $\mathbf{x}$ serves as the state at t_0, as expected. Equation (3-28a) governs the behavior of the state and is called a *state equation.* Equation (3-28b) gives the output and is called an *output equation.* Note that there is no loss of generality in writing the output equation in the form of (3-28b), because by the definition of the state, the knowledge of $\mathbf{x}(t)$ and $\mathbf{u}(t)$ suffices to determine $\mathbf{y}(t)$.

The state space of (3-28) is an n-dimensional real vector space; hence the set of equation (3-28) is called an *n-dimensional dynamical equation.*

Linearity. We use the notation

$$\{\mathbf{u}_{[t_0, \infty)}, \mathbf{x}(t_0)\} \rightarrow \{\mathbf{x}_{[t_0, \infty)}, \mathbf{y}_{[t_0, \infty)}\}$$

to denote that the state $\mathbf{x}(t_0)$ and the input $\mathbf{u}_{[t_0, \infty)}$ excite the output $\mathbf{y}(t)$ and the state $\mathbf{x}(t)$, for $t \geq t_0$, and call it an input-state-output pair. An input-state-output pair of a system is called *admissible* if the system can generate such a pair.

Definition 3-7

A system is said to be *linear* if and only if for any two admissible pairs

$$\{\mathbf{x}^1(t_0), \mathbf{u}^1_{[t_0, \infty)}\} \rightarrow \{\mathbf{x}^1_{[t_0, \infty)}, \mathbf{y}^1_{[t_0, \infty)}\}$$
$$\{\mathbf{x}^2(t_0), \mathbf{u}^2_{[t_0, \infty)}\} \rightarrow \{\mathbf{x}^2_{[t_0, \infty)}, \mathbf{y}^2_{[t_0, \infty)}\}$$

and any real numbers α_1 and α_2, the following pair

$$\{\alpha_1\mathbf{x}^1(t_0)+\alpha_2\mathbf{x}^2(t_0), \alpha_1\mathbf{u}^1_{[t_0,\infty)}+\alpha_2\mathbf{u}^2_{[t_0,\infty)}\}$$
$$\rightarrow\{\alpha_1\mathbf{x}^1_{[t_0,\infty)}+\alpha_2\mathbf{x}^2_{[t_0,\infty)}, \alpha_1\mathbf{y}^1_{[t_0,\infty)}+\alpha_2\mathbf{y}^2_{[t_0,\infty)}\} \tag{3-30}$$

is also admissible. Otherwise, the system is said to be *nonlinear*. ∎

Similar to Equations (3-4) to (3-6), if $\alpha_1=\alpha_2=1$, the relationship in (3-30) is called the property of *additivity*; if $\alpha_2=0$, it is called the property of *homogeneity*. The combination of these two properties is called the *principle of superposition*. In this definition, the superposition property must hold not only at the output but also at all state variables; it must hold for zero initial state as well as nonzero initial state. Hence this definition is much more stringent than the one in Definition 3-2. Consequently, a system may not be linear according to Definition 3-7 but may be linear according to Definition 3-2. For example, the system in Figure 3-8 (a) has a nonlinear capacitor C. If the voltage across the capacitor is zero at $t_0=0$, it will remain to be zero for all $t\geq t_0$ no matter what input waveform is applied. Hence, as far as the behavior at the input and output terminals is concerned, the nonlinear capacitor can be disregarded. Hence the system is linear according to Definition 3-2 but not linear according to Definition 3-7. Consider the network in Figure 3-8(b) with a nonlinear capacitor C and a nonlinear inductor L. Because the L-C loop is in a series connection with the current source, its behavior will not transmit to the output $\mathbf{y}$. Hence the system in Figure 3-8(b) is linear according to Definition 3-2 but not linear according to Definition 3-7.

We discuss now the implication of Definition 3-7. If $\alpha_1=\alpha_2=1$ and if

$$\mathbf{x}^1(t_0)=-\mathbf{x}^2(t_0) \qquad \text{and} \qquad \mathbf{u}^1_{[t_0,\infty)}=-\mathbf{u}^2_{[t_0,\infty)}$$

then the linearity implies that $\{\mathbf{0},\mathbf{0}\}\rightarrow\{\mathbf{0}_{[t_0,\infty)},\mathbf{0}_{[t_0,\infty)}\}$. Hence a necessary condition for a system to be linear is that if $\mathbf{x}(t_0)=\mathbf{0}$ and $\mathbf{u}_{[t_0,\infty)}\equiv\mathbf{0}$, then the responses of the system are identically zero. A very important property of any linear system is that the responses of the system can be decomposed into two parts, as

$$\begin{aligned}&\text{Responses due to } \{\mathbf{x}(t_0), \mathbf{u}_{[t_0,\infty)}\}\\ &\quad=\text{responses due to } \{\mathbf{x}(t_0), \mathbf{0}\}+\text{responses due to } \{\mathbf{0}, \mathbf{u}_{[t_0,\infty)}\}\end{aligned} \tag{3-31}$$

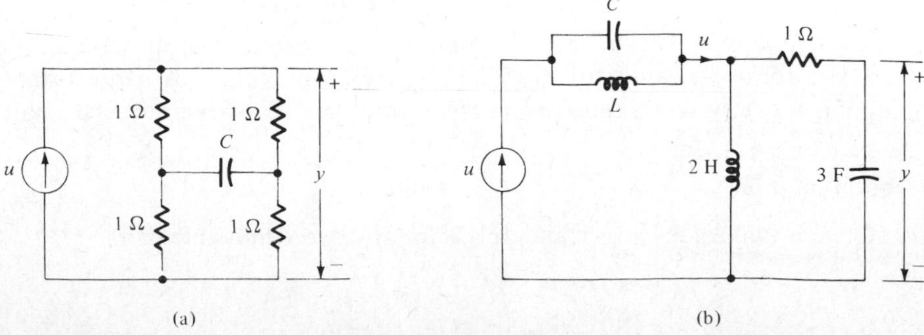

Figure 3-8 Two nonlinear systems which are linear according to Definition 3-2.

The responses due to $\{\mathbf{x}(t_0), \mathbf{0}\}$ are called *zero-input responses*; they are generated exclusively by the nonzero initial state $\mathbf{x}(t_0)$. The responses due to $\{\mathbf{0}, \mathbf{u}_{[t_0, \infty)}\}$ are called *zero-state responses*; they are excited exclusively by the input $\mathbf{u}_{[t_0, \infty)}$. Equation (3-31) follows directly from (3-30) if we choose $\alpha_1 = \alpha_2 = 1$ and

$$\mathbf{x}^1(t_0) = \mathbf{x}(t_0) \qquad \mathbf{u}^1 = \mathbf{0} \qquad \mathbf{x}^2(t_0) = \mathbf{0} \qquad \mathbf{u}^2_{[t_0, \infty)} = \mathbf{u}_{[t_0, \infty)}$$

Hence for linear systems, we may consider the zero-input responses and the zero-state responses independently. The input-output description discussed in the previous section describes only the zero-state responses of linear systems.

If a system is linear, the $\mathbf{h}$ and $\mathbf{g}$ in (3-29) become linear functions of $\mathbf{x}$ and $\mathbf{u}$ as

$$\mathbf{h}(\mathbf{x}(t), \mathbf{u}(t), t) = \mathbf{A}(t)\mathbf{x}(t) + \mathbf{B}(t)\mathbf{u}(t), \mathbf{g}(\mathbf{x}(t), \mathbf{u}(t), t) = \mathbf{C}(t)\mathbf{x}(t) + \mathbf{E}(t)\mathbf{u}(t)$$

where $\mathbf{A}, \mathbf{B}, \mathbf{C}$, and $\mathbf{E}$ are, respectively, $n \times n, n \times p, q \times n$, and $q \times p$ matrices (Problem 3-36). Hence an n-dimensional *linear* dynamical equation is of the form

$$E: \quad \dot{\mathbf{x}}(t) = \mathbf{A}(t)\mathbf{x}(t) + \mathbf{B}(t)\mathbf{u}(t) \qquad \text{(state equation)} \tag{3-32a}$$

$$\mathbf{y}(t) = \mathbf{C}(t)\mathbf{x}(t) + \mathbf{E}(t)\mathbf{u}(t) \qquad \text{(output equation)} \tag{3-32b}$$

A sufficient condition for (3-32) to have a unique solution is that every entry of $\mathbf{A}(\cdot)$ be a continuous function of t defined over $(-\infty, \infty)$. For convenience, the entries of $\mathbf{B}(\cdot), \mathbf{C}(\cdot)$, and $\mathbf{E}(\cdot)$ are also assumed to be continuous in $(-\infty, \infty)$. Since the values of $\mathbf{A}(\cdot), \mathbf{B}(\cdot), \mathbf{C}(\cdot)$, and $\mathbf{E}(\cdot)$ change with time, the dynamical equation E in (3-32) is more suggestively called a *linear time-varying dynamical equation.*

Time invariance. If the characteristics of a system do not change with time, then the system is said to be time invariant. We define it formally in the following. Let Q_α be the shifting operator defined in Figure 3-5.

Definition 3-8

A system is said to be *time invariant* if and only if for any admissible pair

$$\{\mathbf{x}(t_0), \mathbf{u}_{[t_0, \infty)}\} \rightarrow \{\mathbf{x}_{[t_0, \infty)}, \mathbf{y}_{[t_0, \infty)}\}$$

and any real number α, the pair

$$\{Q_\alpha \mathbf{x}(t_0), Q_\alpha \mathbf{u}_{[t_0, \infty)}\} \rightarrow \{Q_\alpha \mathbf{x}_{[t_0, \infty)}, Q_\alpha \mathbf{y}_{[t_0, \infty)}\}$$

is also admissible. Otherwise, the system is said to be *time varying.* ∎

In words, for time-invariant systems, if the initial states are the same and the waveforms of excitations are the same, then the waveform of the responses will always be the same no matter at what instant the excitations are applied. For linear time-invariant systems, the matrices $\mathbf{A}(\cdot), \mathbf{B}(\cdot), \mathbf{C}(\cdot)$, and $\mathbf{E}(\cdot)$ in (3-32) are independent of time, and the equation reduces to

$$FE: \quad \dot{\mathbf{x}}(t) = \mathbf{A}\mathbf{x}(t) + \mathbf{B}\mathbf{u}(t) \tag{3-33a}$$

$$\mathbf{y}(t) = \mathbf{C}\mathbf{x}(t) + \mathbf{E}\mathbf{u}(t) \tag{3-33b}$$

where **A**, **B**, **C** and **E** are, respectively, $n \times n$, $n \times p$, $q \times n$, and $q \times p$ real constant matrices. This set of equations is called a *linear time-invariant n-dimensional* dynamical equation and is denoted by *FE* (fixed equation). For linear time-invariant systems, the responses are independent of the initial time; hence it is always assumed without loss of generality that $t_0 = 0$. The time interval of interest then becomes $[0, \infty)$.

The state space Σ of E or FE is an n-dimensional real vector space $(\mathbb{R}^n, \mathbb{R})$. Hence we can think of the $n \times n$ matrix $\mathbf{A}$ as a linear operator which maps Σ into Σ. As mentioned in the preceding chapter, it is very convenient to introduce the set of the orthonormal vectors $\{\mathbf{n}_1, \mathbf{n}_2, \ldots, \mathbf{n}_n\}$, where $\mathbf{n}_i$ is an $n \times 1$ column vector with 1 at its ith component, and zero elsewhere, as the basis of the state space. In doing so, we may also think of the matrix $\mathbf{A}$ as representing a linear operator with respect to this orthonormal basis. Hence, *unless otherwise stated, the basis of the state space of E or FE is assumed to be the set of the orthonormal vectors*

$$\{\mathbf{n}_1, \mathbf{n}_2, \ldots, \mathbf{n}_n\}$$

The dynamical equations in (3-29), (3-32), and (3-33) can be solved, given $\mathbf{x}(t_0)$ and $\mathbf{u}(\cdot)$, in the direction of positive time or in the direction of negative time. Clearly we are interested in only the direction of positive time. In the positive-time direction, the input $\mathbf{u}(t_1)$ affects only the future responses, the responses for $t \geq t_1$; it does not affect the past responses. Hence the dynamical equations are all causal.

Transfer-function matrix. In the study of linear time-invariant dynamical equations, we may also apply the Laplace transform. Taking the Laplace transform of FE and assuming $\mathbf{x}(0) = \mathbf{x}_0$, we obtain

$$s\hat{\mathbf{x}}(s) - \mathbf{x}_0 = \mathbf{A}\hat{\mathbf{x}}(s) + \mathbf{B}\hat{\mathbf{u}}(s) \tag{3-34a}$$

$$\hat{\mathbf{y}}(s) = \mathbf{C}\hat{\mathbf{x}}(s) + \mathbf{E}\hat{\mathbf{u}}(s) \tag{3-34b}$$

where the circumflex over a variable denotes the Laplace transform of the same variable; for example,

$$\hat{\mathbf{x}}(s) = \int_0^\infty \mathbf{x}(t)e^{-st}\, dt$$

From (3-34), we have

$$\hat{\mathbf{x}}(s) = (s\mathbf{I} - \mathbf{A})^{-1}\mathbf{x}_0 + (s\mathbf{I} - \mathbf{A})^{-1}\mathbf{B}\hat{\mathbf{u}}(s) \tag{3-35a}$$

$$\hat{\mathbf{y}}(s) = \mathbf{C}(s\mathbf{I} - \mathbf{A})^{-1}\mathbf{x}_0 + \mathbf{C}(s\mathbf{I} - \mathbf{A})^{-1}\mathbf{B}\hat{\mathbf{u}}(s) + \mathbf{E}\hat{\mathbf{u}}(s) \tag{3-35b}$$

They are algebraic equations. If $\mathbf{x}_0$ and $\mathbf{u}$ are known, $\hat{\mathbf{x}}(s)$ and $\hat{\mathbf{y}}(s)$ can be computed from (3-35). Note that the determinant of $(s\mathbf{I} - \mathbf{A})$ is different from zero (the zero of the field of rational functions of s)[7]; hence, the inverse of the matrix

[7] That is, the determinant of $(s\mathbf{I} - \mathbf{A})$ is not identically equal to zero. Note that $\det(s\mathbf{I} - \mathbf{A}) = 0$ for some s is permitted.

$(s\mathbf{I}-\mathbf{A})$ always exists. If the initial state $\mathbf{x}_0$ is $\mathbf{0}$—that is, the system is relaxed at $t=0$—then (3-35b) reduces to

$$\hat{\mathbf{y}}(s)=[\mathbf{C}(s\mathbf{I}-\mathbf{A})^{-1}\mathbf{B}+\mathbf{E}]\hat{\mathbf{u}}(s)$$

A comparison of this equation with (3-21) yields

$$\hat{\mathbf{G}}(s)=\mathbf{C}(s\mathbf{I}-\mathbf{A})^{-1}\mathbf{B}+\mathbf{E} \tag{3-36}$$

Hence if a linear time-invariant system is described by the transfer matrix $\hat{\mathbf{G}}(s)$ and the dynamical equation $\{\mathbf{A},\mathbf{B},\mathbf{C},\mathbf{E}\}$, the two descriptions must be related by (3-36). We write (3-36) as

$$\hat{\mathbf{G}}(s)=\frac{1}{\det(s\mathbf{I}-\mathbf{A})}\mathbf{C}[\mathrm{Adj}\,(s\mathbf{I}-\mathbf{A})]\mathbf{B}+\mathbf{E}$$

Every entry of the adjoint of $(s\mathbf{I}-\mathbf{A})$ is a polynomial of degree strictly less than the degree of the determinant of $(s\mathbf{I}-\mathbf{A})$; hence $\mathbf{C}(s\mathbf{I}-\mathbf{A})^{-1}\mathbf{B}$ is a strictly proper rational matrix. If $\mathbf{E}$ is a nonzero matrix, then $\mathbf{C}(s\mathbf{I}-\mathbf{A})^{-1}\mathbf{B}+\mathbf{E}$ is a proper rational matrix. Note that we have

$$\hat{\mathbf{G}}(\infty)=\mathbf{E} \tag{3-37}$$

Analog and digital computer simulations of linear dynamical equations. As will be illustrated in the next section, systems that have a finite number of state variables can always be described by finite-dimensional dynamical equations. We show in this subsection that every finite-dimensional linear dynamical equation can be readily simulated on an analog computer by interconnecting integrators, summers, and amplifiers (or attenuators). The integrator,[8] summer, and amplifier are three basic components of an analog computer, and their functions are illustrated in Figure 3-9. We give in Figure 3-10 a block diagram of analog computer connections of the following two-dimensional time-invariant dynamical equation

$$\begin{bmatrix}\dot{x}_1(t)\\ \dot{x}_2(t)\end{bmatrix}=\begin{bmatrix}a_{11} & a_{12}\\ a_{21} & a_{22}\end{bmatrix}\begin{bmatrix}x_1(t)\\ x_2(t)\end{bmatrix}+\begin{bmatrix}b_{11} & b_{12}\\ b_{21} & b_{22}\end{bmatrix}\begin{bmatrix}u_1(t)\\ u_2(t)\end{bmatrix}$$

$$\begin{bmatrix}y_1(t)\\ y_2(t)\end{bmatrix}=\begin{bmatrix}c_{11} & c_{12}\\ c_{21} & c_{22}\end{bmatrix}\begin{bmatrix}x_1(t)\\ x_2(t)\end{bmatrix}+\begin{bmatrix}e_{11} & e_{12}\\ e_{21} & e_{22}\end{bmatrix}\begin{bmatrix}u_1(t)\\ u_2(t)\end{bmatrix}$$

[8] In practice, pure differentiators are not used for the reason that they will amplify high-frequency noises. On the other hand, integrators will smooth or suppress noises.

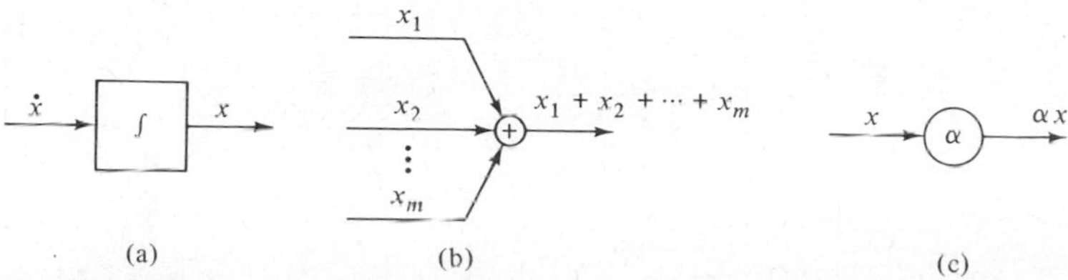

Figure 3-9 Analog computer components. (a) Integrator. (b) Summer. (c) Amplifier or attenuator.

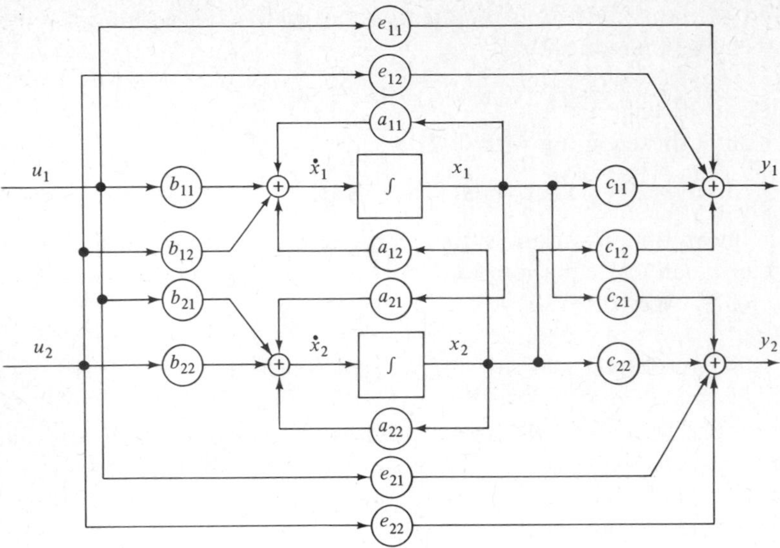

Figure 3-10 Block diagram of two-dimensional dynamical equation.

Note that for a two-dimensional dynamical equation, we need two integrators. *The output of every integrator can be assigned as a state variable.* We see that even for a two-dimensional dynamical equation, the wiring of the block diagram is complicated; hence, for the general case, we usually use a matrix block diagram. The matrix block diagram of the dynamical equation E is shown in Figure 3-11. If E is n-dimensional, the integration block in Figure 3-11 consists of n integrators. The matrix $\mathbf{E}$ represents the *direct transmission part* from the input $\mathbf{u}$ to the output $\mathbf{y}$. If the matrix $\mathbf{A}$ is a zero matrix, then there is no "feedback" in the block diagram. For a discussion of analog computer simulations, see, e.g., Reference S46.

Dynamical equations can be readily simulated on a digital computer. There are many specialized subroutines for solving dynamical equations, such as MIDAS (*M*odified *I*ntegration *D*igital *A*nalog *S*imulator), MIMIC (an improved version of MIDAS), CSMP (*C*ontinuous *S*ystem *M*odeling *P*rogram), TELSIM (*Tel*etype *Sim*ulator), and others; see Reference 26. These programs

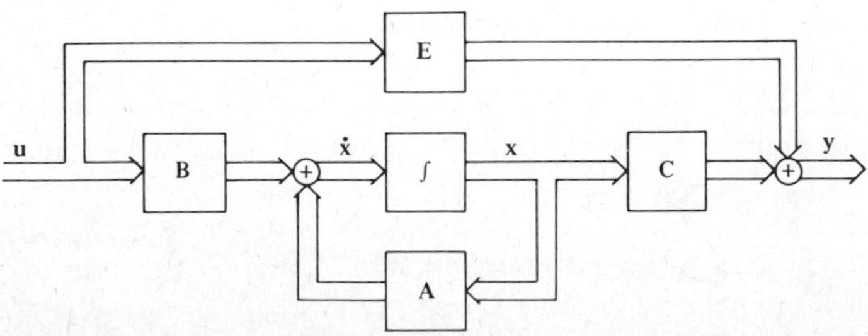

Figure 3-11 Matrix block diagram of the dynamical equation E.

can be easily prepared from the block diagram of a dynamical equation or directly from a dynamical equation.

We give a simple example of the application of System/360 CSMP. Consider the dynamical equation

$$\begin{bmatrix} \dot{x}_1 \\ \dot{x}_2 \end{bmatrix} = \begin{bmatrix} 2 & -1 \\ 1 & 5 \end{bmatrix} \begin{bmatrix} x_1 \\ x_2 \end{bmatrix} + \begin{bmatrix} 0 \\ 1.5 \end{bmatrix} u$$

$$y = [1 \quad 0.6] \begin{bmatrix} x_1 \\ x_2 \end{bmatrix}$$

Find the output y and state variable x_1 from 0 to 20 seconds due to the initial condition $x_1(0) = 1$, $x_2(0) = 0$, and a unit-step-function input.

The CSMP input statements for this problem are listed in the following:

```
DYNAMIC
  PARAMETER U=1.0
        X1DT=2.0*X1-X2
        X2DT=X1+5.0*X2+1.5 *U
        X1=INTGR(1.0,X1DT)
        X2=INTGR(0.0,X2DT)
        Y=X1+0.6 *X2
  TIMER DELT=0.001,FINTIM=20.0,OUTDEL=0.10
  PRTPLT X1,Y
STOP
END
```

The first part of the program is self-explanatory. "DELT" is the integration step size. "FINTIM" is the final time of computation. "OUTDEL" is the interval in which the responses will be printed. In order to have an accurate result, DELT is usually chosen to be very small. It is, however, unnecessary to print out every computed result; therefore the printout interval is chosen much larger than DELT. In employing CSMP the user has to decide the sizes of DELT and OUTDEL. For this program we have asked the computer to *print* as well as *plot* the output y and the state variable x_1.

Comparisons of analog computer simulations and digital computer simulations are in order. On analog computer simulations, the magnitudes of signals are limited to a range, typically ± 10 volts. If the magnitudes go over the range, some components of the analog computer will saturate and the result of the simulation will be erroneous. Hence on analog computer simulations, equations must be properly scaled. This is a difficult problem and usually is carried out by cut and try. On digital computer simulations, because the range of numbers which a digital computer can manage is very large, the problem of magnitude scaling generally does not arise. The accuracy of an analog computer is often limited to 0.1 percent of its full scale; a digital computer may have a precision of eight or more decimal digits. Therefore the result from a digital computer simulation is much more accurate than that from an analog computer. The generation of nonlinear functions is easier on a digital computer than on an analog computer. However, the interaction between an analog computer and the user is very good. By this we mean that the parameters of a simulation can be easily adjusted, and the consequence of the adjustments can be immediately observed. The interaction between a large-frame general-purpose digital computer and the user is generally not very satisfactory, but this interaction has been improving in recent years with the introduction of time sharing and remote terminals. With the increasing speed of computation and graphic display, the interaction between the user and a special-purpose digital computer or a general-purpose mid-frame digital computer is now very good. Furthermore, digital computers are much more versatile; they can be used to carry out the design. Hence digital computer simulations are gaining more popularity in recent years.

3-4 Examples

In this section we shall give some examples to illustrate how the input-output descriptions and state-variable descriptions of linear systems are developed. A system generally consists of many subsystems or components. For example, the network in Figure 3-6 consists of three components: one resistor, one inductor and one capacitor. If any component of the system is nonlinear or time varying, then the overall system is nonlinear or time varying. Hence in order to obtain a linear time-invariant model for a physical system, every component of the system must be modeled as a linear time-invariant element.

Strictly speaking, no physical system is linear and time invariant. A television set, an automobile, or a communication satellite cannot function forever; its performance will deteriorate with time because of aging or other factors. However, if the changes of characteristics are very small in the time interval of interest—say, one year—then these physical systems can be considered as time invariant. Hence over finite time intervals, a great number of physical systems can be modeled by time-invariant systems.

A necessary condition for a system to be linear is that for any admissible pair $\{\mathbf{0}, \mathbf{u}_{[0,\infty)}\} \rightarrow \{\mathbf{x}_{[0,\infty)}, \mathbf{y}_{[0,\infty)}\}$, the pair $\{\mathbf{0}, \alpha\mathbf{u}_{[0,\infty)}\} \rightarrow \{\alpha\mathbf{x}_{[0,\infty)}, \alpha\mathbf{y}_{[0,\infty)}\}$ for any α, even very very large, is also admissible. For any physical system, if the applied signal is larger than certain limit, the system will burn out or saturate. Hence no physical system is linear according to Definition 3-2 or 3-7. However, linear models are often used in practice to represent physical systems. This is possible because most physical systems are designed to operate only in a certain operational ranges. Limited to these ranges, physical systems can often be approximated by linear models. This is accomplished by linearlization or simplification, as will be discussed in the following examples.

Example 1

Consider the mechanical system shown in Figure 3-12. The friction force between the floor and the mass generally consists of three distinct parts: static friction, Coulomb friction, and viscous friction as shown in Figure 3-13. The

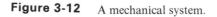

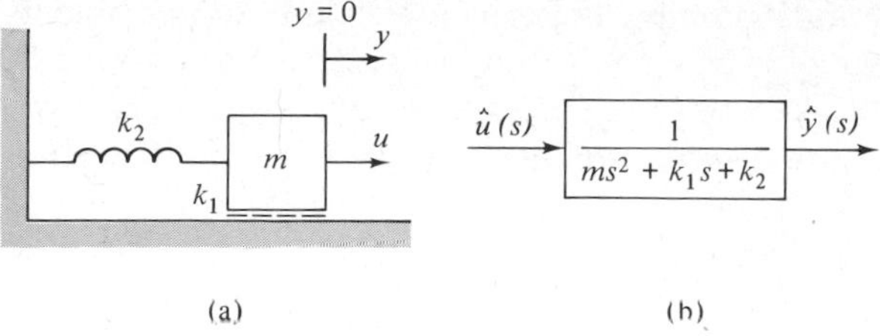

Figure 3-12 A mechanical system.

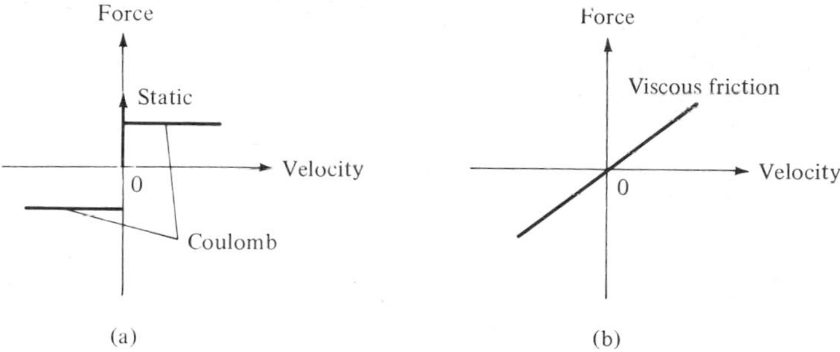

Figure 3-13 (a) Static and Coulomb friction. (b) Viscous friction.

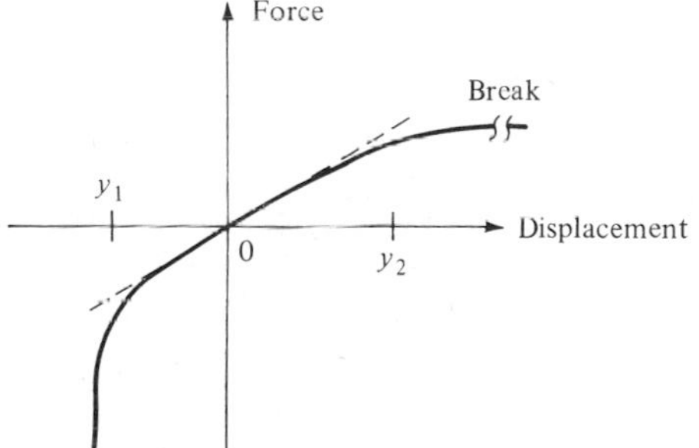

Figure 3-14 Characteristic of the spring.

friction is clearly not a linear function of the velocity. To simplify the analysis, we neglect the static and Coulomb frictions and consider only the viscous friction. Let k_1 be the viscous friction coefficient. Then the friction force f is given by $f = k_1 dy/dt$. The characteristic of the spring is shown in Figure 3-14. It is a nonlinear element. However if the displacement is limited to (y_1, y_2) as shown, then the spring force is equal to $k_2 y$, where k_2 is the spring constant. Hence the linear model in Figure 3-12(a) is obtained by linearization and simplification.

Now we shall develop the input-output description of the system from the external force u (input) to the displacement y (output). The application of Newton's law yields

$$m\frac{d^2y}{dt^2} = u - k_1\frac{dy}{dt} - k_2 y$$

Taking the Laplace transform and assuming the zero initial condition, we obtain

$$\hat{u}(s) = (ms^2 + k_1 s + k_2)\hat{y}(s)$$

Hence the input-output description in the frequency domain of the system is

$$\hat{y}(s) = \frac{1}{ms^2 + k_1 s + k_2}\hat{u}(s)$$

If $m = 1, k_1 = 3, k_2 = 2$, then the impulse response of the system is

$$g(t) = \mathscr{L}^{-1}\left[\frac{1}{s^2 + 3s + 2}\right] = \mathscr{L}^{-1}\left[\frac{1}{(s+1)} - \frac{1}{(s+2)}\right] = e^{-t} - e^{-2t}$$

and the input and output are related by

$$y(t) = \int_0^t g(t-\tau)u(\tau)\,d\tau$$

We next derive the dynamical-equation description of the system. Let the displacement and the velocity of the mass m be the state variables; that is, $x_1 = y$, $x_2 = \dot{y}$. Then we have

$$\dot{x}_1 = x_2 \qquad u = m\dot{x}_2 + k_1 x_2 + k_2 x_1$$

or

$$\begin{bmatrix} \dot{x}_1 \\ \dot{x}_2 \end{bmatrix} = \begin{bmatrix} 0 & 1 \\ -\dfrac{k_2}{m} & -\dfrac{k_1}{m} \end{bmatrix} \begin{bmatrix} x_1 \\ x_2 \end{bmatrix} + \begin{bmatrix} 0 \\ \dfrac{1}{m} \end{bmatrix} u$$

$$y = [1 \quad 0]\begin{bmatrix} x_1 \\ x_2 \end{bmatrix}$$

This is a state-variable description of the system. By choosing a different set of state variables, we may obtain a different state-variable description of the system. ∎

Example 2

Consider a cart with an inverted pendulum hinged on top of it as shown in Figure 3-15. For simplicity, the cart and the pendulum are assumed to move in only one plane, and the friction, the mass of the stick, and the gusts of wind are neglected. The problem is to maintain the pendulum at the vertical position.

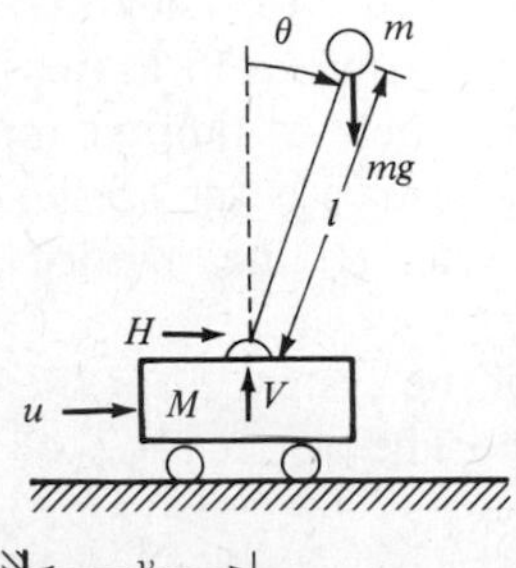

Figure 3-15 A cart with an inverted pendulum.

For example, if the inverted pendulum is falling in the direction shown, the cart is moved to the right and exerts a force, through the hinge, to push the pendulum back to the vertical position. This simple mechanism can be used as a model of a space booster on takeoff.

Let H and V be, respectively, the horizontal and vertical forces exerted by the cart to the pendulum as shown. The application of Newton's law to the linear movements yields

$$M\frac{d^2y}{dt^2}=u-H$$

$$H=m\frac{d^2}{dt^2}(y+l\sin\theta)=m\ddot{y}+ml\cos\theta\,\ddot{\theta}-ml\sin\theta\,(\dot{\theta})^2$$

and

$$mg-V=m\frac{d^2}{dt^2}(l\cos\theta)=ml[-\sin\theta\,\ddot{\theta}-\cos\theta\,(\dot{\theta})^2]$$

The application of Newton's law to the rotational movement of the pendulum yields

$$ml^2\ddot{\theta}=mgl\sin\theta+Vl\sin\theta-Hl\cos\theta$$

These are nonlinear equations. Since the purpose of the problem is to maintain the pendulum at the vertical position, it is reasonable to assume θ and $\dot{\theta}$ to be small. Under this assumption, we may use the approximation $\sin\theta=\theta$ and $\cos\theta=1$. By retaining only the linear terms in θ and $\dot{\theta}$, that is, dropping the terms with θ^2, $\dot{\theta}^2$, $\theta\dot{\theta}$, and $\theta\ddot{\theta}$, we obtain $V=mg$ and

$$M\ddot{y}=u-m\ddot{y}-ml\ddot{\theta}$$
$$ml^2\ddot{\theta}=mgl\theta+mgl\theta-(m\ddot{y}+ml\ddot{\theta})l$$

which imply

$$(M+m)\ddot{y}+ml\ddot{\theta}=u \tag{3-38}$$

and

$$2l\ddot{\theta}-2g\theta+\ddot{y}=0 \tag{3-39}$$

With these linearized equations, we can now derive the input-output and state-variable descriptions of the system. The application of the Laplace transform to (3-38) and (3-39) yields, assuming zero initial conditions,

$$(M+m)s^2\hat{y}(s)+mls^2\hat{\theta}(s)=\hat{u}(s)$$

and

$$(2ls^2-2g)\hat{\theta}(s)+s^2\hat{y}(s)=0$$

From these two equations, the transfer function $\hat{g}_{yu}(s)$ from u to y and the transfer function $\hat{g}_{\theta u}(s)$ from u to θ can be readily obtained as

$$\hat{g}_{yu}(s)=\frac{2ls^2-2g}{s^2[(2M+m)ls^2-2g(M+m)]} \tag{3-40}$$

and

$$\hat{g}_{\theta u}(s)=\frac{-1}{(2M+m)ls^2-2g(M+m)} \tag{3-41}$$

To develop a dynamical equation, we define the state variables as $x_1 = y$, $x_2 = \dot{y}$, $x_3 = \theta$, and $x_4 = \dot{\theta}$. From (3-38) and (3-39), we can solve $\ddot{y}$ and $\ddot{\theta}$ as

$$\ddot{y} = -\frac{2gm}{2M+m}\theta + \frac{2}{2M+m}u$$

and

$$\ddot{\theta} = \frac{2g(M+m)}{(2M+m)l}\theta - \frac{1}{(2M+m)l}u$$

From these two equations and the definition of x_i, the state-variable equation description of the system can be readily obtained as

$$\begin{bmatrix} \dot{x}_1 \\ \dot{x}_2 \\ \dot{x}_3 \\ \dot{x}_4 \end{bmatrix} = \begin{bmatrix} 0 & 1 & 0 & 0 \\ 0 & 0 & \dfrac{-2mg}{2M+m} & 0 \\ 0 & 0 & 0 & 1 \\ 0 & 0 & \dfrac{2g(M+m)}{(2M+m)l} & 0 \end{bmatrix} \begin{bmatrix} x_1 \\ x_2 \\ x_3 \\ x_4 \end{bmatrix} + \begin{bmatrix} 0 \\ \dfrac{2}{2M+m} \\ 0 \\ \dfrac{-1}{(2M+m)l} \end{bmatrix} u \quad \textbf{(3-42)}$$

$$y = [1 \quad 0 \quad 0 \quad 0]\mathbf{x}$$

Note that Equations (3-39) to (3-42) are obtained under simplification and linearization and are applicable only for small θ and $\dot{\theta}$. ∎

Example 3 (Reference S12)

Consider four vehicles moving in a single lane as shown in Figure 3-16. Let y_i, v_i, m_i, and u_i be, respectively, the position, velocity, mass and the applied force of the ith vehicle. Let k be the viscous friction coefficient and be the same for all four vehicles. Then we have, for $i = 1, 2, 3, 4$,

$$v_i = \dot{y}_i \quad \textbf{(3-43)}$$

$$u_i = kv_i + m_i\dot{v}_i \quad \textbf{(3-44)}$$

The purpose of this problem is to maintain the distance between adjacent vehicles at a predetermined value h_0 and to maintain the velocity of each vehicle as close as possible to a desired velocity v_0. Define

$$\bar{y}_{i,i+1}(t) = y_i(t) - y_{i+1}(t) - h_0 \qquad i = 1, 2, 3 \quad \textbf{(3-45)}$$

$$\bar{v}_i(t) = v_i(t) - v_0 \qquad i = 1, 2, 3, 4 \quad \textbf{(3-46)}$$

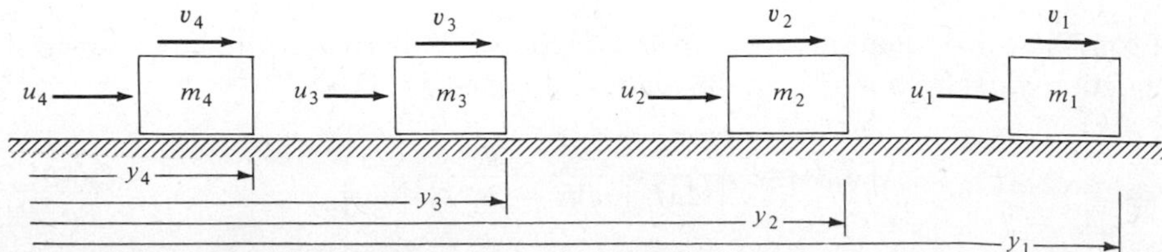

Figure 3-16 Four vehicles moving in a single lane.

and
$$\bar{u}_i(t) = u_i(t) - kv_0 \tag{3-47}$$

The term kv_0 is the force needed to overcome the friction for the vehicles to maintain their velocities at v_0. Now the problem reduces to find $\bar{u}_i(t)$ such that $\bar{y}_{i,i+1}(t)$ and $\bar{v}_i(t)$ are as close as possible to zero for all t. From (3-45), we have, for $i = 1, 2, 3$,

$$\dot{\bar{y}}_{i,i+1}(t) = \dot{y}_i(t) - \dot{y}_{i+1}(t) = \bar{v}_i(t) - \bar{v}_{i+1}(t)$$

From (3-46) and (3-44), we have, for $i = 1, 2, 3, 4$

$$\dot{\bar{v}}_i(t) - \frac{-k}{m_i} v_i + \frac{1}{m_i} u_i = \frac{-k}{m_i} \bar{v}_i(t) + \frac{1}{m_i} \bar{u}_i(t)$$

These equations can be arranged in matrix form as

$$\begin{bmatrix} \dot{\bar{v}}_1(t) \\ \dot{\bar{y}}_{12}(t) \\ \dot{\bar{v}}_2(t) \\ \dot{\bar{y}}_{23}(t) \\ \dot{\bar{v}}_3(t) \\ \dot{\bar{y}}_{34}(t) \\ \dot{\bar{v}}_4(t) \end{bmatrix} = \begin{bmatrix} \frac{-k}{m_1} & 0 & 0 & 0 & 0 & 0 & 0 \\ 1 & 0 & -1 & 0 & 0 & 0 & 0 \\ 0 & 0 & \frac{-k}{m_2} & 0 & 0 & 0 & 0 \\ 0 & 0 & 1 & 0 & -1 & 0 & 0 \\ 0 & 0 & 0 & 0 & \frac{-k}{m_3} & 0 & 0 \\ 0 & 0 & 0 & 0 & 1 & 0 & -1 \\ 0 & 0 & 0 & 0 & 0 & 0 & \frac{-k}{m_4} \end{bmatrix} \begin{bmatrix} \bar{v}_1(t) \\ \bar{y}_{12}(t) \\ \bar{v}_2(t) \\ \bar{y}_{23}(t) \\ \bar{v}_3(t) \\ \bar{y}_{34}(t) \\ \bar{v}_4(t) \end{bmatrix} + \begin{bmatrix} \frac{1}{m_1} & 0 & 0 & 0 \\ 0 & 0 & 0 & 0 \\ 0 & \frac{1}{m_2} & 0 & 0 \\ 0 & 0 & 0 & 0 \\ 0 & 0 & \frac{1}{m_3} & 0 \\ 0 & 0 & 0 & 0 \\ 0 & 0 & 0 & \frac{1}{m_4} \end{bmatrix} \begin{bmatrix} \bar{u}_1(t) \\ \bar{u}_2(t) \\ \bar{u}_3(t) \\ \bar{u}_4(t) \end{bmatrix} \tag{3-48}$$

This is the state equation description of the system. In this problem, we are interested in the distances between adjacent vehicles and their velocities; therefore the absolute distances do not appear in the equations. Depending on what will be considered as outputs, an output equation can be similarly developed.

Example 4 (Reference S96)

Consider a satellite of mass m in earth orbit as shown in Figure 3-17. The altitude of the satellite is specified by $r(t)$, $\theta(t)$, and $\phi(t)$ as shown. The orbit

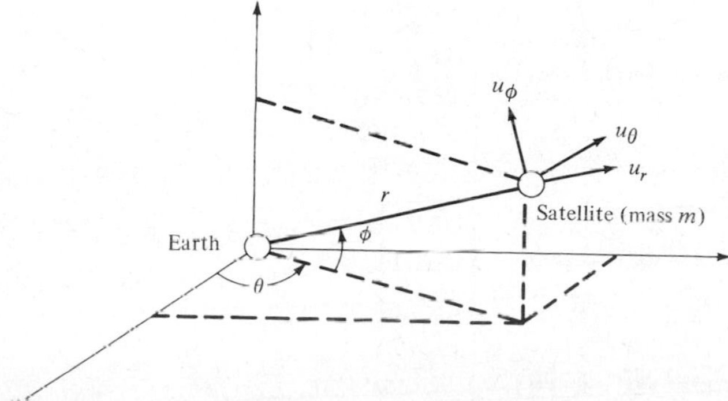

Figure 3-17 Satellite in earth orbit.

can be controlled by three orthogonal thrusts $u_r(t)$, $u_\theta(t)$, and $u_\phi(t)$. The state, input, and output of the system are chosen as

$$\mathbf{x}(t)=\begin{bmatrix} r(t) \\ \dot{r}(t) \\ \theta(t) \\ \dot{\theta}(t) \\ \phi(t) \\ \dot{\phi}(t) \end{bmatrix} \qquad \mathbf{u}(t)=\begin{bmatrix} u_r(t) \\ u_\theta(t) \\ u_\phi(t) \end{bmatrix} \qquad \mathbf{y}(t)=\begin{bmatrix} r(t) \\ \theta(t) \\ \phi(t) \end{bmatrix}$$

Then the system can be shown to be described by

$$\dot{\mathbf{x}}=\mathbf{h}(\mathbf{x},\mathbf{u})=\begin{bmatrix} \dot{r} \\ r\dot{\theta}^2\cos^2\phi + r\dot{\phi}^2 - k/r^2 + u_r/m \\ \dot{\theta} \\ -2\dot{r}\dot{\theta}/r + 2\dot{\theta}\dot{\phi}\sin\phi/\cos\phi + u_\theta/mr\cos\phi \\ \dot{\phi} \\ -\dot{\theta}^2\cos\phi\sin\phi - 2\dot{r}\dot{\phi}/r + u_\phi/mr \end{bmatrix} \tag{3-49a}$$

and

$$\mathbf{y}=\mathbf{C}\mathbf{x}=\begin{bmatrix} 1 & 0 & 0 & 0 & 0 & 0 \\ 0 & 0 & 1 & 0 & 0 & 0 \\ 0 & 0 & 0 & 0 & 1 & 0 \end{bmatrix}\mathbf{x} \tag{3-49b}$$

One solution which corresponds to a circular, equatorial orbit is specified by

$$\mathbf{x}_0(t)=\begin{bmatrix} r_0(t) \\ \dot{r}_0(t) \\ \theta_0(t) \\ \dot{\theta}_0(t) \\ \phi_0(t) \\ \dot{\phi}_0(t) \end{bmatrix}=\begin{bmatrix} r_0 \\ 0 \\ \omega t \\ \omega \\ 0 \\ 0 \end{bmatrix} \qquad u_0(t)=\mathbf{0} \tag{3-50}$$

where r_0 and ω are related by $r_0^3\omega^2=k=$ a known physical constant. Once the satellite reaches this orbit, it will remain in the orbit as long as there are no disturbances. If the satellite deviates from the orbit, thrusts must be applied to push the satellite back to the orbit. Define

$$\mathbf{x}(t)=\mathbf{x}_0(t)+\varepsilon\bar{\mathbf{x}}(t)=\begin{bmatrix} r_0+\varepsilon\bar{x}_1(t) \\ \varepsilon\bar{x}_2(t) \\ \omega t+\varepsilon\bar{x}_3(t) \\ \omega+\varepsilon\bar{x}_4(t) \\ \varepsilon\bar{x}_5(t) \\ \varepsilon\bar{x}_6(t) \end{bmatrix}$$

$$\mathbf{u}(t)=\mathbf{u}_0(t)+\varepsilon\bar{\mathbf{u}}(t)=\begin{bmatrix} \varepsilon\bar{u}_1(t) \\ \varepsilon\bar{u}_2(t) \\ \varepsilon\bar{u}_3(t) \end{bmatrix}$$

and

$$\mathbf{y}(t)=\mathbf{y}_0(t)+\varepsilon\bar{\mathbf{y}}(t)=\begin{bmatrix} r_0+\varepsilon\bar{x}_1(t) \\ \omega t+\varepsilon\bar{x}_3(t) \\ \varepsilon\bar{x}_5(t) \end{bmatrix}$$

If the perturbation is very small, or equivalently, ε is very small, $\mathbf{h}(\mathbf{x}, \mathbf{u})$ in (3-49) can be linearized as

$$\dot{\bar{\mathbf{x}}}(t) = \left[\begin{array}{cccc:cc} 0 & 1 & 0 & 0 & 0 & 0 \\ 3\omega^2 & 0 & 0 & 2\omega r_0 & 0 & 0 \\ 0 & 0 & 0 & 1 & 0 & 0 \\ 0 & \dfrac{-2\omega}{r_0} & 0 & 0 & 0 & 0 \\ \hdashline 0 & 0 & 0 & 0 & 0 & 1 \\ 0 & 0 & 0 & 0 & -\omega^2 & 0 \end{array}\right] \bar{\mathbf{x}}(t) + \left[\begin{array}{cc:c} 0 & 0 & 0 \\ \dfrac{1}{m} & 0 & 0 \\ 0 & 0 & 0 \\ 0 & \dfrac{1}{mr_0} & 0 \\ \hdashline 0 & 0 & 0 \\ 0 & 0 & \dfrac{1}{mr_0} \end{array}\right] \bar{\mathbf{u}}(t) \tag{3-51a}$$

$$\bar{\mathbf{y}}(t) = \left[\begin{array}{cccc:cc} 1 & 0 & 0 & 0 & 0 & 0 \\ 0 & 0 & 1 & 0 & 0 & 0 \\ \hdashline 0 & 0 & 0 & 0 & 1 & 0 \end{array}\right] \bar{\mathbf{x}}(t) \tag{3-51b}$$

where $\mathbf{A} = \partial \mathbf{h}(\mathbf{x}, \mathbf{u})/\partial \mathbf{x}$, $\mathbf{B} = \partial \mathbf{h}(\mathbf{x}, \mathbf{u})/\partial \mathbf{u}$, and computed at the circular orbit at $\mathbf{x}_0$ and $\mathbf{u}_0$. This is a sixth-dimensional linear time-invariant dynamical equation. It can be used to describe and control the satellite so long as the deviation from the circular orbit remains small.

The dashed lines in (3-51) show that the equation can be decomposed into two uncoupled parts, one involving r and θ, the other ϕ. By studying these two parts independently, the analysis and design can be considerably simplified.

***Dynamical equations for RLC networks.** We introduce in this subsection a systematic procedure for assigning state variables and writing dynamical equations for general lumped linear *RLC* networks which may contain independent voltage and current sources. It is well known that if all the inductor currents and the capacitor voltages of an *RLC* networks are known, then the behavior of the network is uniquely determinable for any input. However, it is not necessary to choose all the inductor currents and capacitor voltages as state variables. This can be seen from the simple circuits shown in Figure 3-18. If we assign all the capacitor voltages and inductor currents as state variables as shown, then we see that $x_1(t) = x_2(t)$ for all t. Clearly there is a redundancy

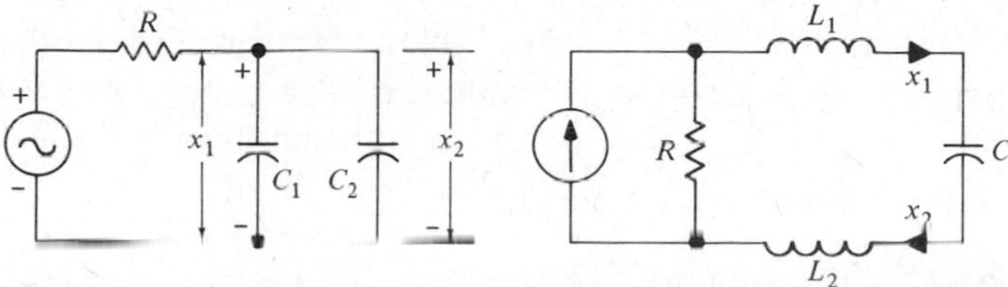

Figure 3-18 Circuits with a loop which consists of capacitors only or a cutset which consists of inductors only.

here. Hence *the state of an RLC network can be chosen to consist of only independent capacitor voltages and independent inductor currents.*

Before proceeding, we review briefly the concepts of tree, link, and cutset of a network. We consider only connected networks. A *tree* of a network is defined as any connected graph (connection of branches) containing all the nodes of the network and not containing any loop. Every branch in a given tree is called a *tree branch.* Every branch not in the tree is called a *link.* A *cutset* of a connected network is any minimal set of branches such that the removal of all the branches in this set causes the remaining network to be unconnected. With respect to any fixed tree, every link and some tree branches form a unique loop called a *fundamental loop*; every tree branch with some links form a unique cutset called a *fundamental cutset.* Hence every fundamental loop includes only one link, and every fundamental cutset includes only one tree branch. With these concepts, we are ready to give a systematic procedure for developing a dynamical-equation description of any *RLC* network that may contain independent voltage sources or current sources[9]:

1. Choose a tree called a *normal tree.* The branches of the normal tree are chosen in the order of voltage sources, capacitors, resistors, inductors, and current sources. Hence, a normal tree consists of all the voltage sources, the maximal number of permissible capacitors (those that do not form a loop), the resistors, and finally the minimal number of inductors. Usually it does not contain any current source.
2. Assign the charges or voltages of *the capacitors in the normal tree* and the flux or current of *the inductors in the links* as state variables. The voltages or charges of the capacitors in the links and the flux or current of the inductors in the normal tree need *not* be chosen as state variables.
3. Express the branch variables (branch voltage and current) of all the resistors, the capacitors in the links, and the inductors in the normal tree in terms of the state variables and the inputs by applying the Kirchhoff voltage or current law to the fundamental loops or cutsets of these branches.
4. Apply the Kirchhoff voltage or current law to the fundamental loop or cutset of every branch that is assigned as a state variable.

Example 5

Consider the linear network shown in Figure 3-19. The normal tree is chosen as shown (heavy lines); it consists of the voltage source, two capacitors, and one resistor. The voltages of the capacitors in the normal tree and the current of the inductor in the link are chosen as state variables. Next we express the variables of resistors ① and ② in terms of the state variables and inputs. By applying the Kirchhoff voltage law (KVL) to the fundamental loop of branch ①, the voltage across ① is found as $(u_1 - x_1)$; hence its current is $(u_1 - x_1)$.

[9] A network with a loop that consists of only voltage sources and capacitors or with a cutset that consists of only current sources and inductors is excluded, because in this case its dynamical-equation description cannot be of the form in (3-33).

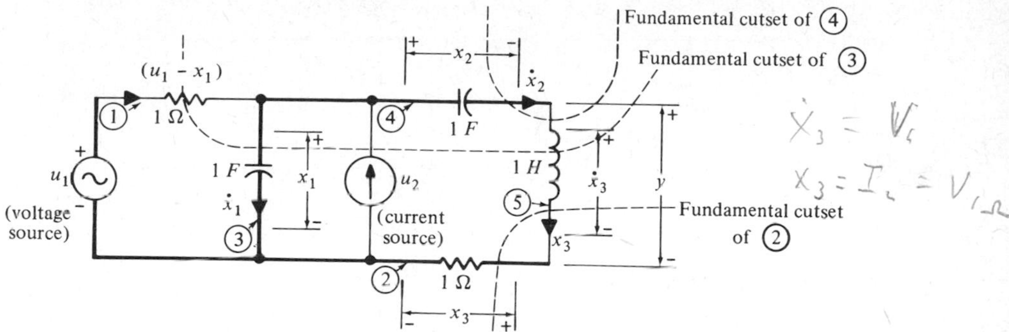

Figure 3-19 A network with voltage and current sources.

By applying the Kirchhoff current law (KCL) to the fundamental cutset of ②, we have immediately that the current through resistor ② is x_3. Consequently the voltage across resistor ② is x_3. Now the characteristics of every branch are expressed in terms of the state variables as shown. If we apply the KCL to the fundamental cutset of branch ③, we have

$$(u_1 - x_1) - \dot{x}_1 + u_2 - x_3 = 0$$

The application of the KCL to the fundamental cutset of ④ yields

$$\dot{x}_2 = x_3$$

This application of the KVL to the fundamental loop of ⑤ yields

$$\dot{x}_3 + x_3 - x_1 + x_2 = 0$$

These three equations can be rearranged in matrix form as

$$\begin{bmatrix} \dot{x}_1 \\ \dot{x}_2 \\ \dot{x}_3 \end{bmatrix} = \begin{bmatrix} -1 & 0 & -1 \\ 0 & 0 & 1 \\ 1 & -1 & -1 \end{bmatrix} \mathbf{x} + \begin{bmatrix} 1 & 1 \\ 0 & 0 \\ 0 & 0 \end{bmatrix} \begin{bmatrix} u_1 \\ u_2 \end{bmatrix}$$

The output equation can be easily found as

$$y = \dot{x}_3 = [1 \quad -1 \quad -1]\mathbf{x}$$

∎

Example 6

Find the input-output description of the network shown in Figure 3-19, or equivalently, the transfer-function matrix of the network. Since there are two inputs, one output, the transfer-function matrix is a 1×2 matrix. Let $\hat{\mathbf{G}}(s) = [\hat{g}_{11}(s) \quad \hat{g}_{12}(s)]$, where $\hat{g}_{11}$ is the transfer function from u_1 to y with $u_2 = 0$, and $\hat{g}_{12}(s)$ is the transfer function from u_2 to y with $u_1 = 0$. With $u_2 = 0$, the network reduces to the one in Figure 3-20(a). By loop analysis, we have

$$\left(1 + \frac{1}{s}\right)\hat{I}_1(s) - \frac{1}{s}\hat{I}_2(s) = \hat{u}_1(s)$$

$$-\frac{1}{s}\hat{I}_1(s) + \left(1 + s + \frac{2}{s}\right)\hat{I}_2(s) = 0$$

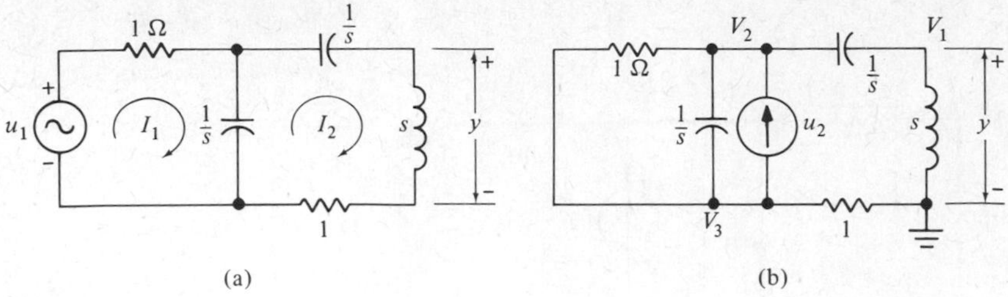

Figure 3-20 Two reduced networks.

Solving for $\hat{I}_2(s)$, we obtain

$$\hat{I}_2(s)=\frac{(1/s)\hat{u}_1(s)}{(1+1/s)(1+s+2/s)-1/s^2}$$

Hence
$$\hat{g}_{11}(s)=\left.\frac{\hat{y}(s)}{\hat{u}_1(s)}\right|_{\substack{u_2=0\\ \text{initially}\\ \text{relaxed}}}=\frac{s\hat{I}_2(s)}{\hat{u}_1(s)}=\frac{s^2}{s^3+2s^2+3s+1}$$

If $u_1=0$, the network in Figure 3-19 reduces to the one in Figure 3-20(b). By node analysis,

$$\left(\frac{1}{s}+s\right)\hat{V}_1(s)-s\hat{V}_2(s)=0$$
$$-s\hat{V}_1(s)+(1+2s)\hat{V}_2(s)-(1+s)\hat{V}_3(s)=\hat{u}_2(s)$$
$$-(1+s)\hat{V}_2(s)+(2+s)\hat{V}_3(s)=-\hat{u}_2(s)$$

Solving for $\hat{V}_1(s)$, we obtain

$$\hat{V}_1(s)=\frac{s^2}{s^3+2s^2+3s+1}\hat{u}_2(s)$$

Hence
$$\hat{g}_{12}(s)=\left.\frac{\hat{V}_1(s)}{\hat{u}_2(s)}\right|_{\substack{u_1=0\\ \text{initially}\\ \text{relaxed}}}=\frac{s^2}{s^3+2s^2+3s+1}$$

Consequently, the input-output description in the frequency domain of the network is given by

$$\hat{y}(s)=\begin{bmatrix}\frac{s^2}{s^3+2s^2+3s+1} & \frac{s^2}{s^3+2s^2+3s+1}\end{bmatrix}\begin{bmatrix}\hat{u}_1(s)\\ \hat{u}_2(s)\end{bmatrix}$$ ∎

Example 7

Consider the network shown in Figure 3-21(a), where T is a tunnel diode with the characteristic shown. Let x_1 be the voltage across the capacitor and x_2 the current through the inductor. Then we have

$$x_2(t)=C\dot{x}_1(t)+i(t)=C\dot{x}_1(t)+h(x_1(t))$$
$$L\dot{x}_2(t)=E-Rx_2(t)-x_1(t)$$

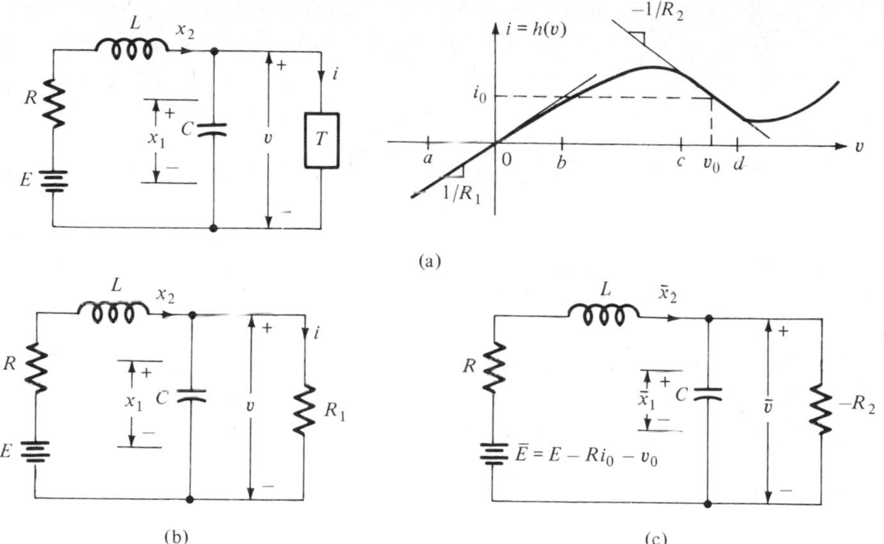

Figure 3-21 Network with a tunnel diode.

These can be rearranged as

$$\begin{aligned}\dot{x}_1(t) &= \frac{-h(x_1(t))}{C} + \frac{x_2(t)}{C} \\ \dot{x}_2(t) &= \frac{-x_1(t) - Rx_2(t)}{L} + \frac{E}{L}\end{aligned} \qquad \textbf{(3-52)}$$

This nonlinear time-invariant dynamical equation describes the network for the general case. Now if $x_1(t)$ is known to operate only inside the range (a, b), then $h(x_1(t))$ can be approximated as $h(x_1(t)) = x_1(t)/R_1$. In this case, the network in Figure 3-21(a) can be reduced to the one in Figure 3-21(b) and is described by the linear time-invariant state equation

$$\begin{bmatrix} \dot{x}_1(t) \\ \dot{x}_2(t) \end{bmatrix} = \begin{bmatrix} -\dfrac{1}{CR_1} & \dfrac{1}{C} \\ -\dfrac{1}{L} & -\dfrac{R}{L} \end{bmatrix} \begin{bmatrix} x_1(t) \\ x_2(t) \end{bmatrix} + \begin{bmatrix} 0 \\ \dfrac{1}{L} \end{bmatrix} E$$

Now if $x_1(t)$ is known to operate only inside the range (c, d), we may introduce the variables $\bar{x}_1(t) = x_1(t) - v_0$, $\bar{x}_2(t) = x_2(t) - i_0$ and approximate $h(x_1(t))$ as $h(x_1(t)) = i_0 - \bar{x}_1(t)/R_2$. The substitution of these into (3-52) yields

$$\begin{bmatrix} \dot{\bar{x}}_1(t) \\ \dot{\bar{x}}_2(t) \end{bmatrix} = \begin{bmatrix} \dfrac{1}{CR_2} & \dfrac{1}{C} \\ -\dfrac{1}{L} & -\dfrac{R}{L} \end{bmatrix} \begin{bmatrix} \bar{x}_1(t) \\ \bar{x}_2(t) \end{bmatrix} + \begin{bmatrix} 0 \\ \dfrac{1}{L} \end{bmatrix} \bar{E}$$

where $\bar{E} = E - v_0 - Ri_0$. This linear equation is applicable only if $v(t)$ is limited inside the range (c, d). An equivalent network of this linear equation is shown in Figure 3-21(c). ∎

3-5 Comparisons of the Input-Output Description and the State Variable Description

We compare in this section the input-output description and the state-variable description of systems.

1. The input-output description of a system describes only the relationship between the input and the output under the assumption that the system is initially relaxed. Hence, if the system is not initially relaxed, the description is not applicable. A more serious problem is that it does not reveal what will happen if the system is not initially relaxed nor does it reveal the behavior inside the system. For example, consider the networks shown in Figures 3-22 and 3-23. The networks are assumed to have a capacitor with -1 farad. If the initial voltage of the capacitor is zero, the current will distribute equally among the two branches. Hence the transfer function of the network in Figure 3-22 is 0.5; the transfer function of the network in Figure 3-23 is 1. Because of the negative capacitance, if the initial voltage across the capacitor is different from zero, then the voltage y_1 in Figure 3-22 will increase with time, whereas the voltage y_2 in Figure 3-23 remains equal to $u(t)$ for all t. It is clear that the network in Figure 3-22 is not satisfactory, because the output will increase without bound if the initial condition is different from zero. Although the output of the network in Figure 3-23 behaves well, the network is still not satisfactory, because the voltages in branches 1 and 2 will increase with time (in different polarity), and the network will eventually burn out. Hence the networks in Figures 3-22 and 3-23 can never function properly. If the internal structure of the network is not known, this fact cannot be detected from its transfer function. Consequently, the input-output description sometimes does not characterize a system completely.

   The dynamical-equation descriptions of the networks in Figures 3-22 and 3-23 (see Problem 3-18) can be found, respectively, as

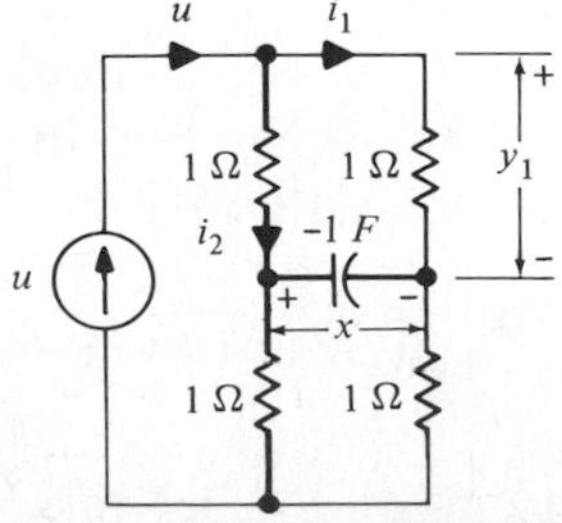

Figure 3-22 A network with linear, time-varying elements.

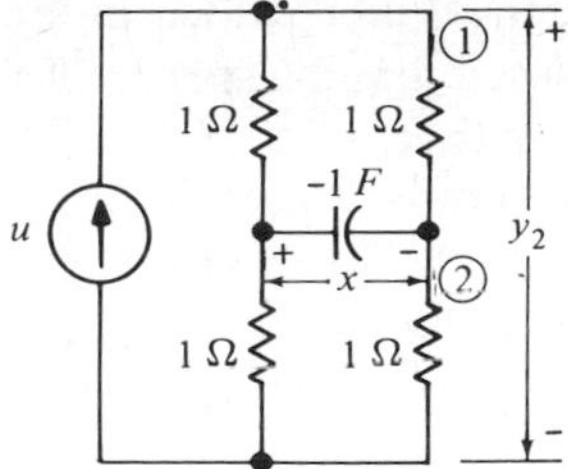

Figure 3-23 Active networks.

$$\begin{cases} \dot{x} = x \\ y_1 = 0.5x + 0.5u \end{cases} \qquad \begin{cases} \dot{x} = x \\ y_2 = u \end{cases}$$

A dynamical equation describes not only the relationship between the input and output but also the behavior inside a system under any initial condition; hence a dynamical equation characterizes a system completely.

2. For extremely complicated linear systems, it might be very involved to find the dynamical-equation descriptions. In these cases, it might be easier to find the input-output descriptions by direct measurements. We apply at each input terminal a very sharp and narrow pulse; the responses of the output terminals give us immediately the impulse-response matrix of the system. In practice, we may have difficulty in generating a very sharp and narrow pulse, or ideally a δ-function; however, this can be avoided by using a unit step function. A *unit step function* $\delta_1(t-t_0)$ is defined as

$$\delta_1(t-t_0) = \begin{cases} 1 & \text{for } t \geq t_0 \\ 0 & \text{for } t < t_0 \end{cases} \tag{3-53}$$

The response of a linear causal system due to a unit-step-function input is given by

$$g_1(t, t_0) = \int_{t_0}^{t} g(t, \tau)\, d\tau \tag{3-54}$$

where $g_1(t, t_0)$ is called the *step response* (due to a unit step function applied at t_0). Differentiating (3-54) with respect to t_0, we obtain

$$g(t, t_0) = -\frac{\partial}{\partial t_0} g_1(t, t_0) \tag{3-55}$$

For the time-invariant case, (3-55) reduces to

$$g(t) = \frac{d}{dt} g_1(t) \tag{3-56}$$

Thus the impulse response can be obtained from the step response by using (3-55) or (3-56). See Problem 3-25.

For linear time-invariant systems, we may measure either transfer functions or impulse responses. The transfer function $\hat{g}(s)$ of a system at frequency $s = j\omega$ can be measured easily and accurately by employing frequency response analyzers. After measuring $\hat{g}(j\omega)$ at a number of frequencies, we can then find a $\hat{g}(s)$ to match the measured $\hat{g}(j\omega)$. This method of determining transfer functions is often used in practice.

3. Prior to 1960, the design of control systems had been mostly carried out by using transfer functions. However, the design had been limited to the single-variable case; its extension to the multivariable case had not been successful. The state-variable approach was developed in the 1960s. In this approach, the formulations in the single- and multivariable cases are the same, and a number of results were established. These results were not available in the transfer-function approach at that time; consequently, interest in this approach was renewed in the 1970s. Now the results in the state-variable approach can also be obtained in the transfer-function approach. It also appears that the latter approach is simpler in concepts and computations.
4. The dynamical equation can be extended to the time-varying case; the extension of the transfer function to the time-varying case has not been successful. In the optimal design, dynamical equations can be used to study finite terminal time problems; this is not possible by using transfer functions.
5. In the study of nonlinear systems, depending on the approach taken, either description can be used. For example, in the study of the stability problem, we use the input-output description in the functional analysis and operator approaches. See References 28, 95, 117, and S79. If Lyapunov's second method is employed, then we must use the dynamical-equation description.
6. If the dynamical-equation description of a system is available, the system can be readily simulated on an analog or a digital computer.

From the foregoing discussion, we see that the input-output and the state-variable descriptions have their own merits. In order to carry out a design efficiently, a designer should make himself familiar with these two mathematical descriptions. In this book, these two descriptions will be developed equally and their relationships will be explored.

3-6 Mathematical Descriptions of Composite Systems[10]

Time-varying case. In engineering, a system is often built by interconnecting a number of subsystems. Such a system is called a *composite system*. Whether or not a system is a composite system depends on what we consider as building blocks. For example, an analog computer can be looked upon as a system, or as a composite system that consists of operational amplifiers, function generators, potentiometers, and others as subsystems. There are many forms of composite systems; however, mostly they are built from the following three basic connections: the parallel, the tandem, and the feedback connections, as shown in Figure 3-24. In this section we shall study the input-output and state-variable descriptions of these three basic composite systems.

We study first the input-output description of composite systems. Consider

[10] The material in this section is not used until Chapter 9; thus its study may be postponed.

two multivariable systems S_i, which are described by

$$\mathbf{y}_i(t) = \int_{-\infty}^{t} \mathbf{G}_i(t, \tau)\mathbf{u}_i(\tau)\, d\tau \qquad i = 1, 2 \tag{3-57}$$

where $\mathbf{u}_i$ and $\mathbf{y}_i$ are the input and the output, $\mathbf{G}_i$ is the impulse-response matrix of the system S_i. Let $\mathbf{u}$, $\mathbf{y}$, and $\mathbf{G}$ be, respectively, the input, the output, and the impulse-response matrix of a composite system. We see from Figure 3-24 that in the parallel connection, we have $\mathbf{u}_1 = \mathbf{u}_2 = \mathbf{u}$, $\mathbf{y} = \mathbf{y}_1 + \mathbf{y}_2$; in the tandem connection, we have $\mathbf{u} = \mathbf{u}_1$, $\mathbf{y}_1 = \mathbf{u}_2$, $\mathbf{y}_2 = \mathbf{y}$; in the feedback connection, we have $\mathbf{u}_1 = \mathbf{u} - \mathbf{y}_2$, $\mathbf{y} = \mathbf{y}_1$. Here we have implicitly assumed that the systems S_1 and S_2 have compatible numbers of input and output; otherwise, the systems cannot be properly connected. It is also assumed that there is no loading effect in the connection; that is, the impulse response matrices $\mathbf{G}_1$, $\mathbf{G}_2$ remain unchanged after connection. It is easy to show that the impulse response matrix of the parallel connection of S_1 and S_2 shown in Figure 3-24(a) is

$$\mathbf{G}(t, \tau) = \mathbf{G}_1(t, \tau) + \mathbf{G}_2(t, \tau) \tag{3-58}$$

For the tandem connection shown in Figure 3-24(b), we have

$$\mathbf{G}(t, \tau) = \int_{\tau}^{t} \mathbf{G}_2(t, v)\mathbf{G}_1(v, \tau)\, dv \tag{3-59}$$

We prove (3-59) for the single-variable case. The impulse response $g(t, \tau)$ is, by definition, the response at y_2 due to a δ-function applied at time τ at u_1. The response at y_1 due to this δ-function is $g_1(t, \tau)$. The output of S_2 due to the input $g_1(t, \tau)$ is

$$\int_{\tau}^{t} g_2(t, v)g_1(v, \tau)\, dv$$

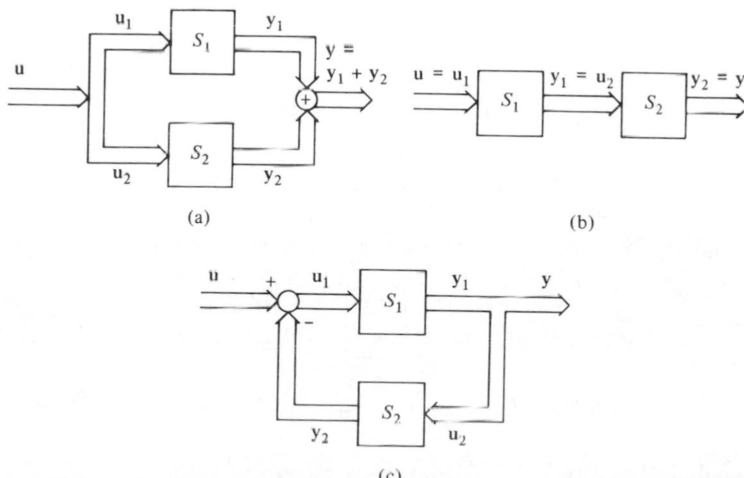

Figure 3.24 Composite connections of two systems. (a) Parallel connection. (b) Tandem connection. (c) Feedback connection.

Hence, $$g(t, \tau) = \int_\tau^t g_2(t, v) g_1(v, \tau)\, dv$$

The same procedure can be used to prove the multivariable case.

For the feedback connection shown in Figure 3-24(c), the impulse-response matrix is the solution of the integral equation

$$\mathbf{G}(t, \tau) = \mathbf{G}_1(t, \tau) - \int_\tau^t \mathbf{G}_1(t, v) \int_\tau^v \mathbf{G}_2(v, s)\mathbf{G}(s, \tau)\, ds\, dv \tag{3-60}$$

where $\mathbf{G}_1$ and $\mathbf{G}_2$ are known and $\mathbf{G}$ is unknown. This equation can be easily verified from the definition of $\mathbf{G}(t, \tau)$ (Problem 3-26). There is a general iterative method for solving Equation (3-60), but it is very involved.

Now we study the state-variable descriptions of composite systems. Let the systems S_1 and S_2 in Figure 3-24 be described by

$$E_i: \quad \dot{\mathbf{x}}_i = \mathbf{A}_i(t)\mathbf{x}_i + \mathbf{B}_i(t)\mathbf{u}_i \qquad i = 1, 2 \tag{3-61a}$$
$$\mathbf{y}_i = \mathbf{C}_i(t)\mathbf{x}_i + \mathbf{E}_i(t)\mathbf{u}_i \tag{3-61b}$$

where $\mathbf{x}_i$ is the state, $\mathbf{u}_i$ is the input, and $\mathbf{y}_i$ is the output; $\mathbf{A}_i, \mathbf{B}_i, \mathbf{C}_i$ and $\mathbf{E}_i$ are matrices of compatible order whose entries are continuous function of t defined over $(-\infty, \infty)$. The state space of S_i is denoted by Σ_i.

Let us introduce the concept of the direct sum of two linear spaces. The linear space Σ is the direct sum of two linear spaces Σ_1 and Σ_2, written as $\Sigma = \Sigma_1 \oplus \Sigma_2$, if every vector in Σ is of the form $[\mathbf{x}_1' \quad \mathbf{x}_2']'$, where $\mathbf{x}_1$ is a vector in Σ_1 and $\mathbf{x}_2$ is a vector in Σ_2. The dimension of Σ is the sum of those of Σ_1 and Σ_2.

It is clear that the composite vector

$$\begin{bmatrix} \mathbf{x}_1 \\ \mathbf{x}_2 \end{bmatrix}$$

qualifies as the state of any composite connection of S_1 and S_2; its state space is the direct sum of the state spaces of S_1 and S_2, $\Sigma_1 \oplus \Sigma_2$. For the parallel connection, we have $\mathbf{u}_1 = \mathbf{u}_2 = \mathbf{u}$, $\mathbf{y} = \mathbf{y}_1 + \mathbf{y}_2$; hence its dynamical equation is

$$\begin{bmatrix} \dot{\mathbf{x}}_1 \\ \dot{\mathbf{x}}_2 \end{bmatrix} = \begin{bmatrix} \mathbf{A}_1(t) & 0 \\ 0 & \mathbf{A}_2(t) \end{bmatrix} \begin{bmatrix} \mathbf{x}_1 \\ \mathbf{x}_2 \end{bmatrix} + \begin{bmatrix} \mathbf{B}_1(t) \\ \mathbf{B}_2(t) \end{bmatrix} \mathbf{u} \tag{3-62a}$$

$$\mathbf{y} = [\mathbf{C}_1(t) \quad \mathbf{C}_2(t)] \begin{bmatrix} \mathbf{x}_1 \\ \mathbf{x}_2 \end{bmatrix} + (\mathbf{E}_1(t) + \mathbf{E}_2(t))\mathbf{u} \tag{3-62b}$$

The dynamical equation of the tandem connection of S_1 and S_2 is given by

$$\begin{bmatrix} \dot{\mathbf{x}}_1 \\ \dot{\mathbf{x}}_2 \end{bmatrix} = \begin{bmatrix} \mathbf{A}_1(t) & \mathbf{0} \\ \mathbf{B}_2(t)\mathbf{C}_1(t) & \mathbf{A}_2(t) \end{bmatrix} \begin{bmatrix} \mathbf{x}_1 \\ \mathbf{x}_2 \end{bmatrix} + \begin{bmatrix} \mathbf{B}_1(t) \\ \mathbf{B}_2(t)\mathbf{E}_1(t) \end{bmatrix} \mathbf{u} \tag{3-63a}$$

$$\mathbf{y} = [\mathbf{E}_2(t)\mathbf{C}_1(t) \quad \mathbf{C}_2(t)] \begin{bmatrix} \mathbf{x}_1 \\ \mathbf{x}_2 \end{bmatrix} + \mathbf{E}_2(t)\mathbf{E}_1(t)\mathbf{u} \tag{3-63b}$$

which can be easily obtained by observing $\mathbf{u}_1 = \mathbf{u}$, $\mathbf{y}_1 = \mathbf{u}_2$, $\mathbf{y} = \mathbf{y}_2$.

For the feedback connection shown in Figure 3-24(c), its dynamical-equation description is

$$\begin{bmatrix} \dot{\mathbf{x}}_1 \\ \dot{\mathbf{x}}_2 \end{bmatrix} = \begin{bmatrix} \mathbf{A}_1(t) - \mathbf{B}_1(t)\mathbf{Y}_2(t)\mathbf{E}_2(t)\mathbf{C}_1(t) & -\mathbf{B}_1(t)\mathbf{Y}_2(t)\mathbf{C}_2(t) \\ \mathbf{B}_2(t)\mathbf{Y}_1(t)\mathbf{C}_1(t) & \mathbf{A}_2(t) - \mathbf{B}_2(t)\mathbf{Y}_1(t)\mathbf{E}_1(t)\mathbf{C}_2(t) \end{bmatrix} \begin{bmatrix} \mathbf{x}_1 \\ \mathbf{x}_2 \end{bmatrix} + \begin{bmatrix} \mathbf{B}_1(t)\mathbf{Y}_2(t) \\ \mathbf{B}_2(t)\mathbf{Y}_1(t)\mathbf{E}_1(t) \end{bmatrix} \mathbf{u} \tag{3-64a}$$

$$\mathbf{y} = [\mathbf{Y}_1(t)\mathbf{C}_1(t) \quad -\mathbf{Y}_1(t)\mathbf{E}_1(t)\mathbf{C}_2(t)] \begin{bmatrix} \mathbf{x}_1 \\ \mathbf{x}_2 \end{bmatrix} + \mathbf{Y}_1(t)\mathbf{E}_1(t)\mathbf{u} \tag{3-64b}$$

where $\mathbf{Y}_1(t) = (\mathbf{I} + \mathbf{E}_1(t)\mathbf{E}_2(t))^{-1}$ and $\mathbf{Y}_2(t) = (\mathbf{I} + \mathbf{E}_2(t)\mathbf{E}_1(t))^{-1}$. It is clear that in order for (3-64) to be defined, we must assume that the inverses of $(\mathbf{I} + \mathbf{E}_1(t)\mathbf{E}_2(t))$ and $(\mathbf{I} + \mathbf{E}_2(t)\mathbf{E}_1(t))$ exist for all t. The dynamical equation (3-64) can be easily verified by observing $\mathbf{u}_1 = \mathbf{u} - \mathbf{y}_2$, $\mathbf{y}_1 = \mathbf{u}_2$, $\mathbf{y} = \mathbf{y}_1$. (See Problem 3-28.)

Time-invariant case. All the results in the preceding subsection can be applied to the time-invariant case without any modification. We shall now discuss the transfer-function matrices of composite systems. Let $\hat{\mathbf{G}}_1(s)$ and $\hat{\mathbf{G}}_2(s)$ be the proper rational transfer-function matrices of S_1 and S_2, respectively; then the transfer-function matrix of the parallel connection of S_1 and S_2 is $\hat{\mathbf{G}}_1(s) + \hat{\mathbf{G}}_2(s)$. The transfer-function matrix of the tandem connection of S_1 followed by S_2 is $\hat{\mathbf{G}}_2(s)\hat{\mathbf{G}}_1(s)$. Note that the order of $\hat{\mathbf{G}}_2(s)\hat{\mathbf{G}}_1(s)$ cannot be reversed. It is clear that if $\hat{\mathbf{G}}_i(s)$, $i = 1, 2$, are proper, so are $\hat{\mathbf{G}}_1(s) + \hat{\mathbf{G}}_2(s)$ and $\hat{\mathbf{G}}_2(s)\hat{\mathbf{G}}_1(s)$. It is implicitly assumed that the orders of $\hat{\mathbf{G}}_1(s)$ and $\hat{\mathbf{G}}_2(s)$ are compatible in both connections.

In order to discuss the transfer-function matrix of the feedback connection shown in Figure 3-24(c), we need some preliminary results.

Theorem 3-2

Let $\hat{\mathbf{G}}_1(s)$ and $\hat{\mathbf{G}}_2(s)$ be, respectively, $q \times p$ and $p \times q$ rational function matrices (not necessarily proper). Then we have

$$\det(\mathbf{I}_p + \hat{\mathbf{G}}_2(s)\hat{\mathbf{G}}_1(s)) = \det(\mathbf{I}_q + \hat{\mathbf{G}}_1(s)\hat{\mathbf{G}}_2(s)) \tag{3-65}$$

∎

Observe that the matrix on the right-hand side of (3-65) is a $q \times q$ matrix, while the matrix on the left-hand side is a $p \times p$ matrix. $\mathbf{I}_m$ is the unit matrix of order m. The elements of these matrices are rational functions of s, and since the rational functions form a field, standard results in matrix theory can be applied.

Proof of Theorem 3-2

It is well known that $\det(\mathbf{NQ}) = \det \mathbf{N} \det \mathbf{Q}$. Hence we have

$$\det(\mathbf{NQP}) = \det \mathbf{N} \det \mathbf{Q} \det \mathbf{P} = \det(\mathbf{PQN}) \tag{3-66}$$

where **N**, **Q**, and **P** are any square matrices of the same order. Let us choose

$$\mathbf{N}=\begin{bmatrix}\mathbf{I}_p & \mathbf{0}\\ -\hat{\mathbf{G}}_1(s) & \mathbf{I}_q\end{bmatrix}\qquad \mathbf{Q}=\begin{bmatrix}\mathbf{I}_p & -\hat{\mathbf{G}}_2(s)\\ \hat{\mathbf{G}}_1(s) & \mathbf{I}_q\end{bmatrix}\qquad \mathbf{P}=\begin{bmatrix}\mathbf{I}_p & \hat{\mathbf{G}}_2(s)\\ \mathbf{0} & \mathbf{I}_q\end{bmatrix}$$

They are square matrices of order $(q+p)$. It is easy to verify that

$$\mathbf{NQP}=\begin{bmatrix}\mathbf{I}_p & \mathbf{0}\\ \mathbf{0} & \mathbf{I}_q+\hat{\mathbf{G}}_1(s)\hat{\mathbf{G}}_2(s)\end{bmatrix}$$

and

$$\mathbf{PQN}=\begin{bmatrix}\mathbf{I}_p+\hat{\mathbf{G}}_2(s)\hat{\mathbf{G}}_1(s) & \mathbf{0}\\ \mathbf{0} & \mathbf{I}_q\end{bmatrix}$$

hence

$$\det(\mathbf{NQP})=\det(\mathbf{I}_q+\hat{\mathbf{G}}_1(s)\hat{\mathbf{G}}_2(s))$$
$$\det(\mathbf{PQN})=\det(\mathbf{I}_p+\hat{\mathbf{G}}_2(s)\hat{\mathbf{G}}_1(s))$$

and the theorem follows immediately from (3-66). For a different proof, see Problem 3-32. Q.E.D.

Theorem 3-3

If $\det(\mathbf{I}_q+\hat{\mathbf{G}}_1(s)\hat{\mathbf{G}}_2(s))\neq 0$, then

$$\hat{\mathbf{G}}_1(s)(\mathbf{I}_p+\hat{\mathbf{G}}_2(s)\hat{\mathbf{G}}_1(s))^{-1}=(\mathbf{I}_q+\hat{\mathbf{G}}_1(s)\hat{\mathbf{G}}_2(s))^{-1}\hat{\mathbf{G}}_1(s)$$

Proof

Note that the zero in $\det(\mathbf{I}_q+\hat{\mathbf{G}}_1(s)\hat{\mathbf{G}}_2(s))\neq 0$ is the zero element in the field of rational functions. Hence, it can be written more suggestively as $\det(\mathbf{I}_q+\hat{\mathbf{G}}_1(s)\hat{\mathbf{G}}_2(s))\not\equiv 0$ for some s. The condition $\det(\mathbf{I}_q+\hat{\mathbf{G}}_1(s)\hat{\mathbf{G}}_2(s))\neq 0$ implies that the inverse of the matrix $(\mathbf{I}_q+\hat{\mathbf{G}}_1(s)\hat{\mathbf{G}}_2(s))$ exists. From Theorem 3-2, we have

$$\det(\mathbf{I}_q+\hat{\mathbf{G}}_1(s)\hat{\mathbf{G}}_2(s))=\det(\mathbf{I}_p+\hat{\mathbf{G}}_2(s)\hat{\mathbf{G}}_1(s))\neq 0$$

Hence, both $(\mathbf{I}_q+\hat{\mathbf{G}}_1(s)\hat{\mathbf{G}}_2(s))^{-1}$ and $(\mathbf{I}_p+\hat{\mathbf{G}}_2(s)\hat{\mathbf{G}}_1(s))^{-1}$ exist. Consider the identity

$$\hat{\mathbf{G}}_1(s)(\mathbf{I}_p+\hat{\mathbf{G}}_2(s)\hat{\mathbf{G}}_1(s))(\mathbf{I}_p+\hat{\mathbf{G}}_2(s)\hat{\mathbf{G}}_1(s))^{-1}=\hat{\mathbf{G}}_1(s)$$

which can be written as

$$(\mathbf{I}_q+\hat{\mathbf{G}}_1(s)\hat{\mathbf{G}}_2(s))\hat{\mathbf{G}}_1(s)(\mathbf{I}_p+\hat{\mathbf{G}}_2(s)\hat{\mathbf{G}}_1(s))^{-1}=\hat{\mathbf{G}}_1(s) \tag{3-67}$$

Premultiplying $(\mathbf{I}_q+\hat{\mathbf{G}}_1(s)\hat{\mathbf{G}}_2(s))^{-1}$ on both sides of (3-67), we obtain the desired equality. Q.E.D.

Knowing Theorem 3-3, we are ready to investigate the transfer-function matrix of the feedback connection of S_1 and S_2.

Corollary 3-3

Consider the feedback system shown in Figure 3-24(c). Let $\hat{\mathbf{G}}_1(s)$ and $\hat{\mathbf{G}}_2(s)$ be $q\times p$ and $p\times q$ proper rational transfer-function matrices of S_1 and S_2, respectively. If $\det(\mathbf{I}_q+\hat{\mathbf{G}}_1(s)\hat{\mathbf{G}}_2(s))\neq 0$, then the transfer-function matrix of the feedback system is given by

$$\hat{\mathbf{G}}(s) = \hat{\mathbf{G}}_1(s)(\mathbf{I}_p + \hat{\mathbf{G}}_2(s)\hat{\mathbf{G}}_1(s))^{-1} = (\mathbf{I}_q + \hat{\mathbf{G}}_1(s)\hat{\mathbf{G}}_2(s))^{-1}\hat{\mathbf{G}}_1(s) \tag{3-68}$$

Proof

From Figure 3-24(c), we have $\hat{\mathbf{G}}_1(s)(\hat{\mathbf{u}}(s) - \hat{\mathbf{G}}_2(s)\hat{\mathbf{y}}(s)) = \hat{\mathbf{y}}(s)$, or

$$(\mathbf{I}_q + \hat{\mathbf{G}}_1(s)\hat{\mathbf{G}}_2(s))\hat{\mathbf{y}}(s) = \hat{\mathbf{G}}_1(s)\hat{\mathbf{u}}(s) \tag{3-69}$$

which, together with Theorem 3-3, implies this corollary. Q.E.D.

Note that the condition $\det(\mathbf{I}_q + \hat{\mathbf{G}}_1(s)\hat{\mathbf{G}}_2(s)) \neq 0$ is essential for a feedback system to be defined. Without this condition, a feedback system may become meaningless in the sense that for certain inputs, there are no outputs satisfying Equation (3-69).

Example 1

Consider a feedback system with

$$\hat{\mathbf{G}}_1(s) = \begin{bmatrix} \dfrac{-s}{s+1} & \dfrac{1}{s+2} \\ \dfrac{1}{s+1} & \dfrac{-s-1}{s+2} \end{bmatrix} \qquad \hat{\mathbf{G}}_2(s) = \begin{bmatrix} 1 & 0 \\ 0 & 1 \end{bmatrix}$$

It is easy to verify that $\det(\mathbf{I}_2 + \hat{\mathbf{G}}_1(s)\hat{\mathbf{G}}_2(s)) = 0$. Let us choose

$$\hat{\mathbf{u}}(s) = \begin{bmatrix} \dfrac{1}{s+2} \\ \dfrac{1}{(s+1)^2} \end{bmatrix}$$

Then (3-69) becomes

$$\begin{bmatrix} \dfrac{1}{s+1} & \dfrac{1}{s+2} \\ \dfrac{1}{s+1} & \dfrac{1}{s+2} \end{bmatrix} \hat{\mathbf{y}}(s) = \begin{bmatrix} \dfrac{1 - s(s+1)}{(s+1)^2(s+2)} \\ 0 \end{bmatrix} \tag{3-70}$$

Obviously there is no $\hat{\mathbf{y}}(s)$ satisfying (3-70). In matrix theory, Equation (3-70) is said to be inconsistent. Hence, in the feedback connection, we require $\det(\mathbf{I}_q + \hat{\mathbf{G}}_1(s)\hat{\mathbf{G}}_2(s)) \neq 0$ for some s. ∎

Recall from (3-64) that the condition for the existence of the dynamical-equation description of the feedback system in Figure 3-24(c) is $\det(\mathbf{I} + \mathbf{E}_1\mathbf{E}_2) \neq 0$ where $\mathbf{E}_i$ is the direct transmission part of the dynamical equation of S_i. Because of $\hat{\mathbf{G}}_i(\infty) = \mathbf{E}_i$ [see Equation (3-37)], if $\det(\mathbf{I} + \mathbf{E}_1\mathbf{E}_2) \neq 0$, then $\det(\mathbf{I} +$

$\hat{\mathbf{G}}_1(s)\hat{\mathbf{G}}_2(s)) \neq 0$. However, the converse is not true; that is, the condition $\det(\mathbf{I}+\hat{\mathbf{G}}_1(s)\hat{\mathbf{G}}_2(s)) \neq 0$ may not imply $\det(\mathbf{I}+\mathbf{E}_1\mathbf{E}_2) \neq 0$. Hence, a feedback system may have the transfer-function matrix description without having the state-variable description. This discrepancy will be resolved in the following subsection.

Well-posedness problem. We showed in Corollary 3-3 that if $\hat{\mathbf{G}}_1(s)$ and $\hat{\mathbf{G}}_2(s)$ are proper rational function matrices and if $\det(\mathbf{I}+\hat{\mathbf{G}}_1(s)\hat{\mathbf{G}}_2(s)) \neq 0$ for some s, then $\hat{\mathbf{G}}(s)=\hat{\mathbf{G}}_1(s)(\mathbf{I}+\hat{\mathbf{G}}_2(s)\hat{\mathbf{G}}_1(s))^{-1}$ is well defined. However, nothing has been said regarding whether or not $\hat{\mathbf{G}}(s)$ is proper. In this subsection, this problem will be studied.

Example 2

Consider the feedback system shown in Figure 3-24(c) with

$$\hat{\mathbf{G}}_1(s)=\begin{bmatrix} -1 & \dfrac{1}{s} \\ \dfrac{1}{s+1} & \dfrac{-s-2}{s+1} \end{bmatrix} \qquad \hat{\mathbf{G}}_2(s)=\mathbf{I}$$

The rational function matrix

$$\mathbf{I}+\hat{\mathbf{G}}_2(s)\hat{\mathbf{G}}_1(s)=\begin{bmatrix} 0 & \dfrac{1}{s} \\ \dfrac{1}{s+1} & \dfrac{-1}{s+1} \end{bmatrix}$$

is clearly nonsingular in the field of rational functions. The overall transfer-function matrix $\hat{\mathbf{G}}(s)$ is given by

$$\hat{\mathbf{G}}(s)=\begin{bmatrix} -1 & \dfrac{1}{s} \\ \dfrac{1}{s+1} & \dfrac{-s-2}{s+1} \end{bmatrix}\begin{bmatrix} 0 & \dfrac{1}{s} \\ \dfrac{1}{s+1} & \dfrac{-1}{s+1} \end{bmatrix}^{-1}=\begin{bmatrix} -s+1 & -s-1 \\ -s & 1 \end{bmatrix}$$

which is not a proper rational function matrix. The block diagram of this feedback system is plotted in Figure 3-25. ∎

For the feedback system in Figure 3-25, if the input signals $\mathbf{u}(t)$ is corrupted by high-frequency noises $\mathbf{n}(t)$, then the noises will be amplified by the improper rational matrix $\hat{\mathbf{G}}(s)$. For example, let $\mathbf{u}(t)=\sin t\,\mathbf{u}_0$ and $\mathbf{n}(t)=0.01\sin 1000t\,\mathbf{n}_0$ where $\mathbf{u}_0$ and $\mathbf{n}_0$ are 2×1 constant vectors. Although the amplitude of the noise is only one hundredth of the amplitude of the signal at the input terminals, the amplitude of the noise at the output terminals is 10 times larger than that

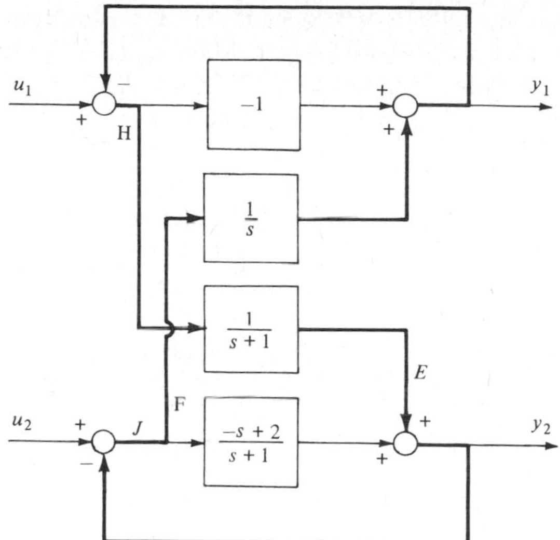

Figure 3-25 Feedback system with an improper transfer-function matrix.

of the signal. Hence, the system in Figure 3.25 is only of limited use in practice, although all of its subsystems have proper rational functions. Thus in the design of feedback systems, we shall require not only that all subsystems have proper transfer functions but also that the resulting overall system has a proper rational matrix. In the remainder of this subsection, we shall study this problem. Before proceeding, we need a preliminary result.

Theorem 3-4

Let $\mathbf{M}(s)$ be a square rational matrix and be decomposed uniquely as

$$\mathbf{M}(s) = \mathbf{M}_p(s) + \mathbf{M}_{sp}(s) \tag{3-71}$$

where $\mathbf{M}_p(s)$ is a polynomial matrix and $\mathbf{M}_{sp}(s)$ is a strictly proper rational matrix. Then $\mathbf{M}^{-1}(s)$ is proper if and only if $\mathbf{M}_p^{-1}(s)$ exists and is proper.[11]

Proof

We first show that if $\mathbf{M}_p^{-1}(s)$ is proper, so is $\mathbf{M}^{-1}(s)$. We write (3-71) as

$$\mathbf{M}^{-1}(s) = \{\mathbf{M}_p(s)[\mathbf{I} + \mathbf{M}_p^{-1}(s)\mathbf{M}_{sp}(s)]\}^{-1} = [\mathbf{I} + \mathbf{M}_p^{-1}(s)\mathbf{M}_{sp}(s)]^{-1}\mathbf{M}_p^{-1}(s)$$

If $\mathbf{M}_p^{-1}(s)$ is proper, then $\mathbf{M}_p^{-1}(\infty)$ is a finite constant matrix. Since $\mathbf{M}_{sp}(s)$ is strictly proper, $\mathbf{M}_{sp}(\infty) = \mathbf{0}$. Hence we have $\mathbf{M}^{-1}(\infty) = \mathbf{M}_p^{-1}(\infty)$, a finite constant matrix. Thus $\mathbf{M}^{-1}(s)$ is proper.

From $\mathbf{M}^{-1}(s) = [\mathbf{M}_p(s) + \mathbf{M}_{sp}(s)]^{-1}$, we have, as $s \to \infty$, $\mathbf{M}^{-1}(s) \to (\mathbf{M}_p(s) + \mathbf{0})^{-1} \to [\text{Adj } \mathbf{M}_p(s)]/\det \mathbf{M}_p(s)$. Hence if $\mathbf{M}_p(s)$ is singular, $\mathbf{M}^{-1}(\infty)$ is

[11] A sufficient condition for $\mathbf{M}_p^{-1}(s)$ to be proper is $\mathbf{M}_p(s)$ column-reduced or row-reduced. See Appendix G. For necessary/and sufficient conditions, see Problem G.17 and Sections 2-4 and 3-4 of Reference S34.

not a finite constant matrix and $\mathbf{M}^{-1}(s)$ is not proper.[12] Now we assume $\mathbf{M}_p(s)$ to be nonsingular and show that if $\mathbf{M}^{-1}(s)$ is proper, so is $\mathbf{M}_p^{-1}(s)$. From $\mathbf{M}(s)=\mathbf{M}_p(s)[\mathbf{I}+\mathbf{M}_p^{-1}(s)\mathbf{M}_{sp}(s)]$, we have

$$\begin{aligned}\mathbf{M}_0(s) \triangleq [\mathbf{I}+\mathbf{M}_p^{-1}(s)\mathbf{M}_{sp}(s)]^{-1} &= \mathbf{M}^{-1}(s)\mathbf{M}_p(s) = \mathbf{M}^{-1}(s)[\mathbf{M}(s)-\mathbf{M}_{sp}(s)] \\ &= \mathbf{I}-\mathbf{M}^{-1}(s)\mathbf{M}_{sp}(s)\end{aligned}$$

which, together with the finiteness of $\mathbf{M}^{-1}(\infty)$ and $\mathbf{M}_{sp}(\infty)=0$, implies $\mathbf{M}_0(\infty)=\mathbf{I}$. Hence we have $\mathbf{M}^{-1}(\infty)=\mathbf{M}_0(\infty)\mathbf{M}_p^{-1}(\infty)=\mathbf{M}_p^{-1}(\infty)$, and $\mathbf{M}_p^{-1}(\infty)$ is a finite constant matrix. Thus $\mathbf{M}_p^{-1}(s)$ is proper. Q.E.D.

This theorem shows that the properness of $\mathbf{M}^{-1}(s)$ depends only on the polynomial part of $\mathbf{M}(s)$. We give a special case in the following.

Corollary 3-4

If $\mathbf{M}(s)$ is a square proper rational matrix, $\mathbf{M}^{-1}(s)$ is proper if and only if $\mathbf{M}(\infty)$ is nonsingular.

Proof

If $\mathbf{M}(s)$ is proper, the polynomial part of $\mathbf{M}(s)$ is $\mathbf{M}_p(s)=\mathbf{M}(\infty)$. Hence $\mathbf{M}^{-1}(s)$ is proper if and only if $\mathbf{M}(\infty)$ is nonsingular. Q.E.D.

Using these results, we are ready to answer the question posed at the beginning of this subsection: the properness of $\hat{\mathbf{G}}(s)$ in (3-68).

Theorem 3-5

Consider the feedback system shown in Figure 3-24. Let $\hat{\mathbf{G}}_1(s)$ and $\hat{\mathbf{G}}_2(s)$ be $q \times p$ and $p \times q$ proper rational transfer matrices of S_1 and S_2. Then the overall transfer matrix

$$\hat{\mathbf{G}}(s)=\hat{\mathbf{G}}_1(s)[\mathbf{I}+\hat{\mathbf{G}}_2(s)\hat{\mathbf{G}}_1(s)]^{-1}$$

is proper if and only if $\mathbf{I}+\hat{\mathbf{G}}_2(\infty)\hat{\mathbf{G}}_1(\infty)$ is nonsingular.

Proof

If $\mathbf{I}+\hat{\mathbf{G}}_2(\infty)\hat{\mathbf{G}}_1(\infty)$ is nonsingular, $[\mathbf{I}+\hat{\mathbf{G}}_2(s)\hat{\mathbf{G}}_1(s)]^{-1}$ is proper. $\hat{\mathbf{G}}(s)$ is the product of two proper rational matrices; hence it is proper.

If $\hat{\mathbf{G}}(s)$ is proper, so are $\hat{\mathbf{G}}_2(s)\hat{\mathbf{G}}(s)$ and $\mathbf{I}-\hat{\mathbf{G}}_2(s)\hat{\mathbf{G}}(s)$. Since

$$[\mathbf{I}+\hat{\mathbf{G}}_2(s)\hat{\mathbf{G}}_1(s)]^{-1}=\mathbf{I}-\hat{\mathbf{G}}_2(s)\hat{\mathbf{G}}_1(s)[\mathbf{I}+\hat{\mathbf{G}}_2(s)\hat{\mathbf{G}}_1(s)]^{-1}=\mathbf{I}-\hat{\mathbf{G}}_2(s)\hat{\mathbf{G}}(s)$$

which can be readily verified by postmultiplying $\mathbf{I}+\hat{\mathbf{G}}_2(s)\hat{\mathbf{G}}_1(s)$, we conclude that if $\hat{\mathbf{G}}(s)$ is proper, so is $[\mathbf{I}+\hat{\mathbf{G}}_2(s)\hat{\mathbf{G}}_1(s)]^{-1}$. Hence $[\mathbf{I}+\hat{\mathbf{G}}_2(\infty)\hat{\mathbf{G}}_1(\infty)]^{-1}$ is finite and $\mathbf{I}+\hat{\mathbf{G}}_2(\infty)\hat{\mathbf{G}}_1(\infty)$ is nonsingular. This establishes the theorem. Q.E.D.

[12] For a different proof of this statement, see Problem 3-44.

The condition $\det[\mathbf{I}+\hat{\mathbf{G}}_1(\infty)\hat{\mathbf{G}}_2(\infty)]=\det[\mathbf{I}+\hat{\mathbf{G}}_2(\infty)\hat{\mathbf{G}}_1(\infty)]\neq 0$ in this theorem is the same as the condition $\det(\mathbf{I}+\mathbf{E}_1\mathbf{E}_2)\neq 0$ for the existence of the dynamical description of the feedback system shown in Figure 3-24(c). This resolves the discrepancy between these two descriptions.

Before discussing the physical implications of the condition $\det[\mathbf{I}+\hat{\mathbf{G}}_2(\infty)\hat{\mathbf{G}}_1(\infty)]=0$, we study first the composite system shown in Figure 3-26. This system is a generalization of the system in Figure 3-24(c) and will be developed in Chapter 7 and then extensively studied in Chapter 9. The transfer matrices $\hat{\mathbf{G}}_i(s)$ in Figure 3-26 are assumed to be proper rational matrices of appropriate orders. Clearly, we have

$$\hat{\mathbf{e}}(s)=\hat{\mathbf{u}}(s)-\hat{\mathbf{G}}_3(s)\hat{\mathbf{G}}_1(s)\hat{\mathbf{e}}(s)-\hat{\mathbf{G}}_4(s)\hat{\mathbf{G}}_2(s)\hat{\mathbf{G}}_1(s)\hat{\mathbf{e}}(s)$$

which implies

$$\hat{\mathbf{e}}(s)=[\mathbf{I}+\hat{\mathbf{G}}_3(s)\hat{\mathbf{G}}_1(s)+\hat{\mathbf{G}}_4(s)\hat{\mathbf{G}}_2(s)\hat{\mathbf{G}}_1(s)]^{-1}\hat{\mathbf{u}}(s) \tag{3-72}$$

Hence the transfer matrix from $\mathbf{u}$ to $\mathbf{y}$ is given by

$$\hat{\mathbf{G}}_f(s)=\hat{\mathbf{G}}_2(s)\hat{\mathbf{G}}_1(s)[\mathbf{I}+\hat{\mathbf{G}}_3(s)\hat{\mathbf{G}}_1(s)+\hat{\mathbf{G}}_4(s)\hat{\mathbf{G}}_2(s)\hat{\mathbf{G}}_1(s)]^{-1} \tag{3-73}$$

In view of Theorem 3-5, one may wonder whether $\hat{\mathbf{G}}_f(s)$ is proper if and only if $\hat{\mathbf{G}}_0(\infty)$ is nonsingular, where

$$\hat{\mathbf{G}}_0(s)\triangleq\mathbf{I}+\hat{\mathbf{G}}_3(s)\hat{\mathbf{G}}_1(s)+\hat{\mathbf{G}}_4(s)\hat{\mathbf{G}}_2(s)\hat{\mathbf{G}}_1(s)$$

If $\hat{\mathbf{G}}_i(s)$ are proper, the polynomial part of $\hat{\mathbf{G}}_0(s)$ is $\hat{\mathbf{G}}_0(\infty)$. Hence, if $\hat{\mathbf{G}}_0(\infty)$ is nonsingular, Corollary 3-4 implies that $\hat{\mathbf{G}}_0^{-1}(s)$ is proper. Consequently, $\hat{\mathbf{G}}_f(s)$ in (3-73) is proper. However, the nonsingularity of $\hat{\mathbf{G}}_0(\infty)$ is not a necessary condition for $\hat{\mathbf{G}}_f(s)$ to be proper. For example, consider

$$\hat{\mathbf{G}}_2(s)=\begin{bmatrix}\frac{1}{s} & 0\\ 0 & \frac{1}{s+1}\end{bmatrix}\qquad \hat{\mathbf{G}}_1(s)=\hat{\mathbf{G}}_4(s)=\mathbf{I}\qquad \hat{\mathbf{G}}_3(s)=-\mathbf{I}$$

Then we have $\hat{\mathbf{G}}_0(\infty)=\mathbf{I}-\mathbf{I}+\mathbf{0}=\mathbf{0}$, which is singular; however, we have

$$\hat{\mathbf{G}}_f(s)=\hat{\mathbf{G}}_2(s)[\mathbf{I}-\mathbf{I}+\hat{\mathbf{G}}_2(s)]^{-1}=\mathbf{I}$$

which is proper.

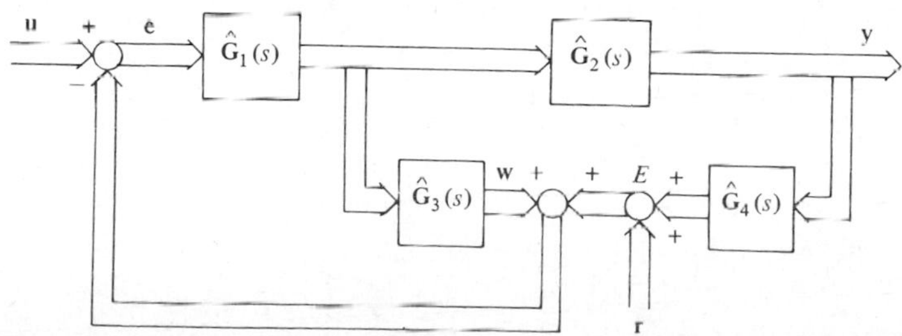

Figure 3-26 A feedback system.

For this system, $\hat{\mathbf{G}}_f(s)$ is proper; whereas $\hat{\mathbf{G}}_0(\infty)$ is singular. Will such a system be acceptable in practice? If $\hat{\mathbf{G}}_0(\infty)$ is singular, the transfer matrix from $\hat{\mathbf{u}}(s)$ to $\hat{\mathbf{e}}(s)$ in (3-72) is improper. In this case, if $\mathbf{u}(t)$ contains a part which has a high-frequency spectrum or is discontinuous, the amplitude of $\mathbf{e}(t)$ will be very large or infinite, and the system may saturate or burn out. Hence, in the design of a feedback system, we shall require not only the overall transfer matrix but also all transfer functions from all possible input-output pairs of the system to be proper. In this case, no signal will be unduly amplified, and in some sense, the smoothness of signals throughout the system will be preserved.

Definition 3-9[13]

Let every subsystem of a composite system be describable by a rational transfer function. Then the composite system is said to be *well posed* if the transfer function of every subsystem is proper and the closed-loop transfer function from any point chosen as an input terminal to every other point along the directed path is well defined and proper. ∎

In this definition, if a point, say E in Figure 3-26, is chosen as an input terminal, then we must add a fictitious input as shown. Then the closed-loop transfer functions from $\mathbf{r}$ to $\mathbf{e}$, $\mathbf{w}$, and $\mathbf{y}$ are, respectively.

$$\hat{\mathbf{e}}(s) = -[\mathbf{I} + \hat{\mathbf{G}}_3(s)\hat{\mathbf{G}}_1(s) + \hat{\mathbf{G}}_4(s)\hat{\mathbf{G}}_2(s)\hat{\mathbf{G}}_1(s)]^{-1}\hat{\mathbf{r}}(s)$$
$$\hat{\mathbf{w}}(s) = -\hat{\mathbf{G}}_3(s)\hat{\mathbf{G}}_1(s)[\mathbf{I} + \hat{\mathbf{G}}_3(s)\hat{\mathbf{G}}_1(s) + \hat{\mathbf{G}}_4(s)\hat{\mathbf{G}}_2(s)\hat{\mathbf{G}}_1(s)]^{-1}\hat{\mathbf{r}}(s) \qquad \textbf{(3-74)}$$
and
$$\hat{\mathbf{y}}(s) = -\hat{\mathbf{G}}_2(s)\hat{\mathbf{G}}_1(s)[\mathbf{I} + \hat{\mathbf{G}}_3(s)\hat{\mathbf{G}}_1(s) + \hat{\mathbf{G}}_4(s)\hat{\mathbf{G}}_2(s)\hat{\mathbf{G}}_1(s)]^{-1}\hat{\mathbf{r}}(s)$$

We note that in computing these transfer functions, no branch in the system should be disconnected. Hence we compute closed-loop transfer functions. If we disconnect any branch, then the system becomes a different system.

Theorem 3-6

The system in Figure 3-26, where $\hat{\mathbf{G}}_i(s)$ are rational transfer matrices of appropriate orders, is well posed if and only if $\hat{\mathbf{G}}_i(s)$, $i = 1, 2, 3, 4$, are proper and the rational matrix

$$\hat{\mathbf{G}}_0^{-1}(s) \triangleq (\mathbf{I} + \hat{\mathbf{G}}_3(s)\hat{\mathbf{G}}_1(s) + \hat{\mathbf{G}}_4(s)\hat{\mathbf{G}}_2(s)\hat{\mathbf{G}}_1(s))^{-1} \qquad \textbf{(3-75)}$$

exists and is proper, or equivalently, the constant matrix

$$\hat{\mathbf{G}}_0(\infty) = \mathbf{I} + \hat{\mathbf{G}}_3(\infty)\hat{\mathbf{G}}_1(\infty) + \hat{\mathbf{G}}_4(\infty)\hat{\mathbf{G}}_2(\infty)\hat{\mathbf{G}}_1(\infty)$$

is nonsingular.

Proof

The transfer matrices in (3-72) to (3-74) are clearly proper if and only if $\hat{\mathbf{G}}_0^{-1}(s)$ is proper. Similarly, it can be shown that if $\hat{\mathbf{G}}_0^{-1}(s)$ is proper, the transfer matrix

[13] This definition is similar to the one in Reference S34 and is applicable only to linear time-invariant lumped systems. For a more general definition and discussion, see References S8, S207 and S214.

from any point to any other point along the directed path is proper. This establishes the first part of the theorem. If all $\hat{\mathbf{G}}_i(s)$, $i=1, 2, 3, 4$, are proper, the polynomial part of $\hat{\mathbf{G}}_0(s)$ is $\hat{\mathbf{G}}_0(\infty)$. Hence $\hat{\mathbf{G}}_0^{-1}(s)$ is, following Corollary 3-4, proper if and only if $\hat{\mathbf{G}}_0(\infty)$ is nonsingular. Q.E.D.

We discuss now the implications of the conditions $\det [\mathbf{I} + \hat{\mathbf{G}}_2(\infty)\hat{\mathbf{G}}_1(\infty)] \neq 0$ and more generally, $\det \hat{\mathbf{G}}_0(\infty) \neq 0$. Before proceeding, we need some concepts. A loop of a block diagram is a closed path which travels from a point along the direction of the path back to the same point and does not pass any point twice. For example, the feedback system in Figure 3-25 has three loops; one of them is indicated by the heavy lines. The *loop gain* of a loop is defined as the product of all transfer functions along the loop including the signs at summing points. In our discussion, we are interested in only the value of the loop gain at $s=\infty$. Clearly if a loop gain is a strictly proper rational function, then its loop gain is zero at $s=\infty$. If all transfer functions of a block diagram are proper, a loop has a nonzero loop gain at $s=\infty$ if and only if all transfer functions along the loop are exactly proper (a rational function is said to be *exactly proper* if the degree of its numerator is equal to that of its denominator). In the following, only loops with nonzero loop gains at $s=\infty$ will come into the discussion.

Consider the feedback systems shown in Figure 3-27(a) and (b). They are not well posed because $\det \hat{\mathbf{G}}_0(\infty)=0$ in both systems. Each system has a loop with loop gain $+1$ at $s=\infty$, which is equivalent to a gain of infinity or an improper transfer function as shown in Figure 3-27(c) and (d). Hence, we may conclude that if a system is not well posed, the system has a loop with loop gain $+1$ at $s=\infty$ which is equivalent to an infinite gain or an improper transfer function.

The system in Figure 3-28(a) has a loop with loop gain 1; however, we have $\hat{G}_0(\infty)=1-1+2\neq 0$, and the system is well posed. This discrepancy arises from the existence of another loop with a nonzero loop gain at $s=\infty$. This loop will offset the loop with loop gain 1, as can be seen from Figure 3-28(c)

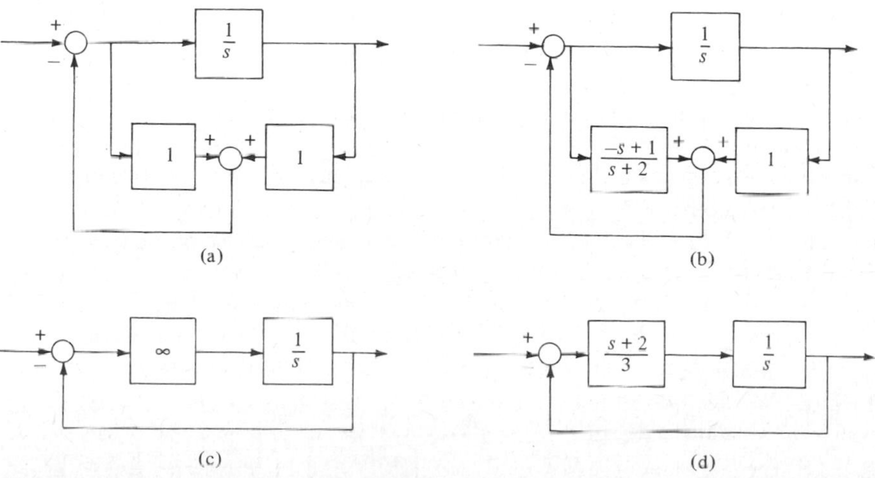

Figure 3-27 Feedback systems which are not well posed.

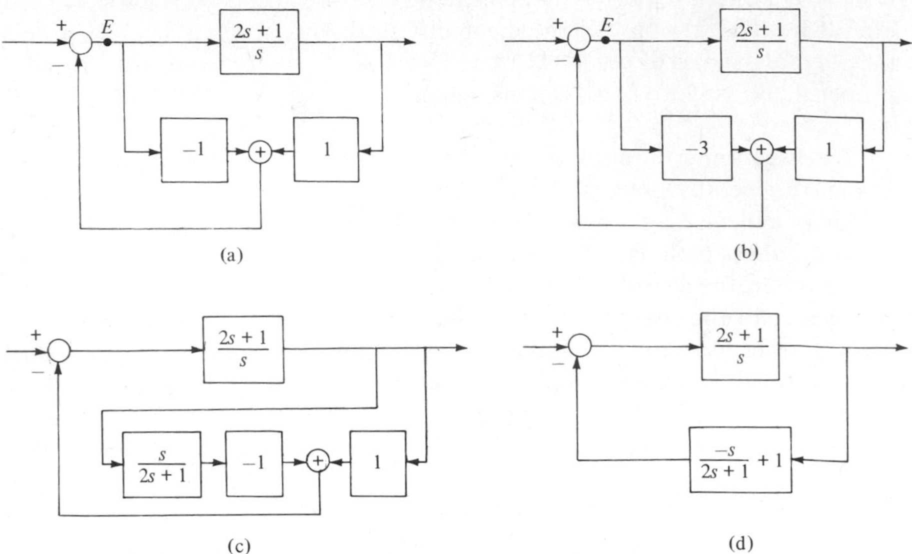

Figure 3-28 (a) A well-posed system. (b) An ill-posed system. (c), (d) Equivalent systems of (a).

and (d). Hence, if there are two or more loops with nonzero loop gains at $s=\infty$ passing the same point, such as the point E shown in Figure 3-28(a), these loops must be grouped into a combined loop with a *net* loop gain equal to the sum of all individual loop gains. For the system in Figure 3-28(a), the combined loop has a net loop gain $1-2=-1$ which is different from $+1$; hence the loop will not cause any problem, and the system is well posed. The system in Figure 3-28(b) has a combined loop with loop gain $3-2=1$; hence the system is not well posed (Problem 3-39).

From these examples, we conclude that if a system has a combined loop with net loop gain 1 at $s=\infty$, the combined loop is equivalent to a gain of infinity or an improper transfer function. Note that what causes the problem is the combined loop instead of individual loops.

To further verify the aforementioned statement, we consider the system shown in Figure 3-25. There are two loops passing through the point H. Along the loop indicated by the heavy lines, there is a loop between points E and F. In computing the loop gain of the heavy-lined loop, the transfer function from E to F should be computed. However because of the two strictly proper transfer functions, the loop gain is zero at $s=\infty$. Hence, the combined loop which passes through the point H has a net loop gain 1 at $s=\infty$. This loop is equivalent to a gain of infinity. Similarly, the combined loop which passes the point J has a net loop gain 1 at $s=\infty$. The transfer function from u_2 to y_2 through the loop is $(s+2)/1$, which is improper. For the system in Figure 3-25, we have $\det[\mathbf{I}+\hat{\mathbf{G}}_2(\infty)\hat{\mathbf{G}}_1(\infty)]=0$.

In conclusion, a system is well posed if and only if the system has no combined loop with net loop gain 1 at $s=\infty$, which is equivalent to an infinite gain or an improper transfer function. In practice, every system should be designed to be well posed.

3-7 Discrete-Time Systems[14]

The inputs and outputs of the systems we studied in the previous sections are defined for all t in $(-\infty, \infty)$ for the time-varying case or in $[0, \infty)$ for the time-invariant case. They are called *continuous-time systems.* In this section we shall study a different class of systems, called *discrete-time systems.* The inputs and outputs of discrete-time systems are defined only at discrete instants of time. For example, a digital computer reads and prints out data that are the values of variables at discrete instants of time; hence it is a discrete-time system. A continuous-time system can also be modeled as a discrete-time system if its responses are of interest or measurable only at certain instants of time.

For convenience, the discrete instants of time at which the input and the output appear will be assumed to be equally spaced by an amount of T. The time interval T is called the *sampling period.* We use $\{\mathbf{u}(k) \triangleq \mathbf{u}(kT)\}$ and $\{\mathbf{y}(k) \triangleq \mathbf{y}(kT)\}$, $k=0, \pm 1, \pm 2, \ldots$, to denote the input and output sequences. In the following, we discuss only single-variable discrete-time systems. As in the continuous-time system, a discrete-time system that is initially relaxed is called a *relaxed discrete-time system.* If the inputs and outputs of a relaxed discrete-time system satisfy the linearity property, then they can be related by

$$y(k) = \sum_{-\infty}^{\infty} g(k, m)u(m) \qquad \textbf{(3-76)}$$

where $g(k, m)$ is called the *weighting sequence* and is the response of the system due to the application of the input

$$u(i) = \begin{cases} 1 & i = m \\ 0 & i \neq m \end{cases}$$

with the system relaxed at time $m-$. If the discrete-time system is causal—that is, the output does not depend on the future values of the input—then $g(k, m) = 0$ for $k < m$. Consequently, if a discrete-time system is causal and is relaxed at k_0, then (3-76) reduces to

$$y(k) = \sum_{m=k_0}^{k} g(k, m)u(m) \qquad \textbf{(3-77)}$$

If a linear causal relaxed discrete-time system is time invariant, then we have $g(k, m) = g(k-m)$ for all $k \geq m$. In this case, the initial time is chosen to be $k_0 = 0$ and the set of time of interest is the set of positive integers. Hence for a linear time-invariant causal relaxed discrete-time system, we have

$$y(k) = \sum_{m=0}^{k} g(k-m)u(m) \qquad k = 0, 1, 2, \ldots \qquad \textbf{(3-78)}$$

---

[14] This section may be skipped without loss in continuity.

Comparing with continuous-time systems, we see that in discrete-time systems, we use summation instead of integration; otherwise, all the concepts are the same. In the continuous-time case, if we apply the Laplace transform, a convolution integral can be transformed into an algebraic equation. We have the same situation here, the transformation which will be used is called the z-transform.

Definition 3-10

The *z-transform* of the sequence $\{u(k), k=0, 1, 2, \ldots\}$ is defined as

$$\hat{u}(z) \triangleq \mathscr{Z}[u(k)] \triangleq \sum_{k=0}^{\infty} u(k)z^{-k}$$

where z is a complex variable. ∎

Example 1

If $u(k)=1$ for $k=0, 1, 2, \ldots$, then

$$\hat{u}(z) = \sum_{k=0}^{\infty} z^{-k} = \frac{1}{1-z^{-1}} = \frac{z}{z-1}$$ ∎

Example 2

If $u(k)=e^{-2k}$ for $k=0, 1, 2, \ldots$, then

$$\hat{u}(z) = \sum_{k=0}^{\infty} e^{-2k}z^{-k} = \frac{1}{1-e^{-2}z^{-1}} = \frac{z}{z-e^{-2}}$$ ∎

Now we shall apply the z-transform to (3-78). For a causal, relaxed system, we have $g(k-m)=0$ for $k<m$; hence we may also write (3-78) as

$$y(k) = \sum_{m=0}^{\infty} g(k-m)u(m)$$

Consequently, we have

$$\begin{aligned} \hat{y}(z) &= \sum_{k=0}^{\infty} y(k)z^{-k} = \sum_{k=0}^{\infty}\sum_{m=0}^{\infty} g(k-m)u(m)z^{-(k-m)}z^{-m} \\ &= \sum_{m=0}^{\infty}\sum_{k=0}^{\infty} g(k-m)z^{-(k-m)}u(m)z^{-m} \\ &= \sum_{k=0}^{\infty} g(k)z^{-k} \sum_{m=0}^{\infty} u(m)z^{-m} = \hat{g}(z)\hat{u}(z) \end{aligned} \tag{3-79}$$

Here we have changed the order of summations and used the fact that $g(k-m)=0$ for $k<m$. The function $\hat{g}(z)$ is the z-transform of the weighting sequence $\{g(k)\}_{k=0}^{\infty}$ and is called the *z-transfer function* or *sampled transfer function.*

The extension of the input-output description of single-variable discrete-time systems to the multivariable case is straightforward and its discussion is omitted. We introduce now the discrete-time dynamical equation. *A linear, time-varying, discrete-time dynamical equation* is defined as

$$DE: \qquad \begin{aligned} \mathbf{x}(k+1) &= \mathbf{A}(k)\mathbf{x}(k) + \mathbf{B}(k)\mathbf{u}(k) \\ \mathbf{y}(k) &= \mathbf{C}(k)\mathbf{x}(k) + \mathbf{E}(k)\mathbf{u}(k) \end{aligned}$$

where $\mathbf{x}$ is the state vector, $\mathbf{u}$ the input, and $\mathbf{y}$ the output. Note that a discrete-time dynamical equation is a set of first-order *difference* equations instead of a set of first-order *differential* equations, as in the continuous-time case.

If $\mathbf{A}(k)$, $\mathbf{B}(k)$, $\mathbf{C}(k)$, and $\mathbf{E}(k)$ are independent of k, then DE reduces to

$$DFE: \qquad \begin{aligned} \mathbf{x}(k+1) &= \mathbf{A}\mathbf{x}(k) + \mathbf{B}\mathbf{u}(k) \\ \mathbf{y}(k) &= \mathbf{C}\mathbf{x}(k) + \mathbf{E}\mathbf{u}(k) \end{aligned} \tag{3-80}$$

which is called a *linear, time-invariant, discrete-time dynamical equation*. Let $\hat{\mathbf{x}}(z)$ be the z-transform of

$$\{\mathbf{x}(k)\}_{k=0}^{\infty}$$

That is,

$$\hat{\mathbf{x}}(z) \triangleq \mathscr{Z}[\mathbf{x}(k)] \triangleq \sum_{k=0}^{\infty} \mathbf{x}(k)z^{-k}$$

Then

$$\begin{aligned} \mathscr{Z}[\mathbf{x}(k+1)] &= \sum_{k=0}^{\infty} \mathbf{x}(k+1)z^{-k} = z \sum_{k=0}^{\infty} \mathbf{x}(k+1)z^{-(k+1)} \\ &= z\left[\sum_{k=-1}^{\infty} \mathbf{x}(k+1)z^{-(k+1)} - \mathbf{x}(0)\right] = z[\hat{\mathbf{x}}(z) - \mathbf{x}(0)] \end{aligned}$$

Hence the application of the z-transform to (3-80) yields

$$\begin{aligned} z\hat{\mathbf{x}}(z) - z\mathbf{x}^0 &= \mathbf{A}\hat{\mathbf{x}}(z) + \mathbf{B}\hat{\mathbf{u}}(z) \\ \hat{\mathbf{y}}(z) &= \mathbf{C}\hat{\mathbf{x}}(z) + \mathbf{E}\hat{\mathbf{u}}(z) \end{aligned} \tag{3-81}$$

where $\mathbf{x}(0) = \mathbf{x}_0$. Equation (3-81) can be arranged as

$$\hat{\mathbf{x}}(z) = (z\mathbf{I} - \mathbf{A})^{-1}z\mathbf{x}_0 + (z\mathbf{I} - \mathbf{A})^{-1}\mathbf{B}\hat{\mathbf{u}}(z) \tag{3-82}$$

$$\hat{\mathbf{y}}(z) = \mathbf{C}(z\mathbf{I} - \mathbf{A})^{-1}z\mathbf{x}_0 + \mathbf{C}(z\mathbf{I} - \mathbf{A})^{-1}\mathbf{B}\hat{\mathbf{u}}(z) + \mathbf{E}\hat{\mathbf{u}}(z) \tag{3-83}$$

They are algebraic equations. The manipulation of these equations is exactly the same as (3-35) and (3-36). If $\mathbf{x}(0) = \mathbf{0}$, then (3-83) reduces to

$$\hat{\mathbf{y}}(z) = [\mathbf{C}(z\mathbf{I} - \mathbf{A})^{-1}\mathbf{B} + \mathbf{E}]\hat{\mathbf{u}}(z) \triangleq \hat{\mathbf{G}}(z)\hat{\mathbf{u}}(z)$$

where

$$\hat{\mathbf{G}}(z) = \mathbf{C}(z\mathbf{I} - \mathbf{A})^{-1}\mathbf{B} + \mathbf{E} \tag{3-84}$$

is called the *sampled transfer-function matrix* of (3-80). We see that (3-84) is identical to the continuous-time case in (3-36) if z is replaced by s.

Similar to the continuous-time case, the sampled transfer functions of practical interest are proper rational functions. Consider the improper transfer function

$$\hat{g}(z) = \frac{z^2+2}{z-1} = z + 1 + 3z^{-1} + 3z^{-2} + 3z^{-3} + \cdots$$

The inverse z-transform of $\hat{g}(z)$ is $g(-1)=1, g(0)=1, g(k)=3, k=1, 2, 3, \ldots$. Its impulse sequence is not zero for $g(k) \neq 0$ for $k < 0$; hence $\{g(k)\}$ is not causal. Consequently, an improper rational function does not describe a causal system. All physical systems are causal; hence we are interested in only proper sampled transfer functions. This reason is different from the continuous-time case where differentiations due to improper transfer functions are to be avoided because of high-frequency noises.

The topics discussed in Sections 3-5 and 3-6 are directly applicable to the discrete-time case without any modification.

3-8 Concluding Remarks

In this chapter we have developed systematically the input-output description and the state-variable description of linear systems. Although the input-output description can be obtained by analysis, it can also be developed from the measurements at the input and output terminals without knowing the internal structure of the system. For linear, time-invariant systems, the transfer-function description can also be used. Whenever transfer functions (including impedances and admittances in network theory) are used, the systems are implicitly assumed to be relaxed at $t=0$.

The condition for a set of first-order differential equations $\dot{\mathbf{x}} = \mathbf{f}(\mathbf{x}, \mathbf{u}, t)$, $\mathbf{y} = \mathbf{g}(\mathbf{x}, \mathbf{u}, t)$—in particular, $\dot{\mathbf{x}} = \mathbf{A}(t)\mathbf{x} + \mathbf{B}(t)\mathbf{u}$, $\mathbf{y} = \mathbf{C}(t)\mathbf{x} + \mathbf{E}(t)\mathbf{u}$—to be qualified as a dynamical equation is that for any initial state $\mathbf{x}_0$ and any input $\mathbf{u}_{[t_0, \infty)}$, there is a *unique* solution $\mathbf{y}_{[t_0, \infty)}$ satisfying the equations. This uniqueness condition is essential in the study of solutions of a dynamical equation. The dynamical equations studied in this book are of the form $\dot{\mathbf{x}} = \mathbf{A}(t)\mathbf{x} + \mathbf{B}(t)\mathbf{u}$, $\mathbf{y} = \mathbf{C}(t)\mathbf{x} + \mathbf{E}(t)\mathbf{u}$. They can be extended to include derivatives of $\mathbf{u}$ in the output equation, such as

$$\mathbf{y}(t) = \mathbf{C}(t)\mathbf{x}(t) + \mathbf{E}(t)\mathbf{u} + \mathbf{E}_1(t)\dot{\mathbf{u}}(t) + \mathbf{E}_2(t)\ddot{\mathbf{u}}(t) + \cdots$$

However, these extensions are of limited interest in practice.

One may wonder why we study a set of first-order differential equations instead of studying higher-order differential equations. The reasons are as follows: (1) every higher-order differential equation can be written as a set of first-order differential equations, (2) the notations used to describe first-order equations are compact and very simple, and (3) first-order differential equations can be readily simulated on an analog or a digital computer.

In addition to transfer functions and dynamical equations, one may also encounter the following description

$$\begin{aligned} \mathbf{P}(s)\hat{\boldsymbol{\xi}}(s) &= \mathbf{Q}(s)\hat{\mathbf{u}}(s) \\ \hat{\mathbf{y}}(s) &= \mathbf{R}(s)\hat{\boldsymbol{\xi}}(s) + \mathbf{W}(s)\hat{\mathbf{u}}(s) \end{aligned}$$

where **P**, **Q**, **R** and **W** are polynomial matrices, in the analyses. This description is called the *polynomial matrix system description* and will be studied in Chapter 6.

Both the input-output description and the state-variable description studied in this chapter are useful in practice. Which description should be used depends on the problem, on the data available, and on the question asked.

Problems

3-1 Consider the memoryless systems with characteristics shown in Figure P3-1, in which u denotes the input and y the output. Which of them is a linear system? Is it possible to introduce a new output so that the system in Figure P3-1(b) is linear?

3-2 We may define the impulse response of a relaxed nonlinear system $g(\cdot, \tau)$ as the response due to a δ-function input applied at time τ. Is this impulse response useful in describing the system? Why?

3-3 The impulse response of a relaxed linear system is found to be $g(t, \tau) = e^{-|t-\tau|}$ for all t and τ. Is this system causal? Is it time invariant?

3-4 The impulse response of an ideal low-pass filter is given by

$$g(t) = 2\omega \frac{\sin 2\omega(t - t_0)}{2\omega(t - t_0)} \qquad \text{for all } t$$

where ω and t_0 are constants. Is the ideal low-pass filter causal? Is it possible to build an ideal low-pass filter in the real world?

3-5 Consider a relaxed system whose input and output are related by

$$\mathbf{y}(t) = (P_\alpha \mathbf{u})(t) \triangleq \begin{cases} \mathbf{u}(t) & \text{for } t \le \alpha \\ \mathbf{0} & \text{for } t > \alpha \end{cases}$$

for any **u**, where α is a fixed constant. In words, this system, called a *truncation operator*, chops off the input after time α. Is this system linear? Is it time invariant? Is it causal?

3-6 Let P_α be the truncation operator defined in Problem 3-5. Show that for any causal relaxed system $\mathbf{y} = H\mathbf{u}$, we have

$$P_\alpha \mathbf{y} = P_\alpha H \mathbf{u} = P_\alpha H P_\alpha \mathbf{u}$$

This fact is often used in stability studies of nonlinear feedback systems.

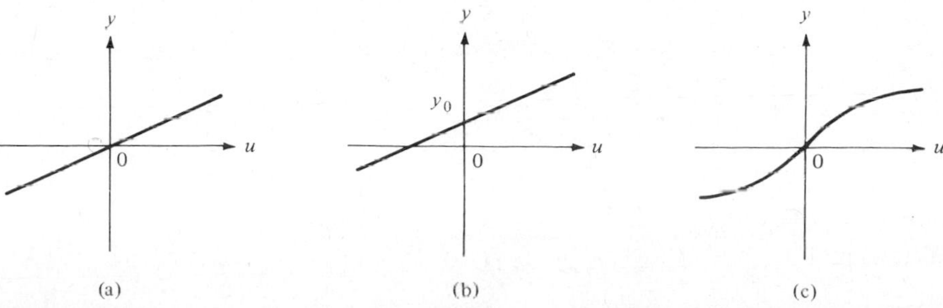

Figure P3-1

3-7 In Problem 3-6, is it true that

$$(P_\alpha \mathbf{y})(t) = (P_\alpha H\mathbf{u})(t) = (HP_\alpha \mathbf{u})(t)$$

for all t in $(-\infty, \infty)$? If not, find the interval of time in which the equation holds.

3-8 Consider a linear system with input u and output y. Three experiments are performed on this system using the inputs $u_1(t)$, $u_2(t)$, and $u_3(t)$ for $t \geq 0$. In each case, the initial state at $t=0$, $\mathbf{x}(0)$, is the same. The corresponding observed outputs are $y_1(t)$, $y_2(t)$, and $y_3(t)$. Which of the following three predictions are true if $\mathbf{x}(0) \neq 0$?

a. If $u_3 = u_1 + u_2$, then $y_3 = y_1 + y_2$.
b. If $u_3 = \frac{1}{2}(u_1 + u_2)$, then $y_3 = \frac{1}{2}(y_1 + y_2)$.
c. If $u_3 = u_1 - u_2$, then $y_3 = y_1 - y_2$.

Which are true if $\mathbf{x}(0) = \mathbf{0}$? (*Answers:* No, yes, no, for $\mathbf{x}(0) \neq \mathbf{0}$; all yes, if $\mathbf{x}(0) = \mathbf{0}$.)

3-9 Show that if $H(\mathbf{u}_1 + \mathbf{u}_2) = H\mathbf{u}_1 + H\mathbf{u}_2$ for any $\mathbf{u}_1$, $\mathbf{u}_2$, then $H\alpha\mathbf{u} = \alpha H\mathbf{u}$ for any rational number α and any $\mathbf{u}$.

3-10 Show that for a fixed α, the shifting operator Q_α defined in Figure 3-5 is a linear time-invariant system. What is its impulse response? What is its transfer function? Is this transfer function a rational function?

3-11 The causality of a relaxed system may also be defined as follows: A relaxed system is causal if and only if $\mathbf{u}_1(t) = \mathbf{u}_2(t)$ for all $t \leq t_0$ implies $(H\mathbf{u}_1)(t) = (H\mathbf{u}_2)(t)$ for all $t \leq t_0$. Show that this definition implies that $\mathbf{y}(t) = H\mathbf{u}_{(-\infty, t]}$, and vice versa.

3-12 Let $g(t, \tau) = g(t + \alpha, \tau + \alpha)$ for all t, τ, and α. Define $x = t + \tau$, $y = t - \tau$, then $g(t, \tau) = g((x+y)/2, (x-y)/2)$. Show that $\partial g(t, \tau)/\partial x = 0$. [From this fact we may conclude that if $g(t, \tau) = g(t + \alpha, \tau + \alpha)$ for all t, τ, and α, then $g(t, \tau)$ depends only on $t - \tau$.]

3-13 Consider a relaxed system that is described by

$$y(t) = \int_0^t g(t-\tau)u(\tau)\, d\tau$$

If the impulse response g is given by Figure P3-13(a), what is the output due to the input shown in Figure P3-13(b)? (Use graphical method.)

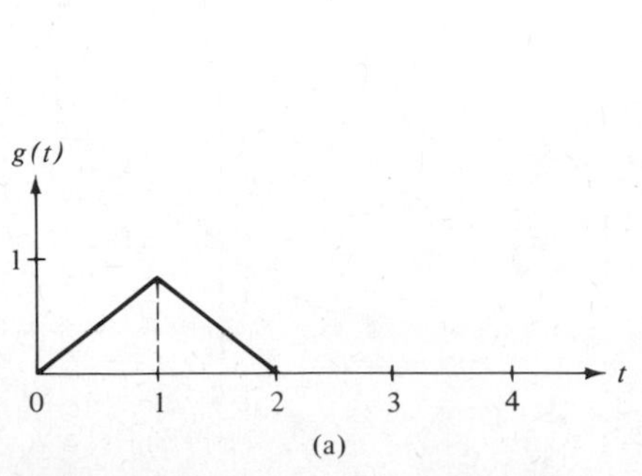

(a)

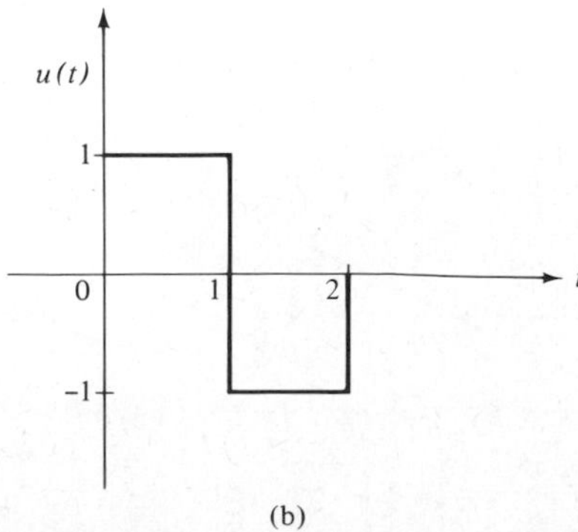

(b)

Figure P3-13

3-14 The input u and the output y of a system is described by

$$\ddot{y}+2\dot{y}+3y=2\dot{u}+u$$

What is the transfer function of the system?

3-15 Consider a multivariable system that is describable by $\hat{\mathbf{y}}(s)=\hat{\mathbf{G}}(s)\hat{\mathbf{u}}(s)$. Show that the ijth element of $\hat{\mathbf{G}}(s)$ can be defined as

$$\hat{g}_{ij}(s)=\left.\frac{\mathscr{L}[y_i(t)]}{\mathscr{L}[u_j(t)]}\right|_{\substack{\text{initially relaxed}\\ \text{and } u_k=0 \text{ for } k\neq j}}$$

where y_i is the ith component of $\mathbf{y}$ and u_j is the jth component of $\mathbf{u}$.

3-16 Consider a multivariable system whose inputs and outputs are described by

$$N_{11}(p)y_1(t)+N_{12}(p)y_2(t)=D_{11}(p)u_1(t)+D_{12}(p)u_2(t)$$
$$N_{21}(p)y_1(t)+N_{22}(p)y_2(t)=D_{21}(p)u_1(t)+D_{22}(p)u_2(t)$$

where the N_{ij}'s and D_{ij}'s are polynomials of $p \triangleq d/dt$. What is the transfer-function matrix of the system?

3-17 Find the dynamical-equation descriptions of the systems shown in Figure P3-17. If θ, θ_1, and θ_2 are very small, can you consider the two systems as linear? Find the transfer functions and dynamical equations to describe the linearized systems.

3-18 Find the dynamical-equation descriptions of the networks shown in Figures 3-22 and 3-23.

3-19 Find the dynamical-equation description and the transfer-function matrix of the network in Figure P3-19.

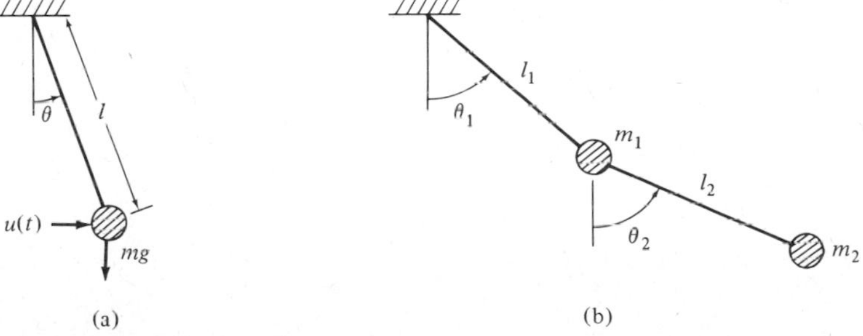

Figure P3-17

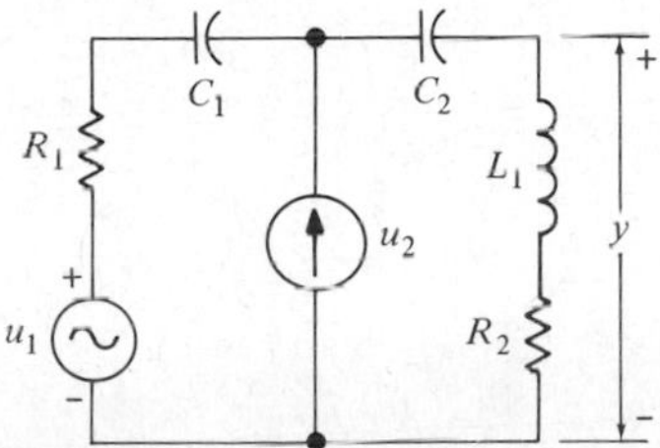

Figure P3-19

3-20 Consider the simplified model of an aircraft shown in Figure P3-20. It is assumed that the aircraft is dynamically equivalent at the pitched angle θ_0, elevator angle u_0, altitude h_0, and cruising speed v_0. It is assumed that small deviations of θ and u from θ_0 and u_0 generate forces $f_1 = k_1\theta$ and $f_2 = k_2 u$, as shown in the figure. Let m be the mass of the aircraft, I the moment of inertia about the center of gravity P, $b\dot{\theta}$ the aerodynamic damping, and h the deviation of the altitude from h_0. Show that the transfer function from u to h is, by neglecting the effect of I,

$$\hat{g}(s) = \frac{\hat{h}(s)}{\hat{u}(s)} = \frac{k_1 k_2 l_2 - k_2 bs}{ms^2(bs + k_1 l_1)}$$

3-21 The soft landing phase of a lunar module descending on the moon can be modeled as shown in Figure P3-21. It is assumed that the thrust generated is proportional to $\dot{m}$, where m is the mass of the module. Then the system can be described by $m\ddot{y} = -k\dot{m} - mg$, where g is the gravity constant on the lunar surface. Define the state variables of the system as $x_1 = y$, $x_2 = \dot{y}$, $x_3 = m$, and $u = \dot{m}$. Find the dynamical-equation description of the system.

3-22 Show that the output of a linear causal relaxed system due to the input $u(t)\delta_1(t - t_0)$ is given by

$$y(t) = u(t_0)g_1(t, t_0) + \int_{t_0}^{t} \dot{u}(\tau)g_1(t, \tau)\,d\tau \qquad \text{for } t \geq t_0$$

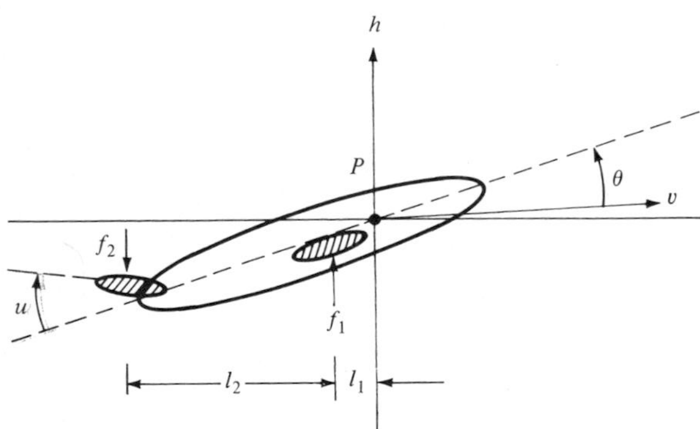

Figure P3-20

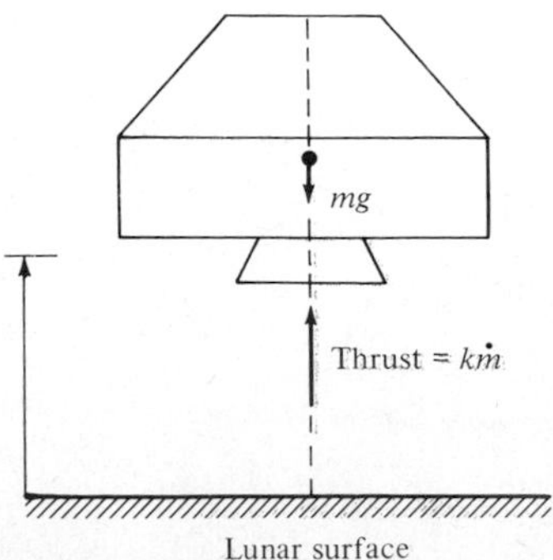

Figure P3-21

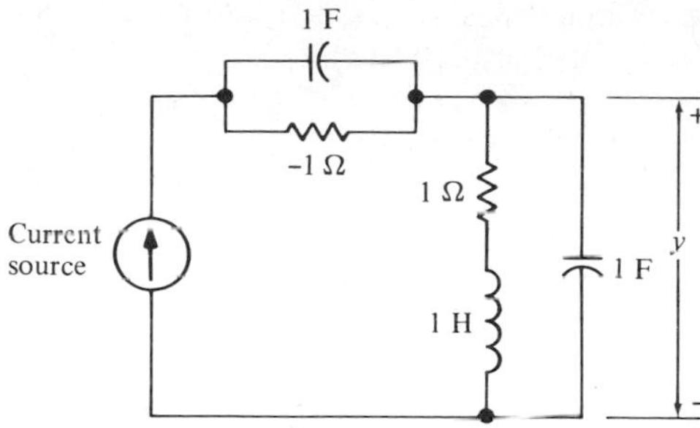

Figure P3-23

where $g_1(t, t_0)$ is the step response and δ_1 is the step function defined in (3-53) and (3-54). [*Hint:* This can be proved either by decomposing u into a sum of step functions or by using (3-55).]

3-23 Find the transfer function and the dynamical-equation description of the network in Figure P3-23. Do you think the transfer function is a good description of this system?

3-24 Show that single-variable, linear, time-invariant, causal, relaxed systems commute in the sense that the order of the tandem connection of two systems is immaterial. Is this true for the time-varying systems?

3-25 The impulse and step responses of a single-variable, linear time-invariant, causal, relaxed system are, by definition, given by $g = H\delta(t)$ and $g_1 = H\delta_1(t)$, where δ is a delta function and δ_1 is a step function. It can be shown that $\delta(t) = \frac{d}{dt}\delta_1(t)$. Verify (3-56) by using the property given in Problem 3-24.

3-26 Verify that the impulse-response matrix of the feedback system in Figure 3-24(c) is given by Equation (3-60).

3-27 Verify the identity

$$\mathbf{I}_p - \mathbf{E}_2(\mathbf{I}_q + \mathbf{E}_1\mathbf{E}_2)^{-1}\mathbf{E}_1 = (\mathbf{I}_p + \mathbf{E}_2\mathbf{E}_1)^{-1}$$

where $\mathbf{E}_1$ is a $q \times p$ matrix and $\mathbf{E}_2$ is a $p \times q$ matrix.

3-28 Verify (3-64) by using the identity in Problem 3-27.

3-29 Show that

$$\det\left(\mathbf{I}_n + \begin{bmatrix} a_1 \\ a_2 \\ \vdots \\ a_n \end{bmatrix} [b_1 \quad b_2 \quad \cdots \quad b_n]\right) = 1 + \sum_{i=1}^{n} a_i b_i$$

Note that the matrix in the left hand side is an $n \times n$ matrix. Let $\mathbf{G}$ be an $n \times n$ matrix of rank 1, show that $\det(\mathbf{I}_n + \mathbf{G}) = 1 + \text{trace } \mathbf{G}$, where trace $\mathbf{G}$ is the sum of all diagonal elements of $\mathbf{G}$. [If $\mathbf{G}$ is not of rank 1, see Equation (9-19) of Chapter 9.]

3-30 Find the transfer-function matrix of the feedback system shown in Figure 3-24(c), where the transfer-function matrices of S_1 and S_2 are, respectively,

$$\hat{\mathbf{G}}_1 = \begin{bmatrix} \frac{1}{s+1} & \frac{1}{s+2} \\ 0 & \frac{s+1}{s+2} \end{bmatrix} \qquad \hat{\mathbf{G}}_2 = \begin{bmatrix} \frac{1}{s+3} & \frac{1}{s+4} \\ \frac{1}{s+1} & 0 \end{bmatrix}$$

3-31 Find the dynamical-equation description of the feedback system in Figure 3-24(c), where S_1 and S_2 are, respectively, described by

$$E^1: \quad \begin{bmatrix} \dot{x}_{11} \\ \dot{x}_{12} \end{bmatrix} = \begin{bmatrix} -2 & 1 \\ 0 & -1 \end{bmatrix} \begin{bmatrix} x_{11} \\ x_{12} \end{bmatrix} + \begin{bmatrix} 4 & 1 \\ -1 & 2 \end{bmatrix} \mathbf{u}_1$$

$$\mathbf{y}_1 = [0 \quad 1]\mathbf{x}_1 + [1 \quad -1]\mathbf{u}_1$$

and

$$E^2: \quad \begin{bmatrix} \dot{x}_{21} \\ \dot{x}_{22} \end{bmatrix} = \begin{bmatrix} 2 \\ 1 \end{bmatrix} \mathbf{u}_2$$

$$\mathbf{y}_2 = \begin{bmatrix} 2 & 0 \\ 1 & -1 \end{bmatrix} \mathbf{x}_2$$

Draw a block diagram of a computer simulation of this feedback system.

3-32 Prove Theorem 3-2 by using det $\mathbf{N}$ = det $\mathbf{P} = 1$ and det $\mathbf{NQ}$ = det $\mathbf{PQ}$.

3-33 Find the overall transfer matrix of the feedback system shown in Figure 3-24(c) with

$$\hat{\mathbf{G}}_1(s) = \begin{bmatrix} \frac{s}{s+1} & \frac{s+1}{s} \\ 0 & \frac{1}{s} \end{bmatrix} \qquad \hat{\mathbf{G}}_2(s) = \begin{bmatrix} -1 & 1 \\ \frac{1}{s+1} & 0 \end{bmatrix}$$

Can you find a combined loop with a net loop gain 1 at $s = \infty$?

3-34 Which of the systems in Figure P3-34 have improper overall transfer functions?

3-35 Can you find a combined loop with a net loop gain 1 in the system shown in Fig. P3-35? What is its overall transfer matrix?

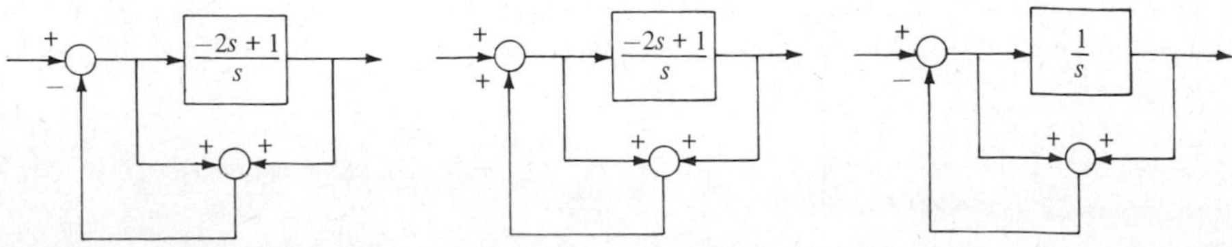

Figure P3-34

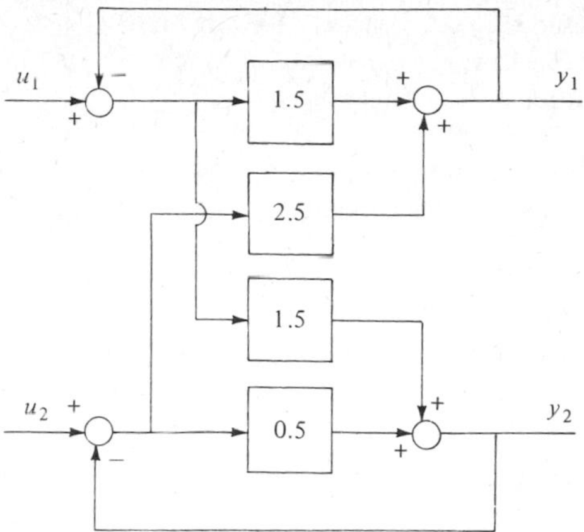

Figure P3-35

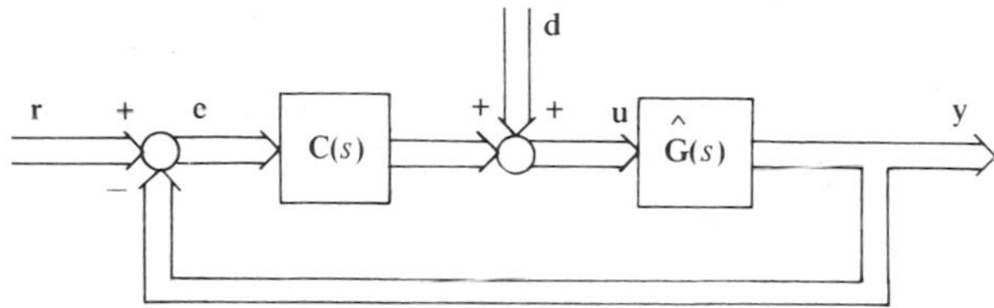

Figure P3-37

3-36 A function $\mathbf{h}(\mathbf{x}(t), \mathbf{u}(t))$ is said to be a linear function of $\mathbf{x}(t)$ and $\mathbf{u}(t)$ if and only if

$$\alpha_1 \mathbf{h}(\mathbf{x}_1(t), \mathbf{u}_1(t)) + \alpha_2 \mathbf{h}(\mathbf{x}_2(t), \mathbf{u}_2(t)) = \mathbf{h}(\alpha_1 \mathbf{x}_1(t) + \alpha_2 \mathbf{x}_2(t), \alpha_1 \mathbf{u}_1(t) + \alpha_2 \mathbf{u}_2(t))$$

for any real numbers α_1, α_2, any $\mathbf{x}_1(t), \mathbf{x}_2(t)$, and any $\mathbf{u}_1(t), \mathbf{u}_2(t)$. Show that $\mathbf{h}(\mathbf{x}(t), \mathbf{u}(t))$ is a linear function of $\mathbf{x}(t)$ and $\mathbf{u}(t)$ if and only if $\mathbf{h}$ is of the form

$$\mathbf{h}(\mathbf{x}(t), \mathbf{u}(t)) = \mathbf{A}(t)\mathbf{x}(t) + \mathbf{B}(t)\mathbf{u}(t)$$

for some $\mathbf{A}$ and $\mathbf{B}$. If $\mathbf{h}, \mathbf{x}$, and $\mathbf{u}$ are square matrix functions instead of vector functions, does the assertion hold? If not, what modification do you need?

3-37 Consider the multivariable feedback system shown in Figure P3-37. Verify the following composite transfer matrix:

$$\begin{bmatrix} \hat{\mathbf{e}}(s) \\ \hat{\mathbf{u}}(s) \end{bmatrix} = \begin{bmatrix} (\mathbf{I} + \hat{\mathbf{G}}(s)\mathbf{C}(s))^{-1} & -\hat{\mathbf{G}}(s)(\mathbf{I} + \mathbf{C}(s)\hat{\mathbf{G}}(s))^{-1} \\ \mathbf{C}(s)(\mathbf{I} + \hat{\mathbf{G}}(s)\mathbf{C}(s))^{-1} & (\mathbf{I} + \mathbf{C}(s)\mathbf{G}(s))^{-1} \end{bmatrix} \begin{bmatrix} \hat{\mathbf{r}}(s) \\ \hat{\mathbf{d}}(s) \end{bmatrix}$$

3-38 The transfer matrix of the system in Figure 3-26 is given in (3-73). Is it possible to have a different form of $\hat{\mathbf{G}}_f(s)$ with $\hat{\mathbf{G}}_1(s)$ or $\hat{\mathbf{G}}_2(s)$ on the right-hand side of the parentheses as in Equation (3-68)?

3-39 Show that the system in Figure 3-28(b) is not well posed by finding an improper transfer function in the system.

3-40 Show that the parallel and tandem connections in Figure 3-24(a) and (b) are well posed if and only if $\hat{\mathbf{G}}_1(s)$ and $\hat{\mathbf{G}}_2(s)$ are proper.

3-41 Let $\hat{\mathbf{G}}(s)=\mathbf{G}_0+\mathbf{G}_1s^{-1}+\mathbf{G}_2s^{-2}+\cdots$ and $\hat{\mathbf{C}}(s)=\mathbf{C}_0+\mathbf{C}_1s^{-1}+\mathbf{C}_2s^{-2}+\cdots$. Show that $(\mathbf{I}+\hat{\mathbf{C}}(s)\hat{\mathbf{G}}(s))^{-1}$ is proper if and only if $(\mathbf{I}+\hat{\mathbf{C}}(\infty)\hat{\mathbf{G}}(\infty))=(\mathbf{I}+\mathbf{C}_0\mathbf{G}_0)$ is nonsingular. Prove it directly by using the power series without using Theorem 3-5.

3-42 A rational matrix $\hat{\mathbf{G}}(s)$ is proper if $\hat{\mathbf{G}}(\infty)$ is a finite constant matrix and improper if $\hat{\mathbf{G}}(\infty)$ is not a finite constant matrix. Are the following statements valid?

1. If $\hat{\mathbf{G}}_1(s)$ and $\hat{\mathbf{G}}_2(s)$ are proper, $\hat{\mathbf{G}}_2(s)\hat{\mathbf{G}}_1(s)$ is proper.
2. If $\hat{\mathbf{G}}_1(s)$ and $\hat{\mathbf{G}}_2(s)$ are improper, $\hat{\mathbf{G}}_2(s)\hat{\mathbf{G}}_1(s)$ is improper.

Answer: Yes; no. Consider $\hat{\mathbf{G}}_2(s)\hat{\mathbf{G}}_1(s)=\hat{\mathbf{G}}_2(s)\mathbf{U}(s)\mathbf{U}^{-1}(s)\hat{\mathbf{G}}_1(s)$. Let $\hat{\mathbf{G}}_i(s)$ be 2×2 proper rational matrices and let

$$\mathbf{U}(s)=\begin{bmatrix} s^n+1 & s^n \\ 1 & 1 \end{bmatrix}$$

3-43 Let $\hat{\mathbf{G}}(s)$ be a $q\times p$ rational matrix, not necessarily proper. Show that the rational matrix

$$\hat{\mathbf{G}}_f(s)=\hat{\mathbf{G}}(s)(\mathbf{I}+\mathbf{K}\hat{\mathbf{G}}(s))^{-1}$$

is proper for almost all $p\times q$ constant matrix $\mathbf{K}$ (or, in mathematical terminology, for a generic $\mathbf{K}$). [*Hint:* Express $\hat{\mathbf{G}}(s)$ as $\hat{\mathbf{G}}(s)=\mathbf{N}(s)\mathbf{D}^{-1}(s)$, where $\mathbf{N}(s)$ and $\mathbf{D}(s)$ are polynomial matrices, and use Theorem G-9. See Reference S196.]

3-44 Let $\mathbf{M}(s)$ be a nonsingular rational matrix decomposed uniquely as

$$\mathbf{M}(s)=\mathbf{M}_p(s)+\mathbf{M}_{sp}(s)$$

where $\mathbf{M}_p(s)$ is a polynomial matrix and $\mathbf{M}_{sp}(s)$ is a strictly proper rational matrix. Show that if $\mathbf{M}_p(s)$ is singular, $\mathbf{M}^{-1}(s)$ is not proper. [*Hint:* Write $\mathbf{M}_{sp}(s)=\mathbf{N}(s)\mathbf{D}^{-1}(s)=(\mathbf{N}(s)\mathbf{U}(s))(\mathbf{D}(s)\mathbf{U}(s))^{-1}$, where $\mathbf{N}(s)$, $\mathbf{D}(s)$, and $\mathbf{U}(s)$ are polynomial matrices, and write

$$\mathbf{M}^{-1}(s)=(\mathbf{D}(s)\mathbf{U}(s))(\mathbf{M}_p(s)\mathbf{D}(s)\mathbf{U}(s)+\mathbf{N}(s)\mathbf{U}(s))^{-1}$$

If $\mathbf{M}_p(s)$ is singular, so is $\mathbf{M}_p(s)\mathbf{D}(s)$. Consequently, there exists a nonsingular $\mathbf{U}(s)$ such that at least one column, say the jth column, of $\mathbf{M}_p(s)\mathbf{D}(s)\mathbf{U}(s)$ is a zero column. Then apply Theorem G-9.]

4 Linear Dynamical Equations and Impulse-Response Matrices

4-1 Introduction

Linear systems can be described, as shown in the previous chapter, by the input-output description and the state-variable description. Once these descriptions are obtained, the next step is naturally to analyze them. There are two types of analyses: qualitative and quantitative. In the qualitative analyses, we are interested in the general properties, such as controllability, observability, and stability, of the equations. These will be discussed in Chapters 5 and 8. In the quantitative analyses, we are interested in the exact responses of equations due to some excitations. Digital computers are now widely available, and they can be used to carry out these analyses (see Section 3-3). Computer solutions however are not in closed forms, and it is difficult to extrapolate any properties from these solutions. In this chapter we shall study the closed-form solutions of the input-output description and the state-variable description. The relationships between these descriptions will also be studied. We note that what will be discussed are applicable to composite systems if their overall mathematical descriptions are first computed.

If the impulse-response matrix $\mathbf{G}(t, \tau)$ of a system is known, then for any input, the output $\mathbf{y}$ can be obtained from

$$\mathbf{y}(t) = \int_{t_0}^{t} \mathbf{G}(t, \tau)\mathbf{u}(\tau)\, d\tau$$

by direct computation or by a graphical method. In the time-invariant case, the equation $\hat{\mathbf{y}}(s) = \hat{\mathbf{G}}(s)\hat{\mathbf{u}}(s)$ can also be used. Unless the transfer-function matrix $\hat{\mathbf{G}}(s)$ is a rational matrix, generally it is easier to compute $\mathbf{y}$ directly in the time domain than from the frequency domain. It is elementary to compute **y** from the input-output description; hence it will not be discussed further.

Solutions of linear dynamical equations are studied in Section 4-2. Solutions are stated in terms of the state transition matrix $\mathbf{\Phi}(t, \tau)$, which is the unique solution of

$$\frac{\partial}{\partial t}\mathbf{\Phi}(t, \tau) = \mathbf{A}(t)\mathbf{\Phi}(t, \tau) \qquad \mathbf{\Phi}(\tau, \tau) = \mathbf{I}$$

In the time-invariant case, we have $\mathbf{\Phi}(t, \tau) = e^{\mathbf{A}(t-\tau)}$. Various methods for the computation of $e^{\mathbf{A}t}$ and $(s\mathbf{I} - \mathbf{A})^{-1}$ are discussed. In Section 4-3 the concept of equivalent dynamical equations is introduced. Equivalent dynamical equations are obtained by changing the basis of the state space. We show that every time-varying linear dynamical equation has an equivalent linear dynamical equation with a constant **A** matrix. We also establish the *theory of Floquet*. In the last section, the relation between linear dynamical equations and impulse-response matrices is studied. The necessary and sufficient condition for an impulse-response matrix to be realizable by a linear dynamical equation is established. We also show that every proper rational matrix has a linear time-invariant dynamical-equation realization.

The references for this chapter are 24, 31, 60, 68, 77, 109, 114, and 116.

4-2 Solutions of a Dynamical Equation

Time-varying case. Consider the n-dimensional linear time-varying dynamical equation

$$E: \quad \dot{\mathbf{x}}(t) = \mathbf{A}(t)\mathbf{x}(t) + \mathbf{B}(t)\mathbf{u}(t) \qquad \text{(state equation)} \qquad \textbf{(4-1a)}$$
$$\mathbf{y}(t) = \mathbf{C}(t)\mathbf{x}(t) + \mathbf{E}(t)\mathbf{u}(t) \qquad \text{(output equation)} \qquad \textbf{(4-1b)}$$

where $\mathbf{A}(\cdot)$, $\mathbf{B}(\cdot)$, $\mathbf{C}(\cdot)$, and $\mathbf{E}(\cdot)$ are $n \times n$, $n \times p$, $q \times n$, and $q \times p$ matrices whose entries are real-valued continuous functions of t defined over $(-\infty, \infty)$. Since $\mathbf{A}(\cdot)$ is assumed to be continuous, for any initial state $\mathbf{x}(t_0)$ and any **u**, there exists a unique solution in the dynamical equation E. This fact will be used frequently in the following development. Before studying the entire dynamical equation E, we study first the solutions of the homogeneous part of E; namely,

$$\dot{\mathbf{x}} = \mathbf{A}(t)\mathbf{x} \qquad \textbf{(4-2)}$$

Solutions of $\dot{\mathbf{x}} = \mathbf{A}(t)\mathbf{x}$

The set of first-order differential equations in (4-2) has a unique solution for every initial state $\mathbf{x}_0$ in $(\mathbb{R}^n, \mathbb{R})$. Since there are infinitely many possible initial states, Equation (4-2) has infinitely many possible solutions. This set of solutions forms a linear space over $\mathbb{R}$. There are only n linearly independent initial states in $(\mathbb{R}^n, \mathbb{R})$; hence the linear space is of dimension n. This fact will be formally established in the following theorem.

Theorem 4-1

The set of all solutions of $\dot{\mathbf{x}}(t)=\mathbf{A}(t)\mathbf{x}(t)$ forms an n-dimensional vector space over the field of real numbers.

Proof

Let $\boldsymbol{\psi}_1$ and $\boldsymbol{\psi}_2$ be two arbitrary solutions of (4-2). Then $\alpha_1\boldsymbol{\psi}_1+\alpha_2\boldsymbol{\psi}_2$ is also a solution of (4-2) for any real α_1 and α_2. We prove this by direct verification:

$$\begin{aligned}\frac{d}{dt}(\alpha_1\boldsymbol{\psi}_1+\alpha_2\boldsymbol{\psi}_2)&=\alpha_1\frac{d}{dt}\boldsymbol{\psi}_1+\alpha_2\frac{d}{dt}\boldsymbol{\psi}_2=\alpha_1\mathbf{A}(t)\boldsymbol{\psi}_1+\alpha_2\mathbf{A}(t)\boldsymbol{\psi}_2\\&=\mathbf{A}(t)(\alpha_1\boldsymbol{\psi}_1+\alpha_2\boldsymbol{\psi}_2)\end{aligned}$$

Hence the set of solutions forms a linear space over $\mathbb{R}$. It is called the *solution space* of (4-2). We next show that the solution space has dimension n. Let $\mathbf{e}_1, \mathbf{e}_2, \ldots, \mathbf{e}_n$ be any linearly independent vectors in $(\mathbb{R}^n, \mathbb{R})$ and $\boldsymbol{\psi}_i$ be the solutions of (4-2) with the initial condition $\boldsymbol{\psi}_i(t_0)=\mathbf{e}_i$, for $i=1, 2, \ldots, n$. If we show that $\boldsymbol{\psi}_i$, for $i=1, 2, \ldots, n$, are linearly independent and that every solution of (4-2) can be written as a linear combination of $\boldsymbol{\psi}_i$, for $i=1, 2, \ldots, n$, then the assertion is proved. We prove by contradiction the fact that the $\boldsymbol{\psi}$'s are linearly independent. Suppose that $\boldsymbol{\psi}_i$, for $i=1, 2, \ldots, n$, are linearly dependent; then, by definition, there exists a nonzero $n \times 1$ real vector $\boldsymbol{\alpha}$ such that

$$[\boldsymbol{\psi}_1 \quad \boldsymbol{\psi}_2 \quad \cdots \quad \boldsymbol{\psi}_n]\boldsymbol{\alpha}=\mathbf{0} \tag{4-3}$$

Note that the $\mathbf{0}$ in the right-hand side of (4-3) is the zero vector of the solution space; therefore, it is more informative to write (4-3) as

$$[\boldsymbol{\psi}_1(t) \quad \boldsymbol{\psi}_2(t) \quad \cdots \quad \boldsymbol{\psi}_n(t)]\boldsymbol{\alpha}=\mathbf{0} \qquad \text{for all } t \text{ in } (-\infty, \infty)$$

In particular, we have

$$[\boldsymbol{\psi}_1(t_0) \quad \boldsymbol{\psi}_2(t_0) \quad \cdots \quad \boldsymbol{\psi}_n(t_0)]\boldsymbol{\alpha}=[\mathbf{e}_1 \quad \mathbf{e}_2 \quad \cdots \quad \mathbf{e}_n]\boldsymbol{\alpha}=\mathbf{0}$$

which implies that $\mathbf{e}_i$, for $i=1, 2, \ldots, n$, are linearly dependent. This contradicts the hypothesis; hence $\boldsymbol{\psi}_i$, for $i=1, 2, \ldots, n$, are linearly independent over $(-\infty, \infty)$.

Let $\boldsymbol{\psi}$ be any solution of (4-2), and let $\boldsymbol{\psi}(t_0)=\mathbf{e}$. Since $\mathbf{e}_1, \mathbf{e}_2, \ldots, \mathbf{e}_n$ are n linearly independent vectors in the n-dimensional vector space $(\mathbb{R}_n, \mathbb{R})$, $\mathbf{e}$ can be written as a unique linear combination of $\mathbf{e}_i$, for $i=1, 2, \ldots, n$—for example, as

$$\mathbf{e}=\sum_{i=1}^{n}\alpha_i\mathbf{e}_i$$

It is clear that

$$\sum_{i=1}^{n}\alpha_i\boldsymbol{\psi}_i(\cdot)$$

is a solution of (4-2) with the initial condition

$$\sum_{i=1}^{n}\alpha_i\boldsymbol{\psi}_i(t_0)=\mathbf{e}$$

Hence, from the uniqueness of the solution, we conclude that

$$\boldsymbol{\psi}(\cdot) = \sum_{i=1}^{n} \alpha_i \boldsymbol{\psi}_i(\cdot)$$

This completes the proof that the solutions of (4-2) form an n-dimensional vector space. Q.E.D.

Definition 4-1

An $n \times n$ matrix function $\boldsymbol{\Psi}$ is said to be a *fundamental matrix* of $\dot{\mathbf{x}} = \mathbf{A}(t)\mathbf{x}$ if and only if the n columns of $\boldsymbol{\Psi}$ consist of n linearly independent solutions of $\dot{\mathbf{x}} = \mathbf{A}(t)\mathbf{x}$. ∎

Example 1

Consider the dynamical equation

$$\dot{\mathbf{x}} = \begin{bmatrix} 0 & 0 \\ t & 0 \end{bmatrix} \mathbf{x}$$

It actually consists of two equations: $\dot{x}_1 = 0$, $\dot{x}_2 = tx_1$. Their solutions are $x_1(t) = x_1(t_0)$ and $x_2(t) = 0.5t^2x_1(t_0) - 0.5t_0^2x_1(t_0) + x_2(t_0)$, which are obtained by first solving for $x_1(t)$ and then substituting $x_1(t)$ into $\dot{x}_2 = tx_1$. Now two linearly independent solutions $\boldsymbol{\psi}_1 = [0 \quad 1]'$ and $\boldsymbol{\psi}_2 = [2 \quad t^2]'$ can be easily obtained by setting $t_0 = 0$, $x_1(t_0) = 0$, $x_2(t_0) = 1$ and $x_1(t_0) = 2$, $x_2(t_0) = 0$. Hence, the matrix

$$\begin{bmatrix} 0 & 2 \\ 1 & t^2 \end{bmatrix}$$

is a fundamental matrix. ∎

Each column of $\boldsymbol{\Psi}$, by definition, satisfies the differential equation $\dot{\mathbf{x}} = \mathbf{A}(t)\mathbf{x}$; hence, it is evident that $\boldsymbol{\Psi}$ satisfies the matrix equation

$$\dot{\boldsymbol{\Psi}} = \mathbf{A}(t)\boldsymbol{\Psi} \tag{4-4}$$

with $\boldsymbol{\Psi}(t_0) = \mathbf{H}$, where $\mathbf{H}$ is some nonsingular real constant matrix. Conversely, if a matrix $\mathbf{M}$ satisfies (4-4) and if $\mathbf{M}(t)$ is nonsingular for some t, then from the proof of Theorem 4-1 we know that all the columns of $\mathbf{M}$ are linearly independent. Hence, the matrix function $\mathbf{M}$ qualifies as a fundamental matrix. Thus we conclude that *a matrix function* $\boldsymbol{\Psi}$ *is a fundamental matrix of* $\dot{\mathbf{x}} = \mathbf{A}(t)\mathbf{x}$ *if and only if* $\boldsymbol{\Psi}$ *satisfies* (4-4) *and* $\boldsymbol{\Psi}(t)$ *is nonsingular for some* t.

An important property of a fundamental matrix $\boldsymbol{\Psi}(\cdot)$ is that the inverse of $\boldsymbol{\Psi}(t)$ exists for each t in $(-\infty, \infty)$. This follows from the following theorem.

Theorem 4-2

Every fundamental matrix $\boldsymbol{\Psi}$ is nonsingular for all t in $(-\infty, \infty)$.

Proof

Before we prove the theorem, we need the following fact: If $\boldsymbol{\psi}(\cdot)$ is a solution of $\dot{\mathbf{x}} = \mathbf{A}(t)\mathbf{x}$ and if $\boldsymbol{\psi}(t_0) = \mathbf{0}$ for some t_0, then the solution $\boldsymbol{\psi}(\cdot)$ is identically zero; that is, $\boldsymbol{\psi}(\cdot) \equiv \mathbf{0}$. It is obvious that $\boldsymbol{\psi}(\cdot) \equiv \mathbf{0}$ is a solution of $\dot{\mathbf{x}} = \mathbf{A}(t)\mathbf{x}$ with $\boldsymbol{\psi}(t_0) = \mathbf{0}$. Again, from the uniqueness of the solution, we conclude that $\boldsymbol{\psi}(\cdot) \equiv \mathbf{0}$ is the only solution with $\boldsymbol{\psi}(t_0) = \mathbf{0}$.

We shall now prove the theorem; we prove it by contradiction. Suppose that $\det \boldsymbol{\Psi}(t_0) = \det [\boldsymbol{\psi}_1(t_0) \quad \boldsymbol{\psi}_2(t_0) \quad \cdots \quad \boldsymbol{\psi}_n(t_0)] = \mathbf{0}$ for some t_0. Then the set of n constant column vectors $\boldsymbol{\psi}_1(t_0), \boldsymbol{\psi}_2(t_0), \ldots, \boldsymbol{\psi}_n(t_0)$ is linearly dependent in $(\mathbb{R}^n, \mathbb{R})$. It follows that there exist real α_i, for $i = 1, 2, \ldots, n$, not all zero, such that

$$\sum_{i=1}^{n} \alpha_i \boldsymbol{\psi}_i(t_0) = \mathbf{0}$$

which, together with the fact that

$$\sum_{i=1}^{n} \alpha_i \boldsymbol{\psi}_i(\cdot)$$

is a solution of $\dot{\mathbf{x}} = \mathbf{A}(t)\mathbf{x}$, implies that

$$\sum_{i=1}^{n} \alpha_i \boldsymbol{\psi}_i(\cdot) \equiv \mathbf{0}$$

This contradicts the assumption that $\boldsymbol{\psi}_i(\cdot)$, for $i = 1, 2, \ldots, n$, are linearly independent. Hence, we conclude that $\det \boldsymbol{\Psi}(t) \neq 0$ for all t in $(-\infty, \infty)$. Q.E.D.

Definition 4-2

Let $\boldsymbol{\Psi}(\cdot)$ be *any* fundamental matrix of $\dot{\mathbf{x}} = \mathbf{A}(t)\mathbf{x}$. Then

$$\boldsymbol{\Phi}(t, t_0) \triangleq \boldsymbol{\Psi}(t)\boldsymbol{\Psi}^{-1}(t_0) \qquad \text{for all } t, t_0 \text{ in } (-\infty, \infty)$$

is said to be *the state transition matrix* of $\dot{\mathbf{x}} = \mathbf{A}(t)\mathbf{x}$. ∎

The physical meaning of $\boldsymbol{\Phi}(t, t_0)$ will be seen later. Since $\boldsymbol{\Psi}(t)$ is nonsingular for all t, its inverse is well defined for each t. From the definition we have immediately the following very important properties of the state transition matrix:

$$\boldsymbol{\Phi}(t, t) = \mathbf{I} \tag{4-5}$$

$$\boldsymbol{\Phi}^{-1}(t, t_0) = \boldsymbol{\Psi}(t_0)\boldsymbol{\Psi}^{-1}(t) = \boldsymbol{\Phi}(t_0, t) \tag{4-6}$$

$$\boldsymbol{\Phi}(t_2, t_0) = \boldsymbol{\Phi}(t_2, t_1)\boldsymbol{\Phi}(t_1, t_0) \tag{4-7}$$

for any t, t_0, t_1, and t_2 in $(-\infty, \infty)$.

Note that $\boldsymbol{\Phi}(t, t_0)$ is uniquely determined by $\mathbf{A}(t)$ and is independent of the particular $\boldsymbol{\Psi}$ chosen. Let $\boldsymbol{\Psi}_1$ and $\boldsymbol{\Psi}_2$ be two different fundamental matrices of $\dot{\mathbf{x}} = \mathbf{A}(t)\mathbf{x}$. Since the columns of $\boldsymbol{\Psi}_1$, as well as the columns of $\boldsymbol{\Psi}_2$, qualify as basis vectors, there exists, as shown in Section 2-3, a nonsingular real constant matrix $\mathbf{P}$ such that $\boldsymbol{\Psi}_2 = \boldsymbol{\Psi}_1\mathbf{P}$. In fact, the ith column of $\mathbf{P}$ is the representation of the ith column of $\boldsymbol{\Psi}_2$ with respect to the basis that consists of the columns of

$\mathbf{\Psi}_1(\cdot)$. By definition, we have

$$\begin{aligned}\mathbf{\Phi}(t, t_0) &= \mathbf{\Psi}_2(t)(\mathbf{\Psi}_2(t_0))^{-1} = \mathbf{\Psi}_1(t)\mathbf{P}\mathbf{P}^{-1}(\mathbf{\Psi}_1(t_0))^{-1} \\ &= \mathbf{\Psi}_1(t)(\mathbf{\Psi}_1(t_0))^{-1}\end{aligned}$$

which shows the uniqueness of $\mathbf{\Phi}(t, t_0)$. From Equation (4-4), it is evident that $\mathbf{\Phi}(t, t_0)$ is the unique solution of the matrix equation

$$\frac{\partial}{\partial t}\mathbf{\Phi}(t, t_0) = \mathbf{A}(t)\mathbf{\Phi}(t, t_0) \tag{4-8}$$

with the initial condition $\mathbf{\Phi}(t_0, t_0) = \mathbf{I}$.

Remarks are in order concerning the solutions of (4-4) and (4-8). If $\mathbf{A}(t)$ is a continuous function of t, then $\mathbf{\Phi}(t, t_0)$ and $\mathbf{\Psi}(t)$ are continuously differentiable[1] in t. More generally, if $\mathbf{A}(t)$ is n times continuously differentiable in t, then $\mathbf{\Phi}(t, t_0)$ and $\mathbf{\Psi}(t)$ are $n+1$ times continuously differential in t; see References 24 and 77.

The computation of the solution of (4-8) in a closed form is generally very difficult, if not impossible, except for some special cases. If $\mathbf{A}(t)$ is a triangular matrix, then its solution can be reduced to solving a set of scalar differential equations, and its closed-form solution can be readily obtained (Problem 4-1). If $\mathbf{A}(t)$ has the following commutative property

$$\mathbf{A}(t)\left(\int_{t_0}^{t} \mathbf{A}(\tau)\, d\tau\right) = \left(\int_{t_0}^{t} \mathbf{A}(\tau)\, d\tau\right)\mathbf{A}(t)$$

for all t and t_0, then the unique solution of (4-8) is given by

$$\mathbf{\Phi}(t, t_0) = \exp\left[\int_{t_0}^{t} \mathbf{A}(\tau)\, d\tau\right] \tag{4-9}$$

(Problem 4-31). If $\mathbf{A}(t)$ is a diagonal matrix or a constant matrix, then it meets the commutative property, and its transition matrix is given by (4-9). For other special cases, see Problems 4-14 and 4-15. See also References S229 to S231.

From the concept of state transition matrix, the solution of $\dot{\mathbf{x}} = \mathbf{A}(t)\mathbf{x}$ follows immediately. To be more informative, we use $\boldsymbol{\phi}(t;\, t_0, \mathbf{x}_0, \mathbf{0})$ to denote the solution of $\dot{\mathbf{x}} = \mathbf{A}(t)\mathbf{x}$ at time t due to the initial condition $\mathbf{x}(t_0) = \mathbf{x}_0$. The fourth argument of $\boldsymbol{\phi}$ denotes the fact that $\mathbf{u} \equiv \mathbf{0}$. The solution of $\dot{\mathbf{x}} = \mathbf{A}(t)\mathbf{x}$ with $\mathbf{x}(t_0) = \mathbf{x}_0$ is given by

$$\mathbf{x}(t) \triangleq \boldsymbol{\phi}(t;\, t_0, \mathbf{x}_0, \mathbf{0}) = \mathbf{\Phi}(t, t_0)\mathbf{x}_0$$

which can be verified by direct substitution. The physical meaning of the state transition matrix $\mathbf{\Phi}(t, t_0)$ is now clear. It governs the motion of the state vector in the time interval in which the input is identically zero. $\mathbf{\Phi}(t, t_0)$ is a linear transformation that maps the state $\mathbf{x}_0$ at t_0 into the state $\mathbf{x}$ at time t.

[1] A function is said to be *continuously differentiable* if its first derivative exists and is continuous.

Solutions of the dynamical equation *E*

We use $\boldsymbol{\phi}(t; t_0, \mathbf{x}_0, \mathbf{u})$ to denote the state resulted at time t due to the initial state $\mathbf{x}(t_0) = \mathbf{x}_0$ and the application of the input $\mathbf{u}$.

Theorem 4-3

The solution of the state equation

$$\dot{\mathbf{x}} = \mathbf{A}(t)\mathbf{x} + \mathbf{B}(t)\mathbf{u} \qquad \mathbf{x}(t_0) = \mathbf{x}_0$$

is given by

$$\mathbf{x}(t) \triangleq \boldsymbol{\phi}(t; t_0, \mathbf{x}_0, \mathbf{u}) = \boldsymbol{\Phi}(t, t_0)\mathbf{x}_0 + \int_{t_0}^{t} \boldsymbol{\Phi}(t, \tau)\mathbf{B}(\tau)\mathbf{u}(\tau)\, d\tau \tag{4-10}$$

$$= \boldsymbol{\Phi}(t, t_0)\left[\mathbf{x}_0 + \int_{t_0}^{t} \boldsymbol{\Phi}(t_0, \tau)\mathbf{B}(\tau)\mathbf{u}(\tau)\, d\tau\right] \tag{4-11}$$

where $\boldsymbol{\Phi}(t, \tau)$ is the state transition matrix of $\dot{\mathbf{x}} = \mathbf{A}(t)\mathbf{x}$; or, equivalently, the unique solution of

$$\frac{\partial}{\partial t}\boldsymbol{\Phi}(t, \tau) = \mathbf{A}(t)\boldsymbol{\Phi}(t, \tau) \qquad \boldsymbol{\Phi}(\tau, \tau) = \mathbf{I}$$ ∎

Proof

Equation (4-11) is obtained from (4-10) by using $\boldsymbol{\Phi}(t, \tau) = \boldsymbol{\Phi}(t, t_0)\boldsymbol{\Phi}(t_0, \tau)$. We first show that (4-10) satisfies the state equation by direct substitution[2]:

$$\begin{aligned}
\frac{d}{dt}\mathbf{x}(t) &= \frac{\partial}{\partial t}\boldsymbol{\Phi}(t, t_0)\mathbf{x}_0 + \frac{\partial}{\partial t}\int_{t_0}^{t} \boldsymbol{\Phi}(t, \tau)\mathbf{B}(\tau)\mathbf{u}(\tau)\, d\tau \\
&= \mathbf{A}(t)\boldsymbol{\Phi}(t, t_0)\mathbf{x}_0 + \boldsymbol{\Phi}(t, t)\mathbf{B}(t)\mathbf{u}(t) + \int_{t_0}^{t} \frac{\partial}{\partial t}\boldsymbol{\Phi}(t, \tau)\mathbf{B}(\tau)\mathbf{u}(\tau)\, d\tau \\
&= \mathbf{A}(t)\left[\boldsymbol{\Phi}(t, t_0)\mathbf{x}_0 + \int_{t_0}^{t} \boldsymbol{\Phi}(t, \tau)\mathbf{B}(\tau)\mathbf{u}(\tau)\, d\tau\right] + \mathbf{B}(t)\mathbf{u}(t) \\
&= \mathbf{A}(t)\mathbf{x}(t) + \mathbf{B}(t)\mathbf{u}(t)
\end{aligned}$$

At $t = t_0$, we have

$$\mathbf{x}(t_0) = \boldsymbol{\Phi}(t_0, t_0)\mathbf{x}_0 + \int_{t_0}^{t_0} \boldsymbol{\Phi}(t_0, \tau)\mathbf{B}(\tau)\mathbf{u}(\tau)\, d\tau = \mathbf{I}\mathbf{x}_0 + \mathbf{0} = \mathbf{x}_0$$

In other words, (4-10) also meets the initial condition. Hence it is the solution. Q.E.D.

We consider again the solution given by Equation (4-10). If $\mathbf{u} \equiv \mathbf{0}$, then (4-10) reduces to

$$\boldsymbol{\phi}(t; t_0, \mathbf{x}_0, \mathbf{0}) = \boldsymbol{\Phi}(t, t_0)\mathbf{x}_0 \tag{4-12}$$

[2] $\dfrac{\partial}{\partial t}\displaystyle\int_{t_0}^{t} f(t, \tau)\, d\tau = f(t, \tau)\Big|_{\tau = t} + \int_{t_0}^{t} \frac{\partial}{\partial t} f(t, \tau)\, d\tau$

If $\mathbf{x}_0 = \mathbf{0}$, Equation (4-10) reduces to

$$\boldsymbol{\phi}(t; t_0, \mathbf{0}, \mathbf{u}) = \int_{t_0}^{t} \boldsymbol{\Phi}(t, \tau)\mathbf{B}(\tau)\mathbf{u}(\tau)\, d\tau \tag{4-13}$$

For obvious reasons, $\boldsymbol{\phi}(t; t_0, \mathbf{x}_0, \mathbf{0})$ is called the *zero-input response*, and $\boldsymbol{\phi}(t; t_0, \mathbf{0}, \mathbf{u})$ is called the *zero-state response* of the state equation. It is clear that $\boldsymbol{\phi}(t; t_0, \mathbf{x}_0, \mathbf{0})$ and $\boldsymbol{\phi}(t; t_0, \mathbf{0}, \mathbf{u})$ are linear functions of $\mathbf{x}_0$ and $\mathbf{u}$, respectively. Using (4-12) and (4-13), the solution given by Equation (4-10) can be written as

$$\boldsymbol{\phi}(t; t_0, \mathbf{x}_0, \mathbf{u}) = \boldsymbol{\phi}(t; t_0, \mathbf{x}_0, \mathbf{0}) + \boldsymbol{\phi}(t; t_0, \mathbf{0}, \mathbf{u}) \tag{4-14}$$

This is a very important property; it says that *the response of a linear state equation can always be decomposed into the zero-state response and the zero-input response.* This is consistent with Equation (3-31).

Note that Equation (4-13) can be derived directly from the fact that it is a linear function of $\mathbf{u}$. The procedure is exactly the same as the one in deriving

$$\int_{t_0}^{t} \mathbf{G}(t, \tau)\mathbf{u}(\tau)\, d\tau$$

in Section 3-2. The response $\boldsymbol{\phi}(t; t_0, \mathbf{0}, \mathbf{u})$ is, by definition, the solution of $\dot{\mathbf{x}} = \mathbf{A}(t)\mathbf{x} + \mathbf{B}(t)\mathbf{u}$ with $\mathbf{0}$ as the initial state. If we cut the input $\mathbf{u}$ into small pulses, say

$$\mathbf{u} = \sum_i \mathbf{u}_{[t_i, t_i + \Delta)}$$

then we have

$$\boldsymbol{\phi}(t; t_0, \mathbf{0}, \mathbf{u}_{[t_i, t_i+\Delta)}) \doteqdot \boldsymbol{\Phi}(t, t_i + \Delta)\mathbf{B}(t_i)\mathbf{u}(t_i)\Delta \tag{4-15}$$

where we have used the fact that if Δ is very small, the solution of $\dot{\mathbf{x}} = \mathbf{A}(t)\mathbf{x} + \mathbf{B}(t)\mathbf{u}$ due to the input $\mathbf{u}_{[t_i, t_i+\Delta)}$ with $\mathbf{0}$ as the initial state is approximately equal to $\mathbf{B}(t_i)\mathbf{u}(t_i)\Delta$. The input $\mathbf{u}_{[t_i, t_i+\Delta)}$ outside the time interval $[t_i, t_i + \Delta)$ is identically zero; hence, the response between $t_i + \Delta$ and t is governed by $\boldsymbol{\Phi}(t, t_i + \Delta)$. Summing up (4-15) for all i and taking the limit $\Delta \to 0$, we immediately obtain the equation

$$\boldsymbol{\phi}(t; t_0, \mathbf{0}, \mathbf{u}) = \int_{t_0}^{t} \boldsymbol{\Phi}(t, \tau)\mathbf{B}(\tau)\mathbf{u}(\tau)\, d\tau$$

We give now the solution of the entire dynamical equation E.

Corollary 4-3

The solution of the dynamical equation E in (4-1) is given by

$$\begin{aligned}\mathbf{y}(t) &= \mathbf{C}(t)\boldsymbol{\Phi}(t, t_0)\mathbf{x}_0 + \mathbf{C}(t)\int_{t_0}^{t} \boldsymbol{\Phi}(t, \tau)\mathbf{B}(\tau)\mathbf{u}(\tau)\, d\tau + \mathbf{E}(t)\mathbf{u}(t) \\ &= \mathbf{C}(t)\boldsymbol{\Phi}(t, t_0)\left[\mathbf{x}_0 + \int_{t_0}^{t} \boldsymbol{\Phi}(t_0, \tau)\mathbf{B}(\tau)\mathbf{u}(\tau)\, d\tau\right] + \mathbf{E}(t)\mathbf{u}(t)\end{aligned} \tag{4-16}$$

∎

By substituting (4-10) and (4-11) into (4-1b), we immediately obtain Corollary 4-3. The output $\mathbf{y}$ can also be decomposed into the zero-state response and the zero-input response. If the dynamical equation is initially in the zero state, Equation (4-16) becomes

$$\mathbf{y}(t) = \int_{t_0}^{t} [\mathbf{C}(t)\mathbf{\Phi}(t, \tau)\mathbf{B}(\tau) + \mathbf{E}(t)\delta(t-\tau)]\mathbf{u}(\tau)\, d\tau$$

$$\triangleq \int_{t_0}^{t} \mathbf{G}(t, \tau)\mathbf{u}(\tau)\, d\tau \tag{4-17}$$

The matrix function

$$\mathbf{G}(t, \tau) \triangleq \mathbf{C}(t)\mathbf{\Phi}(t, \tau)\mathbf{B}(\tau) + \mathbf{E}(t)\delta(t-\tau) \tag{4-18}$$

is called the *impulse-response matrix* of the dynamical equation E. It governs the input-output relation of E if E is initially in the zero state.

We see from (4-16) that if the state transition matrix $\mathbf{\Phi}(t, \tau)$ is known, then the solution of the dynamical equation can be computed. Recall that $\mathbf{\Phi}(t, \tau)$ is the unique solution of

$$\frac{\partial}{\partial t}\mathbf{\Phi}(t, \tau) = \mathbf{A}(t)\mathbf{\Phi}(t, \tau), \quad \mathbf{\Phi}(\tau, \tau) = \mathbf{I}$$

Unfortunately, there is in general no simple relation between $\mathbf{\Phi}(t, \tau)$ and $\mathbf{A}(t)$, and except for very simple cases, state transition matrices $\mathbf{\Phi}(t, \tau)$ cannot be easily found. Therefore, Equations (4-10) and (4-16) are used mainly in the theoretical study of linear system theory. If we are required to find the solution of a dynamical equation due to a given $\mathbf{x}_0$ and $\mathbf{u}$, it is unnecessary to compute $\mathbf{\Phi}(t, \tau)$. The solution can be easily computed by using existing subroutines on a digital computer.

Time-invariant case. In this subsection, we study the linear time-invariant (fixed) dynamical equation

$$FE: \quad \dot{\mathbf{x}} = \mathbf{A}\mathbf{x} + \mathbf{B}\mathbf{u} \tag{4-19a}$$
$$\mathbf{y} = \mathbf{C}\mathbf{x} + \mathbf{E}\mathbf{u} \tag{4-19b}$$

where $\mathbf{A}$, $\mathbf{B}$, $\mathbf{C}$, and $\mathbf{E}$ are $n \times n$, $n \times p$, $q \times n$, and $q \times p$ real constant matrices, respectively. Since the equation FE is a special case of the linear time-varying dynamical equation E, all the results derived in the preceding subsection can be applied here. We have shown in (2-82) that

$$\frac{d}{dt}e^{\mathbf{A}t} = \mathbf{A}e^{\mathbf{A}t}$$

and $e^{\mathbf{A}t}$ is nonsingular at $t=0$; hence $e^{\mathbf{A}t}$ is a fundamental matrix of $\dot{\mathbf{x}} = \mathbf{A}\mathbf{x}$. In fact, $e^{\mathbf{A}t}$ is nonsingular for all t and $(e^{\mathbf{A}t})^{-1} = e^{-\mathbf{A}t}$; therefore, the state transition matrix of $\dot{\mathbf{x}} = \mathbf{A}\mathbf{x}$ is, by the use of (2-81) and (2-79),

$$\mathbf{\Phi}(t, t_0) = e^{\mathbf{A}t}(e^{\mathbf{A}t_0})^{-1} = e^{\mathbf{A}(t-t_0)} = \mathbf{\Phi}(t-t_0)$$

It follows from (4-10) that the solution of (4-19a) is

$$\boldsymbol{\phi}(t; t_0, \mathbf{x}_0, \mathbf{u}) = e^{\mathbf{A}(t-t_0)}\mathbf{x}_0 + \int_{t_0}^{t} e^{\mathbf{A}(t-\tau)}\mathbf{B}\mathbf{u}(\tau)\, d\tau \tag{4-20}$$

If $t_0 = 0$, as is usually assumed in the time-invariant equation, then we have the following theorem.

Theorem 4-4

The solution of the linear time-invariant dynamical equation *FE* given in (4-19) is

$$\mathbf{x}(t) \triangleq \boldsymbol{\phi}(t; 0, \mathbf{x}_0, \mathbf{u}) = e^{\mathbf{A}t}\mathbf{x}_0 + \int_0^t e^{\mathbf{A}(t-\tau)}\mathbf{B}\mathbf{u}(\tau)\, d\tau \tag{4-21}$$

and

$$\mathbf{y}(t) = \mathbf{C}e^{\mathbf{A}t}\mathbf{x}_0 + \mathbf{C}e^{\mathbf{A}t}\int_0^t e^{-\mathbf{A}\tau}\mathbf{B}\mathbf{u}(\tau)\, d\tau + \mathbf{E}\mathbf{u}(t) \tag{4-22}$$

■

The impulse response matrix of *FE* is

$$\mathbf{G}(t, \tau) = \mathbf{G}(t-\tau) = \mathbf{C}e^{\mathbf{A}(t-\tau)}\mathbf{B} + \mathbf{E}\delta(t-\tau)$$

or, as more commonly written,

$$\mathbf{G}(t) = \mathbf{C}e^{\mathbf{A}t}\mathbf{B} + \mathbf{E}\delta(t) \tag{4-23}$$

The solution of a linear time-invariant dynamical equation can also be computed in the frequency domain. Taking the Laplace transform of (4-21) and (4-22), and using $\mathscr{L}[e^{\mathbf{A}t}] = (s\mathbf{I} - \mathbf{A})^{-1}$ [see Equation (2-86)], we obtain

$$\hat{\mathbf{x}}(s) = (s\mathbf{I} - \mathbf{A})^{-1}\mathbf{x}(0) + (s\mathbf{I} - \mathbf{A})^{-1}\mathbf{B}\hat{\mathbf{u}}(s) \tag{4-24}$$

and

$$\hat{\mathbf{y}}(s) = \mathbf{C}(s\mathbf{I} - \mathbf{A})^{-1}\mathbf{x}(0) + \mathbf{C}(s\mathbf{I} - \mathbf{A})^{-1}\mathbf{B}\hat{\mathbf{u}}(s) + \mathbf{E}\hat{\mathbf{u}}(s) \tag{4-25}$$

where the circumflex denotes the Laplace transform of a variable. These equations have been derived in Section 3-3 directly from the dynamical equation. As is defined there, the rational-function matrix

$$\hat{\mathbf{G}}(s) = \mathbf{C}(s\mathbf{I} - \mathbf{A})^{-1}\mathbf{B} + \mathbf{E} \tag{4-26}$$

is called the *transfer-function matrix* of the dynamical equation *FE*. It is the Laplace transform of the impulse-response matrix given in (4-23). The transfer-function matrix governs the zero-state response of the equation *FE*.

We give now some remarks concerning the computation of $e^{\mathbf{A}t}$.[3] We introduced in Section 2-7 three methods of computing functions of a matrix.

---

[3]See also Reference S159.

We may apply them to compute $e^{\mathbf{A}t}$:

1. Using Definition 2-16: First, compute the eigenvalues of $\mathbf{A}$; next, find a polynomial $g(\lambda)$ of degree $n-1$ that is equal to $e^{\lambda t}$ on the spectrum of $\mathbf{A}$; then $e^{\mathbf{A}t}=g(\mathbf{A})$.
2. Using the Jordan canonical form of $\mathbf{A}$: Let $\mathbf{A}=\mathbf{Q}\hat{\mathbf{A}}\mathbf{Q}^{-1}$; then $e^{\mathbf{A}t}=\mathbf{Q}e^{\hat{\mathbf{A}}t}\mathbf{Q}^{-1}$, where $\hat{\mathbf{A}}$ is of the Jordan form. $e^{\hat{\mathbf{A}}t}$ can be obtained immediately by the use of (2-70).
3. Using the infinite series $e^{\mathbf{A}t}=\sum_{k=0}^{\infty} t^k\mathbf{A}^k/k!$: This series will not, generally, give a closed-form solution and is mainly used on digital computer computation.

We introduce one more method of computing $e^{\mathbf{A}t}$. Since $\mathscr{L}[e^{\mathbf{A}t}]=(s\mathbf{I}-\mathbf{A})^{-1}$, we have

$$e^{\mathbf{A}t}=\mathscr{L}^{-1}(s\mathbf{I}-\mathbf{A})^{-1} \tag{4-27}$$

Hence, to compute $e^{\mathbf{A}t}$, we first invert the matrix $(s\mathbf{I}-\mathbf{A})$ and then take the inverse Laplace transform of each element of $(s\mathbf{I}-\mathbf{A})^{-1}$. Computing the inverse of a matrix is generally not an easy job. However, if a matrix is triangular[4] or of order less than 4, its inverse can be easily computed. Note that the inverse of a triangular matrix is again a triangular matrix.

Note that $(s\mathbf{I}-\mathbf{A})^{-1}$ is a function of the matrix $\mathbf{A}$; therefore, again we have many methods to compute it:

1. Taking the inverse of $(s\mathbf{I}-\mathbf{A})$.
2. Using Definition 2-16.
3. Using $(s\mathbf{I}-\mathbf{A})^{-1}=\mathbf{Q}(s\mathbf{I}-\hat{\mathbf{A}})^{-1}\mathbf{Q}^{-1}$ and (2-74).
4. Using Definition 2-17.
5. Taking the Laplace transform of $e^{\mathbf{A}t}$.

In addition, there is an iterative scheme to compute $(s\mathbf{I}-\mathbf{A})^{-1}$ (Problem 2-39).

Example 2

We use methods 1 and 2 to compute $(s\mathbf{I}-\mathbf{A})^{-1}$, where

$$\mathbf{A}=\begin{bmatrix} 0 & -1 \\ 1 & -2 \end{bmatrix}$$

1. $$(s\mathbf{I}-\mathbf{A})^{-1}=\begin{bmatrix} s & 1 \\ -1 & s+2 \end{bmatrix}^{-1}=\frac{1}{s^2+2s+1}\begin{bmatrix} s+2 & -1 \\ 1 & s \end{bmatrix}$$

$$=\begin{bmatrix} \dfrac{s+2}{(s+1)^2} & \dfrac{-1}{(s+1)^2} \\ \dfrac{1}{(s+1)^2} & \dfrac{s}{(s+1)^2} \end{bmatrix}$$

[4] A square matrix is said to be *triangular* if all the elements below or above the main diagonal are zero.

2. The eigenvalues of **A** are -1, -1. Let $g(\lambda)=\alpha_0+\alpha_1\lambda$. If $f(\lambda)\triangleq(s-\lambda)^{-1}=g(\lambda)$ on the spectrum of **A**, then

$$\begin{aligned} f(-1)=g(-1): &\quad (s+1)^{-1}=\alpha_0-\alpha_1 \\ f'(-1)=g'(-1): &\quad (s+1)^{-2}=\alpha_1 \end{aligned}$$

Hence $\qquad g(\lambda)=[(s+1)^{-1}+(s+1)^{-2}]+(s+1)^{-2}\lambda$

and $\quad (s\mathbf{I}-\mathbf{A})^{-1}=g(\mathbf{A})=[(s+1)^{-1}+(s+1)^{-2}]\mathbf{I}+(s+1)^{-2}\mathbf{A}$

$$=\begin{bmatrix} \dfrac{s+2}{(s+1)^2} & \dfrac{-1}{(s+1)^2} \\ \dfrac{1}{(s+1)^2} & \dfrac{s}{(s+1)^2} \end{bmatrix}$$

Example 3

Consider the state equation

$$\begin{bmatrix} \dot{x}_1 \\ \dot{x}_2 \end{bmatrix}=\begin{bmatrix} 0 & -1 \\ 1 & -2 \end{bmatrix}\begin{bmatrix} x_1 \\ x_2 \end{bmatrix}+\begin{bmatrix} 0 \\ 1 \end{bmatrix}u$$

The solution is given by

$$\mathbf{x}(t)=e^{\mathbf{A}t}\mathbf{x}(0)+\int_0^t e^{\mathbf{A}(t-\tau)}\mathbf{B}u(\tau)\,d\tau \tag{4-28}$$

The matrix function $e^{\mathbf{A}t}$ can be obtained by taking the inverse Laplace transform of $(s\mathbf{I}-\mathbf{A})^{-1}$, which is computed in Example 2. Hence

$$e^{\mathbf{A}t}=\mathscr{L}^{-1}\begin{bmatrix} \dfrac{s+2}{(s+1)^2} & \dfrac{-1}{(s+1)^2} \\ \dfrac{1}{(s+1)^2} & \dfrac{s}{(s+1)^2} \end{bmatrix}=\begin{bmatrix} (1+t)e^{-t} & -te^{-t} \\ te^{-t} & (1-t)e^{-t} \end{bmatrix}$$

and

$$\mathbf{x}(t)=\begin{bmatrix} (1+t)e^{-t} & -te^{-t} \\ te^{-t} & (1-t)e^{-t} \end{bmatrix}\mathbf{x}(0)+\int_0^t \begin{bmatrix} -(t-\tau)e^{-(t-\tau)} \\ [1-(t-\tau)]e^{-(t-\tau)} \end{bmatrix}u(\tau)\,d\tau$$

The solutions with $u\equiv 0$ are plotted in Figure 4-1 for the initial states $\mathbf{x}(0)=[8 \quad 8]'$ and $\mathbf{x}(0)=[8 \quad 5]'$. Note that the velocity at each point of the trajectories in Figure 4-1 is equal to $\mathbf{Ax}$. ∎

If the matrix **A** has m distinct eigenvalues λ_i with index $\bar{n}_i$, for $i=1, 2, \ldots, m$ (see Definition 2-15),[5] we claim that every element of $e^{\mathbf{A}t}$ is a linear combination of the factors $t^k e^{\lambda_i t}$, for $k=0, 1, \ldots, \bar{n}_i-1$; $i=1, 2, \ldots, m$. Let $\hat{\mathbf{A}}$ be a Jordan-form representation of **A** and $\mathbf{A}=\mathbf{Q}\hat{\mathbf{A}}\mathbf{Q}^{-1}$. Then $e^{\mathbf{A}t}=\mathbf{Q}e^{\hat{\mathbf{A}}t}\mathbf{Q}^{-1}$. From (2-69) we

[5] If all eigenvalues are distinct, then the index of every eigenvalue is 1. If an eigenvalue has multiplicity k, then its index could be 1, 2, . . . , $k-1$, or k. See Section 2-6 and Definition 2-15.

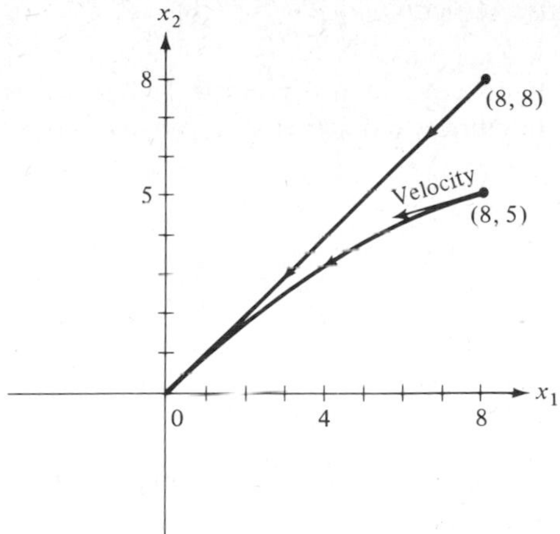

Figure 4-1 Trajectories.

know that every element of $e^{\hat{\mathbf{A}}t}$ is of the form $t^k e^{\lambda_i t}$, for $k=0, 1, \ldots, \bar{n}_i - 1$; $i=1, 2, \ldots, m$. Hence every element of $e^{\mathbf{A}t}$ is a linear combination of these factors. We call $t^k e^{\lambda_i t}$ a *mode* of the dynamical equation FE in (4-19).

The responses of a linear time-invariant dynamical equation are dictated mainly by its modes, or equivalently, the eigenvalues of $\mathbf{A}$. If an eigenvalue has a negative real part, its mode will approach zero exponentially as $t \to \infty$; it approaches zero either monotonically or oscillatorily depending on whether its imaginary part is zero or not. If an eigenvalue has a positive real part, its mode will approach infinity exponentially as $t \to \infty$. If an eigenvalue has a zero real part and has index 1,[5] its mode is a constant or a pure sinusoidal; if its index $\bar{n}_i$ is 2 or higher, then its mode will approach infinity at the rate of $t^{\bar{n}_i - 1}$.

If all eigenvalues of $\mathbf{A}$ are distinct, the response of $\dot{\mathbf{x}} = \mathbf{A}\mathbf{x}$ due to $\mathbf{x}(0) = \mathbf{x}_0$ can be written as, by using Problem 2-44,

$$\hat{\mathbf{x}}(s) = (s\mathbf{I} - \mathbf{A})^{-1}\mathbf{x}_0 = \sum_i \frac{1}{s - \lambda_i} \mathbf{q}_i \mathbf{p}_i \mathbf{x}_0 \tag{4-29}$$

where $\mathbf{q}_i$ and $\mathbf{p}_i$ are, respectively, a right and a left eigenvector of $\mathbf{A}$ associated with λ_i. In the time domain, (4-29) becomes

$$\mathbf{x}(t) = \sum_i (\mathbf{p}_i \mathbf{x}_0) e^{\lambda_i t} \mathbf{q}_i \tag{4-30}$$

If $\mathbf{x}_0$ is chosen so that $\mathbf{p}_i \mathbf{x}_0 = 0$ for all i except $i = j$, then (4-30) reduces to

$$\mathbf{x}(t) = (\mathbf{p}_j \mathbf{x}_0) e^{\lambda_j t} \mathbf{q}_j$$

For this initial state, only the mode $e^{\lambda_j t}$ is excited and $\mathbf{x}(t)$ will travel along the direction of the eigenvector $\mathbf{q}_j$. If $\mathbf{A}$ has eigenvalues with indices 2 or higher, a formula similar to (4-29) can be derived by using (2-74) and the generalized eigenvectors of $\mathbf{A}$. The situation is much more complicated and will not be discussed.

4-3 Equivalent Dynamical Equations

We introduce in this section the concept of equivalent dynamical equations. This concept in the time-invariant case is identical to the one of change of basis introduced in Figure 2-5. Hence, we study first the time-invariant case and then the time-varying case.

Time-invariant case. Consider the linear time-invariant (fixed) dynamical equation

$$FE: \quad \dot{\mathbf{x}} = \mathbf{A}\mathbf{x} + \mathbf{B}\mathbf{u} \tag{4-31a}$$
$$\mathbf{y} = \mathbf{C}\mathbf{x} + \mathbf{E}\mathbf{u} \tag{4-31b}$$

where **A**, **B**, **C**, and **E** are, respectively, $n \times n$, $n \times p$, $q \times n$, and $q \times p$ real constant matrices; **u** is the $p \times 1$ input vector, **y** is the $q \times 1$ output vector, and **x** is the $n \times 1$ state vector. The state space Σ of the dynamical equation is an n-dimensional real vector space and the matrix **A** maps Σ into itself. We have agreed in Section 3-3 to choose the orthonormal vectors $\{\mathbf{n}_1, \mathbf{n}_2, \ldots, \mathbf{n}_n\}$ as the basis vectors of the state space Σ, where $\mathbf{n}_i$ is an $n \times 1$ vector with 1 at its ith component and zeros elsewhere. We study now the effect of changing the basis of the state space. The dynamical equations that result from changing the basis of the state space are called *equivalent dynamical equations.* In order to allow a broader class of equivalent dynamical equations, we shall in this section extend the field of real numbers to the field of complex numbers and consider the state space as an n-dimensional complex vector space. This generalization is needed in order for the dynamical equation FE to have an equivalent Jordan-form dynamical equation.

Definition 4-3

Let **P** be an $n \times n$ nonsingular matrix with coefficients in the field of complex numbers $\mathbb{C}$, and let $\bar{\mathbf{x}} = \mathbf{P}\mathbf{x}$. Then the dynamical equation

$$F\bar{E}: \quad \dot{\bar{\mathbf{x}}} = \bar{\mathbf{A}}\bar{\mathbf{x}} + \bar{\mathbf{B}}\mathbf{u} \tag{4-32a}$$
$$\mathbf{y} = \bar{\mathbf{C}}\bar{\mathbf{x}} + \bar{\mathbf{E}}\mathbf{u} \tag{4-32b}$$

where
$$\bar{\mathbf{A}} = \mathbf{P}\mathbf{A}\mathbf{P}^{-1} \qquad \bar{\mathbf{B}} = \mathbf{P}\mathbf{B} \qquad \bar{\mathbf{C}} = \mathbf{C}\mathbf{P}^{-1} \qquad \bar{\mathbf{E}} = \mathbf{E} \tag{4-33}$$

is said to be *equivalent* to the dynamical equation FE in (4-31), and **P** is said to be an *equivalence transformation.* ∎

The dynamical equation $F\bar{E}$ in (4-32) is obtained from (4-31) by the substitution of $\bar{\mathbf{x}} = \mathbf{P}\mathbf{x}$. In this substitution we have changed the basis vectors of the state space from the orthonormal vectors to the columns of $\mathbf{P}^{-1}$ (see Figure 2-5). Observe that the matrices **A** and $\bar{\mathbf{A}}$ are similar; they are different representations of the same operator. If we let $\mathbf{Q} = \mathbf{P}^{-1} = [\mathbf{q}_1 \quad \mathbf{q}_2 \quad \cdots \quad \mathbf{q}_n]$, then the ith column of $\bar{\mathbf{A}}$ is the representation of $\mathbf{A}\mathbf{q}_i$ with respect to the basis $\{\mathbf{q}_1, \mathbf{q}_2, \ldots, \mathbf{q}_n\}$. From the equation $\bar{\mathbf{B}} = \mathbf{P}\mathbf{B}$ or $\mathbf{B} = \mathbf{P}^{-1}\bar{\mathbf{B}} = [\mathbf{q}_1 \quad \mathbf{q}_2 \quad \cdots \quad \mathbf{q}_n]\bar{\mathbf{B}}$, we see that the ith column of $\bar{\mathbf{B}}$ is the representation of the ith column of **B** with respect to the basis $\{\mathbf{q}_1, \mathbf{q}_2, \ldots, \mathbf{q}_n\}$. The matrix $\bar{\mathbf{C}}$ is to be computed from $\mathbf{C}\mathbf{P}^{-1}$. The matrix $\mathbf{E}$ is the direct transmission part between the input and the output. Since it has nothing to do with the state space, it is not affected by any equivalence transformation.

We explore now the physical meaning of equivalent dynamical equations. Recall that the state of a system is an auxiliary quantity introduced to give a unique relation between the input and the output when the system is not initially relaxed. The choice of the state is not unique; different methods of analysis often lead to different choices of the state.

Example 1

Consider the network shown in Figure 4-2. If the current passing through the inductor x_1 and the voltage across the capacitor x_2 are chosen as the state variables, then the dynamical equation description of the network is

$$\begin{bmatrix} \dot{x}_1 \\ \dot{x}_2 \end{bmatrix} = \begin{bmatrix} 0 & -1 \\ 1 & -1 \end{bmatrix} \begin{bmatrix} x_1 \\ x_2 \end{bmatrix} + \begin{bmatrix} 1 \\ 0 \end{bmatrix} u \qquad \textbf{(4-34a)}$$

$$y = \begin{bmatrix} 0 & 1 \end{bmatrix} \begin{bmatrix} x_1 \\ x_2 \end{bmatrix} \qquad \textbf{(4-34b)}$$

If, instead, the loop currents $\bar{x}_1$ and $\bar{x}_2$ are chosen as the state variables, then the dynamical equation is

$$\begin{bmatrix} \dot{\bar{x}}_1 \\ \dot{\bar{x}}_2 \end{bmatrix} = \begin{bmatrix} -1 & 1 \\ -1 & 0 \end{bmatrix} \begin{bmatrix} \bar{x}_1 \\ \bar{x}_2 \end{bmatrix} + \begin{bmatrix} 1 \\ 1 \end{bmatrix} u \qquad \textbf{(4-35a)}$$

$$y = \begin{bmatrix} 1 & -1 \end{bmatrix} \begin{bmatrix} \bar{x}_1 \\ \bar{x}_2 \end{bmatrix} \qquad \textbf{(435b)}$$

The dynamical equations in (4-34) and (4-35) have the same dimension and describe the same system. Hence they are equivalent. The equivalence transformation between these two equations can be found from Figure 4-2. It is clear that $x_1 = \bar{x}_1$. Since x_2 is equal to the voltage across the 1-ohm resistor, we have $x_2 = (\bar{x}_1 - \bar{x}_2)$. Thus

$$\begin{bmatrix} x_1 \\ x_2 \end{bmatrix} = \begin{bmatrix} 1 & 0 \\ 1 & -1 \end{bmatrix} \begin{bmatrix} \bar{x}_1 \\ \bar{x}_2 \end{bmatrix}$$

or

$$\begin{bmatrix} \bar{x}_1 \\ \bar{x}_2 \end{bmatrix} = \begin{bmatrix} 1 & 0 \\ 1 & -1 \end{bmatrix}^{-1} \begin{bmatrix} x_1 \\ x_2 \end{bmatrix} = \begin{bmatrix} 1 & 0 \\ 1 & -1 \end{bmatrix} \begin{bmatrix} x_1 \\ x_2 \end{bmatrix} \qquad \textbf{(4-36)}$$

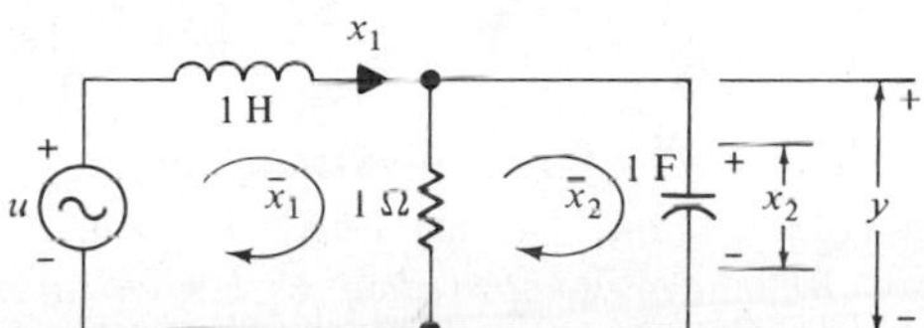

Figure 4-2 A network with two different choices of state variables.

It is easy to verify that the dynamical equations in (4-34) and (4-35) are indeed related by the equivalence transformation (4-36). ∎

Definition 4-4

Two linear dynamical equations are said to be *zero-state equivalent* if and only if they have the same impulse-response matrix or the same transfer-function matrix. Two linear dynamical equations are said to be *zero-input equivalent* if and only if for any initial state in one equation, there exists a state in the other equation, and vice versa, such that the outputs of the two equations due to zero input are identical. ∎

Note that this definition is applicable to linear time-invariant as well as linear time-varying dynamical equations.

Theorem 4-5

Two equivalent linear time-invariant dynamical equations are zero-state equivalent and zero-input equivalent.

Proof

The impulse-response matrix of FE is

$$\mathbf{G}(t)=\mathbf{C}e^{\mathbf{A}t}\mathbf{B}+\mathbf{E}\delta(t)$$

The impulse-response matrix of $F\bar{E}$ is

$$\bar{\mathbf{G}}(t)=\bar{\mathbf{C}}e^{\bar{\mathbf{A}}t}\bar{\mathbf{B}}+\bar{\mathbf{E}}\delta(t)$$

If the equations FE and $F\bar{E}$ are equivalent, we have $\bar{\mathbf{A}}=\mathbf{PAP}^{-1}$, $\bar{\mathbf{B}}=\mathbf{PB}$, $\bar{\mathbf{C}}=\mathbf{CP}^{-1}$, and $\mathbf{E}=\bar{\mathbf{E}}$. Consequently we have

$$e^{\bar{\mathbf{A}}t}=\mathbf{P}e^{\mathbf{A}t}\mathbf{P}^{-1}$$

and $$\bar{\mathbf{C}}e^{\bar{\mathbf{A}}t}\bar{\mathbf{B}}+\bar{\mathbf{E}}\delta(t)=\mathbf{CP}^{-1}\mathbf{P}e^{\mathbf{A}t}\mathbf{P}^{-1}\mathbf{PB}+\mathbf{E}\delta(t)=\mathbf{C}e^{\mathbf{A}t}\mathbf{B}+\mathbf{E}\delta(t)$$

Hence, two equivalent dynamical equations are zero-state equivalent.

The zero-input response of FE is

$$\mathbf{y}(t)=\mathbf{C}e^{\mathbf{A}(t-t_0)}\mathbf{x}(t_0)$$

The zero-input response of $F\bar{E}$ is

$$\bar{\mathbf{y}}(t)=\bar{\mathbf{C}}e^{\bar{\mathbf{A}}(t-t_0)}\bar{\mathbf{x}}(t_0)=\mathbf{C}e^{\mathbf{A}(t-t_0)}\mathbf{P}^{-1}\bar{\mathbf{x}}(t_0)$$

Hence, for any $\mathbf{x}(t_0)$, if we choose $\bar{\mathbf{x}}(t_0)=\mathbf{Px}(t_0)$, then FE and $F\bar{E}$ have the same zero-input response. Q.E.D.

We note that although equivalence implies zero-state equivalence and zero-input equivalence, the converse is not true. That is, *two linear dynamical equations can be zero-state equivalant and zero-input equivalent without being equivalent*. Furthermore, two linear equations can be zero-state equivalent without being zero-input equivalent.

Example 2

Consider the two networks shown in Figure 4-3. If the initial state in the capacitor is zero, the two networks have the same input-output pairs; their impulse responses are all equal to $\delta(t)$, the Dirac δ-function. Therefore, they are zero-state equivalent, or more precisely, their dynamical-equation descriptions are zero-state equivalent. Because of the symmetry of the network, for any initial voltage in the capacitor, the outputs of these two networks due to $u=0$ are all identically zero. Hence, they are zero-input equivalent.

A necessary condition for two dynamical equations to be equivalent is that they have the same dimension. The dynamical-equation description of the network in Figure 4-3(a) has dimension 0; the dynamical-equation description of the network in Figure 4-3(b) has dimension 1. Hence, they are not equivalent. ■

Example 3

The two networks in Figure 4-4 or, more precisely, their dynamical-equation descriptions are zero-state equivalent but not zero-input equivalent. Indeed, if there is a nonzero initial condition in the capacitor, the zero-input response of Figure 4-4(b) is nonzero, whereas the zero-input response of Figure 4-4(a) is identically zero. ■

If the matrix **A** in a dynamical equation is in the Jordan form, then the equation is said to be a Jordan-form dynamical equation. We have shown in Section

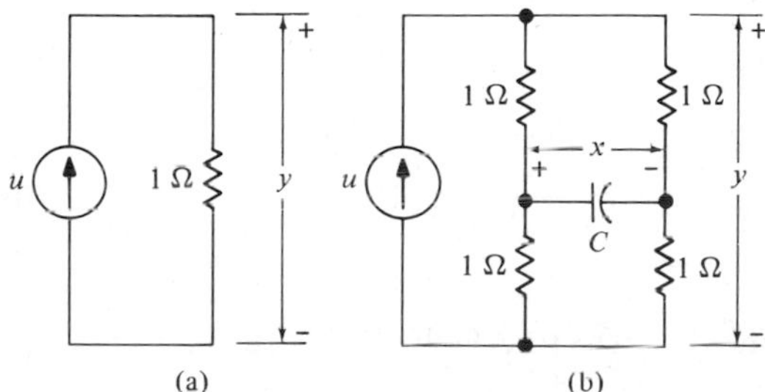

Figure 4-3 Two networks whose dynamical equations are zero-state equivalent and zero-input equivalent without being equivalent.

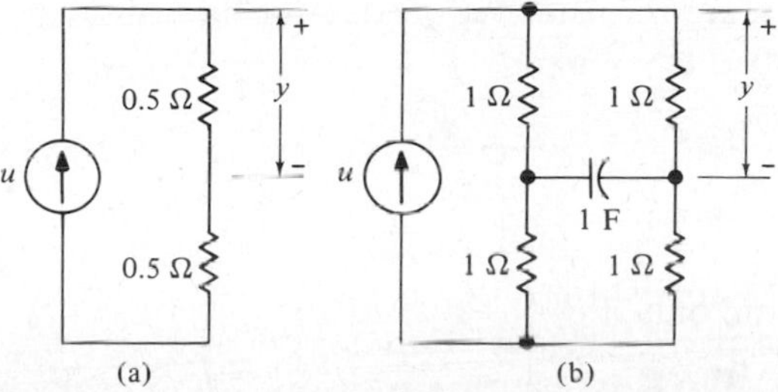

Figure 4-4 Two networks whose dynamical equations are zero-state equivalent.

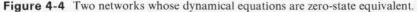

2-6 that every operator that maps $(\mathbb{C}_n, \mathbb{C})$ into itself has a Jordan-form matrix representation. Hence *every linear, time-invariant dynamical equation has an equivalent Jordan-form-dynamical equation.*

Example 4

Consider the dynamical equation

$$FE: \quad \begin{bmatrix} \dot{x}_1 \\ \dot{x}_2 \\ \dot{x}_3 \end{bmatrix} = \begin{bmatrix} 1 & 1 & 2 \\ 0 & 1 & 3 \\ 0 & 0 & 2 \end{bmatrix} \begin{bmatrix} x_1 \\ x_2 \\ x_3 \end{bmatrix} + \begin{bmatrix} 1 & 0 \\ 3 & 2 \\ -1 & -2 \end{bmatrix} \begin{bmatrix} u_1 \\ u_2 \end{bmatrix} \tag{4-37a}$$

$$\begin{bmatrix} y_1 \\ y_2 \end{bmatrix} = \begin{bmatrix} 1 & 0 & 0 \\ 1 & 2 & 2 \end{bmatrix} \begin{bmatrix} x_1 \\ x_2 \\ x_3 \end{bmatrix} \tag{4-37b}$$

It is easy to verify that if we choose

$$\mathbf{P} = \mathbf{Q}^{-1} = \begin{bmatrix} 1 & 0 & 5 \\ 0 & 1 & 3 \\ 0 & 0 & 1 \end{bmatrix}^{-1} = \begin{bmatrix} 1 & 0 & -5 \\ 0 & 1 & -3 \\ 0 & 0 & 1 \end{bmatrix}$$

then the new $\mathbf{A}$ matrix will be in the Jordan form. The method for finding the matrix $\mathbf{P}$ was discussed in Section 2-6. If we substitute $\bar{\mathbf{x}} = \mathbf{P}\mathbf{x}$ into (4-37), we obtain immediately the following equivalent Jordan-form dynamical equation:

$$\begin{bmatrix} \dot{\bar{x}}_1 \\ \dot{\bar{x}}_2 \\ \dot{\bar{x}}_3 \end{bmatrix} = \left[\begin{array}{cc:c} 1 & 1 & 0 \\ 0 & 1 & 0 \\ \hdashline 0 & 0 & 2 \end{array}\right] \begin{bmatrix} \bar{x}_1 \\ \bar{x}_2 \\ \bar{x}_3 \end{bmatrix} + \begin{bmatrix} 6 & 10 \\ 6 & 8 \\ -1 & -2 \end{bmatrix} \begin{bmatrix} u_1 \\ u_2 \end{bmatrix}$$

$$\begin{bmatrix} y_1 \\ y_2 \end{bmatrix} = \begin{bmatrix} 1 & 0 & 5 \\ 1 & 2 & 13 \end{bmatrix} \begin{bmatrix} \bar{x}_1 \\ \bar{x}_2 \\ \bar{x}_3 \end{bmatrix}$$

∎

We give the following important theorem to conclude this subsection.

Theorem 4-6

Two linear time-invariant dynamical equations $\{\mathbf{A}, \mathbf{B}, \mathbf{C}, \mathbf{E}\}$ and $\{\bar{\mathbf{A}}, \bar{\mathbf{B}}, \bar{\mathbf{C}}, \bar{\mathbf{E}}\}$, not necessarily of the same dimension, are zero-state equivalent or have the same transfer-function matrix if and only if $\mathbf{E} = \bar{\mathbf{E}}$ and

$$\mathbf{C}\mathbf{A}^i\mathbf{B} = \bar{\mathbf{C}}\bar{\mathbf{A}}^i\bar{\mathbf{B}} \qquad i = 0, 1, 2, \ldots.$$

Proof

They are zero-state equivalent if and only if

$$\mathbf{E} + \mathbf{C}(s\mathbf{I} - \mathbf{A})^{-1}\mathbf{B} = \bar{\mathbf{E}} + \bar{\mathbf{C}}(s\mathbf{I} - \bar{\mathbf{A}})^{-1}\mathbf{B}$$

This can be written as, by using (2-85),

$$\mathbf{E}+\mathbf{CB}s^{-1}+\mathbf{CAB}s^{-2}+\mathbf{CA}^2\mathbf{B}s^{-3}+\cdots$$
$$=\bar{\mathbf{E}}+\bar{\mathbf{C}}\bar{\mathbf{B}}s^{-1}+\bar{\mathbf{C}}\bar{\mathbf{A}}\bar{\mathbf{B}}s^{-2}+\bar{\mathbf{C}}\bar{\mathbf{A}}^2\bar{\mathbf{B}}s^{-3}+\cdots$$

This equality holds for every s if and only if $\mathbf{E}=\bar{\mathbf{E}}$ and

$$\mathbf{CA}^i\mathbf{B}=\bar{\mathbf{C}}\bar{\mathbf{A}}^i\bar{\mathbf{B}} \qquad i=0,1,2,\ldots$$

This establishes the theorem. Q.E.D.

***Time-varying case.** In this subsection we study equivalent linear time-varying dynamical equations. This is an extension of the time-invariant case. All the discussion and interpretation in the preceding subsection applies here as well. The only conceptual difference is that the basis vectors in the time-invariant case are fixed (independent of time), whereas the basis vectors in the time-varying case may change with time.

Consider the linear time-varying dynamical equation

$$E: \quad \dot{\mathbf{x}}=\mathbf{A}(t)\mathbf{x}+\mathbf{B}(t)\mathbf{u} \tag{4-38a}$$
$$\mathbf{y}=\mathbf{C}(t)\mathbf{x}+\mathbf{E}(t)\mathbf{u} \tag{4-38b}$$

where **A**, **B**, **C**, and **E** are $n\times n$, $n\times p$, $q\times n$, and $q\times p$ matrices whose entries are continuous functions of t.

Definition 4-5[6]

Let $\mathbf{P}(\cdot)$ be an $n\times n$ matrix defined over $(-\infty, \infty)$. It is assumed that $\mathbf{P}(t)$ and $\dot{\mathbf{P}}(t)$ are nonsingular and continuous for all t. Let $\bar{\mathbf{x}}=\mathbf{P}(t)\mathbf{x}$. Then the dynamical equation

$$\bar{E}: \quad \dot{\bar{\mathbf{x}}}=\bar{\mathbf{A}}(t)\bar{\mathbf{x}}+\bar{\mathbf{B}}(t)\mathbf{u} \tag{4-39a}$$
$$\mathbf{y}=\bar{\mathbf{C}}(t)\bar{\mathbf{x}}+\bar{\mathbf{E}}(t)\mathbf{u} \tag{4-39b}$$

where

$$\bar{\mathbf{A}}(t)=(\mathbf{P}(t)\mathbf{A}(t)+\dot{\mathbf{P}}(t))\mathbf{P}^{-1}(t) \tag{4-40a}$$
$$\bar{\mathbf{B}}(t)=\mathbf{P}(t)\mathbf{B}(t) \tag{4-40b}$$
$$\bar{\mathbf{C}}(t)=\mathbf{C}(t)\mathbf{P}^{-1}(t) \tag{4-40c}$$
$$\bar{\mathbf{E}}(t)=\mathbf{E}(t) \tag{4-40d}$$

is said to be *equivalent* to the dynamical equation E in (4-38), and $P(\cdot)$ is said to be an *equivalence transformation*. ∎

The dynamical equation $\bar{E}$ in Equation (4-39) is obtained from (4-38) by the substitution of $\bar{\mathbf{x}}=\mathbf{P}(t)\mathbf{x}$ and $\dot{\bar{\mathbf{x}}}=\dot{\mathbf{P}}(t)\mathbf{x}+\mathbf{P}(t)\dot{\mathbf{x}}$. Let $\boldsymbol{\Psi}$ be a fundamental matrix of E. Then we claim that

$$\bar{\boldsymbol{\Psi}}(t)\triangleq\mathbf{P}(t)\boldsymbol{\Psi}(t) \tag{4-41}$$

is a fundamental matrix of $\bar{E}$. The matrix $\boldsymbol{\Psi}$ is, by assumption, a fundamental

[6]This definition reduces to Definition 4-3 if **P** is independent of time.

matrix of E; hence $\dot{\mathbf{\Psi}}(t)=\mathbf{A}(t)\mathbf{\Psi}(t)$, and $\mathbf{\Psi}(t)$ is nonsingular for all t. Consequently, the matrix $\mathbf{P}(t)\mathbf{\Psi}(t)$ is nonsingular for all t (see Theorem 2-7). Now we show that $\mathbf{P}(t)\mathbf{\Psi}(t)$ satisfies the matrix equation $\dot{\bar{\mathbf{x}}}=\bar{\mathbf{A}}(t)\bar{\mathbf{x}}$. Indeed,

$$\begin{aligned}\frac{d}{dt}(\mathbf{P}(t)\mathbf{\Psi}(t)) &= \dot{\mathbf{P}}(t)\mathbf{\Psi}(t)+\mathbf{P}(t)\dot{\mathbf{\Psi}}(t)\\ &= (\dot{\mathbf{P}}(t)+\mathbf{P}(t)\mathbf{A}(t))\mathbf{P}^{-1}(t)\mathbf{P}(t)\mathbf{\Psi}(t)\\ &= \bar{\mathbf{A}}(t)(\mathbf{P}(t)\mathbf{\Psi}(t))\end{aligned}$$

Hence, $\mathbf{P}(t)\mathbf{\Psi}(t)$ is a fundamental matrix of $\dot{\bar{\mathbf{x}}}=\bar{\mathbf{A}}\bar{\mathbf{x}}$.

Theorem 4-7[7]

Let $\mathbf{A}_0$ be an arbitrary constant matrix. Then the dynamical equation in (4-38) is equivalent to the one in (4-39) with $\bar{\mathbf{A}}(t)=\mathbf{A}_0$.

Proof

Let $\mathbf{\Psi}(t)$ be an fundamental matrix of $\dot{\mathbf{x}}=\mathbf{A}(t)\mathbf{x}$. That is, $\mathbf{\Psi}(t)$ is nonsingular for all t and satisfies $\dot{\mathbf{\Psi}}(t)=\mathbf{A}(t)\mathbf{\Psi}(t)$. The differentiation of $\mathbf{\Psi}^{-1}(t)\mathbf{\Psi}(t)=\mathbf{I}$ yields $\dot{\mathbf{\Psi}}^{-1}(t)\mathbf{\Psi}(t)+\mathbf{\Psi}^{-1}(t)\dot{\mathbf{\Psi}}(t)=0$, which implies $\dot{\mathbf{\Psi}}^{-1}(t)=-\mathbf{\Psi}^{-1}(t)\dot{\mathbf{\Psi}}(t)\mathbf{\Psi}^{-1}(t)=-\mathbf{\Psi}^{-1}(t)\mathbf{A}(t)$. We define, in view of (4-41),

$$\mathbf{P}(t)=e^{\mathbf{A}_0t}\mathbf{\Psi}^{-1}(t)$$

Clearly $\mathbf{P}(t)$ is nonsingular and continuously differentiable for all t and qualifies as an equivalence transformation. We compute

$$\begin{aligned}\bar{\mathbf{A}}(t) &= (\mathbf{P}(t)\mathbf{A}(t)+\dot{\mathbf{P}}(t))\mathbf{P}^{-1}(t)\\ &= (e^{\mathbf{A}_0t}\mathbf{\Psi}^{-1}(t)\mathbf{A}(t)+\mathbf{A}_0e^{\mathbf{A}_0t}\mathbf{\Psi}^{-1}(t)+e^{\mathbf{A}_0t}\dot{\mathbf{\Psi}}^{-1}(t))\mathbf{\Psi}(t)e^{-\mathbf{A}_0t}\\ &= \mathbf{A}_0\end{aligned}$$

This establishes the theorem. Q.E.D.

In this theorem, $\mathbf{\Psi}(\cdot)$ and consequently $\mathbf{P}(\cdot)$ are generally not known, therefore nothing is really gained in this transformation. If $\mathbf{A}_0$ is chosen as zero, then $\mathbf{P}(t)=\mathbf{\Psi}^{-1}(t)$ and (4-40) becomes

$$\bar{\mathbf{A}}(t)=\mathbf{0} \qquad \bar{\mathbf{B}}(t)=\mathbf{\Psi}^{-1}(t)\mathbf{B}(t) \qquad \bar{\mathbf{C}}(t)=\mathbf{C}(t)\mathbf{\Psi}(t) \qquad \bar{\mathbf{E}}(t)=\mathbf{E}(t) \tag{4-42}$$

Its block diagram is plotted in Figure 4-5. Unlike the one in Figure 3-11, there is no feedback in Figure 4-5.

[7] This theorem was pointed out to the author by Professor T. S. Kuo in 1972.

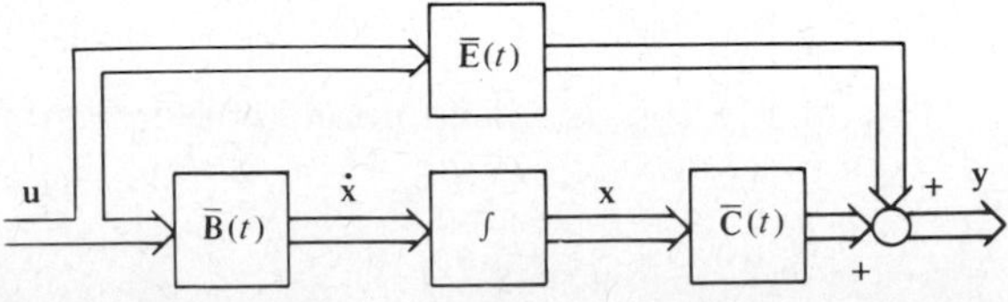

Figure 4-5 Matrix block diagram of the dynamical equation $\bar{E}$ in (4-42).

Definition 4-6

A matrix $\mathbf{P}(\cdot)$ is called a *Lyapunov transformation* if (1) $\mathbf{P}(t)$ and $\dot{\mathbf{P}}(t)$ are continuous and bounded on $[t_0, \infty)$ and (2) there exists a constant m such that

$$0 < m < |\det \mathbf{P}(t)| \qquad \text{for all } t \geq t_0 \tag{4-43}$$

■

Because of (4-43) and the boundedness of $\mathbf{P}(t)$, $\mathbf{P}^{-1}(t)$ is bounded. Because of $\dot{\mathbf{P}}^{-1}(t) = -\mathbf{P}^{-1}(t)\dot{\mathbf{P}}(t)\mathbf{P}^{-1}(t)$ and the boundedness of $\dot{\mathbf{P}}(t)$, $\dot{\mathbf{P}}^{-1}(t)$ is bounded. Consequently, we can show that if $\mathbf{P}(\cdot)$ is a Lyapunov transformation, so is $\mathbf{P}^{-1}(\cdot)$. Clearly a nonsingular constant matrix is a Lyapunov transformation. In Definition 4-5, if $\mathbf{P}(\cdot)$ is a Lyapunov transformation, the dynamical equations are said to be *equivalent in the sense of Lyapunov*. A Lyapunov transformation preserves, as will be discussed in Chapter 8, the stability properties of a dynamical equation, but an equivalence transformation does not. If $\mathbf{P}(\cdot)$ is required to be a Lyapunov transformation, Theorem 4-7 does not hold in general. In other words, not every time-varying dynamical equation can be equivalent in the sense of Lyapunov to a dynamical equation with a constant $\mathbf{A}$. This is possible for the class of time-varying equations with periodic $\mathbf{A}(t)$ as will be discussed in the following.

Linear time-varying dynamical equations with periodic $\mathbf{A}(\cdot)$

Consider the linear time-varying dynamical equation in (4-38). We assume

$$\mathbf{A}(t+T) = \mathbf{A}(t)$$

for all t and for some positive constant T. This means that every element of $\mathbf{A}(\cdot)$ is a periodic function with the same period T. Let $\mathbf{\Psi}(t)$ be a fundamental matrix of $\dot{\mathbf{x}} = \mathbf{A}(t)\mathbf{x}$. Then $\mathbf{\Psi}(t+T)$ is also a fundamental matrix of $\dot{\mathbf{x}} = \mathbf{A}(t)\mathbf{x}$. Indeed we have

$$\dot{\mathbf{\Psi}}(t+T) = \mathbf{A}(t+T)\mathbf{\Psi}(t+T) = \mathbf{A}(t)\mathbf{\Psi}(t+T)$$

The matrix function $\mathbf{\Psi}(t)$ is nonsingular for all t; consequently, so is $\mathbf{\Psi}(t+T)$. Columns of $\mathbf{\Psi}(t)$ and columns of $\mathbf{\Psi}(t+T)$ form two sets of basis vectors in the solution space; hence, there exists a nonsingular constant matrix $\mathbf{Q}$ (see Section 2-3) such that

$$\mathbf{\Psi}(t+T) = \mathbf{\Psi}(t)\mathbf{Q} \tag{4-44}$$

For the nonsingular matrix $\mathbf{Q}$ there exists a constant matrix $\bar{\mathbf{A}}$ such that $e^{\bar{\mathbf{A}}T} = \mathbf{Q}$ (Problem 2-37). Hence, (4-44) can be written as

$$\mathbf{\Psi}(t+T) = \mathbf{\Psi}(t)e^{\bar{\mathbf{A}}T} \tag{4-45}$$

Define

$$\mathbf{P}(t) \triangleq e^{\bar{\mathbf{A}}t}\mathbf{\Psi}^{-1}(t) \tag{4-46}$$

We show that $\mathbf{P}(\cdot)$ is a periodic function with period T:

$$\mathbf{P}(t+T) = e^{\bar{\mathbf{A}}(t+T)}\mathbf{\Psi}^{-1}(t+T) = e^{\bar{\mathbf{A}}t}e^{\bar{\mathbf{A}}T}e^{-\bar{\mathbf{A}}T}\mathbf{\Psi}^{-1}(t) = \mathbf{P}(t)$$

Theorem 4-8

Assume that the matrix **A** in the dynamical equation E in (4-38) is periodic with period T. Let **P** be defined as in (4-46). Then the dynamical equation E in (4-38) and the dynamical equation

$$\bar{E}: \quad \begin{aligned} \dot{\bar{\mathbf{x}}}(t) &= \bar{\mathbf{A}}\bar{\mathbf{x}}(t) + \mathbf{P}(t)\mathbf{B}(t)\mathbf{u}(t) \\ \bar{\mathbf{y}}(t) &= \mathbf{C}(t)\mathbf{P}^{-1}(t)\bar{\mathbf{x}}(t) + \mathbf{E}(t)\mathbf{u}(t) \end{aligned}$$

where $\bar{\mathbf{A}}$ is a constant matrix, are equivalent in the sense of Lyapunov. ∎

The matrix $\mathbf{P}(\cdot)$ in (4-46) is periodic and nonsingular; hence it is bounded. Its derivative is clearly continuous and bounded. Hence, $\mathbf{P}(\cdot)$ is a Lyapunov transformation. The rest of the theorem follows directly from Theorem 4-7.

The homogeneous part of this theorem is the so-called *theory of Floquet*. It states that if $\dot{\mathbf{x}} = \mathbf{A}(t)\mathbf{x}$ and if $\mathbf{A}(t+T) = \mathbf{A}(t)$ for all t, then its fundamental matrix is of the form $\mathbf{P}^{-1}(t)e^{\bar{\mathbf{A}}t}$, where $\mathbf{P}^{-1}(t)$ is a periodic function. Furthermore, $\dot{\mathbf{x}} = \mathbf{A}(t)\mathbf{x}$ is equivalent in the sense of Lyapunov to $\dot{\bar{\mathbf{x}}} = \bar{\mathbf{A}}\bar{\mathbf{x}}$.

4-4 Impulse-Response Matrices and Dynamical Equations

Time-varying case. In this section we shall study the relation between the impulse-response matrix and the dynamical equation. Let the input-output description of a system with p input terminals and q output terminals be

$$\mathbf{y}(t) = \int_{t_0}^{t} \mathbf{G}(t, \tau)\mathbf{u}(\tau)\, d\tau \tag{4-47}$$

where $\mathbf{y}$ is the $q \times 1$ output vector, $\mathbf{u}$ is the $p \times 1$ input vector, and $\mathbf{G}$ is the $q \times p$ impulse-response matrix of the system. We have implicitly assumed in (4-47) that the system is initially relaxed at t_0. The ijth (ith row, jth column) element of $\mathbf{G}(\cdot, \tau)$ is the response at the ith output terminal due to a δ-function input applied at time τ at the jth input terminal. Suppose now that the internal structure of the same system is accessible, and that analysis of this system leads to a dynamical equation of the form

$$E: \quad \dot{\mathbf{x}}(t) = \mathbf{A}(t)\mathbf{x}(t) + \mathbf{B}(t)\mathbf{u}(t) \tag{4-48a}$$

$$\mathbf{y}(t) = \mathbf{C}(t)\mathbf{x}(t) + \mathbf{E}(t)\mathbf{u}(t) \tag{4-48b}$$

where $\mathbf{x}$ is the $n \times 1$ state vector of the system, and **A**, **B**, **C**, and **E** are $n \times n$, $n \times p$, $q \times n$, and $q \times p$ matrices whose entries are continuous functions of t defined over $(-\infty, \infty)$. Since Equations (4-47) and (4-48) are two different descriptions of the same system, they should give the same input-output pairs if the system is initially relaxed. The solution of the dynamical equation E with $\mathbf{x}(t_0) = \mathbf{0}$ is given by

$$\begin{aligned} \mathbf{y}(t) &= \mathbf{C}(t) \int_{t_0}^{t} \mathbf{\Phi}(t, \tau)\mathbf{B}(\tau)\mathbf{u}(\tau)\, d\tau + \mathbf{E}(t)\mathbf{u}(t) \\ &= \int_{t_0}^{t} [\mathbf{C}(t)\mathbf{\Phi}(t, \tau)\mathbf{B}(\tau) + \mathbf{E}(t)\delta(t-\tau)]\mathbf{u}(\tau)\, d\tau \end{aligned} \tag{4-49}$$

where $\mathbf{\Phi}(t, \tau)$ is the state transition matrix of $\dot{\mathbf{x}} = \mathbf{A}(t)\mathbf{x}$. By comparing (4-47) and (4-49), we immediately obtain

$$\mathbf{G}(t, \tau) = \begin{cases} \mathbf{C}(t)\mathbf{\Phi}(t, \tau)\mathbf{B}(\tau) + \mathbf{E}(t)\delta(t-\tau) & \text{for } t \geq \tau \\ \mathbf{0} & \text{for } t < \tau \end{cases} \tag{4-50}$$

That $\mathbf{G}(t, \tau) = \mathbf{0}$ for $t < \tau$ follows from the causality assumption which is implicitly embedded in writing (4-47); that is, the integration is stopped at t.

If the state-variable description of a system is available, the input-output description of the system can be easily obtained from (4-50). The converse problem—to find the state-variable description from the input-output description of a system—is much more complicated, however. It actually consists of two problems: (1) Is it possible at all to obtain the state-variable description from the impulse-response matrix of a system? (2) If yes, how do we obtain the state-variable description from the impulse-response matrix? We shall study the first problem in the remainder of this section. The second problem will be studied in Chapter 6.

Consider a system with the impulse-response matrix $\mathbf{G}(t, \tau)$. If there exists a linear finite-dimensional dynamical equation E that has $\mathbf{G}(t, \tau)$ as its impulse-response matrix, then $\mathbf{G}(t, \tau)$ is said to be *realizable*. We call the dynamical equation E, or more specifically, the matrices $\{\mathbf{A}, \mathbf{B}, \mathbf{C}, \mathbf{E}\}$, a *realization* of $\mathbf{G}(t, \tau)$. The terminology "realization" is justified by the fact that by using the dynamical equation, we can build an operational amplifier circuit that will generate $\mathbf{G}(t, \tau)$. Note that the state of a dynamical-equation realization of the impulse-response matrix of a system is purely an auxiliary variable and it may not have any physical meaning. Note also that *the dynamical-equation realization gives only the same zero-state response of the system.* If the dynamical equation is not in the zero-state, its response may not have any relation to the system.

If the realization of an impulse response $\mathbf{G}(t, \tau)$ is restricted to a finite-dimensional linear dynamical equation of the form (4-48), it is conceivable that not every $\mathbf{G}(t, \tau)$ is realizable. For example, there is no linear equation of the form (4-48) that will generate the impulse response of a unit-time-delay system or the impulse response $1/(t-\tau)$. We give in the following the necessary and sufficient condition for $\mathbf{G}(t, \tau)$ to be realizable.

Theorem 4-9

A $q \times p$ impulse-response matrix $\mathbf{G}(t, \tau)$ is realizable by a finite-dimensional linear dynamical equation of the form (4-48) if and only if $\mathbf{G}(t, \tau)$ can be decomposed into

$$\mathbf{G}(t, \tau) = \mathbf{E}(t)\delta(t-\tau) + \mathbf{M}(t)\mathbf{N}(\tau) \qquad \text{for all } t \geq \tau \tag{4-51}$$

where $\mathbf{E}$ is a $q \times p$ matrix and $\mathbf{M}$ and $\mathbf{N}$ are, respectively, $q \times n$ and $n \times p$ continuous matrices of t.

Proof

Necessity: Suppose the dynamical equation

$$E: \quad \begin{aligned} \dot{\mathbf{x}} &= \mathbf{A}(t)\mathbf{x} + \mathbf{B}(t)\mathbf{u} \\ \mathbf{y} &= \mathbf{C}(t)\mathbf{x} + \mathbf{E}(t)\mathbf{u} \end{aligned}$$

is a realization of $\mathbf{G}(t, \tau)$; then

$$\begin{aligned} \mathbf{G}(t, \tau) &= \mathbf{E}(t)\delta(t-\tau) + \mathbf{C}(t)\mathbf{\Phi}(t, \tau)\mathbf{B}(\tau) \\ &= \mathbf{E}(t)\delta(t-\tau) + \mathbf{C}(t)\mathbf{\Psi}(t)\mathbf{\Psi}^{-1}(\tau)\mathbf{B}(\tau) \end{aligned}$$

where $\mathbf{\Psi}$ is a fundamental matrix of $\dot{\mathbf{x}} = \mathbf{A}(t)\mathbf{x}$. The proof is completed by identifying

$$\mathbf{M}(t) = \mathbf{C}(t)\mathbf{\Psi}(t) \qquad \text{and} \qquad \mathbf{N}(t) = \mathbf{\Psi}^{-1}(t)\mathbf{B}(t)$$

Sufficiency: Let

$$\mathbf{G}(t, \tau) = \mathbf{E}(t)\delta(t-\tau) + \mathbf{M}(t)\mathbf{N}(\tau)$$

where $\mathbf{M}$ and $\mathbf{N}$ are $q \times n$ and $n \times p$ continuous matrices, respectively. Then the following n-dimensional dynamical equation

$$\bar{E}: \quad \dot{\mathbf{x}}(t) = \mathbf{N}(t)\mathbf{u}(t) \tag{4-52a}$$
$$\mathbf{y}(t) = \mathbf{M}(t)\mathbf{x}(t) + \mathbf{E}(t)\mathbf{u}(t) \tag{4-52b}$$

is a realization of $\mathbf{G}(t, \tau)$. Indeed, the state transition matrix of $\bar{E}$ is an $n \times n$ identity matrix; hence,

$$\mathbf{G}(t, \tau) = \mathbf{M}(t)\mathbf{I}\mathbf{N}(\tau) + \mathbf{E}(t)\delta(t-\tau) \qquad \text{Q.E.D.}$$

We note that the dynamical equation in (4-52) can be simulated without using feedback as shown in Figure 4-5 with $\bar{\mathbf{B}}(t) = \mathbf{N}(t)$ and $\bar{\mathbf{C}}(t) = \mathbf{M}(t)$.

Example 1

Consider $g(t, \tau) = g(t-\tau) = (t-\tau)e^{\lambda(t-\tau)}$. It is easy to verify that

$$g(t-\tau) = (t-\tau)e^{\lambda(t-\tau)} = [e^{\lambda t} \quad te^{\lambda t}] \begin{bmatrix} -\tau e^{-\lambda\tau} \\ e^{-\lambda\tau} \end{bmatrix}$$

Hence, the dynamical equation

$$E: \quad \begin{bmatrix} \dot{x}_1 \\ \dot{x}_2 \end{bmatrix} = \begin{bmatrix} -te^{-\lambda t} \\ e^{-\lambda t} \end{bmatrix} u(t)$$
$$y(t) = [e^{\lambda t} \quad te^{\lambda t}] \begin{bmatrix} x_1 \\ x_2 \end{bmatrix}$$

is a realization of $g(t, \tau)$. ∎

All the equivalent dynamical equations have the same impulse-response matrix; hence, if we find a realization of $\mathbf{G}(t, \tau)$, we may obtain different realizations of $\mathbf{G}(t, \tau)$ by applying equivalence transformations. Note that an impulse-response matrix may have different dimensional realizations; for example, the networks in Figure 4-3 are two different dimensional realizations of $g(t, \tau) = \delta(t-\tau)$. For further results in realization, see Reference S128.

Time-invariant case. We shall first apply Theorem 4-9 to the time-invariant case and see what can be established. For the time-invariant case, we have $\mathbf{G}(t, \tau) = \mathbf{G}(t-\tau)$. Consequently, Theorem 4-9 can be read as: An impulse response $\mathbf{G}(t-\tau)$ is realizable by a finite-dimensional (time-varying) dynamical equation if and only if there exist continuous matrices $\mathbf{M}$ and $\mathbf{N}$ such that

$$\mathbf{G}(t-\tau) = \mathbf{M}(t)\mathbf{N}(\tau) + \mathbf{E}(t)\delta(t-\tau) \qquad \text{for all } t \geq \tau$$

There are two objections to using this theorem. First, the condition is stated in terms of $\mathbf{G}(t-\tau)$ instead of $\mathbf{G}(t)$. Second, the condition is given for $\mathbf{G}(t-\tau)$ to be realizable by a linear *time-varying* dynamical equation. What is more desirable is to have conditions on $\mathbf{G}(t)$ under which $\mathbf{G}(t)$ has a linear *time-invariant* dynamical-equation realization. Since, in the time-invariant case, we may also study a system in the frequency domain, we shall first derive the condition of realization in terms of transfer-function matrix, and then state it in terms of $\mathbf{G}(t)$.

Consider a system with the input-output description

$$\mathbf{y}(t) = \int_0^t \mathbf{G}(t-\tau)\mathbf{u}(\tau)\, d\tau$$

or, in the frequency domain,

$$\hat{\mathbf{y}}(s) = \hat{\mathbf{G}}(s)\hat{\mathbf{u}}(s) \tag{4-53}$$

where $\mathbf{G}(\cdot)$ is the impulse-response matrix and $\hat{\mathbf{G}}(s)$ is the transfer-function matrix of the system. Suppose now a state-variable description of the system is found to be

$$FE: \qquad \dot{\mathbf{x}} = \mathbf{A}\mathbf{x} + \mathbf{B}\mathbf{u} \tag{4-54a}$$

$$\mathbf{y} = \mathbf{C}\mathbf{x} + \mathbf{E}\mathbf{u} \tag{4-54b}$$

By taking the Laplace transform and assuming the zero initial state, we obtain

$$\hat{\mathbf{y}}(s) = [\mathbf{C}(s\mathbf{I}-\mathbf{A})^{-1}\mathbf{B} + \mathbf{E}]\hat{\mathbf{u}}(s) \tag{4-55}$$

Since (4-53) and (4-55) describe the same system, we have

$$\hat{\mathbf{G}}(s) = \mathbf{C}(s\mathbf{I}-\mathbf{A})^{-1}\mathbf{B} + \mathbf{E} \tag{4-56}$$

As discussed in (3-36) and (3-37), $\mathbf{C}(s\mathbf{I}-\mathbf{A})^{-1}\mathbf{B}$ is a strictly proper rational matrix, and $\mathbf{C}(s\mathbf{I}-\mathbf{A})^{-1}\mathbf{B} + \mathbf{E}$ is a proper rational matrix. Hence, the transfer function matrix of the dynamical equation in (4-54) is a proper rational matrix.

Theorem 4-10

A transfer-function matrix $\hat{\mathbf{G}}(s)$ is realizable by a finite-dimensional linear time-invariant dynamical equation if and only if $\hat{\mathbf{G}}(s)$ is a proper rational matrix.

Proof

If $\hat{\mathbf{G}}(s)$ is realizable by a finite-dimensional linear time-invariant dynamical equation, then from (4-56), we know that $\hat{\mathbf{G}}(s)$ is a proper rational matrix. Before proving that every proper rational matrix $\hat{\mathbf{G}}(s)$ is realizable, we first prove that every scalar (1×1) proper rational function is realizable. The most general form of a proper rational function is

$$\hat{g}(s) = e + \frac{\beta_1 s^{n-1} + \cdots + \beta_{n-1} s + \beta_n}{s^n + \alpha_1 s^{n-1} + \cdots + \alpha_{n-1} s + \alpha_n} \tag{4-57}$$

We claim that the n-dimensional linear time-invariant dynamical equation

$$\begin{bmatrix} \dot{x}_1 \\ \dot{x}_2 \\ \vdots \\ \dot{x}_{n-1} \\ \dot{x}_n \end{bmatrix} = \begin{bmatrix} 0 & 1 & 0 & \cdots & 0 & 0 \\ 0 & 0 & 1 & \cdots & 0 & 0 \\ \vdots & \vdots & \vdots & & \vdots & \vdots \\ 0 & 0 & 0 & \cdots & 0 & 1 \\ -\alpha_n & -\alpha_{n-1} & -\alpha_{n-2} & \cdots & -\alpha_2 & -\alpha_1 \end{bmatrix} \begin{bmatrix} x_1 \\ x_2 \\ \vdots \\ x_{n-1} \\ x_n \end{bmatrix} + \begin{bmatrix} 0 \\ 0 \\ \vdots \\ 0 \\ 1 \end{bmatrix} u \tag{4-58a}$$

$$y = [\beta_n \quad \beta_{n-1} \quad \beta_{n-2} \quad \cdots \quad \beta_2 \quad \beta_1]\mathbf{x} + eu \tag{4-58b}$$

is a realization of $\hat{g}(s)$. What we have to show is that the transfer function of (4-58) is $\hat{g}(s)$. We shall demonstrate this by using Mason's formula for a signal-flow graph.[8,9] Let us choose $x_1, \dot{x}_1, x_2, \dot{x}_2, \ldots, x_n, \dot{x}_n$ as nodes; then the signal-flow graph of (4-58) is of the form shown in Figure 4-6. There are n loops with loop gain $-\alpha_1/s, -\alpha_2/s^2, \ldots, -\alpha_n/s^n$; and there are, except the direct transmission path e, n forward paths with gains $\beta_1/s, \beta_2/s^2, \ldots, \beta_n/s^n$. Since all the

[8] Mason's gain formula for signal-flow graph is as follows: The transfer function of a signal-flow graph is

$$\frac{\Sigma g_i \Delta_i}{\Delta}$$

where $\Delta = 1 - (\Sigma$ all individual loop gains$) + (\Sigma$ all possible gain products of two nontouching loops$) - \cdots$; g_i = gain of the ith forward path, and Δ_i = the part of Δ not touching the ith forward path. Two loops or two parts of a signal-flow graph are said to be nontouching if they do not have any point in common. See, e.g., Reference S46.

[9] This can also be proved by computing algebraically the transfer function of (4-58). This is done in Chapter 6.

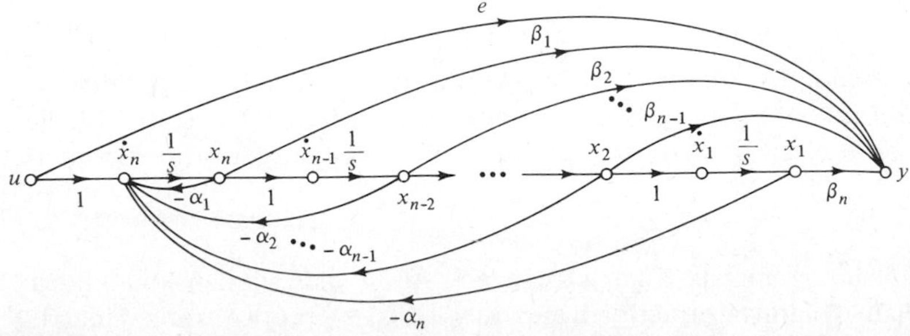

Figure 4-6 Signal-flow graph of the dynamical equation in (4-58).

forward paths and loops have common nodes, we have

$$\Delta = 1 + \frac{\alpha_1}{s} + \frac{\alpha_2}{s^2} + \cdots + \frac{\alpha_n}{s^n} \tag{4-59}$$

and $\Delta_i = 1$, for $i = 1, 2, \ldots, n$. Hence, the transfer function of (4-58) from the input u to the output y is

$$e + \frac{\beta_1/s + \beta_2/s^2 + \cdots + \beta_n/s^n}{1 + \alpha_1/s + \alpha_2/s^2 + \cdots + \alpha_n/s^n} = e + \frac{\beta_1 s^{n-1} + \cdots + \beta_n}{s^n + \alpha_1 s^{n-1} + \cdots + \alpha_n} = \hat{g}(s)$$

This proves the assertion that every scalar proper rational function is realizable.

We are now ready to show that every proper rational matrix is realizable. In order to avoid cumbersome notations, we assume that $\hat{\mathbf{G}}(s)$ is a 2×2 matrix. Let

$$\hat{\mathbf{G}}(s) = \begin{bmatrix} \hat{g}_{11}(s) & \hat{g}_{12}(s) \\ \hat{g}_{21}(s) & \hat{g}_{22}(s) \end{bmatrix} \tag{4-60}$$

and let

$$\dot{\mathbf{x}}_{ij} = \mathbf{A}_{ij}\mathbf{x}_{ij} + \mathbf{b}_{ij}u_j \qquad y_i = \mathbf{c}_{ij}\mathbf{x}_{ij} + e_{ij}u_j$$

be a realization of $\hat{g}_{ij}$, for $i, j = 1, 2$; that is, $\hat{g}_{ij}(s) = \mathbf{c}_{ij}(s\mathbf{I} - \mathbf{A}_{ij})^{-1}\mathbf{b}_{ij} + e_{ij}$. Note that the $\mathbf{b}_{ij}$'s are column vectors and the $\mathbf{c}_{ij}$'s are row vectors. Then the composite dynamical equation

$$\begin{bmatrix} \dot{\mathbf{x}}_{11} \\ \dot{\mathbf{x}}_{12} \\ \dot{\mathbf{x}}_{21} \\ \dot{\mathbf{x}}_{22} \end{bmatrix} = \begin{bmatrix} \mathbf{A}_{11} & \mathbf{0} & \mathbf{0} & \mathbf{0} \\ \mathbf{0} & \mathbf{A}_{12} & \mathbf{0} & \mathbf{0} \\ \mathbf{0} & \mathbf{0} & \mathbf{A}_{21} & \mathbf{0} \\ \mathbf{0} & \mathbf{0} & \mathbf{0} & \mathbf{A}_{22} \end{bmatrix} \begin{bmatrix} \mathbf{x}_{11} \\ \mathbf{x}_{12} \\ \mathbf{x}_{21} \\ \mathbf{x}_{22} \end{bmatrix} + \begin{bmatrix} \mathbf{b}_{11} & \mathbf{0} \\ \mathbf{0} & \mathbf{b}_{12} \\ \mathbf{b}_{21} & \mathbf{0} \\ \mathbf{0} & \mathbf{b}_{22} \end{bmatrix} \begin{bmatrix} u_1 \\ u_2 \end{bmatrix} \tag{4-61a}$$

$$\begin{bmatrix} y_1 \\ y_2 \end{bmatrix} = \begin{bmatrix} \mathbf{c}_{11} & \mathbf{c}_{12} & \mathbf{0} & \mathbf{0} \\ \mathbf{0} & \mathbf{0} & \mathbf{c}_{21} & \mathbf{c}_{22} \end{bmatrix} \begin{bmatrix} \mathbf{x}_{11} \\ \mathbf{x}_{12} \\ \mathbf{x}_{21} \\ \mathbf{x}_{22} \end{bmatrix} + \begin{bmatrix} e_{11} & e_{12} \\ e_{21} & e_{22} \end{bmatrix} \begin{bmatrix} u_1 \\ u_2 \end{bmatrix} \tag{4-61b}$$

is a realization of $\hat{\mathbf{G}}(s)$. Indeed, the transfer-function matrix of (4-61) is

$$\begin{bmatrix} \mathbf{c}_{11} & \mathbf{c}_{12} & \mathbf{0} & \mathbf{0} \\ \mathbf{0} & \mathbf{0} & \mathbf{c}_{21} & \mathbf{c}_{22} \end{bmatrix}$$

$$\cdot \begin{bmatrix} (s\mathbf{I} - \mathbf{A}_{11})^{-1} & \mathbf{0} & \mathbf{0} & \mathbf{0} \\ \mathbf{0} & (s\mathbf{I} - \mathbf{A}_{12})^{-1} & \mathbf{0} & \mathbf{0} \\ \mathbf{0} & \mathbf{0} & (s\mathbf{I} - \mathbf{A}_{21})^{-1} & \mathbf{0} \\ \mathbf{0} & \mathbf{0} & \mathbf{0} & (s\mathbf{I} - \mathbf{A}_{22})^{-1} \end{bmatrix}$$

$$\cdot \begin{bmatrix} \mathbf{b}_{11} & \mathbf{0} \\ \mathbf{0} & \mathbf{b}_{12} \\ \mathbf{b}_{21} & \mathbf{0} \\ \mathbf{0} & \mathbf{b}_{22} \end{bmatrix} + \begin{bmatrix} e_{11} & e_{12} \\ e_{21} & e_{22} \end{bmatrix}$$

$$= \begin{bmatrix} \mathbf{c}_{11}(s\mathbf{I} - \mathbf{A}_{11})^{-1}\mathbf{b}_{11} + e_{11} & \mathbf{c}_{12}(s\mathbf{I} - \mathbf{A}_{12})^{-1}\mathbf{b}_{12} + e_{12} \\ \mathbf{c}_{21}(s\mathbf{I} - \mathbf{A}_{21})^{-1}\mathbf{b}_{21} + e_{21} & \mathbf{c}_{22}(s\mathbf{I} - \mathbf{A}_{22})^{-1}\mathbf{b}_{22} + e_{22} \end{bmatrix} = \hat{\mathbf{G}}(s) \tag{4-62}$$

Thus every proper rational matrix is realizable by a finite-dimensional linear time-invariant dynamical equation. Q.E.D.

The realization procedure discussed in (4-60) and (4-61) is simple and straightforward. We realize every element of a transfer-function matrix independently and then connect them from the input and the output as shown in Figure 4-7. The resulting realization, however, is generally not satisfactory for the following two reasons: First, the realization is internally not coupled as can be seen from Figure 4-7. The coupling or interacting of all variables, however, is a feature of most physical multivariable systems. Second, the dimension of the realization is generally unnecessarily large. These two problems will be resolved in Chapter 6.

The condition of Theorem 4-10 is stated in terms of transfer-function matrices. We may translate it into the time domain as follows:

Corollary 4-10

An impulse-response matrix $\mathbf{G}(t)$ is realizable by a finite-dimensional linear time-invariant dynamical equation if and only if every entry of $\mathbf{G}(t)$ is a linear combination of terms of the form $t^k e^{\lambda_i t}$ (for $k=0, 1, 2, \ldots,$ and $i=1, 2, \ldots$) and possibly contains a δ-function at $t=0$. ■

The impulse-response matrix $\mathbf{G}(t)$ is the inverse Laplace transform of the transfer-function matrix $\hat{\mathbf{G}}(s)$. If a proper rational function is of the form

$$\hat{g}_{ij}(s) = d + \frac{N(s)}{(s-\lambda_1)^{k_1}(s-\lambda_2)^{k_2}\cdots}$$

then its inverse Laplace transform $g_{ij}(t)$ is a linear combination of the terms

$$\delta(t),\ e^{\lambda_1 t},\ te^{\lambda_1 t}, \ldots,\ t^{k_1-1}e^{\lambda_1 t};\ e^{\lambda_2 t},\ te^{\lambda_2 t}, \ldots,\ t^{k_2-1}e^{\lambda_2 t}; \ldots$$

Hence, Corollary 4-10 follows directly from Theorem 4-10.

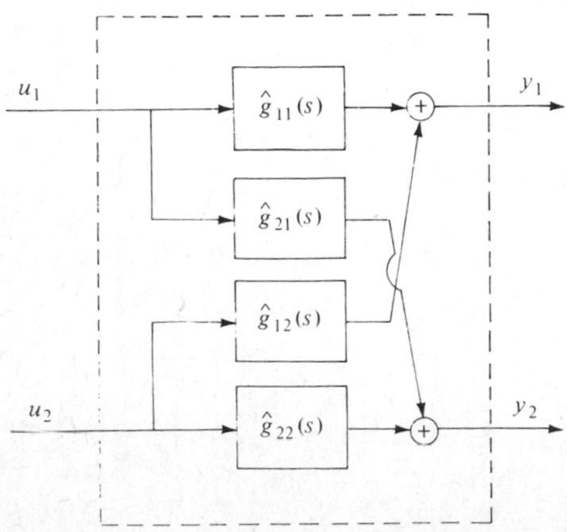

Figure 4-7 Internally uncoupled realization of (4-60).

Since a realizable $\mathbf{G}(t)$ can be decomposed into $\mathbf{G}(t-\tau)=\mathbf{M}(t)\mathbf{N}(\tau)$, and since entries of $\hat{\mathbf{G}}(t)$ are linear combinations of $t^k e^{\lambda_i t}$, for $k=0, 1, \ldots$, and $i=1, 2, \ldots$, the entries of $\mathbf{M}(t)$ and $\mathbf{N}(t)$ must be linear combinations of $t^k e^{\lambda_i t}$. Consequently, the matrices $\mathbf{M}(t)$ and $\mathbf{N}(t)$ are analytic functions of t on the entire real line. As a consequence of this fact, we have the following theorem.

Theorem 4-11

A system that has a proper rational transfer function matrix is relaxed at t_0 if and only if $\mathbf{u}_{[t_0, t_0+\varepsilon]}=\mathbf{0}$ implies $\mathbf{y}_{[t_0, t_0+\varepsilon]}=\mathbf{0}$ for some positive real ε. ∎

This is a restatement of Corollary 3-1. Hence, for the class of systems that have rational transfer-function matrices—or equivalently, are describable by linear time-invariant dynamical equations—if the output is identically zero in an interval (no matter how small), then the system is relaxed at the end of that interval.

4-5 Concluding Remarks

The solutions of linear dynamical equations were studied in this chapter. The solution hinges on the state transition matrix $\mathbf{\Phi}(t, \tau)$, which has the properties $\mathbf{\Phi}(t, t)=\mathbf{I}$, $\mathbf{\Phi}^{-1}(t, \tau)=\mathbf{\Phi}(\tau, t)$, and $\mathbf{\Phi}(t, \tau)\mathbf{\Phi}(\tau, t_0)=\mathbf{\Phi}(t, t_0)$. For the time-varying case, $\mathbf{\Phi}(t, \tau)$ is very difficult to compute; for the time-invariant case, $\mathbf{\Phi}(t, \tau)$ is equal to $e^{\mathbf{A}(t-\tau)}$, which can be computed by using the methods introduced in Section 2-7. In both cases, if only a specific solution is of interest, we may bypass $\mathbf{\Phi}(t, \tau)$ and compute the solution on a digital computer by direct integration.

Different analyses often lead to different dynamical-equation descriptions of a system. Mathematically, it means that dynamical equations depend on the basis chosen for the state space. However, the input-output description has nothing to do with the basis; no matter what analysis is used, it always leads to the same input-output description. This is an advantage of the input-output description over the state-variable description.

Every proper rational function has been shown to be realizable by a finite-dimensional linear time-invariant dynamical equation. This corresponds to the synthesis problem in network theory. A realization of a proper rational matrix was also constructed. However, the realization is not satisfactory, because generally it is possible to construct a lesser-dimensional realization. This will be discussed in Chapter 6 after the introduction of controllability and observability.

To conclude this chapter, we remark briefly the solution of discrete-time equations. The state transition matrix $\mathbf{\Phi}(k, m)$ of $\mathbf{x}(k+1)=\mathbf{A}(k)\mathbf{x}(k)=\mathbf{A}(k)\mathbf{A}(k-1)\mathbf{x}(k-1)=\cdots$ can be readily computed as $\mathbf{\Phi}(k, m)=\mathbf{A}(k-1)\cdots\mathbf{A}(m)$ (Problem 4-27). Unlike the continuous time case, $\mathbf{\Phi}(k, m)$ is easily computable. In terms of $\mathbf{\Phi}(k, m)$, the solution of a discrete-time dynamical equation can be obtained as in Problem 4-29. For the time-invariant case, the equivalent dynamical equations and the realization problem are identical to the continuous-time case. For example, a sampled transfer-function matrix $\hat{\mathbf{G}}(z)$ has a finite-dimensional realization of the form in (3-80) if and only if $\hat{\mathbf{G}}(z)$ is a proper rational matrix in z. The realization procedure in Theorem 4-10 is also directly applicable to the discrete-time case.

Problems

4-1 Find the fundamental matrices and the state transition matrices of the following homogeneous equations:

$$\begin{bmatrix} \dot{x}_1(t) \\ \dot{x}_2(t) \end{bmatrix} = \begin{bmatrix} 0 & 1 \\ 0 & t \end{bmatrix} \begin{bmatrix} x_1(t) \\ x_2(t) \end{bmatrix}$$

and

$$\begin{bmatrix} \dot{x}_1(t) \\ \dot{x}_2(t) \end{bmatrix} = \begin{bmatrix} -1 & e^{2t} \\ 0 & -1 \end{bmatrix} \begin{bmatrix} x_1(t) \\ x_2(t) \end{bmatrix}$$

4-2 Show that $\partial \mathbf{\Phi}(t, \tau)/\partial \tau = -\mathbf{\Phi}(t, \tau)\mathbf{A}(\tau)$.

4-3 Find the solution of

$$\dot{\mathbf{x}} = \begin{bmatrix} 0 & 1 & 0 \\ 0 & 0 & 1 \\ -2 & -4 & -3 \end{bmatrix} \mathbf{x} + \begin{bmatrix} 1 & 0 \\ 0 & 1 \\ -1 & 1 \end{bmatrix} \mathbf{u}$$

$$\mathbf{y} = \begin{bmatrix} 0 & 1 & -1 \\ 1 & 2 & 1 \end{bmatrix} \mathbf{x}$$

with

$$\mathbf{x}(0) = \begin{bmatrix} 1 \\ 0 \\ 0 \end{bmatrix} \qquad \mathbf{u}(t) = \begin{bmatrix} 1 \\ 1 \end{bmatrix} \qquad \text{for } t \geq 0$$

Verify for this problem that $e^{\mathbf{A}(t-\tau)} = e^{\mathbf{A}t}e^{-\mathbf{A}\tau} = e^{-\mathbf{A}\tau}e^{\mathbf{A}t}$.

4-4 Let

$$\mathbf{A} = \begin{bmatrix} 1 & 0 & 1 & 1 \\ 0 & 1 & 0 & 0 \\ 0 & 0 & 1 & -1 \\ 0 & 0 & 0 & 1 \end{bmatrix}$$

Find $e^{\mathbf{A}t}$ by using the formula $\mathscr{L}[e^{\mathbf{A}t}] = (s\mathbf{I} - \mathbf{A})^{-1}$.

4-5 If $\mathbf{T}^{-1}(t)$ exists and is differentiable for all t, show that

$$\frac{d}{dt}[\mathbf{T}^{-1}(t)] = -\mathbf{T}^{-1}(t)\left[\frac{d}{dt}\mathbf{T}(t)\right]\mathbf{T}^{-1}(t)$$

4-6 From $\mathbf{\Phi}(t, \tau)$, show how to compute $\mathbf{A}(t)$.

4-7 Given

$$\mathbf{A}(t)=\begin{bmatrix} a_{11}(t) & a_{12}(t) \\ a_{21}(t) & a_{22}(t) \end{bmatrix}$$

show that

$$\det \mathbf{\Phi}(t, t_0)=\exp\left[\int_{t_0}^{t} (a_{11}(\tau)+a_{22}(\tau))\, d\tau\right]$$

where $\partial\mathbf{\Phi}(t, t_0)/\partial t=\mathbf{A}(t)\mathbf{\Phi}(t, t_0)$ and $\mathbf{\Phi}(t_0, t_0)=\mathbf{I}$. *Hint*: Show that

$$\frac{\partial}{\partial t}\det \mathbf{\Phi}(t, t_0)=(a_{11}(t)+a_{22}(t)) \det \mathbf{\Phi}(t, t_0)$$

4-8 Given $\dot{\mathbf{x}}(t)=\mathbf{A}(t)\mathbf{x}$. The equation $\dot{\mathbf{z}}=-\mathbf{A}^*\mathbf{z}$ is called the *adjoint equation of* $\dot{\mathbf{x}}=\mathbf{A}(t)\mathbf{x}$, where $\mathbf{A}^*$ is the complex conjugate transpose of $\mathbf{A}$. Let $\mathbf{\Phi}(t, t_0)$ and $\mathbf{\Phi}_a(t, t_0)$ be the state transition matrices of $\dot{\mathbf{x}}=\mathbf{A}(t)\mathbf{x}$ and $\dot{\mathbf{z}}=-\mathbf{A}^*\mathbf{z}$, respectively. Verify that

$$\mathbf{\Phi}_a(t, t_0)=\mathbf{\Phi}^*(t_0, t)$$

4-9 Consider

$$\dot{\mathbf{x}}=\mathbf{A}(t)\mathbf{x}+\mathbf{B}(t)\mathbf{u} \qquad \mathbf{y}=\mathbf{C}(t)\mathbf{x}$$

and its adjoint equation

$$\dot{\mathbf{z}}=-\mathbf{A}^*(t)\mathbf{z}+\mathbf{C}^*(t)\mathbf{v} \qquad \mathbf{w}=\mathbf{B}^*(t)\mathbf{z}$$

shown in Figure P4-9. Note the reversal of the flow direction of signals. Let $\mathbf{G}(t, \tau)$ and $\mathbf{G}_a(t, \tau)$ be their impulse response matrices. Show that

$$\mathbf{G}(t, \tau)=\mathbf{G}_a^*(\tau, t)$$

Show that if $\mathbf{A}$, $\mathbf{B}$, and $\mathbf{C}$ are constant matrices and $\hat{\mathbf{G}}(s)$ and $\hat{\mathbf{G}}_a(s)$ are their transfer-function matrices, then

$$\hat{\mathbf{G}}(s)=-\hat{\mathbf{G}}_a^*(-s)$$

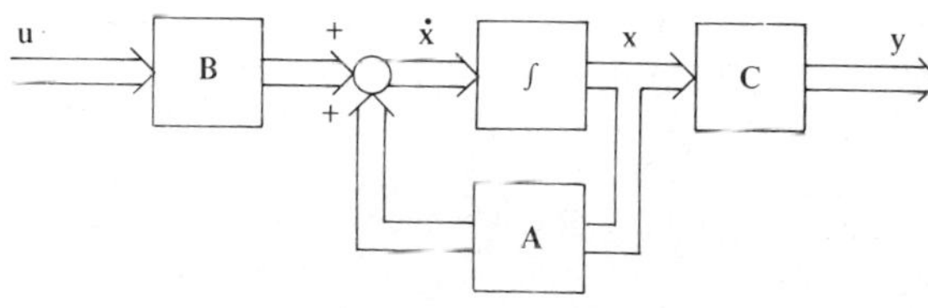

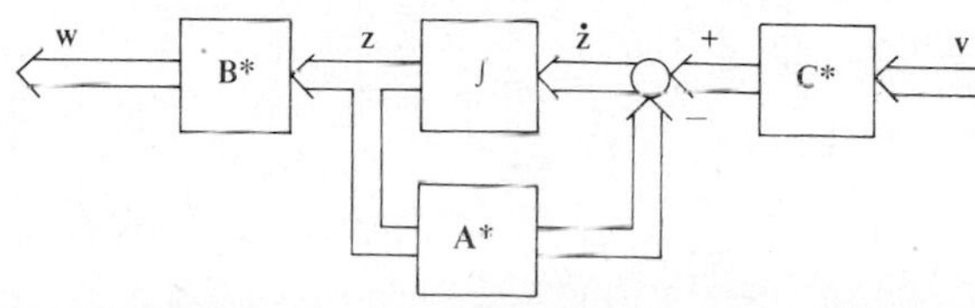

Figure P4-9

4-10 Every element of $\mathbf{\Phi}(t, t_0)$ can be interpreted as the impulse response of some input-output pair. What is the input and the output of the ijth element of $\mathbf{\Phi}(t, t_0)$?

4-11 Let

$$\mathbf{\Phi}(t, t_0) = \begin{bmatrix} \mathbf{\Phi}_{11}(t, t_0) & \mathbf{\Phi}_{12}(t, t_0) \\ \mathbf{\Phi}_{21}(t, t_0) & \mathbf{\Phi}_{22}(t, t_0) \end{bmatrix}$$

be the state transition matrix of

$$\dot{\mathbf{x}} = \begin{bmatrix} \mathbf{A}_{11} & \mathbf{A}_{12} \\ \mathbf{0} & \mathbf{A}_{22} \end{bmatrix} \mathbf{x}$$

Show that $\mathbf{\Phi}_{21}(t, t_0) = \mathbf{0}$ for all t, t_0 and that $(\partial/\partial t)\mathbf{\Phi}_{ii}(t, t_0) = A_{ii}\mathbf{\Phi}_{ii}(t, t_0)$ for $i = 1, 2$.

4-12 Verify that $\mathbf{B}(t) = \mathbf{\Phi}(t, t_0)\mathbf{B}_0\mathbf{\Phi}^*(t, t_0)$ is the solution of

$$\frac{d}{dt}\mathbf{B}(t) = \mathbf{A}(t)\mathbf{B}(t) + \mathbf{B}(t)\mathbf{A}^*(t) \qquad \mathbf{B}(t_0) = \mathbf{B}_0$$

where $\mathbf{\Phi}(t, t_0)$ is the state transition matrix of $\dot{\mathbf{x}} = \mathbf{A}(t)\mathbf{x}$.

4-13 Verify that $\mathbf{X}(t) = e^{\mathbf{A}t}\mathbf{C}e^{\mathbf{B}t}$ is the solution of

$$\frac{d}{dt}\mathbf{X} = \mathbf{AX} + \mathbf{XB} \qquad \mathbf{X}(0) = \mathbf{C}$$

4-14 Show that if $\dot{\mathbf{A}}(t) = \mathbf{A}_1\mathbf{A}(t) - \mathbf{A}(t)\mathbf{A}_1$, then

$$\mathbf{A}(t) = e^{\mathbf{A}_1 t}\mathbf{A}(0)e^{-\mathbf{A}_1 t}$$

Show also that the eigenvalues of $\mathbf{A}(t)$ are independent of t.

4-15 Show that if $\dot{\mathbf{A}}(t) = \mathbf{A}_1\mathbf{A}(t) - \mathbf{A}(t)\mathbf{A}_1$, then a fundamental matrix of $\dot{\mathbf{x}} = \mathbf{A}(t)\mathbf{x}$ is given by

$$\mathbf{\Psi}(t) = e^{\mathbf{A}_1 t}e^{\mathbf{A}_2 t}$$

where $\mathbf{A}_2 \triangleq \mathbf{A}(0) - \mathbf{A}_1$.

4-16 Find the impedance (the transfer function from u to i) of the network in Figure P4-16. If the initial conditions of the inductor and the capacitor are zero and if an input voltage $u(t) = e^{-t}$ is applied, what are $i(t)$, $i_1(t)$ and $i_2(t)$? Note that $i_1(t)$ and $i_2(t)$ contain some exponential functions that do not appear in $i(t)$. How do you explain this?

4-17 Consider the network shown in Figure P4-17. Find the initial inductor current

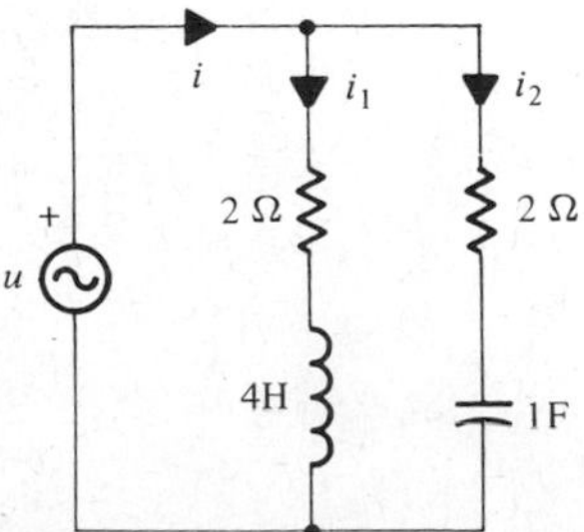

Figure P4-16

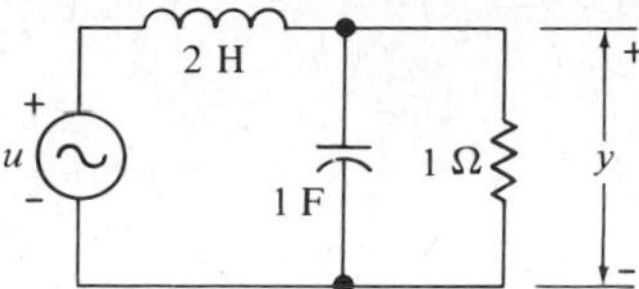

Figure P4-17

and the initial capacitor voltage such that for the input $u(t)=e^{-4t}$ the output $y(t)$ will be immediately of the form e^{-4t} without containing any transient.

4-18 Find an equivalent time-invariant dynamical equation of

$$\dot{x}=(\cos t \sin t)x$$

4-19 Find an equivalent Jordan-canonical-form dynamical equation of

$$\begin{bmatrix}\dot{x}_1\\ \dot{x}_2\\ \dot{x}_3\end{bmatrix}=\begin{bmatrix}0 & 4 & 3\\ 0 & 20 & 16\\ 0 & -25 & -20\end{bmatrix}\begin{bmatrix}x_1\\ x_2\\ x_3\end{bmatrix}+\begin{bmatrix}-1\\ 3\\ 0\end{bmatrix}u$$

$$y=[-1 \quad 3 \quad 0]\begin{bmatrix}x_1\\ x_2\\ x_3\end{bmatrix}+4u$$

4-20 Find an equivalent discrete-time Jordan-canonical-form dynamical equation of

$$\begin{bmatrix}x_1(n+1)\\ x_2(n+1)\\ x_3(n+1)\end{bmatrix}=\begin{bmatrix}0 & 4 & 3\\ 0 & 20 & 16\\ 0 & -25 & -20\end{bmatrix}\begin{bmatrix}x_1(n)\\ x_2(n)\\ x_3(n)\end{bmatrix}+\begin{bmatrix}-1\\ 3\\ 0\end{bmatrix}u(n)$$

$$y(n)=[-1 \quad 3 \quad 0]\begin{bmatrix}x_1(n)\\ x_2(n)\\ x_3(n)\end{bmatrix}+4u(n)$$

4-21 Can you transform a time invariant $\{\mathbf{A}, \mathbf{B}, \mathbf{C}\}$ into $\{\mathbf{0}, \bar{\mathbf{B}}, \bar{\mathbf{C}}\}$ by a time-varying equivalence transformation?

4-22 Find a time-varying dynamical equation realization and a time-invariant dynamical realization of the impulse response $g(t)=t^2e^{\lambda t}$.

4-23 Find a dynamical-equation realization of $g(t, \tau)=\sin te^{-(t-\tau)}\cos \tau$. Is it possible to find a linear time-invariant dynamical-equation realization for it?

4-24 Use a signal-flow graph to show that the transfer function of the following single-variable, linear time-invariant dynamical equation

$$\dot{\mathbf{x}}=\begin{bmatrix}0 & 0 & 0 & \cdots & 0 & -\alpha_n\\ 1 & 0 & 0 & \cdots & 0 & -\alpha_{n-1}\\ 0 & 1 & 0 & \cdots & 0 & -\alpha_{n-2}\\ \vdots & \vdots & \vdots & & \vdots & \vdots\\ 0 & 0 & 0 & \cdots & 1 & -\alpha_1\end{bmatrix}\mathbf{x}+\begin{bmatrix}\beta_n\\ \beta_{n-1}\\ \beta_{n-2}\\ \vdots\\ \beta_1\end{bmatrix}u$$

$$y=[0 \quad 0 \quad 0 \quad \cdots \quad 0 \quad 1]\mathbf{x}+eu$$

is

$$\hat{g}(s)=e+\frac{\beta_1 s^{n-1}+\cdots+\beta_n}{s^n+\alpha_1 s^{n-1}+\cdots+\alpha_{n-1}s+\alpha_n}$$

4-25 Realize the proper rational matrices

$$\begin{bmatrix} \frac{s+2}{s+1} & \frac{1}{s+3} \\ \frac{5}{s+1} & \frac{5s+1}{s+2} \end{bmatrix} \qquad \begin{bmatrix} \frac{z+2}{z+1} & \frac{1}{z+3} \\ \frac{5}{z+1} & \frac{5z+1}{z+2} \end{bmatrix}$$

into continuous-time and discrete-time dynamical equations.

4-26 Consider the equivalent dynamical equations

$$\begin{cases}\dot{\mathbf{x}}=\mathbf{A}\mathbf{x}+\mathbf{B}\mathbf{u}\\ \mathbf{y}=\mathbf{C}\mathbf{x}\end{cases} \qquad \begin{cases}\dot{\bar{\mathbf{x}}}=\bar{\mathbf{A}}\bar{\mathbf{x}}+\bar{\mathbf{B}}\mathbf{u}\\ \mathbf{y}=\bar{\mathbf{C}}\bar{\mathbf{x}}\end{cases}$$

where $\bar{\mathbf{x}}=\mathbf{P}\mathbf{x}$. Their adjoint equations are, respectively,

$$\begin{cases}\dot{\mathbf{z}}=-\mathbf{A}^*\mathbf{z}+\mathbf{C}^*\mathbf{u} & \text{(1a)}\\ \mathbf{y}=\mathbf{B}^*\mathbf{z} & \text{(1b)}\end{cases}$$

$$\begin{cases}\dot{\bar{\mathbf{z}}}=-\bar{\mathbf{A}}^*\bar{\mathbf{z}}+\bar{\mathbf{C}}^*\mathbf{u} & \text{(2a)}\\ \mathbf{y}=\bar{\mathbf{B}}^*\mathbf{z} & \text{(2b)}\end{cases}$$

where $\mathbf{A}^*$ and $\bar{\mathbf{A}}^*$ are the complex conjugate transposes of $\mathbf{A}$ and $\bar{\mathbf{A}}$, respectively. Show that Equations (1) and (2) are equivalent and that they are related by $\bar{\mathbf{z}}=(\mathbf{P}^{-1})^*\mathbf{z}$.

4-27 Consider $\mathbf{x}(k+1)=\mathbf{A}(k)\mathbf{x}(k)$. Define

$$\mathbf{\Phi}(k,m)\triangleq\mathbf{A}(k-1)\mathbf{A}(k-2)\mathbf{A}(k-3)\cdots\mathbf{A}(m) \qquad \text{for } k>m$$
$$\mathbf{\Phi}(m,m)\triangleq\mathbf{I}$$

Show that, given the initial state $\mathbf{x}(m)=\mathbf{x}_0$, the state at time k is given by $\mathbf{x}(k)=\mathbf{\Phi}(k,m)\mathbf{x}_0$. If $\mathbf{A}$ is independent of k, what is $\mathbf{\Phi}(k,m)$?

4-28 For continuous-time dynamical equations, the state transition matrix $\mathbf{\Phi}(t,\tau)$ is defined for all t, τ. However, in discrete-time dynamical equation, $\mathbf{\Phi}(k,m)$ is defined only for $k\geq m$. What condition do we need on $\mathbf{A}(k)$ in order for $\mathbf{\Phi}(k,m)$ to be defined for $k<m$?

4-29 Show that the solution of $\mathbf{x}(k+1)=\mathbf{A}(k)\mathbf{x}(k)+\mathbf{B}(k)\mathbf{u}(k)$ is given by

$$\mathbf{x}(k)=\mathbf{\Phi}(k,m)\mathbf{x}(m)+\sum_{l=m}^{k-1}\mathbf{\Phi}(k,l+1)\mathbf{B}(l)\mathbf{u}(l)$$

[This can be easily verified by considering $\mathbf{B}(l)\mathbf{u}(l)$ as an initial state at time $(l+1)$.] Show that if $\mathbf{A}(k)$ and $\mathbf{B}(k)$ are independent of k, then the solution becomes

$$\mathbf{x}(k)=\mathbf{A}^k\mathbf{x}(0)+\sum_{m=0}^{k-1}\mathbf{A}^{k-1-m}\mathbf{B}\mathbf{u}(m)$$

4-30 Let $\mathbf{A}(t)\triangleq(a_{ij}(t))$. Then, by definition,

$$\frac{d}{dt}\mathbf{A}(t)=\left(\frac{d}{dt}a_{ij}(t)\right)$$

Verify that

$$\frac{d}{dt}(\mathbf{A}(t)\mathbf{B}(t)) = \dot{\mathbf{A}}(t)\mathbf{B}(t) + \mathbf{A}(t)\dot{\mathbf{B}}(t)$$

Verify also that

$$\frac{d}{dt}[\mathbf{A}(t)]^2 \triangleq \frac{d}{dt}(\mathbf{A}(t)\mathbf{A}(t)) = 2\dot{\mathbf{A}}(t)\mathbf{A}(t)$$

if and only if $\dot{\mathbf{A}}(t)$ and $\mathbf{A}(t)$ commute; that is, $\dot{\mathbf{A}}(t)\mathbf{A}(t) = \mathbf{A}(t)\dot{\mathbf{A}}(t)$.

4-31 Show that if $\int_{t_0}^{t} \mathbf{A}(\tau)\, d\tau$ and $\mathbf{A}(t)$ commute for all t, then the unique solution of

$$\frac{\partial}{\partial t}\mathbf{\Phi}(t, t_0) = \mathbf{A}(t)\mathbf{\Phi}(t, t_0) \qquad \mathbf{\Phi}(t_0, t_0) = \mathbf{I}$$

is

$$\mathbf{\Phi}(t, t_0) = \exp \int_{t_0}^{t} \mathbf{A}(\tau)\, d\tau$$

[*Hint:* Use Problem 4-30 and Equation (2-78).]

5
Controllability and Observability of Linear Dynamical Equations

5-1 Introduction

System analyses generally consist of two parts: quantitative and qualitative. In the quantitative study we are interested in the exact response of the system to certain input and initial conditions, as we studied in the preceding chapter. In the qualitative study we are interested in the general properties of a system. In this chapter we shall introduce two qualitative properties of linear dynamical equations: controllability and observability. We shall first give the reader some rough ideas of these two concepts by using the network shown in Figure 5-1. The input u of the network is a current source. It is clear that if the initial voltage in the capacitor C_2 in loop II is zero, no matter what input u is applied, the mode e^{-t} in II can never be excited. Hence the mode e^{-t} in II is said to be not controllable by the input u. On the other hand, the mode e^{-t} in loop I can be excited by the application of the input u; hence the mode in I is controllable by the input u. Although the mode e^{-t} in I can be excited by the input, its presence can never be detected from the output terminal y. Hence it is said to be not

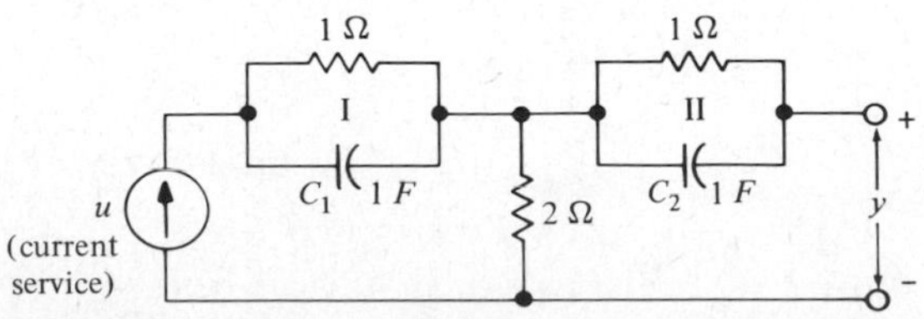

Figure 5-1 A simple network.

observable from the output y. On the other hand, the presence of the mode e^{-t} in II can be detected from the output y; hence the mode is said to be observable. This illustration, though not very accurate, may convey the ideas of the concepts of controllability and observability.

The concepts of controllability and observability are very important in the study of control and filtering problems. As an example, consider the platform system shown in Figure 5-2. The system consists of one platform; both ends of the platform are supported on the ground by means of springs and dashpots. The mass of the platform is, for simplicity, assumed to be zero; hence the movements of the two spring systems are independent. If the initial displacements of both ends of the platform are different from zero, the platform will start to vibrate. If no force is applied, it will take an infinite time for the platform to come back to rest. Now we may ask: For any initial displacements, is it possible to apply a force to bring the platform to rest in a *finite* time? In order to answer this question, the concept of controllability is needed.

This chapter is organized as follows. In Section 5-2 the required mathematical background is introduced. Three theorems that give the conditions for linear independence of a set of vector functions are presented. All the results in controllability and observability follow almost directly from these three theorems. The concept of controllability is introduced in Section 5-3. Necessary and sufficient conditions for linear time-varying dynamical equations and linear time-invariant dynamical equations to be controllable are derived. The concept of observability is introduced in Section 5-4. It is dual to the concept of controllability; hence its discussion is rather brief. Duality theorem is also developed. In Section 5-5, we study dynamical equations which are uncontrollable and/or unobservable. The canonical decomposition theorem is developed. A consequence of this theorem is that the transfer-function matrix of a dynamical equation depends solely on the part of the equation that is controllable and observable. In Section 5-6, we study the controllability and observability of linear time-invariant Jordan-form dynamical equation. Their conditions are very simple and can be checked almost by inspection. In Section 5-7, the concepts of output controllability and output function controllability are introduced. It is shown that they are properties of the input-output description of a system. In the last section, we discuss some computational problems encountered in this chapter.

Although elements of the matrices **A**, **B**, **C**, and **E** are all real-valued, for mathematical convenience they are considered as elements of the field of com

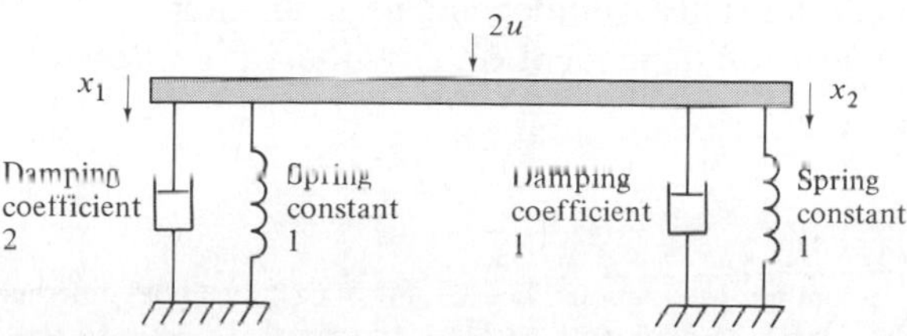

Figure 5-2 A platform system.

plex numbers. Consequently, the state space of an n-dimensional dynamical equation will be taken as an n-dimensional complex vector space $(\mathbb{C}^n, \mathbb{C})$.

The references for this chapter are 2, 8, 11, 13, 14, 20, 21, 48, 55, 56, 60, 61, 69, 71, 98, 103, and 105 to 107.

The reader who is interested in only the time-invariant case may skip Theorems 5-2, 5-5, and 5-6.

5-2 Linear Independence of Time Functions

The concept of linear independence of a set of vectors of a linear space was introduced in Section 2-3. We shall now apply this concept to a set of functions of a real variable. A set of complex-valued functions $f_1, f_2, \dots, f_n$ is said to be linearly dependent on the interval[1] $[t_1, t_2]$ over the field of complex numbers if there exist complex numbers $\alpha_1, \alpha_2, \dots, \alpha_n$, not all zero, such that

$$\alpha_1 f_1(t) + \alpha_2 f_2(t) + \cdots + \alpha_n f_n(t) = 0 \qquad \text{for } all\ t \text{ in } [t_1, t_2] \tag{5-1}$$

Otherwise, the set of functions is said to be linearly independent on $[t_1, t_2]$ over the field of complex numbers. In this definition, the specification of time interval is crucial.

Example 1

Consider the two continuous functions f_1 and f_2, defined by

$$f_1(t) = t \qquad \text{for } t \text{ in } [-1, 1]$$

$$f_2(t) = \begin{cases} t & \text{for } t \text{ in } [0, 1] \\ -t & \text{for } t \text{ in } [-1, 0] \end{cases}$$

It is clear that the functions f_1 and f_2 are linearly dependent on $[0, 1]$, since if we choose $\alpha_1 = 1$, $\alpha_2 = -1$, then $\alpha_1 f_1(t) + \alpha_2 f_2(t) = 0$ for all t in $[0, 1]$. The functions f_1 and f_2 are also linearly dependent on $[-1, 0]$. However, f_1 and f_2 are linearly independent on $[-1, 1]$. ∎

From this example, we see that although a set of functions is linearly independent on an interval, it is not necessary that they are linearly independent on *any* subinterval. However, it is true that there exists a subinterval on which they are linearly independent. For example, in Example 1 the functions f_1 and f_2 are linearly independent on the subinterval $[-\varepsilon, \varepsilon]$ for any positive ε. On the other hand, if a set of functions is linearly independent on an interval $[t_1, t_2]$, then the set of functions is linearly independent on *any* interval that contains $[t_1, t_2]$.

---

[1] The functions we study are mostly continuous functions; hence there is no substantial difference between using the open interval (t_1, t_2) and the closed interval $[t_1, t_2]$. Every $[t_1, t_2]$ is assumed to be a nonzero interval.

The concept of linear independence can be extended to vector-valued functions. Let $\mathbf{f}_i$, for $i=1, 2, \ldots, n$, be $1 \times p$ complex-valued functions of t; then the $1 \times p$ complex-valued functions $\mathbf{f}_1, \mathbf{f}_2, \ldots, \mathbf{f}_n$ are linearly dependent on $[t_1, t_2]$ if there exist complex numbers $\alpha_1, \alpha_2, \ldots, \alpha_n$, not all zero, such that

$$\alpha_1\mathbf{f}_1(t) + \alpha_2\mathbf{f}_2(t) + \cdots + \alpha_n\mathbf{f}_n(t) = \mathbf{0} \qquad \text{for all } t \text{ in } [t_1, t_2] \tag{5-2}$$

Otherwise, the $\mathbf{f}_i$'s are linearly independent on $[t_1, t_2]$. Note that the zero vector in Equation (5-2) is a $1 \times p$ row vector $[0 \quad 0 \quad \cdots \quad 0]$. As in Definition 2-4′, we may also state that $\mathbf{f}_1, \mathbf{f}_2, \ldots, \mathbf{f}_n$ are linearly independent on $[t_1, t_2]$ if and only if

$$[\alpha_1 \quad \alpha_2 \quad \cdots \quad \alpha_n] \begin{bmatrix} \mathbf{f}_1(t) \\ \mathbf{f}_2(t) \\ \vdots \\ \mathbf{f}_n(t) \end{bmatrix} \triangleq \boldsymbol{\alpha}\mathbf{F}(t) = \mathbf{0} \qquad \text{for all } t \text{ in } [t_1, t_2] \tag{5-3}$$

implies $\boldsymbol{\alpha} = \mathbf{0}$, where

$$\boldsymbol{\alpha} \triangleq [\alpha_1 \quad \alpha_2 \quad \cdots \quad \alpha_n] \qquad \mathbf{F} \triangleq \begin{bmatrix} \mathbf{f}_1 \\ \mathbf{f}_2 \\ \vdots \\ \mathbf{f}_n \end{bmatrix}$$

Clearly, $\boldsymbol{\alpha}$ is a constant $1 \times n$ row vector and $\mathbf{F}$ is an $n \times p$ matrix function.

The linear independence of a set of functions is a property associated with an interval; hence in testing for linear independence, we have to consider the entire interval. Let $\mathbf{F}^*(t)$ be the complex conjugate transpose of $\mathbf{F}(t)$.

Theorem 5-1

Let $\mathbf{f}_i$, for $i = 1, 2, \ldots, n$, be $1 \times p$ complex-valued continuous functions defined on $[t_1, t_2]$. Let $\mathbf{F}$ be the $n \times p$ matrix with $\mathbf{f}_i$ as its ith row. Define

$$\mathbf{W}(t_1, t_2) \triangleq \int_{t_1}^{t_2} \mathbf{F}(t)\mathbf{F}^*(t)\, dt$$

Then $\mathbf{f}_1, \mathbf{f}_2, \ldots, \mathbf{f}_n$ are linearly independent on $[t_1, t_2]$ if and only if the $n \times n$ constant matrix $\mathbf{W}(t_1, t_2)$ is nonsingular.[2]

Proof

The proof of this theorem is similar to that of Theorem 2-8. Recall that a matrix is nonsingular if and only if all the columns (rows) are linearly independent. We prove first the necessity of the theorem; we prove it by contradiction. Assume that the $\mathbf{f}_i$'s are linearly independent on $[t_1, t_2]$, but $\mathbf{W}(t_1, t_2)$ is singular Then there exists a nonzero $1 \times n$ row vector $\boldsymbol{\alpha}$ such that $\boldsymbol{\alpha}\mathbf{W}(t_1, t_2) = \mathbf{0}$. This

[2] The matrix $\mathbf{W}(t_1, t_2)$ is in fact positive definite (see Definition 8-6). This property is not needed in this chapter.

implies $\boldsymbol{\alpha}\mathbf{W}(t_1, t_2)\boldsymbol{\alpha}^* = 0$, or

$$\boldsymbol{\alpha}\mathbf{W}(t_1, t_2)\boldsymbol{\alpha}^* = \int_{t_1}^{t_2} (\boldsymbol{\alpha}\mathbf{F}(t))(\boldsymbol{\alpha}\mathbf{F}(t))^* \, dt = 0 \tag{5-4}$$

Since the integrand $(\boldsymbol{\alpha}\mathbf{F}(t))(\boldsymbol{\alpha}\mathbf{F}(t))^*$ is a continuous function and is nonnegative for all t in $[t_1, t_2]$, Equation (5-4) implies that

$$\boldsymbol{\alpha}\mathbf{F}(t) = \mathbf{0} \qquad \text{for all } t \text{ in } [t_1, t_2]$$

This contradicts the linear independence assumption of the set of $\mathbf{f}_i$, $i = 1, 2, \ldots, n$. Hence if the $\mathbf{f}_i$'s are linearly independent on $[t_1, t_2]$, then $\det \mathbf{W}(t_1, t_2) \neq 0$.

We next prove the sufficiency of the theorem. Suppose that $\mathbf{W}(t_1, t_2)$ is nonsingular, but the $\mathbf{f}_i$s are linearly dependent on $[t_1, t_2]$. Then, by definition, there exists a nonzero constant $1 \times n$ row vector $\boldsymbol{\alpha}$ such that $\boldsymbol{\alpha}\mathbf{F}(t) = \mathbf{0}$ for all t in $[t_1, t_2]$. Consequently, we have

$$\boldsymbol{\alpha}\mathbf{W}(t_1, t_2) = \int_{t_1}^{t_2} \boldsymbol{\alpha}\mathbf{F}(t)\mathbf{F}^*(t) \, dt = \mathbf{0} \tag{5-5}$$

which contradicts the assumption that $\mathbf{W}(t_1, t_2)$ is nonsingular. Hence, if $\mathbf{W}(t_1, t_2)$ is nonsingular, then the $\mathbf{f}_i$'s are linearly independent on $[t_1, t_2]$. Q.E.D.

The determinant of $\mathbf{W}(t_1, t_2)$ is called the *Gram determinant* of the $\mathbf{f}_i$'s. In applying Theorem 5-1, the functions $\mathbf{f}_i$, for $i = 1, 2, \ldots, n$, are required to be continuous. If the functions $\mathbf{f}_i$, for $i = 1, 2, \ldots, n$, have continuous derivatives up to order $(n-1)$, then we may use the following theorem.

Theorem 5-2

Assume that the $1 \times p$ complex-valued functions $\mathbf{f}_1, \mathbf{f}_2, \ldots, \mathbf{f}_n$ have continuous derivatives up to order $(n-1)$ on the interval $[t_1, t_2]$. Let $\mathbf{F}$ be the $n \times p$ matrix with $\mathbf{f}_i$ as its ith row, and let $\mathbf{F}^{(k)}$ be the kth derivative of $\mathbf{F}$. If there exists some t_0 in $[t_1, t_2]$ such that the $n \times np$ matrix

$$[\mathbf{F}(t_0) \vdots \mathbf{F}^{(1)}(t_0) \vdots \mathbf{F}^{(2)}(t_0) \vdots \cdots \vdots \mathbf{F}^{(n-1)}(t_0)] \tag{5-6}$$

has rank n, then the $\mathbf{f}_i$'s are linearly independent on $[t_1, t_2]$ over the field of complex numbers.[3]

Proof

We prove the theorem by contradiction. Suppose that there exists some t_0 in $[t_1, t_2]$ such that

$$\rho[\mathbf{F}(t_0) \vdots \mathbf{F}^{(1)}(t_0) \vdots \cdots \vdots \mathbf{F}^{(n-1)}(t_0)] = n$$

[3] If t_0 is at either t_1 or t_2, the end points of an interval, then $\mathbf{F}^{(k)}(t_0)$ is defined as $\mathbf{F}^{(k)}(t)$, with t approaching t_0 from inside the interval.

and the $\mathbf{f}_i$s are linearly dependent on $[t_1, t_2]$. Then by definition, there exists a nonzero $1 \times n$ row vector $\boldsymbol{\alpha}$ such that

$$\boldsymbol{\alpha}\mathbf{F}(t) = \mathbf{0} \qquad \text{for all } t \text{ in } [t_1, t_2]$$

This implies that

$$\boldsymbol{\alpha}\mathbf{F}^{(k)}(t) = \mathbf{0} \qquad \text{for all } t \text{ in } [t_1, t_2] \text{ and } k = 1, 2, \ldots, n-1$$

Hence we have

$$\boldsymbol{\alpha}[\mathbf{F}(t) \vdots \mathbf{F}^{(1)}(t) \vdots \cdots \vdots \mathbf{F}^{(n-1)}(t)] = \mathbf{0} \qquad \text{for all } t \text{ in } [t_1, t_2]$$

in particular,

$$\boldsymbol{\alpha}[\mathbf{F}(t_0) \vdots \mathbf{F}^{(1)}(t_0) \vdots \cdots \vdots \mathbf{F}^{(n-1)}(t_0)] = \mathbf{0}$$

which implies that all the n rows of

$$[\mathbf{F}(t_0) \vdots \mathbf{F}^{(1)}(t_0) \vdots \cdots \vdots \mathbf{F}^{(n-1)}(t_0)]$$

are linearly dependent. This contradicts the hypothesis that

$$[\mathbf{F}(t_0) \vdots \mathbf{F}^{(1)}(t_0) \vdots \cdots \vdots \mathbf{F}^{(n-1)}(t_0)]$$

has rank n. Hence the $\mathbf{f}_i$s are linearly independent on $[t_1, t_2]$. Q.E.D.

The condition of Theorem 5-2 is sufficient but not necessary for a set of functions to be linearly independent. This can be seen from the following example.

Example 2

Consider the two functions

$$f_1(t) = t^3$$
$$f_2(t) = |t^3|$$

defined over $[-1, 1]$. They are linearly independent on $[-1, 1]$; however,

$$\rho\begin{bmatrix} f_1(t) & f_1^{(1)}(t) \\ f_2(t) & f_2^{(1)}(t) \end{bmatrix} = \rho\begin{bmatrix} t^3 & 3t^2 \\ t^3 & 3t^2 \end{bmatrix} = 1 \qquad \text{for all } t \text{ in } (0, 1]$$

$$\rho\begin{bmatrix} f_1(t) & f_1^{(1)}(t) \\ f_2(t) & f_2^{(1)}(t) \end{bmatrix} = \rho\begin{bmatrix} t^3 & 3t^2 \\ -t^3 & -3t^2 \end{bmatrix} = 1 \qquad \text{for all } t \text{ in } [-1, 0)$$

and

$$\rho\begin{bmatrix} f_1(t) & f_1^{(1)}(t) \\ f_2(t) & f_2^{(1)}(t) \end{bmatrix} = 0 \qquad \text{at } t = 0$$

■

To check the linear independence of a set of functions, if the functions are continuous, we can employ Theorem 5-1, which requires an integration over an interval. If the functions are continuously differentiable up to certain order, then Theorem 5-2 can be used. It is clear that Theorem 5-2 is easier to use than Theorem 5-1; however, it gives only sufficient conditions. If the functions are analytic, then we can use Theorem 5-3, which is based on the fact that if a function is analytic on $[t_1, t_2]$, then the function is completely determinable from a point in $[t_1, t_2]$ if all the derivatives of the function at that point are known. (See Appendix B.)

Theorem 5-3

Assume that for each i, $\mathbf{f}_i$ is analytic on $[t_1, t_2]$. Let $\mathbf{F}$ be the $n \times p$ matrix with $\mathbf{f}_i$ as its ith row, and let $\mathbf{F}^{(k)}$ be the kth derivative of $\mathbf{F}$. Let t_0 be any fixed point in $[t_1, t_2]$. Then the $\mathbf{f}_i$'s are linearly independent on $[t_1, t_2]$ if and only if

$$\rho[\mathbf{F}(t_0) \vdots \mathbf{F}^{(1)}(t_0) \vdots \cdots \vdots \mathbf{F}^{(n-1)}(t_0) \vdots \cdots] = n \tag{5-7}$$

Proof

The sufficiency of the theorem can be proved as in Theorem 5-2. Now we prove by contradiction the necessity of the theorem. Suppose that

$$\rho[\mathbf{F}(t_0) \vdots \mathbf{F}^{(1)}(t_0) \vdots \cdots \vdots \mathbf{F}^{(n-1)}(t_0) \vdots \cdots] < n$$

Then the rows of the infinite matrix

$$[\mathbf{F}(t_0) \vdots \mathbf{F}^{(1)}(t_0) \vdots \cdots \vdots \mathbf{F}^{(n-1)}(t_0) \vdots \cdots]$$

are linearly dependent. Consequently, there exists a nonzero $1 \times n$ row vector $\boldsymbol{\alpha}$ such that

$$\boldsymbol{\alpha}[\mathbf{F}(t_0) \vdots \mathbf{F}^{(1)}(t_0) \vdots \cdots \vdots \mathbf{F}^{(n-1)}(t_0) \vdots \cdots] = \mathbf{0} \tag{5-8}$$

The $\mathbf{f}_i$'s are analytic on $[t_1, t_2]$ by assumption; hence there exists an $\varepsilon > 0$ such that, for all t in $[t_0 - \varepsilon, t_0 + \varepsilon]$, $\mathbf{F}(t)$ can be represented as a Taylor series about the point t_0:

$$\mathbf{F}(t) = \sum_{n=0}^{\infty} \frac{(t - t_0)^n}{n!} \mathbf{F}^{(n)}(t_0) \qquad \text{for all } t \text{ in } [t_0 - \varepsilon, t_0 + \varepsilon] \tag{5-9}$$

Premultiplying $\boldsymbol{\alpha}$ on both sides of (5-9) and using (5-8), we obtain

$$\boldsymbol{\alpha}\mathbf{F}(t) = \mathbf{0} \qquad \text{for all } t \text{ in } [t_0 - \varepsilon, t_0 + \varepsilon] \tag{5-10}$$

Since the sum of analytic functions is an analytic function, the analyticity assumption of the $\mathbf{f}_i$'s implies that $\boldsymbol{\alpha}\mathbf{F}(t)$ as a row vector function is analytic over $[t_1, t_2]$. Consequently, Equation (5-10) implies that

$$\boldsymbol{\alpha}\mathbf{F}(t) = \mathbf{0} \qquad \text{for all } t \text{ in } [t_1, t_2]$$

or, equivalently, the $\mathbf{f}_i$'s are linearly dependent on $[t_1, t_2]$. This is a contradiction. Q.E.D.

A direct consequence of this theorem is that if a set of analytic functions is linearly independent on $[t_1, t_2]$, then

$$\rho[\mathbf{F}(t) \vdots \mathbf{F}^{(1)}(t) \vdots \cdots \vdots \mathbf{F}^{(n-1)}(t) \vdots \cdots] = n$$

for *all* t in $[t_1, t_2]$. It follows that *if a set of analytic functions is linearly independent on $[t_1, t_2]$, then the set of analytic functions is linearly independent on* *every subinterval of* $[t_1, t_2]$. In this statement, the analyticity assumption is essential. The statement does not hold without it, as we have seen in Example 1.

Note that Theorem 5-3 is not true if the infinite matrix in (5-7) is replaced by

$$\rho[\mathbf{F}(t_0) \vdots \mathbf{F}^{(1)}(t_0) \vdots \cdots \vdots \mathbf{F}^{(n-1)}(t_0)] = n$$

This can be seen from the following example.

Example 3

Let

$$\mathbf{F}(t) = \begin{bmatrix} \sin 1000t \\ \sin 2000t \end{bmatrix}$$

Then

$$[\mathbf{F}(t) \vdots \mathbf{F}^{(1)}(t)] = \left[\begin{array}{c:c} \sin 1000t & 10^3 \cos 1000t \\ \sin 2000t & 2 \times 10^3 \cos 2000t \end{array}\right]$$

It is easy to verify that $\rho[\mathbf{F}(t) \vdots \mathbf{F}^{(1)}(t)] < 2$ at $t = 0, \pm 10^{-3}\pi, \ldots$. However,

$$\rho[\mathbf{F}(t) \vdots \mathbf{F}^{(1)}(t) \vdots \mathbf{F}^{(2)}(t) \vdots \mathbf{F}^{(3)}(t)]$$

$$= \rho\left[\begin{array}{c:c:c:c} \sin 10^3 t & 10^3 \cos 10^3 t & -10^6 \sin 10^3 t & -10^9 \cos 10^3 t \\ \sin 2 \times 10^3 t & 2 \times 10^3 \cos 2 \times 10^3 t & -4 \times 10^6 \sin 2 \times 10^3 t & -8 \times 10^9 \cos 2 \times 10^3 t \end{array}\right] = 2$$

for all t. ∎

The matrix in (5-7) has n rows but infinitely many columns. However, in many cases, it is not necessary to check all the derivatives of $\mathbf{F}$. For instance, in Example 3, we check only up to $\mathbf{F}^{(3)}$. If we use the matrix

$$[\mathbf{F}(t) \vdots \mathbf{F}^{(1)}(t) \vdots \cdots \vdots \mathbf{F}^{(n-1)}(t)]$$

then we have the following corollary.

Corollary 5-3

Assume that, for each i, $\mathbf{f}_i$ is analytic on $[t_1, t_2]$. Then $\mathbf{f}_1, \mathbf{f}_2, \ldots, \mathbf{f}_n$ are linearly independent on $[t_1, t_2]$ if and only if

$$\rho[\mathbf{F}(t) \vdots \mathbf{F}^{(1)}(t) \vdots \cdots \vdots \mathbf{F}^{(n-1)}(t)] = n$$

for *almost* all t in $[t_1, t_2]$.

This corollary will not be used in this book, and therefore its proof is omitted.

5-3 Controllability of Linear Dynamical Equations

Time-varying case. In this section, we shall introduce the concept of controllability of linear dynamical equations. To be more precise, we study the *state* controllability of linear *state* equations. As will be seen immediately, the state controllability is a property of state equations only; output equations do not play any role here.

Consider the n-dimensional, linear state equation

$$E: \qquad \dot{\mathbf{x}} = \mathbf{A}(t)\mathbf{x}(t) + \mathbf{B}(t)\mathbf{u}(t) \tag{5-11}$$

where $\mathbf{x}$ is the $n \times 1$ state vector, $\mathbf{u}$ is the $p \times 1$ input vector, and $\mathbf{A}$ and $\mathbf{B}$ are, respectively, $n \times n$ and $n \times p$ matrices whose entries are continuous functions of t defined over $(-\infty, \infty)$. The state space of the equation is an n-dimensional complex vector space and is denoted by $\mathbf{\Sigma}$.

Definition 5-1

The state equation E is said to be (state) *controllable*[4] at time t_0, if there exists a finite $t_1 > t_0$ such that for any $\mathbf{x}(t_0)$ in the state space $\mathbf{\Sigma}$ and any $\mathbf{x}_1$ in $\mathbf{\Sigma}$, there exists an input $\mathbf{u}_{[t_0, t_1]}$ that will tıansfer the state $\mathbf{x}(t_0)$ to the state $\mathbf{x}_1$ at time t_1. Otherwise, the state equation is said to be *uncontrollable* at time t_0.

This definition requires only that the input $\mathbf{u}$ be capable of moving any state in the state space to any other state in a *finite* time; what trajectory the state should take is not specified. Furthermore, there is no constraint imposed on the input. Its magnitude can be as large as desired. We give some examples to illustrate this concept.

Example 1

Consider the network shown in Figure 5-3. The state variable x of the system is the voltage across the capacitor. If $x(t_0) = 0$, then $x(t) = 0$ for all $t \geq t_0$ no matter what input is applied. This is due to the symmetry of the network, and the input has no effect on the voltage across the capacitor. Hence the system—or, more

[4] In the literature, if a state can be transferred to the zero state $\mathbf{0}$, the state is said to be *controllable*. If a state can be reached from $\mathbf{0}$, the state is said to be *reachable*. Our definition does not make this distinction to simplify the subsequent presentation. Furthermore, the equation E is said to be, in the literature, *completely controllable*. For conciseness, the adverb "completely" is dropped in this book.

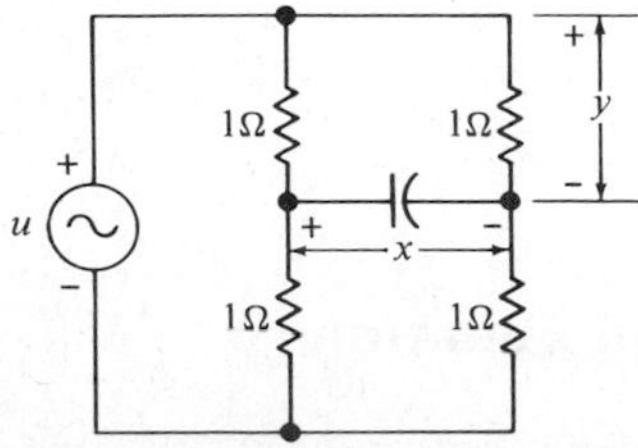

Figure 5-3 An uncontrollable network.

precisely, the dynamical equation that describes the system —is not controllable at any t_0.

Example 2

Consider the system shown in Figure 5-4. There are two state variables x_1 and x_2 in the system. The input can transfer x_1 *or* x_2 to any value; however, it cannot transfer x_1 *and* x_2 to any values. For example, if $x_1(t_0)=0$, $x_2(t_0)=0$, then no matter what input is applied, $x_1(t)$ is always equal to $x_2(t)$, for all $t > t_0$. Hence the equation that describes the system is not controllable at any t_0. ∎

The solution of the state equation E with $\mathbf{x}(t_0)=\mathbf{x}_0$ is given by

$$\mathbf{x}(t)=\boldsymbol{\phi}(t; t_0, \mathbf{x}_0, \mathbf{u})=\mathbf{\Phi}(t, t_0)\mathbf{x}_0+\int_{t_0}^{t}\mathbf{\Phi}(t, \tau)\mathbf{B}(\tau)\mathbf{u}(\tau)\, d\tau$$

$$=\mathbf{\Phi}(t, t_0)\left[\mathbf{x}_0+\int_{t_0}^{t}\mathbf{\Phi}(t_0, \tau)\mathbf{B}(\tau)\mathbf{u}(\tau)\, d\tau\right] \tag{5-12}$$

where $\mathbf{\Phi}(t, t_0)=\mathbf{\Psi}(t)\mathbf{\Psi}^{-1}(t_0)$; $\mathbf{\Psi}$ is a fundamental matrix of $\dot{\mathbf{x}}=\mathbf{A}(t)\mathbf{x}$ and is nonsingular for all t.

Theorem 5-4

The state equation E is controllable at time t_0 if and only if there exists a finite $t_1 > t_0$ such that the n *rows* of the $n \times p$ matrix function $\mathbf{\Phi}(t_0, \cdot)\mathbf{B}(\cdot)$ are linearly independent on $[t_0, t_1]$.

Proof

Sufficiency: If the rows of $\mathbf{\Phi}(t_0, \cdot)\mathbf{B}(\cdot)$ are linearly independent on $[t_0, t_1]$, from Theorem 5-1 the $n \times n$ constant matrix

$$\mathbf{W}(t_0, t_1) \triangleq \int_{t_0}^{t_1}\mathbf{\Phi}(t_0, \tau)\mathbf{B}(\tau)\mathbf{B}^*(\tau)\mathbf{\Phi}^*(t_0, \tau)\, d\tau \tag{5-13}$$

is nonsingular. Given any $\mathbf{x}(t_0)=\mathbf{x}_0$ and any $\mathbf{x}_1$, we claim that the input

$$\mathbf{u}(t)=-\mathbf{B}^*(t)\mathbf{\Phi}^*(t_0, t)\mathbf{W}^{-1}(t_0, t_1)[\mathbf{x}_0-\mathbf{\Phi}(t_0, t_1)\mathbf{x}_1] \tag{5-14}$$

will transfer $\mathbf{x}_0$ to the state $\mathbf{x}_1$ at time t_1. Indeed, by substituting (5-14) into

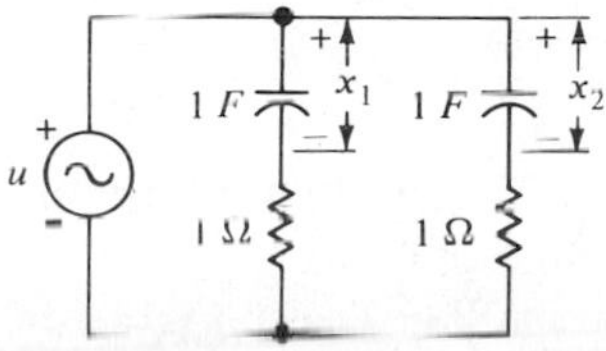

Figure 5-4 An uncontrollable network.

(5-12), we obtain

$$\mathbf{x}(t_1)=\mathbf{\Phi}(t_1, t_0)\left\{\mathbf{x}_0-\int_{t_0}^{t_1}\mathbf{\Phi}(t_0, \tau)\mathbf{B}(\tau)\mathbf{B}^*(\tau)\mathbf{\Phi}^*(t_0, \tau)\, d\tau\, \mathbf{W}^{-1}(t_0, t_1)\cdot[\mathbf{x}_0-\mathbf{\Phi}(t_0, t_1)\mathbf{x}_1]\right\}$$

$$\begin{aligned}&=\mathbf{\Phi}(t_1, t_0)\{\mathbf{x}_0-\mathbf{W}(t_0, t_1)\mathbf{W}^{-1}(t_0, t_1)[\mathbf{x}_0-\mathbf{\Phi}(t_0, t_1)\mathbf{x}_1]\}\\&=\mathbf{\Phi}(t_1, t_0)\mathbf{\Phi}(t_0, t_1)\mathbf{x}_1\\&=\mathbf{x}_1\end{aligned}$$

Thus we conclude that the equation E is controllable. *Necessity*: The proof is by contradiction. Suppose E is controllable at t_0, but the rows of $\mathbf{\Phi}(t_0, \cdot)\mathbf{B}(\cdot)$ are linearly dependent on $[t_0, t_1]$ for all $t_1 > t_0$. Then there exists a nonzero, constant $1 \times n$ row vector $\boldsymbol{\alpha}$ such that

$$\boldsymbol{\alpha}\mathbf{\Phi}(t_0, t)\mathbf{B}(t)=\mathbf{0} \qquad \text{for all } t \text{ in } [t_0, t_1] \tag{5-15}$$

Let us choose $\mathbf{x}(t_0)\triangleq\mathbf{x}_0=\boldsymbol{\alpha}^*$. Then Equation (5-12) becomes

$$\mathbf{\Phi}(t_0, t_1)\mathbf{x}(t_1)=\boldsymbol{\alpha}^*+\int_{t_0}^{t_1}\mathbf{\Phi}(t_0, \tau)\mathbf{B}(\tau)\mathbf{u}(\tau)\, d\tau \tag{5-16}$$

Premultiplying both sides of (5-16) by $\boldsymbol{\alpha}$, we obtain

$$\boldsymbol{\alpha}\mathbf{\Phi}(t_0, t_1)\mathbf{x}(t_1)=\boldsymbol{\alpha}\boldsymbol{\alpha}^*+\int_{t_0}^{t_1}\boldsymbol{\alpha}\mathbf{\Phi}(t_0, \tau)\mathbf{B}(\tau)\mathbf{u}(\tau)\, d\tau \tag{5-17}$$

By hypothesis, E is controllable at t_0; hence for any state—in particular, $\mathbf{x}_1=\mathbf{0}$—there exists $\mathbf{u}_{[t_0, t_1]}$ such that $\mathbf{x}(t_1)=\mathbf{0}$. Since $\boldsymbol{\alpha}\mathbf{\Phi}(t_0, t)\mathbf{B}(t)=\mathbf{0}$, for all t in $[t_0, t_1]$, Equation (5-17) reduces to

$$\boldsymbol{\alpha}\boldsymbol{\alpha}^*=0$$

which, in turn, implies that $\boldsymbol{\alpha}=\mathbf{0}$. This is a contradiction. Q.E.D.

In the proof of this theorem, we also give in (5-14) an input $\mathbf{u}(t)$ that transfers $\mathbf{x}(t_0)$ to $\mathbf{x}_1$ at time t_1. Because of the continuity assumption of $\mathbf{A}$ and $\mathbf{B}$, the input $\mathbf{u}$ in (5-14) is a continuous function of t in $[t_0, t_1]$.

If a linear dynamical equation is controllable, there are generally many different inputs $\mathbf{u}$ that can transfer $\mathbf{x}(t_0)$ to $\mathbf{x}_1$ at time t_1, for the trajectory between $\mathbf{x}(t_0)$ and $\mathbf{x}_1$ is not specified. Among these possible inputs that achieve the same mission, we may ask which input is optimal according to some criterion. If the total energy

$$\int_{t_0}^{t_1}\|\mathbf{u}(t)\|^2\, dt$$

is used as a criterion, the input given in (5-14) will use the minimal energy in transferring $\mathbf{x}(t_0)$ to $\mathbf{x}_1$ at time t_1. This is established in Appendix C.

In order to apply Theorem 5-4, a fundamental matrix $\mathbf{\Psi}$ or the state transition matrix $\mathbf{\Phi}(t, \tau)$ of $\dot{\mathbf{x}}=\mathbf{A}(t)\mathbf{x}$ has to be computed. As we mentioned earlier, this is generally a difficult task. Hence, Theorem 5-4 is not readily applicable.

In the following we shall give a controllability criterion based on the matrices **A** and **B** without solving the state equation. However, in order to do so, we need some additional assumptions on **A** and **B**. Assume that **A** and **B** are $(n-1)$ times continuously differentiable. Define a sequence of $n \times p$ matrices $\mathbf{M}_0(\cdot)$, $\mathbf{M}_1(\cdot), \ldots,$ by the equation

$$\mathbf{M}_{k+1}(t) = -\mathbf{A}(t)\mathbf{M}_k(t) + \frac{d}{dt}\mathbf{M}_k(t) \qquad k=0, 1, 2, \ldots, n-1 \tag{5-18a}$$

with
$$\mathbf{M}_0(t) = \mathbf{B}(t) \tag{5-18b}$$

Observe that

$$\mathbf{\Phi}(t_0, t)\mathbf{B}(t) = \mathbf{\Phi}(t_0, t)\mathbf{M}_0(t) \tag{5-19a}$$

$$\begin{aligned}\frac{\partial}{\partial t}\mathbf{\Phi}(t_0, t)\mathbf{B}(t) &= \mathbf{\Psi}(t_0)\left\{\left[\frac{d}{dt}\mathbf{\Psi}^{-1}(t)\right]\mathbf{B}(t) + \mathbf{\Psi}^{-1}(t)\frac{d}{dt}\mathbf{B}(t)\right\} \\ &= \mathbf{\Psi}(t_0)\mathbf{\Psi}^{-1}(t)\left[-\mathbf{A}(t)\mathbf{B}(t) + \frac{d}{dt}\mathbf{B}(t)\right] \\ &= \mathbf{\Phi}(t_0, t)\mathbf{M}_1(t)\end{aligned} \tag{5-19b}$$

and, in general,

$$\frac{\partial^k}{\partial t^k}\mathbf{\Phi}(t_0, t)\mathbf{B}(t) = \mathbf{\Phi}(t_0, t)\mathbf{M}_k(t) \qquad k=0, 1, 2, \ldots, n-1 \tag{5-19c}$$

Theorem 5-5

Assume that the matrices $\mathbf{A}(\cdot)$ and $\mathbf{B}(\cdot)$ in the n-dimensional state equation E are $n-1$ times continuously differentiable. Then the state equation E is controllable at t_0 if there exists a finite $t_1 > t_0$ such that

$$\rho[\mathbf{M}_0(t_1) \vdots \mathbf{M}_1(t_1) \vdots \cdots \vdots \mathbf{M}_{n-1}(t_1)] = n \tag{5-20}$$

Proof

Define

$$\left.\frac{\partial}{\partial t}\mathbf{\Phi}(t_0, t)\mathbf{B}(t)\right|_{t=t_1} \triangleq \frac{\partial}{\partial t_1}\mathbf{\Phi}(t_0, t_1)\mathbf{B}(t_1)$$

Then, from (5-19) and using (2-2), we have

$$\begin{aligned}&\left[\mathbf{\Phi}(t_0, t_1)\mathbf{B}(t_1) \vdots \frac{\partial}{\partial t_1}\mathbf{\Phi}(t_0, t_1)\mathbf{B}(t_1) \vdots \cdots \vdots \frac{\partial^{n-1}}{\partial t_1^{n-1}}\mathbf{\Phi}(t_0, t_1)\mathbf{B}(t_1)\right] \\ &\qquad = \mathbf{\Phi}(t_0, t_1)[\mathbf{M}_0(t_1) \vdots \mathbf{M}_1(t_1) \vdots \cdots \vdots \mathbf{M}_{n-1}(t_1)]\end{aligned} \tag{5-21}$$

Since $\mathbf{\Phi}(t_0, t_1)$ is nonsingular, the assumption

$$\rho[\mathbf{M}_0(t_1) \vdots \mathbf{M}_1(t_1) \vdots \cdots \vdots \mathbf{M}_{n-1}(t_1)] = n$$

implies

$$\rho\left[\mathbf{\Phi}(t_0, t_1)\mathbf{B}(t_1) \vdots \frac{\partial}{\partial t_1}\mathbf{\Phi}(t_0, t_1)\mathbf{B}(t_1) \vdots \cdots \vdots \frac{\partial^{n-1}}{\partial t_1^{n-1}}\mathbf{\Phi}(t_0, t_1)\mathbf{B}(t_1)\right] = n$$

It follows from Theorem 5-2 that the rows of $\mathbf{\Phi}(t_0, \cdot)\mathbf{B}(\cdot)$ are linearly independent on $[t_0, t_1]$ for any $t_1 > t_0$. Thus, from Theorem 5-4, we conclude that the state equation E is controllable. Q.E.D.

As in Theorem 5-2, the condition of Theorem 5-5 is sufficient but not necessary for the controllability of a state equation.

Example 3

Consider

$$\begin{bmatrix} \dot{x}_1 \\ \dot{x}_2 \\ \dot{x}_3 \end{bmatrix} = \begin{bmatrix} t & 1 & 0 \\ 0 & t & 0 \\ 0 & 0 & t^2 \end{bmatrix} \begin{bmatrix} x_1 \\ x_2 \\ x_3 \end{bmatrix} + \begin{bmatrix} 0 \\ 1 \\ 1 \end{bmatrix} u \tag{5-22}$$

From (5-18), we have

$$\mathbf{M}_0(t) = \begin{bmatrix} 0 \\ 1 \\ 1 \end{bmatrix}$$

$$\mathbf{M}_1(t) = -\mathbf{A}(t)\mathbf{M}_0(t) + \frac{d}{dt}\mathbf{M}_0(t) = \begin{bmatrix} -1 \\ -t \\ -t^2 \end{bmatrix}$$

$$\mathbf{M}_2(t) = -\mathbf{A}(t)\mathbf{M}_1(t) + \frac{d}{dt}\mathbf{M}_1(t) = \begin{bmatrix} 2t \\ t^2 \\ t^4 \end{bmatrix} + \begin{bmatrix} 0 \\ -1 \\ -2t \end{bmatrix} = \begin{bmatrix} 2t \\ t^2 - 1 \\ t^4 - 2t \end{bmatrix}$$

Since the matrix $[\mathbf{M}_0(t) \vdots \mathbf{M}_1(t) \vdots \mathbf{M}_2(t)]$ has rank 3 for all $t \neq 0$, the dynamical equation is controllable at every t. ∎

Differential controllability, instantaneous controllability, and uniform controllability

The remainder of this subsection will be devoted to three distinct types of controllability. This material may be omitted with no loss of continuity.

Definition 5-2

A state equation E is said to be *differentially* (completely) *controllable* at time t_0 if, for any state $\mathbf{x}(t_0)$ in the state space Σ and any state $\mathbf{x}_1$ in Σ, there exists an input $\mathbf{u}$ that will transfer $\mathbf{x}(t_0)$ to the state $\mathbf{x}_1$ in an *arbitrarily* small interval of time. ∎

If every state in Σ can be transferred to any other state in a finite time (no matter how long), the state equation is said to be controllable. If this can be achieved in an arbitrarily small interval of time, then the state equation is said to be differentially controllable. Clearly, differential controllability implies controllability. The condition for differential controllability can be easily obtained by slight modifications of Theorems 5-4 and 5-5. However, if the matrices $\mathbf{A}$ and $\mathbf{B}$ in the state equation E are analytic on $(-\infty, \infty)$, then we have the following theorem.

Theorem 5-6

If the matrices $\mathbf{A}$ and $\mathbf{B}$ are analytic on $(-\infty, \infty)$, then the n-dimensional state equation E is differentially controllable at every t in $(-\infty, \infty)$ if and only if, for any fixed t_0 in $(-\infty, \infty)$,

$$\rho[\mathbf{M}_0(t_0) \vdots \mathbf{M}_1(t_0) \vdots \cdots \vdots \mathbf{M}_{n-1}(t_0) \vdots \cdots] = n$$ ∎

If the matrix $\mathbf{A}$ is analytic on $(-\infty, \infty)$, it can be shown that the state transition matrix $\mathbf{\Phi}(t_0, \cdot)$ of $\dot{\mathbf{x}} = \mathbf{A}(t)\mathbf{x}$ is also analytic on $(-\infty, \infty)$. Since the product of two analytic functions is an analytic function, the assumption of Theorem 5-6 implies that $\mathbf{\Phi}(t_0, \cdot)\mathbf{B}(\cdot)$ is an analytic function. An implication of Theorem 5-3 is that a set of analytic functions is linearly independent on $(-\infty, \infty)$ if and only if the set of analytic functions is linearly independent on every subinterval (no matter how small) of $(-\infty, \infty)$. With this fact, Theorem 5-6 follows immediately from Theorems 5-3 and 5-4. Consequently, if a state equation with analytic $\mathbf{A}$ and $\mathbf{B}$ is controllable at any point at all, it is differentially controllable at every t in $(-\infty, \infty)$.

Definition 5-3

The linear dynamical equation E is said to be *instantaneously controllable*[5] in $(-\infty, \infty)$ if and only if

$$\rho[\mathbf{M}_0(t) \vdots \mathbf{M}_1(t) \vdots \cdots \vdots \mathbf{M}_{n-1}(t)] = n \qquad \text{for all } t \text{ in } (-\infty, \infty)$$

where the $\mathbf{M}_i$'s are as defined in Equation (5-18). ∎

If a dynamical equation is instantaneously controllable, then the transfer of the states can be achieved instantaneously at any time by using an input that consists of δ-functions and their derivatives up to an order of $n-1$. It is clear that instantaneous controllability implies differential controllability. The most important implication of instantaneous controllability is that in the case of a single input, the matrix $[\mathbf{M}_0(t) \quad \mathbf{M}_1(t) \quad \cdots \quad \mathbf{M}_{n-1}(t)]$ qualifies as an equivalence transformation (Definition 4-5). Consequently, many canonical-form equivalent dynamical equations can be obtained for instantaneously controllable dynamical equations. See References 10, 99, 103, and 110.

---

[5] In the engineering literature, it is called *uniform controllability*. However, this terminology was first used in Reference 56 to define a different kind of controllability (see Definition 5-4); hence we adopt the terminology "instantaneous controllability."

Definition 5-4

The dynamical equation E is said to be *uniformly controllable* if and only if there exist a positive σ_c and positive α_i that depend on σ_c such that

$$\mathbf{0} < \alpha_1(\sigma_c)\mathbf{I} \leq \mathbf{W}(t, t + \sigma_c) \leq \alpha_2(\sigma_c)\mathbf{I}$$

and
$$\mathbf{0} < \alpha_3(\sigma_c)\mathbf{I} \leq \mathbf{\Phi}(t + \sigma_c, t)\mathbf{W}(t, t + \sigma_c)\mathbf{\Phi}^*(t + \sigma_c, t) \leq \alpha_4(\sigma_c)\mathbf{I}$$

for all t in $(-\infty, \infty)$, where $\mathbf{\Phi}$ is the state transition matrix and $\mathbf{W}$ is as defined in Equation (5-13). ∎

By $\mathbf{A} > \mathbf{B}$, we mean that the matrix $(\mathbf{A} - \mathbf{B})$ is a positive definite matrix (see Section 8-5). Uniform controllability ensures that the transfer of the states can be achieved in the time interval σ_c. The concept of uniform controllability is needed in the stability study of optimal control systems. See References 56 and 102.

Instantaneous controllability and uniform controllability both imply controllability. However, instantaneous controllability neither implies nor is implied by uniform controllability.

Example 4

Consider the one-dimensional linear dynamical equation

$$\dot{x} = e^{-|t|}u$$

Since $\rho(M_0(t)) = \rho(e^{-|t|}) = 1$ for all t, the dynamical equation is instantaneously controllable at every t. However, the dynamical equation is not uniformly controllable because there exists no α_1 that depends on σ_c but not on t such that

$$W(t, t + \sigma_c) = \int_t^{t + \sigma_c} e^{-2t}\, dt = 0.5e^{-2t}(1 - e^{-2\sigma_c}) > \alpha_1(\sigma_c)$$

for all $t > 0$. ∎

Example 5

Consider the one-dimensional dynamical equation

$$\dot{x} = b(t)u$$

with $b(t)$ defined as in Figure 5-5. The dynamical equation is not instantaneously controllable in the interval $(-1, 1)$. However, it is uniformly controllable in $(-\infty, \infty)$. This can be easily verified by choosing $\sigma_c = 5$. ∎

A remark is in order concerning controllability, differential controllability, and uniform controllability. If a dynamical equation is differentially controllable, a state can be transferred to any other state in an arbitrarily small interval of time. However, the magnitude of the input may become very large; in the extreme case, a delta-function input is required. If a dynamical equation is merely controllable, the transfer of the states may take a very long interval of time. However, if it is uniformly controllable, the transfer of the states can be

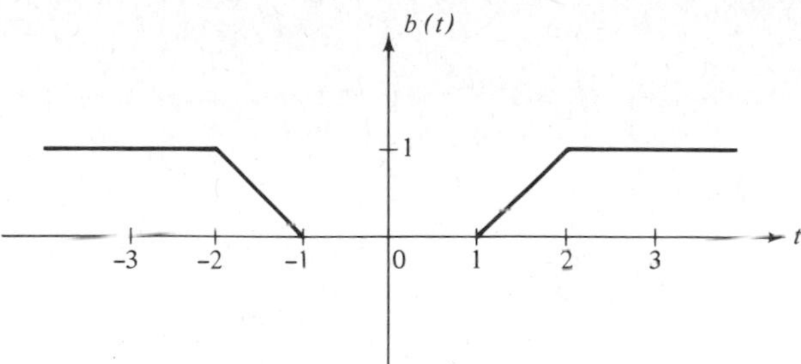

Figure 5-5 The function $b(t)$.

achieved in the time interval σ_c; moreover, the magnitude of the control input will *not* be arbitrarily large [see (5-14) and note that the input is proportional to $\mathbf{W}^{-1}$]. In optimal control theory, the condition of uniform controllability is sometimes required to ensure the stability of an optimal control system.

Time-invariant case. In this subsection, we study the controllability of the n-dimensional linear time-invariant state equation

$$FE: \qquad \dot{\mathbf{x}} = \mathbf{A}\mathbf{x} + \mathbf{B}\mathbf{u} \tag{5-23}$$

where $\mathbf{x}$ is the $n \times 1$ state vector, $\mathbf{u}$ is the $p \times 1$ input vector; $\mathbf{A}$ and $\mathbf{B}$ are $n \times n$ and $n \times p$ real constant matrices, respectively. For time-invariant dynamical equations, the time interval of interest is from the present time to infinity; that is, $[0, \infty)$.

The condition for a time-varying state equation to be controllable at t_0 is that there exists a finite t_1 such that all rows of $\mathbf{\Phi}(t_0, t)\mathbf{B}(t)$ are linearly independent, in t, on $[t_0, t_1]$. In the time-invariant case, we have $\mathbf{\Phi}(t_0, t)\mathbf{B}(t) = e^{\mathbf{A}(t_0 - t)}\mathbf{B}$. As discussed in Chapter 4, all elements of $e^{\mathbf{A}(t_0 - t)}\mathbf{B}$ are linear combinations of terms of form $t^k e^{\lambda t}$; hence they are analytic on $[0, \infty)$ (see Appendix B). Consequently, if its rows are linearly independent on $[0, \infty)$, they are linearly independent on $[t_0, t_1]$ for any t_0 and any $t_1 > t_0$. In other words, if a linear time-invariant state equation is controllable, it is controllable at every $t_0 \geq 0$ and the transfer of any state to any other state can be achieved in any nonzero time interval. Hence the reference of t_0 and t_1 is often dropped in the controllability study of linear time-invariant state equations.

Theorem 5-7

The n-dimensional linear time-invariant state equation in (5-23) is controllable if and only if any of the following equivalent conditions is satisfied:

1. All rows of $e^{-\mathbf{A}t}\mathbf{B}$ (and consequently of $e^{\mathbf{A}t}\mathbf{B}$) are linearly independent on $[0, \infty)$ over $\mathbb{C}$, the field of complex numbers.[6]

1'. All rows of $(s\mathbf{I} - \mathbf{A})^{-1}\mathbf{B}$ are linearly independent over $\mathbb{C}$.

---

[6] Although all the entries of $\mathbf{A}$ and $\mathbf{B}$ are real numbers, we have agreed to consider them as elements of the field of complex numbers.

2. The *controllability grammian*

$$\mathbf{W}_{ct} \triangleq \int_0^t e^{\mathbf{A}\tau}\mathbf{B}\mathbf{B}^* e^{\mathbf{A}^*\tau}\, d\tau$$

is nonsingular for any $t > 0$.[7]

3. The $n \times (np)$ *controllability matrix*

$$\mathbf{U} \triangleq [\mathbf{B} \vdots \mathbf{A}\mathbf{B} \vdots \mathbf{A}^2\mathbf{B} \vdots \cdots \vdots \mathbf{A}^{n-1}\mathbf{B}] \tag{5-24}$$

has rank n.

4. For every eigenvalue λ of $\mathbf{A}$ (and consequently for every λ in $\mathbb{C}$), the $n \times (n+p)$ complex matrix $[\lambda\mathbf{I} - \mathbf{A} \vdots \mathbf{B}]$ has rank n.[8]

Proof

The equivalence of statements 1 and 2 follows directly from Theorems 5-1 and 5-4. Since the entries of $e^{-\mathbf{A}t}\mathbf{B}$ are analytic functions, Theorem 5-3 implies that the rows of $e^{-\mathbf{A}t}\mathbf{B}$ are linearly independent on $[0, \infty)$ if and only if

$$\rho[e^{-\mathbf{A}t}\mathbf{B} \vdots -e^{-\mathbf{A}t}\mathbf{A}\mathbf{B} \vdots \cdots \vdots (-1)^{n-1}e^{-\mathbf{A}t}\mathbf{A}^{n-1}\mathbf{B} \vdots \cdots] = n$$

for *any* t in $[0, \infty)$. Let $t = 0$; then the equation reduces to

$$\rho[\mathbf{B} \vdots -\mathbf{A}\mathbf{B} \vdots \cdots \vdots (-1)^{n-1}\mathbf{A}^{n-1}\mathbf{B} \vdots (-1)^n\mathbf{A}^n\mathbf{B} \vdots \cdots] = n$$

From the Cayley–Hamilton theorem, we know that $\mathbf{A}^m$ with $m \geq n$ can be written as a linear combination of $\mathbf{I}, \mathbf{A}, \ldots, \mathbf{A}^{n-1}$; hence the columns of $\mathbf{A}^m\mathbf{B}$ with $m \geq n$ are linearly dependent on the columns of $\mathbf{B}, \mathbf{A}\mathbf{B}, \ldots, \mathbf{A}^{n-1}\mathbf{B}$. Consequently,

$$\rho[\mathbf{B} \vdots -\mathbf{A}\mathbf{B} \vdots \cdots \vdots (-1)^{n-1}\mathbf{A}^{n-1}\mathbf{B} \vdots \cdots] = \rho[\mathbf{B} \vdots -\mathbf{A}\mathbf{B} \vdots \cdots \vdots (-1)^{n-1}\mathbf{A}^{n-1}\mathbf{B}]$$

Since changing the sign will not change the linear independence, we conclude that the rows of $e^{-\mathbf{A}t}\mathbf{B}$ are linearly independent if and only if $\rho[\mathbf{B} \vdots \mathbf{A}\mathbf{B} \vdots \cdots \vdots \mathbf{A}^{n-1}\mathbf{B}] = n$. This proves the equivalence of statements 1 and 3. In the foregoing argument we also proved that the rows of $e^{-\mathbf{A}t}\mathbf{B}$ are linearly independent if and only if the rows of $e^{\mathbf{A}t}\mathbf{B}$ are linearly independent on $[0, \infty)$ over the field of complex numbers. Next we show the equivalence of statements 1 and 1′. Taking the Laplace transform of $e^{\mathbf{A}t}\mathbf{B}$, we have

$$\mathscr{L}[e^{\mathbf{A}t}\mathbf{B}] = (s\mathbf{I} - \mathbf{A})^{-1}\mathbf{B}$$

Since the Laplace transform is a one-to-one linear operator, if the rows of $e^{\mathbf{A}t}\mathbf{B}$ are linearly independent on $[0, \infty)$ over the field of complex numbers, so are the rows of $(s\mathbf{I} - \mathbf{A})^{-1}\mathbf{B}$, and vice versa.

The proof of statement 4 will be postponed to Section 5-5 (page 206). Q.E.D.

---

[7] The matrix is in fact positive definite. See Problem E-11.

[8] This condition implies that $(s\mathbf{I} - \mathbf{A})$ and $\mathbf{B}$ are left coprime. See Appendix G.

Example 6

Consider the inverted pendulum system studied in Figure 3-15. Its dynamical equation is developed in (3-42). For convenience, we assume $2mg/(2M+m)=1$, $2g(M+m)/(2M+m)l=5$, $2/(2M+m)=1$, and $1/(2M+m)l=2$. Then Equation (3-42) becomes

$$\dot{\mathbf{x}}=\begin{bmatrix}0 & 1 & 0 & 0\\ 0 & 0 & -1 & 0\\ 0 & 0 & 0 & 1\\ 0 & 0 & 5 & 0\end{bmatrix}\mathbf{x}+\begin{bmatrix}0\\ 1\\ 0\\ -2\end{bmatrix}u \qquad \mathbf{y}=[1 \quad 0 \quad 0 \quad 0]\mathbf{x} \tag{5-25}$$

We compute

$$\mathbf{U}=[\mathbf{B} \quad \mathbf{AB} \quad \mathbf{A}^2\mathbf{B} \quad \mathbf{A}^3\mathbf{B}]=\begin{bmatrix}0 & 1 & 0 & 2\\ 1 & 0 & 2 & 0\\ 0 & -2 & 0 & -10\\ -2 & 0 & -10 & 0\end{bmatrix}$$

This matrix can be readily shown to have rank 4. Hence (5-25) is controllable. Thus, if $x_3=\theta$ is different from zero by a small amount, a control u can be found to bring it back to zero. In fact, a control exists to bring $x_1=y$, $x_3=\theta$, and their derivatives back to zero. This is certainly consistent with our experience of balancing a broom on our hand. ∎

Example 7

Consider the platform system shown in Figure 5-2. It is assumed that the mass of the platform is zero and the force is equally divided among the two spring systems. The spring constants are assumed to be 1, and the viscous friction coefficients are assumed to be 2 and 1 as shown. Then we have $x_1+2\dot{x}_1=u$ and $x_2+\dot{x}_2=u$, or

$$\begin{bmatrix}\dot{x}_1\\ \dot{x}_2\end{bmatrix}=\begin{bmatrix}-0.5 & 0\\ 0 & -1\end{bmatrix}\begin{bmatrix}x_1\\ x_2\end{bmatrix}+\begin{bmatrix}0.5\\ 1\end{bmatrix}u \tag{5-26}$$

This is the state-variable description of the system.

Now if the initial displacements $x_1(0)$ and $x_2(0)$ are different from zero, the platform will oscillate, and it will take, theoretically, an infinite time for the platform to come to rest. Now we pose the following problem: If $x_1(0)=10$ and $x_2(0)=-1$, is it possible to apply a force to bring the platform to rest in 2 seconds? The answer does not seem to be obvious because the *same* force is applied to the two spring systems.

For the state equation in (5-26), we compute

$$\rho[\mathbf{B} : \mathbf{AB}]=\rho\begin{bmatrix}0.5 & -0.25\\ 1 & -1\end{bmatrix}=2$$

hence the state equation is controllable. Consequently, the displacements can be brought to zero in 2 seconds by applying a proper input u. Using Equations (5-13) and (5-14), we have

$$\mathbf{W}(0, 2) = \int_0^2 \begin{bmatrix} e^{0.5\tau} & 0 \\ 0 & e^{\tau} \end{bmatrix} \begin{bmatrix} 0.5 \\ 1 \end{bmatrix} [0.5 \quad 1] \begin{bmatrix} e^{0.5\tau} & 0 \\ 0 & e^{\tau} \end{bmatrix} d\tau = \begin{bmatrix} 1.6 & 6.33 \\ 6.33 & 27 \end{bmatrix}$$

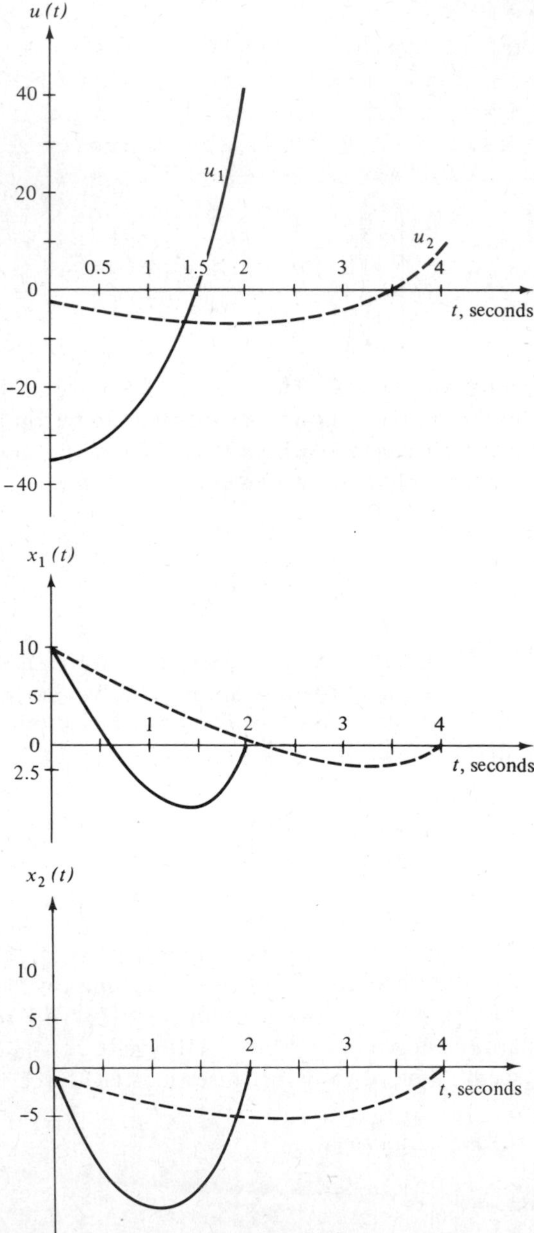

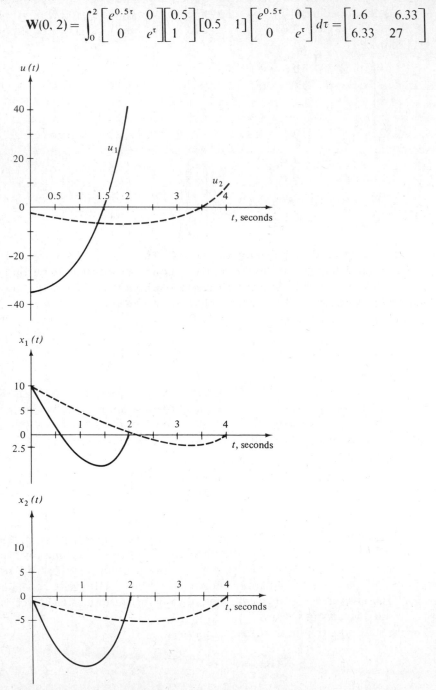

Figure 5-6 Behavior of $x_1(t)$ and $x_2(t)$ and the waveform of u.

and $$u_1(t) = -[0.5 \quad 1]\begin{bmatrix} e^{0.5t} & 0 \\ 0 & e^t \end{bmatrix}\mathbf{W}^{-1}(0, 2)\begin{bmatrix} 10 \\ -1 \end{bmatrix} = -44.1e^{0.5t} + 20.7e^t$$

for t in $[0, 2]$. If a force of the form u_1 is applied, the platform will come to rest at $t = 2$. The behavior of x_1, x_2 and of the input u_1 are plotted by using solid lines in Figure 5-6. ∎

In Figure 5-6 we also plot by using dotted lines the input $u_2(t)$ that transfers $x_1(0) = 10$ and $x_2(0) = -1$ to zero in 4 seconds. We see from Figure 5-6 that, in transferring $\mathbf{x}(0)$ to zero, the smaller the time interval the larger the magnitude of the input. If no restriction is imposed on the input u, then we can transfer $\mathbf{x}(0)$ to zero in an arbitrarily small interval of time; however, the magnitude of the input may become very large. If some restriction on the magnitude of u is imposed, we might not be able to transfer $\mathbf{x}(0)$ to zero in an arbitrarily small interval of time. For example, if we require $|u(t)| \leq 5$ in Example 5, then we might not be able to transfer $\mathbf{x}(0)$ to zero in less than 4 seconds.

Example 8

Consider again the platform system shown in Figure 5-2. Now it is assumed that the viscous friction coefficient and the spring constant of both spring systems are all equal to 1. Then the state-variable description of the platform system is

$$\begin{bmatrix} \dot{x}_1 \\ \dot{x}_2 \end{bmatrix} = \begin{bmatrix} -1 & 0 \\ 0 & -1 \end{bmatrix}\mathbf{x} + \begin{bmatrix} 1 \\ 1 \end{bmatrix}u$$

Clearly we have

$$\rho\mathbf{U} = \rho\begin{bmatrix} 1 & -1 \\ 1 & -1 \end{bmatrix} = 1 < 2$$

and the state equation is not controllable. If $x_1(0) = x_2(0)$, it is possible to find an input to transfer $\mathbf{x}(0)$ to the zero state in a finite time. However, if $x_1(0) \neq x_2(0)$, no input can transfer both $x_1(0)$ and $x_2(0)$ to zero in a finite time. ∎

*Controllability indices

Let $\mathbf{A}$ and $\mathbf{B}$ be $n \times n$ and $n \times p$ constant matrices. Define

$$\mathbf{U}_k = [\mathbf{B} \vdots \mathbf{AB} \vdots \cdots \vdots \mathbf{A}^k\mathbf{B}] \qquad k = 0, 1, 2, \ldots \tag{5-27}$$

It consists of $k+1$ block columns of the form $\mathbf{A}^i\mathbf{B}$ and is of order $n \times (k+1)p$. The matrix $\mathbf{U} \triangleq \mathbf{U}_{n-1}$ is the controllability matrix. If $\{\mathbf{A}, \mathbf{B}\}$ is controllable, then $\mathbf{U}_{n-1}$ has a rank of n and consequently has n linearly independent columns. Note that there are a total of np columns in $\mathbf{U}_{n-1}$; hence there are many possible ways to choose these n linearly independent columns. We discuss in the following the most natural and also the most important way of choosing these columns. Let $\mathbf{b}_i$, $i = 1, 2, \ldots, p$, be the ith column of $\mathbf{B}$. Then the matrix $\mathbf{U}_k$ can be written explicitly as

$$\mathbf{U}_k = [\underbrace{\mathbf{b}_1 \quad \mathbf{b}_2 \quad \cdots \quad \mathbf{b}_p}_{r_0} \ \vdots \ \underbrace{\mathbf{A}\mathbf{b}_1 \quad \cdots \quad \mathbf{A}\mathbf{b}_p}_{r_1} \ \vdots \ \mathbf{A}^2\mathbf{b}_1 \quad \cdots \quad \mathbf{A}^2\mathbf{b}_p \ \vdots \ \cdots$$

$$\vdots \ \underbrace{\mathbf{A}^k\mathbf{b}_1 \quad \cdots \quad \mathbf{A}^k\mathbf{b}_p]}_{r_k \text{ (no. of dependent columns)}} \tag{5-28}$$

Now we search linearly independent columns of $\mathbf{U}_k$ in order from left to right; that is, if a column can be written as a linear combination of its left-hand-side columns, the column is linearly dependent; otherwise, it is linearly independent. This process of searching can be carried out by using, for example, the column-searching algorithm discussed in Appendix A. Let r_i be the number of linearly dependent columns in $\mathbf{A}^i\mathbf{B}$, for $i = 0, 1, \ldots, k$. If $\mathbf{B}$ has a full column rank, then $r_0 = 0$. We note that if a column, say $\mathbf{A}\mathbf{b}_2$, is linearly dependent on its left-hand-side columns, then all $\mathbf{A}^j\mathbf{b}_2$, with $j = 2, 3, \ldots$, will be linearly dependent on its left-hand side columns. Indeed, if

$$\mathbf{A}\mathbf{b}_2 = \alpha_1\mathbf{b}_1 + \alpha_2\mathbf{b}_2 + \cdots + \alpha_p\mathbf{b}_p + \alpha_{p+1}\mathbf{A}\mathbf{b}_1$$

then
$$\mathbf{A}^2\mathbf{b}_2 = \alpha_1\mathbf{A}\mathbf{b}_1 + \alpha_2\mathbf{A}\mathbf{b}_2 + \cdots + \alpha_p\mathbf{A}\mathbf{b}_p + \alpha_{p+1}\mathbf{A}^2\mathbf{b}_1$$

We see that $\mathbf{A}^2\mathbf{b}_2$ is a linear combination of its left-hand-side columns. Proceeding similarly, we can show that $\mathbf{A}^j\mathbf{b}_2$, for $j = 3, 4, \ldots$, are linearly dependent on its left-hand-side columns. Because of this property, we have

$$0 \leq r_0 \leq r_1 \leq r_2 \leq \cdots \leq p$$

Since there are at most n linearly independent columns in $\mathbf{U}_\infty$, there exists an integer μ such that

$$0 \leq r_0 \leq r_1 \leq \cdots \leq r_{\mu-1} < p \tag{5-29a}$$

and

$$r_\mu = r_{\mu+1} = \cdots = p \tag{5-29b}$$

Equivalently, μ is the integer such that

$$\rho\mathbf{U}_0 < \rho\mathbf{U}_1 < \cdots < \rho\mathbf{U}_{\mu-1} = \rho\mathbf{U}_\mu = \rho\mathbf{U}_{\mu+1} = \cdots \tag{5-30}$$

In other words, the rank of $\mathbf{U}_k$ increases monotonically until k reaches $\mu - 1$; thereafter, all p columns of $\mathbf{A}^j\mathbf{B}$ will be linearly dependent of their left-hand-side columns. Hence the controllability of $\{\mathbf{A}, \mathbf{B}\}$ can be checked from $\mathbf{U}_{\mu-1}$ and μ is called the *controllability index*. We claim that the controllability index satisfies the inequalities

$$\frac{n}{p} \leq \mu \leq \min(\bar{n}, n - \bar{p} + 1) \tag{5-31}$$

where $\bar{n}$ is the degree of the minimal polynomial of $\mathbf{A}$ and $\bar{p}$ is the rank of $\mathbf{B}$. From the definition of minimal polynomial, we have

$$\mathbf{A}^{\bar{n}} = \alpha_1\mathbf{A}^{\bar{n}-1} + \alpha_2\mathbf{A}^{\bar{n}-2} + \cdots + \alpha_{\bar{n}}\mathbf{I}$$

for some α_i, which implies

$$\mathbf{A}^{\bar{n}}\mathbf{B} = \alpha_1\mathbf{A}^{\bar{n}-1}\mathbf{B} + \alpha_2\mathbf{A}^{\bar{n}-2}\mathbf{B} + \cdots + \alpha_{\bar{n}}\mathbf{B}$$

Hence, all columns of $\mathbf{A}^{\bar{n}}\mathbf{B}$ are linearly dependent on their left-hand-side columns. Consequently, we have $\mu \leq \bar{n}$. From (5-30), we see that the rank of $\mathbf{U}_k$ must increase by at least 1 as k increased by 1; otherwise, it will cease to increase. Since the rank of $\mathbf{U}_\infty$ is at most n, if the rank of $\mathbf{B}$ is $\bar{p}$, it is sufficient to add at most $n-\bar{p}$ number of $\mathbf{A}^j\mathbf{B}$. Hence, we conclude from (5-30) that $\mu - 1 \leq n - \bar{p}$ or $\mu \leq n - \bar{p} + 1$. The combination of $\mu \leq \bar{n}$ and $\mu \leq n - \bar{p} + 1$ yields the upper bound in (5-31).

In order to have rank n, the matrix $\mathbf{U}_{\mu-1}$ must have more columns than rows. Hence, we need $\mu p \geq n$ or $\mu \geq n/p$. This establishes the lower bound in (5-31).

The degree $\bar{n}$ of the minimal polynomial of $\mathbf{A}$ is not easy to compute; hence $\min(\bar{n}, n-\bar{p}+1)$ in (5-31) is often replaced by $n-\bar{p}+1$. From (5-30) and Theorem 5-7, we have immediately the following corollary.

Corollary 5-7

The state equation FE is controllable if and only if the $n \times (n-\bar{p}+1)p$ matrix

$$\mathbf{U}_{n-\bar{p}} = [\mathbf{B} \quad \mathbf{AB} \quad \cdots \quad \mathbf{A}^{n-\bar{p}}\mathbf{B}]$$

where $\bar{p}$ is the rank of $\mathbf{B}$, has rank n, or the $n \times n$ matrix $\mathbf{U}_{n-\bar{p}}\mathbf{U}^*_{n-\bar{p}}$ is nonsingular. ∎

Example 9

Consider the satellite system studied in Figure 3-17. Its linearized dynamical equation is developed in (3-51). As can be seen from the dotted lines in (3-51), the control of $\bar{x}_i$, $i = 1, 2, 3, 4$, by $\bar{u}_1$ and $\bar{u}_2$ and the control of $\bar{x}_5$ and $\bar{x}_6$ by $\bar{u}_3$ are entirely independent; hence, we may consider only the following subequation of (3-51):

$$\dot{\mathbf{x}} = \begin{bmatrix} 0 & 1 & 0 & 0 \\ 3 & 0 & 0 & 2 \\ 0 & 0 & 0 & 1 \\ 0 & -2 & 0 & 0 \end{bmatrix} \mathbf{x} + \begin{bmatrix} 0 & 0 \\ 1 & 0 \\ 0 & 0 \\ 0 & 1 \end{bmatrix} \mathbf{u} \qquad \textbf{(5-32a)}$$

$$\mathbf{y} = \begin{bmatrix} 1 & 0 & 0 & 0 \\ 0 & 0 & 1 & 0 \end{bmatrix} \mathbf{x} \qquad \textbf{(5-32b)}$$

where we have assumed, for simplicity, $\omega = m = r_0 = 1$ and dropped the overbars. The rank of $\mathbf{B}$ is 2; hence, the state equation is controllable if and only if the matrix

$$[\mathbf{B} \vdots \mathbf{AB} \vdots \mathbf{A}^2\mathbf{B}] = \left[\begin{array}{cc:cc:cc} 0 & 0 & 1 & 0 & 0 & 2 \\ 1 & 0 & 0 & 2 & -1 & 0 \\ 0 & 0 & 0 & 1 & -2 & 0 \\ 0 & 1 & -2 & 0 & 0 & -4 \end{array}\right]$$

has rank 4. This is the case; hence, the state equation in (5-32) is controllable.

For this problem, we can readily show

$$\rho \mathbf{U}_0 = 2 < \rho \mathbf{U}_1 = 4 = \rho \mathbf{U}_2 = \rho \mathbf{U}_3 = \cdots$$

hence, the controllability index of (5-32a) is 2. ∎

We study more the controllability matrix **U**. We assume that the linearly independent columns of **U** in order from left to right have been found. We now rearrange these independent columns as

$$\mathbf{b}_1, \mathbf{A}\mathbf{b}_1, \ldots, \mathbf{A}^{\mu_1 - 1}\mathbf{b}_1, \mathbf{b}_2, \mathbf{A}\mathbf{b}_2, \ldots, \mathbf{A}^{\mu_2 - 1}\mathbf{b}_2, \ldots, \mathbf{b}_p, \mathbf{A}\mathbf{b}_p, \ldots, \mathbf{A}^{\mu_p - 1}\mathbf{b}_p$$

The integer μ_i is the number of linearly independent columns associated with $\mathbf{b}_i$ in the set, or is the length of the *chain* associated with $\mathbf{b}_i$. Clearly we have

$$\mu = \max\{\mu_1, \mu_2, \ldots, \mu_p\}$$

and $\mu_1 + \mu_2 + \cdots + \mu_p \leq n$. The equality holds if $\{\mathbf{A}, \mathbf{B}\}$ is controllable. The set $\{\mu_1, \mu_2, \ldots, \mu_p\}$ will be called the *controllability indices* of $\{\mathbf{A}, \mathbf{B}\}$.

Now we shall establish the relationship between the controllability indices and the r_i's defined in (5-28). In order to visualize the relationship, we use an example. We assume $p = 4$, $\mu_1 = 3$, $\mu_2 = 1$, $\mu_3 = 5$, and $\mu_4 = 3$. These independent columns are arranged in a crate diagram as shown in Figure 5-7. The (i, j)th cell represents the column $\mathbf{A}^{i-1}\mathbf{b}_j$. A column which is linearly independent of its left-hand-side columns in (5-28) is denoted by "x"; otherwise denoted by "0". The search of linearly independent columns in (5-28) from left to right is equivalent to the search from left to right in each row and then to the next row in Figure 5-7. Hence, the number of zeros in the ith row of Figure 5-7 is equal to r_{i-1} as shown. From the crate diagram, we can deduce that r_i is equal to the number of $\{\mathbf{b}_k, k = 1, 2, \ldots, p\}$ with controllability indices equal to or smaller than i. Hence, we conclude that

$$r_i - r_{i-1} = \text{no. of } \{\mathbf{b}_k, k = 1, 2, \ldots, p\} \text{ with controllability index } i \quad \textbf{(5-33)}$$

with $r_{-1} \triangleq 0$. For example, $r_1 - r_0 = 1$, and $\mathbf{b}_2$ has controllability index 1; $r_2 - r_1 = 0$ and no $\mathbf{b}_i$ has controllability index 2; $r_3 - r_2 = 2$, and $\mathbf{b}_1$ and $\mathbf{b}_4$ have controllability index 3; $r_5 - r_4 = 1$, and $\mathbf{b}_3$ has controllability index 5. Hence,

	$\mathbf{b}_1$	$\mathbf{b}_2$	$\mathbf{b}_3$	$\mathbf{b}_4$	
I	x	x	x	x	$r_0 = 0$
A	x	0	x	x	$r_1 = 1$
$\mathbf{A}^2$	x	0	x	x	$r_2 = 1$
$\mathbf{A}^3$	0	0	x	0	$r_3 = 3$
$\mathbf{A}^4$	0	0	x	0	$r_4 = 3$
$\mathbf{A}^5$	0	0	0	0	$r_5 = 4$
μ_i	3	1	5	3	

Figure 5-7 Crate diagram of $\mathbf{A}^{i-1}\mathbf{b}_j$.

the controllability indices of $\{\mathbf{A}, \mathbf{B}\}$ are uniquely determinable from $\{r_i, i=0, 1, \ldots, \mu\}$.

Theorem 5-8

The set of the controllability indices of $\{\mathbf{A}, \mathbf{B}\}$ is invariant under any equivalence transformation and any ordering of the columns of $\mathbf{B}$.

Proof

Define

$$\bar{\mathbf{U}}_k = [\bar{\mathbf{B}} \vdots \bar{\mathbf{A}}\bar{\mathbf{B}} \vdots \cdots \vdots \bar{\mathbf{A}}^k\bar{\mathbf{B}}]$$

where $\bar{\mathbf{A}} = \mathbf{PAP}^{-1}$, $\bar{\mathbf{B}} = \mathbf{PB}$ and $\mathbf{P}$ is any nonsingular matrix. Then it can be easily verified that

$$\bar{\mathbf{U}}_k = \mathbf{PU}_k \qquad \text{for } k=0, 1, 2, \ldots$$

which implies

$$\rho\bar{\mathbf{U}}_k = \rho\mathbf{U}_k \qquad \text{for } k=0, 1, 2, \ldots$$

Hence, the r_i defined in (5-28) and, consequently, the set of controllability indices are invariant under any equivalence transformation.

The rearrangement of the columns of $\mathbf{B}$ can be represented by

$$\tilde{\mathbf{B}} = \mathbf{BM}$$

where $\mathbf{M}$ is a $p \times p$ elementary matrix and is nonsingular. Again it is straightforward to verify that

$$\tilde{\mathbf{U}}_k \triangleq [\tilde{\mathbf{B}} \vdots \mathbf{A}\tilde{\mathbf{B}} \vdots \cdots \vdots \mathbf{A}^k\tilde{\mathbf{B}}] = \mathbf{U}_k \ \mathrm{diag}\{\mathbf{M}, \mathbf{M}, \ldots, \mathbf{M}\}$$

where diag $\{\mathbf{M}, \mathbf{M}, \ldots, \mathbf{M}\}$ consists of $k+1$ number of $\mathbf{M}$, and is clearly nonsingular. Hence, we have

$$\rho\mathbf{U}_k = \rho\tilde{\mathbf{U}}_k \qquad \text{for } k=0, 1, 2, \ldots$$

Consequently, we conclude that the controllability indices are independent of the ordering of the columns of $\mathbf{B}$. Q.E.D.

Now we discuss a different method of searching linearly independent columns in $\mathbf{U} = [\mathbf{B} \quad \mathbf{AB} \quad \cdots \quad \mathbf{A}^{n-1}\mathbf{B}]$. We first rearrange the columns of $\mathbf{U}$ as

$$\mathbf{b}_1, \mathbf{Ab}_1, \mathbf{A}^2\mathbf{b}_1, \ldots, \mathbf{A}^{n-1}\mathbf{b}_1; \mathbf{b}_2, \mathbf{Ab}_2, \ldots, \mathbf{A}^{n-1}\mathbf{b}_2; \ldots; \mathbf{b}_p, \mathbf{Ab}_p, \ldots, \mathbf{A}^{n-1}\mathbf{b}_p \tag{5-34}$$

and then search its linearly independent columns in order from left to right. In terms of the crate diagram in Figure 5-7, the linearly independent columns are searched in order from top to bottom in the first column, then in the second column and so forth. Let

$$\mathbf{b}_1, \mathbf{Ab}_1, \ldots, \mathbf{A}^{\bar{\mu}_1-1}\mathbf{b}_1; \mathbf{b}_2, \mathbf{Ab}_2, \ldots, \mathbf{A}^{\bar{\mu}_2-1}\mathbf{b}_2; \ldots; \mathbf{b}_p, \mathbf{Ab}_p, \ldots, \mathbf{A}^{\bar{\mu}_p-1}\mathbf{b}_p$$

be the resulting linearly independent columns. If $\{\mathbf{A}, \mathbf{B}\}$ is controllable, we have $\bar{\mu}_1 + \bar{\mu}_2 + \cdots + \bar{\mu}_p = n$. The lengths of these chains are $\{\bar{\mu}_1, \bar{\mu}_2, \ldots, \bar{\mu}_p\}$. Unlike the controllability indices, these lengths of chains depend highly on the ordering of $\{\mathbf{b}_i, i = 1, 2, \ldots, p\}$.

Example 10

Consider the state equation in (5-32). If we search linearly independent columns from left to right in

$$\mathbf{b}_1, \mathbf{A}\mathbf{b}_1, \mathbf{A}^2\mathbf{b}_1, \mathbf{A}^3\mathbf{b}_1; \mathbf{b}_2, \mathbf{A}\mathbf{b}_2, \mathbf{A}^2\mathbf{b}_2, \mathbf{A}^3\mathbf{b}_2$$

the resulting linearly independent columns are

$$\mathbf{b}_1, \mathbf{A}\mathbf{b}_1, \mathbf{A}^2\mathbf{b}_1, \mathbf{b}_2$$

Its lengths are $\{3, 1\}$. If we search from left to right in

$$\mathbf{b}_2, \mathbf{A}\mathbf{b}_2, \mathbf{A}^2\mathbf{b}_2, \mathbf{A}^3\mathbf{b}_2; \mathbf{b}_1, \mathbf{A}\mathbf{b}_1, \mathbf{A}^2\mathbf{b}_1, \mathbf{A}^3\mathbf{b}_1$$

the resulting linearly independent columns are

$$\mathbf{b}_2, \mathbf{A}\mathbf{b}_2, \mathbf{A}^2\mathbf{b}_2, \mathbf{A}^3\mathbf{b}_2$$

Its lengths are $\{4, 0\}$. The lengths are indeed different for different ordering of $\mathbf{b}_1$ and $\mathbf{b}_2$. The controllability indices of (5-32) can be computed as $\{2, 2\}$ and are independent of the ordering of $\mathbf{b}_1$ and $\mathbf{b}_2$. ∎

Let $\bar{\mu} = \max\{\bar{\mu}_1, \bar{\mu}_2, \ldots, \bar{\mu}_p\}$. It is clear that $\bar{\mu}$ can never be smaller than μ, the controllability index. Since $\mu = \max\{\mu_1, \mu_2, \ldots, \mu_p\}$, we may conclude that μ is the smallest possible maximum length of chains obtainable in any search of linearly independent columns of $\mathbf{U}$. This controllability index will play an important role in the design of state feedback in Chapter 7 and the design of compensators in Chapter 9.

5-4 Observability of Linear Dynamical Equations

Time-varying case. The concept of observability is dual to that of controllability. Roughly speaking, controllability studies the possibility of steering the state from the input; observability studies the possibility of estimating the state from the output. If a dynamical equation is controllable, all the modes of the equation can be excited from the input; if a dynamical equation is observable, all the modes of the equation can be observed at the output. These two concepts are defined under the assumption that we have the complete knowledge of a dynamical equation; that is, the matrices **A**, **B**, **C**, and **E** are known beforehand. Hence, the problem of observability is different from the problem of realization or identification. The problem of identification is a problem of estimating the matrices **A**, **B**, **C**, and **E** from the information collected at the input and output terminals.

Consider the n-dimensional linear dynamical equation

$$E: \quad \dot{\mathbf{x}} = \mathbf{A}(t)\mathbf{x}(t) + \mathbf{B}(t)\mathbf{u}(t) \tag{5-35a}$$
$$\mathbf{y} = \mathbf{C}(t)\mathbf{x}(t) + \mathbf{E}(t)\mathbf{u}(t) \tag{5-35b}$$

where **A**, **B**, **C**, and **E** are $n \times n$, $n \times p$, $q \times n$, and $q \times p$ matrices whose entries are continuous functions of t defined over $(-\infty, \infty)$.

Definition 5-5

The dynamical equation E is said to be (completely state) *observable* at t_0 if there exists a finite $t_1 > t_0$ such that for any state $\mathbf{x}_0$ at time t_0, the knowledge of the input $\mathbf{u}_{[t_0, t_1]}$ and the output $\mathbf{y}_{[t_0, t_1]}$ over the time interval $[t_0, t_1]$ suffices to determine the state $\mathbf{x}_0$. Otherwise, the dynamical equation E is said to be *unobservable* at t_0. ∎

Example 1

Consider the network shown in Figure 5-8. If the input is zero, no matter what the initial voltage across the capacitor is, in view of the symmetry of the network, the output is identically zero. We know the input and output (both are identically zero), but we are not able to determine the initial condition of the capacitor; hence the system, or more precisely, the dynamical equation that describes the system, is not observable at any t_0.

Example 2

Consider the network shown in Figure 5-9(a). If no input is applied, the network reduces to the one shown in Figure 5-9(b). Clearly the response to the initial

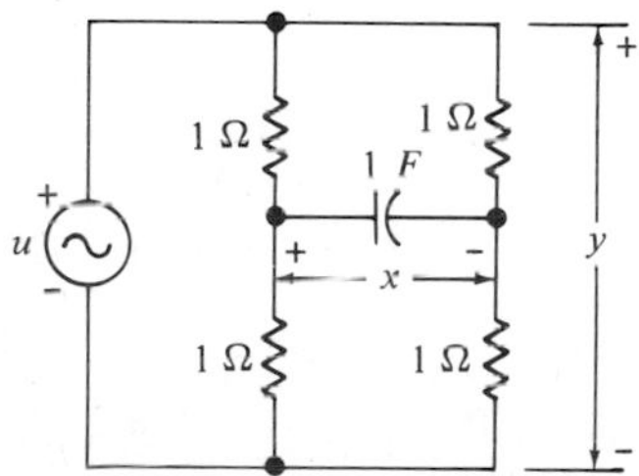

Figure 5-8 An unobservable network.

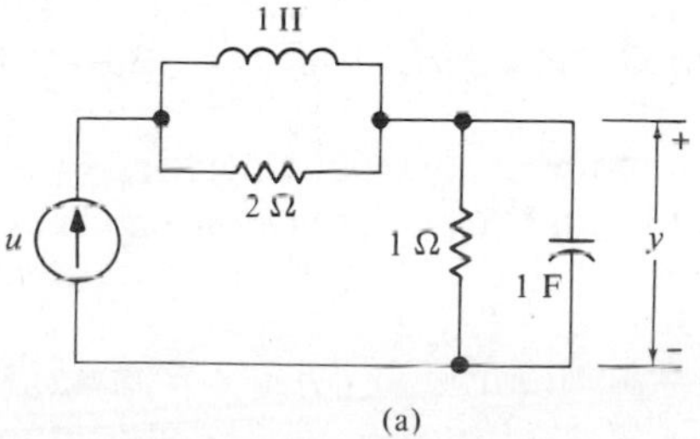

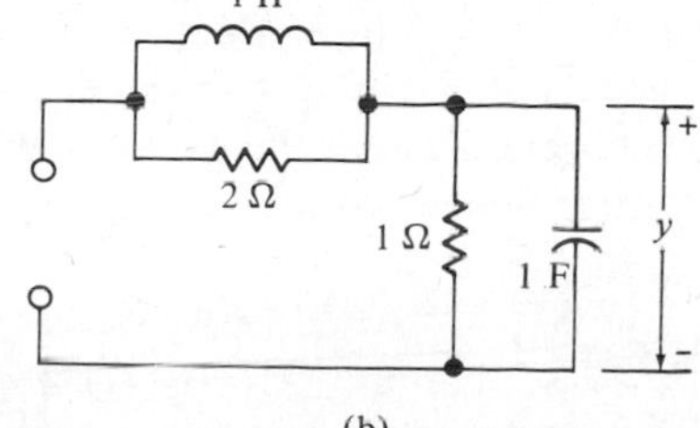

Figure 5-9 An unobservable network.

current in the inductor can never appear at the output terminal. Therefore, there is no way of determining the initial current in the inductor from the input and the output terminals. Hence the system or its dynamical equation is not observable at any t_0. ∎

The response of the dynamical equation (5-35) is given by

$$\mathbf{y}(t) = \mathbf{C}(t)\mathbf{\Phi}(t, t_0)\mathbf{x}(t_0) + \mathbf{C}(t)\int_{t_0}^{t} \mathbf{\Phi}(t, \tau)\mathbf{B}(\tau)\mathbf{u}(\tau)\, d\tau + \mathbf{E}(t)\mathbf{u}(t) \tag{5-36}$$

where $\mathbf{\Phi}(t, \tau)$ is the state transition matrix of $\dot{\mathbf{x}} = \mathbf{A}(t)\mathbf{x}$. In the study of observability, the output $\mathbf{y}$ and the input $\mathbf{u}$ are assumed to be known, the initial state $\mathbf{x}(t_0)$ is the only unknown; hence (5-36) can be written as

$$\bar{\mathbf{y}}(t) = \mathbf{C}(t)\mathbf{\Phi}(t, t_0)\mathbf{x}(t_0) \tag{5-37}$$

where

$$\bar{\mathbf{y}}(t) \triangleq \mathbf{y}(t) - \mathbf{C}(t)\int_{t_0}^{t} \mathbf{\Phi}(t, \tau)\mathbf{B}(\tau)\mathbf{u}(\tau)\, d\tau - \mathbf{E}(t)\mathbf{u}(t) \tag{5-38}$$

is a known function. Consequently, the observability problem is a problem of determining $\mathbf{x}(t_0)$ in (5-37) with the knowledge of $\bar{\mathbf{y}}$, $\mathbf{C}$, and $\mathbf{\Phi}(t, t_0)$. Note that the estimated state $\mathbf{x}(t_0)$ is the state not at time t, but at time t_0. However, if $\mathbf{x}(t_0)$ is known, the state after t_0 can be computed from

$$\mathbf{x}(t) = \mathbf{\Phi}(t, t_0)\mathbf{x}(t_0) + \int_{t_0}^{t} \mathbf{\Phi}(t, \tau)\mathbf{B}(\tau)\mathbf{u}(\tau)\, d\tau \tag{5-39}$$

Theorem 5-9

The dynamical equation E is observable at t_0 if and only if there exists a finite $t_1 > t_0$ such that the *n columns* of the $q \times n$ matrix function $\mathbf{C}(\cdot)\mathbf{\Phi}(\cdot, t_0)$ are linearly independent on $[t_0, t_1]$.

Proof

Sufficiency: Multiplying $\mathbf{\Phi}^*(t, t_0)\mathbf{C}^*(t)$ on both sides of (5-37) and integrating from t_0 *to* t_1, we obtain

$$\int_{t_0}^{t_1} \mathbf{\Phi}^*(t, t_0)\mathbf{C}^*(t)\bar{\mathbf{y}}(t)\, dt = \left[\int_{t_0}^{t_1} \mathbf{\Phi}^*(t, t_0)\mathbf{C}^*(t)\mathbf{C}(t)\mathbf{\Phi}(t, t_0)\, dt\right]\mathbf{x}_0 \triangleq \mathbf{V}(t_0, t_1)\mathbf{x}_0 \tag{5-40}$$

where

$$\mathbf{V}(t_0, t_1) \triangleq \int_{t_0}^{t_1} \mathbf{\Phi}^*(t, t_0)\mathbf{C}^*(t)\mathbf{C}(t)\mathbf{\Phi}(t, t_0)\, dt \tag{5-41}$$

From Theorem 5-1 and the assumption that all the columns of $\mathbf{C}(\cdot)\mathbf{\Phi}(\cdot, t_0)$ are linearly independent on $[t_0, t_1]$, we conclude that $\mathbf{V}(t_0, t_1)$ is nonsingular. Hence, from (5-40) we have

$$\mathbf{x}_0 = \mathbf{V}^{-1}(t_0, t_1)\int_{t_0}^{t_1} \mathbf{\Phi}^*(t, t_0)\mathbf{C}^*(t)\bar{\mathbf{y}}(t)\, dt \tag{5-42}$$

Thus, if the function $\bar{\mathbf{y}}_{[t_0, t_1]}$ is known, $\mathbf{x}_0$ can be computed from (5-42). *Necessity*:

Prove by contradiction. Suppose E is observable at t_0, but there exists *no* $t_1 > t_0$ such that the columns of $\mathbf{C}(\cdot)\mathbf{\Phi}(\cdot, t_0)$ are linearly independent on $[t_0, t_1]$. Then there exists an $n \times 1$ nonzero constant vector $\boldsymbol{\alpha}$ such that

$$\mathbf{C}(t)\mathbf{\Phi}(t, t_0)\boldsymbol{\alpha} = \mathbf{0} \qquad \text{for all } t > t_0$$

Let us choose $\mathbf{x}(t_0) = \boldsymbol{\alpha}$; then

$$\bar{\mathbf{y}}(t) = \mathbf{C}(t)\mathbf{\Phi}(t, t_0)\boldsymbol{\alpha} = \mathbf{0} \qquad \text{for all } t > t_0$$

Hence the initial state $\mathbf{x}(t_0) = \boldsymbol{\alpha}$ cannot be detected. This contradicts the assumption that E is observable. Therefore, if E is observable, there exists a finite $t_1 > t_0$ such that the columns of $\mathbf{C}(\cdot)\mathbf{\Phi}(\cdot, t_0)$ are linearly independent on $[t_0, t_1]$.

Q.E.D.

We see from this theorem that the observability of a linear dynamical equation depends only on $\mathbf{C}(t)$ and $\mathbf{\Phi}(t, t_0)$ or, equivalently, only on $\mathbf{C}$ and $\mathbf{A}$. This can also be deduced from Definition 5-5 by choosing $\mathbf{u} \equiv \mathbf{0}$. Hence in the observability study, it is sometimes convenient to assume $\mathbf{u} \equiv \mathbf{0}$ and study only $\dot{\mathbf{x}} = \mathbf{A}(t)\mathbf{x}$, $\mathbf{y} = \mathbf{C}(t)\mathbf{x}$.

The controllability of a dynamical equation is determined by the linear independence of the *rows* of $\mathbf{\Phi}(t_0, \cdot)\mathbf{B}(\cdot)$, whereas the observability is determined by the linear independence of the *columns* of $\mathbf{C}(\cdot)\mathbf{\Phi}(\cdot, t_0)$. The relationship between these two concepts is established in the following theorem.

Theorem 5-10 (Theorem of duality)

Consider the dynamical equation E in (5-35) and the dynamical equation E^* defined by

$$E^*: \quad \dot{\mathbf{z}} = -\mathbf{A}^*(t)\mathbf{z} + \mathbf{C}^*(t)\mathbf{v} \tag{5-43a}$$

$$\boldsymbol{\gamma} = \mathbf{B}^*(t)\mathbf{z} + \mathbf{E}^*(t)\mathbf{v} \tag{5-43b}$$

where $\mathbf{A}^*$, $\mathbf{B}^*$, $\mathbf{C}^*$, and $\mathbf{E}^*$ are the complex conjugate transposes of $\mathbf{A}$, $\mathbf{B}$, $\mathbf{C}$, and $\mathbf{E}$ in E. The equation E is controllable (observable) at t_0 if and only if the equation E^* is observable (controllable) at t_0.

Proof

From Theorem 5-4, the dynamical equation E is controllable if and only if the *rows* of $\mathbf{\Phi}(t_0, t)\mathbf{B}(t)$ are linearly independent, in t, on $[t_0, t_1]$. From Theorem 5-9, the dynamical equation E^* is observable if and only if the *columns* of $\mathbf{B}^*(t)\mathbf{\Phi}_a(t, t_0)$ are linearly independent, in t, on $[t_0, t_1]$, or equivalently, the *rows* of $[\mathbf{B}^*(t)\mathbf{\Phi}_a(t, t_0)]^* = \mathbf{\Phi}_a^*(t, t_0)\mathbf{B}(t)$ are linearly independent, in t, on $[t_0, t_1]$, where $\mathbf{\Phi}_a$ is the state transition matrix of $\dot{\mathbf{z}} = -\mathbf{A}^*(t)\mathbf{z}$. It is easy to show that $\mathbf{\Phi}_a^*(t, t_0) = \mathbf{\Phi}(t_0, t)$ (see Problem 4-8); hence E is controllable if and only if E^* is observable.

Q.E.D.

We list in the following, for observability, Theorems 5-11 to 5-14 and Definitions 5-6 to 5-8, which are dual to Theorems 5-5 to 5-8 and Definitions 5-2 to 5-4 for controllability. Theorems 5-11 to 5-14 can be proved either directly or by applying Theorem 5-10 to Theorems 5-5 to 5-8. The interpretations in the controllability part also apply to the observability part.

Theorem 5-11

Assume that the matrices $\mathbf{A}(\cdot)$ and $\mathbf{C}(\cdot)$ in the n-dimensional dynamical equation E are $n-1$ times continuously differentiable. Then the dynamical equation E is observable at t_0 if there exists a finite $t_1 > t_0$ such that

$$\rho \begin{bmatrix} \mathbf{N}_0(t_1) \\ \mathbf{N}_1(t_1) \\ \vdots \\ \mathbf{N}_{n-1}(t_1) \end{bmatrix} = n \tag{5-44}$$

where $$\mathbf{N}_{k+1}(t) = \mathbf{N}_k(t)\mathbf{A}(t) + \frac{d}{dt}\mathbf{N}_k(t) \qquad k = 0, 1, 2, \ldots, n-1 \tag{5-45a}$$

with $$\mathbf{N}_0(t) = \mathbf{C}(t) \tag{5-45b}$$

■

*Differential observability, instantaneous observability, and uniform observability

Differential, instantaneous, and uniform observabilities can be defined by using the theorem of duality; for example, we may define $\{\mathbf{A}, \mathbf{C}\}$ to be differentially observable if and only if $\{-\mathbf{A}^*, \mathbf{C}^*\}$ is differentially controllable. However, for ease of reference we shall define them explicitly in the following.

Definition 5-6

The dynamical equation E is said to be *differentially observable* at time t_0 if, for any state $\mathbf{x}(t_0)$ in the state space Σ, the knowledge of the input and the output over an arbitrarily small interval of time suffices to determine $\mathbf{x}(t_0)$. ■

Theorem 5-12

If the matrices $\mathbf{A}$ and $\mathbf{C}$ are analytic on $(-\infty, \infty)$, then the n-dimensional dynamical equation E is differentially observable at every t in $(-\infty, \infty)$ if and only if, for any fixed t_0 in $(-\infty, \infty)$,

$$\rho \begin{bmatrix} \mathbf{N}_0(t_0) \\ \mathbf{N}_1(t_0) \\ \vdots \\ \mathbf{N}_{n-1}(t_0) \\ \vdots \end{bmatrix} = n$$

■

*Sections noted with an asterisk may be skipped without loss of continuity.

Definition 5-7

The linear dynamical equation E is said to be *instantaneously observable* in $(-\infty, \infty)$ if and only if

$$\rho\begin{bmatrix}\mathbf{N}_0(t)\\ \mathbf{N}_1(t)\\ \vdots\\ \mathbf{N}_{n-1}(t)\end{bmatrix}=n \qquad \text{for all } t \text{ in } (-\infty, \infty)$$

where the $\mathbf{N}_i$'s are as defined in (5-45). ∎

Definition 5-8

The linear dynamical equation E is said to be *uniformly observable* in $(-\infty, \infty)$ if and only if there exist a positive σ_o and positive β_i that depends on σ_o such that

$$\mathbf{0}<\beta_1(\sigma_o)\mathbf{I}\le\mathbf{V}(t, t+\sigma_o)\le\beta_2(\sigma_0)\mathbf{I}$$
$$\mathbf{0}<\beta_3(\sigma_o)\mathbf{I}\le\mathbf{\Phi}^*(t, t+\sigma_o)\mathbf{V}(t, t+\sigma_o)\mathbf{\Phi}(t, t+\sigma_o)\le\beta_4(\sigma_o)\mathbf{I}$$

for all t, where $\mathbf{\Phi}$ is the state transition matrix and $\mathbf{V}$ is as defined in (5-41). ∎

Time-invariance case. Consider the linear time-invariant dynamical equation

$$FE: \qquad \dot{\mathbf{x}}=\mathbf{A}\mathbf{x}+\mathbf{B}\mathbf{u} \qquad \textbf{(5-46a)}$$
$$\mathbf{y}=\mathbf{C}\mathbf{x}+\mathbf{E}\mathbf{u} \qquad \textbf{(5-46b)}$$

where $\mathbf{A}$, $\mathbf{B}$, $\mathbf{C}$, and $\mathbf{E}$ are $n\times n$, $n\times p$, $q\times n$, and $q\times p$ constant matrices. The time interval of interest is $[0, \infty)$. Similar to the controllability part, if a linear time-invariant dynamical equation is observable, it is observable at every $t_0\ge 0$, and the determination of the initial state can be achieved in any nonzero time interval. Hence, the reference of t_0 and t_1 is often dropped in the observability study of linear time-invariant dynamical equations.

Theorem 5-13

The n-dimensional linear time-invariant dynamical equation in (5-46) is observable if and only if any of the following equivalent conditions is satisfied:

1. All columns of $\mathbf{C}e^{\mathbf{A}t}$ are linearly independent on $[0, \infty)$ over $\mathbb{C}$, the field of complex numbers.

1'. All columns of $\mathbf{C}(s\mathbf{I}-\mathbf{A})^{-1}$ are linearly independent over $\mathbb{C}$.

2. The *observability grammian*

$$\mathbf{W}_{ot}\triangleq\int_0^t e^{\mathbf{A}^*\tau}\mathbf{C}^*\mathbf{C}e^{\mathbf{A}\tau}\,d\tau$$

is nonsingular for any $t>0$.

3. The *nq × n observability matrix*

$$\mathbf{V} = \begin{bmatrix} \mathbf{C} \\ \mathbf{CA} \\ \mathbf{CA}^2 \\ \vdots \\ \mathbf{CA}^{n-1} \end{bmatrix} \tag{5-47}$$

has rank n.

4. For every eigenvalue λ of $\mathbf{A}$ (and consequently for every λ in $\mathbb{C}$), the $(n+q) \times n$ complex matrix

$$\begin{bmatrix} \lambda\mathbf{I} - \mathbf{A} \\ \mathbf{C} \end{bmatrix}$$

has rank n, or equivalently, $(s\mathbf{I} - \mathbf{A})$ and $\mathbf{C}$ are right coprime.

*Observability Indices

Let $\mathbf{A}$ and $\mathbf{C}$ be $n \times n$ and $q \times n$ constant matrices. Define

$$\mathbf{V}_k = \begin{bmatrix} \mathbf{C} \\ \mathbf{CA} \\ \vdots \\ \mathbf{CA}^k \end{bmatrix} = \begin{bmatrix} \mathbf{c}_1 \\ \mathbf{c}_2 \\ \vdots \\ \mathbf{c}_q \\ \hline \mathbf{c}_1\mathbf{A} \\ \mathbf{c}_2\mathbf{A} \\ \vdots \\ \mathbf{c}_q\mathbf{A} \\ \hline \vdots \\ \hline \mathbf{c}_1\mathbf{A}^k \\ \mathbf{c}_2\mathbf{A}^k \\ \vdots \\ \mathbf{c}_q\mathbf{A}^k \end{bmatrix} \begin{matrix} \left.\begin{matrix} \\ \\ \\ \\ \end{matrix}\right\} r_0 \text{ (no. of dependent rows)} \\ \left.\begin{matrix} \\ \\ \\ \\ \end{matrix}\right\} r_1 \\ \\ \left.\begin{matrix} \\ \\ \\ \\ \end{matrix}\right\} r_k \end{matrix} \tag{5-48}$$

It consists of $k+1$ block rows of the form $\mathbf{CA}^i$ and is of order $(k+1)q \times n$. The matrix $\mathbf{V} = \mathbf{V}_{n-1}$ is the observability matrix. The q rows of $\mathbf{C}$ are denoted by $\mathbf{c}_i$, $i = 1, 2, \ldots, q$. Let us search linearly independent rows of $\mathbf{V}_k$ in order from top to bottom. Let r_i be the number of linearly dependent rows in $\mathbf{CA}^i$, $i = 0, 1, \ldots, k$. Similar to the controllability part, we have

$$r_0 \leq r_1 \leq r_2 \leq \cdots \leq q$$

Since there are at most n linearly independent rows in $\mathbf{V}_\infty$, there exists an integer v such that

$$0 \leq r_0 \leq r_1 \leq \cdots \leq r_{v-1} < q \tag{5-49a}$$

and $$r_\nu = r_{\nu+1} = \cdots = q \tag{5-49b}$$

Equivalently, ν is the integer such that

$$\rho \mathbf{V}_0 < \rho \mathbf{V}_1 < \cdots < \rho \mathbf{V}_{\nu-1} = \rho \mathbf{V}_\nu = \rho \mathbf{V}_{\nu+1} = \cdots \tag{5-50}$$

The integer ν is called the *observability index* of $\{\mathbf{A}, \mathbf{C}\}$. Similar to (5-31), we have

$$\frac{n}{q} \leq \nu \leq \min (\bar{n}, n - \bar{q} + 1) \tag{5-51}$$

where $\bar{n}$ is the degree of the minimal polynomial of $\mathbf{A}$ and $\bar{q}$ is the rank of $\mathbf{C}$.

Corollary 5-13

The dynamical equation *FE* in (5-46) is observable if and only if the matrix $\mathbf{V}_{n-\bar{q}}$, where $\bar{q}$ is the rank of $\mathbf{C}$, is of rank n, or equivalently, the $n \times n$ matrix $\mathbf{V}^*_{n-\bar{q}}\mathbf{V}_{n-\bar{q}}$ is nonsingular. ∎

Consider the matrix $\mathbf{V}_{n-1}$. It is assumed that its linearly independent rows in order from top to bottom have been found. Let ν_i be the number of linearly independent rows associated with $\mathbf{c}_i$. The set of $\{\nu_i, i = 1, 2, \ldots, q\}$ is called the *observability indices* of $\{\mathbf{A}, \mathbf{C}\}$. Clearly we have

$$\nu = \max\{\nu_i, i = 1, 2, \ldots, q\}$$

and $\nu_1 + \nu_2 + \cdots + \nu_q \leq n$. The equality holds if $\{\mathbf{A}, \mathbf{C}\}$ is observable.

Theorem 5-14

The set of observability indices of $\{\mathbf{A}, \mathbf{C}\}$ is invariant under any equivalence transformation and any ordering of the rows of $\mathbf{C}$. ∎

5-5 Canonical Decomposition of a Linear Time-Invariant Dynamical Equation

In the remainder of this chapter, we study exclusively linear time-invariant dynamical equations. Consider the dynamical equation

$$FE: \quad \dot{\mathbf{x}} = \mathbf{A}\mathbf{x} + \mathbf{B}\mathbf{u} \tag{5-52a}$$
$$\mathbf{y} = \mathbf{C}\mathbf{x} + \mathbf{E}\mathbf{u} \tag{5-52b}$$

where $\mathbf{A}$, $\mathbf{B}$, $\mathbf{C}$, and $\mathbf{E}$ are $n \times n$, $n \times p$, $q \times n$, and $q \times p$ real constant matrices. We introduced in the previous sections the concepts of controllability and observability. The conditions for the equation to be controllable and observable are also derived. A question that may be raised at this point is: What can be said if the equation is uncontrollable and/or unobservable? In this section we shall study this problem. Before proceeding, we review briefly the equivalence transformation. Let $\bar{\mathbf{x}} = \mathbf{P}\mathbf{x}$, where $\mathbf{P}$ is a constant nonsingular matrix. The substitution of $\bar{\mathbf{x}} = \mathbf{P}\mathbf{x}$ into (5-52) yields

$$F\bar{E}: \quad \dot{\bar{\mathbf{x}}} = \bar{\mathbf{A}}\bar{\mathbf{x}} + \bar{\mathbf{B}}\mathbf{u} \tag{5-53a}$$

$$\mathbf{y} = \bar{\mathbf{C}}\bar{\mathbf{x}} + \bar{\mathbf{E}}\mathbf{u} \tag{5-53b}$$

where $\bar{\mathbf{A}} = \mathbf{PAP}^{-1}$, $\bar{\mathbf{B}} = \mathbf{PB}$, $\bar{\mathbf{C}} = \mathbf{CP}^{-1}$, and $\bar{\mathbf{E}} = \mathbf{E}$. The dynamical equations FE and $F\bar{E}$ are said to be equivalent, and the matrix $\mathbf{P}$ is called an *equivalence transformation.* Clearly we have

$$\begin{aligned}\bar{\mathbf{U}} \triangleq [\bar{\mathbf{B}} \vdots \bar{\mathbf{A}}\bar{\mathbf{B}} \vdots \cdots \vdots \bar{\mathbf{A}}^{n-1}\bar{\mathbf{B}}] &= [\mathbf{PB} \vdots \mathbf{PAB} \vdots \cdots \vdots \mathbf{PA}^{n-1}\mathbf{B}] \\ &= \mathbf{P}[\mathbf{B} \vdots \mathbf{AB} \vdots \cdots \vdots \mathbf{A}^{n-1}\mathbf{B}] \triangleq \mathbf{PU}\end{aligned}$$

Since the rank of a matrix does not change after multiplication of a nonsingular matrix (Theorem 2-7), we have rank $\mathbf{U} =$ rank $\bar{\mathbf{U}}$. Consequently FE is controllable if and only if $F\bar{E}$ is controllable. A similar statement holds for the observability part.

Theorem 5-15

The controllability and observability of a linear time-invariant dynamical equation are invariant under any equivalence transformation. ∎

This theorem is in fact a special case of Theorems 5-8 and 5-14. For easy reference, we have restated it as a theorem.

In the following, c will be used to stand for controllable, $\bar{c}$ for uncontrollable, o for observable, and $\bar{o}$ for unobservable.

Theorem 5-16

Consider the n-dimensional linear time-invariant dynamical equation FE. If the controllability matrix of FE has rank n_1 (where $n_1 < n$), then there exists an equivalence transformation $\bar{\mathbf{x}} = \mathbf{P}\mathbf{x}$, where $\mathbf{P}$ is a constant nonsingular matrix, which transforms FE into

$$F\bar{E}: \quad \begin{bmatrix} \dot{\bar{\mathbf{x}}}_c \\ \dot{\bar{\mathbf{x}}}_{\bar{c}} \end{bmatrix} = \begin{bmatrix} \bar{\mathbf{A}}_c & \bar{\mathbf{A}}_{12} \\ \mathbf{0} & \bar{\mathbf{A}}_{\bar{c}} \end{bmatrix} \begin{bmatrix} \bar{\mathbf{x}}_c \\ \bar{\mathbf{x}}_{\bar{c}} \end{bmatrix} + \begin{bmatrix} \bar{\mathbf{B}}_c \\ \mathbf{0} \end{bmatrix} \mathbf{u} \tag{5-54a}$$

$$\mathbf{y} = [\bar{\mathbf{C}}_c \quad \bar{\mathbf{C}}_{\bar{c}}] \begin{bmatrix} \bar{\mathbf{x}}_c \\ \bar{\mathbf{x}}_{\bar{c}} \end{bmatrix} + \mathbf{E}\mathbf{u} \tag{5-54b}$$

and the n_1-dimensional subequation of $F\bar{E}$

$$F\bar{E}_c: \quad \dot{\bar{\mathbf{x}}}_c = \bar{\mathbf{A}}_c\bar{\mathbf{x}}_c + \bar{\mathbf{B}}_c\mathbf{u} \tag{5-55a}$$

$$\bar{\mathbf{y}} = \bar{\mathbf{C}}_c\bar{\mathbf{x}}_c + \mathbf{E}\mathbf{u} \tag{5-55b}$$

is controllable[9] and has the same transfer function matrix as FE.

---

[9] It is easy to show that if the equation $F\bar{E}$ is observable, then its subequation $F\bar{E}_c$ is also observable. (Try.)

Proof

If the dynamical equation *FE* is not controllable, then from Theorem 5-7 we have

$$\rho\mathbf{U} \triangleq \rho[\mathbf{B} \vdots \mathbf{AB} \vdots \cdots \vdots \mathbf{A}^{n-1}\mathbf{B}] = n_1 < n$$

Let $\mathbf{q}_1, \mathbf{q}_2, \ldots, \mathbf{q}_{n_1}$ be any n_1 linearly independent columns of $\mathbf{U}$. Note that for each $i = 1, 2, \ldots, n_1$, $\mathbf{Aq}_i$ can be written as a linear combination of $\{\mathbf{q}_1, \mathbf{q}_2, \ldots, \mathbf{q}_{n_1}\}$. (Why?) Define a nonsingular matrix

$$\mathbf{P}^{-1} \triangleq \mathbf{Q} \triangleq [\mathbf{q}_1 \quad \mathbf{q}_2 \quad \cdots \quad \mathbf{q}_{n_1} \quad \cdots \quad \mathbf{q}_n] \tag{5-56}$$

where the last $n - n_1$ columns of $\mathbf{Q}$ are entirely arbitrary so long as the matrix $\mathbf{Q}$ is nonsingular. We claim that the transformation $\bar{\mathbf{x}} = \mathbf{Px}$ will transform *FE* into the form of (5-54). Recall from Figure 2-5 that in the transformation $\bar{\mathbf{x}} = \mathbf{Px}$ we are actually using the columns of $\mathbf{Q} \triangleq \mathbf{P}^{-1}$ as new basis vectors of the state space. The ith column of the new representation $\bar{\mathbf{A}}$ is the representation of $\mathbf{Aq}_i$ with respect to $\{\mathbf{q}_1, \mathbf{q}_2, \ldots, \mathbf{q}_n\}$. Now the vectors $\mathbf{Aq}_i$, for $i = 1, 2, \ldots, n_1$, are linearly dependent on the set $\{\mathbf{q}_1, \mathbf{q}_2, \ldots, \mathbf{q}_{n_1}\}$; hence the matrix $\bar{\mathbf{A}}$ has the form given in (5-54a). The columns of $\bar{\mathbf{B}}$ are the representations of the columns of $\mathbf{B}$ with respect to $\{\mathbf{q}_1, \mathbf{q}_2, \ldots, \mathbf{q}_n\}$. Now the columns of $\mathbf{B}$ depend only on $\{\mathbf{q}_1, \mathbf{q}_2, \ldots, \mathbf{q}_{n_1}\}$; hence $\bar{\mathbf{B}}$ is of the form shown in (5-54a).

Let $\mathbf{U}$ and $\bar{\mathbf{U}}$ be the controllability matrices of *FE* and $F\bar{E}$, respectively. Then we have $\rho\mathbf{U} = \rho\bar{\mathbf{U}} = n_1$ (see Theorem 5-15). It is easy to verify that

$$\bar{\mathbf{U}} = \begin{bmatrix} \bar{\mathbf{B}}_c & \bar{\mathbf{A}}_c\bar{\mathbf{B}}_c & \cdots & \bar{\mathbf{A}}_c^{n-1}\bar{\mathbf{B}}_c \\ \mathbf{0} & \mathbf{0} & & \mathbf{0} \end{bmatrix}$$

$$= \begin{bmatrix} \bar{\mathbf{U}}_c & \bar{\mathbf{A}}_c^{n_1}\bar{\mathbf{B}}_c & \cdots & \bar{\mathbf{A}}_c^{n-1}\bar{\mathbf{B}}_c \\ \mathbf{0} & \mathbf{0} & & \mathbf{0} \end{bmatrix} \begin{matrix} \}n_1 \text{ rows} \\ \}(n-n_1) \text{ rows} \end{matrix}$$

where $\bar{\mathbf{U}}_c$ represents the controllability matrices of $F\bar{E}_c$. Since columns of $\bar{\mathbf{A}}_c^k\mathbf{B}$ with $k \geq n_1$ are linearly dependent on the columns of $\bar{\mathbf{U}}_c$, the condition $\rho\bar{\mathbf{U}} = n_1$ implies $\rho\bar{\mathbf{U}}_c = n_1$. Hence the dynamical equation $F\bar{E}_c$ is controllable.

We show now that the dynamical equations *FE* and $F\bar{E}_c$, or correspondingly, $F\bar{E}$ and $F\bar{E}_c$, have the same transfer-function matrix. It is easy to verify that

$$\begin{bmatrix} s\mathbf{I} - \bar{\mathbf{A}}_c & -\bar{\mathbf{A}}_{12} \\ \mathbf{0} & s\mathbf{I} - \bar{\mathbf{A}}_{\bar{c}} \end{bmatrix} \begin{bmatrix} (s\mathbf{I} - \bar{\mathbf{A}}_c)^{-1} & (s\mathbf{I} - \bar{\mathbf{A}}_c)^{-1}\bar{\mathbf{A}}_{12}(s\mathbf{I} - \bar{\mathbf{A}}_{\bar{c}})^{-1} \\ \mathbf{0} & (s\mathbf{I} - \bar{\mathbf{A}}_{\bar{c}})^{-1} \end{bmatrix} = \mathbf{I} \tag{5-57}$$

Thus the transfer-function matrix of $F\bar{E}$ is

$$[\bar{\mathbf{C}}_c \quad \bar{\mathbf{C}}_{\bar{c}}] \begin{bmatrix} s\mathbf{I} - \bar{\mathbf{A}}_c & -\bar{\mathbf{A}}_{12} \\ \mathbf{0} & s\mathbf{I} - \bar{\mathbf{A}}_{\bar{c}} \end{bmatrix}^{-1} \begin{bmatrix} \bar{\mathbf{B}}_c \\ \mathbf{0} \end{bmatrix} + \mathbf{E}$$

$$= [\bar{\mathbf{C}}_c \quad \bar{\mathbf{C}}_{\bar{c}}] \begin{bmatrix} (s\mathbf{I} - \bar{\mathbf{A}}_c)^{-1} & (s\mathbf{I} - \bar{\mathbf{A}}_c)^{-1}\bar{\mathbf{A}}_{12}(s\mathbf{I} - \bar{\mathbf{A}}_{\bar{c}})^{-1} \\ \mathbf{0} & (s\mathbf{I} - \bar{\mathbf{A}}_{\bar{c}})^{-1} \end{bmatrix} \begin{bmatrix} \bar{\mathbf{B}}_c \\ \mathbf{0} \end{bmatrix} + \mathbf{E}$$

$$= \bar{\mathbf{C}}_c(s\mathbf{I} - \bar{\mathbf{A}}_c)^{-1}\bar{\mathbf{B}}_c + \mathbf{E}$$

which is the transfer-function matrix of $F\bar{E}_c$. Q.E.D.

In the equivalence transformation $\bar{\mathbf{x}} = \mathbf{Px}$, the state space Σ of FE is divided into two subspaces. One is the n_1-dimensional subspace of Σ, denoted by Σ_1, which consists of all the vectors $\begin{bmatrix} \bar{\mathbf{x}}_c \\ \mathbf{0} \end{bmatrix}$; the other is the $(n - n_1)$-dimensional subspace, which consists of all the vectors $\begin{bmatrix} \mathbf{0} \\ \bar{\mathbf{x}}_{\bar{c}} \end{bmatrix}$. Since $F\bar{E}_c$ is controllable, all the vectors $\bar{\mathbf{x}}_c$ in Σ_1 are controllable. Equation (5-54a) shows that the state variables in $\bar{\mathbf{x}}_{\bar{c}}$ are not affected directly by the input $\mathbf{u}$ or indirectly through the state vector $\bar{\mathbf{x}}_c$; therefore, the state vector $\bar{\mathbf{x}}_{\bar{c}}$ is not controllable and is dropped in the reduced equation (5-55). Thus, if a linear time-invariant dynamical equation is not controllable, by a proper choice of a basis, the state vector can be decomposed into two groups: one controllable, the other uncontrollable. By dropping the uncontrollable state vectors, we may obtain a controllable dynamical equation of lesser dimension that is zero-state equivalent to the original equation. See Problems 5-22 and 5-23.

Example 1

Consider the three-dimensional dynamical equation

$$\dot{\mathbf{x}} = \begin{bmatrix} 1 & 1 & 0 \\ 0 & 1 & 0 \\ 0 & 1 & 1 \end{bmatrix} \mathbf{x} + \begin{bmatrix} 0 & 1 \\ 1 & 0 \\ 0 & 1 \end{bmatrix} u \qquad \mathbf{y} = [1 \quad 1 \quad 1]x \tag{5-58}$$

The rank of $\mathbf{B}$ is 2; therefore, we need to check $\mathbf{U}_1 = [\mathbf{B} \vdots \mathbf{AB}]$ in determining the controllability of the equation. Since

$$\rho[\mathbf{B} \vdots \mathbf{AB}] = \rho \begin{bmatrix} 0 & 1 & 1 & 1 \\ 1 & 0 & 1 & 0 \\ 0 & 1 & 1 & 1 \end{bmatrix} = 2 < 3$$

the state equation is not controllable.

Let us choose, as in (5-56),

$$\mathbf{P}^{-1} = \mathbf{Q} = \left[\begin{array}{cc|c} 0 & 1 & 1 \\ 1 & 0 & 0 \\ 0 & 1 & 0 \end{array}\right]$$

The first two columns of $\mathbf{Q}$ are the first two linearly independent columns of $\mathbf{U}_1$; the last column of $\mathbf{Q}$ is chosen arbitrarily to make $\mathbf{Q}$ nonsingular. Let $\bar{\mathbf{x}} = \mathbf{Px}$. We compute

$$\bar{\mathbf{A}} = \mathbf{PAP}^{-1} = \begin{bmatrix} 0 & 1 & 0 \\ 0 & 0 & 1 \\ 1 & 0 & -1 \end{bmatrix} \begin{bmatrix} 1 & 1 & 0 \\ 0 & 1 & 0 \\ 0 & 1 & 1 \end{bmatrix} \begin{bmatrix} 0 & 1 & 1 \\ 1 & 0 & 0 \\ 0 & 1 & 0 \end{bmatrix} = \left[\begin{array}{cc|c} 1 & 0 & 0 \\ 1 & 1 & 0 \\ \hline 0 & 0 & 1 \end{array}\right]$$

$$\bar{\mathbf{B}} = \mathbf{PB} = \begin{bmatrix} 0 & 1 & 0 \\ 0 & 0 & 1 \\ 1 & 0 & -1 \end{bmatrix} \begin{bmatrix} 0 & 1 \\ 1 & 0 \\ 0 & 1 \end{bmatrix} = \left[\begin{array}{cc} 1 & 0 \\ 0 & 1 \\ \hline 0 & 0 \end{array}\right]$$

and $$\bar{\mathbf{C}} = \mathbf{C}\mathbf{P}^{-1} = [1 \quad 1 \quad 1]\begin{bmatrix} 0 & 1 & 1 \\ 1 & 0 & 0 \\ 0 & 1 & 0 \end{bmatrix} = [1 \quad 2 \,\vdots\, 1]$$

Hence, the reduced controllable equation is

$$\dot{\bar{\mathbf{x}}}_c = \begin{bmatrix} 1 & 0 \\ 1 & 1 \end{bmatrix}\bar{\mathbf{x}}_c + \begin{bmatrix} 1 & 0 \\ 0 & 1 \end{bmatrix}\mathbf{u} \qquad \mathbf{y} = [1 \quad 2]\bar{\mathbf{x}}_c \tag{5-59}$$

∎

Dual to Theorem 5-16, we have the following theorem for unobservable dynamical equations.

Theorem 5-17

Consider the n-dimensional linear time-invariant dynamical equation FE. If the observability matrix of FE has rank $n_2 (n_2 < n)$, then there exists an equivalence transformation $\bar{\mathbf{x}} = \mathbf{P}\mathbf{x}$ that transforms FE into

$$F\bar{E}: \quad \begin{bmatrix} \dot{\bar{\mathbf{x}}}_o \\ \dot{\bar{\mathbf{x}}}_{\bar{o}} \end{bmatrix} = \begin{bmatrix} \bar{\mathbf{A}}_o & \mathbf{0} \\ \bar{\mathbf{A}}_{21} & \bar{\mathbf{A}}_{\bar{o}} \end{bmatrix}\begin{bmatrix} \bar{\mathbf{x}}_o \\ \bar{\mathbf{x}}_{\bar{o}} \end{bmatrix} + \begin{bmatrix} \bar{\mathbf{B}}_o \\ \bar{\mathbf{B}}_{\bar{o}} \end{bmatrix}\mathbf{u} \tag{5-60a}$$

$$\mathbf{y} = [\bar{\mathbf{C}}_o \quad \mathbf{0}]\begin{bmatrix} \bar{\mathbf{x}}_o \\ \bar{\mathbf{x}}_{\bar{o}} \end{bmatrix} + \mathbf{E}\mathbf{u} \tag{5-60b}$$

and the n_2-dimensional subequation of $F\bar{E}$

$$F\bar{E}_o: \quad \dot{\bar{\mathbf{x}}}_o = \bar{\mathbf{A}}_o\bar{\mathbf{x}}_o + \bar{\mathbf{B}}_o\mathbf{u} \tag{5-61a}$$

$$\bar{\mathbf{y}} = \bar{\mathbf{C}}_o\bar{\mathbf{x}}_o + \mathbf{E}\mathbf{u} \tag{5-61b}$$

is observable and has the same transfer-function matrix as FE. ∎

This theorem can be readily established by using Theorems 5-16 and 5-10. The first n_2 rows of $\mathbf{P}$ in Theorem 5-17 are any n_2 linearly independent rows of the observability matrix of $\{\mathbf{A}, \mathbf{C}\}$; the remaining $n - n_2$ rows of $\mathbf{P}$ are entirely arbitrary so long as $\mathbf{P}$ is nonsingular. Equation (5-60) shows that the vector $\bar{\mathbf{x}}_{\bar{o}}$ does not appear directly in the output $\mathbf{y}$ or indirectly through $\bar{\mathbf{x}}_o$. Hence the vector $\bar{\mathbf{x}}_{\bar{o}}$ is not observable and is dropped in the reduced equation.

Combining Theorems 5-16 and 5-17, we have the following very important theorem.

Theorem 5-18 (Canonical decomposition theorem)[10]

Consider the linear time-invariant dynamical equation

$$FE: \quad \dot{\mathbf{x}} = \mathbf{A}\mathbf{x} + \mathbf{B}\mathbf{u}$$

$$\mathbf{y} = \mathbf{C}\mathbf{x} + \mathbf{E}\mathbf{u}$$

By equivalence transformations, FE can be transformed into the following

[10] This is a simplified version of the canonical decomposition theorem. For the general form, see References 57, 60, and 116. See also Reference S127.

canonical form

$$F\bar{E}: \quad \begin{bmatrix} \dot{\bar{\mathbf{x}}}_{c\bar{o}} \\ \dot{\bar{\mathbf{x}}}_{co} \\ \dot{\bar{\mathbf{x}}}_{\bar{c}} \end{bmatrix} = \left[\begin{array}{cc:c} \bar{\mathbf{A}}_{c\bar{o}} & \bar{\mathbf{A}}_{12} & \bar{\mathbf{A}}_{13} \\ \mathbf{0} & \bar{\mathbf{A}}_{co} & \bar{\mathbf{A}}_{23} \\ \hdashline \mathbf{0} & \mathbf{0} & \bar{\mathbf{A}}_{\bar{c}} \end{array}\right] \begin{bmatrix} \bar{\mathbf{x}}_{c\bar{o}} \\ \bar{\mathbf{x}}_{co} \\ \bar{\mathbf{x}}_{\bar{c}} \end{bmatrix} + \begin{bmatrix} \bar{\mathbf{B}}_{c\bar{o}} \\ \bar{\mathbf{B}}_{co} \\ \mathbf{0} \end{bmatrix} \mathbf{u} \tag{5-62a}$$

$$\mathbf{y} = [\mathbf{0} \quad \bar{\mathbf{C}}_{co} \ \vdots \ \bar{\mathbf{C}}_{\bar{c}}]\bar{\mathbf{x}} + \mathbf{E}\mathbf{u} \tag{5-62b}$$

where the vector $\bar{\mathbf{x}}_{c\bar{o}}$ is controllable but not observable, $\bar{\mathbf{x}}_{co}$ is controllable and observable, and $\bar{\mathbf{x}}_{\bar{c}}$ is not controllable. Furthermore, the transfer function of FE is

$$\bar{\mathbf{C}}_{co}(s\mathbf{I} - \bar{\mathbf{A}}_{co})^{-1}\bar{\mathbf{B}}_{co} + \mathbf{E}$$

which depends solely on the controllable and observable part of the equation FE.

Proof

If the dynamical equation FE is not controllable, it can be transformed into the form of (5-54). Consider now the dynamical equation $F\bar{E}_c$ which is the controllable part of FE. If $F\bar{E}_c$ is not observable, then $F\bar{E}_c$ can be transformed into the form of (5-60), which can also be written as

$$\begin{bmatrix} \dot{\bar{\mathbf{x}}}_{c\bar{o}} \\ \dot{\bar{\mathbf{x}}}_{co} \end{bmatrix} = \begin{bmatrix} \bar{\mathbf{A}}_{c\bar{o}} & \bar{\mathbf{A}}_{12} \\ \mathbf{0} & \bar{\mathbf{A}}_{co} \end{bmatrix} \begin{bmatrix} \bar{\mathbf{x}}_{c\bar{o}} \\ \bar{\mathbf{x}}_{co} \end{bmatrix} + \begin{bmatrix} \bar{\mathbf{B}}_{c\bar{o}} \\ \bar{\mathbf{B}}_{co} \end{bmatrix} \mathbf{u}$$

$$\mathbf{y} = [\mathbf{0} \quad \bar{\mathbf{C}}_{co}]\bar{\mathbf{x}} + \mathbf{E}\mathbf{u}$$

Combining these two transformations, we immediately obtain Equation (5-62). Following directly from Theorems 5-16 and 5-17, we conclude that the transfer function of FE is given by $\bar{\mathbf{C}}_{co}(s\mathbf{I} - \bar{\mathbf{A}}_{co})^{-1}\bar{\mathbf{B}}_{co} + \mathbf{E}$. Q.E.D.

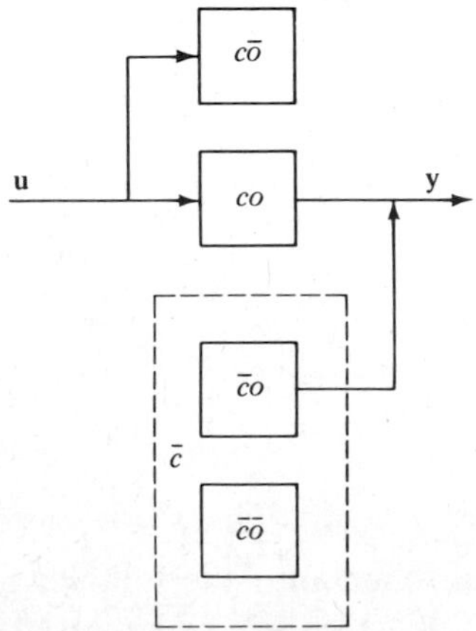

Figure 5-10 Canonical decomposition of a dynamical equation. (c stands for controllable, $\bar{c}$ for uncontrollable, o for observable, and $\bar{o}$ for unobservable.)

This theorem can be illustrated symbolically as shown in Figure 5-10, in which the uncontrollable part is further decomposed into observable and unobservable parts. We see that the transfer-function matrix or the impulse-response matrix of a dynamical equation depends solely on the controllable and observable part of the equation. In other words, *the impulse-response matrix (the input-output description) describes only the part of a system that is controllable and observable.* This is the most important relation between the input-output description and the state-variable description. This theorem tells us why the input-output description is sometimes insufficient to describe a system, for the uncontrollable and/or unobservable parts of the system do not appear in the transfer-function matrix description.

Example 2

Consider the network shown in Figure 5-11(a). Because the input is a current source, the behavior due to the initial conditions in C_1 and L_1 can never be detected from the output. Hence the state variables associated with C_1 and L_1 are not observable (they may be controllable, but we don't care). Similarly the state variable associated with L_2 is not controllable. Because of the symmetry, the state variable associated with C_2 is uncontrollable and unobservable. By dropping the state variables that are either uncontrollable or unobservable, the network in Figure 5-11(a) is reduced to the form in Figure 5-11(b). Hence the transfer function of the network in Figure 5-11(a) is $\hat{g}(s) = 1$. ∎

Before moving to the next topic, we use Theorem 5-16 to prove statement 4 of Theorem 5-7.

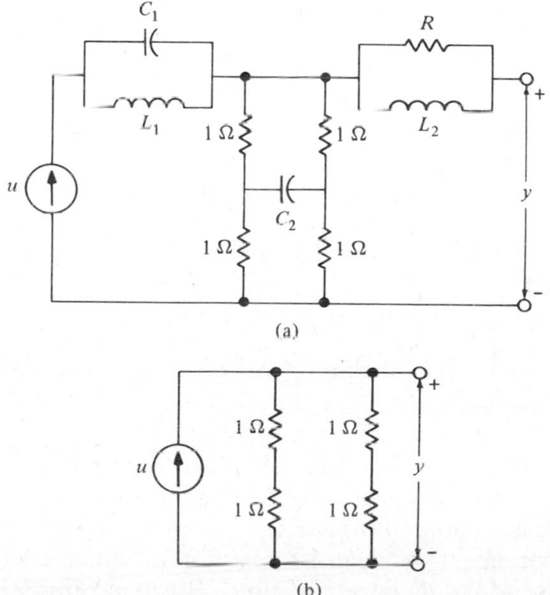

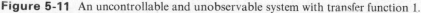

Figure 5-11 An uncontrollable and unobservable system with transfer function 1.

Proof of statement 4 of Theorem 5-7

The matrix $(s\mathbf{I}-\mathbf{A})$ is nonsingular at every s in $\mathbb{C}$ except at the eigenvalues of $\mathbf{A}$; hence the matrix $[s\mathbf{I}-\mathbf{A} \vdots \mathbf{B}]$ has rank n at every s in $\mathbb{C}$ except possibly at the eigenvalues of $\mathbf{A}$. Now we show that if $\{\mathbf{A}, \mathbf{B}\}$ is controllable, then $\rho[s\mathbf{I}-\mathbf{A} \vdots \mathbf{B}]=n$ at every eigenvalue of $\mathbf{A}$. If not, then there exist an eigenvalue λ and a $1 \times n$ vector $\boldsymbol{\alpha} \neq \mathbf{0}$ such that

$$\boldsymbol{\alpha}[\lambda\mathbf{I}-\mathbf{A} \;\vdots\; \mathbf{B}]=\mathbf{0}$$

or

$$\boldsymbol{\alpha}\lambda=\boldsymbol{\alpha}\mathbf{A} \qquad \text{and} \qquad \boldsymbol{\alpha}\mathbf{B}=\mathbf{0}$$

which imply

$$\boldsymbol{\alpha}\mathbf{A}^2=\lambda\boldsymbol{\alpha}\mathbf{A}=\lambda^2\boldsymbol{\alpha}$$

and, in general,

$$\boldsymbol{\alpha}\mathbf{A}^i=\lambda^i\boldsymbol{\alpha} \qquad i=1, 2, \ldots$$

Hence we have

$$\boldsymbol{\alpha}[\mathbf{B} \quad \mathbf{A}\mathbf{B} \quad \cdots \quad \mathbf{A}^{n-1}\mathbf{B}]=[\boldsymbol{\alpha}\mathbf{B} \quad \lambda\boldsymbol{\alpha}\mathbf{B} \quad \cdots \quad \lambda^{n-1}\boldsymbol{\alpha}\mathbf{B}]=\mathbf{0}$$

This contradicts the controllability assumption of $\{\mathbf{A}, \mathbf{B}\}$. Hence, if $\{\mathbf{A}, \mathbf{B}\}$ is controllable, then $\rho[s\mathbf{I}-\mathbf{A} \;\vdots\; \mathbf{B}]=n$ at every eigenvalue of $\mathbf{A}$ and, consequently, at every s in $\mathbb{C}$.

Now we show that if $\{\mathbf{A}, \mathbf{B}\}$ is not controllable, then $\rho[\lambda\mathbf{I}-\mathbf{A} \;\vdots\; \mathbf{B}]<n$ for some eigenvalue λ of $\mathbf{A}$. If $\{\mathbf{A}, \mathbf{B}\}$ is not controllable, there exists an equivalence transformation $\mathbf{P}$ that transforms $\{\mathbf{A}, \mathbf{B}\}$ into $\{\bar{\mathbf{A}}, \bar{\mathbf{B}}\}$ with

$$\bar{\mathbf{A}}=\mathbf{P}\mathbf{A}\mathbf{P}^{-1}=\begin{bmatrix} \bar{\mathbf{A}}_c & \bar{\mathbf{A}}_{12} \\ \mathbf{0} & \bar{\mathbf{A}}_{\bar{c}} \end{bmatrix} \qquad \bar{\mathbf{B}}=\mathbf{P}\mathbf{B}=\begin{bmatrix} \bar{\mathbf{B}}_c \\ \mathbf{0} \end{bmatrix}$$

Let λ be an eigenvalue of $\bar{\mathbf{A}}_{\bar{c}}$. We choose $\boldsymbol{\beta} \neq \mathbf{0}$ such that $\boldsymbol{\beta}\bar{\mathbf{A}}_{\bar{c}}=\lambda\boldsymbol{\beta}$. Now we form a $1 \times n$ vector $\boldsymbol{\alpha}=[\mathbf{0} \quad \boldsymbol{\beta}]$. Then we have

$$\boldsymbol{\alpha}[\lambda\mathbf{I}-\bar{\mathbf{A}} \;\vdots\; \bar{\mathbf{B}}]=[\mathbf{0} \quad \boldsymbol{\beta}]\left[\begin{array}{cc:c} \lambda\mathbf{I}-\bar{\mathbf{A}}_c & -\bar{\mathbf{A}}_{12} & \bar{\mathbf{B}}_c \\ \mathbf{0} & \lambda\mathbf{I}-\bar{\mathbf{A}}_{\bar{c}} & \mathbf{0} \end{array}\right]=\mathbf{0}$$

which implies

$$\mathbf{0}=\boldsymbol{\alpha}[\mathbf{P}(\lambda\mathbf{I}-\mathbf{A})\mathbf{P}^{-1} \vdots \mathbf{P}\mathbf{B}]=\boldsymbol{\alpha}\mathbf{P}[(\lambda\mathbf{I}-\mathbf{A})\mathbf{P}^{-1} \vdots \mathbf{B}]$$

Since $\boldsymbol{\alpha} \neq \mathbf{0}$, we have $\bar{\boldsymbol{\alpha}} \triangleq \boldsymbol{\alpha}\mathbf{P} \neq \mathbf{0}$. Because $\mathbf{P}^{-1}$ is nonsingular, $\bar{\boldsymbol{\alpha}}(\lambda\mathbf{I}-\mathbf{A})\mathbf{P}^{-1}=\mathbf{0}$ implies $\bar{\boldsymbol{\alpha}}(\lambda\mathbf{I}-\mathbf{A})=\mathbf{0}$. Hence we have

$$\bar{\boldsymbol{\alpha}}[\lambda\mathbf{I}-\mathbf{A} \;\vdots\; \mathbf{B}]=\mathbf{0}$$

In other words, if $\{\mathbf{A}, \mathbf{B}\}$ is not controllable, then $[s\mathbf{I}-\mathbf{A} \vdots \mathbf{B}]$ does not have a rank of n for some eigenvalue of $\mathbf{A}$. This completes the proof. Q.E.D.

Irreducible dynamical equation. We have seen from Theorems 5-16 and 5-17 that if a linear time-invariant dynamical equation is either uncontrollable or unobservable, then there exists a dynamical equation of lesser dimension that has the same transfer-function matrix as the original dynamical equation. In other words, if a linear time-invariant dynamical equation is either uncontrollable or unobservable, its dimension can be reduced such that the reduced equation still has the same zero-state response. This fact motivates the following definition.

Definition 5-9

A linear time-invariant dynamical equation FE is said to be *reducible* if and only if there exists a linear time-invariant dynamical equation of lesser dimension that has the same transfer-function matrix as FE or, equivalently, is zero-state equivalent to FE. Otherwise, the equation is said to be *irreducible*.

Theorem 5-19

A linear time-invariant dynamical equation FE is irreducible if and only if FE is controllable and observable.

Proof

If the dynamical equation FE is either uncontrollable or unobservable, then FE is reducible (Theorems 5-16 and 5-17). Hence what we have to prove is that if FE is controllable and observable, then FE is irreducible. We prove this by contradiction. Suppose that the n-dimensional equation FE is controllable and observable and that there exists a linear time-invariant dynamical equation $F\bar{E}$,

$$F\bar{E}: \quad \dot{\bar{\mathbf{x}}} = \bar{\mathbf{A}}\bar{\mathbf{x}} + \bar{\mathbf{B}}\mathbf{u} \tag{5-63a}$$

$$\mathbf{y} = \bar{\mathbf{C}}\bar{\mathbf{x}} + \bar{\mathbf{E}}\mathbf{u} \tag{5-63b}$$

of lesser dimension, say $n_1 < n$, that is zero-state equivalent to FE. Then, from Theorem 4-6, we have $\mathbf{E} = \bar{\mathbf{E}}$ and

$$\mathbf{C}\mathbf{A}^k\mathbf{B} = \bar{\mathbf{C}}\bar{\mathbf{A}}^k\bar{\mathbf{B}} \qquad k = 0, 1, 2, \ldots \tag{5-64}$$

Consider now the product

$$\mathbf{V}\mathbf{U} \triangleq \begin{bmatrix} \mathbf{C} \\ \mathbf{C}\mathbf{A} \\ \vdots \\ \mathbf{C}\mathbf{A}^{n-1} \end{bmatrix} [\mathbf{B} \quad \mathbf{A}\mathbf{B} \quad \cdots \quad \mathbf{A}^{n-1}\mathbf{B}]$$

$$= \begin{bmatrix} \mathbf{C}\mathbf{B} & \mathbf{C}\mathbf{A}\mathbf{B} & \cdots & \mathbf{C}\mathbf{A}^{n-1}\mathbf{B} \\ \mathbf{C}\mathbf{A}\mathbf{B} & \mathbf{C}\mathbf{A}^2\mathbf{B} & \cdots & \mathbf{C}\mathbf{A}^n\mathbf{B} \\ \vdots & \vdots & & \vdots \\ \mathbf{C}\mathbf{A}^{n-1}\mathbf{B} & \mathbf{C}\mathbf{A}^n\mathbf{B} & \cdots & \mathbf{C}\mathbf{A}^{2(n-1)}\mathbf{B} \end{bmatrix} \tag{5-65}$$

By (5-64), we may replace $\mathbf{C}\mathbf{A}^k\mathbf{B}$ in (5-65) by $\bar{\mathbf{C}}\bar{\mathbf{A}}^k\bar{\mathbf{B}}$; consequently, we have

$$\mathbf{V}\mathbf{U} = \bar{\mathbf{V}}_{n-1}\bar{\mathbf{U}}_{n-1} \tag{5-66}$$

where $\bar{\mathbf{V}}_{n-1}$ and $\bar{\mathbf{U}}_{n-1}$ are defined as in (5-48) and (5-27). Since FE is controllable and observable, we have $\rho\mathbf{U} = n$ and $\rho\mathbf{V} = n$. It follows from Theorem 2-6 that $\rho(\mathbf{V}\mathbf{U}) = n$. Now $\bar{\mathbf{V}}_{n-1}$ and $\bar{\mathbf{U}}_{n-1}$ are, respectively, $qn \times n_1$ and $n_1 \times np$ matrices; hence the matrix $\bar{\mathbf{V}}_{n-1}\bar{\mathbf{U}}_{n-1}$ has a rank of at most n_1. However,

(5-66) implies that $\rho(\bar{\mathbf{V}}_{n-1}\bar{\mathbf{U}}_{n-1})=n>n_1$. This is a contradiction. Hence, if FE is controllable and observable, then FE is irreducible. Q.E.D.

Recall from Section 4-4 that if a dynamical equation $\{\mathbf{A}, \mathbf{B}, \mathbf{C}, \mathbf{E}\}$ has a prescribed transfer-function matrix $\hat{\mathbf{G}}(s)$, then the dynamical equation $\{\mathbf{A}, \mathbf{B}, \mathbf{C}, \mathbf{E}\}$ is called a *realization* of $\hat{\mathbf{G}}(s)$. Now if $\{\mathbf{A}, \mathbf{B}, \mathbf{C}, \mathbf{E}\}$ is controllable and observable, then $\{\mathbf{A}, \mathbf{B}, \mathbf{C}, \mathbf{E}\}$ is called an *irreducible realization* of $\hat{\mathbf{G}}(s)$, In the following we shall show that all the irreducible realizations of $\hat{\mathbf{G}}(s)$ are equivalent.

Theorem 5-20

Let the dynamical equation $\{\mathbf{A}, \mathbf{B}, \mathbf{C}, \mathbf{E}\}$ be an irreducible realization of a $q \times p$ proper rational matrix $\hat{\mathbf{G}}(s)$. Then $\{\bar{\mathbf{A}}, \bar{\mathbf{B}}, \bar{\mathbf{C}}, \bar{\mathbf{E}}\}$ is also an irreducible realization of $\hat{\mathbf{G}}(s)$ if and only if $\{\mathbf{A}, \mathbf{B}, \mathbf{C}, \mathbf{E}\}$ and $\{\bar{\mathbf{A}}, \bar{\mathbf{B}}, \bar{\mathbf{C}}, \bar{\mathbf{E}}\}$ are equivalent; that is, there exists a nonsingular constant matrix $\mathbf{P}$ such that $\bar{\mathbf{A}}=\mathbf{PAP}^{-1}$, $\bar{\mathbf{B}}=\mathbf{PB}$, $\bar{\mathbf{C}}=\mathbf{CP}^{-1}$, and $\bar{\mathbf{E}}=\mathbf{E}$.

Proof

The sufficiency follows directly from Theorems 4-6 and 5-15. We show now the necessity of the theorem. Let $\mathbf{U}, \mathbf{V}$ be the controllability and the observability matrices of $\{\mathbf{A}, \mathbf{B}, \mathbf{C}, \mathbf{E}\}$, and let $\bar{\mathbf{U}}, \bar{\mathbf{V}}$ be similarly defined for $\{\bar{\mathbf{A}}, \bar{\mathbf{B}}, \bar{\mathbf{C}}, \bar{\mathbf{E}}\}$. If $\{\mathbf{A}, \mathbf{B}, \mathbf{C}, \mathbf{E}\}$ and $\{\bar{\mathbf{A}}, \bar{\mathbf{B}}, \bar{\mathbf{C}}, \bar{\mathbf{E}}\}$ are realizations of the same $\hat{\mathbf{G}}(s)$, then from (5-64) and (5-65) we have $\mathbf{E}=\bar{\mathbf{E}}$,

$$\mathbf{VU}=\bar{\mathbf{V}}\bar{\mathbf{U}} \tag{5-67}$$

and

$$\mathbf{VAU}=\bar{\mathbf{V}}\bar{\mathbf{A}}\bar{\mathbf{U}} \tag{5-68}$$

The irreducibility assumption implies that $\rho\bar{\mathbf{V}}=n$; hence the matrix $(\bar{\mathbf{V}}^*\bar{\mathbf{V}})$ is nonsingular (Theorem 2-8). Consequently, from (5-67), we have

$$\bar{\mathbf{U}}=(\bar{\mathbf{V}}^*\bar{\mathbf{V}})^{-1}\bar{\mathbf{V}}^*\mathbf{VU} \triangleq \mathbf{PU} \tag{5-69}$$

where $\mathbf{P} \triangleq (\bar{\mathbf{V}}^*\bar{\mathbf{V}})^{-1}\bar{\mathbf{V}}^*\mathbf{V}$. From (5-69) we have $\rho\bar{\mathbf{U}} \leq \min(\rho\mathbf{P}, \rho\mathbf{U})$, which, together with $\rho\bar{\mathbf{U}}=n$, implies that $\rho\mathbf{P}=n$. Hence $\mathbf{P}$ qualifies as an equivalence transformation. The first p columns of (5-69) give $\bar{\mathbf{B}}=\mathbf{PB}$. Since $\rho\mathbf{U}=n$, Equation (5-69) implies that

$$\mathbf{P}=(\bar{\mathbf{U}}\mathbf{U}^*)(\mathbf{UU}^*)^{-1}$$

With $\mathbf{P}=(\bar{\mathbf{V}}^*\bar{\mathbf{V}})^{-1}\bar{\mathbf{V}}^*\mathbf{V}=(\bar{\mathbf{U}}\mathbf{U}^*)(\mathbf{UU}^*)^{-1}$, it is easy to derive from (5-67) and (5-68) that $\mathbf{V}=\bar{\mathbf{V}}\mathbf{P}$ and $\mathbf{PA}=\bar{\mathbf{A}}\mathbf{P}$, which imply that $\mathbf{C}=\bar{\mathbf{C}}\mathbf{P}$ and $\bar{\mathbf{A}}=\mathbf{PAP}^{-1}$. Q.E.D.

This theorem implies that all the irreducible realizations of $\hat{\mathbf{G}}(s)$ have the same dimension. Physically, the dimension of an irreducible dynamical equation is the minimal number of integrators (if we simulate the equation in an analog computer) or the minimal number of energy-storage elements (if the system is an RLC network) required to generate the given transfer-function matrix.

We studied in this section only the canonical decomposition of linear time-invariant dynamical equations. For the time-varying case, the interested reader is referred to References 106 and 108.

*5-6 Controllability and Observability of Jordan-Form Dynamical Equations

The controllability and observability of a linear time-invariant dynamical equation are invariant under any equivalence transformation; hence it is conceivable that we may obtain simpler controllability and observability criteria by transforming the equation into a special form. Indeed, if a dynamical equation is in a Jordan form, the conditions are very simple and can be checked almost by inspection. In this section, we shall derive these conditions.

Consider the n-dimensional linear time-invariant Jordan-form dynamical equation

$$JFE: \quad \dot{\mathbf{x}} = \mathbf{A}\mathbf{x} + \mathbf{B}\mathbf{u} \tag{5-70a}$$

$$\mathbf{y} = \mathbf{C}\mathbf{x} + \mathbf{E}\mathbf{u} \tag{5-70b}$$

where the matrices **A**, **B**, and **C** are assumed of the forms shown in Table 5-1. The $n \times n$ matrix **A** is in the Jordan form, with m distinct eigenvalues $\lambda_1, \lambda_2, \ldots, \lambda_m$. $\mathbf{A}_i$ denotes all the Jordan blocks associated with the eigenvalue λ_i; $r(i)$ is the number of Jordan blocks in $\mathbf{A}_i$; and $\mathbf{A}_{ij}$ is the jth Jordan block in $\mathbf{A}_i$. Clearly,

$$\mathbf{A}_i = \text{diag}\,(\mathbf{A}_{i1}, \mathbf{A}_{i2}, \ldots, \mathbf{A}_{ir(i)}) \qquad \text{and} \qquad \mathbf{A} = \text{diag}\,(\mathbf{A}_1, \mathbf{A}_2, \ldots, \mathbf{A}_m)$$

Table 5-1 Jordan-Form Dynamical Equation

$$\underset{(n \times n)}{\mathbf{A}} = \begin{bmatrix} \mathbf{A}_1 & & & \\ & \mathbf{A}_2 & & \\ & & \ddots & \\ & & & \mathbf{A}_m \end{bmatrix} \qquad \underset{(n \times p)}{\mathbf{B}} = \begin{bmatrix} \mathbf{B}_1 \\ \mathbf{B}_2 \\ \vdots \\ \mathbf{B}_m \end{bmatrix}$$

$$\mathbf{C} = [\mathbf{C}_1 \quad \mathbf{C}_2 \quad \cdots \quad \mathbf{C}_m]$$

$$\underset{(n_i \times n_i)}{\mathbf{A}_i} = \begin{bmatrix} \mathbf{A}_{i1} & & & \\ & \mathbf{A}_{i2} & & \\ & & \ddots & \\ & & & \mathbf{A}_{ir(i)} \end{bmatrix} \qquad \underset{(n_i \times p)}{\mathbf{B}_i} = \begin{bmatrix} \mathbf{B}_{i1} \\ \mathbf{B}_{i2} \\ \vdots \\ \mathbf{B}_{ir(i)} \end{bmatrix}$$

$$\underset{(q \times n_i)}{\mathbf{C}_i} = [\mathbf{C}_{i1} \quad \mathbf{C}_{i2} \quad \cdots \quad \mathbf{C}_{ir(i)}]$$

$$\underset{(n_{ij} \times n_{ij})}{\mathbf{A}_{ij}} = \begin{bmatrix} \lambda_i & 1 & & & \\ & \lambda_i & 1 & & \\ & & \ddots & \ddots & \\ & & & \lambda_i & 1 \\ & & & & \lambda_i \end{bmatrix} \qquad \underset{(n_{ij} \times p)}{\mathbf{B}_{ij}} = \begin{bmatrix} \mathbf{b}_{1ij} \\ \mathbf{b}_{2ij} \\ \vdots \\ \mathbf{b}_{lij} \end{bmatrix}$$

$$\mathbf{C}_{ij} = [\mathbf{c}_{1ij} \quad \mathbf{c}_{2ij} \quad \cdots \quad \mathbf{c}_{lij}]$$

Let n_i and n_{ij} be the order of $\mathbf{A}_i$ and $\mathbf{A}_{ij}$, respectively; then

$$n = \sum_{i=1}^{m} n_i = \sum_{i=1}^{m} \sum_{j=1}^{r(i)} n_{ij}$$

Corresponding to $\mathbf{A}_i$ and $\mathbf{A}_{ij}$, the matrices $\mathbf{B}$ and $\mathbf{C}$ are partitioned as shown. The first *row* and the last row of $\mathbf{B}_{ij}$ are denoted by $\mathbf{b}_{1ij}$ and $\mathbf{b}_{lij}$, respectively. The first *column* and the last column of $\mathbf{C}_{ij}$ are denoted by $\mathbf{c}_{1ij}$ and $\mathbf{c}_{lij}$.

Theorem 5-21

The n-dimensional linear time-invariant Jordan-form dynamical equation JFE is controllable if and only if for each $i = 1, 2, \ldots, m$, the rows of the $r(i) \times p$ matrix

$$\mathbf{B}_i^l \triangleq \begin{bmatrix} \mathbf{b}_{li1} \\ \mathbf{b}_{li2} \\ \vdots \\ \mathbf{b}_{lir(i)} \end{bmatrix} \tag{5-71a}$$

are linearly independent (over the field of complex numbers). JFE is observable if and only if for each $i = 1, 2, \ldots, m$, the columns of the $q \times r(i)$ matrix

$$\mathbf{C}_i^1 = [\mathbf{c}_{1i1} \quad \mathbf{c}_{1i2} \quad \cdots \quad \mathbf{c}_{1ir(i)}] \tag{5-71b}$$

are linearly independent (over the field of complex numbers). ∎

Example 1

Consider the Jordan-form dynamical equation

$$JFE: \quad \dot{\mathbf{x}} = \begin{bmatrix} \lambda_1 & 1 & 0 & 0 & 0 & 0 & 0 \\ 0 & \lambda_1 & 0 & 0 & 0 & 0 & 0 \\ 0 & 0 & \lambda_1 & 0 & 0 & 0 & 0 \\ 0 & 0 & 0 & \lambda_1 & 0 & 0 & 0 \\ 0 & 0 & 0 & 0 & \lambda_2 & 1 & 0 \\ 0 & 0 & 0 & 0 & 0 & \lambda_2 & 1 \\ 0 & 0 & 0 & 0 & 0 & 0 & \lambda_2 \end{bmatrix} \mathbf{x} + \begin{bmatrix} 0 & 0 & 0 \\ 1 & 0 & 0 \\ 0 & 1 & 0 \\ 0 & 0 & 1 \\ 1 & 1 & 2 \\ 0 & 1 & 0 \\ 0 & 0 & 1 \end{bmatrix} \mathbf{u} \begin{matrix} \\ \leftarrow \mathbf{b}_{l11} \\ \leftarrow \mathbf{b}_{l12} \\ \leftarrow \mathbf{b}_{l13} \\ \\ \\ \leftarrow \mathbf{b}_{l21} \end{matrix} \tag{5-72a}$$

$$\mathbf{y} = \begin{bmatrix} 1 & 1 & 2 & 0 & 0 & 2 & 0 \\ 1 & 0 & 1 & 2 & 0 & 1 & 1 \\ 1 & 0 & 2 & 3 & 0 & 2 & 2 \end{bmatrix} \mathbf{x} \tag{5-72b}$$

$$\begin{matrix} \uparrow & & \uparrow & \uparrow & \uparrow \\ \mathbf{c}_{111} & & \mathbf{c}_{112} & \mathbf{c}_{113} & \mathbf{c}_{121} \end{matrix}$$

The matrix $\mathbf{A}$ has two distinct eigenvalues λ_1 and λ_2. There are three Jordan blocks associated with λ_1; hence $r(1) = 3$. There is only one Jordan block associated with λ_2; hence $r(2) = 1$. The conditions for JFE to be controllable are that the set $\{\mathbf{b}_{l11}, \mathbf{b}_{l12}, \mathbf{b}_{l13}\}$ and the set $\{\mathbf{b}_{l21}\}$ be, individually, linearly independent. This is the case; hence JFE is controllable. The conditions for JFE to be observable are that the set $\{\mathbf{c}_{111}, \mathbf{c}_{112}, \mathbf{c}_{113}\}$ and the set $\{\mathbf{c}_{121}\}$ be, individually, linearly independent. Although the set $\{\mathbf{c}_{111}, \mathbf{c}_{112}, \mathbf{c}_{113}\}$ is linearly independent, the set $\{\mathbf{c}_{121}\}$, which consists of a zero vector, is linearly dependent. Hence JFE is not observable. ∎

The conditions for controllability and observability in Theorem 5-21 required that each of the m set of vectors be *individually* tested for linear independence. The linear dependence of one set on the other set is immaterial. Furthermore, the row vectors of $\mathbf{B}$ excluding the $\mathbf{b}_{lij}$'s do not play any role in determining the controllability of the equation.

The physical meaning of the conditions of Theorem 5-21 can be seen from the block diagram of the Jordan-form dynamical equation JFE. Instead of studying the general case, we draw in Figure 5-12 a block diagram for the Jordan-form dynamical equation in (5-72). Observe that each block consists of an integrator and a feedback path, as shown in Figure 5-13. The output of each block, or more precisely the output of each integrator, is assigned as a state variable. Each chain of blocks corresponds to a Jordan block in the equation. Consider the last chain of Figure 5-12. We see that if $\mathbf{b}_{l21} \neq \mathbf{0}$, then all state variables in that chain can be controlled; if $\mathbf{c}_{121} \neq \mathbf{0}$, then all state variables in that chain can be observed. If there are two or more chains associated with the same eigenvalue, then we require the linear independence of the first gain vectors of these chains. The chains associated with different eigenvalues can be studied separately.

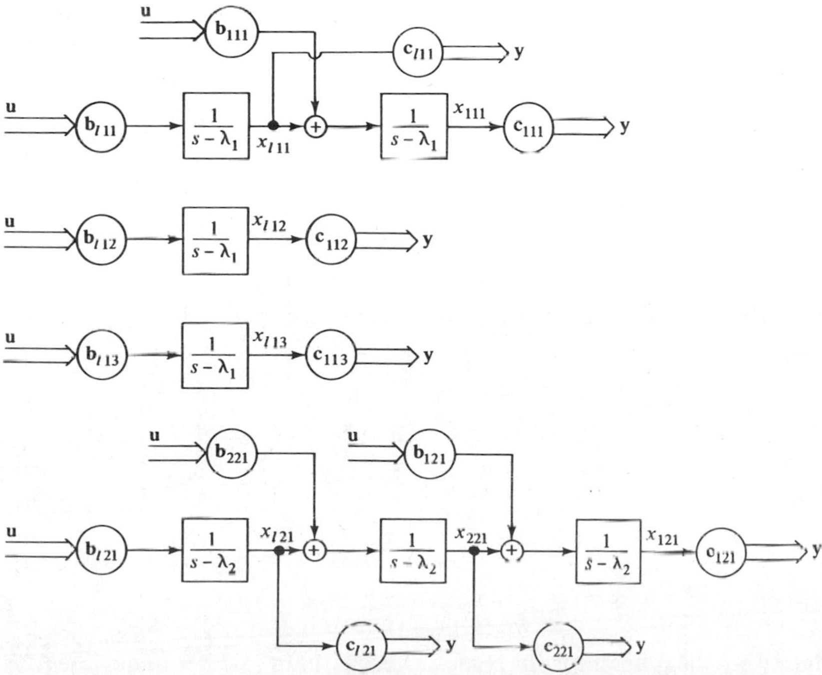

Figure 5-12 Block diagram of the Jordan-form equation (5-72).

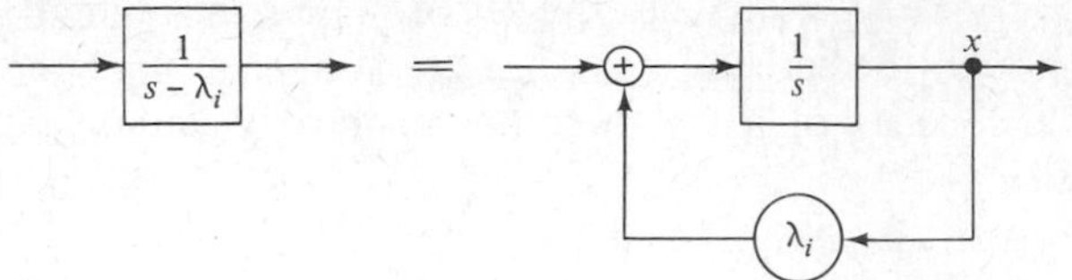

Figure 5-13 Analog computer simulation of $1/(s-\lambda_i)$.

Proof of Theorem 5-21

We use statement 4 of Theorem 5-7 to prove the theorem. In order not to be overwhelmed by notation, we assume $[s\mathbf{I}-\mathbf{A} \;\vdots\; \mathbf{B}]$ to be of the form

$$\left[\begin{array}{ccc|cc|cc|c}
s-\lambda_1 & -1 & 0 & & & & & \mathbf{b}_{111} \\
0 & s-\lambda_1 & -1 & & & & & \mathbf{b}_{211} \\
0 & 0 & s-\lambda_1 & & & & & \mathbf{b}_{l11} \\
\hline
 & & & s-\lambda_1 & -1 & & & \mathbf{b}_{112} \\
 & & & & s-\lambda_1 & & & \mathbf{b}_{l12} \\
\hline
 & & & & & s-\lambda_2 & -1 & \mathbf{b}_{121} \\
 & & & & & & s-\lambda_2 & \mathbf{b}_{l21}
\end{array}\right] \tag{5-73}$$

The matrix $\mathbf{A}$ has two distinct eigenvalues λ_1 and λ_2. There are two Jordan blocks associated with λ_1, one associated with λ_2. If $s=\lambda_1$, (5-73) becomes

$$\left[\begin{array}{ccc|cc|cc|c}
0 & -1 & 0 & & & & & \mathbf{b}_{111} \\
0 & 0 & -1 & & & & & \mathbf{b}_{211} \\
0 & 0 & 0 & & & & & \mathbf{b}_{l11} \\
\hline
 & & & 0 & -1 & & & \mathbf{b}_{112} \\
 & & & 0 & 0 & & & \mathbf{b}_{l12} \\
\hline
 & & & & & \lambda_1-\lambda_2 & -1 & \mathbf{b}_{121} \\
 & & & & & & \lambda_1-\lambda_2 & \mathbf{b}_{l21}
\end{array}\right] \tag{5-74}$$

By a sequence of elementary column operations, the matrix in (5-74) can be transformed into

$$\left[\begin{array}{ccc|cc|cc|c}
0 & -1 & 0 & & & & & \mathbf{0} \\
0 & 0 & -1 & & & & & \mathbf{0} \\
0 & 0 & 0 & & & & & \mathbf{b}_{l11} \\
\hline
 & & & 0 & -1 & & & \mathbf{0} \\
 & & & 0 & 0 & & & \mathbf{b}_{l12} \\
\hline
 & & & & & \lambda_1-\lambda_2 & 0 & \mathbf{0} \\
 & & & & & & \lambda_1-\lambda_2 & \mathbf{0}
\end{array}\right] \tag{5-75}$$

Note that $\lambda_1-\lambda_2$ is different from zero. The matrix in (5-75), or equivalently, the matrix in (5-73) at $s=\lambda_1$, has a full row rank if and only if $\mathbf{b}_{l11}$ and $\mathbf{b}_{l12}$ are linearly independent. By proceeding similarly for each distinct eigenvalue, the theorem can be established. Q.E.D.

Observe that in order for the rows of an $r(i) \times p$ matrix $\mathbf{B}_i^l$ to be linearly independent, it is necessary that $r(i) \leq p$. Hence in the case of single input—that is, $p = 1$— it is necessary to have $r(i) = 1$ in order for the rows of $\mathbf{B}_i^l$ to be linearly independent. In words, a necessary condition for a single-input Jordan-form dynamical equation to be controllable is that there is only one Jordan block associated with each distinct eigenvalue. For $p = 1$, the matrix $\mathbf{B}_i^l$ reduces to a vector. Thus we have the following corollary.

Corollary 5-21

A single-input linear time-invariant Jordan-form dynamical equation is controllable if and only if there is only one Jordan block associated with each distinct eigenvalue and all the components of the column vector $\mathbf{B}$ that correspond to the last row of each Jordan block are different from zero.

A single-output linear time-invariant Jordan-form dynamical equation is observable if and only if there is only one Jordan block associated with each distinct eigenvalue and all the components of the row vector $\mathbf{C}$ that correspond to the first column of each Jordan block are different from zero. ∎

Example 2

Consider the single-variable Jordan-form dynamical equation

$$\dot{\mathbf{x}} = \left[\begin{array}{ccc|c} 0 & 1 & 0 & 0 \\ 0 & 0 & 1 & 0 \\ 0 & 0 & 0 & 0 \\ \hline 1 & 0 & 0 & 1 \end{array}\right] \mathbf{x} + \left[\begin{array}{c} 10 \\ 9 \\ 0 \\ \hline 1 \end{array}\right] u \qquad y = [1 \quad 0 \quad 0 \,\vdots\, 1]\mathbf{x}$$

There are two distinct eigenvalues 0 and 1. The component of $\mathbf{B}$ which corresponds to the last row of the Jordan block associated with eigenvalue 0 is zero; therefore, the equation is not controllable. The two components of $\mathbf{C}$ corresponding to the first column of both Jordan blocks are different from zero; therefore, the equation is observable. ∎

Example 3

Consider the following two Jordan-form state equations:

$$\begin{bmatrix} \dot{x}_1 \\ \dot{x}_2 \end{bmatrix} = \begin{bmatrix} -1 & 0 \\ 0 & -2 \end{bmatrix} \mathbf{x} + \begin{bmatrix} 1 \\ 1 \end{bmatrix} u \tag{5-76}$$

and

$$\begin{bmatrix} \dot{x}_1 \\ \dot{x}_2 \end{bmatrix} = \begin{bmatrix} -1 & 0 \\ 0 & 2 \end{bmatrix} \mathbf{x} + \begin{bmatrix} e^{-t} \\ e^{-2t} \end{bmatrix} u \tag{5-77}$$

That the state equation (5-76) is controllable follows from Corollary 5-21. Equation (5-77) is a time-varying dynamical equation; however, since its $\mathbf{A}$ matrix is in the Jordan form and since the components of $\mathbf{B}$ are different from zero for all t, one might be tempted to conclude that (5-77) is controllable. Let us check this by using Theorem 5-4. For any fixed t_0, we have

$$\boldsymbol{\Phi}(t_0 - t)\mathbf{B}(t) = \begin{bmatrix} e^{-(t_0 - t)} & 0 \\ 0 & e^{-2(t_0 - t)} \end{bmatrix} \begin{bmatrix} e^{-t} \\ e^{-2t} \end{bmatrix} = \begin{bmatrix} e^{-t_0} \\ e^{-2t_0} \end{bmatrix}$$

It is clear that the rows of $\boldsymbol{\Phi}(t_0 - t)\mathbf{B}(t)$ are linearly dependent in t. Hence the state equation (5-77) is *not* controllable at any t_0. ∎

From this example we see that, in applying a theorem, all the conditions should be carefully checked; otherwise, we might obtain an erroneous conclusion.

*5-7 Output Controllability and Output Function Controllability

Similar to the (state) controllability of a dynamical equation, we may define controllability for the output vector of a system. Although these two concepts are the same except that one is defined for the *state* and the other for the *output*, the state controllability is a property of a dynamical equation, whereas the output controllability is a property of the impulse-response matrix of a system.

Consider a system with the input-output description

$$\mathbf{y}(t) = \int_{-\infty}^{t} \mathbf{G}(t, \tau)\mathbf{u}(\tau)\, d\tau$$

where $\mathbf{u}$ is the $p \times 1$ input vector, $\mathbf{y}$ is the $q \times 1$ output vector, $\mathbf{G}(t, \tau)$ is the $q \times p$ impulse-response matrix of the system. We assume for simplicity that $\mathbf{G}(t, \tau)$ does not contain δ-functions and is continuous in t and τ for $t > \tau$.

Definition 5-10

A system with a continuous impulse-response matrix $\mathbf{G}(t, \tau)$ is said to be *output controllable* at time t_0 if, for any $\mathbf{y}_1$, there exist a finite $t_1 > t_0$ and an input $\mathbf{u}_{[t_0, t_1]}$ that transfers the output from $\mathbf{y}(t_0) = \mathbf{0}$ to $\mathbf{y}(t_1) = \mathbf{y}_1$. ∎

Theorem 5-22

A system with a continuous $\mathbf{G}(t, \tau)$ is output controllable at t_0 if and only if there exists a finite $t_1 > t_0$ such that all rows of $\mathbf{G}(t_1, \tau)$ are linearly independent in τ on $[t_0, t_1]$ over the field of complex number. ∎

The proof of this theorem is exactly the same as the one of Theorem 5-4 and is therefore omitted.

We study in the following the class of systems that also have linear time-invariant dynamical-equation descriptions. Consider the system that is describable by

$$FE: \quad \begin{aligned} \dot{\mathbf{x}} &= \mathbf{A}\mathbf{x} + \mathbf{B}\mathbf{u} \\ \mathbf{y} &= \mathbf{C}\mathbf{x} \end{aligned}$$

where **A**, **B**, and **C** are $n \times n$, $n \times p$, $q \times n$ real constant matrices. The impulse-response matrix of the system is

$$\mathbf{G}(t) = \mathbf{C}e^{\mathbf{A}t}\mathbf{B}$$

The transfer-function matrix of the system is

$$\hat{\mathbf{G}}(s) = \mathbf{C}(s\mathbf{I} - \mathbf{A})^{-1}\mathbf{B} \tag{5-78}$$

It is clear that $\hat{\mathbf{G}}(s)$ is a strictly proper rational function matrix.

Corollary 5-22

A system whose transfer function is a strictly proper rational-function matrix is output controllable if and only if all the rows of $\hat{\mathbf{G}}(s)$ are linearly independent over the field of complex numbers or if and only if the $q \times np$ matrix

$$[\mathbf{CB} \vdots \mathbf{CAB} \vdots \cdots \vdots \mathbf{CA}^{n-1}\mathbf{B}] \tag{5-79}$$

has rank q. ∎

The proof of Theorem 5-7 can be applied here with slight modification. A trivial consequence of this corollary is that every single-output system is output controllable. We see that although the condition of output controllability can also be stated in terms of **A**, **B**, and **C**, compared with checking the linear independence of $\hat{\mathbf{G}}(s)$, the condition (5-79) seems more complicated.

The state controllability is defined for the dynamical equation, whereas the output controllability is defined for the input-output description; therefore, these two concepts are not necessarily related.

Example 1

Consider the network shown in Figure 5-14. It is neither state controllable nor observable, but it is output controllable.

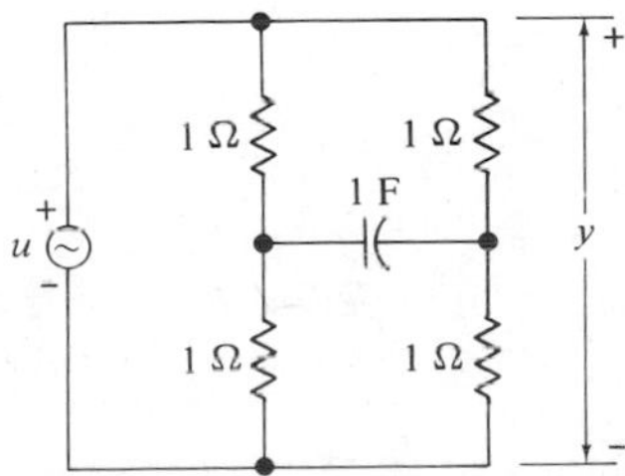

Figure 5-14 A network which is output controllable but neither (state) controllable nor observable.

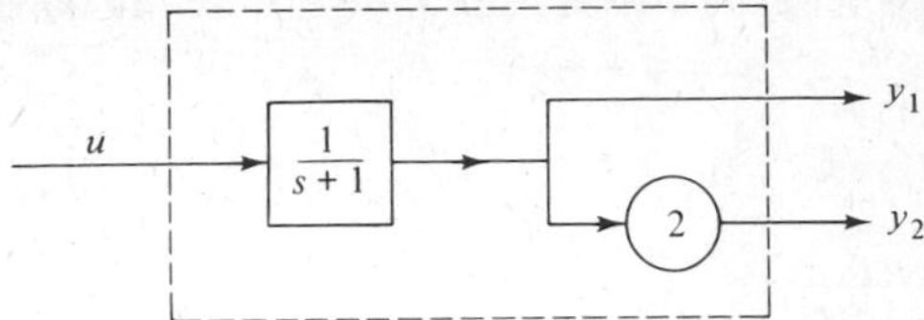

Figure 5-15 A system which is controllable and observable but not output controllable.

Example 2

Consider the system shown in Figure 5-15. The transfer-function matrix of the system is

$$\begin{bmatrix} \dfrac{1}{s+1} \\ \dfrac{2}{s+1} \end{bmatrix}$$

the rows of which are linearly dependent. Hence the system is not output controllable. The dynamical equation of the system is

$$\dot{\mathbf{x}} = -x + u \qquad \mathbf{y} = \begin{bmatrix} 1 \\ 2 \end{bmatrix} x$$

which is controllable and observable. ∎

If a system is output controllable, its output can be transferred to any desired value at certain *instant* of time. A related problem is whether it is possible to steer the output following a preassigned curve over any *interval* of time. A system whose output can be steered over any interval of time is said to be *output function controllable* or *functional reproducible*.

Theorem 5-23

A system with a $q \times p$ proper rational-function matrix $\hat{\mathbf{G}}(s)$ is output function controllable if and only if $\rho\hat{\mathbf{G}}(s) = q$ in $\mathbb{R}(s)$, the field of rational functions with real coefficients.

Proof

If the system is initially relaxed, then we have

$$\hat{\mathbf{y}}(s) = \hat{\mathbf{G}}(s)\hat{\mathbf{u}}(s) \tag{5-80}$$

If $\rho\hat{\mathbf{G}}(s) = q$—that is, all the rows of $\hat{\mathbf{G}}(s)$ are linearly independent over the field of rational functions—then the $q \times q$ matrix $\hat{\mathbf{G}}(s)\hat{\mathbf{G}}^*(s)$ is nonsingular (Theorem 2-8). Consequently, for any $\hat{\mathbf{y}}(s)$, if we choose

$$\hat{\mathbf{u}}(s) = \hat{\mathbf{G}}^*(s)(\hat{\mathbf{G}}(s)\hat{\mathbf{G}}^*(s))^{-1}\hat{\mathbf{y}}(s) \tag{5-81}$$

then Equation (5-80) is satisfied. Consequently, if $\rho\hat{\mathbf{G}}(s) = q$, then the system is output function controllable. If $\rho\hat{\mathbf{G}}(s) < q$, we can always find a $\hat{\mathbf{y}}(s)$, not in the range of $\hat{\mathbf{G}}(s)$, for which there exists no solution $\hat{\mathbf{u}}(s)$ in (5-80) (Theorem 2-4). Q.E.D.

If the input is restricted to the class of piecewise continuous functions of t, then the given output function should be very smooth; otherwise, the input computed from (5-81) will not be piecewise continuous. For example, if the given output has some discontinuity, an input containing δ-functions may be needed to generate the discontinuity.

The condition for output function controllability can also be stated in terms of the matrices **A**, **B**, and **C**. However, it is much more complicated. The interested reader is referred to Reference 8.

Dual to the output function controllability is the input function observability. These problems are intimately related to the *inverse problem*. A system with a $q \times p$ proper rational matrix $\hat{\mathbf{G}}(s)$ is said to have a right (left) inverse if there exists a $p \times q$ rational matrix $\hat{\mathbf{G}}_{RI}(s)$ $[\hat{\mathbf{G}}_{LI}(s)]$ such that

$$\hat{\mathbf{G}}(s)\hat{\mathbf{G}}_{RI}(s) = \mathbf{I}_q \qquad [\hat{\mathbf{G}}_{LI}(s)\hat{\mathbf{G}}(s) = \mathbf{I}_p]$$

A system is said to have an inverse if it has both a right inverse and a left inverse. A necessary and sufficient condition for $\hat{\mathbf{G}}(s)$ to have a right inverse is that $\rho\hat{\mathbf{G}}(s) = q$ in $\mathbb{R}(s)$. This condition is identical to that of the output function controllability. Many questions may be raised regarding a right inverse. Is it unique? Is it a proper rational matrix? What is its minimal degree? Is it stable? What are its equivalent conditions in dynamical equations? These problems will not be studied in this text. The interested reader is referred to References S172, S185, S218, and S239.

*5-8 Computational Problems

In this section, we discuss some computational problems encountered in this chapter. As discussed in Section 2-9, a problem may be well conditioned or ill conditioned; a computational method may be numerically stable or numerically unstable. If we use a numerically stable method to solve a well-conditioned problem, the result will generally be good. If a problem is ill conditioned, even if we use a numerically stable method to solve it, there is no guarantee that the result will be correct. If we use a numerically unstable method to solve a problem, well or ill conditioned, the result must be carefully scrutinized. Given a problem, if we must use an unstable method because of nonexistence of any stable method, the unstable method should be applied at a stage, as late as possible, in the computation.

As discussed in Theorem 5-7, there are several ways of checking the controllability of a state equation. Among them, statements 3 and 4 appear to be more suitable for computer computation. However, they may encounter some computational problems, as will be discussed in the following.

The computation of the controllability matrix $\mathbf{U} = [\mathbf{B} \quad \mathbf{AB} \quad \cdots \quad \mathbf{A}^{n-1}\mathbf{B}]$ is straightforward. Let $\mathbf{K}_0 \triangleq \mathbf{B}$. We compute $\mathbf{K}_i = \mathbf{A}\mathbf{K}_{i-1}$, $i = 1, 2, \ldots, n-1$.

At the end, we have $\mathbf{U} = [\mathbf{K}_0 \quad \mathbf{K}_1 \quad \cdots \quad \mathbf{K}_{n-1}]$. The rank of $\mathbf{U}$ can then be computed by using the singular value decomposition (Appendix E) which is a numerically stable method. If the dimension n of the equation is large, this process requires the computation of $\mathbf{A}^k\mathbf{B}$ for large k and may transform the problem into a *less* well conditioned problem. For convenience in discussion, we assume that all eigenvalues λ_i of $\mathbf{A}$ are distinct and $\mathbf{B}$ is an $n \times 1$ vector. We also arrange λ_i so that $|\lambda_1| \geq |\lambda_2| \geq \cdots \geq |\lambda_n|$. Clearly, we can write $\mathbf{B}$ as

$$\mathbf{B} = \alpha_1 \mathbf{v}_1 + \alpha_2 \mathbf{v}_2 + \cdots + \alpha_n \mathbf{v}_n$$

where $\mathbf{v}_i$ is an eigenvector associated with eigenvalue λ_i; that is, $\mathbf{A}\mathbf{v}_i = \lambda_i \mathbf{v}_i$. It is straightforward to verify

$$\mathbf{A}^k\mathbf{B} = \alpha_1 \lambda_1^k \mathbf{v}_1 + \alpha_2 \lambda_2^k \mathbf{v}_2 + \cdots + \alpha_n \lambda_n^k \mathbf{v}_n$$

If $|\lambda_1|$ is much larger than all other eigenvalues, then we have

$$\mathbf{A}^k\mathbf{B} \rightarrow \alpha_1 \lambda_1^k \mathbf{v}_1 \qquad \text{for } k \text{ large}$$

In other words, $\mathbf{A}^k\mathbf{B}$ tends to approach the same vector, $\mathbf{v}_1$, as k increases. Hence, it will be difficult to check the rank of $\mathbf{U}$ if n is large.

The same conclusion can also be reached by using a different argument. The *condition number* of a matrix $\mathbf{A}$ may be defined as cond $\mathbf{A} \triangleq \|\mathbf{A}\|_2 \|\mathbf{A}^{-1}\|_2 = \sigma_l/\sigma_s$, where σ_l and σ_s are the largest and smallest singular values of $\mathbf{A}$. It can be shown that $\sigma_s \leq |\lambda_n| < |\lambda_1| \leq \sigma_l$. Hence, if $|\lambda_1| \gg |\lambda_n|$, Cond $\mathbf{A}$ is a very large number. In computer computation, the multiplication of a matrix with a large condition number will introduce large computational error and should be avoided. Hence, the use of $[\mathbf{B} \quad \mathbf{AB} \quad \cdots \quad \mathbf{A}^{n-1}\mathbf{B}]$ to check the controllability of $\{\mathbf{A}, \mathbf{B}\}$ is not necessarily a good method.

As an example consider the 10-dimensional state equation (see Reference S169)

$$\dot{\mathbf{x}} = \begin{bmatrix} 1 & & & \mathbf{0} \\ & 2^{-1} & & \\ & & \ddots & \\ \mathbf{0} & & & 2^{-9} \end{bmatrix} \mathbf{x} + \begin{bmatrix} 1 \\ 1 \\ \vdots \\ 1 \end{bmatrix} u$$

We compute $\mathbf{U} = [\mathbf{B} \quad \mathbf{AB} \quad \cdots \quad \mathbf{A}^9\mathbf{B}]$ and then compute its singular values. The three smallest singular values are 0.712×10^{-7}, 0.364×10^{-9}, and 0.613×10^{-12}. If we use a digital computer with a relative precision of 10^{-10} (or a number will be considered as a zero if its absolute value is smaller than 10^{-10}), then the rank of $\mathbf{U}$ is smaller than 10, and we will conclude that the equation is not controllable, although the equation is clearly controllable following Corollary 5-21.

If we use the condition rank $[s\mathbf{I} - \mathbf{A} \quad \mathbf{B}] = n$, for all eigenvalues of $\mathbf{A}$, to check the controllability of $\{\mathbf{A}, \mathbf{B}\}$, we must compute first the eigenvalues of $\mathbf{A}$. The QR method is a stable and reliable method of computing the eigenvalues of a matrix. However, the eigenvalues of some matrices can be very ill conditioned.

For example, consider the matrix

$$\mathbf{A} = \begin{bmatrix} 20 & 20 & & & & \\ & 19 & 20 & & \mathbf{0} & \\ & & 18 & 20 & & \\ & & & \ddots & \ddots & \\ & \mathbf{0} & & & 2 & 20 \\ \varepsilon & & & & & 1 \end{bmatrix}$$

(see Reference S212). The diagonal elements range from 1 to 20; the elements on the superdiagonal are all equal to 20. The rest of the matrix are all zeros except the ε at the (20, 1)th position. The characteristic polynomial of **A** can be computed as

$$\Delta(s) = \prod_{i=1}^{20} (s-i) - (20)^{19}\varepsilon$$

If $\varepsilon = 0$, the eigenvalues of **A** are clearly equal to $\lambda_i = i$, $i = 1, 2, \ldots, 20$. To see the migration of the eigenvalues of **A** as ε increases from 0, we plot, in Figure 5-16, the root locus of $\Delta(s)$ (see Reference S46).[11] The heavy lines with arrows indicate the migration of the eigenvalues as ε increases from 0. We see that all eigenvalues except λ_1 and λ_{20} become complex if ε is sufficiently large (actually

[11] This analysis was suggested by Professor D. Z. Zheng.

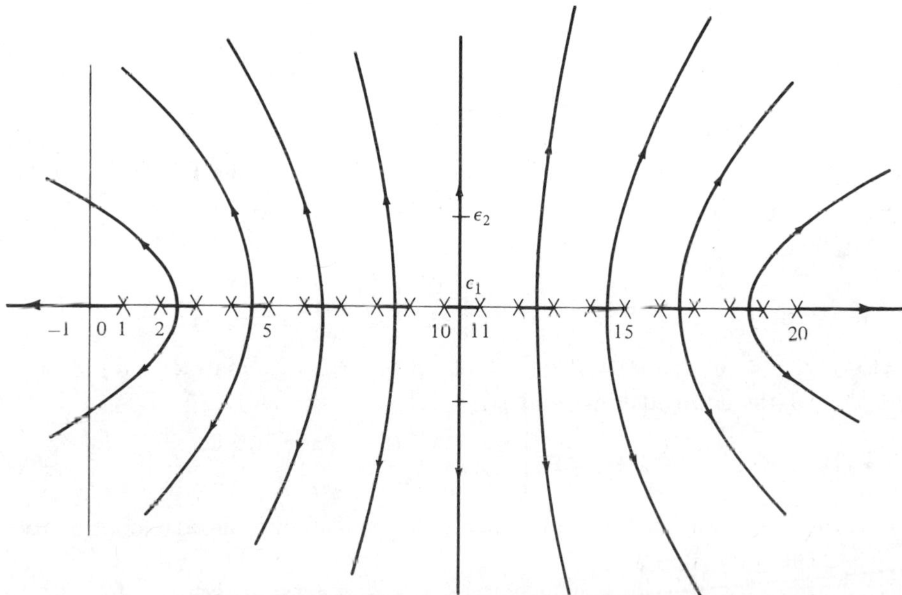

Figure 5-16 The root locus of $\Delta(s)$.

if ε is larger than 7.8×10^{-14}). We list in the following λ_{10} and λ_{11} for some ε:

$$\begin{array}{ll} \varepsilon_1 = 7.8 \times 10^{-14} & 10.5 \pm j0 \\ \varepsilon_2 = 10^{-10} & 10.5 \pm j2.73 \\ \varepsilon_3 = 10^{-5} & 10.5 \pm j8.05 \\ \varepsilon_4 = 1 & 10.5 \pm j16.26 \end{array}$$

They are computed directly from **A** by using the QR method. [The computation of the eigenvalues from $\Delta(s)$ is not advisable because the characteristic polynomial may be more sensitive to ε than the eigenvalues and the roots of a polynomial are very sensitive to the variations of the coefficient.] We see that the eigenvalues are very sensitive to ε. Thus the eigenvalues of **A** are very ill conditioned. For this reason, the use of the criterion rank $[s\mathbf{I} - \mathbf{A} \quad \mathbf{B}] = n$ to check the controllability of $\{\mathbf{A}, \mathbf{B}\}$ may yield an erroneous result.

It turns out that the use of Theorem 5-16 is the best way of checking the controllability of a state equation. Furthermore, if the equation is not controllable, the computation also yields a reduced controllable equation. The proof of Theorem 5-16 provided a procedure of computing the required equivalence transformation. The procedure, however, uses the controllability matrix and is not satisfactory from the computational point of view. In the following, we shall introduce an efficient and numerically stable method to transform a state equation into the form in (5-54). For the single-input case, the method is essentially the procedure of transforming the matrix **A** into the Hessenberg form (see Appendix A). We discuss in the following the general case.

Let $\mathbf{P}_1$ be an orthogonal matrix,[12] that is, $\mathbf{P}_1^{-1} = \mathbf{P}_1'$, such that

$$\mathbf{P}_1\mathbf{B} = \begin{bmatrix} \mathrm{x} & \mathrm{x} & \mathrm{x} & \cdots & \mathrm{x} \\ 0 & \mathrm{x} & \mathrm{x} & \cdots & \mathrm{x} \\ 0 & 0 & \mathrm{x} & \cdots & \mathrm{x} \\ \vdots & \vdots & \vdots & & \vdots \\ 0 & 0 & 0 & \cdots & 0 \end{bmatrix} = \begin{bmatrix} \mathbf{B}_1^{(1)} \\ \mathbf{0} \end{bmatrix} \tag{5-82}$$

where $\mathbf{B}_1^{(1)}$ is an $n_1 \times p$ upper triangular matrix and $n_1 = \text{rank } \mathbf{B}$. This step can be accomplished by a sequence of Householder transformations.[13] We then compute

$$\mathbf{P}_1\mathbf{A}\mathbf{P}_1' = \left[\begin{array}{c|c} \mathbf{A}_{11}^{(1)} & \mathbf{A}_{12}^{(1)} \\ \hline \mathbf{A}_{21}^{(1)} & \mathbf{A}_{22}^{(1)} \end{array}\right] \tag{5-83}$$

where $\mathbf{A}_{11}^{(1)}$ is an $n_1 \times n_1$ matrix, $\mathbf{A}_{21}^{(1)}$ is an $(n - n_1) \times n_1$ matrix, and so forth. If $\mathbf{A}_{21}^{(1)} = \mathbf{0}$, the controllability matrix

$$\mathbf{U} = [\mathbf{B} \quad \mathbf{AB} \vdots \cdots \vdots \mathbf{A}^{n-1}\mathbf{B}] = \mathbf{P}_1 \left[\begin{array}{cccc} \mathbf{B}_1^{(1)} & \mathbf{A}_{11}^{(1)}\mathbf{B}_1^{(1)} & (\mathbf{A}_{11}^{(1)})^2\mathbf{B}_1^{(1)} & \cdots \\ \hline \mathbf{0} & \mathbf{0} & \mathbf{0} & \cdots \end{array}\right] \tag{5-84}$$

has rank $n_1 < n$ and $\{\mathbf{A}, \mathbf{B}\}$ is not controllable and can be reduced to controllable

[12] If $\{\mathbf{A}, \mathbf{B}\}$ has complex elements, we use a unitary matrix, that is, $\mathbf{P}_1^{-1} = \mathbf{P}_1^*$.

[13] The $\mathbf{P}_1$ can also be obtained by using gaussian elimination with partial pivoting. In this case, $\mathbf{P}_1$ is not orthogonal. However, the method is still numerically stable.

$\{\mathbf{A}_{11}^{(1)}, \mathbf{B}_1^{(1)}\}$. If $\mathbf{A}_{21}^{(1)} \neq \mathbf{0}$, we find an orthogonal matrix $\mathbf{P}_2$ such that

$$\mathbf{P}_2\mathbf{A}_{21}^{(1)} = \begin{bmatrix} \mathbf{A}_{21}^{(2)} \\ \mathbf{0} \end{bmatrix} \tag{5-85}$$

where $\mathbf{A}_{21}^{(2)}$ is a $n_2 \times n_1$ matrix with $n_2 = \text{rank } \mathbf{A}_{21}^{(1)}$ and $n_2 \leq n_1$. We compute

$$\begin{bmatrix} \mathbf{I}_{n_1} & \mathbf{0} \\ \mathbf{0} & \mathbf{P}_2 \end{bmatrix} \mathbf{P}_1\mathbf{B} = \begin{bmatrix} \mathbf{B}_1^{(1)} \\ \mathbf{0} \end{bmatrix} \tag{5-86}$$

and

$$\begin{bmatrix} \mathbf{I}_{n_1} & \mathbf{0} \\ \mathbf{0} & \mathbf{P}_2 \end{bmatrix} \mathbf{P}_1\mathbf{A}\mathbf{P}_1' \begin{bmatrix} \mathbf{I}_{n_1} & \mathbf{0} \\ \mathbf{0} & \mathbf{P}_2' \end{bmatrix} = \left[\begin{array}{c|cc} \mathbf{A}_{11}^{(1)} & \multicolumn{2}{c}{\mathbf{A}_{12}^{(2)}} \\ \hline \mathbf{A}_{21}^{(2)} & \mathbf{A}_{22}^{(2)} & \mathbf{A}_{23}^{(2)} \\ \hline \mathbf{0} & \mathbf{A}_{32}^{(2)} & \mathbf{A}_{33}^{(2)} \end{array}\right] \tag{5-87}$$

where $\mathbf{P}_2\mathbf{A}_{22}^{(1)}\mathbf{P}_2'$ has been partitioned according to the dimension of $\mathbf{A}_{21}^{(2)}$. Now if $\mathbf{A}_{32}^{(2)} = \mathbf{0}$, the controllability matrix

$$\mathbf{U} = \begin{bmatrix} \mathbf{I}_{n_1} & \mathbf{0} \\ \mathbf{0} & \mathbf{P}_2 \end{bmatrix} \mathbf{P}_1 \left[\begin{array}{cccccc} \mathbf{B}_1^{(1)} & \mathbf{A}_{11}^{(1)}\mathbf{B}_1^{(1)} & \mathbf{x} & \mathbf{x} & \cdots \\ \hline \mathbf{0} & \mathbf{A}_{21}^{(2)}\mathbf{B}_1^{(1)} & \mathbf{x} & \mathbf{x} & \cdots \\ \hline \mathbf{0} & \mathbf{0} & \mathbf{0} & \mathbf{0} & \cdots \end{array}\right] \tag{5-88}$$

has rank $n_1 + n_2$ (where $\mathbf{x}$ denotes nonzero matrices), and $\{\mathbf{A}, \mathbf{B}\}$ is not controllable and can be reduced. If $\mathbf{A}_{21}^{(2)} \neq \mathbf{0}$, we continue the process until $\{\mathbf{A}, \mathbf{B}\}$ is transformed, by a sequence of orthogonal transformations, into[14]

$$\bar{\mathbf{A}} \triangleq \mathbf{PAP}' = \begin{bmatrix} \mathbf{A}_{11} & \mathbf{A}_{12} & \mathbf{A}_{13} & \cdots & \mathbf{A}_{1,k-1} & \mathbf{A}_{1k} \\ \mathbf{A}_{21} & \mathbf{A}_{22} & \mathbf{A}_{23} & \cdots & \mathbf{A}_{2,k-1} & \mathbf{A}_{2k} \\ \mathbf{0} & \mathbf{A}_{32} & \mathbf{A}_{33} & \cdots & \mathbf{A}_{3,k-1} & \mathbf{A}_{3k} \\ \vdots & \vdots & \vdots & & \vdots & \vdots \\ \mathbf{0} & \mathbf{0} & \mathbf{0} & \cdots & \mathbf{A}_{k,k-1} & \mathbf{A}_{k,k} \end{bmatrix}, \quad \bar{\mathbf{B}} \triangleq \mathbf{PB} = \begin{bmatrix} \mathbf{B}_1^{(1)} \\ \mathbf{0} \\ \mathbf{0} \\ \vdots \\ \mathbf{0} \end{bmatrix} \tag{5-89}$$

$$\bar{\mathbf{C}} = \mathbf{CP}' = [\mathbf{C}_1 \quad \mathbf{C}_2 \quad \mathbf{C}_3 \quad \cdots \quad \mathbf{C}_{k-1} \quad \mathbf{C}_k]$$

where $\rho\mathbf{B}_1^{(1)} = n_1$, $\rho\mathbf{A}_{21} = n_2$, $\rho\mathbf{A}_{32} = n_3, \ldots, \rho\mathbf{A}_{k,k-1} = n_k$. The integer k is the smallest integer such that either $n_k = 0$ or $n_1 + n_2 + \cdots + n_k = n$. Note that $\mathbf{A}_{k,k-1}$ has dimension $n_k \times n_{k-1}$ and has a full row rank. Clearly, we have

$$p \geq n_1 \geq n_2 \geq n_3 \cdots \geq n_k \geq 0$$

If $n_1 + n_2 + \cdots + n_k = n$, $\{\mathbf{A}, \mathbf{B}\}$ is controllable; otherwise, it is not controllable. By this process, the controllability of $\{\mathbf{A}, \mathbf{B}\}$ can be determined.

The $\bar{\mathbf{A}}$ in (5-89) is in the *block Hessenberg form* (see Appendix A). The transformation of $\{\mathbf{A}, \mathbf{B}\}$ into the form in (5-89) can be carried out recursively as follows (see Reference S203):

Step 1. $\mathbf{P} = \mathbf{I}_n$, $\mathbf{A}_0 = \mathbf{A}$, $\mathbf{B}_0 = \mathbf{B}$, $\bar{n} = 0$, $j = 1$.
Step 2. Find an orthogonal transformation $\mathbf{P}_j$ such that

$$\mathbf{P}_j\mathbf{B}_{j-1} = \begin{bmatrix} \mathbf{Z}_j \\ \mathbf{0} \end{bmatrix} \}n_j$$

where rank $\mathbf{Z}_j = n_j$. If $n_j = 0$, go to step 7.

[14] The form of $\bar{\mathbf{A}}$ is identical to the one in (A-17).

Step 3. Compute

$$\mathbf{P}_j\mathbf{A}_{j-1}\mathbf{P}_j' = \left[\begin{array}{c:c} \mathbf{X}_j & \mathbf{Y}_j \\ \hdashline \mathbf{B}_j & \mathbf{A}_j \end{array}\right]$$

where $\mathbf{X}_j$ is a $n_j \times n_j$ matrix.

Step 4. Update

$$\mathbf{P} = \mathbf{P}\begin{bmatrix} \mathbf{I}_{\bar{n}} & \mathbf{0} \\ \mathbf{0} & \mathbf{P}_j \end{bmatrix}$$

Step 5. $\bar{n} = \bar{n} + n_j$. If $\bar{n} = n$, go to step 8.

Step 6. $j = j+1$ and go to step 2.

Step 7. $\{\mathbf{A}, \mathbf{B}\}$ is not controllable and can be reduced to an $\bar{n}$-dimensional controllable equation.

Step 8. $\{\mathbf{A}, \mathbf{B}\}$ is controllable.

This is a numerically stable method of checking the controllability of a state equation. This algorithm was first proposed by Rosenbrock [S185] and then discussed in References S6 and S155. These papers used the gaussian elimination with partial pivoting in finding $\mathbf{P}_j$. Van Dooren [S203] and Patel [S170] suggested to use orthogonal transformations to improve the numerical stability of the algorithm. The singular value decomposition was also suggested in Reference S170 to determine the rank of $\mathbf{B}_i$.

The algorithm actually reveals more than the controllability of $\{\mathbf{A}, \mathbf{B}\}$; it also reveals its controllability indices. A comparison of (5-84) and (5-88) with (5-28) yields immediately

$$r_i = p - n_{i+1} \qquad i = 0, 1, 2, \ldots$$

where r_i is the number of linearly *dependent* columns in $\mathbf{A}^i\mathbf{B}$ and n_{i+1} is the number of linearly *independent* columns in $\mathbf{A}^i\mathbf{B}$. Since the set of controllability indices is uniquely determinable from $\{r_i, i = 0, 1, 2, \ldots\}$ [see Figure 5-7 or Equation (5-33)], it is also uniquely determinable from $\{n_i, i = 1, 2, \ldots\}$.

Actually we can say a little more about the controllability indices. In order not to be overwhelmed by notation, we assume that $\{\mathbf{A}, \mathbf{B}\}$, where $\mathbf{A}$ and $\mathbf{B}$ are respectively 8×8 and 8×3 matrices, has been transformed into

$$\bar{\mathbf{B}} = \left[\begin{array}{ccc} 1 & x & x \\ 0 & 1 & x \\ 0 & 0 & 1 \\ \hdashline 0 & 0 & 0 \\ 0 & 0 & 0 \\ \hdashline 0 & 0 & 0 \\ 0 & 0 & 0 \\ \hdashline 0 & 0 & 0 \end{array}\right] \qquad \bar{\mathbf{A}} = \left[\begin{array}{ccc:cc:cc:c} x & x & x & x & x & x & x & x \\ x & x & x & x & x & x & x & x \\ x & x & x & x & x & x & x & x \\ \hdashline 1 & x & x & x & x & x & x & x \\ 0 & 0 & 1 & x & x & x & x & x \\ \hdashline 0 & 0 & 0 & 1 & x & x & x & x \\ 0 & 0 & 0 & 0 & 1 & b_1 & c_1 & d_1 \\ \hdashline 0 & 0 & 0 & 0 & 0 & 0 & 1 & f_1 \end{array}\right] \tag{5-90}$$

(1st, 2nd, 3rd block column)

Let $\bar{\mathbf{B}} = [\bar{\mathbf{b}}_1 \quad \bar{\mathbf{b}}_2 \quad \bar{\mathbf{b}}_3]$; that is, $\bar{\mathbf{b}}_i$ is the ith column of $\bar{\mathbf{B}}$. Then from the structure of $\bar{\mathbf{A}}$ and $\bar{\mathbf{B}}$ in (5-90), we can readily verify that $\mu_1 = 3$, $\mu_2 = 1$, and $\mu_3 = 4$. These controllability indices can actually be read out directly from $\{\bar{\mathbf{A}}, \bar{\mathbf{B}}\}$ without any computation. The second column of the first block column of $\bar{\mathbf{A}}$ is linearly dependent on its left-hand-side columns. This implies that $\bar{\mathbf{A}}\bar{\mathbf{b}}_2$ is linearly dependent; hence $\mu_2 = 1$. Once a column becomes linearly dependent, it will not appear in the subsequent block columns of $\bar{\mathbf{A}}$. In the second block column of $\bar{\mathbf{A}}$, both columns are linearly independent of their left-hand-side columns of $\bar{\mathbf{A}}$; hence $\bar{\mathbf{A}}^2\bar{\mathbf{b}}_1$ and $\bar{\mathbf{A}}^2\bar{\mathbf{b}}_3$ are linearly independent. In the third block column of $\bar{\mathbf{A}}$, the first column is linearly dependent on its left-hand-side columns; hence $\bar{\mathbf{A}}^3\bar{\mathbf{b}}_1$ is linearly dependent, and $\mu_1 = 3$. The second column of the third block column is linearly independent, and there is no other column of $\bar{\mathbf{A}}$ which is linearly independent of its left-hand-side columns; hence $\mu_3 = 4$. We see that these indices are equal to the numbers of 1 enclosed by circles, squares, and triangles as shown in (5-90).

The $\{\bar{\mathbf{A}}, \bar{\mathbf{B}}\}$ in (5-90) can be further transformed by equivalence transformation into the following form:

$$\tilde{\mathbf{B}} = \left[\begin{array}{ccc} 1 & x & x \\ 0 & 1 & x \\ 0 & 0 & 1 \\ \hline 0 & 0 & 0 \\ 0 & 0 & 0 \\ \hline 0 & 0 & 0 \\ 0 & 0 & 0 \\ \hline 0 & 0 & 0 \end{array}\right] \qquad \tilde{\mathbf{A}} = \left[\begin{array}{ccc|cc|cc|c} x & x & x & x & x & x & x & x \\ x & x & x & x & x & x & x & x \\ x & x & x & x & x & x & x & x \\ \hline 1 & 0 & 0 & 0 & 0 & 0 & 0 & 0 \\ 0 & 0 & 1 & 0 & 0 & 0 & 0 & 0 \\ \hline 0 & 0 & 0 & 1 & 0 & 0 & 0 & 0 \\ 0 & 0 & 0 & 0 & 1 & 0 & 0 & 0 \\ \hline 0 & 0 & 0 & 0 & 0 & 0 & 1 & 0 \end{array}\right] \tag{5-91}$$

The matrix $\tilde{\mathbf{A}}$ is said to be in the *block companion form* or the *block Frobenius form*. This is achieved by choosing

$$\mathbf{P}_1^{-1} = \begin{bmatrix} 1 & & & & & & & \\ & 1 & & & & & & \\ & & 1 & & & & & \\ & & & 1 & & & & \\ & & & & 1 & & & \\ & & & & & 1 & & \\ & & & & & & 1 & -f_1 \\ & & & & & & 0 & 1 \end{bmatrix} \qquad \mathbf{P}_1 = \begin{bmatrix} 1 & & & & & & & \\ & 1 & & & & & & \\ & & 1 & & & & & \\ & & & 1 & & & & \\ & & & & 1 & & & \\ & & & & & 1 & & \\ & & & & & & 1 & f_1 \\ & & & & & & & 1 \end{bmatrix}$$

Then we have

$$\bar{\mathbf{B}}_1 = \mathbf{P}_1\bar{\mathbf{B}} = \bar{\mathbf{B}} \qquad \bar{\mathbf{A}}_1 = \mathbf{P}_1\bar{\mathbf{A}}\mathbf{P}_1^{-1} = \left[\begin{array}{ccc|cc|cc|c} x & x & x & x & x & x & x & x \\ x & x & x & x & x & x & x & x \\ x & x & x & x & x & x & x & x \\ \hline 1 & x & x & x & x & x & x & x \\ 0 & 0 & 1 & x & x & x & x & x \\ \hline 0 & 0 & 0 & 1 & x & x & x & x \\ 0 & 0 & 0 & 0 & 1 & b_2 & c_2 & d_2 \\ \hline 0 & 0 & 0 & 0 & 0 & 0 & 1 & 0 \end{array}\right] \tag{5-92}$$

Proceeding upward, $\{\bar{\mathbf{A}}, \bar{\mathbf{B}}\}$ in (5-90) can be transformed into the form in (5-91). This process can be easily programmed on a digital computer (see Reference S6).

The process of transforming $\{\mathbf{A}, \mathbf{B}\}$ into the block Hessenberg form in (5-89) is numerically stable. The process of transforming the block Hessenberg form into the block companion form in (5-91), however, is *not* numerically stable. The matrix $\mathbf{P}_1^{-1}$ in (5-92) carries out gaussian elimination on columns of $\bar{\mathbf{A}}$ *without* any pivoting. If pivoting is used, the form of $\bar{\mathbf{A}}$ will be altered, and we can never obtain the form in (5-91). Hence, $\mathbf{P}_1^{-1}$ must be chosen without any pivoting and the process is numerically unstable.

The $\{\bar{\mathbf{A}}, \bar{\mathbf{B}}\}$ in (5-90) or, equivalently, the $\{\tilde{\mathbf{A}}, \tilde{\mathbf{B}}\}$ in (5-91) has the controllability indices $\mu_1 = 3$, $\mu_2 = 1$, and $\mu_3 = 4$. Based on these indices, the $\{\tilde{\mathbf{A}}, \tilde{\mathbf{B}}\}$ can be transformed into

$$\hat{\mathbf{B}} = \mathbf{P}\tilde{\mathbf{B}} = \begin{bmatrix} 1 & x & x \\ 0 & 0 & 0 \\ 0 & 0 & 0 \\ \hline 0 & 1 & x \\ \hline 0 & 0 & 1 \\ 0 & 0 & 0 \\ 0 & 0 & 0 \\ 0 & 0 & 0 \end{bmatrix} \qquad \hat{\mathbf{A}} = \mathbf{P}\bar{\mathbf{A}}\mathbf{P}^{-1} = \left[\begin{array}{ccc|c|cccc} x & x & x & x & x & x & x & x \\ 1 & 0 & 0 & 0 & 0 & 0 & 0 & 0 \\ 0 & 1 & 0 & 0 & 0 & 0 & 0 & 0 \\ \hline x & x & x & x & x & x & x & x \\ \hline x & x & x & x & x & x & x & x \\ 0 & 0 & 0 & 0 & 1 & 0 & 0 & 0 \\ 0 & 0 & 0 & 0 & 0 & 1 & 0 & 0 \\ 0 & 0 & 0 & 0 & 0 & 0 & 1 & 0 \end{array}\right] \tag{5-93}$$

by the equivalence transformation[15]

$$\mathbf{P} = \begin{bmatrix} 1 & 0 & 0 & 0 & 0 & 0 & 0 & 0 \\ 0 & 0 & 0 & 1 & 0 & 0 & 0 & 0 \\ 0 & 0 & 0 & 0 & 0 & 1 & 0 & 0 \\ 0 & 1 & 0 & 0 & 0 & 0 & 0 & 0 \\ 0 & 0 & 1 & 0 & 0 & 0 & 0 & 0 \\ 0 & 0 & 0 & 0 & 1 & 0 & 0 & 0 \\ 0 & 0 & 0 & 0 & 0 & 0 & 1 & 0 \\ 0 & 0 & 0 & 0 & 0 & 0 & 0 & 1 \end{bmatrix}$$

The positions of 1 in this matrix are determined from the positions of 1 in (5-90) with the same encirclements. The transformation of $\{\tilde{\mathbf{A}}, \tilde{\mathbf{B}}\}$ into $(\hat{\mathbf{A}}, \hat{\mathbf{B}}\}$ can also be easily programmed without first determining the controllability indices (see Reference S6). The form in (5-93) is said to be in the controllable form and is very useful in the design of state feedback as will be seen in Chapter 7.

We mention one more method of checking controllability to conclude this section. From statement 4 of Theorem 5-7 and Theorem G-8′, we may conclude that $\{\mathbf{A}, \mathbf{B}\}$ is controllable if and only if the polynomial matrices $s\mathbf{I} - \mathbf{A}$ and $\mathbf{B}$

[15] This performs only permutations of columns and rows.

are left coprime. Similar to Theorem G-14, we form the matrix

$$\mathbf{T}_\mu = \left[\begin{array}{cc:cc:c:cc} -\mathbf{A} & \mathbf{B} & \mathbf{0} & \mathbf{0} & & \mathbf{0} & \mathbf{0} \\ \mathbf{I} & \mathbf{0} & -\mathbf{A} & \mathbf{B} & & \mathbf{0} & \mathbf{0} \\ \mathbf{0} & \mathbf{0} & \mathbf{I} & \mathbf{0} & \cdots & \vdots & \vdots \\ \vdots & \vdots & \vdots & \vdots & & -\mathbf{A} & \mathbf{B} \\ \mathbf{0} & \mathbf{0} & \mathbf{0} & \mathbf{0} & & \mathbf{I} & \mathbf{0} \end{array}\right] \tag{5-94}$$

$$\underbrace{\qquad}_{r_0} \quad \underbrace{\qquad}_{r_1} \quad \cdots \quad \underbrace{\qquad}_{r_\mu} \quad \text{(no. of dependent columns)}$$

There are $\mu+1$ block columns in $\mathbf{T}_\mu$; each consists of n columns formed from **A** and **I** and p columns formed from **B**. We call the former A columns, the latter B columns. Now we search linearly independent columns of $\mathbf{T}_\mu$ in order from left to right. Because of the unit matrix **I**, all A columns in $\mathbf{T}_\mu$ are linearly independent of their left-hand-side columns. Let r_i be the number of linearly dependent B columns in the $(i+1)$th block column as shown in (5-94). Because of the structure of $\mathbf{T}_\mu$, we have

$$0 \le r_0 \le r_1 \le \cdots \le r_\mu \cdots \le p$$

Let μ be the least integer such that

$$0 \le r_0 \le r_1 \le \cdots \le r_{\mu-1} < p$$

and

$$r_\mu = r_{\mu+1} = \cdots = p$$

It can be shown directly (Problem 5-35) or deduced from Chapter 6 that the r_i in (5-94) are the same as the r_i in (5-28). Hence we have

$$\begin{aligned} \text{rank } \mathbf{U} = \text{rank } [\mathbf{B} \quad \mathbf{AB} \quad \cdots \quad \mathbf{A}^{\mu-1}\mathbf{B}] &= (p-r_0) + (p-r_1) + \cdots + (p-r_\mu) \\ &= \text{total number of linearly independent B columns in (5-94)} \end{aligned}$$

Consequently, we conclude that $\{\mathbf{A}, \mathbf{B}\}$ is controllable if and only if the total number of linearly independent B columns in (5-94) is n. Note that the linearly independent columns of $\mathbf{T}_\mu$ are to be searched in order from left to right. The position of the rows of $\mathbf{T}_\mu$, however, can be arbitrary altered. Hence, we may apply Householder transformations or gaussian eliminations with partial pivoting on the rows of $\mathbf{T}_\mu$ to transform $\mathbf{T}_\mu$ into an upper triangular form (see Appendix A). Once in this form, the linearly independent B columns can be readily determined.

There are two disadvantages in using $\mathbf{T}_\mu$ to check the controllability of $\{\mathbf{A}, \mathbf{B}\}$. First, the size of $\mathbf{T}_\mu$ is $(2+\mu)n \times (n+p)(\mu+1)$, which is generally much larger than the size of **A**. Second, if $\{\mathbf{A}, \mathbf{B}\}$ is not controllable, its reduced controllable equation cannot be readily obtained. Hence, the transformation of $\{\mathbf{A}, \mathbf{B}\}$ into a block Hessenberg form seems to be a better method of checking the controllability of $\{\mathbf{A}, \mathbf{B}\}$.

The discussion of the observability part is similar to the controllability part and will not be repeated.

5-9 Concluding Remarks

In this chapter we have introduced the concepts of controllability and observability. Various theorems for linear dynamical equations to be controllable and observable were derived. We shall discuss briefly the relations among some of these theorems. We list first those theorems which are dual to each other:

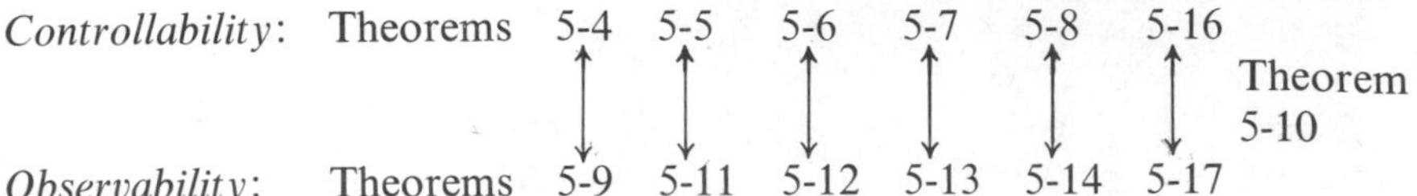

The theorems in the observability part can be easily derived from the controllability part by applying Theorem 5-10 (theorem of duality), and vice versa.

Theorems 5-1 and 5-4 (or 5-9) are two fundamental results of this chapter. They are derived with the least assumption (continuity), and hence they are most widely applicable. If additional assumptions (continuous differentiability) are introduced, then we have Theorems 5-2 and 5-5 (or 5-11), which give only sufficient conditions but are easier to apply. If we have the analyticity assumption (the strongest possible assumption) on time-varying dynamical equations, then we have Theorems 5-3 and 5-6 (or 5-12). Theorem 5-7 (or 5-13), which follows directly from Theorems 5-1, 5-3, and 5-4, gives the necessary and sufficient conditions for a linear time-invariant dynamical equation to be controllable.

The relationship between the transfer-function matrix and the linear time-invariant dynamical equation was established in this chapter. This was achieved by decomposing a dynamical equation into four parts: (1) controllable and observable, (2) controllable but unobservable, (3) uncontrollable and unobservable, and (4) uncontrollable but observable. The transfer-function matrix depends only on the controllable and observable part of the dynamical equation. If a linear time-invariant dynamical equation is not controllable and not observable, it can be reduced to a controllable and observable one.

The concepts of controllability and observability are essential in the study of Chapters 6 to 8. They will be used in the realization of a rational matrix (Chapter 6) and the stability study of linear systems (Chapter 8). Some practical implications of these concepts will be given in Chapter 7.

The computational problems of the various controllability and observability conditions are also discussed. Although the conditions can be stated nicely in terms of the ranks of $[\mathbf{B} \quad \mathbf{AB} \quad \cdots \quad \mathbf{A}^{n-1}\mathbf{B}]$ and $[s\mathbf{I}-\mathbf{A} \quad \mathbf{B}]$ in the controllability case, they are not suitable for computer computations. An efficient and numerically stable method is introduced to transform a dynamical equation into the form in (5-54) or (5-60), and its controllability or observability can then be determined. The algorithm can also be used to reduce a reducible dynamical equation to an irreducible one.

Before concluding this chapter, we remark on the controllability of the n-dimensional linear time-invariant descrete-time equation

$$\mathbf{x}(k+1) = \mathbf{A}\mathbf{x}(k) + \mathbf{B}\mathbf{u}(k)$$

Similar to Definition 5-1, we may define $\{\mathbf{A}, \mathbf{B}\}$ to be controllable if and only if, given any $\mathbf{x}_0$ and any $\mathbf{x}_1$, there exists an input sequence $\{\mathbf{u}(k)\}$ of finite length to transfer $\mathbf{x}_0$ to $\mathbf{x}_1$.[16] The condition of controllability is that rank $[\mathbf{B} \quad \mathbf{AB} \quad \cdots \quad \mathbf{A}^{n-1}\mathbf{B}] = n$ (see Problem 2-20). This condition is identical to statement 3 of Theorem 5-7. Hence, most of the results in the time-invariant continuous-time case are applicable to the discrete-time case without any modification. For the time-varying case, the situation is different but is simpler. This will not be discussed.

Problems

5-1 Which of the following sets are linearly independent over $(-\infty, \infty)$?

a. $\{t, t^2, e^t, e^{2t}, te^t\}$
b. $\{e^t, te^t, t^2e^t, te^{2t}, te^{3t}\}$
c. $\{\sin t, \cos t, \sin 2t\}$

5-2 Check the controllability of the following dynamical equations:

a.
$$\begin{bmatrix} \dot{x}_1 \\ \dot{x}_2 \end{bmatrix} = \begin{bmatrix} 0 & 1 \\ 0 & t \end{bmatrix} \begin{bmatrix} x_1 \\ x_2 \end{bmatrix} + \begin{bmatrix} 0 \\ 1 \end{bmatrix} u$$
$$y = [0 \quad 1] \begin{bmatrix} x_1 \\ x_2 \end{bmatrix}$$

b.
$$\dot{\mathbf{x}} = \begin{bmatrix} 0 & 1 & 0 \\ 0 & 0 & 1 \\ -2 & -4 & -3 \end{bmatrix} \mathbf{x} + \begin{bmatrix} 1 & 0 \\ 0 & 1 \\ -1 & 1 \end{bmatrix} u$$
$$\mathbf{y} = \begin{bmatrix} 0 & 1 & -1 \\ 1 & 2 & 1 \end{bmatrix} \mathbf{x}$$

c.
$$\dot{\mathbf{x}} = \begin{bmatrix} 0 & 4 & 3 \\ 0 & 20 & 16 \\ 0 & -25 & -20 \end{bmatrix} \mathbf{x} + \begin{bmatrix} -1 \\ 3 \\ 0 \end{bmatrix} u$$
$$y = [-1 \quad 3 \quad 0]\mathbf{x}$$

5-3 Show that a linear dynamical equation is controllable at t_0 if and only if there exists a finite $t_1 > t_0$ such that for any $\mathbf{x}_0$, there exists a $\mathbf{u}$ that transfers $\mathbf{x}_0$ to the zero state at time t_1. *Hint:* Use the nonsingularity of the state transition matrix.

5-4 Show that if a linear dynamical equation is controllable at t_0, then it is controllable at any $t < t_0$. Is it true that if a linear dynamical equation is controllable at t_0, then it is controllable at any $t > t_0$? Why?

[16] In the literature, if $\mathbf{x}_0 = \mathbf{0}$, it is called *reachable*; if $\mathbf{x}_1 = \mathbf{0}$, it is called *controllable*. Our definition encompasses both and does not make this distinction. If $\mathbf{A}$ is singular, the condition of reachability and the condition of controllability are slightly different. If $\mathbf{A}$ is nonsingular, they are identical.

5-5 Is it true that $\rho[\mathbf{B} \vdots \mathbf{AB} \vdots \cdots \vdots \mathbf{A}^{n-1}\mathbf{B}] = \rho[\mathbf{AB} \vdots \mathbf{A}^2\mathbf{B} \vdots \cdots \vdots \mathbf{A}^n\mathbf{B}]$? If not, under what condition will it be true?

5-6 Show that if a linear time-invariant dynamical equation is controllable, then it is uniformly controllable.

5-7 Check the observability of the dynamical equations given in Problem 5-2.

5-8 State (without proof) the necessary and sufficient condition for a linear dynamical equation E to be differentially controllable and differentially observable at t_0.

5-9 Check the controllability of the following state equations:

a. $\dot{\mathbf{x}} = \begin{bmatrix} 0 & 0 \\ 0 & 1 \end{bmatrix} \mathbf{x} + \begin{bmatrix} 1 \\ 1 \end{bmatrix} u$

b. $\dot{\mathbf{x}} = \begin{bmatrix} 0 & 0 \\ 0 & 1 \end{bmatrix} \mathbf{x} + \begin{bmatrix} 1 \\ e^{-t} \end{bmatrix} u$

c. $\dot{\mathbf{x}} = \begin{bmatrix} 0 & 0 \\ 0 & 1 \end{bmatrix} \mathbf{x} + \begin{bmatrix} 0 \\ e^{-2t} \end{bmatrix} u$

5-10 Check the controllability and observability of

$$\dot{\mathbf{x}} = \begin{bmatrix} 0 & 1 & 0 \\ 0 & 0 & 1 \\ 0 & 2 & -1 \end{bmatrix} \mathbf{x} + \begin{bmatrix} 0 & 1 \\ 1 & 0 \\ 0 & 0 \end{bmatrix} \mathbf{u}$$

$$\mathbf{y} = [1 \quad 0 \quad 1]\mathbf{x}$$

by using statement 3 of Theorems 5-7 and 5-13.

5-11 What are the controllability indices and the observability index of the equation in Problem 5-10?

5-12 Compute the controllability indices and the controllability index of the state equation in (3-48).

5-13 Given a *controllable* linear time-invariant single-input dynamical equation

$$\dot{\mathbf{x}} = \mathbf{A}\mathbf{x} + \mathbf{b}u$$

where $\mathbf{A}$ is an $n \times n$ matrix and $\mathbf{b}$ is an $n \times 1$ column vector. What is the equivalent dynamical equation if the basis $\{\mathbf{b}, \mathbf{Ab}, \ldots, \mathbf{A}^{n-1}\mathbf{b}\}$ is chosen for the state space? Or, equivalently, if $\bar{\mathbf{x}} = \mathbf{Px}$ and if $\mathbf{P}^{-1} = [\mathbf{b} \quad \mathbf{Ab} \quad \cdots \quad \mathbf{A}^{n-1}\mathbf{b}]$, what is the new state equation?

5-14 Find the dynamical equations for the systems shown in Figures 5-3, 5-4, and 5-11 and check the controllability and observability of these equations.

5-15 Consider the dynamical equation

$$\dot{\mathbf{x}} = \begin{bmatrix} -1 & 1 & 0 \\ 0 & -1 & 0 \\ 0 & 0 & -2 \end{bmatrix} \mathbf{x} + \begin{bmatrix} 0 \\ 1 \\ 1 \end{bmatrix} u$$

$$y = [1 \quad 1 \quad 1]\mathbf{x}$$

Is it possible to choose an initial state at $t=0$ such that the output of the dynamical equation is of the form $y(t)=te^{-t}$ for $t>0$?

5-16 Consider the dynamical equation in Problem 5-15. It is assumed that the initial state of the equation is not known. Is it possible to find an input $u_{[0,\infty)}$ such that the output y is te^{-t} for $t \geq 1$?

5-17 Show that the state of an observable, n-dimensional linear time-invariant dynamical equation can be determined instantaneously from the output and its derivatives up to $n-1$ order. [*Hint*: Compute $y(t), \dot{y}(t), \ldots, y^{(n-1)}(t)$.]

5-18 Show that controllability and observability of linear time-varying dynamical equations are invariant under any equivalence transformation $\bar{\mathbf{x}}=\mathbf{P}(t)\mathbf{x}$, where $\mathbf{P}$ is nonsingular for all t and continuously differentiable in t.

5-19 Reduce the dynamical equation

$$\dot{\mathbf{x}}=\begin{bmatrix}-1 & 4\\ 4 & -1\end{bmatrix}\mathbf{x}+\begin{bmatrix}1\\1\end{bmatrix}u \qquad y=[1 \quad 1]\mathbf{x}$$

to a controllable one.

5-20 Reduce the equation in Problem 5-19 to an observable one.

5-21 Reduce the following dynamical equation

$$\dot{\mathbf{x}}=\begin{bmatrix}\lambda_1 & 1 & 0 & 0 & 0\\ 0 & \lambda_1 & 1 & 0 & 0\\ 0 & 0 & \lambda_1 & 0 & 0\\ 0 & 0 & 0 & \lambda_2 & 1\\ 0 & 0 & 0 & 0 & \lambda_2\end{bmatrix}\mathbf{x}+\begin{bmatrix}0\\1\\0\\0\\1\end{bmatrix}u$$

$$y=[0 \quad 1 \quad 1 \quad 0 \quad 1]\,\mathbf{x}$$

to a controllable and observable equation.

5-22 Consider the n-dimensional linear time-invariant dynamical equation

$$FE: \qquad \begin{aligned}\dot{\mathbf{x}}&=\mathbf{Ax}+\mathbf{Bu}\\ \mathbf{y}&=\mathbf{Cx}+\mathbf{Eu}\end{aligned}$$

The rank of its controllability matrix,

$$\mathbf{U}=[\mathbf{B} \;\vdots\; \mathbf{AB} \;\vdots\; \cdots \;\vdots\; \mathbf{A}^{n-1}\mathbf{B}]$$

is assumed to be n_1 ($<n$). Let $\mathbf{Q}_1$ be an $n \times n_1$ matrix whose columns are any n_1 linearly independent columns of $\mathbf{U}$. Let $\mathbf{P}_1$ be an $n_1 \times n$ matrix such that $\mathbf{P}_1\mathbf{Q}_1=\mathbf{I}_{n_1}$, where $\mathbf{I}_{n_1}$ is the $n_1 \times n_1$ unit matrix. Show that the following n_1-dimensional dynamical equation

$$F\bar{E}: \qquad \begin{aligned}\dot{\bar{\mathbf{x}}}_1&=\mathbf{P}_1\mathbf{AQ}_1\bar{\mathbf{x}}_1+\mathbf{P}_1\mathbf{Bu}\\ \bar{\mathbf{y}}&=\mathbf{CQ}_1\bar{\mathbf{x}}_1+\mathbf{Eu}\end{aligned}$$

is controllable and is zero-state equivalent to FE. In other words, FE is reducible to $F\bar{E}$. *Hint*: Use Theorem 5-16.

5-23 In Problem 5-22, the reduction procedure reduces to solving for $\mathbf{P}_1$ in $\mathbf{P}_1\mathbf{Q}_1=\mathbf{I}_{n_1}$. Find a method to solve $\mathbf{P}_1$ in $\mathbf{P}_1\mathbf{Q}_1=\mathbf{I}_{n_1}$.

5-24 Develop a similar statement as Problem 5-22 for an unobservable linear time-invariant dynamical equation.

5-25 Is the following Jordan-form dynamical equation controllable and observable?

$$\dot{\mathbf{x}} = \begin{bmatrix} 2 & 1 & 0 & 0 & 0 & 0 & 0 \\ 0 & 2 & 0 & 0 & 0 & 0 & 0 \\ 0 & 0 & 2 & 0 & 0 & 0 & 0 \\ 0 & 0 & 0 & 2 & 0 & 0 & 0 \\ 0 & 0 & 0 & 0 & 1 & 1 & 0 \\ 0 & 0 & 0 & 0 & 0 & 1 & 0 \\ 0 & 0 & 0 & 0 & 0 & 0 & 1 \end{bmatrix} \mathbf{x} + \begin{bmatrix} 2 & 1 & 1 \\ 2 & 1 & 1 \\ 1 & 1 & 1 \\ 3 & 2 & 1 \\ -1 & 0 & 0 \\ 1 & 0 & 1 \\ 1 & 0 & 0 \end{bmatrix} \mathbf{u}$$

$$\mathbf{y} = \begin{bmatrix} 2 & 2 & 1 & 3 & -1 & 1 & 1 \\ 1 & 1 & 1 & 2 & 0 & 0 & 0 \\ 1 & 1 & 1 & 1 & 0 & 1 & 0 \end{bmatrix} \mathbf{x}$$

5-26 Is it possible to find a set of b_{ij} and a set of c_{ij} such that the following Jordan-form equation

$$\dot{\mathbf{x}} = \begin{bmatrix} 1 & 1 & 0 & 0 & 0 \\ 0 & 1 & 0 & 0 & 0 \\ 0 & 0 & 1 & 1 & 0 \\ 0 & 0 & 0 & 1 & 0 \\ 0 & 0 & 0 & 0 & 1 \end{bmatrix} \mathbf{x} + \begin{bmatrix} b_{11} & b_{12} \\ b_{21} & b_{22} \\ b_{31} & b_{32} \\ b_{41} & b_{42} \\ b_{51} & b_{52} \end{bmatrix} \mathbf{u}$$

$$\mathbf{y} = \begin{bmatrix} c_{11} & c_{12} & c_{13} & c_{14} & c_{15} \\ c_{21} & c_{22} & c_{23} & c_{24} & c_{25} \\ c_{31} & c_{32} & c_{33} & c_{34} & c_{35} \end{bmatrix} \mathbf{x}$$

is controllable? Observable?

5-27 Show that $\mathbf{A}$ is cyclic (see Problem 2-45) if and only if there exists a vector $\mathbf{b}$ such that $\{\mathbf{A}, \mathbf{b}\}$ is controllable. (*Hint*: Use Corollary 5-21.)

5-28 Show that if $\{\mathbf{A}, \mathbf{B}\}$ is controllable and $\mathbf{A}$ is cyclic, then there exists a $p \times 1$ column vector $\mathbf{r}$ such that $\{\mathbf{A}, \mathbf{Br}\}$ is controllable. (*Hint*: Use Theorem 5-21.)

5-29 Show that a necessary condition for $\{\mathbf{A}, \mathbf{B}\}$ where $\mathbf{B}$ is an $n \times p$ matrix, to be controllable is $p \geq m$, where m is the largest number of Jordan blocks associated with the same eigenvalue of $\mathbf{A}$. (See Problem 5-26 and use Theorem 5-21.)

5-30 In Corollary 5-22 we have that a dynamical equation with $\mathbf{E}=\mathbf{0}$ is output controllable if and only if $\rho[\mathbf{CB} \vdots \mathbf{CAB} \vdots \cdots \vdots \mathbf{CA}^{n-1}\mathbf{B}] = q$. Show that if $\mathbf{E} \neq 0$, then the dynamical equation is output controllable if and only if

$$\rho[\mathbf{CB} \vdots \mathbf{CAB} \vdots \cdots \vdots \mathbf{CA}^{n-1}\mathbf{B} \vdots \mathbf{E}] = q$$

5-31 Consider a linear time-invariant dynamical equation with $\mathbf{E}=\mathbf{0}$. Under what condition on $\mathbf{C}$ will the (state) controllability imply output controllability?

5-32 Consider two systems with the transfer-function matrices

$$\hat{\mathbf{G}}_1 = \begin{bmatrix} \dfrac{1}{s+1} & \dfrac{(s+3)}{(s+2)(s+1)} \\ \dfrac{1}{(s+1)} & \dfrac{(s+2)}{(s+3)(s+1)} \end{bmatrix} \qquad \hat{\mathbf{G}}_2 = \begin{bmatrix} \dfrac{1}{(s+2)} \\ \dfrac{(s+1)}{(s+2)} \end{bmatrix}$$

are they output controllable? Output function controllable?

5-33 A $q \times p$ rational matrix $\hat{\mathbf{G}}(s)$ is said to have a left inverse if there exists a rational matrix $\hat{\mathbf{G}}_{LI}(s)$ such that $\hat{\mathbf{G}}_{LI}(s)\hat{\mathbf{G}}(s) = \mathbf{I}_p$. Show that $\hat{\mathbf{G}}(s)$ has a left inverse if and only if $\rho\hat{\mathbf{G}}(s) = p$ in $\mathbb{R}(s)$.

5-34 Let $\mathbf{P}$ be a nonsingular matrix such that

$$\mathbf{PB} = \begin{bmatrix} \mathbf{B}_1 \\ \mathbf{0} \end{bmatrix} \qquad \mathbf{PAP}^{-1} = \begin{bmatrix} \mathbf{A}_{11} & \mathbf{A}_{12} \\ \mathbf{A}_{21} & \mathbf{A}_{22} \end{bmatrix}$$

where $\mathbf{B}_1$ is an $n_1 \times p$ matrix and $\rho\mathbf{B} = \rho\mathbf{B}_1 = n_1$. The matrices $\mathbf{A}_{21}$ and $\mathbf{A}_{22}$ have dimensions $(n-n_1) \times n_1$ and $(n-n_1) \times (n-n_1)$, respectively. Show that $\{\mathbf{A}, \mathbf{B}\}$ is controllable if and only if $\{\mathbf{A}_{22}, \mathbf{A}_{21}\}$ is controllable.

5-35 Show that

$$\rho\mathbf{U} = \rho[\mathbf{B} \quad \mathbf{AB} \quad \cdots \quad \mathbf{A}^{n-1}\mathbf{B}] = \text{total number of linearly independent B columns of } \mathbf{T}_\mu \text{ in (5-94)}$$

where linearly independent columns of $\mathbf{T}_\mu$ are to be searched in order from left to right. *Hint*: Premultiply $\mathbf{T}_\mu$ by

$$\begin{bmatrix} \mathbf{I} & & & \\ \mathbf{A} & \mathbf{I} & & \\ \mathbf{A}^2 & \mathbf{A} & \mathbf{I} & \\ \vdots & \vdots & \vdots & \ddots \end{bmatrix}$$

5-36 Show that $\{\mathbf{A}, \mathbf{C}\}$ is observable if and only if $\{\mathbf{A}, \mathbf{C}^*\mathbf{C}\}$ is observable, where $\mathbf{A}$ and $\mathbf{C}$ are, respectively, $n \times n$ and $q \times n$ constant matrices and $\mathbf{C}^*$ is the complex conjugate transpose of $\mathbf{C}$.

5-37 Show that $\{\mathbf{A}, \mathbf{B}\}$ is controllable if and only if there exists no left eigenvector of $\mathbf{A}$ that is orthogonal to all columns of $\mathbf{B}$, that is, there exist no eigenvalue λ and nonzero left eigenvector $\boldsymbol{\alpha}$ of $\mathbf{A}$ such that

$$\lambda\boldsymbol{\alpha} = \boldsymbol{\alpha}\mathbf{A} \qquad \text{and} \qquad \boldsymbol{\alpha}\mathbf{B} = \mathbf{0}$$

This is called the *Popov-Belevitch-Hautus test* in Reference S125. (*Hint*: See statement 4 of Theorem 5-7.)

6

Irreducible Realizations, Strict System Equivalence, and Identification

6-1 Introduction

Network synthesis is one of the important disciplines in electrical engineering. It is mainly concerned with determining a passive or an active network that has a prescribed impedance or transfer function. The subject matter we shall introduce in this chapter is along the same line—that is, to determine a linear time-invariant dynamical equation that has a prescribed rational transfer matrix. Hence this chapter might be viewed as a modern version of network synthesis.

A linear time-invariant dynamical equation that has a prescribed transfer matrix $\hat{\mathbf{G}}(s)$ is called a realization of $\hat{\mathbf{G}}(s)$. The term "realization" is justified by the fact that, by using the dynamical equation, the system with the transfer-function matrix $\hat{\mathbf{G}}(s)$ can be built in the real world by using an operational amplifier circuit. Although we have proved in Theorem 4-10 that every proper rational matrix has a finite-dimensional linear time-invariant dynamical-equation realization, there are still many unanswered questions. In this chapter we shall study this and other related problems.

We study the realization problem for the following reasons: First, there are many design techniques and many computational algorithms developed exclusively for dynamical equations. In order to apply these techniques and algorithms, transfer-function matrices must be realized into dynamical equations. Second, in the design of complex system it is always desirable to simulate the system on an analog or a digital computer to check its performance before the system is built. A system cannot be simulated efficiently if the transfer function is the only available description. After a dynamical-equation realization is obtained, by assigning the outputs of integrators as state variables, the system can be readily simulated. The realization can also be built by using operational amplifier circuits. Finally, the results can be used to establish the link between the state-variable approach and the transfer-function approach.

For every realizable transfer-function matrix $\hat{\mathbf{G}}(s)$, there is an unlimited number of linear time-invariant dynamical-equation realizations. Therefore a major problem in the realization is to find a "good" realization. It is clear that a dynamical-equation realization with the least possible dimension is a good realization. We claim that a realization of $\hat{\mathbf{G}}(s)$ with the least possible dimension must be a controllable and observable dynamical equation. Indeed, if a linear time-invariant dynamical-equation realization of $\hat{\mathbf{G}}(s)$ is found, and if the equation is uncontrollable or unobservable, then following from Theorem 5-19 it is possible to reduce the realization to a lesser-dimensional equation that still has $\hat{\mathbf{G}}(s)$ as its transfer-function matrix. This reduction is impossible only if the equation is controllable and observable. Therefore, we conclude that a realization of $\hat{\mathbf{G}}(s)$ with the least possible dimension is a controllable and observable dynamical equation, or equivalently, an irreducible dynamical equation. Such a realization is called a *minimal-dimensional* or *irreducible realization.* In this chapter we study mainly irreducible realizations for the following reasons: (1) A rational transfer-function matrix describes only the controllable and observable part of a dynamical equation; hence a faithful realization should be an irreducible one. (2) When an irreducible realization is used to synthesize a network, the number of integrators needed will be minimal. This is desirable for reasons of economy and sensitivity. Note that if an irreducible realization is found, any other irreducible realization can be obtained by applying an equivalence transformation (Theorem 5-20).

This chapter is organized as follows. In Section 6-2, we introduce the concept of the degree for proper rational matrices. Its significance is demonstrated in Theorem 6-2. In Section 6-3, various realizations are introduced for scalar rational functions. Hankel theorem is also introduced. The realization methods are then extended to vector rational functions in Section 6-4. Two different irreducible realization methods, one based on Hankel matrices and the other on coprime fractions, for proper rational matrices are discussed in Sections 6-5 and 6-6. In Section 6-7 we introduce a new mathematical description, called *polynomial matrix description*, for linear time-invariant systems. Its relationships with transfer functions and dynamical equations are also established. In Section 6-8 the concept of equivalent dynamical equation is extended to strict system equivalence for polynomial matrix description. It is shown that, under the coprimeness assumption, all polynomial matrix descriptions which have the same transfer matrix are strictly system equivalent. In the last section, we study the identification of discrete-time systems from arbitrary input-output pairs. The concept of persistent exciting is introduced.

This chapter is based mainly on references 15, 42, 47, 60, 62, 67, 68, 83, 89, 98, 115, S27, S48, S52, S126, S158, S161, S185, S187, S209, and S218. For the realization of impulse-response matrices $\hat{\mathbf{G}}(t, \tau)$, the interested reader is referred to References 32, 100, 101, 114, and S128.

6-2 The Characteristic Polynomial and the Degree of a Proper Rational Matrix

In this section we shall introduce the concepts of degree and characteristic polynomial for proper rational matrices. These concepts are the extension of the denominator of a proper rational function and its degree to the matrix case. Consider a proper rational function $\hat{g}(s) = N(s)/D(s)$. It is assumed that $N(s)$ and $D(s)$ are coprime (have no nontrivial common factor). Then the denominator of $\hat{g}(s)$ is defined as $D(s)$, and the degree of $\hat{g}(s)$ is defined as the degree of $D(s)$. Without the assumption of coprimeness, the denominator and the degree of $\hat{g}(s)$ are not well defined. In the following, we use det, deg, and dim to stand, respectively, for the determinant, degree, and dimension.

Theorem 6-1

Let the single-variable linear time-invariant dynamical equation

$$FE_1: \qquad \begin{aligned} \dot{\mathbf{x}} &= \mathbf{A}\mathbf{x} + \mathbf{b}u \\ y &= \mathbf{c}\mathbf{x} + eu \end{aligned}$$

be a realization of the proper rational function $\hat{g}(s)$. Then FE_1 is irreducible (controllable and observable) if and only if

$$\det(s\mathbf{I} - \mathbf{A}) = k[\text{denominator of } \hat{g}(s)] \tag{6-1}$$

or

$$\dim \mathbf{A} = \deg \hat{g}(s)$$

where k is a nonzero constant.

Proof

Let $\Delta(s) \triangleq \det(s\mathbf{I} - \mathbf{A})$ and let

$$\mathbf{c}(s\mathbf{I} - \mathbf{A})^{-1}\mathbf{b} \triangleq \frac{N_1(s)}{\Delta(s)} \tag{6-2}$$

First we show that $\{\mathbf{A}, \mathbf{b}, \mathbf{c}\}$ is irreducible if and only if $\Delta(s)$ and $N_1(s)$ are coprime. Indeed, if $\{\mathbf{A}, \mathbf{b}, \mathbf{c}\}$ is not irreducible, then Theorems 5-16 and 5-17 imply the existence of a $\{\bar{\mathbf{A}}, \bar{\mathbf{b}}, \bar{\mathbf{c}}\}$ such that $\dim \bar{\mathbf{A}} < \dim \mathbf{A}$ and

$$\frac{\bar{N}_1(s)}{\bar{\Delta}(s)} \triangleq \bar{\mathbf{c}}(s\mathbf{I} - \bar{\mathbf{A}})^{-1}\bar{\mathbf{b}} = \mathbf{c}(s\mathbf{I} - \mathbf{A})^{-1}\mathbf{b} = \frac{N_1(s)}{\Delta(s)}$$

where $\bar{\Delta}(s) = \det(s\mathbf{I} - \bar{\mathbf{A}})$. Since $\deg \bar{\Delta}(s) = \dim \bar{\mathbf{A}} < \dim \mathbf{A} = \deg \Delta(s)$, we conclude that $\Delta(s)$ and $N_1(s)$ have common factors. Reversing the above argument, we can show that if $\Delta(s)$ and $N_1(s)$ are not coprime, then $\{\mathbf{A}, \mathbf{b}, \mathbf{c}\}$ is not irreducible. Hence we have established that $\{\mathbf{A}, \mathbf{b}, \mathbf{c}\}$ is controllable and observable if and only if $\Delta(s)$ and $N_1(s)$ are coprime.

If FE_1 is a realization of $\hat{g}(s)$, then we have

$$\hat{g}(s) = e + \frac{N_1(s)}{\Delta(s)} = \frac{e\Delta(s) + N_1(s)}{\Delta(s)}$$

It is easy to show that $\Delta(s)$ and $N_1(s)$ are coprime if and only if $\Delta(s)$ and $e\Delta(s) + N_1(s)$ are coprime. Consequently, we have established that FE_1 is irreducible if and only if

$$\text{Denominator of } \hat{g}(s) = \Delta(s) = \det(s\mathbf{I} - \mathbf{A})$$

or

$$\deg \hat{g}(s) = \deg \Delta(s) = \dim \mathbf{A}$$

If the denominator of $\hat{g}(s)$ is not monic (its leading coefficient is not equal to 1), then we need a nonzero constant k in (6-1). Q.E.D.

With this theorem, the irreducibility of a single-variable linear time-invariant dynamical equation can be easily determined from the degree of its transfer function. It is desirable to see whether this is possible for the multivariable case. The answer is affirmative if a "denominator" can be defined for a rational matrix.

Definition 6-1

The *characteristic polynomial* of a proper rational matrix $\hat{\mathbf{G}}(s)$ is defined to be the least common denominator of all minors of $\hat{\mathbf{G}}(s)$. The *degree* of $\hat{\mathbf{G}}(s)$, denoted by $\delta\hat{\mathbf{G}}(s)$, is defined to be the degree of the characteristic polynomial of $\hat{\mathbf{G}}(s)$.[1]

Example 1

Consider the rational-function matrices

$$\hat{\mathbf{G}}_1(s) = \begin{bmatrix} \dfrac{1}{s+1} & \dfrac{1}{s+1} \\ \dfrac{1}{s+1} & \dfrac{1}{s+1} \end{bmatrix} \qquad \hat{\mathbf{G}}_2(s) = \begin{bmatrix} \dfrac{2}{s+1} & \dfrac{1}{s+1} \\ \dfrac{1}{s+1} & \dfrac{1}{s+1} \end{bmatrix}$$

The minors of order 1 of $\hat{\mathbf{G}}_1(s)$ are $1/(s+1)$, $1/(s+1)$, $1/(s+1)$, and $1/(s+1)$. The minor of order 2 of $\hat{\mathbf{G}}_1(s)$ is 0. Hence the characteristic polynomial of $\hat{\mathbf{G}}_1(s)$ is $s+1$ and $\delta\hat{\mathbf{G}}_1(s) = 1$. The minors of order 1 of $\hat{\mathbf{G}}_2(s)$ are $2/(s+1)$, $1/(s+1)$, $1/(s+1)$, and $1/(s+1)$. The minor of order 2 of $\hat{\mathbf{G}}_2(s)$ is $1/(s+1)^2$. Hence the characteristic polynomial of $\hat{\mathbf{G}}_2(s)$ is $(s+1)^2$ and $\delta\hat{\mathbf{G}}_2(s) = 2$. ∎

From this example, we see that the characteristic polynomial of $\hat{\mathbf{G}}(s)$ is in general different from the denominator of the determinant of $\hat{\mathbf{G}}(s)$ [if $\hat{\mathbf{G}}(s)$ is a

[1] It is also called the McMillan degree or the Smith McMillan degree. The definition is applicable only to proper rational matrices. If $\hat{\mathbf{G}}(s)$ is not proper, the definition must be modified to include the poles at $s = \infty$. See References 34, 62, 83, and S185.

square matrix] and different from the least common denominator of all the entries of $\hat{\mathbf{G}}(s)$. If $\hat{\mathbf{G}}(s)$ is scalar (a 1×1 matrix), the characteristic polynomial of $\hat{\mathbf{G}}(s)$ reduces to the denominator of $\hat{\mathbf{G}}(s)$.

Example 2

Consider the 2×3 rational-function matrix

$$\hat{\mathbf{G}}(s)=\begin{bmatrix} \dfrac{s}{s+1} & \dfrac{1}{(s+1)(s+2)} & \dfrac{1}{s+3} \\ \dfrac{-1}{s+1} & \dfrac{1}{(s+1)(s+2)} & \dfrac{1}{s} \end{bmatrix}$$

The minors of order 1 are the entries of $\hat{\mathbf{G}}(s)$. There are three minors of order 2. They are

$$\frac{s}{(s+1)^2(s+2)}+\frac{1}{(s+1)^2(s+2)}=\frac{s+1}{(s+1)^2(s+2)}=\frac{1}{(s+1)(s+2)} \tag{6-3}$$

$$\frac{s}{s+1}\cdot\frac{1}{s}+\frac{1}{(s+1)(s+3)}=\frac{s+4}{(s+1)(s+3)}$$

$$\frac{1}{(s+1)(s+2)s}-\frac{1}{(s+1)(s+2)(s+3)}=\frac{3}{s(s+1)(s+2)(s+3)}$$

Hence the characteristic polynomial of $\hat{\mathbf{G}}(s)$ is $s(s+1)(s+2)(s+3)$ and $\delta\hat{\mathbf{G}}(s)=4$. ∎

Note that in computing the characteristic polynomial of a rational matrix, every minor of the matrix must be reduced to an irreducible one as we did in (6-3); otherwise we will obtain an erroneous result.

In the following, we introduce a different but equivalent definition of the characteristic polynomial and degree of a proper rational matrix. This definition is similar to the scalar case and requires the concepts developed in Appendix G.

*Definition 6-1′

Consider a proper rational matrix $\hat{\mathbf{G}}(s)$ factored as $\hat{\mathbf{G}}(s)=\mathbf{N}_r(s)\mathbf{D}_r^{-1}(s)=\mathbf{D}_l^{-1}(s)\mathbf{N}_l(s)$. It is assumed that $\mathbf{D}_r(s)$ and $\mathbf{N}_r(s)$ are right coprime, and $\mathbf{D}_l(s)$ and $\mathbf{N}_l(s)$ are left coprime. Then the characteristic polynomial of $\hat{\mathbf{G}}(s)$ is defined as

$$\det \mathbf{D}_r(s) \qquad \text{or} \qquad \det \mathbf{D}_l(s)$$

and the degree of $\hat{\mathbf{G}}(s)$ is defined as

$$\deg \hat{\mathbf{G}}(s)=\deg\det \mathbf{D}_r(s)=\deg\det \mathbf{D}_l(s)$$

where deg det stands for the degree of the determinant. ∎

*May be skipped without loss of continuity.

We remark that the polynomials det $\mathbf{D}_r(s)$ and det $\mathbf{D}_l(s)$ differ at most by a nonzero constant [see Equation (6-189a)]; hence either one can be used to define the characteristic polynomial of $\hat{\mathbf{G}}(s)$. The equivalence of Definitions 6-1 and 6-1′ can be established by using the Smith-McMillan form. The interested reader is referred to References 15, S125, and S185.

*Theorem 6-2

Let the multivariable linear time-invariant dynamical equation

$$FE: \quad \begin{aligned} \dot{\mathbf{x}} &= \mathbf{A}\mathbf{x} + \mathbf{B}\mathbf{u} \\ \mathbf{y} &= \mathbf{C}\mathbf{x} + \mathbf{E}\mathbf{u} \end{aligned}$$

be a realization of the proper rational matrix $\hat{\mathbf{G}}(s)$. Then FE is irreducible (controllable and observable) if and only if

$$\det(s\mathbf{I} - \mathbf{A}) = k\,[\text{characteristic polynomial of } \hat{\mathbf{G}}(s)]$$

or

$$\dim \mathbf{A} = \deg \hat{\mathbf{G}}(s)$$

where k is a nonzero constant. ∎

The irreducibility of the realizations in this chapter for the multivariable case will be established by using the conditions of controllability and observability without relying on this theorem. Hence this theorem will not be proved here. In fact, this theorem will be a direct consequence of the irreducible realization discussed in Section 6-6. It is stated in this section because of its analogy to Theorem 6-1 and its bringing out the importance of the concepts of the characteristic polynomial and degree of $\hat{\mathbf{G}}(s)$.

6-3 Irreducible Realizations of Proper Rational Functions

Irreducible realization of $\beta/D(s)$. Before considering the general case, we study first the transfer function

$$\hat{g}(s) = \frac{\beta}{s^n + \alpha_1 s^{n-1} + \cdots + \alpha_{n-1}s + \alpha_n} \triangleq \frac{\beta}{D(s)} \tag{6-4}$$

where β and α_i, for $i = 1, 2, \ldots, n$, are real constants. Note that the leading coefficient of $D(s)$ is 1. Let u and y be the input and output; then we have

$$D(s)\hat{y}(s) = \beta\hat{u}(s) \tag{6-5a}$$

or, in the time domain,

$$(p^n + \alpha_1 p^{n-1} + \cdots + \alpha_n)y(t) = \beta u(t) \tag{6-5b}$$

where p^i stands for d^i/dt^i. This is an nth-order differential equation. It is well known that in an nth-order differential equation, in order to have a unique solution for any u, we need n number of initial conditions. Hence the state vector will consist of n components. In this case the output y and its derivatives up to the $(n-1)$th order qualify as state variables. Define

$$\hat{\mathbf{x}}(s) \triangleq \begin{bmatrix} \hat{x}_1(s) \\ \hat{x}_2(s) \\ \vdots \\ \hat{x}_n(s) \end{bmatrix} \triangleq \begin{bmatrix} \hat{y}(s) \\ s\hat{y}(s) \\ \vdots \\ s^{n-1}\hat{y}(s) \end{bmatrix} = \begin{bmatrix} 1 \\ s \\ \vdots \\ s^{n-1} \end{bmatrix} \hat{y}(s) \tag{6-6a}$$

or, in the time domain,

$$\begin{aligned} x_1(t) &\triangleq y(t) \\ x_2(t) &\triangleq \dot{y}(t) = py(t) = \dot{x}_1(t) \\ x_3(t) &\triangleq \ddot{y}(t) = p^2 y(t) = \dot{x}_2(t) \\ &\cdots\cdots\cdots\cdots\cdots\cdots \\ x_n(t) &\triangleq y^{(n-1)}(t) = p^{n-1} y(t) = \dot{x}_{n-1}(t) \end{aligned} \tag{6-6b}$$

Differentiating $x_n(t)$ once and using (6-5) we obtain

$$\dot{x}_n(t) = p^n y(t) = -\alpha_n x_1 - \alpha_{n-1} x_2 - \cdots - \alpha_1 x_n + \beta u \tag{6-7}$$

These equations can be arranged in matrix form as

$$\dot{\mathbf{x}} = \begin{bmatrix} 0 & 1 & 0 & \cdots & 0 \\ 0 & 0 & 1 & \cdots & 0 \\ 0 & 0 & 0 & \cdots & 0 \\ \vdots & \vdots & \vdots & & \vdots \\ 0 & 0 & 0 & \cdots & 1 \\ -\alpha_n & -\alpha_{n-1} & -\alpha_{n-2} & \cdots & -\alpha_1 \end{bmatrix} \mathbf{x} + \begin{bmatrix} 0 \\ 0 \\ 0 \\ \vdots \\ 0 \\ \beta \end{bmatrix} u \tag{6-8a}$$

$$y = [\,1 \quad 0 \quad 0 \quad \cdots \quad 0\,]\mathbf{x} \tag{6-8b}$$

The first $n-1$ equations of (6-8a) are obtained directly from (6-6b). They are the consequence of the definition of x_i, $i = 1, 2, \ldots, n$, and are independent of the given transfer function. The transfer function $\hat{g}(s)$ comes into (6-8a) only at its last equation through (6-7). We draw in Figure 6-1[2] a block diagram of

[2] Note that if β is moved to the output, then $\mathbf{b}$ and $\mathbf{c}$ in (6-8) become $\mathbf{b}' = [0 \quad 0 \quad \cdots \quad 0 \quad 1]$ and $\mathbf{c} = [\beta \quad 0 \quad 0 \quad \cdots \quad 0]$.

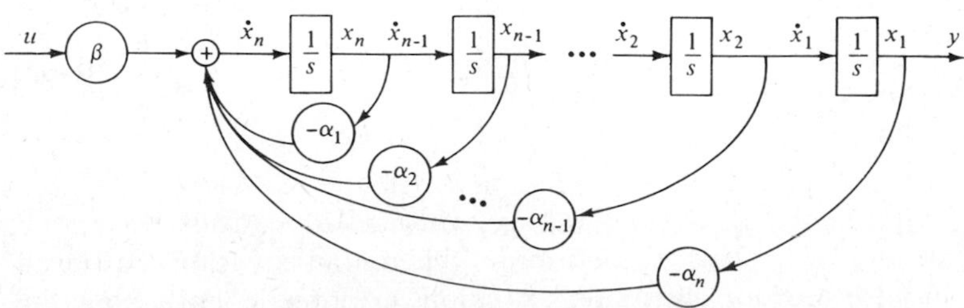

Figure 6-1 The block diagram of Equation (6-8).

(6-8). To show that (6-8) is a realization of $\hat{g}(s)$, we may apply Mason's formula to Figure 6-1 to show, as in Figure 4-6, that the transfer function from u to y is equal to $\hat{g}(s)$. A different way is to verify

$$\hat{g}(s) = \mathbf{c}(s\mathbf{I} - \mathbf{A})^{-1}\mathbf{b}$$

where

$$(s\mathbf{I} - \mathbf{A})^{-1}\mathbf{b} = \begin{bmatrix} s & -1 & \cdots & 0 & 0 \\ 0 & s & \cdots & 0 & 0 \\ \vdots & \vdots & & \vdots & \vdots \\ 0 & 0 & \cdots & s & -1 \\ \alpha_n & \alpha_{n-1} & \cdots & \alpha_2 & s+\alpha_1 \end{bmatrix}^{-1} \begin{bmatrix} 0 \\ 0 \\ \vdots \\ 0 \\ \beta \end{bmatrix} \tag{6-9}$$

It is clear that $(s\mathbf{I}-\mathbf{A})^{-1}\mathbf{b}$ is equal to the last column of $(s\mathbf{I}-\mathbf{A})^{-1}\beta$ or, equivalently, to the cofactors of the last row of $(\beta/\Delta(s))(s\mathbf{I} - \mathbf{A})$, where

$$\Delta(s) \triangleq \det\,(s\mathbf{I} - \mathbf{A}) = s^n + \alpha_1 s^{n-1} + \cdots + \alpha_n$$

(see Problem 2-26). In view of the form of $s\mathbf{I} - \mathbf{A}$, the cofactors of the last row of $(\beta/\Delta(s))(s\mathbf{I} - \mathbf{A})$ can be easily computed as

$$\frac{\beta}{\Delta(s)}[1 \quad s \quad s^2 \quad \cdots \quad s^{n-1}]'$$

Hence,

$$(s\mathbf{I} - \mathbf{A})^{-1}\mathbf{b} = \begin{bmatrix} s & -1 & \cdots & 0 & 0 \\ 0 & s & \cdots & 0 & 0 \\ \vdots & \vdots & & \vdots & \vdots \\ 0 & 0 & \cdots & s & -1 \\ \alpha_n & \alpha_{n-1} & \cdots & \alpha_2 & s+\alpha_1 \end{bmatrix}^{-1} \begin{bmatrix} 0 \\ 0 \\ \vdots \\ 0 \\ \beta \end{bmatrix} = \frac{\beta}{\Delta(s)} \begin{bmatrix} 1 \\ s \\ \vdots \\ s^{n-2} \\ s^{n-1} \end{bmatrix} \tag{6-10}$$

and

$$\hat{g}(s) = c(s\mathbf{I} - \mathbf{A})^{-1}\mathbf{b} = [1 \quad 0 \quad \cdots \quad 0 \quad 0]\frac{\beta}{\Delta(s)} \begin{bmatrix} 1 \\ s \\ \vdots \\ s^{n-2} \\ s^{n-1} \end{bmatrix}$$

$$= \frac{\beta}{\Delta(s)} = \frac{\beta}{s^n + \alpha_1 s^{n-1} + \cdots + \alpha_n} \tag{6-11}$$

This verifies that (6-8) is indeed a realization of $\hat{g}(s)$ in (6-4). Since $\deg \hat{g}(s) = \dim \mathbf{A}$, (6-8) is an irreducible realization (Theorem 6-1). This can also be verified by showing that (6-8) is controllable (except for the trivial case $\beta = 0$) and observable. Note that the realization (6-8) can be obtained directly from the coefficients of $\hat{g}(s)$ in (6-4).

Irreducible realization of $\hat{g}(s)=N(s)/D(s)$. Consider the following scalar proper transfer function

$$\hat{g}_1(s)=\frac{\bar{\beta}_0 s^n+\bar{\beta}_1 s^{n-1}+\cdots+\bar{\beta}_n}{\bar{\alpha}_0 s^n+\bar{\alpha}_1 s^{n-1}+\cdots+\bar{\alpha}_n} \tag{6-12}$$

where $\bar{\alpha}_i$ and $\bar{\beta}_i$, for $i=0, 1, 2, \ldots, n$, are real constants. It is assumed that $\bar{\alpha}_0 \neq 0$. By long division, $\hat{g}_1(s)$ can be written as

$$\hat{g}_1(s)=\frac{\beta_1 s^{n-1}+\beta_2 s^{n-2}+\cdots+\beta_n}{s^n+\alpha_1 s^{n-1}+\cdots+\alpha_{n-1}s+\alpha_n}+e\triangleq\hat{g}(s)+e \tag{6-13}$$

where $e=\hat{g}_1(\infty)=\bar{\beta}_0/\bar{\alpha}_0$. Since the constant e gives immediately the direct transmission part of a realization, we need to consider in the following only the strictly proper rational function

$$\hat{g}(s)\triangleq\frac{N(s)}{D(s)}\triangleq\frac{\beta_1 s^{n-1}+\beta_2 s^{n-2}+\cdots+\beta_n}{s^n+\alpha_1 s^{n-1}+\cdots+\alpha_n} \tag{6-14}$$

Let u and y be the input and output of $\hat{g}(s)$ in (6-14). Then we have

$$D(s)\hat{y}(s)=N(s)\hat{u}(s) \tag{6-15a}$$

or, in the time domain,

$$D(p)y(t)=N(p)u(t) \tag{6-15b}$$

where $D(p)=p^n+\alpha_1 p^{n-1}+\cdots+\alpha_n$, $N(p)=\beta_1 p^{n-1}+\beta_2 p^{n-2}+\cdots+\beta_n$, and p^i stands for d^i/dt^i. Clearly, the transfer function of (6-15) is $\hat{g}(s)$ in (6-14). In the following we introduce several different realizations of (6-15).

Observable canonical-form realization

Consider the nth-order differential equation $D(p)y(t)=N(p)u(t)$. It is well known that if we have n initial conditions—for example, $y(t_0)$, $y^{(1)}(t_0), \ldots, y^{(n-1)}(t_0)$—then for any input $u_{[t_0,t_1]}$, the output $y_{[t_0,t_1]}$ is completely determinable. In this case, however, if we choose $y(t)$, $y^{(1)}(t), \ldots, y^{(n-1)}(t)$ as state variables as we did in (6-6), then we cannot obtain a dynamical equation of the form $\dot{\mathbf{x}}=\mathbf{A}\mathbf{x}+\mathbf{b}u$, $y=\mathbf{c}\mathbf{x}$. Instead, we will obtain an equation of the form $\dot{\mathbf{x}}=\mathbf{A}\mathbf{x}+\mathbf{b}u$, $y=\mathbf{c}\mathbf{x}+e_1u+e_2u^{(1)}+e_3u^{(2)}+\cdots$. Hence in order to realize $N(s)/D(s)$ in the form $\dot{\mathbf{x}}=\mathbf{A}\mathbf{x}+\mathbf{b}u$, $y=\mathbf{c}\mathbf{x}$, a different set of state variables has to be chosen.

Taking the Laplace transform of (6-15) and grouping the terms associated with the same power of s, we finally obtain

$$\begin{aligned}\hat{y}(s)=\frac{N(s)}{D(s)}\hat{u}(s)+\frac{1}{D(s)}\{&y(0)s^{n-1}+[y^{(1)}(0)+\alpha_1 y(0)-\beta_1 u(0)]s^{n-2}+\cdots\\&+[y^{(n-1)}(0)+\alpha_1 y^{(n-2)}(0)-\beta_1 u^{(n-2)}(0)+\alpha_2 y^{(n-3)}(0)\\&-\beta_2 u^{(n-3)}(0)+\cdots+\alpha_{n-1}y(0)-\beta_{n-1}u(0)]\}\end{aligned} \tag{6-16}$$

The term

$$\frac{N(s)}{D(s)}\hat{u}(s)=\hat{g}(s)\hat{u}(s)$$

in the right-hand side of (6-16) gives the response due to the input $\hat{u}(s)$; the remainder gives the response due to the initial conditions. Therefore, if all the coefficients associated with $s^{n-1}, s^{n-2}, \ldots, s^0$ in (6-16) are known, then for any u a unique y can be determined. Consequently, if we choose the state variables as

$$\begin{aligned}
x_n(t) &\triangleq y(t)\\
x_{n-1}(t) &\triangleq y^{(1)}(t)+\alpha_1 y(t)-\beta_1 u(t)\\
x_{n-2}(t) &\triangleq y^{(2)}(t)+\alpha_1 y^{(1)}(t)-\beta_1 u^{(1)}(t)+\alpha_2 y(t)-\beta_2 u(t)\\
&\cdots\cdots\cdots\cdots\cdots\cdots\\
x_1(t) &\triangleq y^{(n-1)}(t)+\alpha_1 y^{(n-2)}(t)-\beta_1 u^{(n-2)}(t)+\cdots+\alpha_{n-1}y(t)-\beta_{n-1}u(t)
\end{aligned} \tag{6-17}$$

then $\mathbf{x}=[x_1 \quad x_2 \quad \cdots \quad x_n]'$ qualifies as the state vector. The set of equations in (6-17) yields

$$\begin{aligned}
y &= x_n\\
x_{n-1} &= \dot{x}_n+\alpha_1 x_n-\beta_1 u\\
x_{n-2} &= \dot{x}_{n-1}+\alpha_2 x_n-\beta_2 u\\
&\cdots\cdots\cdots\cdots\\
x_1 &= \dot{x}_2+\alpha_{n-1}x_n-\beta_{n-1}u
\end{aligned}$$

Differentiating x_1 in (6-17) once and using (6-15), we obtain

$$\dot{x}_1=-\alpha_n x_n+\beta_n u$$

The foregoing equations can be arranged in matrix form as

$$\begin{bmatrix}\dot{x}_1\\ \dot{x}_2\\ \dot{x}_3\\ \vdots\\ \\ \dot{x}_n\end{bmatrix}=\begin{bmatrix}0 & 0 & 0 & \cdots & 0 & -\alpha_n\\ 1 & 0 & 0 & \cdots & 0 & -\alpha_{n-1}\\ 0 & 1 & 0 & \cdots & 0 & -\alpha_{n-2}\\ \vdots & \vdots & \vdots & & \vdots & \vdots\\ 0 & 0 & 0 & \cdots & 0 & -\alpha_2\\ 0 & 0 & 0 & \cdots & 1 & -\alpha_1\end{bmatrix}\begin{bmatrix}x_1\\ x_2\\ x_3\\ \vdots\\ x_{n-1}\\ x_n\end{bmatrix}+\begin{bmatrix}\beta_n\\ \beta_{n-1}\\ \beta_{n-2}\\ \vdots\\ \beta_2\\ \beta_1\end{bmatrix}u \tag{6-18a}$$

$$y=[0 \quad 0 \quad 0 \quad \cdots \quad 0 \quad 1]\mathbf{x} \tag{6-18b}$$

A dynamical equation in the form of (6-18) is said to be in the *observable canonical form.* The block diagram of (6-18) is shown in Figure 6-2. Since (6-18) is derived from (6-15), the transfer function of (6-18) is $\hat{g}(s)$ in (6-14). This can be verified by either applying to Figure 6-2 Mason's formula for signal-flow graph or computing

$$\mathbf{c}(s\mathbf{I}-\mathbf{A})^{-1}\mathbf{b}=[\mathbf{c}(s\mathbf{I}-\mathbf{A})^{-1}\mathbf{b}]'=\mathbf{b}'(s\mathbf{I}-\mathbf{A}')^{-1}\mathbf{c}'$$

with the aid of (6-10), where the "prime" symbol denotes the transpose.

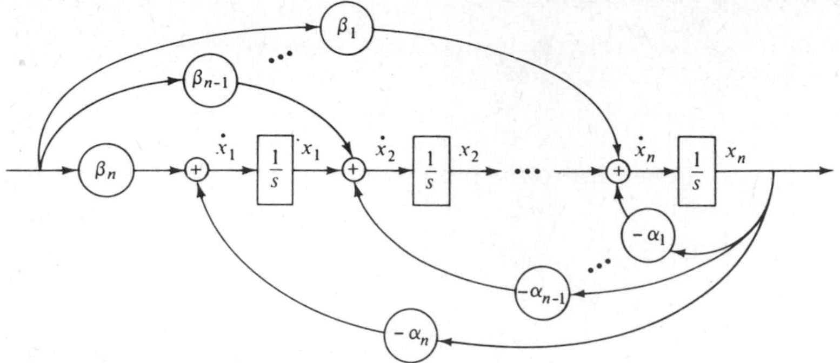

Figure 6-2 Block diagram of the observable canonical-form dynamical equation (6-18).

The observability matrix of (6-18) is

$$\mathbf{V} \triangleq \begin{bmatrix} \mathbf{c} \\ \mathbf{cA} \\ \mathbf{cA}^2 \\ \vdots \\ \\ \mathbf{cA}^{n-1} \end{bmatrix} = \begin{bmatrix} 0 & 0 & 0 & \cdots & 0 & 0 & 1 \\ 0 & 0 & 0 & \cdots & 0 & 1 & -\alpha_1 \\ 0 & 0 & 0 & \cdots & 1 & -\alpha_1 & -\alpha_2+\alpha_1^2 \\ \vdots & \vdots & \vdots & & \vdots & \vdots & \vdots \\ 0 & 1 & x & \cdots & x & x & x \\ 1 & x & x & \cdots & x & x & x \end{bmatrix}$$

where x denotes possible nonzero elements. The matrix $\mathbf{V}$ is nonsingular for any α_i and β_i; hence (6-18) is observable no matter $D(s)$ and $N(s)$ in (6-14) are coprime or not, and is thus called the *observable canonical-form realization.*

The dynamical equation in (6-18) is controllable as well if $D(s)$ and $N(s)$ are coprime. Indeed, if $D(s)$ and $N(s)$ are coprime, then $\deg \hat{g}(s) = \deg D(s) = \dim \mathbf{A}$ and (6-18) is controllable and observable following Theorem 6-1. If $D(s)$ and $N(s)$ are not coprime, the observable dynamical equation (6-18) cannot be controllable as well; otherwise, it would violate Theorem 6-1. In the following we shall show directly that if $D(s)$ and $N(s)$ are not corprime, then (6-18) is not controllable. Let $s-\lambda$ be a common factor of $N(s)$ and $D(s)$; then we have

$$D(\lambda) = \lambda^n + \alpha_1\lambda^{n-1} + \cdots + \alpha_{n-1}\lambda + \alpha_n = 0 \tag{6-19}$$

and

$$N(\lambda) = \beta_1\lambda^{n-1} + \beta_2\lambda^{n-2} + \cdots + \beta_n = 0 \tag{6-20}$$

Define the $1 \times n$ constant vector $\boldsymbol{\alpha}$ as $\boldsymbol{\alpha} \triangleq [1 \quad \lambda \quad \lambda^2 \quad \cdots \quad \lambda^{n-1}]$. Then (6-20) can be written as $\boldsymbol{\alpha}\mathbf{b} = 0$, where $\mathbf{b}$ is defined in (6-18a). Using (6-19), it is easy to verify that $\boldsymbol{\alpha}\mathbf{A} = \lambda\boldsymbol{\alpha}$, $\boldsymbol{\alpha}\mathbf{A}^2 = \lambda^2\boldsymbol{\alpha}, \ldots, \boldsymbol{\alpha}\mathbf{A}^{n-1} = \lambda^{n-1}\boldsymbol{\alpha}$. Hence the identity $\boldsymbol{\alpha}\mathbf{b} = 0$ implies that $\boldsymbol{\alpha}\mathbf{A}^i\mathbf{b} = 0$, for $i = 0, 1, \ldots, n-1$, which can be written as

$$\boldsymbol{\alpha}[\mathbf{b} \quad \mathbf{Ab} \quad \mathbf{A}^2\mathbf{b} \quad \cdots \quad \mathbf{A}^{n-1}\mathbf{b}] = 0$$

which, together with $\boldsymbol{\alpha} \neq \mathbf{0}$, implies that the controllability matrix of (6-18) has a rank less than n. Hence if $D(s)$ and $N(s)$ are not coprime, the realization in (6-18) is not controllable.

Controllable canonical-form realization

We shall now introduce a different realization, called the controllable canonical-form realization, of $\hat{g}(s)=N(s)D^{-1}(s)$ or $\hat{y}(s)=N(s)D^{-1}(s)\hat{u}(s)$. Let us introduce a new variable $v(t)$ defined by $\hat{v}(s)=D^{-1}(s)\hat{u}(s)$. Then we have

$$D(s)\hat{v}(s)=\hat{u}(s) \tag{6-21}$$

and

$$\hat{y}(s)=N(s)\hat{v}(s) \tag{6-22}$$

Equation (6-21) has the same form as (6-5a); hence we may define the state variables as, similar to (6-6a),

$$\hat{\mathbf{x}}(s)\triangleq\begin{bmatrix}\hat{x}_1(s)\\ \hat{x}_2(s)\\ \vdots\\ \hat{x}_n(s)\end{bmatrix}\triangleq\begin{bmatrix}1\\ s\\ \vdots\\ s^{n-1}\end{bmatrix}\hat{v}(s) \quad \text{or} \quad \mathbf{x}(t)\triangleq\begin{bmatrix}x_1(t)\\ x_2(t)\\ \vdots\\ x_n(t)\end{bmatrix}\triangleq\begin{bmatrix}v(t)\\ \dot{v}(t)\\ \vdots\\ v^{(n-1)}(t)\end{bmatrix} \tag{6-23}$$

This definition implies $\dot{x}_1=x_2$, $\dot{x}_2=x_3, \ldots, \dot{x}_{n-1}=x_n$. From (6-21), we have

$$\begin{aligned}sx_n(s)=s^n v(s)&=-\alpha_1 s^{n-1}\hat{v}(s)-\alpha_2 s^{n-2}\hat{v}(s)-\cdots-\alpha_1\hat{v}(s)+\hat{u}(s)\\ &=[-\alpha_n \quad -\alpha_{n-1} \quad \cdots \quad -\alpha_2 \quad -\alpha_1]\hat{\mathbf{x}}(s)+\hat{u}(s)\end{aligned}$$

which becomes, in the time domain,

$$\dot{x}_n(t)=[-\alpha_n \quad -\alpha_{n-1} \quad \cdots \quad -\alpha_2 \quad -\alpha_1]\mathbf{x}(t)+u(t) \tag{6-24}$$

Using (6-23), Equation (6-22) can be written as

$$\hat{y}(s)=[\beta_n \quad \beta_{n-1} \quad \cdots \quad \beta_1]\begin{bmatrix}\hat{v}(s)\\ s\hat{v}(s)\\ \vdots\\ s^{n-1}\hat{v}(s)\end{bmatrix}=[\beta_n \quad \beta_{n-1} \quad \cdots \quad \beta_1]\hat{\mathbf{x}}(s)$$

or, in the time domain,

$$y(t)=[\beta_n \quad \beta_{n-1} \quad \cdots \quad \beta_1]\mathbf{x}(t) \tag{6-25}$$

These equations can be arranged in matrix form as

$$\dot{\mathbf{x}}=\mathbf{A}\mathbf{x}+\mathbf{b}u \qquad y=\mathbf{c}\mathbf{x} \tag{6-26}$$

where

$$\mathbf{A}=\begin{bmatrix}0 & 1 & 0 & \cdots & 0\\ 0 & 0 & 1 & \cdots & 0\\ 0 & 0 & 0 & \cdots & 0\\ \vdots & \vdots & \vdots & \cdots & \vdots\\ 0 & 0 & 0 & \cdots & 1\\ -\alpha_n & -\alpha_{n-1} & -\alpha_{n-2} & \cdots & -\alpha_1\end{bmatrix} \qquad \mathbf{b}=\begin{bmatrix}0\\ 0\\ 0\\ \vdots\\ 0\\ 1\end{bmatrix}$$

$$\mathbf{c}=[\beta_n \quad \beta_{n-1} \quad \beta_{n-2} \quad \cdots \quad \beta_1]$$

This is a realization of $\hat{g}(s)$ in (6-14). Unlike (6-17), there are no simple relationships between x_i, u, and y. The dynamical equation in (6-26) is always controllable no matter whether $D(s)$ and $N(s)$ are coprime or not and is said to be in the *controllable canonical form*. If $D(s)$ and $N(s)$ are coprime, then (6-26) is observable as well; otherwise, it is not observable. This assertion is dual to that of the observable canonical form, and its proof is left as an exercise. A block diagram of (6-26) is drawn in Figure 6-3.

Example 1

Consider the proper irreducible transfer function

$$\hat{g}(s)=\frac{4s^3+25s^2+45s+34}{2s^3+12s^2+20s+16}$$

By long division, $\hat{g}(s)$ can be written as

$$\hat{g}(s)=\frac{0.5s^2+2.5s+1}{s^3+6s^2+10s+8}+2$$

Hence its controllable canonical-form realization is

$$\begin{bmatrix}\dot{x}_1\\ \dot{x}_2\\ \dot{x}_3\end{bmatrix}=\begin{bmatrix}0 & 1 & 0\\ 0 & 0 & 1\\ -8 & -10 & -6\end{bmatrix}\begin{bmatrix}x_1\\ x_2\\ x_3\end{bmatrix}+\begin{bmatrix}0\\ 0\\ 1\end{bmatrix}u$$

$$y=[1 \quad 2.5 \quad 0.5]\,\mathbf{x}+2u$$

Its observable canonical-form realization is

$$\dot{\mathbf{x}}=\begin{bmatrix}0 & 0 & -8\\ 1 & 0 & -10\\ 0 & 1 & -6\end{bmatrix}\mathbf{x}+\begin{bmatrix}1\\ 2.5\\ 0.5\end{bmatrix}u$$

$$y=[0 \quad 0 \quad 1]\,\mathbf{x}+2u$$

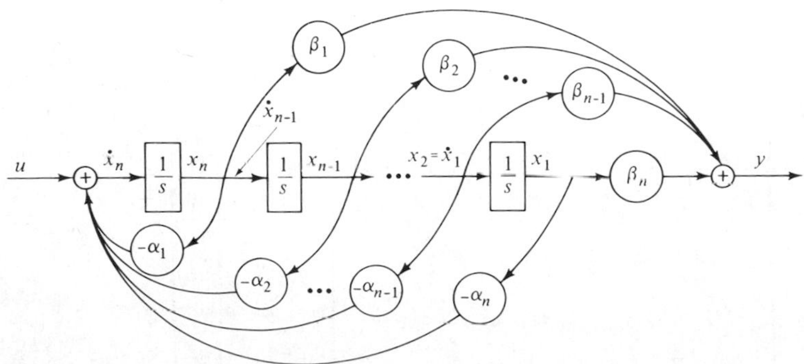

Figure 6-3 Block diagram of the controllable canonical-form dynamical equation (6-26).

Realization from the Hankel matrix

Consider the proper rational function

$$\hat{g}(s) = \frac{\beta_0 s^n + \beta_1 s^{n-1} + \cdots + \beta_n}{s^n + \alpha_1 s^{n-1} + \cdots + \alpha_n} \tag{6-27}$$

We expand it into an infinite power series of descending power of s as

$$\hat{g}(s) = h(0) + h(1)s^{-1} + h(2)s^{-2} + \cdots \tag{6-28}$$

The coefficients $\{h(i),\ i = 0, 1, 2, \ldots\}$ will be called the *Markov parameters*. These parameters can be obtained recursively from α_i and β_i as

$$\begin{aligned} h(0) &= \beta_0 \\ h(1) &= -\alpha_1 h(0) + \beta_1 \\ h(2) &= -\alpha_1 h(1) - \alpha_2 h(0) + \beta_2 \\ &\cdots\cdots\cdots\cdots\cdots\cdots \\ h(n) &= -\alpha_1 h(n-1) - \alpha_2 h(n-2) - \cdots - \alpha_n h(0) + \beta_n \end{aligned} \tag{6-29}$$

$$h(n+i) = -\alpha_1 h(n+i-1) - \alpha_2 h(n+i-2) - \cdots - \alpha_n h(i) \qquad i = 1, 2, \ldots \tag{6-30}$$

These equations are obtained by equating (6-27) and (6-28) as

$$(\beta_0 s^n + \beta_1 s^{n-1} + \cdots + \beta_n) = (s^n + \alpha_1 s^{n-1} + \cdots + \alpha_n)(h(0) + h(1)s^{-1} + h(2)s^{-2} + \cdots)$$

and then equating the coefficients of the same power of s. We form the $\alpha \times \beta$ matrix

$$\mathbf{H}(\alpha, \beta) \triangleq \begin{bmatrix} h(1) & h(2) & h(3) & \cdots & h(\beta) \\ h(2) & h(3) & h(4) & \cdots & h(\beta+1) \\ h(3) & h(4) & h(5) & \cdots & h(\beta+2) \\ \vdots & \vdots & \vdots & & \vdots \\ h(\alpha) & h(\alpha+1) & h(\alpha+2) & \cdots & h(\alpha+\beta-1) \end{bmatrix} \tag{6-31}$$

It is called a *Hankel matrix* of order $\alpha \times \beta$. It is formed from the coefficients $\{h(i),\ i = 1, 2, 3, \ldots\}$. It is important to note that the coefficient $h(0)$ *is not involved in* $\mathbf{H}(\alpha, \beta)$.

Theorem 6-3

The proper transfer function $\hat{g}(s)$ in (6-27) has degree n if and only if

$$\rho\mathbf{H}(n, n) = \rho\mathbf{H}(n+k, n+l) = n \qquad \text{for every } k, l = 1, 2, 3, \ldots \tag{6-32}$$

where ρ denotes the rank.

Proof

We show first that if $\deg \hat{g}(s) = n$, then $\rho\mathbf{H}(n, n) = \rho\mathbf{H}(n+1, \infty) = \rho\mathbf{H}(\infty, \infty) = n$. If $\deg \hat{g}(s) = n$, then (6-30) holds, and n is the smallest integer having this property.

Because of (6-30) the $(n+1)$th row of $\mathbf{H}(n+1, \infty)$ can be written as a linear combination of the n rows of $\mathbf{H}(n, \infty)$. Hence we have $\rho\mathbf{H}(n, \infty) = \rho\mathbf{H}(n+1, \infty)$. Furthermore, we have $\rho\mathbf{H}(n, \infty) = n$; otherwise, there would be an integer $\bar{n}$ smaller than n with the property (6-30). Because of the structure of $\mathbf{H}$, the matrix $\mathbf{H}(n+2, \infty)$ without the first row reduces to the matrix $\mathbf{H}(n+1, \infty)$ without the first column. Hence the $(n+2)$th row of $\mathbf{H}(n+2, \infty)$ is linearly dependent on its previous n rows and, consequently, on the first n row of $\mathbf{H}(n+2, \infty)$. Proceeding in this manner, we can establish $\rho\mathbf{H}(n, \infty) = \rho\mathbf{H}(\infty, \infty) = n$. Again using (6-30), the $(n+1)$th column of $\mathbf{H}(n, \infty)$ is linearly dependent on the columns of $\mathbf{H}(n, n)$. Proceeding similarly, we have $\rho\mathbf{H}(n, n) = \rho\mathbf{H}(n+k, n+l) = n$ for every $k, l = 1, 2, \ldots$.

Now we show that if (6-32) holds, then $\hat{g}(s) = h(0) + h(1)s^{-1} + \cdots$ can be reduced to a proper rational function of degree n. The condition $\rho\mathbf{H}(n, n) = \rho\mathbf{H}(\infty, \infty) = n$ implies the existence of $\{\alpha_i, i = 1, 2, \ldots, n\}$ to meet (6-30). Using (6-29) we can compute $\{\beta_i, i = 0, 1, 2, \ldots, n\}$. Hence we have

$$\hat{g}(s) = h(0) + h(1)s^{-1} + h(2)s^{-2} + \cdots = \frac{\beta_0 s^n + \beta_1 s^{n-1} + \cdots + \beta_n}{s^n + \alpha_1 s^{n-1} + \cdots + \alpha_n}$$

Since the n is the smallest integer having this property, we have $\deg \hat{g}(s) = n$. This completes the proof of this theorem. Q.E.D.

Consider the dynamical equation

$$FE: \quad \dot{\mathbf{x}} = \mathbf{A}\mathbf{x} + \mathbf{b}u$$
$$y = \mathbf{c}\mathbf{x} + eu$$

Its transfer function is clearly equal to

$$\hat{g}(s) = e + \mathbf{c}(s\mathbf{I} - \mathbf{A})^{-1}\mathbf{b} = e + s^{-1}\mathbf{c}(\mathbf{I} - s^{-1}\mathbf{A})^{-1}\mathbf{b}$$

which can be expanded as, by using (2-85),

$$\hat{g}(s) = e + \mathbf{cb}s^{-1} + \mathbf{cAb}s^{-2} + \mathbf{cA}^2\mathbf{b}s^{-3} + \cdots \tag{6-33}$$

From (6-28) and (6-33), we conclude that $\{\mathbf{A}, \mathbf{b}, \mathbf{c}, e\}$ is a realization of $\hat{g}(s)$ in (6-27) if and only if $e = h(0)$ and

$$h(i) = \mathbf{cA}^{i-1}\mathbf{b} \qquad i = 1, 2, 3, \ldots \tag{6-34}$$

With this background, we are ready to introduce a different realization. Consider a proper transfer function $\hat{g}(s) = N(s)/D(s)$ with $\deg D(s) = n$. Here we do not assume that $D(s)$ and $N(s)$ are coprime; hence the degree of $\hat{g}(s)$ may be less than n. We expand $\hat{g}(s)$ as in (6-28) by using the recursive equations in (6-29) and (6-30). We then form the Hankel matrix

$$\mathbf{H}(n+1, n) = \begin{bmatrix} h(1) & h(2) & \cdots & h(n) \\ h(2) & h(3) & \cdots & h(n+1) \\ \vdots & \vdots & & \vdots \\ h(n) & h(n+1) & \cdots & h(2n-1) \\ h(n+1) & h(n+2) & \cdots & h(2n) \end{bmatrix} \tag{6-35}$$

Note that there is one more row than column, and the Markov parameters up to $h(2n)$ are used in forming $\mathbf{H}(n+1, n)$. Now we apply the row-searching algorithm[3] discussed in Appendix A to search the linearly independent rows of $\mathbf{H}$ in (6-35) in order from top to bottom. Let the first σ rows be linearly independent and the $(\sigma+1)$th row of $\mathbf{H}$ be linearly dependent on its previous rows. Then Theorem 6-3 implies that the $(\sigma+k)$th rows, $k=1, 2, 3, \ldots$, are all linearly dependent on their previous rows and the rank of $\mathbf{H}(n+1, n)$ is σ. Hence once a linearly dependent row appears in $\mathbf{H}(n+1, n)$, we may stop the search. We shall call the $(\sigma+1)$th row of $\mathbf{H}(n+1, n)$ the *primary* linearly dependent row; the $(\sigma+k)$th row, $k=2, 3, \ldots$, nonprimary linearly dependent rows. Note that if $D(s)$ and $N(s)$ are coprime, then $\sigma=n$; otherwise, we have $\sigma<n$. The row searching algorithm will also yield $\{a_i, i=1, 2, \ldots,\}$ such that

$$[a_1 \quad a_2 \quad \cdots \quad a_\sigma \quad \boxed{1} \quad 0 \quad \cdots \quad 0]\mathbf{H}(n+1, n)=\mathbf{0} \tag{6-36}$$

This equation expresses the primary linearly dependent row as a unique linear combination of its previous rows. The element $\boxed{1}$ corresponds to the primary dependent row. Note that if $\sigma=n$, then $a_i=\alpha_{n-i}$, $i=1, 2, \ldots, n$. If $\sigma<n$, then we do not have $a_i=\alpha_{n-i}$. We claim that the σ-dimensional dynamical equation

$$\dot{\mathbf{x}}=\mathbf{A}\mathbf{x}+\mathbf{b}u \qquad y=\mathbf{c}\mathbf{x}+eu \tag{6-37}$$

with

$$\mathbf{A}=\begin{bmatrix} 0 & 1 & 0 & \cdots & 0 & 0 \\ 0 & 0 & 1 & \cdots & 0 & 0 \\ \vdots & \vdots & \vdots & & \vdots & \vdots \\ 0 & 0 & 0 & \cdots & 0 & 1 \\ -a_1 & -a_2 & -a_3 & \cdots & -a_{\sigma-1} & -a_\sigma \end{bmatrix} \qquad \mathbf{b}=\begin{bmatrix} h(1) \\ h(2) \\ \vdots \\ h(\sigma-1) \\ h(\sigma) \end{bmatrix} \tag{6-38}$$

$$\mathbf{c}=[\,1 \quad 0 \quad 0 \quad \cdots \quad 0 \quad 0\,] \qquad e=h(0)$$

is a controllable and observable realization of $\hat{g}(s)$. Because of (6-36) and Theorem 6-3, we have

$$h(\sigma+i)=-a_1h(\sigma+i-1)-a_2h(\sigma+i-2)-\cdots-a_\sigma h(i) \quad i=1, 2, 3, \ldots$$

Using this, we can readily show

$$\mathbf{A}\mathbf{b}=\begin{bmatrix} h(2) \\ h(3) \\ \vdots \\ h(\sigma+1) \end{bmatrix}, \quad \mathbf{A}^2\mathbf{b}=\begin{bmatrix} h(3) \\ h(4) \\ \vdots \\ h(\sigma+2) \end{bmatrix}, \quad \ldots, \quad \mathbf{A}^k\mathbf{b}=\begin{bmatrix} h(k+1) \\ h(k+2) \\ \vdots \\ h(k+\sigma) \end{bmatrix}, \quad \ldots \tag{6-39}$$

The effect of the multiplication of $\mathbf{A}$ simply increases the argument i in $h(i)$ by 1, or equivalently, shifts the elements up by one position. Because of the form of $\mathbf{c}$, $\mathbf{c}\mathbf{A}^k\mathbf{b}$ just picks up the first element of $\mathbf{A}^k\mathbf{b}$ as

$$\mathbf{c}\mathbf{b}=h(1), \quad \mathbf{c}\mathbf{A}\mathbf{b}=h(2), \quad \mathbf{c}\mathbf{A}^2\mathbf{b}=h(3), \quad \ldots \tag{6-40}$$

[3] For computer computation, numerically stable methods should be used. See Appendix A.

This shows that (6-37) is indeed a realization of $\hat{g}(s)$. The controllability matrix of (6-37) is

$$[\mathbf{b} \quad \mathbf{Ab} \quad \cdots \quad \mathbf{A}^{\sigma-1}\mathbf{b}] = \mathbf{H}(\sigma, \sigma)$$

The Hankel matrix $\mathbf{H}(\sigma, \sigma)$ has rank σ; hence $\{\mathbf{A}, \mathbf{b}\}$ in (6-37) is controllable. The observability matrix of (6-37) is

$$\begin{bmatrix} \mathbf{c} \\ \mathbf{cA} \\ \mathbf{cA}^2 \\ \vdots \\ \mathbf{cA}^{\sigma-1} \end{bmatrix} = \begin{bmatrix} 1 & 0 & 0 & \cdots & 0 \\ 0 & 1 & 0 & \cdots & 0 \\ 0 & 0 & 1 & \cdots & 0 \\ \vdots & \vdots & \vdots & \cdots & \vdots \\ 0 & 0 & 0 & \cdots & 1 \end{bmatrix}$$

Clearly $\{\mathbf{A}, \mathbf{c}\}$ is observable. Hence (6-37) is an irreducible realization of $\hat{g}(s)$.

Example 2

Consider

$$\hat{g}(s) = \frac{s^4 + s^3 - s - 1}{2s^4 + 2s^3 + 2s^2 + 3s + 1} = \tfrac{1}{2} + 0s^{-1} - \tfrac{1}{2}s^{-2} - \tfrac{3}{4}s^{-3} + \tfrac{1}{2}s^{-4} + s^{-5} - \tfrac{1}{8}s^{-6}$$
$$- \tfrac{5}{4}s^{-7} - \tfrac{3}{8}s^{-8} + \cdots$$

We form the Hankel matrix

$$\mathbf{H}(5, 4) = \begin{bmatrix} 0 & -\frac{1}{2} & -\frac{3}{4} & \frac{1}{2} \\ -\frac{1}{2} & -\frac{3}{4} & \frac{1}{2} & 1 \\ -\frac{3}{4} & \frac{1}{2} & 1 & -\frac{1}{8} \\ \frac{1}{2} & 1 & -\frac{1}{8} & -\frac{5}{4} \\ 1 & -\frac{1}{8} & -\frac{5}{4} & -\frac{3}{8} \end{bmatrix}$$

Using the row searching algorithm discussed in Appendix A, we can readily show that the rank of $\mathbf{H}(5, 4)$ is 3. Hence an irreducible realization of $\hat{g}(s)$ has dimension 3. The row searching algorithm also yields

$$\mathbf{kH}(5, 4) \triangleq [\underbrace{0.5 \quad 1 \quad 0}\ \ \boxed{1}\ \ 0]\mathbf{H}(5, 4) = \mathbf{0}$$

Hence an irreducible realization of $\hat{g}(s)$ is

$$\dot{\mathbf{x}} = \begin{bmatrix} 0 & 1 & 0 \\ 0 & 0 & 1 \\ -0.5 & -1 & 0 \end{bmatrix} \mathbf{x} + \begin{bmatrix} 0 \\ -\frac{1}{2} \\ -\frac{3}{4} \end{bmatrix} u$$
$$y = [\ 1 \quad 0 \quad 0]\ \mathbf{x} + 0.5u$$

The last row of the companion-form matrix $\mathbf{A}$ consists of the first three elements of $\mathbf{k}$ with the signs reversed. The $\mathbf{b}$ vector consists of the first three Markov parameters of $\hat{g}(s)$ [excluding $h(0)$]. The form of $\mathbf{c}$ is fixed and is independent of $\hat{g}(s)$. ∎

We note that this procedure also reveals whether the numerator and denominator of $\hat{g}(s)$ are coprime. If the rank of the Hankel matrix of $\hat{g}(s)$ is smaller than the degree of its denominator, then $\hat{g}(s)$ is not irreducible.

Dual to the introduced procedure, we may also search the linearly independent columns of $\mathbf{H}(n, n+1)$ in order from left to right and obtain a different irreducible realization. The procedure will not be repeated.

*Jordan-canonical-form realization

We use an example to illustrate the procedure to realize a transfer function into a Jordan-form dynamical equation. The idea can be easily extended to the general case. Assume that $D(s)$ consists of three distinct roots λ_1, λ_2, and λ_3, and assume that $D(s)$ can be factored as $D(s)=(s-\lambda_1)^3(s-\lambda_2)(s-\lambda_3)$. We also assume that $\hat{g}(s)$ can be expanded by partial fraction expansion into

$$\hat{g}(s)=\frac{e_{11}}{(s-\lambda_1)^3}+\frac{e_{12}}{(s-\lambda_1)^2}+\frac{e_{13}}{(s-\lambda_1)}+\frac{e_2}{(s-\lambda_2)}+\frac{e_3}{(s-\lambda_3)} \tag{6-41}$$

The block diagrams of $\hat{g}(s)$ are given in Figure 6-4. In Figure 6-4(a), the coefficients e_{11}, e_{12}, e_{13}, e_2, and e_3 are associated with the output. In Figure 6-4(b),

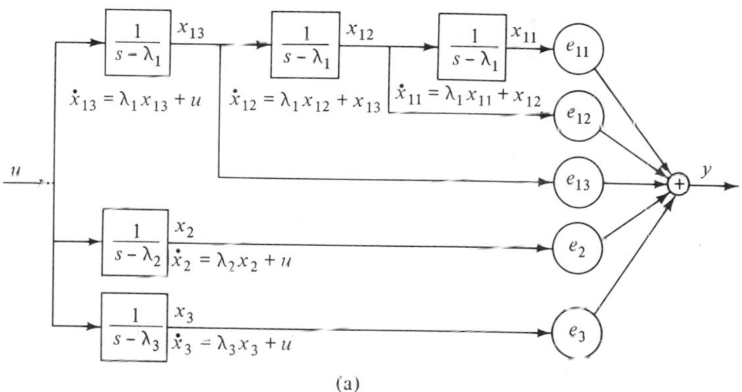

(a)

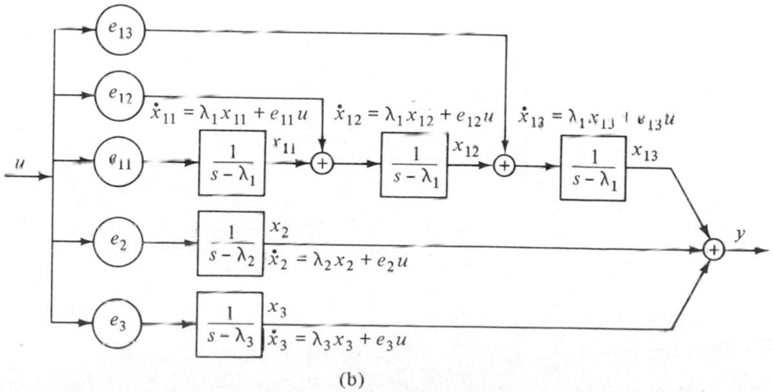

(b)

Figure 6-4 Two block diagrams of $\hat{g}(s)$ in Equation (6-41).

they are associated with the input. We note that every block in Figure 6-4 can be viewed as consisting of an integrator, as shown in Figure 6-5. Hence the output of each block qualifies as a state variable. By assigning the output of each block as a state variable, and referring to Figure 6-5, we can readily obtain the dynamical equation for each block of Figure 6-4 as shown. By grouping the equations in Figure 6-4(a), we obtain

$$\begin{bmatrix} \dot{x}_{11} \\ \dot{x}_{12} \\ \dot{x}_{13} \\ \dot{x}_2 \\ \dot{x}_3 \end{bmatrix} = \left[\begin{array}{ccc:c:c} \lambda_1 & 1 & 0 & 0 & 0 \\ 0 & \lambda_1 & 1 & 0 & 0 \\ 0 & 0 & \lambda_1 & 0 & 0 \\ \hdashline 0 & 0 & 0 & \lambda_2 & 0 \\ \hdashline 0 & 0 & 0 & 0 & \lambda_3 \end{array}\right] \mathbf{x} + \begin{bmatrix} 0 \\ 0 \\ 1 \\ 1 \\ 1 \end{bmatrix} u \tag{6-42a}$$

$$y = [e_{11} \quad e_{12} \quad e_{13} \quad e_2 \quad e_3]\mathbf{x} \tag{6-42b}$$

Equation (6-42) is in the Jordan canonical form. There is one Jordan block associated with each eigenvalue. The equation is clearly controllable (Corollary (5-21); it is also observable, except for the trivial cases $e_{11}=0$, $e_2=0$, or $e_3=0$. Therefore the dynamical equation (6-42) is an irreducible realization of the $\hat{g}(s)$ in (6-41).

If the block diagram in Figure 6-4(b) is used and if the state variables are chosen as shown, then the dynamical equation is

$$\begin{bmatrix} \dot{x}_{11} \\ \dot{x}_{12} \\ \dot{x}_{13} \\ \dot{x}_2 \\ \dot{x}_3 \end{bmatrix} = \left[\begin{array}{ccc:c:c} \lambda_1 & 0 & 0 & 0 & 0 \\ 1 & \lambda_1 & 0 & 0 & 0 \\ 0 & 1 & \lambda_1 & 0 & 0 \\ \hdashline 0 & 0 & 0 & \lambda_2 & 0 \\ \hdashline 0 & 0 & 0 & 0 & \lambda_3 \end{array}\right] \mathbf{x} + \begin{bmatrix} e_{11} \\ e_{12} \\ e_{13} \\ e_2 \\ e_3 \end{bmatrix} u$$

$$y = [0 \quad 0 \quad 1 \quad 1 \quad 1]\mathbf{x}$$

which is another irreducible Jordan-form realization of $\hat{g}(s)$. Note the differences in the assignment of the state variables in Figures 6-4(a) and 6-4(b).

There are two difficulties in realizing a transfer function into a Jordan canonical form. First, the denominator of the transfer function must be factored, or correspondingly, the poles of the transfer function $\hat{g}(s)$ must be computed. This is a difficult task if the degree of $\hat{g}(s)$ is larger than 3. Second, if the transfer function has complex poles, the matrices **A**, **b**, and **c** will consist of complex

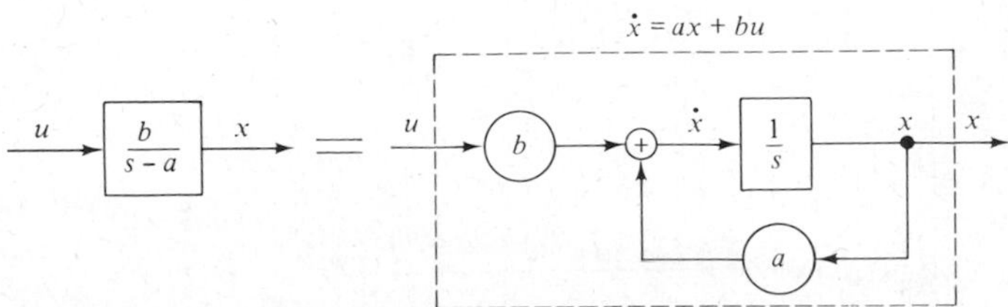

Figure 6-5 Internal structure of block $b/(s-a)$.

numbers. In this case, the equation cannot be simulated on an analog computer, for complex numbers cannot be generated in the real world. However, this can be taken care of by introducing some equivalence transformation, as will be demonstrated in the following. Since all the coefficients of $\hat{g}(s)$ are assumed to be real, if a complex number λ is a pole of $\hat{g}(s)$, its complex conjugate $\bar{\lambda}$, is also a pole of $\hat{g}(s)$. Hence in the Jordan-form realization of $\hat{g}(s)$, we have the following subequation:

$$\begin{bmatrix} \dot{\mathbf{x}}_1 \\ \dot{\mathbf{x}}_2 \end{bmatrix} = \begin{bmatrix} \mathbf{A}_1 & \mathbf{0} \\ \mathbf{0} & \bar{\mathbf{A}}_1 \end{bmatrix} \begin{bmatrix} \mathbf{x}_1 \\ \mathbf{x}_2 \end{bmatrix} + \begin{bmatrix} \mathbf{b}_1 \\ \bar{\mathbf{b}}_1 \end{bmatrix} u \tag{6-43a}$$

$$y = [\mathbf{c}_1 \quad \bar{\mathbf{c}}_1] \begin{bmatrix} \mathbf{x}_1 \\ \mathbf{x}_2 \end{bmatrix} \tag{6-43b}$$

where $\mathbf{A}_1$ is the Jordan block associated with λ and $\bar{\mathbf{A}}_1$ is the complex conjugate (no transpose) of $\mathbf{A}_1$. Clearly, $\bar{\mathbf{A}}_1$ is the Jordan block associated with $\bar{\lambda}$. Let us introduce the equivalence transformation $\bar{\mathbf{x}} = \mathbf{P}\mathbf{x}$, where

$$\mathbf{P} = \begin{bmatrix} \mathbf{I} & \mathbf{I} \\ i\mathbf{I} & -i\mathbf{I} \end{bmatrix} \qquad (i^2 = -1)$$

and

$$\mathbf{P}^{-1} = \frac{1}{2} \begin{bmatrix} \mathbf{I} & -i\mathbf{I} \\ \mathbf{I} & i\mathbf{I} \end{bmatrix}$$

Then it can be easily verified that the dynamical equation in (6-43) can be transformed into

$$\begin{bmatrix} \dot{\bar{\mathbf{x}}}_1 \\ \dot{\bar{\mathbf{x}}}_2 \end{bmatrix} = \begin{bmatrix} \operatorname{Re} \mathbf{A}_1 & \operatorname{Im} \mathbf{A}_1 \\ -\operatorname{Im} \mathbf{A}_1 & \operatorname{Re} \mathbf{A}_1 \end{bmatrix} \begin{bmatrix} \bar{\mathbf{x}}_1 \\ \bar{\mathbf{x}}_2 \end{bmatrix} + \begin{bmatrix} 2 \operatorname{Re} \mathbf{b}_1 \\ -2 \operatorname{Im} \mathbf{b}_1 \end{bmatrix} u \tag{6-44a}$$

$$y = [\operatorname{Re} \mathbf{c}_1 \quad \operatorname{Im} \mathbf{c}_1] \begin{bmatrix} \bar{\mathbf{x}}_1 \\ \bar{\mathbf{x}}_2 \end{bmatrix} \tag{6-44b}$$

where Re $\mathbf{A}$ and Im $\mathbf{A}$ denote the real part and the imaginary part of $\mathbf{A}$, respectively. Since all the coefficients in (6-44) are real, this equation can be used on analog computer simulations. Another convenient way to transform a Jordan-form dynamical equation into an equation with real coefficients is to use the transformation introduced in Problems 6-14 and 6-15.

Example 3

Consider the Jordan-form equation with complex eigenvalues.

$$\dot{\mathbf{x}} = \left[\begin{array}{cc:cc:c} 1+2i & 1 & 0 & 0 & 0 \\ 0 & 1+2i & 0 & 0 & 0 \\ \hdashline 0 & 0 & 1-2i & 1 & 0 \\ 0 & 0 & 0 & 1-2i & 0 \\ \hdashline 0 & 0 & 0 & 0 & 2 \end{array}\right] \mathbf{x} + \left[\begin{array}{c} 2-3i \\ 1 \\ \hdashline 2+3i \\ 1 \\ \hdashline 2 \end{array}\right] u \tag{6-45}$$

$$y = [\,1 \quad -i \;\vdots\; 1 \quad i \;\vdots\; 2\,]\,\mathbf{x}$$

Let $\bar{\mathbf{x}} = \mathbf{P}\mathbf{x}$, where

$$\mathbf{P} = \left[\begin{array}{cc:cc:c} 1 & 0 & 1 & 0 & 0 \\ 0 & 1 & 0 & 1 & 0 \\ \hdashline i & 0 & -i & 0 & 0 \\ 0 & i & 0 & -i & 0 \\ \hdashline 0 & 0 & 0 & 0 & 1 \end{array}\right]$$

Then (6-45) can be transformed into

$$\dot{\bar{\mathbf{x}}} = \left[\begin{array}{cc:cc:c} 1 & 1 & 2 & 0 & 0 \\ 0 & 1 & 0 & 2 & 0 \\ \hdashline -2 & 0 & 1 & 1 & 0 \\ 0 & -2 & 0 & 1 & 0 \\ \hdashline 0 & 0 & 0 & 0 & 2 \end{array}\right] \bar{\mathbf{x}} + \left[\begin{array}{c} 4 \\ 2 \\ \hdashline 6 \\ 0 \\ \hdashline 2 \end{array}\right] u$$

$$y = \left[\begin{array}{cc:cc:c} 1 & 0 & 0 & -1 & 2 \end{array}\right] \bar{\mathbf{x}}$$

whose coefficients are all real. ∎

A remark is in order regarding these realizations of $\hat{g}(s)$. Clearly, the controllable canonical form and observable canonical form realizations are the easiest to obtain. These realizations, however, are not necessarily irreducible unless the given $\hat{g}(s) = N(s)/D(s)$ is known to be irreducible [$N(s)$ and $D(s)$ are coprime]. The realization obtained from the Markov parameters and the Jordan form realization are always controllable and observable no matter the given $\hat{g}(s)$ is irreducible or not. Because of the requirement of computing the poles of $\hat{g}(s)$, the Jordan-form realization is generally more difficult to compute.

The Jordan-form realization however is, as discussed in References 23, 81 and 87, least sensitive to parameter variations among all realizations. The *sensitivity* is defined as the shifting of the eigenvalues due to parameter variations. In practice, in order to reduce sensitivity, a transfer function $\hat{g}(s)$ is factored as

$$\hat{g}(s) = \hat{g}_1(s)\hat{g}_2(s)\cdots$$

or

$$\hat{g}(s) = \hat{\hat{g}}_1(s) + \hat{\hat{g}}_2(s) + \cdots$$

where $\hat{g}_i(s)$ and $\hat{\hat{g}}_i(s)$ are transfer functions of degree 1 or 2. We then realize each $\hat{g}_i(s)$ and $\hat{\hat{g}}_i(s)$, and then connect them together. The first one is called a *tandem realization*; the second one, a *parallel realization*. This type of realization is often used in the design of digital filters (see Reference S47).

***Realization of linear time-varying differential equations.** Before concluding this section, we shall briefly discuss the setup of dynamical equations for linear, time-varying differential equations. If an nth-order, linear, time-varying differential equation is of the form

$$(p^n + \alpha_1(t)p^{n-1} + \cdots + \alpha_n(t))y(t) = \beta(t)u(t) \tag{6-46}$$

by choosing $y(t), \dot{y}(t), \ldots, y^{(n-1)}(t)$ as state variables, a dynamical equation of exactly the same form as (6-8) can be set up. However, if the right-hand side of (6-46) consists of the derivatives of u, although it can still be realized into a dynamical equation, the situation becomes very involved. Instead of giving a general formula, we give an example to illustrate the procedure.

Example 4

Consider the following second-order time-varying differential equation

$$[p^2 + \alpha_1(t)p + \alpha_2(t)]y(t) = [\beta_0(t)p^2 + \beta_1(t)p + \beta_2(t)]u(t) \tag{6-47}$$

The procedure of formulating a dynamical equation for (6-47) is as follows: We first assume that (6-47) can be set into the form

$$\begin{bmatrix} \dot{x}_1 \\ \dot{x}_2 \end{bmatrix} = \begin{bmatrix} 0 & 1 \\ -\alpha_2(t) & -\alpha_1(t) \end{bmatrix} \begin{bmatrix} x_1 \\ x_2 \end{bmatrix} + \begin{bmatrix} b_1(t) \\ b_2(t) \end{bmatrix} u \tag{6-48a}$$

$$y = [\,1 \quad 0\,]\mathbf{x} + e(t)u \tag{6-48b}$$

and then verify this by computing the unknown time functions b_1, b_2, and e in terms of the coefficients of (6-47). Differentiating (6-48b) and using (6-48a), we obtain

$$\dot{y} = py = x_2 + b_1(t)u(t) + \dot{e}(t)u(t) + e(t)\dot{u}(t)$$
$$\ddot{y} = p^2 y = -\alpha_2 x_1 - \alpha_1 x_2 + b_2 u + \dot{b}_1 u + b_1 \dot{u} + \ddot{e}u + 2\dot{e}\dot{u} + e\ddot{u}$$

Substituting these into (6-47) and equating the coefficients of u, $\dot{u}$, and $\ddot{u}$, we obtain

$$\begin{aligned} e(t) &= \beta_0(t) \\ b_1(t) &= \beta_1(t) - \alpha_1(t)\beta_0(t) - 2\dot{\beta}_0(t) \\ b_2(t) &= \beta_2(t) - \dot{b}_1(t) - \alpha_1(t)b_1(t) - \alpha_1(t)\dot{\beta}_0(t) - \alpha_2(t)\beta_0(t) - \ddot{\beta}_0 \end{aligned} \tag{6-49}$$

Since the time functions b_1, b_2 and d can be solved from (6-49), we conclude that the differential equation (6-47) can be transformed into a dynamical equation of the form in (6-48). We see that even for a second-order differential equation, the relations between $\mathbf{b}$, e, and the α_i's and β_i's are very complicated. ∎

6-4 Realizations of Vector Proper Rational Transfer Functions

In this section realizations of vector proper rational transfer functions will be studied. By a vector rational function we mean either a $1 \times p$ or a $q \times 1$ rational-function matrix. Consider the $q \times 1$ proper rational-function matrix

$$\hat{\mathbf{G}}(s) = \begin{bmatrix} \hat{g}'_1(s) \\ \hat{g}'_2(s) \\ \vdots \\ \hat{g}'_q(s) \end{bmatrix} \tag{6-50}$$

It is assumed that every $\hat{g}_i'(s)$ is irreducible. We first expand $\hat{\mathbf{G}}$ into

$$\hat{\mathbf{G}}(s)=\begin{bmatrix} e_1 \\ e_2 \\ \vdots \\ e_q \end{bmatrix}+\begin{bmatrix} \hat{g}_1(s) \\ \hat{g}_2(s) \\ \vdots \\ \hat{g}_q(s) \end{bmatrix} \tag{6-51}$$

where $e_i=\hat{g}_i'(\infty)$, and $\hat{g}_i(s)\triangleq\hat{g}_i'(s)-e_i$ is a strictly proper rational function. We compute the least common denominator of $\hat{g}_i$, for $i=1, 2, \ldots, q$, say $s^n+\alpha_1 s^{n-1}+\cdots+\alpha_n$, and then express $\hat{\mathbf{G}}(s)$ as

$$\hat{\mathbf{G}}(s)=\begin{bmatrix} e_1 \\ e_2 \\ \vdots \\ e_q \end{bmatrix}+\frac{1}{s^n+\alpha_1 s^{n-1}+\cdots+\alpha_n}\begin{bmatrix} \beta_{11}s^{n-1}+\cdots+\beta_{1n} \\ \beta_{21}s^{n-1}+\cdots+\beta_{2n} \\ \vdots \\ \beta_{q1}s^{n-1}+\cdots+\beta_{qn} \end{bmatrix} \tag{6-52}$$

It is claimed that the dynamical equation

$$\dot{\mathbf{x}}=\begin{bmatrix} 0 & 1 & 0 & \cdots & 0 \\ 0 & 0 & 1 & \cdots & 0 \\ \vdots & \vdots & \vdots & & \vdots \\ 0 & 0 & 0 & \cdots & 1 \\ -\alpha_n & -\alpha_{n-1} & -\alpha_{n-2} & \cdots & -\alpha_1 \end{bmatrix}\mathbf{x}+\begin{bmatrix} 0 \\ 0 \\ \vdots \\ 0 \\ 1 \end{bmatrix}u \tag{6-53a}$$

$$\begin{bmatrix} y_1 \\ y_2 \\ \vdots \\ y_q \end{bmatrix}=\begin{bmatrix} \beta_{1n} & \beta_{1(n-1)} & \cdots & \beta_{11} \\ \beta_{2n} & \beta_{2(n-1)} & \cdots & \beta_{21} \\ \vdots & \vdots & & \vdots \\ \beta_{qn} & \beta_{q(n-1)} & \cdots & \beta_{qn} \end{bmatrix}\mathbf{x}+\begin{bmatrix} e_1 \\ e_2 \\ \vdots \\ e_q \end{bmatrix}u \tag{6-53b}$$

is a realization of (6-52). This can be proved by using the controllable-form realization of $\hat{g}(s)$ in (6-14). By comparing (6-53) with (6-26), we see that the transfer function from u to y_i is equal to

$$e_i+\frac{\beta_{i1}s^{n-1}+\cdots+\beta_{in}}{s^n+\alpha_1 s^{n-1}+\cdots+\alpha_n}$$

which is the ith component of $\hat{\mathbf{G}}(s)$. This proves the assertion. Since $\hat{g}_i'(s)$ for $i=1, 2, \ldots, q$ are assumed to be irreducible, the degree of $\hat{\mathbf{G}}(s)$ is equal to n. The dynamical equation (6-53) has dimension n; hence it is a minimal-dimensional realization of $\hat{\mathbf{G}}(s)$ in (6-52). We note that if some or all of $\hat{g}_i'(s)$ are not irreducible, then (6-53) is not observable, although it remains to be controllable.

For single-input, single-output transfer functions, we have both controllable-form and the observable-form realizations. But for column rational functions it is *not* possible to have the observable-form realization.

Example 1

Consider

$$\hat{\mathbf{G}}(s)=\begin{bmatrix}\dfrac{s+3}{(s+1)(s+2)}\\[2ex] \dfrac{s+4}{s+3}\end{bmatrix}=\begin{bmatrix}0\\1\end{bmatrix}+\begin{bmatrix}\dfrac{s+3}{(s+1)(s+2)}\\[2ex] \dfrac{1}{s+3}\end{bmatrix}$$

$$=\begin{bmatrix}0\\1\end{bmatrix}+\frac{1}{(s+1)(s+2)(s+3)}\begin{bmatrix}(s+3)^2\\(s+1)(s+2)\end{bmatrix}$$

$$=\begin{bmatrix}0\\1\end{bmatrix}+\frac{1}{s^3+6s^2+11s+6}\begin{bmatrix}s^2+6s+9\\s^2+3s+2\end{bmatrix}$$

Hence a minimal-dimensional realization of $\hat{\mathbf{G}}(s)$ is given by

$$\begin{bmatrix}\dot{x}_1\\\dot{x}_2\\\dot{x}_3\end{bmatrix}=\begin{bmatrix}0&1&0\\0&0&1\\-6&-11&-6\end{bmatrix}\mathbf{x}+\begin{bmatrix}0\\0\\1\end{bmatrix}u$$

$$y=\begin{bmatrix}9&6&1\\2&3&1\end{bmatrix}\mathbf{x}+\begin{bmatrix}0\\1\end{bmatrix}u \tag{6-54}$$

■

We study now the realizations of $1\times p$ proper rational-function matrices. Since its development is similar to the one for the $q\times 1$ case, we present only the result. Consider the $1\times p$ proper rational matrix

$$\begin{aligned}\hat{\mathbf{G}}(s)&=[\hat{g}'_1(s)\ \vdots\ \hat{g}'_2(s)\ \vdots\ \cdots\ \vdots\ \hat{g}'_p(s)]\\&=[e_1\ \vdots\ e_2\ \vdots\ \cdots\ \vdots\ e_p]+[\hat{g}_1(s)\ \vdots\ \hat{g}_2(s)\ \vdots\ \cdots\ \vdots\ \hat{g}_p(s)]\\&=[e_1\ \vdots\ e_2\ \vdots\ \cdots\ \vdots\ e_p]+\frac{1}{s^n+\alpha_1 s^{n-1}+\cdots+\alpha_n}\\&\quad\times[\beta_{11}s^{n-1}+\beta_{12}s^{n-2}+\cdots+\beta_{1n}\ \vdots\ \beta_{21}s^{n-1}+\beta_{22}s^{n-2}+\cdots\\&\qquad+\beta_{2n}\ \vdots\ \cdots\ \vdots\ \beta_{p1}s^{n-1}+\beta_{p2}s^{n-2}+\cdots+\beta_{pn}]\end{aligned} \tag{6-55}$$

Then the dynamical equation

$$\begin{bmatrix}\dot{x}_1\\\dot{x}_2\\\dot{x}_3\\\vdots\\\dot{x}_n\end{bmatrix}=\begin{bmatrix}0&0&\cdots&0&-\alpha_n\\1&0&\cdots&0&-\alpha_{n-1}\\0&1&\cdots&0&-\alpha_{n-2}\\\vdots&\vdots&&\vdots&\vdots\\0&0&\cdots&1&-\alpha_1\end{bmatrix}\mathbf{x}+\begin{bmatrix}\beta_{1n}&\beta_{2n}&\cdots&\beta_{pn}\\\beta_{1(n-1)}&\beta_{2(n-1)}&\cdots&\beta_{p(n-1)}\\\beta_{1(n-2)}&\beta_{2(n-2)}&\cdots&\beta_{p(n-2)}\\\vdots&\vdots&&\vdots\\\beta_{11}&\beta_{21}&\cdots&\beta_{p1}\end{bmatrix}\mathbf{u}$$

$$y=[0\quad 0\quad\cdots\quad 0\quad 1]\mathbf{x}+[e_1\quad e_2\quad\cdots\quad e_p]\mathbf{u} \tag{6-56}$$

is a realization of (6-55). The realization is always observable whether $\hat{g}_i'(s)$, $i = 1, 2, \ldots, p$, are irreducible or not. If they are all irreducible, then the realization is controllable as well.

It is also possible to find Jordan-form realizations for vector proper rational functions. The procedure is similar to the one for the scalar case. We use an example to illustrate the procedure.

*Example 2

Find a Jordan-form realization of the 2 × 1 rational function

$$\hat{\mathbf{G}}(s) = \begin{bmatrix} \dfrac{s+3}{(s+1)(s+2)} \\ \dfrac{s^2+4}{(s+1)^2} \end{bmatrix} = \begin{bmatrix} \dfrac{2}{s+1} - \dfrac{1}{s+2} \\ 1 - \dfrac{2}{s+1} + \dfrac{5}{(s+1)^2} \end{bmatrix}$$

$$= \begin{bmatrix} 0 \\ 1 \end{bmatrix} + \frac{1}{s+1}\begin{bmatrix} 2 \\ -2 \end{bmatrix} + \frac{1}{s+2}\begin{bmatrix} -1 \\ 0 \end{bmatrix} + \frac{1}{(s+1)^2}\begin{bmatrix} 0 \\ 5 \end{bmatrix} \tag{6-57}$$

we draw in Figure 6-6 a block diagram of (6-57). With the state variables chosen as shown, we can obtain the Jordan-form equation

$$\begin{bmatrix} \dot{x}_1 \\ \dot{x}_2 \\ \dot{x}_3 \end{bmatrix} = \left[\begin{array}{cc:c} -1 & 1 & 0 \\ 0 & -1 & 0 \\ \hdashline 0 & 0 & -2 \end{array}\right] \mathbf{x} + \left[\begin{array}{c} 0 \\ 1 \\ \hdashline 1 \end{array}\right] u$$

$$y = \left[\begin{array}{cc:c} 0 & 2 & -1 \\ 5 & -2 & 0 \end{array}\right] \mathbf{x} + \begin{bmatrix} 0 \\ 1 \end{bmatrix} u$$

to describe the block diagram. The equation can be readily shown to be controllable and observable by using Theorem 5-21. Hence it is an irreducible Jordan-form realization of (6-57).

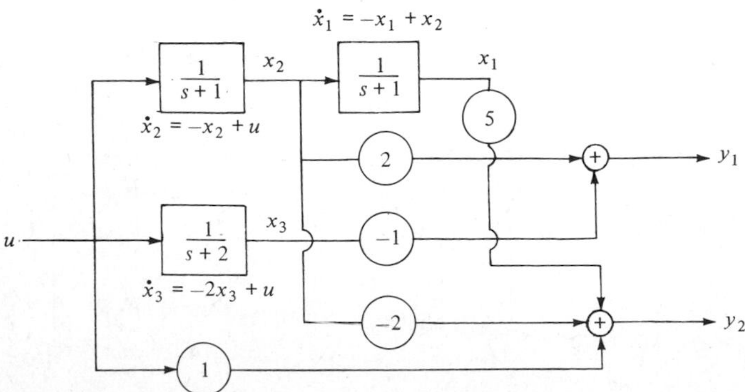

Figure 6-6 Block diagram of Equation (6-57).

Realization from the Hankel matrix

Consider a $q \times 1$ proper rational matrix $\hat{\mathbf{G}}(s)$ expanded as

$$\hat{\mathbf{G}}(s) = \begin{bmatrix} \hat{g}_1(s) \\ \hat{g}_2(s) \\ \vdots \\ \hat{g}_q(s) \end{bmatrix} = \begin{bmatrix} h_1(0) + h_1(1)s^{-1} + h_1(2)s^{-2} + \cdots \\ h_2(0) + h_2(1)s^{-1} + h_2(2)s^{-2} + \cdots \\ \vdots \\ h_q(0) + h_q(1)s^{-1} + h_q(2)s^{-2} + \cdots \end{bmatrix} \qquad \textbf{(6-58)}$$

The Markov parameters $h_i(j)$ can be obtained recursively as in (6-29) and (6-30). For each $\hat{g}_i(s)$, we form a Hankel matrix $\mathbf{H}_i$ defined in (6-31). We then form the composite matrix

$$\mathbf{T} \triangleq \begin{bmatrix} \mathbf{H}_1(\alpha_1 + 1, \beta) \\ \hdashline \mathbf{H}_2(\alpha_2 + 1, \beta) \\ \hdashline \vdots \\ \hdashline \mathbf{H}_q(\alpha_q + 1, \beta) \end{bmatrix} \begin{matrix} \} \sigma_1 \text{ (no. of linearly independent rows)} \\ \} \sigma_2 \\ \\ \} \sigma_q \end{matrix} \qquad \textbf{(6-59)}$$

Note that $\mathbf{H}_i$, $i = 1, 2, \ldots, q$, have the same number of columns but different number of rows. The integer α_i is the degree of the denominator of $\hat{g}_i(s)$ and β is equal to or larger than the degree of the least common denominator of $\{\hat{g}_i(s), i = 1, 2, \ldots, q\}$. If the least common denominator of $\{\hat{g}_i(s), i = 1, 2, 3, \ldots, q\}$ is not available, we may choose β to be equal to the sum of the degrees of the denominator of $\hat{g}_i(s)$, $i = 1, 2, \ldots, q$. Note that in this method, $\hat{g}_i(s)$ need not be irreducible.

Now we shall apply the row searching algorithm to search the linear independent rows of $\mathbf{T}$ in order from top to bottom. Because of the structure of $\mathbf{T}$ and Theorem 6-3, if one row in $\mathbf{H}_i$ is linearly dependent on its previous rows of $\mathbf{T}$ in (6-59), then all subsequent rows in $\mathbf{H}_i$ will also be linearly dependent. Let σ_i be the number of linearly independent rows in $\mathbf{H}_i$. If all $\hat{g}_i(s)$ are irreducible, then the first α_i rows of $\mathbf{H}_i$ will be linearly independent in $\mathbf{H}_i$ (Theorem 6-3). However these linear independent rows of $\mathbf{H}_i$ may not be all linearly independent in $\mathbf{T}$ because they may become dependent on the rows of $\mathbf{H}_j$ for $j < i$; hence we have $\sigma_i \leq \alpha_i$, $i = 2, 3, \ldots, q$. Note that if $\hat{g}_1(s)$ is irreducible, then we do have $\sigma_1 = \alpha_1$. The row of $\mathbf{H}_i$ which first becomes linearly dependent on its previous rows of $\mathbf{T}$ is called the *primary* linearly dependent row of $\mathbf{H}_i$. If σ_i is the number of linearly independent rows in $\mathbf{H}_i$, then the $(\sigma_i + 1)$th row of $\mathbf{H}_i$ is the primary linearly dependent row. Corresponding to these q primary dependent rows, the row-searching algorithm will yield

$$\begin{aligned}
\mathbf{k}_1 &= [\overbrace{\underbrace{a_{11}(1)\, a_{11}(2) \cdots a_{11}(\sigma_1)}_{\sigma_1}\ 1}^{\alpha_1 + 1} \vdots \overbrace{0 \quad \cdots 0 \qquad 0 \cdots 0\ 0}^{\alpha_2 + 1} \vdots \cdots \vdots \overbrace{0 \quad \cdots 0 \qquad 0 \cdots 0}^{\alpha_q + 1}] \\
\mathbf{k}_2 &= [a_{21}(1)\, a_{21}(2) \cdots a_{21}(\sigma_1)\ 0 \vdots a_{22}(1) \cdots a_{22}(\sigma_2)\ 1 \cdots 0\ 0 \vdots \cdots \vdots 0 \quad \cdots 0 \qquad 0 \cdots 0] \\
&\ \ \vdots \\
\mathbf{k}_q &= [a_{q1}(1)\, a_{q1}(2) \cdots a_{q1}(\sigma_1)\ 0 \vdots \underbrace{a_{q2}(1) \cdots a_{q2}(\sigma_2)}_{\sigma_2}\ 0 \cdots 0\ 0 \vdots \cdots \vdots \underbrace{a_{qq}(1) \cdots a_{qq}(\sigma_q)}_{\sigma_q}\ 1 \cdots 0]
\end{aligned}$$

(6-60)

such that $\mathbf{k}_i\mathbf{T}=\mathbf{0}$. Note that the row vector $\mathbf{k}_j$ has, except element 1, only $(\sigma_1+\sigma_2+\cdots+\sigma_j)$ possible nonzero elements. This is a consequence of the row searching algorithm. The $\mathbf{k}_i$ expresses the primary dependent row of $\mathbf{H}_i$ as a unique linear combination of its previous linearly independent rows of $\mathbf{T}$. See Appendix A.

Now we claim that $\hat{\mathbf{G}}(s)$ in (6-58) has the following irreducible realization:

$$\dot{\mathbf{x}}=\mathbf{A}\mathbf{x}+\mathbf{b}u \tag{6-61a}$$

$$\mathbf{y}=\mathbf{C}\mathbf{x}+\mathbf{e}u \tag{6-61b}$$

with

$$\mathbf{A}=\begin{bmatrix}\mathbf{A}_{11} & \mathbf{0} & \cdots & \mathbf{0}\\ \mathbf{A}_{21} & \mathbf{A}_{22} & \cdots & \mathbf{0}\\ \vdots & \vdots & & \vdots\\ \mathbf{A}_{q1} & \mathbf{A}_{q2} & \cdots & \mathbf{A}_{qq}\end{bmatrix},\qquad \mathbf{b}=\begin{bmatrix}\mathbf{b}_1\\ \mathbf{b}_2\\ \vdots\\ \mathbf{b}_q\end{bmatrix},\qquad \mathbf{e}=\begin{bmatrix}h_1(0)\\ h_2(0)\\ \vdots\\ h_q(0)\end{bmatrix}$$

where, for $i=1,2,\ldots,q$,

$$\mathbf{A}_{ii}=\begin{bmatrix}0 & 1 & 0 & \cdots & 0\\ 0 & 0 & 1 & \cdots & 0\\ \vdots & \vdots & \vdots & & \vdots\\ 0 & 0 & 0 & \cdots & 1\\ -a_{ii}(1) & -a_{ii}(2) & -a_{ii}(3) & \cdots & -a_{ii}(\sigma_i)\end{bmatrix}\quad (\sigma_i\times\sigma_i)\text{ matrix}$$

for $i>j$,

$$\mathbf{A}_{ij}=\begin{bmatrix}0 & 0 & \cdots & 0\\ 0 & 0 & \cdots & 0\\ \vdots & \vdots & & \vdots\\ 0 & 0 & \cdots & 0\\ -a_{ij}(1) & -a_{ij}(2) & \cdots & -a_{ij}(\sigma_j)\end{bmatrix}\quad (\sigma_i\times\sigma_j)\text{ matrix}$$

and, for $i=1,2,\ldots,q$,

$$\mathbf{b}_i=\begin{bmatrix}h_{i1}(1)\\ h_{i1}(2)\\ \vdots\\ h_{i1}(\sigma_i)\end{bmatrix}$$

The matrix $\mathbf{C}$ is given by, if $\sigma_i\neq 0$, for $i=1,2,\ldots,q$,

$$\mathbf{C}=\left.\left[\begin{array}{cccc|cccc|c|cccc}1 & 0 & \cdots & 0 & 0 & 0 & \cdots & 0 & & 0 & 0 & \cdots & 0\\ 0 & 0 & \cdots & 0 & 1 & 0 & \cdots & 0 & \cdots & 0 & 0 & \cdots & 0\\ \vdots & \vdots & & \vdots & \vdots & \vdots & & \vdots & & \vdots & \vdots & & \vdots\\ 0 & 0 & \cdots & 0 & 0 & 0 & \cdots & 0 & & 1 & 0 & \cdots & 0\end{array}\right]\right\}q \tag{6-62a}$$

with the three column blocks of widths σ_1, σ_2, …, σ_q.

If $\sigma_i = 0$ for some i, say $i = 3$ for convenience of illustration, then

$$\mathbf{C} = \left[\begin{array}{ccc:ccc:cccc:cccc} 1 & \cdots & 0 & 0 & \cdots & 0 & 0 & 0 & \cdots & 0 & 0 & 0 & \cdots & 0 \\ 0 & \cdots & 0 & 1 & \cdots & 0 & 0 & 0 & \cdots & 0 & 0 & 0 & \cdots & 0 \\ -a_{31}(1) & \cdots & -a_{31}(\sigma_1) & -a_{32}(1) & \cdots & -a_{32}(\sigma_2) & 0 & 0 & \cdots & 0 & 0 & 0 & \cdots & 0 \\ 0 & \cdots & 0 & 0 & \cdots & 0 & 1 & 0 & \cdots & 0 & 0 & 0 & \cdots & 0 \\ \vdots & & \vdots & \vdots & & \vdots & \vdots & \vdots & & \vdots & \vdots & \vdots & & \vdots \\ 0 & \cdots & 0 & 0 & \cdots & 0 & 0 & 0 & \cdots & 0 & 1 & 0 & \cdots & 0 \end{array}\right]$$
$$\qquad\qquad \underbrace{\qquad\qquad}_{\sigma_1} \qquad\qquad \underbrace{\qquad\qquad}_{\sigma_2} \qquad\qquad \underbrace{\qquad\quad}_{\sigma_4} \qquad\qquad \underbrace{\qquad\quad}_{\sigma_q} \tag{6-62b}$$

The dimension of this realization is $n = \sigma_1 + \sigma_2 + \cdots + \sigma_q$ which is equal to the rank of $\mathbf{H}$.

The assertion that (6-61) is a realization of $\hat{\mathbf{G}}(s)$ in (6-58) can be established by using the procedure used in the scalar case. Similar to (6-39), it is straightforward, though tedious, to verify

$$\mathbf{A}^k\mathbf{b} = \left[\begin{array}{c} h_1(k+1) \\ h_1(k+2) \\ \vdots \\ h_1(k+\sigma_1) \\ \hdashline \vdots \\ \hdashline h_q(k+1) \\ h_q(k+2) \\ \vdots \\ h_q(k+\sigma_q) \end{array}\right] \qquad k = 0, 1, 2, \ldots \tag{6-63}$$

Let $\mathbf{c}_i$ be the ith row of $\mathbf{C}$. Then we have

$$\mathbf{c}_i\mathbf{A}^k\mathbf{b} = h_i(k+1), \qquad k = 0, 1, 2, \ldots; i = 1, 2, \ldots, q$$

This establishes, following (6-34), the assertion.

We show next that the realization in (6-61) is irreducible. From (6-63), we see that $[\mathbf{b} \quad \mathbf{Ab} \quad \cdots \quad \mathbf{A}^{n-1}\mathbf{b}]$ consists of all linear independent rows of $\mathbf{H}$ in (6-59); hence its rank is n and $\{\mathbf{A}, \mathbf{b}\}$ is controllable. Similarly it can be verified that the matrix

$$\left[\begin{array}{c} \mathbf{c}_1 \\ \mathbf{c}_1\mathbf{A} \\ \vdots \\ \mathbf{c}_1\mathbf{A}^{\sigma_1 - 1} \\ \vdots \\ \mathbf{c}_q \\ \vdots \\ \mathbf{c}_q\mathbf{A}^{\sigma_q - 1} \end{array}\right]$$

is the identity matrix of order n; hence $\{\mathbf{A}, \mathbf{C}\}$ is observable. This completes the proof that (6-61) is an irreducible realization of $\hat{\mathbf{G}}(s)$ in (6-58).

Example 3

Consider

$$\hat{\mathbf{G}}(s)=\begin{bmatrix}\dfrac{s+3}{s^2(s+1)^2}\\[2ex] \dfrac{1}{s^3(s+1)}\end{bmatrix}=\begin{bmatrix}s^{-3}+s^{-4}-3s^{-5}+5s^{-6}-7s^{-7}+9s^{-8}-11s^{-9}+\cdots\\ s^{-4}-s^{-5}+s^{-6}-s^{-7}+s^{-8}-\cdots\end{bmatrix}$$

We form the Hankel matrix, with $\alpha_1=4$, $\alpha_2=4$, and $\beta=5$,

$$\mathbf{T}=\begin{bmatrix}\mathbf{H}_1(5,5)\\ \mathbf{H}_2(5,5)\end{bmatrix}=\begin{bmatrix}0 & 0 & 1 & 1 & -3\\ 0 & 1 & 1 & -3 & 5\\ 1 & 1 & -3 & 5 & -7\\ 1 & -3 & 5 & -7 & 9\\ -3 & 5 & -7 & 9 & -11\\ \hline 0 & 0 & 0 & 1 & -1\\ 0 & 0 & 1 & -1 & 1\\ 0 & 1 & -1 & 1 & -1\\ 1 & -1 & 1 & -1 & 1\\ -1 & 1 & -1 & 1 & -1\end{bmatrix}$$

If we apply the row-searching algorithm to $\mathbf{T}$, we will finally obtain

$$\begin{bmatrix}1 &&&&&&&&&\\ -1 & 1 &&&&&&&&\\ 3 & -1 & 1 &&&&&&&\\ -5 & 3 & -1 & 1 &&&&&&\\ 7 & -5 & 3 & 2 & 1 &&&&&\\ \hline 0 & 0 & 0 & \frac{1}{36} & 0 & 1 &&&&\\ -1 & 0 & 0 & -\frac{2}{36} & 0 & 0 & 1 &&&\\ \hline 1 & -1 & 0 & \frac{6}{36} & 0 & 0 & 0 & 1 &&\\ -1 & 1 & -1 & -\frac{1}{2} & 0 & 0 & 0 & 0 & 1 &\\ 1 & -1 & 1 & \frac{1}{2} & 0 & 0 & 0 & 0 & 0 & 1\end{bmatrix} * \mathbf{T}$$

$$=\begin{bmatrix}0 & 0 & \textcircled{1} & 1 & -3\\ 0 & \textcircled{1} & 0 & -4 & 8\\ \textcircled{1} & 0 & 0 & 12 & -24\\ 0 & 0 & 0 & -36 & 72\\ 0 & 0 & 0 & 0 & 0\\ \hline 0 & 0 & 0 & 0 & \textcircled{1}\\ 0 & 0 & 0 & 0 & 0\\ 0 & 0 & 0 & 0 & 0\\ 0 & 0 & 0 & 0 & 0\\ 0 & 0 & 0 & 0 & 0\end{bmatrix}\begin{matrix}\left.\vphantom{\begin{matrix}0\\0\\0\\0\end{matrix}}\right\}\sigma_1=4\\ \\ \sigma_2=1\\ \\ \\ \\ \\ \end{matrix}$$

The leftmost matrix is the matrix **F** defined in (F-9). From the rightmost matrix, we have $\sigma_1 = 4$ and $\sigma_2 = 1$. The fifth and seventh rows of **K** can be obtained by using the formula in (F-11) as

$$\begin{aligned}\mathbf{k}_1 &= [\underbrace{\begin{matrix}0 & 0 & 1 & 2\end{matrix}}_{\sigma_1=4} \quad \textcircled{1} \;\vdots\; \underbrace{0}_{\sigma_2=1} \quad 0 \quad 0 \quad 0 \quad 0] \\ \mathbf{k}_2 &= [\begin{matrix}-\frac{1}{3} & -\frac{2}{9} & \frac{1}{18} & -\frac{1}{18}\end{matrix} \quad 0 \;\vdots\; 0 \quad \textcircled{1} \quad 0 \quad 0 \quad 0]\end{aligned}$$

The elements $\textcircled{1}$ correspond to the primary linearly dependent rows. Hence an irreducible realization of $\hat{\mathbf{G}}(s)$ is given by

$$\dot{\mathbf{x}} = \left[\begin{array}{cccc:c} 0 & 1 & 0 & 0 & 0 \\ 0 & 0 & 1 & 0 & 0 \\ 0 & 0 & 0 & 1 & 0 \\ 0 & 0 & -1 & -2 & 0 \\ \hdashline \frac{1}{3} & \frac{2}{9} & -\frac{1}{18} & \frac{1}{18} & 0 \end{array}\right]\mathbf{x} + \left[\begin{array}{c} 0 \\ 0 \\ 1 \\ 1 \\ \hdashline 0 \end{array}\right] u$$

$$\mathbf{y} = \left[\begin{array}{cccc:c} 1 & 0 & 0 & 0 & 0 \\ 0 & 0 & 0 & 0 & 1 \end{array}\right]\mathbf{x}$$

The $(\sigma_1 = 4)$th and $(\sigma_1 + \sigma_2 = 5)$th rows of **A** are taken from the $\mathbf{k}_1$ and $\mathbf{k}_2$ rows with the signs reversed. The vector **b** consists of the first σ_1 Markov parameters of $\hat{g}_1(s)$ and the first σ_2 Markov parameter of $\hat{g}_2(s)$. The form of **C** is fixed and depends only on σ_1 and σ_2.

It is possible to search the linearly independent rows of **T** in a different order. First we expand $\hat{\mathbf{G}}(s)$ as

$$\hat{\mathbf{G}}(s) = \mathbf{H}(0) + \mathbf{H}(1)s^{-1} + \mathbf{H}(2)s^{-2} + \mathbf{H}(3)s^{-3} + \cdots$$

where $\mathbf{H}(i)$ have the same order as $\hat{\mathbf{G}}(s)$. Let $\alpha = \max\{\alpha_i, i = 1, 2, \ldots, q\}$, where α_i are defined in (6-59). We then form the Hankel matrix

$$\bar{\mathbf{T}} = \left[\begin{array}{ccccc} \mathbf{H}(1) & \mathbf{H}(2) & \mathbf{H}(3) & \cdots & \mathbf{H}(\beta) \\ \hdashline \mathbf{H}(2) & \mathbf{H}(3) & \mathbf{H}(4) & \cdots & \mathbf{H}(\beta+1) \\ \vdots & \vdots & \vdots & & \vdots \\ \mathbf{H}(\alpha) & \mathbf{H}(\alpha+1) & \mathbf{H}(\alpha+2) & \cdots & \mathbf{H}(\alpha+\beta-1) \\ \mathbf{H}(\alpha+1) & \mathbf{H}(\alpha+2) & \mathbf{H}(\alpha+3) & \cdots & \mathbf{H}(\alpha+\beta) \end{array}\right] \begin{array}{l}\}\text{one block row} \\ \\ \\ \\ \\ \end{array} \tag{6-64}$$

There are $\alpha + 1$ block rows in $\bar{\mathbf{T}}$; each block has q rows. The ith row of every block row is associated with $\hat{g}_i(s)$. Except from having more rows and different ordering of rows, the Hankel matrix in (6-64) is basically the same as the one in (6-59). Now we shall apply the row searching algorithm to search the linearly independent rows of $\bar{\mathbf{T}}$ in order from top to bottom. Clearly, if the ith row of a block row is linearly dependent on its previous rows, then all the ith rows in the subsequent block rows are linearly dependent on their previous rows.[4] After

[4] See the discussion in the subsections on the controllability indices and observability indices in Chapter 5. See also Schemes 1 and 2 of the search of linearly independent columns of the controllability matrix in the second half of Section 7-2.

the completion of this search, we then rearrange the rows of $\bar{\mathbf{T}}$ into the form of (6-59) as

$$\hat{\mathbf{T}} = \begin{bmatrix} \mathbf{H}_1(\alpha+1, \beta) \\ \mathbf{H}_2(\alpha+1, \beta) \\ \vdots \\ \mathbf{H}_q(\alpha+1, \beta) \end{bmatrix} \begin{matrix} \} \; v_1 \text{ (no. of linearly independent rows)} \\ \\ \\ \} \; v_q \end{matrix} \tag{6-65}$$

Clearly the total numbers of independent rows of $\bar{\mathbf{T}}$ in (6-64), $\hat{\mathbf{T}}$ in (6-65), and $\mathbf{T}$ in (6-59) are all the same. Let v_i be the number of linearly independent rows of $\mathbf{H}_i(\alpha+1, \beta)$ in (6-65). Then we have $v_1 + v_2 + \cdots + v_q = \sigma_1 + \sigma_2 + \cdots + \sigma_q$, where σ_i are defined in (6-59). Note that generally we have $\sigma_i \neq v_i$, for some or all i. Clearly the primary linearly dependent row of $\mathbf{H}_i$ in $\hat{\mathbf{T}}$ is its (v_i+1)th row. The major difference between $\mathbf{T}$ in (6-59) and $\hat{\mathbf{T}}$ in (6-65) is the orders in which the linearly independent rows are searched. In $\hat{\mathbf{T}}$, the linearly independent rows are searched in the order of the first row of $\mathbf{H}_i$, $i = 1, 2, \ldots, q$; the second row of $\mathbf{H}_i$, $i = 1, 2, \ldots, q$, and so forth. For convenience of searching, we search the rows of $\bar{\mathbf{T}}$ in order from top to bottom and then rearrange it as $\hat{\mathbf{T}}$.

Now from the row searching algorithm and the rearrangement of the coefficients of combinations according to the rearrangements from $\bar{\mathbf{T}}$ to $\hat{\mathbf{T}}$, we can obtain, similar to (6-60),

$$\begin{aligned}
\hat{\mathbf{k}}_1 &= [\overbrace{\underbrace{a_{11}(1) \; a_{11}(2) \; \cdots \; a_{11}(v_1)}_{v_1} \; 1 \; 0 \cdots 0}^{\alpha+1} \;\vdots\; \overbrace{\underbrace{a_{12}(1) \cdots a_{12}(v_2)}_{v_2} \; 0 \; 0 \cdots 0}^{\alpha+1} \;\vdots\; \cdots \;\vdots\; \overbrace{\underbrace{a_{1q}(1) \cdots a_{1q}(v_q)}_{v_q} \; 0 \; 0 \cdots 0}^{\alpha+1}] \\
\hat{\mathbf{k}}_2 &= [a_{21}(1) \; a_{21}(2) \; \cdots \; a_{21}(v_1) \; 0 \; 0 \cdots 0 \;\vdots\; a_{22}(1) \cdots a_{22}(v_2) \; 1 \; 0 \cdots 0 \;\vdots\; \cdots \;\vdots\; a_{2q}(1) \cdots a_{2q}(v_q) \; 0 \; 0 \cdots 0] \\
&\;\;\vdots \\
\hat{\mathbf{k}}_q &= [a_{q1}(1) \; a_{q1}(2) \; \cdots \; a_{q1}(v_1) \; 0 \; 0 \cdots 0 \;\vdots\; a_{q2}(1) \cdots a_{q2}(v_2) \; 0 \; 0 \cdots 0 \;\vdots\; \cdots \;\vdots\; a_{qq}(1) \cdots a_{qq}(v_q) \; 1 \; 0 \cdots 0]
\end{aligned} \tag{6-66}$$

such that $\hat{\mathbf{k}}_i \hat{\mathbf{T}} = \mathbf{0}$. The major difference between (6-66) and (6-60) is that $a_{ij}(k)$, for $j > i$, are generally different from zeros in (6-66). In other words, the (v_1+1)th row in $\mathbf{H}_1$ in (6-65) depends not only on the first v_1 rows of $\mathbf{H}_1$ but also on the first v_1 rows of $\mathbf{H}_2, \mathbf{H}_3, \ldots, \mathbf{H}_q$ in (6-65); whereas the (σ_1+1)th row of $\mathbf{H}_1$ in (6-59) depends only on the first σ_1 rows of $\mathbf{H}_1$. Similar remarks apply to the (v_i+1)th row of $\mathbf{H}_i$ in (6-65). Now we claim that the dynamical equation

$$\dot{\mathbf{x}} = \mathbf{A}\mathbf{x} + \mathbf{B}\mathbf{u} \qquad \mathbf{y} = \mathbf{C}\mathbf{x} \tag{6-67}$$

with

$$\mathbf{A} = \begin{bmatrix} \mathbf{A}_{11} & \mathbf{A}_{12} & \cdots & \mathbf{A}_{1q} \\ \mathbf{A}_{21} & \mathbf{A}_{22} & \cdots & \mathbf{A}_{2q} \\ \vdots & \vdots & & \vdots \\ \mathbf{A}_{q1} & \mathbf{A}_{q2} & \cdots & \mathbf{A}_{qq} \end{bmatrix}$$

where $\mathbf{A}_{ij}$, $\mathbf{B}$, and $\mathbf{C}$ are given as in (6-61) and (6-62) with σ_i replaced by v_i and $a_{ij}(k)$ taken from (6-66). The proof of this statement is similar to the one in (6-60); it is more tedious, however, because of the necessity of tracking the rearrangements of coefficients. The proof does not involve any new idea and will be omitted. Instead we give an example to illustrate the procedure.

Example 4

Consider the transfer-function matrix $\hat{\mathbf{G}}(s)$ in Example 3. We write

$$\hat{\mathbf{G}}(s)=\begin{bmatrix}1\\0\end{bmatrix}s^{-3}+\begin{bmatrix}1\\1\end{bmatrix}s^{-4}+\begin{bmatrix}-3\\-1\end{bmatrix}s^{-5}+\begin{bmatrix}5\\1\end{bmatrix}s^{-6}+\begin{bmatrix}-7\\-1\end{bmatrix}s^{-7}+\begin{bmatrix}9\\1\end{bmatrix}s^{-8}+\cdots$$

We form $\bar{\mathbf{T}}$ with $\alpha=\max\{4, 4\}=4$ and $\beta=5$, and then search its linearly independent rows as

$$\begin{bmatrix}1&&&&&&&&&\\0&1&&&&&&&&\\-1&4&1&&&&&&&\\-1&2&0&1&&&&&&\\3&-8&-1&6&1&&&&&\\1&-2&-1&3&0&1&&&&\\-5&12&3&-12&-1&0&1&&&\\-1&2&1&-3&-1&0&0&1&&\\7&-16&-5&18&3&0&0&0&1&\\1&-2&-1&3&1&0&0&0&0&1\end{bmatrix}\left[\begin{array}{ccccc}0&0&1&1&-3\\0&0&0&1&-1\\ \hdashline 0&1&1&-3&5\\0&0&1&-1&1\\ \hdashline 1&1&-3&5&-7\\0&1&-1&1&-1\\ \hdashline 1&3&5&-7&9\\1&-1&1&-1&1\\ \hdashline -3&5&-7&9&-11\\-1&1&-1&1&-1\end{array}\right]$$

$$=\left[\begin{array}{ccccc}0&0&1&1&-3\\0&0&0&1&-1\\ \hdashline 0&1&0&0&4\\0&0&0&0&2\\ \hdashline 1&0&0&0&0\\0&0&0&0&0\\ \hdashline 0&0&0&0&0\\0&0&0&0&0\\ \hdashline 0&0&0&0&0\\0&0&0&0&0\end{array}\right]\begin{array}{l}\\ \\ \\ \\ \\ \leftarrow\\ \leftarrow\\ \\ \\ \\ \end{array}$$

The rows of $\mathbf{K}$ corresponding to the zero rows indicated by arrows can be obtained, by using the formula in (F-11), as

$$\begin{aligned}\bar{\mathbf{k}}_1&=[\;\;6\;\;0\;\vdots\;\;4\;\;-18\;\vdots\;-1\;\;0\;\vdots\;1\;\;0\;\vdots\;0\;\;0]\\ \bar{\mathbf{k}}_2&=[-1\;\;0\;\vdots\;-1\;\;3\;\vdots\;\;0\;\;1\;\vdots\;0\;\;0\;\vdots\;0\;\;0]\end{aligned}$$

Note that the $\bar{\mathbf{k}}_1$ row corresponds to the first row of $\mathbf{H}(k)$ which first becomes linearly dependent or, equivalently, the primary dependent row of $\mathbf{H}_1$, and the $\bar{\mathbf{k}}_2$ row corresponds to the primary dependent row of $\mathbf{H}_2$. Now we rearrange $\bar{\mathbf{T}}$ and $\bar{\mathbf{k}}_i$ as

$$\begin{bmatrix}\hat{\mathbf{k}}_1\\ \hat{\mathbf{k}}_2\end{bmatrix}\hat{\mathbf{T}}\triangleq\left[\begin{array}{ccccc:ccccc}6&4&-1&1&0&0&-18&0&0&0\\ \underbrace{-1\;\;-1\;\;\;0}_{\nu_1}&&&0&0&\underbrace{0\;\;\;3}_{\nu_2}&&1&0&0\end{array}\right]$$

$$\times\begin{bmatrix} 0 & 0 & 1 & 1 & -3 \\ 0 & 1 & 1 & -3 & 5 \\ 1 & 1 & -3 & 5 & -7 \\ 1 & -3 & 5 & -7 & 9 \\ -3 & 5 & -7 & 9 & -11 \\ \hdashline 0 & 0 & 0 & 1 & -1 \\ 0 & 0 & 1 & -1 & 1 \\ 0 & 1 & -1 & 1 & -1 \\ 1 & -1 & 1 & -1 & 1 \\ -1 & 1 & -1 & 1 & -1 \end{bmatrix} = \begin{bmatrix} \mathbf{0} \\ \mathbf{0} \end{bmatrix} \begin{matrix} \\ \\ \\ \leftarrow \text{Primary dependent row of } \mathbf{H}_1 \\ \\ \\ \\ \leftarrow \text{Primary dependent row of } \mathbf{H}_2 \\ \\ \\ \end{matrix}$$

Hence an irreducible realization of $\hat{\mathbf{G}}(s)$ is given by

$$\dot{\mathbf{x}} = \left[\begin{array}{ccc:cc} 0 & 1 & 0 & 0 & 0 \\ 0 & 0 & 1 & 0 & 0 \\ -6 & -4 & 1 & 0 & 18 \\ \hdashline 0 & 0 & 0 & 0 & 1 \\ 1 & 1 & 0 & 0 & -3 \end{array}\right]\mathbf{x} + \begin{bmatrix} 0 \\ 0 \\ 1 \\ \hdashline 0 \\ 0 \end{bmatrix} u$$

$$\mathbf{y} = \left[\begin{array}{ccc:cc} 1 & 0 & 0 & 0 & 0 \\ 0 & 0 & 0 & 1 & 0 \end{array}\right]\mathbf{x}$$

The v_1th and $(v_1 + v_2)$th rows of $\mathbf{A}$ are taken from the $\hat{\mathbf{k}}_1$ and $\hat{\mathbf{k}}_2$ rows with the signs reversed. The vector $\mathbf{b}$ consists of the first v_1 and v_2 Markov parameters of $\hat{g}_1$ and $\hat{g}_2$. The form of $\mathbf{C}$ is fixed and depends only on v_1 and v_2. ∎

Remarks are in order regarding the use of the Hankel matrices in (6-59) and (6-64) in the realizations. First we show that the v_i in (6-65), which are obtained by searching the linearly independent rows of $\bar{\mathbf{T}}$ in order from top to bottom, are the observability indices of any irreducible realization of $\hat{\mathbf{G}}(s)$. Indeed, from (5-64) and $\mathbf{H}(k) = \mathbf{CA}^{k-1}\mathbf{B}$, we have

$$\bar{\mathbf{T}} = \mathbf{V}_\alpha \mathbf{U}_{\beta-1}$$

where $\mathbf{V}_k$ and $\mathbf{U}_k$ are defined in (5-48) and (5-27). The postmultiplication of $\mathbf{U}_{\beta-1}$ on $\mathbf{V}_\alpha$ operates only on the columns of $\mathbf{V}_\alpha$, hence the linearly independent rows of $\bar{\mathbf{T}}$ in order from top to bottom are the same as the linearly independent rows of $\mathbf{V}_\alpha$ in order from top to bottom. Consequently, we conclude that the v_i, $i = 1, 2, \ldots, q$, are the observability indices of any irreducible realization of $\hat{\mathbf{G}}(s)$. This can also be easily verified for the $\hat{\mathbf{G}}(s)$ in Example 4.

Because v_i, $i = 1, 2, \ldots, q$, are the observability indices, we have

$$v \triangleq \max\{v_i, i = 1, 2, \ldots, q\} \leq \sigma = \max\{\sigma_i, i = 1, 2, \ldots, q\}$$

Hence the largest order of $\mathbf{A}_{ii}$ in (6-67) is smaller than or equal to the one in (6-60). This implies that the data up to $\mathbf{CA}^{v+\beta+1}\mathbf{B}$ are used in the realization in (6-67); whereas, the data up to $\mathbf{CA}^{\sigma+\beta+1}\mathbf{B}$ are used in the realization in (6-60). As discussed in Section 5-8, the larger the power of $\mathbf{A}$, the more errors may be introduced in $\mathbf{CA}^k\mathbf{B}$. Hence it is more difficult to determine the correct σ_i, $i=1, 2, \ldots, q$, on digital computer computation. We note that the order of $\bar{\mathbf{T}}$, which is $q(\alpha+1)\times\beta$, is larger than the order of $\mathbf{T}$, which is $(\alpha_1+\alpha_2+\cdots+\alpha_q)\times\beta$; however, in the search of linearly independent rows of $\bar{\mathbf{T}}$, once a dependent row, say the ith row, in a block row is found, all the ith rows in the subsequent block rows can be skipped. Hence, although the order of $\bar{\mathbf{T}}$ in (6-64) is larger than the $\mathbf{T}$ in (6-59), the highest power of $\mathbf{CA}^k\mathbf{B}$ used in the search of linearly independent rows in $\bar{\mathbf{T}}$ is smaller. As a result, it is easier to determine v_i than σ_i on computer computation, and $\bar{\mathbf{T}}$ should be used.

We discuss now an irreducible realization for a $1\times p$ proper rational matrix. Consider

$$\hat{\mathbf{G}}(s)=[\hat{g}_1(s)\quad \hat{g}_2(s)\quad \cdots\quad \hat{g}_p(s)]=\mathbf{H}(0)+\mathbf{H}(1)s^{-1}+\mathbf{H}(2)s^{-2}+\cdots$$

where $\mathbf{H}(k)$ are $1\times p$ constant matrices. Let α be the degree of the least common denominator of all elements of $\hat{\mathbf{G}}(s)$ and let β_i be the degree of the denominator of $\hat{g}_i(s)$. Define $\beta=\max\{\beta_i, i=1, 2, \ldots, p\}$. We form the Hankel matrix

$$\mathbf{T}=\begin{bmatrix}\mathbf{H}(1) & \mathbf{H}(2) & \cdots & \mathbf{H}(\beta)\\ \mathbf{H}(2) & \mathbf{H}(3) & \cdots & \mathbf{H}(\beta+1)\\ \vdots & \vdots & & \vdots\\ \mathbf{H}(\alpha+1) & \mathbf{H}(\alpha+2) & \cdots & \mathbf{H}(\alpha+\beta)\end{bmatrix}$$

It is a $(\alpha+1)\times\beta p$ matrix. Now we search the linearly independent rows of $\mathbf{T}$ in order from top to bottom. It is assumed that the first v rows are linearly independent and the $(v+1)$th row is linearly dependent on its previous rows. Then there exist $a(i)$, $i=1, 2, \ldots, v$, such that

$$[a(1)\quad a(2)\quad \cdots\quad a(v)\quad 1\quad 0\quad \cdots\quad 0]\mathbf{T}=\mathbf{0}$$

It is claimed that the dynamical equation

$$\dot{\mathbf{x}}=\begin{bmatrix}0 & 1 & 0 & \cdots & 0\\ 0 & 0 & 1 & \cdots & 0\\ \vdots & \vdots & \vdots & & \vdots\\ 0 & 0 & 0 & & 1\\ -a(1) & -a(2) & -a(3) & \cdots & -a(v)\end{bmatrix}\mathbf{x}+\begin{bmatrix}\mathbf{H}(1)\\ \mathbf{H}(2)\\ \vdots\\ \mathbf{H}(v-1)\\ \mathbf{H}(v)\end{bmatrix}\mathbf{u} \tag{6-68}$$
$$y=[\,1\quad 0\quad 0\quad \cdots\quad 0\,]\mathbf{x}+\mathbf{H}(0)\mathbf{u}$$

is an irreducible realization of the $1\times p$ proper rational transfer matrix. The proof of this statement is similar to the column case and is left as an exercise.

*6-5 Irreducible Realizations of Proper Rational Matrices: Hankel Methods

There are many approaches to find irreducible realizations for $q\times p$ proper rational matrices. One approach is to first find a reducible realization and then apply the reduction procedure discussed in Section 5-8 to reduce it to an irreducible one. We discuss this approach first.

Given a $q \times p$ proper rational matrix $\hat{\mathbf{G}}(s) = (\hat{g}_{ij}(s))$, if we first find an irreducible realization for every element $\hat{g}_{ij}(s)$ of $\hat{\mathbf{G}}(s)$, and then combine them together as in (4-61) or Figure 4-7, then the resulting realization is generally not controllable and not observable. To reduce this realization requires the application of the reduction procedure twice. If we find the controllable canonical-form realization for the ith column, $\hat{\mathbf{G}}_i(s)$, of $\hat{\mathbf{G}}(s)$, say

$$\dot{\mathbf{x}}_i = \mathbf{A}_i \mathbf{x}_i + \mathbf{b}_i u_i \qquad \mathbf{y}_i = \mathbf{C}_i \mathbf{x}_i + \mathbf{e}_i u_i$$

where $\mathbf{A}_i$, $\mathbf{b}_i$, $\mathbf{C}_i$, and $\mathbf{e}_i$ are of the form shown in (6-53), u_i is the ith component of $\mathbf{u}$ and $\mathbf{y}_i$ is the $q \times 1$ output vector due to the input u_i, then the composite dynamical equation

$$\begin{bmatrix} \dot{\mathbf{x}}_1 \\ \dot{\mathbf{x}}_2 \\ \vdots \\ \dot{\mathbf{x}}_p \end{bmatrix} = \begin{bmatrix} \mathbf{A}_1 & \mathbf{0} & \cdots & \mathbf{0} \\ \mathbf{0} & \mathbf{A}_2 & \cdots & \mathbf{0} \\ \vdots & \vdots & & \vdots \\ \mathbf{0} & \mathbf{0} & \cdots & \mathbf{A}_p \end{bmatrix} \begin{bmatrix} \mathbf{x}_1 \\ \mathbf{x}_2 \\ \vdots \\ \mathbf{x}_p \end{bmatrix} + \begin{bmatrix} \mathbf{b}_1 & \mathbf{0} & \cdots & \mathbf{0} \\ \mathbf{0} & \mathbf{b}_2 & \cdots & \mathbf{0} \\ \vdots & \vdots & & \vdots \\ \mathbf{0} & \mathbf{0} & \cdots & \mathbf{b}_p \end{bmatrix} \begin{bmatrix} u_1 \\ u_2 \\ \vdots \\ u_p \end{bmatrix}$$

$$\mathbf{y} = [\mathbf{C}_1 \quad \mathbf{C}_2 \quad \cdots \quad \mathbf{C}_p]\ \mathbf{x} + [\mathbf{e}_1 \quad \mathbf{e}_2 \quad \cdots \quad \mathbf{e}_p]\ \mathbf{u}$$

is a realization of $\hat{\mathbf{G}}(s)$. (Prove.) Because of the structure of $\mathbf{A}_i$, $\mathbf{b}_i$ it can be readily verified that the realization is always controllable. It is however generally not observable. To reduce the realization to an irreducible one requires the application of the reduction procedure only once.

It is possible to obtain different controllable realizations of a proper $\hat{\mathbf{G}}(s)$. Let $\hat{\mathbf{G}}(s) = \hat{\mathbf{G}}(s) + \hat{\mathbf{G}}(\infty)$, where $\hat{\mathbf{G}}(s)$ is strictly proper. Let $\psi(s)$ be the monic least common denominator of $\hat{\mathbf{G}}(s)$ and of the form

$$\psi(s) = s^m + \alpha_1 s^{m-1} + \alpha_2 s^{m-2} + \cdots + \alpha_m \tag{6-69}$$

Then we can write $\hat{\mathbf{G}}(s)$ as

$$\hat{\mathbf{G}}(s) = \frac{1}{\psi(s)} [\mathbf{R}_1 s^{m-1} + \mathbf{R}_2 s^{m-2} + \cdots + \mathbf{R}_m] \tag{6-70}$$

where $\mathbf{R}_i$ are $q \times p$ constant matrices. Let $\mathbf{I}_p$ be the $p \times p$ unit matrix and $\mathbf{0}_p$ be the $p \times p$ zero matrix. Then the dynamical equation

$$\dot{\mathbf{x}} = \begin{bmatrix} \mathbf{0}_p & \mathbf{I}_p & \mathbf{0}_p & \cdots & \mathbf{0}_p \\ \mathbf{0}_p & \mathbf{0}_p & \mathbf{I}_p & \cdots & \mathbf{0}_p \\ \vdots & \vdots & \vdots & & \vdots \\ \mathbf{0}_p & \mathbf{0}_p & \mathbf{0}_p & \cdots & \mathbf{I}_p \\ -\alpha_m \mathbf{I}_p & -\alpha_{m-1}\mathbf{I}_p & -\alpha_{m-2}\mathbf{I}_p & \cdots & -\alpha_1 \mathbf{I}_p \end{bmatrix} \mathbf{x} + \begin{bmatrix} \mathbf{0}_p \\ \mathbf{0}_p \\ \vdots \\ \mathbf{0}_p \\ \mathbf{I}_p \end{bmatrix} \mathbf{u} \tag{6-71a}$$

$$\mathbf{y} = [\mathbf{R}_m \quad \mathbf{R}_{m-1} \quad \mathbf{R}_{m-2} \quad \cdots \quad \mathbf{R}_1]\mathbf{x} + \hat{\mathbf{G}}(\infty)\mathbf{u} \tag{6-71b}$$

is a realization of $\hat{\mathbf{G}}(s)$. To show this, it is sufficient to show $\mathbf{C}(s\mathbf{I} - \mathbf{A})^{-1}\mathbf{B} = \hat{\mathbf{G}}(s)$. Define $\mathbf{V}(s) \triangleq (s\mathbf{I} - \mathbf{A})^{-1}\mathbf{B}$ or $(s\mathbf{I} - \mathbf{A})\mathbf{V}(s) = \mathbf{B}$. $\mathbf{V}(s)$ is a $mp \times p$ matrix. If we partition it as $\mathbf{V}'(s) = [\mathbf{V}'_1(s) \quad \mathbf{V}'_2(s) \quad \cdots \quad \mathbf{V}'_m(s)]$, where the prime denotes the transpose and $\mathbf{V}_i(s)$ is a $p \times p$ matrix, then $(s\mathbf{I} - \mathbf{A})\mathbf{V}(s) = \mathbf{B}$ or $s\mathbf{V}(s) = \mathbf{A}\mathbf{V}(s)$

$+\mathbf{B}$ implies

$$\begin{aligned} s\mathbf{V}_1(s) &= \mathbf{V}_2(s) \\ s\mathbf{V}_2(s) &= \mathbf{V}_3(s) = s^2\mathbf{V}_1(s) \\ &\cdots\cdots\cdots \\ s\mathbf{V}_{m-1}(s) &= \mathbf{V}_m(s) = s^{m-1}\mathbf{V}_1(s) \end{aligned}$$

and

$$s\mathbf{V}_m(s) = -\alpha_m\mathbf{V}_1(s) - \alpha_{m-1}\mathbf{V}_2(s) - \cdots - \alpha_1\mathbf{V}_m(s) + \mathbf{I}_p$$

These equations imply

$$(s^m + \alpha_1 s^{m-1} + \cdots + \alpha_m)\mathbf{V}_1(s) = \psi(s)\mathbf{V}_1(s) = \mathbf{I}_p$$

and

$$\mathbf{V}_i(s) = \frac{s^{i-1}\mathbf{I}_p}{\psi(s)} \qquad i = 1, 2, \ldots, m \tag{6-72}$$

Consider

$$\mathbf{C}(s\mathbf{I} - \mathbf{A})^{-1}\mathbf{B} = \mathbf{C}\mathbf{V}(s) = \mathbf{R}_m\mathbf{V}_1(s) + \mathbf{R}_{m-1}\mathbf{V}_2(s) + \cdots + \mathbf{R}_1\mathbf{V}_m(s)$$

which becomes, after the substitution of (6-72),

$$\mathbf{C}(s\mathbf{I} - \mathbf{A})^{-1}\mathbf{B} = \frac{\mathbf{R}_m + s\mathbf{R}_{m-1} + \cdots + s^{m-1}\mathbf{R}_1}{\psi(s)} = \hat{\mathbf{G}}(s)$$

This shows that Equation (6-71) is a realization of $\hat{\mathbf{G}}(s)$. Because of the forms of $\mathbf{A}$ and $\mathbf{B}$, it is easy to verify that the realization is controllable. It is, however, generally not observable.

We note that the realization in (6-71) is a generalization of the controllable-form realization for a scalar transfer function shown in (6-26). It is also possible to find a generalization of the one in (6-38). Let

$$\hat{\mathbf{G}}(s) = \mathbf{H}(0) + \mathbf{H}(1)s^{-1} + \mathbf{H}(2)s^{-2} + \cdots \tag{6-73}$$

where $\mathbf{H}(i)$ are $q \times p$ constant matrices. Let $\psi(s)$ be the monic least common denominator of $\hat{\mathbf{G}}(s)$ and of the form shown in (6-69). Then similar to (6-30), we can show

$$\mathbf{H}(m+i) = -\alpha_1\mathbf{H}(m+i-1) - \alpha_2\mathbf{H}(m+i-2) - \cdots - \alpha_m\mathbf{H}(i), \qquad i = 1, 2, \ldots \tag{6-74}$$

This is a key equation in the following development.

Let $\{\mathbf{A}, \mathbf{B}, \mathbf{C}, \mathbf{E}\}$ be a realization of $\hat{\mathbf{G}}(s)$ in (6-73). Then we have, similar to (6-33),

$$\hat{\mathbf{G}}(s) = \mathbf{E} + \mathbf{C}(s\mathbf{I} - \mathbf{A})^{-1}\mathbf{B} = \mathbf{E} + \mathbf{C}\mathbf{B}s^{-1} + \mathbf{C}\mathbf{A}\mathbf{B}s^{-2} + \mathbf{C}\mathbf{A}^2\mathbf{B}s^{-3} + \cdots \tag{6-75}$$

From (6-73) and (6-75), we may conclude, similar to (6-34), that $\{\mathbf{A}, \mathbf{B}, \mathbf{C}, \mathbf{E}\}$ is a realization of $\hat{\mathbf{G}}(s)$ in (6-73) if and only if $\mathbf{E} = \mathbf{H}(0)$ and

$$\mathbf{H}(i+1) = \mathbf{C}\mathbf{A}^i\mathbf{B}, \qquad i = 0, 1, 2, \ldots \tag{6-76}$$

Now we claim that the dynamical equation

$$\dot{\mathbf{x}} = \begin{bmatrix} \mathbf{0}_q & \mathbf{I}_q & \mathbf{0}_q & \cdots & \mathbf{0}_q \\ \mathbf{0}_q & \mathbf{0} & \mathbf{I}_q & \cdots & \mathbf{0}_q \\ \vdots & \vdots & \vdots & & \vdots \\ \mathbf{0}_q & \mathbf{0}_q & \mathbf{0}_q & \cdots & \mathbf{I}_q \\ -\alpha_m \mathbf{I}_q & -\alpha_{m-1}\mathbf{I}_q & -\alpha_{m-2}\mathbf{I}_q & \cdots & -\alpha_1 \mathbf{I}_q \end{bmatrix} \mathbf{x} + \begin{bmatrix} \mathbf{H}(1) \\ \mathbf{H}(2) \\ \vdots \\ \mathbf{H}(m-1) \\ \mathbf{H}(m) \end{bmatrix} \mathbf{u} \qquad \text{(6-77a)}$$

$$\mathbf{y} = [\ \mathbf{I}_q \quad \mathbf{0} \quad \mathbf{0} \quad \cdots \quad \mathbf{0}\ ]\ \mathbf{x} + \mathbf{H}(0)\mathbf{u} \qquad \text{(6-77b)}$$

is a qm-dimensional realization of $\hat{\mathbf{G}}(s)$. Indeed, using (6-74), we can readily verify that

$$\mathbf{AB} = \begin{bmatrix} \mathbf{H}(2) \\ \mathbf{H}(3) \\ \vdots \\ \mathbf{H}(m+1) \end{bmatrix}, \quad \mathbf{A}^2\mathbf{B} = \begin{bmatrix} \mathbf{H}(3) \\ \mathbf{H}(4) \\ \vdots \\ \mathbf{H}(m+2) \end{bmatrix}, \quad \ldots, \quad \mathbf{A}^i\mathbf{B} = \begin{bmatrix} \mathbf{H}(i+1) \\ \mathbf{H}(i+2) \\ \vdots \\ \mathbf{H}(i+m) \end{bmatrix} \qquad \text{(6-78)}$$

Consequently we have $\mathbf{CA}^i\mathbf{B} = \mathbf{H}(i+1)$. This establishes the assertion. The observability matrix of (6-77) is the unit matrix of order qm; hence (6-77) is always observable. It is, however, generally not controllable.

It is possible to introduce other either controllable or observable realizations. Their introduction however would not introduce any new concept and will be skipped. Instead, we shall discuss in the following two methods which will yield directly irreducible realizations. Both methods are based on the Hankel matrices.

Method I. Singular value decomposition. Consider the $q \times p$ proper rational matrix $\hat{\mathbf{G}}(s)$ given in (6-73). Let $\psi(s) = s^m + \alpha_1 s^{m-1} + \cdots + \alpha_m$ be the least common denominator of all elements of $\hat{\mathbf{G}}(s)$. Define the $qm \times qm$ and $pm \times pm$ matrices

$$\mathbf{M} \triangleq \begin{bmatrix} \mathbf{0} & \mathbf{I}_q & \mathbf{0} & \cdots & \mathbf{0} \\ \mathbf{0} & \mathbf{0} & \mathbf{I}_q & \cdots & \mathbf{0} \\ \vdots & \vdots & \vdots & & \vdots \\ \mathbf{0} & \mathbf{0} & \mathbf{0} & \cdots & \mathbf{I}_q \\ -\alpha_m \mathbf{I}_q & -\alpha_{m-1}\mathbf{I}_q & -\alpha_{m-2}\mathbf{I}_q & \cdots & -\alpha_1 \mathbf{I}_q \end{bmatrix} \qquad \text{(6-79a)}$$

$$\mathbf{N} \triangleq \begin{bmatrix} \mathbf{0} & \mathbf{0} & \cdots & \mathbf{0} & -\alpha_m \mathbf{I}_p \\ \mathbf{I}_p & \mathbf{0} & \cdots & \mathbf{0} & -\alpha_{m-1}\mathbf{I}_p \\ \mathbf{0} & \mathbf{I}_p & \cdots & \mathbf{0} & -\alpha_{m-2}\mathbf{I}_p \\ \vdots & \vdots & & \vdots & \vdots \\ \mathbf{0} & \mathbf{0} & \cdots & \mathbf{I}_p & -\alpha_1 \mathbf{I}_p \end{bmatrix} \qquad \text{(6-79b)}$$

where $\mathbf{I}_n$ denotes the $n \times n$ unit matrix. We also define the following two

Hankel matrices

$$\mathbf{T} \triangleq \begin{bmatrix} \mathbf{H}(1) & \mathbf{H}(2) & \cdots & \mathbf{H}(m) \\ \mathbf{H}(2) & \mathbf{H}(3) & \cdots & \mathbf{H}(m+1) \\ \vdots & \vdots & & \vdots \\ \mathbf{H}(m) & \mathbf{H}(m+1) & \cdots & \mathbf{H}(2m-1) \end{bmatrix} \tag{6-80}$$

and

$$\tilde{\mathbf{T}} \triangleq \begin{bmatrix} \mathbf{H}(2) & \mathbf{H}(3) & \cdots & \mathbf{H}(m+1) \\ \mathbf{H}(3) & \mathbf{H}(4) & \cdots & \mathbf{H}(m+2) \\ \vdots & \vdots & & \vdots \\ \mathbf{H}(m+1) & \mathbf{H}(m+2) & \cdots & \mathbf{H}(2m) \end{bmatrix} \tag{6-81}$$

Since $\mathbf{T}$ and $\tilde{\mathbf{T}}$ consist of m block rows and m block columns, and since $\mathbf{H}(i)$ are $q \times p$ matrices, $\mathbf{T}$ and $\tilde{\mathbf{T}}$ are of order $qm \times pm$. Using (6-74), it can be readily verified that

$$\tilde{\mathbf{T}} = \mathbf{MT} = \mathbf{TN} \tag{6-82}$$

and, in general,

$$\mathbf{M}^i\mathbf{T} = \mathbf{TN}^i \qquad i = 0, 1, 2, \ldots \tag{6-83}$$

Note that the left-upper-corner element of $\mathbf{M}^i\mathbf{T} = \mathbf{TN}^i$ is $\mathbf{H}(i+1)$. Let $\mathbf{I}_{k,l}$ be a $k \times l\,(l > k)$ constant matrix of the form $\mathbf{I}_{k,l} = [\mathbf{I}_k \quad \mathbf{0}]$, where $\mathbf{I}_k$ is the unit matrix of order k, and $\mathbf{0}$ is the $k \times (l-k)$ zero matrix. Then the corner element $\mathbf{H}(i+1)$ can be removed from $\mathbf{M}^i\mathbf{T} = \mathbf{TN}^i$ as

$$\mathbf{H}(i+1) = \mathbf{I}_{q,qm}\mathbf{M}^i\mathbf{T}\mathbf{I}'_{p,pm} = \mathbf{I}_{q,qm}\mathbf{T}\mathbf{N}^i\mathbf{I}'_{p,pm} \qquad i = 0, 1, 2, \ldots \tag{6-84}$$

where the prime denotes the transpose. From this equation and (6-76), we conclude that $\{\mathbf{A} = \mathbf{M},\ \mathbf{B} = \mathbf{T}\mathbf{I}'_{p,pm},\ \mathbf{C} = \mathbf{I}_{q,qm},\ \mathbf{E} = \mathbf{H}(0)\}$ is a qm-dimensional realization of $\hat{\mathbf{G}}(s)$ in (6-73). Note that this realization is the one in (6-77) and is observable but not necessarily controllable. Similarly, the dynamical equation

$$\begin{aligned} \mathbf{C} &= \mathbf{I}_{q,qm}\mathbf{T} = [\mathbf{H}(1) \quad \mathbf{H}(2) \quad \cdots \quad \mathbf{H}(m)] \\ \mathbf{A} &= \mathbf{N} \\ \mathbf{B} &= \mathbf{I}'_{p,pm} = [\mathbf{I}_p \quad \mathbf{0} \quad \cdots \quad \mathbf{0}]' \\ \mathbf{E} &= \mathbf{H}(0) \end{aligned} \tag{6-85}$$

is a pm-dimensional realization of $\hat{\mathbf{G}}(s)$. The controllability matrix of this realization is a unit matrix; hence the realization is always controllable. The realization however is generally not observable.

Now we shall use Theorem E-5, the singular value decomposition, to find an irreducible realization directly from $\mathbf{T}$ and $\tilde{\mathbf{T}}$. Theorem E-5 implies the existence of $qm \times qm$ and $pm \times pm$ unitary matrices $\mathbf{K}$ and $\mathbf{L}$ such that

$$\mathbf{T} = \mathbf{K}\begin{bmatrix} \mathbf{\Sigma} & \mathbf{0} \\ \mathbf{0} & \mathbf{0} \end{bmatrix}\mathbf{L} \tag{6-86}$$

where $\mathbf{\Sigma} = \text{diag}\,\{\lambda_1, \lambda_2, \ldots, \lambda_n\}$ and λ_i, $i = 1, 2, \ldots, n$, are the positive square roots of the eigenvalues of $\mathbf{T}^*\mathbf{T}$, where the asterisk stands for the complex conjugate transpose. Clearly n is the rank of $\mathbf{T}$. Let $\mathbf{K}_1$ denote the first n columns of $\mathbf{K}$ and $\mathbf{L}_1$ denote the first n rows of $\mathbf{L}$. Then we can write $\mathbf{T}$ as

$$\mathbf{T} = \mathbf{K}_1 \mathbf{\Sigma} \mathbf{L}_1 = \mathbf{K}_1 \mathbf{\Sigma}^{1/2} \mathbf{\Sigma}^{1/2} \mathbf{L}_1 \triangleq \mathbf{V}\mathbf{U} \tag{6-87}$$

where $\mathbf{\Sigma}^{1/2} = \text{diag}\{\sqrt{\lambda_1}, \sqrt{\lambda_2}, \ldots, \sqrt{\lambda_n}\}$, $\mathbf{V} \triangleq \mathbf{K}_1 \mathbf{\Sigma}^{1/2}$ is a $qm \times n$ matrix, and $\mathbf{U} = \mathbf{\Sigma}^{1/2}\mathbf{L}_1$ is an $n \times pm$ matrix. Define[5]

$$\mathbf{V}^+ = \mathbf{\Sigma}^{-1/2}\mathbf{K}_1^* \qquad \text{and} \qquad \mathbf{U}^+ = \mathbf{L}_1^* \mathbf{\Sigma}^{-1/2} \tag{6-88}$$

Because of $\mathbf{K}_1^* \mathbf{K}_1 = \mathbf{I}_n$, $\mathbf{L}_1 \mathbf{L}_1^* = \mathbf{I}_n$ (see Problem E-7), we have

$$\mathbf{V}^+ \mathbf{V} = \mathbf{\Sigma}^{-1/2} \mathbf{K}_1^* \mathbf{K}_1 \mathbf{\Sigma}^{1/2} = \mathbf{I}_n \tag{6-89}$$

and

$$\mathbf{U}\mathbf{U}^+ = \mathbf{\Sigma}^{1/2} \mathbf{L}_1 \mathbf{L}_1^* \mathbf{\Sigma}^{-1/2} = \mathbf{I}_n \tag{6-90}$$

With these preliminaries we can now establish the following theorem.

Theorem 6-4

Consider a $q \times p$ proper rational matrix $\hat{\mathbf{G}}(s)$ expanded as $\hat{\mathbf{G}}(s) = \sum_{i=0}^{\infty} \mathbf{H}(i)s^{-i}$. We form $\mathbf{T}$ and $\tilde{\mathbf{T}}$ as in (6-80) and (6-81), and factor $\mathbf{T}$ as $\mathbf{T} = \mathbf{V}\mathbf{U}$, and $\rho\mathbf{T} = \rho\mathbf{V} = \rho\mathbf{U}$, where ρ denotes the rank, by using the singular value decomposition. Then the $\{\mathbf{A}, \mathbf{B}, \mathbf{C}, \mathbf{E}\}$ defined by

$$\mathbf{A} = \mathbf{V}^+ \tilde{\mathbf{T}} \mathbf{U}^+ \tag{6-91a}$$
$$\mathbf{B} = \text{first } p \text{ columns of } \mathbf{U} = \mathbf{U}\mathbf{I}'_{p,pm} \tag{6-91b}$$
$$\mathbf{C} = \text{first } q \text{ rows of } \mathbf{V} = \mathbf{I}_{q,qm}\mathbf{V} \tag{6-91c}$$
$$\mathbf{E} = \mathbf{H}(0)$$

is an irreducible realization of $\hat{\mathbf{G}}(s)$. ∎

We give first a justification of (6-91) before giving a formal proof. We see from (5-65) that $\mathbf{V}$ and $\mathbf{U}$ are, respectively, the observability matrix and the controllability matrix of $\{\mathbf{A}, \mathbf{B}, \mathbf{C}\}$. Consequently, we have (6-91b) and (6-91c). Again from (5-65), we can easily verify that $\tilde{\mathbf{T}} = \mathbf{V}\mathbf{A}\mathbf{U}$. The pre- and postmultiplication of $\mathbf{V}^+$ and $\mathbf{U}^+$ to $\tilde{\mathbf{T}} = \mathbf{V}\mathbf{A}\mathbf{U}$ and the use of (6-89) and (6-90) yield (6-91a). Since the dimension of $\mathbf{A}$ is equal to the rank of $\mathbf{T}$, the irreducibility of (6-91a) follows from $\rho\mathbf{V} = \rho\mathbf{U} = \dim \mathbf{A}$.

Now we give a formal proof of the theorem.

Proof of Theorem 6-4

Define

$$\mathbf{T}^+ \triangleq \mathbf{U}^+ \mathbf{V}^+ \tag{6-92}$$

Then we have, by using (6-87), (6-89), and (6-90),

$$\mathbf{T}\mathbf{T}^+\mathbf{T} = \mathbf{V}\mathbf{U}\mathbf{U}^+\mathbf{V}^+\mathbf{V}\mathbf{U} = \mathbf{V}\mathbf{U} = \mathbf{T} \tag{6-93}$$

---

[5] $\mathbf{V}^+$ is called the *pseudoinverse* of $\mathbf{V}$. See Reference 116.

Consider

$$\begin{aligned}
\mathbf{A}^2 &= (\mathbf{V}^+\tilde{\mathbf{T}}\mathbf{U}^+)^2 = (\mathbf{V}^+\mathbf{M}\mathbf{T}\mathbf{U}^+)^2 && \text{[using (6-82)]}\\
&= (\mathbf{V}^+\mathbf{M}\mathbf{T}\mathbf{U}^+)(\mathbf{V}^+\mathbf{M}\mathbf{T}\mathbf{U}^+)\\
&= \mathbf{V}^+\mathbf{M}\mathbf{T}\mathbf{T}^+\mathbf{T}\mathbf{N}\mathbf{U}^+ && \text{[using (6-92) and (6-83)]}\\
&= \mathbf{V}^+\mathbf{M}\mathbf{T}\mathbf{N}\mathbf{U}^+ && \text{[using (6-93)]}\\
&= \mathbf{V}^+\mathbf{M}\mathbf{M}\mathbf{T}\mathbf{U}^+ && \text{[using (6-83)]}\\
&= \mathbf{V}^+\mathbf{M}^2\mathbf{T}\mathbf{U}^+
\end{aligned}$$

Repeating the process we can show

$$\mathbf{A}^i = \mathbf{V}^+\mathbf{M}^i\mathbf{T}\mathbf{U}^+ \qquad i = 1, 2, 3, \ldots$$

Consider (6-84):

$$\begin{aligned}
\mathbf{H}(i+1) &= \mathbf{I}_{q,qm}\mathbf{M}^i\mathbf{T}\mathbf{I}'_{p,pm}\\
&= \mathbf{I}_{q,qm}\mathbf{M}^i\mathbf{T}\mathbf{T}^+\mathbf{T}\mathbf{I}'_{p,pm} && \text{[using (6-93)]}\\
&= \mathbf{I}_{q,qm}\mathbf{T}\mathbf{N}^i\mathbf{T}^+\mathbf{T}\mathbf{I}'_{p,pm} && \text{[using (6-83)]}\\
&= \mathbf{I}_{q,qm}\mathbf{T}\mathbf{T}^+\mathbf{T}\mathbf{N}^i\mathbf{T}^+\mathbf{T}\mathbf{I}'_{p,pm} && \text{[using (6-93)]}\\
&= \underbrace{\mathbf{I}_{q,qm}\mathbf{V}}_{\mathbf{C}}\underbrace{\mathbf{U}\mathbf{U}^+}_{\mathbf{I}_n}\underbrace{\mathbf{V}^+\mathbf{M}^i\mathbf{T}\mathbf{U}^+}_{\mathbf{A}^i}\underbrace{\mathbf{V}^+\mathbf{V}}_{\mathbf{I}_n}\underbrace{\mathbf{U}\mathbf{I}'_{p,pm}}_{\mathbf{B}} && \text{[using (6-87), (6-92), and (6-83)]}\\
&= \mathbf{C}\mathbf{A}^i\mathbf{B}, \qquad i = 0, 1, 2, \ldots
\end{aligned}$$

This shows that (6-91) is a realization of $\hat{\mathbf{G}}(s)$.

The dimension of $\mathbf{A}$ is equal to n, the rank of $\mathbf{T}$. Let $\bar{\mathbf{U}} = [\mathbf{B} \quad \mathbf{AB} \quad \cdots \quad \mathbf{A}^{m-1}\mathbf{B}]$ and $\bar{\mathbf{V}}' = [\mathbf{C}' \quad \mathbf{A}'\mathbf{C}' \quad \cdots \quad (\mathbf{A}')^{m-1}\mathbf{C}']$. Then from (5-65), we have $\mathbf{T} = \bar{\mathbf{V}}\bar{\mathbf{U}}$ and $n = \mathbf{T} \leq \min(\rho\bar{\mathbf{V}}, \rho\bar{\mathbf{U}})$. Since $\bar{\mathbf{U}}$ is a $n \times mp$ matrix and $\bar{\mathbf{V}}$ is a $mq \times n$ matrix, we conclude $\rho\bar{\mathbf{U}} = n$ and $\rho\bar{\mathbf{V}} = n$. Hence the realization in (6-91) is controllable and observable. This completes the proof of the theorem. Q.E.D.

The crux of this realization is the singular value decomposition of $\mathbf{T}$. Numerically stable computer programs are available for this computation. Hence this realization can be readily carried out on a digital computer.

The decomposition of $\mathbf{T} = \mathbf{VU}$ in (6-87) is not unique. For example, we may choose $\mathbf{V} = \mathbf{K}_1$ and $\mathbf{U} = \mathbf{\Sigma}\mathbf{L}_1$, and the subsequent derivation still applies. The $\mathbf{V}$ and $\mathbf{U}$ defined in (6-87), however, have the property

$$\mathbf{V}^*\mathbf{V} = \mathbf{\Sigma}^{1/2}\mathbf{K}_1^*\mathbf{K}_1\mathbf{\Sigma}^{1/2} = \mathbf{\Sigma} = \mathbf{\Sigma}^{1/2}\mathbf{L}_1\mathbf{L}_1^*\mathbf{\Sigma}^{1/2} = \mathbf{U}\mathbf{U}^*$$

A realization whose controllability and observability matrices having the property $\mathbf{V}^*\mathbf{V} = \mathbf{U}\mathbf{U}^*$ is called an *internally balanced realization.* Roughly speaking, the signal transfer effect from the input to the state and that from the state to the output are similar or balanced in an internally balanced realization (see Problems 6-22 and 6-23).

In the application, it is often desirable to find a simplified or reduced-order model for a given system. This realization procedure can be directly applied to this problem. For example, consider

$$\mathbf{T} = \mathbf{K}\begin{bmatrix} \mathbf{\Sigma} & \mathbf{0} \\ \mathbf{0} & \mathbf{0} \end{bmatrix}\mathbf{L}$$

where $\mathbf{\Sigma} = \text{diag}\{\lambda_1, \lambda_2, \ldots \lambda_n\}$ and $\lambda_1 \geq \lambda_2 \geq \cdots \geq \lambda_n$. If the system is to be approximated by an m-dimensional model with $m < n$, then the reduced model can be obtained from

$$\mathbf{T} = \mathbf{K}_1 \text{ diag}\{\lambda_1, \lambda_2, \ldots, \lambda_m\}\mathbf{L}_1$$

where $\mathbf{K}_1$ is the first m columns of $\mathbf{K}$ and $\mathbf{L}_1$ is the first m rows of $\mathbf{L}$. See References S141 and S162.

To conclude this subsection, we remark that the dimension of the irreducible realization is equal to the rank of $\mathbf{T}$. Because of Theorem 6-2, we have deg $\hat{\mathbf{G}}(s) = \text{rank } \mathbf{T}$. Consequently, the degree of $\hat{\mathbf{G}}(s)$ can also be computed from $\mathbf{T}$. If the degree of $\hat{\mathbf{G}}$ is known a priori, then the m in (6-80) can be chosen to be the least integer such that the rank of $\mathbf{T}$ is n. Because of $\mathbf{T} = \mathbf{VU}$, from the definitions of controllability index μ and the observability index v, we may conclude that the least integer to have $\rho\mathbf{T} = n$ is equal to

$$\max\{\mu, v\}$$

Note that μ and v can also be obtained directly from the column and row degrees of coprime fractions of $\hat{\mathbf{G}}(s)$. See Theorem 6-6 on page 284.

Method II. Row Searching method. In this subsection we shall search the linear independent rows of a Hankel matrix of $\hat{\mathbf{G}}(s)$ to find an irreducible realization. This method is a direct extension of the vector case discussed in Section 6-4 and the reader is advised to review the section before proceeding.

Let $\hat{\mathbf{G}}(s)$ be a $q \times p$ proper rational matrix. Let α_i be the degree of the least common denominator of the ith row of $\hat{\mathbf{G}}(s)$ and β_j be the degree of the least common denominator of the jth column of $\hat{\mathbf{G}}(s)$. If their computations are complicated, they may simply be chosen as the sums of the denominator degrees of the ith row and jth column of $\hat{\mathbf{G}}(s)$. Define $\alpha = \max\{\alpha_i, i = 1, 2, \ldots, q\}$ and $\beta = \max\{\beta_j, j = 1, 2, \ldots, p\}$. Similar to (6-59), we may form the Hankel matrix

$$\mathbf{T} = \begin{bmatrix} \mathbf{H}_{11}(\alpha_1+1, \beta_1) & \mathbf{H}_{12}(\alpha_1+1, \beta_2) & \cdots & \mathbf{H}_{1p}(\alpha_1+1, \beta_p) \\ \mathbf{H}_{21}(\alpha_2+1, \beta_1) & \mathbf{H}_{22}(\alpha_2+1, \beta_2) & \cdots & \mathbf{H}_{2p}(\alpha_2+1, \beta_p) \\ \vdots & \vdots & & \vdots \\ \mathbf{H}_{q1}(\alpha_q+1, \beta_1) & \mathbf{H}_{q2}(\alpha_q+1, \beta_2) & \cdots & \mathbf{H}_{qp}(\alpha_q+1, \beta_p) \end{bmatrix} \begin{matrix} \} \sigma_1 \text{ (no. of} \\ \} \sigma_2 \text{ linearly} \\ \text{independent} \\ \} \sigma_p \text{ rows)} \end{matrix} \tag{6-94}$$

where $\mathbf{H}_{ij}(\alpha_i+1, \beta_j)$ is the Hankel matrix of $\hat{g}_{ij}(s)$ defined as in (6-35). Now we search the linearly independent rows of $\mathbf{T}$ in order from top to bottom. Then corresponding to the primary linearly dependent rows of $[\mathbf{H}_{i1} \quad \mathbf{H}_{i2} \quad \cdots \quad \mathbf{H}_{ip}]$, $i = 1, 2, \ldots, q$, we may obtain an irreducible realization similar to the one in (6-61). Another way is to form the Hankel matrix

$$\bar{\mathbf{T}} = \begin{bmatrix} \mathbf{H}(1) & \mathbf{H}(2) & \cdots & \mathbf{H}(\beta) \\ \mathbf{H}(2) & \mathbf{H}(3) & \cdots & \mathbf{H}(\beta+1) \\ \vdots & \vdots & & \vdots \\ \mathbf{H}(\alpha+1) & \mathbf{H}(\alpha+2) & \cdots & \mathbf{H}(\alpha+\beta) \end{bmatrix} \tag{6-95}$$

as in (6-64). We then search its linearly independent rows in order from top to bottom. After the completion of the search, we then rearrange the rows of $\mathbf{T}$ into the form

$$\hat{\mathbf{T}} = \begin{bmatrix} \mathbf{H}_{11}(\alpha+1, \beta) & \mathbf{H}_{12}(\alpha+1, \beta) & \cdots & \mathbf{H}_{1p}(\alpha+1, \beta) \\ \mathbf{H}_{21}(\alpha+1, \beta) & \mathbf{H}_{22}(\alpha+1, \beta) & \cdots & \mathbf{H}_{2p}(\alpha+1, \beta) \\ \vdots & \vdots & & \vdots \\ \mathbf{H}_{q1}(\alpha+1, \beta) & \mathbf{H}_{q2}(\alpha+1, \beta) & \cdots & \mathbf{H}_{qp}(\alpha+1, \beta) \end{bmatrix} \begin{matrix} \} \, v_1 \text{ (no. of linearly independent rows)} \\ \\ \\ \} \, v_q \end{matrix} \tag{6-96}$$

as in (6-65). Using the coefficients of linear combinations corresponding to the q primary linearly dependent rows of $\hat{\mathbf{T}}$, an irreducible realization similar to (6-67) can be readily obtained. Similar to (6-65), the v_i in (6-96) yield the observability indices of any irreducible realization of $\hat{\mathbf{G}}(s)$. We use an example to illustrate the procedures.

Example 1

Consider the proper rational matrix

$$\hat{\mathbf{G}}(s) = \begin{bmatrix} \dfrac{-2s^2-3s-2}{(s+1)^2} & \dfrac{1}{s^2} \\ \dfrac{4s+5}{s+1} & \dfrac{-3s-5}{s+2} \end{bmatrix} \tag{6-97}$$

We compute the Markov parameters:

$$\frac{-2s^2-3s-2}{(s+1)^2} = -2 + s^{-1} - 2s^{-2} + 3s^{-3} - 4s^{-4} + 5s^{-5} - 6s^{-6} + \cdots$$

$$\frac{1}{s^2} = 0 + 0s^{-1} + s^{-2} + 0s^{-3} + \cdots$$

$$\frac{4s+5}{s+1} = 4 + s^{-1} - s^{-2} + s^{-3} - s^{-4} + \cdots$$

and

$$\frac{-3s-5}{s+2} = -3 + s^{-1} - 2s^{-2} + 4s^{-3} - 8s^{-4} + 16s^{-5} - \cdots$$

It is clear that $\alpha_1 = 4$, $\alpha_2 = 2$, $\beta_1 = 2$ and $\beta_2 = 3$. We form

$$\mathbf{T} = \begin{bmatrix} \mathbf{H}_{11}(\alpha_1+1, \beta_1) & \mathbf{H}_{12}(\alpha_1+1, \beta_2) \\ \mathbf{H}_{21}(\alpha_2+1, \beta_1) & \mathbf{H}_{22}(\alpha_2+1, \beta_2) \end{bmatrix}$$

$$= \left[\begin{array}{rr|rrr} 1 & -2 & 0 & 1 & 0 \\ -2 & 3 & 1 & 0 & 0 \\ 3 & -4 & 0 & 0 & 0 \\ -4 & 5 & 0 & 0 & 0 \\ 5 & -6 & 0 & 0 & 0 \\ \hline 1 & -1 & 1 & -2 & 4 \\ -1 & 1 & -2 & 4 & -8 \\ 1 & -1 & 4 & -8 & 16 \end{array}\right]$$

We note that $h_{ij}(0)$ are not used in forming **T**. The application of the row-searching algorithm to **T** yields

$$\begin{bmatrix} 1 & & & & & & & \\ 0 & 1 & & & & & & \\ 0 & 0 & 1 & & & & & \\ 0 & 0 & \frac{4}{3} & 1 & & & & \\ 0 & 0 & -\frac{5}{3} & 2 & 1 & & & \\ \hdashline 2 & -1 & -\frac{5}{3} & -4 & 0 & 1 & & \\ -4 & 2 & 3 & 9 & 0 & 2 & 1 & 0 \\ \hdashline 8 & -4 & -\frac{17}{3} & -19 & 0 & -4 & 0 & 1 \end{bmatrix} \mathbf{T} = \begin{bmatrix} 1 & -2 & 0 & 1 & 0 \\ -2 & 3 & 1 & 0 & 0 \\ 3 & -4 & 0 & 0 & 0 \\ 0 & -\frac{1}{3} & 0 & 0 & 0 \\ 0 & 0 & 0 & 0 & 0 \\ \hdashline 0 & 0 & 0 & 0 & 4 \\ 0 & 0 & 0 & 0 & 0 \\ 0 & 0 & 0 & 0 & 0 \end{bmatrix} \begin{matrix} \left.\vphantom{\begin{matrix}1\\1\\1\\1\end{matrix}}\right\} \sigma_1 = 4 \\ \\ \}\, \sigma_2 = 1 \\ \\ \\ \end{matrix} \tag{6-98}$$

Clearly we have $\sigma_1 = 4$ and $\sigma_2 = 1$. Hence the dimension of any irreducible realization of $\hat{\mathbf{G}}(s)$ is 5. Corresponding to the fifth and seventh row of the right-hand-side matrix of (6-98), we can compute, by using (F-11), from the leftmost matrix in (6-98):

$$\mathbf{k}_1\mathbf{T} = [0 \quad 0 \quad 1 \quad 2 \quad \boxed{1} \;\vdots\; 0 \quad 0 \quad 0]\mathbf{T} = \mathbf{0}$$

and

$$\mathbf{k}_2\mathbf{T} = [\underbrace{0 \quad 0 \quad 1 \quad 1}_{} \quad 0 \;\vdots\; \underbrace{2}_{} \quad \boxed{1} \quad 0]\mathbf{T} = \mathbf{0}$$

Hence an irreducible realization of $\hat{\mathbf{G}}(s)$ is given by

$$\dot{\mathbf{x}} = \left[\begin{array}{cccc:c} 0 & 1 & 0 & 0 & 0 \\ 0 & 0 & 1 & 0 & 0 \\ 0 & 0 & 0 & 1 & 0 \\ 0 & 0 & -1 & -2 & 0 \\ \hdashline 0 & 0 & -1 & -1 & -2 \end{array}\right] \mathbf{x} + \begin{bmatrix} 1 & 0 \\ -2 & 1 \\ 3 & 0 \\ -4 & 0 \\ \hdashline 1 & 1 \end{bmatrix} \mathbf{u}$$

$$\mathbf{y} = \left[\begin{array}{cccc:c} 1 & 0 & 0 & 0 & 0 \\ 0 & 0 & 0 & 0 & 1 \end{array}\right] \mathbf{x} + \begin{bmatrix} -2 & 0 \\ 4 & -3 \end{bmatrix} \mathbf{u}$$

The σ_1th and $(\sigma_1 + \sigma_2)$th rows of **A** are taken from $\mathbf{k}_1$ and $\mathbf{k}_2$ with the signs reversed. The matrix **B** consists of the first σ_1 Markov parameters of $\hat{g}_{11}(s)$ and $\hat{g}_{12}(s)$ and the first σ_2 Markov parameters of $\hat{g}_{21}(s)$ and $\hat{g}_{22}(s)$. The form of **C** is fixed and depends only on σ_1 and σ_2.

Next we use $\bar{\mathbf{T}}$ and $\hat{\mathbf{T}}$ to find a different irreducible realization. We write

$$\hat{\mathbf{G}}(s)=\begin{bmatrix}-2 & 0\\ 4 & -3\end{bmatrix}+\begin{bmatrix}1 & 0\\ 1 & 1\end{bmatrix}s^{-1}+\begin{bmatrix}-2 & 1\\ -1 & -2\end{bmatrix}s^{-2}$$
$$+\begin{bmatrix}3 & 0\\ 1 & 4\end{bmatrix}s^{-3}+\begin{bmatrix}-4 & 0\\ -1 & -8\end{bmatrix}s^{-4}+\cdots$$

We have $\alpha=\max\{\alpha_1, \alpha_2\}=4$ and $\beta=\max\{\beta_1, \beta_2\}=3$. We form the Hankel matrix $\bar{\mathbf{T}}$ and apply the row-searching algorithm to yield

$$\begin{bmatrix}
1 \\
2 & 1 \\
0 & -1 & 1 \\
-4 & 2 & 0 & 1 \\
0 & 0 & 0 & -3 & 1 \\
8 & -4 & 0 & 3 & 0 & 1 \\
0 & 0 & 0 & 4 & 1 & 0 & 1 \\
-16 & 8 & 0 & -7 & 0 & 0 & 0 & 1 \\
0 & 0 & 0 & -5 & -1 & 0 & 0 & 0 & 1 \\
32 & -16 & 0 & 15 & 0 & 0 & 0 & 0 & 0 & 1
\end{bmatrix}$$

$$**\begin{bmatrix}
1 & 0 & -2 & 1 & 3 & 0\\
1 & 1 & -1 & -2 & 1 & 4\\
\hline
-2 & 1 & 3 & 0 & -4 & 0\\
-1 & -2 & 1 & 4 & -1 & -8\\
\hline
3 & 0 & -4 & 0 & 5 & 0\\
1 & 4 & -1 & -8 & 1 & 16\\
\hline
-4 & 0 & 5 & 0 & -6 & 0\\
-1 & -8 & 1 & 16 & -1 & -32\\
\hline
5 & 0 & -6 & 0 & 7 & 0\\
1 & 16 & -1 & -32 & 1 & 64
\end{bmatrix}=\begin{bmatrix}
1 & 0 & -2 & 1 & 3 & 0\\
3 & 1 & -5 & 0 & 7 & 4\\
\hline
-5 & 0 & 8 & 0 & -11 & -4\\
1 & 0 & -1 & 0 & 1 & 0\\
\hline
0 & 0 & -1 & 0 & 2 & 0\\
0 & 0 & 0 & 0 & 0 & 0\\
\hline
0 & 0 & 0 & 0 & 0 & 0\\
0 & 0 & 0 & 0 & 0 & 0\\
\hline
0 & 0 & 0 & 0 & 0 & 0\\
0 & 0 & 0 & 0 & 0 & 0
\end{bmatrix}\begin{matrix}\\ \\ \\ \\ \\ \leftarrow \\ \leftarrow \\ \\ \\ \\ \end{matrix}$$

Corresponding to the primary linearly dependent rows indicated by the arrows as shown, we can compute, by using (F-11), the coefficients of combinations as

$$\begin{aligned}
\bar{\mathbf{k}}_1&=[0 \quad 2 \;\vdots\; 0 \quad 1 \;\vdots\; 1 \quad 0 \;\vdots\; \textcircled{1} \quad 0 \;\vdots\; 0 \quad 0]\\
\bar{\mathbf{k}}_2&=[0 \quad 2 \;\vdots\; 0 \quad 3 \;\vdots\; 0 \quad \textcircled{1} \;\vdots\; 0 \quad 0 \;\vdots\; 0 \quad 0]
\end{aligned} \tag{6-99}$$

The elements $\textcircled{1}$ correspond to the primary linearly dependent rows. Now we rearrange $\bar{\mathbf{T}}$ into $\hat{\mathbf{T}}$ and rearrange $\bar{k}_i$ accordingly to yield

$$\begin{bmatrix}\hat{\mathbf{k}}_1\\ \hat{\mathbf{k}}_2\end{bmatrix}\hat{\mathbf{T}}\triangleq\begin{bmatrix}0 & 0 & 1 & \textcircled{1} & 0 & \vdots & 2 & 1 & 0 & 0 & 0\\ 0 & 0 & 0 & 0 & 0 & \vdots & 2 & 3 & \textcircled{1} & 0 & 0\end{bmatrix}$$

$$\underbrace{\qquad\qquad}_{v_1=3}\qquad\qquad\underbrace{\qquad}_{v_2=2}$$

$$
\times \left[\begin{array}{rrr|rrr}
1 & -2 & 3 & 0 & 1 & 0 \\
-2 & 3 & -4 & 1 & 0 & 0 \\
3 & -4 & 5 & 0 & 0 & 0 \\
-4 & 5 & -6 & 0 & 0 & 0 \\
5 & -6 & 7 & 0 & 0 & 0 \\
\hline
1 & -1 & 1 & 1 & -2 & 4 \\
-1 & 1 & -1 & -2 & 4 & -8 \\
1 & -1 & 1 & 4 & -8 & 16 \\
-1 & 1 & -1 & -8 & 16 & -32 \\
1 & -1 & 1 & 16 & -32 & 64
\end{array}\right]
\begin{array}{l} \\ \\ \\ \leftarrow \\ \\ \\ \\ \leftarrow \\ \\ \\ \end{array}
= \begin{bmatrix} \mathbf{0} \\ \mathbf{0} \end{bmatrix}
$$

Hence a different irreducible realization of $\hat{\mathbf{G}}(s)$ is

$$
\dot{\mathbf{x}} = \left[\begin{array}{rrr|rr}
0 & 1 & 0 & 0 & 0 \\
0 & 0 & 1 & 0 & 0 \\
0 & 0 & -1 & -2 & -1 \\
\hline
0 & 0 & 0 & 0 & 1 \\
0 & 0 & 0 & -2 & -3
\end{array}\right] \mathbf{x} + \left[\begin{array}{rr}
1 & 0 \\
-2 & 1 \\
3 & 0 \\
\hline
1 & 1 \\
-1 & -2
\end{array}\right] \mathbf{u}
$$

$$
\mathbf{y} = \left[\begin{array}{rrr|rr}
1 & 0 & 0 & 0 & 0 \\
0 & 0 & 0 & 1 & 0
\end{array}\right] \mathbf{x} + \begin{bmatrix} -2 & 0 \\ 4 & -3 \end{bmatrix} \mathbf{u}
$$

The v_1th and $(v_1 + v_2)$th rows of $\mathbf{A}$ are taken from $\hat{\mathbf{k}}_1$ and $\hat{\mathbf{k}}_2$ with the signs reversed. The matrix $\mathbf{B}$ consists of the first v_1 Markov parameters (excluding $h_{ij}(0)$) of $\hat{g}_{11}(s)$ and $\hat{g}_{12}(s)$ and the first v_2 Markov parameters of $\hat{g}_{21}(s)$ and $\hat{g}_{22}(s)$. The form of $\mathbf{C}$ is fixed and the positions of 1 depend on v_1 and v_2. The observability indices of this realization are v_1 and v_2 (Verify them.)

*6-6 Irreducible Realizations of $\hat{\mathbf{G}}(s)$: Coprime Fraction Method

Controllable-form realization. Consider a $q \times p$ proper rational matrix $\hat{\hat{\mathbf{G}}}(s)$. By long division of each element of $\hat{\hat{\mathbf{G}}}(s)$, we can write $\hat{\hat{\mathbf{G}}}(s)$ as

$$\hat{\hat{\mathbf{G}}}(s) = \hat{\hat{\mathbf{G}}}(\infty) + \hat{\mathbf{G}}(s)$$

where $\hat{\mathbf{G}}(s)$ is strictly proper. Since $\hat{\hat{\mathbf{G}}}(\infty)$ gives the direct transmission part of a realization, we may exclude it in the following discussion. Using the procedure discussed in Appendix G, we can factor $\hat{\mathbf{G}}(s)$ as

$$\hat{\mathbf{G}}(s) = \mathbf{N}(s)\mathbf{D}^{-1}(s)$$

where $\mathbf{N}(s)$ and $\mathbf{D}(s)$ are $q \times p$ and $p \times p$ polynomial matrices and are right coprime. Furthermore, $\mathbf{D}(s)$ is column reduced and $\delta_{ci}\mathbf{N}(s) < \delta_{ci}\mathbf{D}(s) = \mu_i$, for $i = 1, 2, \ldots, p$. Define

$$n = \mu_1 + \mu_2 + \cdots + \mu_p$$

That is, n is the sum of all the column degrees of $\mathbf{D}(s)$. Let us write

$$\mathbf{D}(s) = \mathbf{D}_{hc}\mathbf{H}(s) + \mathbf{D}_{lc}\mathbf{L}(s) \tag{6-100}$$

with

$$\mathbf{H}(s) = \begin{bmatrix} s^{\mu_1} & 0 & 0 & \cdots & 0 \\ 0 & s^{\mu_2} & 0 & \cdots & 0 \\ 0 & 0 & s^{\mu_3} & \cdots & 0 \\ \vdots & \vdots & \vdots & & \vdots \\ 0 & 0 & 0 & \cdots & s^{\mu_p} \end{bmatrix} \quad \mathbf{L}(s) = \left[\begin{array}{c:c:c:c} 1 & 0 & & 0 \\ s & 0 & & 0 \\ \vdots & \vdots & & \vdots \\ s^{\mu_1-1} & 0 & & 0 \\ \hdashline 0 & 1 & & 0 \\ \vdots & \vdots & \cdots & \vdots \\ 0 & s^{\mu_2-1} & & 0 \\ \hdashline & \vdots & \ddots & \\ \hdashline 0 & 0 & & 1 \\ \vdots & \vdots & & \vdots \\ 0 & 0 & & s^{\mu_p-1} \end{array}\right] \tag{6-101}$$

The matrix $\mathbf{H}(s)$ is a $p \times p$ diagonal matrix with s to the powers of column degrees on the diagonal. The matrix $\mathbf{L}(s)$ is an $n \times p$ matrix; if μ_i is zero, then the corresponding column is a zero column. The matrices $\mathbf{D}_{hc}$ and $\mathbf{D}_{lc}$ are, respectively, $p \times p$ and $p \times n$ constant matrices. Because of $\delta_{ci}\mathbf{N}(s) < \delta_{ci}\mathbf{D}(s)$, $\mathbf{N}(s)$ can be expressed as

$$\mathbf{N}(s) = \mathbf{N}_{lc}\mathbf{L}(s) \tag{6-102}$$

where $\mathbf{N}_{lc}$ is a $q \times n$ constant matrix.

The procedure to be discussed follows the one for developing the controllable canonical-form realization discussed from (6-21) to (6-26). The only difference is that the $\mathbf{D}(s)$ in the scalar case is normalized to be monic, but not here. Let $\mathbf{v}$ be a $p \times 1$ variable defined by $\hat{\mathbf{v}}(s) = \mathbf{D}^{-1}(s)\hat{\mathbf{u}}(s)$. Then we have

$$\mathbf{D}(s)\hat{\mathbf{v}}(s) = \hat{\mathbf{u}}(s) \tag{6-103}$$

and

$$\hat{\mathbf{y}}(s) = \mathbf{N}(s)\hat{\mathbf{v}}(s) \tag{6-104}$$

The substitution of (6-100) and (6-102) into (6-103) and (6-104) yields

$$(\mathbf{D}_{hc}\mathbf{H}(s) + \mathbf{D}_{lc}\mathbf{L}(s))\hat{\mathbf{v}}(s) = \hat{\mathbf{u}}(s) \tag{6-105}$$

and

$$\hat{\mathbf{y}}(s) = \mathbf{N}_{lc}\mathbf{L}(s)\hat{\mathbf{v}}(s) \tag{6-106}$$

Since $\mathbf{D}(s)$ is column reduced, the constant matrix $\mathbf{D}_{hc}$ is nonsingular. Hence (6-105) can be written as[6]

$$\mathbf{H}(s)\mathbf{v}(s) = -\mathbf{D}_{hc}^{-1}\mathbf{D}_{lc}\mathbf{L}(s)\hat{\mathbf{v}}(s) + \mathbf{D}_{hc}^{-1}\hat{\mathbf{u}}(s) \tag{6-107}$$

Let us define

$$\hat{\mathbf{x}}(s) = \mathbf{L}(s)\hat{\mathbf{v}}(s) \tag{6-108}$$

[6] This step does not appear in the scalar case because $D(s)$ is normalized to be monic before the realization.

We express this explicitly in the time domain:

$$\begin{bmatrix} x_{11} \\ x_{12} \\ \vdots \\ x_{1\mu_1} \\ \hdashline x_{21} \\ \vdots \\ x_{2\mu_2} \\ \hdashline \vdots \\ \hdashline x_{p1} \\ \vdots \\ x_{p\mu_p} \end{bmatrix} = \begin{bmatrix} v_1 \\ \dot{v}_1 \\ \vdots \\ v_1^{(\mu_1-1)} \\ \hdashline v_2 \\ \vdots \\ v_2^{(\mu_2-1)} \\ \hdashline \vdots \\ \hdashline v_p \\ \vdots \\ v_p^{(\mu_p-1)} \end{bmatrix} \tag{6-109}$$

where v_i is the ith component of $\mathbf{v}$, and $v_i^{(k)} = d^k v_i/dt^k$. From this definition we have

$$\dot{x}_{ik} = x_{i(k+1)} \qquad i = 1, 2, \ldots, p;\ k = 1, 2, \ldots, \mu_i - 1 \tag{6-110}$$

From the definition of $\mathbf{H}(s)$ and (6-107), we have

$$\begin{bmatrix} \dot{x}_{1\mu_1} \\ \dot{x}_{2\mu_2} \\ \vdots \\ \dot{x}_{p\mu_p} \end{bmatrix} = -\mathbf{D}_{hc}^{-1}\mathbf{D}_{lc}\mathbf{x}(t) + \mathbf{D}_{hc}^{-1}\mathbf{u}(t) \triangleq \begin{bmatrix} -\mathbf{a}_{1\mu_1} \\ -\mathbf{a}_{2\mu_2} \\ \vdots \\ -\mathbf{a}_{p\mu_p} \end{bmatrix} \mathbf{x}(t) + \begin{bmatrix} \mathbf{b}_{1\mu_1} \\ \mathbf{b}_{2\mu_2} \\ \vdots \\ \mathbf{b}_{p\mu_p} \end{bmatrix} \mathbf{u}(t) \tag{6-111}$$

where $\mathbf{a}_{i\mu_i}$ denotes the ith row of $\mathbf{D}_{hc}^{-1}\mathbf{D}_{lc}$, and $\mathbf{b}_{i\mu_i}$ denotes the ith row of $\mathbf{D}_{hc}^{-1}$. A block diagram of (6-110) and (6-111) is shown in Figure 6-7. It consists of p chains of integrators. The output of each integrator is chosen as a state variable as shown. This block diagram is a generalization of the one in Figure 6-3.

From this block diagram or, equivalently, from (6-110) and (6-111), we can readily obtain the following dynamical equation.

$$\begin{bmatrix} \dot{x}_{11} \\ \dot{x}_{12} \\ \vdots \\ \dot{x}_{1\mu_1} \\ \hdashline \dot{x}_{21} \\ \dot{x}_{22} \\ \vdots \\ \dot{x}_{2\mu_2} \\ \hdashline \vdots \\ \hdashline \dot{x}_{p1} \\ \vdots \\ \dot{x}_{p\mu_p} \end{bmatrix} = \left[\begin{array}{ccccc:ccccc:c:ccccc}
0 & 1 & 0 & \cdots & 0 & & & & & & & & & & & \\
0 & 0 & 1 & \cdots & 0 & & & 0 & & & & & & 0 & & \\
\vdots & \vdots & \vdots & & \vdots & & & & & & & & & & & \\
0 & 0 & 0 & \cdots & 1 & & & & & & & & & & & \\
x & x & x & \cdots & x & x & \cdots & (-\mathbf{a}_{1\mu_1}) & x & x & & x & x & x & \cdots & x \\ \hdashline
 & & & & & 0 & 1 & 0 & \cdots & 0 & & & & & & \\
 & & & & & 0 & 0 & 1 & \cdots & 0 & & & & & & \\
 & & 0 & & & \vdots & \vdots & \vdots & & \vdots & \cdots & & & 0 & & \\
 & & & & & 0 & 0 & 0 & \cdots & 1 & & & & & & \\
x & x & x & \cdots & x & x & \cdots & (-\mathbf{a}_{2\mu_2}) & \cdots & x & & x & x & x & \cdots & x \\ \hdashline
 & & \vdots & & & & & & & & \ddots & & & \vdots & & \\ \hdashline
 & & & & & & & & & & & 0 & 1 & 0 & \cdots & 0 \\
 & & & & & & & & & & & 0 & 0 & 1 & \cdots & 0 \\
 & & 0 & & & & & 0 & & & & \vdots & \vdots & \vdots & & \vdots \\
 & & & & & & & & & & & 0 & 0 & 0 & \cdots & 1 \\
x & x & x & \cdots & x & x & \cdots & (-\mathbf{a}_{p\mu_p}) & \cdots & x & & x & x & x & \cdots & x
\end{array}\right] \mathbf{x}(t) + \begin{bmatrix} 0 \\ 0 \\ \vdots \\ 0 \\ \mathbf{b}_{1\mu_1} \\ \hdashline 0 \\ 0 \\ \vdots \\ 0 \\ \mathbf{b}_{2\mu_2} \\ \hdashline \vdots \\ \hdashline 0 \\ 0 \\ \vdots \\ \mathbf{b}_{p\mu_p} \end{bmatrix} \mathbf{u}(t) \tag{6-112a}$$

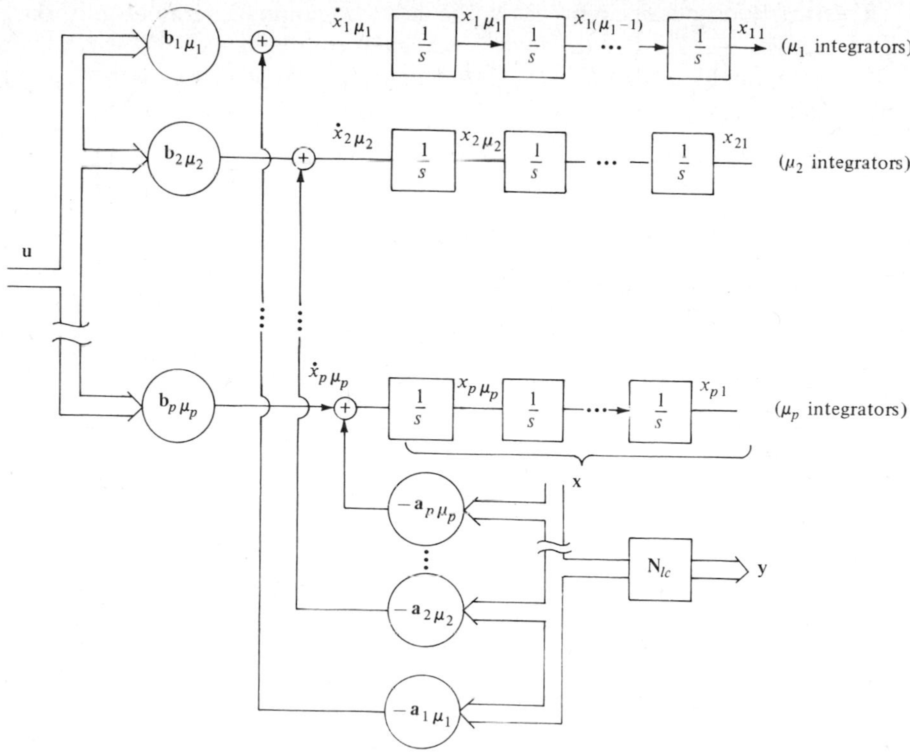

Figure 6-7 Block diagram of $\hat{\mathbf{G}}(s) = \mathbf{N}(s)\mathbf{D}^{-1}(s)$.

$$\mathbf{y}(t) = \mathbf{C}\mathbf{x}(t) = \mathbf{N}_{lc}\mathbf{x}(t) \tag{6-112b}$$

where "x" denotes possible nonzero elements. The μ_1th row of $\mathbf{A}$ is equal to $-\mathbf{a}_{1\mu_1}$, the $(\mu_1+\mu_2)$th row of $\mathbf{A}$ is equal to $-\mathbf{a}_{2\mu_2}$ and so forth. Equation (6-112b) follows directly from (6-106) by the substitution of $\hat{\mathbf{x}}(s) = \mathbf{L}(s)\hat{\mathbf{v}}(s)$. We note that if $\mu_i = 0$, then the ith chain of integrators will not appear in Fig. 6-7 and the ith row of $\mathbf{D}_{hc}^{-1}\mathbf{D}_{lc}$ and $\mathbf{D}_{hc}^{-1}$ will not appear in Equation (6-112).

Example 1

Consider the strictly proper rational matrix

$$\hat{\mathbf{G}}(s) = \begin{bmatrix} \dfrac{-3s^2-6s-2}{(s+1)^3} & \dfrac{s^3-3s-1}{(s-2)(s+1)^3} & \dfrac{1}{(s-2)(s+1)^2} \\ \dfrac{s}{(s+1)^3} & \dfrac{s}{(s-2)(s+1)^3} & \dfrac{s}{(s-2)(s+1)^2} \end{bmatrix}$$
$$= \begin{bmatrix} (s+1)^3(s-2) & 0 \\ 0 & (s+1)^3(s-2) \end{bmatrix}^{-1} \begin{bmatrix} (-3s^2-6s-2)(s-2) & s^3-3s-1 & s+1 \\ s(s-2) & s & s(s+1) \end{bmatrix} \tag{6-113}$$

This fraction is obtained by using the least common denominator of each row and is generally not irreducible. From the coefficients of this fraction, we can obtain, by using the procedure discussed in Appendix G, the following irreducible fraction:

$$\hat{\mathbf{G}}(s)=\begin{bmatrix}-3s^2-6s-2 & 1 & 0\\ s & 0 & 0\end{bmatrix}\begin{bmatrix}s^3+3s^2+3s+1 & -1 & -1\\ 0 & s-2 & -3\\ 0 & 0 & 1\end{bmatrix}^{-1}$$

Hence we have $\mu_1=3$, $\mu_2=1$, and $\mu_3=0$ and

$$n=\mu_1+\mu_2+\mu_3=4$$

We write $\mathbf{D}(s)$ and $\mathbf{N}(s)$ in the form of (6-100) and (6-102) as

$$\mathbf{D}(s)=\begin{bmatrix}1 & 0 & -1\\ 0 & 1 & -3\\ 0 & 0 & 1\end{bmatrix}\begin{bmatrix}s^3 & 0 & 0\\ 0 & s & 0\\ 0 & 0 & 1\end{bmatrix}+\left[\begin{array}{ccc:c}1 & 3 & 3 & -1\\ 0 & 0 & 0 & -2\\ 0 & 0 & 0 & 0\end{array}\right]\left[\begin{array}{c:c:c}1 & 0 & 0\\ s & 0 & 0\\ s^2 & 0 & 0\\ \hdashline 0 & 1 & 0\end{array}\right]=\mathbf{D}_{hc}\mathbf{H}(s)+\mathbf{D}_{lc}\mathbf{L}(s)$$

$$\mathbf{N}(s)=\left[\begin{array}{ccc:c}-2 & -6 & -3 & 1\\ 0 & 1 & 0 & 0\end{array}\right]\left[\begin{array}{c:c:c}1 & 0 & 0\\ s & 0 & 0\\ s^2 & 0 & 0\\ \hdashline 0 & 1 & 0\end{array}\right]=\mathbf{N}_{lc}\mathbf{L}(s)$$

Note that, because of $\mu_3=0$, the last column of $\mathbf{L}(s)$ is a zero column. We compute

$$\mathbf{D}_{hc}^{-1}=\begin{bmatrix}1 & 0 & 1\\ 0 & 1 & 3\\ 0 & 0 & 1\end{bmatrix}$$

and $$\mathbf{D}_{hc}^{-1}\mathbf{D}_{lc}=\begin{bmatrix}1 & 0 & 1\\ 0 & 1 & 3\\ 0 & 0 & 1\end{bmatrix}\begin{bmatrix}1 & 3 & 3 & -1\\ 0 & 0 & 0 & -2\\ 0 & 0 & 0 & 0\end{bmatrix}=\begin{bmatrix}1 & 3 & 3 & -1\\ 0 & 0 & 0 & -2\\ 0 & 0 & 0 & 0\end{bmatrix}$$

Hence a realization of $\hat{\mathbf{G}}(s)$ is given by

$$\begin{bmatrix}\dot{x}_{11}\\ \dot{x}_{12}\\ \dot{x}_{13}\\ \dot{x}_{21}\end{bmatrix}=\left[\begin{array}{ccc:c}0 & 1 & 0 & 0\\ 0 & 0 & 1 & 0\\ -1 & -3 & -3 & 1\\ \hdashline 0 & 0 & 0 & 2\end{array}\right]\mathbf{x}+\left[\begin{array}{ccc}0 & 0 & 0\\ 0 & 0 & 0\\ 1 & 0 & 1\\ \hdashline 0 & 1 & 3\end{array}\right]\mathbf{u} \qquad \textbf{(6-114a)}$$

$$\mathbf{y}=\left[\begin{array}{ccc:c}-2 & -6 & -3 & 1\\ 0 & 1 & 0 & 0\end{array}\right]\mathbf{x} \qquad \textbf{(6-114b)}$$

Note that, because of $\mu_3=0$, the third row of $\mathbf{D}_{hc}^{-1}$ and $\mathbf{D}_{hc}^{-1}\mathbf{D}_{lc}$ do not appear in (6-114).

The dynamical equation in (6-112) is developed from the block diagram in Figure 6-7 which, in turn, is developed from $\hat{\mathbf{G}}(s)=\mathbf{N}(s)\mathbf{D}^{-1}(s)$; hence (6-112) is a realization of $\hat{\mathbf{G}}(s) = \mathbf{N}(s)\mathbf{D}^{-1}(s)$. We shall now formally establish this by showing that the **A**, **B**, and **C** in (6-112) satisfies

$$\mathbf{C}(s\mathbf{I}-\mathbf{A})^{-1}\mathbf{B} = \mathbf{N}_{lc}(s\mathbf{I}-\mathbf{A})^{-1}\mathbf{B} = \mathbf{N}(s)\mathbf{D}^{-1}(s) = \mathbf{N}_{lc}\mathbf{L}(s)\mathbf{D}^{-1}(s)$$

or equivalently,

$$(s\mathbf{I}-\mathbf{A})^{-1}\mathbf{B} = \mathbf{L}(s)\mathbf{D}^{-1}(s) = \mathbf{L}(s)(\mathbf{D}_{hc}\mathbf{H}(s) + \mathbf{D}_{lc}\mathbf{L}(s))^{-1} \tag{6-115}$$

This equation implies

$$\mathbf{B}\mathbf{D}(s) = \mathbf{B}\mathbf{D}_{hc}(\mathbf{H}(s) + \mathbf{D}_{hc}^{-1}\mathbf{D}_{lc}\mathbf{L}(s)) = (s\mathbf{I}-\mathbf{A})\mathbf{L}(s) \tag{6-116}$$

From (6-112), we see that all rows of **B**, except the $(\sum_{m=1}^{i} \mu_m)$th rows, $i = 1, 2, \ldots, p$, are zero rows. Furthermore, these nonzero rows are equal to $\mathbf{D}_{hc}^{-1}$. Hence the $(\sum_{m=1}^{i} \mu_m)$th row of

$$\mathbf{B}\mathbf{D}_{hc}(\mathbf{H}(s) + \mathbf{D}_{hc}^{-1}\mathbf{D}_{lc}\mathbf{L}(s))$$

is equal to the ith row of $\mathbf{H}(s) + \mathbf{D}_{hc}^{-1}\mathbf{D}_{lc}\mathbf{L}(s)$ for $i = 1, 2, \ldots, p$, and the remainder are all zero rows. To establish the equality in (6-116), we write

$$(s\mathbf{I}-\mathbf{A})\mathbf{L}(s) = \left[\begin{array}{ccccc|cccc|cccc}
s & -1 & 0 & \cdots & 0 & & & & & & & & \\
0 & s & -1 & \cdots & 0 & & & & & & & & \\
\vdots & \vdots & \vdots & & \vdots & & & & & & & & \\
0 & 0 & 0 & \cdots & -1 & & & & & & & & \\
-x & -x & -x & \cdots & s-x & -x & -x & \cdots & -x & -x & -x & \cdots & -x \\
\hline
 & & & & & s & -1 & \cdots & 0 & & & & \\
 & & & & & 0 & s & \cdots & 0 & & & & \\
 & & & & & \vdots & \vdots & & \vdots & & & & \\
 & & & & & 0 & 0 & & -1 & & & & \\
-x & -x & -x & \cdots & -x & -x & -x & \cdots & s-x & -x & -x & \cdots & -x \\
\hline
 & & & \vdots & & & & & & & & & \\
\hline
 & & & & & & & & & s & -1 & \cdots & 0 \\
 & & & & & & & & & 0 & s & & 0 \\
 & & & & & & & & & \vdots & \vdots & & \vdots \\
-x & -x & -x & \cdots & -x & -x & -x & \cdots & -x & -x & -x & \cdots & s-x
\end{array}\right]$$

$$\times \left[\begin{array}{c|c|c|c}
1 & 0 & & 0 \\
s & 0 & & 0 \\
\vdots & \vdots & & \vdots \\
s^{\mu_1-1} & 0 & & 0 \\
\hline
0 & 1 & & 0 \\
0 & s & & 0 \\
\vdots & \vdots & & \vdots \\
0 & s^{\mu_2-1} & & 0 \\
\hline
 & & \ddots & \\
\hline
0 & 0 & & 1 \\
0 & 0 & & s \\
\vdots & \vdots & & \vdots \\
0 & 0 & & s^{\mu_p-1}
\end{array}\right] \tag{6-117}$$

From this equation, we can readily verify that all rows, except the $(\sum_{m=1}^{i} \mu_m)$th row, for $i = 1, 2, \ldots, p$, of $(s\mathbf{I} - \mathbf{A})\mathbf{L}(s)$ are zero rows. Because the $(\sum_{m=1}^{i} \mu_m)$th row of $\mathbf{A}$ is equal to the ith row of $-\mathbf{D}_{hc}^{-1}\mathbf{D}_{lc}$, as can be seen from (6-111), the $(\sum_{m=1}^{i} \mu_m)$th row of $(s\mathbf{I} - \mathbf{A})\mathbf{L}(s) = s\mathbf{L}(s) - \mathbf{A}\mathbf{L}(s)$ is equal to the ith row of $\mathbf{H}(s) + \mathbf{D}_{hc}^{-1}\mathbf{D}_{lc}\mathbf{L}(s)$. This establishes the equality in (6-116). Hence the dynamical equation in (6-112) is a realization of $\hat{\mathbf{G}}(s) = \mathbf{N}(s)\mathbf{D}^{-1}(s)$.

We discuss now the relationship between $\det(s\mathbf{I} - \mathbf{A})$ and $\det \mathbf{D}(s)$. From $(s\mathbf{I} - \mathbf{A})^{-1}\mathbf{B} = \mathbf{L}(s)\mathbf{D}^{-1}(s)$ in (6-115), we have

$$\frac{1}{\det (s\mathbf{I} - \mathbf{A})} (\text{Adj } (s\mathbf{I} - \mathbf{A}))\mathbf{B} = \frac{1}{\det \mathbf{D}(s)} \mathbf{L}(s) \text{ Adj } \mathbf{D}(s)$$

where "det" stands for the determant and "Adj" the adjoint. Since $(\text{Adj } (s\mathbf{I} - \mathbf{A}))\mathbf{B}$ and $\mathbf{L}(s) \text{ Adj } \mathbf{D}(s)$ are polynomial matrices and since $\det (s\mathbf{I} - \mathbf{A})$ and $\det \mathbf{D}(s)$ are polynomial of degree n, we must have

$$\det(s\mathbf{I} - \mathbf{A}) = k \det \mathbf{D}(s)$$

for some constant k. Det $(s\mathbf{I} - \mathbf{A})$ has a leading coefficient of 1, and $\det \mathbf{D}(s)$ has a leading coefficient of $\det \mathbf{D}_{hc}$, hence we have $k = (\det \mathbf{D}_{hc})^{-1}$ and

$$\det(s\mathbf{I} - \mathbf{A}) = (\det \mathbf{D}_{hc})^{-1} \det \mathbf{D}(s) \tag{6-118}$$

Realization of $\mathbf{N}(s)\mathbf{D}^{-1}(s)$ where $\mathbf{D}(s)$ and $\mathbf{N}(s)$ are not right coprime

The realization procedure discussed from (6-100) to (6-112) is applicable whether $\hat{\mathbf{G}}(s) = \mathbf{N}(s)\mathbf{D}^{-1}(s)$ is a coprime fraction or not as long as $\mathbf{D}(s)$ is column reduced and $\delta_{ci}\mathbf{N}(s) < \delta_{ci}\mathbf{D}(s)$. Now we shall show that the dynamical equation in (6-112) is controllable whether $\mathbf{D}(s)$ and $\mathbf{N}(s)$ are right coprime or not. This can be established by employing the condition rank $[\mathbf{B} \quad \mathbf{AB} \quad \cdots \quad \mathbf{A}^{n-1}\mathbf{B}] = n$ or the left coprimeness of $(s\mathbf{I} - \mathbf{A})$ and $\mathbf{B}$. For this problem, it is easier to employ the latter. From (6-115), we have

$$(s\mathbf{I} - \mathbf{A})^{-1}\mathbf{B} = \mathbf{L}(s)\mathbf{D}^{-1}(s) \tag{6-119}$$

Since $\mathbf{L}(s)$ consists of, as can be seen from (6-101), a unit matrix of order p as a submatrix, we have

$$\text{rank} \begin{bmatrix} \mathbf{L}(s) \\ \mathbf{D}(s) \end{bmatrix} = p \qquad \text{for every } s \text{ in } \mathbb{C} \tag{6-120}$$

Hence $\mathbf{D}(s)$ and $\mathbf{L}(s)$ are right coprime (Theorem G-8). From the realization procedure, we have $\dim \mathbf{A} = \deg \det(s\mathbf{I} - \mathbf{A}) = \deg \det \mathbf{D}(s)$. Hence Corollary G-8 implies that $s\mathbf{I} - \mathbf{A}$ and $\mathbf{B}$ are left coprime or rank $[s\mathbf{I} - \mathbf{A} \quad \mathbf{B}] = n$, for every s in $\mathbb{C}$. Consequently, $\{\mathbf{A}, \mathbf{B}\}$ is controllable (Theorem 5-7). Up to this point, the coprimeness of $\mathbf{D}(s)$ and $\mathbf{N}(s)$ has not been used.

Theorem 6-5

The controllable realization of $\hat{\mathbf{G}}(s) = \mathbf{N}(s)\mathbf{D}^{-1}(s)$ in (6-112) is observable if and only if $\mathbf{D}(s)$ and $\mathbf{N}(s)$ are right coprime.

Proof

We first show that if $\mathbf{D}(s)$ and $\mathbf{N}(s)$ are not right coprime, then (6-112) is not observable. Under the premise, we can find a $\bar{\mathbf{D}}(s)$ with deg det $\bar{\mathbf{D}}(s) <$ deg det $\mathbf{D}(s)$ such that $\hat{\mathbf{G}}(s) = \bar{\mathbf{N}}(s)\bar{\mathbf{D}}^{-1}(s) = \mathbf{N}(s)\mathbf{D}^{-1}(s)$. Using $\bar{\mathbf{N}}(s)\bar{\mathbf{D}}^{-1}(s)$, we can find a realization of $\hat{\mathbf{G}}(s)$ with a dimension smaller than that of (6-112). Hence (6-112) is not irreducible (controllable and observable) (Theorem 5-19). However, (6-112) is known to be controllable; hence (6-112) is not observable. We shall establish this statement once again by using a different argument. If $\mathbf{D}(s)$ and $\mathbf{N}(s)$ are not right coprime, then there exists at least one s, say $s = \lambda$, such that

$$\text{rank} \begin{bmatrix} \mathbf{D}(\lambda) \\ \mathbf{N}(\lambda) \end{bmatrix} < p$$

which implies the existence of a $p \times 1$ nonzero constant vector $\boldsymbol{\alpha}$ such that

$$\mathbf{D}(\lambda)\boldsymbol{\alpha} = \mathbf{0}$$

and

$$\mathbf{N}(\lambda)\boldsymbol{\alpha} = \mathbf{0}$$

Because of (6-116), (6-102) and (6-112b), we have

$$(\lambda\mathbf{I} - \mathbf{A})\mathbf{L}(\lambda)\boldsymbol{\alpha} = \mathbf{0} \qquad \mathbf{C}\mathbf{L}(\lambda)\boldsymbol{\alpha} = \mathbf{0}$$

or

$$\begin{bmatrix} \lambda\mathbf{I} - \mathbf{A} \\ \mathbf{C} \end{bmatrix} \mathbf{L}(\lambda)\boldsymbol{\alpha} = \mathbf{0} \tag{6-121}$$

Since $\mathbf{L}(\lambda)$ consists of a unit matrix of order p as a submatrix, if $\boldsymbol{\alpha}$ is nonzero, so is the $n \times 1$ vector $\mathbf{L}(\lambda)\boldsymbol{\alpha}$. Hence we conclude from (6-121) that $[(s\mathbf{I} - \mathbf{A})' \quad \mathbf{C}']'$ does not have a full rank at $s = \lambda$, and $\{\mathbf{A}, \mathbf{C}\}$ is not observable.

Now we show that if $\{\mathbf{A}, \mathbf{C}\}$ is not observable, then $\mathbf{D}(s)$ and $\mathbf{N}(s)$ are not right coprime. Under the premise, there exists, from the dual of Problem 5-37 on page 231, an eigenvalue λ of $\mathbf{A}$ and its $n \times 1$ eigenvector $\mathbf{e}$ such that

$$\begin{bmatrix} \lambda\mathbf{I} - \mathbf{A} \\ \mathbf{C} \end{bmatrix} \mathbf{e} = \mathbf{0} \tag{6-122}$$

Now we show that there exists a $p \times 1$ vector $\boldsymbol{\alpha}$ such that

$$\mathbf{e} = \mathbf{L}(\lambda)\boldsymbol{\alpha} \qquad \text{and} \qquad \mathbf{D}(\lambda)\boldsymbol{\alpha} = \mathbf{0} \tag{6-123}$$

To show this, consider (6-116) or

$$[\lambda\mathbf{I} - \mathbf{A} \quad -\mathbf{B}] \begin{bmatrix} \mathbf{L}(\lambda) \\ \mathbf{D}(\lambda) \end{bmatrix} = \mathbf{0} \tag{6-124}$$

The composite matrix $[\lambda\mathbf{I} - \mathbf{A} \quad -\mathbf{B}]$ is of dimension $n \times (n+p)$; hence Theorem 2-5 implies

$$\text{rank}\,[\lambda\mathbf{I} - \mathbf{A} \quad -\mathbf{B}] + \text{nullity}\,[\lambda\mathbf{I} - \mathbf{A} \quad -\mathbf{B}] = n + p$$

which implies, because $\{\mathbf{A}, \mathbf{B}\}$ is controllable,

$$\text{nullity}\,[\lambda\mathbf{I} - \mathbf{A} \quad -\mathbf{B}] = p$$

Consequently, the p linear independent columns of $[\mathbf{L}'(\lambda) \quad \mathbf{D}'(\lambda)]'$ form a basis of the null space of $[\lambda\mathbf{I}-\mathbf{A} \quad -\mathbf{B}]$. Because of

$$[\lambda\mathbf{I}-\mathbf{A} \quad -\mathbf{B}]\begin{bmatrix}\mathbf{e}\\ \mathbf{0}\end{bmatrix}=\mathbf{0}$$

the vector $[\mathbf{e}' \quad \mathbf{0}']'$ is in the null space. Hence there exists a $p \times 1$ vector $\boldsymbol{\alpha}$ such that

$$\begin{bmatrix}\mathbf{e}\\ \mathbf{0}\end{bmatrix}=\begin{bmatrix}\mathbf{L}(\lambda)\\ \mathbf{D}(\lambda)\end{bmatrix}\boldsymbol{\alpha}$$

This establishes (6-123). The substitution of $\mathbf{e}=\mathbf{L}(\lambda)\boldsymbol{\alpha}$ into (6-122) yields (6-121) which implies, by using (6-116), (6-102), and (6-112b),

$$\begin{bmatrix}\mathbf{D}(\lambda)\\ \mathbf{N}(\lambda)\end{bmatrix}\boldsymbol{\alpha}=\mathbf{0}$$

Hence $\mathbf{D}(s)$ and $\mathbf{N}(s)$ are not coprime. This completes the proof of this theorem.
Q.E.D.

Theorem 6-5 implies that the realization in (6-112) is irreducible if and only if $\hat{\mathbf{G}}(s)=\mathbf{N}(s)\mathbf{D}^{-1}(s)$ is a right-coprime fraction. If $\hat{\mathbf{G}}(s)=\mathbf{N}(s)\mathbf{D}^{-1}(s)$ is a right-coprime fraction, then we have, according to Definition 6-1,

$$\deg \hat{\mathbf{G}}(s)=\deg \det \mathbf{D}(s)$$

Consequently, the realization in (6-112) is irreducible if and only if $\deg \hat{\mathbf{G}}(s)= \dim \mathbf{A}$. Since all irreducible realizations of $\hat{\mathbf{G}}(s)$ are equivalent (Theorem 5-20), we conclude that a realization of $\hat{\mathbf{G}}(s)$ is irreducible if and only if its dimension is equal to the degree of $\hat{\mathbf{G}}(s)$. This establishes essentially Theorem 6-2.

The realization in (6-112) is a generalization of the controllable *canonical-*form realization discussed in (6-26). Hence we shall call (6-112) a multivariable controllable-form realization. We note that the realization depends on the fraction $\mathbf{N}(s)\mathbf{D}^{-1}(s)$ used. For example, if we use $(\mathbf{N}(s)\mathbf{U}(s))(\mathbf{D}(s)\mathbf{U}(s))^{-1}$, where $\mathbf{U}(s)$ is a unimodular matrix, and $\mathbf{D}(s)\mathbf{U}(s)$ remains to be column reduced, then we will obtain a different multivariable controllable-form realization. Hence, unlike the scalar case, the controllable-form realization in the multivariable case is not unique. Consequently, the adjective "canonical" is not used.

Column degrees and controllability indices

Before proceeding, we give a relationship between column degrees and controllability indices.

Theorem 6-6

The set of the controllability indices of any irreducible realization of a strictly proper rational matrix $\hat{\mathbf{G}}(s)$ is equal to the set of the column degrees of $\mathbf{D}(s)$ in any column-reduced right coprime fraction $\hat{\mathbf{G}}(s)=\mathbf{N}(s)\mathbf{D}^{-1}(s)$.

Proof

Controllability indices are, as shown in Theorem 5-8, invariant under any similarity transformation and any rearrangement of input vectors by a nonsingular matrix. Column degrees are, as discussed in Theorem G-15′, intrinsic properties of $\hat{\mathbf{G}}(s)$. Hence we may use the realization in (6-112) to establish this theorem. If we rearrange the input vector $\mathbf{u}(t)$ by

$$\bar{\mathbf{u}}(t) = \mathbf{D}_{hc}\mathbf{u}(t)$$

where $\mathbf{D}_{hc}$ is nonsingular by assumption, then the new $\mathbf{B}$ matrix in (6-112) becomes

$$\bar{\mathbf{B}}' = \left[\begin{array}{ccccc|ccccc|c|ccccc} 0 & \cdots & 0 & 1 & & 0 & \cdots & 0 & 0 & & & 0 & \cdots & 0 & 0 \\ 0 & \cdots & 0 & 0 & & 0 & \cdots & 0 & 1 & & \cdots & 0 & \cdots & 0 & 0 \\ \vdots & & \vdots & \vdots & & \vdots & & \vdots & \vdots & & & \vdots & & \vdots & \vdots \\ 0 & \cdots & 0 & 0 & & 0 & \cdots & 0 & 0 & & & 0 & \cdots & 0 & 1 \end{array}\right]$$
$$\qquad\underbrace{\qquad\qquad}_{\mu_1}\qquad\underbrace{\qquad\qquad}_{\mu_2}\qquad\qquad\underbrace{\qquad\qquad}_{\mu_p}$$

where the prime denotes the transpose. Using this $\bar{\mathbf{B}}$ and the $\mathbf{A}$ in (6-112), we can readily show that the controllability indices are $\mu_1, \mu_2, \ldots, \mu_p$, which are the column degrees of $\mathbf{D}(s)$. This establishes the theorem. Q.E.D.

Observable-Form Realization. Similar to the observable canonical-form realization discussed in (6-18) for scalar transfer functions, we can develop observable form realizations for transfer matrices. This realization will be used extensively in the next section.

Consider a strictly proper $q \times p$ rational matrix $\hat{\mathbf{G}}(s)$. We factor it as

$$\hat{\mathbf{G}}(s) = \mathbf{D}^{-1}(s)\mathbf{N}(s) \tag{6-125}$$

where the polynomial matrices $\mathbf{D}(s)$ and $\mathbf{N}(s)$ are left coprime and $\mathbf{D}(s)$ is row reduced. We take the transpose of $\hat{\mathbf{G}}(s)$:

$$\hat{\mathbf{G}}'(s) = \mathbf{N}'(s)(\mathbf{D}^{-1}(s))' = \mathbf{N}'(s)(\mathbf{D}'(s))^{-1} \tag{6-126}$$

The rows of $\mathbf{D}(s)$ become the columns of $\mathbf{D}'(s)$. Hence $\mathbf{D}'(s)$ is column reduced. Furthermore, $\mathbf{N}'(s)$ and $\mathbf{D}'(s)$ are right coprime. Hence an irreducible realization

$$\dot{\mathbf{z}} = \mathbf{A}\mathbf{z} + \mathbf{B}\mathbf{v} \qquad \mathbf{w} = \mathbf{C}\mathbf{z}$$

of the form in (6-112) can be developed for $\mathbf{N}'(s)(\mathbf{D}'(s))^{-1}$. Consequently, the equation

$$\dot{\mathbf{x}} = \mathbf{A}'\mathbf{x} + \mathbf{C}'\mathbf{u} \qquad \mathbf{y} = \mathbf{B}'\mathbf{x}$$

is an irreducible realization of $\mathbf{D}^{-1}(s)\mathbf{N}(s)$, and the equation is said to be in a multivariable observable form.

Of course, an observable form realization can also be obtained directly from (6-125). Let the row degrees of $\mathbf{D}(s)$ be v_i, that is, $\delta_{ri}\mathbf{D}(s) = v_i$, $i = 1, 2, \ldots, q$, and let

$$n = v_1 + v_2 + \cdots + v_q \tag{6-127}$$

We write

$$\mathbf{D}(s) = \bar{\mathbf{H}}(s)\mathbf{D}_{hr} + \bar{\mathbf{L}}(s)\mathbf{D}_{lr} = (\bar{\mathbf{H}}(s) + \mathbf{L}(s)\mathbf{D}_{lr}\mathbf{D}_{hr}^{-1})\mathbf{D}_{hr} \tag{6-128}$$

where

$$\bar{\mathbf{H}}(s) = \begin{bmatrix} s^{v_1} & 0 & \cdots & 0 \\ 0 & s^{v_2} & \cdots & 0 \\ \vdots & \vdots & & \vdots \\ 0 & 0 & \cdots & s^{v_q} \end{bmatrix}$$

$$\bar{\mathbf{L}}(s) = \left[\begin{array}{c|c|c|c} 1 \;\; s \;\; \cdots \;\; s^{v_1-1} & \mathbf{0} & & \mathbf{0} \\ \hline \mathbf{0} & 1 \;\; s \;\; \cdots \;\; s^{v_2-1} & & \mathbf{0} \\ \hline \vdots & \vdots & \ddots & \\ \hline \mathbf{0} & \mathbf{0} & & 1 \;\; s \;\; \cdots \;\; s^{v_q-1} \end{array}\right] \tag{6-129}$$

and

$$\mathbf{N}(s) = \bar{\mathbf{L}}(s)\mathbf{N}_{lc} \tag{6-130}$$

Note that $\mathbf{D}_{hr}$ is a $q \times q$ matrix and is nonsingular, for $\mathbf{D}(s)$ is row reduced. The matrix $\bar{\mathbf{L}}(s)$ is a $q \times n$ matrix. $\mathbf{D}_{lr}$ and $\mathbf{D}_{lr}\mathbf{D}_{hr}^{-1}$ are $n \times q$ matrices. Thus a realization of (6-125) is given by

$$\dot{\mathbf{x}}(t) = \left[\begin{array}{ccccc|ccccc|c|ccccc} 0 & 0 & \cdots & 0 & x & & & & & x & & & & & & x \\ 1 & 0 & & 0 & x & & & & & x & & & & & & x \\ 0 & 1 & \cdots & 0 & x & & & \mathbf{0} & & x & & & & \mathbf{0} & & x \\ \vdots & \vdots & & \vdots & \vdots & & & & & \vdots & & & & & & \vdots \\ 0 & 0 & \cdots & 1 & x & & & & & x & & & & & & x \\ \hline & & & & x & 0 & 0 & \cdots & 0 & x & & & & & & x \\ & & & & x & 1 & 0 & & 0 & x & & & & & & x \\ & & \mathbf{0} & & x & 0 & 1 & & 0 & x & & & & \mathbf{0} & & x \\ & & & & \vdots & \vdots & \vdots & & \vdots & \vdots & & & & & & \vdots \\ & & & & x & 0 & 0 & \cdots & 1 & x & & & & & & x \\ \hline & & \vdots & & & & & & & & \ddots & & & \vdots & & \\ \hline & & & & x & & & & & x & & 0 & 0 & \cdots & 0 & x \\ & & & & x & & & & & x & & 1 & 0 & \cdots & 0 & x \\ & & \mathbf{0} & & x & & & \mathbf{0} & & x & & 0 & 1 & \cdots & 0 & x \\ & & & & \vdots & & & & & \vdots & & \vdots & \vdots & & \vdots & \vdots \\ & & & & x & & & & & x & & 0 & 0 & \cdots & 1 & x \end{array}\right]\mathbf{x}(t) + \mathbf{N}_{lc}\mathbf{u}(t) \tag{6-131a}$$

$$\mathbf{y} = [\mathbf{0} \;\; \mathbf{0} \;\; \cdots \;\; \mathbf{0} \;\; \mathbf{c}_{1v_1} \,|\, \mathbf{0} \;\; \mathbf{0} \;\; \cdots \;\; \mathbf{0} \;\; \mathbf{c}_{2v_2} \,|\, \cdots \,|\, \mathbf{0} \;\; \mathbf{0} \;\; \cdots \;\; \mathbf{0} \;\; \mathbf{c}_{qv_q}]\mathbf{x} \tag{6-131b}$$

where $\mathbf{c}_{iv_i}$ is the ith column of $\mathbf{D}_{hr}^{-1}$ and the $(\sum_{m=1}^{i} v_m)$th column of the matrix $\mathbf{A}$ is the ith column of $-\mathbf{D}_{lr}\mathbf{D}_{hr}^{-1}$. Since (6-131) is a realization of (6-125), we have

$$\begin{aligned}\mathbf{C}(s\mathbf{I}-\mathbf{A})^{-1}\mathbf{B} &= \mathbf{C}(s\mathbf{I}-\mathbf{A})^{-1}\mathbf{N}_{lc} \\ &= \mathbf{D}^{-1}(s)\mathbf{N}(s) = \mathbf{D}^{-1}(s)\bar{\mathbf{L}}(s)\mathbf{N}_{lc}\end{aligned} \tag{6-132}$$

which implies

$$\mathbf{D}^{-1}(s)\bar{\mathbf{L}}(s) = \mathbf{C}(s\mathbf{I}-\mathbf{A})^{-1} \tag{6-133}$$

or

$$\bar{\mathbf{L}}(s)(s\mathbf{I}-\mathbf{A}) = \mathbf{D}(s)\mathbf{C} \tag{6-134}$$

This is a key equation in this realization, and will be used in the next section. Equation (6-131) will be called a multivariable observable-form realization of $\hat{\mathbf{G}}(s)$.

Dual to a multivariable controllable-form realization, a multivariable observable-form realization of $\hat{\mathbf{G}}(s) = \mathbf{D}^{-1}(s)\mathbf{N}(s)$ is always observable no matter whether $\mathbf{D}(s)$ and $\mathbf{N}(s)$ are left coprime or not. If $\mathbf{D}(s)$ and $\mathbf{N}(s)$ are left coprime, then the realization is controllable as well; otherwise, it is not controllable.

*6-7 Polynomial Matrix Description

Consider the network shown in Figure 6-8. If we assign the capacitor voltages and inductor currents as state variables and use the procedure discussed in Section 3-4, then a dynamical equation can be developed to describe the network. There are, however, other methods available to develop mathematical equations to describe the network. For example, if we assign the loop currents $\xi_1(t)$ and $\xi_2(t)$ as shown and use the loop analysis, then we can obtain

$$\left(2s+\frac{1}{3s}\right)\hat{\xi}_1(s) - \frac{1}{3s}\hat{\xi}_2(s) = \hat{u}(s)$$

$$-\frac{1}{3s}\hat{\xi}_1(s) + \left(\frac{1}{3s}+\frac{1}{s}+2s+1\right)\hat{\xi}_2(s) = 0$$

or

$$\begin{bmatrix} 6s^2+1 & -1 \\ -1 & 6s^2+3s+4 \end{bmatrix}\begin{bmatrix}\hat{\xi}_1(s) \\ \hat{\xi}_2(s)\end{bmatrix} = \begin{bmatrix} 3s \\ 0 \end{bmatrix}\hat{u}(s)$$

to describe the network. The output y is equal to $\hat{y}(s) = 2s\hat{\xi}_2(s)$ or

$$\hat{y}(s) = [0 \quad 2s]\begin{bmatrix}\hat{\xi}_1(s) \\ \hat{\xi}_2(s)\end{bmatrix} + 0\cdot\hat{u}(s)$$

This set of equations can be written in a general form as

$$\mathbf{P}(s)\hat{\boldsymbol{\xi}}(s) = \mathbf{Q}(s)\hat{\mathbf{u}}(s) \tag{6-135a}$$

$$\hat{\mathbf{y}}(s) = \mathbf{R}(s)\hat{\boldsymbol{\xi}}(s) + \mathbf{W}(s)\hat{\mathbf{u}}(s) \tag{6-135b}$$

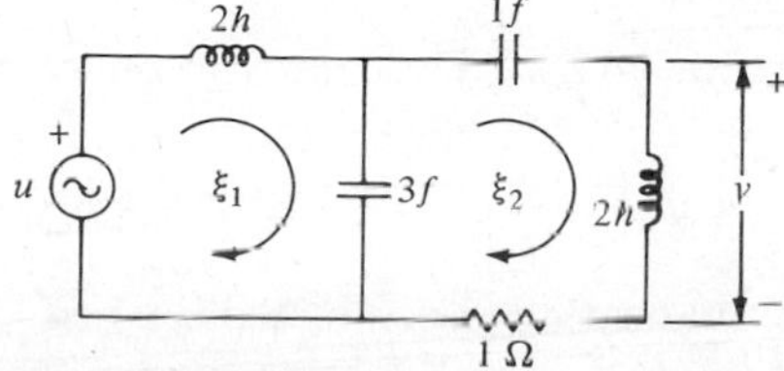

Figure 6-8 A network.

where $\mathbf{P}(s)$, $\mathbf{Q}(s)$, $\mathbf{R}(s)$, and $\mathbf{W}(s)$ are, respectively, $m \times m$, $m \times p$, $q \times m$, and $q \times p$ polynomial matrices; $\hat{\mathbf{u}}(s)$ is the $p \times 1$ input vector, and $\hat{\mathbf{y}}(s)$ is the $q \times 1$ output vector. The variable $\hat{\boldsymbol{\xi}}(s)$ is a $m \times 1$ vector and will be called the *pseudostate*. In order to insure a unique solution in (6-135), the square matrix $\mathbf{P}(s)$ is assumed to be nonsingular. The set of equations in (6-135) is called the polynomial matrix description of a system. In the time domain, this set of equations becomes

$$\mathbf{P}(p)\boldsymbol{\xi}(t) = \mathbf{Q}(p)\mathbf{u}(t) \tag{6-136a}$$

$$\mathbf{y}(t) = \mathbf{R}(p)\boldsymbol{\xi}(t) + \mathbf{W}(p)\mathbf{u}(t) \tag{6-136b}$$

where p stands for d/dt and p^i for d^i/dt^i. From (6-135a) we have $\hat{\boldsymbol{\xi}}(s) = \mathbf{P}^{-1}(s)\mathbf{Q}(s)\hat{\mathbf{u}}(s)$. The substitution of $\hat{\boldsymbol{\xi}}(s)$ into (6-135b) yields

$$\hat{\mathbf{y}}(s) = [\mathbf{R}(s)\mathbf{P}^{-1}(s)\mathbf{Q}(s) + \mathbf{W}(s)]\hat{\mathbf{u}}(s)$$

Hence the transfer matrix of (6-135) from $\mathbf{u}$ to $\mathbf{y}$ is equal to

$$\hat{\mathbf{G}}(s) = \mathbf{R}(s)\mathbf{P}^{-1}(s)\mathbf{Q}(s) + \mathbf{W}(s) \tag{6-137}$$

From this, we can see that the state-variable equation is a special case of (6-135) if we identify $\mathbf{R}(s) = \mathbf{C}$, $\mathbf{P}(s) = s\mathbf{I} - \mathbf{A}$, $\mathbf{Q}(s) = \mathbf{B}$ and $\mathbf{W}(s) = \mathbf{E}$. The fraction $\hat{\mathbf{G}}(s) = \mathbf{N}_r(s)\mathbf{D}_r^{-1}(s)$ is a special case of (6-135) if we identify $\mathbf{R}(s) = \mathbf{N}_r(s)$, $\mathbf{P}(s) = \mathbf{D}_r(s)$, $\mathbf{Q}(s) = \mathbf{I}$, and $\mathbf{W}(s) = \mathbf{0}$. The fraction $\hat{\mathbf{G}}(s) = \mathbf{D}_l^{-1}(s)\mathbf{N}_l(s)$ is also a special case of (6-135) if we identify $\mathbf{R}(s) = \mathbf{I}$, $\mathbf{P}(s) = \mathbf{D}_l(s)$, $\mathbf{Q}(s) = \mathbf{N}_l(s)$, and $\mathbf{W}(s) = \mathbf{0}$. Hence this description is the most general one.

Given a polynomial matrix description (6-135), its transfer function matrix $\hat{\mathbf{G}}(s)$ from $\mathbf{u}$ to $\mathbf{y}$ can be readily computed by using (6-137). Note that the computed $\hat{\mathbf{G}}(s)$ is not necessarily proper. In the following we discuss a method of finding a dynamical equation for (6-135). We consider first (6-135a). If $\mathbf{P}(s)$ is not row reduced, we can find an $m \times m$ unimodular matrix $\mathbf{M}(s)$ such that $\mathbf{M}(s)\mathbf{P}(s)$ is row reduced (Theorem G-11). The premultiplication of $\mathbf{M}(s)$ on both sides of (6-135) yields,

$$\mathbf{M}(s)\mathbf{P}(s)\hat{\boldsymbol{\xi}}(s) = \mathbf{M}(s)\mathbf{Q}(s)\hat{\mathbf{u}}(s) \tag{6-138}$$

which implies

$$\hat{\boldsymbol{\xi}}(s) = (\mathbf{M}(s)\mathbf{P}(s))^{-1}\mathbf{M}(s)\mathbf{Q}(s)\hat{\mathbf{u}}(s) \triangleq \mathbf{P}_r^{-1}(s)\mathbf{Q}_r(s)\hat{\mathbf{u}}(s) \tag{6-139}$$

where $\mathbf{P}_r \triangleq \mathbf{M}(s)\mathbf{P}(s)$ and $\mathbf{Q}_r(s) \triangleq \mathbf{M}(s)\mathbf{Q}(s)$. Note that deg det $\mathbf{P}_r(s) =$ deg det $\mathbf{P}(s)$. In general $\mathbf{P}_r^{-1}(s)\mathbf{Q}_r(s)$ is not strictly proper. If so, we carry out the division:

$$\mathbf{Q}_r(s) = \mathbf{P}_r(s)\mathbf{Y}(s) + \bar{\mathbf{Q}}_r(s) \tag{6-140}$$

so that $\mathbf{P}_r^{-1}(s)\bar{\mathbf{Q}}_r(s)$ is strictly proper (Theorem G-12′). The substitution of (6-140) into (6-139) yields

$$\hat{\boldsymbol{\xi}}(s) = \mathbf{P}_r^{-1}(s)\bar{\mathbf{Q}}_r(s)\hat{\mathbf{u}}(s) + \mathbf{Y}(s)\hat{\mathbf{u}}(s) \tag{6-141}$$

Now $\mathbf{P}_r^{-1}(s)\bar{\mathbf{Q}}_r(s)$ is strictly proper; hence the realization procedure discussed from (6-125) to (6-131) can be applied to find $\{\mathbf{A}, \mathbf{B}, \mathbf{C}_0\}$ of the form shown in

(6-131) such that

$$\mathbf{C}_0(s\mathbf{I}-\mathbf{A})^{-1}\mathbf{B}=\mathbf{P}_r^{-1}(s)\bar{\mathbf{Q}}_r(s) \tag{6-142}$$

If deg det $\mathbf{P}_r(s)=n$, then $\mathbf{A}$, $\mathbf{B}$, and $\mathbf{C}_0$ are, respectively, $n\times n$, $n\times p$, and $m\times n$ matrices. Similar to (6-130), (6-133), and (6-134), we have

$$\bar{\mathbf{Q}}_r(s)=\bar{\mathbf{L}}(s)\mathbf{B} \tag{6-143}$$

and

$$\mathbf{C}_0(s\mathbf{I}-\mathbf{A})^{-1}=\mathbf{P}_r^{-1}(s)\bar{\mathbf{L}}(s) \tag{6-144}$$

or

$$\mathbf{P}_r(s)\mathbf{C}_0=\bar{\mathbf{L}}(s)(s\mathbf{I}-\mathbf{A}) \tag{6-145}$$

where $\bar{\mathbf{L}}(s)$ is defined as in (6-129) and is a $m\times n$ matrix. Since $\bar{\mathbf{L}}(s)$ contains a unit matrix of order m, the rank of $[\bar{\mathbf{L}}(s) \ \vdots\ \mathbf{P}_r(s)]$ is m for every s in $\mathbb{C}$. Hence $\mathbf{P}_r(s)$ and $\bar{\mathbf{L}}(s)$ are left coprime (Theorem G-8′). Because dim $\mathbf{A}=$ deg det $\mathbf{P}_r(s)$, Corollary G-8 implies that $\mathbf{C}_0$ and $(s\mathbf{I}-\mathbf{A})$ are right coprime, and consequently, $\{\mathbf{A}, \mathbf{C}_0\}$ is observable. Note that no assumption of the coprimeness of $\{\mathbf{P}(s), \mathbf{Q}(s)\}$ or $\{\mathbf{P}(s), \mathbf{R}(s)\}$ is used here. The substitution of (6-141) and (6-142) into (6-135b) yields

$$\hat{\mathbf{y}}(s)=\mathbf{R}(s)\mathbf{C}_0(s\mathbf{I}-\mathbf{A})^{-1}\mathbf{B}\hat{\mathbf{u}}(s)+(\mathbf{R}(s)\mathbf{Y}(s)+\mathbf{W}(s))\hat{\mathbf{u}}(s) \tag{6-146}$$

This is not exactly in the state variable form because the matrix $\mathbf{R}(s)\mathbf{C}_0(s\mathbf{I}-\mathbf{A})^{-1}\mathbf{B}$ may not be strictly proper. If so, we can apply Corollary G-12 to $\mathbf{R}(s)\mathbf{C}_0$ to obtain

$$\mathbf{R}(s)\mathbf{C}_0=\mathbf{X}(s)(s\mathbf{I}-\mathbf{A})+\mathbf{C} \tag{6-147}$$

where $\mathbf{X}(s)$ is a polynomial matrix and $\mathbf{C}=\mathbf{R}_r(\mathbf{A})\mathbf{C}_0$ is a $q\times n$ constant matrix defined as in Corollary G-12. Using (6-147), Equation (6-146) becomes

$$\hat{\mathbf{y}}(s)=\mathbf{C}(s\mathbf{I}-\mathbf{A})^{-1}\mathbf{B}\hat{\mathbf{u}}(s)+(\mathbf{X}(s)\mathbf{B}+\mathbf{R}(s)\mathbf{Y}(s)+\mathbf{W}(s))\hat{\mathbf{u}}(s) \tag{6-148}$$

in which $\mathbf{C}(s\mathbf{I}-\mathbf{A})^{-1}\mathbf{B}$ is a $q\times p$ strictly proper rational matrix and

$$\mathbf{E}(s)\triangleq\mathbf{X}(s)\mathbf{B}+\mathbf{R}(s)\mathbf{Y}(s)+\mathbf{W}(s) \tag{6-149}$$

is a $q\times p$ polynomial matrix. From this development, we conclude that the dynamical equation

$$\dot{\mathbf{x}}(t)=\mathbf{A}\mathbf{x}(t)+\mathbf{B}\mathbf{u}(t) \tag{6-150a}$$

$$\mathbf{y}(t)=\mathbf{C}\mathbf{x}(t)+\mathbf{E}(p)\mathbf{u}(t) \tag{6-150b}$$

where $p=d/dt$, is a realization of the polynomial matrix description in (6-135) or the transfer matrix $\hat{\mathbf{G}}(s)$ in (6-137). Note that $\mathbf{W}(s)=\mathbf{0}$ does not imply that $\hat{\mathbf{G}}(s)$ in (6-137) is strictly proper. Whether or not $\hat{\mathbf{G}}(s)$ is proper is determined by $\mathbf{E}(s)$ defined in (6-149). If $\mathbf{E}(p)=\mathbf{E}_0+\mathbf{E}_1p+\mathbf{E}_2p^2+\cdots$, then (6-150b) can be written explicitly as

$$\mathbf{y}(t)=\mathbf{C}\mathbf{x}(t)+\mathbf{E}_0\mathbf{u}(t)+\mathbf{E}_1\dot{\mathbf{u}}(t)+\mathbf{E}_2\ddot{\mathbf{u}}(t)+\cdots$$

From (6-144) and $\mathbf{P}_r(s)=\mathbf{M}(s)\mathbf{P}(s)$, where $\mathbf{M}(s)$ is a unimodular matrix, we have

$$\det(s\mathbf{I}-\mathbf{A})=k\det\mathbf{P}(s)$$

for some nonzero constant k. This relation is a consequence of the realization procedure and is independent of the coprimeness among $\mathbf{P}(s)$, $\mathbf{R}(s)$, and $\mathbf{Q}(s)$. The $\{\mathbf{A}, \mathbf{C}_0\}$ in (6-144) is of the multivariable observable form and is always observable. The observability of $\{\mathbf{A}, \mathbf{C}_0\}$, however, does not imply the observability of $\{\mathbf{A}, \mathbf{C}\}$ in (6-150). In the following, we shall establish

$$\{\mathbf{P}(s), \mathbf{Q}(s) \text{ left coprime}\} \quad \text{if and only if} \quad \{\mathbf{A}, \mathbf{B} \text{ controllable}\} \tag{6-151}$$

$$\{\mathbf{P}(s), \mathbf{R}(s) \text{ right coprime}\} \quad \text{if and only if} \quad \{\mathbf{A}, \mathbf{C} \text{ observable}\} \tag{6-152}$$

Before doing so, we need some preliminary development. We combine (6-145), (6-140), (6-143), (6-147), and (6-149) to form

$$\begin{bmatrix} \bar{\mathbf{L}}(s) & \mathbf{0} \\ -\mathbf{X}(s) & \mathbf{I}_q \end{bmatrix}\begin{bmatrix} s\mathbf{I}-\mathbf{A} & \mathbf{B} \\ -\mathbf{C} & \mathbf{E}(s) \end{bmatrix} = \begin{bmatrix} \mathbf{P}_r(s) & \mathbf{Q}_r(s) \\ -\mathbf{R}(s) & \mathbf{W}(s) \end{bmatrix}\begin{bmatrix} \mathbf{C}_0 & -\mathbf{Y}(s) \\ \mathbf{0} & \mathbf{I}_p \end{bmatrix} \tag{6-153}$$

They are, respectively, $(m+q)\times(n+q)$, $(n+q)\times(n+p)$, $(m+q)\times(m+p)$, and $(m+p)\times(n+p)$ matrices. Because of the right coprimeness of $\{\mathbf{P}_r(s), \bar{\mathbf{L}}(s)\}$ and the left coprimeness of $\{s\mathbf{I}-\mathbf{A}, \mathbf{C}_0\}$, (6-145) and Problem G-11 imply the existence of $\mathbf{U}_{11}(s)$ and $\mathbf{U}_{12}(s)$ such that

$$\underbrace{\begin{bmatrix} -\mathbf{U}_{11}(s) & \mathbf{U}_{12}(s) \\ \mathbf{P}_r(s) & \bar{\mathbf{L}}(s) \end{bmatrix}}_{\text{unimodular}}\begin{bmatrix} -\mathbf{C}_0 \\ s\mathbf{I}-\mathbf{A} \end{bmatrix} = \begin{bmatrix} \mathbf{I}_n \\ \mathbf{0} \end{bmatrix} \tag{6-154}$$

where the leftmost matrix is a unimodular matrix. Using this matrix, we expand (6-153) as

$$\left[\begin{array}{cc:c} -\mathbf{U}_{11}(s) & \mathbf{U}_{12}(s) & \mathbf{0} \\ \mathbf{P}_r(s) & \bar{\mathbf{L}}(s) & \mathbf{0} \\ \hdashline \mathbf{0} & -\mathbf{X}(s) & \mathbf{I}_q \end{array}\right]\left[\begin{array}{c:cc} \mathbf{I}_m & \mathbf{0} & \mathbf{0} \\ \hdashline \mathbf{0} & (s\mathbf{I}-\mathbf{A}) & \mathbf{B} \\ \mathbf{0} & -\mathbf{C} & \mathbf{E}(s) \end{array}\right]$$

$$= \left[\begin{array}{c:cc} \mathbf{I}_n & \mathbf{0} & \mathbf{0} \\ \hdashline \mathbf{0} & \mathbf{P}_r(s) & \mathbf{Q}_r(s) \\ \mathbf{0} & -\mathbf{R}(s) & \mathbf{W}(s) \end{array}\right]\left[\begin{array}{cc:c} -\mathbf{U}_{11}(s) & \mathbf{U}_{12}(s)(s\mathbf{I}-\mathbf{A}) & \mathbf{U}_{12}(s)\mathbf{B} \\ \mathbf{I}_m & \mathbf{C}_0 & -\mathbf{Y}(s) \\ \hdashline \mathbf{0} & \mathbf{0} & \mathbf{I}_p \end{array}\right] \tag{6-155}$$

This is obtained by augmenting the first three matrices in (6-153) as shown in (6-155) and then searching the fourth matrix in (6-155) to complete the equality. Since the left-upper-corner matrix in the leftmost matrix in (6-155) is unimodular, so is the entire leftmost matrix in (6-155). From (6-154), we have $\mathbf{U}_{12}(s)(s\mathbf{I}-\mathbf{A}) = \mathbf{I}_n - \mathbf{U}_{11}(s)\mathbf{C}_0$ and

$$\begin{bmatrix} -\mathbf{U}_{11}(s) & \mathbf{U}_{12}(s)(s\mathbf{I}-\mathbf{A}) \\ \mathbf{I}_m & \mathbf{C}_0 \end{bmatrix} = \begin{bmatrix} -\mathbf{U}_{11}(s) & \mathbf{I}_n - \mathbf{U}_{11}(s)\mathbf{C}_0 \\ \mathbf{I}_m & \mathbf{C}_0 \end{bmatrix}$$

$$= \begin{bmatrix} \mathbf{I}_n & -\mathbf{U}_{11}(s) \\ \mathbf{0} & \mathbf{I}_m \end{bmatrix}\begin{bmatrix} \mathbf{0} & \mathbf{I}_n \\ \mathbf{I}_m & \mathbf{C}_0 \end{bmatrix} \tag{6-156}$$

Since the two matrices after the last equality are unimodular, so is the first matrix in (6-156). Consequently, the rightmost matrix in (6-155) is also unimodular.

As the last step, we shall replace $\mathbf{P}_r(s)$ and $\mathbf{Q}_r(s)$ in (6-155) by $\mathbf{P}(s)$ and $\mathbf{Q}(s)$.

Recall that $\mathbf{P}_r(s)=\mathbf{M}(s)\mathbf{P}(s)$ and $\mathbf{Q}_r(s)=\mathbf{M}(s)\mathbf{Q}(s)$, where $\mathbf{M}(s)$ is unimodular. The premultiplication of diag $\{\mathbf{I}_n, \mathbf{M}^{-1}(s), \mathbf{I}_q\}$, a unimodular matrix, to (6-155) yields

$$\left[\begin{array}{cc|c} -\mathbf{U}_{11}(s) & \mathbf{U}_{12}(s) & \mathbf{0} \\ \mathbf{P}(s) & \mathbf{M}^{-1}(s)\bar{\mathbf{L}}(s) & \mathbf{0} \\ \hline \mathbf{0} & -\mathbf{X}(s) & \mathbf{I}_q \end{array}\right]\left[\begin{array}{c|cc} \mathbf{I}_m & \mathbf{0} & \mathbf{0} \\ \hline \mathbf{0} & (s\mathbf{I}-\mathbf{A}) & \mathbf{B} \\ \mathbf{0} & -\mathbf{C} & \mathbf{E}(s) \end{array}\right]$$

$$=\left[\begin{array}{c|cc} \mathbf{I}_n & \mathbf{0} & \mathbf{0} \\ \hline \mathbf{0} & \mathbf{P}(s) & \mathbf{Q}(s) \\ \mathbf{0} & -\mathbf{R}(s) & \mathbf{W}(s) \end{array}\right]\left[\begin{array}{cc|c} -\mathbf{U}_{11}(s) & \mathbf{U}_{12}(s)(s\mathbf{I}-\mathbf{A}) & \mathbf{U}_{12}(s)\mathbf{B} \\ \mathbf{I}_m & \mathbf{C}_0 & -\mathbf{Y}(s) \\ \hline \mathbf{0} & \mathbf{0} & \mathbf{I}_p \end{array}\right] \qquad \textbf{(6-157)}$$

where the leftmost and rightmost matrices remain to be unimodular. With this equation, we are ready to establish (6-151) and (6-152).

From the first two block rows of (6-157), we can readily obtain

$$\begin{bmatrix} -\mathbf{U}_{11}(s) & \mathbf{U}_{12}(s) \\ \mathbf{P}(s) & \mathbf{M}^{-1}(s)\bar{\mathbf{L}}(s) \end{bmatrix}\begin{bmatrix} \mathbf{I}_m & \mathbf{0} & \mathbf{0} \\ \mathbf{0} & (s\mathbf{I}-\mathbf{A}) & \mathbf{B} \end{bmatrix}=\begin{bmatrix} \mathbf{I}_n & \mathbf{0} & \mathbf{0} \\ \mathbf{0} & \mathbf{P}(s) & \mathbf{Q}(s) \end{bmatrix}\mathbf{S}(s) \qquad \textbf{(6-158)}$$

where $\mathbf{S}(s)$ denotes the rightmost matrix in (6-157). Since the leftmost and rightmost matrices in (6-158) are unimodular, we conclude that

$$\begin{bmatrix} \mathbf{I}_m & \mathbf{0} & \mathbf{0} \\ \mathbf{0} & (s\mathbf{I}-\mathbf{A}) & \mathbf{B} \end{bmatrix} \text{has rank } n+m \text{ for every } s \text{ in } \mathbb{C}$$

if and only if

$$\begin{bmatrix} \mathbf{I}_n & \mathbf{0} & \mathbf{0} \\ \mathbf{0} & \mathbf{P}(s) & \mathbf{Q}(s) \end{bmatrix} \text{has rank } n+m \text{ for every } s \text{ in } \mathbb{C}$$

which implies that

$$[(s\mathbf{I}-\mathbf{A}) \;\vdots\; \mathbf{B}] \text{ has rank } n \text{ for every } s \text{ in } \mathbb{C}$$

$$\text{if and only if } [\mathbf{P}(s) \;\vdots\; \mathbf{Q}(s)] \text{ has rank } m \text{ for every } s \text{ in } \mathbb{C}$$

or equivalently, $\{\mathbf{A}, \mathbf{B}\}$ is controllable if and only if $\mathbf{P}(s)$ and $\mathbf{Q}(s)$ are left coprime. Similarly, we can establish that $\{\mathbf{A}, \mathbf{C}\}$ is observable if and only if $\mathbf{P}(s)$ and $\mathbf{R}(s)$ are right coprime.

Definition 6-2

The polynomial matrix description in (6-135) is said to be irreducible if and only if $\mathbf{P}(s)$ and $\mathbf{Q}(s)$ are left coprime and $\mathbf{P}(s)$ and $\mathbf{R}(s)$ are right coprime.

Theorem 6-7

Consider the polynomial matrix description in (6-135) with $n=\deg\det \mathbf{P}(s)$. Then an n-dimensional dynamical equation realization of (6-135) is irreducible or, equivalently, controllable and observable if and only if (6-135) is irreducible. ∎

This theorem follows directly from Definition 6-2 and the statements in (6-151) and (6-152). To conclude this section, we discuss the situation where (6-135) is not irreducible. If $\mathbf{P}(s)$ and $\mathbf{Q}(s)$ are not left coprime, then there exists a $m \times m$ polynomial matrix $\mathbf{H}(s)$ with deg det $\mathbf{H}(s) > 1$ such that

$$\mathbf{P}(s) = \mathbf{H}(s)\bar{\mathbf{P}}(s) \qquad \text{and} \qquad \mathbf{Q}(s) = \mathbf{H}(s)\bar{\mathbf{Q}}(s)$$

Consequently, at the roots, λ, of det $\mathbf{H}(s) = 0$, we have

$$\rho[\mathbf{P}(\lambda) \quad \mathbf{Q}(\lambda)] = \rho\mathbf{H}(\lambda)[\bar{\mathbf{P}}(\lambda) \quad \bar{\mathbf{Q}}(\lambda)] < m$$

where ρ stands for the rank. If an n-dimensional dynamical equation is developed for $\{\mathbf{P}(s), \mathbf{Q}(s), \mathbf{R}(s), \mathbf{W}(s)\}$ with $n = \deg \det \mathbf{P}(s)$, then this n-dimensional state equation will not be controllable. If this equation is decomposed into the form in (5-54), then the eigenvalues associated with $\bar{\mathbf{A}}_{\bar{c}}$, the uncontrollable part, will be equal to the roots of det $\mathbf{H}(s)$. Hence the roots of det $\mathbf{H}(s)$ or, equivalently, those λ in $\mathbb{C}$ with $\rho[\mathbf{P}(\lambda) \quad \mathbf{Q}(\lambda)] < m$ will be called the uncontrollable mode of $\{\mathbf{P}(s), \mathbf{Q}(s)\}$. These uncontrollable modes are called the input-decoupling zeros in Reference S185.

Similarly, if $\mathbf{P}(s)$ and $\mathbf{R}(s)$ are not right coprime, then the roots of the determinant of their greatest common right divisor will be called the unobservable modes of $\{\mathbf{P}(s), \mathbf{R}(s)\}$. These roots are called the output-decoupling zeros in Reference S185. Hence an irreducible $\{\mathbf{P}(s), \mathbf{Q}(s), \mathbf{R}(s), \mathbf{W}(s)\}$ does not have any uncontrollable or unobservable modes.

The discussion of the polynomial matrix description in this section is not complete. We discuss only its realization problem. For a more detailed discussion, see References S34 and S218. In S218, the description is called the differential operator description.

*6-8 Strict System Equivalence

In this text, we have introduced three types of mathematical descriptions for linear time-invariant multivariable systems. They are state-variable equation

$$\dot{\mathbf{x}} = \mathbf{A}\mathbf{x} + \mathbf{B}\mathbf{u} \qquad \mathbf{y} = \mathbf{C}\mathbf{x} + \mathbf{E}\mathbf{u} + \mathbf{E}_1\dot{\mathbf{u}} + \cdots \tag{6-159}$$

transfer matrix in fractional forms

$$\hat{\mathbf{y}}(s) = \hat{\mathbf{G}}(s)\hat{\mathbf{u}}(s) = \mathbf{N}_r(s)\mathbf{D}_r^{-1}(s)\hat{\mathbf{u}}(s) = \mathbf{D}_l^{-1}(s)\mathbf{N}_l(s)\hat{\mathbf{u}}(s) \tag{6-160}$$

and polynomial matrix description

$$\begin{aligned} \mathbf{P}(s)\hat{\boldsymbol{\xi}}(s) &= \mathbf{Q}(s)\hat{\mathbf{u}}(s) \\ \hat{\mathbf{y}}(s) &= \mathbf{R}(s)\hat{\boldsymbol{\xi}}(s) + \mathbf{W}(s)\hat{\mathbf{u}}(s) \end{aligned} \tag{6-161}$$

Equation (6-159) is more general than the one studied in earlier sections by including derivatives of $\mathbf{u}$ in $\mathbf{y}$. By so doing, its transfer matrix can be extended to include improper case. Given a state-variable equation, the transfer function description can be obtained as $\hat{\mathbf{G}}(s) = \mathbf{C}(s\mathbf{I} - \mathbf{A})^{-1}\mathbf{B} + (\mathbf{E} + \mathbf{E}_1 s + \cdots)$. If $\hat{\mathbf{G}}(s)$ is factored as $\hat{\mathbf{G}}(s) = \mathbf{N}_r(s)\mathbf{D}_r^{-1}(s)$, then a polynomial matrix description can be obtained as $\mathbf{Q}(s) = \mathbf{I}$, $\mathbf{P}(s) = \mathbf{D}_r(s)$, $\mathbf{R}(s) = \mathbf{N}_r(s)$, and $\mathbf{W}(s) = \mathbf{0}$. Conversely, given a transfer matrix or polynomial matrix description, we may use the procedures in Section 6-7 to develop a state variable description. Hence the relationships among them have been essentially established. Even so, some questions can still be posed regarding these descriptions. For example, consider a transfer matrix $\hat{\mathbf{G}}(s)$ which is not necessarily proper. It can be factored as $\hat{\mathbf{G}}(s) = \mathbf{N}_r(s)\mathbf{D}_r^{-1}(s)$ or it can be decomposed as $\hat{\mathbf{G}}(s) = \hat{\mathbf{G}}_1(s) + \mathbf{E}(s)$, where $\hat{\mathbf{G}}_1$ is strictly proper and $\mathbf{E}(s)$ is a polynominal matrix and then factored as $\hat{\mathbf{G}}(s) = \mathbf{N}_1(s)\mathbf{D}_1^{-1}(s) + \mathbf{E}(s)$. The question is then: what is the relationship between $\{\mathbf{N}_r(s), \mathbf{D}_r(s)\}$ and $\{\mathbf{N}_1(s), \mathbf{D}_1(s), \mathbf{E}(s)\}$? To answer this and other related questions, we rewrite (6-161) as

$$\begin{bmatrix} \mathbf{P}(s) & \mathbf{Q}(s) \\ -\mathbf{R}(s) & \mathbf{W}(s) \end{bmatrix} \begin{bmatrix} \hat{\boldsymbol{\xi}}(s) \\ -\hat{\mathbf{u}}(s) \end{bmatrix} = \begin{bmatrix} \mathbf{0} \\ -\hat{\mathbf{y}}(s) \end{bmatrix} \tag{6-162}$$

where $\boldsymbol{\xi}$ is called the *pseudostate*, and the matrix

$$\mathbf{S}(s) = \begin{bmatrix} \mathbf{P}(s) & \mathbf{Q}(s) \\ -\mathbf{R}(s) & \mathbf{W}(s) \end{bmatrix} \tag{6-163}$$

is called the *system matrix*. Its transfer function from $\mathbf{u}$ to $\mathbf{y}$ is

$$\hat{\mathbf{G}}(s) = \mathbf{R}(s)\mathbf{P}^{-1}(s)\mathbf{Q}(s) + \mathbf{W}(s) \tag{6-164}$$

If we identify $\mathbf{R}(s) = \mathbf{C}$, $\mathbf{P}(s) = (s\mathbf{I} - \mathbf{A})$, $\mathbf{Q}(s) = \mathbf{B}$, and $\mathbf{W}(s) = \mathbf{E} + \mathbf{E}_1 s + \cdots$, or identify $\mathbf{R}(s) = \mathbf{N}_r(s)$, $\mathbf{P}(s) = \mathbf{D}_r(s)$, $\mathbf{Q}(s) = \mathbf{I}$, and $\mathbf{W}(s) = \mathbf{0}$, then the system matrix includes (6-159) and (6-160) as special cases. Hence the system matrix $\mathbf{S}(s)$ can be used to describe any of the three descriptions in (6-159)–(6-161).

Consider the system matrix $\mathbf{S}(s)$ in (6-163). We extend it to

$$\mathbf{S}_e(s) = \left[\begin{array}{cc:c} \mathbf{I} & \mathbf{0} & \mathbf{0} \\ \mathbf{0} & \mathbf{P}(s) & \mathbf{Q}(s) \\ \hdashline \mathbf{0} & -\mathbf{R}(s) & \mathbf{W}(s) \end{array}\right] \triangleq \begin{bmatrix} \mathbf{P}_e(s) & \mathbf{Q}_e(s) \\ -\mathbf{R}_e(s) & \mathbf{W}(s) \end{bmatrix} \tag{6-165}$$

where $\mathbf{I}$ is a unit matrix of any order so that the order of $\mathbf{P}_e(s)$ is equal to or larger than deg det $\mathbf{P}(s)$. It is clear that

$$\det \mathbf{P}_e(s) = \det \mathbf{P}(s)$$

and $$\hat{\mathbf{G}}_e(s) = \mathbf{R}_e(s)\mathbf{P}_e^{-1}(s)\mathbf{Q}_e(s) + \mathbf{W}(s) = \mathbf{R}(s)\mathbf{P}^{-1}(s)\mathbf{Q}(s) + \mathbf{W} = \hat{\mathbf{G}}(s)$$

Hence the input-output behavior of $\mathbf{S}(s)$ and that of $\mathbf{S}_e(s)$ are identical. In fact, their entire dynamical behaviors are, as will be shown later, equivalent.

Consider two system matrices

$$\mathbf{S}_i(s) = \begin{bmatrix} \mathbf{P}_i(s) & \mathbf{Q}_i(s) \\ -\mathbf{R}_i(s) & \mathbf{W}_i(s) \end{bmatrix} \qquad i = 1, 2 \tag{6-166}$$

where $\mathbf{P}_i(s)$, $\mathbf{Q}_i(s)$, $\mathbf{R}_i(s)$, and $\mathbf{W}_i(s)$ are respectively $m_i \times m_i$, $m_i \times p$, $q \times m_i$, and $q \times p$ polynomial matrices. Note that m_1 is not necessarily equal to m_2. However, because of the discussion in (6-165), we may extend either m_1 or m_2 or both to make them equal and require $m = m_1 = m_2 \geq \deg \det \mathbf{P}_i(s)$, $i = 1, 2$. Without this requirement, the subsequent discussion may not hold. See References S125, S186, and S187.

Definition 6-3

Two system matrices $\mathbf{S}_1(s)$ and $\mathbf{S}_2(s)$ are said to be *strictly system equivalent* if and only if there exist $m \times m$ unimodular polynomial matrices $\mathbf{U}(s)$ and $\mathbf{V}(s)$ and $q \times m$ and $m \times p$ polynomial matrices $\mathbf{X}(s)$ and $\mathbf{Y}(s)$ such that

$$\begin{bmatrix} \mathbf{U}(s) & \mathbf{0} \\ \mathbf{X}(s) & \mathbf{I}_q \end{bmatrix} \begin{bmatrix} \mathbf{P}_1(s) & \mathbf{Q}_1(s) \\ -\mathbf{R}_1(s) & \mathbf{W}_1(s) \end{bmatrix} \begin{bmatrix} \mathbf{V}(s) & \mathbf{Y}(s) \\ \mathbf{0} & \mathbf{I}_p \end{bmatrix} = \begin{bmatrix} \mathbf{P}_2(s) & \mathbf{Q}_2(s) \\ -\mathbf{R}_2(s) & \mathbf{W}_2(s) \end{bmatrix} \tag{6-167}$$

∎

Since $\mathbf{U}(s)$ and $\mathbf{V}(s)$ are unimodulars, so are

$$\begin{bmatrix} \mathbf{U}(s) & \mathbf{0} \\ \mathbf{X}(s) & \mathbf{I}_q \end{bmatrix}, \begin{bmatrix} \mathbf{V}(s) & \mathbf{Y}(s) \\ \mathbf{0} & \mathbf{I}_p \end{bmatrix}$$

and their inverses. Using this property, it can be readily verified that this equivalence relation has the symmetry property [if $S_1(s) \sim S_2(s)$, then $S_2(s) \sim S_1(s)$], the reflexitivity property [$S_1(s) \sim S_1(s)$] and the transitivity property [if $S_1(s) \sim S_2(s)$ and $S_2(s) \sim S_3(s)$, then $S_1(s) \sim S_3(s)$, where $\sim$ denotes strict system equivalence].

The reason of using extended $\mathbf{S}_i(s)$ in Definition 6-3 can be seen from the realization developed in Section 6-7, especially Equations (6-153) and (6-155). Without extending $\mathbf{P}_i(s)$, the matrices $\mathbf{U}(s)$ and $\mathbf{V}(s)$ in (6-167) may not be square and, certainly, may not be unimodular. Using the unimodularity property, we can write (6-167) as

$$\begin{bmatrix} \mathbf{U}(s) & \mathbf{0} \\ \mathbf{X}(s) & \mathbf{I} \end{bmatrix} \begin{bmatrix} \mathbf{P}_1(s) & \mathbf{Q}_1(s) \\ -\mathbf{R}_1(s) & \mathbf{W}_1(s) \end{bmatrix} = \begin{bmatrix} \mathbf{P}_2(s) & \mathbf{Q}_2(s) \\ -\mathbf{R}_2(s) & \mathbf{W}_2(s) \end{bmatrix} \begin{bmatrix} \bar{\mathbf{V}}(s) & \bar{\mathbf{Y}}(s) \\ \mathbf{0} & \mathbf{I}_p \end{bmatrix} \tag{6-168}$$

where $\bar{\mathbf{V}}(s) = \mathbf{V}^{-1}(s)$ and $\bar{\mathbf{Y}}(s) = -\mathbf{V}^{-1}(s)\mathbf{Y}(s)$. Clearly $\bar{\mathbf{V}}(s)$ is unimodular and $\bar{\mathbf{Y}}(s)$ is a polynomial matrix. A comparison of (6-168) and (6-155) reveals immediately that $\{\mathbf{P}(s), \mathbf{Q}(s), \mathbf{R}(s), \mathbf{W}(s)\}$ and its realization $\{\mathbf{A}, \mathbf{B}, \mathbf{C}, \mathbf{D}\}$ developed in Section 6-7 are strictly system equivalent.

Mathematical descriptions which are strictly system equivalent are equivalent in every sense. They have the same transfer matrix; their pseudostates are related by an invertible transformation. Their dynamical-equation realizations developed by using the procedure discussed in Section 6-7 are zero-input and zero-state equivalent (see Section 4-3). If one realization is controllable (observable), so is the other, and conversely. These properties will be established in the following.

Theorem 6-8

Two system matrices which are strictly system equivalent have the same transfer matrix and $\det \mathbf{P}_1(s) = k \det \mathbf{P}_2(s)$, where k is a nonzero constant.

Proof

We multiply (6-167) out to yield

$$\begin{bmatrix} \mathbf{UP}_1\mathbf{V} & \mathbf{UP}_1\mathbf{Y} + \mathbf{UQ}_1 \\ -(\mathbf{R}_1 - \mathbf{XP}_1)\mathbf{V} & (\mathbf{XP}_1 - \mathbf{R}_1)\mathbf{Y} + (\mathbf{XQ}_1 + \mathbf{W}_1) \end{bmatrix} = \begin{bmatrix} \mathbf{P}_2 & \mathbf{Q}_2 \\ -\mathbf{R}_2 & \mathbf{W}_2 \end{bmatrix}$$

Hence we have

$$\mathbf{U}(s)\mathbf{P}_1(s)\mathbf{V}(s)=\mathbf{P}_2(s)$$

which, together with the unimodularity of $\mathbf{U}(s)$ and $\mathbf{V}(s)$, implies

$$k \det \mathbf{P}_1(s)=\det \mathbf{P}_2(s) \tag{6-169}$$

where k is a nonzero constant. Hence we have $\deg \det \mathbf{P}_1(s)=\deg \det \mathbf{P}_2(s)$. We compute

$$\begin{aligned}\hat{\mathbf{G}}_2(s)&=\mathbf{R}_2\mathbf{P}_2^{-1}\mathbf{Q}_2+\mathbf{W}_2\\&=(\mathbf{R}_1-\mathbf{X}\mathbf{P}_1)\mathbf{V}(\mathbf{U}\mathbf{P}_1\mathbf{V})^{-1}\mathbf{U}(\mathbf{P}_1\mathbf{Y}+\mathbf{Q}_1)+(\mathbf{X}\mathbf{P}_1-\mathbf{R}_1)\mathbf{Y}+(\mathbf{X}\mathbf{Q}_1+\mathbf{W}_1)\end{aligned}$$

which, by simple manipulation, becomes

$$\hat{\mathbf{G}}_2(s)=\mathbf{R}_2\mathbf{P}_2^{-1}\mathbf{Q}_2+\mathbf{W}_2=\mathbf{R}_1\mathbf{P}_1^{-1}\mathbf{Q}_1+\mathbf{W}_1=\hat{\mathbf{G}}_1(s)$$

This completes the proof of this theorem. Q.E.D.

In order to establish the relationship between the pseudostates $\boldsymbol{\xi}_1$ and $\boldsymbol{\xi}_2$, we use (6-162) to write $\mathbf{S}_i$ as

$$\begin{bmatrix}\mathbf{P}_1(s) & \mathbf{Q}_1(s)\\ -\mathbf{R}_1(s) & \mathbf{W}_1(s)\end{bmatrix}\begin{bmatrix}\hat{\boldsymbol{\xi}}_1(s)\\ -\hat{\mathbf{u}}(s)\end{bmatrix}=\begin{bmatrix}\mathbf{0}\\ -\hat{\mathbf{y}}(s)\end{bmatrix} \tag{6-170}$$

and

$$\begin{bmatrix}\mathbf{P}_2(s) & \mathbf{Q}_2(s)\\ -\mathbf{R}_2(s) & \mathbf{W}_2(s)\end{bmatrix}\begin{bmatrix}\hat{\boldsymbol{\xi}}_2(s)\\ -\hat{\mathbf{u}}(s)\end{bmatrix}=\begin{bmatrix}\mathbf{0}\\ -\hat{\mathbf{y}}(s)\end{bmatrix} \tag{6-171}$$

Note that $\mathbf{S}_1(s)$ and $\mathbf{S}_2(s)$ are two different descriptions of the same system, hence $\mathbf{u}$ and $\mathbf{y}$ in (6-170) and (6-171) are the same. However, their pseudostates $\boldsymbol{\xi}_1$ and $\boldsymbol{\xi}_2$ are different in general. From (6-168), we have

$$\begin{bmatrix}\mathbf{U} & \mathbf{0}\\ \mathbf{X} & \mathbf{I}\end{bmatrix}\begin{bmatrix}\mathbf{P}_1 & \mathbf{Q}_1\\ -\mathbf{R}_1 & \mathbf{W}_1\end{bmatrix}\begin{bmatrix}\hat{\boldsymbol{\xi}}_1\\ -\hat{\mathbf{u}}\end{bmatrix}=\begin{bmatrix}\mathbf{P}_2 & \mathbf{Q}_2\\ -\mathbf{R}_2 & \mathbf{W}_2\end{bmatrix}\begin{bmatrix}\bar{\mathbf{V}} & \bar{\mathbf{Y}}\\ \mathbf{0} & \mathbf{I}\end{bmatrix}\begin{bmatrix}\hat{\boldsymbol{\xi}}_1\\ -\hat{\mathbf{u}}\end{bmatrix}$$

which can be written as, by the substitution of (6-170),

$$\begin{bmatrix}\mathbf{U} & \mathbf{0}\\ \mathbf{X} & \mathbf{I}\end{bmatrix}\begin{bmatrix}\mathbf{0}\\ -\hat{\mathbf{y}}\end{bmatrix}=\begin{bmatrix}\mathbf{0}\\ -\hat{\mathbf{y}}\end{bmatrix}=\begin{bmatrix}\mathbf{P}_2 & \mathbf{Q}_2\\ -\mathbf{R}_2 & \mathbf{W}_2\end{bmatrix}\begin{bmatrix}\bar{\mathbf{V}}\hat{\boldsymbol{\xi}}_1-\bar{\mathbf{Y}}\hat{\mathbf{u}}\\ -\hat{\mathbf{u}}\end{bmatrix} \tag{6-172}$$

A comparison of (6-171) and (6-172) yields

$$\hat{\boldsymbol{\xi}}_2(s)=\bar{\mathbf{V}}(s)\hat{\boldsymbol{\xi}}_1(s)-\bar{\mathbf{Y}}(s)\hat{\mathbf{u}}(s) \tag{6-173}$$

where $\bar{\mathbf{V}}(s)$ is unimodular, and

$$\begin{aligned}\hat{\boldsymbol{\xi}}_1(s)&=\bar{\mathbf{V}}^{-1}(s)\hat{\boldsymbol{\xi}}_2(s)+\bar{\mathbf{V}}^{-1}(s)\bar{\mathbf{Y}}(s)\hat{\mathbf{u}}(s)\\&=\mathbf{V}(s)\hat{\boldsymbol{\xi}}_2(s)-\mathbf{Y}(s)\hat{\mathbf{u}}(s)\end{aligned} \tag{6-174}$$

We see that $\hat{\boldsymbol{\xi}}_1$ and $\hat{\boldsymbol{\xi}}_2$ are related by the invertible transformation pair in (6-173) and (6-174). Hence, if $\mathbf{S}_1(s)$ and $\mathbf{S}_2(s)$ are strictly system equivalent, then for any input $\mathbf{u}(t)$, and any set of initial conditions in $\mathbf{S}_1$, there exists a unique set of initial conditions in $\mathbf{S}_2$, and vice versa, such that the outputs of $\mathbf{S}_1$ and $\mathbf{S}_2$ are identical and their pseudostates are related by (6-173) and (6-174). In short, if

$\mathbf{S}_i$, $i=1,2$, are strictly system equivalent, there is no difference in their dynamical behaviors.

Consider two system matrices $\mathbf{S}_1(s)$ and $\mathbf{S}_2(s)$. If they are realized by using the procedure developed in Section 6-7, then the realizations $\{\mathbf{A}_i, \mathbf{B}_i, \mathbf{C}_i, \mathbf{E}_i\}$ have the property

$$\det(s\mathbf{I}-\mathbf{A}_i)=k_i \det \mathbf{P}_i(s) \qquad i=1,2$$

If $\mathbf{S}_1(s)$ and $\mathbf{S}_2(s)$ are strictly system equivalent, then $\det \mathbf{P}_1(s)=k \det \mathbf{P}_2(s)$. Hence the dynamical-equation realizations of $\mathbf{S}_1(s)$ and $\mathbf{S}_2(s)$ have the same dimension and the same characteristic polynomial.

Theorem 6-9

Coprimeness, controllability, and observability are invariant under the transformation of strict system equivalence.

Proof

Consider (6-167). Because of the presence of the two zero matrices, the first block row of (6-167) can be written as

$$\mathbf{U}(s)[\mathbf{P}_1(s) \quad \mathbf{Q}_1(s)]\begin{bmatrix} \mathbf{V}(s) & \mathbf{Y}(s) \\ \mathbf{0} & \mathbf{I}_p \end{bmatrix}=[\mathbf{P}_2(s) \quad \mathbf{Q}_2(s)] \tag{6-175}$$

and the first block column of (6-167) can be written as

$$\begin{bmatrix} \mathbf{U}(s) & \mathbf{0} \\ \mathbf{X}(s) & \mathbf{I}_q \end{bmatrix}\begin{bmatrix} \mathbf{P}_1(s) \\ \mathbf{R}_1(s) \end{bmatrix}\mathbf{V}(s)=\begin{bmatrix} \mathbf{P}_2(s) \\ -\mathbf{R}_2(s) \end{bmatrix} \tag{6-176}$$

Since $\mathbf{U}(s)$ and $\mathbf{V}(s)$ are unimodular, we have, for every s in $\mathbb{C}$,

$$\rho[\mathbf{P}_1(s) \quad \mathbf{Q}_1(s)]=\rho[\mathbf{P}_2(s) \quad \mathbf{Q}_2(s)]$$

and

$$\rho\begin{bmatrix} \mathbf{P}_1(s) \\ -\mathbf{R}_1(s) \end{bmatrix}=\rho\begin{bmatrix} \mathbf{P}_2(s) \\ -\mathbf{R}_2(s) \end{bmatrix}$$

where ρ denotes the rank in the field of complex numbers. Hence $\mathbf{P}_1(s)$ and $\mathbf{Q}_1(s)$ are left coprime if and only if $\mathbf{P}_2(s)$ and $\mathbf{Q}_2(s)$ are left coprime. If $\mathbf{P}_2(s)= s\mathbf{I}-\mathbf{A}$ and $\mathbf{Q}_2(s)=\mathbf{B}$, then $\mathbf{P}_1(s)$ and $\mathbf{Q}_1(s)$ are left coprime if and only if $\{\mathbf{A}, \mathbf{B}\}$ is controllable [see Equation (6-151)]. The rest of the theorem can be similarly proved. Q.E.D.

In Section 4-3, we introduced the concept of equivalent dynamical equations. Now we shall show that strict system equivalence is a generalization of this concept.

Theorem 6-10

Two dynamical equations $\{\mathbf{A}, \mathbf{B}, \mathbf{C}, \mathbf{E}\}$ and $\{\bar{\mathbf{A}}, \bar{\mathbf{B}}, \bar{\mathbf{C}}, \bar{\mathbf{E}}\}$ are equivalent if and only if their system matrices are strictly system equivalent.

Proof

If $\{\mathbf{A}, \mathbf{B}, \mathbf{C}, \mathbf{E}\}$ and $\{\bar{\mathbf{A}}, \bar{\mathbf{B}}, \bar{\mathbf{C}}, \bar{\mathbf{E}}\}$ are equivalent, then there exists a nonsingular constant matrix $\mathbf{P}$ such that $\bar{\mathbf{A}} = \mathbf{PAP}^{-1}$, $\bar{\mathbf{B}} = \mathbf{PB}$, $\bar{\mathbf{C}} = \mathbf{CP}^{-1}$ and $\bar{\mathbf{E}} = \mathbf{E}$. With these, it is straightforward to verify

$$\begin{bmatrix} \mathbf{P} & \mathbf{0} \\ \mathbf{0} & \mathbf{I} \end{bmatrix} \begin{bmatrix} s\mathbf{I} - \mathbf{A} & \mathbf{B} \\ -\mathbf{C} & \mathbf{E} \end{bmatrix} \begin{bmatrix} \mathbf{P}^{-1} & \mathbf{0} \\ \mathbf{0} & \mathbf{I} \end{bmatrix} = \begin{bmatrix} s\mathbf{I} - \bar{\mathbf{A}} & \bar{\mathbf{B}} \\ -\bar{\mathbf{C}} & \bar{\mathbf{E}} \end{bmatrix}$$

Hence their system matrices are strictly system equivalent.

Now we assume that $\{\mathbf{A}, \mathbf{B}, \mathbf{C}, \mathbf{D}\}$ and $\{\bar{\mathbf{A}}, \bar{\mathbf{B}}, \bar{\mathbf{C}}, \bar{\mathbf{D}}\}$ are strictly system equivalent. Then from (6-168), we have

$$\begin{bmatrix} \mathbf{U}(s) & \mathbf{0} \\ \mathbf{X}(s) & \mathbf{I}_q \end{bmatrix} \begin{bmatrix} s\mathbf{I} - \mathbf{A} & \mathbf{B} \\ -\mathbf{C} & \mathbf{E} \end{bmatrix} = \begin{bmatrix} s\mathbf{I} - \bar{\mathbf{A}} & \bar{\mathbf{B}} \\ -\bar{\mathbf{C}} & \bar{\mathbf{E}} \end{bmatrix} \begin{bmatrix} \bar{\mathbf{V}}(s) & \bar{\mathbf{Y}}(s) \\ \mathbf{0} & \mathbf{I}_p \end{bmatrix} \tag{6-177}$$

for some unimodular $\mathbf{U}(s)$ and $\bar{\mathbf{V}}(s)$. Since $\{\mathbf{A}, \mathbf{B}, \mathbf{C}, \mathbf{E}\}$ and $\{\bar{\mathbf{A}}, \bar{\mathbf{B}}, \bar{\mathbf{C}}, \bar{\mathbf{E}}\}$ have the same transfer matrix (Theorem 6-8), we have $\mathbf{E} = \bar{\mathbf{E}}$. Equation (6-177) implies

$$\mathbf{U}(s)(s\mathbf{I} - \mathbf{A}) = (s\mathbf{I} - \bar{\mathbf{A}})\bar{\mathbf{V}}(s) \tag{6-178}$$

Using Corollary G-12, we can write $\mathbf{U}(s)$ as

$$\mathbf{U}(s) = (s\mathbf{I} - \bar{\mathbf{A}})\bar{\mathbf{U}}(s) + \mathbf{P} \tag{6-179}$$

where $\mathbf{P}$ is a constant matrix. The substitution of (6-179) into (6-178) yields

$$\begin{aligned} \mathbf{P}(s\mathbf{I} - \mathbf{A}) &= (s\mathbf{I} - \bar{\mathbf{A}})[\bar{\mathbf{V}}(s) - \bar{\mathbf{U}}(s)(s\mathbf{I} - \mathbf{A})] \\ &\triangleq (s\mathbf{I} - \bar{\mathbf{A}})\bar{\mathbf{P}}(s) \end{aligned} \tag{6-180}$$

which implies $\bar{\mathbf{P}}(s)(s\mathbf{I} - \mathbf{A})^{-1} = (s\mathbf{I} - \bar{\mathbf{A}})^{-1}\mathbf{P}$. Since $(s\mathbf{I} - \bar{\mathbf{A}})^{-1}\mathbf{P}$ is a strictly proper rational matrix, so must be $\bar{\mathbf{P}}(s)(s\mathbf{I} - \mathbf{A})^{-1}$. This is possible only if $\bar{\mathbf{P}}(s)$ is a constant matrix. Hence we may replace $\bar{\mathbf{P}}(s)$ in (6-180) by $\bar{\mathbf{P}}(s) = \mathbf{P}_0$ to yield

$$s\mathbf{P} - \mathbf{PA} = s\mathbf{P}_0 - \bar{\mathbf{A}}\mathbf{P}_0$$

which implies $\mathbf{P} = \mathbf{P}_0$ and $\mathbf{PA} = \bar{\mathbf{A}}\mathbf{P}_0$. Hence we have

$$\mathbf{P}(s\mathbf{I} - \mathbf{A}) = (s\mathbf{I} - \bar{\mathbf{A}})\mathbf{P} \qquad \text{and} \qquad \mathbf{PA} = \bar{\mathbf{A}}\mathbf{P} \tag{6-181}$$

Now we shall show that the $\mathbf{P}$ in (6-179) is nonsingular. Since $\mathbf{U}(s)$ is unimodular, $\mathbf{U}^{-1}(s)$ is also unimodular and can be expressed as

$$\mathbf{U}^{-1}(s) = (s\mathbf{I} - \mathbf{A})\mathbf{U}_1(s) + \mathbf{P}_1 \tag{6-182}$$

for some polynomial matrix $\mathbf{U}_1(s)$ and some constant matrix $\mathbf{P}_1$. The multiplication of (6-179) and (6-182) yields

$$\mathbf{U}(s)\mathbf{U}^{-1}(s) = (s\mathbf{I} - \bar{\mathbf{A}})\bar{\mathbf{U}}(s)(s\mathbf{I} - \mathbf{A})\mathbf{U}_1(s) + (s\mathbf{I} - \bar{\mathbf{A}})\bar{\mathbf{U}}(s)\mathbf{P}_1 + \mathbf{P}(s\mathbf{I} - \mathbf{A})\mathbf{U}_1(s) + \mathbf{PP}_1$$

which becomes, because of (6-181) and $\mathbf{U}(s)\mathbf{U}^{-1}(s) = \mathbf{I}$,

$$\mathbf{I} - \mathbf{PP}_1 = (s\mathbf{I} - \bar{\mathbf{A}})[\bar{\mathbf{U}}(s)(s\mathbf{I} - \mathbf{A})\mathbf{U}_1(s) + \bar{\mathbf{U}}(s)\mathbf{P}_1 + \mathbf{PU}_1(s)]$$

or

$$(s\mathbf{I}-\bar{\mathbf{A}})^{-1}(\mathbf{I}-\mathbf{P}\mathbf{P}_1)=\bar{\mathbf{U}}(s)(s\mathbf{I}-\mathbf{A})\mathbf{U}_1(s)+\bar{\mathbf{U}}(s)\mathbf{P}_1+\mathbf{P}\mathbf{U}_1(s) \qquad \textbf{(6-183)}$$

Since its left-hand side is a strictly proper rational matrix, whereas its right-hand side is a polynomial matrix, (6-183) holds only if both sides are identically equal to zero. Hence we have $\mathbf{I}-\mathbf{PP}_1=\mathbf{0}$ or $\mathbf{PP}_1=\mathbf{I}$. Consequently, $\mathbf{P}$ is nonsingular and is qualified as an equivalence transformation.

To complete the proof of this theorem, we must show $\bar{\mathbf{B}}=\mathbf{PB}$ and $\bar{\mathbf{C}}=\mathbf{CP}^{-1}$ From (6-177), we have

$$\mathbf{U}(s)\mathbf{B}=(s\mathbf{I}-\bar{\mathbf{A}})\bar{\mathbf{Y}}(s)+\bar{\mathbf{B}}$$

which becomes, by using (6-179),

$$(s\mathbf{I}-\bar{\mathbf{A}})\bar{\mathbf{U}}(s)\mathbf{B}+\mathbf{PB}=(s\mathbf{I}-\bar{\mathbf{A}})\bar{\mathbf{Y}}(s)+\bar{\mathbf{B}}$$

or

$$\bar{\mathbf{U}}(s)\mathbf{B}-\bar{\mathbf{Y}}(s)=(s\mathbf{I}-\bar{\mathbf{A}})^{-1}(\bar{\mathbf{B}}-\mathbf{PB})$$

Its left-hand side is a polynomial matrix, whereas its right-hand side is a rational matrix. This is possible only if $\bar{\mathbf{B}}-\mathbf{PB}=\mathbf{0}$. Hence we conclude $\bar{\mathbf{B}}=\mathbf{PB}$. Similarly, we can show $\bar{\mathbf{C}}=\mathbf{CP}^{-1}$. This completes the proof of this theorem.
Q.E.D.

System matrices which are strictly system equivalent have the same transfer matrix. System matrices which have the same transfer matrix, however, are not necessarily strictly system equivalent. Example 2 of Section 4-3 is an example of this statement. A different example will be $\hat{\mathbf{G}}(s)=\mathbf{N}(s)\mathbf{D}^{-1}(s)=\bar{\mathbf{N}}(s)\bar{\mathbf{D}}^{-1}(s)$, where $\mathbf{N}(s)$ and $\mathbf{D}(s)$ are right coprime, but $\bar{\mathbf{N}}(s)$ and $\bar{\mathbf{D}}(s)$ are *not* right coprime. Clearly the two system matrices $\{\mathbf{D}(s), \mathbf{I}, \mathbf{N}(s), \mathbf{0}\}$ and $\{\bar{\mathbf{D}}(s), \mathbf{I}, \bar{\mathbf{N}}(s), \mathbf{0}\}$ have the same transfer matrix. However, because of $\det \mathbf{D}(s) \neq \det \bar{\mathbf{D}}(s)$, they are not strictly system equivalent (Theorem 6-8).

Although system matrices $\{\mathbf{P}_i(s), \mathbf{Q}_i(s), \mathbf{R}_i(s), \mathbf{W}_i(s)\}$, which have the same transfer matrix, are generally not strictly system equivalent, they become strictly system equivalent if the system matrices are irreducible; that is, $\{\mathbf{P}_i(s), \mathbf{Q}_i(s)\}$ are left coprime and $\{\mathbf{P}_i(s), \mathbf{R}_i(s)\}$ are right coprime. We establish first a special case.

Theorem 6-11

All coprime fractions of $\hat{\mathbf{G}}(s)$ are strictly system equivalent where $\hat{\mathbf{G}}(s)$ is a rational matrix, not necessarily proper.

Proof

Consider the two right coprime fractions $\hat{\mathbf{G}}(s)=\mathbf{N}_1(s)\mathbf{D}_1^{-1}(s)=\mathbf{N}_2(s)\mathbf{D}_2^{-1}(s)$. Theorem G-13 implies the existence of a unimodular matrix $\mathbf{T}(s)$ such that

$\mathbf{D}_2(s) = \mathbf{D}_1(s)\mathbf{T}(s)$ and $\mathbf{N}_2(s) = \mathbf{N}_1(s)\mathbf{T}(s)$.[7] Hence we have,

$$\begin{bmatrix} \mathbf{I} & \mathbf{0} \\ \mathbf{0} & \mathbf{I} \end{bmatrix} \begin{bmatrix} \mathbf{D}_1(s) & \mathbf{I} \\ -\mathbf{N}_1(s) & \mathbf{0} \end{bmatrix} \begin{bmatrix} \mathbf{T}(s) & \mathbf{0} \\ \mathbf{0} & \mathbf{I} \end{bmatrix} = \begin{bmatrix} \mathbf{D}_2(s) & \mathbf{I} \\ -\mathbf{N}_2(s) & \mathbf{0} \end{bmatrix} \tag{6-184}$$

which implies the strict system equivalence of $\{\mathbf{D}_1, \mathbf{I}, \mathbf{N}_1, \mathbf{0}\}$ and $\{\mathbf{D}_2, \mathbf{I}, \mathbf{N}_2, \mathbf{0}\}$.

Now if $\hat{\mathbf{G}}(s)$ is not strictly proper, then we may also factor $\hat{\mathbf{G}}(s)$ as

$$\hat{\mathbf{G}}(s) = \bar{\mathbf{N}}(s)\bar{\mathbf{D}}^{-1}(s) + \mathbf{E}(s) = (\bar{\mathbf{N}}(s) + \mathbf{E}(s)\bar{\mathbf{D}}(s))\bar{\mathbf{D}}^{-1}(s) \triangleq \mathbf{N}(s)\bar{\mathbf{D}}^{-1}(s)$$

where $\bar{\mathbf{N}}(s)\bar{\mathbf{D}}^{-1}(s)$ is strictly proper and coprime; $\mathbf{N}(s) = \bar{\mathbf{N}}(s) + \mathbf{E}(s)\bar{\mathbf{D}}(s)$. Because of

$$\begin{bmatrix} \bar{\mathbf{D}}(s) & \mathbf{I} \\ -\mathbf{N}(s) & \mathbf{0} \end{bmatrix} = \begin{bmatrix} \mathbf{I} & \mathbf{0} \\ -\mathbf{E}(s) & \mathbf{I} \end{bmatrix} \begin{bmatrix} \bar{\mathbf{D}}(s) & \mathbf{I} \\ -\bar{\mathbf{N}}(s) & \mathbf{E}(s) \end{bmatrix} \begin{bmatrix} \mathbf{I} & \mathbf{0} \\ \mathbf{0} & \mathbf{I} \end{bmatrix} \tag{6-185}$$

We conclude that $\{\bar{\mathbf{D}}, \mathbf{I}, \bar{\mathbf{N}}, \mathbf{E}\}$ and $\{\bar{\mathbf{D}}, \mathbf{I}, \mathbf{N}, \mathbf{0}\}$ are strictly system equivalent. Similarly we can show that all left-coprime fractions are strictly system equivalent.

What remains to be shown[8] is the strict system equivalence of right- and left-coprime fractions of $\hat{\mathbf{G}}(s)$. Consider the coprime fractions $\hat{\mathbf{G}}(s) = \mathbf{N}_r\mathbf{D}_r^{-1} = \mathbf{D}_l^{-1}\mathbf{N}_l$. Then there exist $\mathbf{U}_{11}$ and $\mathbf{U}_{12}$ such that

$$\begin{bmatrix} \mathbf{U}_{11} & \mathbf{U}_{12} \\ \mathbf{D}_l & \mathbf{N}_l \end{bmatrix} \begin{bmatrix} -\mathbf{N}_r \\ \mathbf{D}_r \end{bmatrix} = \begin{bmatrix} \mathbf{I} \\ \mathbf{0} \end{bmatrix} \tag{6-186}$$

where the leftmost matrix is a unimodular matrix (Problem G-11). Now using $\mathbf{D}_l\mathbf{N}_r = \mathbf{N}_l\mathbf{D}_r$, we form the identity

$$\left[\begin{array}{cc|c} \mathbf{U}_{11} & \mathbf{U}_{12} & \mathbf{0} \\ \mathbf{D}_l & \mathbf{N}_l & \mathbf{0} \\ \hline -\mathbf{I} & \mathbf{0} & \mathbf{I} \end{array}\right] \left[\begin{array}{c|cc} \mathbf{I} & \mathbf{0} & \mathbf{0} \\ \hline \mathbf{0} & \mathbf{D}_r & \mathbf{I} \\ \mathbf{0} & -\mathbf{N}_r & \mathbf{0} \end{array}\right] = \left[\begin{array}{c|cc} \mathbf{I} & \mathbf{0} & \mathbf{0} \\ \hline \mathbf{0} & \mathbf{D}_l & \mathbf{N}_l \\ \mathbf{0} & -\mathbf{I} & \mathbf{0} \end{array}\right] \left[\begin{array}{cc|c} \mathbf{U}_{11} & \mathbf{U}_{12}\mathbf{D}_r & \mathbf{U}_{12} \\ \mathbf{I} & \mathbf{N}_r & \mathbf{0} \\ \hline \mathbf{0} & \mathbf{0} & \mathbf{I} \end{array}\right] \tag{6-187}$$

The second and third matrices from the left in (6-187) are extended system matrices of $\{\mathbf{D}_r, \mathbf{I}, \mathbf{N}_r, \mathbf{0}\}$ and $\{\mathbf{D}_l, \mathbf{N}_l, \mathbf{I}, \mathbf{0}\}$. The leftmost matrix in (6-187) is unimodular because its left-upper-corner matrix is unimodular following from (6-186). We show in the following that the rightmost matrix in (6-187) is also unimodular. From (6-186), we have $-\mathbf{U}_{11}\mathbf{N}_r + \mathbf{U}_{12}\mathbf{D}_r = \mathbf{I}$ and

$$\begin{bmatrix} \mathbf{U}_{11} & \mathbf{U}_{12}\mathbf{D}_r \\ \mathbf{I} & \mathbf{N}_r \end{bmatrix} = \begin{bmatrix} \mathbf{U}_{11} & \mathbf{U}_{11}\mathbf{N}_r + \mathbf{I} \\ \mathbf{I} & \mathbf{N}_r \end{bmatrix} = \begin{bmatrix} \mathbf{U}_{11} & \mathbf{I} \\ \mathbf{I} & \mathbf{0} \end{bmatrix} \begin{bmatrix} \mathbf{I} & \mathbf{N}_r \\ \mathbf{0} & \mathbf{I} \end{bmatrix} \tag{6-188}$$

Since the two matrices after the last equality in (6-188) are unimodular, so is the first matrix in (6-188). This implies the unimodularity of the rightmost matrix in (6-187). Consequently, we conclude from Definition 6-3 that $\{\mathbf{D}_r, \mathbf{I}, \mathbf{N}_r, \mathbf{0}\}$ and $\{\mathbf{D}_l, \mathbf{N}_l, \mathbf{I}, \mathbf{0}\}$ are strictly system equivalent. This completes the proof of this theorem.
Q.E.D.

---

[7] The proof of Theorem G-13 does not use the properness of $\hat{\mathbf{G}}(s)$; hence the theorem is applicable to improper $\hat{\mathbf{G}}(s)$.

[8] The procedure is similar to the one from (6-154) to (6-156).

Theorem 6-12

All irreducible polynomial matrix descriptions $\{\mathbf{P}_i(s), \mathbf{Q}_i(s), \mathbf{R}_i(s), \mathbf{W}_i(s)\}$ which have the same transfer matrix are strictly system equivalent.

Proof

Every irreducible $\{\mathbf{P}_i, \mathbf{Q}_i, \mathbf{R}_i, \mathbf{W}_i\}$ has an irreducible realization $\{\mathbf{A}, \mathbf{B}, \mathbf{C}, \mathbf{E}(p)\}$ with dim $\mathbf{A} = \deg \det \mathbf{P}_i(s)$. All irreducible $\{\mathbf{A}, \mathbf{B}, \mathbf{C}, \mathbf{E}(p)\}$ which have the same transfer matrix are equivalent (Theorem 5-20) and, consequently, strictly system equivalent (Theorem 6-10). Hence by the transitivity property, we conclude that all irreducible $\{\mathbf{P}_i \ \mathbf{Q}_i \ \mathbf{R}_i \ \mathbf{W}_i\}$ of the same transfer matrix are strictly system equivalent. Q.E.D.

All irreducible dynamical equations, all coprime fractions and all irreducible polynomial matrix descriptions which have the same transfer matrix are strictly system equivalent and consequently have, following Theorems 6-2 and 6-8 and Definition 6-1′, the following properties:

$$\Delta(\hat{\mathbf{G}}(s)) \sim \det(s\mathbf{I} - \mathbf{A}) \sim \det \mathbf{D}(s) \sim \det \mathbf{P}(s) \tag{6-189}$$

where $\Delta(\hat{\mathbf{G}}(s))$ is the characteristic polynomial of $\hat{\mathbf{G}}(s)$ and $\sim$ denotes the equality of two polynomials modulo a nonzero constant factor. Conversely, if all descriptions have the same transfer matrix and satisfy (6-189), then they must be all irreducible. Under the irreducibility assumption, any one of the descriptions can be used, without loss of any essential information, to study and design a system.

To conclude this section, we mention that system matrices which are strictly system equivalent to

$$\mathbf{S}(s) = \begin{bmatrix} \mathbf{P}(s) & \mathbf{Q}(s) \\ -\mathbf{R}(s) & \mathbf{W}(s) \end{bmatrix}$$

where $\mathbf{P}$, $\mathbf{Q}$, $\mathbf{R}$, and $\mathbf{W}$ are, respectively, $m \times m$, $m \times p$, $q \times m$, and $q \times p$ polynomial matrices, can be generated by the following elementary operations:

1. Multiplication of any of the first m rows or columns by a nonzero constant.
2. Interchange of any two of the first m rows or columns.
3. Addition of the multiple of any of the first m rows (columns) by a polynomial to any of the $m+q$ rows (the $m+p$ columns).

These operations can be readily derived from the unimodular matrices

$$\begin{bmatrix} \mathbf{U}(s) & \mathbf{0} \\ \mathbf{X}(s) & \mathbf{I}_q \end{bmatrix}, \begin{bmatrix} \mathbf{V}(s) & \mathbf{Y}(s) \\ \mathbf{0} & \mathbf{I}_p \end{bmatrix}$$

used in the definition of strict system equivalence.

*6-9 Identification of Discrete-Time Systems from Noise-Free Data

In the previous sections, we introduced various realization methods for continuous-time systems described by transfer matrices. We also introduced polynomial matrix description and the concept of strict system equivalence. These results are directly applicable to the discrete-time systems if we carry out the following transformations:

	Continuous-time systems	Discrete-time systems
Transfer matrix	$\hat{\mathbf{G}}(s)=\mathbf{N}_r(s)\mathbf{D}_r^{-1}(s)$ $=\mathbf{D}_l^{-1}(s)\mathbf{N}_l(s)$	$\hat{\mathbf{G}}(z)=\mathbf{N}_r(z)\mathbf{D}_r^{-1}(z)$ $=\mathbf{D}_l^{-1}(z)\mathbf{N}_l(z)$
Dynamical equation	$\dot{\mathbf{x}}(t)=\mathbf{A}\mathbf{x}(t)+\mathbf{B}\mathbf{u}(t)$ $\mathbf{y}(t)=\mathbf{C}\mathbf{x}(t)+\mathbf{E}\mathbf{u}(t)$ $\hat{\mathbf{G}}(s)=\mathbf{E}+\mathbf{C}(s\mathbf{I}-\mathbf{A})^{-1}\mathbf{B}$	$\mathbf{x}(k+1)=\mathbf{A}\mathbf{x}(k)+\mathbf{B}\mathbf{u}(k)$ $\mathbf{y}(k)=\mathbf{C}\mathbf{x}(k)+\mathbf{E}\mathbf{u}(k)$ $\hat{\mathbf{G}}(z)=\mathbf{E}+\mathbf{C}(z\mathbf{I}-\mathbf{A})^{-1}\mathbf{B}$
Polynomial matrix description	$\mathbf{P}(s)\hat{\boldsymbol{\xi}}(s)=\mathbf{Q}(s)\hat{\mathbf{u}}(s)$ $\hat{\mathbf{y}}(s)=\mathbf{R}(s)\hat{\boldsymbol{\xi}}(s)+\mathbf{W}(s)\hat{\mathbf{u}}(s)$	$\mathbf{P}(z)\hat{\boldsymbol{\xi}}(z)=\mathbf{Q}(z)\hat{\mathbf{u}}(z)$ $\hat{\mathbf{y}}(z)=\mathbf{R}(z)\hat{\boldsymbol{\xi}}(z)+\mathbf{W}(z)\hat{\mathbf{u}}(z)$
System matrix	$\mathbf{S}(s)=\begin{bmatrix}\mathbf{P}(s) & \mathbf{Q}(s)\\ -\mathbf{R}(s) & \mathbf{W}(s)\end{bmatrix}$	$\mathbf{S}(z)=\begin{bmatrix}\mathbf{P}(z) & \mathbf{Q}(z)\\ -\mathbf{R}(z) & \mathbf{W}(z)\end{bmatrix}$

Hence the discussion of these problems will not be repeated for the discrete-time case. Note that all block diagrams in this chapter are also applicable to the discrete-time case if every integrator is replaced by a unit-delay element or, equivalently, s^{-1} is replaced by z^{-1}.

There is, however, one problem in the discrete-time case which deserves special discussion. Consider a sampled transfer function, $\hat{g}(z)$, of a discrete-time system expanded as

$$\hat{g}(z)=h(0)+h(1)z^{-1}+h(2)z^{-2}+\cdots$$

As in the continuous-time case, we shall call $\{h(i),\ i=0, 1, 2, \ldots\}$ the Markov parameters of $\hat{g}(z)$. In the continuous-time case, the Markov parameters must be computed from $d^i g(t)/dt^i$, $i=0, 1, 2, \ldots$, at $t=0$, where $g(t)$ is the impulse response of the system or the inverse Laplace transform of $\hat{g}(s)$. The generating of an impulse as an input is not possible in practice; repetitive differentiations of $g(t)$ are again impractical. Hence the Markov parameters in the continuous-time case are not really available.[9] In the discrete-time case, the situation is entirely different. If we apply the input $\{u(0)=1,\ u(i)=0,\ i=1, 2, \ldots\}$ to an initially relaxed linear time-invariant discrete-time system, then the measured data at the output terminal are the Markov parameters, that is, $y(i)=h(i)$, $i=0, 1, 2, \ldots$. Hence the realization from the Markov parameters in the discrete-time case can be considered as an identification problem—a problem of determining a mathematical description of a system from the data obtained by direct measurement at the input and output terminals. In actual measurement, all data will be corrupted by noises. A study of the identification problem with noisy data requires the concepts of probability and statistics and is outside the scope of this

[9] Consequently methods are developed to find realizations by using the moments defined by $M_k=\int_0^\infty t^k g(t)\,dt$, $k=0, 1, 2, \ldots$ (see References S28 and S146).

text. Hence we assume in this text that all data are free of noise and call the problem the *deterministic identification.*

Consider a single-variable linear time-invariant discrete-time system with transfer function $\hat{g}(z)$. If we apply the impulse sequence $\{u(0)=1,\ u(i)=0,\ i=1, 2, 3, \ldots\}$, then the zero-state response yields the Markov parameters $h(i)=y(i),\ i=0, 1, 2, \ldots$. Clearly the z-transform of $\{h(i)\}$ yields the transfer function $\hat{g}(z)$. However, this approach requires the use of an infinite number of $\{h(i)\}$. If we form a Hankel matrix from $\{h(i)\}$, and if the system is known to have a degree bounded by N,[10] then we need only $2N+1$ of the $h(i),\ i=0, 1, \ldots, 2N$. From the Hankel matrix, we can readily obtain a dynamical-equation description of the system as shown in (6-35) to (6-38). If the transfer function description is desired, we have

$$\hat{g}(z)=\frac{\beta_0 z^\sigma+\beta_1 z^{\sigma-1}+\cdots+\beta_\sigma}{z^\sigma+\alpha_1 z^{\sigma-1}+\cdots+\alpha_\sigma}$$

where α_i are obtained from the Hankel matrix as in (6-36) and β_i can be computed from

$$\begin{bmatrix} h(0) & 0 & \cdots & 0 \\ h(1) & h(0) & \cdots & 0 \\ \vdots & \vdots & & \vdots \\ h(\sigma) & h(\sigma-1) & \cdots & h(0) \end{bmatrix}\begin{bmatrix} 1 \\ \alpha_1 \\ \vdots \\ \alpha_\sigma \end{bmatrix}=\begin{bmatrix} \beta_0 \\ \beta_1 \\ \vdots \\ \beta_\sigma \end{bmatrix}$$

This matrix equation is just the set of equations in (6-29) arranged in matrix form. Consequently, the transfer function description of a system can be obtained from a finite number of Markov parameters. This assertion also applies to the multivariable case.

In order to obtain Markov parameters, the system must be initially relaxed. In a multivariable system with p inputs and q outputs, we apply an impulse sequence to the first input terminal and no input to all other input terminals, then the responses at the outputs yield $\{h_{k1}(i), k=1, 2, \ldots, q; i=0, 1, 2, \ldots\}$. After the system is at rest again [in theory, after an infinite time; in practice, after $h(i)$ is practically zero or becomes almost periodic[11]], we then repeat the process for the second input terminal and so forth. Hence the measurement of Markov parameters is possible only if the system is at our disposal. If a system is in continuous operation, its Markov parameters cannot be measured.

[10] If no bound of the degree of a system is available, then it is theoretically impossible to identify the system. For example, if we have $\{h(0)=0, h(i)=1, i=1, 2, \ldots, 200\}$ and if the degree of the system is bounded by 10, then we have $g(z)=1/(z-1)$. However the system $1/(z-1)+1/z^{1000}$ may also generate the given sequence. Hence if no bound is available, there is no way to identify a system from a finite sequence of $\{h(i), i=0, 1, 2, \ldots, N\}$. The problem of finding a transfer function to match a finite sequence of Markov parameters is called the *partial realization problem* (see References 68, S126 and S239).

[11] If a system is not BIBO stable (see Chapter 8), then $h(i)$ will approach infinity or remain oscillatory (including approach a nonzero constant). In the former case, the system will saturate or burn out and the linear model is no longer applicable. In the latter case, the system can be brought to rest by resetting. In theory, the realization or identification is applicable no matter the system is stable or not so long as the data are available.

In the following, we shall discuss a method of identifying a system from an arbitrary input-output pair. The concept of persistently exciting will be introduced. The problem of nonzero initial conditions will also be discussed.

Consider a linear time-invariant discrete-time system with $q \times p$ proper transfer matrix $\hat{\mathbf{G}}(z)$. Let $\hat{\mathbf{G}}(z)$ be factored as

$$\hat{\mathbf{G}}(z) = \mathbf{D}^{-1}(z)\mathbf{N}(z) \tag{6-190}$$

where $\mathbf{D}(z)$ and $\mathbf{N}(z)$ are, respectively, $q \times q$ and $q \times p$ polynomial matrices. Let the highest degree of all entires of $\mathbf{D}(z)$ be v. Then $\mathbf{D}(z)$ and $\mathbf{N}(z)$ can be expressed as

$$\mathbf{D}(z) = \mathbf{D}_0 + \mathbf{D}_1 z + \cdots + \mathbf{D}_v z^v \tag{6-191}$$

and

$$\mathbf{N}(z) = \mathbf{N}_0 + \mathbf{N}_1 z + \cdots + \mathbf{N}_v z^v \tag{6-192}$$

where $\mathbf{D}_i$, $\mathbf{N}_i$, $i = 0, 1, \ldots, v$ are $q \times q$ and $q \times p$ real constant matrices. If we apply the $p \times 1$ input sequence

$$\hat{\mathbf{u}}(z) = \mathbf{u}(0) + \mathbf{u}(1)z^{-1} + \mathbf{u}(2)z^{-2} + \cdots \tag{6-193}$$

to the initially relaxed system, then the output is a $q \times 1$ sequence given by

$$\hat{\mathbf{y}}(z) = \hat{\mathbf{G}}(z)\hat{\mathbf{u}}(z) = \mathbf{y}(0) + \mathbf{y}(1)z^{-1} + \mathbf{y}(2)z^{-2} + \cdots \tag{6-194}$$

The substitution of (6-190) into (6-194) yields $\mathbf{D}(z)\hat{\mathbf{y}}(z) = \mathbf{N}(z)\hat{\mathbf{u}}(z)$ or

$$-\mathbf{N}(z)\hat{\mathbf{u}}(z) + \mathbf{D}(z)\hat{\mathbf{y}}(z) = 0 \tag{6-195}$$

Equating the coefficient of z^i, $i = v, v-1, \ldots, -\infty$, to zero yields

$$[-\mathbf{N}_0 \quad \mathbf{D}_0 \quad -\mathbf{N}_1 \quad \mathbf{D}_1 \quad \cdots \quad -\mathbf{N}_v \quad \mathbf{D}_v] \begin{bmatrix} \mathbf{0} & \cdots & \mathbf{0} & \mathbf{u}(0) & \mathbf{u}(1) & \cdots \\ \mathbf{0} & \cdots & \mathbf{0} & \mathbf{y}(0) & \mathbf{y}(1) & \cdots \\ \mathbf{0} & \cdots & \mathbf{u}(0) & \mathbf{u}(1) & \mathbf{u}(2) & \cdots \\ \mathbf{0} & \cdots & \mathbf{y}(0) & \mathbf{y}(1) & \mathbf{y}(2) & \cdots \\ \vdots & & \vdots & \vdots & & \\ \mathbf{u}(0) & \cdots & \mathbf{u}(v-1) & \mathbf{u}(v) & \mathbf{u}(v+1) & \cdots \\ \mathbf{y}(0) & \cdots & \mathbf{y}(v-1) & \mathbf{y}(v) & \mathbf{y}(v+1) & \cdots \end{bmatrix}$$

$$\triangleq \mathbf{M}\mathbf{S}_v(-v, \infty) = \mathbf{0} \tag{6-196}$$

where $\mathbf{M} = [-\mathbf{N}_0 \quad \mathbf{D}_0 \quad \cdots \quad -\mathbf{N}_v \quad \mathbf{D}_v]$ is a $q \times (p+q)(v+1)$ matrix and the matrix $\mathbf{S}_v(-v, \infty)$ has $(p+q)(v+1)$ number of rows and an infinite number of columns. This equation is applicable no matter what input sequence is applied. Hence, given an arbitrary input-output pair, the identification problem reduces to the search of $\mathbf{M}$ to meet (6-196). There are q rows of $\mathbf{M}$. In order to have q rows of nontrivial solutions in (6-196), we need q linearly dependent rows in $\mathbf{S}_v(-v, \infty)$. Since it is desirable to have v, the degree of $\mathbf{D}(z)$ and $\mathbf{N}(z)$, as small as possible, we search, roughly speaking, the first q linearly dependent rows of $\mathbf{S}_v(-v, \infty)$. Hence the identification problem reduces to the search of linearly dependent rows of $\mathbf{S}_v(-v, \infty)$ in order from top to bottom. This problem is similar to the coprime fraction problem discussed in Section G-4.

The systems to be identified are assumed to be causal and to have proper rational transfer matrices. Hence if $\mathbf{u}(k) = \mathbf{0}$ for $k < 0$, then we have $\mathbf{y}(k) = \mathbf{0}$ for $k < 0$. Define

$$\bar{\mathbf{u}}(k, l) \triangleq [\mathbf{u}(k) \quad \mathbf{u}(k+1) \quad \cdots \quad \mathbf{u}(l)] \tag{6-197}$$
$$\bar{\mathbf{y}}(k, l) \triangleq [\mathbf{y}(k) \quad \mathbf{y}(k+1) \quad \cdots \quad \mathbf{y}(l)] \tag{6-198}$$

and define

$$\mathbf{S}_\alpha(k, l) \triangleq \begin{bmatrix} \bar{\mathbf{u}}(k, l) \\ \bar{\mathbf{y}}(k, l) \\ \hdashline \bar{\mathbf{u}}(k+1, l+1) \\ \bar{\mathbf{y}}(k+1, l+1) \\ \hdashline \vdots \\ \hdashline \bar{\mathbf{u}}(k+\alpha, l+\alpha) \\ \bar{\mathbf{y}}(k+\alpha, l+\alpha) \end{bmatrix} \qquad \alpha = 0, 1, 2 \ldots \tag{6-199}$$

There are $\alpha + 1$ block rows in $\mathbf{S}_\alpha$. Each block row consists of p rows of $\bar{\mathbf{u}}$ and q rows of $\bar{\mathbf{y}}$, hence $\mathbf{S}_\alpha(k, l)$ is of order $(p+q)(\alpha+1) \times (l-k+1)$. We call the rows formed from $\bar{\mathbf{u}}$ u rows and the rows formed from $\bar{\mathbf{y}}$ y rows.

There are infinitely many equations in (6-196). It is much more than necessary to solve for $\mathbf{M}$. In fact, $\mathbf{M}$ can often be solved from $\mathbf{MS}_\nu(0, l) = \mathbf{0}$, with l chosen so that $\mathbf{S}_\nu(0, l)$ has more columns than rows. The range of data $\{k, l\}$ used in the identification is not critical if the system is initially relaxed. However, it becomes critical, as will be discussed later, if the nonzero initial conditions are to be identified as well.

Definition 6-4

An input sequence $\{\mathbf{u}(n)\}$ is called *persistently exciting* if every u row in (6-199) is linearly independent of its previous rows in $\mathbf{S}_\alpha(k, l)$. ∎

This definition, as such, is not well defined because of its dependence on k, l, and α. The integer k is usually chosen as 0, although other value is also permitted. In theory, the integer l should be infinity; in practice, it is chosen so that $\mathbf{S}_\alpha(k, l)$ has more columns than rows. The integer α should be equal to or larger than, as will be discussed later, the observability index of any irreducible realization of $\hat{\mathbf{G}}(z)$ or the largest row degree of row reduced $\mathbf{D}(z)$ in any left-coprime fraction of $\hat{\mathbf{G}}(z) = \mathbf{D}^{-1}(z)\mathbf{N}(z)$.

The persistent exciting of an input sequence is not defined solely on the input signal; it also depends on the output it generates. In other words, whether or not an input is persistently exciting depends on the system to be identified. An input sequence which is persistently exciting to a system may not be so to a different system. Roughly speaking, an input sequence must be sufficiently random or rich in order to identify a system. Since the space spanned by the rows of $\mathbf{S}_\alpha(k, l)$ is of a finite dimension, whereas all possible input sequences form an space of an infinite dimension, almost all input sequences are persistently exciting. It means that if an input is generated randomly, the probability for the input to be persistently exciting is almost 1.

Consider $\mathbf{S}_0(k, \infty)$.[12] We use, for example, the row searching algorithm to search the linearly dependent rows of $\mathbf{S}_0(k, \infty)$ in order from top to bottom. Because $\bar{\mathbf{u}}$ is persistently exciting by assumption, all dependent rows will appear in $\bar{\mathbf{y}}(k, \infty)$. Let r_0 be the number of dependent rows in $\bar{\mathbf{y}}(k, \infty)$. Clearly $r_0 \leq q$, and $q - r_0$ is the number of linearly independent rows in $\bar{\mathbf{y}}(k, \infty)$. Next, we apply the row-searching algorithm to $\mathbf{S}_1(k, \infty)$. Let r_1 be the number of linearly dependent rows in $\bar{\mathbf{y}}(k+1, \infty)$. Let $\bar{\mathbf{y}}_i(k, \infty)$ be the ith row of $\bar{\mathbf{y}}(k, \infty)$. If $\bar{\mathbf{y}}_i(k, \infty)$ is linearly dependent in $\mathbf{S}_0(k, \infty)$, then $\bar{\mathbf{y}}_i(k+1, \infty)$ will also be linearly dependent in $\mathbf{S}_1(k, \infty)$. This follows from the fact that $\bar{\mathbf{u}}(k, \infty) = [\mathbf{u}(k) \vdots \bar{\mathbf{u}}(k+1, \infty)]$ and $\bar{\mathbf{y}}(k, \infty) = [\mathbf{y}(k) \vdots \bar{\mathbf{y}}(k+1, \infty)]$. Because of this property, we have $r_1 \geq r_0$. We continue this process until $r_v = q$ as shown.

$$
\mathbf{S}_v(k, \infty) = \begin{bmatrix} \bar{\mathbf{u}}(k, \infty) \\ \bar{\mathbf{y}}(k, \infty) \\ \hdashline \bar{\mathbf{u}}(k+1, \infty) \\ \bar{\mathbf{y}}(k+1, \infty) \\ \hdashline \vdots \\ \hdashline \bar{\mathbf{u}}(k+v-1, \infty) \\ \bar{\mathbf{y}}(k+v-1, \infty) \\ \hdashline \bar{\mathbf{u}}(k+v, \infty) \\ \bar{\mathbf{y}}(k+v, \infty) \end{bmatrix}
\begin{matrix} \\ \} \; r_0 \; [\text{no. of dependent rows in } \bar{\mathbf{y}}(k, \infty)] \\ \\ \} \; r_1 \\ \\ \\ \} \; r_{v-1} \\ \\ \} \; r_v = q \end{matrix}
\tag{6-200}
$$

with $0 \leq r_0 \leq r_1 \leq \cdots \leq r_{v-1} \leq r_v = q$. Note that if $r_v = q$, then $r_j = q$ for all $j \geq v$. Note also that all the $\bar{\mathbf{u}}$ rows in $\mathbf{S}_v(k, \infty)$ are linearly independent by the assumption of persistently exciting. The number of linearly independent y rows in $\mathbf{S}_v(k, \infty)$ is clearly equal to

$$
\begin{aligned}
n &\triangleq \text{number of linearly independent } y \text{ rows} \\
&= (q - r_0) + (q - r_1) + \cdots + (q - r_{v-1}) = vq - \sum_{i=0}^{v-1} r_i
\end{aligned}
\tag{6-201}
$$

Let $\bar{\mathbf{y}}_i$, $i = 1, 2, \ldots, q$, be the ith row of $\bar{\mathbf{y}}$. Let v_i be the number of linearly independent $\bar{\mathbf{y}}_i$ in $\mathbf{S}_v(k, \infty)$. By this, we mean that the rows $\bar{\mathbf{y}}_i(k, \infty)$, $\bar{\mathbf{y}}_i(k+1, \infty)$, $\ldots$, $\bar{\mathbf{y}}_i(k+v_i-1, \infty)$ are linearly independent of their previous rows in $\mathbf{S}_v(k, \infty)$, and $\bar{\mathbf{y}}_i(k+l, \infty)$, $l = v_i, v_i+1, \ldots, v-1$ are linearly dependent in $\mathbf{S}_v(k, \infty)$. Then $\bar{\mathbf{y}}_i$ is said to have *index* v_i. Clearly, we have $v_i \leq v$, $v = \max\{v_i, i = 1, 2, \ldots, q\}$, and

$$
n = v_1 + v_2 + \cdots + v_q \tag{6-202}
$$

---

[12] The subsequent analysis is similar to the one in Section G-4.

As an example, suppose $q=3$ and

$$\mathbf{K}\begin{bmatrix}\bar{\mathbf{u}}(k,\infty)\\ \bar{\mathbf{y}}_1(k,\infty)\\ \bar{\mathbf{y}}_2(k,\infty)\\ \bar{\mathbf{y}}_3(k,\infty)\\ \hdashline \bar{\mathbf{u}}(k+1,\infty)\\ \bar{\mathbf{y}}_1(k+1,\infty)\\ \bar{\mathbf{y}}_2(k+1,\infty)\\ \bar{\mathbf{y}}_3(k+1,\infty)\\ \hdashline \mathbf{u}(k+2,\infty)\\ \bar{\mathbf{y}}_1(k+2,\infty)\\ \bar{\mathbf{y}}_2(k+2,\infty)\\ \bar{\mathbf{y}}_3(k+2,\infty)\end{bmatrix}=\begin{bmatrix}x\\ x\\ \mathbf{0}\\ x\\ \hdashline x\\ x\\ \mathbf{0}\\ x\\ \hdashline x\\ \mathbf{0}\\ \mathbf{0}\\ \mathbf{0}\end{bmatrix}\begin{matrix}\\ \left.\begin{matrix}\\ \\ \\ \end{matrix}\right\} r_0=1\\ \\ \left.\begin{matrix}\\ \\ \\ \end{matrix}\right\} r_1=1\\ \\ \left.\begin{matrix}\\ \\ \\ \end{matrix}\right\} r_2=3=q\end{matrix} \tag{6-203}$$

where $\mathbf{K}$ is a lower triangular matrix with 1 on the diagonal, as discussed in (A-7) of Appendix A; and x denotes nonzero rows. Then we have $v_1=2$, $v_2=0$, and $v_3=2$, and $\bar{\mathbf{y}}_1$ has index 2, $\bar{\mathbf{y}}_2$ has index 0, and $\bar{\mathbf{y}}_3$ has index 2.

The $\bar{\mathbf{y}}_i$, $i=1,2,\ldots,q$, which first becomes linearly dependent on its previous rows in $\mathbf{S}_v$ are called the *primary dependent rows*. For example, the $\bar{\mathbf{y}}_2$ in the first block row of (6-203) is the primary dependent row; the $\bar{\mathbf{y}}_2$ in the second and third block rows are not. The $\bar{\mathbf{y}}_1$ and $\bar{\mathbf{y}}_3$ in the third block-row of (6-203) are the primary dependent rows. The positions of the primary dependent rows are clearly determinable from the index of $\bar{\mathbf{y}}_i$. The primary dependent $\bar{\mathbf{y}}_i$ row appears in the (v_i+1)th block row of $\mathbf{S}_v$.

Theorem 6-13

Consider an initially relaxed system excited by a persistently exciting input sequence. We form $\mathbf{S}_\alpha(k,\infty)$ and search its linearly independent rows in order from top to bottom to yield $\mathbf{K}\mathbf{S}_v=\bar{\mathbf{S}}_v$, where $\mathbf{K}$ is a lower triangular matrix with 1 on the diagonal, and v is the first integer such that all y-rows in the last block row of $\mathbf{S}_v$ are linearly dependent. Let $[-\mathbf{N}_0 \quad \mathbf{D}_0 \vdots \cdots \vdots -\mathbf{N}_v \quad \mathbf{D}_v]$ be the q rows of $\mathbf{K}$ corresponding to the q primary dependent rows of $\mathbf{S}_v$. Then the transfer matrix of the system is given by $\hat{\mathbf{G}}(z)=\mathbf{D}^{-1}(z)\mathbf{N}(z)$, where

$$\mathbf{D}(z)=\mathbf{D}_0+\mathbf{D}_1 z+\cdots+\mathbf{D}_v z^v \tag{6-204a}$$

and

$$\mathbf{N}(z)=\mathbf{N}_0+\mathbf{N}_1 z+\cdots+\mathbf{N}_v z^v \tag{6-204b}$$

Furthermore, $\mathbf{D}(z)$ and $\mathbf{N}(z)$ are left coprime, deg $\hat{\mathbf{G}}(z)=$ deg det $\mathbf{D}(z)$ and $\mathbf{D}(z)$ is row reduced, column reduced, and actually in the polynomial echelon form (see Appendix G).

Proof[13]

In order not to be overwhelmed by notations, we use an example to prove the

[13] The proof is identical to the proof of Theorem G-14.

theorem. Let $q=3$ and $\nu_1=3$, $\nu_2=1$ and $\nu_3=3$. Then we have

$$[-\mathbf{N}_0 \quad \mathbf{D}_0 \vdots -\mathbf{N}_1 \quad \mathbf{D}_1 \vdots -\mathbf{N}_2 \quad \mathbf{D}_2 \vdots -\mathbf{N}_3 \quad \mathbf{D}_3]$$
$$=\left[\begin{array}{cccc|cccc|cccc|cccc} & d_{11}^0 & d_{12}^0 & d_{13}^0 & & d_{11}^1 & d_{12}^1 & d_{13}^1 & & d_{11}^2 & d_{12}^2 & d_{13}^2 & & \boxed{d_{11}^3} & d_{12}^3 & d_{13}^3 \\ -\mathbf{N}_0 & d_{21}^0 & d_{22}^0 & d_{23}^0 & -\mathbf{N}_1 & d_{21}^1 & \boxed{d_{22}^1} & d_{23}^1 & -\mathbf{N}_2 & d_{21}^2 & d_{22}^2 & d_{23}^2 & -\mathbf{N}_3 & d_{21}^3 & d_{22}^3 & d_{23}^3 \\ & d_{31}^0 & d_{32}^0 & d_{33}^0 & & d_{31}^1 & d_{32}^1 & d_{33}^1 & & d_{31}^2 & d_{32}^2 & d_{33}^2 & & d_{31}^3 & d_{32}^3 & \boxed{d_{33}^3} \end{array}\right] \tag{6-205}$$

Because of the assumption $\nu_1=3$, $\nu_2=1$, and $\nu_3=3$, we have $d_{11}^3=1$, $d_{22}^1=1$, $d_{33}^3=1$, and all elements in (6-205) on the right-hand sides of d_{11}^3, d_{22}^1, and d_{33}^3 are zeros. In addition, the columns associated with d_{11}^{3+l}, d_{22}^{1+l}, d_{33}^{3+l} are zero for $l=1, 2, 3, \ldots$. Hence the $\mathbf{D}_i$ reduce to[14]

$$[\mathbf{D}_0 \quad \mathbf{D}_1 \quad \mathbf{D}_2 \quad \mathbf{D}_3] \doteq \left[\begin{array}{ccc|ccc|ccc|ccc} d_{11}^0 & d_{12}^0 & d_{13}^0 & d_{11}^1 & 0 & d_{13}^1 & d_{11}^2 & 0 & d_{13}^2 & \boxed{1} & 0 & 0 \\ d_{21}^0 & d_{22}^0 & d_{23}^0 & d_{21}^1 & \boxed{1} & 0 & 0 & 0 & 0 & 0 & 0 & 0 \\ d_{31}^0 & d_{32}^0 & d_{33}^0 & d_{31}^1 & 0 & d_{33}^1 & d_{31}^2 & 0 & d_{33}^2 & 0 & 0 & \boxed{1} \end{array}\right]$$

and

$$\mathbf{D}(z)=\begin{bmatrix} d_{11}^0+d_{11}^1z+d_{11}^2z^2+z^3 & d_{12}^0 & d_{13}^0+d_{13}^1z+d_{13}^2z^2 \\ d_{21}^0+d_{21}^1z & d_{22}^0+z & d_{23}^0 \\ d_{31}^0+d_{31}^1z+d_{31}^2z^2 & d_{32}^0 & d_{33}^0+d_{33}^1z+d_{33}^2z^2+z^3 \end{bmatrix} \tag{6-206}$$

This $\mathbf{D}(z)$ is clearly row reduced. It is also column reduced. (This property, however, is not needed here.) Therefore we have

$$\deg \det \mathbf{D}(z)=\nu_1+\nu_2+\nu_3=n \tag{6-207}$$

Because all elements on the right-hand side of d_{22}^1, d_{11}^3, and d_{33}^3 in (6-205) are zeros, the row degree of $\mathbf{N}(z)$ is at most equal to the corresponding row degree of $\mathbf{D}(z)$. Hence $\mathbf{D}^{-1}(z)\mathbf{N}(z)$ is a proper transfer matrix. We claim that $\mathbf{N}(z)$ and $\mathbf{D}(z)$ are left coprime. Suppose not, then there exists a $q \times q$ polynomial matrix $\mathbf{R}(z)$ with $\deg \det \mathbf{R}(z)>0$ such that

$$\mathbf{N}(z)=\mathbf{R}(z)\bar{\mathbf{N}}(z) \qquad \mathbf{D}(z)=\mathbf{R}(z)\bar{\mathbf{D}}(z) \tag{6-208}$$

This implies $\deg \det \mathbf{D}(z)>\deg \det \bar{\mathbf{D}}(z)$. However, this is not possible because the $n=\sum \nu_i$ computed in the algorithm is unique and smallest possible. Therefore we conclude that $\mathbf{N}(z)$ and $\mathbf{D}(z)$ are left coprime, and the degree of the transfer matrix is equal to $\deg \det \mathbf{D}(z)=\sum_i \nu_i$.

That $\mathbf{D}(z)$ is in the polynomial echelon form follows from the definition given in Appendix G. This completes the proof of this theorem. Q.E.D.

An example will be given, after the discussion of nonzero initial conditions, to illustrate the identification procedure. We note that once the transfer matrix, in the coprime fractional form, of a system is identified, a dynamical-equation description can be readily obtained by using the procedure discussed in Section 6-6.

Persistently exciting input sequences. The identification of a multivariable system can be achieved by searching the linearly dependent rows of

[14] The matrix is essentially in the echelon form. See Appendix G-3.

$\mathbf{S}_v(k, \infty)$. In practice, we use the matrix $\mathbf{S}_v(k, l)$ for a finite l. Clearly, l must be larger than $(v+1)(p+q)$ to ensure that there are more columns than rows. In actual computation, once $\bar{\mathbf{y}}_i(k+v_i, l)$ is linearly dependent in $\mathbf{S}_{v_i}(k, l)$, then $\bar{\mathbf{y}}_i$ may be deleted in forming $\mathbf{S}_{v_i+j}$, $j=1, 2, \ldots$.

If $\mathbf{u}$ is persistently exciting, then the $\mathbf{N}_i$ and $\mathbf{D}_i$ computed by using the row searching algorithm have the properties that $\mathbf{D}(z)$ is row reduced and, consequently, nonsingular and $\mathbf{D}^{-1}(z)\mathbf{N}(z)$ is proper. If $\bar{\mathbf{u}}$ is not persistently exciting, then the computed $\mathbf{D}^{-1}(z)\mathbf{N}(z)$ may not be proper. For example, consider $\hat{g}(z)=1/(z+1)$. If we apply $\hat{u}(z)=1+z^{-1}$, then the output $\hat{y}(z)$ is equal to z^{-1}. We form

$$\mathbf{S}_1=\begin{bmatrix} \bar{\mathbf{u}}(0,\infty) \\ \bar{\mathbf{y}}(0,\infty) \\ \hdashline \bar{\mathbf{u}}(1,\infty) \\ \bar{\mathbf{y}}(1,\infty) \end{bmatrix}=\begin{bmatrix} 1 & 1 & 0 & 0 & 0 & \cdots \\ 0 & 1 & 0 & 0 & 0 & \cdots \\ \hdashline 1 & 0 & 0 & 0 & 0 & \cdots \\ 1 & 0 & 0 & 0 & 0 & \cdots \end{bmatrix} \tag{6-209}$$

Clearly $\bar{\mathbf{u}}(1, \infty)$ is linearly dependent on its previous rows; hence $\hat{u}(z)$ is not persistently exciting. If we solve (6-209) by using the first linearly dependent row of $\mathbf{S}_1$, then the solution is

$$[-1 \quad 1 \,\vdots\, 1 \quad 0]\mathbf{S}_1=\mathbf{0}$$

and

$$\hat{y}(z)=\frac{1-z}{1}\,\hat{u}(z) \tag{6-210}$$

We see that this $\hat{g}(z)$ is not proper and is erroneous. Fortunately, this problem can be automatically detected in the search of linearly independent rows of $\mathbf{S}_i$.

It is possible to obtain a different but equivalent equation of (6-196) by grouping all u rows at the upper half of $\mathbf{S}_v$ and all y rows at the lower half. For this example, the equation becomes

$$[-N_0 \quad -N_1 \,\vdots\, D_0 \quad D_1]\begin{bmatrix} \bar{\mathbf{u}}(0,\infty) \\ \bar{\mathbf{u}}(1,\infty) \\ \hdashline \bar{\mathbf{y}}(0,\infty) \\ \bar{\mathbf{y}}(1,\infty) \end{bmatrix} \triangleq [-1 \quad 1 \,\vdots\, 1 \quad 0]\tilde{\mathbf{S}}_1=\mathbf{0} \tag{6-211}$$

and the solution is also equal to $\hat{g}(z)=(1-z)/1$. Since all u rows of $\mathbf{S}_1$ in (6-211) are linearly independent by themselves, there is no way to check from $\mathbf{S}_1$ that $\hat{g}(z)$ is erroneous. Hence the persistent exciting of $\{\mathbf{u}(n)\}$ cannot be defined solely on $\{\mathbf{u}(n)\}$.

A necessary condition for an input sequence to be persistently exciting is that the matrix

$$\mathbf{U}(k,\infty)=\begin{bmatrix} \mathbf{u}(k,\infty) \\ \mathbf{u}(k+1,\infty) \\ \vdots \\ \mathbf{u}(k+v,\infty) \end{bmatrix} \tag{6-212}$$

has a full row rank. We give some sufficient condition for $\mathbf{U}(k, \infty)$ to have a full row rank. The condition depends on the value of k. If $k=1$, then $\mathbf{U}(1, \infty)$ is the Hankel matrix of $\hat{\mathbf{u}}(z)$. Hence, if $\hat{\mathbf{u}}(z)$ is a rational vector, then the conditions for all rows of $\mathbf{U}(1, \infty)$ to be linearly independent are, as can be deduced from the results in Section 6-4,

$$\delta\hat{\mathbf{u}}(z) \geq p(v+1) \tag{6-213a}$$

and

$$\delta\hat{u}_i(z) \geq v+1 \qquad i=1, 2, \ldots, p \tag{6-213b}$$

where $\hat{u}_i(z)$ is the ith component of $\hat{\mathbf{u}}(z)$ and δ denotes the degree. These conditions state that the input signals must be more complicated than the system if we use $\mathbf{S}_v(1, \infty)$ in the identification. If $k=-v$, then the matrix $\mathbf{U}(-v, \infty)$ is the Hankel matrix of $\mathbf{u}(z)/z^{v+1}$. Therefore the conditions for all rows of (6-212) to be linearly independent are

$$\delta[\hat{\mathbf{u}}(z)/z^{v+1}] \geq p(v+1) \tag{6-214a}$$

and

$$\delta[\hat{u}_i(z)/z^{v+1}] \geq v+1 \qquad i=1, 2, \ldots, p \tag{6-214b}$$

The conditions in (6-214) are less stringent than the ones in (6-213). For example, one of $\hat{u}_i(z)$ may have degree 0 and satisfies (6-214), but will not satisfy (6-213). See also Problem 6-26. In practice, we do not have v a priori. In this case, we may choose an upper bound $v^* \geq v$ and use v^* in (6-213) or (6-214). If the chosen or given input sequence is not persistently exciting, then $\bar{\mathbf{u}}$ will appear as linear dependent rows in $\mathbf{S}_\alpha$. Therefore the procedure will check automatically whether or not the input sequence is persistently exciting.

Nonzero initial conditions. In this subsection we study the identification of systems which are not necessarily initially relaxed. Consider the controllable and observable discrete-time equation

$$\mathbf{x}(k+1) = \mathbf{A}\mathbf{x}(k) + \mathbf{B}\mathbf{u}(k) \tag{6-215a}$$

$$\mathbf{y}(k) = \mathbf{C}\mathbf{x}(k) + \mathbf{E}\mathbf{u}(k) \tag{6-215b}$$

The application of the z-transform to (6-215) yields

$$\hat{\mathbf{y}}(z) = \mathbf{C}(z\mathbf{I}-\mathbf{A})^{-1}z\mathbf{x}(0) + [\mathbf{C}(z\mathbf{I}-\mathbf{A})^{-1}\mathbf{B} + \mathbf{E}]\hat{\mathbf{u}}(z) \tag{6-216}$$

which can be written as

$$\hat{\mathbf{y}}(z) = \mathbf{D}^{-1}(z)\mathbf{R}(z) + \mathbf{D}^{-1}(z)\mathbf{N}(z)\hat{\mathbf{u}}(z) \tag{6-217}$$

where $\mathbf{R}(z)$ is a $q \times 1$ polynomial matrix with row degree smaller than or equal to the corresponding row degree of $\mathbf{D}(z)$. Clearly, $\mathbf{R}(z)$ is dependent on $\mathbf{x}(0)$. We rewrite (6-217) as

$$[-\mathbf{R}(z) \quad -\mathbf{N}(z)]\begin{bmatrix} 1 \\ \hat{\mathbf{u}}(z) \end{bmatrix} + \mathbf{D}(z)\hat{\mathbf{y}}(z) = 0 \tag{6-218}$$

This equation is similar to (6-195) and implies

$$[-\mathbf{R}_0 \quad -\mathbf{N}_0 \quad \mathbf{D}_0 \vdots \cdots \vdots -\mathbf{R}_v \quad -\mathbf{N}_v \quad \mathbf{D}_v] \left[\begin{array}{cccccc} 0 & \cdots & 0 & 1 & 0 & \cdots \\ 0 & \cdots & 0 & \mathbf{u}(0) & \mathbf{u}(1) & \cdots \\ 0 & \cdots & 0 & \mathbf{y}(0) & \mathbf{y}(1) & \cdots \\ \hline 0 & \cdots & 1 & 0 & 0 & \\ 0 & \cdots & \mathbf{u}(0) & \mathbf{u}(1) & \mathbf{u}(2) & \cdots \\ 0 & \cdots & \mathbf{y}(0) & \mathbf{y}(1) & \mathbf{y}(2) & \cdots \\ \hline & & & \vdots & & \\ \hline 1 & \cdots & 0 & 0 & 0 & \cdots \\ \mathbf{u}(0) & \cdots & \mathbf{u}(v-1) & \mathbf{u}(v) & \mathbf{u}(v+1) & \cdots \\ \mathbf{y}(0) & \cdots & \mathbf{y}(v-1) & \mathbf{y}(v) & \mathbf{y}(v+1) & \cdots \end{array}\right]$$

$$= \hat{\mathbf{M}}\hat{\mathbf{S}}_v(-v, \infty) = \mathbf{0} \qquad \textbf{(6-219)}$$

where $\hat{\mathbf{M}} = [-\mathbf{R}_0 \quad -\mathbf{N}_0 \quad \mathbf{D}_0 \vdots \cdots \vdots -\mathbf{R}_v \quad -\mathbf{N}_v \quad \mathbf{D}_v]$, and $\hat{\mathbf{S}}_v(-v, \infty)$ denotes the $v(p+1+q) \times \infty$ matrix in (6-219). Now we require the input $[1 \quad \hat{\mathbf{u}}'(z)]'$, where the prime denotes the transpose, to be persistently exciting. This input $[1 \quad \hat{\mathbf{u}}'(z)]'$ violates the conditions in (6-213b); hence we cannot use $\hat{\mathbf{S}}_v(k, l)$ with $k \geq 1$, in the identification. This new input however may meet the conditions in (6-214), hence by using $\hat{\mathbf{S}}_v(-v, l)$, the system with its initial conditions can be identified.

Is it really necessary to identify the initial conditions? In the design, it seems that initial conditions are never required. What is essential is the transfer function or dynamical equation. Furthermore, if the system is not relaxed at $n=0$, and if the system is controllable and observable, then we can always find an input sequence $\{\mathbf{u}(n), n = -n_0, -n_0+1, \ldots, -1\}$ so that the system is relaxed at $n = -n_0$ and has the given initial state at $n=0$. Since the initial time is not critical in the time-invariant case, a system with nonzero initial conditions at $n=0$ can be viewed as a system with zero initial conditions at $n = -n_0$. Consequently, (6-196) with a proper modification [since $\mathbf{u}(n)$ and $\mathbf{y}(n)$ are no longer zero for $n<0$] can be used in the identification of the system. Since (6-196) contains more than enough equations and since we do not have the information of $\mathbf{u}(n)$ and $\mathbf{y}(n)$ for $n<0$, we simply use $\mathbf{MS}_v(k, \infty)=0$, for any $k \geq 0$, in the identification. Consequently, given a system and given $\{\mathbf{u}(n), \mathbf{y}(n), i=0, 1, 2, \ldots\}$, no matter the initial conditions of the system are zero or not, its transfer matrix can be obtained by solving $\mathbf{MS}_v(0, \infty)=\mathbf{0}$.

Theorem 6-14

Given a linear time-invariant system and given $\{\mathbf{u}(n), \mathbf{y}(n), n=0, 1, 2, \ldots\}$, the transfer matrix of the system obtained from $\mathbf{S}_v(0, \infty)$ by assuming zero initial condition and the one obtained from $\hat{\mathbf{S}}_v(-v, \infty)$ without the assumption of zero initial conditions are the same.

Proof

Consider (6-196) and (6-219) with (6-196) modified as $\mathbf{MS}_v(0, \infty)=\mathbf{0}$. First we note that, by deleting the zero rows, the matrix $\hat{\mathbf{S}}_v(1, \infty)$ reduces to $\mathbf{S}_v(1, \infty)$;

hence the $\mathbf{N}_i$, $\mathbf{D}_i$ computed from (6-219) will also satisfy (6-196). Conversely, with the $\mathbf{N}_i$, $\mathbf{D}_i$ computed from (6-196), a set of $\mathbf{R}_i$, $i=0, 1, \ldots, \nu$, can be computed by solving successively the first $\nu+1$ column equations of $\hat{\mathbf{M}}\hat{\mathbf{S}}_\nu(-\nu, \infty)=\mathbf{0}$. In other words the solution $\mathbf{M}$ of $\mathbf{MS}_\nu(0, \infty)=\mathbf{0}$ is also a part of the solution of $\hat{\mathbf{M}}\hat{\mathbf{S}}_\nu(-\nu, \infty)=\mathbf{0}$. Hence the $\mathbf{D}(z)$ and $\mathbf{N}(z)$ computed from $\mathbf{MS}_\nu(0, \infty)=\mathbf{0}$, and the ones computed from $\hat{\mathbf{M}}\hat{\mathbf{S}}_\nu(-\nu, \infty)=\mathbf{0}$ will differ at most by a nonsingular polynomial matrix (dual of Theorem G-13). Consequently, their transfer matrices are the same. This completes the proof of the theorem. Q.E.D.

Example 1

Consider a system with transfer matrix

$$\hat{\mathbf{G}}(z)=\begin{bmatrix} \dfrac{z-0.5}{z^2-1} & \dfrac{1}{z-1} \\ \dfrac{z^2}{z^2-1} & \dfrac{2z}{z-1} \end{bmatrix} \tag{6-220}$$

An irreducible realization can be found as

$$\mathbf{x}(k+1)=\begin{bmatrix} 0 & 1 \\ 1 & 0 \end{bmatrix}\mathbf{x}(k)+\begin{bmatrix} 1 & 1 \\ -0.5 & 1 \end{bmatrix}\mathbf{u}(k) \tag{6-221}$$

$$\mathbf{y}(k)=\begin{bmatrix} 1 & 0 \\ \frac{2}{3} & \frac{4}{3} \end{bmatrix}\mathbf{x}(k)+\begin{bmatrix} 0 & 0 \\ 1 & 2 \end{bmatrix}\mathbf{u}(k)$$

With the initial state $\mathbf{x}(0)=[0.5 \quad 3.5]'$, we can obtain the following input-output pair

k	0	1	2	3	4	5	6	7	8	9	10	11
$u_1(k)$	0	0	−1	1	0	0	0	1	−1	0	0	0
$u_2(k)$	0	0	1	0	0	0	1	0	0	0	0	1
$y_1(k)$	0.5	3.5	0.5	3.5	3	3	3	4	5	2.5	5.5	2.5
$y_2(k)$	5	3	6	6	6	6	8	9	7	9	7	11

Now we shall try to identify the system from the input-output sequence $\{\mathbf{u}(n), \mathbf{y}(n)\}$. First we assume that the system is initially relaxed and apply the row-searching algorithm to $\mathbf{S}_2(0, 11)$ as

$$\begin{bmatrix} 1 \\ 0 & 1 \\ -\frac{7}{2} & -3 & 1 \\ -6 & -8 & -2 & 1 \\ 0 & -1 & 0 & 0 & 1 \\ 0 & 0 & 0 & 0 & 0 & 1 \\ -3 & -4 & -1 & -\frac{3}{4} & 0 & 0 & 1 \\ -6 & -9 & -2 & -\frac{1}{2} & -1 & -2 & 0 & 1 \end{bmatrix}$$

$$
\begin{bmatrix}
0 & 0 & -1 & \textcircled{1} & 0 & 0 & 0 & 1 & -1 & 0 & 0 \\
0 & 0 & 1 & 0 & 0 & 0 & \textcircled{1} & 0 & 0 & 0 & 0 \\
0.5 & 3.5 & 0.5 & 3.5 & \textcircled{3} & 3 & 3 & 4 & 5 & 2.5 & 5.5 \\
\textcircled{5} & 3 & 6 & 6 & 6 & 6 & 8 & 9 & 7 & 9 & 7 \\
\hline
0 & -1 & 1 & 0 & 0 & 0 & 1 & \textcircled{-1} & 0 & 0 & 0 \\
0 & \textcircled{1} & 0 & 0 & 0 & 1 & 0 & 0 & 0 & 0 & 1 \\
3.5 & 0.5 & 3.5 & 3 & 3 & 3 & 4 & 5 & 2.5 & 5.5 & 2.5 \\
3 & 6 & 6 & 6 & 6 & 8 & 9 & 7 & 9 & 7 & 11
\end{bmatrix}
=
\begin{bmatrix}
x \\ x \\ x \\ x \\ \hline x \\ x \\ \mathbf{0} \\ \mathbf{0}
\end{bmatrix}
\quad (*\ *)
$$

where the leftmost matrix is the $\mathbf{F}$ defined in (A-9). Its first column is chosen to make the fourth column, except the first element, of $\mathbf{S}_2(0, 11)$ a zero column. Its second column will make the seventh column, except the first two elements, of $\mathbf{K}_1\mathbf{S}_2(0, 11)$ a zero column. Note that the location of the pivot element in each row is encircled. Since the computation is carried out by hand, the pivot element is chosen for convenience in computation. For this problem, we have $r_0 = 0$, $r_1 = 2 = q$, and $v_1 = v_2 = 1$. The last two rows of $\mathbf{K}$, which correspond to the linearly dependent rows of $\mathbf{S}_2(0, 11)$, can be readily computed by using the formula in (F-11) as

$$
[-\mathbf{N}_0 \quad \mathbf{D}_0 \vdots -\mathbf{N}_1 \quad \mathbf{D}_1] = \left[\begin{array}{cc:cc:cc:cc} -\frac{1}{4} & \frac{1}{2} & \frac{1}{2} & -\frac{3}{4} & 0 & 0 & 1 & 0 \\ \frac{1}{2} & -1 & -1 & -\frac{1}{2} & -1 & -2 & 0 & 1 \end{array}\right]
$$

Hence we have

$$
\mathbf{N}(z) = \begin{bmatrix} \frac{1}{4} & -\frac{1}{2} \\ z - \frac{1}{2} & 2z + 1 \end{bmatrix} \qquad \mathbf{D}(z) = \begin{bmatrix} z + \frac{1}{2} & -\frac{3}{4} \\ -1 & z - \frac{1}{2} \end{bmatrix} \tag{6-222}
$$

It can be readily verified that $\mathbf{D}^{-1}(z)\mathbf{N}(z) = \hat{\mathbf{G}}(z)$.

Now we assume that the system is not initially relaxed. In this case we must use (6-219). The application of the row-searching algorithm yields

$$
\begin{bmatrix}
1 \\
0 & 1 \\
0 & 0 & 1 \\
-0.5 & -\frac{7}{2} & -3 & 1 \\
-5 & -6 & -8 & -2 & 1 \\
0 & 0 & 0 & 0 & 0 & 1 \\
0 & 0 & -1 & 0 & 0 & 0 & 1 \\
0 & 0 & 0 & 0 & 0 & 0 & 1 & 1 \\
-3.5 & -3 & -4 & -1 & -\frac{3}{4} & -\frac{1}{2} & 0 & 0 & 1 \\
-3 & -6 & -9 & -2 & -\frac{1}{2} & -5 & 1 & -2 & 0 & 1
\end{bmatrix}
$$

$$
\begin{array}{c}
\\ \\ \\ \\ * \\ * \\ \\ \\ \\ \\
\end{array}
\begin{bmatrix}
0 & 1 & 0 & 0 & 0 & 0 & 0 & 0 & 0 & 0 & 0 & 0 \\
0 & 0 & 0 & -1 & 1 & 0 & 0 & 0 & 1 & -1 & 0 & 0 \\
0 & 0 & 0 & 1 & 0 & 0 & 0 & 1 & 0 & 0 & 0 & 0 \\
0 & 0.5 & 3.5 & 0.5 & 3.5 & 3 & 3 & 3 & 4 & 5 & 2.5 & 5.5 \\
0 & 5 & 3 & 6 & 6 & 6 & 6 & 8 & 9 & 7 & 9 & 7 \\
\hline
1 & 0 & 0 & 0 & 0 & 0 & 0 & 0 & 0 & 0 & 0 & 0 \\
0 & 0 & -1 & 1 & 0 & 0 & 0 & 1 & 1 & 0 & 0 & 0 \\
0 & 0 & 1 & 0 & 0 & 0 & 1 & 0 & 0 & 0 & 0 & 1 \\
0.5 & 3.5 & 0.5 & 3.5 & 3 & 3 & 3 & 4 & 5 & 2.5 & 5.5 & 2.5 \\
5 & 3 & 6 & 6 & 6 & 6 & 8 & 9 & 7 & 9 & 7 & 11
\end{bmatrix}
=
\begin{bmatrix}
x \\ x \\ x \\ x \\ x \\ \hline x \\ x \\ x \\ \mathbf{0} \\ \mathbf{0}
\end{bmatrix}
$$

The last two rows of $\mathbf{K}$ can be computed as

$$
\begin{aligned}
&[-\mathbf{R}_0 \quad -\mathbf{N}_0 \quad \mathbf{D}_0 \,\vdots\, -\mathbf{R}_1 \quad -\mathbf{N}_1 \quad \mathbf{D}_1] \\
&\qquad = \left[\begin{array}{c:cc:cc:c:cc:cc}
0 & -\frac{1}{4} & \frac{1}{2} & \frac{1}{2} & -\frac{3}{4} & -\frac{1}{2} & 0 & 0 & 1 & 0 \\
0 & \frac{1}{2} & -1 & -1 & -\frac{1}{2} & -5 & -1 & -2 & 0 & 1
\end{array}\right]
\end{aligned}
$$

Hence we have

$$
\mathbf{R}(z) = \begin{bmatrix} \dfrac{z}{2} \\ 5z \end{bmatrix} \qquad \mathbf{N}(z) = \begin{bmatrix} \frac{1}{4} & -\frac{1}{2} \\ z - \frac{1}{2} & 2z + 1 \end{bmatrix} \qquad \mathbf{D}(z) = \begin{bmatrix} z + \frac{1}{2} & -\frac{3}{4} \\ -1 & z - \frac{1}{2} \end{bmatrix}
$$

which are the same as (6-222) obtained by disregarding the initial conditions. For this example, we can also verify that

$$
\mathbf{D}^{-1}(z)\mathbf{R}(z) = \mathbf{C}(z\mathbf{I} - \mathbf{A})^{-1} z\mathbf{x}(0)
$$

Hence the result is correct. ∎

6-10 Concluding Remarks

In this chapter, we discussed three approaches to find irreducible realizations for proper rational matrices. The first approach is to find a reducible one and then use the procedure in Section 5-8 to reduce it to an irreducible one. The second approach computes the Markov parameters by using the recursive formula in (6-29) and (6-30) and form the Hankel matrix. We can then use the singular value decomposition or the row searching algorithm to find irreducible realizations. In the last approach, irreducible realizations are obtained from coprime fractions of strictly proper rational matrices. The major effort in this approach is the computation of coprime fractions and the procedure in Appendix G can be employed. Although we have introduced a number of irreducible realization methods, the treatment is not exhaustive. For example, the Jordan-form realization is not discussed for the general case. The interested reader is referred to References S38, S59, and S145. For other methods, see, for example, References, S62, S63, and S191.

Now we compare briefly the numerical efficiencies of these three approaches. Given a $q \times p$ strictly proper rational matrix. Let $\alpha_i(s)$ and $\beta_j(s)$ be respectively the (least) common denominators of the ith row and jth column of $\hat{\mathbf{G}}(s)$. Because the computation of the least common denominator is not a simple task, we may simply multiply all denominators of the ith row and the jth column of $\hat{\mathbf{G}}(s)$ to yield $\alpha_i(s)$ and $\beta_j(s)$. For simplicity, we assume $p = q$ and $\alpha = \deg \alpha_i(s) = \deg \beta_j(s)$, for all i, j. Then we can find a controllable but not necessarily observable (or vice versa) realization of dimension αp. To reduce this reducible realization, we must append $\mathbf{B}$ and $\mathbf{C}$ to $\mathbf{A}$; hence the matrix under operation is roughly of order $(\alpha + 1)p$. In the Hankel matrix, the order of the Hankel matrix is $(\alpha + 1)p \times \alpha p$. In the coprime fraction method, if we use the method discussed in Appendix G to find a coprime fraction, we must form a generalized resultant. Each block row has $(p + q) = 2p$ rows, and we need roughly $(\alpha + 1)$ block rows. Hence the generalized resultant has $2(\alpha + 1)p$ number of rows. Thus we conclude that the matrices used in the first two approaches are roughly of the same order; the matrix used in the third approach is about twice the sizes of those used in the first two approaches.

The operation in the first approach requires the triangularization of a matrix by similarity transformations (both column and row operations). In the second and third approaches, we require the triangularization of a matrix by either row or column operations. Hence the first approach requires twice the computations of the second and third approaches. Thus we conclude that the Hankel method is probably the most efficient. For a more detailed comparison, see Reference S202. We note that all three approaches can be implemented by using numerically stable methods.

It is of interest to compare the singular value decomposition and the row-searching method discussed in Section 6-5. They are implemented on UNIVAC 1110 at Stony Brook in Reference S202. The row searching algorithm is carried out by applying Householder transformations and the gaussian elimination with partial pivoting on the columns of Hankel matrices and then computing the coefficients of combinations by back substitution. The scaling problem is included in the program. The examples in Reference S202 do not show any substantial difference between Householder transformations and the gaussian elimination with partial pivoting. The singular value decomposition is most reliable. However, its reliability over the row searching method is not overwhelming in the examples in Reference S202. This seems to be consistent with the remarks in Reference S82, p. 11.23.

The identification of linear time-invariant discrete-time systems is studied. The identification is carried out from a set of arbitrary input-output pair so long as the input sequence is persistently exciting. Whether the system is initially relaxed or not, the transfer function matrix in coprime fractional form can be obtained by solving $\mathbf{MS}_v(0, \infty) = \mathbf{0}$. The realization problem can be considered as a special case of the identification problem in which the system is to be identified from a particular input-output pair, the impulse response or the Markov parameters. The identification method introduced is applicable to continuous-time systems if they are first discretized. This, however, will introduce errors due to discretization. This is outside the scope of this text and will not be discussed.

In this chapter, we also introduced the polynomial matrix description and established its relationships with dynamical equations and transfer matrices. We showed that if the three descriptions have the same transfer matrix and are irreducible, then they are all strictly system equivalent. In this case, any one of them can be used in the analysis and design without loss of any essential information.

The degree of a proper rational matrix is introduced in this chapter. The degree can be computed by using any of the following methods:

1. Compute the least common denominator, $\Delta(s)$, of all minors of $\hat{\mathbf{G}}(s)$. Then we have $\deg \hat{\mathbf{G}}(s) = \deg \Delta(s)$.
2. Compute the Markov parameters of $\hat{\mathbf{G}}(s)$ and form the Hankel matrix $\mathbf{T}$ shown in (6-80). Then we have $\deg \hat{\mathbf{G}}(s) = \text{rank } \mathbf{T}$. The rank of $\mathbf{T}$ can be computed by using the singular value decomposition.
3. Find a right fraction $\hat{\mathbf{G}}(s) = \hat{\mathbf{N}}(s)\mathbf{D}^{-1}(s)$. If the fraction is right coprime, then $\deg \hat{\mathbf{G}}(s) = \deg \det \mathbf{D}(s)$. If the fraction is not right coprime, we use the coefficient matrices of $\mathbf{D}(s)$ and $\mathbf{N}(s)$ to form the matrix $\mathbf{S}_k$ shown in (G-67). We then search the linear independent rows of $\mathbf{S}_k$ in order from top to bottom. Then we have

$$\deg \hat{\mathbf{G}}(s) = \text{total number of linear independent } N \text{ rows in } \mathbf{S}_\infty.$$

   Similar results can be stated for a left fraction of $\hat{\mathbf{G}}(s)$.
4. Consider the realization of $\hat{\mathbf{G}}(s)$ shown in Equation (6-71). The realization is controllable but not necessarily observable. Define

$$\mathbf{Q}_{10} = \mathbf{0}, \qquad \mathbf{Q}_{1j} = \mathbf{R}_{m-j+1} \qquad j = 1, 2, \ldots, m$$
$$\mathbf{Q}_{i0} = \mathbf{0}, \mathbf{Q}_{ij} = \mathbf{Q}_{(i-1)(j-1)} - \alpha_{m-j+1}\mathbf{Q}_{(i-1)m} \qquad i = 2, 3, \ldots, m; j = 1, 2, \ldots, m$$

   Using $\mathbf{Q}_{ij}$, the observability matrix of (6-71) can be computed as

$$\mathbf{V} \triangleq \begin{bmatrix} \mathbf{Q}_{11} & \mathbf{Q}_{12} & \cdots & \mathbf{Q}_{1m} \\ \mathbf{Q}_{21} & \mathbf{Q}_{22} & \cdots & \mathbf{Q}_{2m} \\ \vdots & \vdots & & \vdots \\ \mathbf{Q}_{m1} & \mathbf{Q}_{m2} & \cdots & \mathbf{Q}_{mm} \end{bmatrix}$$

   Then we have $\deg \hat{\mathbf{G}}(s) = \text{rank } \mathbf{V}$ (why?).
5. Find a realization of $\hat{\mathbf{G}}(s)$ and reduce it, by using the method discussed in Section 5-8, to an irreducible one. Then we have $\deg \hat{\mathbf{G}}(s) = \dim \mathbf{A}$.

These methods can be readily programmed on a digital computer. A comparison of these methods in terms of computational efficiency and numerical stability seems to be unavailable at present.

To conclude this chapter, we give a different derivation of the observable realization in Equation (6-131). It will be derived from three maps developed in algebraic system theory. We discuss the discrete-time case. Consider the $q \times p$ strictly proper rational matrix $\hat{\mathbf{G}}(z) = \mathbf{D}^{-1}(z)\mathbf{N}(z)$, where $\mathbf{D}(z)$ and $\mathbf{N}(z)$

are respectively $q \times q$ and $q \times p$ real polynomial matrices. It is assumed that $\mathbf{D}(z)$ is row-reduced and its row degree is v_i, that is, $\delta_{ri}\mathbf{D}(s) = v_i$. Clearly we have $\delta_{ri}\mathbf{N}(z) < v_i$.

Let $\mathbb{K}_D$ denote the set of all $q \times 1$ real polynomial vectors with row degrees smaller than v_i. It can be readily verified that $(\mathbb{K}_D, \mathbb{R})$ is a linear space over the field of real numbers $\mathbb{R}$. The dimension of $(\mathbb{K}_D, \mathbb{R})$ is $n \triangleq v_1 + v_2 + \cdots + v_q$. (See Example 5, on page 16.) In this polynomial vector space, we shall choose the n columns of the $q \times n$ matrix:

$$\bar{\mathbf{L}}(z) = \begin{bmatrix} 1 & z & \cdots & z^{v_1-1} & 0 & 0 & \cdots & 0 & \cdots & 0 & 0 & \cdots & 0 \\ 0 & 0 & \cdots & 0 & 1 & z & \cdots & z^{v_2-1} & \cdots & 0 & 0 & \cdots & 0 \\ \vdots & \vdots & & \vdots & \vdots & \vdots & & \vdots & & \vdots & \vdots & & \vdots \\ 0 & 0 & \cdots & 0 & 0 & 0 & \cdots & 0 & \cdots & 1 & z & \cdots & z^{v_q-1} \end{bmatrix} \tag{6-223}$$

as a basis. Its columns will be denoted by $\mathbf{q}_{ij}$, $i = 1, 2, \ldots, q$; $j = 1, 2, \ldots, v_i$.

Let $\mathbf{f}(z)$ be a $q \times 1$ rational vector. Define the operator Π as

$$\Pi(\mathbf{f}(z)) = \text{strictly proper part of } \mathbf{f}(z) \tag{6-224}$$

If $\mathbf{f}(z)$ is a polynomial vector, then $\Pi(\mathbf{f}(z)) = \mathbf{0}$; if $\mathbf{f}(z)$ is strictly proper, then $\Pi(\mathbf{f}(z)) = \mathbf{f}(z)$. It is a linear operator which maps $(\mathbb{R}^q(z), \mathbb{R})$ into itself. Next we define, for any $q \times 1$ real polynomial,

$$\Pi_D(\mathbf{h}(z)) = \mathbf{D}(z)\Pi(\mathbf{D}^{-1}(z)\mathbf{h}(z)) \tag{6-225}$$

It is a linear operator which maps $(\mathbb{R}^q[z], \mathbb{R})$ into itself. Because the ith row degree of $\Pi_D(\mathbf{h}(z))$ is at most $v_i - 1$, the range space of $\Pi_D(\mathbf{h}(z))$ is $(\mathbb{K}_D, \mathbb{R})$.

In algebraic system theory, a realization can be expressed by the following three maps:

$$\begin{aligned} &\mathbf{B}_D : \mathbb{R}^p \to \mathbb{K}_D;\ \mathbf{u} \mapsto \mathbf{N}(z)\mathbf{u} \\ &\mathbf{A}_D : \mathbb{K}_D \to \mathbb{K}_D;\ \mathbf{x} \mapsto \Pi_D(z\mathbf{x}) \\ &\mathbf{C}_D : \mathbb{K}_D \to \mathbb{R}^q;\ x \to (\mathbf{D}^{-1}(z)\mathbf{x})_{-1} \end{aligned} \tag{6-226}$$

where, if we expand $\mathbf{D}^{-1}(z)\mathbf{x} = \boldsymbol{\alpha}_1 z^{-1} + \boldsymbol{\alpha}_2 z^{-2} + \cdots$, then $(\mathbf{D}^{-1}(z)\mathbf{x})_{-1} \triangleq \boldsymbol{\alpha}_1$. If the columns $\mathbf{q}_{ij}$ in (6-223) are chosen as a basis of $\mathbb{K}_D$, then the map $\mathbf{B}_D$ is the representation of $\mathbf{N}(z)$ with respect to the basis, that is,

$$\mathbf{N}(z) = \bar{\mathbf{L}}(z)\mathbf{B}$$

as shown in (6-130). The ith column of $\mathbf{A}$ is the representation of $\Pi_D(z\mathbf{q}_{ij})$ with respect to the basis in (6-223). Indeed, if we rewrite (6-128) here as

$$\mathbf{D}(z) = \bar{\mathbf{H}}(z)\mathbf{D}_{hr} + \bar{\mathbf{L}}(z)\mathbf{D}_{lr} \tag{6-227}$$

Then we have

$$\bar{\mathbf{H}}(z) = \mathbf{D}(z)\mathbf{D}_{hr}^{-1} - \bar{\mathbf{L}}(z)\mathbf{D}_{lr}\mathbf{D}_{hr}^{-1}$$

Consider

$$\Pi_D(z\mathbf{q}_{ij}) = \mathbf{D}(z)\Pi(\mathbf{D}^{-1}(z)z\mathbf{q}_{ij})$$

If $i=1, j=1$, then $\mathbf{D}^{-1}(s)z\mathbf{q}_{11}$ is strictly proper and

$$\Pi_D(z\mathbf{q}_{11})=\mathbf{D}(s)\mathbf{D}^{-1}(s)zq_{11}=z\mathbf{q}_{11}=\mathbf{q}_{12}=\bar{\mathbf{L}}(s)\begin{bmatrix}0\\1\\0\\\vdots\\0\end{bmatrix}$$

This is the first column of $\mathbf{A}$ as shown in (6-131a). Proceeding similarly, all columns of $\mathbf{A}$ except the $(\sum_{m=1}^{i} v_m)$th columns can be obtained as shown in (6-131a). Consider now

$$\Pi_D(\bar{\mathbf{H}}(z))=\mathbf{D}(z)\Pi(\mathbf{D}^{-1}(z)\bar{\mathbf{H}}(z))=\mathbf{D}(z)\Pi[\mathbf{D}_{hr}^{-1}-\mathbf{D}^{-1}(z)\bar{\mathbf{L}}(z)\mathbf{D}_{lr}\mathbf{D}_{hr}^{-1}]$$

Because $\mathbf{D}_{hr}^{-1}$ is a constant matrix (a polynomial of degree 0), we have $\Pi(\mathbf{D}_{hr}^{-1})=\mathbf{0}$. Because $\mathbf{D}^{-1}(z)\bar{\mathbf{L}}(z)\mathbf{D}_{lr}\mathbf{D}_{hr}^{-1}$ is strictly proper, $\Pi[\mathbf{D}^{-1}(z)\bar{\mathbf{L}}(z)\mathbf{D}_{lr}\mathbf{D}_{hr}^{-1}]=\mathbf{D}^{-1}(z)\bar{\mathbf{L}}(z)\mathbf{D}_{lr}\mathbf{D}_{hr}^{-1}$. Hence we have

$$\Pi_D(\bar{\mathbf{H}}(z))=-\mathbf{D}(z)\mathbf{D}^{-1}(z)\bar{\mathbf{L}}(z)\mathbf{D}_{lr}\mathbf{D}_{hr}^{-1}=-\bar{\mathbf{L}}(z)\mathbf{D}_{lr}\mathbf{D}_{hr}^{-1}$$

Note that the ith column of $\bar{\mathbf{H}}(z)$ is zq_{iv_i}. Hence the $(\sum_{m=1}^{i})$th column of $\mathbf{A}$ is given by the ith column of $-\mathbf{D}_{lr}\mathbf{D}_{hr}^{-1}$, as shown in (6-131a).

Now we show that the ijth column of $\mathbf{C}$ in (6-131a) is equal to $(\mathbf{D}^{-1}(z)\mathbf{q}_{ij})_{-1}$. From (6-226), we have

$$\begin{aligned}\mathbf{D}^{-1}(z)&=\mathbf{D}_{hr}^{-1}[\bar{\mathbf{H}}(z)(\mathbf{I}+\bar{\mathbf{H}}^{-1}(z)\bar{\mathbf{L}}(z)\mathbf{D}_{lr})]^{-1}=\mathbf{D}_{hr}^{-1}(\mathbf{I}+\bar{\mathbf{H}}^{-1}(z)\bar{\mathbf{L}}(z)\mathbf{D}_{lr})^{-1}\bar{\mathbf{H}}^{-1}(z)\\&=\mathbf{D}_{hr}^{-1}(\mathbf{I}+\mathbf{E}_1z^{-1}+\mathbf{E}_2z^{-2}+\cdots)\bar{\mathbf{H}}^{-1}(z)\end{aligned}$$

for some constant $\mathbf{E}_i$, $i=1,2,\ldots$. From this equation, we can readily show that if $j<v_i$, then $(\mathbf{D}^{-1}(z)\mathbf{q}_{ij})_{-1}=\mathbf{0}$ and $(\mathbf{D}^{-1}(z)\mathbf{q}_{iv_i})=$the ith column of $\mathbf{D}_{hr}^{-1}$. This completes the derivation of the observable realization in (6-131) from the three maps in (6-226). For a derivation of (6-226), see Reference S102. The maps in (6-226) are useful in solving polynomial matrix equations; see Reference S85 and Section 9-8.

Problems

6-1 Find the degrees and the characteristic polynomials of the following proper rational matrices.

a. $$\begin{bmatrix}\dfrac{1}{(s+1)^2} & \dfrac{s+3}{s+2} & \dfrac{1}{s+5}\\[2ex] \dfrac{1}{(s+3)^2} & \dfrac{s+1}{s+4} & \dfrac{1}{s}\end{bmatrix}$$

b. $$\begin{bmatrix}\dfrac{1}{(s+1)^2} & \dfrac{1}{(s+1)(s+2)}\\[2ex] \dfrac{1}{(s+2)} & \dfrac{1}{(s+1)(s+2)}\end{bmatrix}$$

c. $\begin{bmatrix} \dfrac{1}{s} & \dfrac{s+3}{s+1} \\ \dfrac{1}{s+3} & \dfrac{s}{s+1} \end{bmatrix}$

6-2 Find the dynamical-equation description of the block diagram with state variables chosen as shown in Figure P6-2.

Figure P6-2

6-3 Find the dynamical-equation realizations of the transfer functions

a. $\dfrac{s^4+1}{4s^4+2s^3+2s+1}$

b. $\dfrac{s^2-s+1}{s^5-s^4+s^3-s^2+s-1}$

Are these realizations irreducible? Find block diagrams for analog computer simulations of these transfer functions.

6-4 Find Jordan-canonical-form dynamical-equation realizations of the transfer functions

a. $\dfrac{s^2+1}{(s+1)(s+2)(s+3)}$

b. $\dfrac{s^2+1}{(s+2)^3}$

c. $\dfrac{s^2+1}{s^2+2s+2}$

If the Jordan-form realizations consist of complex numbers, find their equivalent dynamical equations that do not contain any complex numbers.

6-5 Write a dynamical equation for the feedback system shown in Figure P6-5. First find an overall transfer function and then realize it. Second, realize the open-loop transfer function and then make the necessary connection. Which realization is more convenient in computer simulations if the gain k is to be varied?

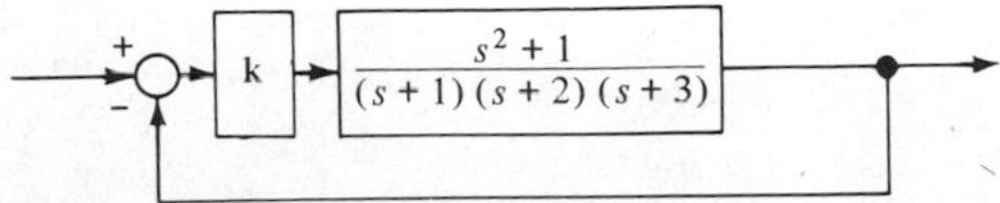

Figure P6-5

6-6 Find an irreducible realization, an uncontrollable dynamical-equation realization, an unobservable dynamical-equation realization, and an unobservable and uncontrollable dynamical-equation realization of $1/(s^3+1)$.

6-7 Find the controllable canonical-form, the observable canonical-form, and the Jordan canonical-form dynamical-equation realizations of $1/s^4$.

6-8 Set up a linear time-varying dynamical equation for the differential equation

$$(p^3+\alpha_1(t)p^2+\alpha_2(t)p+\alpha_3(t))y(t)=(\beta_0(t)p^2+\beta_1(t)p+\beta_2(t))u(t)$$

where $p^i \triangleq d^i/dt^i$.

6-9 Find irreducible controllable or observable canonical-form realizations for the matrices

a. $$\begin{bmatrix} \dfrac{2s}{(s+1)(s+2)(s+3)} \\ \dfrac{s^2+2s+2}{s(s+1)^2(s+4)} \end{bmatrix}$$

b. $$\begin{bmatrix} \dfrac{2s+3}{(s+1)^2(s+2)} & \dfrac{s^2+2s+2}{s(s+1)^3} \end{bmatrix}$$

6-10 Find irreducible realizations of the rational matrix

$$\begin{bmatrix} \dfrac{s+2}{s+1} & \dfrac{1}{s+3} \\ \dfrac{s}{s+1} & \dfrac{s+1}{s+2} \end{bmatrix}$$

Use two different methods.

6-11 Find irreducible realizations of the rational matrix

$$\begin{bmatrix} \dfrac{s^2+1}{s^3} & \dfrac{2s+1}{s^2} \\ \dfrac{s+3}{s^2} & \dfrac{2}{s} \end{bmatrix}$$

Use two different methods.

6-12 Find a linear time-invariant discrete-time dynamical equation whose sampled transfer function is

$$\frac{z^4+1}{4z^4+2z^3+2z+1}$$

[See Problem 6-3(a).]

6-13 Find an irreducible, discrete-time dynamical-equation realization of the sampled transfer-function matrix

$$\begin{bmatrix} \dfrac{z+2}{z+1} & \dfrac{1}{z+3} \\ \dfrac{z}{z+1} & \dfrac{z+1}{z+2} \end{bmatrix}$$

6-14 Consider

$$\dot{\mathbf{x}} = \begin{bmatrix} \lambda & 0 \\ 0 & \bar{\lambda} \end{bmatrix} \mathbf{x} + \begin{bmatrix} b_1 \\ \bar{b}_1 \end{bmatrix} u \qquad y = [c_1 \quad \bar{c}_1]\mathbf{x}$$

where the overbar denotes the complex conjugate. Verify that by using the transformation $\mathbf{x} = \mathbf{Q}_1 \tilde{\mathbf{x}}$, where

$$\mathbf{Q}_1 = \begin{bmatrix} -\bar{\lambda} b_1 & b_1 \\ -\lambda \bar{b}_1 & \bar{b}_1 \end{bmatrix}$$

the equation can be transformed into

$$\dot{\tilde{\mathbf{x}}} = \tilde{\mathbf{A}} \tilde{\mathbf{x}} + \tilde{\mathbf{b}} u \qquad y = \tilde{\mathbf{c}}_1 \tilde{\mathbf{x}}$$

where

$$\tilde{\mathbf{A}} = \begin{bmatrix} 0 & 1 \\ -\lambda\bar{\lambda} & \lambda + \bar{\lambda} \end{bmatrix} \qquad \tilde{\mathbf{b}} = \begin{bmatrix} 0 \\ 1 \end{bmatrix} \qquad \tilde{\mathbf{c}}_1 = [-2 \operatorname{Re}(\bar{\lambda} b_1 c_1) \quad 2 \operatorname{Re}(b_1 c_1)]$$

6-15 Verify that the Jordan-form dynamical equation

$$\dot{\mathbf{x}} = \begin{bmatrix} \lambda & 1 & 0 & 0 & 0 & 0 \\ 0 & \lambda & 1 & 0 & 0 & 0 \\ 0 & 0 & \lambda & 0 & 0 & 0 \\ 0 & 0 & 0 & \bar{\lambda} & 1 & 0 \\ 0 & 0 & 0 & 0 & \bar{\lambda} & 1 \\ 0 & 0 & 0 & 0 & 0 & \bar{\lambda} \end{bmatrix} \mathbf{x} + \begin{bmatrix} b_1 \\ b_2 \\ b_3 \\ \bar{b}_1 \\ \bar{b}_2 \\ \bar{b}_3 \end{bmatrix} u$$

$$y = [c_1 \quad c_2 \quad c_3 \quad \bar{c}_1 \quad \bar{c}_2 \quad \bar{c}_3]\mathbf{x}$$

can be transformed into

$$\dot{\tilde{\mathbf{x}}} = \begin{bmatrix} \tilde{\mathbf{A}} & \mathbf{I}_2 & 0 \\ 0 & \tilde{\mathbf{A}} & \mathbf{I}_2 \\ 0 & 0 & \tilde{\mathbf{A}} \end{bmatrix} \tilde{\mathbf{x}} + \begin{bmatrix} \tilde{\mathbf{b}} \\ \tilde{\mathbf{b}} \\ \tilde{\mathbf{b}} \end{bmatrix} u \qquad y = [\tilde{\mathbf{c}}_1 \quad \tilde{\mathbf{c}}_2 \quad \tilde{\mathbf{c}}_3]\tilde{\mathbf{x}}$$

where $\tilde{\mathbf{A}}$, $\tilde{\mathbf{b}}$, and $\tilde{\mathbf{c}}_i$ are defined in Problem 6-14 and $\mathbf{I}_2$ is the unit matrix of order 2. [*Hint*: Change the order of state variables from $[x_1 \; x_2 \; x_3 \; x_4 \; x_5 \; x_6]'$ to $[x_1 \; x_4 \; x_2 \; x_5 \; x_3 \; x_6]'$ and then apply the equivalence transformation $\mathbf{x} = \mathbf{Q}\tilde{\mathbf{x}}$, where $\mathbf{Q} = \operatorname{diag}(\mathbf{Q}_1, \mathbf{Q}_2, \mathbf{Q}_3)$.]

6-16 Write an irreducible dynamical equation for the following simultaneous differential equation:

$$\begin{aligned} 2(p+1)y_1 + (p+1)y_2 &= pu_1 + u_2 \\ (p+1)y_1 + (p+1)y_2 &= (p-1)u_1 \end{aligned}$$

where $p \triangleq d/dt$.

6-17 Show that (6-77) is a realization of $\hat{\mathbf{G}}(s)$ in (6-73) by establishing $\mathbf{H}(k+1)=\mathbf{CA}^k\mathbf{B}$, $k=0, 1, 2, \ldots$.

6-18 Show that (6-77) is always observable but not necessarily controllable.

6-19 Consider the $q \times p$ proper rational matrix

$$\hat{\mathbf{G}}(s)=\mathbf{H}(0)+\mathbf{H}(1)s^{-1}+\mathbf{H}(2)s^{-2}+\cdots$$

It is assumed that there exist $q \times q$ constant matrices $\mathbf{K}_i$, $i=0, 1, 2, \ldots, v$, such that

$$\mathbf{K}_0\mathbf{H}(v+i)=-\mathbf{K}_1\mathbf{H}(v+i-1)-\mathbf{K}_2\mathbf{H}(v+i-2)-\cdots-\mathbf{K}_v\mathbf{H}(i) \qquad i=1, 2, 3, \ldots$$

or

$$[\mathbf{K}_1 \quad \mathbf{K}_2 \quad \cdots \quad \mathbf{K}_v \quad \mathbf{K}_0 \quad 0 \quad \cdots \quad 0]\bar{\mathbf{T}}=\mathbf{0}$$

where $\bar{\mathbf{T}}$ is defined as in (6-95). If $\mathbf{K}_0$ is nonsingular, show that the vq-dimensional dynamical equation

$$\dot{\mathbf{x}}=\begin{bmatrix} -\mathbf{K}_v\mathbf{K}_0^{-1} & \mathbf{I} & \mathbf{0} & \cdots & \mathbf{0} \\ -\mathbf{K}_{v-1}\mathbf{K}_0^{-1} & \mathbf{0} & \mathbf{I} & \cdots & \mathbf{0} \\ \vdots & \vdots & \vdots & & \vdots \\ -\mathbf{K}_2\mathbf{K}_0^{-1} & \mathbf{0} & \mathbf{0} & \cdots & \mathbf{I} \\ -\mathbf{K}_1\mathbf{K}_0^{-1} & \mathbf{0} & \mathbf{0} & \cdots & \mathbf{0} \end{bmatrix}\mathbf{x}+\begin{bmatrix} \mathbf{K}_0 & \mathbf{0} & \cdots & \mathbf{0} & \mathbf{0} \\ \mathbf{K}_v & \mathbf{K}_0 & \cdots & \mathbf{0} & \mathbf{0} \\ \vdots & \vdots & & \vdots & \vdots \\ \mathbf{K}_3 & \mathbf{K}_4 & \cdots & \mathbf{K}_0 & \mathbf{0} \\ \mathbf{K}_2 & \mathbf{K}_3 & \cdots & \mathbf{K}_v & \mathbf{K}_0 \end{bmatrix}\begin{bmatrix} \mathbf{H}(1) \\ \mathbf{H}(2) \\ \vdots \\ \mathbf{H}(v-1) \\ \mathbf{H}(v) \end{bmatrix}\mathbf{u}$$

$$\mathbf{y}=[\mathbf{K}_0^{-1} \quad \mathbf{0} \quad \mathbf{0} \quad \cdots \quad \mathbf{0}]\mathbf{x}+\mathbf{H}(0)\mathbf{u}$$

is a realization of $\hat{\mathbf{G}}(s)$. Show that the realization is always observable but not necessarily controllable.

6-20 Use Problem 6-19 and (6-99) to find a realization of $\hat{\mathbf{G}}(s)$ in (6-97). Reduce it to an irreducible one. (*Hint:* Shift $\bar{\mathbf{k}}_2$ in (6-99) to right to form

$$\left[\begin{array}{cc:cc:cc:cc} 0 & 2 & 0 & 1 & 1 & 0 & 1 & 0 \\ 0 & 0 & 0 & 2 & 0 & 3 & 0 & 1 \end{array}\right]=[\mathbf{K}_1 \quad \mathbf{K}_2 \quad \mathbf{K}_3 \quad \mathbf{K}_0]$$

and then proceed. This procedure can be used to find an irreducible realization of the form in (6-131) except the rearrangements of state variables as discussed in Section 5-8. See References S63 and S201.)

6-21 Let $\mathbf{N}(s)\mathbf{D}^{-1}(s)$ be a $q \times p$ strictly proper rational matrix and let

$$\begin{aligned} \mathbf{N}(s)&=\mathbf{N}_0+\mathbf{N}_1 s+\cdots+\mathbf{N}_{\mu-1}s^{\mu-1} \\ \mathbf{D}(s)&=\mathbf{D}_0+\mathbf{D}_1 s+\cdots+\mathbf{D}_\mu s^\mu=\mathbf{D}_\mu(\bar{\mathbf{D}}_0+\bar{\mathbf{D}}_1 s+\cdots+\mathbf{I}s^\mu) \end{aligned}$$

where $\mathbf{D}_\mu$ is assumed to be nonsingular and $\bar{\mathbf{D}}_i=\mathbf{D}_\mu^{-1}\mathbf{D}_i$. Show that the $p\mu$-dimensional dynamical equation

$$\dot{\mathbf{x}}=\begin{bmatrix} \mathbf{0} & \mathbf{I} & \mathbf{0} & \cdots & \mathbf{0} \\ \mathbf{0} & \mathbf{0} & \mathbf{I} & \cdots & \mathbf{0} \\ \vdots & \vdots & \vdots & & \vdots \\ \mathbf{0} & \mathbf{0} & \mathbf{0} & \cdots & \mathbf{I} \\ -\bar{\mathbf{D}}_0 & -\bar{\mathbf{D}}_1 & -\bar{\mathbf{D}}_2 & \cdots & -\bar{\mathbf{D}}_{\mu-1} \end{bmatrix}\mathbf{x}+\begin{bmatrix} \mathbf{0} \\ \mathbf{0} \\ \vdots \\ \mathbf{0} \\ \mathbf{D}_\mu^{-1} \end{bmatrix}\mathbf{u}$$

$$\mathbf{y}=[\mathbf{N}_0 \quad \mathbf{N}_1 \quad \mathbf{N}_2 \quad \cdots \quad \mathbf{N}_{\mu-1}]\mathbf{x}$$

is a controllable realization of $\mathbf{N}(s)\mathbf{D}^{-1}(s)$.

6-22 Consider the equivalent time-invariant dynamical equations $\{\mathbf{A}, \mathbf{B}, \mathbf{C}\}$ and $\{\bar{\mathbf{A}}, \bar{\mathbf{B}}, \bar{\mathbf{C}}\}$ with $\bar{\mathbf{A}} = \mathbf{PAP}^{-1}$, $\bar{\mathbf{B}} = \mathbf{PB}$, and $\bar{\mathbf{C}} = \mathbf{CP}^{-1}$. Let $\mathbf{W}_{ct}$ and $\mathbf{W}_{ot}$ be the controllability and observability grammians defined in Theorems 5-7 and 5-13. Show that

$$\bar{\mathbf{W}}_{ct} = \mathbf{P}\mathbf{W}_{ct}\mathbf{P}^*$$

and

$$\bar{\mathbf{W}}_{ot} = (\mathbf{P}^{-1})^*\mathbf{W}_{ot}\mathbf{P}^{-1}$$

6-23 Consider the irreducible dynamical equation $\{\mathbf{A}, \mathbf{B}, \mathbf{C}\}$. Using Theorem E-4 to write its grammians as

$$\mathbf{W}_{ct} = \mathbf{R}_c\mathbf{\Sigma}_c^2\mathbf{R}_c^* \qquad \text{and} \qquad \mathbf{W}_{ot} = \mathbf{R}_0\mathbf{\Sigma}_o^2\mathbf{R}_o^*$$

where $\mathbf{R}^*\mathbf{R} = \mathbf{I}$ and $\mathbf{\Sigma}^2 = \text{diag}\,\{\lambda_1^2, \lambda_2^2, \ldots, \lambda_n^2\}$. Define

$$\mathbf{H} \triangleq \mathbf{\Sigma}_o^*\mathbf{R}_o^*\mathbf{R}_c\mathbf{\Sigma}_c$$

Using the singular value decomposition (Theorem E-5), we write

$$\mathbf{H} = \mathbf{R}_H\mathbf{\Sigma}_H\mathbf{Q}_H$$

where $\mathbf{R}_H^*\mathbf{R}_H = \mathbf{I}$ and $\mathbf{Q}_H\mathbf{Q}_H^* = \mathbf{I}$. Show that if $\mathbf{P}$ in Problem 6-22 is chosen as

i. $\mathbf{P}_{in} = \mathbf{Q}_H\mathbf{\Sigma}_c^{-1}\mathbf{R}_c^*$
ii. $\mathbf{P}_{out} = \mathbf{R}_H^*\mathbf{\Sigma}_o^*\mathbf{R}_o^*$
iii. $\mathbf{P}_{ib} = \mathbf{P}_{in}\mathbf{\Sigma}_H^{1/2} = \mathbf{\Sigma}_H^{1/2}\mathbf{P}_{out}$

Then its equivalent dynamical equation $\{\bar{\mathbf{A}}, \bar{\mathbf{B}}, \bar{\mathbf{C}}\}$ has the following corresponding grammians:

i. $\bar{\mathbf{W}}_{ct} = \mathbf{I}$, $\bar{\mathbf{W}}_{ot} = \mathbf{\Sigma}_H^2$
ii. $\bar{\mathbf{W}}_{ct} = \mathbf{\Sigma}_H^2$, $\bar{\mathbf{W}}_{ot} = \mathbf{I}$
iii. $\bar{\mathbf{W}}_{ct} = \bar{\mathbf{W}}_{ot} = \mathbf{\Sigma}_H$

They are called, respectively, *input-normal*, *output-normal*, and *internally balanced* on $[0, t]$ (see References S161 and S162).

6-24 Consider the irreducible dynamical equation

$$\dot{\mathbf{x}} = \mathbf{A}\mathbf{x} + \mathbf{B}\mathbf{u} \qquad \mathbf{y} = \mathbf{C}\mathbf{x}$$

It is assumed that all the eigenvalues of $\mathbf{A}$ have negative real parts and that the equation is internally balanced, that is,

$$\mathbf{W} = \mathbf{W}_{c\infty} = \mathbf{W}_{o\infty} = \mathbf{\Sigma}_H$$

where $\mathbf{W}$ is a diagonal matrix with positive diagonal entries. Show that $\mathbf{W}$ is the unique solution of

$$\mathbf{AW} + \mathbf{WA}^* = -\mathbf{BB}^* \qquad \text{or} \qquad \mathbf{WA} + \mathbf{A}^*\mathbf{W} = -\mathbf{C}^*\mathbf{C}$$

(*Hint*: Use Corollary F-1b.)

6-25 Consider the internally balanced system partitioned as

$$\begin{bmatrix} \dot{\mathbf{x}}_1 \\ \dot{\mathbf{x}}_2 \end{bmatrix} = \begin{bmatrix} \mathbf{A}_{11} & \mathbf{A}_{12} \\ \mathbf{A}_{21} & \mathbf{A}_{22} \end{bmatrix} \begin{bmatrix} \mathbf{x}_1 \\ \mathbf{x}_2 \end{bmatrix} + \begin{bmatrix} \mathbf{B}_1 \\ \mathbf{B}_2 \end{bmatrix} \mathbf{u}$$

$$\mathbf{y} = [\mathbf{C}_1 \quad \mathbf{C}_2] \begin{bmatrix} \mathbf{x}_1 \\ \mathbf{x}_2 \end{bmatrix}$$

It is assumed that all eigenvalues of $\mathbf{A}$ have negative real parts. Show that the reduced system

$$\dot{\bar{\mathbf{x}}} = \bar{\mathbf{A}}\bar{\mathbf{x}} + \bar{\mathbf{B}}\mathbf{u}, \qquad \bar{\mathbf{y}} = \bar{\mathbf{C}}\bar{\mathbf{x}}$$

where

$$\begin{aligned}\bar{\mathbf{A}} &= \mathbf{A}_{11} - \mathbf{A}_{12}\mathbf{A}_{22}^{-1}\mathbf{A}_{21} \\ \bar{\mathbf{B}} &= \mathbf{B}_1 - \mathbf{A}_{12}\mathbf{A}_{22}^{-1}\mathbf{B}_2 \\ \bar{\mathbf{C}} &= \mathbf{C}_1 - \mathbf{C}_2\mathbf{A}_{22}^{-1}\mathbf{A}_{21}\end{aligned}$$

is also internally balanced. This result is useful in system reduction. See Reference S93. (*Hint*: Write $\mathbf{W} = \text{diag}\,\{\mathbf{W}_1, \mathbf{W}_2\}$ and use Problem 6-24.)

6-26 Show that for any single-variable system with a proper transfer function, if $\mathbf{S}_v(-v, \infty)$ is used in the identification, the impulse sequence $\{u(0) = 1, u(i) = 0, i = 1, 2, 3, \ldots\}$ or any nonzero input sequence is always persistently exciting. Is the statement still true if $\mathbf{S}_v(0, \infty)$ is used?

6-27 Identify a linear time-invariant system from the following input-output sequence:

k	0	1	2	3	4	5	6	7	8	9	10	11	12	13	14 ···
$u_1(k)$	1	1	−1	0	0	1	0	−1	0	1	0	−1	−1	−2	1
$u_2(k)$	1	0	0	−1	0	0	0	1	−1	0	−1	1	0	1	−2
$y_1(k)$	−1	1	2	1	0	0	1	1	1	0	1	0	0	−1	−2
$y_2(k)$	−2	−0.5	1	−0.5	−0.5	−0.5	1	1	−0.5	0.5	1	−1	−2.5	−4	−7

First identify the system by disregarding the initial conditions and then identify the system as well as its initial conditions.

6-28. Let $\{\mathbf{A}, \mathbf{B}, \mathbf{C}\}$ be irreducible and let $\hat{\mathbf{G}}(s) = \mathbf{C}(s\mathbf{I} - \mathbf{A})^{-1}\mathbf{B} = \mathbf{N}(s)\mathbf{D}^{-1}(s)$, where $\mathbf{D}(s)$ and $\mathbf{N}(s)$ are right coprime. Show that for any $\mathbf{C}_1$, there exists a polynomial matrix $\mathbf{N}_1(s)$ such that $\mathbf{C}_1(s\mathbf{I} - \mathbf{A})^{-1}\mathbf{B} = \mathbf{N}_1(s)\mathbf{D}^{-1}(s)$. Show also that, conversely, for any strictly proper $\mathbf{N}_1(s)\mathbf{D}^{-1}(s)$, there exists a $\mathbf{C}_1$ such that $\mathbf{N}_1(s)\mathbf{D}^{-1}(s) = \mathbf{C}_1(s\mathbf{I} - \mathbf{A})^{-1}\mathbf{B}$.

7

State Feedback and State Estimators

7-1 Introduction

In engineering, design techniques are often developed from qualitative analyses of systems. For example, the design of feedback control systems by using Bode's plot was developed from the stability study of feedback systems. In Chapter 5 we introduced two qualitative properties of dynamical equations: controllability and observability. In this chapter we shall study their practical implications and develop some design techniques from them.

If an n-dimensional linear time-invariant dynamical equation is controllable, the controllability matrix $[\mathbf{B} \quad \mathbf{AB} \quad \cdots \quad \mathbf{A}^{n-1}\mathbf{B}]$ has n linearly independent columns. By using these independent columns or their linear combinations as basis vectors of the state space, various canonical forms can be obtained. We introduce in Section 7-2 the most useful one: the controllable canonical form. We also introduce there the observable canonical form dynamical equation. These two canonical forms are very useful in the designs of state feedback and state estimators.

Consider a system, called a *plant*, and a desired or reference signal; the control problem is to find a control signal or an actuating signal so that the output of the plant will be as close as possible to the reference signal. If a control signal is predetermined and is independent of the actual response of the plant, the control is called an *open-loop control*. This type of control is not satisfactory if there are disturbances or changes in the system. If a control signal depends on the actual response of the system, it is called a *feedback* or *closed-loop control*. Since the state of a system contains all the essential information of the system, if a control signal is designed to be a function of the state and the reference signal, a reasonably good control can be achieved. In Section 7-3, we study the effects of introducing a linear state feedback of the form $\mathbf{u}=\mathbf{r}+\mathbf{Kx}$ on a dynamical equation. We show what can be achieved by the linear state feedback under the assumption of controllability. Stabilization of an uncontrollable dynamical equation is also discussed.

In state feedback, all the state variables are assumed to be available as outputs. This assumption generally does not hold in practice. Therefore if we want to introduce state feedback, the state vector has to be generated or estimated from the available information. In Section 7-4 we introduce asymptotic state estimators under the assumption of observability. The outputs of an asymptotic state estimator give an estimate of the state of the original equation. Both full-dimensional and reduced-dimensional state estimators are introduced.

In Section 7-5 we apply the state feedback to the output of a state estimator. We shall establish the separation property which shows that the state feedback and state estimator can be designed independently and their connection will not affect their intended designs. We also show that the state estimator will not appear in the transfer-function matrix from the reference input to the plant output.

The inputs and outputs of a multivariable system are generally coupled; that is, every input controls more than one output, and every output is controlled by more than one input. If a compensator can be found such that every input controls one and only one output, then the multivariable system is said to be *decoupled*. We study in Section 7-6 the decoupling of a multivariable system by state feedback. The necessary and sufficient condition under which a system can be decoupled by linear state feedback is derived.

We study in this chapter only linear time-invariant dynamical equations. The material is based in part on References 1, 4, 7, 17, 18, 36, 41, 46, 51, 52, 68, 78 to 80, 84, 99, 111, 112, S17, S23, S69, and S110. The results in this chapter can be extended to linear time-varying dynamical equations if they are instantaneously controllable (see Definition 5-3). The interested reader is referred to References 10, 98, 99, 110, and S119.

7-2 Canonical-Form Dynamical Equations

Single-variable case. Consider the n-dimensional linear time-invariant, single-variable dynamical equation:

$$FE_1: \qquad \dot{\mathbf{x}} = \mathbf{Ax} + \mathbf{b}u \qquad \text{(7-1a)}$$
$$y = \mathbf{cx} + eu \qquad \text{(7-1b)}$$

where $\mathbf{A}$, $\mathbf{b}$, $\mathbf{c}$, and e are, respectively, $n \times n$, $n \times 1$, $1 \times n$, and 1×1 real constant matrices. Note that the subscript 1 in FE_1 stands for single variableness. In this section we shall derive various equivalent dynamical equations of FE_1 under the controllability or observability assumption.

Let the following dynamical equation $F\bar{E}_1$

$$F\bar{E}_1: \qquad \dot{\bar{\mathbf{x}}} = \bar{\mathbf{A}}\bar{\mathbf{x}} + \bar{\mathbf{b}}u \qquad \text{(7-2a)}$$
$$y = \bar{\mathbf{c}}\bar{\mathbf{x}} + eu \qquad \text{(7-2b)}$$

be an equivalent dynamical equation of FE_1, which is obtained by introducing $\bar{\mathbf{x}} = \mathbf{P}\mathbf{x} = \mathbf{Q}^{-1}\mathbf{x}$, where $\mathbf{P} \triangleq \mathbf{Q}^{-1}$ is a nonsingular constant matrix. Then from (4-33) we have

$$\bar{\mathbf{A}} = \mathbf{P}\mathbf{A}\mathbf{P}^{-1} \qquad \bar{\mathbf{b}} = \mathbf{P}\mathbf{b} \qquad \bar{\mathbf{c}} = \mathbf{c}\mathbf{P}^{-1}$$

The controllability matrices of FE_1 and $F\bar{E}_1$ are, respectively,

$$\mathbf{U} \triangleq [\mathbf{b} \quad \mathbf{A}\mathbf{b} \quad \cdots \quad \mathbf{A}^{n-1}\mathbf{b}] \tag{7-3}$$

$$\begin{aligned} \bar{\mathbf{U}} &\triangleq [\bar{\mathbf{b}} \quad \bar{\mathbf{A}}\bar{\mathbf{b}} \quad \cdots \quad \bar{\mathbf{A}}^{n-1}\bar{\mathbf{b}}] \\ &= \mathbf{P}[\mathbf{b} \quad \mathbf{A}\mathbf{b} \quad \cdots \quad \mathbf{A}^{n-1}\mathbf{b}] = \mathbf{P}\mathbf{U} = \mathbf{Q}^{-1}\mathbf{U} \end{aligned} \tag{7-4}$$

Now if the dynamical equations FE_1 and, consequently, $F\bar{E}_1$ are controllable,[1] then the controllability matrices $\mathbf{U}$ and $\bar{\mathbf{U}}$ are nonsingular. Hence, from (7-4) we have

$$\mathbf{P} = \bar{\mathbf{U}}\mathbf{U}^{-1} \tag{7-5a}$$

or

$$\mathbf{Q} = \mathbf{U}\bar{\mathbf{U}}^{-1} \tag{7-5b}$$

Similarly, if $\mathbf{V}$ and $\bar{\mathbf{V}}$ are the observability matrices of FE_1 and $F\bar{E}_1$, that is,

$$\mathbf{V} = \begin{bmatrix} \mathbf{c} \\ \mathbf{c}\mathbf{A} \\ \vdots \\ \mathbf{c}\mathbf{A}^{n-1} \end{bmatrix} \qquad \bar{\mathbf{V}} = \begin{bmatrix} \bar{\mathbf{c}} \\ \bar{\mathbf{c}}\bar{\mathbf{A}} \\ \vdots \\ \bar{\mathbf{c}}\bar{\mathbf{A}}^{n-1} \end{bmatrix} = \mathbf{V}\mathbf{P}^{-1} = \mathbf{V}\mathbf{Q}$$

and if FE_1 and $F\bar{E}_1$ are observable, then

$$\mathbf{P} = \bar{\mathbf{V}}^{-1}\mathbf{V} \qquad \text{or} \qquad \mathbf{Q} = \mathbf{V}^{-1}\bar{\mathbf{V}} \tag{7-6}$$

Thus, if two same-dimensional linear time-invariant, single-variable dynamical equations are known to be equivalent and if they are either controllable or observable, then the equivalence transformation $\mathbf{P}$ between them can be computed by the use of either (7-5) or (7-6). This is also true for multivariable controllable or observable time-invariant dynamical equations. For multivariable equivalent equations, the relation $\bar{\mathbf{U}} = \mathbf{P}\mathbf{U}$ still holds. In this case $\mathbf{U}$ is not a square matrix; however, from Theorem 2-8, we have $\bar{\mathbf{U}}\mathbf{U}^* = \mathbf{P}\mathbf{U}\mathbf{U}^*$ and $\mathbf{P} = \bar{\mathbf{U}}\mathbf{U}^*(\mathbf{U}\mathbf{U}^*)^{-1}$.

Let the characteristic polynomial of the matrix $\mathbf{A}$ in (7-1a) be

$$\Delta(\lambda) = \det(\lambda\mathbf{I} - \mathbf{A}) = \lambda^n + \alpha_1\lambda^{n-1} + \cdots + \alpha_{n-1}\lambda + \alpha_n$$

Theorem 7-1

If the n-dimensional linear time-invariant, single-variable dynamical equation FE_1 is controllable, then it can be transformed, by an equivalence transformation, into the form

[1] See Theorem 5-15.

CFE_1:

$$\dot{\bar{\mathbf{x}}} = \begin{bmatrix} 0 & 1 & 0 & \cdots & 0 & 0 \\ 0 & 0 & 1 & \cdots & 0 & 0 \\ 0 & 0 & 0 & \cdots & 0 & 0 \\ \vdots & \vdots & \vdots & & \vdots & \vdots \\ 0 & 0 & 0 & \cdots & 0 & 1 \\ -\alpha_n & -\alpha_{n-1} & -\alpha_{n-2} & \cdots & -\alpha_2 & -\alpha_1 \end{bmatrix} \bar{\mathbf{x}} + \begin{bmatrix} 0 \\ 0 \\ 0 \\ \vdots \\ 0 \\ 1 \end{bmatrix} u \tag{7-7a}$$

$$\mathbf{y} = [\ \beta_n \quad \beta_{n-1} \quad \beta_{n-2} \quad \cdots \quad \beta_2 \quad \beta_1] \bar{\mathbf{x}} + eu \tag{7-7b}$$

where $\alpha_1, \alpha_2, \ldots, \alpha_n$ are the coefficients of the characteristic polynomial of $\mathbf{A}$, and the β_i's are to be computed from FE_1. The dynamical equation (7-7) is said to be in the *controllable canonical form.* The transfer function of FE_1 is

$$\hat{g}(s) = \frac{\beta_1 s^{n-1} + \beta_2 s^{n-2} + \cdots + \beta_n}{s^n + \alpha_1 s^{n-1} + \cdots + \alpha_{n-1} s + \alpha_n} + e \tag{7-8}$$

Proof

The dynamical equation FE_1 is controllable by assumption; hence the set of $n \times 1$ column vectors $\mathbf{b}, \mathbf{Ab}, \ldots, \mathbf{A}^{n-1}\mathbf{b}$ is linearly independent. Consequently, the following set of $n \times 1$ vectors

$$\begin{aligned} \mathbf{q}_n &\triangleq \mathbf{b} \\ \mathbf{q}_{n-1} &\triangleq \mathbf{A}\mathbf{q}_n + \alpha_1 \mathbf{q}_n = \mathbf{Ab} + \alpha_1 \mathbf{b} \\ \mathbf{q}_{n-2} &\triangleq \mathbf{A}\mathbf{q}_{n-1} + \alpha_2 \mathbf{q}_n = \mathbf{A}^2\mathbf{b} + \alpha_1 \mathbf{Ab} + \alpha_2 \mathbf{b} \\ &\cdots\cdots\cdots\cdots \\ \mathbf{q}_1 &\triangleq \mathbf{A}\mathbf{q}_2 + \alpha_{n-1}\mathbf{q}_n = \mathbf{A}^{n-1}\mathbf{b} + \alpha_1 \mathbf{A}^{n-2}\mathbf{b} + \cdots + \alpha_{n-1}\mathbf{b} \end{aligned} \tag{7-9}$$

is linearly independent and qualifies as a basis of the state space of FE_1. Recall from Figure 2-5 that if the vectors $\{\mathbf{q}_1, \mathbf{q}_2, \ldots, \mathbf{q}_n\}$ are used as a basis, then the ith column of the new representation $\bar{\mathbf{A}}$ is the representation of $\mathbf{A}\mathbf{q}_i$ with respect to the basis $\{\mathbf{q}_1, \mathbf{q}_2, \ldots, \mathbf{q}_n\}$. Observe that

$$\mathbf{A}\mathbf{q}_1 = (\mathbf{A}^n + \alpha_1 \mathbf{A}^{n-1} + \cdots + \alpha_{n-1}\mathbf{A} + \alpha_n \mathbf{I})\mathbf{b} - \alpha_n \mathbf{b}$$

$$= -\alpha_n \mathbf{b} = -\alpha_n \mathbf{q}_n = [\mathbf{q}_1 \quad \mathbf{q}_2 \quad \cdots \quad \mathbf{q}_n] \begin{bmatrix} 0 \\ 0 \\ \vdots \\ 0 \\ -\alpha_n \end{bmatrix}$$

$$\mathbf{A}\mathbf{q}_2 = \mathbf{q}_1 - \alpha_{n-1}\mathbf{q}_n = [\mathbf{q}_1 \quad \mathbf{q}_2 \quad \cdots \quad \mathbf{q}_n] \begin{bmatrix} 1 \\ 0 \\ \vdots \\ 0 \\ -\alpha_{n-1} \end{bmatrix}$$

$$\cdots\cdots\cdots\cdots\cdots\cdots\cdots\cdots$$

$$\mathbf{A}\mathbf{q}_n = \mathbf{q}_{n-1} - \alpha_1 \mathbf{q}_n - [\mathbf{q}_1 \quad \mathbf{q}_2 \quad \cdots \quad \mathbf{q}_n] \begin{bmatrix} 0 \\ 0 \\ \vdots \\ 1 \\ -\alpha_1 \end{bmatrix}$$

Hence if we choose $\{\mathbf{q}_1, \mathbf{q}_2, \ldots, \mathbf{q}_n\}$ as a new basis of the state space, then $\mathbf{A}$ and $\mathbf{b}$ have new representations of the form

$$\bar{\mathbf{A}} = \begin{bmatrix} 0 & 1 & 0 & \cdots & 0 & 0 \\ 0 & 0 & 1 & \cdots & 0 & 0 \\ 0 & 0 & 0 & \cdots & 0 & 0 \\ \vdots & \vdots & \vdots & & \vdots & \vdots \\ 0 & 0 & 0 & \cdots & 0 & 1 \\ -\alpha_n & -\alpha_{n-1} & -\alpha_{n-2} & \cdots & -\alpha_2 & -\alpha_1 \end{bmatrix} \qquad \bar{\mathbf{b}} = \begin{bmatrix} 0 \\ 0 \\ 0 \\ \vdots \\ 0 \\ 1 \end{bmatrix} \tag{7-10}$$

The matrices $\bar{\mathbf{A}}$ and $\bar{\mathbf{b}}$ can also be obtained by using an equivalence transformation. Let $\mathbf{Q} \triangleq [\mathbf{q}_1 \quad \mathbf{q}_2 \quad \cdots \quad \mathbf{q}_n] \triangleq \mathbf{P}^{-1}$, and let $\bar{\mathbf{x}} = \mathbf{P}\mathbf{x}$ or $\mathbf{x} = \mathbf{Q}\bar{\mathbf{x}}$, then the dynamical equation FE_1 can be transformed into

$$\begin{aligned} \dot{\bar{\mathbf{x}}} &= \mathbf{Q}^{-1}\mathbf{A}\mathbf{Q}\bar{\mathbf{x}} + \mathbf{Q}^{-1}\mathbf{b}u \\ y &= \mathbf{c}\mathbf{Q}\bar{\mathbf{x}} + eu \end{aligned}$$

The reader is advised to verify that $\mathbf{Q}^{-1}\mathbf{A}\mathbf{Q} = \bar{\mathbf{A}}$ or $\mathbf{A}\mathbf{Q} = \mathbf{Q}\bar{\mathbf{A}}$ and $\mathbf{Q}^{-1}\mathbf{b} = \bar{\mathbf{b}}$. The vector $\bar{\mathbf{c}}$ is to be computed from $\mathbf{c}\mathbf{Q}$ as

$$\bar{\mathbf{c}} = \mathbf{c}\mathbf{Q} \triangleq [\beta_n \quad \beta_{n-1} \quad \beta_{n-2} \quad \cdots \quad \beta_1] \tag{7-11}$$

Hence the controllable dynamical equation FE_1 has the equivalent controllable canonical-form dynamical equation CFE_1. This proves the first part of the theorem.

The dynamical equations FE_1 and CFE_1 are equivalent; hence they have the same transfer function. It has been shown in Sections 4-4 and 6-3 that the transfer function of CFE_1 is equal to $\hat{g}(s)$ in Equation (7-8). Q.E.D.

One may wonder how we obtain the set of basis vectors given in (7-9). This is derived in the following. Let $\bar{\mathbf{U}}$ be the controllability matrix of the controllable canonical-form dynamical equation, then

$$\bar{\mathbf{U}} \triangleq [\bar{\mathbf{b}} \quad \bar{\mathbf{A}}\bar{\mathbf{b}} \quad \cdots \quad \bar{\mathbf{A}}^{n-1}\bar{\mathbf{b}}] = \begin{bmatrix} 0 & 0 & 0 & \cdots & 1 \\ 0 & 0 & 0 & \cdots & e_1 \\ \vdots & \vdots & \vdots & & \vdots \\ 0 & 0 & 1 & \cdots & e_{n-3} \\ 0 & 1 & e_1 & \cdots & e_{n-2} \\ 1 & e_1 & e_2 & \cdots & e_{n-1} \end{bmatrix} \tag{7-12}$$

where

$$e_k = -\sum_{i=0}^{k-1} \alpha_{i+1} e_{k-i-1} \qquad k = 1, 2, \ldots, n-1;\ e_0 = 1$$

The controllability matrix $\bar{\mathbf{U}}$ is nonsingular for any $\alpha_1, \alpha_2, \ldots, \alpha_n$. Therefore the controllable canonical-form dynamical equation is always controllable, as one might expect. The inverse of $\bar{\mathbf{U}}$ has the following very simple form:

$$\bar{\mathbf{U}}^{-1}=\begin{bmatrix} \alpha_{n-1} & \alpha_{n-2} & \cdots & \alpha_1 & 1 \\ \alpha_{n-2} & \alpha_{n-3} & \cdots & 1 & 0 \\ \vdots & \vdots & & \vdots & \vdots \\ \alpha_1 & 1 & \cdots & 0 & 0 \\ 1 & 0 & \cdots & 0 & 0 \end{bmatrix} \triangleq \mathbf{\Lambda} \tag{7-13}$$

This can be directly verified by showing that $\bar{\mathbf{U}}\bar{\mathbf{U}}^{-1}=\mathbf{I}$. The dynamical equations FE_1 and CFE_1 are related by the equivalence transformation $\mathbf{x}=\mathbf{Q}\bar{\mathbf{x}}$; hence $\mathbf{Q}=\mathbf{U}\bar{\mathbf{U}}^{-1}$. In the equivalence transformation $\mathbf{x}=\mathbf{Q}\bar{\mathbf{x}}$, we use the columns of $\mathbf{Q}$ as new basis vectors of the state space (see Figure 2-5). By computing $\mathbf{U}\bar{\mathbf{U}}^{-1}$, we see that the columns of $\mathbf{Q}$ are indeed those given in (7-9); that is,

$$\mathbf{Q}=[\mathbf{q}_1 \quad \mathbf{q}_2 \quad \cdots \quad \mathbf{q}_n]=[\mathbf{b} \quad \mathbf{Ab} \quad \cdots \quad \mathbf{A}^{n-1}\mathbf{b}]\bar{\mathbf{U}}^{-1}=\mathbf{U\Lambda} \tag{7-14}$$

We have the following theorem, which is similar to Theorem 7-1, for an observable dynamical equation.

Theorem 7-2

If the n-dimensional linear time-invariant, single-variable dynamical equation FE_1 is observable, then it can be transformed, by an equivalence transformation, into the form

OFE_1:

$$\dot{\bar{\mathbf{x}}}=\begin{bmatrix} 0 & 0 & \cdots & 0 & -\alpha_n \\ 1 & 0 & \cdots & 0 & -\alpha_{n-1} \\ 0 & 1 & \cdots & 0 & -\alpha_{n-2} \\ \vdots & \vdots & & \vdots & \vdots \\ 0 & 0 & \cdots & 1 & -\alpha_1 \end{bmatrix}\bar{\mathbf{x}}+\begin{bmatrix} \beta_n \\ \beta_{n-1} \\ \beta_{n-2} \\ \vdots \\ \beta_1 \end{bmatrix}u \tag{7-15a}$$

$$y=[0 \quad 0 \quad \cdots \quad 0 \quad 1]\bar{\mathbf{x}}+eu \tag{7-15b}$$

The dynamical equation (7-15) is said to be in the *observable canonical form*; moreover, the transfer function of FE_1 is

$$\hat{g}(s)=\frac{\beta_1 s^{n-1}+\beta_2 s^{n-2}+\cdots+\beta_{n-1}s+\beta_n}{s^n+\alpha_1 s^{n-1}+\cdots+\alpha_{n-1}s+\alpha_n}+e \qquad \blacksquare$$

This theorem can be proved either by a direct verification or by using the theorem of duality (Theorem 5-10). Its proof is left as an exercise. The equivalence transformation $\bar{\mathbf{x}}=\mathbf{Px}$ between (7-1) and (7-15) can be obtained by using Equation (7-6): $\mathbf{P}=\bar{\mathbf{V}}^{-1}\mathbf{V}$. It is easy to verify that the observability matrix $\bar{\mathbf{V}}$ of the observable canonical-form dynamical equation OFE_1 is the same as the matrix in (7-12). Hence the equivalence transformation $\bar{\mathbf{x}}=\mathbf{Px}$ between (7-1) and (7-15) is given by

$$\mathbf{P}=\begin{bmatrix} \alpha_{n-1} & \alpha_{n-2} & \cdots & \alpha_1 & 1 \\ \alpha_{n-2} & \alpha_{n-3} & \cdots & 1 & 0 \\ \vdots & \vdots & & \vdots & \vdots \\ \alpha_1 & 1 & \cdots & 0 & 0 \\ 1 & 0 & \cdots & 0 & 0 \end{bmatrix}\begin{bmatrix} \mathbf{c} \\ \mathbf{cA} \\ \vdots \\ \mathbf{cA}^{n-2} \\ \mathbf{cA}^{n-1} \end{bmatrix}=\mathbf{\Lambda V} \tag{7-16}$$

We see from (7-9) that the basis of the controllable canonical-form dynamical equation is obtained from a linear combination of the vectors $\{\mathbf{b}, \mathbf{Ab}, \ldots, \mathbf{A}^{n-1}\mathbf{b}\}$. One may wonder what form we shall obtain if we choose $\{\mathbf{b}, \mathbf{Ab}, \ldots, \mathbf{A}^{n-1}\mathbf{b}\}$ as the basis. Define $\mathbf{q}_1 \triangleq \mathbf{b}$, $\mathbf{q}_2 \triangleq \mathbf{Ab}, \ldots, \mathbf{q}_n \triangleq \mathbf{A}^{n-1}\mathbf{b}$ and let $\bar{\mathbf{x}} = \mathbf{Q}^{-1}\mathbf{x}$, where $\mathbf{Q} \triangleq [\mathbf{q}_1 \quad \mathbf{q}_2 \quad \cdots \quad \mathbf{q}_n]$; then we can obtain the following new representation (see Problem 7-2):

$$\dot{\bar{\mathbf{x}}} = \begin{bmatrix} 0 & 0 & \cdots & 0 & -\alpha_n \\ 1 & 0 & \cdots & 0 & -\alpha_{n-1} \\ 0 & 1 & \cdots & 0 & -\alpha_{n-2} \\ \vdots & \vdots & & \vdots & \vdots \\ 0 & 0 & \cdots & 1 & -\alpha_1 \end{bmatrix} \bar{\mathbf{x}} + \begin{bmatrix} 1 \\ 0 \\ 0 \\ \vdots \\ 0 \end{bmatrix} u$$

$$y = \mathbf{cQ}\bar{\mathbf{x}} + eu$$

This equation has the same $\mathbf{A}$ matrix as the one in (7-15) and is easier to obtain because its equivalence transformation $\mathbf{Q}$ is simpler. However, the usefulness of an equation of this form is not known at present.

Example 1

Transform the following controllable and observable single-variable dynamical equation

$$\dot{\mathbf{x}} = \begin{bmatrix} 1 & 2 & 0 \\ 3 & -1 & 1 \\ 0 & 2 & 0 \end{bmatrix} \mathbf{x} + \begin{bmatrix} 2 \\ 1 \\ 1 \end{bmatrix} u \qquad \textbf{(7-17a)}$$

$$y = [0 \quad 0 \quad 1]\mathbf{x} \qquad \textbf{(7-17b)}$$

into the controllable and observable canonical-form dynamical equations.

The characteristic polynomial of the matrix $\mathbf{A}$ in (7-17) is

$$\Delta(\lambda) = \det \begin{bmatrix} \lambda - 1 & -2 & 0 \\ -3 & \lambda + 1 & -1 \\ 0 & -2 & \lambda \end{bmatrix} = \lambda^3 - 9\lambda + 2$$

Hence $\alpha_3 = 2$, $\alpha_2 = -9$, and $\alpha_1 = 0$. The controllability matrix $\mathbf{U}$ is

$$\mathbf{U} = \begin{bmatrix} 2 & 4 & 16 \\ 1 & 6 & 8 \\ 1 & 2 & 12 \end{bmatrix}$$

From Equations (7-5b) and (7-13), we have

$$\mathbf{Q} = \mathbf{U\Lambda} = \begin{bmatrix} 2 & 4 & 16 \\ 1 & 6 & 8 \\ 1 & 2 & 12 \end{bmatrix} \begin{bmatrix} -9 & 0 & 1 \\ 0 & 1 & 0 \\ 1 & 0 & 0 \end{bmatrix} = \begin{bmatrix} -2 & 4 & 2 \\ -1 & 6 & 1 \\ 3 & 2 & 1 \end{bmatrix}$$

The matrix $\mathbf{Q}$ can also be obtained from (7-9). From (7-11), we have

$$[\beta_3 \quad \beta_2 \quad \beta_1] = \mathbf{cQ} = [3 \quad 2 \quad 1]$$

Hence the equivalent controllable canonical-form and observable canonical-form dynamical equations are, respectively,

$$CFE_1: \quad \dot{\bar{\mathbf{x}}} = \begin{bmatrix} 0 & 1 & 0 \\ 0 & 0 & 1 \\ -2 & 9 & 0 \end{bmatrix} \bar{\mathbf{x}} + \begin{bmatrix} 0 \\ 0 \\ 1 \end{bmatrix} u$$

$$y = [\ 3 \quad 2 \quad 1]\,\bar{\mathbf{x}}$$

$$OFE_1: \quad \dot{\bar{\mathbf{x}}} = \begin{bmatrix} 0 & 0 & -2 \\ 1 & 0 & 9 \\ 0 & 1 & 0 \end{bmatrix} \bar{\mathbf{x}} + \begin{bmatrix} 3 \\ 2 \\ 1 \end{bmatrix} u$$

$$y = [0 \quad 0 \quad 1]\,\bar{\mathbf{x}}$$

The transfer function of (7-17) is thus equal to

$$\hat{g}(s) = \frac{s^2 + 2s + 3}{s^3 - 9s + 2}$$

The controllable or observable canonical-form dynamical equation can also be obtained by first computing the transfer function of the dynamical equation. The coefficients of the transfer function give immediately the canonical-form equations. ∎

The controllable canonical-form and the observable canonical-form dynamical equations are useful in the study of state feedback and state estimator. They are also useful in the simulation of a dynamical equation on an analog computer. For example, in simulating the dynamical equation FE_1, we generally need n^2 amplifiers and attenuators (potentiometers) to simulate $\mathbf{A}$, and $2n$ amplifiers and attenuators to simulate $\mathbf{b}$ and $\mathbf{c}$. However, if FE_1 is transformed into the controllable or observable canonical form or into the form of (7-16), then the number of components required in the simulation is reduced from $n^2 + 2n$ to $2n$.

***Multivariable case.** Consider the n-dimensional linear time-invariant, multivariable dynamical equation

$$FE: \quad \dot{\mathbf{x}} = \mathbf{A}\mathbf{x} + \mathbf{B}\mathbf{u} \qquad \textbf{(7-18a)}$$

$$\mathbf{y} = \mathbf{C}\mathbf{x} + \mathbf{E}\mathbf{u} \qquad \textbf{(7-18b)}$$

where $\mathbf{A}, \mathbf{B}, \mathbf{C}$, and $\mathbf{E}$ are $n \times n$, $n \times p$, $q \times n$, and $q \times p$ real constant matrices, respectively. Let $\mathbf{b}_i$ be the ith column of $\mathbf{B}$; that is, $\mathbf{B} = [\mathbf{b}_1 \quad \mathbf{b}_2 \quad \cdots \quad \mathbf{b}_p]$.

If the dynamical equation FE is controllable, then the controllability matrix

$$\mathbf{U} = [\mathbf{b}_1 \quad \mathbf{b}_2 \quad \cdots \quad \mathbf{b}_p \quad \mathbf{A}\mathbf{b}_1 \quad \cdots \quad \mathbf{A}\mathbf{b}_p \quad \cdots \quad \mathbf{A}^{n-1}\mathbf{b}_1 \quad \mathbf{A}^{n-1}\mathbf{b}_2 \quad \cdots \quad \mathbf{A}^{n-1}\mathbf{b}_p] \qquad \textbf{(7-19)}$$

has rank n. Consequently, there are n linearly independent column vectors in $\mathbf{U}$. There are many ways to choose n linearly independent column vectors from the $n \times np$ composite matrix $\mathbf{U}$. In the following, we shall give two schemes for choosing n linearly independent column vectors to form new bases for the state space. These two schemes were first discussed in Section 5-3.

Scheme 1

We start with the vector $\mathbf{b}_1$ and then proceed to $\mathbf{Ab}_1$, $\mathbf{A}^2\mathbf{b}_1$, up to $\mathbf{A}^{\bar{\mu}_1-1}\mathbf{b}_1$ until the vector $\mathbf{A}^{\bar{\mu}_1}\mathbf{b}_1$ can be expressed as a linear combination of $\{\mathbf{b}_1, \ldots, \mathbf{A}^{\bar{\mu}_1-1}\mathbf{b}_1\}$. If $\bar{\mu}_1 = n$, the equation can be controlled by the first column of $\mathbf{B}$ alone. If $\bar{\mu}_1 < n$, we select $\mathbf{b}_2$, $\mathbf{Ab}_2$, up to $\mathbf{A}^{\bar{\mu}_2-1}\mathbf{b}_2$, until the vector $\mathbf{A}^{\bar{\mu}_2}\mathbf{b}_2$ can be expressed as a linear combination of $\{\mathbf{b}_1, \ldots, \mathbf{A}^{\bar{\mu}_1-1}\mathbf{b}_1, \mathbf{b}_2, \ldots, \mathbf{A}^{\bar{\mu}_2-1}\mathbf{b}_2\}$ (see Problem 7-5). If $\bar{\mu}_1 + \bar{\mu}_2 < n$, we proceed to $\mathbf{b}_3, \mathbf{Ab}_3, \ldots, \mathbf{A}^{\bar{\mu}_3-1}\mathbf{b}_3$ and so forth. Assume that $\bar{\mu}_1 + \bar{\mu}_2 + \bar{\mu}_3 = n$, and the n vectors

$$\{\mathbf{b}_1, \mathbf{Ab}_1, \ldots, \mathbf{A}^{\bar{\mu}_1-1}\mathbf{b}_1; \mathbf{b}_2, \mathbf{Ab}_2, \ldots, \mathbf{A}^{\bar{\mu}_2-1}\mathbf{b}_2; \mathbf{b}_3, \mathbf{Ab}_3, \ldots, \mathbf{A}^{\bar{\mu}_3-1}\mathbf{b}_3\} \tag{7-20}$$

are linearly independent. An important property of this set is that the vector $\mathbf{A}^{\bar{\mu}_i}\mathbf{b}_i$ can be expressed as a linear combination of the preceding vectors; for example, $\mathbf{A}^{\bar{\mu}_2}\mathbf{b}_2$ can be expressed as a linear combination of $\{\mathbf{b}_1, \mathbf{Ab}_1, \ldots, \mathbf{A}^{\bar{\mu}_1-1}\mathbf{b}_1, \mathbf{b}_2, \mathbf{Ab}_2, \ldots, \mathbf{A}^{\bar{\mu}_2-1}\mathbf{b}_2\}$.

Scheme 2

The linearly independent vectors are selected in the order of (7-19); that is, we start from $\mathbf{b}_1, \mathbf{b}_2, \ldots, \mathbf{b}_p$ and then $\mathbf{Ab}_1, \mathbf{Ab}_2, \ldots, \mathbf{Ab}_p$, and then $\mathbf{A}^2\mathbf{b}_1, \mathbf{A}^2\mathbf{b}_2$, and so forth, until we obtain n linearly independent vectors. Note that if a vector, say $\mathbf{Ab}_2$, is skipped because of linear dependence on the vectors $\{\mathbf{b}_1, \mathbf{b}_2, \ldots, \mathbf{b}_p, \mathbf{Ab}_1\}$, then all vectors of the form $\mathbf{A}^k\mathbf{b}_2$, for $k \geq 1$, can also be skipped because they must also be dependent on the previous columns. After choosing n linearly independent vectors in this order, we *rearrange* them as

$$\{\mathbf{b}_1, \ldots, \mathbf{A}^{\mu_1-1}\mathbf{b}_1; \mathbf{b}_2, \ldots, \mathbf{A}^{\mu_2-1}\mathbf{b}_2; \ldots; \mathbf{b}_p, \ldots, \mathbf{A}^{\mu_p-1}\mathbf{b}_p\} \tag{7-21}$$

where $\mu_1 + \mu_2 + \cdots + \mu_p = n$. Note that the main difference between this scheme and Scheme 1 is that in Equation (7-20) $\mathbf{A}^{\bar{\mu}_1}\mathbf{b}_1$ can be expressed as a linear combination of $\{\mathbf{b}_1, \mathbf{Ab}_1, \ldots, \mathbf{A}^{\bar{\mu}_1-1}\mathbf{b}_1\}$, whereas the vector $\mathbf{A}^{\mu_1}\mathbf{b}_1$ in (7-21) cannot be expressed as a linear combination of $\{\mathbf{b}_1, \ldots, \mathbf{A}^{\mu_1-1}\mathbf{b}_1\}$; $\mathbf{A}^{\mu_1}\mathbf{b}_1$ is generally linearly dependent on all vectors in (7-21).[2] Similar remarks apply to $\mathbf{A}^{\mu_i}\mathbf{b}_i$, for $i = 1, 2, \ldots, p$.

Now if the set of vectors in (7-20) is chosen as a basis of the state space of *FE* or, equivalently, let $\bar{\mathbf{x}} = \mathbf{Q}^{-1}\mathbf{x}$ where

$$\mathbf{Q} \triangleq [\mathbf{q}_1 \quad \mathbf{q}_2 \quad \cdots \quad \mathbf{q}_n] \triangleq [\mathbf{b}_1 \quad \cdots \quad \mathbf{A}^{\bar{\mu}_1-1}\mathbf{b}_1 \quad \mathbf{b}_2 \quad \cdots \quad \mathbf{A}^{\bar{\mu}_2-1}\mathbf{b}_2 \quad \mathbf{b}_3 \quad \cdots \quad \mathbf{A}^{\bar{\mu}_3-1}\mathbf{b}_3]$$

then the matrices $\bar{\mathbf{A}}$ and $\bar{\mathbf{B}}$ will be of the forms

[2] More can be said by using the relative sizes of μ_i. For example, if $\mu_1 \leq \mu_2 \leq \cdots \leq \mu_p$, then $\mathbf{A}^{\mu_1}\mathbf{b}_1$ depends on $\{\mathbf{A}^k\mathbf{b}_i, i = 1, 2, \ldots, p; k = 0, 1, \ldots, \mu_1 - 1\}$; $\mathbf{A}^{\mu_2}\mathbf{b}_2$ depends on $\{\mathbf{A}^k\mathbf{b}_1, k = 0, 1, \ldots, \mu_1 - 1; \mathbf{A}^k\mathbf{b}_i, i = 2, 3, \ldots, p; k = 0, 1, \ldots, \mu_2 - 1\}$ and so forth.

$$\bar{\mathbf{A}} = \left[\begin{array}{cccccc:cccccc:cccccc}
0 & 0 & 0 & \cdots & 0 & x & & & & & & x & & & & & & x \\
1 & 0 & 0 & \cdots & 0 & x & & & & & & x & & & & & & x \\
0 & 1 & 0 & \cdots & 0 & x & & & & & & x & & & & & & x \\
\vdots & \vdots & \vdots & & \vdots & \vdots & & & & & & \vdots & & & & & & \vdots \\
0 & 0 & 0 & \cdots & 1 & x & & & & & & x & & & & & & x \\
\hdashline
\multicolumn{6}{c:}{(\bar{\mu}_1 \times \bar{\mu}_1)} & 0 & 0 & 0 & \cdots & 0 & x & & & & & & x \\
& & & & & & 1 & 0 & 0 & \cdots & 0 & x & & & & & & x \\
& & & & & & 0 & 1 & 0 & \cdots & 0 & x & & & & & & x \\
& & & & & & \vdots & \vdots & \vdots & & \vdots & \vdots & & & & & & \vdots \\
& & & & & & 0 & 0 & 0 & \cdots & 1 & x & & & & & & x \\
\hdashline
& & & & & & \multicolumn{6}{c:}{(\bar{\mu}_2 \times \bar{\mu}_2)} & 0 & 0 & 0 & \cdots & 0 & x \\
& & & & & & & & & & & & 1 & 0 & 0 & \cdots & 0 & x \\
& & & & & & & & & & & & 0 & 1 & 0 & \cdots & 0 & x \\
& & & & & & & & & & & & \vdots & \vdots & \vdots & & \vdots & \vdots \\
& & & & & & & & & & & & 0 & 0 & 0 & \cdots & 1 & x \\
& & & & & & & & & & & & \multicolumn{6}{c}{(\bar{\mu}_3 \times \bar{\mu}_3)}
\end{array}\right]
\qquad
\bar{\mathbf{B}} = \left[\begin{array}{ccc}
1 & 0 & 0 \\
0 & 0 & 0 \\
0 & 0 & 0 \\
\vdots & \vdots & \vdots \\
0 & 0 & 0 \\
\hdashline
0 & 1 & 0 \\
0 & 0 & 0 \\
0 & 0 & 0 \\
\vdots & \vdots & \vdots \\
0 & 0 & 0 \\
\hdashline
0 & 0 & 1 \\
0 & 0 & 0 \\
0 & 0 & 0 \\
\vdots & \vdots & \vdots \\
0 & 0 & 0
\end{array} \quad \bar{\mathbf{b}}_4 \cdots \bar{\mathbf{b}}_p \right]
\tag{7-22}$$

where the X's denote possible nonzero elements and unfilled positions are all zeros. This can be easily verified by observing that the ith column of $\bar{\mathbf{A}}$ is the representation of $\mathbf{A}\mathbf{q}_i$ with respect to the basis vectors $[\mathbf{q}_1 \quad \mathbf{q}_2 \quad \cdots \quad \mathbf{q}_n]$.

If the set of vectors in (7-21) is used as a basis, then the new matrices $\tilde{\mathbf{A}}$ and $\tilde{\mathbf{B}}$ will be of the form[3]

$$\tilde{\mathbf{A}} = \left[\begin{array}{cccc:cccc:c:cccc}
0 & 0 & \cdots & x & & & & x & & & & & x \\
1 & 0 & \cdots & x & & & & x & & & & & x \\
0 & 1 & \cdots & x & & & & x & & & & & x \\
\vdots & \vdots & & \vdots & & & & \vdots & & & & & \vdots \\
0 & 0 & \cdots & x & & & & x & & & & & x \\
\hdashline
\multicolumn{3}{c}{(\mu_1 \times \mu_1)} & x & 0 & 0 & \cdots & x & & & & & x \\
& & & x & 1 & 0 & \cdots & x & & & & & x \\
& & & x & 0 & 1 & \cdots & x & & & & & x \\
& & & \vdots & \vdots & \vdots & & \vdots & & & & & \vdots \\
& & & x & 0 & 0 & \cdots & x & & & & & x \\
\hdashline
& & & & \multicolumn{4}{c:}{(\mu_2 \times \mu_2)} & & & & & \\
& & & \vdots & & & & \vdots & \ddots & & & & \vdots \\
& & & & & & & & & \multicolumn{4}{c}{(\mu_p \times \mu_p)} \\
\hdashline
& & & x & & & & x & & 0 & 0 & \cdots & x \\
& & & x & & & & x & & 1 & 0 & \cdots & x \\
& & & x & & & & x & & 0 & 1 & \cdots & x \\
& & & \vdots & & & & \vdots & & \vdots & \vdots & & \vdots \\
& & & x & & & & x & & 0 & 0 & \cdots & x
\end{array}\right]
\qquad
\tilde{\mathbf{B}} = \left[\begin{array}{cccc}
1 & 0 & \cdots & 0 \\
0 & 0 & \cdots & 0 \\
0 & 0 & \cdots & 0 \\
\vdots & \vdots & & \vdots \\
0 & 0 & \cdots & 0 \\
\hdashline
0 & 1 & \cdots & 0 \\
0 & 0 & \cdots & 0 \\
0 & 0 & \cdots & 0 \\
\vdots & \vdots & & \vdots \\
0 & 0 & \cdots & 0 \\
\hdashline
 & & \vdots & \\
\hdashline
0 & 0 & \cdots & 1 \\
0 & 0 & \cdots & 0 \\
0 & 0 & \cdots & 0 \\
\vdots & \vdots & & \vdots \\
0 & 0 & \cdots & 0
\end{array}\right]
\tag{7-23}$$

[3] Using footnote 2, some elements denoted by X can be replaced by zeros.

where the X's again denote possible nonzero elements and unfilled positions are all zeros. The matrices $\tilde{\mathbf{A}}$ and $\tilde{\mathbf{B}}$ can be verified by inspection. The matrix $\mathbf{C}$ in both cases are to be computed from $\mathbf{CQ}$.

By comparing (7-22) with (7-23), we see immediately the differences in the matrices $\mathbf{A}$ and $\mathbf{B}$ due to the different choices of basic vectors according to Schemes 1 and 2. The matrix $\bar{\mathbf{A}}$ has three blocks in the diagonal, whereas the matrix $\tilde{\mathbf{A}}$ has p blocks. The first three columns of $\bar{\mathbf{B}}$ are very simple, whereas every column of $\tilde{\mathbf{B}}$ consists of only one nonzero element.

The usefulness of the forms in (7-22) and (7-23) is not known at present. The purpose of introducing these two forms is to show that there is no additional conceptual difficulty in developing canonical forms for multivariable dynamical equations. By rearranging the vectors in (7-20) or (7-21), different dynamical equations can be obtained. One of them will be discussed in Section 7-3. For a survey of various canonical forms, see References S154.

7-3 State Feedback

In this section we study the effect of the linear state feedback $\mathbf{u} = \mathbf{r} + \mathbf{Kx}$ on a linear time-invariant dynamical equation. The variable $\mathbf{r}$ denotes the reference input and the $\mathbf{K}$ is the feedback gain matrix and is required to be real. In this study, we implicitly assume that all state variables are available for the feedback.

Single-variable case. Consider the single-variable, linear time-invariant dynamical equation

$$FE_1: \qquad \dot{\mathbf{x}} = \mathbf{Ax} + \mathbf{b}u \tag{7-24a}$$
$$y = \mathbf{cx} + eu \tag{7-24b}$$

where $\mathbf{x}$ is the $n \times 1$ state vector, u is the scalar input, y is the scalar output, $\mathbf{A}$ is an $n \times n$ real constant matrix, $\mathbf{b}$ is an $n \times 1$ real constant column vector, and $\mathbf{c}$ is a $1 \times n$ real constant row vector. In state feedback, every state variable is multiplied by a gain and fed back into the input terminal. Let the gain between the ith state variable and the input be k_i. Define $\mathbf{k} \triangleq [k_1 \quad k_2 \quad \cdots \quad k_n]$. Then the dynamical equation of the state-feedback system shown in Figure 7-1 is

$$FE_1^f: \qquad \dot{\mathbf{x}} = (\mathbf{A} + \mathbf{bk})\mathbf{x} + \mathbf{b}r \tag{7-25a}$$
$$y = (\mathbf{c} + e\mathbf{k})\mathbf{x} + er \tag{7-25b}$$

which is obtained by replacing u in (7-24) by $r + \mathbf{kx}$, where r is the reference input. Note that the dynamical equations (7-24) and (7-25) have the same dimension and the same state space. Now we shall show that the controllability of a linear time-invariant dynamical equation is invariant under any linear state feedback.

Theorem 7-3

The state feedback dynamical equation FE_1^f in (7-25) is controllable for any $1 \times n$ real vector $\mathbf{k}$ if and only if the dynamical equation FE_1 in (7-24) is controllable.

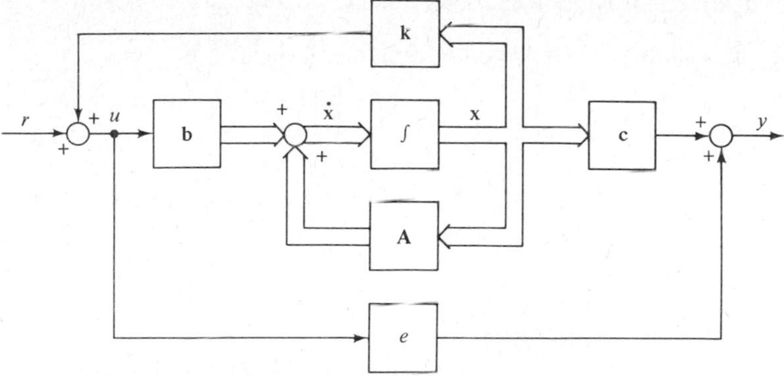

Figure 7-1 A state feedback system.

Proof

First we show that the controllability of FE_1 implies the controllability of FE_1^f. Let $\mathbf{x}_0$ and $\mathbf{x}_1$ be two arbitrary states in the state space Σ. By the controllability assumption of FE_1, there exists an input u that will transfer $\mathbf{x}_0$ to $\mathbf{x}_1$ in a finite time. Now for the state-feedback dynamical equation, if we choose $r(t)=u(t)-\mathbf{k}\mathbf{x}(t)$, then the input r will transfer $\mathbf{x}_0$ to $\mathbf{x}_1$. Therefore we conclude that FE_1^f is controllable.

We see from Figure 7-1 that the input r does not control the state $\mathbf{x}$ directly, it generates u to control $\mathbf{x}$. Therefore, if u cannot control $\mathbf{x}$, neither can r. In other words, if FE_1 is not controllable, neither is FE_1^f. Q.E.D.

We see that in the proof, the assumptions of single-variableness and time-invariance are not used. Therefore we have the following corollary.

Corollary 7-3

The controllability of a multivariable linear time-varying dynamical equation is invariant under any state feedback of the form $\mathbf{u}(t)=\mathbf{r}(t)+\mathbf{K}(t)\mathbf{x}(t)$. ∎

Note that Theorem 7-3 can also be proved by showing that

$$\rho[\mathbf{b} \quad \mathbf{A}\mathbf{b} \quad \cdots \quad \mathbf{A}^{n-1}\mathbf{b}]=\rho[\mathbf{b} \quad (\mathbf{A}+\mathbf{b}\mathbf{k})\mathbf{b} \quad \cdots \quad (\mathbf{A}+\mathbf{b}\mathbf{k})^{n-1}\mathbf{b}] \qquad \textbf{(7-26)}$$

for any $1\times n$ real constant vector $\mathbf{k}$ (see Problem 7-10).

Although state feedback preserves the controllability of a dynamical equation, it is always possible to destroy the observability property of a dynamical equation by some choice of $\mathbf{k}$. For example, if $e\neq 0$ and if $\mathbf{k}=(-1/e)\mathbf{c}$, then the state-feedback equation (7-25) is not observable even if FE_1 is observable. If $e=0$, it is still possible to choose some $\mathbf{k}$ such that the state-feedback dynamical equation will not preserve the observability property (see Problem 7-14).

Example 1

Consider the controllable and observable dynamical equation

$$FE_1: \qquad \dot{\mathbf{x}} = \begin{bmatrix} 1 & 2 \\ 3 & 1 \end{bmatrix} \mathbf{x} + \begin{bmatrix} 0 \\ 1 \end{bmatrix} u$$

$$y = [1 \quad 2]\,\mathbf{x}$$

If we introduce the state feedback

$$u = r + [-3 \quad -1]\mathbf{x}$$

then the state-feedback equation is

$$FE_1^f: \qquad \dot{\mathbf{x}} = \begin{bmatrix} 1 & 2 \\ 0 & 0 \end{bmatrix} \mathbf{x} + \begin{bmatrix} 0 \\ 1 \end{bmatrix} r$$

$$y = [1 \quad 2]\,\mathbf{x}$$

which is controllable but not observable. ∎

An important property of state feedback is that it can be used to control the eigenvalues of a dynamical equation.

Theorem 7-4

If the single-variable dynamical equation FE_1 given in (7-24) is controllable, then by the state feedback $u = r + \mathbf{kx}$, where $\mathbf{k}$ is a $1 \times n$ real vector, the eigenvalues of $(\mathbf{A} + \mathbf{bk})$ can be arbitrarily assigned, provided that complex conjugate eigenvalues appear in pair.

Proof

If the dynamical equation FE_1 is controllable, by an equivalence transformation $\bar{\mathbf{x}} = \mathbf{Px}$, FE_1 can be transformed into the following controllable canonical form (Theorem 7-1):

$$CFE_1: \quad \dot{\bar{\mathbf{x}}} = \begin{bmatrix} 0 & 1 & 0 & \cdots & 0 & 0 \\ 0 & 0 & 1 & \cdots & 0 & 0 \\ 0 & 0 & 0 & \cdots & 0 & 0 \\ \vdots & \vdots & \vdots & & \vdots & \vdots \\ 0 & 0 & 0 & \cdots & 0 & 1 \\ -\alpha_n & -\alpha_{n-1} & -\alpha_{n-2} & \cdots & -\alpha_2 & -\alpha_1 \end{bmatrix} \bar{\mathbf{x}} + \begin{bmatrix} 0 \\ 0 \\ 0 \\ \vdots \\ 0 \\ 1 \end{bmatrix} u \tag{7-27a}$$

$$y = [\beta_n \quad \beta_{n-1} \quad \beta_{n-2} \quad \cdots \quad \beta_2 \quad \beta_1]\bar{\mathbf{x}} + eu \tag{7-27b}$$

Let $\bar{\mathbf{A}}$ and $\bar{\mathbf{b}}$ denote the matrices in (7-27a), then $\bar{\mathbf{A}} = \mathbf{PAP}^{-1}$, $\bar{\mathbf{b}} = \mathbf{Pb}$. Because of the equivalence transformation, the state feedback becomes

$$u = r + \mathbf{kx} = r + \mathbf{kP}^{-1}\bar{\mathbf{x}} \triangleq r + \bar{\mathbf{k}}\bar{\mathbf{x}} \tag{7-28}$$

where $\bar{\mathbf{k}} \triangleq \mathbf{kP}^{-1}$. It is easy to see that the set of the eigenvalues of $(\mathbf{A} + \mathbf{bk})$ is equal to the set of the eigenvalues of $(\bar{\mathbf{A}} + \bar{\mathbf{b}}\bar{\mathbf{k}})$. Let the characteristic polynomial of the matrix $(\mathbf{A}+\mathbf{bk})$ or, correspondingly, of $(\bar{\mathbf{A}}+\bar{\mathbf{b}}\bar{\mathbf{k}})$ with desired eigenvalues be

$$s^n+\bar{\alpha}_1 s^{n-1}+\cdots+\bar{\alpha}_n$$

If $\bar{\mathbf{k}}$ is chosen as

$$\bar{\mathbf{k}}=[\alpha_n-\bar{\alpha}_n \quad \alpha_{n-1}-\bar{\alpha}_{n-1} \quad \cdots \quad \alpha_1-\bar{\alpha}_1] \tag{7-29}$$

then the state-feedback dynamical equation becomes

$$CFE_1^f:\quad \dot{\bar{\mathbf{x}}}=\begin{bmatrix} 0 & 1 & 0 & \cdots & 0 & 0 \\ 0 & 0 & 1 & \cdots & 0 & 0 \\ 0 & 0 & 0 & \cdots & 0 & 0 \\ \vdots & \vdots & \vdots & & \vdots & \vdots \\ 0 & 0 & 0 & \cdots & 0 & 1 \\ -\bar{\alpha}_n & -\bar{\alpha}_{n-1} & -\bar{\alpha}_{n-2} & \cdots & -\bar{\alpha}_2 & -\bar{\alpha}_1 \end{bmatrix}\bar{\mathbf{x}}+\begin{bmatrix} 0 \\ 0 \\ 0 \\ \vdots \\ 0 \\ 1 \end{bmatrix} r \tag{7-30a}$$

$$y=[\beta_n+e(\alpha_n-\bar{\alpha}_n) \quad \beta_{n-1}+e(\alpha_{n-1}-\bar{\alpha}_{n-1}) \quad \cdots \quad \beta_1+e(\alpha_1-\bar{\alpha}_1)]\bar{\mathbf{x}}+er \tag{7-30b}$$

Since the characteristic polynomial of the $\mathbf{A}$ matrix in (7-30) is $s^n+\bar{\alpha}_1 s^{n-1}+\cdots+\bar{\alpha}_n$, we conclude that the state-feedback equation has the desired eigenvalues. Q.E.D.

The gain vector $\bar{\mathbf{k}}$ in (7-29) is chosen with respect to the state $\bar{\mathbf{x}}$; that is, $u=r+\bar{\mathbf{k}}\bar{\mathbf{x}}$. Therefore, with respect to the original state $\mathbf{x}$, we have to use $\mathbf{k}=\bar{\mathbf{k}}\mathbf{P}$, where the columns of $\mathbf{P}^{-1}=\mathbf{Q}$ are those vectors introduced in (7-9). The matrices $\mathbf{A}$, $\mathbf{b}$, and $\mathbf{k}$ are assumed to be real; hence, if a complex eigenvalue is assigned to the matrix $(\mathbf{A}+\mathbf{bk})$, its complex conjugate must also be assigned. The procedure of choosing $\mathbf{k}$ has nothing to do with the number of outputs; therefore it can be applied to any controllable single-input, multiple-output, linear time-invariant dynamical equation. We summarize the procedure of choosing $\mathbf{k}$ in the following.

Algorithm

Given a controllable $\{\mathbf{A}, \mathbf{b}\}$ and a set of eigenvalues $\bar{\lambda}_1, \bar{\lambda}_2, \ldots, \bar{\lambda}_n$. Find the $1\times n$ real vector $\mathbf{k}$ such that the matrix $(\mathbf{A}+\mathbf{bk})$ has the set $\{\bar{\lambda}_1, \bar{\lambda}_2, \ldots, \bar{\lambda}_n\}$ as its eigenvalues.

1. Find the characteristic polynomial of $\mathbf{A}$: $\det(s\mathbf{I}-\mathbf{A})=s^n+\alpha_1 s^{n-1}+\cdots+\alpha_n$.
2. Compute $(s-\bar{\lambda}_1)(s-\bar{\lambda}_2)\cdots(s-\bar{\lambda}_n)=s_n+\bar{\alpha}_1 s^{n-1}+\cdots+\bar{\alpha}_n$.
3. Compute $\bar{\mathbf{k}}=[\alpha_n-\bar{\alpha}_n \quad \alpha_{n-1}-\bar{\alpha}_{n-1} \quad \cdots \quad \alpha_1-\bar{\alpha}_1]$.
4. Compute $\mathbf{q}_{n-i}=\mathbf{A}\mathbf{q}_{n-i+1}+\alpha_i\mathbf{q}_n$, for $i=1, 2, \ldots, (n-1)$, with $\mathbf{q}_n=\mathbf{b}$.
5. Form $\mathbf{Q}=[\mathbf{q}_1 \quad \mathbf{q}_2 \quad \cdots \quad \mathbf{q}_n]$.
6. Find $\mathbf{P}\triangleq\mathbf{Q}^{-1}$.
7. $\mathbf{k}=\bar{\mathbf{k}}\mathbf{P}$. ∎

This algorithm can also be developed directly without transforming the equation explicitly into the controllable canonical form. Let $\bar{\Delta}(s)$ be the characteristic polynomial of $\mathbf{A}+\mathbf{bk}$. Then we have

$$\begin{aligned}\bar{\Delta}(s) &= \det(s\mathbf{I}-\mathbf{A}-\mathbf{bk}) = \det[(s\mathbf{I}-\mathbf{A})(\mathbf{I}-(s\mathbf{I}-\mathbf{A})^{-1}\mathbf{bk})] \\ &= \det(s\mathbf{I}-\mathbf{A})\det[\mathbf{I}-(s\mathbf{I}-\mathbf{A})^{-1}\mathbf{bk}]\end{aligned}$$

This can be written as, by using (3-65) and defining $\Delta(s)=\det(s\mathbf{I}-\mathbf{A})$,

$$\bar{\Delta}(s) = \Delta(s)(1-\mathbf{k}(s\mathbf{I}-\mathbf{A})^{-1}\mathbf{b})$$

or

$$\Delta(s)-\bar{\Delta}(s) = \Delta(s)\mathbf{k}(s\mathbf{I}-\mathbf{A})^{-1}\mathbf{b} \tag{7-31}$$

From this equation and using Problem 2-39, Equations (7-13) and (7-14), we can readily establish $\bar{\mathbf{k}}=\mathbf{k}\mathbf{U}\mathbf{\Lambda}=\mathbf{kQ}$ or $\mathbf{k}=\bar{\mathbf{k}}\mathbf{P}$.

Example 2

Consider the inverted pendulum problem studied in Example 4 on page 185. Its dynamical equation is, as derived in (5-25),

$$\dot{\mathbf{x}} = \begin{bmatrix} 0 & 1 & 0 & 0 \\ 0 & 0 & -1 & 0 \\ 0 & 0 & 0 & 1 \\ 0 & 0 & 5 & 0 \end{bmatrix}\mathbf{x} + \begin{bmatrix} 0 \\ 1 \\ 0 \\ -2 \end{bmatrix} u$$

$$y = \begin{bmatrix} 1 & 0 & 0 & 0 \end{bmatrix}\mathbf{x}$$

It is controllable; hence its eigenvalues can be arbitrarily assigned. Let the desired eigenvalues be $-1, -2, -1\pm j$. Then we have

$$\bar{\Delta}(s) = (s+1)(s+2)(s+1+j)(s+1-j) = s^4+5s^3+10s^2+10s+4$$

The characteristic polynomial of

$$\mathbf{A}+\mathbf{bk} = \begin{bmatrix} 0 & 1 & 0 & 0 \\ k_1 & k_2 & k_3-1 & k_4 \\ 0 & 0 & 0 & 1 \\ -2k_1 & -2k_2 & 5-2k_3 & -2k_4 \end{bmatrix}$$

where $\mathbf{k}=[k_1 \quad k_2 \quad k_3 \quad k_4]$ can be computed as

$$\det(s\mathbf{I}-\mathbf{A}-\mathbf{bk}) = s^4+(2k_4-k_2)s^3+(2k_3-k_1-5)s^2+3k_2 s+3k_1$$

The comparison of the coefficients of $\det(s\mathbf{I}-\mathbf{A}-\mathbf{bk})$ and those of $\bar{\Delta}(s)$ yields

$$k_1=\tfrac{3}{4} \qquad k_2=\tfrac{10}{3} \qquad k_3=\tfrac{63}{8} \qquad k_4=\tfrac{25}{6}$$

Hence the introduction of the state feedback

$$u(t)=r(t)+[\tfrac{3}{4} \quad \tfrac{10}{3} \quad \tfrac{63}{8} \quad \tfrac{25}{6}]\mathbf{x}$$

will place the eigenvalues of the resulting equation at $-1, -2, -1\pm j$. ∎

Even for this simple $\{\mathbf{A}, \mathbf{b}\}$, direct computation of $\det(s\mathbf{I}-\mathbf{A}-\mathbf{bk})$ is very complicated and is not suitable for computer computation. Hence, for large n, the use of the algorithm is much more convenient. In the multivariable case, a different method of computing $\mathbf{k}$ by solving a Lyapunov equation (Appendix F) will be introduced. The method is directly applicable to the single-variable case. To avoid repetition, it will not be discussed here.

Stabilization

If a dynamical equation is controllable, then all the eigenvalues can be arbitrarily assigned by the introduction of state feedback. If a dynamical equation is not controllable, one may wonder how many eigenvalues can be controlled. It is shown in Theorem 5-16 that if a linear time-invariant dynamical equation is not controllable, by a proper choice of basis vectors, the state equation can be transformed into

$$\dot{\bar{\mathbf{x}}} = \bar{\mathbf{A}}\bar{\mathbf{x}} + \bar{\mathbf{b}}u \tag{7-32}$$

where

$$\bar{\mathbf{A}} = \begin{bmatrix} \bar{\mathbf{A}}_{11} & \bar{\mathbf{A}}_{12} \\ \mathbf{0} & \bar{\mathbf{A}}_{22} \end{bmatrix} \qquad \bar{\mathbf{b}} = \begin{bmatrix} \bar{\mathbf{b}}_1 \\ \mathbf{0} \end{bmatrix} \tag{7-33}$$

and the reduced equation $\dot{\bar{\mathbf{x}}}_1 = \bar{\mathbf{A}}_{11}\bar{\mathbf{x}}_1 + \bar{\mathbf{b}}_1 u$ is controllable. Because of the form of $\bar{\mathbf{A}}$, the set of eigenvalues of $\bar{\mathbf{A}}$ is the union of the sets of the eigenvalues of $\bar{\mathbf{A}}_{11}$ and of $\bar{\mathbf{A}}_{22}$. In view of the form of $\bar{\mathbf{b}}$, it is easy to see that the matrix $\bar{\mathbf{A}}_{22}$ is not affected by the introduction of any state feedback of the form $u = r + \bar{\mathbf{k}}\bar{\mathbf{x}}$. Therefore all the eigenvalues of $\bar{\mathbf{A}}_{22}$ cannot be controlled. On the other hand, all the eigenvalues of $\bar{\mathbf{A}}_{11}$ can be arbitrarily assigned because $\{\bar{\mathbf{A}}_{11}, \bar{\mathbf{b}}_1\}$ is controllable. Hence we conclude that the eigenvalues of $(\mathbf{A} + \mathbf{bk})$ can be arbitrarily assigned if and only if $\{\mathbf{A}, \mathbf{b}\}$ is controllable. Since $\mathbf{k}$ is required to be a real vector, complex conjugate eigenvalues must be assigned in pairs.

In the design of a system, sometimes it is required only to change unstable eigenvalues (the eigenvalues with nonnegative real parts) to stable eigenvalues (the eigenvalues with negative real parts). This is called *stabilization*. From the above discussion we see that if the matrix $\bar{\mathbf{A}}_{22}$ has unstable eigenvalues, then the equation cannot be stabilized. Hence we may conclude that if $\{\mathbf{A}, \mathbf{b}\}$ is transformed into the form in (7-33), and if $\{\bar{\mathbf{A}}_{11}, \bar{\mathbf{b}}_1\}$ is controllable and all the eigenvalues of $\bar{\mathbf{A}}_{22}$ have negative real parts, then $\{\mathbf{A}, \mathbf{b}\}$ is stabilizable. Whether a dynamical equation is stabilizable can also be seen from its equivalent Jordan-form dynamical equation. If all the Jordan blocks associated with unstable eigenvalues are controllable, then the equation is stabilizable. From the proof of Statement 4 of Theorem 5-7 on page 206, we can conclude that λ is in $\bar{\mathbf{A}}_{22}$ if and only if the matrix $[\lambda\mathbf{I} - \mathbf{A} \quad \mathbf{B}]$ does not have a full row rank. We note that $\bar{\mathbf{A}}_{11}$ and $\bar{\mathbf{A}}_{22}$ in (7-33) may have same eigenvalues.

Effect on the numerator of $\hat{g}(s)$

Consider the dynamical equation (7-24). Its transfer function is $\hat{g}(s) = \mathbf{c}(s\mathbf{I} - \mathbf{A})^{-1}\mathbf{b} + e$. Since every pole of $\hat{g}(s)$ is an eigenvalue of $\mathbf{A}$, from Theorem 7-4 we conclude that the poles of $\hat{g}(s)$ can be arbitrarily assigned by the introduction of state feedback. It is of interest to note that although the poles of $\hat{g}(s)$ are shifted by state feedback, the zeros of $\hat{g}(s)$ are not affected. That is, the zeros of $\hat{g}(s)$ remain unchanged after the introduction of state feedback. We prove this by showing that the numerator of the transfer function of FE_1 in (7-24) is equal to the numerator of the transfer function of CFE_1 in (7-30). The transfer function of FE_1 is

$$\hat{g}(s)=\frac{\beta_1 s^{n-1}+\beta_2 s^{n-2}+\cdots+\beta_n}{s^n+\alpha_1 s^{n-1}+\cdots+\alpha_n}+e$$

$$=\frac{es^n+(\beta_1+e\alpha_1)s^{n-1}+\cdots+(\beta_n+e\alpha_n)}{s^n+\alpha_1 s^{n-1}+\cdots+\alpha_n} \tag{7-34}$$

The transfer function of (7-30) is

$$\hat{g}_f(s)=\frac{[\beta_1+e(\alpha_1-\bar{\alpha}_1)]s^{n-1}+\cdots+[\beta_n+e(\alpha_n-\bar{\alpha}_n)]}{s^n+\bar{\alpha}_1 s^{n-1}+\cdots+\bar{\alpha}_n}+e$$

$$=\frac{es^n+(\beta_1+e\alpha_1)s^{n-1}+\cdots+(\beta_n+e\alpha_n)}{s^n+\bar{\alpha}_1 s^{n-1}+\cdots+\bar{\alpha}_n} \tag{7-35}$$

which has the same numerator as $\hat{g}(s)$ in (7-34). This proves that the zeros of $\hat{g}(s)$ remain unchanged after the introduction of state feedback.

This property can be used to explain why a state feedback may alter the observability property of an equation. If some of the poles are shifted to coincide with zeros of $\hat{g}(s)$, then the degree of $\hat{g}_f(s)$ in (7-35) is less than n, and the dynamical equation in (7-25) is not irreducible. In this case, (7-25) must be unobservable because it is controllable (Theorem 7-3).

Asymptotic tracking problem—nonzero set point

In this subsection we use state feedback to design a system so that its output will track the step reference input $r(t)=r_d$, for $t\geq 0$. If $r_d=0$, the problem is called a *regulator problem*; if $r_d\neq 0$, it is a special case of the *asymptotic tracking problem*. A step reference input can be set by the position of a potentiometer and is often referred to as set point. In this problem, the input u is chosen as

$$u(t)=pr(t)+\mathbf{k}\mathbf{x}(t) \tag{7-36}$$

where p is a constant gain and $\mathbf{k}$ is the feedback gain vector. The substitution of (7-36) into (7-24) and then the application of the Laplace transform to the resulting equation yields

$$\hat{y}(s)=(\mathbf{c}+e\mathbf{k})[(s\mathbf{I}-\mathbf{A}-\mathbf{b}\mathbf{k})^{-1}\mathbf{x}(0)+(s\mathbf{I}-\mathbf{A}-\mathbf{b}\mathbf{k})^{-1}\mathbf{b}p\hat{r}(s)]+ep\hat{r}(s) \tag{7-37}$$

If $r(t)=r_d$, for $t\geq 0$, then $\hat{r}(s)=r_d/s$. The substitution of the identity

$$(s\mathbf{I}-\mathbf{A}-\mathbf{b}\mathbf{k})^{-1}s^{-1}=(-\mathbf{A}-\mathbf{b}\mathbf{k})^{-1}s^{-1}+(s\mathbf{I}-\mathbf{A}-\mathbf{b}\mathbf{k})^{-1}(\mathbf{A}+\mathbf{b}\mathbf{k})^{-1}$$

[see also Equation (H-1)] and

$$\hat{g}_f(s)=(\mathbf{c}+e\mathbf{k})(s\mathbf{I}-\mathbf{A}-\mathbf{b}\mathbf{k})^{-1}\mathbf{b}+e \tag{7-38}$$

into (7-37) yields

$$\hat{y}(s)=(\mathbf{c}+e\mathbf{k})(s\mathbf{I}-\mathbf{A}-\mathbf{b}\mathbf{k})^{-1}[\mathbf{x}(0)\mathbf{I}+(\mathbf{A}+\mathbf{b}\mathbf{k})^{-1}\mathbf{b}pr_d]+\hat{g}_f(0)pr_d s^{-1}$$

which becomes, in the time domain,

$$y(t) = (\mathbf{c} + e\mathbf{k})e^{(\mathbf{A}+\mathbf{bk})t}\,[\mathbf{x}(0)\mathbf{I} + (\mathbf{A}+\mathbf{bk})^{-1}\mathbf{b}pr_d] + \hat{g}_f(0)pr_d \tag{7-39}$$

for $t \geq 0$. If all eigenvalues of $(\mathbf{A}+\mathbf{bk})$ have negative real parts, then $e^{(\mathbf{A}+\mathbf{bk})t}$ approaches zero as $t \to \infty$ and consequently $y(t)$ approaches $\hat{g}_f(0)pr_d$ as $t \to \infty$. If $\{\mathbf{A}, \mathbf{b}\}$ is controllable, all eigenvalues of $(\mathbf{A}+\mathbf{bk})$ can be arbitrarily assigned by a proper choice of $\mathbf{k}$. Clearly, if the eigenvalues of $(\mathbf{A}+\mathbf{bk})$ are shifted deeper inside the left-half s plane, the faster the output approaches $\hat{g}_f(0)pr_d$. However, in this case the required feedback gain will be larger and may cause saturation. Furthermore, the system may be more susceptible to noises. Thus the choice of proper eigenvalues is not a simple task and is generally reached by a compromise between conflicting criteria.

As discussed in the previous subsection, the zeros of $\hat{g}(s)$ are not affected by state feedback; hence we have $\hat{g}_f(0) \neq 0$ if and only if $\hat{g}(0) \neq 0$. If $\hat{g}(0) \neq 0$, we may choose $p = 1/\hat{g}_f(0)$. In this case, we have

$$y(t) \to r_d \qquad \text{as } t \to \infty$$

and the design is completed. We note that the design can also be achieved if $\{\mathbf{A}, \mathbf{b}\}$ is not controllable but is stabilizable. In this case, however, we do not have complete control over the rate of convergence of $y(t)$ to r_d.

Much more can be said regarding the design of asymptotic tracking problem. It appears that the design can be carried out more easily by using the transfer function and will be discussed in Section 9-6. Hence this problem will not be discussed further in this chapter.

***Multivariable case.** Consider the n-dimensional linear time-invariant, multivariable dynamical equation

$$FE: \quad \dot{\mathbf{x}} = \mathbf{Ax} + \mathbf{Bu} \tag{7-40a}$$
$$\mathbf{y} = \mathbf{Cx} + \mathbf{Eu} \tag{7-40b}$$

where $\mathbf{A}$, $\mathbf{B}$, $\mathbf{C}$, and $\mathbf{E}$ are, respectively, $n \times n$, $n \times p$, $q \times n$, and $q \times p$ constant real matrices. In state feedback, the input $\mathbf{u}$ in FE is replaced by

$$\mathbf{u} = \mathbf{r} + \mathbf{Kx}$$

where $\mathbf{r}$ stands for a reference input vector and $\mathbf{K}$ is a $p \times n$ real constant matrix, called the feedback gain matrix; and Equation (7-40) becomes

$$FE^f: \quad \dot{\mathbf{x}} = (\mathbf{A} + \mathbf{BK})\mathbf{x} + \mathbf{Br} \tag{7-41a}$$
$$\mathbf{y} = (\mathbf{C} + \mathbf{EK})\mathbf{x} + \mathbf{Er} \tag{7-41b}$$

In the following, we shall show that if the dynamical equation FE is controllable, then the eigenvalues of $(\mathbf{A}+\mathbf{BK})$ can be arbitrarily assigned by a proper choice of $\mathbf{K}$. This will be established by using three different methods.

Method 1

In this method we change the multivariable problem into a single-variable problem and then apply the result in the previous subsection.

A matrix **A** is called *cyclic* if its characteristic polynomial is equal to its minimal polynomial. From Theorem 2-12 or Example 1 on page 48, we can conclude immediately that **A** is cyclic if and only if the Jordan canonical form of **A** has one and only one Jordan block associated with each distinct eigenvalue (Problem 2-45). The term "cyclicity" arises from the property that if **A** is cyclic, then there exists a vector **b** such that the vectors **b**, **Ab**, ..., $\mathbf{A}^{n-1}\mathbf{b}$ span the n-dimensional real space or, equivalently, {**A**, **b**} is controllable. This property can be easily deduced from Corollary 5-21.

Theorem 7-5

If {**A**, **B**} is controllable and if **A** is cyclic, then for almost any $p \times 1$ real vector **v**, the single-input pair {**A**, **Bv**} is controllable.

Proof

Controllability is invariant under any equivalence transformation, hence we may assume **A** to be in the Jordan canonical form. To see the basic idea, we use the following example:

$$\mathbf{A} = \left[\begin{array}{ccc:cc} 2 & 1 & 0 & 0 & 0 \\ 0 & 2 & 1 & 0 & 0 \\ 0 & 0 & 2 & 0 & 0 \\ \hdashline 0 & 0 & 0 & 1 & 1 \\ 0 & 0 & 0 & 0 & 1 \end{array}\right] \qquad \mathbf{B} = \begin{bmatrix} 0 & 1 \\ 0 & 0 \\ 1 & 2 \\ 4 & 9 \\ 1 & 0 \end{bmatrix} \qquad \mathbf{Bv} = \mathbf{B}\begin{bmatrix} v_1 \\ v_2 \end{bmatrix} = \begin{bmatrix} x \\ x \\ \alpha \\ x \\ \beta \end{bmatrix} \tag{7-42}$$

There is only one Jordan block associated with each distinct eigenvalue; hence **A** is cyclic. The condition for {**A**, **B**} to be controllable is that the two rows in **B** encircled by dotted lines in (7-42) are nonzeros (Theorem 5-21).

The necessary and sufficient condition for {**A**, **Bv**} to be controllable is that $\alpha \neq 0$ and $\beta \neq 0$ in (7-42). $\alpha = 0$ and $\beta = 0$ if and only if $v_1 = 0$ and $v_1/v_2 = -2/1$. Hence any **v** other than $v_1 = 0$ and $v_1/v_2 = -2$ will make {**A**, **Bv**} controllable. This establishes the theorem. Q.E.D.

The cyclicity assumption in this theorem is essential. Without this assumption, the theorem does not hold. For example, the {**A**, **B**} in

$$\mathbf{A} = \left[\begin{array}{cc:c} 2 & 1 & 0 \\ 0 & 2 & 0 \\ \hdashline 0 & 0 & 2 \end{array}\right] \qquad \mathbf{B} = \left[\begin{array}{cc} 1 & 3 \\ 1 & 0 \\ \hdashline 0 & 1 \end{array}\right]$$

is controllable. However, there is no **v** such that {**A**, **Bv**} is controllable.

If all the eigenvalues of **A** are distinct, then there is only one Jordan block associated with each eigenvalue. Hence a sufficient condition for **A** to be cyclic is that all the eigenvalues of **A** are distinct.

Theorem 7-6[4]

If $\{\mathbf{A}, \mathbf{B}\}$ is controllable, then for almost any $p \times n$ real constant matrix $\mathbf{K}$, all the eigenvalues of $\mathbf{A} + \mathbf{BK}$ are distinct and consequently $(\mathbf{A} + \mathbf{BK})$ is cyclic.

Proof

Let the characteristic polynomial of $\mathbf{A} + \mathbf{BK}$ be

$$\bar{\Delta}(s) = s^n + a_1 s^{n-1} + \cdots + a_{n-1}s + a_n$$

where a_i, $i = 1, 2, \ldots, n$, are functions of all elements, k_{ij}, of $\mathbf{K}$. The differentiation of $\bar{\Delta}(s)$ with respect to s yields

$$\bar{\Delta}'(s) = ns^{n-1} + (n-1)a_1 s^{n-2} + \cdots + a_{n-1}$$

If $\bar{\Delta}(s)$ has repeated roots, then $\bar{\Delta}(s)$ and $\bar{\Delta}'(s)$ are not coprime. A necessary and sufficient condition for $\bar{\Delta}(s)$ and $\bar{\Delta}'(s)$ to be not coprime is

$$\det \begin{bmatrix} a_n & a_{n-1} & \cdots & a_1 & 1 & 0 & \cdots & 0 \\ 0 & a_n & \cdots & a_2 & a_1 & 1 & \cdots & 0 \\ \vdots & \vdots & & \vdots & \vdots & \vdots & & \vdots \\ 0 & 0 & \cdots & a_n & a_{n-1} & \cdots & \cdots & 1 \\ \hdashline a_{n-1} & 2a_{n-2} & \cdots & n & 0 & 0 & \cdots & 0 \\ 0 & a_{n-1} & \cdots & (n-1)a_1 & n & 0 & \cdots & 0 \\ \vdots & \vdots & & \vdots & \vdots & \vdots & & \vdots \\ 0 & 0 & \cdots & 0 & a_{n-1} & \cdots & \cdots & n \end{bmatrix} \triangleq \gamma(k_{ij}) = 0$$

(See Appendix G). We note that $\gamma(k_{ij})$ is generally a nonhomogeneous polynomial of k_{ij}. There is a total of $p \times n$ number of k_{ij} in $\mathbf{K}$. Hence we may think of $\{k_{ij}\}$ as a vector in the $(p \times n)$-dimensional real vector space $(\mathbb{R}^{p \times n}, \mathbb{R})$. It is clear that the solution, k_{ij}, of $\gamma(k_{ij}) = 0$ is a subset of $(\mathbb{R}^{p \times n}, \mathbb{R})$. Hence for almost all $\mathbf{K}$, we have $\gamma(k_{ij}) \neq 0$ and all the roots of $\bar{\Delta}(s)$ are distinct. This establishes the theorem. See Problem 7-8. Q.E.D.

We are now ready to establish the following main theorem.

Theorem 7-7

If the dynamical equation FE in (7-40) is controllable, by a linear state feedback of the form $\mathbf{u} = \mathbf{r} + \mathbf{Kx}$, where $\mathbf{K}$ is a $p \times n$ real constant matrix, the eigenvalues of $(\mathbf{A} + \mathbf{BK})$ can be arbitrarily assigned provided complex conjugate eigenvalues appear in pairs.

---

[4] This theorem can be extended as follows: If $\{\mathbf{A}, \mathbf{B}, \mathbf{C}\}$ is irreducible, then for almost any $p \times q$ real constant matrix $\mathbf{H}$, the eigenvalues of $\mathbf{A} + \mathbf{BHC}$ are distinct and consequently $(\mathbf{A} + \mathbf{BHC})$ is cyclic. The matrix $\mathbf{A} + \mathbf{BHC}$ results from the constant output feedback $\mathbf{u} = \mathbf{r} + \mathbf{Hy}$. The proof of Theorem 7-6 is equally applicable here.

Proof

If $\mathbf{A}$ is not cyclic, we introduce $\mathbf{u}=\mathbf{w}+\mathbf{K}_1\mathbf{x}$ such that $\bar{\mathbf{A}} \triangleq \mathbf{A}+\mathbf{BK}_1$ in

$$\dot{\mathbf{x}}=(\mathbf{A}+\mathbf{BK}_1)\mathbf{x}+\mathbf{Bw} \tag{7-43}$$

is cyclic. Since $\{\mathbf{A}, \mathbf{B}\}$ is controllable, so is $\{\bar{\mathbf{A}}, \mathbf{B}\}$ (Corollary 7-3). Hence there exists a $p \times 1$ real vector $\mathbf{v}$ such that $\{\bar{\mathbf{A}}, \mathbf{Bv}\}$ is controllable.[5] Now we introduce another state feedback $\mathbf{w}=\mathbf{r}+\mathbf{K}_2\mathbf{x}$, as shown in Figure 7-2, with $\mathbf{K}_2$ chosen of the form

$$\mathbf{K}_2=\mathbf{vk}$$

where $\mathbf{k}$ is a $1 \times n$ real vector, then (7-43) becomes

$$\dot{\mathbf{x}}=(\bar{\mathbf{A}}+\mathbf{BK}_2)\mathbf{x}+\mathbf{Br}=(\bar{\mathbf{A}}+\mathbf{Bvk})\mathbf{x}+\mathbf{Br}$$

Since $\{\bar{\mathbf{A}}, \mathbf{Bv}\}$ is controllable, the eigenvalues of $(\bar{\mathbf{A}}+\mathbf{Bvk})$ can be arbitrarily assigned by a proper choice of $\mathbf{k}$ (Theorem 7-4). By combining the state feedback $\mathbf{u}=\mathbf{w}+\mathbf{K}_1\mathbf{x}$ and the state feedback $\mathbf{w}=\mathbf{r}+\mathbf{K}_2\mathbf{x}$ as

$$\mathbf{u}=\mathbf{r}+(\mathbf{K}_1+\mathbf{K}_2)\mathbf{x} \triangleq \mathbf{r}+\mathbf{Kx}$$

the theorem is proved. Q.E.D.

[5] The choices of $\mathbf{K}_1$ and $\mathbf{v}$ are not unique. They can be chosen arbitrarily, and the probability is almost 1 that they will meet the requirements. In Theorem 7-5 of Reference S38, a procedure is given to choose $\mathbf{K}_1$ and $\mathbf{v}$ without any uncertainty. The computation required however is very complicated.

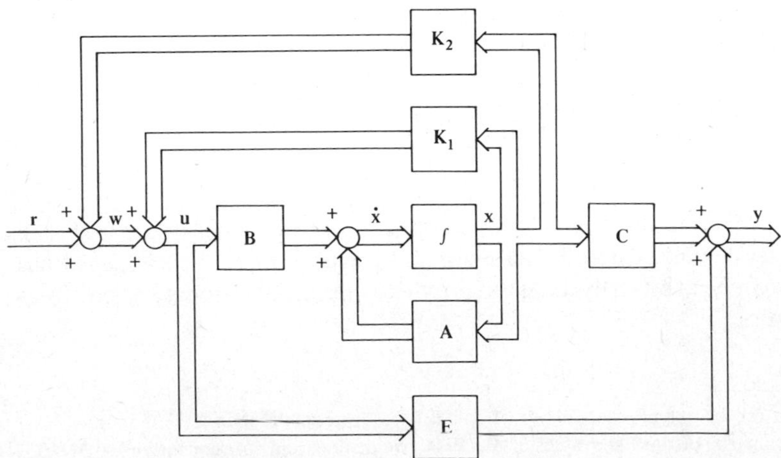

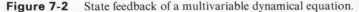

Figure 7-2 State feedback of a multivariable dynamical equation.

If $\{\mathbf{A}, \mathbf{B}\}$ is not controllable, they can be transformed into

$$\begin{bmatrix} \bar{\mathbf{A}}_{11} & \bar{\mathbf{A}}_{12} \\ 0 & \bar{\mathbf{A}}_{22} \end{bmatrix} \quad \begin{bmatrix} \bar{\mathbf{B}}_1 \\ 0 \end{bmatrix}$$

and the eigenvalues of $\bar{\mathbf{A}}_{22}$ are not affected by any state feedback. Hence we conclude that the eigenvalues of $(\mathbf{A}+\mathbf{BK})$ can be arbitrarily assigned if and only if $\{\mathbf{A}, \mathbf{B}\}$ is controllable.

Method II

In this method, we shall transform $\{\mathbf{A}, \mathbf{B}\}$ into the controllable form discussed in (5-93) and (6-112) and then compute the required $\mathbf{K}$. The procedure is similar to the single-input case. In order not to be overwhelmed by notations, we assume $n=9$ and $p=3$. It is also assumed that the controllability indices are $\mu_1=3$, $\mu_2=2$, and $\mu_3=4$. Then the matrix

$$\mathbf{M}=[\mathbf{b}_1 \quad \mathbf{A}\mathbf{b}_1 \quad \mathbf{A}^2\mathbf{b}_1 \quad \mathbf{b}_2 \quad \mathbf{A}\mathbf{b}_2 \quad \mathbf{b}_3 \quad \mathbf{A}\mathbf{b}_3 \quad \mathbf{A}^2\mathbf{b}_3 \quad \mathbf{A}^3\mathbf{b}_3] \qquad \text{(7-44)}$$

(see Section 5-3) is nonsingular. We compute $\mathbf{M}^{-1}$ and name its *rows* as

$$\mathbf{M}^{-1}=\begin{bmatrix} \mathbf{e}_{11} \\ \mathbf{e}_{12} \\ \mathbf{e}_{13} \\ \hdashline \mathbf{e}_{21} \\ \mathbf{e}_{22} \\ \hdashline \mathbf{e}_{31} \\ \mathbf{e}_{32} \\ \mathbf{e}_{33} \\ \mathbf{e}_{34} \end{bmatrix} \begin{matrix} \\ \\ \leftarrow e_{1\mu_1} \\ \\ \leftarrow e_{2\mu_2} \\ \\ \\ \\ \leftarrow e_{3\mu_3} \end{matrix}$$

From $\mathbf{M}^{-1}\mathbf{M}=\mathbf{I}$, we can obtain some relationships such as $\mathbf{e}_{13}\mathbf{A}^k\mathbf{b}_1=0$, $k=0, 1$; $\mathbf{e}_{13}\mathbf{A}^2\mathbf{b}_1=1$; $\mathbf{e}_{13}\mathbf{A}^k\mathbf{b}_2=0, k=0, 1$; $\mathbf{e}_{13}\mathbf{A}^k\mathbf{b}_3=0, k=0, 1, 2, 3$. Note that we do *not* have $\mathbf{e}_{13}\mathbf{A}^2\mathbf{b}_2=0$ because $\mathbf{A}^2\mathbf{b}_2$ is a linear combination of $\{\mathbf{b}_j, \mathbf{A}\mathbf{b}_j, j=1, 2, 3$ and $\mathbf{A}^2\mathbf{b}_1\}$ and $\mathbf{e}_{13}\mathbf{A}^2\mathbf{b}_1=1$. Now we use $\mathbf{e}_{i\mu_i}$, $i=1, 2, 3$, to form the square matrix[6]

$$\mathbf{P}=\begin{bmatrix} \mathbf{e}_{13} \\ \mathbf{e}_{13}\mathbf{A} \\ \mathbf{e}_{13}\mathbf{A}^2 \\ \hdashline \mathbf{e}_{22} \\ \mathbf{e}_{22}\mathbf{A} \\ \hdashline \mathbf{e}_{34} \\ \mathbf{e}_{34}\mathbf{A} \\ \mathbf{e}_{34}\mathbf{A}^2 \\ \mathbf{e}_{34}\mathbf{A}^3 \end{bmatrix} \begin{matrix} \\ \\ \leftarrow e_{1\mu_1}\mathbf{A}^{\mu_1-1} \\ \\ \\ \\ \\ \\ \\ \end{matrix} \qquad \text{(7-45)}$$

[6] This matrix can be obtained directly from the coefficients of linear combinations of $\mathbf{A}^{\mu_i}\mathbf{b}_1$ without computing explicitly $\mathbf{M}^{-1}$. See Reference S178.

Using the aforementioned relationship $\mathbf{e}_{ij}\mathbf{A}^k\mathbf{b}=0$ or 1, it is straightforward, though very tedious, to verify that $\mathbf{P}$ is nonsingular and[7]

$$\bar{\mathbf{A}}=\mathbf{PAP}^{-1}=\left[\begin{array}{ccc|cc|cccc} 0 & 1 & 0 & & & & & & \\ 0 & 0 & 1 & & & & & & \\ x & x & x & x & x & x & x & x & x \\ \hline & & & 0 & 1 & & & & \\ x & x & x & x & x & x & x & x & x \\ \hline & & & & & 0 & 1 & 0 & 0 \\ & & & & & 0 & 0 & 1 & 0 \\ & & & & & 0 & 0 & 0 & 1 \\ x & x & x & x & x & x & x & x & x \end{array}\right] \qquad \bar{\mathbf{B}}=\mathbf{PB}=\left[\begin{array}{ccc} 0 & 0 & 0 \\ 0 & 0 & 0 \\ 1 & x & 0 \\ \hline 0 & 0 & 0 \\ 0 & 1 & 0 \\ \hline 0 & 0 & 0 \\ 0 & 0 & 0 \\ 0 & 0 & 0 \\ 0 & 0 & 1 \end{array}\right] \tag{7-46}$$

where the unfilled positions are all zeros. The pair $\{\bar{\mathbf{A}}, \bar{\mathbf{B}}\}$ is said to be in a multivariable controllable form. The introduction of $\mathbf{u}=\mathbf{r}+\bar{\mathbf{K}}\bar{\mathbf{x}}$, where $\bar{\mathbf{K}}$ is a 3×9 real constant matrix, yields

$$\dot{\bar{\mathbf{x}}}=(\bar{\mathbf{A}}+\bar{\mathbf{B}}\bar{\mathbf{K}})\bar{\mathbf{x}}+\bar{\mathbf{B}}\mathbf{r}$$

Because of the form of $\bar{\mathbf{B}}$, all rows of $\bar{\mathbf{A}}$ except the three rows denoted by strings of x are not affected by the state feedback. Because the three nonzero rows of $\bar{\mathbf{B}}$ are linearly independent, the three rows of $\bar{\mathbf{A}}$ denoted by strings of x can be arbitrarily assigned. For example, we may choose $\bar{\mathbf{K}}$ so that $(\bar{\mathbf{A}}+\bar{\mathbf{B}}\bar{\mathbf{K}})$ is of the form

$$\left[\begin{array}{ccc|cc|cccc} 0 & 1 & 0 & & & & & & \\ 0 & 0 & 1 & & & & & & \\ a_1 & a_2 & a_3 & & & & & & \\ \hline & & & 0 & 1 & & & & \\ & & & b_1 & b_2 & & & & \\ \hline & & & & & 0 & 1 & 0 & 0 \\ & & & & & 0 & 0 & 1 & 0 \\ & & & & & 0 & 0 & 0 & 1 \\ & & & & & c_1 & c_2 & c_3 & c_4 \end{array}\right] \quad \text{or} \quad \left[\begin{array}{ccc|cc|cccc} 0 & 1 & 0 & 0 & 0 & 0 & 0 & 0 & 0 \\ 0 & 0 & 1 & 0 & 0 & 0 & 0 & 0 & 0 \\ 0 & 0 & 0 & 1 & 0 & 0 & 0 & 0 & 0 \\ \hline 0 & 0 & 0 & 0 & 1 & 0 & 0 & 0 & 0 \\ 0 & 0 & 0 & 0 & 0 & 1 & 0 & 0 & 0 \\ \hline 0 & 0 & 0 & 0 & 0 & 0 & 1 & 0 & 0 \\ 0 & 0 & 0 & 0 & 0 & 0 & 0 & 1 & 0 \\ 0 & 0 & 0 & 0 & 0 & 0 & 0 & 0 & 1 \\ d_1 & d_2 & d_3 & d_4 & d_5 & d_6 & d_7 & d_8 & d_9 \end{array}\right] \tag{7-47}$$

The left-hand-side matrix has three blocks of companion form on the diagonal; the right-hand-side matrix has only one companion form. Of course, it is

---

[7] If $\mu_1>\mu_2>\mu_3$, the three nonzero rows of $\bar{\mathbf{B}}$ become $\begin{bmatrix} 1 & x & x \\ 0 & 1 & x \\ 0 & 0 & 1 \end{bmatrix}$. If $\mu_1\le\mu_2\le\mu_3$, they become $\begin{bmatrix} 1 & 0 & 0 \\ 0 & 1 & 0 \\ 0 & 0 & 1 \end{bmatrix}$ In our example, $\mu_1=3$, $\mu_2=2$, and $\mu_3=4$, and the three nonzero rows of $\bar{\mathbf{B}}$ are $\begin{bmatrix} 1 & x & 0 \\ 0 & 1 & 0 \\ 0 & 0 & 1 \end{bmatrix}$. A simple pattern can be deduced from these examples.

possible to obtain some other forms, for example, two blocks of companion form on the diagonal or a block triangular form with two or three blocks of companion form on the diagonal. The characteristic polynomials of the $(\bar{\mathbf{A}} + \bar{\mathbf{B}}\bar{\mathbf{K}})$ in (7-47) are

$$(s^3 - a_3 s^2 - a_2 s - a_1)(s^2 - b_2 s - b_1)(s^4 - c_4 s^3 - c_3 s^2 - c_2 s - c_1)$$

and
$$s^9 - d_9 s^8 - d_8 s^7 - \cdots - d_2 s - d_1$$

Since a_i, b_i, c_i, and d_i can be arbitrarily assigned, we have established once again Theorem 7-7.

Method III

We introduce a method of computing the feedback gain matrix without transforming **A** into a controllable form. It will be achieved by solving a Lyapunov equation. See Appendix F.

Algorithm

Consider a controllable $\{\mathbf{A}, \mathbf{B}\}$, where **A** and **B** are, respectively, $n \times n$ and $n \times p$ constant matrices. Find a **K** so that $\mathbf{A} + \mathbf{B}\mathbf{K}$ has a set of desired eigenvalues.

1. Choose an arbitrary $n \times n$ matrix **F** which has no eigenvalues in common with those of **A**.
2. Choose an arbitrary $p \times n$ matrix $\bar{\mathbf{K}}$ such that $\{\mathbf{F}, \bar{\mathbf{K}}\}$ is observable.
3. Solve the unique **T** in the Lyapunov equation $\mathbf{AT} - \mathbf{TF} = -\mathbf{B}\bar{\mathbf{K}}$.
4. If **T** is nonsingular, then we have $\mathbf{K} = \bar{\mathbf{K}}\mathbf{T}^{-1}$, and $\mathbf{A} + \mathbf{BK}$ has the same eigenvalues as those of **F**. If **T** is singular, choose a different **F** or a different $\bar{\mathbf{K}}$ and repeat the process. ∎

We justify first the algorithm. If **T** is nonsingular, the Lyapunov equation $\mathbf{AT} - \mathbf{TF} = -\mathbf{B}\bar{\mathbf{K}}$ implies

$$(\mathbf{A} + \mathbf{B}\bar{\mathbf{K}}\mathbf{T}^{-1})\mathbf{T} = \mathbf{TF} \qquad \text{or} \qquad \mathbf{A} + \mathbf{BK} = \mathbf{TFT}^{-1}$$

Hence $\mathbf{A} + \mathbf{BK}$ and **F** are similar and have the same set of eigenvalues. Since **F** can be arbitrarily chosen so long as its eigenvalues are distinct from those of **A**, the eigenvalues of $\mathbf{A} + \mathbf{BK}$ can almost be arbitrarily assigned.

As discussed in Appendix F, if **A** and **F** have no common eigenvalues,[8] a solution **T** always exists in $\mathbf{AT} - \mathbf{TF} = -\mathbf{B}\bar{\mathbf{K}}$ for any $\bar{\mathbf{K}}$ and is unique. If **A** and **F** have common eigenvalues, a solution **T** may or may not exist depending on $\mathbf{B}\bar{\mathbf{K}}$. To remove this uncertainty, we require **A** and **F** to have no common eigenvalues. A necessary condition for **T** to be nonsingular is that $\{\mathbf{A}, \mathbf{B}\}$ controllable and $\{\mathbf{F}, \bar{\mathbf{K}}\}$ observable. The condition becomes sufficient as well for the single-variable case ($p = 1$). The dual of this assertion will be proved in Theorem 7-10; the proof will again be used in the proof of Theorem 7-11. Hence the assertion will not be proved here.

[8] Note that we have $\mathscr{A}(\mathbf{T}) = \mathbf{AT} - \mathbf{TF}$ rather than $\mathscr{A}(\mathbf{T}) = \mathbf{AT} + \mathbf{TF}$. Hence the eigenvalues of $\mathscr{A}(\mathbf{T})$ are $\lambda_i(\mathbf{A}) - \mu_j(\mathbf{F})$.

Nonuniqueness of feedback gain matrix K

From the design procedures discussed in the previous subsections, we see that many different feedback gain matrices can yield the same eigenvalues of $(\mathbf{A}+\mathbf{BK})$. Hence it is natural to ask which $\mathbf{K}$ should we choose? Before discussing this problem, we give an example.

Example 3

Consider the dynamical equation

$$\dot{\bar{\mathbf{x}}} = \left[\begin{array}{ccc:cc} 0 & 1 & 0 & 0 & 0 \\ 0 & 0 & 1 & 0 & 0 \\ 2 & 0 & 0 & 1 & 1 \\ \hdashline 0 & 0 & 0 & 0 & 1 \\ 0 & 0 & 0 & -1 & -2 \end{array}\right] \bar{\mathbf{x}} + \left[\begin{array}{cc} 0 & 0 \\ 0 & 0 \\ 1 & 0 \\ \hdashline 0 & 0 \\ 0 & 1 \end{array}\right] \mathbf{u} \qquad \textbf{(7-48)}$$

$$y = \left[\begin{array}{ccc:cc} 1 & -1 & 3 & 2 & 0 \end{array}\right] \bar{\mathbf{x}}$$

The equation is in a multivariable controllable form. The problem is to find a $\bar{\mathbf{K}}$ so that the eigenvalues of the resulting matrix are -1, $-2 \pm j$, and $-1 \pm 2j$.

We compute

$$\bar{\Delta}(s) = s^5 + 7s^4 + 24s^3 + 48s^2 + 55s + 25$$

If we choose $\bar{\mathbf{K}}$ as

$$\bar{\mathbf{K}}_1 = \begin{bmatrix} -2 & 0 & 0 & -1 & -1 \\ -25 & -55 & -48 & -23 & -5 \end{bmatrix}$$

then we have

$$\bar{\mathbf{A}} + \bar{\mathbf{B}}\bar{\mathbf{K}}_1 = \begin{bmatrix} 0 & 1 & 0 & 0 & 0 \\ 0 & 0 & 1 & 0 & 0 \\ 0 & 0 & 0 & 1 & 0 \\ 0 & 0 & 0 & 0 & 1 \\ -25 & -55 & -48 & -24 & -7 \end{bmatrix}$$

If we choose $\bar{\mathbf{K}}$ as

$$\bar{\mathbf{K}}_2 = \begin{bmatrix} -7 & -9 & -5 & -1 & -1 \\ 0 & 0 & 0 & -4 & 0 \end{bmatrix}$$

then we have

$$\bar{\mathbf{A}} + \bar{\mathbf{B}}\bar{\mathbf{K}}_2 = \left[\begin{array}{ccc:cc} 0 & 1 & 0 & 0 & 0 \\ 0 & 0 & 1 & 0 & 0 \\ -5 & -9 & -5 & 0 & 0 \\ \hdashline 0 & 0 & 0 & 0 & 1 \\ 0 & 0 & 0 & -5 & -2 \end{array}\right]$$

This matrix has two blocks of companion form. One has eigenvalues -1, $-2 \pm j$; the other has $-1 \pm 2j$. These two blocks are noninteracting.

Now we are ready to compare these two $\bar{\mathbf{K}}$'s. The elements of $\bar{\mathbf{K}}_1$, which yields one block of companion form, are much larger in magnitude than the elements of $\bar{\mathbf{K}}_2$, which preserves the original blocks of companion form. If most of the assigned eigenvalues have magnitudes larger than one, then it is clear that the larger the order of a companion-form block, the larger the magnitudes of the feedback gains.

Next we compare the responses of the two systems. Although $\bar{\mathbf{K}}_1$ and $\bar{\mathbf{K}}_2$ yield the same eigenvalues, the transfer matrix $\bar{\mathbf{C}}(s\mathbf{I}-\bar{\mathbf{A}}-\bar{\mathbf{B}}\bar{\mathbf{K}}_1)^{-1}\bar{\mathbf{B}}$ and the transfer matrix $\bar{\mathbf{C}}(s\mathbf{I}-\bar{\mathbf{A}}-\bar{\mathbf{B}}\bar{\mathbf{K}}_2)^{-1}\bar{\mathbf{B}}$ are generally different. Thus, their zero-input and zero-state responses are generally different. For convenience, we compare only their zero-input responses. We plot in Figure 7-3(a) and (b) the responses of $\dot{\bar{\mathbf{x}}}=(\bar{\mathbf{A}}+\bar{\mathbf{B}}\bar{\mathbf{K}}_1)\bar{\mathbf{x}}$ and $\dot{\bar{\mathbf{x}}}=(\bar{\mathbf{A}}+\bar{\mathbf{B}}\bar{\mathbf{K}}_2)\bar{\mathbf{x}}$ due to $\bar{\mathbf{x}}'(0)=[2 \quad 1 \quad 0 \quad -1 \quad -2]$. We see that the largest magnitude in transient in $\bar{\mathbf{K}}_1$ is roughly three times larger than the one in $\bar{\mathbf{K}}_2$. Hence $\bar{\mathbf{K}}_2$ has smaller gains and yields smaller magnitudes in transient and is preferred. ∎

For a companion-form matrix of order m, if all its eigenvalues are distinct, it is shown in Reference S17 that the largest magnitude in transient is roughly proportional to

$$(|\bar{\lambda}|_{\max})^{m-1}$$

where $|\bar{\lambda}|_{\max}$ is its largest eigenvalue in magnitude. The magnitudes of feedback gains are also proportional to m. Hence, in order to have small feedback gains

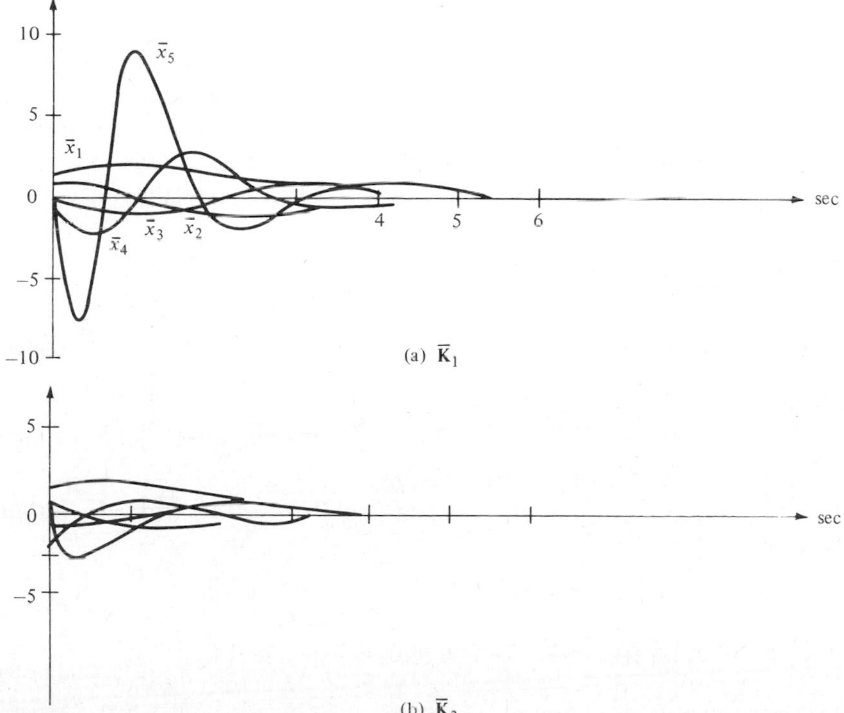

Figure 7-3 Transient responses.

and small transient, the order of the largest companion-form block should be kept as small as possible.

The orders of companion-form blocks are determined by the lengths of chains of vectors $\mathbf{b}_i, \mathbf{A}\mathbf{b}_i, \ldots, \mathbf{A}^k\mathbf{b}_i$ in (7-44). If they are chosen by using Scheme 2 discussed in (7-21), the orders of companion-form blocks are equal to the controllability indices of $\{\mathbf{A}, \mathbf{B}\}$. In this case, the largest order is equal to the controllability index μ. If we use Scheme 1 discussed in (7-20) or any other scheme, the length of the longest chain, and consequently, the order of the largest companion-form block will be larger than, at best equal to, μ. Hence in order to have small feedback gains and small transient, we shall use Scheme 2 to search the linearly independent columns in $[\mathbf{B} \quad \mathbf{AB} \quad \cdots \quad \mathbf{A}^{n-1}\mathbf{B}]$ and the orders of the resulting companion-form blocks are $\mu_1, \mu_2, \ldots, \mu_p$, the controllability indices.

The introduction of state feedback can increase the order of a companion-form block, as shown in Example 3, but can never decrease it. Hence in finding a $\bar{\mathbf{K}}$, we should preserve the orders of the original companion-form blocks. Even so, there are still a lot of flexibility in choosing $\bar{\mathbf{K}}$. We may choose $\bar{\mathbf{K}}$ so that the resulting matrix is in block diagonal form or in block triangular forms. Furthermore different grouping of desired eigenvalues will yield different $\bar{\mathbf{K}}$. If we require the resulting matrix to be block diagonal and group the desired eigenvalues to minimize

$$\text{mas}\ \{(|\bar{\lambda}_1|_{\max})^{\mu_1-1}, (|\bar{\lambda}_2|_{\max})^{\mu_2-1}, \ldots, (|\bar{\lambda}_p|_{\max})^{\mu_p-1}\}$$

where $|\bar{\lambda}_i|_{\max}$ is the largest eigenvalue in magnitude in the ith companion-form block of order μ_i, then the flexibility of choosing $\bar{\mathbf{K}}$ is considerably reduced.

A problem may sometimes arise in preserving the orders of companion-form blocks. For example, let $\mu_i = 3, i = 1, 2$, and let the desired eigenvalues be $-3 \pm 3j$, $-4 \pm 2j$, and $-3 \pm j$. In this case, it is not possible to choose a *real* $\bar{\mathbf{K}}$ so that $\bar{\mathbf{A}} + \bar{\mathbf{B}}\bar{\mathbf{K}}$ preserve μ_i and has a block diagonal or a block triangular form. If we combine the two companion-form blocks into one, then the resulting $\bar{\mathbf{A}} + \bar{\mathbf{B}}\bar{\mathbf{K}}$ is

$$\begin{bmatrix} 0 & 1 & 0 & 0 & 0 & 0 \\ 0 & 0 & 1 & 0 & 0 & 0 \\ 0 & 0 & 0 & 1 & 0 & 0 \\ 0 & 0 & 0 & 0 & 1 & 0 \\ 0 & 0 & 0 & 0 & 0 & 1 \\ -3600 & -4800 & -2804 & -920 & -180 & -20 \end{bmatrix} \tag{7-49}$$

We see that the feedback gains are very large.

Now we use the procedure discussed in Problem 7-20 to find a $\bar{\mathbf{K}}$ so that the resulting $\bar{\mathbf{A}} + \bar{\mathbf{B}}\bar{\mathbf{K}}$ is of the form

$$\left[\begin{array}{ccc|ccc} 0 & 1 & 0 & 0 & 0 & 0 \\ 0 & 0 & 1 & 0 & 0 & 0 \\ -54 & -36 & -9 & -20 & -8 & -1 \\ \hline 0 & 0 & 0 & 0 & 1 & 0 \\ 0 & 0 & 0 & 0 & 0 & 1 \\ 18 & 6 & 1 & -60 & -44 & -11 \end{array}\right] \tag{7-50}$$

and of eigenvalues $-3 \pm 3j$, $-4 \pm 2j$, and $-3 \pm j$. We see that the feedback gains required in (7-50) are much smaller than those in (7-49). It can also be shown by a computer simulation that the largest magnitude in the transient of (7-50) is much much smaller than the one in (7-49). See Reference S17.

We recapitulate the preceding discussion in the following. In the transformation of $\{\mathbf{A}, \mathbf{B}\}$ into a multivariable controllable form, we should use Scheme 2 discussed in (7-21) to search the linearly independent columns in $[\mathbf{B} \quad \mathbf{AB} \quad \cdots \quad \mathbf{A}^{n-1}\mathbf{B}]$. The orders of companion-form blocks in the resulting $\{\bar{\mathbf{A}}, \bar{\mathbf{B}}\}$ are equal to the controllability indices, μ_i, 1, 2, ..., p, of $\{\mathbf{A}, \mathbf{B}\}$ and the order of the largest companion-form block in $\{\bar{\mathbf{A}}, \bar{\mathbf{B}}\}$ is the smallest possible among all transformations. We then choose a $\bar{\mathbf{K}}$ so that $(\bar{\mathbf{A}} + \bar{\mathbf{B}}\bar{\mathbf{K}})$ is of block diagonal form with companion-form blocks of order μ_i on the diagonal. If this is not possible due to complex eigenvalues, we may use the procedure in Problem 7-20 to choose a $\bar{\mathbf{K}}$. This process of choosing $\bar{\mathbf{K}}$ and consequently $\mathbf{K} = \bar{\mathbf{K}}\mathbf{P}$ may yield a system with small feedback gains and small transient.

The design procedure discussed in Method I transforms the problem into a single-input system; hence the order of the companion-form block is n. Thus the method should not be used to avoid large feedback gains and large transient. However, the method may require a smaller-dimensional state estimator as will be discussed in Sections 7-5 and 9-5.

Up to this point, we have not yet discussed how to choose a set of desired eigenvalues. This depends highly on the performance criteria of the design such as the rise time, settling time, overshoot, largest magnitude of the actuating signals, and so forth. Even if these criteria are precisely specified, there is no simple answer to the posed problem. One way to proceed is by computer simulations. Of course, the set of eigenvalues obtained by this process will not be unique. The only known systematic way of finding a unique $\mathbf{K}$ and, consequently, a unique set of eigenvalues is by minimizing the quadratic performance index

$$J = \int_0^\infty [\mathbf{x}^*(t)\mathbf{Q}\mathbf{x}(t) + \mathbf{u}^*(t)\mathbf{R}\mathbf{u}(t)]\, dt$$

The $\mathbf{K}$ can be uniquely determined by solving an algebraic Riccati equation. This is outside the scope of this text and the reader is referred to References 3 and S4, or any other book on optimal control.

Assignment of eigenvalues and eigenvectors

In this subsection, we discuss the assignment of eigenvalues as well as the eigenvectors of the resulting $\mathbf{A} + \mathbf{BK}$. For convenience, we assume that the assigned eigenvalues, $\bar{\lambda}_i$, are all distinct. Let $\mathbf{e}_i$ be an eigenvector of $(\mathbf{A} + \mathbf{BK})$ associated with $\bar{\lambda}_i$, that is

$$(\mathbf{A} + \mathbf{BK})\mathbf{e}_i = \bar{\lambda}_i \mathbf{e}_i \qquad i = 1, 2, \ldots, n$$

or

$$\mathbf{BK}\mathbf{e}_i = (\bar{\lambda}_i \mathbf{I} - \mathbf{A})\mathbf{e}_i \triangleq \mathbf{f}_i$$

which implies that

$$\mathbf{BK}[\mathbf{e}_1 \quad \mathbf{e}_2 \quad \cdots \quad \mathbf{e}_n] = [\mathbf{f}_1 \quad \mathbf{f}_2 \quad \cdots \quad \mathbf{f}_n]$$

Now if $\{\mathbf{e}_i, i=1,2,\ldots,n\}$ are linearly independent, then we have

$$\mathbf{BK}=[\mathbf{f}_1 \quad \mathbf{f}_2 \quad \cdots \quad \mathbf{f}_n][\mathbf{e}_1 \quad \mathbf{e}_2 \quad \cdots \quad \mathbf{e}_n]^{-1} \tag{7-51}$$

Unless all columns of the matrix on the right-hand side of (7-51) are inside the range space of $\mathbf{B}$, no $\mathbf{K}$ will satisfy the equation. Hence the assignment of eigenvectors cannot be entirely independent from the assignment of eigenvalues.

For a general matrix $\mathbf{A}$, there is no way to predict what form an eigenvector will assume. However if a matrix is of the companion forms shown in (7-47), then its eigenvectors are of the form $[1 \ \ \bar{\lambda}_i \ \ \bar{\lambda}_i^2 \ \ 0 \ \ 0 \ \ 0 \ \ 0 \ \ 0 \ \ 0]'$, $[0 \ \ 0 \ \ 0 \ \ 1 \ \ \bar{\lambda}_i \ \ 0 \ \ 0 \ \ 0 \ \ 0]'$, or $[0 \ \ 0 \ \ 0 \ \ 0 \ \ 0 \ \ 1 \ \ \bar{\lambda}_i \ \ \bar{\lambda}_i^2 \ \ \bar{\lambda}_i^3]'$ for the left-hand-side matrix in (7-47) and $[1 \ \ \bar{\lambda}_i \ \ \bar{\lambda}_i^2 \ \ \bar{\lambda}_i^3 \ \ \bar{\lambda}_i^4 \ \ \bar{\lambda}_i^5 \ \ \bar{\lambda}_i^6 \ \ \bar{\lambda}_i^7 \ \ \bar{\lambda}_i^8]'$ for the right-hand-side matrix in (7-47) (see Problem 2-26). Hence for the $\{\bar{\mathbf{A}}, \bar{\mathbf{B}}\}$ in (7-46), after the eigenvalues are chosen, if the eigenvectors are chosen as discussed, then a solution $\bar{\mathbf{K}}$ exists in (7-51). In fact, the solution is unique and can be solved by using the three nonzero rows of $\bar{\mathbf{B}}$ in (7-46).

This process of choosing eigenvectors is far from an arbitrary assignment. Once we choose the structure of $\bar{\mathbf{A}}+\bar{\mathbf{B}}\bar{\mathbf{K}}$ (one, two, or three companion-form blocks), and once we assign the eigenvalues for each block, the eigenvectors are practically uniquely determined. Hence if the eigenvectors are chosen as discussed, there is no essential difference between the assignment of eigenvectors and the assignment of structure of $(\bar{\mathbf{A}}+\bar{\mathbf{B}}\bar{\mathbf{K}})$ discussed in the previous subsection. For further discussion of the assignment of eigenvectors, the reader is referred to References S91, S125, S160.

Effect on the Numerator matrix of $\hat{\mathbf{G}}(s)$

In the single-variable case, we showed that the numerator of $\hat{g}(s)$ is not affected by any state feedback, whereas its denominator can be arbitrarily assigned. Now we establish a similar statement for the multivariable case. Before proceeding, the reader should review the material in Section 6-6, especially Equations (6-100) through (6-117).

Once $\{\mathbf{A}, \mathbf{B}, \mathbf{C}\}$ is transformed into the form in (7-46), by reversing the procedure from (6-100) to (6-117), we can obtain

$$\hat{\mathbf{G}}(s)=\mathbf{C}(s\mathbf{I}-\mathbf{A})^{-1}\mathbf{B}+\mathbf{E}=\mathbf{N}(s)\mathbf{D}^{-1}(s)+\mathbf{E}=(\mathbf{N}(s)+\mathbf{E}\mathbf{D}(s))\mathbf{D}^{-1}(s)$$

with

$$\mathbf{D}(s)=\mathbf{D}_{hc}\mathbf{H}(s)+\mathbf{D}_{lc}\mathbf{L}(s)$$
$$\mathbf{N}(s)=\mathbf{C}\mathbf{L}(s)$$

where $\mathbf{H}(s)$ and $\mathbf{L}(s)$ are defined as in (6-101). We call $\mathbf{N}(s)+\mathbf{E}\mathbf{D}(s)$ the numerator matrix and $\mathbf{D}(s)$ denominator matrix of $\hat{\mathbf{G}}(s)$. If we plot $\hat{\mathbf{G}}(s)$ as in Figure 7-4 and call the output of $\mathbf{D}^{-1}(s)$ the pseudostate $\hat{\boldsymbol{\xi}}(s)$, then the state $\hat{\mathbf{x}}(s)$ is, as established in (6-108), equal to $\hat{\mathbf{x}}(s)=\mathbf{L}(s)\hat{\boldsymbol{\xi}}(s)$. Hence the state feedback $\hat{\mathbf{u}}(s)=\hat{\mathbf{r}}(s)+\mathbf{K}\hat{\mathbf{x}}(s)$ can be achieved by applying $\mathbf{K}$ as shown in Figure 7-4. From Figure 7-4 we have

$$\hat{\mathbf{u}}(s)=\hat{\mathbf{r}}(s)+\mathbf{K}\mathbf{L}(s)\hat{\boldsymbol{\xi}}(s)=\hat{\mathbf{r}}(s)+\mathbf{K}\mathbf{L}(s)\mathbf{D}^{-1}(s)\hat{\mathbf{u}}(s)$$

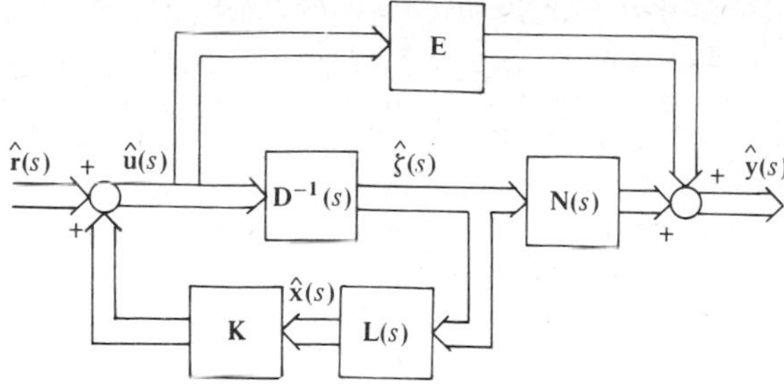

Figure 7-4 The effect of state feedback.

which implies

$$\hat{\mathbf{u}}(s) = (\mathbf{I} - \mathbf{K}\mathbf{L}(s)\mathbf{D}^{-1}(s))^{-1}\hat{\mathbf{r}}(s) = \mathbf{D}(s)(\mathbf{D}(s) - \mathbf{K}\mathbf{L}(s))^{-1}\hat{\mathbf{r}}(s) \tag{7-52}$$

The substitution of (7-52) into $\hat{\mathbf{y}}(s) = \hat{\mathbf{G}}(s)\hat{\mathbf{u}}(s) = (\mathbf{N}(s) + \mathbf{E}\mathbf{D}(s))\mathbf{D}^{-1}(s)\hat{\mathbf{u}}(s)$ yields

$$\hat{\mathbf{y}}(s) = (\mathbf{N}(s) + \mathbf{E}\mathbf{D}(s))\mathbf{D}^{-1}(s)\mathbf{D}(s)(\mathbf{D}(s) - \mathbf{K}\mathbf{L}(s))^{-1}\hat{\mathbf{r}}(s)$$

Hence the transfer matrix from $\hat{\mathbf{r}}(s)$ to $\hat{\mathbf{y}}(s)$ is

$$\begin{aligned}\hat{\mathbf{G}}_f(s) &= (\mathbf{N}(s) + \mathbf{E}\mathbf{D}(s))(\mathbf{D}(s) - \mathbf{K}\mathbf{L}(s))^{-1}\\ &= (\mathbf{N}(s) + \mathbf{E}\mathbf{D}(s))[\mathbf{D}_{hc}\mathbf{H}(s) + (\mathbf{D}_{lc} - \mathbf{K})\mathbf{L}(s)]^{-1}\end{aligned} \tag{7-53}$$

We see that the numerator matrix $\mathbf{N}(s) + \mathbf{E}\mathbf{D}(s)$ of $\hat{\mathbf{G}}_f(s)$ is not affected by any state feedback. The column degrees of $\mathbf{H}(s)$ and the column-degree-coefficient matrix of the denominator matrix $\mathbf{D}_f(s) \triangleq \mathbf{D}(s) - \mathbf{K}\mathbf{L}(s)$ are not affected by any state feedback. The lower degree part of $\mathbf{D}_f(s)$ however can be arbitrarily assigned by a proper choice of $\mathbf{K}$. This result is similar to the single-variable case.

If $\hat{\mathbf{G}}(s)$ is factored as $\mathbf{N}(s)\mathbf{D}^{-1}(s)$, the numerator matrix $\mathbf{N}(s)$ will not be affected by any state feedback. This, however, does not imply that the numerator of every element of $\hat{\mathbf{G}}(s)$ will be unaffected. In fact, zeros of element of $\hat{\mathbf{G}}(s)$ will be affected by state feedback. This is one of the reasons why different feedback gain matrices $\mathbf{K}$, which yield the same set of poles, may yield vastly different transient responses.[9]

Computational problems

In this subsection, we discuss some computational problems in Methods II and III for state feedback. The major step in Method II is the transformation of $\{\mathbf{A}, \mathbf{B}\}$ into a multivariable controllable form. The equivalence transformations to achieve this are given in (7-14) and (7-45). They require the computation of $\mathbf{A}^k\mathbf{B}$ for $k = 1, 2, \ldots, n-1$ and may change the problem into a less well-conditioned problem, as discussed in Section 5-8. Hence this procedure is generally not satisfactory from a computer computational point of view.

[9] This observation was provided by Professor D. Z. Zheng of Tsinghua University, Beijing, China.

In Section 5-8, we discussed a different method of transforming an equation into a controllable form without explicitly computing $\mathbf{A}^k\mathbf{B}$. The method is first to transform $\mathbf{A}$ into a Hessenberg form as in (5-89), and then transform it into a controllable form as in (5-93). (Note the different ordering of the state variables.) The first step is carried out by a sequence of orthogonal transformations or gaussian eliminations with partial pivoting and is numerically stable. The second step must be carried out by gaussian eliminations without any pivoting and is numerically unstable. Hence this transformation method is again not satisfactory.

In Section 6-6, we introduced an irreducible realization of $\hat{\mathbf{G}}(s) = \mathbf{N}(s)\mathbf{D}^{-1}(s)$ in a controllable form. Hence, if $\mathbf{N}(s)\mathbf{D}^{-1}(s)$ can be computed from $\{\mathbf{A}, \mathbf{B}, \mathbf{C}\}$, then a controllable form dynamical equation can be obtained. In Reference S171, an algorithm is presented to compute a coprime $\mathbf{N}(s)\mathbf{D}^{-1}(s)$ from an Hessenberg-form dynamical equation. However in the design, we still need the equivalence transformation. Thus it is not clear whether this design procedure will be satisfactory. In References S156 and S157 a dynamical equation is first transformed into the Hessenberg form, the pole assignment is then carried out by using the QR algorithm. The methods used in References S156 and S157 are all numerically stable. Once a dynamical equation is transformed into a Hessenberg form, the Hyman method in Reference S212 can also be employed to carry out the pole-assignment. For the 20×20 matrix on page 219 with $\varepsilon = 0$, the Hyman method yields a comparable result as the QR method. See Reference S202.

The procedure in Method III does not require the transformation of an equation into a controllable form. It requires merely the solution of a Lyapunov equation. The solution of Lyapunov equations has been extensively studied in the literature, see, for example, References S9, S18, and S107. The algorithm in Reference S107 is claimed to be efficient and numerically stable and may be employed in our design.

A limited comparison between Method II and Method III in terms of operational efficiency and numerical stability was carried out in Reference S202. Because of the complexities of the problem, no clear-cut conclusion was reached in Reference S202.

7-4 State Estimators

In the previous section we introduced state feedback under the assumption that all the state variables are available as outputs. This assumption often does not hold in practice, however, either because the state variables are not accessible for direct measurement or because the number of measuring devices is limited. Thus, in order to apply state feedback to stabilize, to optimize, or to decouple (see Section 7-6) a system, a reasonable substitute for the state vector often has to be found. In this section we shall show how the available inputs and outputs of a dynamical equation can be used to drive a device so that the outputs of the device will approximate the state vector. The device that constructs an approximation of the state vector is called a *state estimator* or a *state observer.* In this section we shall introduce two types of state estimators: full-dimensional and reduced-dimensional. The design procedure is basically dual to the design of state feedback; hence its discussion will be brief. The single-variable case will not be discussed separately and will be included as a special case of the multivariable case.

In this section we use the circumflex over a variable to denote an estimate of the variable. For example, $\hat{\mathbf{x}}$ is an estimate of $\mathbf{x}$; $\hat{\bar{\mathbf{x}}}$ is an estimate of $\bar{\mathbf{x}}$.

Full-dimensional state estimator. Consider the n-dimensional linear time-invariant dynamical equation

$$FE: \qquad \dot{\mathbf{x}} = \mathbf{A}\mathbf{x} + \mathbf{B}\mathbf{u} \qquad \textbf{(7-54a)}$$

$$\mathbf{y} = \mathbf{C}\mathbf{x} \qquad \textbf{(7-54b)}$$

where $\mathbf{A}$, $\mathbf{B}$, and $\mathbf{C}$ are, respectively, $n \times n$, $n \times p$, and $q \times n$ real constant matrices. For simplicity, the direct transmission part has been assumed to be zero. We assume now that the state variables are not accessible. Note that although the state variables are not accessible, the matrices $\mathbf{A}$, $\mathbf{B}$, and $\mathbf{C}$ are assumed to be completely known. Hence the problem is that of estimating or generating $\mathbf{x}(t)$ from the available input $\mathbf{u}$ and the output $\mathbf{y}$ with the knowledge of the matrices $\mathbf{A}$, $\mathbf{B}$, and $\mathbf{C}$. If we know the matrices $\mathbf{A}$ and $\mathbf{B}$, we can duplicate the original system as shown in Figure 7-5. We called the system an *open-loop estimator.* Now if the original equation FE and the estimator have the same initial state and are driven by the same input, the output $\hat{\mathbf{x}}(t)$ of the estimator will be equal to the actual state $\mathbf{x}(t)$ for all t. Therefore, the remaining question is how to find the initial state of FE and set the initial state of the estimator to that state. This problem was solved in Section 5-4. It is shown there that if the dynamical equation FE is observable, the initial state of FE can be computed from its input and output. Consequently, if the equation FE is observable, an open-loop estimator can be used to generate the state vector.

There are, however, two disadvantages in using an open-loop estimator. First, the initial state must be computed and set each time we use the estimator.

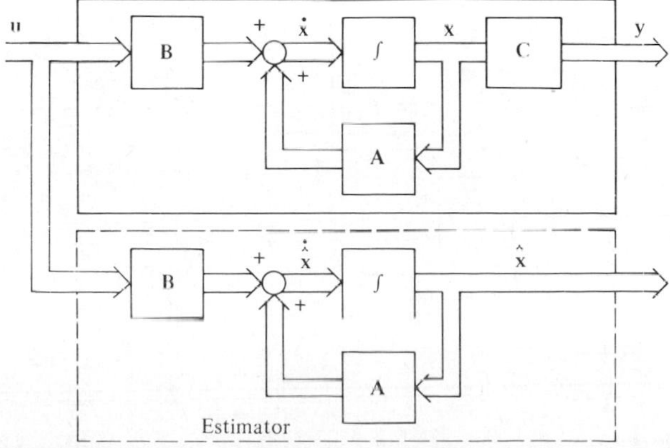

Figure 7-5 An open-loop state estimator.

This is very inconvenient. Second, and more seriously, if the matrix **A** has eigenvalues with positive real parts, then even for a very small difference between $\mathbf{x}(t_0)$ and $\hat{\mathbf{x}}(t_0)$ at some t_0, which may be caused by disturbance or incorrect estimation of the initial state, the difference between the actual $\mathbf{x}(t)$ and the estimated $\hat{\mathbf{x}}(t)$ will increase with time. Therefore an open-loop estimator is, in general, not satisfactory.

Another possible way to generate the n-dimensional state vector is to differentiate the output and the input $n-1$ times. If the dynamical equation is observable, then from $\mathbf{u}(t)$, $\mathbf{y}(t)$, and their derivatives, the state vector can be computed (see Problem 5-17). However, pure differentiators are not easy to build. Furthermore, the estimated state might be severely distorted by noises if pure differentiators are used.

We see from Figure 7-5 that although the input and the output of FE are available, we use only the input in the open-loop estimator. It is conceivable that if both the output and input are utilized, the performance of an estimator can be improved.

Consider the state estimator shown in Figure 7-6. The estimator is driven by the input as well as the output of the original system. The output of FE, $\mathbf{y}=\mathbf{C}\mathbf{x}$, is compared with $\hat{\mathbf{y}} \triangleq \mathbf{C}\hat{\mathbf{x}}$, and their difference is used to serve as a correcting term. The difference of $\mathbf{y}$ and $\mathbf{C}\hat{\mathbf{x}}$, $\mathbf{y}-\mathbf{C}\hat{\mathbf{x}}$, is multiplied by an $n \times q$ real constant matrix **L** and fed into the input of the integrators of the estimator. This estimator will be called an *asymptotic state estimator*, for a reason to be seen later.

The dynamical equation of the asymptotic state estimator shown in Figure 7-6 is given by

$$\dot{\hat{\mathbf{x}}} = \mathbf{A}\hat{\mathbf{x}} + \mathbf{B}\mathbf{u} + \mathbf{L}(\mathbf{y} - \mathbf{C}\hat{\mathbf{x}}) \tag{7-55}$$

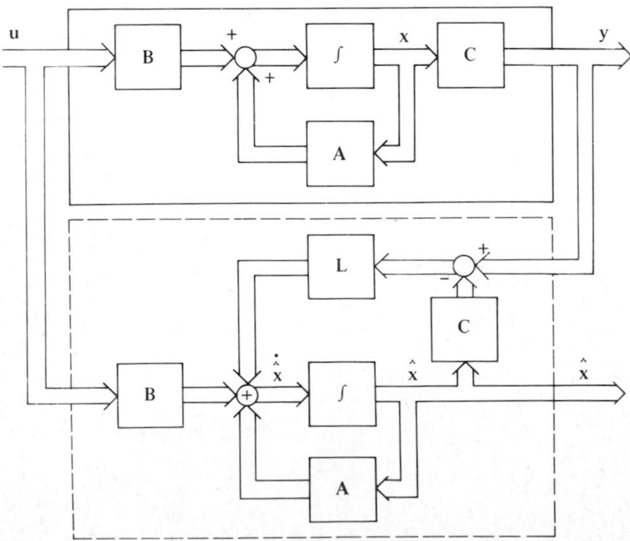

Figure 7-6 An asymptotic state estimator.

which can be written as

$$\dot{\hat{\mathbf{x}}} = (\mathbf{A} - \mathbf{LC})\hat{\mathbf{x}} + \mathbf{Ly} + \mathbf{Bu} \tag{7-56}$$

The asymptotic estimator in Figure 7-6 can be redrawn as in Figure 7-7 either by a block diagram manipulation or from (7-56). Define

$$\tilde{\mathbf{x}} \triangleq \mathbf{x} - \hat{\mathbf{x}}$$

Clearly $\tilde{\mathbf{x}}$ is the error between the actual state and the estimated state. Subtracting (7-56) from (7-54), we obtain

$$\dot{\tilde{\mathbf{x}}} = (\mathbf{A} - \mathbf{LC})\tilde{\mathbf{x}} \tag{7-57}$$

If the eigenvalues of $(\mathbf{A} - \mathbf{LC})$ can be chosen arbitrarily, then the behavior of the error $\tilde{\mathbf{x}}$ can be controlled. For example, if all the eigenvalues of $(\mathbf{A} - \mathbf{LC})$ have negative real parts smaller than $-\sigma$, then all the elements of $\tilde{\mathbf{x}}$ will approach zero at rates faster than $e^{-\sigma t}$. Consequently, even if there is a large error between $\hat{\mathbf{x}}(t_0)$ and $\mathbf{x}(t_0)$ at initial time t_0, the vector $\hat{\mathbf{x}}$ will approach $\mathbf{x}$ rapidly. Thus, if the eigenvalues of $\mathbf{A} - \mathbf{LC}$ can be chosen properly, an asymptotic state estimator is much more desirable than an open-loop estimator.

In the following we discuss two different methods of designing a state estimator.

Method I

In this method, we apply the design procedures for state feedback to design state estimators.

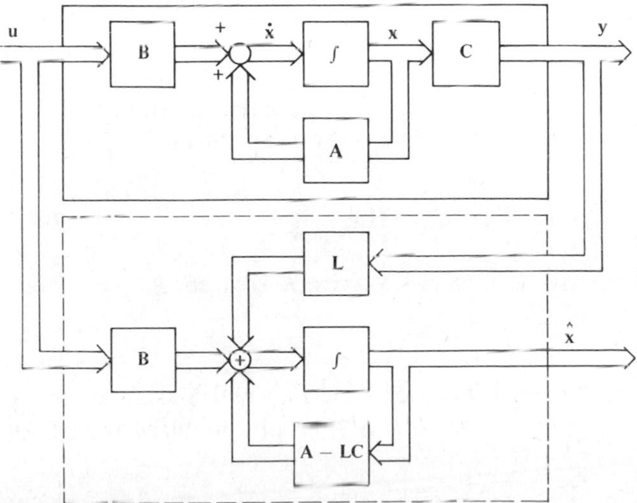

Figure 7-7 An asymptotic state estimator.

Theorem 7-8

If the n-dimensional dynamical equation FE in (7-54) is observable, its state can be estimated by using the n-dimensional state estimator

$$\dot{\hat{\mathbf{x}}} = (\mathbf{A} - \mathbf{LC})\hat{\mathbf{x}} + \mathbf{Ly} + \mathbf{Bu}$$

with the error $\tilde{\mathbf{x}} = \mathbf{x} - \hat{\mathbf{x}}$ governed by

$$\dot{\tilde{\mathbf{x}}} = (\mathbf{A} - \mathbf{LC})\tilde{\mathbf{x}}$$

and all the eigenvalues of $\mathbf{A} - \mathbf{LC}$ can be arbitrarily assigned, provided complex conjugate eigenvalues appear in pairs. ∎

To show the theorem, it is sufficient to show that if $\{\mathbf{A}, \mathbf{C}\}$ is observable, then all the eigenvalues of $\mathbf{A} - \mathbf{LC}$ can be arbitrarily assigned. Indeed, if $\{\mathbf{A}, \mathbf{C}\}$ is observable, then $\{\mathbf{A}', \mathbf{C}'\}$, where the prime denotes the transpose, is controllable (Theorem 5-10). Hence for any set of n eigenvalues, we can find a real $\mathbf{K}$ so that $\mathbf{A}' + \mathbf{C}'\mathbf{K}$ has the set as its eigenvalues (Theorem 7-7). Since the eigenvalues of $\mathbf{A} + \mathbf{K}'\mathbf{C}$ and those of $\mathbf{A}' + \mathbf{C}'\mathbf{K}$ are the same, the theorem is established by choosing $\mathbf{L} = -\mathbf{K}'$.

It is clear that all the design procedures for state feedback can be used to design state estimators. Hence their discussions will not be repeated. We mention only that, in order to have small gains in $\mathbf{L}$ and small transient in $\dot{\tilde{\mathbf{x}}} = (\mathbf{A} - \mathbf{LC})\tilde{\mathbf{x}}$, the largest order of companion-form blocks in the transformed $\mathbf{A}$ matrix should be as small as possible. Dual to the state-feedback part, the smallest such order is the observability index of $\{\mathbf{A}, \mathbf{C}\}$.

Method II

Consider the n-dimensional dynamical equation

$$FE: \quad \dot{\mathbf{x}} = \mathbf{Ax} + \mathbf{Bu} \qquad \textbf{(7-58a)}$$
$$\mathbf{y} = \mathbf{Cx} \qquad \textbf{(7-58b)}$$

where $\mathbf{A}$, $\mathbf{B}$, and $\mathbf{C}$ are, respectively, $n \times n$, $n \times p$, and $q \times n$ real constant matrices. It is assumed that FE is irreducible. Define the n-dimensional dynamical equation

$$\dot{\mathbf{z}} = \mathbf{Fz} + \mathbf{Gy} + \mathbf{Hu} \qquad \textbf{(7-59)}$$

where $\mathbf{F}$, $\mathbf{G}$, and $\mathbf{H}$ are, respectively, $n \times n$, $n \times q$, and $n \times p$ real constant matrices.

Theorem 7-9

The state $\mathbf{z}(t)$ in (7-59) is an estimate of $\mathbf{Tx}(t)$ for some $n \times n$ real constant matrix $\mathbf{T}$ in the sense that $\mathbf{z}(t) - \mathbf{Tx}(t) \to \mathbf{0}$ as $t \to \infty$, for any $\mathbf{x}(0)$, $\mathbf{z}(0)$, and $\mathbf{u}(t)$ if and only if

1. $\mathbf{TA} - \mathbf{FT} = \mathbf{GC}$
2. $\mathbf{H} = \mathbf{TB}$ **(7-60)**
3. All the eigenvalues of $\mathbf{F}$ have negative real parts.

Proof

Define $$\mathbf{e} \triangleq \mathbf{z} - \mathbf{Tx}$$

Then we have

$$\begin{aligned}\dot{\mathbf{e}} &= \dot{\mathbf{z}} - \mathbf{T}\dot{\mathbf{x}} = \mathbf{Fz} + \mathbf{Gy} + \mathbf{Hu} - \mathbf{TAx} - \mathbf{TBu} \\ &= \mathbf{Fe} + (\mathbf{FT} - \mathbf{TA} + \mathbf{GC})\mathbf{x} + (\mathbf{H} - \mathbf{TB})\mathbf{u}\end{aligned}$$

If the three conditions in (7-60) are met, then $\mathbf{e}(t) = e^{\mathbf{F}t}\mathbf{e}(0) \to \mathbf{0}$ for any $\mathbf{x}(0)$, $\mathbf{z}(0)$, and $\mathbf{u}(t)$. Hence $\mathbf{z}(t)$ is an estimate of $\mathbf{Tx}(t)$.

Now we show the necessity of the conditions. If 3 is not met, then for $\mathbf{x}(0) = \mathbf{0}$ and $\mathbf{u}(t) \equiv \mathbf{0}$, we have $\mathbf{e}(t) = e^{\mathbf{F}t}\mathbf{z}(0) \not\to 0$ as $t \to \infty$. If $\mathbf{H} \neq \mathbf{TB}$, we can find a $\mathbf{u}(t)$ to make $\mathbf{e}(t) \not\to 0$ as $t \to \infty$. If $\mathbf{TA} - \mathbf{FT} \neq \mathbf{GC}$ and if $\{\mathbf{A}, \mathbf{B}\}$ is controllable, we can find a $\mathbf{u}(t)$ to generate a $\mathbf{x}(t)$ which makes $\mathbf{e}(t) \not\to 0$ as $t \to \infty$. This establishes the necessity part of the theorem. Q.E.D.

With this theorem, we can now propose a design algorithm.

Algorithm

1. Choose an $\mathbf{F}$ so that all of its eigenvalues have negative real parts and are disjoint from those of $\mathbf{A}$.
2. Choose a $\mathbf{G}$ so that $\{\mathbf{F}, \mathbf{G}\}$ is controllable.
3. Solve the unique $\mathbf{T}$ in $\mathbf{TA} - \mathbf{FT} = \mathbf{GC}$.
4. If $\mathbf{T}$ is nonsingular, compute $\mathbf{H} = \mathbf{TB}$. The equation in (7-59) with these $\mathbf{F}$, $\mathbf{G}$, and $\mathbf{H}$ is an estimate of $\mathbf{Tx}(t)$ or $\hat{\mathbf{x}}(t) = \mathbf{T}^{-1}\mathbf{z}(t)$. If $\mathbf{T}$ is singular, choose different $\mathbf{F}$ and/or $\mathbf{G}$ and repeat the process. ■

We give some remarks regarding this algorithm. If $\mathbf{A}$ and $\mathbf{F}$ have no common eigenvalue, then a solution $\mathbf{T}$ always exists in $\mathbf{TA} - \mathbf{FT} = \mathbf{GC}$ and is unique (Appendix F). If $\mathbf{A}$ and $\mathbf{F}$ have common eigenvalues, then a solution $\mathbf{T}$ may or may not exist in $\mathbf{TA} - \mathbf{FT} = \mathbf{GC}$. To remove this uncertainty, we require $\mathbf{F}$ and $\mathbf{A}$ to have no common eigenvalues.

The controllability of $\{\mathbf{F}, \mathbf{G}\}$ is needed for the existence of a nonsingular $\mathbf{T}$, as will be established in the following theorem.

Theorem 7-10

If $\mathbf{A}$ and $\mathbf{F}$ have no common eigenvalues, necessary conditions for the existence of a nonsingular $\mathbf{T}$ in $\mathbf{TA} - \mathbf{FT} = \mathbf{GC}$ are $\{\mathbf{A}, \mathbf{C}\}$ observable and $\{\mathbf{F}, \mathbf{G}\}$ controllable. For the single-output case ($q = 1$), the conditions are sufficient as well

Proof

Let the characteristic polynomial of $\mathbf{A}$ be

$$\Delta(s) = \det(s\mathbf{I} - \mathbf{A}) = s^n + \alpha_1 s^{n-1} + \alpha_2 s^{n-2} + \cdots + \alpha_n$$

Clearly we have $\Delta(\mathbf{A}) = \mathbf{0}$ (Cayley-Hamilton theorem). If λ_i is an eigenvalue of $\mathbf{F}$, then $\Delta(\lambda_i)$ is an eigenvalue of $\Delta(\mathbf{F})$ (Problem 2-32). Since $\mathbf{A}$ and $\mathbf{F}$ have no common eigenvalue, we have $\Delta(\lambda_i) \neq 0$ for all eigenvalues λ_i of $\mathbf{F}$ and

$$\det \Delta(\mathbf{F}) = \prod_{i=1}^{n} \Delta(\lambda_i) \neq 0$$

(Problem 2-22). Hence $\Delta(\mathbf{F})$ is nonsingular.

The substitution of $\mathbf{TA} = \mathbf{FT} + \mathbf{GC}$ into $\mathbf{TA}^2 - \mathbf{F}^2\mathbf{T}$ yields

$$\mathbf{TA}^2 - \mathbf{F}^2\mathbf{T} = (\mathbf{FT} + \mathbf{GC})\mathbf{A} - \mathbf{F}^2\mathbf{T} = \mathbf{F}(\mathbf{TA} - \mathbf{FT}) + \mathbf{GCA} = \mathbf{FGC} + \mathbf{GCA}$$

Proceeding similarly, we can obtain the following set of equalities:

$$\begin{aligned}
\mathbf{TI} - \mathbf{IT} &= \mathbf{0} \\
\mathbf{TA} - \mathbf{FT} &= \mathbf{GC} \\
\mathbf{TA}^2 - \mathbf{F}^2\mathbf{T} &= \mathbf{GCA} + \mathbf{FGC} \\
\mathbf{TA}^3 - \mathbf{F}^3\mathbf{T} &= \mathbf{GCA}^2 + \mathbf{FGCA} + \mathbf{F}^2\mathbf{GC} \\
&\cdots\cdots\cdots\cdots \\
\mathbf{TA}^n - \mathbf{F}^n\mathbf{T} &= \mathbf{GCA}^{n-1} + \mathbf{FGCA}^{n-2} + \cdots + \mathbf{F}^{n-2}\mathbf{GCA} + \mathbf{F}^{n-1}\mathbf{GC}
\end{aligned}$$

We multiply the first equation by α_n, the second equation by $\alpha_{n-1}, \ldots$, the last equation by 1, and then sum them up. After some manipulation, we finally obtain

$$\mathbf{T}\Delta(\mathbf{A}) - \Delta(\mathbf{F})\mathbf{T} = [\mathbf{G} \;\; \mathbf{FG} \;\; \cdots \;\; \mathbf{F}^{n-1}\mathbf{G}] \begin{bmatrix} \alpha_{n-1}\mathbf{I}_q & \alpha_{n-2}\mathbf{I}_q & \cdots & \alpha_1\mathbf{I}_q & \mathbf{I}_q \\ \alpha_{n-2}\mathbf{I}_q & \alpha_{n-3}\mathbf{I}_q & \cdots & \mathbf{I}_q & \mathbf{0} \\ \vdots & \vdots & & \vdots & \vdots \\ \alpha_1\mathbf{I}_q & \mathbf{I}_q & \cdots & \mathbf{0} & \mathbf{0} \\ \mathbf{I}_q & \mathbf{0} & \cdots & \mathbf{0} & \mathbf{0} \end{bmatrix} \begin{bmatrix} \mathbf{C} \\ \mathbf{CA} \\ \vdots \\ \mathbf{CA}^{n-2} \\ \mathbf{CA}^{n-1} \end{bmatrix}$$

$$\triangleq \mathbf{U}_F \mathbf{\Lambda}_q \mathbf{V}_A \tag{7-61}$$

where $\mathbf{U}_F$ is the controllability matrix of $\{\mathbf{F}, \mathbf{G}\}$, and $\mathbf{V}_A$ is the observability matrix of $\{\mathbf{A}, \mathbf{C}\}$. The matrix $\mathbf{\Lambda}_q$ is an $nq \times nq$ matrix and is similar to the one in (7-13). Since $\Delta(\mathbf{A}) = \mathbf{0}$ and $\Delta(\mathbf{F})$ is nonsingular, Equation (7-61) implies

$$\mathbf{T} = -\Delta^{-1}(\mathbf{F})\mathbf{U}_F \mathbf{\Lambda}_q \mathbf{V}_A \tag{7-62}$$

From this equation we conclude that if rank $\mathbf{U}_F < n$ or rank $\mathbf{V}_A < n$, then rank $\mathbf{T} < n$ (Theorem 2-6) and $\mathbf{T}$ is singular. However, rank $\mathbf{U}_F = n$ and rank $\mathbf{V}_A = n$ do not imply the nonsingularity of $\mathbf{T}$. Hence they are only necessary conditions for $\mathbf{T}$ to be nonsingular.

If $q = 1$, $\mathbf{U}_F$, $\mathbf{\Lambda}_q$, and $\mathbf{V}_A$ are $n \times n$ square matrices. $\mathbf{\Lambda}_q$ is always nonsingular. Hence $\mathbf{T}$ is nonsingular if and only if $\mathbf{U}_F$ and $\mathbf{V}_A$ are nonsingular or, equivalently, $\{\mathbf{F}, \mathbf{G}\}$ is controllable and $\{\mathbf{A}, \mathbf{C}\}$ is observable. This establishes the theorem. Q.E.D.

Although the unique solution $\mathbf{T}$ in $\mathbf{TA} - \mathbf{FT} = \mathbf{GC}$ can be computed from (7-62), the method may not be desirable because it requires the computation of $\mathbf{U}_F$ and $\mathbf{V}_A$ (see Section 5-8). The computational problem of $\mathbf{TA} - \mathbf{FT} = \mathbf{GC}$ has been studied extensively in the literature; see, for example, References S9, S18, and S107.

The design procedure in Method I requires the transformation of a dynamical equation into a multivariable observable form. This step is by no means simple numerically. See the discussions in the state-feedback part. It is not clear at present which method, I or II, is better in terms of computational efficiency and numerical stability.

Reduced-dimensional state estimator

Method I

Consider the n-dimensional dynamical equation

$$\dot{\mathbf{x}} = \mathbf{A}\mathbf{x} + \mathbf{B}\mathbf{u} \tag{7-63a}$$
$$\mathbf{y} = \mathbf{C}\mathbf{x} \tag{7-63b}$$

where $\mathbf{A}$, $\mathbf{B}$, and $\mathbf{C}$ are, respectively, $n \times n$, $n \times p$, and $q \times n$ real constant matrices. In this section we assume that $\mathbf{C}$ has a full row rank, that is, rank $\mathbf{C} = q$. Define

$$\mathbf{P} \triangleq \begin{bmatrix} \mathbf{C} \\ \mathbf{R} \end{bmatrix} \tag{7-64}$$

where $\mathbf{R}$ is an $(n-q) \times n$ real constant matrix and is entirely arbitrary so long as $\mathbf{P}$ is nonsingular. We compute the inverse of $\mathbf{P}$ as

$$\mathbf{Q} \triangleq \mathbf{P}^{-1} \triangleq [\mathbf{Q}_1 \vdots \mathbf{Q}_2] \tag{7-65}$$

where $\mathbf{Q}_1$ and $\mathbf{Q}_2$ are $n \times q$ and $n \times (n-q)$ matrices. Clearly we have

$$\mathbf{I}_n = \mathbf{P}\mathbf{Q} = \begin{bmatrix} \mathbf{C} \\ \mathbf{R} \end{bmatrix} [\mathbf{Q}_1 \quad \mathbf{Q}_2] = \begin{bmatrix} \mathbf{C}\mathbf{Q}_1 & \mathbf{C}\mathbf{Q}_2 \\ \mathbf{R}\mathbf{Q}_1 & \mathbf{R}\mathbf{Q}_2 \end{bmatrix} = \begin{bmatrix} \mathbf{I}_q & \mathbf{0} \\ \mathbf{0} & \mathbf{I}_{n-q} \end{bmatrix} \tag{7-66}$$

Now we transform Equation (7-63) into, by the equivalence transformation $\bar{\mathbf{x}} = \mathbf{P}\mathbf{x}$,

$$\dot{\bar{\mathbf{x}}} = \mathbf{P}\mathbf{A}\mathbf{P}^{-1}\bar{\mathbf{x}} + \mathbf{P}\mathbf{B}\mathbf{u}$$
$$\mathbf{y} = \mathbf{C}\mathbf{P}^{-1}\bar{\mathbf{x}} = \mathbf{C}\mathbf{Q}\bar{\mathbf{x}} = [\mathbf{I}_q \quad \mathbf{0}]\bar{\mathbf{x}}$$

which can be partitioned as

$$\begin{bmatrix} \dot{\bar{\mathbf{x}}}_1 \\ \dot{\bar{\mathbf{x}}}_2 \end{bmatrix} = \begin{bmatrix} \bar{\mathbf{A}}_{11} & \bar{\mathbf{A}}_{12} \\ \bar{\mathbf{A}}_{21} & \bar{\mathbf{A}}_{22} \end{bmatrix} \begin{bmatrix} \bar{\mathbf{x}}_1 \\ \bar{\mathbf{x}}_2 \end{bmatrix} + \begin{bmatrix} \bar{\mathbf{B}}_1 \\ \bar{\mathbf{B}}_2 \end{bmatrix} \mathbf{u} \tag{7-67a}$$

$$\mathbf{y} = [\mathbf{I}_q \quad \mathbf{0}]\bar{\mathbf{x}} = \bar{\mathbf{x}}_1 \tag{7-67b}$$

where $\bar{\mathbf{x}}_1$ consists of the first q elements of $\bar{\mathbf{x}}$ and $\bar{\mathbf{x}}_2$ the remainder of $\bar{\mathbf{x}}$; $\bar{\mathbf{A}}_{11}$, $\bar{\mathbf{A}}_{12}$, $\bar{\mathbf{A}}_{21}$, and $\bar{\mathbf{A}}_{22}$ are, respectively, $q \times q$, $q \times (n-q)$, $(n-q) \times q$, and $(n-q) \times (n-q)$ matrices; $\bar{\mathbf{B}}_1$ and $\bar{\mathbf{B}}_2$ are partitioned accordingly. We see from (7-67b) that $\mathbf{y} = \bar{\mathbf{x}}_1$. Hence only the last $n-q$ elements of $\bar{\mathbf{x}}$ need to be estimated. Consequently we need only an $(n-q)$-dimensional state estimator rather than an n-dimensional estimator as discussed in the previous subsection.

Using $\bar{\mathbf{x}}_1 = \mathbf{y}$, we write (7-67a) as

$$\dot{\mathbf{y}} = \bar{\mathbf{A}}_{11}\mathbf{y} + \bar{\mathbf{A}}_{12}\bar{\mathbf{x}}_2 + \bar{\mathbf{B}}_1\mathbf{u}$$
$$\dot{\bar{\mathbf{x}}}_2 = \bar{\mathbf{A}}_{22}\bar{\mathbf{x}}_2 + \bar{\mathbf{A}}_{21}\mathbf{y} + \bar{\mathbf{B}}_2\mathbf{u} \tag{7-68a}$$

which become, by defining $\bar{\mathbf{u}} = \bar{\mathbf{A}}_{21}\mathbf{y} + \bar{\mathbf{B}}_2\mathbf{u}$ and $\mathbf{w} = \dot{\mathbf{y}} - \bar{\mathbf{A}}_{11}\mathbf{y} - \bar{\mathbf{B}}_1\mathbf{u}$,

$$\begin{aligned} \dot{\bar{\mathbf{x}}}_2 &= \bar{\mathbf{A}}_{22}\bar{\mathbf{x}}_2 + \bar{\mathbf{u}} \\ \mathbf{w} &= \bar{\mathbf{A}}_{12}\bar{\mathbf{x}}_2 \end{aligned} \tag{7-68b}$$

We note that $\bar{\mathbf{u}}$ and $\mathbf{w}$ are functions of known signals $\mathbf{u}$ and $\mathbf{y}$. Now if the dynamical equation in (7-68) is observable, an estimator of $\bar{\mathbf{x}}_2$ can be constructed.

Theorem 7-11

The pair $\{\mathbf{A}, \mathbf{C}\}$ in Equation (7-63) or, equivalently, the pair $\{\bar{\mathbf{A}}, \bar{\mathbf{C}}\}$ in Equation (7-67) is observable if and only if the pair $\{\bar{\mathbf{A}}_{22}, \bar{\mathbf{A}}_{12}\}$ in (7-68) is observable. ∎

The controllability part of this theorem was implicitly established in Section 5-8. See also Problem 5-34. Thus the proof of this theorem is left as an exercise.

If $\{\mathbf{A}, \mathbf{C}\}$ is observable, then $\{\bar{\mathbf{A}}_{22}, \bar{\mathbf{A}}_{12}\}$ is observable. Consequently, there exists an $(n-q)$-dimensional state estimator of $\bar{\mathbf{x}}_2$ of the form

$$\dot{\hat{\bar{\mathbf{x}}}}_2 = (\bar{\mathbf{A}}_{22} - \bar{\mathbf{L}}\bar{\mathbf{A}}_{12})\hat{\bar{\mathbf{x}}}_2 + \bar{\mathbf{L}}\mathbf{w} + \bar{\mathbf{u}} \tag{7-69}$$

such that the eigenvalues of $(\bar{\mathbf{A}}_{22} - \bar{\mathbf{L}}\bar{\mathbf{A}}_{12})$ can be arbitrarily assigned by a proper choice of $\bar{\mathbf{L}}$ (Theorem 7-8). The substitution of $\bar{\mathbf{u}}$ and $\mathbf{w}$ into (7-69) yields

$$\dot{\hat{\bar{\mathbf{x}}}}_2 = (\bar{\mathbf{A}}_{22} - \bar{\mathbf{L}}\bar{\mathbf{A}}_{12})\hat{\bar{\mathbf{x}}}_2 + \bar{\mathbf{L}}(\dot{\mathbf{y}} - \bar{\mathbf{A}}_{11}\mathbf{y} - \bar{\mathbf{B}}_1\mathbf{u}) + (\bar{\mathbf{A}}_{21}\mathbf{y} + \bar{\mathbf{B}}_2\mathbf{u}) \tag{7-70}$$

This equation contains the derivative of $\mathbf{y}$. This can be eliminated by defining

$$\mathbf{z} = \hat{\bar{\mathbf{x}}}_2 - \bar{\mathbf{L}}\mathbf{y} \tag{7-71}$$

Then (7-70) becomes

$$\begin{aligned} \dot{\mathbf{z}} &= (\bar{\mathbf{A}}_{22} - \bar{\mathbf{L}}\bar{\mathbf{A}}_{12})(\mathbf{z} + \bar{\mathbf{L}}\mathbf{y}) + (\bar{\mathbf{A}}_{21} - \bar{\mathbf{L}}\bar{\mathbf{A}}_{11})\mathbf{y} + (\bar{\mathbf{B}}_2 - \bar{\mathbf{L}}\bar{\mathbf{B}}_1)\mathbf{u} \\ &= (\bar{\mathbf{A}}_{22} - \bar{\mathbf{L}}\bar{\mathbf{A}}_{12})\mathbf{z} + [(\bar{\mathbf{A}}_{22} - \bar{\mathbf{L}}\bar{\mathbf{A}}_{12})\bar{\mathbf{L}} + (\bar{\mathbf{A}}_{21} - \bar{\mathbf{L}}\bar{\mathbf{A}}_{11})]\mathbf{y} + (\bar{\mathbf{B}}_2 - \bar{\mathbf{L}}\bar{\mathbf{B}}_1)\mathbf{u} \end{aligned} \tag{7-72}$$

This is an $(n-q)$-dimensional dynamical equation with $\mathbf{u}$ and $\mathbf{y}$ as the inputs and can be readily constructed by using, for example, operational amplifier circuits. From (7-71), we see that $\mathbf{z} + \bar{\mathbf{L}}\mathbf{y}$ is an estimate of $\bar{\mathbf{x}}_2$. Indeed, if we define $\mathbf{e} = \bar{\mathbf{x}}_2 - (\mathbf{z} + \bar{\mathbf{L}}\mathbf{y})$, then we have

$$\begin{aligned} \dot{\mathbf{e}} &= \dot{\bar{\mathbf{x}}}_2 - (\dot{\mathbf{z}} + \bar{\mathbf{L}}\dot{\bar{\mathbf{x}}}_1) = \bar{\mathbf{A}}_{21}\bar{\mathbf{x}}_1 + \bar{\mathbf{A}}_{22}\bar{\mathbf{x}}_2 + \bar{\mathbf{B}}_2\mathbf{u} - (\bar{\mathbf{A}}_{22} - \bar{\mathbf{L}}\bar{\mathbf{A}}_{12})(\mathbf{z} + \bar{\mathbf{L}}\bar{\mathbf{x}}_1) \\ &\quad - (\bar{\mathbf{A}}_{21} - \bar{\mathbf{L}}\bar{\mathbf{A}}_{11})\bar{\mathbf{x}}_1 - (\bar{\mathbf{B}}_2 - \bar{\mathbf{L}}\bar{\mathbf{B}}_1)\mathbf{u} - \bar{\mathbf{L}}\bar{\mathbf{A}}_{11}\bar{\mathbf{x}}_1 - \bar{\mathbf{L}}\bar{\mathbf{A}}_{12}\bar{\mathbf{x}}_2 - \bar{\mathbf{L}}\bar{\mathbf{B}}_1\mathbf{u} \\ &= (\bar{\mathbf{A}}_{22} - \bar{\mathbf{L}}\bar{\mathbf{A}}_{12})(\bar{\mathbf{x}}_2 - \mathbf{z} - \bar{\mathbf{L}}\bar{\mathbf{x}}_1) \end{aligned}$$

or

$$\dot{\mathbf{e}} = (\bar{\mathbf{A}}_{22} - \bar{\mathbf{L}}\bar{\mathbf{A}}_{12})\mathbf{e} \tag{7-73}$$

Since the eigenvalues of $(\bar{\mathbf{A}}_{22} - \bar{\mathbf{L}}\bar{\mathbf{A}}_{12})$ can be arbitrarily assigned, the rate of $\mathbf{e}(t)$ approaching zero or, equivalently, the rate of $(\mathbf{z} + \bar{\mathbf{L}}\mathbf{y})$ approaching $\bar{\mathbf{x}}_2$ can be determined by the designer. Hence $\mathbf{z} + \bar{\mathbf{L}}\mathbf{y}$ yields an estimate of $\bar{\mathbf{x}}_2$.

Now we combine $\bar{\mathbf{x}}_1 = \mathbf{y} = \hat{\bar{\mathbf{x}}}_1$ with $\hat{\bar{\mathbf{x}}}_2 = \mathbf{z} + \bar{\mathbf{L}}\mathbf{y}$ to form

$$\hat{\bar{\mathbf{x}}} = \begin{bmatrix} \hat{\bar{\mathbf{x}}}_1 \\ \hat{\bar{\mathbf{x}}}_2 \end{bmatrix} = \begin{bmatrix} \mathbf{y} \\ \bar{\mathbf{L}}\mathbf{y} + \mathbf{z} \end{bmatrix}$$

Since $\bar{\mathbf{x}} = \mathbf{P}\mathbf{x}$, we have $\mathbf{x} = \mathbf{P}^{-1}\bar{\mathbf{x}} = \mathbf{Q}\bar{\mathbf{x}}$ or

$$\hat{\mathbf{x}} = \mathbf{Q}\hat{\bar{\mathbf{x}}} = [\mathbf{Q}_1 \quad \mathbf{Q}_2]\begin{bmatrix} \mathbf{y} \\ \bar{\mathbf{L}}\mathbf{y} + \mathbf{z} \end{bmatrix} = [\mathbf{Q}_1 \quad \mathbf{Q}_2]\begin{bmatrix} \mathbf{I}_q & \mathbf{0} \\ \bar{\mathbf{L}} & \mathbf{I}_{n-q} \end{bmatrix}\begin{bmatrix} \mathbf{y} \\ \mathbf{z} \end{bmatrix} \tag{7-74}$$

This gives an estimate of the original n-dimensional state vector $\mathbf{x}$. A block diagram of the $(n-q)$-dimensional state estimator in (7-72) and (7-74) is plotted in Figure 7-8.

A comparison between an $(n-q)$-dimensional and an n-dimensional state estimators is in order. The reduced-dimensional estimator requires the equivalence transformation discussed in (7-64). Excluding this step, the computation required in the reduced-dimensional case is clearly less than the one in the full-dimensional case. In the implementation, the former also requires less number of integrators. In the reduced-dimensional case, the signal $\mathbf{y}$ is fed through the constant matrix $\mathbf{Q}_1$ to the output of the estimator. Hence, if $\mathbf{y}$ is corrupted with noises, the noises will appear in the output of the estimator. In the full-dimensional estimator, the signal $\mathbf{y}$ is integrated or filtered; hence high-frequency noises in $\mathbf{y}$ will be suppressed in the full-dimensional estimator.

Method II

Consider the n-dimensional dynamical equation

$$FE: \qquad \dot{\mathbf{x}} = \mathbf{A}\mathbf{x} + \mathbf{B}\mathbf{u} \tag{7-75a}$$

$$\mathbf{y} = \mathbf{C}\mathbf{x} \tag{7-75b}$$

where $\mathbf{A}$, $\mathbf{B}$, and $\mathbf{C}$ are, respectively, $n \times n$, $n \times p$, and $q \times n$ real constant matrices. It is assumed that FE is irreducible and rank $\mathbf{C} = q$. Let

$$\dot{\mathbf{z}} = \mathbf{F}\mathbf{z} + \mathbf{G}\mathbf{y} + \mathbf{H}\mathbf{u} \tag{7-76}$$

be an $(n-q)$-dimensional dynamical equation. $\mathbf{F}$, $\mathbf{G}$, and $\mathbf{H}$ are, respectively, $(n-q) \times (n-q)$, $(n-q) \times q$, and $(n-q) \times p$ real constant matrices to be designed.

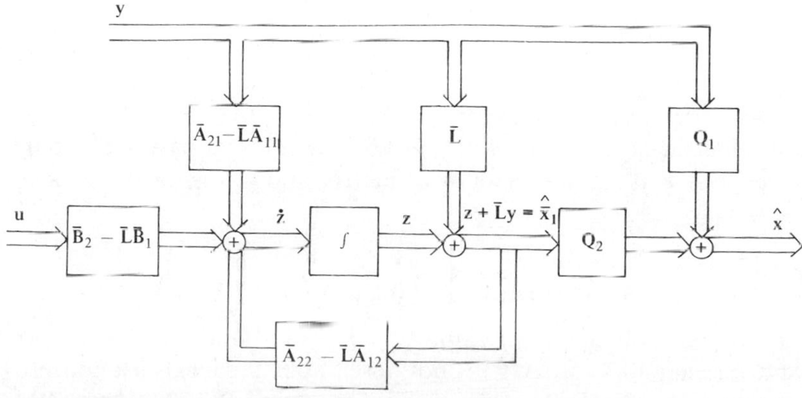

Figure 7-8 An $(n-q)$-dimensional state estimator.

The procedure of computing **F**, **G**, and **H** is similar to Method II of the full-dimensional case.

Algorithm

1. Choose an $(n-q)\times(n-q)$ real constant matrix **F** so that all of its eigenvalues have negative real parts and are distinct from those of **A**.
2. Choose a **G** so that $\{\mathbf{F}, \mathbf{G}\}$ is controllable, that is

$$\text{rank}\,[\mathbf{G}\quad \mathbf{FG}\quad \cdots\quad \mathbf{F}^{n-q-1}\mathbf{G}]=n-q$$

3. Solve the unique **T** in $\mathbf{TA}-\mathbf{FT}=\mathbf{GC}$. Note that **T** is an $(n-q)\times n$ matrix.
4. If the square matrix of order n

$$\mathbf{P}=\begin{bmatrix}\mathbf{C}\\ \mathbf{T}\end{bmatrix} \tag{7-77}$$

is singular, go back to step 1 and/or step 2 and repeat the process. If **P** is nonsingular, compute $\mathbf{H}=\mathbf{TB}$. Then the **z** in Equation (7-76) is an estimate of **Tx** and the original state **x** can be estimated by

$$\hat{\mathbf{x}}=\begin{bmatrix}\mathbf{C}\\ \mathbf{T}\end{bmatrix}^{-1}\begin{bmatrix}\mathbf{y}\\ \mathbf{z}\end{bmatrix} \tag{7-78}$$

∎

The proof of Theorem 7-9 can be used to establish that **z** is an estimate of **Tx**. The combination of $\mathbf{y}=\mathbf{C}\hat{\mathbf{x}}$ and $\mathbf{z}=\mathbf{T}\hat{\mathbf{x}}$ yields immediately (7-78).

Theorem 7-12

If **A** and **F** have no common eigenvalues, necessary conditions for the existence of a full rank **T** in $\mathbf{TA}-\mathbf{FT}=\mathbf{GC}$ such that

$$\mathbf{P}=\begin{bmatrix}\mathbf{C}\\ \mathbf{T}\end{bmatrix}$$

is nonsingular are $\{\mathbf{A}, \mathbf{C}\}$ observable and $\{\mathbf{F}, \mathbf{G}\}$ controllable. For the single-output case ($q=1$), the conditions are sufficient as well.

Proof

First we note that Equation (7-62) still holds for the present case. The only differences are that **F** is an $(n-q)\times(n-q)$ matrix and $\mathbf{U}_F$ is an $(n-q)\times nq$ matrix. We write

$$\mathbf{P}=\begin{bmatrix}\mathbf{C}\\ \mathbf{T}\end{bmatrix}=\begin{bmatrix}\mathbf{C}\\ -\Delta^{-1}(\mathbf{F})\mathbf{U}_F\mathbf{\Lambda}_q\mathbf{V}_A\end{bmatrix}=\begin{bmatrix}\mathbf{I} & \mathbf{0}\\ \mathbf{0} & -\Delta^{-1}(\mathbf{F})\end{bmatrix}\begin{bmatrix}\mathbf{C}\\ \mathbf{U}_F\mathbf{\Lambda}_q\mathbf{V}_A\end{bmatrix}$$

If $\{\mathbf{F}, \mathbf{G}\}$ is not controllable, then rank $\mathbf{U}_F<n-q$. Hence we have rank $\mathbf{T}<n-q$ and **P** is singular. If $\{\mathbf{A}, \mathbf{C}\}$ is not observable, there exists a nonzero $n\times 1$ vector **r** such that $\mathbf{V}_A\mathbf{r}=\mathbf{0}$. This implies $\mathbf{Cr}=\mathbf{0}$ and $\mathbf{Pr}=\mathbf{0}$. Hence **P** is singular. This establishes the necessity of the theorem.

Now we consider the single-output case. We show the sufficiency of the theorem by showing that $\mathbf{r}=\mathbf{0}$ is the only possible solution in

$$\begin{bmatrix} \mathbf{C} \\ \mathbf{U}_F\mathbf{\Lambda}_1\mathbf{V}_A \end{bmatrix}\mathbf{r}=\mathbf{0} \tag{7-79}$$

Define $\boldsymbol{\beta}=\mathbf{\Lambda}_1\mathbf{V}_A\mathbf{r}=[\beta_1 \quad \beta_2 \quad \cdots \quad \beta_n]'$, or

$$\begin{bmatrix} \beta_1 \\ \beta_2 \\ \vdots \\ \beta_n \end{bmatrix}=\begin{bmatrix} \alpha_{n-1} & \alpha_{n-2} & \cdots & \alpha_1 & 1 \\ \alpha_{n-2} & \alpha_{n-3} & \cdots & 1 & 0 \\ \vdots & \vdots & & \vdots & \vdots \\ 1 & 0 & \cdots & 0 & 0 \end{bmatrix}\begin{bmatrix} \mathbf{C} \\ \mathbf{CA} \\ \vdots \\ \mathbf{CA}^{n-1} \end{bmatrix}\mathbf{r}=\begin{bmatrix} x \\ x \\ \vdots \\ \mathbf{Cr} \end{bmatrix}$$

where x denotes nonzero elements. From this equation, we see that if $\mathbf{Cr}=0$, then we have $\beta_n=0$. From $\mathbf{U}_F\mathbf{\Lambda}_1\mathbf{V}_A\mathbf{r}=\mathbf{U}_F\boldsymbol{\beta}=\mathbf{0}$, we have, by using $\beta_n=0$,

$$\mathbf{U}_F\mathbf{\Lambda}_1\mathbf{V}_A\mathbf{r}=\sum_{i=1}^{n}\beta_i\mathbf{F}^{i-1}\mathbf{G}=\sum_{i=1}^{n-1}\beta_i\mathbf{F}^{i-1}\mathbf{G}=\mathbf{0} \tag{7-80}$$

which together with the controllability of $\{\mathbf{F}, \mathbf{G}\}$ imply $\beta_i=0, i=1, 2, \ldots, n-1$. Hence (7-79) and rank $\mathbf{U}_F=n-1$ imply $\boldsymbol{\beta}=\mathbf{0}$.

Because $\mathbf{\Lambda}_1$ is nonsingular, $\boldsymbol{\beta}=\mathbf{\Lambda}_1\mathbf{V}_A\mathbf{r}=\mathbf{0}$ implies $\mathbf{V}_A\mathbf{r}=\mathbf{0}$, which together with the observability of $\{\mathbf{A}, \mathbf{C}\}$ imply $\mathbf{r}=\mathbf{0}$. This establishes that if $\{\mathbf{A}, \mathbf{C}\}$ is observable and $\{\mathbf{F}, \mathbf{G}\}$ is controllable, then $\mathbf{P}$ is nonsingular. This completes the proof of this theorem. Q.E.D.

The discussion in Method II of the full-dimensional case regarding the computational problem and the discussion in Method I of the reduced-dimensional case regarding the comparison with the full-dimensional case are equally applicable here; hence they will not be repeated.

To conclude this section, we summarize the preceding results as a theorem.

Theorem 7-13

If the n-dimensional dynamical equation FE is observable, an $(n-q)$-dimensional state estimator, as given in (7-72) and (7-74) or in (7-76) and (7-78), with any desired eigenvalues (provided complex conjugate eigenvalues appear in pair) can be constructed, where q is the rank of matrix $\mathbf{C}$ in FE. ∎

7-5 Connection of State Feedback and State Estimator

Consider the n-dimensional dynamical equation

$$FE: \quad \dot{\mathbf{x}}=\mathbf{Ax}+\mathbf{Bu} \tag{7-81a}$$
$$\mathbf{y}=\mathbf{Cx} \tag{7-81b}$$

where $\mathbf{A}$, $\mathbf{B}$, and $\mathbf{C}$ are, respectively, $n \times n$, $n \times p$, and $q \times n$ real constant matrices. It was shown in Section 7-3 that if $\{\mathbf{A}, \mathbf{B}\}$ is controllable, the introduction of the state feedback $\mathbf{u}=\mathbf{r}+\mathbf{Kx}$ can place the eigenvalues of $(\mathbf{A}+\mathbf{BK})$ in any desired positions. It was shown in Section 7-4 that if $\{\mathbf{A}, \mathbf{C}\}$ is observable, a state estimator, with arbitrary eigenvalues, of dimension n or $n-q$ can be constructed by using **u** and **y** as the inputs. In the following, we consider only the reduced-dimensional case; the full-dimensional case is left as an exercise. Consider the $(n-q)$-dimensional state estimator developed in (7-76) and (7-78):

$$\dot{\mathbf{z}} = \mathbf{F}\mathbf{z} + \mathbf{G}\mathbf{y} + \mathbf{H}\mathbf{u} \tag{7-82a}$$

$$\hat{\mathbf{x}} = \begin{bmatrix} \mathbf{C} \\ \mathbf{T} \end{bmatrix}^{-1} \begin{bmatrix} \mathbf{y} \\ \mathbf{z} \end{bmatrix} = [\bar{\mathbf{Q}}_1 \quad \bar{\mathbf{Q}}_2] \begin{bmatrix} \mathbf{y} \\ \mathbf{z} \end{bmatrix} \tag{7-82b}$$

with $\mathbf{TA} - \mathbf{FT} = \mathbf{GC}$, $\mathbf{H} = \mathbf{TB}$ and

$$[\bar{\mathbf{Q}}_1 \quad \bar{\mathbf{Q}}_2] \begin{bmatrix} \mathbf{C} \\ \mathbf{T} \end{bmatrix} = \bar{\mathbf{Q}}_1\mathbf{C} + \bar{\mathbf{Q}}_2\mathbf{T} = \mathbf{I} \tag{7-83}$$

If the eigenvalues of **F** are chosen to have negative real parts, the estimated state $\hat{\mathbf{x}}(t)$ will approach the actual state $\mathbf{x}(t)$ exponentially.

The state feedback is designed with respect to the actual state of FE. If the actual state $\mathbf{x}(t)$ is not available for feedback, it is natural to apply the feedback gain **K** to the estimated state $\hat{\mathbf{x}}(t)$, that is,

$$\mathbf{u} = \mathbf{r} + \mathbf{K}\hat{\mathbf{x}} \tag{7-84}$$

as shown in Figure 7-9. Three questions may be raised in this connection: (1) In the state feedback $\mathbf{u} = \mathbf{r} + \mathbf{K}\mathbf{x}$, the eigenvalues of the resulting equation are given by the eigenvalues of $\mathbf{A} + \mathbf{BK}$. In the estimated state feedback $\mathbf{u} = \mathbf{r} + \mathbf{K}\hat{\mathbf{x}}$, do we still have the same set of eigenvalues? (2) Will the eigenvalues of the state estimator be affected by the feedback $\mathbf{u} = \mathbf{r} + \mathbf{K}\hat{\mathbf{x}}$? (3) What is the effect of the estimator on the transfer-function matrix from **r** to **y**? To answer these questions, we form the composite dynamical equation of the system in Figure 7-9 as

$$\begin{bmatrix} \dot{\mathbf{x}} \\ \dot{\mathbf{z}} \end{bmatrix} = \begin{bmatrix} \mathbf{A} + \mathbf{BK}\bar{\mathbf{Q}}_1\mathbf{C} & \mathbf{BK}\bar{\mathbf{Q}}_2 \\ \mathbf{GC} + \mathbf{HK}\bar{\mathbf{Q}}_1\mathbf{C} & \mathbf{F} + \mathbf{HK}\bar{\mathbf{Q}}_2 \end{bmatrix} \begin{bmatrix} \mathbf{x} \\ \mathbf{z} \end{bmatrix} + \begin{bmatrix} \mathbf{B} \\ \mathbf{H} \end{bmatrix} \mathbf{r} \tag{7-85a}$$

$$\mathbf{y} = [\mathbf{C} \quad \mathbf{0}] \begin{bmatrix} \mathbf{x} \\ \mathbf{z} \end{bmatrix} \tag{7-85b}$$

This is obtained by the substitution of (7-84) and (7-82b) into (7-81) and (7-82a).

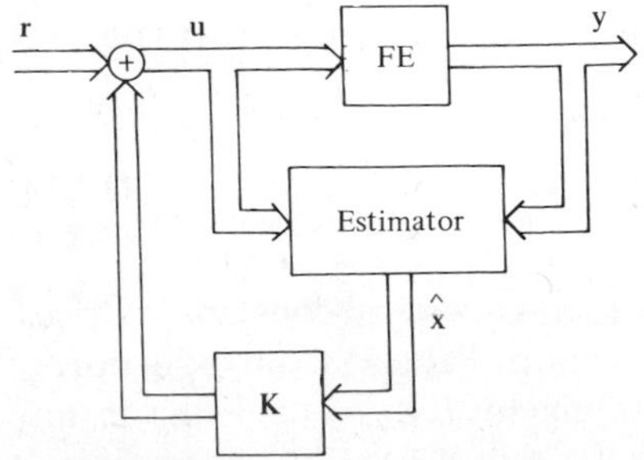

Figure 7-9 Feedback from the estimated state.

To simplify (7-85), we introduce the following equivalence transformation

$$\begin{bmatrix} \mathbf{x} \\ \mathbf{e} \end{bmatrix} = \begin{bmatrix} \mathbf{x} \\ \mathbf{z} - \mathbf{T}\mathbf{x} \end{bmatrix} = \begin{bmatrix} \mathbf{I} & \mathbf{0} \\ -\mathbf{T} & \mathbf{I} \end{bmatrix} \begin{bmatrix} \mathbf{x} \\ \mathbf{z} \end{bmatrix}$$

After some manipulation and using $\mathbf{TA} - \mathbf{FT} = \mathbf{GC}, \mathbf{H} = \mathbf{TB}$ and (7-83), we finally obtain the following equivalent dynamical equation

$$\begin{bmatrix} \dot{\mathbf{x}} \\ \dot{\mathbf{e}} \end{bmatrix} = \begin{bmatrix} \mathbf{A} + \mathbf{BK} & \mathbf{BK}\bar{\mathbf{Q}}_2 \\ \mathbf{0} & \mathbf{F} \end{bmatrix} \begin{bmatrix} \mathbf{x} \\ \mathbf{e} \end{bmatrix} + \begin{bmatrix} \mathbf{B} \\ \mathbf{0} \end{bmatrix} \mathbf{r} \tag{7-86a}$$

$$\mathbf{y} = [\mathbf{C} \quad \mathbf{0}] \begin{bmatrix} \mathbf{x} \\ \mathbf{e} \end{bmatrix} \tag{7-86b}$$

The eigenvalues of a dynamical equation are invariant under any equivalence transformation; hence we conclude from (7-86a) that the eigenvalues of the system in Figure 7-9 are the union of those of $\mathbf{A} + \mathbf{BK}$ and those of $\mathbf{F}$. We see that the eigenvalues of the state estimator are not affected by the feedback and, as far as the eigenvalues are concerned, there is no difference in state feedback from the estimated state $\hat{\mathbf{x}}$ or from the actual state $\mathbf{x}$. Consequently, the design of state feedback and the design of a state estimator can be carried out independently and the eigenvalues of the entire system are the union of those of state feedback and those of the state estimator. This property is often called the *separation property*.

We discuss now the transfer matrix from $\mathbf{r}$ to $\mathbf{y}$. Equation (7-86) is of the form shown in (5-54). Hence the transfer matrix of (7-86) or of the system in Figure 7-9 can be computed from

$$\dot{\mathbf{x}} = (\mathbf{A} + \mathbf{BK})\mathbf{x} + \mathbf{Br} \qquad \mathbf{y} = \mathbf{Cx}$$

as

$$\hat{\mathbf{G}}_f(s) = \mathbf{C}(s\mathbf{I} - \mathbf{A} - \mathbf{BK})^{-1}\mathbf{B}$$

(Theorem 5-16). This is the transfer matrix of the perfect state feedback without the use of a state estimator. In other words, the estimator is completely canceled and does not appear in the transfer matrix from $\mathbf{r}$ to $\mathbf{y}$. This has a simple explanation. In the computation of transfer matrix, all initial states are assumed to be zero. Consequently, we have $\hat{\mathbf{x}}(0) = \mathbf{x}(0) = \mathbf{0}$, which implies $\hat{\mathbf{x}}(t) = \mathbf{x}(t)$ for all t. Hence, as far as the transfer matrix from $\mathbf{r}$ to $\mathbf{y}$ is concerned, there is no difference whether a state estimator is employed or not.

If $\hat{\mathbf{x}}(0) \neq \mathbf{x}(0)$, the rate of $\hat{\mathbf{x}}(t)$ approaching $\mathbf{x}(t)$ is determined by the eigenvalues of the estimator. Clearly, the larger the magnitudes of the negative real parts of these eigenvalues, the faster $\hat{\mathbf{x}}(t)$ approaches $\mathbf{x}(t)$. However, these will cause larger gains in $\mathbf{L}$ and generate larger magnitudes in transient, as in the state feedback case. As the eigenvalues approach negative infinity, the estimator will act as a differentiator and will be susceptible to noises. Large gains and large transient will also cause the estimator to saturate. Hence there is no simple answer to the question of what are the best eigenvalues. It has been suggested in the literature that the eigenvalues of $\mathbf{A} - \mathbf{LC}$ or $\mathbf{F}$ be chosen to be two or three times faster than the eigenvalues of $\mathbf{A} + \mathbf{BK}$, that is, the magnitudes of the negative real parts of the eigenvalues of $\mathbf{A} - \mathbf{LC}$ or $\mathbf{F}$ be two or three times of those of $\mathbf{A} + \mathbf{BK}$. This seems to be a simple and reasonable guideline. If a dynamical equation is corrupted by noises and modeled as a stochastic process, the matrix $\mathbf{L}$ can be computed by minimizing the mean square error. The interested reader is referred to References S4 and S10.

Example 1

Consider the inverted pendulum problem studied in Exmple 2 on page 338. Its dynamical equation is

$$\dot{\mathbf{x}} = \left[\begin{array}{c:ccc} 0 & 1 & 0 & 0 \\ \hdashline 0 & 0 & -1 & 0 \\ 0 & 0 & 0 & 1 \\ 0 & 0 & 5 & 0 \end{array}\right]\mathbf{x} + \left[\begin{array}{c} 0 \\ \hdashline 1 \\ 0 \\ -2 \end{array}\right]u$$
$$\mathbf{y} = \left[\begin{array}{c:ccc} 1 & 0 & 0 & 0 \end{array}\right]\mathbf{x}$$

As discussed there, if we introduce the state feedback

$$u(t) = r(t) + \left[\tfrac{3}{4} \quad \tfrac{10}{3} \quad \tfrac{63}{8} \quad \tfrac{25}{6}\right]\mathbf{x}$$

then the resulting equation has eigenvalues -1, -2, and $-1 \pm j$.

Now if the state $\mathbf{x}$ is not available for feedback, we must design a state estimator. We use Method I of the reduced-dimensional state estimator to design a three-dimensional estimator. Since the equation is already in the form of (7-67), the step of the equivalence transformation in (7-64) can be skipped. Clearly we have

$$\bar{\mathbf{A}}_{22} = \begin{bmatrix} 0 & -1 & 0 \\ 0 & 0 & 1 \\ 0 & 5 & 0 \end{bmatrix} \qquad \bar{\mathbf{A}}_{12} = [1 \quad 0 \quad 0]$$

Let us choose arbitrarily the eigenvalues of the estimator as -3 and $-3 \pm 2j$. Then we have to find a $\bar{\mathbf{L}}$ such that the eigenvalues of $\bar{\mathbf{A}}_{22} - \bar{\mathbf{L}}\bar{\mathbf{A}}_{12}$ are -3 and $-3 \pm 2j$. For this problem, we shall solve $\bar{\mathbf{L}}$ by brute force. Let $\bar{\mathbf{L}}' = [l_1 \quad l_2 \quad l_3]$, where the prime denotes the transpose. Then we have

$$\bar{\bar{\mathbf{A}}} \triangleq \bar{\mathbf{A}}_{22} - \bar{\mathbf{L}}\bar{\mathbf{A}}_{12} = \begin{bmatrix} -l_1 & -1 & 0 \\ -l_2 & 0 & 1 \\ -l_3 & 5 & 0 \end{bmatrix}$$

and $$\det(s\mathbf{I} - \bar{\bar{\mathbf{A}}}) = s^3 + l_1 s^2 - (5 + l_2)s - (l_3 + 5l_1)$$

By equating its coefficients with those of

$$(s+3)(s+3+2j)(s+3-2j) = s^3 + 9s^2 + 31s + 39$$

we obtain

$$l_1 = 9 \qquad l_2 = -36 \qquad l_3 = -84$$

Hence a three-dimensional state estimator is, following (7-72) and (7-74),

$$\dot{\mathbf{z}} = \begin{bmatrix} -9 & -1 & 0 \\ 36 & 0 & 1 \\ 84 & 5 & 0 \end{bmatrix}\begin{bmatrix} z_1 \\ z_2 \\ z_3 \end{bmatrix} + \begin{bmatrix} -45 \\ 240 \\ 576 \end{bmatrix} y + \begin{bmatrix} 1 \\ 0 \\ -2 \end{bmatrix} u$$

$$\hat{\mathbf{x}} = \begin{bmatrix} 1 & 0 & 0 & 0 \\ 9 & 1 & 0 & 0 \\ -36 & 0 & 1 & 0 \\ -84 & 0 & 0 & 1 \end{bmatrix} \begin{bmatrix} y \\ z_1 \\ z_2 \\ z_3 \end{bmatrix}$$

Now we may apply the state feedback from $\hat{\mathbf{x}}$ as

$$u = r + \begin{bmatrix} \frac{3}{4} & \frac{10}{3} & \frac{63}{8} & \frac{25}{6} \end{bmatrix} \hat{\mathbf{x}}$$

This completes the design. ∎

Functional estimators. If all the n state variables are to be reconstructed, an estimator of dimension at least $n-q$ is required. In many cases, however, it is not necessary to reconstruct all the state variables; what is needed is to generate some functions of the state, for example, to generate $\mathbf{Kx}$ in the state feedback. In these cases, the dimensions of the estimators may be reduced considerably. In the following, we discuss a method of designing an estimator to generate $\mathbf{kx}$, where $\mathbf{k}$ is an arbitrary $1 \times n$ constant vector. Such an estimator is called a *functional estimator* or *observer.*

Consider the dynamical equation in (7-63) with rank $\mathbf{C} = q$. The problem is to design an estimator so that its output will approach $\mathbf{kx}$ as $t \to \infty$. First we transform (7-63) into the form in (7-67), rewritten here for convenience,[10]

$$\begin{bmatrix} \dot{\bar{\mathbf{x}}}_1 \\ \dot{\bar{\mathbf{x}}}_2 \end{bmatrix} = \begin{bmatrix} \bar{\mathbf{A}}_{11} & \bar{\mathbf{A}}_{12} \\ \bar{\mathbf{A}}_{21} & \bar{\mathbf{A}}_{22} \end{bmatrix} \begin{bmatrix} \bar{\mathbf{x}}_1 \\ \bar{\mathbf{x}}_2 \end{bmatrix} + \begin{bmatrix} \bar{\mathbf{B}}_1 \\ \bar{\mathbf{B}}_2 \end{bmatrix} \mathbf{u} \triangleq \bar{\mathbf{A}}\bar{\mathbf{x}} + \bar{\mathbf{B}}\mathbf{u}$$
$$\mathbf{y} = [\mathbf{I}_q \quad \mathbf{0}]\bar{\mathbf{x}} \triangleq \bar{\mathbf{C}}\bar{\mathbf{x}}$$

where $\bar{\mathbf{x}} = \mathbf{Px}$ with $\mathbf{P}$ given in (7-64). The functional estimator with a single output $w(t)$ will be chosen of the form

$$\dot{\mathbf{z}} = \mathbf{Fz} + \mathbf{Gy} + \mathbf{Hu} \tag{7-87a}$$
$$w = \mathbf{Mz} + \mathbf{Ny} \tag{7-87b}$$

where $\mathbf{F}$ is a $m \times m$ matrix; $\mathbf{G}, m \times q$; $\mathbf{H}, m \times p$; $\mathbf{M}, 1 \times m$ and $\mathbf{N}, 1 \times q$. The problem is to design such an estimator of a dimension as small as possible so that $w(t)$ will approach $\mathbf{kx} = \mathbf{kP}^{-1}\bar{\mathbf{x}} = \bar{\mathbf{k}}\bar{\mathbf{x}}$, where $\bar{\mathbf{k}} = \mathbf{kP}^{-1}$. Following Theorem 7-9, we can show that if

$$\mathbf{T}\bar{\mathbf{A}} - \mathbf{FT} = \mathbf{G}\bar{\mathbf{C}} \tag{7-88a}$$
$$\mathbf{H} = \mathbf{T}\bar{\mathbf{B}} \tag{7-88b}$$

and all eigenvalues of $\mathbf{F}$ have negative real parts, where $\mathbf{T}$ is a $m \times n$ real constant matrix, then $\mathbf{z}(t)$ approaches $\mathbf{T}\bar{\mathbf{x}}(t)$ as $t \to \infty$. Hence, if $\mathbf{M}$ and $\mathbf{N}$ are designed to meet

$$\mathbf{MT} + \mathbf{N}\bar{\mathbf{C}} = \bar{\mathbf{k}} \tag{7-88c}$$

then $w(t)$ approaches $\mathbf{kx}(t)$ as $t \to \infty$.

---

[10] $[\mathbf{I}_q \quad \mathbf{0}]$ can be replaced by $[\bar{\mathbf{C}}_1 \quad \mathbf{0}]$, where $\bar{\mathbf{C}}_1$ is a $q \times q$ nonsingular matrix, and the subsequent development, with slight modification, still holds. This subsection follows closely Reference S110.

Now we partition $\mathbf{T}$ as $\mathbf{T}=[\mathbf{T}_1 \quad \mathbf{T}_2]$, where $\mathbf{T}_1$ and $\mathbf{T}_2$ are, respectively, $m \times q$ and $m \times (n-q)$ matrices. Then (7-88a) can be written as, using $\bar{\mathbf{C}}=[\mathbf{I}_q \quad \mathbf{0}]$,

$$\mathbf{T}_1\bar{\mathbf{A}}_{11}+\mathbf{T}_2\bar{\mathbf{A}}_{21}-\mathbf{F}\mathbf{T}_1=\mathbf{G} \tag{7-89a}$$
$$\mathbf{T}_1\bar{\mathbf{A}}_{12}+\mathbf{T}_2\bar{\mathbf{A}}_{22}-\mathbf{F}\mathbf{T}_2=\mathbf{0} \tag{7-89b}$$

If we also partition $\bar{\mathbf{k}}$ as $\bar{\mathbf{k}}=[\bar{\mathbf{k}}_1 \quad \bar{\mathbf{k}}_2]$, where $\bar{\mathbf{k}}_1$ and $\bar{\mathbf{k}}_2$ are, respectively, $1 \times q$ and $1 \times (n-q)$ vectors, then (7-88c) becomes

$$\mathbf{M}\mathbf{T}_1+\mathbf{N}=\bar{\mathbf{k}}_1 \tag{7-89c}$$

and

$$\mathbf{M}\mathbf{T}_2=\bar{\mathbf{k}}_2 \tag{7-89d}$$

Now we choose $\mathbf{F}$ and $\mathbf{M}$ to be of the forms

$$\mathbf{F}=\begin{bmatrix} 0 & 1 & 0 & \cdots & 0 \\ 0 & 0 & 1 & \cdots & 0 \\ \vdots & \vdots & \vdots & & \vdots \\ 0 & 0 & 0 & \cdots & 1 \\ -\alpha_m & -\alpha_{m-1} & -\alpha_{m-2} & \cdots & -\alpha_1 \end{bmatrix}$$
$$\mathbf{M}=[\,1 \quad 0 \quad 0 \quad \cdots \quad 0\,]$$

where α_i can be arbitrarily assigned. Note that $\{\mathbf{F}, \mathbf{M}\}$ is observable. Let $\mathbf{t}_{ij}$, $j=1, 2, \ldots, m$, be the jth row of $\mathbf{T}_i$, $i=1, 2$. Then because of the forms of $\mathbf{F}$ and $\mathbf{M}$, (7-89d) and (7-89b) imply

$$\begin{aligned} \mathbf{t}_{21}&=\bar{\mathbf{k}}_2 \\ \mathbf{t}_{11}\bar{\mathbf{A}}_{12}+\mathbf{t}_{21}\bar{\mathbf{A}}_{22}&=\mathbf{t}_{22} \\ \mathbf{t}_{12}\bar{\mathbf{A}}_{12}+\mathbf{t}_{22}\bar{\mathbf{A}}_{22}&=\mathbf{t}_{23} \\ &\cdots\cdots \\ \mathbf{t}_{1m}\bar{\mathbf{A}}_{12}+\mathbf{t}_{2m}\bar{\mathbf{A}}_{22}&=-\alpha_m\mathbf{t}_{21}-\alpha_{m-1}\mathbf{t}_{22}-\cdots-\alpha_1\mathbf{t}_{2m} \end{aligned} \tag{7-90a}$$

These equations, except the last, can be written as

$$\begin{aligned} \mathbf{t}_{21}&=\bar{\mathbf{k}}_2 \\ \mathbf{t}_{22}&=\bar{\mathbf{k}}_2\bar{\mathbf{A}}_{22}+\mathbf{t}_{11}\bar{\mathbf{A}}_{12} \\ \mathbf{t}_{23}&=\bar{\mathbf{k}}_2\bar{\mathbf{A}}_{22}^2+\mathbf{t}_{11}\bar{\mathbf{A}}_{12}\bar{\mathbf{A}}_{22}+\mathbf{t}_{12}\bar{\mathbf{A}}_{12} \\ &\cdots\cdots \\ \mathbf{t}_{2m}&=\bar{\mathbf{k}}_2\bar{\mathbf{A}}_{22}^{m-1}+\mathbf{t}_{11}\bar{\mathbf{A}}_{12}\bar{\mathbf{A}}_{22}^{m-2}+\cdots+\mathbf{t}_{1,m-2}\bar{\mathbf{A}}_{12}\bar{\mathbf{A}}_{22}+\mathbf{t}_{1,m-1}\bar{\mathbf{A}}_{12} \end{aligned} \tag{7-90b}$$

The substitution of these $\mathbf{t}_{2j}$, $j=1, 2, \ldots, m$, into the last equation of (7-90a) yields, after some straightforward manipulation,

$$\begin{aligned} \tilde{\mathbf{k}} &\triangleq \bar{\mathbf{k}}_2\bar{\mathbf{A}}_{22}^m+\alpha_1\bar{\mathbf{k}}_2\bar{\mathbf{A}}_{22}^{m-1}+\cdots+\alpha_{m-1}\bar{\mathbf{k}}_2\bar{\mathbf{A}}_{22}+\alpha_m\bar{\mathbf{k}}_2 \\ &=-[\mathbf{t}_{1m} \quad \mathbf{t}_{1,m-1} \quad \cdots \quad \mathbf{t}_{12} \quad \mathbf{t}_{11}] \begin{bmatrix} \mathbf{I} & & & & \\ \alpha_1\mathbf{I} & \mathbf{I} & & & \\ \alpha_2\mathbf{I} & \alpha_1\mathbf{I} & \mathbf{I} & & \\ \vdots & & & \ddots & \\ \alpha_{m-1}\mathbf{I} & & \cdots & \alpha_1\mathbf{I} & \mathbf{I} \end{bmatrix} \begin{bmatrix} \bar{\mathbf{A}}_{12} \\ \bar{\mathbf{A}}_{12}\bar{\mathbf{A}}_{22} \\ \bar{\mathbf{A}}_{12}\bar{\mathbf{A}}_{22}^2 \\ \vdots \\ \bar{\mathbf{A}}_{12}\bar{\mathbf{A}}_{22}^{m-1} \end{bmatrix} \end{aligned} \tag{7-91}$$

The vector $\tilde{\mathbf{k}}$ is a $1 \times (n-q)$ vector and is known once $\mathbf{k}$ is given and α_i are chosen.

The $mq \times mq$ matrix in (7-91) is nonsingular for any α_i. Thus we conclude from Theorem 2-4 that, for any $\tilde{\mathbf{k}}$, a solution $[\mathbf{t}_{1m} \quad \mathbf{t}_{1,m-1} \quad \cdots \quad \mathbf{t}_{11}]$ exists in (7-91) if and only if the $mq \times (n-q)$ matrix

$$\mathbf{V} = \begin{bmatrix} \bar{\mathbf{A}}_{12} \\ \bar{\mathbf{A}}_{12}\bar{\mathbf{A}}_{22} \\ \vdots \\ \bar{\mathbf{A}}_{12}\bar{\mathbf{A}}_{22}^{m-1} \end{bmatrix}$$

has rank $n-q$. Now $\{\mathbf{A}, \mathbf{C}\}$ in (7-63) is observable if and only if $\{\bar{\mathbf{A}}_{22}, \bar{\mathbf{A}}_{12}\}$ is observable. Furthermore, the observability index of $\{\bar{\mathbf{A}}_{22}, \bar{\mathbf{A}}_{12}\}$ is $v-1$, where v is the observability index of $\{\mathbf{A}, \mathbf{C}\}$ (Problem 7-27). Hence, if $m = v-1$, then $\mathbf{V}$ has rank $n-q$, and for any $\bar{\mathbf{k}}$ and any α_i, a solution $\mathbf{T}_1$ exists in (7-91). Once $\mathbf{T}_1$ is known, we compute $\mathbf{T}_2$ from (7-90b) and then compute $\mathbf{G}$, $\mathbf{H}$, and $\mathbf{N}$ from (7-89a), (7-88b), and (7-89c). The resulting estimator is a functional estimator of $\mathbf{kx}$. This is stated as a theorem.

Theorem 7-14

If the dynamical equation in (7-63) is observable with observability index v, then for any $1 \times n$ real constant vector $\mathbf{k}$, there exists a $(v-1)$-dimensional functional estimator with arbitrarily assignable eigenvalues so that its output approaches $\mathbf{kx}(t)$ exponentially. ■

The dimension $(v-1)$ could be much smaller than $n-q$ for large q. For example, if $n=12$ and $q=4$, then we have $n-q=8$; whereas the observability index v could be $\frac{12}{4}=3$ [see Equation (5-51)], and we have $v-1=2$. If $q=1$, then $n-q=v-1$.

If a $p \times n$ feedback gain matrix $\mathbf{K}$ is of rank 1, it can be written as $\mathbf{K} = \mathbf{vk}$, where $\mathbf{v}$ is a $p \times 1$ vector and $\mathbf{k}$ is a $1 \times n$ vector as the $\mathbf{K}_2$ in the design in Figure 7-2, then $\mathbf{Kx}$ can also be generated by using a $(v-1)$-dimensional functional estimator. For a general $\mathbf{K}$, the situation is much more complicated. The eigenvalues of the functional estimators discussed above are permitted to be arbitrarily chosen. For a given set of eigenvalues, if they are chosen properly, it may be possible to further reduce the dimension of a functional estimator. In the design, if we augument the matrix $\mathbf{C}$ in (7-63b) as $[\mathbf{C}' \quad \mathbf{K}']'$, then the dimension of estimators may also be reduced. See References S92, S97, S110, S131 and S184.

*7-6 Decoupling by State Feedback

Consider a p-input p-output system with the linear, time-invariant, dynamical-equation description

$$FE: \qquad \dot{\mathbf{x}} = \mathbf{Ax} + \mathbf{Bu} \qquad \textbf{(7-92a)}$$
$$\mathbf{y} = \mathbf{Cx} \qquad \textbf{(7-92b)}$$

where $\mathbf{u}$ is the $p \times 1$ input vector, $\mathbf{y}$ is the $p \times 1$ output vector; $\mathbf{A}$, $\mathbf{B}$, and $\mathbf{C}$ are

$n \times n$, $n \times p$, and $p \times n$ real constant matrices, respectively. It is assumed that $p \leq n$. The transfer function of the system is

$$\hat{\mathbf{G}}(s) = \mathbf{C}(s\mathbf{I} - \mathbf{A})^{-1}\mathbf{B} \tag{7-93}$$

Clearly $\hat{\mathbf{G}}(s)$ is a $p \times p$ rational-function matrix. If the system is initially in the zero state, its inputs and outputs are related by

$$\begin{aligned} \hat{y}_1(s) &= \hat{g}_{11}(s)\hat{u}_1(s) + \hat{g}_{12}(s)\hat{u}_2(s) + \cdots + \hat{g}_{1p}(s)\hat{u}_p(s) \\ \hat{y}_2(s) &= \hat{g}_{21}(s)\hat{u}_1(s) + \hat{g}_{22}(s)\hat{u}_2(s) + \cdots + \hat{g}_{2p}(s)\hat{u}_p(s) \\ &\cdots\cdots\cdots\cdots\cdots\cdots\cdots\cdots\cdots\cdots \\ \hat{y}_p(s) &= \hat{g}_{p1}(s)\hat{u}_1(s) + \hat{g}_{p2}(s)\hat{u}_2(s) + \cdots + \hat{g}_{pp}(s)\hat{u}_p(s) \end{aligned} \tag{7-94}$$

where $\hat{g}_{ij}$ is the ijth element of $\hat{\mathbf{G}}(s)$. We see from (7-94) that every input controls more than one output and that every output is controlled by more than one input. Because of this phenomenon, which is called *coupling* or *interacting*, it is generally very difficult to control a multivariable system. For example, suppose we wish to control $\hat{y}_1(s)$ without affecting $\hat{y}_2(s), \hat{y}_3(s), \ldots, \hat{y}_p(s)$; the required inputs $\hat{u}_1(s), \hat{u}_2(s), \ldots, \hat{u}_p(s)$ cannot be readily found. Therefore, in some cases we like to introduce some compensator so that a coupled multivariable system may become decoupled in the sense that every input controls only one output and every output is controlled by only one input. Consequently, a decoupled system can be considered as consisting of a set of independent single-variable systems. It is clear that if the transfer-function matrix of a multivariable system is diagonal, then the system is decoupled.

Definition 7-1

A multivariable system is said to be *decoupled* if its transfer-function matrix is diagonal and nonsingular. ∎

In this section we shall study the problem of the decoupling of multivariable systems by linear state feedback of the form

$$\mathbf{u}(t) = \mathbf{K}\mathbf{x} + \mathbf{H}\mathbf{r} \tag{7-95}$$

where $\mathbf{K}$ is a $p \times n$ real constant matrix, $\mathbf{H}$ is a $p \times p$ real constant nonsingular matrix, and $\mathbf{r}$ denotes the $p \times 1$ reference input. The state feedback is shown in Figure 7-10. Substituting (7-95) into (7-92), we obtain

$$FE^f: \quad \dot{\mathbf{x}} = (\mathbf{A} + \mathbf{B}\mathbf{K})\mathbf{x} + \mathbf{B}\mathbf{H}\mathbf{r} \tag{7-96a}$$

$$\mathbf{y} = \mathbf{C}\mathbf{x} \tag{7-96b}$$

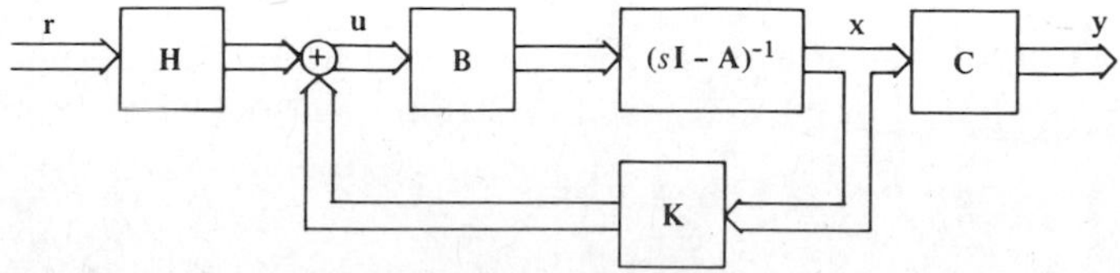

Figure 7-10 Decoupling by state feedback.

The transfer function of the state-feedback system is

$$\hat{\mathbf{G}}_f(s, \mathbf{K}, \mathbf{H}) \triangleq \mathbf{C}(s\mathbf{I} - \mathbf{A} - \mathbf{B}\mathbf{K})^{-1}\mathbf{B}\mathbf{H} \tag{7-97}$$

We shall derive in the following the condition on $\hat{\mathbf{G}}(s)$ under which the system can be decoupled by state feedback. Define the nonnegative integer d_i as

$d_i \triangleq$ min (the difference of the degree in s of the denominator and the numerator of each entry of the ith row of $\hat{\mathbf{G}}(s)$) -1

and the $1 \times p$ constant row vector $\mathbf{E}_i$ as

$$\mathbf{E}_i \triangleq \lim_{s \to \infty} s^{d_i+1}\hat{\mathbf{G}}_i(s)$$

where $\hat{\mathbf{G}}_i(s)$ is the ith row of $\hat{\mathbf{G}}(s)$.

Example 1

Consider

$$\hat{\mathbf{G}}(s) = \begin{bmatrix} \dfrac{s+2}{s^2+s+1} & \dfrac{1}{s^2+s+2} \\ \dfrac{1}{s^2+2s+1} & \dfrac{3}{s^2+s+4} \end{bmatrix}$$

The differences in degree of the first row of $\hat{\mathbf{G}}(s)$ are 1 and 2; hence $d_1 = 0$ and

$$\mathbf{E}_1 = \lim_{s \to \infty} s \begin{bmatrix} \dfrac{s+2}{s^2+s+1} & \dfrac{1}{s^2+s+2} \end{bmatrix} = [1 \quad 0]$$

The differences in degree of the second row of $\hat{\mathbf{G}}(s)$ are 2 and 2, hence $d_2 = 1$ and

$$\mathbf{E}_2 = \lim_{s \to \infty} s^2 \begin{bmatrix} \dfrac{1}{s^2+2s+1} & \dfrac{3}{s^2+s+4} \end{bmatrix} = [1 \quad 3]$$ ∎

Theorem 7-15

A system with the transfer-function matrix $\hat{\mathbf{G}}(s)$ can be decoupled by state feedback of the form $\mathbf{u} = \mathbf{K}\mathbf{x} + \mathbf{H}\mathbf{r}$ if and only if the constant matrix

$$\mathbf{E} = \begin{bmatrix} \mathbf{E}_1 \\ \mathbf{E}_2 \\ \vdots \\ \mathbf{E}_p \end{bmatrix}$$

is nonsingular. ∎

We see from this theorem that whether or not a system can be decoupled is a property of its transfer-function matrix. The dynamical-equation description comes into play a role only when the gain matrix $\mathbf{K}$ is to be found. Therefore the controllability and observability of the dynamical-equation description of the system are immaterial here. Let $\hat{\mathbf{G}}_i(s)$ and $\mathbf{C}_i$ be the ith row of $\hat{\mathbf{G}}(s)$ and $\mathbf{C}$, respectively. Then, from (7-93) and using (2-3), we have

$$\hat{\mathbf{G}}_i(s) = \mathbf{C}_i(s\mathbf{I} - \mathbf{A})^{-1}\mathbf{B} \tag{7-98}$$

In the following, we shall establish the relations between the integer d_i, the vector $\mathbf{E}_i$, and the matrices $\mathbf{C}_i$, $\mathbf{A}$, and $\mathbf{B}$. First, we expand (7-98) into (see Problem 2-39)

$$\begin{aligned}\hat{\mathbf{G}}_i(s) &= \frac{1}{\Delta(s)}\mathbf{C}_i[s^{n-1}\mathbf{I} + \mathbf{R}_1 s^{n-2} + \cdots + \mathbf{R}_{n-1}]\mathbf{B} \\ &= \frac{1}{\Delta(s)}[\mathbf{C}_i\mathbf{B}s^{n-1} + \mathbf{C}_i\mathbf{R}_1\mathbf{B}s^{n-2} + \cdots + \mathbf{C}_i\mathbf{R}_{d_i}\mathbf{B}s^{n-d_i-1} + \cdots + \mathbf{C}_i\mathbf{R}_{n-1}\mathbf{B}]\end{aligned} \tag{7-99}$$

where

$$\Delta(s) \triangleq \det(s\mathbf{I} - \mathbf{A}) \triangleq s^n + \alpha_1 s^{n-1} + \cdots + \alpha_n \tag{7-100}$$

and

$$\begin{aligned}\mathbf{R}_1 &= \mathbf{A} + \alpha_1\mathbf{I} \\ \mathbf{R}_2 &= \mathbf{A}\mathbf{R}_1 + \alpha_2\mathbf{I} = \mathbf{A}^2 + \alpha_1\mathbf{A} + \alpha_2\mathbf{I} \\ &\cdots\cdots\cdots\cdots\cdots\cdots\cdots\cdots \\ \mathbf{R}_{n-1} &= \mathbf{A}\mathbf{R}_{n-2} + \alpha_{n-1}\mathbf{I} = \mathbf{A}^{n-1} + \alpha_1\mathbf{A}^{n-2} + \cdots + \alpha_{n-1}\mathbf{I}\end{aligned} \tag{7-101}$$

Since $d_i + 1$ is the smallest difference in degree between the numerator and the denominator of each entry of $\hat{\mathbf{G}}_i(s)$, we conclude from (7-99) that the coefficients associated with $s^{n-d_i}, s^{n-d_i+1}, \ldots,$ and s^{n-1} vanish, but the coefficient matrix associated with s^{n-d_i-1} is different from zero. Hence we have

$$\mathbf{C}_i\mathbf{B} = \mathbf{0}, \quad \mathbf{C}_i\mathbf{R}_1\mathbf{B} = \mathbf{0}, \quad \ldots, \quad \mathbf{C}_i\mathbf{R}_{d_i-1}\mathbf{B} = \mathbf{0} \tag{7-102}$$

and[11]

$$\mathbf{E}_i = \mathbf{C}_i\mathbf{R}_{d_i}\mathbf{B} \neq \mathbf{0} \tag{7-103}$$

From (7-101) it is easy to see that conditions (7-102) imply that

$$\mathbf{C}_i\mathbf{B} = \mathbf{0}, \quad \mathbf{C}_i\mathbf{A}\mathbf{B} = \mathbf{0}, \quad \ldots, \quad \mathbf{C}_i\mathbf{A}^{d_i-1}\mathbf{B} = \mathbf{0} \tag{7-104}$$

and Equation (7-103) becomes

$$\mathbf{E}_i = \mathbf{C}_i\mathbf{A}^{d_i}\mathbf{B} \neq \mathbf{0} \tag{7-105}$$

Therefore, d_i can also be defined from a dynamical equation as the smallest integer such that $\mathbf{C}_i\mathbf{A}^{d_i}\mathbf{B} \neq \mathbf{0}$. If $\mathbf{C}_i\mathbf{A}^k\mathbf{B} = \mathbf{0}$, for all $k \leq n$, then $d_i \triangleq n - 1$. (See Problem 7-28.)

Similarly, we may define $\bar{d}_i$ and $\bar{\mathbf{E}}_i$ for the transfer-function matrix $\hat{\mathbf{G}}_f$ in (7-97). If we expand $\hat{\mathbf{G}}_{fi}$ into

$$\hat{\mathbf{G}}_{fi} = \frac{1}{\bar{\Delta}(s)}[\mathbf{C}_i\mathbf{B}\mathbf{H}s^{n-1} + \mathbf{C}_i\bar{\mathbf{R}}_1\mathbf{B}\mathbf{H}s^{n-2} + \cdots + \mathbf{C}_i\bar{\mathbf{R}}_{n-1}\mathbf{B}\mathbf{H}] \tag{7-106}$$

where

$$\bar{\Delta}(s) \triangleq \det(s\mathbf{I} - \mathbf{A} - \mathbf{B}\mathbf{K}) \triangleq s^n + \bar{\alpha}_1 s^{n-1} + \cdots + \bar{\alpha}_n \tag{7-107}$$

---

[11] $\mathbf{E}_i$ and $\mathbf{C}_i\mathbf{R}_{di}\mathbf{B}$ may differ by a constant because $\Delta(s)$ is factored out in (7-99). For convenience, this difference, if any, is neglected.

and
$$\begin{aligned}\bar{\mathbf{R}}_1 &= (\mathbf{A}+\mathbf{BK})+\bar{\alpha}_1\mathbf{I}\\ \bar{\mathbf{R}}_2 &= (\mathbf{A}+\mathbf{BK})^2+\bar{\alpha}_1(\mathbf{A}+\mathbf{BK})+\bar{\alpha}_2\mathbf{I}\\ &\cdots\cdots\cdots\cdots\cdots\cdots\cdots\cdots\cdots\cdots\\ \bar{\mathbf{R}}_{n-1} &= (\mathbf{A}+\mathbf{BK})^{n-1}+\bar{\alpha}_1(\mathbf{A}+\mathbf{BK})^{n-2}+\cdots+\bar{\alpha}_{n-1}\mathbf{I}\end{aligned} \tag{7-108}$$

then

$$\mathbf{CBH}=\mathbf{0},\quad \mathbf{C}(\mathbf{A}+\mathbf{BK})\mathbf{BH}=\mathbf{0},\quad \ldots,\quad \mathbf{C}(\mathbf{A}+\mathbf{BK})^{\bar{d}_i-1}\mathbf{BH}=\mathbf{0} \tag{7-109}$$

and
$$\bar{\mathbf{E}}=\mathbf{C}(\mathbf{A}+\mathbf{BK})^{\bar{d}_i}\mathbf{BH}\neq\mathbf{0} \tag{7-110}$$

We establish in the following the relationships between d_i and $\bar{d}_i$ and between $\mathbf{E}_i$ and $\bar{\mathbf{E}}_i$.

Theorem 7-16

For any $\mathbf{K}$ and any nonsingular $\mathbf{H}$, we have $\bar{d}_i=d_i$ and $\bar{\mathbf{E}}_i=\mathbf{E}_i\mathbf{H}$.

Proof

It is easy to verify that, for each i, the conditions $\mathbf{C}_i\mathbf{B}=\mathbf{0}$, $\mathbf{C}_i\mathbf{AB}=\mathbf{0}$, . . . , $\mathbf{C}_i\mathbf{A}^{d_i-1}\mathbf{B}=\mathbf{0}$ imply that

$$\mathbf{C}_i(\mathbf{A}+\mathbf{BK})^k=\mathbf{C}_i\mathbf{A}^k \qquad k=0,1,2,\ldots,d_i \tag{7-111}$$

and
$$\mathbf{C}_i(\mathbf{A}+\mathbf{BK})^k=\mathbf{C}_i\mathbf{A}^{d_i}(\mathbf{A}+\mathbf{BK})^{k-d_i} \qquad k=d_i+1,d_i+2,\ldots \tag{7-112}$$

Consequently, we have

$$\mathbf{C}_i(\mathbf{A}+\mathbf{BK})^k\mathbf{BH}=\mathbf{0} \qquad k=0,1,\ldots,d_i-1$$

and
$$\mathbf{C}_i(\mathbf{A}+\mathbf{BK})^{d_i}\mathbf{BH}=\mathbf{C}_i\mathbf{A}^{d_i}\mathbf{BH}=\mathbf{E}_i\mathbf{H}$$

Since $\mathbf{H}$ is nonsingular by assumption, if $\mathbf{E}_i$ is nonzero, so is $\mathbf{E}_i\mathbf{H}$. Therefore we conclude that $\bar{d}_i=d_i$ and $\bar{\mathbf{E}}_i=\mathbf{E}_i\mathbf{H}$. Q.E.D.

With these preliminaries, we are ready to prove Theorem 7-15.

Proof of Theorem 7-15

Necessity: Suppose that there exists $\mathbf{K}$ and $\mathbf{H}$ such that $\hat{\mathbf{G}}_f(s,\mathbf{K},\mathbf{H})$ is diagonal and nonsingular. Then

$$\bar{\mathbf{E}}=\begin{bmatrix}\bar{\mathbf{E}}_1\\ \bar{\mathbf{E}}_2\\ \vdots\\ \bar{\mathbf{E}}_p\end{bmatrix}$$

is a diagonal constant matrix and nonsingular. Since $\bar{\mathbf{E}}=\mathbf{EH}$ and since $\mathbf{H}$ is nonsingular by assumption, we conclude that $\mathbf{E}$ is nonsingular.

Sufficiency: If the matrix $\mathbf{E}$ is nonsingular, then the system can be decoupled Define

$$\mathbf{F}\triangleq\begin{bmatrix}\mathbf{C}_1\mathbf{A}^{d_1+1}\\ \mathbf{C}_2\mathbf{A}^{d_2+1}\\ \vdots\\ \mathbf{C}_p\mathbf{A}^{d_p+1}\end{bmatrix} \tag{7-113}$$

We show that if $\mathbf{K} = -\mathbf{E}^{-1}\mathbf{F}$ and $\mathbf{H} = \mathbf{E}^{-1}$, then the system can be decoupled and the transfer function of the decoupled system is

$$\hat{\mathbf{G}}_f(s, -\mathbf{E}^{-1}\mathbf{F}, \mathbf{E}^{-1}) = \begin{bmatrix} 1/s^{(d_1+1)} & 0 & \cdots & 0 \\ 0 & 1/s^{(d_2+1)} & \cdots & 0 \\ \vdots & \vdots & & \vdots \\ 0 & 0 & \cdots & 1/s^{(d_p+1)} \end{bmatrix}$$

or, equivalently,

$$\mathbf{C}_i(s\mathbf{I} - \mathbf{A} - \mathbf{BK})^{-1}\mathbf{BH} = [1/s^{(d_i+1)}]\mathbf{e}_i \tag{7-114}$$

where $\mathbf{e}_i$ is a row vector with 1 in the ith place and zeros elsewhere. First we show that $\mathbf{C}_i(\mathbf{A} + \mathbf{BK})^{d_i+1} = \mathbf{0}$. From (7-105), (7-112) and (7-113) and using $\mathbf{K} = -\mathbf{E}^{-1}\mathbf{F}$ and $\mathbf{E}_i\mathbf{E}^{-1} = \mathbf{e}_i$, we obtain

$$\begin{aligned} \mathbf{C}_i(\mathbf{A} + \mathbf{BK})^{d_i+1} &= \mathbf{C}_i\mathbf{A}^{d_i}(\mathbf{A} + \mathbf{BK}) = \mathbf{C}_i\mathbf{A}^{d_i+1} + \mathbf{C}_i\mathbf{A}^{d_i}\mathbf{BK} \\ &= \mathbf{F}_i - \mathbf{E}_i\mathbf{E}^{-1}\mathbf{F} = \mathbf{F}_i - \mathbf{e}_i\mathbf{F} = \mathbf{0} \end{aligned}$$

where $\mathbf{F}_i$ and $\mathbf{E}_i$ are the ith row of $\mathbf{F}$ and $\mathbf{E}$, respectively. Hence we conclude that

$$\mathbf{C}_i(\mathbf{A} + \mathbf{BK})^{d_i+k} = \mathbf{0} \qquad \text{for any positive integer } k \tag{7-115}$$

Since $\bar{d}_i = d_i$, Equation (7-106) reduces to

$$\begin{aligned} \mathbf{C}_i(s\mathbf{I} - \mathbf{A} - \mathbf{BK})^{-1}\mathbf{BH} = \frac{1}{\bar{\Delta}(s)}[\mathbf{C}_i\bar{\mathbf{R}}_{d_i}\mathbf{B}s^{n-d_i-1} + \mathbf{C}_i\bar{\mathbf{R}}_{d_i+1}\mathbf{B}s^{n-d_i-2} + \cdots \\ + \mathbf{C}_i\bar{\mathbf{R}}_{n-1}\mathbf{B}]\mathbf{H} \end{aligned} \tag{7-116}$$

Now, from (7-105), (7-108), (7-112), (7-115), and the fact $\mathbf{C}_i(\mathbf{A} + \mathbf{BK})^k\mathbf{B} = \mathbf{0}$, for $k = 0, 1, \ldots, \bar{d}_i - 1$, which follows from (7-109) and the nonsingularity of $\mathbf{H}$, it is straightforward to verify that

$$\begin{aligned} \mathbf{C}_i\bar{\mathbf{R}}_{d_i}\mathbf{B} &= \mathbf{C}_i(\mathbf{A} + \mathbf{BK})^{d_i}\mathbf{B} = \mathbf{C}_i\mathbf{A}^{d_i}\mathbf{B} = \mathbf{E}_i \\ \mathbf{C}_i\bar{\mathbf{R}}_{d_i+1}\mathbf{B} &= \mathbf{C}_i[(\mathbf{A} + \mathbf{BK})^{d_i+1} + \bar{\alpha}_1(\mathbf{A} + \mathbf{BK})^{d_i}]\mathbf{B} = \bar{\alpha}_1\mathbf{E}_i \\ &\cdots\cdots\cdots \\ \mathbf{C}_i\bar{\mathbf{R}}_{n-1}\mathbf{B} &= \bar{\alpha}_{n-1-d_i}\mathbf{E}_i \end{aligned}$$

Consequently, Equation (7-116) becomes

$$\mathbf{C}_i(s\mathbf{I} - \mathbf{A} - \mathbf{BK})^{-1}\mathbf{BH} = \frac{1}{\bar{\Delta}(s)}[s^{n-d_i-1} + \bar{\alpha}_1 s^{n-d_i-2} + \cdots + \bar{\alpha}_{n-1-d_i}]\mathbf{E}_i\mathbf{H} \tag{7-117}$$

What is left to be shown is that

$$\begin{aligned} \bar{\Delta}(s) &= s^n + \bar{\alpha}_1 s^{n-1} + \bar{\alpha}_2 s^{n-2} + \cdots + \bar{\alpha}_n \\ &= s^{d_i+1}(s^{n-d_i-1} + \bar{\alpha}_1 s^{n-d_i-2} + \cdots + \bar{\alpha}_{n-d_i-1}) \end{aligned} \tag{7-118}$$

From the Cayley-Hamilton theorem, we have

$$(\mathbf{A} + \mathbf{BK})^n + \bar{\alpha}_1(\mathbf{A} + \mathbf{BK})^{n-1} + \cdots + \bar{\alpha}_n\mathbf{I} = \mathbf{0} \tag{7-119}$$

By multiplying $\mathbf{C}_i(\mathbf{A} + \mathbf{BK})^{d_i}$ to (7-119) and using (7-115), we can obtain $\bar{\alpha}_n\mathbf{C}_i(\mathbf{A} + \mathbf{BK})^{d_i} = \mathbf{0}$ which implies that $\bar{\alpha}_n = 0$. Next by multiplying $\mathbf{C}_i(\mathbf{A} + \mathbf{BK})^{d_i-1}$, we can show that $\bar{\alpha}_{n-1} = 0$. Proceeding similarly, we can prove Equation (7-118). By substituting (7-118) into (7-117) and using $\mathbf{E}_i\mathbf{H} = \mathbf{E}_i\mathbf{E}^{-1} =$

$\mathbf{e}_i$, we immediately obtain Equation (7-114). Consequently, the system can be decoupled by using $\mathbf{K} = -\mathbf{E}^{-1}\mathbf{F}$ and $\mathbf{H} = \mathbf{E}^{-1}$. Q.E.D.

Although a system can be decoupled by using $\mathbf{K} = -\mathbf{E}^{-1}\mathbf{F}$ and $\mathbf{H} = \mathbf{E}^{-1}$, the resulting system is not satisfactory because all the poles of the decoupled system are at the origin. However, we may introduce additional state feedback to move these poles to the desired location.

If the dynamical equation is controllable, then we can control all the eigenvalues by state feedback. Now, if in addition we like to decouple the system, the number of eigenvalues that can be controlled by state feedback will be reduced. For a complete discussion of this problem and others, such as how to compute **K** and **H** to have a decoupled system with desired poles, see References 36, 41, 112, S218, and S227.

7-7 Concluding Remarks

In this chapter we studied the practical implications of controllability and observability. We showed that if a linear time-invariant dynamical equation is controllable, we can, by introducing state feedback, arbitrarily assign the eigenvalues of the resulting dynamical equation. This was achieved by the use of controllable canonical-form dynamical equations. When we introduce state feedback, all the state variables must be available. If they are not, an asymptotic state estimator has to be constructed. If a dynamical equation is observable, a state estimator with a set of arbitrary eigenvalues can be constructed. The construction of state estimators is achieved by the use of observable canonical-form dynamical equation or by the solution of the equation $\mathbf{TA} - \mathbf{FT} = \mathbf{GC}$.

We studied in this chapter only the time-invariant case. The results in Section 7-2, 7-3, and 7-4 can be extended to the linear time-varying case if the equation is instantaneously controllable since, under the instantaneous controllability assumption, the controllability matrix of the linear time-varying dynamical equation has a full rank for all t. Hence by using it in the equivalence transformation, a controllable canonical-form dynamical equation can be obtained. Consequently the state feedback can be introduced. Similar remarks apply to the observability part.

We also studied the decoupling of a transfer function matrix by introducing state feedback. The concepts of controllability and observability are not required in this part of study. It was discussed in this chapter because the decoupling also used the state feedback as in Section 7-3.

The combination of state feedback and state estimator will again be studied in Section 9-5 directly from the transfer-function matrix. Comparisons between the state-variable approach and the transfer-function approach will also be discussed there.

The introduction of constant gain feedback from the output is called the *constant gain output feedback*. Unlike the state feedback, the output feedback cannot arbitrarily assign all the eigenvalues of the resulting equation. Roughly speaking, the number of eigenvalues which a constant gain output feedback can assign is equal to min $\{n, p+q-1\}$, where $p = \text{rank } \mathbf{B}$ and $q = \text{rank } \mathbf{C}$. In this assignment, no repeated eigenvalues are permitted. Furthermore, the assignment cannot be exact but only arbitrarily close. See References S31, S71, S129, S130, and S131. See also Reference S165.

To conclude this chapter, we mention that all the results in this chapter can be applied directly, without any modification, to the discrete-time case. We mention that the discrete-time state estimator should be modeled as

$$\hat{\mathbf{x}}(k+1) = \mathbf{A}\hat{\mathbf{x}}(k) + \mathbf{B}\mathbf{u}(k) + \mathbf{L}(\mathbf{y}(k) - \mathbf{C}\mathbf{x}(k))$$

In other words, the estimated $\hat{\mathbf{x}}(k+1)$ is reconstructed from $\{\mathbf{y}(0), \mathbf{y}(1), \ldots, \mathbf{y}(k)\}$. If $\hat{\mathbf{x}}(k+1)$ is reconstructed from $\{\mathbf{y}(0), \mathbf{y}(1), \ldots, \mathbf{y}(k), \mathbf{y}(k+1)\}$ and modeled as

$$\hat{\mathbf{x}}(k+1) = \mathbf{A}\hat{\mathbf{x}}(k) + \mathbf{B}\mathbf{u}(k) + \mathbf{L}(\mathbf{y}(k+1) - \mathbf{C}\mathbf{x}(k+1))$$

then the situation will be different. The reader is referred to Reference S215.

Problems

7-1 The equivalence transformation $\mathbf{Q}$ of $\bar{\mathbf{x}} = \mathbf{P}\mathbf{x} = \mathbf{Q}^{-1}\mathbf{x}$ in Theorem 7-1 can be obtained either by using $\mathbf{Q} = [\mathbf{q}_1 \quad \mathbf{q}_2 \quad \cdots \quad \mathbf{q}_n]$, where the $\mathbf{q}_i$'s are given in (7-9), or by using $\mathbf{Q} = \mathbf{U}\bar{\mathbf{U}}^{-1}$, where $\mathbf{U}$ is the controllability matrix of FE_1 and $\bar{\mathbf{U}}^{-1}$ is given in (7-13). From a computational point of view, which one is easier to use?

7-2 Consider the controllable dynamical equation

$$FE_1: \quad \begin{aligned} \dot{\mathbf{x}} &= \mathbf{A}\mathbf{x} + \mathbf{b}u \\ y &= \mathbf{c}\mathbf{x} \end{aligned}$$

If the vectors $\{\mathbf{b}, \mathbf{A}\mathbf{b}, \ldots, \mathbf{A}^{n-1}\mathbf{b}\}$ are used as basis vectors, what is its equivalent dynamical equation? Does the new $\bar{\mathbf{c}}$ bear a simple relation to the coefficients of the transfer function of FE_1 as it does in the controllable canonical form? (*Hint*: See Problem 2-15.)

7-3 Find the matrix $\mathbf{Q}$ which transforms (7-17) into the observable canonical-form dynamical equation.

7-4 Transform the equation

$$\dot{\mathbf{x}} = \begin{bmatrix} -1 & -2 & -2 \\ 0 & -1 & 1 \\ 1 & 0 & -1 \end{bmatrix} \mathbf{x} + \begin{bmatrix} 2 \\ 0 \\ 1 \end{bmatrix} u$$

$$y = [1 \quad 1 \quad 0]\mathbf{x}$$

into the controllable canonical-form dynamical equation. What is its transfer function?

7-5 Let μ_1 be the largest integer such that $\{\mathbf{b}_1, \mathbf{A}\mathbf{b}_1, \ldots, \mathbf{A}^{\mu_1-1}\mathbf{b}_1\}$ is a linearly independent set. Let μ_2 be the largest integer such that $\{\mathbf{b}_1, \mathbf{A}\mathbf{b}_1, \ldots, \mathbf{A}^{\mu_1-1}\mathbf{b}_1, \mathbf{b}_2, \mathbf{A}\mathbf{b}_2, \ldots, \mathbf{A}^{\mu_2-1}\mathbf{b}_2\}$ is a linearly independent set. Show that $\mathbf{A}^n\mathbf{b}_2$ for all $n \geq \mu_2$ is linearly dependent on $\{\mathbf{b}_1, \ldots, \mathbf{A}^{\mu_1-1}\mathbf{b}_1, \mathbf{b}_2, \ldots, \mathbf{A}^{\mu_2-1}\mathbf{b}_2\}$.

7-6 Transform the dynamical equation in Problem 7-4 into the observable canonical-form dynamical equation.

7-7 Consider the state feedback and the output feedback systems shown in Figure P7-7. Show that for any constant matrix **H**, there exists a constant matrix **K** such that $\mathbf{Kx}=\mathbf{HCx}$. Under what condition on **C** will there exist a matrix **H** such that $\mathbf{K}=\mathbf{HC}$ for any **K**? (**C**, **K** and **H** are $q\times n$, $p\times n$, *and* $p\times q$ constant matrices, respectively. It is generally assumed that $n\geq q$.) (*Answer*: A solution **H** exists in $\mathbf{K}=\mathbf{HC}$ for a given **K** if and only if $\rho\begin{bmatrix}\mathbf{C}\\ \mathbf{K}\end{bmatrix}=\rho\mathbf{C}$. Consequently, a solution **H** exists in $\mathbf{K}=\mathbf{HC}$ for *any* **K** if and only if **C** is square and nonsingular.)

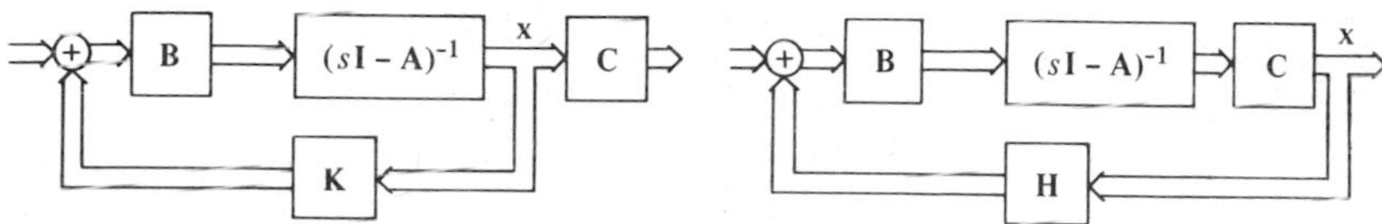

Figure P7-7

7-8 Let

$$\mathbf{A}=\begin{bmatrix}1 & 1\\ 0 & 1\end{bmatrix}\qquad \mathbf{b}=\begin{bmatrix}0\\ 1\end{bmatrix}\qquad \mathbf{c}=[1\quad 0]$$

Find the values of $\mathbf{k}=[k_1\quad k_2]$ such that the matrix $(\mathbf{A}+\mathbf{bk})$ has repeated eigenvalues. Find the values of h such that the matrix $(\mathbf{A}+\mathbf{b}h\mathbf{c})$ has repeated eigenvalues. Can you conclude that for almost all **k** and h, the two matrices have distinct eigenvalues?

7-9 Use state feedback to transfer the eigenvalues of the dynamical equation in Problem 7-4 to -1, -2, and -2. Draw a block diagram for the dynamical equation in Problem 7-4 and then add the required state feedback.

7-10 Show that

$$\rho[\mathbf{b}\quad \mathbf{Ab}\quad \cdots\quad \mathbf{A}^{n-1}\mathbf{b}]=\rho[\mathbf{b}\quad (\mathbf{A}-\mathbf{bk})\mathbf{b}\quad \cdots\quad (\mathbf{A}-\mathbf{bk})^{n-1}\mathbf{b}]$$

for any $1\times n$ constant vector **k**.

7-11 Consider the Jordan-form dynamical equation

$$\dot{\mathbf{x}}=\left[\begin{array}{cc|ccc}-2 & 1 & & & \\ 0 & -2 & & & \\ \hline & & 1 & 1 & 0\\ & & 0 & 1 & 1\\ & & 0 & 0 & 1\end{array}\right]\mathbf{x}+\left[\begin{array}{c}1\\ 0\\ \hline 0\\ 1\\ 1\end{array}\right]u$$

which has an unstable eigenvalue 1. The dynamical equation is not controllable, but the subequation associated with eigenvalue 1 is controllable. Do you think it is possible to stabilize the equation by using state feedback? If yes, find the gain vector **k** such that the closed-loop equation has eigenvalues -1, -1, -2, -2, and -2.

7-12 Given

$$\dot{\mathbf{x}}=\begin{bmatrix}2 & 1\\ -1 & 1\end{bmatrix}\mathbf{x}+\begin{bmatrix}1\\ 2\end{bmatrix}u$$

Find the gain vector $[k_1 \quad k_2]$ such that the state-feedback system has -1 and -2 as its eigenvalues. Compute k_1, k_2 directly without using any equivalence transformation.

7-13 Consider the uncontrollable state equation

$$\dot{\mathbf{x}} = \begin{bmatrix} 2 & 1 & 0 & 0 \\ 0 & 2 & 0 & 0 \\ 0 & 0 & -1 & 0 \\ 0 & 0 & 0 & -1 \end{bmatrix} \mathbf{x} + \begin{bmatrix} 0 \\ 1 \\ 1 \\ 1 \end{bmatrix} u$$

Is it possible to find a gain vector $\mathbf{k}$ such that the equation with state feedback $u = \mathbf{kx} + r$ has eigenvalues $-2, -2, -1, -1$? Is it possible to have eigenvalues $-2, -2, -2, -1$? How about $-2, -2, -2, -2$? (*Answers:* Yes; yes; no.)

7-14 The observability of a dynamical equation is not invariant under any state feedback. Where does the argument fail if the argument used in the proof of Theorem 7-3 is used to prove the observability part?

7-15 Find a three-dimensional state estimator with eigenvalues -2, -2, and -3 for the dynamical equation in Problem 7-4. Use two different methods.

7-16 Find a two-dimensional state estimator with eigenvalues -2 and -3 for the dynamical equation in Problem 7-4. Use two different methods.

7-17 Consider a system with the transfer function

$$\frac{(s-1)(s+2)}{(s+1)(s-2)(s+3)}$$

Is it possible to change the transfer function to

$$\frac{s-1}{(s+2)(s+3)}$$

by state feedback? If yes, how?

7-18 In Problem 7-17, if the state is not available for feedback, can you carry out the design by using a state estimator? Choose the eigenvalues of the estimator as -4 and -4. What is its overall transfer function?

7-19 Prove Theorem 7-11.

7-20 Consider the matrix in Problem 2-30. Show that the eigenvalues of the matrix are $\lambda_1, \lambda_2, \ldots, \lambda_{n_1-1}, \alpha \pm j\beta, \gamma_1, \gamma_2, \ldots, \gamma_{n_2-1}$ if a_{ij} are chosen as

$$\Delta_{11}(s) = s^{n_1} + a_{111}s^{n_1-1} + \cdots + a_{11n_1} = (s-\alpha)\prod_{i=1}^{n_1-1}(s-\lambda_i)$$

$$\Delta_{21}(s) = a_{211}s^{n_1-1} + \cdots + a_{21n_1} = -\beta\prod_{i=1}^{n_1-1}(s-\lambda_i)$$

$$\Delta_{22}(s) = s^{n_2} + a_{221}s^{n_2-1} + \cdots + a_{22n_2} = (s-\alpha)\prod_{i=1}^{n_2-1}(s-\gamma_i)$$

$$\Delta_{12}(s) = a_{121}s^{n_2-1} + \cdots + a_{12n_2} = \beta\prod_{i=1}^{n_2-1}(s-\gamma_i)$$

7-21 Verify that the matrix in (7-50) has eigenvalues $-3 \pm 3j$, $-3 \pm j$ and $-4 \pm 2j$.

7-22 Let

$$\mathbf{A} = \begin{bmatrix} 0 & 1 & 0 & 0 \\ 0 & 0 & 1 & 0 \\ -3 & 1 & 2 & 3 \\ 2 & 1 & 0 & 0 \end{bmatrix} \qquad \mathbf{B} = \begin{bmatrix} 0 & 0 \\ 0 & 0 \\ 1 & 2 \\ 0 & 2 \end{bmatrix}$$

Find two different real constant 2×4 matrices $\mathbf{K}$ such that the matrix $(\mathbf{A} + \mathbf{BK})$ has eigenvalues $-4 \pm 3j$ and $-5 \pm 4j$.

7-23 Show that the state estimator in Theorem 7-8 meets Theorem 7-9.

7-24 Show that the $(n-q)$-dimensional state estimator in (7-72) and (7-74) meets Theorem 7-9.

7-25 Establish the separation property by using a full-dimensional state estimator.

7-26 Consider the dynamical equation

$$\dot{\mathbf{x}} = \begin{bmatrix} 0 & 0 & 1 & 0 & 0 \\ 1 & 0 & 0 & 0 & 0 \\ 0 & 1 & 0 & 1 & -1 \\ 0 & 1 & 1 & 1 & 0 \\ 0 & 0 & 1 & 0 & 0 \end{bmatrix} \mathbf{x} + \begin{bmatrix} 1 \\ 2 \\ 1 \\ 0 \\ 1 \end{bmatrix} u$$

$$\mathbf{y} = \begin{bmatrix} 0 & 0 & 0 & 1 & 0 \\ 0 & 0 & 0 & 0 & 1 \end{bmatrix} \mathbf{x}$$

Find a full-dimensional estimator, a reduced-dimensional estimator, and a functional estimator for $\mathbf{kx}(t)$ with

$$\mathbf{k} = [0 \quad 1 \quad 1 \quad 2 \quad -1]$$

7-27 Prove Theorem 7-11. Let v_i be the observability indices of $\{\mathbf{A}, \mathbf{C}\}$. Show that the observability indices of $\{\bar{\mathbf{A}}_{22}, \bar{\mathbf{A}}_{12}\}$ are $v_i - 1$. (*Hint:* Compute

$$\begin{bmatrix} \mathbf{I} & & & \\ & \mathbf{I} & & \\ & & \mathbf{I} & \\ & & -\bar{\mathbf{A}}_{11} & \mathbf{I} \end{bmatrix} \begin{bmatrix} \mathbf{I} & & & \\ & \mathbf{I} & & \\ & -\bar{\mathbf{A}}_{11} & \mathbf{I} & \\ & & -\bar{\mathbf{A}}_{11} & \mathbf{I} \end{bmatrix} \begin{bmatrix} \bar{\mathbf{C}} \\ \bar{\mathbf{C}}\bar{\mathbf{A}} \\ \bar{\mathbf{C}}\bar{\mathbf{A}}^2 \\ \bar{\mathbf{C}}\bar{\mathbf{A}}^3 \end{bmatrix} = \left[\begin{array}{c|c} \mathbf{I}_q & \mathbf{0} \\ \hline x & \bar{\mathbf{A}}_{12} \\ x & \bar{\mathbf{A}}_{12}\bar{\mathbf{A}}_{22} \\ x & \bar{\mathbf{A}}_{12}\bar{\mathbf{A}}_{22}^2 \end{array}\right]$$

and note that if linearly independent rows of the observability matrices are searched in order from top to bottom, the locations of the linearly independent rows of both observability matrices are identical.)

7-28 Consider the irreducible single-variable dynamical equation

$$\dot{\mathbf{x}} = \mathbf{Ax} + \mathbf{b}u \qquad y = \mathbf{cx}$$

where $\mathbf{A}$ is an $n \times n$ matrix, $\mathbf{b}$ is an $n \times 1$ vector, and $\mathbf{c}$ is a $1 \times n$ vector. Let $\hat{g}$ be its transfer function. Show that $\hat{g}(s)$ has m zeros—in other words, that the numerator of $\hat{g}(s)$ has degree m—if and only if

$$\mathbf{cA}^i\mathbf{b}=\mathbf{0} \qquad \text{for } i=0, 1, 2, \ldots, n-m-2$$

and $\mathbf{cA}^{n-m-1}\mathbf{b}\neq\mathbf{0}$. Or equivalently, the difference between the degree of the denominator and the degree of the numerator of $\hat{g}(s)$ is $\bar{d}=n-m$ if and only if

$$\mathbf{cA}^{\bar{d}-1}\mathbf{b}\neq\mathbf{0} \qquad \mathbf{cA}^i\mathbf{b}=\mathbf{0} \qquad \text{for } i=0, 1, 2, \ldots, \bar{d}-2$$

(*Hint*: Show that $\mathbf{cA}^i\mathbf{b}$ is invariant under any equivalence transformation and then use the controllable canonical-form dynamical equation.)

7-29 a. Can a multivariable system with the transfer function

$$\begin{bmatrix} \dfrac{1}{s^2+1} & \dfrac{2}{s^2+1} \\ \dfrac{2s+1}{s^3+s+1} & \dfrac{1}{s} \end{bmatrix}$$

be decoupled by state feedback?

b. Can a system with the dynamical-equation description

$$\dot{\mathbf{x}}=\begin{bmatrix} 3 & 1 & 0 \\ 0 & 0 & -1 \\ 0 & 1 & -1 \end{bmatrix}\mathbf{x}+\begin{bmatrix} 0 & 0 \\ 1 & 0 \\ 0 & 1 \end{bmatrix}u$$

$$\mathbf{y}=\begin{bmatrix} 2 & -1 & 1 \\ 0 & 2 & 1 \end{bmatrix}\mathbf{x}$$

be decoupled by state feedback?

7-30 Consider the linear time-invariant discrete-time equation

$$\mathbf{x}(k+1)=\begin{bmatrix} -1 & -2 & -2 \\ 0 & -1 & 1 \\ 1 & 0 & -1 \end{bmatrix}\mathbf{x}(k)+\begin{bmatrix} 2 \\ 0 \\ 1 \end{bmatrix}u(k)$$

$$\mathbf{y}(k)=[\,1 \quad 1 \quad 0\,]\mathbf{x}(k)$$

Find a feedback gain $\mathbf{k}$ so that all the eigenvalues of the resulting equation are zeros. Show that, for any initial state, the zero-input response of the equation becomes identically zero for $k\geq 3$. This is called a *dead beat* control. See Problem 2-50.

7-31 Establish a counterpart of Theorem 7-9 for the discrete-time case.

7-32 Show that the controllability and observability indices of $\{\mathbf{A}+\mathbf{BHC}, \mathbf{B}, \mathbf{C}\}$ are the same as those of $\{\mathbf{A}, \mathbf{B}, \mathbf{C}\}$ for any $\mathbf{H}$. Consequently, the controllability and observability indices are invariant under any constant gain output feedback.

7-33 Given

$$\mathbf{A}=\begin{bmatrix} 2 & 1 & 0 \\ 0 & 1 & 0 \\ 1 & 0 & 1 \end{bmatrix} \qquad \mathbf{b}=\begin{bmatrix} 0 \\ 1 \\ 0 \end{bmatrix}$$

Find a **k** so that (**A** + **bk**) is similar to

$$\mathbf{F} = \begin{bmatrix} -1 & 0 & 0 \\ 0 & -2 & 0 \\ 0 & 0 & -3 \end{bmatrix}$$

Hint: Use the method on page 347. Choose $\bar{\mathbf{k}} = [6 \quad 12 \quad 20]$, then

$$\mathbf{T} = \begin{bmatrix} 1 & 1 & 1 \\ -3 & -4 & -5 \\ -1/2 & -1/3 & -1/4 \end{bmatrix}$$

8

Stability of Linear Systems

8-1 Introduction

Controllability and observability, introduced in Chapter 5, are two important qualitative properties of linear systems. In this chapter we shall introduce another qualitative property of systems—namely, stability. The concept of stability is extremely important, because almost every workable system is designed to be stable. If a system is not stable, it is usually of no use in practice.

We have introduced the input-output description and the dynamical-equation description of systems; hence it is natural to study stability in terms of these two descriptions separately. In Section 8-2 the bounded-input–bounded-output (BIBO) stability of systems is introduced in terms of the input-output description. We show that if the impulse response of a system is absolutely integrable, then the system is BIBO stable. The stability condition in terms of rational transfer functions is also given in this section. In Section 8-3 we introduce the Routh-Hurwitz criterion which can be used to check whether or not all the roots of a polynomial have negative real parts. We study in Section 8-4 the stability of a system in terms of the state-variable description. We introduce there the concepts of equilibrium state, stability in the sense of Lyapunov, asymptotic stability, and total stability. The relationships between these concepts are established. In Section 8-5, we introduce a Lyapunov theorem and then use it to establish the Routh-Hurwitz criterion. In the last section, the stability of linear discrete-time systems is discussed.

The references for this chapter are 58, 59, 65, 76, 90, 102, 116, S79, S111, S135, and S153.

8-2 Stability Criteria in Terms of the Input-Output Description

Time-varying case. We consider first single-variable systems, that is, systems with only one input terminal and only one output terminal. It was shown in Chapter 3 that if the input and output of an initially relaxed system satisfy the homogeneity and additivity properties, and if the system is causal, then the input u and the output y of the system can be related by

$$y(t) = \int_{-\infty}^{t} g(t, \tau)u(\tau)\, d\tau \qquad \text{for all } t \text{ in } (-\infty, \infty) \tag{8-1}$$

where $g(t, \tau)$ is the impulse response of the system and is, by definition, the output measured at time t due to an impulse function input applied at time τ.

In the qualitative study of a system from the input and output terminals, perhaps the only question which can be asked is that if the input has certain properties, under what condition will the output have the same properties? For example, if the input u is bounded, that is,

$$|u(t)| \leq k_1 < \infty \qquad \text{for all } t \text{ in } (-\infty, \infty) \tag{8-2}$$

under what condition on the system does there exist a constant k_2 such that the output y satisfies

$$|y(t)| \leq k_2 < \infty$$

for all t in $(-\infty, \infty)$? If the input is of finite energy, that is,

$$\left(\int_{-\infty}^{\infty} |u(t)|^2\, dt \right)^{1/2} \leq k_3 < \infty \tag{8-3}$$

does there exist a constant k_4 such that

$$\left(\int_{-\infty}^{\infty} |y(t)|^2\, dt \right)^{1/2} \leq k_4 < \infty\,?$$

If the input approaches a periodic function, under what condition will the output approach another periodic function with the same period? If the input approaches a constant, will the output approach some constant? According to these various properties, we may introduce different stability definitions for the system. We shall introduce here only the one which is most commonly used in linear systems: the bounded-input–bounded-output stability.

Recall that the input-output description of a system is applicable only when the system is initially relaxed. A system that is initially relaxed is called a *relaxed system.* Hence the stability that is defined in terms of the input-output description is applicable only to relaxed systems.

Definition 8-1

A relaxed system is said to be BIBO (bounded-input–bounded-output) stable if and only if for any bounded input, the output is bounded. ∎

We illustrate the importance of the qualification "relaxedness" by showing that a system, being BIBO stable under the relaxedness assumption, may not be BIBO stable if it is not initially relaxed.

Example 1

Consider the network shown in Figure 8-1. If the system is initially relaxed, that is, the initial voltage across the capacitor is zero, then $y(t)=u(t)/2$ for all t. Therefore, for any bounded input, the output is also bounded. However, if the initial voltage across the capacitor is not zero, because of the negative capacitance, the output will increase to infinity even if no input is applied. ∎

Theorem 8-1

A relaxed single-variable system that is described by

$$y(t)=\int_{-\infty}^{t} g(t, \tau)u(\tau)\, d\tau$$

is BIBO stable if and only if there exists a finite number k such that

$$\int_{-\infty}^{t} |g(t, \tau)|\, d\tau \leq k < \infty$$

for all t in $(-\infty, \infty)$.

Proof

Sufficiency: Let u be an arbitrary input and let $|u(t)| \leq k_1$ for all t in $(-\infty, \infty)$. Then

$$|y(t)|=\left|\int_{-\infty}^{t} g(t, \tau)u(\tau)\, d\tau\right| \leq \int_{-\infty}^{t} |g(t, \tau)|\,|u(\tau)|\, d\tau \leq k_1 \int_{-\infty}^{t} |g(t, \tau)|\, d\tau \leq kk_1$$

for all t in $(-\infty, \infty)$. *Necessity:* A rigorous proof of this part is rather involved. We shall exhibit the basic idea by showing that if

$$\int_{-\infty}^{t} |g(t, \tau)|\, d\tau=\infty$$

for some t, say t_1, then we can find a bounded input that excites an unbounded

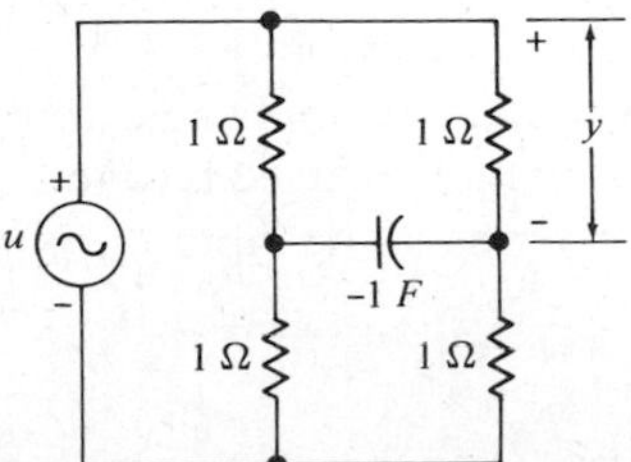

Figure 8-1 A system whose output is bounded for any bounded input if the initial voltage across the capacitor is zero.

output. Let us choose

$$u(t) = \operatorname{sgn}\,[g(t_1, t)] \tag{8-4}$$

where

$$\operatorname{sgn} x = \begin{cases} 0 & \text{if } x = 0 \\ 1 & \text{if } x > 0 \\ -1 & \text{if } x < 0 \end{cases}$$

Clearly u is bounded. However, the output excited by this input,

$$y(t_1) = \int_{-\infty}^{t_1} g(t_1, \tau)u(\tau)\,d\tau = \int_{-\infty}^{t_1} |g(t_1, \tau)|\,d\tau = \infty$$

is not bounded. Q.E.D.

We consider now multivariable systems. Consider a system with p input terminals and q output terminals, which is described by

$$\mathbf{y}(t) = \int_{-\infty}^{t} \mathbf{G}(t, \tau)\mathbf{u}(\tau)\,d\tau \tag{8-5}$$

where $\mathbf{u}$ is the $p \times 1$ input vector, $\mathbf{y}$ is the $q \times 1$ output vector, and $\mathbf{G}(t, \tau)$ is the $q \times p$ impulse-response matrix of the system. Let

$$\mathbf{G}(t, \tau) = \begin{bmatrix} g_{11}(t, \tau) & g_{12}(t, \tau) & \cdots & g_{1p}(t, \tau) \\ g_{21}(t, \tau) & g_{22}(t, \tau) & \cdots & g_{2p}(t, \tau) \\ \vdots & \vdots & & \vdots \\ g_{q1}(t, \tau) & g_{q2}(t, \tau) & \cdots & g_{qp}(t, \tau) \end{bmatrix} \tag{8-6}$$

Then g_{ij} is the impulse response between the jth input terminal and the ith output terminal. Similar to single-variable systems, a relaxed multivariable system is defined to be BIBO stable if and only if for any bounded-input vector, the output vector is bounded. By a bounded vector, we mean that every component of the vector is bounded. Applying Theorem 8-1 to every possible pair of input and output terminals, and using the fact that the sum of a finite number of bounded functions is bounded, we have immediately the following theorem.

Theorem 8-2

A relaxed multivariable system that is described by

$$\mathbf{y}(t) = \int_{-\infty}^{t} \mathbf{G}(t, \tau)\mathbf{u}(\tau)\,d\tau$$

is BIBO stable if and only if there exists a finite number k such that, for every entry of $\mathbf{G}$,

$$\int_{-\infty}^{t} |g_{ij}(t, \tau)|\,d\tau \le k < \infty$$

for all t in $(-\infty, \infty)$. ∎

Time-invariant case. Consider a single-variable system with the following input-output description:

$$y(t)=\int_0^t g(t-\tau)u(\tau)\,d\tau=\int_0^t g(\tau)u(t-\tau)\,d\tau \tag{8-7}$$

where $g(t)$ is the impulse response of the system. Recall that in order to have a description of the form (8-7), the input-output pairs of the system must satisfy linearity, causality, and time-invariance properties. In addition, the system is assumed to be relaxed at $t=0$.

Corollary 8-1

A relaxed single-variable system which is described by

$$y(t)=\int_0^t g(t-\tau)u(\tau)\,d\tau$$

is BIBO stable if and only if

$$\int_0^\infty |g(t)|\,dt\le k<\infty$$

for some constant k.

Proof

This follows directly from Theorem 8-1 by observing that

$$\int_0^t |g(t,\tau)|\,d\tau=\int_0^t |g(t-\tau)|\,d\tau=\int_0^t |g(\alpha)|\,d\alpha\le\int_0^\infty |g(\alpha)|\,d\alpha \tag{8-8}$$

For the time-invariant case, the initial time is chosen at $t=0$; hence the integration in (8-8) starts from 0 instead of $-\infty$. Q.E.D.

A function g is said to be absolutely integrable on $[0, \infty)$ if

$$\int_0^\infty |g(t)|\,dt\le k<\infty$$

Graphically, it says that the total area under $|g|$ is finite. The fact that g is absolutely integrable does not imply that g is bounded on $[0, \infty)$ nor that $g(t)$ approaches zero as $t\to\infty$. Indeed, consider the function defined by

$$f(t-n)=\begin{cases} n+(t-n)n^4 & \text{for } \dfrac{-1}{n^3}<(t-n)\le 0 \\[2ex] n-(t-n)n^4 & \text{for } 0\le(t-n)\le\dfrac{1}{n^3} \end{cases}$$

for $n=2, 3, 4, \ldots$. The function f is depicted in Figure 8-2. It is easy to verify that f is absolutely integrable; however, f is neither bounded on $[0, \infty)$ nor approaching zero as $t\to\infty$.

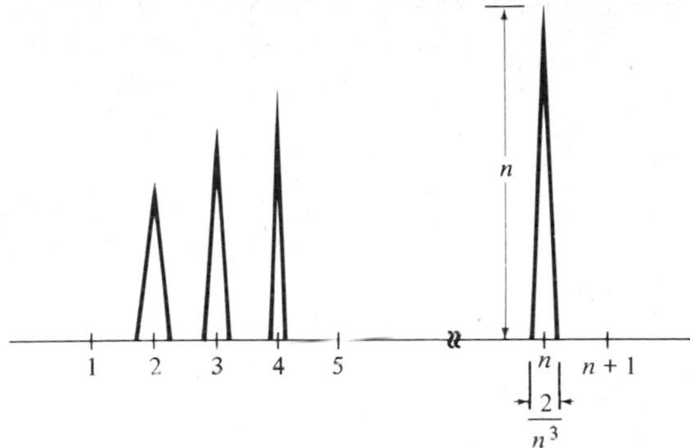

Figure 8-2 An absolutely integrable function that is neither bounded nor tending to zero as $t \to \infty$.

If g is absolutely integrable on $[0, \infty)$, then[1]

$$\int_{\alpha}^{\infty} |g(t)|\, dt \to 0 \tag{8-9}$$

as $\alpha \to \infty$. We shall use this fact in the proof of the following theorem.

Theorem 8-3

Consider a relaxed single-variable system whose input u and output y are related by

$$y(t) = \int_0^t g(t-\tau)u(\tau)\, d\tau$$

[1] Since

$$\int_0^{\infty} |g(t)|\, dt = \int_0^{\alpha} |g(t)|\, dt + \int_{\alpha}^{\infty} |g(t)|\, dt$$

and since

$$\int_0^{\alpha} |g(t)|\, dt$$

is nondecreasing as α increases,

$$\lim_{\alpha \to \infty} \int_0^{\alpha} |g(t)|\, dt \to \int_0^{\infty} |g(t)|\, dt$$

Consequently,

$$\lim_{\alpha \to \infty} \int_{\alpha}^{\infty} |g(t)|\, dt \to 0$$

If

$$\int_0^\infty |g(t)|\, dt \le k < \infty$$

for some constant k, then we have the following:

1. If u is a periodic function with period T—that is, $u(t) = u(t+T)$ for all $t \ge 0$—then the output y tends to a periodic function with the same period (not necessarily of the same waveform).
2. If u is bounded and tends to a constant, then the output will tend to a constant.
3.[2] If u is of finite energy, that is,

$$\left(\int_0^\infty |u(t)|^2\, dt \right)^{1/2} \le k_1 < \infty$$

then the output is also of finite energy; that is, there exists a finite k_2 that depends on k_1 such that

$$\left(\int_0^\infty |y(t)|^2\, dt \right)^{1/2} \le k_2 < \infty$$

Proof

1. We shall show that if $u(t) = u(t+T)$ for all $t \ge 0$, then $y(t) \to y(t+T)$ as $t \to \infty$. It is clear that

$$y(t) = \int_0^t g(\tau) u(t-\tau)\, d\tau \tag{8-10}$$

and

$$y(t+T) = \int_0^{t+T} g(\tau) u(t+T-\tau)\, d\tau = \int_0^{t+T} g(\tau) u(t-\tau)\, d\tau \tag{8-11}$$

Subtracting (8-10) from (8-11), we obtain

$$\begin{aligned} |y(t+T) - y(t)| &= \left| \int_t^{t+T} g(\tau) u(t-\tau)\, d\tau \right| \\ &\le \int_t^{t+T} |g(\tau)|\, |u(t-\tau)|\, d\tau \le u_M \int_t^{t+T} |g(\tau)|\, d\tau \end{aligned}$$

where $u_M \triangleq \max_{0 \le t \le T} |u(t)|$. It follows from (8-9) that $|y(t) - y(t+T)| \to 0$ as $t \to \infty$, or $y(t) \to y(t+T)$ as $t \to \infty$.

---

[2] This can be extended to as follows: For any real number p in $[1, \infty]$, if

$$\left(\int_0^\infty |u(t)|^p\, dt \right)^{1/p} \le k_1 < \infty$$

then there exists a finite k_2 such that

$$\left(\int_0^\infty |y(t)|^p\, dt \right)^{1/p} \le k_2 < \infty$$

and the system is said to be *L_p-stable*.

2. We prove this part by an intuitive argument. Consider

$$y(t) = \int_0^{t_1} g(\tau)u(t-\tau)\,d\tau + \int_{t_1}^{t} g(\tau)u(t-\tau)\,d\tau \tag{8-12}$$

Let $u_M \triangleq \max_t |u(t)|$. Then we have, with $t > t_1$,

$$\left| \int_{t_1}^{t} g(\tau)u(t-\tau)\,d\tau \right| \leq \int_{t_1}^{t} |g(\tau)|\,|u(t-\tau)|\,d\tau \leq u_M \int_{t_1}^{t} |g(\tau)|\,d\tau$$

which approaches zero as $t_1 \to \infty$, following from (8-9). Hence if t_1 is sufficiently large, (8-12) can be approximated by

$$y(t) \doteq \int_0^{t_1} g(\tau)u(t-\tau)\,d\tau \tag{8-13}$$

for all $t \geq t_1$. As can be seen from Figure 8-3, if t is much larger than t_1, $u(t-\tau)$ is approximately equal to a constant, say α, for all τ in $[0, t_1]$. Hence (8-13) becomes, for all $t \gg t_1 \gg 0$,

$$y(t) \doteq \alpha \int_0^{t_1} g(\tau)\,d\tau$$

which is independent of t. This completes the proof.

3. It can be shown that the set of all real-valued functions defined over $[0, \infty)$ with the property

$$\left(\int_0^{\infty} |f(t)|^2\,dt \right)^{1/2} \leq \bar{k} < \infty$$

for some constant $\bar{k}$ forms a linear space over $\mathbb{R}$, It is called the L_2 function space. In this space we may define a norm and an inner product as follows (see Problem 2-49):

$$\|f\| \triangleq \left(\int_0^{\infty} |f(t)|^2\,dt \right)^{1/2} \qquad \langle f, g \rangle \triangleq \int_0^{\infty} f(t)g(t)\,dt$$

The Schwartz inequality (Theorem 2-14) reads as

$$\left| \int_0^{\infty} f(t)g(t)\,dt \right| \leq \left(\int_0^{\infty} |f(t)|^2\,dt \right)^{1/2} \left(\int_0^{\infty} |g(t)|^2\,dt \right)^{1/2}$$

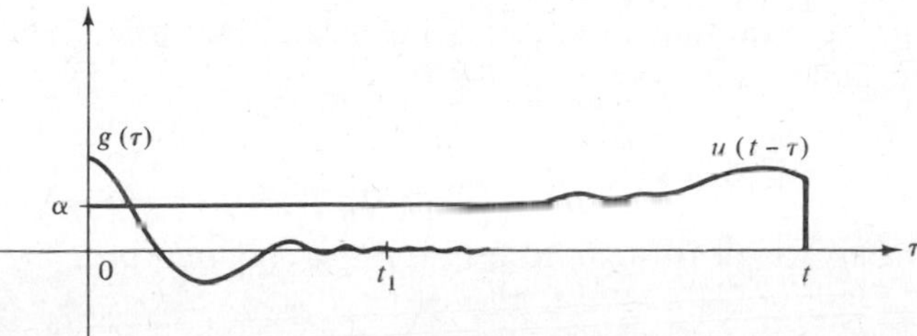

Figure 8-3 The convolution of g and u.

With these preliminaries, we are ready to proceed with the proof. Consider

$$|y(t)| = \left| \int_0^t g(\tau)u(t-\tau)\, d\tau \right| \leq \int_0^\infty |g(\tau)|\, |u(t-\tau)|\, d\tau$$

which can be written as

$$|y(t)| \leq \int_0^\infty (|g(\tau)|^{1/2})(|g(\tau)|^{1/2}|u(t-\tau)|)\, d\tau \tag{8-14}$$

Applying the Schwartz inequality, (8-14) becomes

$$|y(t)|^2 \leq \left(\int_0^\infty |g(\tau)|\, d\tau \right)\left(\int_0^\infty |g(\tau)|\, |u(t-\tau)|^2\, d\tau \right) \leq k \int_0^\infty |g(\tau)|\, |u(t-\tau)|^2\, d\tau$$

Consider now

$$\begin{aligned} \int_0^\infty |y(t)|^2\, dt &\leq k \int_0^\infty \int_0^\infty |g(\tau)|\, |u(t-\tau)|^2\, d\tau\, dt \\ &= k \int_0^\infty \left(\int_0^\infty |u(t-\tau)|^2\, dt \right) |g(\tau)|\, d\tau \end{aligned} \tag{8-15}$$

where we have changed the order of integration. By assumption, u is of finite energy; hence there exists a finite k_1 such that

$$\int_0^\infty |u(t-\tau)|^2\, dt \leq k_1$$

Hence (8-15) implies that

$$\int_0^\infty |y(t)|^2\, dt \leq kk_1 \int_0^\infty |g(\tau)|\, d\tau = k^2 k_1$$

In other words, if g is absolutely integrable on $[0, \infty)$ and if u is of finite energy, then the output y is also of finite energy. Q.E.D.

We see from Corollary 8-1 and Theorem 8-3 that if a system that is describable by

$$y(t) = \int_0^t g(t-\tau)u(\tau)\, d\tau$$

is BIBO stable, then if the input has certain property, the output will have the same property. For the time-varying case, this is not necessarily true. The following is a very important corollary of Theorem 8-3.

Corollary 8-3

Consider a relaxed single-variable system whose input u and output y are related by

$$y(t) = \int_0^t g(t-\tau)u(\tau)\, d\tau$$

If

$$\int_0^\infty |g(t)|\, dt \le k < \infty$$

for some k and if $u(t) = \sin \omega t$, for $t \ge 0$, then

$$y(t) \to |\hat{g}(i\omega)| \sin(\omega t + \theta) \qquad \text{as } t \to \infty \tag{8-16}$$

where $\theta = \tan^{-1}(\operatorname{Im} \hat{g}(i\omega)/\operatorname{Re} \hat{g}(i\omega))$ and $\hat{g}(s)$ is the Laplace transform of $g(\cdot)$; Re and Im denote the real part and the imaginary part, respectively.

Proof

Since $\sin(t - \tau) = \sin t \cos \tau - \cos t \sin \tau$, we have

$$y(t) = \int_0^t g(\tau) u(t-\tau)\, d\tau = \int_0^t g(\tau)[\sin \omega t \cos \omega\tau - \cos \omega t \sin \omega\tau]\, d\tau$$

$$= \sin \omega t \int_0^\infty g(\tau) \cos \omega\tau\, d\tau - \cos \omega t \int_0^\infty g(\tau) \sin \omega\tau\, d\tau - \int_t^\infty g(\tau) \sin \omega(t-\tau)\, d\tau$$

It is clear that

$$\left| \int_t^\infty g(\tau) \sin \omega(t-\tau)\, d\tau \right| \le \int_t^\infty |g(\tau)|\, |\sin \omega(t-\tau)|\, d\tau \le \int_t^\infty |g(\tau)|\, d\tau$$

Hence, from (8-9) we conclude that as $t \to \infty$, we obtain

$$y(t) \to \sin \omega t \int_0^\infty g(\tau) \cos \omega\tau\, d\tau - \cos \omega t \int_0^\infty g(\tau) \sin \omega\tau\, d\tau \tag{8-17}$$

By definition, $\hat{g}(s)$ is the Laplace transform of $g(\cdot)$; that is,

$$\hat{g}(s) = \int_0^\infty g(t) e^{-st}\, dt$$

which implies that

$$\hat{g}(i\omega) = \int_0^\infty g(t) e^{-i\omega t}\, dt = \int_0^\infty g(t) \cos \omega t\, dt - i \int_0^\infty g(t) \sin \omega t\, dt$$

Since $g(t)$ is real-valued, we have

$$\operatorname{Re} \hat{g}(i\omega) = \int_0^\infty g(t) \cos \omega t\, dt \tag{8-18a}$$

$$\operatorname{Im} \hat{g}(i\omega) = -\int_0^\infty g(t) \sin \omega t\, dt \tag{8-18b}$$

Hence (8-17) becomes

$$y(t) \to \sin \omega t\, (\operatorname{Re} \hat{g}(i\omega)) + \cos \omega t\, (\operatorname{Im} \hat{g}(i\omega)) = |\hat{g}(i\omega)| \sin(\omega t + \theta)$$

where $\theta = \tan^{-1}(\operatorname{Im} \hat{g}(i\omega)/\operatorname{Re} \hat{g}(i\omega))$. Q.E.D.

This corollary shows that for a BIBO-stable, linear time-invariant relaxed system, if the input is a sinusoidal function, after the transient dies out, the output is also a sinusoidal function. Furthermore, from this sinusoidal output, the magnitude and phase of the transfer function at that frequency can be read out directly. This fact is often used in practice to measure the transfer function of a linear time-invariant relaxed system.

Linear time-invariant systems are often described by transfer functions; hence it is useful to study the stability conditions in terms of transfer functions. If $\hat{g}(s)$ is a proper rational function of s, the stability condition can be easily stated in terms of $\hat{g}(s)$. If $\hat{g}(s)$ is not a rational function, then the situation is much more complicated; see Reference S32.

Theorem 8-4

A relaxed single-variable system that is described by a proper rational function $\hat{g}(s)$ is BIBO stable if and only if all the poles of $\hat{g}(s)$ are in the open left-half s plane or, equivalently, all the poles of $\hat{g}(s)$ have negative real parts. ■

By the open left-half s plane, we mean the left-half s plane excluding the imaginary axis. On the other hand, the closed left-half s plane is the left-half s plane including the imaginary axis. Note that stability of a system is independent of the zeros of $\hat{g}(s)$.

Proof of Theorem 8-4

If $\hat{g}(s)$ is a proper rational function, it can be expanded by partial fraction expansion into a sum of finite number of terms of the form

$$\frac{\beta}{(s-\lambda_i)^k}$$

and possibly of a constant, where λ_i is a pole of $\hat{g}(s)$. Consequently, $g(t)$ is a sum of finite number of the term $t^{k-1}e^{\lambda_i t}$ and possibly of a δ-function. It is easy to show that $t^{k-1}e^{\lambda_i t}$ is absolutely integrable if and only if λ_i has a negative real part. Hence we conclude that the relaxed system is BIBO stable if and only if all the poles of $\hat{g}(s)$ have negative real parts. Q.E.D.

Example 2

Consider the system with a transfer function $\hat{g}(s)=1/s$. The pole of $\hat{g}(s)$ is on the imaginary axis. Hence the system is not BIBO stable. This can also be shown from the definition. Let the input be a unit step function, then $\hat{u}(s)=1/s$. Corresponding to this bounded input, the output is $\mathscr{L}^{-1}[\hat{g}(s)\hat{u}(s)]=\mathscr{L}^{-1}[1/s^2]=t$, which is not bounded. Hence the system is not BIBO stable. ■

For multivariable systems, we have the following results.

Corollary 8-2

A relaxed multivariable system that is described by

$$\mathbf{y}(t)=\int_0^t \mathbf{G}(t-\tau)\mathbf{u}(\tau)\,d\tau$$

is BIBO stable if and only if there exists a finite number k such that, for every entry of $\mathbf{G}$,

$$\int_0^\infty |g_{ij}(t)|\, dt \leq k < \infty \qquad \blacksquare$$

If the Laplace transform of $\mathbf{G}(t)$, denoted by $\hat{\mathbf{G}}(s)$, is a proper rational-function matrix, then the BIBO stability condition can also be stated in terms of $\hat{\mathbf{G}}(s)$.

Theorem 8-5

A relaxed multivariable system that is described by $\hat{\mathbf{y}}(s) = \hat{\mathbf{G}}(s)\hat{\mathbf{u}}(s)$, where $\hat{\mathbf{G}}(s)$ is a proper rational-function matrix, is BIBO stable if and only if all the poles of every entry of $\hat{\mathbf{G}}(s)$ have negative real parts. $\blacksquare$

Corollary 8-2 and Theorem 8-5 follow immediately from Theorems 8-2 and 8-4 if we consider every entry of $\mathbf{G}$ as the impulse response of a certain input-output pair.

8-3 Routh-Hurwitz Criterion

If the transfer function of a system is a rational function of s, then the BIBO stability of the system is completely determined by the poles of $\hat{g}(s)$. If $\hat{g}(s)$ is irreducible, that is, there is no nontrivial common factor between its denominator and numerator, then the poles of $\hat{g}(s)$ are equal to the roots of the denominator of $\hat{g}(s)$. Hence under the irreducibility of $\hat{g}(s)$, the BIBO stability of a system is determined by the roots of a polynomial, the denominator of $\hat{g}(s)$. A polynomial is called a *Hurwitz polynomial* if all of its roots have negative real parts. Hence a system is BIBO stable if and only if the denominator of its irreducible transfer function is a Hurwitz polynomial (Theorem 8-4). Whether a polynomial is Hurwitz or not can be readily determined once all of its roots are computed. However, if the degree of the polynomial is three or higher, the computation of the roots is not a simple task. Furthermore, the knowledge of the exact locations of the roots is not needed in determining the stability. Therefore it is desirable to have some method of determining the stability without solving for the roots. In this section we shall introduce one such method: the Routh-Hurwitz criterion.

Consider the polynomial

$$D(s) = a_0 s^n + a_1 s^{n-1} + a_2 s^{n-2} + \cdots + a_{n-1} s + a_n \qquad a_0 > 0 \tag{8-19}$$

where the a_i's are real numbers. Before developing the Routh-Hurwitz criterion, we shall give some necessary condition for $D(s)$ to be Hurwitz. If $D(s)$ is a Hurwitz polynomial—that is, if all the roots of $D(s)$ have negative real parts—then $D(s)$ can be factored as

$$\begin{aligned} D(s) &= a_0 \prod_k (s + \alpha_k) \prod_j (s + \beta_j + i\gamma_j)(s + \beta_j - i\gamma_j) \\ &= a_0 \prod_k (s + \alpha_k) \prod_j (s^2 + 2\beta_j s + \beta_j^2 + \gamma_j^2) \end{aligned} \tag{8-20}$$

where $\alpha_k > 0$, $\beta_j > 0$, and $i^2 = -1$. Since all the coefficients of the factors in the right-hand side of (8-20) are positive, we conclude that if $D(s)$ is a Hurwitz polynomial, its coefficients a_i, $i = 1, 2, \ldots, n$ must be all positive. Hence *given a polynomial with a positive leading coefficient, if some of its coefficients are negative or zero then the polynomial is not a Hurwitz polynomial.* The condition that all coefficients of a polynomial be positive is only a necessary condition for the polynomial to be Hurwitz. A polynomial with positive coefficients may still not be a Hurwitz polynomial; for example, the polynomial with positive coefficients

$$s^3 + s^2 + 11s + 51 = (s+3)(s-1+4i)(s-1-4i)$$

is not a Hurwitz polynomial.

Consider the polynomial $D(s)$ given in (8-19). We form the following polynomials

$$D_0(s) = a_0 s^n + a_2 s^{n-2} + \cdots \tag{8-21a}$$

$$D_1(s) = a_1 s^{n-1} + a_3 s^{n-3} + \cdots \tag{8-21b}$$

that is, if n is even, $D_0(s)$ consists of the even part of $D(s)$ and $D_1(s)$ consists of the odd part of $D(s)$; if n is odd, $D_0(s)$ consists of the odd part of $D(s)$ and $D_1(s)$ consists of the even part of $D(s)$. Observe that the degree of $D_0(s)$ is always one degree higher than that of $D_1(s)$. Now we expand $D_0(s)/D_1(s)$ in the following Stieljes continued fraction expansion:

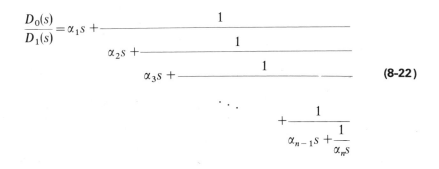

$$\frac{D_0(s)}{D_1(s)} = \alpha_1 s + \cfrac{1}{\alpha_2 s + \cfrac{1}{\alpha_3 s + \cfrac{1}{\ddots + \cfrac{1}{\alpha_{n-1} s + \cfrac{1}{\alpha_n s}}}}} \tag{8-22}$$

Theorem 8-6

The polynomial $D(s)$ in (8-19) is a Hurwitz polynomial if and only if the n numbers $\alpha_1, \alpha_2, \ldots, \alpha_n$ in (8-22) are all positive. ∎

This theorem will be proved in Section 8-5. Here we give an example to illustrate its application.

Example 1

Consider $D(s) = s^4 + 2s^3 + 6s^2 + 4s + 1$. We have $D_0(s) = s^4 + 6s^2 + 1$, $D_1(s) = 2s^3 + 4s$, and

$$\frac{D_0(s)}{D_1(s)}=\frac{1}{2}s+\cfrac{1}{\cfrac{2s^3+4s}{4s^2+1}}=\tfrac{1}{2}s+\cfrac{1}{\tfrac{1}{2}s+\cfrac{1}{\tfrac{8}{7}s+\cfrac{1}{\tfrac{7}{2}s}}} \tag{8-23}$$

Since the four numbers $\alpha_1=\frac{1}{2}, \alpha_2=\frac{1}{2}, \alpha_3=\frac{8}{7}$, and $\alpha_4=\frac{7}{2}$ are all positive, we conclude from Theorem 8-6 that the polynomial $D(s)$ is a Hurwitz polynomial. ∎

In order to obtain $\alpha_1, \alpha_2, \ldots, \alpha_n$, a series of long division must be performed. These processes can be carried out in tabular form. First we assume that n is an even number and rename the coefficients of $D_0(s)$ and $D_1(s)$ in (8-21) as, with $n' \triangleq n/2$,

$$\begin{aligned} D_0(s)&=a_0^{(0)}s^n+a_1^{(0)}s^{n-2}+\cdots+a_{n'-1}^{(0)}s^2+a_{n'}^{(0)} \\ D_1(s)&=a_0^{(1)}s^{n-1}+a_1^{(1)}s^{n-3}+\cdots+a_{n'-1}^{(1)}s \end{aligned} \tag{8-24}$$

Note that the number of coefficients in $D_1(s)$ is one less than that of $D_0(s)$ if n is even; they are equal if n is odd. Using these coefficients we form the table shown in Table 8-1. It is called the Routh table. The coefficients in the first two rows are just the coefficients of $D_0(s)$ and $D_1(s)$. The coefficients in the remaining rows are obtained from its previous two rows by using the formula

$$a_i^{(k+2)}=\frac{a_0^{(k+1)}a_{i+1}^{(k)}-a_0^{(k)}a_{i+1}^{(k+1)}}{a_0^{(k+1)}}=a_{i+1}^{(k)}-\alpha_{k+1}a_{i+1}^{(k+1)} \tag{8-25}$$

Table 8.1 Continuous-Time Stability Table (for n even)†

s^n	$a_0^{(0)}$	$a_1^{(0)}$	$a_2^{(0)}$	$\cdots$		$a_{n'-1}^{(0)}$	$a_{n'}^{(0)}$	
s^{n-1}	$a_0^{(1)}$	$a_1^{(1)}$	$a_2^{(1)}$	$\cdots$		$a_{n'-1}^{(1)}$		$\alpha_1=\dfrac{a_0^{(0)}}{a_0^{(1)}}$
s^{n-2}	$a_0^{(2)}$	$a_1^{(2)}$	$a_2^{(2)}$	$\cdots$		$a_{n'-1}^{(2)}$		$\alpha_2=\dfrac{a_0^{(1)}}{a_0^{(2)}}$
s^{n-3}	$a_0^{(3)}$	$a_1^{(3)}$	$a_2^{(3)}$	$\cdots$	$a_{n'-2}^{(3)}$			$\alpha_3=\dfrac{a_0^{(2)}}{a_0^{(3)}}$
$\vdots$	$\vdots$	$\vdots$	$\vdots$		$\vdots$			$\vdots$
s^2	$a_0^{(n-2)}$	$a_1^{(n-2)}$						
s^1	$a_0^{(n-1)}$							$\alpha_{n-1}=\dfrac{a_0^{(n-2)}}{a_0^{(n-1)}}$
s^0	$a_0^{(n)}$							$\alpha_n=\dfrac{a_0^{(n-1)}}{a_0^{(n)}}$

† If n is odd, the first two rows have the same number of coefficients. Otherwise, the pattern is identical.

where
$$\alpha_{k+1}=\frac{a_0^{(k)}}{a_0^{(k+1)}} \tag{8-26}$$

There is a simple pattern in (8-25). Using this pattern, the Routh table can be readily computed. We note that the number of coefficients will decrease by one at every odd power of s at the leftmost column of Table 8-1. The α_i are defined as the ratios of two subsequent coefficients in the first column of the table. We shall show that these α_i are equal to the α_i in (8-22). In order to do so, we use the coefficients in Table 8-1 to form the polynomials, for $k=0, 1, \ldots, n$,

$$D_k(s)=a_0^{(k)}s^{n-k}+a_1^{(k)}s^{n-k-2}+\cdots+a_{k'}^{(k)}s^{n-2k'}$$

where
$$k'=\begin{cases}\dfrac{n-k}{2} & \text{if } n-k \text{ is even}\\[2ex] \dfrac{n-k-1}{2} & \text{if } n-k \text{ is odd}\end{cases}$$

Then it is straightforward to verify, by using (8-25) and (8-26),

$$D_{k+2}(s)=D_k(s)-\alpha_{k+1}sD_{k+1}(s)$$

which implies

$$\frac{D_k(s)}{D_{k+1}(s)}=\alpha_{k+1}s+\frac{D_{k+2}(s)}{D_{k+1}(s)}=\alpha_{k+1}s+\cfrac{1}{\cfrac{D_{k+1}(s)}{D_{k+2}(s)}} \tag{8-27}$$

This equation holds for $k=0, 1, \ldots, n-2$. For $k=n-1$, we have $D_n(s)/D_{n-1}(s)=a_0^{(n)}/a_0^{(n-1)}s=1/\alpha_n s$. Using (8-27), we can write $D_0(s)/D_1(s)$ into the form in (8-22). This establishes that the α_i in (8-22) can indeed be computed from the coefficients in the first column of Table 8-1 as shown.

Theorem 8-7 (Routh-Hurwitz criterion)

The polynomial $D(s)$ of degree n in (8-19) is a Hurwitz polynomial if and only if the n terms $a_0^{(i)}$, $i=1, 2, \ldots, n$, in the first column of Table 8-1 are all positive or if and only if all the coefficients $a_j^{(i)}$ in Table 8-1 are positive.

Proof

The assumption $a_0=a_0^{(0)}>0$ in (8-19) and the relations $\alpha_i=a_0^{(i-1)}/a_0^{(i)}$, $i=1, 2, \ldots,$ n, imply $\alpha_i>0$, $i=1, 2, \ldots, n$ if and only if $a_0^{(i)}>0$, $i=1, 2, \ldots, n$. Hence the first part of the theorem follows directly from Theorem 8-6. To show the second part, we show that $a_0^{(i)}>0$, $i=1, 2, \ldots, n$ if and only if all the coefficients in Table 8-1 are positive. The sufficient part is obvious. To show the necessary part, we write (8-25) as

$$a_{i+1}^{(k)}=a_i^{(k+2)}+\alpha_{k+1}a_{i+1}^{(k+1)}$$

Using this equation we can show that if $a_0^{(n)} > 0$, $a_0^{(n-1)} > 0$ and $\alpha_{n-1} > 0$, then $a_1^{(n-2)} > 0$. Similarly we can show that the coefficients associated with s^3 in Table 8-1 are all positive. Proceeding upward, we can show that if $a_0^{(i)} > 0$, $i = 1, 2, \ldots, n$, then all the coefficients in Table 8-1 are positive. This establishes the theorem.

Q.E.D.

In forming the Routh table for a polynomial, if a negative number or a zero appears, we may stop the computation and conclude that the polynomial is not a Hurwitz polynomial. However, to conclude that a polynomial is a Hurwitz polynomial, we must complete the table and then check whether all coefficients are positive.

Example 2

Consider the polynomial in Example 1. We form

$$
\begin{array}{llll}
s^4 & 1 & 6 & 1 \\
s^3 & 2 & 4 & \\
s^2 & 4 & 1 & \\
s^1 & 3.5 & & \\
s^0 & 1 & &
\end{array}
$$

All the coefficients are positive; hence the polynomial is Hurwitz. ∎

Example 3

Consider the polynomial $D(s) = 3s^7 + 2s^6 + 2s^5 + s^4 + 3s^3 + s^2 + 1.5s + 1$. We form

$$
\begin{array}{lllll}
s^7 & 3 & 2 & 3 & 1.5 \\
s^6 & 2 & 1 & 1 & 1 \\
s^5 & \frac{1}{2} & \frac{3}{2} & 0 &
\end{array}
$$

A zero appears in the table; hence the polynomial is not Hurwitz. ∎

Example 4

Consider $2s^4 + 2s^3 + s^2 + 3s + 2$. We have

$$
\begin{array}{llll}
s^4 & 2 & 1 & 2 \\
s^3 & 2 & 3 & \\
s^2 & -2 & &
\end{array}
$$

A negative number appears in the table; hence the polynomial is not a Hurwitz polynomial. ∎

We note that in computing Table 8-1, the signs of the numbers in the table are not affected if a row is multiplied by a positive number. By using this fact, the computation of Table 8-1 may often be simplified.

Example 5

Consider $D(s) = 2s^4 + 5s^3 + 5s^2 + 2s + 1$. We form the following table:

s^4	2	5	1	
s^3	5	2		
s^2	21	5		(after the multiplication of 5)
s^1	17			(after the multiplication of 21)
s^0	5			

There are four positive numbers in the first column (excluding the first number); hence $D(s)$ is a Hurwitz polynomial. ∎

To conclude this section, we mention that the Routh table can also be used to determine the number of roots in the right-half s plane. If $a_0^{(i)} \neq 0$, $i = 1, 2, \ldots, n$, then the number of roots in the right-half s plane is equal to the number of sign changes in $\{a_0^{(1)}, a_0^{(2)}, \ldots, a_0^{(n)}\}$. See, for example, References S41 and S44. If $a_0^{(i)} = 0$ for some i, the situation becomes complicated. See Reference S37 and a number of papers discussing this topic in Reference S239.

8-4 Stability of Linear Dynamical Equations

Time-varying case. Consider the n-dimensional linear time-varying dynamical equation

$$E: \quad \dot{\mathbf{x}} = \mathbf{A}(t)\mathbf{x} + \mathbf{B}(t)\mathbf{u} \tag{8-28a}$$

$$\mathbf{y} = \mathbf{C}(t)\mathbf{x} \tag{8-28b}$$

where $\mathbf{x}$ is the $n \times 1$ state vector, $\mathbf{u}$ is the $p \times 1$ input vector, $\mathbf{y}$ is the $q \times 1$ output vector; and $\mathbf{A}$, $\mathbf{B}$, and $\mathbf{C}$ are $n \times n$, $n \times p$, and $q \times n$ matrices, respectively. It is assumed that every entry of $\mathbf{A}$, $\mathbf{B}$, and $\mathbf{C}$ are continuous functions of t in $(-\infty, \infty)$. No assumption as to boundedness of these entries is made. We have assumed that there is no direct transmission part in (8-28) (that is, $\mathbf{E} = \mathbf{0}$) because it does not play any role in the stability study.

The response of the state equation (8-28a) can always be decomposed into the zero-input response and the zero-state response as

$$\mathbf{x}(t) = \boldsymbol{\phi}(t;\, t_0, \mathbf{x}_0, \mathbf{u}) = \boldsymbol{\phi}(t;\, t_0, \mathbf{x}_0, \mathbf{0}) + \boldsymbol{\phi}(t;\, t_0, \mathbf{0}, \mathbf{u})$$

Hence it is very convenient to study the stabilities of the zero-input response and of the zero-state response separately. Combining these results, we shall immediately obtain the stability properties of the entire dynamical equation.

First we consider the stability of the zero-state response. The response of E with $\mathbf{0}$ as the initial state at time t_0 is given by

$$\begin{aligned} \mathbf{y}(t) &= \int_{t_0}^{t} \mathbf{C}(t)\boldsymbol{\Phi}(t, \tau)\mathbf{B}(\tau)\mathbf{u}(\tau)\, d\tau \\ &= \int_{t_0}^{t} \mathbf{G}(t, \tau)\mathbf{u}(\tau)\, d\tau \end{aligned}$$

where $\mathbf{G}(t, \tau) \triangleq \mathbf{C}(t)\mathbf{\Phi}(t, \tau)\mathbf{B}(\tau)$ is, by definition, the impulse-response matrix of E. What we have discussed in Section 8-2 (in particular, Theorem 8-2) can be applied here. We shall rephrase Theorem 8-2 by using the notion of norm (see Section 2-8).

Theorem 8-8

The zero-state response of the dynamical equation E is BIBO stable if and only if there exists a finite number k such that

$$\int_{t_0}^{t} \|\mathbf{C}(t)\mathbf{\Phi}(t, \tau)\mathbf{B}(\tau)\| \, d\tau \leq k < \infty \tag{8-29}$$

for any t_0 and for all $t \geq t_0$. ∎

The norm used in (8-29) is defined in terms of the norm of $\mathbf{u}$. At any instant of time, $\|\mathbf{u}(t)\|$ can be chosen as $\sum_i |u_i(t)|$, $\max_i |u_i(t)|$, or $(\sum_i |u_i(t)|^2)^{1/2}$. Corresponding to these different norms of $\mathbf{u}$,

$$\|\mathbf{C}(t)\mathbf{\Phi}(t, \tau)\mathbf{B}(\tau)\|$$

has different values. However, as far as stability is concerned, any norm can be used.

Next we study the stability of the zero-input response. More specifically, we study the response of

$$\dot{\mathbf{x}} = \mathbf{A}(t)\mathbf{x}$$

due to any initial state. The response of $\dot{\mathbf{x}} = \mathbf{A}(t)\mathbf{x}(t)$ due to the initial state $\mathbf{x}(t_0) = \mathbf{x}_0$ is given by

$$\mathbf{x}(t) \triangleq \boldsymbol{\phi}(t; t_0, \mathbf{x}_0, \mathbf{0}) = \mathbf{\Phi}(t, t_0)\mathbf{x}_0$$

It is clear that the BIBO stability can no longer be applied here. Hence we must introduce different kinds of stability. Before doing so we need the concept of equilibrium state.

Definition 8-2

A state $\mathbf{x}_e$ of a dynamical equation is said to be an *equilibrium state* at t_0 if and only if

$$\mathbf{x}_e = \boldsymbol{\phi}(t; t_0, \mathbf{x}_e, \mathbf{0})$$

for all $t \geq t_0$. ∎

We see from this definition that if a trajectory reaches an equilibrium state and if no input is applied, the trajectory will stay at the equilibrium state forever. Hence at any equilibrium state, $\dot{\mathbf{x}}_e(t) = \mathbf{0}$ for all $t \geq t_0$. Consequently, an equilibrium state of $\dot{\mathbf{x}} = \mathbf{A}(t)\mathbf{x}$ is a solution of $\mathbf{A}(t)\mathbf{x} = \mathbf{0}$ for all $t \geq t_0$. Or from definition, an equilibrium state of $\dot{\mathbf{x}} = \mathbf{A}(t)\mathbf{x}$ is a solution of

$$\mathbf{x}_e = \mathbf{\Phi}(t, t_0)\mathbf{x}_e \qquad \text{or} \qquad (\mathbf{\Phi}(t, t_0) - \mathbf{I})\mathbf{x}_e = \mathbf{0}$$

for all $t \geq t_0$. Clearly, the zero state, $\mathbf{0}$, is always an equilibrium state of $\dot{\mathbf{x}} = \mathbf{A}(t)\mathbf{x}$.

We shall now define the stability of an equilibrium state in terms of the zero-input response.

Definition 8-3

An equilibrium state $\mathbf{x}_e$ is said to be *stable in the sense of Lyapunov* at t_0 if and only if for every $\varepsilon > 0$, there exists a positive number δ which depends on ε and t_0 such that if $\|\mathbf{x}_0 - \mathbf{x}_e\| \leq \delta$, then

$$\|\boldsymbol{\phi}(t; t_0, \mathbf{x}_0, \mathbf{0}) - \mathbf{x}_e\| \leq \varepsilon$$

for all $t \geq t_0$. It is said to be *uniformly stable i.s.L.* (in the sense of Lyapunov) over $[t_0, \infty)$ if and only if for every $\varepsilon > 0$, there exists a positive δ which depends on ε but not on t_0 such that if $\|\mathbf{x}_0 - \mathbf{x}_e\| \leq \delta$, then

$$\|\boldsymbol{\phi}(t; t_1, \mathbf{x}_0, \mathbf{0}) - \mathbf{x}_e\| \leq \varepsilon$$

for any $t_1 \geq t_0$ and for all $t \geq t_1$. ∎

If an equilibrium state is uniformly stable i.s.L., then it is stable i.s.L. However, the converse may not be true. For example, the zero state of

$$\dot{\mathbf{x}}(t) = (6t \sin t - 2t)\mathbf{x}(t)$$

is stable i.s.L., but not uniformly stable i.s.L. See Reference S206. Roughly speaking, an equilibrium state $\mathbf{x}_e$ is stable i.s.L. if the response due to any initial state that is sufficiently near to $\mathbf{x}_e$ will not move far away from $\mathbf{x}_e$. For time-invariant systems, there is no difference between stability and uniform stability.

Example 1

Consider the pendulum system shown in Figure P3-17 (page 127). The application of Newton's law yields

$$u(t) \cos \theta - mg \sin \theta = ml\ddot{\theta}$$

Let $x_1 = \theta$, and $x_2 = \dot{\theta}$. Then we have

$$\dot{x}_1 = x_2$$

$$\dot{x}_2 = \left(-\frac{g}{l}\right) \sin x_1 + \frac{\cos x_1}{ml} u$$

Its equilibrium states are the solutions of $\dot{x}_1 = 0$ and $\dot{x}_2 = 0$ with $u(t) = 0$ which yield

$$\mathbf{x}_e = \begin{bmatrix} k\pi \\ 0 \end{bmatrix} \qquad k = 0, \pm 1, \pm 2, \ldots$$

The equilibrium states $[k\pi \quad 0]'$, $k = 0, \pm 2, \pm 4, \ldots$, are uniformly stable i.s.L. The equilibrium states $[k\pi \quad 0]'$, $k = \pm 1, \pm 3, \ldots$, however, are not stable i.s.L. (Why?)

Definition 8-4

An equilibrium state $\mathbf{x}_e$ is said to be *asymptotically stable* at t_0 if it is stable i.s.L. at t_0 and if every motion starting sufficiently near $\mathbf{x}_e$ converges to $\mathbf{x}_e$ as $t \to \infty$. More precisely, there is some $\gamma > 0$ such that if $\|\mathbf{x}(t_1) - \mathbf{x}_e\| \leq \gamma$, then for any $\bar{\varepsilon} > 0$, there exists a positive T which depends on $\bar{\varepsilon}$, γ, and t_1 such that

$$\|\boldsymbol{\phi}(t; t_1, \mathbf{x}(t_1), 0) - \mathbf{x}_e\| \leq \bar{\varepsilon}$$

for all $t \geq t_1 + T$. If it is uniformly stable i.s.L. over $[t_0, \infty)$ and if T is independent of t_1, the equilibrium state is said to be *uniformly asymptotically stable* over $[t_0, \infty)$. ∎

The concept of asymptotic stability is illustrated in Figure 8-4. We see that the stabilities defined in Definitions 8-3 and 8-4 are local properties, because we do not know how small δ in Definition 8-3 and γ in Definition 8-4 should be chosen. However, for linear systems, because of the homogeneity property, δ and γ can be extended to the entire state space. Hence, if an equilibrium state of a linear equation is stable at all, it will be *globally* stable or stable *in the large*.

Theorem 8-9

Every equilibrium state of $\dot{\mathbf{x}} = \mathbf{A}(t)\mathbf{x}$ is stable in the sense of Lyapunov at t_0 if and only if there exists some constant k which depends on t_0 such that

$$\|\boldsymbol{\Phi}(t, t_0)\| \leq k < \infty$$

for all $t \geq t_0$. If k is independent of t_0, it is uniformly stable i.s.L.

Proof

Sufficiency: Let $\mathbf{x}_e$ be an equilibrium state of $\dot{\mathbf{x}} = \mathbf{A}(t)\mathbf{x}$; that is

$$\mathbf{x}_e = \boldsymbol{\Phi}(t, t_0)\mathbf{x}_e \qquad \text{for all } t \geq t_0$$

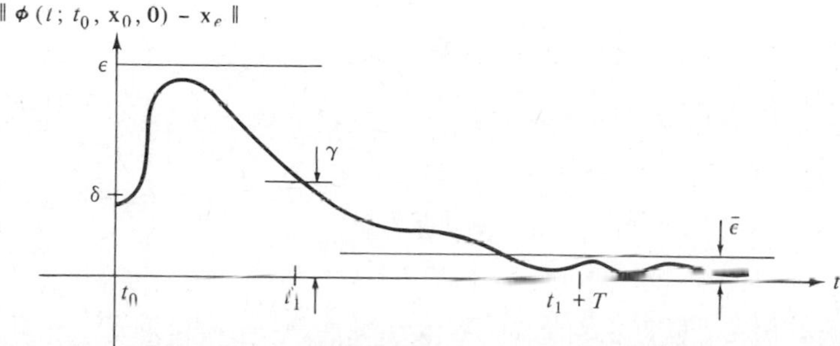

Figure 8-4 Asymptotic stability.

Then we have

$$\mathbf{x}(t) - \mathbf{x}_e = \mathbf{\Phi}(t, t_0)\mathbf{x}_0 - \mathbf{x}_e = \mathbf{\Phi}(t, t_0)(\mathbf{x}_0 - \mathbf{x}_e) \tag{8-30}$$

It follows from (2-90) that

$$\|\mathbf{x}(t) - \mathbf{x}_e\| \le \|\mathbf{\Phi}(t, t_0)\| \, \|\mathbf{x}_0 - \mathbf{x}_e\| \le k\|\mathbf{x}_0 - \mathbf{x}_e\|$$

for all t. Hence, for any ε, if we choose $\delta = \varepsilon/k$, then $\|\mathbf{x}_0 - \mathbf{x}_e\| \le \delta$ implies that

$$\|\mathbf{x}(t) - \mathbf{x}_e\| \le \varepsilon \qquad \text{for all } t \ge t_0$$

Necessity: If $\mathbf{x}_e$ is stable i.s.L., then $\mathbf{\Phi}(t, t_0)$ is bounded. We prove this by contradiction. Suppose that $\mathbf{x}_e$ is stable i.s.L. and that $\mathbf{\Phi}(t, t_0)$ is not bounded, then for some t_0, say $\tilde{t}_0$, at least one element of $\mathbf{\Phi}(t, \tilde{t}_0)$, say $\phi_{ij}(t, \tilde{t}_0)$, becomes arbitrarily large as $t \to \infty$. Let us choose $\mathbf{x}_0$ at time $\tilde{t}_0$ such that all the components of $(\mathbf{x}_0 - \mathbf{x}_e)$ are zero except the jth component, which is equal to α. Then the ith component of $(\mathbf{x}(t) - \mathbf{x}_e)$ is equal to $\phi_{ij}(t, \tilde{t}_0)\cdot\alpha$, which becomes arbitrarily large as $t \to \infty$, no matter how small α is. Hence $\mathbf{x}_e$ is not stable i.s.L. This is a contradiction. Hence, if $\mathbf{x}_e$ is stable i.s.L., then $\mathbf{\Phi}(t, t_0)$ is bounded.
Q.E.D.

The zero state is, as mentioned earlier, an equilibrium state. If *the zero state is asymptotically stable, then the zero state is the only equilibrium state of* $\dot{\mathbf{x}} = \mathbf{A}(t)\mathbf{x}$. Indeed, if there were another equilibrium state different from the zero state, then by choosing that equilibrium state as an initial state, the response would not approach the zero state. Hence, if $\dot{\mathbf{x}} = \mathbf{A}(t)\mathbf{x}$ is asymptotically stable, the zero state is the only equilibrium state.

Theorem 8-10

The zero state of $\dot{\mathbf{x}} = \mathbf{A}(t)\mathbf{x}$ is asymptotically stable at t_0 if and only if $\|\mathbf{\Phi}(t, t_0)\| \le k(t_0) < \infty$ and $\|\mathbf{\Phi}(t, t_0)\| \to 0$ as $t \to \infty$. The zero state is uniformly asymptotically stable over $[0, \infty)$ if and only if there exist positive numbers k_1 and k_2 such that

$$\|\mathbf{\Phi}(t, t_0)\| \le k_1 e^{-k_2(t - t_0)} \tag{8-31}$$

for any $t_0 \ge 0$ and for all $t \ge t_0$.

Proof

We prove only the second part of the theorem.

Sufficiency: If

$$\|\mathbf{\Phi}(t, t_0)\| \le k_1 e^{-k_2(t - t_0)}$$

then

$$\|\mathbf{x}(t)\| = \|\mathbf{\Phi}(t, t_0)\mathbf{x}(t_0)\| \le \|\mathbf{\Phi}(t, t_0)\| \, \|\mathbf{x}_0\| \le k_1 e^{-k_2(t - t_0)}\|\mathbf{x}_0\|$$

which implies that $\|\mathbf{x}(t)\| \to 0$ at $t \to \infty$, uniformly in t_0. *Necessity*: If the zero state is asymptotically stable, then by definition it is stable in the sense of

Lyapunov. Consequently, there exists a finite number k_3 such that $\|\Phi(t, t_0)\| \leq k_3$ for any t_0 and all $t \geq t_0$ (Theorem 8-9). From Definition 8-4, there is some $\gamma > 0$, and for every $\bar{\varepsilon} > 0$ there exists a positive T such that

$$\|\mathbf{x}(t_0 + T)\| = \|\Phi(t_0 + T, t_0)\mathbf{x}_0\| \leq \bar{\varepsilon} \tag{8-32}$$

for all $\|\mathbf{x}_0\| \leq \gamma$, and for any t_0. Now choose an $\mathbf{x}_0$ such that $\|\mathbf{x}_0\| = \gamma$ and $\|\Phi(t_0 + T, t_0)\mathbf{x}_0\| = \|\Phi(t_0 + T, t_0)\| \, \|\mathbf{x}_0\|$ and choose $\bar{\varepsilon} = \gamma/2$; then (8-32) implies that

$$\|\Phi(t_0 + T, t_0)\| \leq \tfrac{1}{2} \qquad \text{for any } t_0$$

This condition and $\|\Phi(t, t_0)\| \leq k_3$ imply that

$$\|\Phi(t, t_0)\| \leq k_3 \qquad \text{for all } t \text{ in } [t_0, t_0 + T)$$

$$\|\Phi(t, t_0)\| = \|\Phi(t, t_0 + T)\Phi(t_0 + T, t_0)\| \leq \|\Phi(t, t_0 + T)\| \, \|\Phi(t_0 + T, t_0)\|$$

$$\leq \frac{k_3}{2} \qquad \text{for all } t \text{ in } [t_0 + T, t_0 + 2T)$$

$$\|\Phi(t, t_0)\| \leq \|\Phi(t, t_0 + 2T)\| \, \|\Phi(t_0 + 2T, t_0 + T)\| \, \|\Phi(t_0 + T, t_0)\|$$

$$\leq \frac{k_3}{2^2} \qquad \text{for all } t \text{ in } [t_0 + 2T, t_0 + 3T)$$

and so forth, as shown in Figure 8-5. Let us choose k_2 such that $e^{-k_2 T} = \frac{1}{2}$ and let $k_1 = 2k_3$. Then from Figure 8-5, we see immediately that

$$\|\Phi(t, t_0)\| \leq k_1 e^{-k_2(t - t_0)}$$

for any t_0 and for all $t \geq t_0$. Q.E.D.

If the zero state of $\dot{\mathbf{x}} = \mathbf{A}(t)\mathbf{x}$ is uniformly asymptotically stable, then for any initial state, the zero-input response will tend to zero exponentially. Hence, for linear equations, if the zero state is uniformly asymptotically stable, it is also said to be *exponentially stable*.

We have discussed separately the stability of the zero-input response and the stability of the zero-state response. By combining these results, we may give

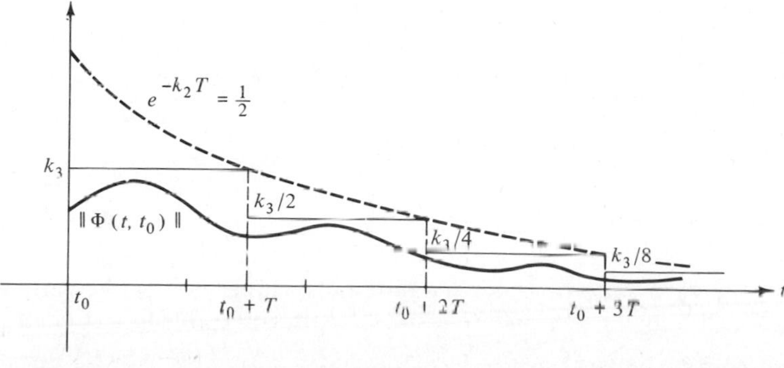

Figure 8-5 $\|\Phi(t, t_0)\|$ bounded by an exponentially decreasing function.

various definitions and theorems for the stability of the entire dynamical equation E. Before doing so, we shall discuss the relation between the stability of the zero-state response and the stability of the zero-input response.

The necessary and sufficient condition for the zero-state response of E to be BIBO stable is that, for some finite k,

$$\int_{t_0}^{t} \|\mathbf{C}(t)\mathbf{\Phi}(t,\tau)\mathbf{B}(\tau)\|\, d\tau \leq k < \infty$$

for any t_0 and for all $t \geq t_0$. It has been shown by the function given in Figure 8-2 that an absolutely integrable function is not necessarily bounded. Conversely, a bounded function need not be absolutely integrable. Hence, the stability i.s.L. of an equilibrium state, in general, does not imply nor is implied by the BIBO stability of the zero-state response. A function that approaches zero as $t \to \infty$ may not be absolutely integrable; hence asymptotic stability may not imply BIBO stability. If a system is uniformly asymptotically stable, then $\mathbf{\Phi}(t, \tau)$ is bounded and absolutely integrable as implied by (8-31); hence with some conditions on **B** and **C**, uniformly asymptotic stability may imply BIBO stability.

Theorem 8-11

Consider the dynamical equation E given in Equation (8-28). If the matrices **B** and **C** are bounded on $(-\infty, \infty)$, then the uniformly asymptotic stability of the zero state implies the BIBO stability of the zero-state response. ∎

Proof

This theorem follows directly from the fact that

$$\int \|\mathbf{C}(t)\mathbf{\Phi}(t,\tau)\mathbf{B}(\tau)\|\, d\tau \leq \int \|\mathbf{C}(t)\|\, \|\mathbf{\Phi}(t,\tau)\|\, \|\mathbf{B}(\tau)\|\, d\tau \leq k_1 k_2 \int \|\mathbf{\Phi}(t,\tau)\|\, d\tau$$

where $\|\mathbf{B}(t)\| \leq k_1$, $\|\mathbf{C}(t)\| \leq k_2$ for all t. Q.E.D.

The converse problem of Theorem 8-11—that of determining the conditions under which the BIBO stability of the zero-state response implies the asymptotic stability of the zero state—is much more difficult. In order to solve this problem, the concepts of uniform controllability and uniform observability (Definitions 5-4 and 5-8) are needed. We state only the result; its proof can be found in Reference 102.

*Theorem 8-12

Consider the dynamical equation E given in (8-28). If the matrices **A**, **B**, and **C** are bounded on $(-\infty, \infty)$ and if E is uniformly controllable and uniformly observable, then the zero state of E is asymptotically stable (under the zero-input response) if and only if its zero-state response is BIBO stable. ∎

We have studied the stability of the zero-input response and the stability of the zero-state response. Their relations are also established. We shall now study the entire response.

Definition 8-5

A linear dynamical equation is said to be *totally stable*, or *T-stable* for short, if and only if for any initial state and for any bounded input, the output as well as all the state variables are bounded. ■

We see that the conditions of T-stability are more stringent than those of BIBO stability; they require not only the boundedness of the output but also of all state variables; the boundedness must hold not only for the zero state but also for any initial state. A system that is BIBO stable sometimes cannot function properly, because some of the state variables might increase with time, and the system will burn out or at least be saturated. Therefore, in practice every system is required to be T-stable.

Theorem 8-13

A system that is described by the linear dynamical equation E in (8-28) is totally stable if and only if $\mathbf{C}(t_0)$ and $\mathbf{\Phi}(t, t_0)$ are bounded and

$$\int_{t_0}^{t} \|\mathbf{\Phi}(t, \tau)\mathbf{B}(\tau)\| \, d\tau \leq k < \infty$$

for any t_0 and for all $t \geq t_0$.

Proof

The response of (8-28a) is

$$\mathbf{x}(t) = \mathbf{\Phi}(t, t_0)\mathbf{x}_0 + \int_{t_0}^{t} \mathbf{\Phi}(t, \tau)\mathbf{B}(\tau)\mathbf{u}(\tau) \, d\tau$$

Hence we conclude from Theorems 8-8 and 8-9 that $\mathbf{x}$ is bounded for any initial state and any bounded input if and only if $\mathbf{\Phi}(t, t_0)$ is bounded and $\mathbf{\Phi}(t, \tau)\mathbf{B}(\tau)$ is absolutely integrable. From $\mathbf{y}(t) = \mathbf{C}(t)\mathbf{x}(t)$, we conclude that $\mathbf{y}$ is bounded if and only if $\mathbf{C}(t)$ is bounded. Q.E.D.

Total stability clearly implies BIBO stability; the converse however is not true. A system may be totally stable without being asymptotically stable. If $\mathbf{B}$ and $\mathbf{C}$ are bounded, then uniformly asymptotic stability implies total stability. The situation is similar to Theorem 8-11.

Time-invariant case. Although various stability conditions have been obtained for linear, time-varying dynamical equations, they can hardly be used, because all the conditions are stated in terms of state transition matrices, which are very difficult, if not impossible, to obtain. In the stability study of linear time-invariant dynamical equations, the knowledge of the state transition matrix is, however, not needed. The stability can be determined directly from the matrix $\mathbf{A}$.

Consider the n-dimensional linear time-invariant dynamical equation

$$FE: \quad \dot{\mathbf{x}} = \mathbf{A}\mathbf{x} + \mathbf{B}\mathbf{u} \tag{8-33a}$$
$$\mathbf{y} = \mathbf{C}\mathbf{x} \tag{8-33b}$$

where $\mathbf{A}$, $\mathbf{B}$, $\mathbf{C}$ are $n \times n$, $n \times p$, $q \times n$ real constant matrices, respectively. As in the time-varying case, we study first the zero-state response and the zero-input response and then the entire response. The zero-state response of FE is characterized by

$$\hat{\mathbf{G}}(s) = \mathbf{C}(s\mathbf{I} - \mathbf{A})^{-1}\mathbf{B}$$

From Theorem 8-5, *the zero-state response of FE is BIBO stable if and only if all the poles of every entry of* $\hat{\mathbf{G}}(s)$ *have negative real parts.* The zero-input response of FE is governed by $\dot{\mathbf{x}} = \mathbf{A}\mathbf{x}$ or $\mathbf{x}(t) = e^{\mathbf{A}t}\mathbf{x}_0$. Recall that an equilibrium state of $\dot{\mathbf{x}} = \mathbf{A}\mathbf{x}$ is a solution of $\dot{\mathbf{x}} = \mathbf{0}$ or $\mathbf{A}\mathbf{x} = \mathbf{0}$, and has the property $\mathbf{x}_e = e^{\mathbf{A}t}\mathbf{x}_e$ for all $t \geq 0$. Note also that stability implies uniform stability in the time-invariant case.

Theorem 8-14

Every equilibrium state of $\dot{\mathbf{x}} = \mathbf{A}\mathbf{x}$ is stable in the sense of Lyapunov if and only if all the eigenvalues of $\mathbf{A}$ have nonpositive (negative or zero) real parts and those with zero real parts are distinct roots of the minimal polynomial of $\mathbf{A}$.[3]

Proof

Let $\mathbf{x}_e$ be an equilibrium state of $\dot{\mathbf{x}} = \mathbf{A}\mathbf{x}$. Then $\mathbf{x}(t) - \mathbf{x}_e = e^{\mathbf{A}t}(\mathbf{x}_0 - \mathbf{x}_e)$. Hence every equilibrium state is stable i.s.L. if and only if there is a constant k such that $\|e^{\mathbf{A}t}\| \leq k < \infty$ for all $t \geq 0$. Let $\mathbf{P}$ be the nonsingular matrix such that $\hat{\mathbf{A}} = \mathbf{P}\mathbf{A}\mathbf{P}^{-1}$ and $\hat{\mathbf{A}}$ is in the Jordan form. Since $e^{\hat{\mathbf{A}}t} = \mathbf{P}e^{\mathbf{A}t}\mathbf{P}^{-1}$, then

$$\|e^{\hat{\mathbf{A}}t}\| \leq \|\mathbf{P}\| \, \|e^{\mathbf{A}t}\| \, \|\mathbf{P}^{-1}\|$$

Consequently, if $\|e^{\mathbf{A}t}\|$ is bounded, so is $\|e^{\hat{\mathbf{A}}t}\|$. Conversely, from the equation $e^{\mathbf{A}t} = \mathbf{P}^{-1}e^{\hat{\mathbf{A}}t}\mathbf{P}$, we see that if $\|e^{\hat{\mathbf{A}}t}\|$ is bounded, so is $\|e^{\mathbf{A}t}\|$. Hence we conclude that every equilibrium state is stable i.s.L. if and only if $\|e^{\hat{\mathbf{A}}t}\|$ is bounded on $[0, \infty)$. Now $\|e^{\hat{\mathbf{A}}t}\|$ is bounded if and only if every entry of $e^{\hat{\mathbf{A}}t}$ is bounded. Since $\hat{\mathbf{A}}$ is in the Jordan form, every entry of $e^{\hat{\mathbf{A}}t}$ is of the form $t^k e^{\alpha_j t + i\omega_j t}$, where $\alpha_j + i\omega_j$ is an eigenvalue of $\hat{\mathbf{A}}$ (see Section 2-7). If α_j is negative, it is easy to see that $t^k e^{\alpha_j t + i\omega_j t}$ is bounded on $[0, \infty)$ for any integer k. If $\alpha_j = 0$, the function $t^k e^{i\omega_j t}$ is bounded if and only if $k = 0$—that is, the order of the Jordan block associated with the eigenvalue with $\alpha_j = 0$ is 1. Q.E.D.

Example 2

Consider

[3] Equivalently, if $\mathbf{A}$ is transformed into the Jordan form, the order of every Jordan blocks associated with eigenvalues with zero real parts is 1. Note that this condition does not imply that the eigenvalues of $\mathbf{A}$ with zero real parts are distinct.

$$\dot{\mathbf{x}}(t) = \begin{bmatrix} 0 & 0 & 0 \\ 0 & 0 & 0 \\ 0 & 0 & -1 \end{bmatrix} \begin{bmatrix} x_1 \\ x_2 \\ x_3 \end{bmatrix}$$

Its equilibrium states are $[x_{1e} \quad x_{2e} \quad 0]'$, for any $x_{1e} \neq 0, x_{2e} \neq 0$. In other words, every point in the $x_1 - x_2$ plane is an equilibrium state. The eigenvalues of the matrix are $-1, 0$, and 0. Its minimal polynomial is $s(s+1)$. The eigenvalue 0, which has a zero real part, is a distinct root of the minimal polynomial. Hence every equilibrium state is stable i.s.L. ∎

Example 3

Consider

$$\dot{\mathbf{x}}(t) = \begin{bmatrix} 0 & 1 & 0 \\ 0 & 0 & 0 \\ 0 & 0 & -1 \end{bmatrix} \mathbf{x}(t)$$

It has eigenvalues $-1, 0$, and 0. Its minimal polynomial is $s^2(s+1)$. The eigenvalue 0 is not a distinct root of the minimal polynomial. Hence the equilibrium states of the equation are not stable i.s.L. ∎

Theorem 8-15

The zero state of $\dot{\mathbf{x}} = \mathbf{A}\mathbf{x}$ is asymptotically stable if and only if all the eigenvalues of $\mathbf{A}$ have negative real parts.

Proof

In order for the zero state to be asymptotically stable, in addition to the boundedness of $\|e^{\mathbf{A}t}\|$, it is required that $\|e^{\mathbf{A}t}\|$ tends to zero as $t \to \infty$, or equivalently, that $\|e^{\hat{\mathbf{A}}t}\| \to 0$ as $t \to \infty$. Since every entry of $e^{\hat{\mathbf{A}}t}$ is of the form $t^k e^{\alpha_j t + i\omega_j t}$, we conclude that $\|e^{\hat{\mathbf{A}}t}\| \to 0$ as $t \to \infty$ if and only if all the eigenvalues of $\hat{\mathbf{A}}$, and consequently of $\mathbf{A}$ have negative real parts. Q.E.D.

If a linear time-invariant system is asymptotically stable, its zero input response will approach zero exponentially; thus it is also said to be *exponentially stable*. This is consistent with Theorem 8-10 because asymptotic stability in the time-invariant case implies uniformly asymptotic stability.

The eigenvalues of $\mathbf{A}$ are the roots of the characteristic equation of $\mathbf{A}$, $\det(s\mathbf{I} - \mathbf{A}) = 0$. We have introduced in the previous section the Routh-Hurwitz criterion to check whether or not all the roots of a polynomial have negative real parts. Hence the asymptotic stability of the zero-input response of $\dot{\mathbf{x}} = \mathbf{A}\mathbf{x}$ can be easily determined by first forming the characteristic polynomial of $\mathbf{A}$ and then applying the Routh-Hurwitz criterion.

The BIBO stability of the linear time-invariant dynamical equation *FE* is determined by the poles of $\hat{\mathbf{G}}(s)$. Since

$$\hat{\mathbf{G}}(s) = \mathbf{C}(s\mathbf{I} - \mathbf{A})^{-1}\mathbf{B} = \frac{1}{\det(s\mathbf{I} - \mathbf{A})} \mathbf{C}[\operatorname{Adj}(s\mathbf{I} - \mathbf{A})]\mathbf{B}$$

every pole of $\hat{\mathbf{G}}(s)$ is an eigenvalue of $\mathbf{A}$ (the converse is not true). Consequently, if the zero state of *FE* is asymptotically stable, the zero-state response of *FE* will also be BIBO stable. (This fact can also be deduced directly from Theorem 8-11.) Conversely, the BIBO stability of the zero-state response in general does not imply the asymptotic stability of the zero state, because the zero-state response is determined by the transfer function, which, however, describes only the controllable and observable part of a dynamical equation.

Example 4

Consider a system with the following dynamical equation description:

$$\dot{\mathbf{x}} = \begin{bmatrix} 1 & 0 \\ 1 & -1 \end{bmatrix} \mathbf{x} + \begin{bmatrix} 0 \\ 1 \end{bmatrix} u$$
$$\mathbf{y} = [1 \quad 1]\,\mathbf{x}$$

Its transfer function is

$$\hat{g}(s) = [1 \quad 1] \begin{bmatrix} s-1 & 0 \\ -1 & s+1 \end{bmatrix}^{-1} \begin{bmatrix} 0 \\ 1 \end{bmatrix} = \frac{1}{s+1}$$

Hence the zero-state response of the dynamical equation is BIBO stable; however, the zero state is not asymptotically stable, because there is a positive eigenvalue. ∎

Theorem 8-16

Let

$$\dot{\bar{\mathbf{x}}} = \begin{bmatrix} \bar{\mathbf{A}}_c & \bar{\mathbf{A}}_{12} \\ \mathbf{0} & \bar{\mathbf{A}}_{\bar{c}} \end{bmatrix} \bar{\mathbf{x}} + \begin{bmatrix} \bar{\mathbf{B}}_c \\ \mathbf{0} \end{bmatrix} \mathbf{u} \qquad \textbf{(8-34a)}$$
$$\mathbf{y} = [\bar{\mathbf{C}}_c \quad \bar{\mathbf{C}}_{\bar{c}}]\,\bar{\mathbf{x}} \qquad \textbf{(8-34b)}$$

be an equivalent dynamical equation of the dynamical equation in (8-33) with $\{\bar{\mathbf{A}}_c, \bar{\mathbf{B}}_c\}$ controllable. Then the dynamical equation in (8-33) is totally stable if and only if all the eigenvalues of $\bar{\mathbf{A}}_c$ have negative real parts and all the eigenvalues of $\bar{\mathbf{A}}_{\bar{c}}$ have negative or zero real parts and those with zero real parts are distinct roots of the minimal polynomial of $\bar{\mathbf{A}}_{\bar{c}}$.

Proof

The application of the Laplace transform to (8-34a) yields

$$\hat{\bar{\mathbf{x}}}(s) = \begin{bmatrix} s\mathbf{I} - \bar{\mathbf{A}}_c & -\bar{\mathbf{A}}_{12} \\ \mathbf{0} & s\mathbf{I} - \bar{\mathbf{A}}_{\bar{c}} \end{bmatrix}^{-1} \mathbf{x}(0) + \begin{bmatrix} s\mathbf{I} - \bar{\mathbf{A}}_c & -\bar{\mathbf{A}}_{12} \\ \mathbf{0} & s\mathbf{I} - \bar{\mathbf{A}}_{\bar{c}} \end{bmatrix}^{-1} \begin{bmatrix} \bar{\mathbf{B}}_c \\ \mathbf{0} \end{bmatrix} \hat{\mathbf{u}}(s)$$
$$= \begin{bmatrix} (s\mathbf{I} - \bar{\mathbf{A}}_c)^{-1} & \mathbf{M} \\ \mathbf{0} & (s\mathbf{I} - \bar{\mathbf{A}}_{\bar{c}})^{-1} \end{bmatrix} \mathbf{x}(0) + \begin{bmatrix} (s\mathbf{I} - \bar{\mathbf{A}}_c)^{-1}\bar{\mathbf{B}}_c \\ \mathbf{0} \end{bmatrix} \hat{\mathbf{u}}(s)$$

where $\mathbf{M} = (s\mathbf{I} - \bar{\mathbf{A}}_c)^{-1}\bar{\mathbf{A}}_{12}(s\mathbf{I} - \bar{\mathbf{A}}_{\bar{c}})^{-1}$. From this equation, we may conclude that $\mathbf{x}$ is bounded for any initial state and any bounded $\mathbf{u}$ if and only if the conditions in the theorem hold. If $\mathbf{x}$ is bounded, so is $\mathbf{y}$. This establishes the theorem. Q.E.D.

It is clear that total stability implies BIBO stability. BIBO stability however may not imply total stability because no condition is imposed on $\bar{\mathbf{A}}_{\bar{c}}$ in the BIBO stability. A comparison of Theorems 8-15 and 8-16 yields that asymptotic stability implies total stability but not conversely. Hence asymptotic stability is the most stringent among these three different stabilities.

If a linear time-invariant dynamical equation is controllable and observable, then the characteristic polynomial of $\mathbf{A}$ is equal to the characteristic polynomial of $\hat{\mathbf{G}}(s)$ (Theorem 6-2). This implies that every eigenvalue of $\mathbf{A}$ is a pole of $\hat{\mathbf{G}}(s)$, and every pole of $\hat{\mathbf{G}}(s)$ is an eigenvalue of $\mathbf{A}$. Consequently, we have the following theorem.

Theorem 8-17

If a linear time-invariant dynamical equation *FE* is controllable and observable, then the following statements are equivalent:

1. The dynamical equation is totally stable.
2. The zero-state response of *FE* is BIBO stable.
3. The zero state of *FE* is asymptotically stable (under the zero-input response).
4. All the poles of the transfer function matrix of *FE* have negative real parts.
5. All the eigenvalues of the matrix $\mathbf{A}$ of *FE* have negative real parts. ■

A system is said to be completely or faithfully characterized by its transfer-function matrix if the dynamical-equation description of the system is controllable and observable. We see from Theorem 8-17 that *if a system is completely characterized by its transfer-function matrix, then asymptotic stability of the system can be determined from its transfer-function matrix alone* with no need of considering the dynamical-equation description of the system.

A remark is in order concerning the stability of $\dot{\mathbf{x}} = \mathbf{A}(t)\mathbf{x}$. If the matrix $\mathbf{A}$ is independent of t and if all the eigenvalues of $\mathbf{A}$ have negative real parts, then the zero state is asymptotically stable. Hence one might be tempted to suggest that if for each t, all the eigenvalues of $\mathbf{A}(t)$ have negative real parts, then the zero state of $\dot{\mathbf{x}} = \mathbf{A}(t)\mathbf{x}$ is asymptotically stable. This is not so, as can be seen from the following example.

Example 5

Consider the linear time-varying equation

$$\dot{\mathbf{x}} = \begin{bmatrix} -1 & e^{2t} \\ 0 & -1 \end{bmatrix} \mathbf{x}$$

The characteristic polynomial of the matrix $\mathbf{A}$ at each t is given by

$$\det\,[\lambda\mathbf{I} - \mathbf{A}] = \det \begin{bmatrix} \lambda + 1 & -e^{2t} \\ 0 & \lambda + 1 \end{bmatrix} = (\lambda + 1)^2$$

Hence the eigenvalues of $\mathbf{A}$ are -1 and -1 for all t. However, the zero state of the equation is neither asymptotically stable nor stable i.s.L., because the state transition matrix of the equation is

$$\Phi(t, 0) = \begin{bmatrix} e^{-t} & (e^{t} - e^{-t})/2 \\ 0 & e^{-t} \end{bmatrix}$$

(as in Problem 4-1 or by direct verification), whose norm tends to infinity as $t \to \infty$. ∎

*8-5 Lyapunov Theorem

The asymptotic stability of $\dot{\mathbf{x}} = \mathbf{A}\mathbf{x}$ can be determined by first computing the characteristic polynomial of $\mathbf{A}$ and then applying the Routh-Hurwitz criterion. If all the roots of the characteristic polynomial have negative real parts, then the zero state of $\dot{\mathbf{x}} = \mathbf{A}\mathbf{x}$ is asymptotically stable. There is one more method of checking the asymptotic stability of $\dot{\mathbf{x}} = \mathbf{A}\mathbf{x}$ without computing explicitly the eigenvalues of $\mathbf{A}$. We shall discuss such a method in this section and then apply it to establish the Routh-Hurwitz criterion.

Before proceeding, we need the concept of positive definite and positive semidefinite matrices. An $n \times n$ matrix $\mathbf{M}$ with elements in the field of complex numbers is said to be a *hermitian matrix* if $\mathbf{M}^* = \mathbf{M}$, where $\mathbf{M}^*$ is the complex conjugate transpose of $\mathbf{M}$. If $\mathbf{M}$ is a real matrix, $\mathbf{M}$ is said to be *symmetric*. The matrix $\mathbf{M}$ can be considered as an operator that maps $(\mathbb{C}^n, \mathbb{C})$ into itself. It is shown in Theorem E-1 that all the eigenvalues of a hermitian matrix are real, and that there exists a nonsingular matrix $\mathbf{P}$, called a unitary matrix, such that $\mathbf{P}^{-1} = \mathbf{P}^*$ and $\hat{\mathbf{M}} = \mathbf{P}\mathbf{M}\mathbf{P}^*$, where $\hat{\mathbf{M}}$ is a diagonal matrix with eigenvalues on the diagonal (Theorem E-4). We shall use this fact to establish the following theorem.

Theorem 8-18

Let $\mathbf{M}$ be a hermitian matrix and let $\lambda_{\min}$ and $\lambda_{\max}$ be the smallest and largest eigenvalues of $\mathbf{M}$, respectively. Then

$$\lambda_{\min} \|\mathbf{x}\|^2 \leq \mathbf{x}^*\mathbf{M}\mathbf{x} \leq \lambda_{\max} \|\mathbf{x}\|^2 \tag{8-35}$$

for any $\mathbf{x}$ in the n-dimensional complex vector space $\mathbb{C}^n$, where

$$\|\mathbf{x}\|^2 \triangleq \langle \mathbf{x}, \mathbf{x} \rangle \triangleq \mathbf{x}^*\mathbf{x} = \sum_{i=1}^{n} |x_i|^2$$

and x_i is the ith component of $\mathbf{x}$.

Proof

Note that $\mathbf{x}^*\mathbf{M}\mathbf{x}$ is a real number for any $\mathbf{x}$ in $\mathbb{C}^n$. Let $\mathbf{P}$ be the nonsingular matrix such that $\mathbf{P}^{-1} = \mathbf{P}^*$ and $\hat{\mathbf{M}} = \mathbf{P}\mathbf{M}\mathbf{P}^*$, where $\hat{\mathbf{M}}$ is a diagonal matrix with eigenvalues of $\mathbf{M}$ on the diagonal. Let $\bar{\mathbf{x}} = \mathbf{P}\mathbf{x}$ or $\mathbf{x} = \mathbf{P}^{-1}\bar{\mathbf{x}} = \mathbf{P}^*\bar{\mathbf{x}}$, then

$$\mathbf{x}^*\mathbf{M}\mathbf{x} = \bar{\mathbf{x}}^*\mathbf{P}\mathbf{M}\mathbf{P}^*\bar{\mathbf{x}} = \bar{\mathbf{x}}^*\hat{\mathbf{M}}\bar{\mathbf{x}} = \sum_{i=1}^{n} \lambda_i |\bar{x}_i|^2$$

where the λ_i's are the eigenvalues of $\mathbf{M}$. It follows that

$$\lambda_{\min} \sum_{i=1}^{n} |\bar{x}_i|^2 \leq \mathbf{x}^*\mathbf{M}\mathbf{x} = \bar{\mathbf{x}}^*\hat{\mathbf{M}}\bar{\mathbf{x}} = \sum_{i=1}^{n} \lambda_i |\bar{x}_i|^2 \leq \lambda_{\max} \sum_{i=1}^{n} |\bar{x}_i|^2 \tag{8-36}$$

The fact that $\mathbf{P}^{-1} = \mathbf{P}^*$ implies that

$$||\mathbf{x}||^2 = \mathbf{x}^*\mathbf{x} = \bar{\mathbf{x}}^*\bar{\mathbf{x}} = \sum_{i=1}^{n} |\bar{x}_i|^2$$

Hence, the inequality (8-36) implies (8-35). Q.E.D.

Definition 8-6

A hermitian matrix $\mathbf{M}$ is said to be *positive definite* if and only if $\mathbf{x}^*\mathbf{M}\mathbf{x} > 0$ for all nonzero $\mathbf{x}$ in $\mathbb{C}^n$. A hermitian matrix $\mathbf{M}$ is said to be *positive semidefinite* or *nonnegative definite* if and only if $\mathbf{x}^*\mathbf{M}\mathbf{x} \geq 0$ for all $\mathbf{x}$ in $\mathbb{C}^n$, and the equality holds for some nonzero $\mathbf{x}$ in $\mathbb{C}^n$. ■

Theorem 8-19

A hermitian matrix $\mathbf{M}$ is positive definite (positive semidefinite) if and only if any one of the following conditions holds:

1. All the eigenvalues of $\mathbf{M}$ are positive (nonnegative).
2. All the leading principal minors[4] of $\mathbf{M}$ are positive (all the principal minors of $\mathbf{M}$ are nonnegative).[5]
3. There exists a nonsingular matrix $\mathbf{N}$ (a singular matrix $\mathbf{N}$) such that $\mathbf{M} = \mathbf{N}^*\mathbf{N}$.[6]

---

[4] The *principal minors* of the matrix

$$\mathbf{M} = \begin{bmatrix} m_{11} & m_{12} & m_{13} \\ m_{21} & m_{22} & m_{23} \\ m_{31} & m_{32} & m_{33} \end{bmatrix}$$

are m_{11}, m_{22}, m_{33}, $\det \begin{bmatrix} m_{11} & m_{12} \\ m_{21} & m_{22} \end{bmatrix}$, $\det \begin{bmatrix} m_{11} & m_{13} \\ m_{31} & m_{33} \end{bmatrix}$, $\det \begin{bmatrix} m_{22} & m_{23} \\ m_{32} & m_{33} \end{bmatrix}$, and $\det \mathbf{M}$, that is, the minors whose diagonal elements are also diagonal elements of the matrix. The *leading principal minors* of $\mathbf{M}$ are m_{11}, $\det \begin{bmatrix} m_{11} & m_{12} \\ m_{21} & m_{22} \end{bmatrix}$, and $\det \mathbf{M}$, that is, the minors obtained by deleting the last k columns and the last k rows, for $k = 2, 1$, and 0.

[5] It is shown in Reference 39 that if all the leading principal minors of a matrix are positive, then all the principal minors are positive. However, it is *not* true that if all the leading principal minors are nonnegative, then all the principal minors are nonnegative. For a counterexample, try Problem 8-32b.

[6] If $\mathbf{N}$ is an upper triangular matrix, it is called the *Cholesky decomposition*. Subroutines are available in LINPACK and IBM Scientific Subroutine Package to carry out this decomposition.

Proof

Condition 1 follows directly from Theorem 8-18. A proof of condition 2 can be found, for example, in References 5 and 39. For a proof of condition 3, see Problem 8-24. Q.E.D.

With these preliminaries, we are ready to introduce the Lyapunov theorem and its extension. They will be used to prove the Routh-Hurwitz criterion.

Theorem 8-20 (Lyapunov theorem)

All the eigenvalues of **A** have negative real parts or, equivalently, the zero state of $\dot{\mathbf{x}} = \mathbf{A}\mathbf{x}$ is asymptotically stable if and only if for any given positive definite hermitian matrix **N**, the matrix equation

$$\mathbf{A}^*\mathbf{M} + \mathbf{M}\mathbf{A} = -\mathbf{N} \tag{8-37}$$

has a unique hermitian solution **M** and **M** is positive definite. ∎

Corollary 8-20

All the eigenvalues of **A** have negative real parts, or equivalently, the zero state of $\dot{\mathbf{x}} = \mathbf{A}\mathbf{x}$ is asymptotically stable, if and only if for any given positive semidefinite hermitian matrix **N** with the property $\{\mathbf{A}, \mathbf{N}\}$ observable, the matrix equation

$$\mathbf{A}^*\mathbf{M} + \mathbf{M}\mathbf{A} = -\mathbf{N}$$

has a unique hermitian solution **M** and **M** is positive definite. ∎

The implication of Theorem 8-20 and Corollary 8-20 is that if **A** is asymptotically stable and if **N** is positive definite or positive semidefinite, then the solution **M** of (8-37) must be positive definite. However, it does *not* say that if **A** is asymptotically stable and if **M** is positive definite, then the matrix **N** computed from (8-37) is positive definite or positive semidefinite.

Before proving the Lyapunov theorem, we make a few comments. Since Theorem 8-20 holds for any positive definite hermitian matrix **N**, the matrix **N** in (8-37) is often chosen to be a unit matrix. Since **M** is a hermitian matrix, there are n^2 unknown numbers in **M** to be solved. If **M** is a real symmetric matrix there are $n(n+1)/2$ unknown numbers in **M** to be solved. Hence the matrix equation (8-37) actually consists of n^2 linear algebraic equations. To apply Theorem 8-20, we first solve these n^2 equations for **M**, and then check whether or not **M** is positive definite. This is not an easy task. Hence Theorem 8-20 and its corollary are generally not used in determining the stability of $\dot{\mathbf{x}} = \mathbf{A}\mathbf{x}$. However, they are very important in the stability study of nonlinear time-varying systems by using the so-called second method of Lyapunov. Furthermore, we shall use it to prove the Routh-Hurwitz criterion.

We give now a physical interpretation of the Lyapunov theorem. If the hermitian matrix **M** is positive definite, the plot of $V(\mathbf{x})$

$$V(\mathbf{x}) \triangleq \mathbf{x}^*\mathbf{M}\mathbf{x} \tag{8-38}$$

will be bowl shaped, as shown in Figure 8-6. Consider now the successive values taken by V along a trajectory of $\dot{\mathbf{x}} = \mathbf{A}\mathbf{x}$. We like to know whether the value of V will increase or decrease with time as the state moving along the trajectory. Taking the derivative of V with respect to t along any trajectory of $\dot{\mathbf{x}} = \mathbf{A}\mathbf{x}$, we obtain

$$\begin{aligned}\frac{d}{dt} V(\mathbf{x}(t)) &= \frac{d}{dt}(\mathbf{x}^*(t)\mathbf{M}\mathbf{x}(t)) = \left(\frac{d}{dt}\mathbf{x}^*(t)\right)\mathbf{M}\mathbf{x}(t) + \mathbf{x}^*(t)\mathbf{M}\left(\frac{d}{dt}\mathbf{x}(t)\right) \\ &= \mathbf{x}^*(t)\mathbf{A}^*\mathbf{M}\mathbf{x}(t) + \mathbf{x}^*(t)\mathbf{M}\mathbf{A}\mathbf{x}(t) = \mathbf{x}^*(t)(\mathbf{A}^*\mathbf{M} + \mathbf{M}\mathbf{A})\mathbf{x}(t) \\ &= -\mathbf{x}^*(t)\mathbf{N}\mathbf{x}(t)\end{aligned} \tag{8-39}$$

where $\mathbf{N} \triangleq -(\mathbf{A}^*\mathbf{M} + \mathbf{M}\mathbf{A})$. This equation gives the rate of change of $V(\mathbf{x})$ along any trajectory of $\dot{\mathbf{x}} = \mathbf{A}\mathbf{x}$. Now if $\mathbf{N}$ is positive definite, the function $-\mathbf{x}^*(t)\mathbf{N}\mathbf{x}(t)$ is always negative. This implies that $V(\mathbf{x}(t))$ decreases monotonically with time along any trajectory of $\dot{\mathbf{x}} = \mathbf{A}\mathbf{x}$; hence $V(\mathbf{x}(t))$ will eventually approach zero as $t \to \infty$. Now since $V(\mathbf{x})$ is positive definite, we have $V(\mathbf{x}) = 0$ only at $\mathbf{x} = \mathbf{0}$; hence we conclude that if we can find positive definite matrices $\mathbf{M}$ and $\mathbf{N}$ that are related by (8-37), then every trajectory of $\dot{\mathbf{x}} = \mathbf{A}\mathbf{x}$ will approach the zero state as $t \to \infty$. The function $V(\mathbf{x})$ is called a *Lyapunov function* of $\dot{\mathbf{x}} = \mathbf{A}\mathbf{x}$. A Lyapunov function can be considered as a generalization of the concept of distance or energy. If the "distance" of the state along any trajectory of $\dot{\mathbf{x}} = \mathbf{A}\mathbf{x}$ decreases with time, then $\mathbf{x}(t)$ must tend to $\mathbf{0}$ as $t \to \infty$.

Proof of Theorem 8-20

Sufficiency: Consider $V(\mathbf{x}) = \mathbf{x}^*\mathbf{M}\mathbf{x}$. Then we have

$$\dot{V}(\mathbf{x}) \triangleq \frac{d}{dt} V(\mathbf{x}) = -\mathbf{x}^*\mathbf{N}\mathbf{x}$$

along any trajectory of $\dot{\mathbf{x}} = \mathbf{A}\mathbf{x}$. From Theorem 8-18, we have

$$\frac{\dot{V}}{V} = -\frac{\mathbf{x}^*\mathbf{N}\mathbf{x}}{\mathbf{x}^*\mathbf{M}\mathbf{x}} \leq -\frac{(\lambda_{\mathbf{N}})_{\min}}{(\lambda_{\mathbf{M}})_{\max}} \tag{8-40}$$

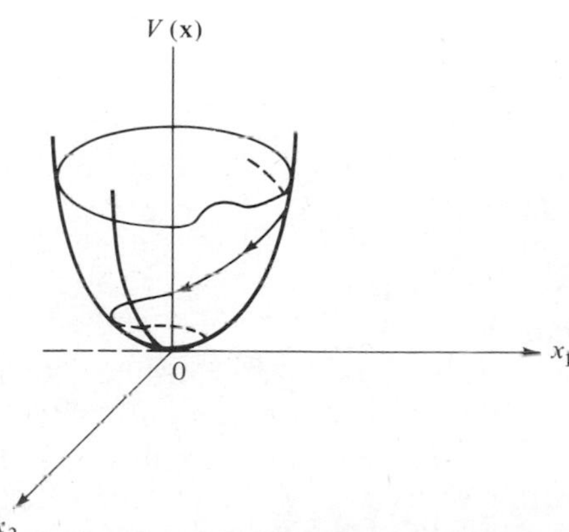

Figure 8-6 A Lyapunov function $V(\mathbf{x})$.

where $(\lambda_N)_{min}$ is the smallest eigenvalue of **N** and $(\lambda_M)_{max}$ is the largest eigenvalue of **M**. From Theorem 8-19 and from the assumption that the matrices **M** and **N** are positive definite, we have $(\lambda_N)_{min} > 0$ and $(\lambda_M)_{max} > 0$. If we define

$$\alpha \triangleq \frac{(\lambda_N)_{min}}{(\lambda_M)_{max}}$$

then inequality (8-40) becomes $\dot{V} \leq -\alpha V$, which implies that $V(t) \leq e^{-\alpha t} V(0)$. It is clear that $\alpha > 0$; hence V decreases exponentially to zero on every trajectory of $\dot{\mathbf{x}} = \mathbf{A}\mathbf{x}$. Now $V(\mathbf{x}) = 0$ only at $\mathbf{x} = \mathbf{0}$; hence we conclude that the response of $\dot{\mathbf{x}} = \mathbf{A}\mathbf{x}$ due to any initial state $\mathbf{x}_0$ tends to $\mathbf{0}$ as $t \to \infty$. This proves that the zero state of $\dot{\mathbf{x}} = \mathbf{A}\mathbf{x}$ is asymptotically stable. *Necessity:* If the zero state of $\dot{\mathbf{x}} = \mathbf{A}\mathbf{x}$ is asymptotically stable, then all the eigenvalues of **A** have negative real parts. Consequently, for any **N**, there exists a unique matrix **M** satisfying

$$\mathbf{A}^*\mathbf{M} + \mathbf{M}\mathbf{A} = -\mathbf{N}$$

and **M** can be expressed as

$$\mathbf{M} = \int_0^\infty e^{\mathbf{A}^* t} \mathbf{N} e^{\mathbf{A}t}\, dt \tag{8-41}$$

(see Appendix F). Now we show that if **N** is positive definite, so is **M**. Let **H** be a nonsingular matrix such that $\mathbf{N} = \mathbf{H}^*\mathbf{H}$ (Theorem 8-19). Consider

$$\mathbf{x}_0^*\mathbf{M}\mathbf{x}_0 = \int_0^\infty \mathbf{x}_0^* e^{\mathbf{A}^* t} \mathbf{H}^*\mathbf{H} e^{\mathbf{A}t} \mathbf{x}_0\, dt = \int_0^\infty \|\mathbf{H} e^{\mathbf{A}t}\mathbf{x}_0\|^2\, dt \tag{8-42}$$

Since **H** is nonsingular and $e^{\mathbf{A}t}$ is nonsingular for all t, we have $\mathbf{H}e^{\mathbf{A}t}\mathbf{x}_0 \neq \mathbf{0}$ for all t unless $\mathbf{x}_0 = \mathbf{0}$. Hence we conclude that $\mathbf{x}_0^*\mathbf{M}\mathbf{x}_0 > 0$ for all $\mathbf{x}_0 \neq \mathbf{0}$, and **M** is positive definite. This completes the proof of this theorem. Q.E.D.

In order to establish Corollary 8-20, we show that if **N** is positive semidefinite and if $\{\mathbf{A}, \mathbf{N}\}$ is observable, then $\mathbf{x}^*(t)\mathbf{N}\mathbf{x}(t)$ cannot be identically zero along any nontrivial trajectory of $\dot{\mathbf{x}} = \mathbf{A}\mathbf{x}$ (any solution due to any nonzero initial state $\mathbf{x}_0$). First we use Theorem 8-19 to write **N** as $\mathbf{N} = \mathbf{H}^*\mathbf{H}$. Then the observability of $\{\mathbf{A}, \mathbf{N}\}$ implies the observability of $\{\mathbf{A}, \mathbf{H}\}$ (Problem 5-36). Consider

$$\mathbf{x}^*(t)\mathbf{N}\mathbf{x}(t) = \mathbf{x}_0^* e^{\mathbf{A}^* t}\mathbf{H}^*\mathbf{H}e^{\mathbf{A}t}\mathbf{x}_0 = \|\mathbf{H}e^{\mathbf{A}t}\mathbf{x}_0\|^2$$

Since $\{\mathbf{A}, \mathbf{H}\}$ is observable, all rows of $\mathbf{H}e^{\mathbf{A}t}$ are linearly independent on $[0, \infty)$. Hence we have that $\mathbf{H}e^{\mathbf{A}t}\mathbf{x}_0 = \mathbf{0}$ for all t if and only if $\mathbf{x}_0 = \mathbf{0}$. Because $e^{\mathbf{A}t}$ is analytic over $[0, \infty)$, we conclude that for any $\mathbf{x}_0 \neq \mathbf{0}$, $\mathbf{H}e^{\mathbf{A}t}\mathbf{x}_0$ can never be identically zero over any finite interval, no matter how small; otherwise it would be identically zero over $[0, \infty)$. See Theorem B-1. Note that $\mathbf{H}e^{\mathbf{A}t}\mathbf{x}_0 = \mathbf{0}$, at some discrete instants of time, is permitted.

With the preceding discussion, we are ready to establish Corollary 8-20. Consider the Lyapunov function $V(\mathbf{x})$ defined in (8-38) and $dV(\mathbf{x})/dt = -\mathbf{x}^*(t)\mathbf{N}\mathbf{x}(t)$ in (8-39). If $\mathbf{x}_0 \neq \mathbf{0}$, $dV(\mathbf{x})/dt \leq 0$ and the equality holds only at some discrete instants of time; hence $V(\mathbf{x}(t))$ will decrease with time, not necessarily monotonic at every instant of time, and will eventually approach zero as

$t \to \infty$. This shows the sufficiency of the corollary. The necessary part can be similarly proved as in Theorem 8-20 by using (8-42).

Theorem 8-20 and its corollary are generally not used in checking the asymptotic stability of $\dot{\mathbf{x}} = \mathbf{A}\mathbf{x}$. However, they are important by their own right and are basic in the stability study of nonlinear systems. They also provide a simple proof of the Routh-Hurwitz criterion.

A proof of the Routh-Hurwitz criterion. Consider the polynomial

$$D(s) = a_0 s^n + a_1 s^{n-1} + a_2 s^{n-2} + \cdots + a_n \qquad a_0 > 0$$

with real coefficients a_i, $i = 0, 1, 2, \ldots, n$. We form the polynomials

$$\begin{aligned} D_0(s) &= a_0 s^n + a_2 s^{n-2} + \cdots \\ D_1(s) &= a_1 s^{n-1} + a_3 s^{n-3} + \cdots \end{aligned}$$

and compute

$$\frac{D_0(s)}{D_1(s)} = \alpha_1 s + \cfrac{1}{\alpha_2 s + \cfrac{1}{\alpha_3 s + \cfrac{1}{\ddots + \cfrac{1}{\alpha_{n-1} s + \cfrac{1}{\alpha_n s}}}}}$$

For convenience, we shall restate the theorem here.

Theorem 8-6

The polynomial $D(s)$ is a Hurwitz polynomial if and only if all the n numbers $\alpha_1, \alpha_2, \ldots, \alpha_n$ are positive.

Proof[7]

First we assume that all the α_i's are different from zero. Consider the rational function

$$\hat{g}(s) \triangleq \frac{D_1(s)}{D(s)} = \frac{D_1(s)}{D_0(s) + D_1(s)} = \frac{1}{1 + D_0(s)/D_1(s)}$$

The assumption $\alpha_i \neq 0$, for $i = 1, 2, \ldots, n$, implies that there is no common factor between $D_0(s)$ and $D_1(s)$. Consequently, there is no common factor between $D_1(s)$ and $D(s)$; in other words, $\hat{g}(s)$ is irreducible.

Consider the block diagram shown in Figure 8-7. We show that the transfer function from u to y is $\hat{g}(s)$. Let $\hat{h}_1(s)$ be the transfer function from x_n to x_{n-1} or equivalently, from the terminal E to the terminal F, as shown in Figure 8-7.

---

[7] This follows Reference 90.

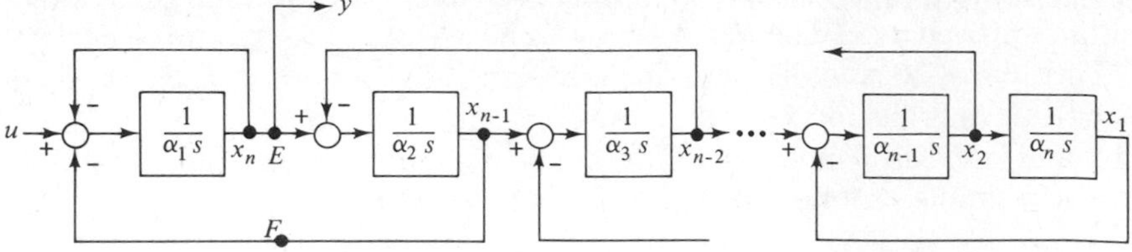

Figure 8-7 A block diagram of $\hat{g}(s)$.

Then

$$\frac{\hat{y}(s)}{\hat{u}(s)}=\frac{1/\alpha_1 s}{1+[1+\hat{h}_1(s)]/\alpha_1 s}=\frac{1}{1+\alpha_1 s+\hat{h}_1(s)}$$

Let $\hat{h}_2(s)$ be the transfer function from x_{n-1} to x_{n-2}, then $\hat{h}_1(s)$ can be written as

$$\hat{h}_1(s)=\frac{1/\alpha_2 s}{1+\hat{h}_2(s)/\alpha_2 s}=\frac{1}{\alpha_2 s+\hat{h}_2(s)}$$

Proceeding forward, we can show easily that the transfer function from u to y is indeed $\hat{g}(s)$. With the state variables chosen as shown, we can readily write the dynamical equation of the block diagram as

$$\begin{bmatrix}\dot{x}_1\\ \dot{x}_2\\ \dot{x}_3\\ \vdots\\ \dot{x}_{n-1}\\ \dot{x}_n\end{bmatrix}=\begin{bmatrix}0 & \frac{1}{\alpha_n} & 0 & \cdots & 0 & 0 & 0\\ \frac{-1}{\alpha_{n-1}} & 0 & \frac{1}{\alpha_{n-1}} & \cdots & 0 & 0 & 0\\ 0 & \frac{-1}{\alpha_{n-2}} & 0 & \cdots & 0 & 0 & 0\\ \vdots & \vdots & \vdots & & \vdots & \vdots & \vdots\\ 0 & 0 & 0 & \cdots & \frac{-1}{\alpha_2} & 0 & \frac{1}{\alpha_2}\\ 0 & 0 & 0 & \cdots & 0 & \frac{-1}{\alpha_1} & \frac{-1}{\alpha_1}\end{bmatrix}\begin{bmatrix}x_1\\ x_2\\ x_3\\ \vdots\\ x_{n-1}\\ x_n\end{bmatrix}+\begin{bmatrix}0\\ 0\\ 0\\ \vdots\\ 0\\ 1\end{bmatrix}u \tag{8-43}$$

$$y=[\,0\quad 0\quad 0\quad \cdots\quad 0\quad 0\quad 1\,]\mathbf{x}$$

Irreducibility of (8-43) can be verified either by showing that it is controllable and observable or by the fact that its dimension is equal to the degree of the denominator of $\hat{g}(s)$. Consequently, the characteristic polynomial of the matrix $\mathbf{A}$ in (8-43) is equal to the denominator of $\hat{g}(s)$ (Theorem 6-2). Now we shall derive the condition for the zero state of (8-43) to be asymptotically stable.

Let the **M** matrix in Corollary 8-20 be chosen as

$$\mathbf{M} = \begin{bmatrix} \alpha_n & 0 & \cdots & 0 & 0 \\ 0 & \alpha_{n-1} & \cdots & 0 & 0 \\ \vdots & \vdots & & \vdots & \vdots \\ 0 & 0 & \cdots & \alpha_2 & 0 \\ 0 & 0 & \cdots & 0 & \alpha_1 \end{bmatrix} \tag{8-44}$$

Then it is easy to verify that

$$\mathbf{A}^*\mathbf{M} + \mathbf{MA} = -\begin{bmatrix} 0 & 0 & 0 & \cdots & 0 & 0 \\ 0 & 0 & 0 & \cdots & 0 & 0 \\ 0 & 0 & 0 & \cdots & 0 & 0 \\ \vdots & \vdots & \vdots & & \vdots & \vdots \\ 0 & 0 & 0 & \cdots & 0 & 0 \\ 0 & 0 & 0 & \cdots & 0 & 2 \end{bmatrix} \triangleq -\mathbf{N} \tag{8-45}$$

It is clear that **N** is a positive semidefinite matrix. It is straightforward to verify that $\{\mathbf{A}, \mathbf{N}\}$ is observable. Hence from Corollary 8-20 we conclude that the zero state of $\dot{\mathbf{x}} = \mathbf{Ax}$ is asymptotically stable if and only if the matrix **M** is positive definite, or equivalently, the n numbers $\alpha_1, \alpha_2, \ldots, \alpha_n$ are positive. Now the zero state of $\dot{\mathbf{x}} = \mathbf{Ax}$ is asymptotically stable if and only if all the eigenvalues of **A**, or equivalently all the roots of $D(s)$, have negative real parts. In other words, $D(s)$ is a Hurwitz polynomial if and only if all the n numbers $\alpha_1, \alpha_2, \ldots, \alpha_n$ are positive.

Consider now the case in which not all the α_i's are different from zero. In other words, some of the coefficients in the first column of Table 8-1 are equal to zero. Suppose $a_0^{(2)} = 0$. If all the coefficients in the s^{n-2} row of Table 8-1 are equal to zero, it implies that $D_0(s)$ and $D_1(s)$ have at least one common factor. The common factor is clearly either an even or an odd function of s, say $f(s)$. Then $D(s)$ can be factored as $f(s)\bar{D}(s)$. Since not all the roots of an even function or an odd function can have negative real parts, $D(s)$ is not a Hurwitz polynomial. If not all the coefficients in the s^{n-2} row are equal to zero, we may replace $a_0^{(2)}$ by a very small positive number ε and continue to complete Table 8-1. In this case it can be seen that some α_i will be negative. If some α_i is negative, at least one root of the modified $D(s)$ (since $a_0^{(2)}$ is replaced by ε) has a positive real part. Now the roots of a polynomial are continuous functions of its coefficients. Hence, as $\varepsilon \to 0$, at least one root of $D(s)$ has a positive or zero real part. Q.E.D.

*8-6 Linear Time-Invariant Discrete-Time systems

The stability concepts introduced for the continuous-time systems are directly applicable to the discrete-time case. However, the conditions of stability are quite different. In this section we shall discuss some of these conditions.

Consider a relaxed linear time-invariant discrete-time system described by

$$y(k)=\sum_{m=0}^{k} g(k-m)u(m) \tag{8-46}$$

Then for any bounded-input sequence $\{u(k)\}$, (that is, there exists a finite h such that $|u(k)|<h$ for $k=0, 1, 2, \ldots$), the output sequence $\{y(k)\}$ is bounded, if and only if

$$\sum_{k=0}^{\infty} |g(k)|<\infty \tag{8-47}$$

that is, $\{g(k)\}$ is absolutely summerable.[8] The proof of (8-47) is similar to the continuous-time case and is left as an exercise. The z-transform of (8-46) yields

$$\hat{y}(z)=\hat{g}(z)\hat{u}(z)$$

If $\hat{g}(z)$ is a rational function of z, then the system is BIBO stable if and only if all the poles of $\hat{g}(z)$ have magnitudes less than 1, or equivalently, all the poles of $\hat{g}(z)$ lie inside the unit circle of the z plane. This can be readily proved by noting the z-transform pair

$$g(k)=b^k \qquad k=0, 1, 2, \ldots \Leftrightarrow \hat{g}(z)=\mathscr{Z}[g(k)]=\frac{z}{z-b}$$

where b is a real or a complex number. If $|b|<1$, then

$$\sum_{k=0}^{\infty} |b|^k<\infty$$

Otherwise, it diverges.

If $\hat{g}(z)$ is irreducible, the poles of $\hat{g}(z)$ are equal to the roots of its denominator. If the degree of the denominator is three or higher, the computation of the roots is complicated. We introduce in the following a method of checking whether or not all the roots of a polynomial are inside the unit circle without computing explicitly the roots. The method is a counterpart of the Routh-Hurwitz criterion.

Consider the polynomial with real coefficients

$$D(z)=a_0z^n+a_1z^{n-1}+\cdots+a_{n-1}z+a_n \qquad a_0>0 \tag{8-48}$$

We define $a_i^{(0)}=a_i$, $i=0, 1, \ldots, n$, and form the table in Table 8-2. The first row is just the coefficients of $D(z)$. The constant k_0 is the quotient of its last and first elements. The second row is obtained by multiplying k_0 on the first row, except the first element, and then reversing its order. The third row

---

[8] An absolutely integrable function is neither necessarily bounded nor necessarily approaches zero as $t\to\infty$ as shown in Figure 8-2. An absolutely summerable sequence however is always bounded and approaches zero as $k\to\infty$. Thus the stability problem in the discrete-time case is simpler than the one in the continuing time case.

Table 8.2 Discrete-time Stability Table

$a_0^{(0)}$	$a_1^{(0)}$	$\cdots$	$a_{n-2}^{(0)}$	$a_{n-1}^{(0)}$	$a_n^{(0)}$	$k_0 = a_n^{(0)}/a_0^{(0)}$
$-)\ k_0 a_n^{(0)}$	$k_0 a_{n-1}^{(0)}$	$\cdots$	$k_0 a_2^{(0)}$	$k_0 a_1^{(0)}$		
$a_0^{(1)}$	$a_1^{(1)}$	$\cdots$	$a_{n-2}^{(1)}$	$a_{n-1}^{(1)}$		$k_1 = a_{n-1}^{(1)}/a_0^{(1)}$
$-)\ k_1 a_{n-1}^{(1)}$	$k_1 a_{n-2}^{(1)}$	$\cdots$	$k_1 a_1^{(1)}$			
$a_0^{(2)}$	$a_1^{(2)}$	$\cdots$	$a_{n-2}^{(2)}$			$k_2 = a_{n-2}^{(2)}/a_0^{(2)}$
	$\vdots$					
$a_0^{(n-1)}$	$a_1^{(n-1)}$					$k_{n-1} = a_1^{(n-1)}/a_0^{(n-1)}$
$-)\ k_{n-1} a_1^{(n-1)}$						
$a_0^{(n)}$						

is the difference of its two previous rows. The remainder of the table is obtained by the same procedure until n numbers $\{a_0^{(1)}, a_0^{(2)}, \ldots, a_0^{(n)}\}$ are obtained. We define $\alpha_i = a_0^{(i)}$, $i = 0, 1, \ldots, n$.

Theorem 8-21

All the roots of $D(z)$ in (8-48) have magnitudes less than 1 if and only if the n numbers $\alpha_i \triangleq a_0^{(i)}$, $i = 1, 2, \ldots, n$, computed in Table 8-2 are all positive. ∎

We shall prove this after the establishment of the Lyapunov theorem for the discrete-time systems. Consider the linear time-invariant discrete-time dynamical equation

$$\mathbf{x}(k+1) = \mathbf{A}\mathbf{x}(k) + \mathbf{B}\mathbf{u}(k)$$
$$\mathbf{y}(k) = \mathbf{C}\mathbf{x}(k)$$

The concepts of equilibrium state, stability in the sense of Lyapunov and asymptotic stability are identical to the continuous-time case. Similar to Theorem 8-14, every equilibrium state of $\mathbf{x}(k+1) = \mathbf{A}\mathbf{x}(k)$ is stable i.s.L. if and only if all the eigenvalues of $\mathbf{A}$ have magnitudes equal to or less than 1 and those with magnitudes equal to 1 are distinct roots of the minimal polynomial of $\mathbf{A}$. Similar to Theorem 8-15, the zero state of $\mathbf{x}(k+1) = \mathbf{A}\mathbf{x}(k)$ is asymptotically stable if and only if all the eigenvalues of $\mathbf{A}$ have magnitudes less than 1. The Lyapunov theorem for the discrete-time case reads as:

Theorem 8-22

All the eigenvalues of $\mathbf{A}$ have magnitudes less than 1 if and only if for any given positive definite hermitian matrix $\mathbf{N}$ or for any given positive semidefinite hermitian matrix $\mathbf{N}$ with the property $\{\mathbf{A}, \mathbf{N}\}$ observable, the matrix equation

$$\mathbf{A}^*\mathbf{M}\mathbf{A} - \mathbf{M} = -\mathbf{N}$$

has a unique hermitian solution $\mathbf{M}$ and $\mathbf{M}$ is positive definite. ∎

This theorem can be proved by defining

$$V(\mathbf{x}(k)) = \mathbf{x}^*(k)\mathbf{M}\mathbf{x}(k)$$

and computing

$$\begin{aligned}\Delta V(\mathbf{x}(k)) \triangleq V(\mathbf{x}(k+1)) - V(\mathbf{x}(k)) &= \mathbf{x}^*(k)\mathbf{A}^*\mathbf{M}\mathbf{A}\mathbf{x}(k) - \mathbf{x}^*(k)\mathbf{M}\mathbf{x}(k)\\ &= \mathbf{x}^*(k)(\mathbf{A}^*\mathbf{M}\mathbf{A} - \mathbf{M})\mathbf{x}(k)\\ &= -\mathbf{x}^*(k)\mathbf{N}\mathbf{x}(k)\end{aligned}$$

and is left as an exercise (Problem 8-36).

A Proof of Theorem 8-21[9]

Now we shall use Theorem 8-22 to prove Theorem 8-21. Define, for $i = 0, 1, \ldots, n$.

$$D_i(z) = a_0^{(i)}z^{n-i} + a_1^{(i)}z^{n-i-1} + \cdots + a_{n-1}^{(i)} \qquad k_i = a_{n-i}^{(i)}/a_0^{(i)} \tag{8-49}$$

and
$$\bar{D}_i(z) = a_{n-i}^{(i)}z^{n-i} + a_{n-i-1}^{(i)}z^{n-i-1} + \cdots + a_1^{(i)}z + a_0^{(i)} \tag{8-50}$$

where $\bar{D}_i(z)$ is the reciprocal of $D_i(z)$. These polynomials can be defined recursively by

$$D_i(z) = \frac{D_{i-1}(z) - k_{i-1}\bar{D}_{i-1}(z)}{z} \qquad i = 1, 2, \ldots, n \tag{8-51}$$

with $D_0(z) = D(z)$. Note that the coefficients of $D_i(z)$ are the $a_j^{(i)}$, $j = 0, 1, \ldots, n-i$ defined in Table 8-2. It can be verified that the reciprocal of $D_i(z)$ can be expressed as

$$\bar{D}_i(z) = \bar{D}_{i-1}(z) - k_{i-1}D_{i-1}(z) \qquad i = 1, 2, \ldots, n \tag{8-52}$$

From Table 8-2, we have $a_0^{(i+1)} = a_0^{(i)} - k_i a_{n-i}^{(i)} = a_0^{(i)}(1 - k_i^2)$ or

$$k_i^2 = 1 - \frac{a_0^{(i+1)}}{a_0^{(i)}} = 1 - \frac{\alpha_{i+1}}{\alpha_i} \tag{8-53}$$

Simple manipulation among (8-51), (8-52), and (8-53) yields

$$D_{i-1}(z) = (zD_i(z) + k_{i-1}\bar{D}_i(z))\frac{\alpha_{i-1}}{\alpha_i} \tag{8-54}$$

$$\bar{D}_{i-1}(z) = (k_{i-1}zD_i(z) + \bar{D}_i(z))\frac{\alpha_{i-1}}{\alpha_i} \tag{8-55}$$

These recursive equations are valid for $i = 1, 2, \ldots, n$. Note that $D_n(z) = \bar{D}_n(z) = a_0^{(n)}$. Consider now the transfer function

$$\frac{\bar{D}_{i-1}(z)}{D_{i-1}(z)} = \frac{k_{i-1}zD_i(z) + \bar{D}_i(z)}{zD_i(z) + k_{i-1}\bar{D}_i(z)} = k_{i-1} + \frac{(1 - k_{i-1}^2)\bar{D}_i(z)}{zD_i(z) + k_{i-1}\bar{D}_i(z)}$$

which becomes, by using (8-53),

$$\frac{\bar{D}_{i-1}(z)}{D_{i-1}(z)} = k_{i-1} + \frac{\alpha_i\bar{D}_i(z)/\alpha_{i-1}D_i(z)}{z + k_{i-1}\bar{D}_i(z)/D_i(z)} \tag{8-56}$$

[9] This follows Reference S111.

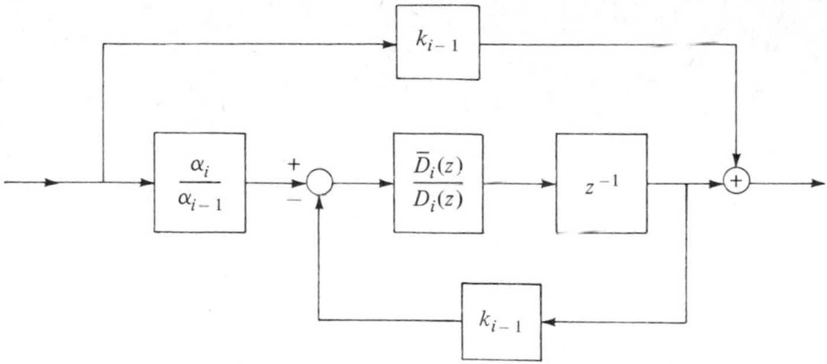

Figure 8-8 A block diagram of $\bar{D}_{i-1}(z)/D_{i-1}(z)$.

for $i=1, 2, \ldots, n$. The block diagram of (8-56) is shown in Figure 8-8. If we apply this diagram repetitively, we can finally obtain the block diagram of $\bar{D}_0(z)/D_0(z)$ shown in Figure 8-9. Note that $\bar{D}_n(z)/D_n(z)=1$. In this block diagram, there is a total of n unit delay elements. If we assign the output of z^{-1} as a state variable as shown, then we will obtain the following state equation

$$\mathbf{x}(k+1)=\mathbf{A}\mathbf{x}(k)+\mathbf{b}u(k)$$
$$y(k)=\mathbf{c}\mathbf{x}(k)+k_0u(k)$$

with

$$\mathbf{c}=[1 \quad 0 \quad 0 \quad \cdots \quad 0 \quad 0]$$

$$\mathbf{b}=\begin{bmatrix} \dfrac{\alpha_1 k_1}{\alpha_0} & \dfrac{\alpha_2 k_2}{\alpha_0} & \cdots & \dfrac{\alpha_{n-1} k_{n-1}}{\alpha_0} & \dfrac{\alpha_n}{\alpha_0} \end{bmatrix}'$$

and

$$\mathbf{A}=\begin{bmatrix} -k_0k_1 & 1 & 0 & \cdots & 0 & 0 \\ -k_0k_2\alpha_2/\alpha_1 & -k_1k_2 & 1 & \cdots & 0 & 0 \\ -k_0k_3\alpha_3/\alpha_1 & -k_1k_3\alpha_3/\alpha_2 & -k_2k_3 & \cdots & 0 & 0 \\ \vdots & \vdots & \vdots & & \vdots & \vdots \\ -k_0k_{n-1}\alpha_{n-1}/\alpha_1 & -k_1k_{n-1}\alpha_{n-1}/\alpha_2 & -k_2k_{n-1}\alpha_{n-1}/\alpha_3 & \cdots & -k_{n-2}k_{n-1} & 1 \\ -k_0\alpha_n/\alpha_1 & -k_1\alpha_n/\alpha_2 & -k_2\alpha_n/\alpha_3 & \cdots & -k_{n-2}\alpha_n/\alpha_{n-1} & -k_{n-1} \end{bmatrix} \tag{8-57}$$

This state equation is obtained from $\bar{D}(z)/D(z)$, and the degree of $D(z)$ is equal to the dimension of $\mathbf{A}$; hence the roots of $D(z)$ are identical to the eigenvalues of $\mathbf{A}$. For this matrix $\mathbf{A}$, if we choose $\mathbf{M}$ as

$$\mathbf{M}=\text{diag}\left\{\frac{1}{\alpha_1}, \frac{1}{\alpha_2}, \ldots, \frac{1}{\alpha_n}\right\}$$

then it is straightforward to verify

$$\mathbf{A}^*\mathbf{M}\mathbf{A}-\mathbf{M}=-\mathbf{N}=-\text{diag}\left\{\frac{1}{a_0}, 0, 0, \ldots, 0\right\}$$

By assumption, we have $a_0>0$; hence $\mathbf{N}$ is positive semidefinite. Furthermore it has the property that $\{\mathbf{A}, \mathbf{N}\}$ is observable. Hence we conclude that all eigenvalues of $\mathbf{A}$ and, consequently, all roots of $D(z)$ have magnitudes less than 1 if and only if $\mathbf{M}$ is positive definite or, equivalently, $\alpha_i>0$, $i=1, 2, \ldots, n$. Q.E.D.

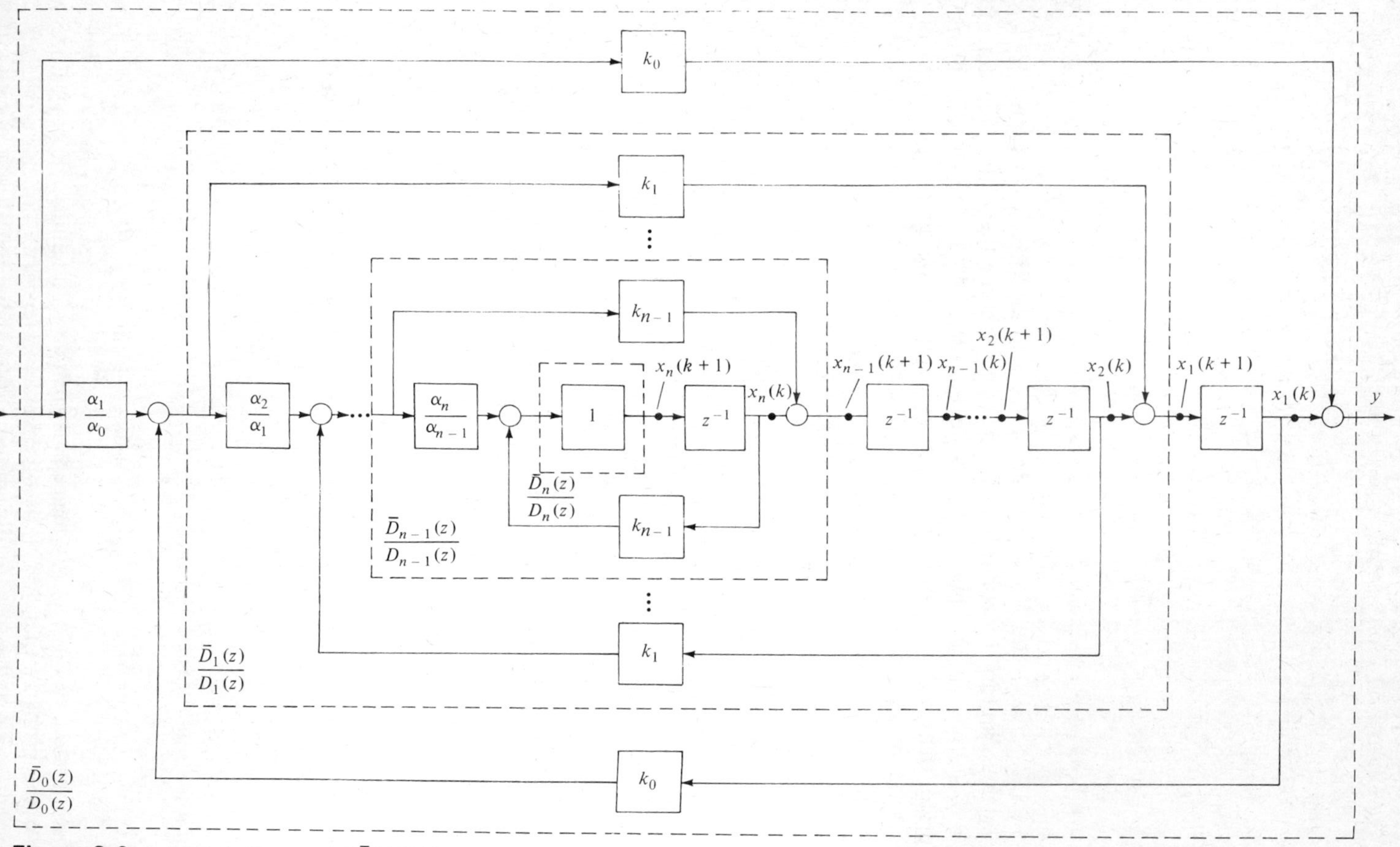

Figure 8-9 A block diagram of $\bar{D}(z)/D(z)$.

8-7 Concluding Remarks

In this chapter we introduced the BIBO stability for the zero-state response and the stability i.s.L. and asymptotic stability for the equilibrium state of the zero-input response. For the time varying case, a system may be stable without being uniformly stable. For the time-invariant case, there is, however, no distinction between uniform stabilities and (nonuniform) stabilities. Although necessary and sufficient conditions are established for the time-varying case, they can hardly be employed because state transition matrices are generally not available.

For the time-invariant case, the stability can be checked from the poles of the transfer function or from the eigenvalues of the matrix **A**. Whether or not all the eigenvalues of **A** have negative real parts can be checked by applying the Routh-Hurwitz criterion to the characteristic polynomial of **A** or by applying the Lyapunov theorem. The characteristic polynomial of **A** can be computed by using the Leverrier algorithm (Problem 2-39), which however is sensitive to computational errors. If the matrix **A** is first transformed into a Hessenberg form by a numerically stable method, the characteristic polynomial of **A** can then be more easily computed. See References S90 and S212. A Hessenberg form is also used in the efficient method of solving the Lyapunov equation in Reference S107. Once **A** is transformed into a Hessenberg form, the computation needed in the Routh-Hurwitz method is much less than that in solving the Lyapunov equation and checking the positive definiteness of **M**. Hence the Routh-Hurwitz method is simpler computationally than the Lyapunov method in checking the stability of **A**. Although no comparison has been carried out, the former may also be more stable numerically than the latter.

The concepts of stability are equally applicable to the discrete-time case. The stability conditions, however, are different. If all the poles of a transfer function lie inside the open left-half s plane, then the continuous-time system is BIBO stable; whereas if all the poles of a transfer function lie inside the open unit circle of the z plane, then the discrete-time system is BIBO stable. Their relationship can be established by using the bilinear transformation

$$s = \frac{z-1}{z+1}$$

which maps the left-half s plane into the unit circle in the z plane. Although we may transform a discrete-time problem into a continuous-time problem by using the bilinear transformation, it is simpler to check the stability directly on discrete-time equations.

Problems

8-1 Is a system with the impulse responses $g(t, \tau) = e^{-2|t| - |\tau|}$, for $t \geq \tau$, BIBO stable? How about $g(t, \tau) = \sin t \, e^{-(t-\tau)} \cos \tau$?

8-2 Is the network shown in Figure P8-2 BIBO stable? If not, find a bounded input that will excite an unbounded output.

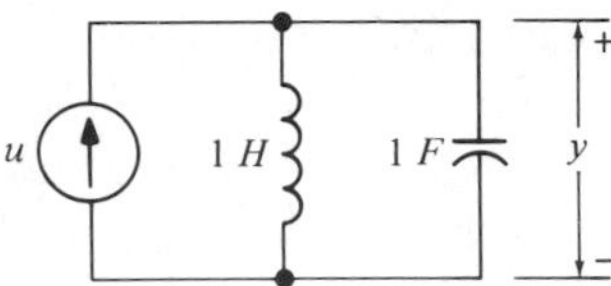

Figure P8-2

8-3 Consider a system with the transfer function $\hat{g}(s)$ that is not necessarily a rational function of s. Show that a necessary condition for the system to be BIBO stable is that $|\hat{g}(s)|$ is finite for all Re $s \geq 0$.

8-4 Consider a system with the impulse response shown in Figure P8-4. If the input $u(t) = \sin 2\pi t$, for $t \geq 0$, is applied, what is the waveform of the output? After how many seconds will the output reach its steady state?

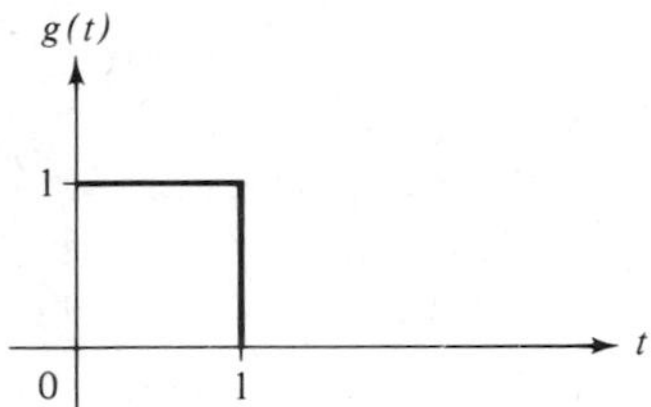

Figure P8-4

8-5 Is a system with the impulse response $g(t) = 1/(1+t)$ BIBO stable?

8-6 Is a system with the transfer function $\hat{g}(s) = e^{-s}/(s+1)$ BIBO stable?

8-7 Use the Routh-Hurwitz criterion to determine which of the following polynomials are Hurwitz polynomials.

a. $s^5 + 4s^4 + 10s^3 + 2s^2 + 5s + 6$
b. $s^5 + s^4 + 2s^3 + 2s^2 + 5s + 5$
c. $-2s^4 - 7s^3 - 4s^2 - 5s - 10$

8-8 Can you determine without solving the roots that the real parts of all the roots of $s^4 + 14s^3 + 71s^2 + 154s + 120$ are smaller than -1? (*Hint*: Let $s = s' - 1$.)

8-9 Give the necessary and sufficient conditions for the following polynomials to be Hurwitz polynomials:

a. $a_0 s^2 + a_1 s + a_2$
b. $a_0 s^3 + a_1 s^2 + a_2 s + a_3$

8-10 Find the dynamical-equation description of the network shown in Problem 8-2. Find the equilibrium states of the equation. Is the equilibrium state stable in the sense of Lyapunov? Is it asymptotically stable?

8-11 Prove parts 1 and 2 of Theorem 8-3 by using the Laplace transform for the class of systems describable by rational transfer functions.

8-12 Consider the dynamical equation

$$\dot{\mathbf{x}} = \begin{bmatrix} 0 & 1 & 0 & 0 & 0 \\ 0 & 0 & 1 & 0 & 0 \\ 0 & 0 & 0 & 1 & 0 \\ 0 & 0 & 0 & 0 & 1 \\ -1 & -1 & -2 & -10 & -4 \end{bmatrix} \mathbf{x} + \begin{bmatrix} 0 \\ 0 \\ 0 \\ 0 \\ 1 \end{bmatrix} u$$

$$y = [\ 0 \quad 0 \quad 0 \quad 0 \quad 0]\mathbf{x}$$

Is the zero state asymptotically stable (for the case $u \equiv 0$)? Is its zero-state response BIBO stable? Is the equation totally stable?

8-13 Consider

$$\begin{bmatrix} \dot{x}_1 \\ \dot{x}_2 \\ \dot{x}_3 \end{bmatrix} = \begin{bmatrix} 0 & 0 & 0 \\ 0 & 0 & 0 \\ 0 & 0 & 0 \end{bmatrix} \begin{bmatrix} x_1 \\ x_2 \\ x_3 \end{bmatrix} + \begin{bmatrix} 1 \\ 0 \\ 0 \end{bmatrix} u$$

$$y = [1 \quad 1 \quad 1]\mathbf{x}$$

Find all the equilibrium states of the equation. Is every equilibrium state stable i.s.L.? Is it asymptotically stable? Is its zero-state response BIBO stable? Is the equation totally stable?

8-14 Check the BIBO stability of a system with the transfer function

$$\frac{2s^2 - 1}{s^5 + 2s^4 + 3s^3 + 5s^2 + 2s + 2}$$

8-15 Find the ranges of k_1 and k_2 such that the system with the transfer function

$$\frac{s + k_1}{s^3 + 2s^2 + k_2 s + 4}$$

is BIBO stable.

8-16 It is known that the dynamical equation

$$\dot{\mathbf{x}} = \begin{bmatrix} \lambda & 1 & 0 \\ 0 & \lambda & 0 \\ 0 & 0 & \lambda \end{bmatrix} \mathbf{x} + \begin{bmatrix} 0 & 0 \\ 1 & 0 \\ 0 & 1 \end{bmatrix} \mathbf{u}$$

$$\mathbf{y} = \begin{bmatrix} 1 & 2 & 0 \\ 0 & 1 & 1 \end{bmatrix} \mathbf{x}$$

is BIBO stable. Can we conclude that the real part of λ is negative? Why?

8-17 Show that $\mathbf{x} = \mathbf{0}$ is the only solution satisfying $(\mathbf{\Phi}(t, t_0) - \mathbf{I})\mathbf{x} \equiv \mathbf{0}$ if $\|\mathbf{\Phi}(t, t_0)\| \to 0$ as $t \to \infty$.

8-18 Consider the following linear time-varying dynamical equation:

$$E: \quad \dot{x} = 2tx + u$$
$$y = x$$

Show that the zero state of E is not stable i.s.L.(under the zero-input response).

8-19 Consider the equivalent equation of E in Problem 8-18 obtained by the equivalence transformation $\bar{x} = P(t)x$, where $P(t) = e^{-t^2}$:

$$\bar{E}: \quad \dot{\bar{x}} = (2te^{-t^2} - 2te^{-t^2})e^{t^2}\bar{x} + e^{-t^2}u = 0 + e^{-t^2}u$$
$$y = e^{t^2}\bar{x}$$

Show that the zero state of $\bar{E}$ is stable i.s.L. (under the zero-input response). From Problems 8-18 and 8-19, we conclude that *an equivalence transformation need not preserve the stability of the zero state.* Does an equivalence transformation preserve the BIBO stability of the zero-state response?

8-20 Show that stability i.s.L. and asymptotic stability of the zero state of $\dot{\mathbf{x}} = \mathbf{A}(t)\mathbf{x}$ are invariant under any Lyapunov transformation (see Definition 4-6). Is the transformation $P(t) = e^{-t^2}$ in Problem 8-19 a Lyapunov transformation?

8-21 Show that if $\dot{\mathbf{x}} = \mathbf{A}(t)\mathbf{x}$ is stable i.s.L. at t_0, then it is stable i.s.L. at every $t_1 \geq t_0$. [*Hint*: Use $\mathbf{\Phi}(t, t_1) = \mathbf{\Phi}(t, t_0)\mathbf{\Phi}^{-1}(t_1, t_0)$ and note the boundedness of $\mathbf{\Phi}^{-1}(t_1, t_0)$ for any finite t_1 and t_0.]

8-22 Consider a system with the following dynamical-equation description:

$$\dot{\mathbf{x}} = \begin{bmatrix} -1 & 1 & 0 \\ 0 & -1 & 0 \\ 0 & 0 & 0 \end{bmatrix} \mathbf{x} + \begin{bmatrix} 1 \\ 2 \\ 0 \end{bmatrix} u$$
$$y = [\,2 \quad 3 \quad 1\,]\,\mathbf{x}$$

Is the zero state asymptotically stable? Is the zero-state response BIBO stable? Is the system T-stable?

8-23 Prove Theorem 8-13.

8-24 Prove condition 3 of Theorem 8-19. *Hint:* Use $\mathbf{M} = \mathbf{P}\hat{\mathbf{M}}\mathbf{P}^*$ and $\hat{\mathbf{M}} = (\hat{\mathbf{M}})^{1/2}(\hat{\mathbf{M}})^{1/2}$.

8-25 Consider the linear time-invariant controllable state equation

$$\dot{\mathbf{x}} = \mathbf{A}\mathbf{x} + \mathbf{B}\mathbf{u}$$

Show that if $\mathbf{u} = -\mathbf{B}^*\mathbf{W}^{-1}(T)\mathbf{x}$, where

$$\mathbf{W}(T) = \int_0^T e^{-\mathbf{A}\tau}\mathbf{B}\mathbf{B}^* e^{\mathbf{A}^*\tau}\, d\tau \qquad T \text{ is an arbitrary positive number}$$

then the overall system is asymptotically stable. Furthermore, $V(\mathbf{x}(t)) = \mathbf{x}^*(t)\mathbf{W}^{-1}(T)\mathbf{x}(t)$ is a suitable Lyapunov function for the closed-loop system.

8-26 Are the networks shown in Figure P8-26 totally stable? *Answers*: No; yes.

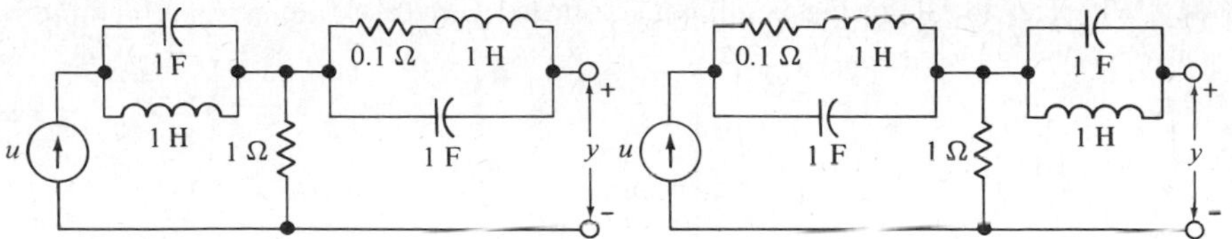

Figure P8-26

8-27 Consider a discrete-time system that is described by

$$y(n) = \sum_{m=-\infty}^{n} g(n, m)u(m)$$

Show that any bounded-input sequence $\{u(n)\}$ excites a bounded-output sequence $\{y(n)\}$ if and only if

$$\sum_{m=-\infty}^{n} |g(n, m)| \leq k \leq \infty \qquad \text{for all } n$$

8-28 Consider a discrete-time system that is described by

$$y(n) = \sum_{m=0}^{n} g(n-m)u(m)$$

Show that any bounded-input sequence $\{u(n)\}$ excites a bounded-output sequence $\{y(n)\}$ if and only if

$$\sum_{m=0}^{\infty} |g(m)| \leq k < \infty$$

8-29 Consider a system with the impulse response

$$g(t) = g_1(t) + \sum_{i=0}^{\infty} a_i \delta(t - \tau_i)$$

Show that the system is BIBO stable if and only if

$$\int_0^{\infty} |g_1(t)|\, dt \leq k_1 < \infty$$

and

$$\sum_{i=0}^{\infty} |a_i| \leq k_2 < \infty$$

8-30 Prove Corollary 8-3 by using partial fraction expansion for the class of systems that have proper rational transfer-function descriptions.

8-31 Is the function

$$[x_1 \quad x_2 \quad x_3] \begin{bmatrix} 1 & 2 & 2 \\ 0 & 1 & 1 \\ 2 & 2 & 6 \end{bmatrix} \begin{bmatrix} x_1 \\ x_2 \\ x_3 \end{bmatrix}$$

positive definite or semidefinite? [*Hint*: Use Equation (E-2) in Appendix E.]

8-32 Which of the following hermitian (symmetric) matrices are positive definite or positive semidefinite?

a. $\begin{bmatrix} 2 & 3 & 2 \\ 3 & 1 & 0 \\ 2 & 0 & 2 \end{bmatrix}$ **b.** $\begin{bmatrix} 0 & 0 & 1 \\ 0 & 0 & 0 \\ 1 & 0 & 2 \end{bmatrix}$

c. $\begin{bmatrix} 0 & 0 & 0 \\ 0 & 1 & 0 \\ 0 & 0 & 0 \end{bmatrix}$ **d.** $\begin{bmatrix} a_1a_1 & a_1a_2 & a_1a_3 \\ a_1a_2 & a_2a_2 & a_2a_3 \\ a_1a_3 & a_2a_3 & a_3a_3 \end{bmatrix}$

where a_i, $i=1, 2, 3$ are any real numbers.

8-33 Let $\lambda_1 = -1$, $\lambda_2 = -2$, $\lambda_3 = -3$ and let a_1, a_2, a_3 be arbitrary real numbers, not zero. Prove by using Corollary 8-20 that the matrix

$$\mathbf{M} = \begin{bmatrix} -\dfrac{a_1^2}{2\lambda_1} & -\dfrac{a_1a_2}{\lambda_1+\lambda_2} & -\dfrac{a_1a_3}{\lambda_1+\lambda_3} \\ -\dfrac{a_2a_1}{\lambda_2+\lambda_1} & -\dfrac{a_2^2}{2\lambda_2} & -\dfrac{a_2a_3}{\lambda_2+\lambda_3} \\ -\dfrac{a_3a_1}{\lambda_1+\lambda_3} & -\dfrac{a_3a_2}{\lambda_2+\lambda_3} & -\dfrac{a_3^2}{2\lambda_3} \end{bmatrix}$$

is a positive definite matrix. [*Hint*: Let $\mathbf{A} = \text{diag}(\lambda_1, \lambda_2, \lambda_2)$.]

8-34 A real matrix $\mathbf{M}$ (not necessarily symmetric) is defined to be, as in Definition 8-6, positive definite if $\mathbf{x}'\mathbf{M}\mathbf{x} > 0$ for all nonzero $\mathbf{x}$ in $\mathbb{R}^n$. Is it true that the matrix $\mathbf{M}$ is positive definite if all the eigenvalues of $\mathbf{M}$ are positive real or if all the leading principal minors are positive? If not, how do you check its positive definiteness? *Hint:* Try

$$\begin{bmatrix} 0 & 1 \\ -2 & 3 \end{bmatrix} \quad \begin{bmatrix} 2 & 1 \\ 1.9 & 1 \end{bmatrix}$$

8-35 Let $\mathbf{M}$ be a hermitian matrix of order n. Let $\mathbf{e}_i$, $i=1, 2, \ldots, n$, be a set of linearly independent vectors. Is it true that if $\mathbf{e}_i^*\mathbf{M}\mathbf{e}_i > 0$ for $i=1, 2, \ldots, n$, then $\mathbf{M}$ is positive definite? (*Answer*: False.)

8-36 Prove Theorem 8-22.

8-37 Determine the asymptotic stability of the discrete-time equation

$$\mathbf{x}(k+1) = \begin{bmatrix} 0 & 1 & 0 & 0 \\ 0 & 0 & 1 & 0 \\ 0 & 0 & 0 & 1 \\ 0.008 & 0.008 & -0.79 & -0.8 \end{bmatrix} \mathbf{x}(k)$$

(1) by computing its characteristic polynomial and then applying Theorem 8-21 and (2) by solving the Lyapunov equation in Theorem 8-22.

8-38 Consider the internally balanced system in Problems 6-23 to 6-25 partitioned as

$$\begin{bmatrix} \dot{\mathbf{x}}_1 \\ \dot{\mathbf{x}}_2 \end{bmatrix} = \begin{bmatrix} \mathbf{A}_{11} & \mathbf{A}_{12} \\ \mathbf{A}_{21} & \mathbf{A}_{22} \end{bmatrix} \begin{bmatrix} \mathbf{x}_1 \\ \mathbf{x}_2 \end{bmatrix} + \begin{bmatrix} \mathbf{B}_1 \\ \mathbf{B}_2 \end{bmatrix} \mathbf{u}$$

$$\mathbf{y} = [\mathbf{C}_1 \quad \mathbf{C}_2] \begin{bmatrix} \mathbf{x}_1 \\ \mathbf{x}_2 \end{bmatrix}$$

and $\mathbf{W} = \text{diag}\,\{\mathbf{W}_1, \mathbf{W}_2\}$. Show that if the equation is asymptotically stable and if $\mathbf{W}_1$ and $\mathbf{W}_2$ have no diagonal entries in common, then the subsystems

$$\dot{\mathbf{x}}_i = \mathbf{A}_{ii}\mathbf{x}_i + \mathbf{B}_i\mathbf{u}$$
$$\mathbf{y}_i = \mathbf{C}_i\mathbf{x}_i$$

for $i = 1, 2$, are asymptotically stable. This result is useful in system reduction. See References S162 and S176.

8-39 Consider the polynomial in (8-48) with the following companion matrix

$$\mathbf{A}_c = \begin{bmatrix} 0 & 1 & 0 & \cdots & 0 \\ 0 & 0 & 1 & \cdots & 0 \\ \vdots & \vdots & \vdots & & \vdots \\ 0 & 0 & 0 & \cdots & 1 \\ -a_n & -a_{n-1} & -a_{n-2} & \cdots & -a_1 \end{bmatrix}$$

and the matrix $\mathbf{A}$ in (8-57) obtained from the discrete-time stability table in Table 8-2. Verify that $\mathbf{A}_c = \mathbf{Q}^{-1}\mathbf{A}\mathbf{Q}$ or $\mathbf{Q}\mathbf{A}_c = \mathbf{A}\mathbf{Q}$, where

$$\mathbf{Q} = \begin{bmatrix} a_0^{(1)} & a_1^{(1)} & a_2^{(1)} & \cdots & a_{n-2}^{(1)} & a_{n-1}^{(1)} \\ 0 & a_0^{(2)} & a_1^{(2)} & \cdots & a_{n-3}^{(2)} & a_{n-2}^{(2)} \\ \vdots & \vdots & \vdots & & \vdots & \vdots \\ 0 & 0 & 0 & \cdots & a_0^{(n-1)} & a_1^{(n-1)} \\ 0 & 0 & 0 & \cdots & 0 & a_0^{(n)} \end{bmatrix}$$

(Due to Y. P. Harn.)

8-40 Consider the system shown in Figure P8-40, where S_i is described by $\dot{\mathbf{x}}_i = \mathbf{A}_i\mathbf{x}_i + \mathbf{B}_i\mathbf{u}_i$, $\mathbf{y}_i = \mathbf{C}_i\mathbf{x}_i + \mathbf{E}_i\mathbf{u}_i$. Use the composite state $\mathbf{x}' = [\mathbf{x}_1' \quad \mathbf{x}_2']$ to develop a state variable description of the system with $[\mathbf{r}_1' \quad \mathbf{r}_2']'$ as the input and $[\mathbf{y}_1' \quad \mathbf{y}_2' \quad \mathbf{u}_1' \quad \mathbf{u}_2']'$ as the output. Is the matrix $\mathbf{A}$ of this description the same as the one in the time-invariant case of (3-64)? What is the condition for the system to be asymptotically stable? Will the condition ensure that the system be BIBO stable from any input-output pair?

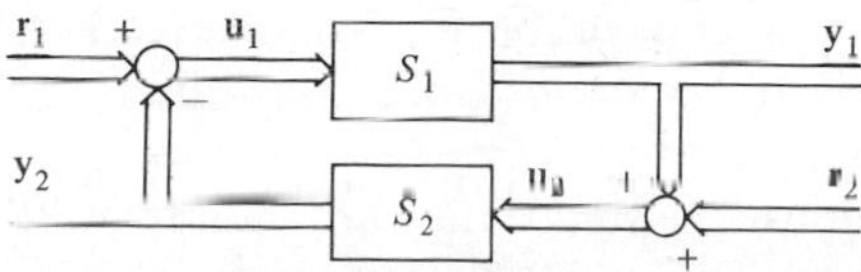

Figure P8-40

9

Linear Time-Invariant Composite Systems: Characterization, Stability, and Designs

9-1 Introduction

The design of control systems can be formulated as follows. Given a plant, design an overall system to meet certain design objectives. Because of the presence of noises, the compensators are required to have proper rational functions or matrices. In order to reduce sensitivity due to the plant variations and load disturbances, the configuration of the overall system must be of feedback or closed-loop type. There are many possible feedback configurations. In this chapter we study only the two configurations shown in Figure 9-1. The one in Figure 9-1(a) will be called the *unity feedback system*. The one in Figure 9-1(b) will be called the *plant input-output feedback system* or, simply, *the input-output feedback system*,[1] for the feedbacks are introduced from both the input and output of the plant $\hat{g}(s)$. This configuration arises from the combination of the state-feedback and state-estimator studied in Chapter 7. Before proceeding, the reader should review the time-invariant part of Section 3-6.

A control system is, as can be seen from Figure 9-1, basically a composite system. By a composite system, we mean that the system consists of two or more subsystems. Rather than plunging directly into the design problem, we study first some basic problems associated with composite systems. We assume that there is no loading effect in any connection of two subsystems; that is, the transfer function of each subsystem remains unchanged after the connection (see Reference S46).

Although there are many forms of composite systems, they are mainly built

[1] This terminology is by no means universal. Another possible name is the Luenberger- or state-estimator-type configuration.

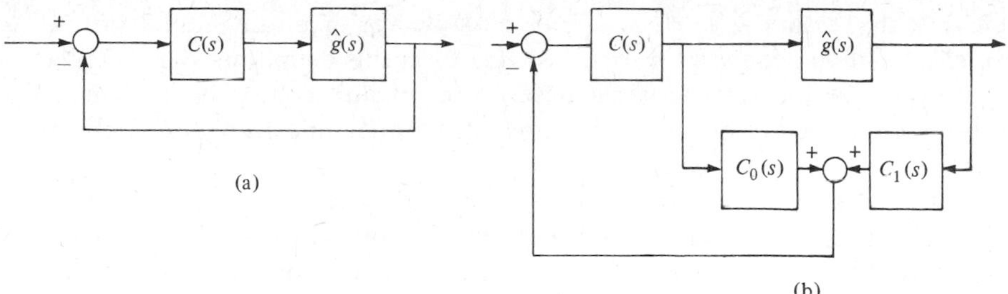

Figure 9-1 (a) Unity feedback system. (b) Input-output feedback system.

up from three basic connections: parallel, tandem, and feedback connections. Hence we shall restrict ourselves to the studies of these three connections. If all the subsystems that form a composite system are all linear and time-invariant, it is easy to show that the composite system is again a linear time-invariant system. For linear composite systems, all the results in the previous chapters can be directly applied. For example, consider the feedback connection of two linear time-invariant systems with a transfer function $\hat{g}_1(s)$ in the forward path and a transfer function $\hat{g}_2(s)$ in the feedback path. If the transfer function of the feedback connection, $\hat{g}_f(s) = (1 + \hat{g}_1(s)\hat{g}_2(s))^{-1}\hat{g}_1(s)$, is computed, then the input-output stability of the feedback system can be determined from $\hat{g}_f(s)$. However, many questions can still be raised regarding the composite system. For example, what is the implication if there are pole-zero cancellations between $\hat{g}_1(s)$ and $\hat{g}_2(s)$? Is it possible to determine the stability of the feedback system from $\hat{g}_1$ and $\hat{g}_2$ without computing $\hat{g}_f(s)$? In the first part of this chapter, these questions for the single-variable as well as multivariable systems will be answered. Before proceeding, we introduce a definition.

Definition 9-1

A system is said to be *completely characterized* by its rational transfer-function matrix if and only if the dynamical-equation description of the system is controllable and observable. ∎

The motivation of this definition is as follows. If the dynamical-equation description of a system is uncontrollable and/or unobservable, then the transfer-function matrix of the system describes only the part of the system which is controllable from the input and observable at the output; hence the transfer function does not describe the system fully. On the other hand, if the dynamical-equation description of a system is controllable and observable, then the information obtained from the dynamical equation and the one from the transfer function of the system will be essentially the same. Hence, in this case the system is said to be completely characterized by its transfer-function matrix.

Every dynamical equation can be reduced to a controllable and observable one; hence when we apply Definition 9-1 to a composite system, we must clarify what its dynamical-equation description is. Let S_i, $i = 1, 2$, be two subsystems with state vectors $\mathbf{x}_i$, $i=1,2$. Then the state vector of any connection of S_1 and S_2 will be defined as $\mathbf{x}' = [\mathbf{x}_1' \quad \mathbf{x}_2']$. With this definition, there will be no confusion in the state-variable description of any composite system. The state-variable descriptions of the tandem, parallel, and feedback connections of two systems are derived in (3-62) to (3-64).

All the subsystems that form a composite system will be assumed to be completely characterized by their transfer-function matrices. This assumption, however, does not imply that a composite system is completely characterized by its composite transfer-function matrix. In Section 9-2 we study the conditions of complete characterization of composite systems. For single-variable systems, the conditions are very simple; if there is no common pole in the parallel connection, or no pole-zero cancellation in the tandem and the feedback connections, then the composite system is completely characterized by its transfer function. In Section 9-3 we extend the results of Section 9-2 to the multivariable case, with the condition of pole-zero cancellation replaced by the condition of coprimeness. In Section 9-4 we study the stability problem of composite systems. The stability conditions are stated in terms of the transfer matrices of the subsystems. The remainder of the chapter is devoted to the design problem. In Section 9-5 we study the design of compensators in the unity feedback system to achieve arbitrary pole placement and arbitrary denominator matrix. The problem of pole placement and that of arbitrary denominator are identical in the single-variable case. They are, however, different in the multivariable case; the compensator required for the latter is much more complicated than that required for the former. In Section 9-6 we design robust control systems to achieve asymptotic tracking and disturbance rejection. The static decoupling problem is also discussed; both robust and nonrobust designs are considered. In the last section, we study the design of compensators in the input-output feedback system. The results are more general than those obtained in the state-variable approach discussed in Chapter 7.

The references for this chapter are S2, S11, S19, S34, S35, S40, S49 to S51, S54, S55, S64 to S66, S75, S81, S85, S93, S94, S98, S174, S185, S199, S218, S237, and S238.

9-2 Complete Characterization of Single-Variable Composite Systems

Consider two systems S_i, for $i=1,2$, with the dynamical-equation descriptions

$$FE^i: \quad \dot{\mathbf{x}}_i = \mathbf{A}_i\mathbf{x}_i + \mathbf{B}_i\mathbf{u}_i \tag{9-1a}$$

$$\mathbf{y}_i = \mathbf{C}_i\mathbf{x}_i + \mathbf{E}_i\mathbf{u}_i \tag{9-1b}$$

where $\mathbf{x}_i$, $\mathbf{u}_i$, and $\mathbf{y}_i$ are, respectively, the state, the input, and the output of the system S_i. $\mathbf{A}_i$, $\mathbf{B}_i$, $\mathbf{C}_i$, and $\mathbf{E}_i$ are real constant matrices. The transfer-function matrix of S_i is

$$\hat{\mathbf{G}}_i(s) = \mathbf{C}_i(s\mathbf{I} - \mathbf{A}_i)^{-1}\mathbf{B}_i + \mathbf{E}_i \tag{9-2}$$

It is assumed that the systems S_1 and S_2 are completely characterized by their transfer-function matrices $\hat{\mathbf{G}}_1(s)$ and $\hat{\mathbf{G}}_2(s)$; or, equivalently, the dynamical equations (9-1) are controllable and observable. It was shown in Section 3-6 that the transfer-function matrix of the parallel connection of S_1 and S_2 is $\hat{\mathbf{G}}_1(s)+\hat{\mathbf{G}}_2(s)$; the transfer-function matrix of the tandem connection of S_1 followed by S_2 is $\hat{\mathbf{G}}_2(s)\hat{\mathbf{G}}_1(s)$; the transfer-function matrix of the feedback connection of S_1 with S_2 in the feedback path is $\hat{\mathbf{G}}_1(s)(\mathbf{I}+\hat{\mathbf{G}}_2(s)\hat{\mathbf{G}}_1(s))^{-1}=(\mathbf{I}+\hat{\mathbf{G}}_1(s)\hat{\mathbf{G}}_2(s))^{-1}\hat{\mathbf{G}}_1(s)$. Although $\hat{\mathbf{G}}_1(s)$ and $\hat{\mathbf{G}}_2(s)$ completely characterize the systems S_1 and S_2, respectively, it does not follow that a composite transfer function $\hat{\mathbf{G}}(s)$ completely characterizes a composite system.

Example 1

Consider the parallel connection of two single-variable systems S_1 and S_2 whose dynamical-equation descriptions are, respectively,

$$FE_1^1: \quad \dot{x}_1 = x_1 + u_1$$
$$y_1 = x_1 + u_1$$

and

$$FE_1^2: \quad \dot{x}_2 = x_2 - u_2$$
$$y_2 = x_2$$

Their transfer functions are

$$\hat{g}_1(s)=\frac{s}{s-1} \qquad \text{and} \qquad \hat{g}_2(s)=\frac{-1}{s-1}$$

The composite transfer function of the parallel connection of S_1 and S_2 is

$$g(s)=\hat{g}_1(s)+\hat{g}_2(s)=\frac{s}{s-1}+\frac{-1}{s-1}=1$$

It is clear that $\hat{g}(s)=1$ does not characterize completely the composite system, because $\hat{g}(s)$ does not reveal the unstable mode e^t in the system. This can also be checked from the composite dynamical equation. In the parallel connection, we have $u_1=u_2=u$ and $y=y_1+y_2$; hence the composite dynamical equation is

$$\begin{bmatrix} \dot{x}_1 \\ \dot{x}_2 \end{bmatrix} = \begin{bmatrix} 1 & 0 \\ 0 & 1 \end{bmatrix} \begin{bmatrix} x_1 \\ x_2 \end{bmatrix} + \begin{bmatrix} 1 \\ -1 \end{bmatrix} u$$
$$y = [1 \quad 1]\mathbf{x} + u$$

It is easy to check that the composite equation is not controllable and not observable; hence from Definition 9-1, the composite system is not completely characterized by its composite transfer function. ∎

In the following we shall study the conditions under which composite transfer functions completely describe composite systems. We study this problem directly from transfer-function matrices without looking into dynamical equations. If a system is completely characterized by its transfer-function matrix, then the dimension of the dynamical-equation description of the system is equal to the degree of its transfer-function matrix (Theorem 6-2).

Therefore, whether or not a system is completely characterized by its transfer-function matrix can be checked from the number of state variables of the system. If a system is an *RLC* network,[2] then the number of state variables is equal to the number of energy-storage elements (inductors and capacitors); hence an *RLC network*[2] *is completely characterized by its transfer-function matrix if and only if the number of energy storage elements is equal to the degree of its transfer-function matrix.* Consider now two *RLC* networks S_1 and S_2, which are completely characterized by their transfer-function matrices $\hat{\mathbf{G}}_1(s)$ and $\hat{\mathbf{G}}_2(s)$, respectively. The number of energy-storage elements in any composite connection of S_1 and S_2 is clearly equal to $\delta\hat{\mathbf{G}}_1(s)+\delta\hat{\mathbf{G}}_2(s)$. Let $\hat{\mathbf{G}}(s)$ be the transfer-function matrix of the composite connection of S_1 and S_2. Now the composite system consists of $(\delta\hat{\mathbf{G}}_1(s)+\delta\hat{\mathbf{G}}_2(s))$ energy-storage elements; hence, in order for $\hat{\mathbf{G}}(s)$ to characterize the composite system completely, it is necessary and sufficient to have $\delta\hat{\mathbf{G}}(s)=\delta\hat{\mathbf{G}}_1(s)+\delta\hat{\mathbf{G}}_2(s)$. This is stated as a theorem.

Theorem 9-1

Consider two systems S_1 and S_2, which are completely characterized by their proper transfer-function matrices $\hat{\mathbf{G}}_1(s)$ and $\hat{\mathbf{G}}_2(s)$, respectively. Any composite connection of S_1 and S_2 is completely characterized by its composite transfer-function matrix $\hat{\mathbf{G}}(s)$ if and only if

$$\delta\hat{\mathbf{G}}(s)=\delta\hat{\mathbf{G}}_1(s)+\delta\hat{\mathbf{G}}_2(s) \qquad \blacksquare$$

This theorem can also be verified from the dynamical-equation descriptions of systems. Recall from Section 3-6 that the state space of any composite connection of S_1 and S_2 is chosen to be the direct sum of the state spaces of S_1 and S_2; consequently, the dimension of the composite dynamical equation is the sum of the dimensions of the dynamical-equation descriptions of S_1 and S_2. Hence Theorem 9-1 follows directly from Definition 9-1 and Theorem 6-2.

In order to apply Theorem 9-1 we must first compute the transfer-function matrix of a composite system. This is not desirable, particularly in the design of feedback control systems. Hence the conditions in terms of $\hat{\mathbf{G}}_1$ and $\hat{\mathbf{G}}_2$ for $\hat{\mathbf{G}}$ to characterize completely the composite connections of S_1 and S_2 will be studied. We study in this section only single-variable systems. The multivariable systems will be studied in the next section.

The transfer function of a single-variable system is a scalar, and its degree is just the degree of its denominator if the transfer function is irreducible. We assume in this section that all transfer functions are irreducible; that is, their denominators and numerators are coprime.

Theorem 9-2

Consider two single-variable systems S_1 and S_2, which are completely characterized by their proper rational transfer functions $\hat{g}_1(s)$ and $\hat{g}_2(s)$.

[2] We assume that there are no capacitors-only loops and inductors-only cutsets in the network.

1. The parallel connection of S_1 and S_2 is completely characterized by $\hat{g}(s) = \hat{g}_1(s) + \hat{g}_2(s)$ if and only if $\hat{g}_1(s)$ and $\hat{g}_2(s)$ do not have any pole in common.
2. The tandem connection of S_1 and S_2 is completely characterized by $\hat{g}(s) = \hat{g}_2(s)\hat{g}_1(s)$ if and only if there is no pole-zero cancellation between $\hat{g}_1$ and $\hat{g}_2$.
3. The feedback connection of S_1 and S_2 shown in Figure 9-2 is completely characterized by $\hat{g}(s) = (1 + \hat{g}_1\hat{g}_2)^{-1}\hat{g}_1$ if and only if there is no pole of $\hat{g}_2(s)$ canceled by any zero of $\hat{g}_1(s)$.

Proof

1. It is obvious that if $\hat{g}_1$ and $\hat{g}_2$ have at least one pole in common, then $\delta\hat{g} < \delta\hat{g}_1 + \delta\hat{g}_2$. Let $\hat{g}_i = N_i/D_i$, for $i = 1, 2$; then
$$\hat{g} = \hat{g}_1 + \hat{g}_2 = \frac{N_1D_2 + N_2D_1}{D_1D_2}$$
We show now that if $\hat{g}_1$ and $\hat{g}_2$ do not have any pole in common, then $\delta\hat{g} = \delta\hat{g}_1 + \delta\hat{g}_2$. We prove this by contradiction. Suppose $\delta\hat{g} < \delta\hat{g}_1 + \delta\hat{g}_2$, then there is at least one common factor between $N_1D_2 + N_2D_1$ and D_1D_2. If there is a common factor, say, between $N_1D_2 + N_2D_1$ and D_1, then the assumption that there is no common factor between D_1 and D_2 implies that there is a common factor between N_1 and D_1. This contradicts the assumption that $\hat{g}_1$ is irreducible. Hence we conclude that if $\hat{g}_1$ and $\hat{g}_2$ have no pole in common, then $\hat{g}(s) = \hat{g}_1(s) + \hat{g}_2(s)$ characterizes completely the parallel connection of S_1 and S_2.
2. The proof of this part is obvious and is omitted.
3. The transfer function of the feedback system shown in Figure 9-2 is
$$\hat{g}(s) = \frac{\hat{g}_1(s)}{1 + \hat{g}_1(s)\hat{g}_2(s)} = \frac{D_2N_1}{D_1D_2 + N_1N_2}$$
By the irreducibility assumption, D_1 and N_1 have no common factor, nor have D_2 and N_2. Hence D_2N_1 and $D_1D_2 + N_1N_2$ have common factors if and only if D_2 and N_1 have common factors. Q.E.D.

Example 2

Consider the tandem connection of S_1 and S_2 with transfer functions

$$\frac{1}{s-1} \quad \text{and} \quad \frac{s-1}{s+1}$$

as shown in Figure 9-3(a). There is a pole-zero cancellation between $\hat{g}_1$ and $\hat{g}_2$.

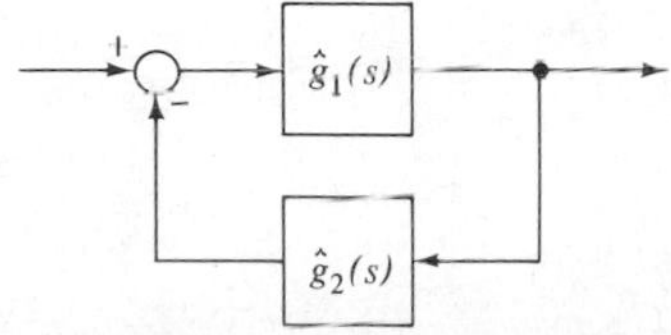

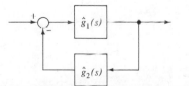

Figure 9-2 A single-variable feedback system.

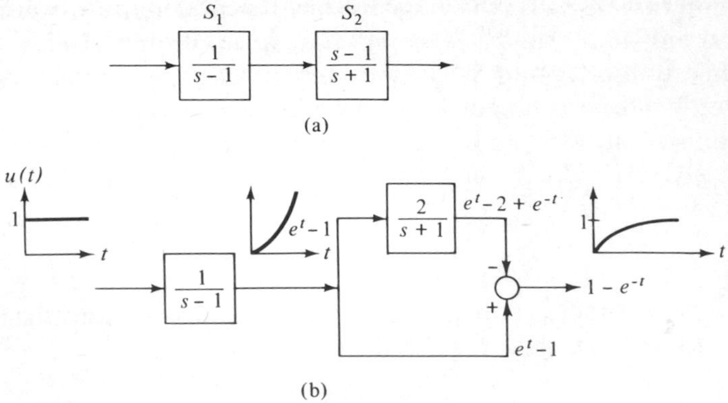

Figure 9-3 Tandem connection of S_1 and S_2, which is not characterized completely by $\hat{g}(s)=\hat{g}_2\hat{g}_1=1/(s+1)$.

Hence the composite transfer function

$$\hat{g}(s)=\hat{g}_2(s)\hat{g}_1(s)=\frac{1}{s+1}$$

does not completely characterize the tandem connection. This can be seen by applying a unit step input to the composite system; although the output of the tandem connection is bounded, the output of S_1 increases exponentially with time, as shown in Figure 9-3(b). ∎

Example 3

Consider the feedback connections shown in Figure 9-4. In Figure 9-4(a), the pole of the transfer function in the feedback path is canceled by the zero of the transfer function in the forward path. Hence the transfer function of the feedback system does not completely describe the feedback system. Indeed, its transfer function is

$$\hat{g}(s)=\frac{\dfrac{s-1}{s+1}}{1+\dfrac{s-1}{s+1}\dfrac{1}{s-1}}=\frac{s-1}{s+2}$$

Its degree is smaller than 2.

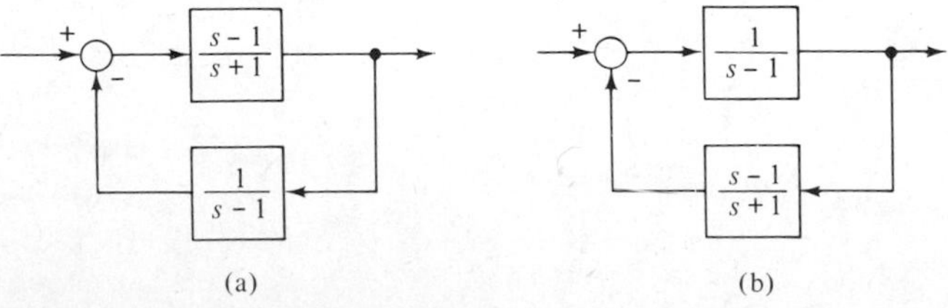

Figure 9-4 Feedback systems.

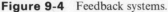

On the other hand, although the pole of the transfer function in the forward path of Figure 9-4(b) is canceled by the zero of the transfer function in the feedback path, the transfer function of the feedback system still completely characterizes the feedback system. Its transfer function is

$$\hat{g}(s) = \frac{s+1}{(s-1)(s+2)}$$ ∎

Consider a special case of the feedback system shown in Figure 9-2 with $\hat{g}_2(s) = k$, where k is a real constant. Since there is no pole in $\hat{g}_2(s)$ to be canceled by $\hat{g}_1(s)$, the transfer function

$$\hat{g}(s) = [1 + k\hat{g}_1(s)]^{-1}\hat{g}_1(s) = \frac{N_1}{D_1 + kN_1}$$

always characterizes the feedback system completely.

9-3 Controllability and Observability of Composite Systems

Let $\hat{g}_i(s) = N_i(s)D_i^{-1}(s)$, $i = 1, 2$, and let $N_i(s)$ and $D_i(s)$ be coprime. In terms of N_i and D_i, Theorem 9-2 can be stated as follows: The parallel connection of S_1 and S_2 is completely characterized by $\hat{g}_1(s) + \hat{g}_2(s)$ if and only if $D_1(s)$ and $D_2(s)$ are coprime. The tandem connection is completely characterized by $\hat{g}_2(s)\hat{g}_1(s)$ if and only if $D_1(s)$ and $N_2(s)$ are coprime and $D_2(s)$ and $N_1(s)$ are coprime. In this section we shall extend these conditions to the multivariable composite systems. Before proceeding, the reader is advised to review Section 6-8, on strict system equivalence.

A system can be described by a dynamical equation $\{\mathbf{A}, \mathbf{B}, \mathbf{C}, \mathbf{E}\}$, a transfer matrix in fractional forms $\hat{\mathbf{G}}(s) = \mathbf{N}_r(s)\mathbf{D}_r^{-1}(s) = \mathbf{D}_l^{-1}(s)\mathbf{N}_l(s)$, or a system matrix

$$\begin{bmatrix} \mathbf{P}(s) & \mathbf{Q}(s) \\ -\mathbf{R}(s) & \mathbf{W}(s) \end{bmatrix} \begin{bmatrix} \hat{\boldsymbol{\xi}}(s) \\ -\hat{\mathbf{u}}(s) \end{bmatrix} = \begin{bmatrix} \mathbf{0} \\ -\hat{\mathbf{y}}(s) \end{bmatrix}$$

where $\mathbf{u}$ and $\mathbf{y}$ are the input and output and $\boldsymbol{\xi}$ is the pseudostate. If these descriptions describe the same system, they all have the same transfer matrix. If $\{\mathbf{A}, \mathbf{B}, \mathbf{C}\}$ is not irreducible (not controllable or not observable) or if $\{\mathbf{D}_r, \mathbf{N}_r\}$, $\{\mathbf{D}_l, \mathbf{N}_l\}$, and $\{\mathbf{P}(s), \mathbf{Q}(s), \mathbf{R}(s)\}$ are not coprime, then they are generally not strictly system equivalent. However, if they are all irreducible, they are strictly system equivalent, and any one of them can be used in the analysis and design. For our problem, it turns out that the use of the system matrix is the most convenient. Hence the system matrix will be used extensively in this section. We recall from (6-151) and (6-152) that $\{\mathbf{A}, \mathbf{B}\}$ is controllable if and only if $\{\mathbf{P}(s), \mathbf{Q}(s)\}$ is left coprime; $\{\mathbf{A}, \mathbf{C}\}$ is observable if and only if $\{\mathbf{P}(s), \mathbf{R}(s)\}$ is right coprime. These properties will be used to establish the conditions for complete characterization for the multivariable case.

Let $\hat{\mathbf{G}}_i(s)=\mathbf{N}_{ri}(s)\mathbf{D}_{ri}^{-1}(s)=\mathbf{D}_{li}^{-1}(s)\mathbf{N}_{li}(s)$ be, respectively, right and left coprime fraction of $\hat{\mathbf{G}}_i(s)$. Then system S_i can also be described by

$$\begin{bmatrix} \mathbf{D}_{ri}(s) & \mathbf{I} \\ -\mathbf{N}_{ri}(s) & \mathbf{0} \end{bmatrix}\begin{bmatrix} \hat{\boldsymbol{\xi}}_i(s) \\ -\hat{\mathbf{u}}_i(s) \end{bmatrix}=\begin{bmatrix} \mathbf{0} \\ -\hat{\mathbf{y}}_i(s) \end{bmatrix} \qquad i=1,2 \tag{9-3}$$

where $\hat{\boldsymbol{\xi}}_i(s)$, $\hat{\mathbf{u}}_i(s)$ and $\hat{\mathbf{y}}_i(s)$ are, respectively, $p_i \times 1$, $p_i \times 1$, and $q_i \times 1$ vectors, or

$$\begin{bmatrix} \mathbf{D}_{li}(s) & \mathbf{N}_{li}(s) \\ -\mathbf{I} & \mathbf{0} \end{bmatrix}\begin{bmatrix} \hat{\bar{\boldsymbol{\xi}}}_i(s) \\ -\hat{\mathbf{u}}_i(s) \end{bmatrix}=\begin{bmatrix} \mathbf{0} \\ -\hat{\mathbf{y}}_i(s) \end{bmatrix} \qquad i=1,2 \tag{9-4}$$

where $\hat{\bar{\boldsymbol{\xi}}}_i(s)$, $\hat{\mathbf{u}}_i(s)$, and $\mathbf{y}_i(s)$ are, respectively, $q_i \times 1$, $p_i \times 1$, and $q_i \times 1$ vectors.

Parallel connection. For the parallel connection shown in Figure 9-5, we have $p_1=p_2$, $q_1=q_2$, $\mathbf{u}_1=\mathbf{u}_2=\mathbf{u}$, and $\mathbf{y}=\mathbf{y}_1+\mathbf{y}_2$, where $\mathbf{u}$ and $\mathbf{y}$ denote the input and output of the overall system. Using these equations and (9-3), we can obtain

$$\left[\begin{array}{cc:c} \mathbf{D}_{r1}(s) & \mathbf{0} & \mathbf{I} \\ \mathbf{0} & \mathbf{D}_{r2}(s) & \mathbf{I} \\ \hdashline -\mathbf{N}_{r1}(s) & -\mathbf{N}_{r2}(s) & \mathbf{0} \end{array}\right]\left[\begin{array}{c} \hat{\boldsymbol{\xi}}_1(s) \\ \hat{\boldsymbol{\xi}}_2(s) \\ \hdashline -\hat{\mathbf{u}}(s) \end{array}\right]=\left[\begin{array}{c} \mathbf{0} \\ \mathbf{0} \\ \hdashline -\hat{\mathbf{y}}(s) \end{array}\right] \tag{9-5}$$

to describe the parallel connection. Although (9-3) are irreducible by assumption, the system matrix in (9-5) is not necessarily irreducible. In order to obtain a simpler result, we carry out the following strict system equivalence transformation:

$$\left[\begin{array}{cc:c} \mathbf{I} & -\mathbf{I} & \mathbf{0} \\ \mathbf{0} & \mathbf{I} & \mathbf{0} \\ \hdashline \mathbf{0} & \mathbf{0} & \mathbf{I} \end{array}\right]\left[\begin{array}{cc:c} \mathbf{D}_{r1}(s) & \mathbf{0} & \mathbf{I} \\ \mathbf{0} & \mathbf{D}_{r2}(s) & \mathbf{I} \\ \hdashline -\mathbf{N}_{r1}(s) & -\mathbf{N}_{r2}(s) & \mathbf{0} \end{array}\right]\left[\begin{array}{cc:c} \mathbf{I} & \mathbf{0} & \mathbf{0} \\ \mathbf{0} & \mathbf{I} & \mathbf{0} \\ \hdashline \mathbf{0} & \mathbf{0} & \mathbf{I} \end{array}\right]=\left[\begin{array}{cc:c} \mathbf{D}_{r1}(s) & -\mathbf{D}_{r2}(s) & \mathbf{0} \\ \mathbf{0} & \mathbf{D}_{r2}(s) & \mathbf{I} \\ \hdashline -\mathbf{N}_{r1}(s) & -\mathbf{N}_{r2}(s) & \mathbf{0} \end{array}\right]$$

This transformation merely substracts the second block row of the system matrix in (9-5) from the first block row. Since controllability and coprimeness are invariant under the transformation of strict system equivalence (Theorem 6-9), we conclude that the dynamical equation description of the parallel connection is controllable if and only if

$$\left[\begin{array}{cc:c} \mathbf{D}_{r1}(s) & \mathbf{0} & \mathbf{I} \\ \mathbf{0} & \mathbf{D}_{r2}(s) & \mathbf{I} \end{array}\right] \quad \text{or} \quad \left[\begin{array}{cc:c} \mathbf{D}_{r1}(s) & -\mathbf{D}_{r2}(s) & \mathbf{0} \\ \hdashline \mathbf{0} & \mathbf{D}_{r2}(s) & \mathbf{I} \end{array}\right] \tag{9-6}$$

has a full rank for every s in $\mathbb{C}$. Because of the block triangular form, the second matrix has a full rank for every s in $\mathbb{C}$ if and only if the matrix $[\mathbf{D}_{r1}(s) \quad -\mathbf{D}_{r2}(s)]$ has a full rank for every s in $\mathbb{C}$ or, following Theorem G-8′,

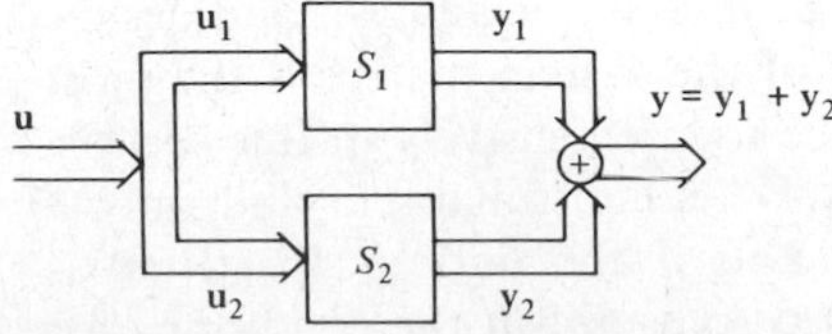

Figure 9-5 The parallel connection of S_1 and S_2.

$\mathbf{D}_{r1}(s)$ and $\mathbf{D}_{r2}(s)$ are left coprime. Hence we conclude that the parallel connection is controllable if and only if $\mathbf{D}_{r1}(s)$ and $\mathbf{D}_{r2}(s)$ are left coprime.

If we use (9-4), then the system matrix of the parallel connection is given by

$$\left[\begin{array}{cc|c} \mathbf{D}_{l1}(s) & \mathbf{0} & \mathbf{N}_{l1}(s) \\ \mathbf{0} & \mathbf{D}_{l2}(s) & \mathbf{N}_{l2}(s) \\ \hline -\mathbf{I} & -\mathbf{I} & \mathbf{0} \end{array}\right] \tag{9-7}$$

By a similar argument, it can be shown that the parallel connection is observable if and only if $\mathbf{D}_{l1}(s)$ and $\mathbf{D}_{l2}(s)$ are right coprime. We recapitulate the above results as a theorem.

Theorem 9-3

Consider two systems which are completely characterized by their transfer matrices $\hat{\mathbf{G}}_1(s)$ and $\hat{\mathbf{G}}_2(s)$. Let $\hat{\mathbf{G}}_i(s) = \mathbf{D}_{li}^{-1}(s)\mathbf{N}_{li}(s) = \mathbf{N}_{ri}(s)\mathbf{D}_{ri}^{-1}(s)$ be coprime fractions of $\hat{\mathbf{G}}_i(s)$. Then the parallel connection of these two systems is controllable if and only if $\mathbf{D}_{r1}(s)$ and $\mathbf{D}_{r2}(s)$ are left coprime. The parallel connection is observable if and only if $\mathbf{D}_{l1}(s)$ and $\mathbf{D}_{l2}(s)$ are right coprime. ∎

The roots of det $\mathbf{D}_{ri}(s)$ or det $\mathbf{D}_{li}(s)$ are called the *poles* of $\hat{\mathbf{G}}_i(s)$ (see Appendix H). If the poles of $\hat{\mathbf{G}}_1(s)$ and those of $\hat{\mathbf{G}}_2(s)$ are disjoint, then the matrix $[\mathbf{D}_{r1}(s) \quad \mathbf{D}_{r2}(s)]$ and the matrix $[\mathbf{D}'_{l1}(s) \quad \mathbf{D}'_{l2}(s)]'$ have a full rank for every s in $\mathbb{C}$ (why?). Hence a sufficient condition for the parallel connection to be controllable and observable is that $\hat{\mathbf{G}}_1(s)$ and $\hat{\mathbf{G}}_2(s)$ have no pole in common. This condition, however, is not a necessary condition (see Problem 9-4).

If $\hat{\mathbf{G}}_1(s)$ and $\hat{\mathbf{G}}_2(s)$ are 1×1 rational functions, then this theorem reduces to the following: The parallel connection is controllable and observable if and only if their denominators have no roots in common. This provides a different proof of statement 1 of Theorem 9-2. By combining the conditions in Theorem 9-3, we have the necessary and sufficient conditions for $\hat{\mathbf{G}}_1(s) + \hat{\mathbf{G}}_2(s)$ to characterize completely the parallel connection of the two systems.

Tandem connection. For the tandem connection of S_1 followed by S_2 shown in Figure 9-6, we have $q_1 = p_2$, $\hat{\mathbf{u}}(s) = \hat{\mathbf{u}}_1(s)$, $\hat{\mathbf{u}}_2(s) = \hat{\mathbf{y}}_1(s)$, and $\hat{\mathbf{y}}(s) = \hat{\mathbf{y}}_2(s)$. Using these relations and (9-3), we can obtain

$$\left[\begin{array}{ccc|c} \mathbf{D}_{r1}(s) & \mathbf{0} & \mathbf{0} & \mathbf{I} \\ \mathbf{0} & \mathbf{D}_{r2}(s) & \mathbf{I} & \mathbf{0} \\ -\mathbf{N}_{r1}(s) & \mathbf{0} & -\mathbf{I} & \mathbf{0} \\ \hline \mathbf{0} & -\mathbf{N}_{r2}(s) & \mathbf{0} & \mathbf{0} \end{array}\right] \left[\begin{array}{c} \hat{\boldsymbol{\xi}}_1(s) \\ \hat{\boldsymbol{\xi}}_2(s) \\ \hat{\mathbf{y}}_1(s) \\ \hline -\hat{\mathbf{u}}(s) \end{array}\right] = \left[\begin{array}{c} \mathbf{0} \\ \mathbf{0} \\ \mathbf{0} \\ \hline -\hat{\mathbf{y}}(s) \end{array}\right] \tag{9-8}$$

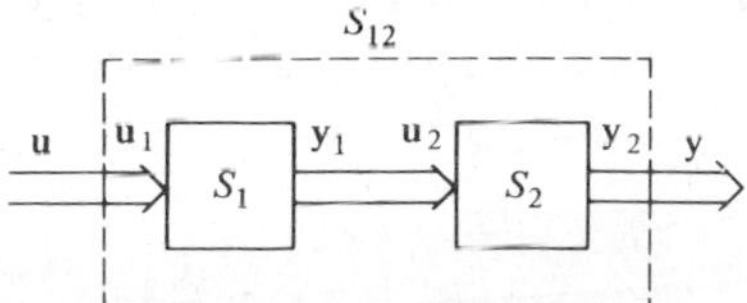

Figure 9-6 S_{12}, the tandem connection of S_1 followed by S_2.

to describe the tandem connection. Note that $\hat{\mathbf{y}}_1(s)=\hat{\mathbf{u}}_2(s)$ is a part of the pseudostate of the tandem connection. It is clear that the tandem connection is controllable if and only if, for every s in $\mathbb{C}$, the matrix

$$\begin{bmatrix} \mathbf{D}_{r1}(s) & \mathbf{0} & \mathbf{0} & \vdots & \mathbf{I} \\ \mathbf{0} & \mathbf{D}_{r2}(s) & \mathbf{I} & \vdots & \mathbf{0} \\ -\mathbf{N}_{r1}(s) & \mathbf{0} & -\mathbf{I} & \vdots & \mathbf{0} \end{bmatrix} \tag{9-9}$$

has a full rank for every s in $\mathbb{C}$. By adding the second block row to the third block row,[3] (9-9) becomes

$$\begin{bmatrix} \mathbf{D}_{r1}(s) & \mathbf{0} & \vdots & \mathbf{0} & \mathbf{I} \\ \mathbf{0} & \mathbf{D}_{r2}(s) & \vdots & \mathbf{I} & \mathbf{0} \\ \hline -\mathbf{N}_{r1}(s) & \mathbf{D}_{r2}(s) & \vdots & \mathbf{0} & \mathbf{0} \end{bmatrix}$$

which implies that the matrix in (9-9) has a full rank for every s in $\mathbb{C}$ if and only if $[-\mathbf{N}_{r1}(s) \quad \mathbf{D}_{r2}(s)]$ has a full rank for every s in $\mathbb{C}$. Hence we conclude that the tandem connection is controllable if and only if $\mathbf{D}_{r2}(s)$ and $\mathbf{N}_{r1}(s)$ are, following Theorem G-8′, left coprime.

If we use (9-3) to describe S_1 and (9-4) to describe S_2, then the tandem connection is described by

$$\begin{bmatrix} \mathbf{D}_{r1}(s) & \mathbf{0} & \mathbf{0} & \vdots & \mathbf{I} \\ -\mathbf{N}_{r1}(s) & \mathbf{0} & -\mathbf{I} & \vdots & \mathbf{0} \\ \mathbf{0} & \mathbf{D}_{l2}(s) & \mathbf{N}_{l2}(s) & \vdots & \mathbf{0} \\ \hline \mathbf{0} & \mathbf{I} & \mathbf{0} & \vdots & \mathbf{0} \end{bmatrix} \begin{bmatrix} \hat{\boldsymbol{\xi}}_1(s) \\ \hat{\boldsymbol{\xi}}_2(s) \\ -\hat{\mathbf{y}}_1(s) \\ \hline -\hat{\mathbf{u}}(s) \end{bmatrix} = \begin{bmatrix} \mathbf{0} \\ \mathbf{0} \\ \mathbf{0} \\ \hline -\hat{\mathbf{y}}(s) \end{bmatrix} \tag{9-10}$$

If we add the product of the second block row and $\mathbf{N}_{l2}$ to the third block row,[3] then the system matrix in (9-10) becomes

$$\begin{bmatrix} \mathbf{D}_{r1}(s) & \mathbf{0} & \mathbf{0} & \vdots & \mathbf{I} \\ -\mathbf{N}_{r1}(s) & \mathbf{0} & -\mathbf{I} & \vdots & \mathbf{0} \\ -\mathbf{N}_{l2}(s)\mathbf{N}_{r1}(s) & \mathbf{D}_{l2}(s) & \mathbf{0} & \vdots & \mathbf{0} \\ \hline \mathbf{0} & \mathbf{I} & \mathbf{0} & \vdots & \mathbf{0} \end{bmatrix}$$

Hence we conclude that the tandem connection is controllable if and only if $\mathbf{N}_{l2}(s)\mathbf{N}_{r1}(s)$ and $\mathbf{D}_{l2}(s)$ are left coprime. By a similar argument, we can establish the following theorem.

Theorem 9-4

Consider two systems S_i which are completely characterized by their transfer matrices $\hat{\mathbf{G}}_i(s)$, $i=1, 2$. Let $\hat{\mathbf{G}}_i(s)=\mathbf{D}_{li}^{-1}(s)\mathbf{N}_{li}(s)=\mathbf{N}_{ri}(s)\mathbf{D}_{ri}^{-1}(s)$ be coprime fractions of $\hat{\mathbf{G}}_i(s)$. Then the tandem connection of S_1 followed by S_2 is controllable if and only if any one of the following three pairs of polynomial matrices, $\mathbf{D}_{r2}(s)$ and $\mathbf{N}_{r1}(s)$, $\mathbf{D}_{l1}(s)\mathbf{D}_{r2}(s)$ and $\mathbf{N}_{l1}(s)$, or $\mathbf{D}_{l2}(s)$ and $\mathbf{N}_{l2}(s)\mathbf{N}_{r1}(s)$, are left coprime. The tandem connection is observable if and only if any one of the

[3] This is a transformation of strict system equivalence. See the statement at the end of Section 6-8.

following three pairs of polynomial matrices, $\mathbf{D}_{l1}(s)$ and $\mathbf{N}_{l2}(s)$, $\mathbf{D}_{l1}(s)\mathbf{D}_{r2}(s)$ and $\mathbf{N}_{r2}(s)$, or $\mathbf{D}_{r1}(s)$ and $\mathbf{N}_{l2}(s)\mathbf{N}_{r1}(s)$, are right coprime. ∎

Let $\hat{\mathbf{G}}_i(s)$ be a $q_i \times p_i$ rational matrix and have the coprime fractions $\hat{\mathbf{G}}_i(s) = \mathbf{N}_{ri}(s)\mathbf{D}_{ri}^{-1}(s) = \mathbf{D}_{li}^{-1}(s)\mathbf{N}_{li}(s)$. Then the roots of det $\mathbf{D}_{ri}(s)$ or det $\mathbf{D}_{li}(s)$ are called the *poles* of $\hat{\mathbf{G}}_i(s)$ and those s for which $\rho\mathbf{N}_{ri}(s) < \min(p_i, q_i)$ or $\rho\mathbf{N}_{li}(s) < \min(p_i, q_i)$, where ρ stands for the rank, are called the *transmission zeros* of $\hat{\mathbf{G}}_i(s)$ (see Appendix H).

Corollary 9-4

A sufficient condition for the tandem connection of S_1 followed by S_2 with $p_1 \geq q_1$, $q_2 \geq p_2$, and $q_1 = p_2$ to be controllable (observable) is that no pole of $\hat{\mathbf{G}}_2(s)$ is a transmission zero of $\hat{\mathbf{G}}_1(s)$ [no pole of $\hat{\mathbf{G}}_1(s)$ is a transmission zero of $\hat{\mathbf{G}}_2(s)$].

Proof

Let $\hat{\mathbf{G}}_1(s) = \mathbf{N}_{r1}(s)\mathbf{D}_{r1}^{-1}(s)$ and $\hat{\mathbf{G}}_2(s) = \mathbf{N}_{r2}(s)\mathbf{D}_{r2}^{-1}(s)$, where $\mathbf{N}_{r1}(s)$ is $q_1 \times p_1$ polynomial matrix and $\mathbf{D}_{r2}(s)$ is a $p_2 \times p_2$ polynomial matrix. $\mathbf{D}_{r2}(s)$ has rank p_2 at every s except the roots of det $\mathbf{D}_{r2}(s)$. If no pole of $\hat{\mathbf{G}}_2(s)$ is a transmission zero of $\hat{\mathbf{G}}_1(s)$, we have $\rho\mathbf{N}_{r1}(s) = q_1 = p_2$ at the roots of det $\mathbf{D}_{r2}(s)$. Hence the matrix $[\mathbf{D}_{r2}(s) \quad \mathbf{N}_{r1}(s)]$ has rank p_2 at every s in $\mathbb{C}$ and, consequently, is left coprime. Hence if no pole of $\hat{\mathbf{G}}_2(s)$ is a transmission zero of $\hat{\mathbf{G}}_1(s)$, the tandem connection of S_1 followed by S_2 is, following Theorem 9-4, controllable. The observability part can be similarly proved. Q.E.D.

The combination of the controllability and observability conditions in Theorem 9-4 yields the condition for $\hat{\mathbf{G}}_2(s)\hat{\mathbf{G}}_1(s)$ to completely characterize the tandem connection. If $\hat{\mathbf{G}}_1(s)$ and $\hat{\mathbf{G}}_2(s)$ are 1×1 rational functions, then Theorem 9-4 reduces to that the tandem connection of $\hat{g}_1(s)$ followed by $\hat{g}_2(s)$ is controllable (observable) if and only if no pole of $\hat{g}_2(s)$ is canceled by any zero of $\hat{g}_1(s)$ [no pole of $\hat{g}_1(s)$ is canceled by any zero of $\hat{g}_2(s)$]. Hence $\hat{g}_2(s)\hat{g}_1(s)$ characterizes completely the tandem connection of $\hat{g}_1(s)$ followed by $\hat{g}_2(s)$ if and only if there is no pole-zero cancellation between $\hat{g}_1(s)$ and $\hat{g}_2(s)$. This provides a different proof of statement 2 of Theorem 9-2.

The concept of pole-zero cancellation in the scalar case can be extended to the matrix case if it is carefully defined. If $\hat{\mathbf{G}}_2(s)\hat{\mathbf{G}}_1(s)$ characterizes completely the tandem connection in Figure 9-6, then we have, following Theorem 9-1, $\delta\hat{\mathbf{G}}(s) = \delta(\hat{\mathbf{G}}_2(s)\hat{\mathbf{G}}_1(s)) = \delta\hat{\mathbf{G}}_2(s) + \delta\hat{\mathbf{G}}_1(s)$, where δ denotes the degree. If the tandem connection is either uncontrollable or unobservable, then we have

$$\delta\hat{\mathbf{G}}(s) = \delta(\hat{\mathbf{G}}_2(s)\hat{\mathbf{G}}_1(s)) < \delta\hat{\mathbf{G}}_2(s) + \delta\hat{\mathbf{G}}_1(s)$$

In this case, we may define that there are pole-zero cancellations in $\hat{\mathbf{G}}_2(s)\hat{\mathbf{G}}_1(s)$. For example, if

$$\hat{\mathbf{G}}_2(s) = [1 \quad -1] \qquad \hat{\mathbf{G}}_1(s) = \begin{bmatrix} \dfrac{s}{s-1} \\ \dfrac{1}{s-1} \end{bmatrix}$$

then $\delta(\hat{\mathbf{G}}_2(s)\hat{\mathbf{G}}_1(s))=0<\delta\hat{\mathbf{G}}_2(s)+\delta\hat{\mathbf{G}}_1(s)=1$ and there is one pole-zero cancellation in $\hat{\mathbf{G}}_2(s)\hat{\mathbf{G}}_1(s)$. Clearly, the existence of pole-zero cancellations in $\hat{\mathbf{G}}_2(s)\hat{\mathbf{G}}_1(s)$ does not imply the existence of pole-zero cancellations in $\hat{\mathbf{G}}_1(s)\hat{\mathbf{G}}_2(s)$, as can be verified from the example. Unlike the scalar case, pole-zero cancellations in the multivariable case may not involve actual cancellations of poles and zeros.

Let $\Delta(s)$ and $\Delta_i(s)$ be, respectively, the characteristic polynomials of $\hat{\mathbf{G}}(s)$ and $\hat{\mathbf{G}}_i(s)$. If $\hat{\mathbf{G}}(s)=\hat{\mathbf{G}}_2(s)\hat{\mathbf{G}}_1(s)$ has no pole-zero cancellation, then we have $\Delta(s)=\Delta_1(s)\Delta_2(s)$, as can be easily seen from Equation (3-63a). If $\hat{\mathbf{G}}_2(s)\hat{\mathbf{G}}_1(s)$ has pole-zero cancellations, then $\deg \Delta(s)<\deg \Delta_1(s)+\deg \Delta_2(s)$, and the roots of $\Delta_1(s)\Delta_2(s)/\Delta(s)$ are called the cancelled poles. If λ is a cancelled pole, then from Theorem 9-4, we have the following equivalent conditions:

1. $[\mathbf{D}_{r2}(\lambda) \quad \mathbf{N}_{r1}(\lambda)]$ and/or $\begin{bmatrix}\mathbf{D}_{l1}(\lambda)\\ \mathbf{N}_{l2}(\lambda)\end{bmatrix}$ have no full rank.
2. $[\mathbf{D}_{l1}(\lambda)\mathbf{D}_{r2}(\lambda) \quad \mathbf{N}_{l1}(\lambda)]$ and/or $\begin{bmatrix}\mathbf{D}_{l1}(\lambda)\mathbf{D}_{r2}(\lambda)\\ \mathbf{N}_{r2}(\lambda)\end{bmatrix}$ have no full rank.
3. $[\mathbf{D}_{l2}(\lambda) \quad \mathbf{N}_{l2}(\lambda)\mathbf{N}_{r1}(\lambda)]$ and/or $\begin{bmatrix}\mathbf{D}_{r1}(\lambda)\\ \mathbf{N}_{l2}(\lambda)\mathbf{N}_{r1}(\lambda)\end{bmatrix}$ have no full rank.

For a direct proof of this statement, see Reference S2. See also the discussion of the input-decoupling zero on page 292. If all cancelled poles have negative real parts, then $\hat{\mathbf{G}}_2(s)\hat{\mathbf{G}}_1(s)$ is said to have no unstable pole-zero cancellation.

Feedback connection. Consider the feedback connection of S_1 and S_2 shown in Figure 9-7. It is assumed that S_i is completely characterized by $\hat{\mathbf{G}}_i(s)$. It is also assumed that $\det(\mathbf{I}+\hat{\mathbf{G}}_1(s)\hat{\mathbf{G}}_2(s))\neq 0$ at $s=\infty$ to ensure that the feedback transfer matrix is well defined and proper (Theorem 3-5). Let S_{12} denote the tandem connection of S_1 followed by S_2; S_{21}, the tandem connection of S_2 followed by S_1; and S_f, the feedback system in Figure 9-7.

Theorem 9-5

Consider two systems S_i which are completely characterized by their transfer matrices $\hat{\mathbf{G}}_i(s)$, $i=1, 2$. It is assumed that $\det(\mathbf{I}+\hat{\mathbf{G}}_1(\infty)\hat{\mathbf{G}}_2(\infty))\neq 0$. Then the feedback system S_f is controllable (observable) if and only if S_{12} is controllable (S_{21} is observable).

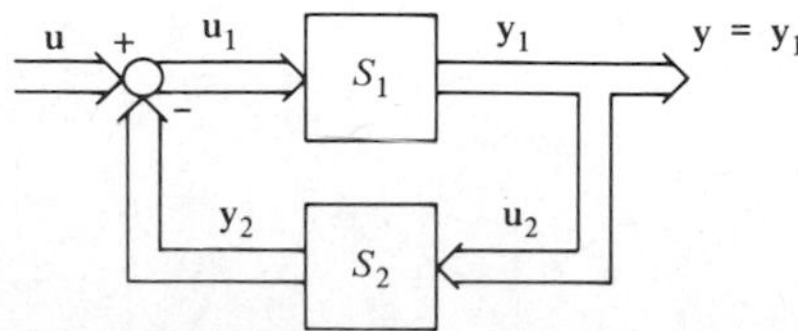

Figure 9-7 Feedback connection of S_1 and S_2.

Proof

We shall prove this theorem from the definition of controllability and observability. Let $\bar{\mathbf{x}}$ be the composite state of S_1 and S_2; that is, $\bar{\mathbf{x}} = [\mathbf{x}_1' \quad \mathbf{x}_2']'$, where $\mathbf{x}_i$ is the state of S_i. If S_{12} is controllable, then for any $\bar{\mathbf{x}}_0$ and any $\bar{\mathbf{x}}_1$ in the composite state space, an input $\mathbf{u}_1$ to S_{12} exists to transfer $\bar{\mathbf{x}}_0$ to $\bar{\mathbf{x}}_1$ in a finite time. Let $\mathbf{y}_2$ be the output of S_{12} due to $\mathbf{u}_1$. Now if we choose $\mathbf{u} = \mathbf{u}_1 + \mathbf{y}_2$ for S_f, this input will transfer $\bar{\mathbf{x}}_0$ to $\bar{\mathbf{x}}_1$. This proves that if S_{12} is controllable, so is S_f. Conversely, if S_f is controllable, corresponding to any $\mathbf{u}$, we choose $\mathbf{u}_1 = \mathbf{u} - \mathbf{y}$. Hence S_{12} is controllable.

Now we consider the observability part. Let the zero-input response of S_f due to some unknown state $\bar{\mathbf{x}}_0$ be $\bar{\mathbf{y}}_0$. If S_{21} is in the same initial state and if $-\bar{\mathbf{y}}_0$ is applied to S_{21}, then from Figure 9-7, we can see that the output of S_{21} is $\bar{\mathbf{y}}_0$. If S_{21} is observable, from the knowledge of the input $-\bar{\mathbf{y}}_0$ and the output $\bar{\mathbf{y}}_0$, the initial state $\bar{\mathbf{x}}_0$ can be determined. Consequently, S_f is observable. If S_{21} is not observable, we cannot determine $\bar{\mathbf{x}}_0$ from its input $-\bar{\mathbf{y}}_0$ and output $\bar{\mathbf{y}}_0$. Consequently, we cannot determine the initial state $\bar{\mathbf{x}}_0$ of S_f from its zero-input response, and S_f is not observable. This completes the proof of this theorem. Q.E.D.

In general, the condition of controllability of S_{12} is different from the condition of observability of S_{21}. However, in the single-variable case the two conditions are identical. Indeed if no pole of $\hat{g}_2(s)$ is canceled by any zero of $\hat{g}_1(s)$, then $\hat{g}_1(s)$ followed by $\hat{g}_2(s)$ is controllable, and $\hat{g}_2(s)$ followed by $\hat{g}_1(s)$ is observable. Hence we conclude that, for the single-variable case, the feedback connection in Figure 9-7 is controllable and observable if and only if S_{12} is controllable or, equivalently, no pole of $\hat{g}_2(s)$ is canceled by any zero of $\hat{g}_1(s)$. This checks with what we have in part 3 of Theorem 9-2.

We consider a special case of Figure 9-7 in which $\hat{\mathbf{G}}_2(s)$ is a constant matrix $\mathbf{K}$. We call this a *constant output feedback system*. Since $\mathbf{K}$ does not introduce any new state variable, if S_1 is controllable, so is S_1 followed by $\mathbf{K}$. If S_1 is observable, so is $\mathbf{K}$ followed by S_1. Hence the feedback connection of S_1 with a constant gain $\mathbf{K}$ in the feedback path is controllable and observable if and only if S_1 is controllable and observable. Hence controllability and observability are invariant under any constant output feedback. This contrasts with the constant state feedback discussed in Chapter 7 where controllability is preserved under constant state feedback, but observability is generally not preserved.

We discuss further the feedback system in Figure 9-7. Let the system S_i be described by irreducible $\hat{\mathbf{G}}_i(s) = \mathbf{N}_{ri}(s)\mathbf{D}_{ri}^{-1}(s) = \mathbf{D}_{li}^{-1}(s)\mathbf{N}_{li}(s)$ and irreducible $\{\mathbf{A}_i, \mathbf{B}_i, \mathbf{C}_i, \mathbf{E}_i\}$, where $\hat{\mathbf{G}}_1(s)$ and $\hat{\mathbf{G}}_2(s)$ are, respectively, $q \times p$ and $p \times q$ proper rational matrices. By assumption, we have

$$\Delta[\hat{\mathbf{G}}_i(s)] \sim \det \mathbf{D}_{ri}(s) \sim \det \mathbf{D}_{li}(s) \sim \det (s\mathbf{I} - \mathbf{A}_i)$$

and
$$\deg \hat{\mathbf{G}}_i(s) = \deg \det \mathbf{D}_{ri}(s) = \deg \det \mathbf{D}_{li}(s) = \dim \mathbf{A}_i = n_i$$

where $\Delta[\cdot]$ denotes the characteristic polynomial of a rational matrix and $\sim$ denotes equality of polynomials modulo a nonzero constant factor. The overall transfer matrix of the feedback system in Figure 9-7 is, as derived in (3-68),

$$\begin{aligned}\hat{\mathbf{G}}_f(s) &= \hat{\mathbf{G}}_1(s)(\mathbf{I}+\hat{\mathbf{G}}_2(s)\hat{\mathbf{G}}_1(s))^{-1}\\ &= \mathbf{N}_{r1}(s)\mathbf{D}_{r1}^{-1}(s)[\mathbf{I}+\mathbf{D}_{l2}^{-1}(s)\mathbf{N}_{l2}(s)\mathbf{N}_{r1}(s)\mathbf{D}_{r1}^{-1}(s)]^{-1}\\ &= \mathbf{N}_{r1}(s)[\mathbf{D}_{l2}(s)\mathbf{D}_{r1}(s)+\mathbf{N}_{l2}(s)\mathbf{N}_{r1}(s)]^{-1}\mathbf{D}_{l2}(s)\end{aligned} \tag{9-11}$$

The matrix **A** of its state-variable description is, as derived in (3-64),

$$\mathbf{A}_f = \begin{bmatrix} \mathbf{A}_1 - \mathbf{B}_1\mathbf{Y}_2\mathbf{E}_2\mathbf{C}_1 & -\mathbf{B}_1\mathbf{Y}_2\mathbf{C}_2 \\ \mathbf{B}_2\mathbf{Y}_1\mathbf{C}_1 & \mathbf{A}_2 - \mathbf{B}_2\mathbf{Y}_1\mathbf{E}_1\mathbf{C}_2 \end{bmatrix} \tag{9-12}$$

with $\mathbf{Y}_1 = (\mathbf{I}+\mathbf{E}_1\mathbf{E}_2)^{-1}$ and $\mathbf{Y}_2 = (\mathbf{I}+\mathbf{E}_2\mathbf{E}_1)^{-1}$. If the feedback system is not completely characterized by $\hat{\mathbf{G}}_f(s)$, then

$$\Delta[\hat{\mathbf{G}}_f(s)] \not\sim \det(s\mathbf{I}-\mathbf{A}_f)$$

and

$$\deg \hat{\mathbf{G}}_f(s) < \dim \mathbf{A}_f = n_1 + n_2$$

where $\not\sim$ denotes not equal modulo a nonzero constant factor. Conversely, if the equalities hold, then the (n_1+n_2)-dimensional dynamical equation description of the feedback system in Figure 9-7 is controllable and observable. Note that we have

$$\Delta[\hat{\mathbf{G}}_f(s)] \not\sim \det[\mathbf{D}_{l2}(s)\mathbf{D}_{r1}(s)+\mathbf{N}_{l2}(s)\mathbf{N}_{r1}(s)]$$

because the factorization in (9-11) is not necessarily irreducible.

The controllability and observability of a system depends on the assignment of the input and output. Although the system in Figure 9-7 may be uncontrollable or unobservable, if we assign the additional input and output as shown in Figure 9-8, then it is easy to show, by using the argument in the proof of Theorem 9-5, that the feedback system is always controllable and observable from the input $[\mathbf{r}_1' \quad \mathbf{r}_2']'$ and the output $[\mathbf{y}_1' \quad \mathbf{y}_2']'$. Similarly, the system is always controllable and observable from the input $[\mathbf{r}_1' \quad \mathbf{r}_2']'$ and the output $[\mathbf{u}_1' \quad \mathbf{u}_2']'$. We study now the transfer matrix $\mathbf{H}(s)$ from $[\mathbf{r}_1' \quad \mathbf{r}_2']'$ to $[\mathbf{u}_1' \quad \mathbf{u}_2']'$. From $\hat{\mathbf{u}}_1(s) = \hat{\mathbf{r}}_1(s) - \hat{\mathbf{y}}_2(s) = \hat{\mathbf{r}}_1(s) - \hat{\mathbf{G}}_2(s)\hat{\mathbf{u}}_2(s)$ and $\hat{\mathbf{u}}_2(s) = \hat{\mathbf{r}}_2(s) + \hat{\mathbf{y}}_1(s) = \hat{\mathbf{r}}_2(s) + \hat{\mathbf{G}}_1(s)\hat{\mathbf{u}}_1(s)$, we have

$$\begin{bmatrix}\hat{\mathbf{r}}_1(s)\\ \hat{\mathbf{r}}_2(s)\end{bmatrix} = \begin{bmatrix}\mathbf{I}_p & \hat{\mathbf{G}}_2(s)\\ -\hat{\mathbf{G}}_1(s) & \mathbf{I}_q\end{bmatrix}\begin{bmatrix}\hat{\mathbf{u}}_1(s)\\ \hat{\mathbf{u}}_2(s)\end{bmatrix}$$

By definition,

$$\begin{bmatrix}\hat{\mathbf{u}}_1(s)\\ \hat{\mathbf{u}}_2(s)\end{bmatrix} = \mathbf{H}(s)\begin{bmatrix}\hat{\mathbf{r}}_1(s)\\ \hat{\mathbf{r}}_2(s)\end{bmatrix}$$

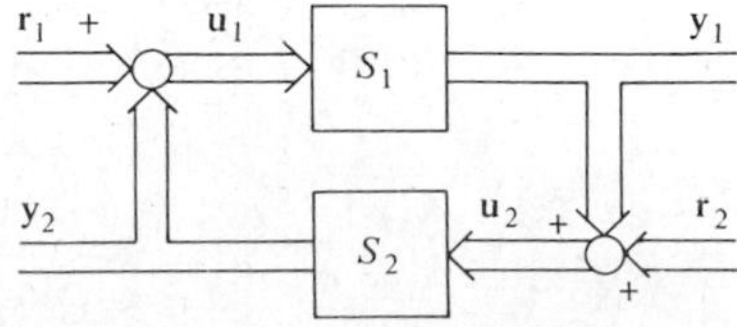

Figure 9-8 Feedback system with additional input and output.

hence, we have

$$\mathbf{H}(s)=\begin{bmatrix} \mathbf{I}_p & \hat{\mathbf{G}}_2(s) \\ -\hat{\mathbf{G}}_1(s) & \mathbf{I}_q \end{bmatrix}^{-1}$$

The state equation of the system in Figure 9-8 can be readily computed as

$$\begin{bmatrix} \dot{\mathbf{x}}_1 \\ \dot{\mathbf{x}}_2 \end{bmatrix}=\begin{bmatrix} \mathbf{A}_1-\mathbf{B}_1\mathbf{Y}_2\mathbf{E}_2\mathbf{C}_1 & -\mathbf{B}_1\mathbf{Y}_2\mathbf{C}_2 \\ \mathbf{B}_2\mathbf{Y}_1\mathbf{C}_1 & \mathbf{A}_2-\mathbf{B}_2\mathbf{Y}_1\mathbf{E}_1\mathbf{C}_2 \end{bmatrix}\begin{bmatrix} \mathbf{x}_1 \\ \mathbf{x}_2 \end{bmatrix} + \begin{bmatrix} \mathbf{B}_1\mathbf{Y}_2 & -\mathbf{B}_1\mathbf{Y}_2\mathbf{E}_2 \\ \mathbf{B}_2\mathbf{Y}_1\mathbf{E}_1 & \mathbf{B}_2\mathbf{Y}_1 \end{bmatrix}\begin{bmatrix} \mathbf{r}_1 \\ \mathbf{r}_2 \end{bmatrix}$$

Its $\mathbf{A}$ matrix is identical to the $\mathbf{A}_f$ in (9-12).

Theorem 9-6

Consider the system in Figure 9-8 with the assumption $\det[\mathbf{I}+\hat{\mathbf{G}}_1(\infty)\hat{\mathbf{G}}_2(\infty)]\neq 0$. Then we have

$$\Delta(\mathbf{H}(s)) \sim \det[\mathbf{D}_{l1}(s)\mathbf{D}_{r2}(s)+\mathbf{N}_{l1}(s)\mathbf{N}_{r2}(s)] \quad \textbf{(9-13a)}$$

$$\sim \det[\mathbf{D}_{l2}(s)\mathbf{D}_{r1}(s)+\mathbf{N}_{l2}(s)\mathbf{N}_{r1}(s)] \quad \textbf{(9-13b)}$$

$$\sim \det\begin{bmatrix} \mathbf{D}_{r1}(s) & \mathbf{N}_{r2}(s) \\ -\mathbf{N}_{r1}(s) & \mathbf{D}_{r2}(s) \end{bmatrix} \quad \textbf{(9-13c)}$$

$$\sim \det\begin{bmatrix} \mathbf{D}_{l2}(s) & \mathbf{N}_{l2}(s) \\ -\mathbf{N}_{l1}(s) & \mathbf{D}_{l1}(s) \end{bmatrix} \quad \textbf{(9-13d)}$$

$$\sim \det(s\mathbf{I}-\mathbf{A}_f) \quad \textbf{(9-13e)}$$

$$\sim \Delta[\mathbf{G}_1(s)]\Delta[\mathbf{G}_2(s)]\det[\mathbf{I}+\hat{\mathbf{G}}_1(s)\hat{\mathbf{G}}_2(s)] \quad \textbf{(9-13f)}$$

where $\sim$ denotes the equality of polynomials modulo a nonzero constant factor.

Proof

We remark first that the condition $\det[\mathbf{I}+\hat{\mathbf{G}}_1(\infty)\hat{\mathbf{G}}_2(\infty)]\neq 0$ ensures, as shown in Section 3-6, the properness of all elements of $\mathbf{H}(s)$. Without the assumption, $\mathbf{H}(s)$ may have poles at $s=\infty$ and the theorem may not hold. Consider the right fraction

$$\mathbf{H}(s)=\begin{bmatrix} \mathbf{I} & \hat{\mathbf{G}}_2(s) \\ -\hat{\mathbf{G}}_1(s) & \mathbf{I} \end{bmatrix}^{-1}=\begin{bmatrix} \mathbf{I} & \mathbf{N}_{r2}(s)\mathbf{D}_{r2}^{-1}(s) \\ -\mathbf{N}_{r1}(s)\mathbf{D}_{r1}^{-1}(s) & \mathbf{I} \end{bmatrix}^{-1}$$

$$=\begin{bmatrix} \mathbf{D}_{r1}(s) & \mathbf{0} \\ \mathbf{0} & \mathbf{D}_{r2}(s) \end{bmatrix}\begin{bmatrix} \mathbf{D}_{r1}(s) & \mathbf{N}_{r2}(s) \\ -\mathbf{N}_{r1}(s) & \mathbf{D}_{r2}(s) \end{bmatrix}^{-1}$$

which can be easily verified. The fraction is right coprime as can be seen from, by carrying out elementary row operations,

$$\begin{bmatrix} \mathbf{D}_{r1}(s) & \mathbf{N}_{r2}(s) \\ -\mathbf{N}_{r1}(s) & \mathbf{D}_{r2}(s) \\ \mathbf{D}_{r1}(s) & \mathbf{0} \\ \mathbf{0} & \mathbf{D}_{r2}(s) \end{bmatrix} \rightarrow \begin{bmatrix} \mathbf{0} & \mathbf{N}_{r2}(s) \\ -\mathbf{N}_{r1}(s) & \mathbf{0} \\ \mathbf{D}_{r1}(s) & \mathbf{0} \\ \mathbf{0} & \mathbf{D}_{r2}(s) \end{bmatrix}$$

and the right coprime assumption of $\mathbf{N}_{ri}(s)\mathbf{D}_{ri}^{-1}(s)$ (Theorem G-8). Thus, we have established, following (6-189), the equality in (9-13c).

Again by direct verification, we have

$$\begin{aligned} \mathbf{H}(s) &= \begin{bmatrix} \mathbf{I} & \mathbf{N}_{r2}(s)\mathbf{D}_{r2}^{-1}(s) \\ -\mathbf{D}_{l1}^{-1}(s)\mathbf{N}_{l1}(s) & \mathbf{I} \end{bmatrix}^{-1} \\ &= \begin{bmatrix} \mathbf{I} & 0 \\ \mathbf{0} & \mathbf{D}_{r2}(s) \end{bmatrix} \begin{bmatrix} \mathbf{I} & \mathbf{N}_{r2}(s) \\ -\mathbf{N}_{l1}(s) & \mathbf{D}_{l1}(s)\mathbf{D}_{r2}(s) \end{bmatrix}^{-1} \begin{bmatrix} \mathbf{I} & 0 \\ \mathbf{0} & \mathbf{D}_{l1}(s) \end{bmatrix} \end{aligned}$$

It is easy to show that the factorization is irreducible. Hence we have

$$\Delta(\mathbf{H}(s)) = \det \begin{bmatrix} \mathbf{I} & \mathbf{N}_{r2}(s) \\ -\mathbf{N}_{l1}(s) & \mathbf{D}_{l1}(s)\mathbf{D}_{r2}(s) \end{bmatrix} = \det\left[\mathbf{D}_{l1}(s)\mathbf{D}_{r2}(s) + \mathbf{N}_{l1}(s)\mathbf{N}_{r2}(s)\right]$$

where we have used, in the last step, the formula

$$\begin{aligned} \det \begin{bmatrix} \mathbf{P} & \mathbf{U} \\ \mathbf{Q} & \mathbf{V} \end{bmatrix} &= \det \begin{bmatrix} \mathbf{I} & \mathbf{0} \\ -\mathbf{Q}\mathbf{P}^{-1} & \mathbf{I} \end{bmatrix} \begin{bmatrix} \mathbf{P} & \mathbf{U} \\ \mathbf{Q} & \mathbf{V} \end{bmatrix} = \det \begin{bmatrix} \mathbf{P} & \mathbf{U} \\ \mathbf{0} & \mathbf{V} - \mathbf{Q}\mathbf{P}^{-1}\mathbf{U} \end{bmatrix} \\ &= \det \mathbf{P} \det (\mathbf{V} - \mathbf{Q}\mathbf{P}^{-1}\mathbf{U}) \end{aligned}$$

This establishes (9-13a). The equality $\Delta[\mathbf{H}(s)] \sim \det(s\mathbf{I} - \mathbf{A}_f)$ follows directly from (6-189) and the fact that the state-variable equation is controllable and observable from the input $[\mathbf{r}_1' \quad \mathbf{r}_2']'$ and the output $[\mathbf{u}_1' \quad \mathbf{u}_2']'$.

From (9-13a), we have

$$\begin{aligned} &\det\{\mathbf{D}_{l1}[\mathbf{I} + \mathbf{D}_{l1}^{-1}(s)\mathbf{N}_{l1}(s)\mathbf{N}_{r2}(s)\mathbf{D}_{r2}^{-1}(s)]\mathbf{D}_{r2}(s)\} \\ &\quad = \det \mathbf{D}_{l1}(s) \det[\mathbf{I} + \hat{\mathbf{G}}_1(s)\hat{\mathbf{G}}_2(s)] \det \mathbf{D}_{r2}(s) \end{aligned}$$

This implies (9-13f). Q.E.D.

This theorem provides five formulas for computing $\det(s\mathbf{I} - \mathbf{A}_f)$ and is useful in stability study. In the single-variable case, there is no distinction between right and left fractions, and the theorem reduces to

$$\det(s\mathbf{I} - \mathbf{A}_f) = D_1(s)D_2(s) + N_1(s)N_2(s)$$

where $\hat{g}_i(s) = N_i(s)/D_i(s)$, $i = 1, 2$.

9-4 Stability of Feedback Systems

Every control system is designed to be stable. Therefore it is important to study the stability problem of composite systems. We recall from Chapter 8 that a system or, more precisely, the zero-input response of a system is asymptotically stable if the response due to every initial state approaches the zero state as $t \to \infty$. A system is bounded-input–bounded-output (BIBO) stable if the zero-state response due to every bounded input is bounded. If a system is asymptotically stable, then the system is also BIBO stable, but not conversely. However, if a system is completely characterized by its transfer matrix, then asymptotic stability implies and is implied by BIBO stability.

Consider two linear time-invariant systems S_1 and S_2 which are completely characterized by their transfer matrices $\hat{\mathbf{G}}_1(s)$ and $\hat{\mathbf{G}}_2(s)$. If the composite transfer matrix $\hat{\mathbf{G}}(s)$ of any connection of S_1 and S_2 is computed, the BIBO stability of the composite system can be determined from $\hat{\mathbf{G}}(s)$. However, its asymptotic stability cannot be determined from $\hat{\mathbf{G}}(s)$ unless the composite system is known to be completely characterized by $\hat{\mathbf{G}}(s)$. It is not a simple task to compute composite transfer matrices, especially, in the feedback connection. Even if the composite transfer function is obtained, it is still quite tedious to check whether or not the composite system is completely characterized. Therefore, it is desirable to be able to determine the stability of composite systems from $\hat{\mathbf{G}}_1(s)$ and $\hat{\mathbf{G}}_2(s)$ without computing composite-transfer-function matrices. We shall study this problem in this section.

For the parallel and tandem connections of S_1 and S_2, the problem is very simple. The parallel or tandem connection of S_1 and S_2 is asymptotically stable if and only if S_1 and S_2 are asymptotically stable. The condition for asymptotic stability is that all eigenvalues of $\mathbf{A}$ have negative real parts. From (3-62) and (3-63), we see that the eigenvalues of the parallel or tandem connection of S_1 and S_2 are the union of those of S_1 and those of S_2. Hence the assertion follows.

For the feedback system, the situation is much more complicated. A feedback system may be stable with unstable subsystems; conversely, a feedback system may be unstable with stable subsystems. We study first the single-variable case and then the multivariable case.

Single-variable feedback systems. Consider the single-variable feedback system shown in Figure 9-9. The transfer function of the feedback system is

$$\hat{g}_f(s) = \frac{\hat{g}_1(s)}{1 + \hat{g}_1(s)\hat{g}_2(s)} = \frac{N_1(s)D_2(s)}{D_1(s)D_2(s) + N_1(s)N_2(s)} \triangleq \frac{N_f(s)}{D_f(s)}$$

where $\hat{g}_i(s) \triangleq N_i(s)/D_i(s)$ for $i = 1, 2$ and N_f and D_f are assumed to be coprime. Recall that every transfer function is assumed to be irreducible; therefore, when we speak of $\hat{g}_f(s)$, we mean $\hat{g}_f(s) = N_f(s)/D_f(s)$. It was shown in Theorem 9-2 that if N_1 and D_2 have no common factor, then $D_f = D_1D_2 + N_1N_2$, and the system is completely characterized by $\hat{g}_f(s)$. If there are common factors between N_1 and D_2, then D_f consists of only a part of $D_1D_2 + N_1N_2$, and the system is not completely characterized by $\hat{g}_f(s)$.

Theorem 9-7

Consider the feedback system shown in Figure 9-9. It is assumed that S_1 and S_2 are completely characterized by their proper transfer functions $\hat{g}_1(s)$ and

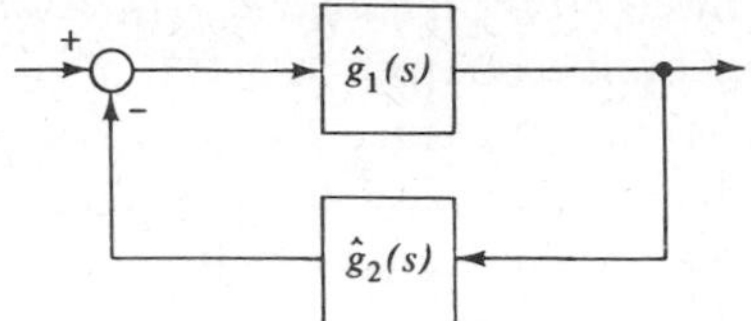

Figure 9-9 A feedback system.

$\hat{g}_2(s)$. It is also assumed that $1+\hat{g}_1(\infty)\hat{g}_2(\infty)\neq 0$. Let $\hat{g}_i(s)=N_i(s)/D_i(s)$. Then the feedback system is asymptotically stable if and only if all the roots of $D_1(s)D_2(s)+N_1(s)N_2(s)$ have negative real parts. The condition is sufficient but not necessary for the system to be BIBO stable.

Proof

First we note that the condition $1+\hat{g}_1(\infty)\hat{g}_2(\infty)\neq 0$ will ensure the properness of $\hat{g}_f(s)$. Let $\mathbf{x}=[\mathbf{x}_1' \quad \mathbf{x}_2']'$ be the composite state vector of the feedback system, where $\mathbf{x}_i$ is the state vector of S_i, and let $\dot{\mathbf{x}}=\mathbf{A}_f\mathbf{x}+\mathbf{b}_f u$, $y=\mathbf{c}_f\mathbf{x}+e_f u$ be the dynamical equation describing the feedback system [see Equation (3-64)]. Then the characteristic polynomial, $\det(s\mathbf{I}-\mathbf{A}_f)$, of $\mathbf{A}_f$ is a polynomial of degree n_1+n_2, where $\deg D_i(s)=n_i$, $i=1, 2$.

Now we show that if $1+\hat{g}_1(\infty)\hat{g}_2(\infty)\neq 0$, then $\deg[D_1(s)D_2(s)+N_1(s)N_2(s)]=n_1+n_2$. In other words, the term $s^{n_1+n_2}$ in $D_1(s)D_2(s)$ will not be canceled by any term in $N_1(s)N_2(s)$. Let

$$\hat{g}_i(s)=\frac{b_{i0}s^{n_i}+b_{i1}s^{n_i-1}+\cdots}{a_{i0}s^{n_i}+a_{i1}s^{n_i-1}+\cdots}$$

Then we have

$$\lim_{s\to\infty}[1+\hat{g}_1(s)\hat{g}_2(s)]=1+\frac{b_{10}b_{20}}{a_{10}a_{20}}=\frac{a_{10}a_{20}+b_{10}b_{20}}{a_{10}a_{20}} \tag{9-14}$$

and

$$\begin{aligned} D_1(s)D_2(s)+N_1(s)N_2(s)&=a_{10}a_{20}s^{n_1+n_2}+\cdots+b_{10}b_{20}s^{n_1+n_2}+\cdots \\ &=(a_{10}a_{20}+b_{10}b_{20})s^{n_1+n_2} \\ &\quad +\text{terms of lower degrees} \end{aligned} \tag{9-15}$$

From (9-14) and (9-15) we see immediately that if $1+\hat{g}_1(\infty)\hat{g}_2(\infty)\neq 0$, then $\deg[D_1(s)D_2(s)+N_1(s)N_2(s)]=n_1+n_2=\deg\det(s\mathbf{I}-\mathbf{A}_f)$. Using this fact and the relation

$$\begin{aligned} \hat{g}_f(s)&=\mathbf{c}_f(s\mathbf{I}-\mathbf{A}_f)^{-1}\mathbf{b}_f+e_f=\frac{1}{\det(s\mathbf{I}-\mathbf{A}_f)}[\mathbf{c}_f\operatorname{Adj}(s\mathbf{I}-\mathbf{A}_f)\,\mathbf{b}_f]+e_f \\ &=\frac{N_1(s)D_2(s)}{D_1(s)D_2(s)+N_1(s)N_2(s)} \end{aligned}$$

we conclude that

$$\det(s\mathbf{I}-\mathbf{A}_f)=k(D_1(s)D_2(s)+N_1(s)N_2(s)) \tag{9-16}$$

for some constant k. In other words, the set of the eigenvalues of $\mathbf{A}_f$ and the set of the roots of $D_1(s)D_2(s)+N_1(s)N_2(s)$ are the same. Hence the feedback system is asymptotically stable if and only if all the roots of $D_1(s)D_2(s) + N_1(s)N_2(s)$ have negative real parts.

Because of possible cancellations between $N_1(s)$ and $D_2(s)$, some roots of $D_1(s)D_2(s) + N_1(s)N_2(s)$ may not appear as poles of $\hat{g}_f(s)$; hence the condition in the theorem is only a sufficient condition for the system to be BIBO stable. Consequently, the system in Figure 9-9 can be BIBO stable without being asymptotically stable. Q.E.D.

This theorem can in fact be deduced directly from Theorem 9-6. It is proved here, however, by using explicitly the condition $1+\hat{g}_1(\infty)\hat{g}_2(\infty)\neq 0$. From (9-15) we see that (9-16) holds if and only if $1+\hat{g}_1(\infty)\hat{g}_2(\infty)\neq 0$; it is independent of whether the (n_1+n_2)-dimensional dynamical equation description of the feedback system is irreducible or not. It is also independent of whether the feedback system is completely characterized by $\hat{g}_f(s)$. If $1+\hat{g}_1(\infty)\hat{g}_2(\infty)=0$, even if all roots of $D_1(s)D_2(s)+N_1(s)N_2(s)$ have negative real parts, the theorem still does not hold. For example, if $\hat{g}_1(s)=(-s^2+s+1)/s^2$, $\hat{g}_2(s)=1$, then we have $1+\hat{g}_1(\infty)\hat{g}_2(\infty)=0$ and $\hat{g}_f(s)=(-s^2+s+1)/(s+1)$ which is improper. Although the root of $D_1(s)D_2(s)+N_1(s)N_2(s)=(s+1)$ has a negative real part, the application of the bounded input $\sin t^2$ will excite a term of $2t\cos t^2$, which is not bounded, at the output of the feedback system. Thus without imposing the condition $1+\hat{g}_1(\infty)\hat{g}_2(\infty)\neq 0$, Theorem 9-7 does not hold in general.

If $\hat{g}_2(s)$ in Figure 9-9 is a constant k, the feedback system reduces to a constant output feedback system and is, as discussed in the previous section, always controllable and observable. In this case, asymptotic stability implies and is implied by BIBO stability.

Corollary 9-7

Consider the feedback system shown in Figure 9-9 with $\hat{g}_2(s)=k$. It is assumed that S_1 is completely characterized by its proper transfer function $\hat{g}_1(s)$ and $1+k\hat{g}_1(\infty)\neq 0$. Then the feedback system is BIBO stable and asymptotically stable if and only if all zeros of $1+k\hat{g}_1(s)$ or, equivalently, all roots of $D_1(s)+kN_1(s)$ have negative real parts.

Multivariable feedback systems. The situation in the multivariable case is much more complex, hence we study first the simplest possible case. Consider the multivariable feedback system shown in Figure 9-10. The system S_1 is assumed to be completely characterized by its $p\times p$ proper rational matrix $\hat{\mathbf{G}}_1(s)$. The assumption that $\hat{\mathbf{G}}_1(s)$ is square is necessary for the proper feedback connection. Let $\hat{\mathbf{G}}_f(s)$ be the transfer matrix of the feedback system. Then

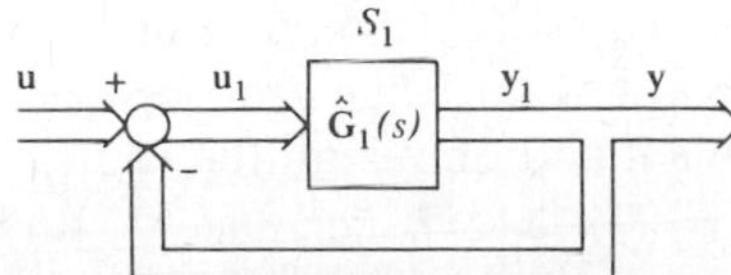

Figure 9-10 A multivariable feedback system.

we have

$$\hat{\mathbf{G}}_f(s) = [\mathbf{I} + \hat{\mathbf{G}}_1(s)]^{-1}\hat{\mathbf{G}}_1(s) = \frac{1}{\det[\mathbf{I} + \hat{\mathbf{G}}_1(s)]}\{\text{Adj}[\mathbf{I} + \hat{\mathbf{G}}_1(s)]\}\hat{\mathbf{G}}_1(s) \quad \textbf{(9-17)}$$

where Adj(·) denotes the adjoint of a matrix. Since S_1 is completely characterized by $\hat{\mathbf{G}}_1(s)$, it follows from Theorem 9-5 that the feedback system in Figure 9-10 is completely characterized by $\hat{\mathbf{G}}_f(s)$. If the order of $\hat{\mathbf{G}}_1(s)$ reduces to 1×1, then $\det[\mathbf{I} + \hat{\mathbf{G}}_1(s)]$ becomes $1 + \hat{g}_1(s)$ and the stability depends only on the zeros of $1 + \hat{g}_1(s)$. Hence, one might be tempted to assert that the stability of multivariable feedback systems depends only on the zeros of $\det[\mathbf{I} + \hat{\mathbf{G}}_1(s)]$.

Example 1

Consider the multivariable feedback system shown in Figure 9-10 with

$$\hat{\mathbf{G}}_1(s) = \begin{bmatrix} \dfrac{-s}{s-1} & \dfrac{s}{s+1} \\ 1 & \dfrac{-2}{s+1} \end{bmatrix}$$

It can be easily verified that

$$\hat{\mathbf{G}}_f(s) = \begin{bmatrix} \dfrac{2s}{s+1} & \dfrac{-s}{s+1} \\ -1 & \dfrac{s-2}{s-1} \end{bmatrix}$$

which is unstable. However, we have $\det[\mathbf{I} + \hat{\mathbf{G}}_1(s)] = -1$, which does not have any right-half-plane zeros. Hence the stability of the feedback system cannot be determined solely from the zeros of $\det[\mathbf{I} + \hat{\mathbf{G}}_1(s)]$. ∎

In order to study the stability of multivariable feedback systems, the concept of the characteristic polynomial of a proper rational matrix is needed. Recall from Definition 6-1 that the characteristic polynomial of $\hat{\mathbf{G}}_1(s)$, denoted by $\Delta_1(s)$, is the least common denominator of all the minors of $\hat{\mathbf{G}}_1(s)$. If $\hat{\mathbf{G}}_1(s)$ has the coprime fraction $\mathbf{N}_1(s)\mathbf{D}_1^{-1}(s)$, then we also have $\Delta_1(s) = k \det D_1(s)$, for some nonzero constant k.

Theorem 9-8

Consider the feedback system shown in Figure 9-10. It is assumed that S_1 is completely characterized by its proper rational matrix $\hat{\mathbf{G}}_1(s)$ and that $\det[\mathbf{I} + \hat{\mathbf{G}}_1(\infty)] \neq 0$. Let $\hat{\mathbf{G}}_1(s) = \mathbf{N}_{r1}(s)\mathbf{D}_{r1}^{-1}(s) = \mathbf{D}_{l1}^{-1}(s)\mathbf{N}_{l1}(s)$ be coprime fractions and $\Delta_1(s)$ be the characteristic polynomial of $\hat{\mathbf{G}}_1(s)$. Then the feedback system is asymptotically stable and BIBO stable if and only if all the roots of any of the three polynomials $\Delta_1(s)\det[\mathbf{I} + \hat{\mathbf{G}}_1(s)]$, $\det[\mathbf{D}_{r1}(s) + \mathbf{N}_{r1}(s)]$ and $\det[\mathbf{D}_{l1}(s) + \mathbf{N}_{l1}(s)]$ have negative real parts. ∎

First we note that $\hat{\mathbf{G}}_f(s)$ is proper because of the assumption $\det[\mathbf{I}+\hat{\mathbf{G}}_1(\infty)]\neq 0$ (Theorem 3-5). Before proving the theorem, we show that $\Delta_1(s)\det[\mathbf{I}+\hat{\mathbf{G}}_1(s)]$ is a polynomial. Define

$$\det[\mathbf{I}+\hat{\mathbf{G}}_1(s)] \triangleq \frac{E(s)}{F(s)} \tag{9-18}$$

where $E(s)$ and $F(s)$ are assumed to be coprime. Then we have

$$\Delta_1(s)\det[\mathbf{I}+\hat{\mathbf{G}}_1(s)] = \frac{\Delta_1(s)E(s)}{F(s)}$$

To show that $\Delta_1(s)\det[\mathbf{I}+\hat{\mathbf{G}}_1(s)]$ is a polynomial is the same as showing that $\Delta_1(s)$ is divisible without remainder by $F(s)$. This will be shown by using the following identity:

$$\det[\mathbf{I}+\hat{\mathbf{G}}_1(s)] = \sum_{i=1}^{p} \alpha_i(s) + 1 \tag{9-19}$$

where p is the order of $\hat{\mathbf{G}}_1(s)$ and α_i is the sum of all the principal minors of $\hat{\mathbf{G}}_1(s)$ of order i; for example, $\alpha_p(s)=\det\hat{\mathbf{G}}_1(s)$, $\alpha_1(s)=$ sum of the diagonal elements of $\hat{\mathbf{G}}_1(s)$. The verification of (9-19) is straightforward and is omitted. Now $\Delta_1(s)$ is the least common denominator of all the minors of $\hat{\mathbf{G}}_1(s)$, whereas $F(s)$ is at most equal to the least common denominator of all the principal minors of $\hat{\mathbf{G}}_1(s)$ following (9-19); hence we conclude that $\Delta_1(s)$ is divisible without remainder by $F(s)$ and that $\Delta_1(s)\det[\mathbf{I}+\hat{\mathbf{G}}_1(s)]$ is a polynomial. Theorem 9-8 can be deduced from Theorem 9-6; however, because of its importance, we shall prove it directly in the following.

Proof of Theorem 9-8

We shall use the dynamical-equation description of the system to prove the first part of the theorem. Let

$$FE: \quad \dot{\mathbf{x}}_1 = \mathbf{A}_1\mathbf{x}_1 + \mathbf{B}_1\mathbf{u}_1 \tag{9-20a}$$
$$\mathbf{y}_1 = \mathbf{C}_1\mathbf{x}_1 + \mathbf{E}_1\mathbf{u}_1 \tag{9-20b}$$

be the dynamical-equation description of the system S_1. It is clear that

$$\hat{\mathbf{G}}_1(s) = \mathbf{C}_1(s\mathbf{I}-\mathbf{A}_1)^{-1}\mathbf{B}_1 + \mathbf{E}_1 \tag{9-21}$$

Note that the dynamical equation FE is used only implicitly in the proof; its knowledge is not required in the application of the theorem. Now the assumption that S_1 is completely characterized by $\hat{\mathbf{G}}_1(s)$ implies that FE is controllable and observable. Consequently, the characteristic polynomial $\Delta_1(s)$ of $\hat{\mathbf{G}}_1(s)$ is equal to the characteristic polynomial of $\mathbf{A}_1$; that is,

$$\Delta_1(s) \triangleq \Delta[\hat{\mathbf{G}}_1(s)] = \det(s\mathbf{I}-\mathbf{A}_1) \tag{9-22}$$

By the substitution of $\mathbf{u}_1=\mathbf{u}-\mathbf{y}_1$, $\mathbf{y}=\mathbf{y}_1$ and the use of the identity

$$\mathbf{I}-(\mathbf{I}+\mathbf{E}_1)^{-1}\mathbf{E}_1 = (\mathbf{I}+\mathbf{E}_1)^{-1}$$

the dynamical-equation description of the feedback system can be obtained as

$$FE_f: \quad \dot{\mathbf{x}}_1 = [\mathbf{A}_1 - \mathbf{B}_1(\mathbf{I}+\mathbf{E}_1)^{-1}\mathbf{C}_1]\mathbf{x}_1 + \mathbf{B}_1(\mathbf{I}+\mathbf{E}_1)^{-1}\mathbf{u} \quad \text{(9-23a)}$$

$$\mathbf{y} = (\mathbf{I}+\mathbf{E}_1)^{-1}\mathbf{C}_1\mathbf{x}_1 + (\mathbf{I}+\mathbf{E}_1)^{-1}\mathbf{E}_1\mathbf{u} \quad \text{(9-23b)}$$

The dynamical equation FE_f is controllable and observable following Theorem 9-5. Hence we conclude that the feedback system is BIBO and asymptotically stable if and only if all the eigenvalues of $\mathbf{A}_1 - \mathbf{B}_1(\mathbf{I}+\mathbf{E}_1)^{-1}\mathbf{C}_1$ have negative real parts. We show in the following that the set of the eigenvalues of $\mathbf{A}_1 - \mathbf{B}_1(\mathbf{I}+\mathbf{E}_1)^{-1}\mathbf{C}_1$ is equal to the set of the zeros of $\Delta_1(s)\det[\mathbf{I}+\hat{\mathbf{G}}_1(s)]$. Consider

$$\Delta_f(s) \triangleq \det[s\mathbf{I} - \mathbf{A}_1 + \mathbf{B}_1(\mathbf{I}+\mathbf{E}_1)^{-1}\mathbf{C}_1] = \det[(s\mathbf{I} - \mathbf{A}_1) + \mathbf{B}_1(\mathbf{I}+\mathbf{E}_1)^{-1}\mathbf{C}_1]$$

$$= \det\{(s\mathbf{I} - \mathbf{A}_1)[\mathbf{I} + (s\mathbf{I} - \mathbf{A}_1)^{-1}\mathbf{B}_1(\mathbf{I}+\mathbf{E}_1)^{-1}\mathbf{C}_1]\} \quad \text{(9-24)}$$

It is well known that $\det \mathbf{AB} = \det \mathbf{A} \det \mathbf{B}$; hence (9-24) becomes

$$\Delta_f(s) = \det(s\mathbf{I} - \mathbf{A}_1)\det[\mathbf{I} + (s\mathbf{I} - \mathbf{A}_1)^{-1}\mathbf{B}_1(\mathbf{I}+\mathbf{E}_1)^{-1}\mathbf{C}_1] \quad \text{(9-25)}$$

It is shown in Theorem 3-2 that $\det[\mathbf{I}+\hat{\mathbf{G}}_1(s)\hat{\mathbf{G}}_2(s)] = \det[\mathbf{I}+\hat{\mathbf{G}}_2(s)\hat{\mathbf{G}}_1(s)]$ and, since $\Delta_1(s) = \det(s\mathbf{I} - \mathbf{A}_1)$, Equation (9-25) implies that

$$\begin{aligned}\Delta_f(s) &= \Delta_1(s)\det[\mathbf{I} + \mathbf{C}_1(s\mathbf{I} - \mathbf{A}_1)^{-1}\mathbf{B}_1(\mathbf{I}+\mathbf{E}_1)^{-1}] \\ &= \Delta_1(s)\det\{[(\mathbf{I}+\mathbf{E}_1) + \mathbf{C}_1(s\mathbf{I} - \mathbf{A}_1)^{-1}\mathbf{B}_1](\mathbf{I}+\mathbf{E}_1)^{-1}\} \\ &= \Delta_1(s)\det[\mathbf{I}+\hat{\mathbf{G}}_1(s)]\det(\mathbf{I}+\mathbf{E}_1)^{-1}\end{aligned} \quad \text{(9-26)}$$

Since $\det(\mathbf{I}+\mathbf{E}_1)^{-1} = \det[\mathbf{I}+\hat{\mathbf{G}}_1(\infty)]^{-1} \neq 0$, we conclude from (9-26) that the set of the eigenvalues of $\mathbf{A}_1 - \mathbf{B}_1(\mathbf{I}+\mathbf{E}_1)^{-1}\mathbf{C}_1$ is equal to the set of the zeros of $\Delta_1(s)\det[\mathbf{I}+\hat{\mathbf{G}}_1(s)]$. This proves the first part of the theorem.

Now we show the second part. Let $\hat{\mathbf{G}}_1(s) = \mathbf{N}_{r1}(s)\mathbf{D}_{r1}^{-1}(s)$. Then we have

$$\begin{aligned}\hat{\mathbf{G}}_f(s) = \hat{\mathbf{G}}_1(s)[\mathbf{I}+\hat{\mathbf{G}}_1(s)]^{-1} &= \mathbf{N}_{r1}(s)\mathbf{D}_{r1}^{-1}(s)[\mathbf{I}+\mathbf{N}_{r1}(s)\mathbf{D}_{r1}^{-1}(s)]^{-1} \\ &= \mathbf{N}_{r1}(s)\mathbf{D}_{r1}^{-1}(s)\{[\mathbf{D}_{r1}(s)+\mathbf{N}_{r1}(s)]\mathbf{D}_{r1}^{-1}(s)\}^{-1} \\ &= \mathbf{N}_{r1}(s)[\mathbf{D}_{r1}(s)+\mathbf{N}_{r1}(s)]^{-1}\end{aligned} \quad \text{(9-27)}$$

We show that if $\mathbf{D}_{r1}(s)$ and $\mathbf{N}_{r1}(s)$ are right coprime, so are $\mathbf{D}_{r1}(s)+\mathbf{N}_{r1}(s)$ and $\mathbf{N}_{r1}(s)$. The coprimeness of $\mathbf{D}_{r1}(s)$ and $\mathbf{N}_{r1}(s)$ implies the existence of polynomial matrices $\mathbf{X}(s)$ and $\mathbf{Y}(s)$ such that

$$\mathbf{X}(s)\mathbf{D}_{r1}(s) + \mathbf{Y}(s)\mathbf{N}_{r1}(s) = \mathbf{I}$$

(Theorem G-8). This equation implies

$$\mathbf{X}(s)[\mathbf{D}_{r1}(s)+\mathbf{N}_{r1}(s)] + [\mathbf{Y}(s)-\mathbf{X}(s)]\mathbf{N}_{r1}(s) = \mathbf{I}$$

Since $\mathbf{X}(s)$ and $\mathbf{Y}(s)-\mathbf{X}(s)$ are both polynomial matrices, Theorem G-8 implies the right coprimeness of $\mathbf{D}_{r1}(s)+\mathbf{N}_{r1}(s)$ and $\mathbf{N}_{r1}(s)$. Consequently, $\hat{\mathbf{G}}_f(s) = \mathbf{N}_{r1}(s)[\mathbf{D}_{r1}(s)+\mathbf{N}_{r1}(s)]^{-1}$ is a coprime fraction and $\Delta[\hat{\mathbf{G}}_f(s)] \sim \det[\mathbf{D}_{r1}(s)+\mathbf{N}_{r1}(s)] \sim \Delta_f(s)$. Hence the feedback system is BIBO and asymptotically stable if and only if all the roots of $\det[\mathbf{D}_{r1}(s)+\mathbf{N}_{r1}(s)]$ have negative real parts.

Q.E.D.

Example 2

Determine the stability of the feedback system in Figure 9-10, with

$$\hat{\mathbf{G}}_1(s) = \begin{bmatrix} \dfrac{-s}{s-1} & \dfrac{s}{s+1} \\ 1 & \dfrac{-2}{s+1} \end{bmatrix}$$

The characteristic polynomial of $\mathbf{G}_1(s)$ is $\Delta_1(s) = (s-1)(s+1)$. It is easy to verify that $\det[\mathbf{I} + \hat{\mathbf{G}}_1(s)] = -1$. Hence the feedback system is stable if and only if all the roots of $\Delta_1(s)\det[\mathbf{I} + \hat{\mathbf{G}}_1(s)] = -(s-1)(s+1)$ have negative real parts. This is not the case; hence the feedback system is neither BIBO stable nor asymptotically stable. ∎

It is of interest to give some interpretation of the roots of the polynomial

$$\Delta_1(s)\det[\mathbf{I} + \hat{\mathbf{G}}_1(s)] = \Delta_1(s)\frac{E(s)}{F(s)}$$

where $E(s)$ and $F(s)$ have no common factor. The roots of $E(s)$ are poles of the feedback system introduced by feedback, whereas the roots of $\Delta_1(s)/F(s)$ are poles which are possessed by the open-loop system S_1 as well as by the feedback system. For example, in Example 1 we have $E(s) = -1$, $F(s) = 1$, $\Delta_1(s)/F(s) = (s+1)(s-1)$; hence no new pole is introduced in the feedback system $\hat{\mathbf{G}}_f(s)$; $\hat{\mathbf{G}}_f(s)$ and $\hat{\mathbf{G}}_1(s)$ have the same poles 1 and -1.

Roughly speaking, the polynomial $\Delta_1(s)/F(s)$ takes care of the missing zeros of $\det[\mathbf{I} + \hat{\mathbf{G}}_1(s)]$. Therefore one might suggest that if we do not cancel out any common factors in $\det[\mathbf{I} + \hat{\mathbf{G}}_1(s)]$, then the stability of the feedback system might be determinable from the zeros of $\det[\mathbf{I} + \hat{\mathbf{G}}_1(s)]$. This is however not necessarily correct, as can be seen from the following example.

Example 3

Consider the feedback system shown in Figure 9-10 with

$$\hat{\mathbf{G}}_1(s) = \begin{bmatrix} \dfrac{1}{(s+1.5)(s-0.5)} & \dfrac{1}{s-0.5} \\ \dfrac{s+0.5}{(s+1)(s-0.5)} & \dfrac{1}{s-0.5} \end{bmatrix}$$

It is straightforward to verify that

$$\det[\mathbf{I} + \hat{\mathbf{G}}_1(s)] = \frac{(s+0.5)(s^2+s+0.25)}{(s+1.5)(s-0.5)^2} - \frac{s+0.5}{(s+1.5)(s-0.5)^2}$$

$$-\frac{(s+0.5)(s-0.5)(s+1.5)}{(s-0.5)(s-0.5)(s+1.5)}$$

If we do not cancel out the common factor, then $\det[\mathbf{I} + \hat{\mathbf{G}}_1(s)]$ has a right-half-plane zero. Thus we may suggest erroneously that the feedback system is not stable.

The determinant of $\hat{\mathbf{G}}_1(s)$ is given by

$$\det \hat{\mathbf{G}}_1(s) = \frac{1-(s+0.5)}{(s+1.5)(s-0.5)^2} = \frac{-1}{(s+1.5)(s-0.5)}$$

Hence the characteristic polynomial, Δ_1, of $\hat{\mathbf{G}}_1(s)$ is

$$\Delta_1(s) = (s+1.5)(s-0.5)$$

Since we have $E(s)/F(s) = (s+0.5)/(s-0.5)$ and $\Delta_1(s)E(s)/F(s) = (s+1.5)(s+0.5)$, the feedback system is asymptotically stable and BIBO stable. ∎

We consider now the case where a dynamical system appears in the feedback path as shown in Figure 9-11. This feedback system is generally not completely characterized by its composite transfer matrix; hence the situation here is more complicated than the unity feedback system shown in Figure 9-10.

Theorem 9-9

Consider the feedback system shown in Figure 9-11. It is assumed that S_1 and S_2 are completely characterized by their $q \times p$ and $p \times q$ proper transfer matrices $\hat{\mathbf{G}}_1(s)$ and $\hat{\mathbf{G}}_2(s)$. It is also assumed that $\det[\mathbf{I} + \hat{\mathbf{G}}_1(\infty)\hat{\mathbf{G}}_2(\infty)] \neq 0$. Let $\hat{\mathbf{G}}_i(s) = \mathbf{N}_{ri}(s)\mathbf{D}_{ri}^{-1}(s) = \mathbf{D}_{li}^{-1}(s)\mathbf{N}_{li}(s)$, $i = 1, 2$, be coprime fractions and $\Delta_i(s)$ be the characteristic polynomial of $\hat{\mathbf{G}}_i(s)$. Then the feedback system is asymptotically stable if and only if all the roots of $\Delta_1(s)\Delta_2(s) \det[\mathbf{I} + \hat{\mathbf{G}}_1(s)\hat{\mathbf{G}}_2(s)]$, or $\mathbf{D}_{l1}(s)\mathbf{D}_{r2}(s) + \mathbf{N}_{l1}(s)\mathbf{N}_{r2}(s)$, or any other of (9-13) have negative real parts. The condition is sufficient but not necessarily for the system to be BIBO stable. If $\hat{\mathbf{G}}_2(s) = \mathbf{K}$, a $p \times q$ constant matrix, the condition is necessary as well for the system to be BIBO stable. ∎

This theorem can be deduced directly from Theorem 9-6. It can also be proved by using its dynamical equation description to establish, as in (9-26),

$$\Delta_f(s) \triangleq \det(s\mathbf{I} - \mathbf{A}_f) = \Delta_1(s)\Delta_2(s) \frac{\det[\mathbf{I} + \hat{\mathbf{G}}_1(s)\hat{\mathbf{G}}_2(s)]}{\det[\mathbf{I} + \hat{\mathbf{G}}_1(\infty)\hat{\mathbf{G}}_2(\infty)]} \tag{9-28}$$

where $\mathbf{A}_f$ is given as in (9-12) and $\Delta_i(s) \triangleq \Delta[\hat{\mathbf{G}}_i(s)]$. The interested reader is referred to Reference 49. Similar to the proof in (9-14) and (9-15), it is possible to establish, by assuming the column reducedness of $\mathbf{D}_{r2}(s)$ and the row reducedness of $\mathbf{D}_{l1}(s)$, that

$$\deg \det[\mathbf{D}_{l1}(s)\mathbf{D}_{r2}(s) + \mathbf{N}_{l1}(s)\mathbf{N}_{r2}(s)] = \deg \hat{\mathbf{G}}_1(s) + \deg \hat{\mathbf{G}}_2(s) = n_1 + n_2$$

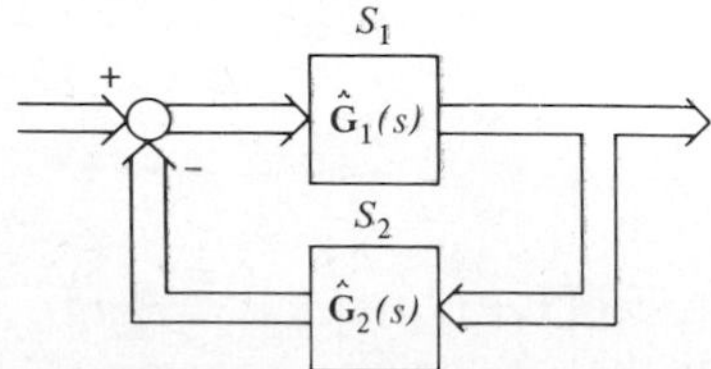

Figure 9-11 A multivariable feedback system.

and consequently,

$$\det(s\mathbf{I}-\mathbf{A}_f) \sim \det[\mathbf{D}_{l1}(s)\mathbf{D}_{r2}(s)+\mathbf{N}_{l1}(s)\mathbf{N}_{r2}(s)] \tag{9-29}$$

if and only if

$$\det[\mathbf{I}+\hat{\mathbf{G}}_1(\infty)\hat{\mathbf{G}}_2(\infty)] \neq 0$$

To save space, this assertion will not be proved here. We note that the validity of (9-29) is independent of whether or not the feedback system is completely characterized by

$$\begin{aligned}\hat{\mathbf{G}}_f(s) &= [\mathbf{I}+\hat{\mathbf{G}}_1(s)\hat{\mathbf{G}}_2(s)]^{-1}\hat{\mathbf{G}}_1(s)\\ &= [\mathbf{I}+\mathbf{D}_{l1}^{-1}(s)\mathbf{N}_{l1}(s)\mathbf{N}_{r2}(s)\mathbf{D}_{r2}^{-1}(s)]^{-1}\mathbf{D}_{l1}^{-1}(s)\mathbf{N}_{l1}(s)\\ &= \mathbf{D}_{r2}(s)[\mathbf{D}_{l1}(s)\mathbf{D}_{r2}(s)+\mathbf{N}_{l1}(s)\mathbf{N}_{r2}(s)]^{-1}\mathbf{N}_{l1}(s)\end{aligned} \tag{9-30}$$

The factorization in (9-30) is not necessarily irreducible; hence generally we have

$$\Delta[\mathbf{G}_f(s)] \not\sim \det[\mathbf{D}_{l1}(s)\mathbf{D}_{r2}(s)+\mathbf{N}_{l1}(s)\mathbf{N}_{r2}(s)] \sim \det(s\mathbf{I}-\mathbf{A}_f) \tag{9-31}$$

For this reason, the condition in Theorem 9-9 is only sufficient for $\hat{\mathbf{G}}_f(s)$ or the system in Figure 9-11 to be BIBO stable. If $\hat{\mathbf{G}}_2(s)=\mathbf{K}$, a $p \times q$ constant matrix, the feedback system is, as discussed in the previous section, completely characterized by $\hat{\mathbf{G}}_f(s)$ and $\Delta[\hat{\mathbf{G}}_f(s)] \sim \det[\mathbf{D}_{l1}(s)+\mathbf{N}_{l1}(s)\mathbf{K}]$; hence the condition in Theorem 9-9 becomes necessary as well for $\hat{\mathbf{G}}_f(s)$ to be BIBO stable.

Now we consider the same system in Figure 9-11 but with additional input and output as shown in Figure 9-8. As derived in (9-12), we have

$$\begin{bmatrix}\hat{\mathbf{u}}_1(s)\\ \hat{\mathbf{u}}_2(s)\end{bmatrix} = \mathbf{H}(s)\begin{bmatrix}\hat{\mathbf{r}}_1(s)\\ \hat{\mathbf{r}}_2(s)\end{bmatrix} = \begin{bmatrix}\mathbf{I}_p & \hat{\mathbf{G}}_2(s)\\ -\hat{\mathbf{G}}_1(s) & \mathbf{I}_q\end{bmatrix}^{-1}\begin{bmatrix}\hat{\mathbf{r}}_1(s)\\ \hat{\mathbf{r}}_2(s)\end{bmatrix}$$

By direct computation, we have

$$\begin{aligned}\mathbf{H}(s) &= \begin{bmatrix}\mathbf{I}_p & \hat{\mathbf{G}}_2(s)\\ -\hat{\mathbf{G}}_1(s) & \mathbf{I}_q\end{bmatrix}^{-1}\\ &= \begin{bmatrix}[\mathbf{I}_p+\hat{\mathbf{G}}_2(s)\hat{\mathbf{G}}_1(s)]^{-1} & -\hat{\mathbf{G}}_2(s)[\mathbf{I}_q+\hat{\mathbf{G}}_1(s)\hat{\mathbf{G}}_2(s)]^{-1}\\ \hat{\mathbf{G}}_1(s)[\mathbf{I}_p+\hat{\mathbf{G}}_2(s)\hat{\mathbf{G}}_1(s)]^{-1} & [\mathbf{I}_q+\hat{\mathbf{G}}_1(s)\hat{\mathbf{G}}_2(s)]^{-1}\end{bmatrix}\end{aligned} \tag{9-32}$$

Because the characteristic polynomial of $\mathbf{H}(s)$ is, as developed in Theorem 9-6, equal to $\det(s\mathbf{I}-\mathbf{A}_f)$, the condition in Theorem 9-9 is now necessary and sufficient for the system in Figure 9-8 to be BIBO stable. It is important to note that there exist $\hat{\mathbf{G}}_1(s)$ and $\hat{\mathbf{G}}_2(s)$ so that any three entries of $\mathbf{H}(s)$ can be BIBO stable but the fourth entry is not. For all possible combinations of stability and instability of the entries of $\mathbf{H}(s)$, see Reference S75.

In the design of feedback systems, we shall require, as discussed in the subsection of well-posedness, every transfer function from all possible input-output pairs to be proper. Similarly we shall require a system to be BIBO stable from all possible input-output pairs. For example, if the transfer matrix from $\mathbf{r}_2$ to $\mathbf{u}_1$ is not BIBO stable, any load disturbances which can be modeled as a nonzero input at $\mathbf{r}_2$ will cause the system in Figure 9-8 to saturate. Hence it is reasonable in practice to require every feedback system to be BIBO stable from all possible input-output pairs.

Consider again the system in Figure 9-8 with $[\mathbf{r}_1' \quad \mathbf{r}_2']'$ as the input and $[\mathbf{y}_1' \quad \mathbf{y}_2']'$ as the output. Clearly, we have

$$\begin{bmatrix} \hat{\mathbf{y}}_1(s) \\ \hat{\mathbf{y}}_2(s) \end{bmatrix} \triangleq \mathbf{H}_y(s) \begin{bmatrix} \hat{\mathbf{r}}_1(s) \\ \hat{\mathbf{r}}_2(s) \end{bmatrix} = \begin{bmatrix} \hat{\mathbf{G}}_1(s)\hat{\mathbf{u}}_1(s) \\ \hat{\mathbf{G}}_2(s)\hat{\mathbf{u}}_2(s) \end{bmatrix} = \begin{bmatrix} \hat{\mathbf{G}}_1(s) & \mathbf{0} \\ \mathbf{0} & \hat{\mathbf{G}}_2(s) \end{bmatrix} \mathbf{H}(s) \begin{bmatrix} \hat{\mathbf{r}}_1(s) \\ \hat{\mathbf{r}}_2(s) \end{bmatrix}$$

which implies, by using (9-32),

$$\mathbf{H}_y(s) = \begin{bmatrix} \hat{\mathbf{G}}_1(s)[\mathbf{I}_p + \hat{\mathbf{G}}_2(s)\hat{\mathbf{G}}_1(s)]^{-1} & -\hat{\mathbf{G}}_1(s)\hat{\mathbf{G}}_2(s)[\mathbf{I}_q + \hat{\mathbf{G}}_1(s)\hat{\mathbf{G}}_2(s)]^{-1} \\ \hat{\mathbf{G}}_2(s)\hat{\mathbf{G}}_1(s)[\mathbf{I}_p + \hat{\mathbf{G}}_2(s)\hat{\mathbf{G}}_1(s)]^{-1} & \hat{\mathbf{G}}_2(s)[\mathbf{I}_q + \hat{\mathbf{G}}_1(s)\hat{\mathbf{G}}_2(s)]^{-1} \end{bmatrix}$$

Using the identity

$$\mathbf{I} - \hat{\mathbf{G}}_i(s)\hat{\mathbf{G}}_j(s)[\mathbf{I} + \hat{\mathbf{G}}_i(s)\hat{\mathbf{G}}_j(s)]^{-1} = [\mathbf{I} + \hat{\mathbf{G}}_i(s)\hat{\mathbf{G}}_j(s)]^{-1} \qquad \text{for } i, j = 1, 2 \tag{9-34}$$

and defining

$$\mathbf{J} = \begin{bmatrix} \mathbf{0} & \mathbf{I}_p \\ -\mathbf{I}_q & \mathbf{0} \end{bmatrix} \tag{9-35}$$

we can readily establish

$$\mathbf{H}_y(s) = \mathbf{J}[\mathbf{H}(s) - \mathbf{I}_{p+q}] \qquad \mathbf{H}(s) = \mathbf{I}_{p+q} - \mathbf{J}\mathbf{H}_y(s) \tag{9-36}$$

From this relationship, we see that, because of the nonsingularity of $\mathbf{J}$, every pole of $\mathbf{H}(s)$ is a pole of $\mathbf{H}_y(s)$ and vice versa. Hence $\mathbf{H}(s)$ is BIBO stable if and only if $\mathbf{H}_y(s)$ is BIBO stable. This fact can also be seen from Figure 9-8. Because of $\mathbf{y}_2 = \mathbf{r}_1 - \mathbf{u}_1$ and $\mathbf{y}_1 = \mathbf{u}_2 - \mathbf{r}_2$, if $\mathbf{r}_i(t)$ and $\mathbf{u}_i(t)$ are bounded, so are $\mathbf{y}_i(t)$, $i = 1, 2$.

Corollary 9-9

The feedback system in Figure 9-8 is BIBO stable if and only if the feedback system in Figure 9-11 is asymptotically stable. ∎

This corollary can also be concluded from Problem 8-40. Thus, if a feedback system is designed to be asymptotically stable, then the system is BIBO stable from all possible input-output pairs. On the other hand, if a system is BIBO stable from one input-output pair and if there is no unstable hidden mode or, roughly speaking, no unstable pole-zero cancellation, then the system is asymptotically stable. In practice, every control system should be designed to be asymptotically stable.

9-5 Design of Compensators: Unity Feedback Systems

Single-variable case. In Chapter 7, we studied the problem of state feedback and state estimator and showed that the eigenvalues of the original system can be arbitrarily assigned without affecting its numerator. In this section, we shall study a similar problem by using transfer functions in fractional forms. We study only unity feedback systems in this section; the configuration developed in Chapter 7 will be studied in Section 9-7.

Consider the unity feedback system shown in Figure 9-12. The compen-

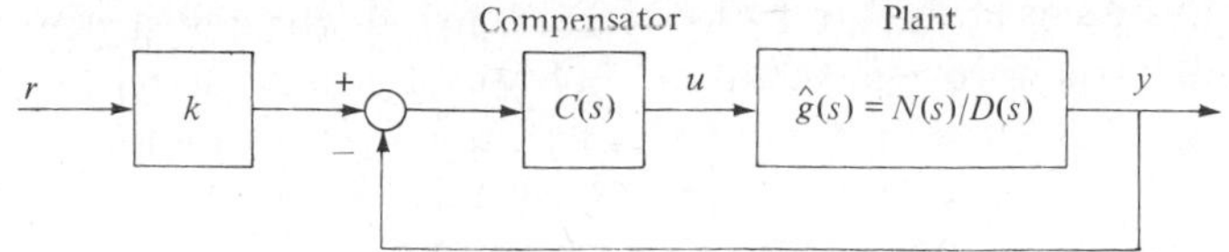

Figure 9-12 Unity feedback system.

sator $C(s) = N_c(s)/D_c(s)$ is required to be a proper rational function. The gain k is a real constant. We shall study what can be achieved in this configuration without imposing any constraint on $C(s)$ so long as it is proper. The properness of $C(s)$ is imposed to avoid amplification of high-frequency noises.

Let $\hat{g}_f(s)$ be the transfer function from r to y in Figure 9-12. Then we have

$$\hat{g}_f(s) = \frac{kC(s)\hat{g}(s)}{1 + C(s)\hat{g}(s)}$$

From $\hat{g}_f(s) + \hat{g}_f(s)C(s)\hat{g}(s) = kC(s)\hat{g}(s)$, and using $\hat{g}_f(s) = N_f(s)/D_f(s)$ and $\hat{g}(s) = N(s)/D(s)$, we can write $C(s)$ as

$$C(s) \triangleq \frac{N_c(s)}{D_c(s)} = \frac{\hat{g}_f(s)}{\hat{g}(s)(k - \hat{g}_f(s))} = \frac{D(s)N_f(s)}{N(s)(kD_f(s) - N_f(s))} \tag{9-37}$$

Now we claim that for any $\hat{g}_f(s) = N_f(s)/D_f(s)$ meeting the pole-zero excess inequality

$$\delta D_f(s) - \delta N_f(s) \geq \delta D(s) - \delta N(s) \tag{9-38}$$

where δ denotes the degree, a constant gain k and a proper compensator $C(s)$ exist to achieve the design. Indeed, by a proper choice of k, we have

$$\delta[kD_f(s) - N_f(s)] = \delta D_f(s)$$

which, together with (9-37) and (9-38), imply

$$\delta D_c(s) - \delta N_c(s) = [\delta D_f(s) - \delta N_f(s)] - [\delta D(s) - \delta N(s)] \geq 0$$

Hence $C(s)$ is proper. Thus for any $\hat{g}_f(s)$ meeting the pole-zero excess inequality, a proper compensator exists in Figure 9-12 to meet the design. We note that not only the poles but also the zeros of the overall system can be arbitrarily assigned.

There is, however, a very serious problem in the design. From (9-37), we see that $C(s)$ contains the factor $1/\hat{g}(s) = D(s)/N(s)$. Hence the tandem connection of $C(s)$ followed by $\hat{g}(s)$ always involves pole-zero cancellations. Furthermore, the canceled poles are dictated by the given plant; the designer has no control over these poles. If the given $\hat{g}(s)$ has poles or zeros in the open right-half s plane, then the design will contain *unstable* pole-zero cancellations. Hence this design is not always permissible in practice. In other words, if undesirable pole-zero cancellations are not permitted, arbitrary assignments of poles and zeros are not always possible in Figure 9-12 by computing $C(s)$ from (9-37).

In the following, we discuss a method of computing $C(s)$ which will not involve any undesirable pole-zero cancellations. However, we must pay a price for achieving this. The zeros of $N_f(s)$ of the resulting system cannot be arbitrarily chosen; they are dictated by the zeros of $C(s)$ and $\hat{g}(s)$. If we wish to control the poles as well as the zeros of $\hat{g}_f(s)$, then we must choose a different configuration (see Problem 9-10 and Reference S34).

The substitution of $\hat{g}_f = N_f/D_f$, $\hat{g} = N/D$, and $C = N_c/D_c$ into $\hat{g}_f(s) = kC(s)\hat{g}(s)/[1 + C(s)\hat{g}(s)]$ yields

$$\frac{N_f(s)}{D_f(s)} = \frac{kN_c(s)N(s)}{D_c(s)D(s) + N_c(s)N(s)} \tag{9-39}$$

In the arbitrary pole placement, we study only the polynomial equation

$$D_f(s) = D_c(s)D(s) + N_c(s)N(s) \tag{9-40}$$

In the following, we shall study the condition on $D(s)$ and $N(s)$ under which a set of solution $\{D_c(s), N_c(s)\}$ exists in (9-40) for any $D_f(s)$. We shall also study the degrees of $D_c(s)$ and $N_c(s)$ in order to achieve arbitrary pole placement.[4] By arbitrary pole placement, we always assume implicitly that complex conjugate poles appear in pair. Otherwise, the compensator will have complex coefficients and cannot be implemented in practice.

The polynomial equation in (9-40) is called the *Diophantine equation* in honor of Diophantus of the third century in Reference S139. It is called the compensator equation in Reference S34. This is the most important equation in the remainder of this chapter. Instead of studying it directly (see Problem G-14), we shall translate it into a set of linear algebraic equations. From the set of algebraic equations, we will then develop all properties of the Diophantine equation. The set of algebraic equations can also be used to compute the compensator $C(s)$. Let us define

$$D(s) = D_0 + D_1 s + \cdots + D_n s^n \qquad D_n \neq 0 \tag{9-41a}$$

$$N(s) = N_0 + N_1 s + \cdots + N_n s^n \tag{9-41b}$$

$$D_c(s) = D_{c0} + D_{c1} s + \cdots + D_{cm} s^m \tag{9-42a}$$

$$N_c(s) = N_{c0} + N_{c1} s + \cdots + N_{cm} s^m \tag{9-42b}$$

and define

$$D_f(s) = F_0 + F_1 s + F_2 s^2 + \cdots + F_{n+m} s^{n+m} \tag{9-43}$$

where D_i, N_i, D_{ci}, N_{ci}, and F_i are real constant, not necessarily all nonzero. The substitution of (9-41) to (9-43) into (9-40) yields

$$\begin{aligned} F_0 + F_1 s + \cdots + F_{n+m} s^{n+m} = {} & (D_{c0} + D_{c1} s + \cdots + D_{cm} s^m)(D_0 + D_1 s + \cdots + D_n s^n) \\ & + (N_{c0} + N_{c1} s + \cdots + N_{cm} s^m)(N_0 + N_1 s + \cdots + N_n s^n) \end{aligned}$$

[4] Mathematically, the finding of polynomial solutions in (9-40) is equivalent to the finding of integer solutions x, y in $ax + by = f$, where a, b, and f are given integers. This is a topic in number theory or continued fractions.

Equating the coefficients of the same power of s yields

$$[D_{c0}N_{c0} \vdots D_{c1}N_{c1} \vdots \cdots \vdots D_{cm}N_{cm}]\mathbf{S}_m = [F_0 \quad F_1 \quad F_2 \quad \cdots \quad F_{n+m}] \triangleq \mathbf{F} \tag{9-44}$$

with

$$\mathbf{S}_m = \left[\begin{array}{cccccccc} D_0 & D_1 & \cdots & D_{n-1} & D_n & 0 & 0 & \cdots & 0 \\ N_0 & N_1 & \cdots & N_{n-1} & N_n & 0 & 0 & \cdots & 0 \\ \hline 0 & D_0 & \cdots & D_{n-2} & D_{n-1} & D_n & 0 & \cdots & 0 \\ 0 & N_0 & \cdots & N_{n-2} & N_{n-1} & N_n & 0 & \cdots & 0 \\ \hline & & \vdots & & & & & \vdots & \\ \hline 0 & 0 & \cdots & 0 & D_0 & D_1 & D_2 & \cdots & D_n \\ 0 & 0 & \cdots & 0 & N_0 & N_1 & N_2 & \cdots & N_n \end{array}\right] \begin{array}{l} \left.\right\} \text{1st block row} \\ \\ \\ \\ \\ \left.\right\} (m+1)\text{th block row} \end{array} \tag{9-45}$$

This is a set of linear algebraic equations. There is a one-to-one correspondence between the polynomial equation in (9-40) and the algebraic equations in (9-44); hence it is permissible to study the former by using the latter. The matrix $\mathbf{S}_m$ consists of $m+1$ block rows; each block row has two rows formed from the coefficients of $\hat{g}(s)$ and can be obtained by shifting its previous block row to the right by one column. It is clear that $\mathbf{S}_m$ is a $2(m+1) \times (n+m+1)$ matrix. The application of Theorem 2-4 to the transpose of (9-44) reveals that for every $\mathbf{F}$, (9-44) has a solution $\{D_{ci}, N_{ci}, i=0, 1, \ldots, m\}$ if and only if $\rho\mathbf{S}_m = n+m+1$ or $\mathbf{S}_m$ has a full column rank. A necessary condition for $\mathbf{S}_m$ to have a full column rank is that $2(m+1) \geq n+m+1$, or $m \geq n-1$. Hence the smallest degree of a compensator to achieve arbitrary pole placement is $n-1$. From Corollary G-4, we have that $\mathbf{S}_{n-1}$ is nonsingular if and only if $D(s)$ and $N(s)$ are coprime. In fact, from the proof of Corollary G-4, we have that $\mathbf{S}_m$ has a full column rank for $m \geq n-1$ if and only if $D(s)$ and $N(s)$ are coprime. Hence we have established the following theorem.[5]

Theorem 9-10

Consider the polynomial equation

$$D_f(s) = D_c(s)D(s) + N_c(s)N(s) \tag{9-46}$$

with $\deg N(s) \leq \deg D(s) = n$. Let $D_c(s)$ and $N_c(s)$ be of degree m or less. Then for every $D_f(s)$ of degree $n+m$ or less, there exist $D_c(s)$ and $N_c(s)$ to meet (9-46) if and only if $D(s)$ and $N(s)$ are coprime and $m \geq n-1$. ∎

This theorem states only the conditions for the existence of $D_c(s)$ and $N_c(s)$ to meet (9-46). Nothing has been said regarding the properness of $N_c(s)/D_c(s)$.

[5] If $D(s)$ and $N(s)$ are not coprime, then a solution $\{D_c(s), N_c(s)\}$ exists in (9-46) for any $D_f(s)$ if and only if the greatest common divisor of $D(s)$ and $N(s)$ is a divisor of $D_f(s)$. See Problem G-14.

This question will be answered in the following two theorems by removing the condition $\deg D_f(s) < n+m$ from Theorem 9-10.

Theorem 9-11

Consider the feedback system shown in Figure 9-12 with $\hat{g}(s) = N(s)/D(s)$ and $\deg N(s) < \deg D(s) = n$. Then for any $D_f(s)$ of degree $n+m$, a proper compensator $C(s)$ of degree m exists so that the feedback system has transfer function $N(s)D_f^{-1}(s)N_c(s)$ if and only if $D(s)$ and $N(s)$ are coprime and $m \geq n-1$.

Proof

If we show that D_{cm} in (9-44) is different from zero, then this theorem follows directly from Theorem 9-10. If $\hat{g}(s)$ is strictly proper, then $N_n = 0$. Consequently, from the last column of (9-44), we have $D_{cm} = F_{n+m}/D_n$. Hence if F_{n+m} is different from zero, so is D_{cm}. This proves that the compensator $C(s)$ is proper. Q.E.D.

Theorem 9-11′

Consider the feedback system shown in Figure 9-12 with $\hat{g}(s) = N(s)/D(s)$ and $\deg N(s) \leq \deg D(s) = n$. Then for any $D_f(s)$ of degree $n+m$, a strictly proper compensator $C(s)$ of degree m exists so that the feedback system has transfer function $N(s)D_f^{-1}(s)N_c(s)$ if and only if $D(s)$ and $N(s)$ are coprime and $m \geq n$.

Proof

For a proper $\hat{g}(s)$, the solution $\{D_{ci}, N_{ci}\}$ in (9-44) exists if and only if $N(s)$ and $D(s)$ are coprime and $m \geq n-1$. However, if $m = n-1$, the solution is unique, and there is no guarantee that $D_{cm} \neq 0$, and the compensator may become improper. If $m \geq n$, the number of unknown $\{D_{ci}, N_{ci}\}$ in (9-44) is larger than the number of equations. Since the last row of $\mathbf{S}_m$, with $m \geq n$, is linearly dependent on its previous rows (see Section G-2), we may choose $N_{cm} = 0$. For this choice, we have $D_{cm} = F_{n+m}/D_n$. Hence the compensator is strictly proper. This establishes the theorem. Q.E.D.

In Theorem 9-11, if $m = n-1$ and $D_f(s) = P(s)D(s)$, where $P(s)$ is an arbitrary polynomial of degree m, then the unique solution of (9-46) is $D_c(s) = P(s)$ and $N_c(s) = 0$. In this case, no pole of the plant is to be altered, and the compensator is zero. This is a degenerated case. Similarly, in Theorem 9-11′, if $m = n$ and $D_f(s) = P(s)D(s)$, and if the compensator is required to be strictly proper, then $N_c(s) = 0$, and the compensator is zero. Clearly, Theorem 9-11′ still holds if the compensator is permitted to be proper. As discussed in Section 3-6, the well-posedness problem of feedback systems should be checked in the design. In Theorem 9-11, $\hat{g}(s)$ is strictly proper and $C(s)$ is proper; hence $1 + C(\infty)\hat{g}(\infty) \neq 0$, and the feedback system is well posed. In Theorem 9-11′, $\hat{g}(s)$ is proper, but $C(s)$ is strictly proper; hence $1 + C(\infty)\hat{g}(\infty) \neq 0$, and the feedback system is again well posed. The design may involve polezero cancellations between $D_f(s)$ and $N(s)N_c(s)$. However these poles are all assignable by the designer; hence they will not cause any problem in the design.

The employment of Theorems 9-11 and 9-11′ is very simple. We form the set of linear algebraic equations in (9-44) with $m=n-1$ or $m=n$, depending on whether $\hat{g}(s)$ is strictly proper or proper. Its solution yields immediately the required compensator. This is illustrated by an example.

Example 1

Consider a plant with transfer function

$$\hat{g}(s)=\frac{s^2+1}{s^2+2s-2}$$

Let $C(s)$ be chosen as

$$C(s)=\frac{N_{c0}+N_{c1}s}{D_{c0}+D_{c1}s} \tag{9-47}$$

Then the equation in (9-44) becomes

$$[D_{c0} \quad N_{c0} \;\vdots\; D_{c1} \quad N_{c1}]\left[\begin{array}{cccc} -2 & 2 & 1 & 0 \\ 1 & 0 & 1 & 0 \\ \hdashline 0 & -2 & 2 & 1 \\ 0 & 1 & 0 & 1 \end{array}\right]=[F_0 \quad F_1 \quad F_2 \quad F_3] \tag{9-48}$$

Since the square matrix in (9-48) is nonsingular, for any $F_i, i=0, 1, 2, 3$, a solution exists in (9-48). In other words, for any three arbitrary poles, a compensator of degree 1 always exists. However, for some set of chosen poles, the D_{c1} in (9-48) may become zero. For example, if $D_f(s)=2+2s+0.5s^2+3s^3$, then the solution in (9-48) can be computed as $D_{c0}=-0.5, D_{c1}=0, N_{c0}=1, N_{c1}=3$, or $C(s)=(3s+1)/(-0.5)$. This compensator does not have a proper transfer function.

Now we choose

$$C(s)=\frac{N_{c0}+N_{c1}s+N_{c2}s^2}{D_{c0}+D_{c1}s+D_{c2}s^2}$$

In this case, (9-44) becomes

$$[D_{c0} \quad N_{c0} \;\vdots\; D_{c1} \quad N_{c1} \;\vdots\; D_{c2} \quad N_{c2}]\left[\begin{array}{ccccc} -2 & 2 & 1 & 0 & 0 \\ 1 & 0 & 1 & 0 & 0 \\ \hdashline 0 & -2 & 2 & 1 & 0 \\ 0 & 1 & 0 & 1 & 0 \\ \hdashline 0 & 0 & -2 & 2 & 1 \\ 0 & 0 & 1 & 0 & 1 \end{array}\right]$$

$$=[F_0 \quad F_1 \quad F_2 \quad F_3 \quad F_4] \tag{9-49}$$

The first five rows of the 6×5 matrix in (9-49) are linearly independent and have a full column rank; hence for any $F_i, i=0, 1, \ldots, 4$, a solution $\{D_{ci}, N_{ci}$ with $D_{c2}=F_4$ and $N_{c2}=0\}$ exists in (9-49). In other words, the compensator has a strictly proper transfer function. This confirms with Theorem 9-11′. ∎

In Theorem 9-11, $n-1$ is the smallest degree which a compensator must have in order to achieve arbitrary pole placement. In other words, for *every* $D_f(s)$, a compensator $C(s)$ of degree $n-1$ always exists so that the unity feedback system in Figure 9-12 has $D_f(s)$ as its denominator. If $D_f(s)$ is prechosen, then it may be possible to find a compensator of degree smaller than $n-1$ to achieve the design. For example, if $D_f(s)$ is chosen as

$$D_f(s) = 4s^2 + 2s + 1$$

in the example, then we have

$$[D_{c0} \quad N_{c0}]\mathbf{S}_0 = [1 \quad 2 \quad 4]$$

and $D_{c0} = 1, N_{c0} = 3$. In other words, we can find a compensator $C(s) = \frac{3}{1}$ of degree 0 to achieve this particular pole placement. Hence, if the degree of a compensator is extremely important, we may proceed as follows: First, we form $\mathbf{S}_0$ and check whether the chosen $\bar{D}_f(s)$ of degree n is in the row space of $\mathbf{S}_0$. If yes, the design can be completed by using a compensator of degree 0. If not, we form $\mathbf{S}_1$ and increase the degree of $D_f(s)$ by one to $D_f(s) = \bar{D}_f(s)(s+k)$. If $D_f(s)$ is in the row space of $\mathbf{S}_1$ for some k, and if the pole $s+k$ of this k is acceptable, then the design can be accomplished by using a compensator of degree 1. If $D_f(s)$ is not in the row space of $\mathbf{S}_1$, then we must increase m by 1 and repeat the process. For a study of this problem, the reader is referred to References S97, S184, S210, and S223.

The larger the degree of a compensator, the larger the number of parameters available for adjustment to achieve design purposes. If the number of parameters is larger than the minimum required for arbitrary pole placement, the spared parameters can be used to achieve other design objectives such as the assignment of zeros or the minimization of sensitivity functions. The interested reader is referred to References S25 and S83 and Problem 9-29.

In Theorems 9-11 and 9-11′, if $D_f(s)$ is chosen as a Hurwitz polynomial, then the feedback system is stable. Hence we conclude that every plant with a proper transfer function can be stabilized by using a compensator with a proper transfer function in the unity feedback configuration shown in Figure 9-12.

Single-input or single-output case. In this section we discuss the design of compensators to achieve pole placement for single-input multiple-output and multiple-input single-output systems. The general case will be postponed to the next subsection.

Consider the feedback system shown in Figure 9-13(a). The plant is de-

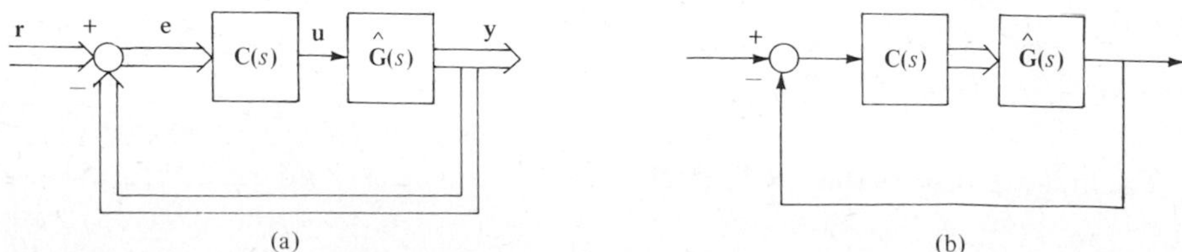

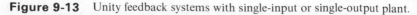

Figure 9-13 Unity feedback systems with single-input or single-output plant.

scribed by the $q \times 1$ proper rational matrix

$$\hat{\mathbf{G}}(s) = \begin{bmatrix} \dfrac{N_1'(s)}{D_1'(s)} \\ \dfrac{N_2'(s)}{D_2'(s)} \\ \vdots \\ \dfrac{N_q'(s)}{D_q'(s)} \end{bmatrix} = \frac{1}{D(s)} \begin{bmatrix} N_1(s) \\ N_2(s) \\ \vdots \\ N_q(s) \end{bmatrix} \triangleq \mathbf{N}(s)D^{-1}(s) \tag{9-50}$$

where $D(s)$ the least common denominator of all elements of $\hat{\mathbf{G}}(s)$. We assume

$$\begin{aligned} D(s) &= D_0 + D_1 s + D_2 s^2 + \cdots + D_n s^n \qquad D_n \neq 0 \\ \mathbf{N}(s) &= \mathbf{N}_0 + \mathbf{N}_1 s + \mathbf{N}_2 s^2 + \cdots + \mathbf{N}_n s^n \end{aligned} \tag{9-51}$$

where D_i are constants and $\mathbf{N}_i$ are $q \times 1$ constant vectors. The problem is to find a compensator with a proper transfer matrix of degree m so that $n+m$ number of poles of the feedback system in Figure 9-13(a) can be arbitrarily assigned. Furthermore, the degree m of the compensator is required to be as small as possible. From Figure 9-13(a) we have

$$\hat{u}(s) = (1 + \mathbf{C}(s)\hat{\mathbf{G}}(s))^{-1}\mathbf{C}(s)\hat{\mathbf{r}}(s)$$

Hence the transfer matrix of the overall feedback system in Figure 9-13(a) is equal to

$$\hat{\mathbf{G}}_f(s) = \hat{\mathbf{G}}(s)[1 + \mathbf{C}(s)\hat{\mathbf{G}}(s)]^{-1}\mathbf{C}(s) \tag{9-52}$$

Let us write the compensator $\mathbf{C}(s)$ as

$$\mathbf{C}(s) = \frac{1}{D_c(s)}[N_{c1}(s) \quad N_{c2}(s) \quad \cdots \quad N_{cq}(s)] = D_c^{-1}(s)\mathbf{N}_c(s) \tag{9-53}$$

with

$$\begin{aligned} D_c(s) &= D_{c0} + D_{c1}s + \cdots + D_{cm}s^m \\ \mathbf{N}_c(s) &= \mathbf{N}_{c0} + \mathbf{N}_{c1}s + \cdots + \mathbf{N}_{cm}s^m \end{aligned} \tag{9-54}$$

where D_i are scalars and $\mathbf{N}_{ci}$ are $1 \times q$ constant vectors. The substitution of (9-50) and (9-53) into (9-52) yields

$$\begin{aligned} \hat{\mathbf{G}}_f(s) &= \mathbf{N}(s)D^{-1}(s)[1 + D_c^{-1}(s)\mathbf{N}_c(s)\mathbf{N}(s)D^{-1}(s)]^{-1}D_c^{-1}(s)\mathbf{N}_c(s) \\ &= [D_c(s)D(s) + \mathbf{N}_c(s)\mathbf{N}(s)]^{-1}\mathbf{N}(s)\mathbf{N}_c(s) \end{aligned} \tag{9-55}$$

Because $\mathbf{N}(s)$ and $\mathbf{N}_c(s)$ are $q \times 1$ and $1 \times q$ vectors, $\mathbf{N}_c(s)\mathbf{N}(s)$ is a 1×1 matrix and $\mathbf{N}(s)\mathbf{N}_c(s)$ is a $q \times q$ matrix. Hence $\hat{\mathbf{G}}_f(s)$ is a $q \times q$ rational matrix. Define[6]

$$D_f(s) = D_c(s)D(s) + \mathbf{N}_c(s)\mathbf{N}(s) \tag{9-56}$$

Hence the problem of pole placement reduces to the solving of Equation (9-56). This equation is a generalization of the Diophantine equation in (9-40). Similar to (9-40), we shall translate it into a set of linear algebraic equations. Let

$$D_f(s) = F_0 + F_1 s + F_2 s^2 + \cdots + F_{n+m}s^{n+m} \tag{9-57}$$

[6] Note that, because of possible pole-zero cancellations, not all roots of $D_f(s)$ are poles of $\hat{\mathbf{G}}_f(s)$.

The substitution of (9-51), (9-54), and (9-57) into (9-56) and equating the coefficients of the same power of s yield

$$[D_{c0} \quad \mathbf{N}_{c0} \,\vdots\, D_{c1} \quad \mathbf{N}_{c1} \,\vdots\, \cdots \,\vdots\, D_{cm} \quad \mathbf{N}_{cm}]\mathbf{S}_m = [F_0 \quad F_1 \quad \cdots \quad F_{n+m}] \triangleq \mathbf{F} \tag{9-58}$$

with

$$\mathbf{S}_m = \left[\begin{array}{ccccccc} D_0 & D_1 & \cdots & D_n & 0 & \cdots & 0 \\ \mathbf{N}_0 & \mathbf{N}_1 & \cdots & \mathbf{N}_n & 0 & \cdots & 0 \\ \hdashline 0 & D_0 & \cdots & D_{n-1} & D_n & \cdots & 0 \\ 0 & \mathbf{N}_0 & \cdots & \mathbf{N}_{n-1} & \mathbf{N}_n & \cdots & 0 \\ \hdashline & & & \vdots & & & \\ \hdashline 0 & 0 & 0 & D_0 & D_1 & \cdots & D_n \\ 0 & 0 & 0 & \mathbf{N}_0 & \mathbf{N}_1 & \cdots & \mathbf{N}_n \end{array}\right] \begin{array}{l} \Big\} \text{one block row} \\ \\ \\ \\ \\ \\ \\ \end{array} \tag{9-59}$$

We call the rows formed from $\{D_i\}$ D rows and the rows formed from $\{\mathbf{N}_i\}$ N rows. Then every block row in (9-59) has one D row and q N rows. The matrix $\mathbf{S}_m$ consists of $m+1$ block rows; every block row is the shifting to the right by one column of its previous block row. The matrix $\mathbf{S}_m$ is clearly a $(1+q)(m+1)\times(n+1+m)$ matrix. Now we search the linearly independent rows of $\mathbf{S}_m$ in order from top to bottom by using, for example, the row-searching algorithm discussed in Appendix A. Then as discussed in Theorem G-13, all D rows in $\mathbf{S}_m$ will be linearly independent, and some N rows will be linearly dependent on their previous rows in $\mathbf{S}_m$. Let r_i be the number of linearly dependent N rows in the $i+1$ block row of $\mathbf{S}_m$. Then because of the structure of $\mathbf{S}_m$, we have $0 \le r_0 \le r_1 \le \cdots \le q$. Let v be the integer such that $r_0 \le r_1 \le \cdots \le r_{v-1} < q$ and $r_v = r_{v+1} = \cdots = q$. We call v the row index of $\hat{\mathbf{G}}(s)$. Then v is the largest row degree of $\mathbf{A}(s)$ in any left-coprime fraction of $\hat{\mathbf{G}}(s) = \mathbf{A}^{-1}(s)\mathbf{B}(s)$ with $\mathbf{A}(s)$ row reduced [see (G-80) and (G-81)] or the observability index of any irreducible realization of $\hat{\mathbf{G}}(s)$ (dual of Theorem 6-6).

Theorem 9-12[7]

Consider the feedback system shown in Figure 9-13(a) with the plant described by a $q \times 1$ strictly proper (proper) rational matrix $\hat{\mathbf{G}}(s) = \mathbf{N}(s)D^{-1}(s)$ with $\deg D(s) = n$. Then for any $D_f(s)$ of degree $n+m$, there exists a $1 \times q$ proper (strictly proper) compensator $C(s) = D_c^{-1}(s)\mathbf{N}_c(s)$ with $\deg D_c(s) = m$ so that the feedback system has $q \times q$ transfer matrix $\mathbf{N}(s)D_f^{-1}(s)\mathbf{N}_c(s)$ if and only if $D(s)$ and $\mathbf{N}(s)$ are right coprime and $m \ge v-1$ ($m \ge v$), where v is the row index of $\hat{\mathbf{G}}(s)$, or the observability index of any irreducible realization of $\hat{\mathbf{G}}(s)$, or the largest row degree of $\mathbf{A}(s)$ in any left-coprime fraction of $\hat{\mathbf{G}}(s) = \mathbf{A}^{-1}(s)\mathbf{B}(s)$ with $\mathbf{A}(s)$ row reduced.

---

[7] If $q = 1$, this theorem reduces to Theorems 9-11 and 9-11′ and v is equal to n.

Proof

The design problem is equivalent to the solving of Equation (9-58). The application of Theorem 2-4 to the transpose of (9-58) yields that for every $\mathbf{F}$, there exists a set of $\{D_{ci}, \mathbf{N}_{ci}\}$ or, equivalently, a compensator $\mathbf{C}(s)$ if and only if the matrix $\mathbf{S}_m$ has a full column rank, that is, $\rho\mathbf{S}_m = n + m + 1$. From the definition of r_i and the linear independence of all D rows, we have

$$\rho\mathbf{S}_m = (m+1) + \sum_{i=0}^{m} (q - r_i) \tag{9-60}$$

Hence the condition for $\mathbf{S}_m$ to have a rank of $n+m+1$ is $n = \sum_{i=0}^{m} (q - r_i)$. This is the case if and only if $D(s)$ and $\mathbf{N}(s)$ are right coprime and $m \geq v - 1$ (Corollary G-14). This establishes the existence of D_{ci} and $\mathbf{N}_{ci}$ in (9-58) for any $\{F_i\}$.

If $\hat{\mathbf{G}}(s)$ is strictly proper, then $\mathbf{N}_n = \mathbf{0}$. Consequently, D_{cm} in (9-58) can be solved as $D_{cm} = F_{n+m}/D_n$, and the compensator $\mathbf{C}(s)$ in (9-53) is proper. If $\hat{\mathbf{G}}(s)$ is proper and if $m = v - 1$, then D_{cm} in (9-58) may become zero. However, if $m \geq v$, then the last q rows of $\mathbf{S}_m$ are linearly dependent on their previous rows. Consequently, we may choose $\mathbf{N}_{cm} = \mathbf{0}$. For this choice, D_{cm} becomes $D_{cm} = F_{n+m}/D_n$. Hence the compensator $\mathbf{C}(s)$ is strictly proper. This completes the proof of this theorem. Q.E.D.

The application of this theorem is very simple. If $D(s), N_1(s), N_2(s), \ldots, N_q(s)$ in (9-50) have no common roots, then $D(s)$ and $\mathbf{N}(s)$ are right coprime. We form $\mathbf{S}_m$ by using the coefficients of $D(s)$ and $\mathbf{N}(s)$, and then search its linearly independent rows in order from top to bottom by using the row-searching algorithm. Once v is found, we choose $n + v - 1$ poles for the overall systems if $\hat{\mathbf{G}}(s)$ is strictly proper. From these poles, we form Equation (9-58) with $m = v - 1$. Its solution yields the compensator. We note that in hand calculation, the solution of (9-58) can be obtained by continuing the row searching to

$$\begin{bmatrix} \mathbf{S}_{v-1} \\ -\mathbf{F} \end{bmatrix}$$

We emphasize that although Theorem 9-12 is stated in terms of the observability index, the concept is not needed in applying the theorem. The theorem hinges on the search of m so that $\mathbf{S}_m$ in (9-59) has a full column rank. Since $\mathbf{S}_m$ is a $(1+q)(m+1) \times (n+m+1)$ matrix, in order to have a full column rank, we need $(1+q)(m+1) \geq (n+m+1)$ or $m \geq (n/q) - 1$. Hence in searching m, we may start from the smallest integer $m \geq (n/q) - 1$ rather than from $m = 0$. We note that all remarks for the single-variable case are equally applicable here. For example, for a given set of poles, it may be possible to find a compensator of a degree smaller than $v - 1$ to achieve pole placement. By increasing the degree of a compensator, it is possible to achieve, in addition to the pole placement, other design objectives.

Dual to Theorem 9-12, we have the following theorem for the feedback system shown in Figure 9-13(b).

Theorem 9-12′

Consider the feedback system shown in Figure 9-13(b) with the plant described by a strictly proper (proper) $1 \times p$ rational matrix $\hat{\mathbf{G}}(s) = D^{-1}(s)\mathbf{N}(s)$ with $\deg D(s) = n$. Then for any $D_f(s)$ of degree $n+m$, there exists a $p \times 1$ proper (strictly proper) compensator $\mathbf{C}(s) = \mathbf{N}_c(s)D_c^{-1}(s)$ with $\deg D_c(s) = m$ so that the feedback system has 1×1 transfer function $\mathbf{N}(s)D_f^{-1}(s)\mathbf{N}_c(s)$ if and only if $D(s)$ and $\mathbf{N}(s)$ are left coprime and $m \geq \mu - 1 (m \geq \mu)$, where μ is the column index of $\hat{\mathbf{G}}(s)$, or the controllability index of any irreducible realization of $\hat{\mathbf{G}}(s)$, or the largest column degree of $\mathbf{A}(s)$ in any right-coprime factorization of $\hat{\mathbf{G}}(s) = \mathbf{B}(s)\mathbf{A}^{-1}(s)$ with $\mathbf{A}(s)$ column reduced. ∎

The polynomial equation arises in this theorem is of the form

$$D_f(s) = D(s)D_c(s) + \mathbf{N}(s)\mathbf{N}_c(s) \tag{9-61}$$

Note that D_c and $\mathbf{N}_c$ are on the right-hand side of $D(s)$ and $\mathbf{N}(s)$, rather than the left-hand side as in (9-56). Equation (9-61) can be solved indirectly by taking its transpose to become the form of (9-56) or solved directly as follows: Using the coefficient matrices of $D(s)$, $\mathbf{N}(s)$, $D_c(s)$, $\mathbf{N}_c(s)$, and $D_f(s)$, we form the linear algebraic equation

$$\mathbf{T}_m \begin{bmatrix} D_{c0} \\ \mathbf{N}_{c0} \\ D_{c1} \\ \mathbf{N}_{c1} \\ \vdots \\ D_{cm} \\ \mathbf{N}_{cm} \end{bmatrix} \triangleq \underbrace{\left[\begin{array}{cc|cc|c|cc} D_0 & \mathbf{N}_0 & 0 & \mathbf{0} & & 0 & \mathbf{0} \\ D_1 & \mathbf{N}_1 & D_0 & \mathbf{N}_0 & & 0 & \mathbf{0} \\ & & & & & \vdots & \vdots \\ \vdots & \vdots & \vdots & \vdots & & 0 & \mathbf{0} \\ D_n & \mathbf{N}_n & D_{n-1} & \mathbf{N}_{n-1} & \cdots & D_0 & \mathbf{N}_0 \\ 0 & \mathbf{0} & D_n & \mathbf{N}_n & & D_1 & \mathbf{N}_1 \\ \vdots & \vdots & \vdots & \vdots & & \vdots & \vdots \\ 0 & \mathbf{0} & 0 & \mathbf{0} & & D_n & \mathbf{N}_n \end{array}\right]}_{m+1 \text{ block columns}} \begin{bmatrix} D_{c0} \\ \mathbf{N}_{c0} \\ D_{c1} \\ \mathbf{N}_{c1} \\ \vdots \\ D_{cm} \\ \mathbf{N}_{cm} \end{bmatrix} = \begin{bmatrix} F_0 \\ F_1 \\ F_2 \\ \\ \vdots \\ \\ F_{n+m} \end{bmatrix} \tag{9-62}$$

This equation is essentially the transpose of (9-58) (see Theorems G-14 and G-14′). We then search linearly independent columns of T_m in order from left to right. Let μ be the least integer such that the last p N columns of T_μ are all linearly dependent of their left-hand-side columns. The μ will be called the *column index* of $\hat{\mathbf{G}}(s)$. It is equal to the controllability index of any irreducible realization of $\hat{\mathbf{G}}(s)$ or the largest column degree of the column-reduced $\mathbf{A}(s)$ in any right-coprime factorization of $\hat{\mathbf{G}}(s) = \mathbf{B}(s)\mathbf{A}^{-1}(s)$. The proof of Theorem 9-12′ is similar to the one of Theorem 9-12 and will not be repeated.

Multivariable case: Arbitrary pole assignment. In this section, the design technique developed in the previous subsection will be extended to general proper rational matrices. We extend it first to a special class of rational matrices, called cyclic rational matrices, and then to the general case.

Consider a $q \times p$ proper rational matrix $\hat{\mathbf{G}}(s)$. Let $\psi(s)$ be the least common denominator of all elements of $\hat{\mathbf{G}}(s)$ or, equivalently, of all minors of order 1 of $\hat{\mathbf{G}}(s)$.[8] Let $\Delta(s)$ be the characteristic polynomial of $\hat{\mathbf{G}}(s)$ defined as the least common denominator of all minors of orders $1, 2, \ldots, \min(p, q)$, of $\hat{\mathbf{G}}(s)$ (see Definition 6-1). Clearly, in general, we have $\Delta(s) = \psi(s)h(s)$ for some polynomial $h(s)$. If $\Delta(s) = \psi(s)\,k$ for some constant k, then $G(s)$ is called a *cyclic* rational matrix. For example, if

$$\hat{\mathbf{G}}_1(s) = \begin{bmatrix} \dfrac{1}{s+1} & \dfrac{2}{s+1} \\ \dfrac{1}{s+1} & 0 \end{bmatrix} \qquad \hat{\mathbf{G}}_2(s) = \begin{bmatrix} \dfrac{1}{s+1} & \dfrac{1}{s+1} \\ \dfrac{1}{s+1} & \dfrac{s+2}{s+1} \end{bmatrix} \qquad \hat{\mathbf{G}}_3(s) = \begin{bmatrix} \dfrac{1}{s+1} & \dfrac{1}{s} \\ \dfrac{s+1}{s+2} & \dfrac{1}{s+2} \end{bmatrix}$$

then it can be readily verified that $\hat{\mathbf{G}}_2(s)$ and $\hat{\mathbf{G}}_3(s)$ are *cyclic*, but $\hat{\mathbf{G}}_1(s)$ is not. Every $1 \times p$ or $q \times 1$ proper rational matrix is cyclic. If $\hat{\mathbf{G}}(s)$ can be expressed as $\hat{\mathbf{G}}(s) = \psi^{-1}(s)\mathbf{N}(s)\mathbf{N}_c(s)$, where $\mathbf{N}(s)$ and $\mathbf{N}_c(s)$ are $q \times 1$ and $1 \times p$ polynomial matrices, then $\hat{\mathbf{G}}(s)$ is cyclic (why?). If no λ appears as a pole of two or more elements $\hat{g}_{ij}(s)$ of $\hat{\mathbf{G}}(s)$, then $\hat{\mathbf{G}}(s)$ is cyclic (Problem 9-14).

Consider the cyclic rational matrix $\hat{\mathbf{G}}_2(s)$. We form

$$\hat{\mathbf{G}}_2(s)\mathbf{a} \triangleq \hat{\mathbf{G}}_2(s)\begin{bmatrix} a_1 \\ a_2 \end{bmatrix} = \begin{bmatrix} \dfrac{a_1 + a_2}{s+1} \\ \dfrac{a_2 s + (2a_2 + a_1)}{s+1} \end{bmatrix}$$

We see that if $a_1 = -a_2$, then $\hat{\mathbf{G}}_2(s)\mathbf{a} = [0 \quad -a_2]'$ and the characteristic polynomials of $\hat{\mathbf{G}}_2(s)$ and $\hat{\mathbf{G}}_2(s)\mathbf{a}$ are different. For $\mathbf{a}$ with $a_1 \neq -a_2$, their characteristic polynomials, however, are equal. Recall that for cyclic rational matrices, the characteristic polynomial is equal to the minimal polynomial. This example shows that in any linear combination of the columns of $\hat{\mathbf{G}}_2(s)$, the chance of canceling a pole is very small. This is true in general as state in the following theorem.

Theorem 9-13

Consider a $q \times p$ cyclic proper rational matrix $\hat{\mathbf{G}}(s)$. Then for almost all $p \times 1$ and $1 \times q$ real constant vectors $\mathbf{t}_1$ and $\mathbf{t}_2$, we have

$$\Delta[\hat{\mathbf{G}}(s)] = \Delta[\hat{\mathbf{G}}(s)\mathbf{t}_1] = \Delta[\mathbf{t}_2\hat{\mathbf{G}}(s)] \tag{9-63}$$

where $\Delta(\cdot)$ denotes the characteristic polynomial of a rational matrix.[9]

---

[8] Let $\{\mathbf{A}, \mathbf{B}, \mathbf{C}, \mathbf{E}\}$ be an irreducible realization of $\hat{\mathbf{G}}(s)$. Then it can be shown that $\psi(s)$ is the minimal polynomial of $\mathbf{A}$. Compare the definition of cyclic rational matrix with the one for a constant $\mathbf{A}$ matrix in Problem 2-45.

[9] Compare this theorem with Theorem 7-5.

Proof

Let us write

$$\hat{\mathbf{G}}(s) = \frac{1}{\psi(s)}\mathbf{N}(s) = \frac{1}{\psi(s)}\begin{bmatrix} \mathbf{N}_1(s) \\ \mathbf{N}_2(s) \\ \vdots \\ \mathbf{N}_q(s) \end{bmatrix}$$

where $\psi(s)$ is the least common denominator of all elements of $\hat{\mathbf{G}}(s)$; $\mathbf{N}(s)$ is a $q \times p$ polynomial matrix, and $\mathbf{N}_i(s)$ is the ith row of $\mathbf{N}(s)$. We assume that every element of $\hat{\mathbf{G}}(s)$ is irreducible. The cyclicity assumption of $\hat{\mathbf{G}}(s)$ implies $\Delta[\hat{\mathbf{G}}(s)] = \psi(s)$. Let λ_i, $i = 1, 2, \ldots, m$, be the distinct roots of $\psi(s)$. First we assume that λ_i, $i = 1, 2, \ldots, m$, are real. Then we have

$$\mathbf{N}(\lambda_i) \neq \mathbf{0} \qquad \text{for } i = 1, 2, \ldots, m$$

Hence the rank of $\mathbf{N}(\lambda_i)$ is at least 1. Consequently, the null space of $\mathbf{N}(\lambda_i)$ has a dimension of at most $p-1$ (Theorem 2-5). In other words, the set of vectors $\mathbf{t}$ in the p-dimensional real vector space $\mathbb{R}^p$ which satisfy $\mathbf{N}(\lambda_i)\mathbf{t} = \mathbf{0}$ is a linear space of dimension at most $p-1$. To help to visualize what will be discussed, we assume $p = 2$. Then those $\mathbf{t}$ satisfying $\mathbf{N}(\lambda_i)\mathbf{t} = \mathbf{0}$, for each i, is at most a straight line in $\mathbb{R}^2$. Consequently, there are at most m straight lines on which $\mathbf{N}(\lambda_i)\mathbf{t} = \mathbf{0}$ for some i. The $\mathbf{t}_1$ not on these straight lines have the property

$$\mathbf{N}(\lambda_i)\mathbf{t}_1 \neq \mathbf{0} \qquad \text{for all } i = 1, 2, \ldots, m$$

This implies that $\psi(s)$ and $\mathbf{N}_i(s)\mathbf{t}_1$, $i = 1, 2, \ldots, q$, in

$$\hat{\mathbf{G}}(s)\mathbf{t}_1 = \frac{1}{\psi(s)}\begin{bmatrix} \mathbf{N}_1(s)\mathbf{t}_1 \\ \mathbf{N}_2(s)\mathbf{t}_1 \\ \vdots \\ \mathbf{N}_q(s)\mathbf{t}_1 \end{bmatrix}$$

have no common factor. Hence $\psi(s)$ is the characteristic polynomial of $\hat{\mathbf{G}}(s)\mathbf{t}_1$. This establishes $\Delta[\hat{\mathbf{G}}(s)] = \Delta[\hat{\mathbf{G}}(s)\mathbf{t}_1]$ for the case where all roots of $\psi(s)$ are real. If some of λ_i, $i = 1, 2, \ldots, m$, are not real, the preceding proof still holds if we use $\mathbf{N}(\lambda_i) + \mathbf{N}(\lambda_i^*)$. This completes the proof of $\Delta[\hat{\mathbf{G}}(s)] = \Delta[\hat{\mathbf{G}}(s)\mathbf{t}_1]$. The other part of the theorem can be similarly proved.

Now we show that (9-63) holds for almost all $\mathbf{t}_1$ in $\mathbb{R}^p$. For easy visualization, we assume $p = 2$ and consider only $\mathbf{t}_1$ with $\|\mathbf{t}_1\|_2 = 1$. Then from the preceding proof, we see that at most $2m$ points on the unit circle of $\mathbb{R}^2$ will not meet (9-63), the rest of the unit circle (infinitely many points) will meet (9-63). Hence almost every vector $\mathbf{t}_1$ in $\mathbb{R}^2$ has the property $\Delta[\hat{\mathbf{G}}(s)] = \Delta[\hat{\mathbf{G}}(s)\mathbf{t}_1]$. Hence if a vector $\mathbf{t}_1$ is chosen arbitrarily or generated randomly, the probability of having $\Delta[\hat{\mathbf{G}}(s)] = \Delta[\hat{\mathbf{G}}(s)\mathbf{t}_1]$ is almost equal to 1. This establishes the theorem. Q.E.D.

The cyclicity of $\hat{\mathbf{G}}(s)$ is essential in this theorem. If $\hat{\mathbf{G}}(s)$ is not cyclic, the theorem does not hold in general (try $\hat{\mathbf{G}}_4(s)$ in Problem 9-13). Using Theorem 9-13, we can now extend the design procedure in Theorems 9-12 and 9-12′ to cyclic rational matrices.

Theorem 9-14

Consider the feedback system shown in Figure 9-14 with the plant described by a $q \times p$ cyclic strictly proper (proper) rational matrix $\hat{\mathbf{G}}(s)$ of degree n. The compensator is assumed to have a $p \times q$ proper (strictly proper) rational matrix $\mathbf{C}(s)$ of degree m. If $m \geq \min(\mu - 1, v - 1)$ $[m \geq \min(\mu, v)]$, then all $n + m$ poles of the unity feedback system can be arbitrarily assigned, where μ and v are, respectively, the controllability and observability indices of any irreducible realization of $\hat{\mathbf{G}}(s)$ or the column index and row index of $\hat{\mathbf{G}}(s)$.

Proof

Since $\hat{\mathbf{G}}(s)$ is cyclic by assumption, there exists a $p \times 1$ constant vector $\mathbf{t}_1$ such that $\Delta[\hat{\mathbf{G}}(s)] = \Delta[\hat{\mathbf{G}}(s)\mathbf{t}_1]$. Let us write the $q \times 1$ rational matrix $\hat{\mathbf{G}}(s)\mathbf{t}_1$ as

$$\hat{\mathbf{G}}(s)\mathbf{t}_1 = \mathbf{N}(s)D^{-1}(s)$$

Then Theorem 9-12 implies the existence of a $1 \times q$ proper rational matrix $\bar{\mathbf{C}}(s) = D_c^{-1}(s)\mathbf{N}_c(s)$ with $\deg \bar{\mathbf{C}}(s) = m \geq v - 1$ if $\hat{\mathbf{G}}(s)$ is strictly proper, such that all $n + m$ roots of

$$D_f(s) = D_c(s)D(s) + \mathbf{N}_c(s)\mathbf{N}(s)$$

can be arbitrary assigned. Now we show that the roots of $D_f(s)$ give the poles of the feedback system in Figure 9-14(a). Indeed, from Figure 9-14(a), we have

$$\hat{h}(s) = \bar{\mathbf{C}}(s)\hat{\mathbf{e}}(s) = \bar{\mathbf{C}}(s)[\hat{\mathbf{r}}(s) - \hat{\mathbf{G}}(s)\mathbf{t}_1\hat{h}(s)]$$

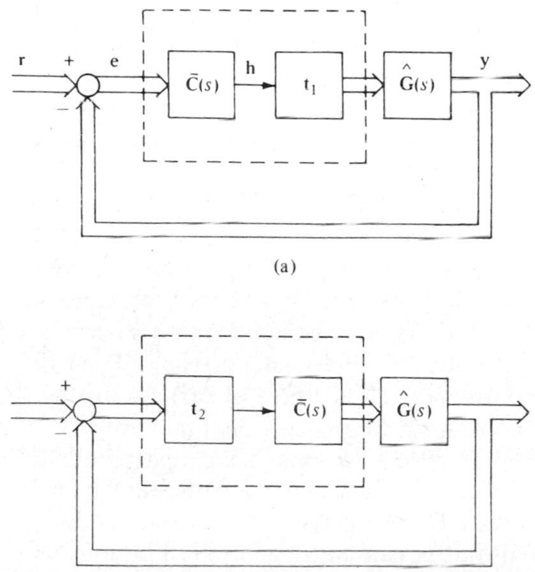

Figure 9-14 Design of compensators for plant with cyclic proper rational matrix.

which implies

$$\hat{h}(s)=[1+\bar{\mathbf{C}}(s)\hat{\mathbf{G}}(s)\mathbf{t}_1]^{-1}\bar{\mathbf{C}}(s)\hat{\mathbf{r}}(s)$$

and

$$\hat{\mathbf{G}}_f(s)=\hat{\mathbf{G}}(s)\mathbf{t}_1[1+\bar{\mathbf{C}}(s)\hat{\mathbf{G}}(s)\mathbf{t}_1]^{-1}\bar{\mathbf{C}}(s) \tag{9-64}$$

The substitution of $\hat{\mathbf{G}}(s)\mathbf{t}_1=\mathbf{N}(s)D^{-1}(s)$ and $\bar{\mathbf{C}}(s)=D_c^{-1}(s)\mathbf{N}_c(s)$ into (9-64) yields

$$\begin{aligned}\hat{\mathbf{G}}_f(s)&=\mathbf{N}(s)D^{-1}(s)[1+D_c^{-1}(s)\mathbf{N}_c(s)\mathbf{N}(s)D^{-1}(s)]^{-1}D_c^{-1}(s)\mathbf{N}_c(s)\\&=\mathbf{N}(s)[D_c(s)D(s)+\mathbf{N}_c(s)\mathbf{N}(s)]^{-1}\mathbf{N}_c(s)\end{aligned} \tag{9-65}$$

where we have used the fact that $D_c(s)$ and $D(s)$ are 1×1 polynomial matrices. From (9-65) we conclude that the roots of $D_f(s)$ give the poles of (9-65). Hence the $q\times p$ compensator defined by $\mathbf{C}(s)=\mathbf{t}_1\bar{\mathbf{C}}(s)=D_c^{-1}(s)\mathbf{t}_1\mathbf{N}_c(s)$ can achieve arbitrary pole placement.

Now we show that the observability index of $\hat{\mathbf{G}}(s)\mathbf{t}_1$ is equal to the one of $\hat{\mathbf{G}}(s)$. If $\hat{\mathbf{G}}(s)$ is factored as $\hat{\mathbf{G}}(s)=\mathbf{A}^{-1}(s)\mathbf{B}(s)$, where $\mathbf{A}(s)$ and $\mathbf{B}(s)$ are left coprime and $\mathbf{A}(s)$ is row reduced, then the observability index of $\hat{\mathbf{G}}(s)$ is equal to the largest row degree of $\mathbf{A}(s)$ (see Theorem 6-6). Consider $\hat{\mathbf{G}}(s)\mathbf{t}_1=\mathbf{A}^{-1}(s)\mathbf{B}(s)\mathbf{t}_1$. The condition $\Delta[\hat{\mathbf{G}}(s)]=\Delta[\hat{\mathbf{G}}(s)\mathbf{t}_1]=\det \mathbf{A}(s)$ implies that $\mathbf{A}(s)$ and $\mathbf{B}(s)\mathbf{t}_1$ are left coprime. Hence the observability index of $\hat{\mathbf{G}}(s)\mathbf{t}_1$ is also equal to the largest row degree of $\mathbf{A}(s)$. This establishes that the observability index of $\hat{\mathbf{G}}(s)$ and that of $\hat{\mathbf{G}}(s)\mathbf{t}_1$ are the same. Hence we have $\deg \bar{\mathbf{C}}(s)\geq v-1$, where v is the observability index of $\hat{\mathbf{G}}(s)$ or $\hat{\mathbf{G}}(s)\mathbf{t}_1$. Since $\mathbf{C}(s)=\mathbf{t}_1\bar{\mathbf{C}}(s)$ is cyclic, we have $\deg \mathbf{C}(s)=\deg \bar{\mathbf{C}}(s)\geq v-1$. This completes the proof of one part of the theorem. The rest can be similarly proved. Q.E.D.

From (9-65), we see that the transfer matrix from $\mathbf{r}$ to $\mathbf{y}$ is of the form $\mathbf{N}(s)D_c^{-1}(s)\mathbf{N}_c(s)$, as in the vector case. However, the $\mathbf{N}(s)$ and $\mathbf{N}_c(s)$ in this design are not unique; they depend on the choice of $\mathbf{t}_1$. Although the degree of compensators in Theorems 9-12 and 9-12′ are minimal to achieve pole placement, the degree in Theorem 9-14 may not be minimal. In other words, it may be possible to design a compensator of degree less than $\min(\mu-1, v-1)$ to achieve arbitrary pole placement for a $q\times p$ cyclic proper rational matrix. What is the minimum degree seems to be a difficult problem.

With Theorem 9-14, we can now discuss the design of compensators for general proper rational matrices. The procedure consists of two steps: First change a noncyclic rational matrix into a cyclic one and then apply Theorem 9-14. Consider a proper rational matrix $\hat{\mathbf{G}}(s)$. Let $\Delta(s)$ be its characteristic polynomial. We claim that if all roots of $\Delta(s)$ are distinct, then $\hat{\mathbf{G}}(s)$ is cyclic. Let $\psi(s)$ be the least common denominator of all elements of $\hat{\mathbf{G}}(s)$. Then we have $\Delta(s)=\psi(s)h(s)$ for some polynomial $h(s)$. If all roots of $\Delta(s)$ are distinct, then we have $\Delta(s)=\psi(s)k$ for some constant k, for $\psi(s)$ must contain every root of $\hat{\mathbf{G}}(s)$. Hence $\hat{\mathbf{G}}(s)$ is cyclic. Note that a cyclic $\hat{\mathbf{G}}(s)$ may have repeated poles. Hence the condition that all roots of $\Delta(s)$ are distinct is a sufficient but not necessary condition for $\hat{\mathbf{G}}(s)$ to be cyclic. This property will be used to establish that every noncyclic proper rational matrix can be transformed into a cyclic one by introducing a constant gain feedback from the output to the input, as stated in the following theorem.

Theorem 9-15

Consider a $q \times p$ proper (strictly proper) rational matrix $\hat{\mathbf{G}}(s)$. Then for almost every $p \times q$ constant matrix $\mathbf{K}$, the $q \times p$ rational matrix

$$\hat{\hat{\mathbf{G}}}(s) = [\mathbf{I} + \hat{\mathbf{G}}(s)\mathbf{K}]^{-1}\hat{\mathbf{G}}(s) = \hat{\mathbf{G}}(s)[\mathbf{I} + \mathbf{K}\hat{\mathbf{G}}(s)]^{-1}$$

is proper (strictly proper) and cyclic.

Proof[10]

We show that the roots of the characteristic polynomial, $\bar{\Delta}(s)$, of $\hat{\hat{\mathbf{G}}}(s)$ are all distinct for almost all $\mathbf{K}$. Let

$$\bar{\Delta}(s) = a_0 s^n + a_1 s^{n-1} + \cdots + a_n$$

where a_i, $i = 1, 2, \ldots, n$, are functions of all elements, k_{ij}, of $\mathbf{K}$. The differentiation of $\bar{\Delta}(s)$ with respect to s yields

$$\bar{\Delta}'(s) = n a_0 s^{n-1} + (n-1) a_1 s^{n-2} + \cdots + a_{n-1}$$

If $\Delta(s)$ has repeated roots, then $\Delta(s)$ and $\Delta'(s)$ are not coprime. A necessary and sufficient condition for $\Delta(s)$ and $\Delta'(s)$ to be not coprime is

$$\det \left[\begin{array}{ccccccc} a_n & a_{n-1} & \cdots & a_1 & a_0 & 0 \;\cdots & 0 \\ 0 & a_n & \cdots & a_2 & a_1 & a_0 \;\cdots & 0 \\ & & \vdots & & & & \\ 0 & 0 & \cdots & a_n & a_{n-1} & \cdots & a_0 \\ \hline a_{n-1} & 2a_{n-2} & \cdots & n a_0 & 0 & \cdots & 0 \\ & & \vdots & & & & \\ 0 & 0 & \cdots & a_{n-1} & 2a_{n-2} & \cdots & n a_0 \end{array}\right] = \gamma(k_{ij}) = 0$$

(See Appendix G). We note that $\gamma(k_{ij})$ is generally a nonhomogeneous polynomial of k_{ij}. There is a total of $p \times q$ number of k_{ij} in $\mathbf{K}$. Hence we may think of $\{k_{ij}\}$ as a vector in the $(p \times q)$-dimensional real vector space $\mathbb{R}^{p \times q}$. It is clear that the solution, k_{ij}, of $\gamma(k_{ij}) = 0$ is a subset in $\mathbb{R}^{p \times q}$. In other words, for almost every k_{ij}, we have $\gamma(k_{ij}) \neq 0$. Consequently, all roots of $\bar{\Delta}(s)$ are distinct. (This is similar to that for almost all α_1, α_2, and α_3, the roots of $\alpha_1 s^2 + \alpha_2 s + \alpha_3$ are distinct.) Hence $\hat{\hat{\mathbf{G}}}(s)$ is cyclic.

If $\hat{\mathbf{G}}(s)$ is strictly proper, so is $\hat{\hat{\mathbf{G}}}(s)$ for every $\mathbf{K}$. (Prove it.) If $\hat{\mathbf{G}}(s)$ is proper, the condition for $\hat{\hat{\mathbf{G}}}(s)$ to be proper, as discussed in Section 3-6, is $\det(\mathbf{I} + \hat{\mathbf{G}}(\infty)\mathbf{K}) \neq 0$. Almost all $\mathbf{K}$ satisfy this condition. Hence if $\hat{\mathbf{G}}(s)$ is proper, so is $\hat{\hat{\mathbf{G}}}(s)$ for almost all $\mathbf{K}$. This completes the proof of this theorem.

Q.E.D.

With this theorem, the design of a compensator to achieve arbitrary pole placement for general $\hat{\mathbf{G}}(s)$ becomes obvious. We first introduce a constant gain output feedback to make $\hat{\hat{\mathbf{G}}}(s) = [\mathbf{I} + \hat{\mathbf{G}}(s)\mathbf{K}]^{-1}\hat{\mathbf{G}}(s)$ cyclic. We then apply

[10] The proof is identical to the one of Theorem 7-6.

Theorem 9-14 to design a compensator $\mathbf{C}(s)$ of degree $m \geq \min(\bar{\mu}-1, \bar{\nu}-1)$ or $m \geq \min(\bar{\mu}, \bar{\nu})$ depending on whether $\hat{\mathbf{G}}(s)$ is strictly proper or proper, where $\bar{\mu}$ and $\bar{\nu}$ are, respectively, controllability and observability indices of $\hat{\bar{\mathbf{G}}}(s)$. Hence all poles of the feedback system in Figure 9-15(a) can be arbitrarily assigned.

The feedback configuration in Figure 9-15(a) is not exactly a unity feedback system. However, if we are concerned only with the poles of $\hat{\mathbf{G}}_f(s)$, not the structure of $\hat{\mathbf{G}}_f(s)$, we may combine the parallel connection of $\mathbf{C}(s)$ and $\mathbf{K}$ into $\tilde{\mathbf{C}}(s)$ and the configuration reduces to the unity feedback system shown in Figure 9-15(b).

Theorem 9-16

Consider the feedback system shown in Figure 9-15(b) with the plant described by a $q \times p$ strictly proper (proper) rational matrix $\hat{\mathbf{G}}(s)$ of degree n. The compensator $\tilde{\mathbf{C}}(s)$ is assumed to have a $p \times q$ proper (strictly proper) rational matrix of degree m. If $m \geq \min(\mu-1, \nu-1)[m \geq \min(\mu, \nu)]$, then all $n+m$ poles[11] of the unity feedback system in Figure 9-15(b) can be arbitrarily assigned, where μ and ν are, respectively, the controllability and observability indices of any irreducible realization of $\hat{\mathbf{G}}(s)$, or the column index and row index of $\hat{\mathbf{G}}(s)$.

Proof

First we show that if $\tilde{\mathbf{C}}(s) = \mathbf{C}(s) + \mathbf{K}$ and if $\mathbf{C}(s)$ can be written as $\mathbf{C}(s) = D_c^{-1}(s)\mathbf{t}_1\mathbf{N}_c(s)$, where $\mathbf{t}_1$ is a $p \times 1$ constant vector and $\mathbf{N}_c(s)$ is a $1 \times q$ polynomial matrix, then the poles of the system in Figure 9-15(a) and those in Figure 9-15(b) are the same. The transfer matrix of the feedback system in Figure 9-15(a) is

[11] Because of possible pole-zero cancellations, not all $n+m$ of these roots will be the poles of the resulting system. However, for convenience, we shall call all of them poles. The strictly proper part of this theorem was first established in Reference S26 in the state variable approach and in Reference S51 in the transfer-function approach.

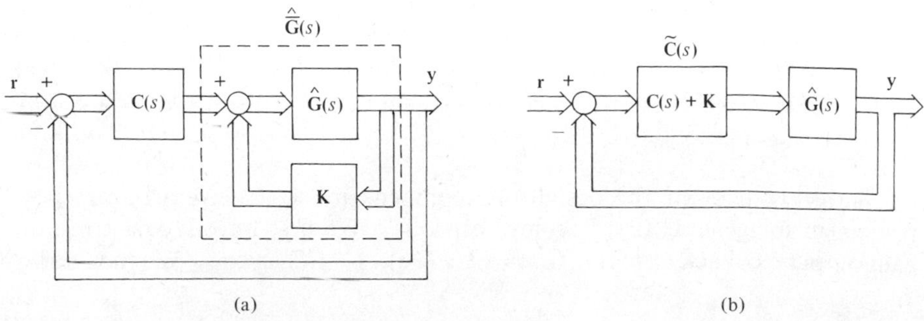

Figure 9-15 Design of compensator.

given by

$$\begin{aligned}\hat{\mathbf{G}}_{f1}(s) &= [\mathbf{I} + \hat{\bar{\mathbf{G}}}(s)\mathbf{C}(s)]^{-1}\hat{\bar{\mathbf{G}}}(s)\mathbf{C}(s)\\ &= [\mathbf{I} + (\mathbf{I} + \hat{\mathbf{G}}(s)\mathbf{K})^{-1}\hat{\mathbf{G}}(s)\mathbf{C}(s)]^{-1}(\mathbf{I} + \hat{\mathbf{G}}(s)\mathbf{K})^{-1}\hat{\mathbf{G}}(s)\mathbf{C}(s)\\ &= [\mathbf{I} + \hat{\mathbf{G}}(s)\mathbf{K} + \hat{\mathbf{G}}(s)\mathbf{C}(s)]^{-1}\hat{\mathbf{G}}(s)\mathbf{C}(s)\end{aligned}$$

The substitution of $\hat{\mathbf{G}}(s) = \mathbf{D}^{-1}(s)\mathbf{N}(s)$ and $\mathbf{C}(s) = D_c^{-1}(s)\mathbf{t}_1\mathbf{N}_c(s)$ yields

$$\begin{aligned}\hat{\mathbf{G}}_{f1}(s) &= \{\mathbf{D}^{-1}(s)[\mathbf{D}(s)D_c(s) + \mathbf{N}(s)\mathbf{K}D_c(s) + \mathbf{N}(s)\mathbf{t}_1\mathbf{N}_c(s)]D_c^{-1}(s)\}^{-1}\\ &\quad\times \mathbf{D}^{-1}(s)\mathbf{N}(s)D_c^{-1}\mathbf{t}_1\mathbf{N}_c(s)\\ &= [\mathbf{D}(s)D_c(s) + \mathbf{N}(s)\mathbf{K}D_c(s) + \mathbf{N}(s)\mathbf{t}_1\mathbf{N}_c(s)]^{-1}\mathbf{N}(s)\mathbf{t}_1\mathbf{N}_c(s)\end{aligned} \tag{9-66}$$

By a similar manipulation, the transfer matrix of the system in Figure 9-15(b) can be computed as

$$\hat{\mathbf{G}}_f(s) = [\mathbf{D}(s)D_c(s) + \mathbf{N}(s)\mathbf{K}D_c(s) + \mathbf{N}(s)\mathbf{t}_1\mathbf{N}_c(s)]^{-1}\mathbf{N}(s)[\mathbf{t}_1\mathbf{N}_c(s) + \mathbf{K}D_c(s)] \tag{9-67}$$

From (9-66) and (9-67), we conclude that the systems in Figure 9-15(a) and (b) have the same set of poles.[11] Since the poles in Figure 9-15(a) can be arbitrarily assigned by a proper choice of $\mathbf{C}(s)$ and $\mathbf{K}$, we conclude that by choosing $\tilde{\mathbf{C}}(s) = \mathbf{C}(s) + \mathbf{K}$, the poles of the system in Figure 9-15(b) can be arbitrarily assigned.

Now we claim that deg det $\mathbf{C}(s) =$ deg det $\tilde{\mathbf{C}}(s)$. Let $\dot{\mathbf{x}} = \mathbf{A}\mathbf{x} + \mathbf{B}\mathbf{u}$, $\mathbf{y} = \mathbf{C}\mathbf{x} + \mathbf{E}\mathbf{u}$ be an irreducible realization of $\mathbf{C}(s)$; then we have deg det $\mathbf{C}(s) =$ dim $\mathbf{A}$. Clearly $\dot{\mathbf{x}} = \mathbf{A}\mathbf{x} + \mathbf{B}\mathbf{u}$, $\mathbf{y} = \mathbf{C}\mathbf{x} + (\mathbf{E} + \mathbf{K})\mathbf{u}$ is an irreducible realization of $\tilde{\mathbf{C}}(s) = \mathbf{C}(s) + \mathbf{K}$. Hence we have deg det $\mathbf{C}(s) =$ dim $\mathbf{A} =$ deg det $\tilde{\mathbf{C}}(s)$. This fact can also be proved by using Theorem 9-3. If we write $\mathbf{K} = \mathbf{K}\mathbf{I}_q^{-1} = \mathbf{I}_p^{-1}\mathbf{K}$, then Theorem 9-3 implies that the parallel connection of $\mathbf{C}(s)$ and $\mathbf{K}$ is controllable and observable. Hence we have deg det $\tilde{\mathbf{C}}(s) =$ deg det $\mathbf{C}(s) +$ deg det $\mathbf{K} =$ deg det $\mathbf{C}(s)$.

What remains to be proved is that the controllability and observability indices of $\hat{\mathbf{G}}(s)$ and $\hat{\bar{\mathbf{G}}}(s) = [\mathbf{I} + \hat{\mathbf{G}}(s)\mathbf{K}]^{-1}\hat{\mathbf{G}}(s) = \hat{\mathbf{G}}(s)[\mathbf{I} + \mathbf{K}\hat{\mathbf{G}}(s)]^{-1}$ are the same for almost every constant $\mathbf{K}$. If $\hat{\mathbf{G}}(s)$ is factored as $\hat{\mathbf{G}}(s) = \mathbf{N}(s)\mathbf{D}^{-1}(s)$, where $\mathbf{N}(s)$ and $\mathbf{D}(s)$ are right coprime and $\mathbf{D}(s)$ is column reduced, then the controllability index of any irreducible realization of $\hat{\mathbf{G}}(s)$ is equal to the largest column degree of $\mathbf{D}(s)$ (see Theorem 6-6). Using this fraction, $\hat{\bar{\mathbf{G}}}(s)$ becomes

$$\hat{\bar{\mathbf{G}}}(s) = \mathbf{N}(s)\mathbf{D}^{-1}(s)[\mathbf{I} + \mathbf{K}\mathbf{N}(s)\mathbf{D}^{-1}(s)]^{-1} = \mathbf{N}(s)[\mathbf{D}(s) + \mathbf{K}\mathbf{N}(s)]^{-1}$$

Since $\mathbf{N}(s)$ and $\mathbf{D}(s)$ are right coprime, there exist polynomial matrices $\mathbf{X}(s)$ and $\mathbf{Y}(s)$ such that $\mathbf{X}(s)\mathbf{D}(s) + \mathbf{Y}(s)\mathbf{N}(s) = \mathbf{I}$ (Theorem G-8). We modify it as

$$\mathbf{X}(s)[\mathbf{D}(s) + \mathbf{K}\mathbf{N}(s)] + [\mathbf{Y}(s) - \mathbf{X}(s)\mathbf{K}]\mathbf{N}(s) = \mathbf{I}$$

Hence Theorem G-8 implies that $\mathbf{N}(s)$ and $\mathbf{D}(s) + \mathbf{K}\mathbf{N}(s)$ are right coprime. The column degrees of $\mathbf{D}(s)$ are clearly equal to the column degrees of $\mathbf{D}(s) + \mathbf{K}\mathbf{N}(s)$ for all $\mathbf{K}$ if $\mathbf{N}(s)\mathbf{D}^{-1}(s)$ is strictly proper, and for almost all $\mathbf{K}$ if $\mathbf{N}(s)\mathbf{D}^{-1}(s)$ is proper. Hence the controllability indices of $\hat{\mathbf{G}}(s)$ and $\hat{\bar{\mathbf{G}}}(s)$ are the same. This fact can also be proved by using dynamical equations. The observability part can be similarly proved. This completes the proof of this theorem. Q.E.D.

The transfer matrix $\hat{\mathbf{G}}_f(s)$ of the resulting feedback system is given in (9-67). Unlike the cases of single input, single output, and cyclic plant, $\hat{\mathbf{G}}_f(s)$ is not in the form of $\mathbf{N}(s)D_f^{-1}(s)\mathbf{N}_c(s)$ with scalar $D_f(s)$. However, from the design procedure, we may conclude that $\hat{\mathbf{G}}_f(s)$ in (9-67) is cyclic. See also Problem 9-16.

We recapture the design procedure in the following:

Step 1. Find a $\mathbf{K}$ such that $\hat{\bar{\mathbf{G}}}(s)=(\mathbf{I}+\hat{\mathbf{G}}(s)\mathbf{K})^{-1}\mathbf{G}(s)$ is cyclic.
Step 2. Find a $\mathbf{t}_1$ so that $\Delta[\hat{\bar{\mathbf{G}}}(s)]=\Delta[\hat{\bar{\mathbf{G}}}(s)\mathbf{t}_1]$.
Step 3. Write $\hat{\bar{\mathbf{G}}}(s)\mathbf{t}_1=\mathbf{N}(s)D^{-1}(s)=(\mathbf{N}_0+\mathbf{N}_1s+\cdots+\mathbf{N}_ns^n)(D_0+D_1s+\cdots+D_ns^n)^{-1}$ and form the $\mathbf{S}_m$ in (9-59). Find the least integer v such that $\mathbf{S}_{v-1}$ has a full column rank. The integer v is called the *row index* or the *observability index*.
Step 4. Choose $n+v-1$ number of poles and compute

$$\Delta_f(s)=F_0+F_1s+F_2s^2+\cdots+F_{n+v-1}s^{n+v-1}$$

Step 5. Solve D_{ci} and N_{ci} from

$$[D_{c0}\mathbf{N}_{c0} \vdots \cdots \vdots D_{c(v-1)}\mathbf{N}_{c(v-1)}]\mathbf{S}_{v-1}=[F_0 \quad F_1 \quad F_2 \quad \cdots \quad F_{n+v-1}]$$

Then the compensator is given by

$$\tilde{\mathbf{C}}(s)=\mathbf{t}_1\mathbf{C}(s)+\mathbf{K}$$

where $\mathbf{C}(s)=(D_{c0}+D_{c1}s+\cdots+D_{c(v-1)}s^{v-1})^{-1}[\mathbf{N}_{c0}+\mathbf{N}_{c1}s+\cdots+\mathbf{N}_{c(v-1)}s^{v-1}]$

∎

In step 2, if we find a $\mathbf{t}_2$ such that $\Delta[\hat{\bar{\mathbf{G}}}(s)]=\Delta[\mathbf{t}_2\hat{\bar{\mathbf{G}}}(s)]$, then we must modify steps 3 and 5. In step 3, we form $\mathbf{T}_m$ as in (9-62) and search its linearly independent columns in order from left to right. Let μ be the least integer such that $\mathbf{T}_{\mu-1}$ has a full row rank. The integer μ is called the column index or the controllability index. In step 5, we solve $\mathbf{C}(s)$ from the linear algebraic equation in (9-62). Then the compensator is given by $\tilde{\mathbf{C}}(s)=\mathbf{C}(s)\mathbf{t}_2+\mathbf{K}$.

Consider a $q\times p$ proper rational matrix $\hat{\mathbf{G}}(s)$ of degree n. If its controllability and observability indices are roughly of equal lengths, then we have $\mu\approx n/p$ and $v\approx n/q$ (see Equations (5-31) and (5-51)). Hence if $p\geq q$, we use $\mathbf{t}_2$; otherwise, use $\mathbf{t}_1$. By so choosing, we may achieve min $\{\mu-1, v-1\}$ without computing explicitly both μ and v.

Example 2

Consider

$$\hat{\mathbf{G}}(s)=\begin{bmatrix}\frac{1}{s^2} & \frac{1}{s} & 0\\ 0 & 0 & \frac{1}{s}\end{bmatrix}$$

It is a noncyclic rational matrix of degree 3. Rather arbitrarily, we choose the constant output feedback gain as

$$\mathbf{K}=\begin{bmatrix}1 & -1\\ -1 & 0\\ 2 & 1\end{bmatrix}$$

We compute

$$\hat{\bar{\mathbf{G}}}(s)=[\mathbf{I}+\hat{\mathbf{G}}(s)\mathbf{K}]^{-1}\mathbf{G}(s)=\frac{1}{s^3+3}\begin{bmatrix} s+1 & s(s+1) & 1 \\ -2 & -2s & s^2-s+1 \end{bmatrix}$$

The minimal polynomial of $\hat{\bar{\mathbf{G}}}(s)$ is clearly equal to s^3+3. To find the characteristic polynomial of $\hat{\bar{\mathbf{G}}}(s)$, we compute its minors of order 2 as

$$\frac{1}{(s^3+3)^2}[(s+1)(-2s)-s(s+1)(-2)]=0$$

$$\frac{1}{(s^3+3)^2}[(s+1)(s^2-s+1)+2]=\frac{1}{(s^3+3)^2}(s^3+3)=\frac{1}{(s^3+3)}$$

$$\frac{1}{(s^3+3)^2}[s(s+1)(s^2-s+1)+2s]=\frac{1}{(s^3+3)^2}s(s^3+3)=\frac{s}{(s^3+3)}$$

Hence the characteristic polynomial of $\hat{\bar{\mathbf{G}}}(s)$ is s^3+3, which is equal to the minimal polynomial. Hence $\bar{\mathbf{G}}(s)$ is cyclic. The system has three inputs and two outputs; hence the controllability index is probably smaller than the observability index. We choose $\mathbf{t}_2=[1 \quad 0]$. Then we have

$$\mathbf{t}_2\hat{\bar{\mathbf{G}}}(s)=\frac{1}{s^3+3}[s+1 \quad s(s+1) \quad 1]$$

and $\Delta[\hat{\bar{\mathbf{G}}}(s)]=\Delta[\mathbf{t}_2\hat{\bar{\mathbf{G}}}(s)]$. We form, as in (9-62),

$$\mathbf{T}_0=\left[\begin{array}{c:ccc} 3 & 1 & 0 & 1 \\ 0 & 1 & 1 & 0 \\ 0 & 0 & 1 & 0 \\ 1 & 0 & 0 & 0 \end{array}\right]$$

The matrix $\mathbf{T}_0$ has a full row rank; hence $\mu=1$, and the degree of compensator is $\mu-1=0$. Let

$$\mathbf{C}(s)=\frac{1}{D_{c0}}\begin{bmatrix} N_{c0}^1 \\ N_{c0}^2 \\ N_{c0}^3 \end{bmatrix}$$

where the superscripts denote the components of $\mathbf{N}_c(s)$. The $n+(\mu-1)=3$ poles are chosen arbitrarily as -1, -1, and -2. Hence we have

$$\Delta_f(s)=(s+1)^2(s+2)=2+5s+4s^2+s^3$$

The solutions of

$$\mathbf{T}_0\begin{bmatrix} D_{c0} \\ N_{c0}^1 \\ N_{c0}^2 \\ N_{c0}^3 \end{bmatrix}=\begin{bmatrix} 2 \\ 5 \\ 4 \\ 1 \end{bmatrix}$$

are $D_{c0}=1$, $N_{c0}^1=1$, $N_{c0}^2=4$, and $N_{c0}^3=-2$. Hence the compensator in Figure 9-15(b) is given by

$$\tilde{\mathbf{C}}(s)=\mathbf{C}(s)\mathbf{t}_2+\mathbf{K}=\begin{bmatrix} 1 \\ 4 \\ -2 \end{bmatrix}[1 \quad 0]+\begin{bmatrix} 1 & -1 \\ -1 & 0 \\ 2 & 1 \end{bmatrix}=\begin{bmatrix} 2 & -1 \\ 3 & 0 \\ 0 & 1 \end{bmatrix}$$

It is a compensator of degree 0. As a check, we compute

$$\hat{\mathbf{G}}_f(s) = [\mathbf{I} + \hat{\mathbf{G}}(s)\tilde{\mathbf{C}}(s)]^{-1}\hat{\mathbf{G}}(s)\tilde{\mathbf{C}}(s) = \frac{1}{(s+1)^2(s+2)}\begin{bmatrix} (s+1)(3s+2) & 1 \\ 0 & (s+1)(s+2) \end{bmatrix}$$

Its poles are -1, -1, and -2. We note that $\hat{\mathbf{G}}_f$ is cyclic (verify it). ∎

To conclude this subsection, we mention that although we can place arbitrary the poles, the resulting $\hat{\mathbf{G}}_f(s)$ is always cyclic.[12] In other words, the structure of $\hat{\mathbf{G}}_f(s)$ is restricted; we have no control over it. This restriction will be removed in the next subsection.

Multivariable case: Arbitrary denominator matrix assignment.[13] In this subsection, we study the design of compensators to achieve arbitrary denominator matrices. If we can assign an entire denominator matrix, then certainly we can achieve arbitrary pole assignment. Hence this problem accomplishes more than the pole assignment discussed in the previous subsection. Consequently, the degrees of compensators required for arbitrary denominator matrix are generally much larger than the ones required for pole assignment.

Consider the unity feedback system in Figure 9-16. The plant is described by a $q \times p$ proper rational matrix $\hat{\mathbf{G}}(s)$. The compensator to be designed is required to have a $p \times q$ proper rational matrix $\mathbf{C}(s)$. Let $\hat{\mathbf{G}}_f(s)$ be the transfer matrix of the overall system. Then we have

$$\begin{aligned}\hat{\mathbf{G}}_f(s) &= [\mathbf{I} + \hat{\mathbf{G}}(s)\mathbf{C}(s)]^{-1}\hat{\mathbf{G}}(s)\mathbf{C}(s) \\ &= \hat{\mathbf{G}}(s)\mathbf{C}(s)[\mathbf{I} + \hat{\mathbf{G}}(s)\mathbf{C}(s)]^{-1} \\ &= \hat{\mathbf{G}}(s)[\mathbf{I} + \mathbf{C}(s)\hat{\mathbf{G}}(s)]^{-1}\mathbf{C}(s)\end{aligned} \tag{9-68}$$

The first equality is obtained from $\hat{\mathbf{y}}(s) = \hat{\mathbf{G}}(s)\mathbf{C}(s)[\hat{\mathbf{r}}(s) - \hat{\mathbf{y}}(s)]$; the second one from $\hat{\mathbf{e}}(s) = \hat{\mathbf{r}}(s) - \hat{\mathbf{G}}(s)\mathbf{C}(s)\mathbf{e}(s)$; and the third one from $\hat{\mathbf{u}}(s) = \mathbf{C}(s)[\hat{\mathbf{r}}(s) - \hat{\mathbf{G}}(s)\mathbf{u}(s)]$. (Verify them. Compare also with Theorem 3-3.) In the single-variable case, if we assign both poles and zeros of the unity feedback system, then the design will generally involve undesirable pole-zero cancellations. In order to avoid these cancellations, we assign only the poles and leave the zeros unspecified.

[12] Compare with Method I of the multivariable case in the design of state feedback.
[13] Follows closely References S49 and S237.

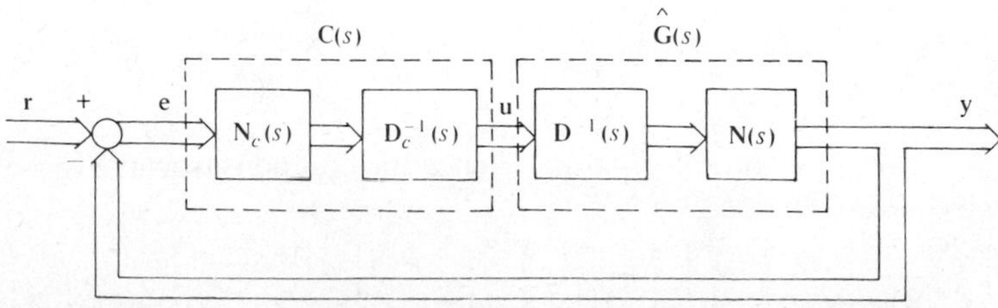

Figure 9-16 Multivariable feedback system.

We shall do the same for the multivariable case.

Let $\hat{\mathbf{G}}(s)=\mathbf{N}(s)\mathbf{D}^{-1}(s)$ and let $\mathbf{C}(s)=\mathbf{D}_c^{-1}(s)\mathbf{N}_c(s)$. Then (9-68) implies

$$\begin{aligned}\hat{\mathbf{G}}_f(s)&=\mathbf{N}(s)\mathbf{D}^{-1}(s)[\mathbf{I}+\mathbf{D}_c^{-1}(s)\mathbf{N}_c(s)\mathbf{N}(s)\mathbf{D}^{-1}(s)]^{-1}\mathbf{D}_c^{-1}(s)\mathbf{N}_c(s)\\&=\mathbf{N}(s)[\mathbf{D}_c(s)\mathbf{D}(s)+\mathbf{N}_c(s)\mathbf{N}(s)]^{-1}\mathbf{N}_c(s)\end{aligned} \tag{9-69}$$

Define the polynomial matrix

$$\mathbf{D}_f(s)=\mathbf{D}_c(s)\mathbf{D}(s)+\mathbf{N}_c(s)\mathbf{N}(s) \tag{9-70}$$

Then we have $\hat{\mathbf{G}}_f(s)=\mathbf{N}(s)\mathbf{D}_f^{-1}(s)\mathbf{N}_c(s)$. Hence the design problem becomes: Given $\mathbf{D}(s)$ and $\mathbf{N}(s)$ and an arbitrary $\mathbf{D}_f(s)$, find $\mathbf{D}_c(s)$ and $\mathbf{N}_c(s)$ to meet (9-70). This is the matrix version of the Diophantine equation in (9-40). Similar to (9-40), we shall translate it into a set of linear algebraic equations. Let us write[14]

$$\mathbf{D}(s)=\mathbf{D}_0+\mathbf{D}_1s+\cdots+\mathbf{D}_\mu s^\mu \tag{9-71a}$$
$$\mathbf{N}(s)=\mathbf{N}_0+\mathbf{N}_1s+\cdots+\mathbf{N}_\mu s^\mu \tag{9-71b}$$

We also write

$$\mathbf{D}_c(s)=\mathbf{D}_{c0}+\mathbf{D}_{c1}s+\cdots+\mathbf{D}_{cm}s^m \tag{9-72a}$$
$$\mathbf{N}_c(s)=\mathbf{N}_{c0}+\mathbf{N}_{c1}s+\cdots+\mathbf{N}_{cm}s^m \tag{9-72b}$$

and

$$\mathbf{D}_f(s)=\mathbf{F}_0+\mathbf{F}_1s+\mathbf{F}_2s^2+\cdots+\mathbf{F}_{m+\mu}s^{m+\mu} \tag{9-73}$$

The substitution of these into (9-70) yields

$$[\mathbf{D}_{c0}\quad \mathbf{N}_{c0}\ \vdots\ \mathbf{D}_{c1}\quad \mathbf{N}_{c1}\ \vdots\cdots\vdots\ \mathbf{D}_{cm}\quad \mathbf{N}_{cm}]\mathbf{S}_m=[\mathbf{F}_0\quad \mathbf{F}_1\quad \mathbf{F}_2\quad\cdots\quad \mathbf{F}_{m+\mu}]\triangleq\mathbf{F} \tag{9-74}$$

where

$$\mathbf{S}_m=\left[\begin{array}{cccccccc}\mathbf{D}_0 & \mathbf{D}_1 & \cdots & \mathbf{D}_\mu & \mathbf{0} & \mathbf{0} & \cdots & \mathbf{0}\\ \mathbf{N}_0 & \mathbf{N}_1 & \cdots & \mathbf{N}_\mu & \mathbf{0} & \mathbf{0} & \cdots & \mathbf{0}\\ \hline \mathbf{0} & \mathbf{D}_0 & \cdots & \mathbf{D}_{\mu-1} & \mathbf{D}_\mu & \mathbf{0} & \cdots & \mathbf{0}\\ \mathbf{0} & \mathbf{N}_0 & \cdots & \mathbf{N}_{\mu-1} & \mathbf{N}_\mu & \mathbf{0} & \cdots & \mathbf{0}\\ \hline & & \vdots & & & & & \\ \hline \mathbf{0} & \cdots & \mathbf{0} & \mathbf{D}_0 & \cdots & \cdots & \mathbf{D}_{\mu-1} & \mathbf{D}_\mu\\ \mathbf{0} & \cdots & \mathbf{0} & \mathbf{N}_0 & \cdots & \cdots & \mathbf{N}_{\mu-1} & \mathbf{N}_\mu\end{array}\right]\begin{array}{l}\\ \}r_0\ \text{(number of dependent rows)}\\ \\ \}r_1\\ \\ \\ \\ \}r_m\end{array} \tag{9-75}$$

The matrix $\mathbf{S}_m$ has $m+1$ block rows; each block row consists of p rows formed from $\mathbf{D}_i$ and q rows formed from $\mathbf{N}_i$. We call the former D rows and the latter N rows. This matrix $\mathbf{S}_m$ is studied in Appendix G. It is shown in Theorem G-13 that if $\mathbf{N}(s)\mathbf{D}^{-1}(s)$ is proper, then all D rows in $\mathbf{S}_m$, $m=0, 1, 2, \ldots$, are linearly independent of their previous rows. Some N rows in each block, however, may be linearly dependent on their previous rows. Let r_i be the number of linearly dependent N rows in the $(i+1)$th block. Then because of the structure of $\mathbf{S}_m$, we have $r_0\le r_1\le\cdots\le r_m\le q$. Let v be the least integer such that $r_v=q$.

[14] The integer μ is the largest column degree of $\mathbf{D}(s)$. It is different from deg det $\mathbf{D}(s)=n$. In the single-input case, however we have $\mu=n$.

See Equation (G-70). Then we have

$$\operatorname{rank} \mathbf{S}_m = (m+1)p + \sum_{j=0}^{m} (q - r_j) \qquad \text{for } m < \nu - 1$$

and

$$\operatorname{rank} \mathbf{S}_m = (m+1)p + \sum_{j=0}^{\nu-1} (q - r_j) = (m+1)p + n \qquad \text{for } m \geq \nu - 1$$

(9-76)

where $n \triangleq \sum_{j=0}^{\nu-1} (q - r_j)$ is the degree of $\hat{\mathbf{G}}(s)$. We call ν the row index of $\hat{\mathbf{G}}(s)$. It is the largest row degree of $\mathbf{A}(s)$ in any left coprime fraction of $\hat{\mathbf{G}}(s) = \mathbf{A}^{-1}(s)\mathbf{B}(s)$ with $\mathbf{A}(s)$ row reduced. It is also equal to the observability index of any irreducible realization of $\hat{\mathbf{G}}(s)$. See Equation (G-81) and Theorem 6-6.

Theorem 9-17

Consider a $q \times p$ proper rational matrix with the fraction $\hat{\mathbf{G}}(s) = \mathbf{N}(s)\mathbf{D}^{-1}(s)$. Let μ_i, $i = 1, 2, \ldots, p$, be the column degrees of $\mathbf{D}(s)$, and let ν be the row index of $\hat{\mathbf{G}}(s)$. If $m \geq \nu - 1$, then for any $\mathbf{D}_f(s)$ with column degrees $m + \mu_i$, $i = 1, 2, \ldots, p$, or less, there exist $\mathbf{D}_c(s)$ and $\mathbf{N}_c(s)$ of row degree m or less to meet

$$\mathbf{D}_f(s) = \mathbf{D}_c(s)\mathbf{D}(s) + \mathbf{N}_c(s)\mathbf{N}(s)$$

if and only if $\mathbf{D}(s)$ and $\mathbf{N}(s)$ are right coprime and $\mathbf{D}(s)$ is column reduced.[15]

Proof

Let $\mu = \max \{\mu_i, i = 1, 2, \ldots, p\}$. Since $\hat{\mathbf{G}}(s)$ is proper, the column degrees of $\mathbf{N}(s)$ are equal to or smaller than the corresponding column degrees of $\mathbf{D}(s)$. Consequently, the matrix

$$\mathbf{S}_0 = \begin{bmatrix} \mathbf{D}_0 & \mathbf{D}_1 & \cdots & \mathbf{D}_{\mu-1} & \mathbf{D}_\mu \\ \mathbf{N}_0 & \mathbf{N}_1 & \cdots & \mathbf{N}_{\mu-1} & \mathbf{N}_\mu \end{bmatrix}$$

has at least a total of $\sum_{i=1}^{p} (\mu - \mu_i)$ zero columns. In the matrix $\mathbf{S}_1$, some new zero column will be created in the rightmost block column; however, some zero columns in $\mathbf{S}_0$ will disappear from $\mathbf{S}_1$. Hence the number of zero columns in $\mathbf{S}_1$ remains to be

$$\sum_{i=1}^{p} (\mu - \mu_i) = p\mu - \sum_{i=1}^{p} \mu_i$$

In fact, this is the minimum number of zero columns in S_i, $i = 2, 3, \ldots$. Let $\tilde{\mathbf{S}}_{\nu-1}$ be the matrix $\mathbf{S}_{\nu-1}$ after deleting these zero columns. Since the number of columns in $\mathbf{S}_m$ is $(\mu + 1 + m)p$, the number of columns in $\tilde{\mathbf{S}}_{\nu-1}$ is equal to

$$(\mu + \nu)p - \left(p\mu - \sum_{i=1}^{p} \mu_i\right) = \nu p + \sum_{i=1}^{p} \mu_i \qquad \textbf{(9-77)}$$

[15] This theorem reduces to Theorem 9-10 for the single variable case. In the single-variable case, we have $\mu = \nu = \deg D(s) = n$, and the search of the row index becomes unnecessary.

The rank of $\tilde{\mathbf{S}}_{v-1}$ is clearly equal to the rank of $\mathbf{S}_{v-1}$. Hence we have, from (9-76),

$$\operatorname{rank} \tilde{\mathbf{S}}_{v-1} = \operatorname{rank} \mathbf{S}_{v-1} = vp + n \tag{9-78}$$

A comparison of (9-78) with (9-77) reveals immediately that $\tilde{\mathbf{S}}_{v-1}$ has a full column rank if and only if

$$\sum_{i=1}^{p} \mu_i = n$$

This equality holds if and only if, following Corollary G-14 and Definition G-4, $\mathbf{D}(s)$ and $\mathbf{N}(s)$ are right coprime and $\mathbf{D}(s)$ is column reduced. Since both the rank of $\tilde{\mathbf{S}}_k$ and the number of columns in $\tilde{\mathbf{S}}_k$ increase by p as k increases by 1 from $v-1$, we conclude that $\tilde{\mathbf{S}}_m$, for $m \geq v-1$, has a full column rank if and only if $\mathbf{D}(s)$ and $\mathbf{N}(s)$ are right coprime and $\mathbf{D}(s)$ is column reduced.

If $\mathbf{D}_f(s)$ is of column degree $m+\mu_i$, then there are at least $\sum_{i=1}^{p} (\mu - \mu_i)$ zero columns in the $\mathbf{F}$ matrix in (9-74). Furthermore, the positions of these zero columns coincide with those of $\mathbf{S}_m$. Since $\tilde{\mathbf{S}}_m$ has a full column rank, the $\mathbf{F}$ must be inside the row space of $\mathbf{S}_m$. Hence a set of solutions $\{\mathbf{D}_{ci}, \mathbf{N}_{ci}\}$ exists in (9-74), or equivalently, a set of solutions $\{\mathbf{D}_c(s), \mathbf{N}_c(s)\}$ exists in (9-70). This completes the proof of the theorem. Q.E.D.

This theorem states the condition for the existence of $\mathbf{D}_c(s)$ and $\mathbf{N}_c(s)$ to meet (9-70), but states nothing regarding whether $\mathbf{D}_c^{-1}(s)\mathbf{N}_c(s)$ is proper or not. To study this question, we consider separately the case where $\hat{\mathbf{G}}(s)$ is strictly proper or proper. The situation is similar to the single-variable case (Theorems 9-11 and 9-11′). However, the proof must be modified because the leading coefficient matrix $\mathbf{D}_\mu$ of $\mathbf{D}(s)$ is generally not nonsingular.

Before proceeding, we define

$$\mathbf{H}(s) = \operatorname{diag}\{s^{\mu_1}, s^{\mu_2}, \ldots, s^{\mu_p}\} \tag{9-79}$$

and

$$\mathbf{H}_c(s) = \operatorname{diag}\{s^{m_1}, s^{m_2}, \ldots, s^{m_p}\} \tag{9-80}$$

Theorem 9-18

Consider a $q \times p$ strictly proper (proper) rational matrix $\hat{\mathbf{G}}(s)$ with the fraction $\hat{\mathbf{G}}(s) = \mathbf{N}(s)\mathbf{D}^{-1}(s)$. Let μ_i, $i = 1, 2, \ldots, p$, be the column degrees of $\mathbf{D}(s)$ and let v be the row index of $\hat{\mathbf{G}}(s)$. Let m_i be the row degrees of $\mathbf{D}_c(s)$. If $m_i \geq v-1$ ($m_i \geq v$) for all i, then for any $\mathbf{D}_f(s)$ with the property that

$$\lim_{s\to\infty} \mathbf{H}_c^{-1}(s)\mathbf{D}_f(s)\mathbf{H}^{-1}(s) = \mathbf{J} \tag{9-81}$$

exists and is nonsingular, there exists proper (strictly proper) $\mathbf{D}_c^{-1}(s)\mathbf{N}_c(s)$ to meet

$$\mathbf{D}_f(s) = \mathbf{D}_c(s)\mathbf{D}(s) + \mathbf{N}_c(s)\mathbf{N}(s) \tag{9-82}$$

if and only if $\mathbf{D}(s)$ and $\mathbf{N}(s)$ are right coprime and $\mathbf{D}(s)$ is column reduced.

Proof

Let $m = \max\{m_i\}$ and $\mu = \max\{\mu_i\}$. Consider the ith row equation of (9-74):

$$[\mathbf{D}_{ic0} \quad \mathbf{N}_{ic0} \quad \cdots \quad \mathbf{D}_{icm_i} \quad \mathbf{N}_{icm_i}]\mathbf{S}_{m_i} = [\mathbf{F}_{i0} \quad \mathbf{F}_{i1} \quad \cdots \quad \mathbf{F}_{i(m_i+\mu)}] \triangleq \mathbf{F}_i$$

where $\mathbf{D}_{icj}$ denotes the ith row of $\mathbf{D}_{cj}$ and so forth. Since $m_i \geq v-1$, the resultant $\mathbf{S}_{m_i}$, excluding $\sum(\mu-\mu_i)$ number of zero columns, has a full column rank. The assumption of (9-81) implies that the ith row of $\mathbf{D}_f(s)$ has column degrees at most $m_i + \mu_j$. Hence $\mathbf{F}_i$ has $\sum(\mu-\mu_j)$ number of zero elements whose positions coincide with those of $\mathbf{S}_{m_i}$. Hence we conclude that for any $\mathbf{D}_f(s)$ meeting (9-81), solutions $\mathbf{D}_c(s)$ and $\mathbf{N}_c(s)$ of row degrees at most m_i exist in (9-82).

We write (G-57) as $\mathbf{M}(s) = [\mathbf{M}_{hc} + \mathbf{M}_l(s)]\mathbf{H}_c(s)$, where $\mathbf{M}_l(s) \triangleq \mathbf{M}_{lc}(s)\mathbf{H}_c^{-1}(s)$ is a strictly proper rational matrix and $\mathbf{M}_l(\infty) = \mathbf{0}$. Similarly, we write

$$\begin{aligned}
\mathbf{D}(s) &= [\mathbf{D}_h + \mathbf{D}_l(s)]\mathbf{H}(s) \\
\mathbf{N}(s) &= [\mathbf{N}_h + \mathbf{N}_l(s)]\mathbf{H}(s) \\
\mathbf{D}_c(s) &= \mathbf{H}_c(s)[\mathbf{D}_{ch} + \mathbf{D}_{cl}(s)] \\
\mathbf{N}_c(s) &= \mathbf{H}_c(s)[\mathbf{N}_{ch} + \mathbf{N}_{cl}(s)] \\
\mathbf{D}_f(s) &= \mathbf{H}_c(s)[\mathbf{D}_{fh} + \mathbf{D}_{fl}(s)]\mathbf{H}(s)
\end{aligned} \tag{9-83}$$

where $\mathbf{D}_l(s)$, $\mathbf{D}_{cl}(s)$, $\mathbf{N}_l(s)$, $\mathbf{N}_{cl}(s)$ and $\mathbf{D}_{fl}(s)$ are strictly proper rational matrices. The constant matrix $\mathbf{D}_h$ is the column degree coefficient matrix of $\mathbf{D}(s)$ and is nonsingular by assumption. The substitution of (9-83) into (9-82) yields, at $s = \infty$,

$$\mathbf{D}_{ch}\mathbf{D}_h + \mathbf{N}_{ch}\mathbf{N}_h = \mathbf{D}_{fh} \tag{9-84}$$

If $\hat{\mathbf{G}}(s)$ is strictly proper, $\mathbf{N}_h = \mathbf{0}$ and $\mathbf{D}_{ch} = \mathbf{D}_{fh}\mathbf{D}_h^{-1}$. By assumption, $\mathbf{D}_{fh} = \mathbf{J}$ is nonsingular; hence $\mathbf{D}_c(s)$ is row reduced and $\mathbf{D}_c^{-1}(s)\mathbf{N}_c(s)$ is proper. If $\hat{\mathbf{G}}(s)$ is proper, we have $\mathbf{N}_h \neq \mathbf{0}$. However, the row of $\mathbf{S}_m$ corresponding to $\mathbf{N}_{ich}$ are linearly dependent if $m \geq v$. Hence we may choose $\mathbf{N}_{ich} = \mathbf{0}$. For this choice, we have $\mathbf{D}_{ich} = \mathbf{D}_{ifh}\mathbf{D}_h^{-1}$ and $\mathbf{D}_{ch} = \mathbf{D}_{fh}\mathbf{D}_h^{-1}$. Hence $\mathbf{D}_c(s)$ is row reduced. Because of $\mathbf{N}_{ich} = \mathbf{0}$, $\mathbf{D}_c^{-1}(s)\mathbf{N}_c(s)$ is strictly proper. Q.E.D.

It is important to note that if $m_i \geq v$ for some i, the solutions of (9-74) or (9-82) are not unique and there may be improper solutions. However, there exists, under the assumption of (9-81), at least one set of proper or strictly proper solutions. If $\mathbf{D}_f(s)$ meets (9-81), it is said to be row-column-reduced in Reference S34. Every $\mathbf{D}_f(s)$ can be transformed, by elementary operations, into a row-column-reduced one.

Now we may apply Theorem 9-18 to the design problem.

Theorem 9-19

Consider a $q \times p$ strictly proper (proper) rational matrix $\hat{\mathbf{G}}(s)$ with the fraction $\hat{\mathbf{G}}(s) = \mathbf{N}(s)\mathbf{D}^{-1}(s)$. Let μ_i, $i = 1, 2, \ldots, p$ be the column degrees of $\mathbf{D}(s)$ and let v be the row index of $\hat{\mathbf{G}}(s)$. Let the row degrees of $\mathbf{D}_c(s)$ be m_i, $i = 1, 2, \ldots, p$. If $m_i \geq v-1$ ($m_i \geq v$) for all i, then for any $\mathbf{D}_f(s)$ with the property that

$$\lim_{s \to \infty} \mathbf{H}_c^{-1}(s)\mathbf{D}_f(s)\mathbf{H}^{-1}(s) = \mathbf{J}$$

exists and is nonsingular, there exists a compensator with a $p \times q$ proper (strictly proper) rational matrix $\mathbf{D}_c^{-1}(s)\mathbf{N}_c(s)$ such that the transfer matrix of the unity feedback system in Figure 9-16 is $\mathbf{N}(s)\mathbf{D}_f^{-1}(s)\mathbf{N}_c(s)$ if and only if $\mathbf{D}(s)$ and $\mathbf{N}(s)$ are right coprime and $\mathbf{D}(s)$ is column reduced. ∎

This theorem follows directly from Theorem 9-18 and Equations (9-69) and (9-70). Several remarks are in order regarding this theorem. We discuss only the case where $\hat{\mathbf{G}}(s)$ is strictly proper. If $m_i = v - 1$, for all i, and if we choose $\mathbf{D}_f(s) = \mathbf{P}(s)\mathbf{D}(s)$, then the unique solution of (9-70) is $\mathbf{D}_c(s) = \mathbf{P}(s)$ and $\mathbf{N}_c(s) = \mathbf{0}$. In other words, if the denominator matrix of the plant is not to be altered, the compensator $\mathbf{C}(s) = \mathbf{D}_c^{-1}(s)\mathbf{N}_c(s)$ is $\mathbf{0}$. This is a degenerated case. The second remark concerns the possible existence of common divisors between $\mathbf{N}(s)$ and $\mathbf{D}_f(s)$ and between $\mathbf{D}_f(s)$ and $\mathbf{N}_c(s)$ in $\hat{\mathbf{G}}_f(s) = \mathbf{N}(s)\mathbf{D}_f^{-1}(s)\mathbf{N}_c(s)$; and between $\mathbf{N}_c(s)$ and $\mathbf{D}_c(s)$ in $\mathbf{C}(s) = \mathbf{D}_c^{-1}(s)\mathbf{N}_c(s)$. In the first case, the pole-zero cancellations involve the poles which the designer has the freedom in placing, therefore these cancellations are permitted in practice. From (9-70), we see that if $\mathbf{D}_c(s)$ and $\mathbf{N}_c(s)$ have a common left divisor, then it is also a left divisor of $\mathbf{D}_f(s)$. Hence the pole-zero cancellations between $\mathbf{D}_c(s)$ and $\mathbf{N}_c(s)$ involve again only assignable poles. The final remark concerns the well posedness of the feedback system. Since $\hat{\mathbf{G}}(s)$ is strictly proper and $\mathbf{C}(s)$ is proper, we have $\mathbf{I} + \hat{\mathbf{G}}(\infty)\mathbf{C}(\infty) = \mathbf{I}$. Hence the unity feedback system in Theorem 9-18 is well posed. Note that if $\mathbf{D}_c(s)$ and $\mathbf{N}_c(s)$ are left coprime, the degree of the compensator is $\sum m_i \geq p(v-1)$, which is much larger than the one required for arbitrary pole assignment.

We state the dual of Theorem 9-19 as a corollary.

Corollary 9-19

Consider a $q \times p$ strictly proper (proper) rational matrix $\hat{\mathbf{G}}(s)$ with the fraction $\hat{\mathbf{G}}(s) = \mathbf{D}^{-1}(s)\mathbf{N}(s)$. Let v_i, $i = 1, 2, \ldots, q$, be the row degrees of $\mathbf{D}(s)$ and let μ be the column index of $\hat{\mathbf{G}}(s)$. Let the column degrees of $\mathbf{D}_c(s)$ be m_i, $i = 1, 2, \ldots, q$. If $m_i \geq \mu - 1$ ($m_i \geq \mu$) for all i, then for any $\mathbf{D}_f(s)$ with the property that

$$\lim_{s \to \infty} \operatorname{diag}\{s^{-v_1}, s^{-v_2}, \ldots, s^{-v_q}\}\mathbf{D}_f(s)\operatorname{diag}\{s^{-m_1}, s^{-m_2}, \ldots, s^{-m_q}\} = \mathbf{J}$$

exists and is nonsingular, there exists a compensator with a $q \times p$ proper (strictly proper) rational matrix $\mathbf{N}_c(s)\mathbf{D}_c^{-1}(s)$ such that the transfer matrix from $\mathbf{r}$ to $\mathbf{y}$ in Figure 9-17 is equal to $\mathbf{I} - \mathbf{D}_c(s)\mathbf{D}_f^{-1}(s)\mathbf{D}(s)$ if and only if $\mathbf{D}(s)$ and $\mathbf{N}(s)$ are left coprime and $\mathbf{D}(s)$ is row reduced.

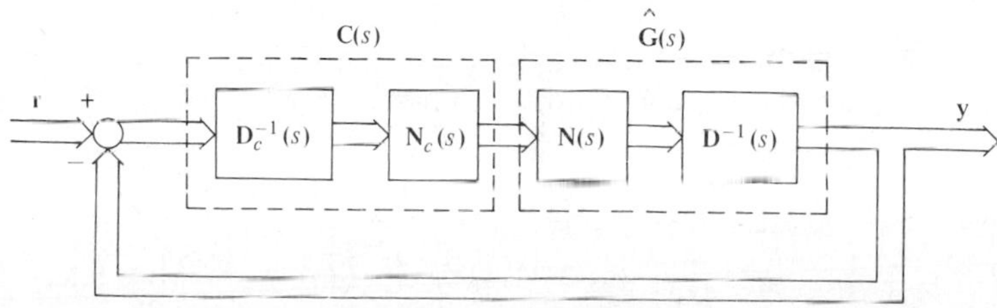

Figure 9-17 Unity feedback system.

The substitution of $\hat{\mathbf{G}}(s)=\mathbf{D}^{-1}(s)\mathbf{N}(s)$ and $\mathbf{C}(s)=\mathbf{N}_c(s)\mathbf{D}_c^{-1}(s)$ into the first equality of (9-68) yields

$$\begin{aligned}\hat{\mathbf{G}}_f(s)&=[\mathbf{I}+\mathbf{D}^{-1}(s)\mathbf{N}(s)\mathbf{N}_c(s)\mathbf{D}_c^{-1}(s)]^{-1}\mathbf{D}^{-1}(s)\mathbf{N}(s)\mathbf{N}_c(s)\mathbf{D}_c^{-1}(s)\\&=\mathbf{D}_c(s)[\mathbf{D}(s)\mathbf{D}_c(s)+\mathbf{N}(s)\mathbf{N}_c(s)]^{-1}\mathbf{N}(s)\mathbf{N}_c(s)\mathbf{D}_c^{-1}(s)\end{aligned} \tag{9-85}$$

Define

$$\mathbf{D}_f(s)=\mathbf{D}(s)\mathbf{D}_c(s)+\mathbf{N}(s)\mathbf{N}_c(s) \tag{9-86}$$

Then $\hat{\mathbf{G}}_f(s)$ becomes

$$\begin{aligned}\hat{\mathbf{G}}_f(s)&=\mathbf{D}_c(s)\mathbf{D}_f^{-1}(s)[\mathbf{D}_f(s)-\mathbf{D}(s)\mathbf{D}_c(s)]\mathbf{D}_c^{-1}(s)\\&=\mathbf{I}-\mathbf{D}_c(s)\mathbf{D}_f^{-1}(s)\mathbf{D}(s)\end{aligned}$$

This shows that the transfer matrix from $\mathbf{r}$ to $\mathbf{y}$ in Figure 9-17 is equal to $[\mathbf{I}-\mathbf{D}_c(s)\mathbf{D}_f^{-1}(s)\mathbf{D}(s)]$. The design in Corollary 9-19 hinges on solving (9-86). Note that the transpose of (9-86) becomes (9-70); left coprime and row reduced become right coprime and column reduced. Hence Theorem 9-18 can be applied directly to the transpose of (9-86). Of course Equation (9-86) can also be solved directly. We use the coefficient matrices of $\mathbf{D}(s)$ and $\mathbf{N}(s)$ to form $\mathbf{T}_k$ as shown in Equation (9-62) and then search linearly independent columns in order from left to right. The least integer μ such that all N columns in the last block column of T_μ are linearly dependent, is called the column index of $\hat{\mathbf{G}}(s)$. Dual to the row index, the column index is equal to the largest column degree of the column reduced $\mathbf{D}_r(s)$ in any right coprime fraction of $\hat{\mathbf{G}}(s)=\mathbf{N}_r(s)\mathbf{D}_r^{-1}(s)$. It is also equal to the controllability index of any irreducible realization of $\hat{\mathbf{G}}(s)$. The proof of Corollary 9-19 is similar to the one of Theorem 9-19 and will not be repeated.

Example 3

We give an example to illustrate the application of Theorem 9-19. Consider the proper rational matrix

$$\hat{\mathbf{G}}(s)=\mathbf{N}(s)\mathbf{D}^{-1}(s)=\begin{bmatrix}s^2+1 & s\\ 0 & s^2+s+1\end{bmatrix}\begin{bmatrix}s^2-1 & 0\\ 0 & s^2-1\end{bmatrix}^{-1}$$

We form $\mathbf{S}_0, \mathbf{S}_1, \ldots,$ and search their linearly dependent rows in order from top to bottom. For this example, we can readily obtain that $v=2$. Clearly we have $\mu_1=\mu_2=2$. Let $m_1=m_2=v-1=1$. We choose

$$\mathbf{D}_f(s)=\begin{bmatrix}(s+1)^3 & 0\\ 0 & (s+1)(s^2+s+1)\end{bmatrix}=\begin{bmatrix}1&0\\0&1\end{bmatrix}+\begin{bmatrix}3&0\\0&2\end{bmatrix}s+\begin{bmatrix}3&0\\0&2\end{bmatrix}s^2+\begin{bmatrix}1&0\\0&1\end{bmatrix}s^3$$

Then the compensator is the solution of

$$[\mathbf{D}_{c0} \quad \mathbf{N}_{c0} \vdots \mathbf{D}_{c1} \quad \mathbf{N}_{c1}]\left[\begin{array}{cccccc:cc} -1 & 0 & 0 & 0 & 1 & 0 & 0 & 0 \\ 0 & -1 & 0 & 0 & 0 & 1 & 0 & 0 \\ 1 & 0 & 0 & 1 & 1 & 0 & 0 & 0 \\ 0 & 1 & 0 & 1 & 0 & 1 & 0 & 0 \\ \hdashline 0 & 0 & -1 & 0 & 0 & 0 & 1 & 0 \\ 0 & 0 & 0 & 1 & 0 & 0 & 0 & 1 \\ 0 & 0 & 1 & 0 & 0 & 1 & 1 & 0 \\ 0 & 0 & 0 & 1 & 0 & 1 & 0 & 1 \end{array}\right]$$

$$= \begin{bmatrix} 1 & 0 & 3 & 0 & 3 & 0 & 1 & 0 \\ 0 & 1 & 0 & 2 & 0 & 2 & 0 & 1 \end{bmatrix}$$

which yields

$$\mathbf{D}_c(s) = \begin{bmatrix} -s+1 & \dfrac{2s-2}{3} \\ 0 & 0 \end{bmatrix} \qquad \mathbf{N}_c(s) = \begin{bmatrix} 2(s+1) & -\dfrac{2s+2}{3} \\ 0 & s+1 \end{bmatrix}$$

We see that $\mathbf{D}_c(s)$ is singular and the compensator $\mathbf{D}_c^{-1}(s)\mathbf{N}_c(s)$ is not defined. Hence if $\hat{\mathbf{G}}(s)$ is proper, the choice of $m_i = v-1$ may not yield the required compensator.

If we choose $m_i = v$ and choose

$$\mathbf{D}_f(s) = \begin{bmatrix} (s+1)^4 & 0 \\ 0 & (s+1)^2(s^2+s+1) \end{bmatrix}$$

then the compensator can be computed as

$$\mathbf{D}_c(s) = \begin{bmatrix} s^2+3 & 4(s-1)/3 \\ 0 & s^2+s+1 \end{bmatrix} \qquad \mathbf{N}_c(s) = \begin{bmatrix} 4s+4 & -4(s+1)/3 \\ 0 & 2(s+1) \end{bmatrix}$$

and $\mathbf{D}_c^{-1}(s)\mathbf{N}_c(s)$ is strictly proper. The degree of $\mathbf{D}_c(s)$ is equal to 4. ∎

Remarks are in order regarding the design of arbitrary assignment of poles and the design of arbitrary assignment of denominator matrices. If $\hat{\mathbf{G}}(s)$ is strictly proper, the minimal degree of compensator for the former is $\min(\mu-1, v-1)$ (Theorem 9-16); whereas the minimal degree for the latter is $p(v-1)$ (Theorem 9-18) or $q(\mu-1)$ (Theorem 9-19). The design of pole assignment assigns only the poles or the determinant of the denominator matrix and always yields a cyclic overall transfer matrix. The design of denominator matrix yields generally a noncyclic overall transfer matrix. If the discussion of the cyclic and noncyclic designs in the multivariable case of Section 7-3 is applicable here, the transient responses of the system obtained by the pole-placement design are probably worse than those of the system obtained by the denominator matrix design with the same set of poles. A detailed comparison of these two designs is not available at present.

Decoupling. The results in Theorem 9-19 may be used to design a unity feedback system so that its transfer matrix $\hat{\mathbf{G}}_f(s)$ is diagonal and nonsingular. Such a system is said to be *decoupled* (see Definition 7-1). If the plant transfer matrix $\hat{\mathbf{G}}(s)=\mathbf{N}(s)\mathbf{D}^{-1}(s)$ is square and nonsingular, we may choose $\mathbf{D}_f(s)=\bar{\mathbf{D}}_f(s)\mathbf{N}(s)$, where $\bar{\mathbf{D}}_f(s)$ is diagonal, in Theorem 9-19. Then the overall transfer matrix in Figure 9-16 is

$$\hat{\mathbf{G}}_f(s)=\mathbf{N}(s)\mathbf{D}_f^{-1}(s)\mathbf{N}_c(s)=\mathbf{N}(s)(\bar{\mathbf{D}}_f(s)\mathbf{N}(s))^{-1}\mathbf{N}_c(s)=\bar{\mathbf{D}}_f^{-1}(s)\mathbf{N}_c(s)$$

Now if the degree of compensator is sufficiently large, we may be able to choose a $\mathbf{N}_c(s)$ which is diagonal. In this case, $\hat{\mathbf{G}}_f(s)$ becomes diagonal and the system is decoupled. This design is achieved by canceling $\mathbf{N}(s)$ and may involve undesirable pole-zero cancellations. Hence this decoupling method is not always satisfactory.

In the following, we discuss a different method of achieving decoupling.[16] The method involves direct cancellations and is applicable only to stable plants. Consider a $p \times p$ proper rational matrix $\hat{\mathbf{G}}(s)=\mathbf{N}(s)\mathbf{D}^{-1}(s)$. It is assumed that $\hat{\mathbf{G}}(s)$ is nonsingular and det $\mathbf{D}(s)$ is a Hurwitz polynomial. Let $\mathbf{P}(s)$ in Figure 9-18 be a diagonal rational matrix of the form

$$\mathbf{P}(s)=\operatorname{diag}\left\{\frac{\beta_1(s)}{\alpha_1(s)}, \frac{\beta_2(s)}{\alpha_2(s)}, \ldots, \frac{\beta_p(s)}{\alpha_p(s)}\right\}$$

We see that the transfer matrix of the tandem connection of the compensator followed by the plant is $\hat{\mathbf{G}}(s)\mathbf{C}(s)=\mathbf{N}(s)\mathbf{D}^{-1}(s)\mathbf{D}(s)\mathbf{N}^{-1}(s)\mathbf{P}(s)=\mathbf{P}(s)$. Because $\hat{\mathbf{G}}(s)\mathbf{C}(s)=\mathbf{P}(s)$ is diagonal, the feedback system in Figure 9-18 reduces to a set of p number of decoupled single-loop systems. Hence the system is decoupled.

Now the matrix $\mathbf{P}(s)$ is to be chosen sc that the compensator

$$\mathbf{C}(s)=\mathbf{D}(s)\mathbf{N}^{-1}(s)\mathbf{P}(s)=\hat{\mathbf{G}}^{-1}(s)\mathbf{P}(s) \tag{9-87}$$

is a proper rational matrix and the overall transfer matrix

$$\hat{\mathbf{G}}_f(s)=\operatorname{diag}\left\{\frac{\beta_1(s)}{\alpha_1(s)+\beta_1(s)}, \frac{\beta_2(s)}{\alpha_2(s)+\beta_2(s)}, \ldots, \frac{\beta_p(s)}{\alpha_p(s)+\beta_p(s)}\right\}$$

is BIBO stable. If det $\mathbf{N}(s)$ is Hurwitz, the design in Figure 9-18 will not involve

[16] Follows Reference S76.

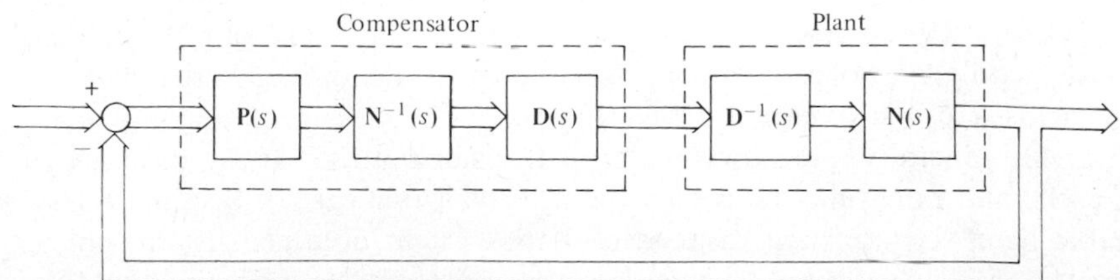

Figure 9-18 Decoupling of a plant.

any unstable pole-zero cancellations. In this case, we may choose $\beta_i(s)=1$, $i=1,2,\ldots,p$, and the poles of $\hat{\mathbf{G}}_f(s)$ becomes $\alpha_i(s)+1$. The degree of $\alpha_i(s)$ and consequently the number of poles of $\hat{\mathbf{G}}_f(s)$ are to be chosen so that $\mathbf{C}(s)$ in (9-87) is proper. Once the poles of $\mathbf{G}_f(s)$ are specified, $\alpha_i(s)$ can be readily computed. In this design, it is also possible to design a stable compensator by requiring both $\alpha_i(s)$ and $(\alpha_i(s)+1)$ to be Hurwitz.

If det $\mathbf{N}(s)$ is not Hurwitz, then the preceding design will involve unstable pole-zero cancellations. These cancellations arise from the unstable poles of $\mathbf{N}^{-1}(s)$. [Recall that $\hat{\mathbf{G}}(s)$ is assumed to be stable, hence there are no unstable poles in $\mathbf{D}^{-1}(s)$.] To avoid these, we choose $\beta_i(s)$ so that the rational matrix

$$\mathbf{N}^{-1}(s)\,\mathrm{diag}\,\{\beta_1(s), \beta_2(s), \ldots, \beta_p(s)\} \tag{9-88}$$

has no open right-half s plane poles. This is accomplished by choosing $\beta_i(s)$ to be the least common denominator of the unstable poles of the ith column of $\mathbf{N}^{-1}(s)$. Once $\beta_i(s)$ is chosen, the degree of $\alpha_i(s)$ is to be chosen so that $\mathbf{C}(s)$ in (9-87) is proper. Note that although $\hat{\mathbf{G}}(s)$ is proper, $\hat{\mathbf{G}}^{-1}(s)$ is generally not proper. Let $\hat{\mathbf{G}}^{-1}(s) \triangleq (\bar{n}_{ij}(s)/\bar{d}_{ij}(s))$. Then from (9-87), we can readily verify that if

$$\deg \alpha_i(s) - \deg \beta_i(s) \geq \max_j\,[\deg \bar{n}_{ij}(s) - \deg \bar{d}_{ij}(s)] \tag{9-89}$$

for $i=1,2,\ldots,p$, then $\mathbf{C}(s)$ is proper. That is, if the pole-zero excess[17] of the ith column of $\mathbf{P}(s)$ is equal to or larger than the largest *zero-pole* excess of the ith column of $\hat{\mathbf{G}}^{-1}(s)$, then $\mathbf{C}(s)$ is proper. Note that the left-hand side of (9-89) is also equal to the pole-zero excess of $\hat{\mathbf{G}}_f(s)$. If $p=1$, the zero-pole excess of $\hat{\mathbf{G}}^{-1}(s)$ is equal to the pole-zero excess of $\hat{\mathbf{G}}(s)$, and (9-89) reduces to the pole-zero excess inequality discussed in (9-38) for the single-variable case.

The poles of $\hat{\mathbf{G}}_f(s)$ are the zeros of $\alpha_i(s)+\beta_i(s)$. Once $\beta_i(s)$ and the degrees of $\alpha_i(s)$ are determined from (9-88) and (9-89), from the assignment of the poles of $\hat{\mathbf{G}}_f(s)$, we can readily compute $\alpha_i(s)$. Using these $\alpha_i(s)$, the compensator $\mathbf{C}(s)$ in (9-87) is proper and the unity feedback system $\hat{\mathbf{G}}_f(s)$ in Figure 9-18 is decoupled. The poles of $\hat{\mathbf{G}}_f(s)$ are assignable by the designer. The zeros, $\beta_i(s)$, of $\hat{\mathbf{G}}_f(s)$ are dictated by the closed right-half plane roots of det $\mathbf{N}(s)$, called the *nonminimum-phase zeros* of $\hat{\mathbf{G}}(s)$. They are chosen as in (9-88) to avoid unstable pole-zero cancellations. In other words, nonminimum phase zeros of $\hat{\mathbf{G}}(s)$ should not be canceled and should be retained in $\hat{\mathbf{G}}_f(s)$.

If a plant is not stable, it must be stabilized before the application of the decoupling procedure. We see that the decoupling is achieved by exact cancellations, although all stable ones. If there are any perturbations in $\hat{\mathbf{G}}(s)$ and $\mathbf{C}(s)$, the property of decoupling will be destroyed. Furthermore, the degree of compensator is usually very large for an exact decoupling. Hence decoupling is very sensitive to parameter variations and is expensive to implement.

[17] Let $\hat{g}(s)=N(s)/D(s)$. We call $\deg D(s)-\deg N(s)$ the pole-zero excess and $\deg N(s)-\deg D(s)$ the zero-pole excess of $\hat{g}(s)$.

9-6 Asymptotic Tracking and Disturbance Rejection

Single-variable case. Consider the design of the control system shown in Figure 9-19(a). The plant with transfer function $\hat{g}(s)$ is given, the problem is to find a compensator with a proper transfer function $C(s)$ so that the feedback system is asymptotically stable and meets some other specifications. One of the important specifications is to require the output of the plant $y(t)$ to track the reference signal $r(t)$. Because of physical limitations, it is not possible to design a feedback system so that $y(t)=r(t)$ for all t (see Reference S46). The best we can achieve is that

$$\lim_{t\to\infty} e(t) = \lim_{t\to\infty} [r(t)-y(t)] = 0$$

This is called the *asymptotic tracking.* It is well known that if $r(t)$ is a step function, or $\hat{r}(s)=\mathscr{L}[r(t)]=1/s$ and if $C(s)\hat{g}(s)$ is a type 1 system, that is, $C(s)\hat{g}(s)$ has one pole at the origin, and if the feedback system in Figure 9-19(a) is asymptotically stable, then the output $y(t)$ will track $r(t)$ asymptotically. In this section, this statement and its generalization will be established.

Consider the feedback system shown in Figure 9-19(b). The plant is written as $\hat{g}(s)=D^{-1}(s)N(s)$. At the input of $D^{-1}(s)$, a signal $w(t)$ is injected into the plant. The signal $w(t)$ will be called the *disturbance signal.* Now in addition to asymptotic tracking, we also require that the affect of $w(t)$ at the output approaches zero as $t\to\infty$; that is, $\lim y_w(t)\to 0$ as $t\to\infty$, where $y_w(t)$ is the output of the feedback system in Figure 9-19(b) due to the application of $w(t)$ and $r(t)\equiv 0$. This is called the *disturbance rejection.* Hence, if we succeed in finding a compensator $C(s)=N_c(s)/D_c(s)$ in Figure 9-19(b) so that for any $r(t)$ and any $w(t)$, we have

$$\lim_{t\to\infty} e(t) = \lim_{t\to\infty} [r(t)-y(t)] = 0 \qquad \textbf{(9-90)}$$

then the feedback system achieves asymptotic tracking and disturbance rejection. In this section we shall study the design of this problem.

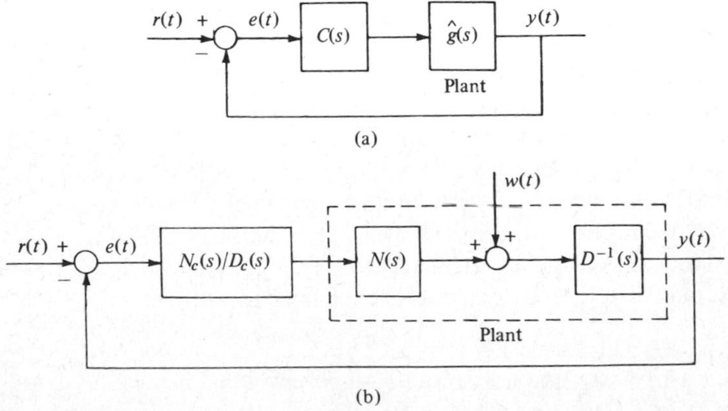

Figure 9-19 Design of control systems.

Before proceeding, we discuss first the nature of the signals $r(t)$ and $w(t)$. If $r(t)$ and $w(t)$ both go to zero as $t \to \infty$, then (9-90) will be automatically met if the feedback system in Figure 9-19(b) is asymptotically stable. If either $r(t)$ or $w(t)$ does not go to zero, and if we have no knowledge whatsoever about its nature, then it is not possible to achieve asymptotic tracking and disturbance rejection. Hence we need some information of $r(t)$ and $w(t)$ before carrying out the design. We assume that the Laplace transforms of $r(t)$ and $w(t)$ are given by

$$\hat{r}(s) = \mathscr{L}[r(t)] = \frac{N_r(s)}{D_r(s)} \tag{9-91}$$

and

$$\hat{w}(s) = \mathscr{L}[w(t)] = \frac{N_w(s)}{D_w(s)} \tag{9-92}$$

where the polynomials $D_r(s)$ and $D_w(s)$ are known and the polynomials $N_r(s)$ and $N_w(s)$ are however arbitrary so long as $\hat{r}(s)$ and $\hat{w}(s)$ are proper. This is equivalent to the assumption that $r(t)$ and $w(t)$ are generated by

$$\dot{\mathbf{x}}_r = \mathbf{A}_r \mathbf{x}_r \tag{9-93a}$$

$$r(t) = \mathbf{c}_r \mathbf{x}_r \tag{9-93b}$$

and

$$\dot{\mathbf{x}}_w(t) = \mathbf{A}_w \mathbf{x}_w \tag{9-94a}$$

$$w(t) = \mathbf{c}_w \mathbf{x}_w \tag{9-94b}$$

with some unknown initial states $\mathbf{x}_r(0)$ and $\mathbf{x}_w(0)$. The minimal polynomials of $\mathbf{A}_r$ and $\mathbf{A}_w$ are $D_r(s)$ and $D_w(s)$. The parts of $r(t)$ and $w(t)$ which go to zero as $t \to \infty$ have no effect on y as $t \to \infty$, hence we assume that some roots of $D_r(s)$ and $D_w(s)$ have zero or positive real parts. Let $\phi(s)$ be the least common denominator of the unstable poles of $\hat{r}(s)$ and $\hat{w}(s)$. Then all roots of $\phi(s)$ have zero or positive real parts.

Theorem 9-20

Consider the feedback system shown in Figure 9-19(b), where the plant is completely characterized by its proper transfer function $\hat{g}(s)$. The reference signal $r(t)$ and disturbance signal $w(t)$ are modeled as $\hat{r}(s) = N_r(s)/D_r(s)$ and $\hat{w}(s) = N_w(s)/D_w(s)$. Let $\phi(s)$ be the least common denominator of the unstable poles of $\hat{r}(s)$ and $\hat{w}(s)$. If no root of $\phi(s)$ is a zero of $\hat{g}(s)$, there exists a compensator with a proper transfer function so that the unity feedback system is asymptotically stable and achieves asymptotic tracking and disturbance rejection.

Proof

If no root of $\phi(s)$ is a zero of $\hat{g}(s)$, the tandem connection of the system with transfer function $1/\phi(s)$ followed by $\hat{g}(s) = N(s)/D(s)$ is controllable and observable (Theorem 9-2). Consequently, the polynomials $N(s)$ and $D(s)\phi(s)$ are coprime, and there exists a compensator $\bar{C}(s) = N_c(s)/D_c(s)$ such that the feedback system shown in Figure 9-20 is asymptotically stable (Theorems 9-11 and

9-11′) or, equivalently, all roots of

$$D_f(s) \triangleq D_c(s)D(s)\phi(s) + N_c(s)N(s)$$

have negative real parts.

Now we claim that the feedback system in Figure 9-20 with $C(s) = N_c(s)/D_c(s)\phi(s)$ will achieve asymptotic tracking and disturbance rejection. Indeed, the output $y(t)$ excited by $w(t)$ and $r(t) \equiv 0$ is equal to

$$\begin{aligned}\hat{y}_w(s) = -\hat{e}_w(s) &= \frac{D^{-1}(s)}{1 + N_c(s)N(s)/D_c(s)\phi(s)D(s)}\hat{w}(s) \\ &= \frac{D_c(s)\phi(s)}{D_c(s)\phi(s)D(s) + N_c(s)N(s)}\frac{N_w(s)}{D_w(s)} \\ &= \frac{D_c(s)N_w(s)}{D_c(s)D(s)\phi(s) + N_c(s)N(s)}\frac{\phi(s)}{D_w(s)}\end{aligned} \tag{9-95}$$

Since all unstable roots of $D_w(s)$ are canceled by $\phi(s)$, all the poles of $\hat{y}_w(s)$ have negative real parts. Hence we have $y_w(t) = -e_w(t) \to 0$ as $t \to \infty$. We see that even though $w(t)$ does not go to zero as $t \to \infty$, its effect on $y(t)$ diminishes as $t \to \infty$.

Let $y_r(t)$ be the output excited exclusively by $r(t)$. Then we have

$$\begin{aligned}\hat{r}(s) - \hat{y}_r(s) &= \left(1 - \frac{N_c(s)N(s)/D_c(s)D(s)\phi(s)}{1 + N_c(s)N(s)/D_c(s)D(s)\phi(s)}\right)\hat{r}(s) \\ &= \frac{D_c(s)D(s)\phi(s)}{D_c(s)D(s)\phi(s) + N_c(s)N(s)}\frac{N_r(s)}{D_r(s)} \\ &= \frac{D_c(s)D(s)N_r(s)}{D_c(s)D(s)\phi(s) + N_c(s)N(s)}\frac{\phi(s)}{D_r(s)}\end{aligned} \tag{9-96}$$

Again, all the poles of $\hat{r}(s) - \hat{y}_r(s)$ have negative real parts; hence we have $r(t) - y_r(t) \to 0$ as $t \to \infty$. Because of linearity, we have $y(t) = y_w(t) + y_r(t)$ and $e(t) = r(t) - y(t) \to 0$ as $t \to \infty$. This establishes the theorem. Q.E.D.

The design procedure developed in this proof consists of two steps: introduction of $1/\phi(s)$, a model of the reference and disturbance signals, inside the loop, and the stabilization of the feedback system by introducing the compensator $N_c(s)/D_c(s)$. The duplication of the dynamic or model, $1/\phi(s)$, inside the loop is often referred as the *internal model principle*. As will be discussed later,

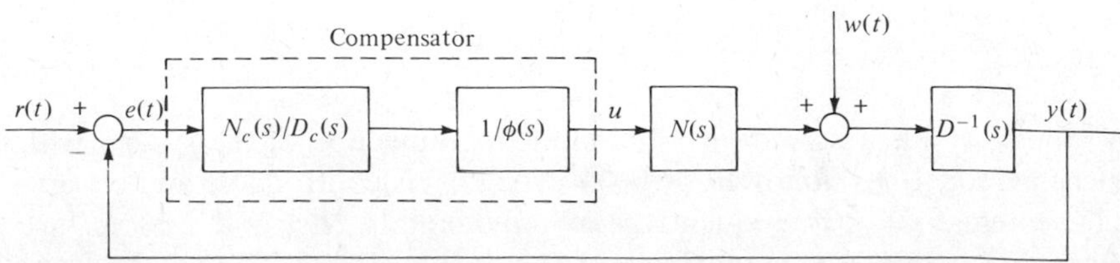

Figure 9-20 Asymptotic tracking and disturbance rejection.

for some reference input and $w(t)=0$, it is possible to achieve asymptotic tracking without introducing an internal model. However, the employment of the internal model will make the feedback system insensitive to parameter variations of $\hat{g}(s)=N(s)/D(s)$ and $N_c(s)/D_c(s)$. Hence this design is said to be *robust*. This will be expanded in the following remarks:

1. It is well known in the design of unity feedback systems that, in order to have a zero steady-state error for a step reference input (a ramp reference input), the plant must be of type 1 (type 2) transfer function, that is, $\hat{g}(s)$ has one pole (two poles) at the origin. If $r(t)$ is a step function (a ramp function), then we have $\phi(s)=s$ $[\phi(s)=s^2]$ and $(1/\phi(s))\hat{g}(s)$ is of type 1 (type 2) transfer function. Hence the well-known result is a special case of this design.
2. From (9-95) and (9-96), we see that asymptotic tracking and disturbance rejection are achieved by exact cancellations of the unstable modes of the reference and disturbance signals. This is accomplished by duplicating those modes $1/\phi(s)$ inside the loop so that $\phi(s)$ appears as numerators of the transfer functions from r to e and from w to y. In other words, the employment of the internal model is to create the required numerator to cancel the undesirable modes.
3. The location of the internal model $1/\phi(s)$ is not critical in the single-variable case so long as $1/\phi(s)$ does not appear in the forward paths from r to e and from w to y. If it appears in the forward path from, say, w to y, then $\phi(s)$ will not appear in the numerator of the transfer function from w to y.
4. Because of aging or the variation of the load, the plant transfer function $\hat{g}(s)$ or, equivalently, the coefficients of $N(s)$ and $D(s)$ may change with time. This is called the *parameter perturbation*. In this design, the parameter perturbation, even large perturbation, of $N(s)$, $D(s)$, $N_c(s)$, and $D_c(s)$ are permitted so long as all roots of

$$D_c(s)D(s)\phi(s)+N_c(s)N(s)$$

remain to have negative real parts. [Note that the perturbation of $\phi(s)$ is not permitted.] Hence this design is insensitive to parameter perturbation and is said to be *robust*.

The robustness is due to the presence of the internal model. We use the case where $\hat{r}(s)=1/s$ and $\hat{w}(s)=0$ to illustrate this point. Consider a plant with transfer function

$$\hat{g}(s)=\frac{N_n s^n+N_{n-1}s^{n-1}+\cdots+N_1 s+N_0}{D_n s^n+D_{n-1}s^{n-1}+\cdots+D_1 s+D_0}$$

and consider the feedback system shown in Figure 9-21, where a constant

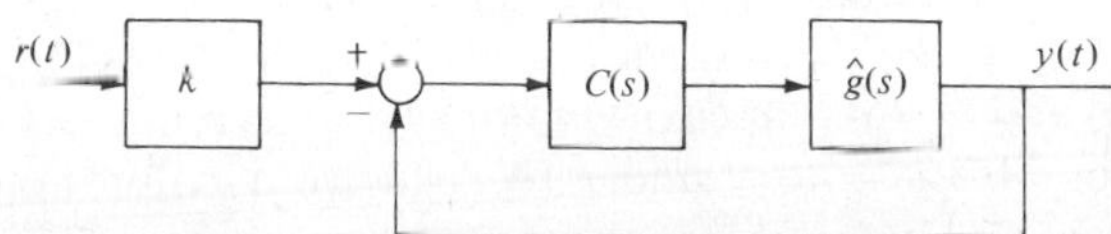

Figure 9-21 Asymptotic tracking with and without an internal model.

gain k is placed at the input. Let

$$C(s)=\frac{N_{cm}s^m+\cdots+N_{c1}s+N_{c0}}{D_{cm}s^m+\cdots+D_{c1}s+D_{c0}}$$

Then the transfer function, $\hat{g}_f(s)$, from r to y is given by

$$\hat{g}_f(s)=\frac{k\hat{g}(s)C(s)}{1+\hat{g}(s)C(s)}=\frac{k(N_nN_{cm}s^{n+m}+\cdots+N_0N_{c0})}{(D_nD_{cm}+N_nN_{cm})s^{n+m}+\cdots+(D_0D_{c0}+N_0N_{c0})}$$

It is assumed that $\hat{g}_f(s)$ is asymptotically stable. If $\hat{r}(s)=1/s$, the application of the final-value theorem yields

$$\lim_{t\to\infty}(r(t)-y(t))=\lim_{s\to 0}s(\hat{r}(s)-\hat{g}_f(s)\hat{r}(s))=1-\hat{g}_f(0)$$

$$=1-\frac{kN_0N_{c0}}{D_0D_{c0}+N_0N_{c0}} \tag{9-97}$$

If $C(s)$ contains the internal model $1/s$ and if s is not a zero of $\hat{g}(s)$, then we have $D_{c0}=0$, $N_{c0}\neq 0$ and $N_0\neq 0$. In this case, if $k=1$, then $r(t)-y(t)\to 0$ as $t\to\infty$. As long as $D_{c0}=0$, $N_{c0}\neq 0$ and $\hat{g}_f(s)$ remains to be asymptotically stable, we always have $r(t)-y(t)\to 0$ as $t\to\infty$ for all parameter perturbations of $\hat{g}(s)$ and $C(s)$. Hence the design is robust.

Now we assume that $\hat{g}(s)C(s)$ does not contain the internal model $1/s$. Then we have $D_0\neq 0$ and $D_{c0}\neq 0$. In this case, if we choose k in (9-97) as

$$k=\frac{D_0D_{c0}+N_0N_{c0}}{N_0N_{c0}}=1+\frac{D_0D_{c0}}{N_0N_{c0}}$$

then we have $r(t)-y(t)\to 0$ as $t\to\infty$. In order to have a finite k, we need $N_0\neq 0$ and $N_{c0}\neq 0$. In the design of compensators, it is always possible to find a $C(s)$ with $N_{c0}\neq 0$. If $D_0\neq 0$, we have $N_0\neq 0$ if and only if $\phi(s)=s$ is not a zero of $\hat{g}(s)$. Hence we conclude that, under the condition stated in Theorem 9-20, if $\hat{r}(s)=1/s$, it is also possible to design a feedback system, without introducing the internal model, to achieve asymptotic tracking. This design, however, is not robust. If there are perturbations in any of N_0, D_0, N_{c0}, and D_{c0}, we do not have $r(t)-y(t)\to 0$ as $t\to\infty$. Hence the design which does not employ the internal model principle is not robust. We emphasize that the condition that no root of $\phi(s)$ is a zero of $\hat{g}(s)$ is needed in both robust and nonrobust designs.

5. From the discussion in item 4, we may conclude that the condition stated in Theorem 9-20 is necessary as well. If we introduce the internal model and if any root of $\phi(s)$ is a zero of $\hat{g}(s)$, then the root, which is unstable, becomes a hidden mode and will not be affected by any compensation. Hence the unity feedback system can never be asymptotically stable. If no internal model is employed, even though $\hat{g}(s)$ is of the form $\phi(s)\bar{N}(s)/D(s)$, it is straightforward to show from (9-95) and (9-96) that $\phi(s)$ will not appear as a numerator of the transfer functions from w to y and r to e; hence asymptotic tracking and disturbance rejection cannot be achieved.

6. Unlike $D_c(s)$, $N_c(s)$, $D(s)$, and $N(s)$, the variation of the coefficients of $\phi(s)$ is not permitted in robust design, because tracking and disturbance rejection are achieved by exact cancellation of the unstable modes of $r(t)$ and $w(t)$ by the roots of $\phi(s)$. In practice, exact cancellations are very difficult and expensive to achieve; hence inexact cancellations often occur. We study in the following the effect of inexact cancellation. For simplicity, we assume

$$\hat{r}(s) = \frac{1}{s-a}\hat{r}_0(s)$$

and

$$\phi(s) = (s - a + \varepsilon)\phi_0(s)$$

where $a \geq 0$ and ε is a small real number which denotes the amount of perturbation or inexact implementation of $\phi(s)$. The transfer function from r to e in Figure 9-20 then has the form

$$\hat{g}_r(s) = (s - a + \varepsilon)\hat{g}_{r0}(s)$$

where all poles of $\hat{g}_{r0}(s)$ have negative real parts. The signal $\hat{e}(s)$ due to $\hat{r}(s)$ and $\hat{w}(s) = 0$ is equal to

$$\hat{e}(s) = (s - a + \varepsilon)\hat{g}_{r0}(s)\frac{1}{s-a}\hat{r}_0(s)$$

$$= \frac{k_1}{s-a} + \text{terms due to the poles of } \hat{g}_{r0}(s) \text{ and } \hat{r}_0(s)$$

with

$$k_1 = \frac{(s-a+\varepsilon)}{s-a}\hat{g}_{r0}(s)\hat{r}_0(s)(s-a)\bigg|_{s=a} = \varepsilon\hat{g}_{r0}(a)\hat{r}_0(a)$$

We see that k_1 is proportional to the deviation ε. If the cancellation is exact, then k_1 is zero; otherwise, k_1 is nonzero.

If k_1 is nonzero, even though very small, then $e(t)$ contains the term $k_1 e^{at}$. If a is positive, this term will approach infinity as $t \to \infty$, and the output $y(t)$ will not track $r(t)$ asymptotically. If a is zero, then $y(t)$ will track $r(t)$ but with a finite deviation. If $\hat{r}(s)$ has a repeated pole $(s-a)^m$, with $m > 1$, and if cancellations are not exact, then asymptotic tracking is again not possible.

This, however, does not mean that Theorem 9-20 is only of theoretical interest. In the design, if the internal model principle is not employed, the result will be even worse. In practice, most $r(t)$ and $w(t)$ are bounded. In this case, if the internal model principle is employed, even though there are errors in implementing $\phi(t)$, the output will still track the reference signal but with a finite steady-state error. The more accurate the implementation of $\phi(s)$, the smaller the error.

7. The design of $N_c(s)/D_c(s)$ to stabilize the feedback system requires the solving of the polynomial equation

$$D_f(s) = D_c(s)D(s)\phi(s) + N_c(s)N(s)$$

Under the condition of Theorem 9-20, $D(s)\phi(s)$ and $N(s)$ are coprime. If we use Theorem 9-10, the degree of $D_c(s)$ is deg $D(s)$ + deg $\phi(s) - 1$ and the total degree of the compensator $N_c(s)/D_c(s)\phi(s)$ is deg $D(s) + 2$ deg $\phi(s) - 1$. A different procedure is to solve the equation

$$D_f(s) = \bar{D}_c(s)D(s) + \bar{N}_c(s)N(s)$$

with the constraints that $\bar{D}_c(s)$ contains the factor $\phi(s)$ and $\bar{N}_c(s)/\bar{D}_c(s)$ is proper. By so doing, the degree of compensator can be considerably reduced. See References S34 and S238.

8. We give a state-variable interpretation of the disturbance $w(t)$ in the plant shown in Figure 9-20. The output $\hat{y}(s)$ is equal to

$$\hat{y}(s) = D^{-1}(s)\hat{w}(s) + D^{-1}(s)N(s)\hat{u}(s)$$

Let $D(s) = s^n + D_{n-1}s^{n-1} + \cdots + D_0$ and $N(s) = N_{n-1}s^{n-1} + N_{n-2}s^{n-2} + \cdots + N_0$. Then $\hat{y}_u(s) = N(s)D^{-1}(s)\hat{u}(s)$ can be realized as

$$\dot{\mathbf{x}} = \mathbf{A}\mathbf{x} + \mathbf{b}_u u(t) \qquad y_u = \mathbf{c}\mathbf{x} \tag{9-98}$$

with $\mathbf{A}$ and $\mathbf{c}$ of the forms shown in (6-38). Similarly, $y_w(s) = D^{-1}(s)w(s)$ can be realized as, by using (6-8),

$$\dot{\mathbf{x}} = \mathbf{A}\mathbf{x} + \mathbf{b}_w w(t) \qquad y = \mathbf{c}\mathbf{x} \tag{9-99}$$

where $\mathbf{A}$ and $\mathbf{c}$ are identical to the $\mathbf{A}$, $\mathbf{c}$ in (9-98). Hence we may combine (9-98) and (9-99) to yield

$$\dot{\mathbf{x}} = \mathbf{A}\mathbf{x} + \mathbf{b}_u u(t) + \mathbf{b}_w w(t) \qquad y = \mathbf{c}\mathbf{x}$$

From the equation, we see that the disturbance in Figure 9-19 can be considered as imposed on the state of the system. This equation can be generalized to

$$\dot{\mathbf{x}} = \mathbf{A}\mathbf{x} + \mathbf{b}_u u(t) + \mathbf{b}_w w(t) \qquad y = \mathbf{c}\mathbf{x} + \mathbf{e}_u u(t) + \mathbf{e}_w w(t)$$

and $w(t)$ is called an *additive disturbance*. This is the type of disturbance most often studied in dynamical equations.

9. There are two types of specifications in the design of control systems. One is called the steady-state performance, the other the transient performance. The steady-state performance is specified for the response as $t \to \infty$. Hence asymptotic tracking and disturbance rejection belong to this type of specification. The transient performance is specified for the response right after the application of the reference signal and disturbance. Typical specifications are rise time, settling time, and overshoot. They are governed mainly by the location of the poles of the overall systems or, equivalently, the roots of $D_c(s)D(s)\phi(s) + N_c(s)N(s)$. The relationship between these poles and the transient performance is generally complicated. For a discussion of this problem, see Reference S46.

10. To conclude this subsection, we remark that disturbances can be roughly classified as noise-type and waveform-structured disturbances. The former requires the statistical description and is studied in stochastic control theory. See, for example, Reference S10. The latter is describable by differential equations such as the ones in (9-93) and (9-94). For an extensive discussion of this type of disturbances, see References S121 and S122.

***Multivariable case.** Consider the design of the multivariable control system shown in Figure 9-22. The plant is described by a $q \times p$ proper rational matrix $\hat{\mathbf{G}}(s)$ factored as $\hat{\mathbf{G}}(s) = \mathbf{D}^{-1}(s)\mathbf{N}(s)$, where $\mathbf{D}(s)$ and $\mathbf{N}(s)$ are, respectively, $q \times q$ and $q \times p$ polynomial matrices. It is assumed that the $q \times 1$ reference signal $\mathbf{r}(t)$ and the $q \times 1$ disturbance signal $\mathbf{w}(t)$ are modeled as

$$\hat{\mathbf{r}}(s) = \mathbf{D}_r^{-1}(s)\mathbf{N}_r(s) \tag{9-100}$$

and

$$\hat{\mathbf{w}}(s) = \mathbf{D}_w^{-1}(s)\mathbf{N}_w(s) \tag{9-101}$$

where $\mathbf{D}_r(s)$ and $\mathbf{D}_w(s)$ are $q \times q$ polynomial matrices and $\mathbf{N}_r(s)$ and $\mathbf{N}_w(s)$ are $q \times 1$ polynomial matrices. The problem is to find a compensator so that, for any $\mathbf{N}_r(s)$ and any $\mathbf{N}_w(s)$,

$$\mathbf{e}(t) = \lim_{t \to \infty} (\mathbf{r}(t) - \mathbf{y}(t)) = \mathbf{0} \tag{9-102}$$

This is the problem of asymptotic tracking and disturbance rejection.

Following the single-variable case, the design will consist of two steps: introduction of an internal model and stabilization of the feedback system. There is, however, one important difference: the location of the internal model is critical in the multivariable case. To illustrate this point, we consider only $\mathbf{w}(t)$ by assuming $\mathbf{r}(t) = \mathbf{0}$. The model $\mathbf{D}_w^{-1}(s)$ of the disturbance $\mathbf{w}(t)$ is placed as shown in Figure 9-23. Let $\hat{\mathbf{e}}_w(s)$ be the input of $\mathbf{D}^{-1}(s)$ as shown. Then we have

$$\hat{\mathbf{e}}_w(s) = \mathbf{w}(s) - \mathbf{N}(s)\mathbf{D}_w^{-1}(s)\mathbf{N}_c(s)\mathbf{D}_c^{-1}(s)\mathbf{D}^{-1}(s)\hat{\mathbf{e}}_w(s)$$

or

$$\hat{\mathbf{e}}_w(s) = [\mathbf{I} + \mathbf{N}(s)\mathbf{D}_w^{-1}(s)\mathbf{N}_c(s)\mathbf{D}_c^{-1}(s)\mathbf{D}^{-1}(s)]^{-1}\hat{\mathbf{w}}(s)$$

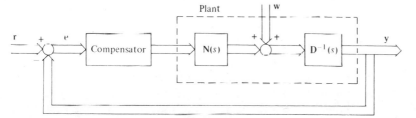

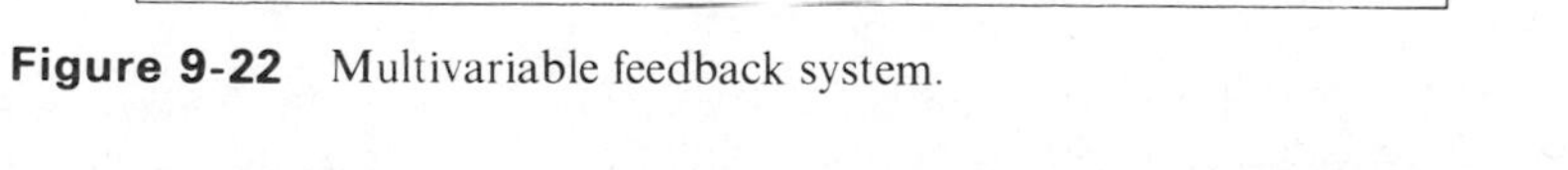

Figure 9-22 Multivariable feedback system.

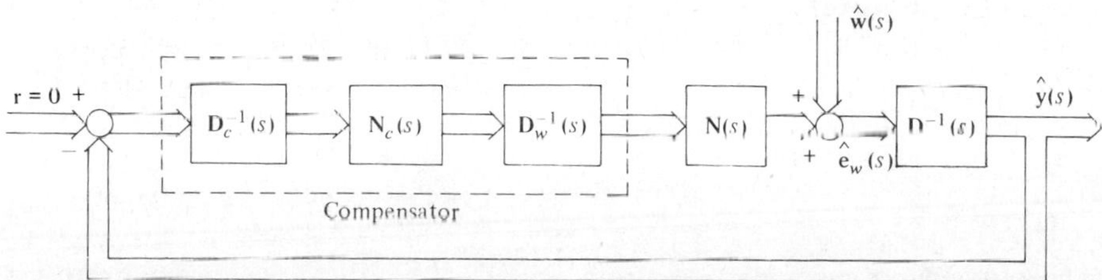

Figure 9-23 Placement of internal model for disturbance rejection.

This equation and (9-101) imply

$$\begin{aligned}\hat{\mathbf{y}}_w(s) &= \mathbf{D}^{-1}(s)[\mathbf{I} + \mathbf{N}(s)\mathbf{D}_w^{-1}(s)\mathbf{N}_c(s)\mathbf{D}_c^{-1}(s)\mathbf{D}^{-1}(s)]^{-1}\mathbf{D}_w^{-1}(s)\mathbf{N}_w(s) \\ &= \mathbf{D}_c(s)[\mathbf{D}(s)\mathbf{D}_c(s) + \mathbf{N}(s)\mathbf{D}_w^{-1}(s)\mathbf{N}_c(s)]^{-1}\mathbf{D}_w^{-1}(s)\mathbf{N}_w(s) \qquad \textbf{(9-103)}\end{aligned}$$

Although there is $\mathbf{D}_w^{-1}(s)$ inside the parentheses, because of the noncommutative of $\mathbf{N}(s)$ and $\mathbf{D}_w^{-1}(s)$, it cannot be brought out of the parentheses to cancel $\mathbf{D}_w^{-1}(s)$. Hence there is no guarantee that the internal model $\mathbf{D}_w^{-1}$ located as shown will accomplish the disturbance rejection.

Now we shall place the internal model $\mathbf{D}_w^{-1}(s)$ at the input of the summing point where $\mathbf{w}(t)$ enters the system. In other words, we interchange the positions of the blocks $\mathbf{D}_w^{-1}(s)$ and $\mathbf{N}(s)$ in Figure 9-23, or replace $\mathbf{N}(s)\mathbf{D}_w^{-1}(s)$ by $\mathbf{D}_w^{-1}(s)\mathbf{N}(s)$ in (9-103). Then we have

$$\begin{aligned}\hat{\mathbf{y}}_w(s) &= \mathbf{D}_c(s)[\mathbf{D}(s)\mathbf{D}_c(s) + \mathbf{D}_w^{-1}(s)\mathbf{N}(s)\mathbf{N}_c(s)]^{-1}\mathbf{D}_w^{-1}(s)\mathbf{N}_w(s) \\ &= \mathbf{D}_c(s)\{\mathbf{D}_w^{-1}(s)[\mathbf{D}_w(s)\mathbf{D}(s)\mathbf{D}_c(s) + \mathbf{N}(s)\mathbf{N}_c(s)]\}^{-1}\mathbf{D}_w^{-1}(s)\mathbf{N}_w(s) \\ &= \mathbf{D}_c(s)[\mathbf{D}_w(s)\mathbf{D}(s)\mathbf{D}_c(s) + \mathbf{N}(s)\mathbf{N}_c(s)]^{-1}\mathbf{D}_w(s)\mathbf{D}_w^{-1}(s)\mathbf{N}_w(s) \qquad \textbf{(9-104)}\end{aligned}$$

We see that if there is no cancellation between $\mathbf{D}_w(s)\mathbf{D}(s)\mathbf{D}_c(s) + \mathbf{N}(s)\mathbf{N}_c(s)$ and $\mathbf{D}_w(s)$, then the modes of $\mathbf{w}(t)$ are completely canceled by the internal model. Therefore, in the multivariable case, the position of the internal model is very important; it must be placed at the summation junction where the disturbance signal enters the loop. In the scalar case, because of the commutative property, the position of the internal model is not critical, so long as it is not in the forward paths from r and e and from w to y.

By a similar argument, in order to achieve asymptotic tracking, we must place $\mathbf{D}_r^{-1}(s)$ at the input of the summing point where the reference signal $\mathbf{r}$ enters the loop. In other words, we must place $\mathbf{D}_r^{-1}(s)$ after $\mathbf{D}^{-1}(s)$ or in the feedback path. If the internal model is required to locate inside the compensator shown in Figure 9-22, then it must commute with $\mathbf{N}(s)$ to achieve disturbance rejection and commute with $\mathbf{N}(s)$ and $\mathbf{D}^{-1}(s)$ to achieve asymptotic tracking. For a general $q \times p$ rational matrix $\hat{\mathbf{G}}(s) = \mathbf{D}^{-1}(s)\mathbf{N}(s)$, the internal model which has these commutative properties must be very restricted. If the internal model is chosen of the form $a(s)\mathbf{I}$, where $a(s)$ is a polynomial and $\mathbf{I}$ is a unit matrix, then it has all the commutative properties. Furthermore, these commutative properties hold even if there are parameter perturbations in $\hat{\mathbf{G}}(s)$. Hence, by choosing the internal model of the form $a(s)\mathbf{I}$, the design will be robust.

Consider the reference signal and disturbance signal given in (9-100) and (9-101). Let $\phi(s)$ be the least common denominator of the unstable poles of every element of $\mathbf{D}_r^{-1}(s)$ and $\mathbf{D}_w^{-1}(s)$. Let $\hat{\mathbf{G}}(s) = \mathbf{D}_l^{-1}(s)\mathbf{N}_l(s) = \mathbf{N}(s)\mathbf{D}^{-1}(s)$, where $\mathbf{D}_l(s)$ and $\mathbf{N}_l(s)$ are left coprime and $\mathbf{N}(s)$ and $\mathbf{D}(s)$ are right coprime. Then λ is called a *transmission zero* of $\hat{\mathbf{G}}(s)$ if rank $\mathbf{N}(\lambda) < \min(p, q)$ or rank $\mathbf{N}_l(\lambda) < \min(p, q)$. See Appendix H.

Theorem 9-21

The tandem connection of $\phi^{-1}(s)\mathbf{I}_p$ followed by the $q \times p$ proper rational matrix $\hat{\mathbf{G}}(s)$($\hat{\mathbf{G}}(s)$ followed by $\phi^{-1}(s)\mathbf{I}_q$) is controllable and observable if and only if no root of $\phi(s)$ is a transmission zero of $\hat{\mathbf{G}}(s)$ and $q \geq p$ ($p \geq q$) or, equivalently,

$$\text{rank}\begin{bmatrix} \lambda\mathbf{I}-\mathbf{A} & \mathbf{B} \\ -\mathbf{C} & \mathbf{E} \end{bmatrix} = n+p \quad (n+q) \qquad \text{for every root } \lambda \text{ of } \phi(s) \tag{9-105}$$

where $\{\mathbf{A}, \mathbf{B}, \mathbf{C}, \mathbf{E}\}$ is any irreducible realization of $\hat{\mathbf{G}}(s)$ and n is the dimension of $\mathbf{A}$ or the degree of $\hat{\mathbf{G}}(s)$.

Proof

Using Theorem 9-4, the tandem connection of $\mathbf{I}_p(\phi(s)\mathbf{I}_p)^{-1} = (\phi(s)\mathbf{I}_p)^{-1}\mathbf{I}_p$ followed by $\hat{\mathbf{G}}(s) = \mathbf{D}_l^{-1}(s)\mathbf{N}_l(s)$, where $\mathbf{D}_l(s)$ and $\mathbf{N}_l(s)$ are left coprime, is controllable if and only if $\mathbf{D}_{l2}(s) \triangleq \mathbf{D}_l(s)$ and $\mathbf{N}_{l2}(s)\mathbf{N}_{r1} \triangleq \mathbf{N}_l(s)\mathbf{I}_p$ are left coprime. Since $\mathbf{D}_l(s)$ and $\mathbf{N}_l(s)$ are left coprime by assumption, the tandem connection is always controllable.

The tandem connection is observable if and only if $\mathbf{D}_{r1}(s) \triangleq (\phi(s)\mathbf{I}_p)$ and $\mathbf{N}_{l2}(s) \triangleq \mathbf{N}_l(s)$ are right coprime or, equivalently, the $(p+q) \times p$ polynomial matrix

$$\mathbf{M}(s) \triangleq \begin{bmatrix} \phi(s)\mathbf{I}_p \\ \mathbf{N}_l(s) \end{bmatrix} \tag{9-106}$$

has rank p for every s in $\mathbb{C}$ (Theorem G-8). If λ is not a root of $\phi(s)=0$, then $\phi(\lambda) \neq 0$, and rank $\mathbf{M}(\lambda) = p$. If λ is a root of $\phi(s)=0$, then $\phi(\lambda)=0$ and rank $\mathbf{M}(\lambda) =$ rank $\mathbf{N}_l(\lambda)$. Hence the tandem connection is observable if and only if rank $\mathbf{N}_l(\lambda) = p$. By Definition H-3, rank $\mathbf{N}_l(\lambda) = p$ if and only if λ is not a transmission zero of $\hat{\mathbf{G}}(s)$. This together with Theorem H-6 completes the proof of this part of the theorem. The proof of the tandem connection of $\hat{\mathbf{G}}(s)$ followed by $[\phi(s)\mathbf{I}_q]^{-1}$ is similar and is omitted. Q.E.D.

With this result, we are ready to develop the condition for the existence of compensators to achieve tracking and disturbance rejection. The condition is a generalization of the scalar case in Theorem 9-20.

Theorem 9-22

Consider the feedback system shown in Figure 9-24 where the plant is completely characterized by its $q \times p$ proper rational matrix $\hat{\mathbf{G}}(s)$. It is assumed that the reference signal $\mathbf{r}(t)$ and the disturbance signal $\mathbf{w}(t)$ are modeled as

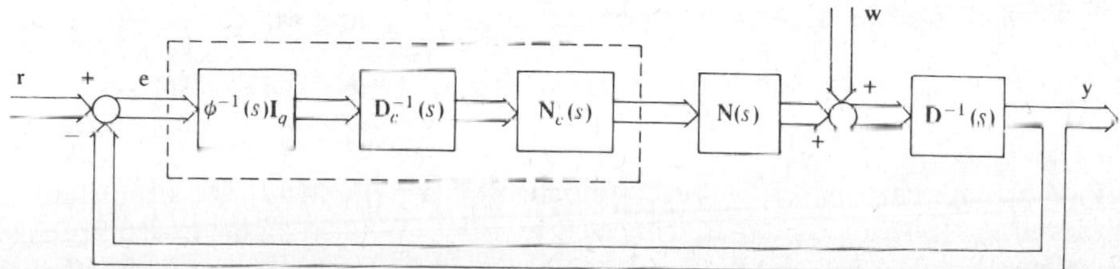

Figure 9-24 Multivariable feedback system.

$\hat{\mathbf{r}}(s)=\mathbf{D}_r^{-1}(s)\mathbf{N}_r(s)$ and $\hat{\mathbf{w}}(s)=\mathbf{D}_w^{-1}(s)\mathbf{N}_w(s)$. Let $\phi(s)$ be the least common denominator of the unstable poles of every element of $\mathbf{D}_r^{-1}(s)$ and $\mathbf{D}_w^{-1}(s)$. If no root of $\phi(s)$ is a transmission zero of $\hat{\mathbf{G}}(s)$ and $p \geq q$ or, equivalently,

$$\text{rank}\begin{bmatrix} \lambda\mathbf{I}-\mathbf{A} & \mathbf{B} \\ -\mathbf{C} & \mathbf{E} \end{bmatrix} = n+q \qquad \text{for every root } \lambda \text{ of } \phi(s)$$

there exists a compensator with a $p \times q$ proper rational matrix such that the feedback system is asymptotically stable and achieves asymptotic tracking and disturbance rejection.

Proof

Let $\hat{\mathbf{G}}(s)=\mathbf{D}^{-1}(s)\mathbf{N}(s)$ be a left-coprime fraction and let $\phi^{-1}(s)\mathbf{I}_q$ be the internal model. The compensator $\mathbf{N}_c(s)\mathbf{D}_c^{-1}(s)$ is to be designed to stabilize the unity feedback system. From Figure 9-24, we have

$$\hat{\mathbf{e}}(s)=\hat{\mathbf{r}}(s)-\mathbf{D}^{-1}(s)\mathbf{N}(s)\mathbf{N}_c(s)\mathbf{D}_c^{-1}(s)(\phi(s)\mathbf{I}_q)^{-1}\hat{\mathbf{e}}(s)$$

Hence the transfer matrix from $\mathbf{r}$ to $\mathbf{e}$ is equal to

$$\begin{aligned}\hat{\mathbf{G}}_{er}(s)&=[\mathbf{I}+\mathbf{D}^{-1}(s)\mathbf{N}(s)\mathbf{N}_c(s)\mathbf{D}_c^{-1}(s)(\phi(s)\mathbf{I}_q)^{-1}]^{-1}\\ &=\{\mathbf{D}^{-1}(s)[\mathbf{D}(s)(\phi(s)\mathbf{I}_q)\mathbf{D}_c(s)+\mathbf{N}(s)\mathbf{N}_c(s)]\mathbf{D}_c^{-1}(s)(\phi(s)\mathbf{I}_q)^{-1}\}^{-1} \\ &=\phi(s)\mathbf{D}_c(s)[\phi(s)\mathbf{D}(s)\mathbf{D}_c(s)+\mathbf{N}(s)\mathbf{N}_c(s)]^{-1}\mathbf{D}(s)\end{aligned} \qquad \textbf{(9-107)}$$

Similarly, the transfer matrix from $\mathbf{w}$ to $\mathbf{y}$ can be computed as

$$\hat{\mathbf{G}}_{yw}(s)=\phi(s)\mathbf{D}_c(s)[\phi(s)\mathbf{D}(s)\mathbf{D}_c(s)+\mathbf{N}(s)\mathbf{N}_c(s)]^{-1} \qquad \textbf{(9-108)}$$

If $p \geq q$ and if no root of $\phi(s)$ is a transmission zero of $\hat{\mathbf{G}}(s)$, then $\hat{\mathbf{G}}(s)=\mathbf{D}^{-1}(s)\mathbf{N}(s)$ followed by $\phi^{-1}(s)\mathbf{I}_q$ is controllable and observable (Theorem 9-21). Hence we have $\deg[\phi^{-1}(s)\mathbf{I}_q\mathbf{D}^{-1}(s)\mathbf{N}(s)]=\deg[\phi^{-1}(s)\mathbf{I}_q]+\deg\det\mathbf{D}(s)$(Theorem 9-1) which implies that $\phi(s)\mathbf{D}(s)$ and $\mathbf{N}(s)$ are left coprime. Consequently the roots of the determinant of

$$\mathbf{D}_f(s) \triangleq \phi(s)\mathbf{D}(s)\mathbf{D}_c(s)+\mathbf{N}(s)\mathbf{N}_c(s)$$

can be arbitrarily placed, in particular, placed in the open left-half plane by a proper choice of $\mathbf{D}_c(s)$ and $\mathbf{N}_c(s)$ (Theorem 9-16 or 9-19). Hence the unity feedback system is asymptotically stable.

All roots of $\phi(s)$ have zero or positive real parts, hence no cancellation between the roots of $\phi(s)\mathbf{D}(s)\mathbf{D}_c(s)+\mathbf{N}(s)\mathbf{N}_c(s)$ and the roots of $\phi(s)$ will occur. Consequently, $\phi(s)$ will appear as a numerator of every element of $\hat{\mathbf{G}}_{er}(s)$ and $\hat{\mathbf{G}}_{yw}(s)$. Note that we have $\hat{\mathbf{G}}_{ew}(s)=-\hat{\mathbf{G}}_{yw}(s)$.

With $\hat{\mathbf{G}}_{er}(s)$ and $\hat{\mathbf{G}}_{ew}(s)$, the response $\mathbf{e}(t)$ due to $\mathbf{r}(t)$ and $\mathbf{w}(t)$ can be written as

$$\hat{\mathbf{e}}(s)=\hat{\mathbf{G}}_{er}(s)\hat{\mathbf{r}}(s)+\hat{\mathbf{G}}_{ew}(s)\hat{\mathbf{w}}(s)=\hat{\mathbf{G}}_{er}(s)\mathbf{D}_r^{-1}(s)\mathbf{N}_r(s)+\hat{\mathbf{G}}_{ew}(s)\mathbf{D}_w^{-1}(s)\mathbf{N}_w(s) \qquad \textbf{(9-109)}$$

Since $\phi(s)$ appears as zeros of every element of $\hat{\mathbf{G}}_{er}(s)$ and $\hat{\mathbf{G}}_{ew}(s)$, all unstable poles of $\mathbf{D}_r^{-1}(s)$ and $\mathbf{D}_w^{-1}(s)$ are canceled by $\phi(s)$. What remains in $\hat{\mathbf{e}}(s)$ are all stable poles. Hence, for any $\mathbf{N}_w(s)$ and $\mathbf{N}_r(s)$, we have $\mathbf{e}(t)\to\mathbf{0}$ as $t\to\infty$. This proves the theorem. Q.E.D.

The design consists of two steps: introduction of the internal model $\phi^{-1}(s)\mathbf{I}_q$ and stabilization of the feedback system by the compensator $\mathbf{N}_c(s)\mathbf{D}_c^{-1}(s)$. Hence the total compensator is given by $\mathbf{N}_c(s)(\mathbf{D}_c(s)\phi(s))^{-1}$. Because the internal model $\phi(s)$ appears as zeros of every element of $\hat{\mathbf{G}}_{er}(s)$ and $\hat{\mathbf{G}}_{ew}(s)$, these zeros are called, according to Definition H-4, the *blocking zeros*. If the unstable modes of $\mathbf{r}(t)$ and $\mathbf{w}(t)$ are the blocking zeros of $\hat{\mathbf{G}}_{er}(s)$ and $\hat{\mathbf{G}}_{ew}(s)$, then, as established in Theorem H-7, for any initial state these unstable modes will not appear in $\mathbf{e}(t)$. In creating these blocking zeros if any of them are a transmission zero of the plant $\hat{\mathbf{G}}(s)$, then this zero will become an uncontrollable or unobservable mode or a hidden mode of the system. This unstable hidden mode will make the system useless in practice. Hence, in order to achieve asymptotic tracking and disturbance rejection, no root of $\phi(s)$ can be a transmission zero of $\hat{\mathbf{G}}(s)$ and all of $\phi(s)$ must be the blocking zeros of $\hat{\mathbf{G}}_{er}(s)$ and $\hat{\mathbf{G}}_{ew}(s)$.

We note that the internal model consists of q copies $\phi^{-1}(s)\mathbf{I}_q$ of the dynamic of the reference and disturbance signals. Because of these q copies, $\phi(s)$ can be brought out of the parentheses, as shown in (9-107) and (9-108), and become the blocking zeros of $\hat{\mathbf{G}}_{er}(s)$ and $\hat{\mathbf{G}}_{yw}(s)$. Consequently, the perturbations of $\mathbf{D}(s)$, $\mathbf{D}_c(s)$, $\mathbf{N}(s)$ and $\mathbf{N}_c(s)$ will not affect these blocking zeros as can be seen from (9-107) and (9-108). Hence the design is robust. For a more detailed discussion of robustness, see Reference S81.

We discuss in the following that the design can also be achieved by using a single copy of the reference signal as an internal model. The design, however, will be generally not robust. Consider Figure 9-22. Define $\hat{\hat{\mathbf{G}}}(s) = \hat{\mathbf{G}}(s)\mathbf{C}(s) = \mathbf{D}^{-1}(s)\mathbf{N}(s)\mathbf{N}_c(s)\mathbf{D}_c^{-1}(s) \triangleq \bar{\mathbf{D}}^{-1}(s)\bar{\mathbf{N}}(s)$, where $\bar{\mathbf{D}}(s)$ and $\bar{\mathbf{N}}(s)$ are left coprime. Then we have

$$\hat{\mathbf{e}}(s) = [\mathbf{I} + \hat{\hat{\mathbf{G}}}(s)]^{-1}\hat{\mathbf{r}}(s) = [\bar{\mathbf{D}}(s) + \bar{\mathbf{N}}(s)]^{-1}\bar{\mathbf{D}}(s)\mathbf{D}_r^{-1}(s)\mathbf{N}_r(s)$$

We assume, without loss of generality, that all poles of $\mathbf{D}_r^{-1}(s)$ have nonnegative real parts. Since all poles of $(\bar{\mathbf{D}}(s) + \bar{\mathbf{N}}(s))^{-1}$ have negative real parts, all poles of $\hat{\mathbf{e}}(s)$ have negative real parts if and only if $\bar{\mathbf{D}}(s)\mathbf{D}_r^{-1}(s)\mathbf{N}_r(s)$ is a polynomial matrix. If $\mathbf{D}_r(s)$ and $\mathbf{N}_r(s)$ are left coprime, then $\bar{\mathbf{D}}(s)\mathbf{D}_r^{-1}(s)\mathbf{N}_r(s)$ is a polynomial matrix if and only if $\mathbf{D}_r(s)$ is a right divisor of $\bar{\mathbf{D}}(s)$, or there exists a $\tilde{\mathbf{D}}(s)$ such that $\bar{\mathbf{D}}(s) = \tilde{\mathbf{D}}(s)\mathbf{D}_r(s)$ (see Problems 9-19 to 9-21). Hence asymptotic tracking is possible if and only if a compensator $\mathbf{C}(s)$ can be found such that

$$\begin{aligned}\hat{\mathbf{G}}(s)\mathbf{C}(s) &= \mathbf{D}^{-1}(s)\mathbf{N}(s)\mathbf{N}_c(s)\mathbf{D}_c^{-1}(s) = (\tilde{\mathbf{D}}(s)\mathbf{D}_r(s))^{-1}\bar{\mathbf{N}}(s) \\ &= \mathbf{D}_r^{-1}(s)\tilde{\mathbf{D}}^{-1}(s)\bar{\mathbf{N}}(s)\end{aligned} \tag{9-110}$$

This is consistent with the statements that the internal model, $\mathbf{D}_r^{-1}(s)$, must be placed at the point where the reference signal enters the loop. We see that, in this design, we require only one copy of the dynamic of the reference signal. However, the computing of $\mathbf{C}(s)$ to meet (9-110) is complicated. Even if such a $\mathbf{C}(s)$ is found for a given $\hat{\mathbf{G}}(s)$, there is no guarantee that $\mathbf{D}_r^{-1}(s)$ can still be subtracted and moved to the leftmost position as in (9-110) if there are perturbations in $\hat{\mathbf{G}}(s)$. Hence this design will not be robust. In contrast with this design, the design procedure of employing $\phi^{-1}(s)\mathbf{I}_q$ as an internal model is straightforward and the design is robust. However, the degree of compensators is much larger.

Similar to the single-variable case, the perturbation of the internal model is not permitted in the robust design. All other remarks in the single-variable case are equally applicable here and will not be repeated.

Before moving to the next topic, we remark on the necessity of the condition $p \geq q$; that is, the number of plant inputs be larger than or equal to the number of plant outputs. If $q > p$, the tandem condition of $\phi^{-1}(s)\mathbf{I}_p$ followed by $\hat{\mathbf{G}}(s)$ is controllable and observable, although $\hat{\mathbf{G}}(s)$ followed by $\phi^{-1}(s)\mathbf{I}_q$ is not (Theorem 9-21). Consequently, we may introduce the internal model as shown in Figure 9-25. The transfer matrix from **r** to **e** in Figure 9-25 can be computed as

$$\hat{\mathbf{G}}_{er}(s) = [\mathbf{I} + \mathbf{D}^{-1}(s)\mathbf{N}(s)\phi^{-1}(s)\mathbf{I}_p\mathbf{N}_c(s)\mathbf{D}_c^{-1}(s)]^{-1} \qquad \textbf{(9-111)}$$

If we write $\mathbf{D}^{-1}(s)\mathbf{N}(s)\phi^{-1}(s)\mathbf{I}_p = \mathbf{D}^{-1}(s)\phi^{-1}(s)\mathbf{N}(s) = (\phi(s)\mathbf{D}(s))^{-1}\mathbf{N}(s)$, then (9-111) can be simplified as

$$\hat{\mathbf{G}}_{er}(s) = \mathbf{D}_c(s)[\phi(s)\mathbf{D}(s)\mathbf{D}_c(s) + \mathbf{N}(s)\mathbf{N}_c(s)]^{-1}\mathbf{D}(s)\phi(s) \qquad \textbf{(9-112)}$$

This equation is similar to (9-107). Although $\phi(s)$ in (9-107) may become blocking zeros of $\hat{\mathbf{G}}_{er}(s)$, the $\phi(s)$ in (9-112) can never be all blocking zeros of $\hat{\mathbf{G}}_{er}(s)$. This can be seen from the step of writing

$$\mathbf{D}^{-1}(s)\mathbf{N}(s)\phi^{-1}(s)\mathbf{I}_p = \mathbf{D}^{-1}(s)\phi^{-1}(s)\mathbf{I}_q\mathbf{N}(s) = [\phi(s)\mathbf{D}(s)]^{-1}\mathbf{N}(s)$$

We see that p copies of $\phi(s)$ is increased to q copies. Because $\phi^{-1}(s)\mathbf{I}_p$ followed by $\hat{\mathbf{G}}(s)$ is controllable and observable, we have $\deg(\mathbf{D}^{-1}(s)\mathbf{N}(s)\phi^{-1}(s)\mathbf{I}_p) = \deg\det\mathbf{D}(s) + p\deg\phi(s)$ (Theorem 9-1). The degree of $\det\phi(s)\mathbf{D}(s)$ is equal to $q\deg\phi(s) + \deg\det\mathbf{D}(s)$, which is, because $q > p$, larger than $\deg((\phi(s)\mathbf{D}(s))^{-1}\mathbf{N}(s))$. Hence $\phi(s)\mathbf{D}(s)$ and $\mathbf{N}(s)$ are not left coprime. Consequently, not all the roots of the determinant of

$$\phi(s)\mathbf{D}(s)\mathbf{D}_c(s) + \mathbf{N}(s)\mathbf{N}_c(s)$$

can be arbitrarily assigned (Corollary 9-19). Furthermore some roots of $\phi(s)$ will appear as unassignable roots, and there is always cancellation between $\phi(s)\mathbf{D}(s)\mathbf{D}_c(s) + \mathbf{N}(s)\mathbf{N}_c(s)$ and $\phi(s)$. Hence not all roots of $\phi(s)$ will be the blocking zeros of $\hat{\mathbf{G}}_{er}(s)$ in (9-112). Consequently, if $q > p$, asymptotic tracking is not always possible. A similar remark applies to disturbance rejection.

We use a different argument to establish once again the necessity of $p \geq q$.[18]

[18] This argument was provided to the author by Professor C. A. Desoer.

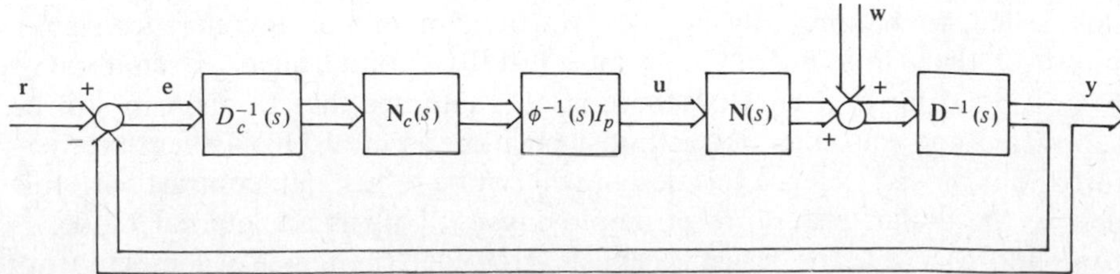

Figure 9-25 Multivariable feedback system with $q > p$.

Consider the feedback system shown in Figure 9-25 with

$$\hat{\mathbf{G}}(s)=\begin{bmatrix} \dfrac{s-1}{s+2} \\ \dfrac{-s}{s+3} \end{bmatrix}$$

It is a plant with one input and two outputs and is BIBO stable. Let $\hat{\mathbf{r}}(s)=\mathbf{r}_0/s$. No matter how the feedback system is designed, in order to have step functions at the outputs, the input to the plant must be of the form

$$\hat{u}(s)=\frac{k}{s}+(\text{terms with poles inside the open left-half } s \text{ plane})$$

Hence we have, by using the final-value theorem,

$$\lim_{t\to\infty}\mathbf{y}(t)=\lim_{s\to 0} s\hat{\mathbf{G}}(s)\hat{u}(s)=\hat{\mathbf{G}}(0)k=\begin{bmatrix} -0.5 \\ 0 \end{bmatrix}k$$

We see that, unless $\mathbf{r}_0$ is inside the range space of $\hat{\mathbf{G}}(0)$, it is not possible to achieve asymptotic tracking. Hence for any $\mathbf{r}_0$ in $\mathbb{R}^q$, in order to achieve asymptotic tracking, it is necessary to have the range space, $\mathscr{R}(\hat{\mathbf{G}}(0))$, of $\hat{\mathbf{G}}(0)$ equal to the q-dimensional real space $(\mathbb{R}^q, \mathbb{R})$ or rank $\hat{\mathbf{G}}(0)=q$. Since $\hat{\mathbf{G}}(s)$ is a $q\times p$ matrix, in order to have rank $\hat{\mathbf{G}}(0)=q$, we need $p\geq q$; that is, the number of plant inputs must be greater than or equal to the number of plant outputs.

Using this argument, we can now establish intuitively the conditions in Theorem 9-22. If $\hat{\mathbf{r}}(s)=\mathbf{r}_0/(s-\lambda)$ with $\lambda\geq 0$, in order to achieve asymptotic tracking, we need rank $\hat{\mathbf{G}}(\lambda)=q$. For the general case, we require rank $\hat{\mathbf{G}}(\lambda)=q$ for every root λ of $\phi(s)$. If we use the coprime fraction $\hat{\mathbf{G}}(s)=\mathbf{N}(s)\mathbf{D}^{-1}(s)$ and if λ is not a pole of $\hat{\mathbf{G}}(s)$, then rank $\hat{\mathbf{G}}(\lambda)=q$ if and only if rank $\mathbf{N}(\lambda)=q$, or λ is not a transmission zero of $\hat{\mathbf{G}}(s)$. This is essentially Theorem 9-22.

Static decoupling: Robust and nonrobust designs. Consider the feedback system shown in Figure 9-26 where the plant is completely characterized by its $q\times p$ proper rational matrix $\hat{\mathbf{G}}(s)$. Let the reference signal $\hat{\mathbf{r}}(s)$ be of the form $\hat{\mathbf{r}}(s)=\mathbf{d}s^{-1}$, where $\mathbf{d}$ is an arbitrary $q\times 1$ constant vector; that is, the reference signals are step functions of various magnitudes. The response at the output as $t\to\infty$ is called the *steady-state response*. The response right after the application of $\mathbf{r}(t)$ is called the *transient response*. Let the overall transfer matrix from $\mathbf{r}$ to $\mathbf{y}$ be $\hat{\mathbf{G}}_f(s)$. If $\hat{\mathbf{G}}_f(s)$ is BIBO stable, then the steady-state

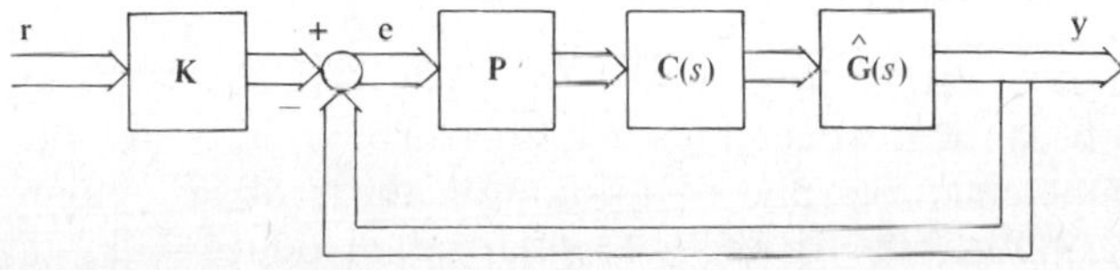

Figure 9-26 Design of unity feedback system.

response due to $\hat{\mathbf{r}}(s) = \mathbf{d}s^{-1}$ can be computed as, by using the final-value theorem,

$$\lim_{t\to\infty} \mathbf{y}(t) = \lim_{s\to 0} s\hat{\mathbf{G}}_f(s)\hat{\mathbf{r}}(s) = \lim_{s\to 0} s\hat{\mathbf{G}}_f(s)\mathbf{d}s^{-1} = \hat{\mathbf{G}}_f(0)\mathbf{d} \tag{9-113}$$

Now if $\hat{\mathbf{G}}_f(0)$ is diagonal and nonsingular, in particular, a unit matrix, then the feedback system is said to be *statically decoupled.* Indeed, if $\hat{\mathbf{G}}_f(0) = \text{diag}\{h_1, h_2, \ldots, h_q\}$, then (9-113) implies

$$\lim_{t\to\infty} y_i(t) = h_i d_i \qquad i = 1, 2, \ldots, q$$

where y_i and d_i are the ith components of $\mathbf{y}$ and $\mathbf{d}$. Hence the steady-state response at the ith output of a statically decoupled system depends solely on the ith reference input and is decoupled from the other inputs. If there is a change in the magnitude of the ith reference step input, it will cause responses at all output terminals. However, as time approaches infinity or as the transient dies out, it will cause only a change at the ith output and no change at other outputs. Hence in a statically decoupled system, only the steady-state responses are decoupled, but not the transient responses. This differs from the decoupled system discussed in Section 7-6 where all the responses, transient as well as steady state, are decoupled. Furthermore, decoupling is defined for any reference signal; whereas, static decoupling is defined only for step reference inputs. The class of step reference signals, however, is very important and is often encountered in practice. For example, temperature and humidity controls of a room are this type of reference signals. Maintaining an aircraft at a fixed altitude is another example.

Asymptotic tracking actually achieves decoupling as $t \to \infty$; the steady state of $y_i(t)$ tracks $r_i(t)$ and is independent of $r_j(t)$, for $j \neq i$. Hence the design for asymptotic tracking can be directly applied to static decoupling. In this case, we have $\phi(s) = s$. Let $K = \mathbf{I}_q$ and $\mathbf{P} = \phi^{-1}(s)\mathbf{I}_q = s^{-1}\mathbf{I}_q$ in Figure 9-26. We then design a compensator $\mathbf{C}(s)$ in Figure 9-26 to stabilize the feedback system. As shown in Theorem 9-22, if $p \geq q$ and if s is not a transmission zero of $\hat{\mathbf{G}}(s)$, then s will appear as a zero of every element of $\hat{\mathbf{G}}_{er}(s)$, the transfer matrix from $\mathbf{r}$ to $\mathbf{e}$. Hence we have $\hat{\mathbf{G}}_{er}(0) = \mathbf{0}$ and

$$\lim_{t\to\infty} \mathbf{e}(t) = \lim_{t\to\infty} [\mathbf{r}(t) - \mathbf{y}(t)] = \lim_{s\to 0} s\hat{\mathbf{G}}_{er}(s)\mathbf{d}s^{-1} = \hat{\mathbf{G}}_{er}(0)\mathbf{d} = \mathbf{0}$$

Because of $\hat{\mathbf{e}}(s) = \hat{\mathbf{r}}(s) - \hat{\mathbf{y}}(s)$ or $\hat{\mathbf{y}}(s) = \hat{\mathbf{r}}(s) - \hat{\mathbf{G}}_{er}(s)\hat{\mathbf{r}}(s)$, the transfer matrix, $\hat{\mathbf{G}}_f(s)$, from $\mathbf{r}$ to $\mathbf{y}$ is equal to

$$\hat{\mathbf{G}}_f(s) = \mathbf{I} - \hat{\mathbf{G}}_{er}(s)$$

Consequently, we have $\hat{\mathbf{G}}_f(0) = \mathbf{I}$ and the feedback system is statically decoupled. Note that every element of $\hat{\mathbf{G}}_f(s)$, except those on the diagonal, has s as a zero in its numerator.

Because of the presence of the internal model, the design is robust. That is, the system remains to be statically decoupled with perturbations, even large perturbations, of the parameters of $\mathbf{C}(s)$ and $\hat{\mathbf{G}}(s)$, so long as the feedback system remains to be asymptotically stable. In the following, we introduce a design which is not robust. The design is an extension of the single-variable case discussed in Figure 9-21. In this design, no internal model will be employed; hence we set $\mathbf{P}=\mathbf{I}$ in Figure 9-26. Let $\hat{\mathbf{G}}(s)=\mathbf{N}(s)\mathbf{D}^{-1}(s)$ and $\mathbf{C}(s)=\mathbf{D}_c^{-1}(s)\mathbf{N}_c(s)$. We find a $\mathbf{C}(s)$ so that all the roots of the determinant of the polynomial matrix

$$\mathbf{D}_f(s)=\mathbf{D}_c(s)\mathbf{D}(s)+\mathbf{N}_c(s)\mathbf{N}(s)$$

have negative real parts. The transfer matrix from $\mathbf{r}$ to $\mathbf{y}$ in Figure 9-26 can be computed, by using (9-68), as

$$\begin{aligned}\hat{\mathbf{G}}_f(s)&=\mathbf{G}(s)[\mathbf{I}+\mathbf{C}(s)\hat{\mathbf{G}}(s)]^{-1}\mathbf{C}(s)\mathbf{K}\\&=\mathbf{N}(s)[\mathbf{D}_c(s)\mathbf{D}(s)+\mathbf{N}_c(s)\mathbf{N}(s)]^{-1}\mathbf{N}_c(s)\mathbf{K}=\mathbf{N}(s)\mathbf{D}_f^{-1}(s)\mathbf{N}_c(s)\mathbf{K}\end{aligned}$$

Hence we have

$$\hat{\mathbf{G}}_f(0)=\mathbf{N}(0)\mathbf{D}_f^{-1}(0)\mathbf{N}_c(0)\mathbf{K} \tag{9-114}$$

Since all the poles of $\mathbf{D}_f^{-1}(s)$ have negative real parts, the constant matrix $\mathbf{D}_f(0)$ is nonsingular. If s is not a transmission zero of $\hat{\mathbf{G}}(s)$, then rank $\mathbf{N}(0)=q$. Now we may design a compensator $\mathbf{C}(s)$ so that rank $\mathbf{N}_c(0)=q$ and the $q\times q$ constant matrix $\mathbf{N}(0)\mathbf{D}_f^{-1}(0)\mathbf{N}_c(0)$ is nonsingular. Under these assumptions, we may choose

$$\mathbf{K}=[\mathbf{N}(0)\mathbf{D}_f^{-1}(0)\mathbf{N}_c(0)]^{-1} \tag{9-115}$$

and $\hat{\mathbf{G}}_f(0)$ becomes

$$\hat{\mathbf{G}}_f(0)=\mathbf{I}$$

Hence the system in Figure 9-26 is statically decoupled. This design is, similar to the one in Figure 9-21, not robust. If there are any perturbations in $\mathbf{N}(0)$, $\mathbf{N}_c(0)$ and $\mathbf{D}_f(0)$, then the system will not be statically decoupled. We summarize the preceding results as a corollary.

Corollary 9-22

Consider the feedback system shown in Figure 9-26 where the plant is completely characterized by its $q\times p$ proper rational matrix $\hat{\mathbf{G}}(s)$. If s is not a transmission zero of $\hat{\mathbf{G}}(s)$ and $p\geq q$ or, equivalently,

$$\text{rank}\begin{bmatrix}\mathbf{A} & \mathbf{B}\\ \mathbf{C} & \mathbf{E}\end{bmatrix}-n+q$$

then there exists a compensator with a $q\times p$ proper rational matrix such that the feedback system is asymptotically stable and statically decoupled, where $\{\mathbf{A},\mathbf{B},\mathbf{C},\mathbf{E}\}$ is any irreducible realization of $\hat{\mathbf{G}}(s)$. ∎

We remark once again that the design can be robust by introducing an internal model or nonrobust without introducing an internal model. Similar to the remark in (5) on the single-variable case, the condition is necessary as well. To check the condition in Corollary 9-22, we must find a coprime fraction of $\hat{\mathbf{G}}(s)$ and is complicated. If $\hat{\mathbf{G}}(s)$ has no pole at $s=0$, the condition in Corollary 9-22 is equivalent to rank $\hat{\mathbf{G}}(0)=q$, which can be easily checked.

State-variable approach. In this subsection, we shall discuss the design of robust control systems to achieve asymptotic tracking and disturbance rejection by using state-variable equations. The discussion will be brief because the procedure is similar to the transfer function approach. Consider a plant described by

$$\dot{\mathbf{x}} = \mathbf{A}\mathbf{x} + \mathbf{B}\mathbf{u} + \mathbf{B}_w\mathbf{w}(t)$$
$$\mathbf{y} = \mathbf{C}\mathbf{x} + \mathbf{E}\mathbf{u} + \mathbf{E}_w\mathbf{w}(t)$$

where $\mathbf{A}, \mathbf{B}, \mathbf{C}, \mathbf{E}, \mathbf{B}_w$, and $\mathbf{E}_w$ are, respectively, $n \times n$, $n \times p$, $q \times n$, $q \times p$, $n \times q$, and $q \times q$ constant matrices. It is assumed that $\{\mathbf{A}, \mathbf{B}\}$ is controllable and $\{\mathbf{A}, \mathbf{C}\}$ is observable. It is also assumed that the disturbance signal $\mathbf{w}(t)$ is generated by

$$\dot{\mathbf{x}}_w(t) = \mathbf{A}_w\mathbf{x}_w(t) \qquad \mathbf{w}(t) = \mathbf{C}_w\mathbf{x}_w(t)$$

with some unknown initial state. The problem is to design a robust control system so that the output of the plant will track asymptotically the reference signal $\mathbf{r}(t)$ generated by

$$\dot{\mathbf{x}}_r(t) = \mathbf{A}_r\mathbf{x}_r(t) \qquad \mathbf{r}(t) = \mathbf{C}_r\mathbf{x}_r(t)$$

with some unknown initial state. Let $\phi_w(s)$ and $\phi_r(s)$ be the minimal polynomials of $\mathbf{A}_w$ and $\mathbf{A}_r$, respectively, and let

$$\phi(s) = s^m + \alpha_1 s^{m-1} + \alpha_2 s^{m-2} + \cdots + \alpha_m$$

be the least common multiple of the closed right-half s plane roots of $\phi_w(s)$ and $\phi_r(s)$. Thus all roots of $\phi(s)$ have nonnegative real parts. The internal model $\phi^{-1}(s)\mathbf{I}_q$ can be realized as

$$FE_c: \qquad \dot{\mathbf{x}}_c = \mathbf{A}_c\mathbf{x}_c + \mathbf{B}_c\mathbf{e}$$
$$\mathbf{y}_c = \mathbf{x}_c$$

where

$$\mathbf{A}_c = \text{block diag}\,\underbrace{\{\boldsymbol{\Gamma}, \boldsymbol{\Gamma}, \ldots, \boldsymbol{\Gamma}\}}_{q\text{-tuple}}$$

$$\mathbf{B}_c = \text{block diag}\,\overbrace{\{\boldsymbol{\tau}, \boldsymbol{\tau}, \ldots, \boldsymbol{\tau}\}}$$

with

$$\boldsymbol{\Gamma} = \begin{bmatrix} 0 & 1 & 0 & \cdots & 0 \\ 0 & 0 & 1 & \cdots & 0 \\ \vdots & \vdots & \vdots & & \vdots \\ 0 & 0 & 0 & \cdots & 1 \\ -\alpha_m & -\alpha_{m-1} & -\alpha_{m-2} & \cdots & -\alpha_1 \end{bmatrix} \qquad \boldsymbol{\tau} = \begin{bmatrix} 0 \\ 0 \\ \vdots \\ 0 \\ 1 \end{bmatrix}$$

and $\mathbf{e} = \mathbf{r} - \mathbf{y}$ as shown in Figure 9-27. FE_c is an mq-dimensional equation. This internal model is called the *servocompensator* in References S64 to S66. Note that the output of the servocompensator consists of all qm numbers of state variables. This is possible if the servocompensator is implemented by using operational amplifiers, resistors, and capacitors. Now consider the tandem connection of the plant followed by the servocompensator. Its composite dynamical equation is, as derived in (3-63),

$$\begin{bmatrix} \dot{\mathbf{x}} \\ \dot{\mathbf{x}}_c \end{bmatrix} = \begin{bmatrix} \mathbf{A} & \mathbf{0} \\ -\mathbf{B}_c\mathbf{C} & \mathbf{A}_c \end{bmatrix} \begin{bmatrix} \mathbf{x} \\ \mathbf{x}_c \end{bmatrix} + \begin{bmatrix} \mathbf{B} \\ -\mathbf{B}_c\mathbf{E} \end{bmatrix} \mathbf{u} \tag{9-116}$$

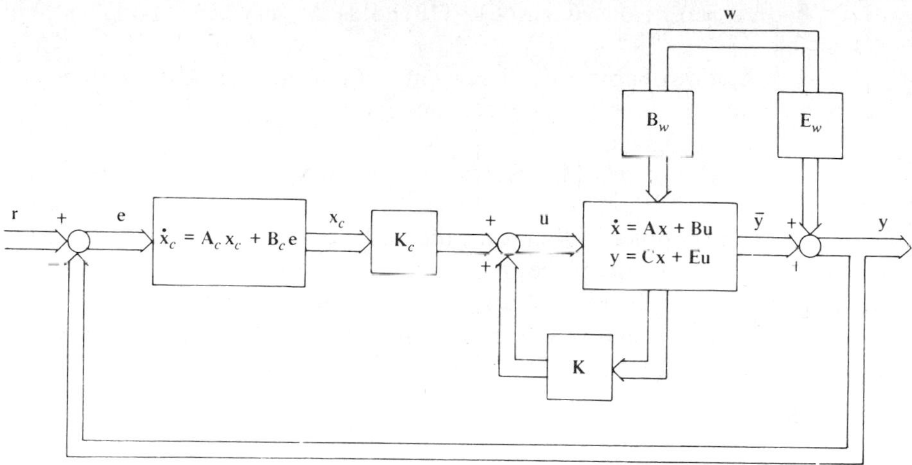

Figure 9-27 Design of robust system to achieve tracking and disturbance rejection.

This connection was shown in Theorem 9-21 to be controllable and observable if and only if $p \geq q$ and no root of $\phi(s)$ is a transmission zero of the plant. Now we reestablish directly that (9-116) is controllable if

$$\text{rank}\begin{bmatrix} \lambda\mathbf{I}-\mathbf{A} & \mathbf{B} \\ -\mathbf{C} & \mathbf{E} \end{bmatrix} = n+q \qquad \text{for every root } \lambda \text{ of } \phi(s) \tag{9-117}$$

Theorem 5-7 implies that (9-116) is controllable if and only if

$$\text{rank } \mathbf{V}(s) \triangleq \text{rank}\left[\begin{array}{cc|c} s\mathbf{I}-\mathbf{A} & 0 & \mathbf{B} \\ \mathbf{B}_c\mathbf{C} & s\mathbf{I}-\mathbf{A}_c & -\mathbf{B}_c\mathbf{E} \end{array}\right] = n+mq$$

for every s in $\mathbb{C}$. If $\{\mathbf{A}, \mathbf{B}\}$ is controllable, we have rank $[s\mathbf{I}-\mathbf{A} \quad \mathbf{B}] = n$ for every s in $\mathbb{C}$. If s is not an eigenvalue of $\mathbf{A}_c$ or, equivalently, a root of $\phi(\lambda)$, then rank $(s\mathbf{I}-\mathbf{A}_c) = mq$. Hence we conclude from the structure of $\mathbf{V}(s)$ that if s is not a root of $\phi(s)$, then we have

$$\text{rank } \mathbf{V}(s) = n+mq \tag{9-118}$$

Next we show that under the condition of (9-117), Equation (9-118) still holds at the roots of $\phi(s)$. We write

$$\mathbf{V}(s) = \begin{bmatrix} \mathbf{I}_n & \mathbf{0} & \mathbf{0} \\ \mathbf{0} & \mathbf{B}_c & s\mathbf{I}-\mathbf{A}_c \end{bmatrix} \left.\begin{bmatrix} s\mathbf{I}-\mathbf{A} & \mathbf{0} & \mathbf{B} \\ \mathbf{C} & \mathbf{0} & -\mathbf{E} \\ \mathbf{0} & \mathbf{I}_{mq} & \mathbf{0} \end{bmatrix}\right\} n+q+mq$$

The first factor has rank $n+mq$ by the irreducible realization of $\phi^{-1}(s)\mathbf{I}_q$. The second factor has rank $n+q+mq$ at every root of $\phi(s)$, by the assumption of (9-117). Hence Sylvester's inequality (Theorem 2-6) implies

$$\text{rank } \mathbf{V}(s) \geq (n+mq) + (n+q+mq) - (n+q+mq) = n+mq$$

Since $\mathbf{V}(s)$ is an $(n+mq) \times (n+mq+q)$ matrix, we conclude that

$$\text{rank } \mathbf{V}(s) = n+mq$$

at every root of $\phi(s)$. Consequently, (9-118) holds at every s in $\mathbb{C}$ and (9-116) is controllable.

If (9-116) is controllable, then the eigenvalues of the composite system can be arbitrarily assigned by the state feedback

$$\mathbf{u} = [\mathbf{K} \quad \mathbf{K}_c] \begin{bmatrix} \mathbf{x} \\ \mathbf{x}_c \end{bmatrix} = \mathbf{K}\mathbf{x} + \mathbf{K}_c\mathbf{x}_c$$

Hence by a proper choice of $\mathbf{K}$ and $\mathbf{K}_c$, the feedback system in Figure 9-27 can be stabilized. If the state of the plant is not available, we can design a state estimator and then apply the feedback gain $\mathbf{K}$ at the output of the state estimator. This completes the design of a robust control system to achieve asymptotic tracking and disturbance rejection. For a more detailed discussion, see Reference S81.

9-7 Design of Compensators: Input-Output Feedback Systems

Single-variable case. In this section, we study the design problem in Section 9-5 for a different configuration. Specifically, we study the condition on the plant for the existence of proper compensators in the input-output feedback system shown in Figure 9-1(b), to achieve arbitrary denominator matrix. We also study the minimum degrees of compensators to achieve the design. We note that the input-output feedback configuration is developed from the design of state feedback and state estimator discussed in Chapter 7.

Consider the system in Figure 9-1(b). We study first the case $C(s) = 1$. The compensators $C_0(s)$ and $C_1(s)$ are required to have the same denominator as

$$C_0(s) = \frac{L(s)}{D_c(s)} \tag{9-119}$$

and

$$C_1(s) = \frac{M(s)}{D_c(s)} \tag{9-120}$$

Because of this restriction, the feedback system in Figure 9-1(b) can be redrawn as shown in Figure 9-28(a). If $\deg D_c(s) = m$, the 1×2 rational matrix $[C_0(s) \quad C_1(s)]$ is of degree m and can be realized by using m integrators. If the configuration in Figure 9-28(a) is reduced to the single-loop system shown in Figure 9-28(b), then the denominators of the two compensators are different. Hence their implementations require twice as many integrators as the one in Figure 9-28(a). Hence, although the two configurations in Figure 9-28 are equivalent mathematically, they are different in actual implementation.

The transfer function from r to y in Figure 9-28(a) can be computed as

$$\hat{g}_f(s) = \frac{\hat{g}(s)}{1 + C_0(s) + C_1(s)\hat{g}(s)} = \frac{N(s)D_c(s)}{D_c(s)D(s) + L(s)D(s) + M(s)N(s)} \tag{9-121}$$

Define

$$D_f(s) = D_c(s)D(s) + L(s)D(s) + M(s)N(s)$$

or

$$D_f(s) - D_c(s)D(s) = L(s)D(s) + M(s)N(s) \tag{9-122}$$

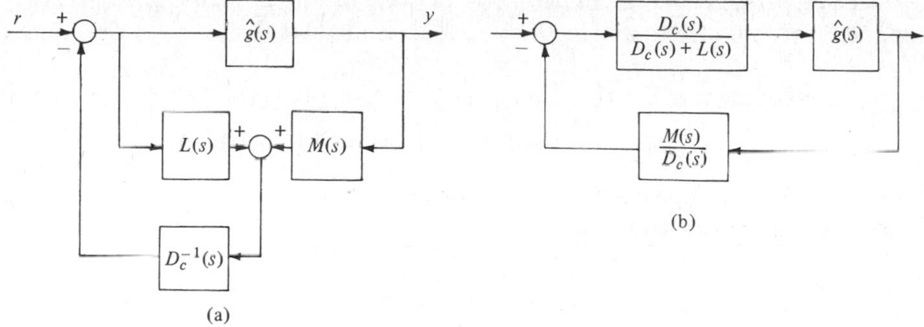

Figure 9-28 Input-output feedback system.

Then $\hat{g}_f(s)$ becomes

$$\hat{g}_f(s) = N(s)D_f^{-1}(s)D_c(s) \tag{9-123}$$

Theorem 9-23′

Consider a plant with transfer function $\hat{g}(s) = N(s)/D(s)$ and deg $N(s) \leq$ deg $D(s) = n$. For any $D_c(s)$ of degree m and any $D_f(s)$ of degree $n+m$ or less, there exist proper compensators $L(s)/D_c(s)$ and $M(s)/D_c(s)$ so that the feedback system in Figure 9-28(a) has transfer function $N(s)D_f^{-1}(s)D_c(s)$ from r to y if and only if $D(s)$ and $N(s)$ are coprime and $m \geq n-1$.

Proof

For any $D_c(s)$ of degree m and any $D_f(s)$ of degree $n+m$ or less, the polynomial $D_f(s) - D_c(s)D(s)$ is of degree $n+m$ or less. The application of Theorem 9-10 to (9-122) yields that (9-122) has solutions $L(s)$ and $M(s)$ of degrees m or less if and only if $D(s)$ and $N(s)$ are coprime and $m \geq n-1$. This proves the theorem. Q.E.D.

We compare first this theorem with Theorems 9-11 and 9-11′. First, in order to ensure the properness of the compensators, we must consider separately in Theorems 9-11 and 9-11′ the cases where the plant is proper or strictly proper and require the degree of $\mathbf{D}_f(s)$ to be exactly equal to $n+m$. Since the $D_c(s)$ in Theorem 9-23′ is chosen a priori to have degree m, the compensators are always proper whether $D_f(s)$ is of degree $n+m$ or not. Second, the overall transfer function of the unity feedback system in Figure 9-12 is $N(s)D_f^{-1}(s)N_c(s)$, where we can control only $D_f(s)$. The overall transfer function of the input-output feedback system in Figure 9-28(a) is $N(s)D_f^{-1}(s)D_c(s)$, where we can control $D_f(s)$ as well as $D_c(s)$. The reason for having this extra freedom can be seen from Figures 9-12 and 9-28(a). Although the compensator $D_c^{-1}(s)N_c(s)$ in Figure 9-12 and the compensator $D_c^{-1}(s)[L(s) \quad M(s)]$ in Figure 9-28(a) have the same degree, the former has two sets of parameters D_{ci}, N_{ci}, whereas the latter has three sets of parameters D_{ci}, L_i, M_i. Since it requires only two sets of parameters to meet (9-40) and (9-122), either $D_c(s)$ or $L(s)$ in (9-122) can be arbitrarily assigned. We choose to assign $D_c(s)$ because it dictates the poles of the compensator.

The employment of Theorem 9-23′ requires the solving of (9-122) or

$$E(s) \triangleq D_f(s) - D_c(s)D(s) = L(s)D(s) + M(s)N(s) \tag{9-124}$$

This is the Diophantine equation in (9-40) and, hence, can be translated into a set of linear algebraic equations. Let

$$\begin{aligned} D(s) &= D_0 + D_1 s + \cdots + D_n s^n \\ N(s) &= N_0 + N_1 s + \cdots + N_n s^n \\ D_c(s) &= D_{c0} + D_{c1} s + \cdots + D_{cm} s^m \\ L(s) &= L_0 + L_1 s + \cdots + L_m s^m \\ M(s) &= M_0 + M_1 s + \cdots + M_m s^m \\ D_f(s) &= F_0 + F_1 s + \cdots + F_{n+m} s^{n+m} \\ \text{and} \quad E(s) &= E_0 + E_1 s + \cdots + E_{n+m} s^{n+m} \end{aligned} \tag{9-125}$$

Then Equation (9-124) is equivalent to the algebraic equation

$$[L_0 \quad M_0 \,\vdots\, L_1 \quad M_1 \,\vdots \cdots \vdots\, L_m \quad M_m]\mathbf{S}_m = [E_0 \quad E_1 \quad \cdots \quad E_{n+m}] \triangleq \mathbf{E} \tag{9-126}$$

where $\mathbf{S}_m$ is defined as in (9-45). If $m \geq n-1$, then $\mathbf{S}_m$ has a full column rank. Hence for any $\mathbf{E}$, solutions $\{L_i, M_i\}$ exist in (9-126). The solutions yield immediately the proper compensators $C_0(s)$ and $C_1(s)$.

Example 1

Consider a plant with transfer function $\hat{g}(s) = N(s)/D(s) = (s-1)/s(s-2)$. Clearly, we have $n = 2$ and $m \geq n - 1 = 1$. Let us choose arbitrarily $D_c(s) = s + 1$, and $D_f(s) = s^2 + 2s + 2$. Note that the degree of $D_f(s)$ is smaller than $n + m = 3$. We compute

$$\begin{aligned} E(s) = D_f(s) - D_c(s)D(s) &= s^2 + 2s + 2 - (s+1)s(s-2) \\ &= 2 + 4s + 2s^2 - s^3 \end{aligned}$$

and form

$$[L_0 \quad M_0 \,\vdots\, L_1 \quad M_1] \left[\begin{array}{ccc:c} 0 & -2 & 1 & 0 \\ -1 & 1 & 0 & 0 \\ \hdashline 0 & 0 & -2 & 1 \\ 0 & -1 & 1 & 0 \end{array}\right] = [2 \quad 4 \quad 2 \quad -1]$$

Its solutions are $L_0 = -6$, $L_1 = -1$, $M_0 = -2$ and $M_1 = 6$. Hence the compensators are

$$C_0(s) = \frac{L(s)}{D_c(s)} = \frac{-6 - s}{1 + s}$$

and

$$C_1(s) = \frac{M(s)}{D_c(s)} = \frac{-2 + 6s}{1 + s}$$

They are all proper rational functions. The block diagram of the feedback system is shown in Figure 9-29. ∎

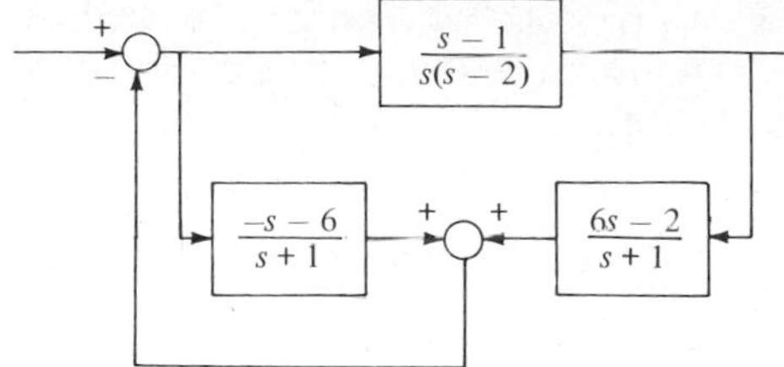

Figure 9-29 An input-output feedback system which is not well posed.

Although all compensators in Figure 9-29 are proper, the system has one serious problem. The system has a loop which yields an improper transfer function and is, as discussed in Section 3-6, not well-posed. In the design of feedback systems, we require not only all compensators to be proper but also the resulting system to be well-posed. Theorem 9-23′ may not yield a well-posed system and should be replaced by the following theorem.

Theorem 9-23

Consider a plant with transfer function $\hat{g}(s)=N(s)/D(s)$ and $\deg N(s)\leq \deg D(s)=n$. For any $D_c(s)$ of degree m and any $D_f(s)$ of degree $n+m$, there exist proper compensators $L(s)/D_c(s)$ and $M(s)/D_c(s)$ such that the feedback system in Figure 9-28(a) is well posed and has a transfer function $N(s)D_f^{-1}(s)D_c(s)$ from r to y if and only if $D(s)$ and $N(s)$ are coprime and $m\geq n-1$. ∎

We see that by removing $\deg D_f(s)<n+m$ from Theorem 9-23′, we can then assert the well posedness in Theorem 9-23. We show that the input-output feedback system is well posed by showing that $1+C_0(s)+C_1(s)\hat{g}(s)$ is different from 0 at $s=\infty$ if $\deg D_f(s)=n+m$. From (9-122), we have

$$\begin{aligned} D_c^{-1}(s)D_f(s)D^{-1}(s) &= 1+D_c^{-1}(s)L(s)+D_c^{-1}(s)M(s)N(s)D^{-1}(s) \\ &= 1+C_0(s)+C_1(s)\hat{g}(s) \end{aligned} \tag{9-127}$$

Hence the well posedness of the input-output feedback system depends on the value of $D_c^{-1}(s)D_f(s)D^{-1}(s)$ at $s=\infty$. Using the notations of (9-125), we have, for $s\rightarrow\infty$,

$$D_c^{-1}(s)D_f(s)D^{-1}(s)\rightarrow D_{cm}^{-1}s^{-m}(F_{n+m}s^{n+m}+F_{n+m-1}s^{n+m-1}+\cdots)s^{-n}D_n^{-1}$$

Hence if $F_{n+m}=0$ or $\deg D_f(s)<n+m$, then $1+C_0(\infty)+C_1(\infty)\hat{g}(\infty)=0$ and the system is not well posed. However, if $F_{n+m}\neq 0$ or $\deg D_f(s)=n+m$, then $1+C_0(\infty)+C_1(\infty)\hat{g}(\infty)\neq 0$ and the system is well posed. It is of interest to note that if $\deg D_f(s)<n+m$, the pole-zero excess inequality in (9-38) is violated.

We consider now a special case of Theorem 9-23.

Corollary 9-23

Consider a plant with transfer function $\hat{g}(s)=N(s)/D(s)$ and $\deg N(s)\leq \deg D(s)=n$. For any $\bar{D}_f(s)$ of degree n, there exist proper compensators $L(s)/D_c(s)$ and

$M(s)/D_c(s)$ of degree $n-1$ and with arbitrarily assignable poles such that the feedback system in Figure 9-28(a) is well posed and has transfer function $N(s)\bar{D}_f^{-1}(s)$ if and only if $D(s)$ and $N(s)$ are coprime. ∎

This corollary follows directly from Theorem 9-23 by choosing $D_f(s) = D_c(s)\bar{D}_f(s)$ and noting

$$\hat{g}_f(s) = N(s)D_f^{-1}(s)D_c(s) = N(s)\bar{D}_f^{-1}(s)D_c^{-1}(s)D_c(s) = N(s)\bar{D}_f^{-1}(s)$$

This design always involves the cancellation of $D_c(s)$, which can be chosen by the designer, however. The degree of $D_c(s)$ in the corollary can be larger than $n-1$; however, it does not seem to serve any design purpose because $D_c(s)$ is completely canceled in the design.

We compare now Corollary 9-23 with the result obtained in the state-variable approach. By state feedback, we can assign the eigenvalues of $\mathbf{A}$ or the poles of $\hat{g}(s) = N(s)/D(s)$ as the roots of $\bar{D}_f(s)$ without affecting $N(s)$. An $(n-1)$-dimensional state estimator with arbitrary eigenvalues can be constructed to generate an estimate of the state. The connection of the feedback gain from the output of the estimator yields the overall transfer function $N(s)/\bar{D}_f(s)$ (see Section 7-5). Hence Corollary 9-23 establishes essentially the result of state feedback and state estimator. However, the result in Corollary 9-23 is slightly more general: the plant is permitted to have a proper transfer function. In Chapter 7, we design state estimators only for strictly proper plants or dynamical equations with the direct transmission parts equal to zero. In Corollary 9-23, we require $\deg \bar{D}_f(s) = \deg D(s)$; in the design of state feedback, we require $\deg \bar{D}_f(s) = \deg D(s)$ and $\bar{D}_{fn} = D_n$ (that is, their leading coefficients are equal). If $\deg \bar{D}_f(s) = \deg D(s)$ and $\bar{D}_{fn} = D_n$, the compensator $C_0(s)$ is always strictly proper for $\hat{g}(s)$ strictly proper. This can be verified from the last column equation of (9-126) (Problem 9-28). Hence we always have $1 + C_0(\infty) + C_1(\infty)\hat{g}(\infty) \neq 0$ for the class of systems studied in the state-variable approach, and consequently, the well-posedness problem does not arise in the approach. Hence the result in Corollary 9-23 is more general than the result of state feedback and state estimator.

In the transfer-function approach, we require only the concept of coprimeness. In the state-variable approach, we require the concepts of controllability and observability. In the former approach, the design consists of forming a linear algebraic equation and its solutions yield immediately the required compensators. In the latter approach, the design requires one similarity transformation to compute the feedback gain, and requires one similarity transformation or one solution of a Lyapunov matrix equation to find a state estimator. Hence for the single-variable case, it appears that the design in the transfer-function approach is simpler conceptually and computationally than the one in the state-variable approach.

To conclude this subsection, we remark that Corollary 9-23 can be used in the design of optimal systems. Consider a plant with transfer function $\hat{g}(s) = N(s)/D(s)$ with input $u(t)$ and output $y(t)$. It is required to design an overall system to minimize the quadratic performance index

$$J = \int_0^\infty \{q[y(t) - r(t)]^2 + u^2(t)\}\, dt \tag{9-128}$$

where $q > 0$ is a weighting factor and $r(t)$ is the reference signal. If $r(t)$ is a step function, then the optimal system which has the smallest J is of the form

$$\hat{g}_f(s) = \frac{N(s)}{\bar{D}_f(s)}$$

where $\bar{D}_f(s)$ is a Hurwitz polynomial and has the same degree as $D(s)$. The $N(s)$ in the numerator of $\hat{g}_f(s)$ is the same as the numerator of the plant transfer function $\hat{g}(s)$. See Reference S46. Hence the design of the optimal system requires the solution of Corollary 9-23.

Multivariable case. In this subsection, the results in the single-variable case will be extended to the multivariable case. Consider the input-output feedback system shown in Figure 9-30. The plant is described by the $q \times p$ proper rational matrix $\hat{\mathbf{G}}(s) = \mathbf{N}(s)\mathbf{D}^{-1}(s)$. The compensators are denoted by the $p \times p$ proper rational matrix $\mathbf{C}_0(s) = \mathbf{D}_c^{-1}(s)\mathbf{L}(s)$ and the $p \times q$ proper rational matrix $\mathbf{C}_1(s) = \mathbf{D}_c^{-1}(s)\mathbf{M}(s)$. The transfer matrix from $\mathbf{r}$ to $\mathbf{y}$ in Figure 9-30 can be computed as

$$\begin{aligned}\hat{\mathbf{G}}_f(s) &= \hat{\mathbf{G}}(s)[\mathbf{I} + \mathbf{C}_0(s) + \mathbf{C}_1(s)\hat{\mathbf{G}}(s)]^{-1} \\ &= \mathbf{N}(s)\mathbf{D}^{-1}(s)[\mathbf{I} + \mathbf{D}_c^{-1}(s)\mathbf{L}(s) + \mathbf{D}_c^{-1}(s)\mathbf{M}(s)\mathbf{N}(s)\mathbf{D}^{-1}(s)]^{-1} \\ &= \mathbf{N}(s)[\mathbf{D}_c(s)\mathbf{D}(s) + \mathbf{L}(s)\mathbf{D}(s) + \mathbf{M}(s)\mathbf{N}(s)]^{-1}\mathbf{D}_c(s)\end{aligned} \tag{9-129}$$

Define

$$\mathbf{D}_f(s) = \mathbf{D}_c(s)\mathbf{D}(s) + \mathbf{L}(s)\mathbf{D}(s) + \mathbf{M}(s)\mathbf{N}(s) \tag{9-130}$$

or

$$\mathbf{E}(s) \triangleq \mathbf{D}_f(s) - \mathbf{D}_c(s)\mathbf{D}(s) = \mathbf{L}(s)\mathbf{D}(s) + \mathbf{M}(s)\mathbf{N}(s) \tag{9-131}$$

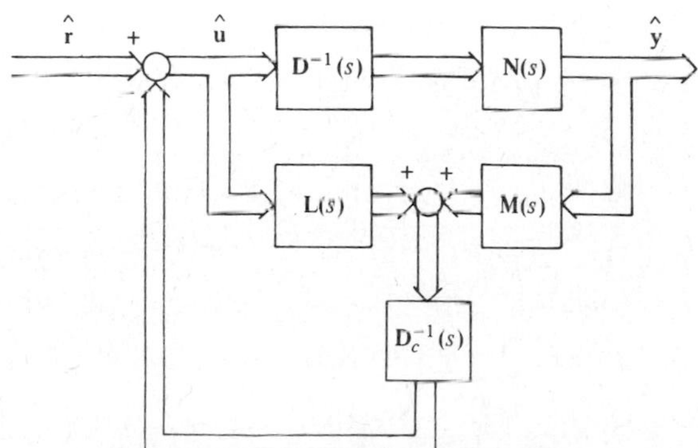

Figure 9-30 Input-output feedback system.

Then $\hat{\mathbf{G}}_f(s)$ becomes

$$\hat{\mathbf{G}}_f(s) = \mathbf{N}(s)\mathbf{D}_f^{-1}(s)\mathbf{D}_c(s) \tag{9-132}$$

Define

$$\mathbf{H}(s) = \text{diag}\{s^{\mu_1}, s^{\mu_2}, \ldots, s^{\mu_p}\} \tag{9-133}$$
$$\mathbf{H}_c(s) = \text{diag}\{s^{m_1}, s^{m_2}, \ldots, s^{m_p}\} \tag{9-134}$$

Theorem 9-24

Consider a plant with $q \times p$ proper rational matrix $\hat{\mathbf{G}}(s) = \mathbf{N}(s)\mathbf{D}^{-1}(s)$. Let μ_i, $i = 1, 2, \ldots, p$, be the column degrees of $\mathbf{D}(s)$, and let v be the row index of $\hat{\mathbf{G}}(s)$. Let $m_i \geq v - 1$, for $i = 1, 2, \ldots, p$. Then for any $\mathbf{D}_c(s)$ of row degrees m_i and row reduced, and any $\mathbf{D}_f(s)$ of the property that

$$\lim_{s \to \infty} \mathbf{H}_c^{-1}(s)\mathbf{D}_f(s)\mathbf{H}^{-1}(s) = \mathbf{J} \tag{9-135}$$

exists and is nonsingular, there exist compensators with proper rational matrices $\mathbf{D}_c^{-1}(s)\mathbf{L}(s)$ and $\mathbf{D}_c^{-1}(s)\mathbf{M}(s)$ such that the feedback system in Figure 9-30 is well posed and has transfer matrix $\mathbf{N}(s)\mathbf{D}_f^{-1}(s)\mathbf{D}_c(s)$ if and only if $\mathbf{D}(s)$ and $\mathbf{N}(s)$ are right coprime and $\mathbf{D}(s)$ is column reduced.

Proof

Consider the $\mathbf{E}(s)$ defined in (9-131). Clearly we have

$$\lim_{s \to \infty} \mathbf{H}_c^{-1}(s)\mathbf{E}(s)\mathbf{H}^{-1}(s) = \mathbf{J} - \lim_{s \to \infty} \mathbf{H}_c^{-1}(s)\mathbf{D}_c(s)\mathbf{D}(s)\mathbf{H}^{-1}(s) = \mathbf{J} - \mathbf{D}_{ch}\mathbf{D}_h \tag{9-136}$$

where $\mathbf{D}_{ch}$ is the row degree coefficient matrix of $\mathbf{D}_c(s)$ and $\mathbf{D}_h$ is the column degree coefficient matrix of $\mathbf{D}(s)$. Because $\mathbf{J} - \mathbf{D}_{ch}\mathbf{D}_h$ exists, we conclude from the proof of Theorem 9-18 that solutions $\mathbf{L}(s)$ and $\mathbf{M}(s)$ of row degrees at most m_i exist in (9-131). Consequently $\mathbf{D}_c^{-1}(s)\mathbf{L}(s)$ and $\mathbf{D}_c^{-1}(s)\mathbf{M}(s)$ are proper following the assumption of $\mathbf{D}_c(s)$. Note that whether $\mathbf{J} - \mathbf{D}_{ch}\mathbf{D}_h$ is nonsingular or not is immaterial here.

We show that the system is well posed by showing that $\mathbf{I} + \mathbf{C}_0(\infty) + \mathbf{C}_1(\infty)\hat{\mathbf{G}}(\infty)$ is nonsingular. From (9-130), we have, similar to (9-127),

$$\begin{aligned}\mathbf{D}_c^{-1}(s)\mathbf{D}_f(s)\mathbf{D}^{-1}(s) &= \mathbf{I} + \mathbf{D}_c^{-1}(s)\mathbf{L}(s) + \mathbf{D}_c^{-1}(s)\mathbf{M}(s)\mathbf{N}(s)\mathbf{D}^{-1}(s) \\ &= \mathbf{I} + \mathbf{C}_0(s) + \mathbf{C}_1(s)\hat{\mathbf{G}}(s)\end{aligned} \tag{9-137}$$

We write, similar to (9-83),

$$\mathbf{D}(s) = [\mathbf{D}_h + \mathbf{D}_l(s)]\mathbf{H}(s) \tag{9-138}$$
$$\mathbf{D}_c(s) = \mathbf{H}_c(s)[\mathbf{D}_{ch} + \mathbf{D}_{cl}(s)] \tag{9-139}$$

where $\mathbf{D}_l(s)$ and $\mathbf{D}_{cl}(s)$ are strictly proper. Then we have

$$\begin{aligned}\lim_{s \to \infty} \mathbf{D}_c^{-1}(s)\mathbf{D}_f(s)\mathbf{D}(s) &= \lim_{s \to \infty} [\mathbf{D}_{ch} + \mathbf{D}_{cl}(s)]^{-1}\mathbf{H}_c^{-1}(s)\mathbf{D}_f(s)\mathbf{H}^{-1}(s)[\mathbf{D}_h + \mathbf{D}_l(s)]^{-1} \\ &= \mathbf{D}_{ch}^{-1}\mathbf{J}\mathbf{D}_h^{-1}\end{aligned} \tag{9-140}$$

which is nonsingular by the assumptions of $\mathbf{D}_c(s)$, $\mathbf{D}(s)$, and (9-135). Hence the input-output feedback system in Figure 9-30 is, following Theorem 3-6, well posed.

Q.E.D.

In the design, the denominator matrix $\mathbf{D}_c(s)$ of the compensators can be chosen as

$$\mathbf{D}_c(s) = \text{diag}\,\{d_{c1}(s), d_{c2}(s), \dots, d_{cp}(s)\} \tag{9-141}$$

where $d_{ci}(s)$ are arbitrary Hurwitz polynomials of degrees m_i. If some or all m_i are odd integers, then we may not be able to assign complex conjugate roots. In this case, if we choose $D_c(s)$ to be of the form shown in Problem 2-29, then the difficulty of assigning complex conjugate roots will not arise.

Corollary 9-24

Consider a plant with $q \times p$ proper rational matrix $\hat{\mathbf{G}}(s) = \mathbf{N}(s)\mathbf{D}^{-1}(s)$. Let $\mu_i, i = 1, 2, \dots, p$, be the column degrees of $\mathbf{D}(s)$ and let v be the row index of $\hat{\mathbf{G}}(s)$. Then for any $\mathbf{D}_c(s)$ of row degrees all equal to $v-1$ and row reduced, and any $\bar{\mathbf{D}}_f(s)$ of column degrees $\mu_i, i = 1, 2, \dots, p$, and column reduced, there exist compensators with proper rational matrices $\mathbf{D}_c^{-1}(s)\mathbf{L}(s)$ and $\mathbf{D}_c^{-1}(s)\mathbf{M}(s)$ such that the feedback system in Figure 9-30 is well posed and has transfer matrix $\mathbf{N}(s)\bar{\mathbf{D}}_f^{-1}(s)$ if and only if $\mathbf{D}(s)$ and $\mathbf{N}(s)$ are right coprime and $\mathbf{D}(s)$ is column reduced.

Proof

Let $\mathbf{D}_f(s) = \mathbf{D}_c(s)\bar{\mathbf{D}}_f(s)$. Clearly the degree requirements of $\mathbf{D}_f(s)$ in Theorem 9-24 are met under the assumptions of $\mathbf{D}_c(s)$ and $\bar{\mathbf{D}}_f(s)$. With this $\mathbf{D}_f(s)$, (9-131) and $\hat{\mathbf{G}}_f(s)$ become

$$\mathbf{D}_c(s)[\bar{\mathbf{D}}_f(s) - \mathbf{D}(s)] = \mathbf{L}(s)\mathbf{D}(s) + \mathbf{M}(s)\mathbf{N}(s) \tag{9-142}$$

and

$$\hat{\mathbf{G}}_f(s) = \mathbf{N}(s)[\mathbf{D}_c(s)\bar{\mathbf{D}}_f(s)]^{-1}\mathbf{D}_c(s) = \mathbf{N}(s)\bar{\mathbf{D}}_f^{-1}(s) \tag{9-143}$$

This establishes the corollary.

Q.E.D.

The application of Theorem 9-24 and its corollary is straightforward. First we use the coefficient matrices of $\mathbf{D}(s)$ and $\mathbf{N}(s)$ to form the matrix $\mathbf{S}_m$ shown in (9-75). We then search the linearly independent rows of $\mathbf{S}_m$ in order from top to bottom. Let v be the least integer such that all N rows in the last block row of $\mathbf{S}_v$ are linearly dependent. This v is the row index of $\hat{\mathbf{G}}(s)$. For convenience, we assume $m_i = v-1$ for all i in Theorem 9-24. Let

$$\mathbf{L}(s) = \mathbf{L}_0 + \mathbf{L}_1 s + \cdots + \mathbf{L}_{v-1}s^{v-1} \tag{9-144}$$

$$\mathbf{M}(s) = \mathbf{M}_0 + \mathbf{M}_1 s + \cdots + \mathbf{M}_{v-1}s^{v-1} \tag{9-145}$$

and

$$\mathbf{E}(s) = \mathbf{D}_f(s) - \mathbf{D}_c(s)\mathbf{D}(s) = \mathbf{E}_0 + \mathbf{E}_1 s + \cdots + \mathbf{E}_{\mu+v-1}s^{\mu+v-1} \tag{9-146}$$

where $\mu = \max\,\{\mu_i, i = 1, 2, \dots, p\}$. The substitution of (9-144) to (9-146) into (9-131) yields

$$[\mathbf{L}_0 \quad \mathbf{M}_0 \,\vdots\, \mathbf{L}_1 \quad \mathbf{M}_1 \,\vdots\, \cdots \,\vdots\, \mathbf{L}_{v-1} \quad \mathbf{M}_{v-1}]\mathbf{S}_{v-1} = [\mathbf{E}_0 \quad \mathbf{E}_1 \quad \cdots \quad \mathbf{E}_{\mu+v-1}] \tag{9-147}$$

The solution of this set of linear algebraic equations yields the required compensators.

It is of interest to compare the result in Corollary 9-24 with the one developed in the state-variable approach. If $\mathbf{D}(s)$ and $\mathbf{N}(s)$ are right coprime and $\mathbf{D}(s)$ is column reduced, then an irreducible realization can be readily found for $\hat{\mathbf{G}}(s) = \mathbf{N}(s)\mathbf{D}^{-1}(s)$ (Section 6-6). By state feedback, we can achieve $\mathbf{N}(s)\bar{\mathbf{D}}_f^{-1}(s)$, where $\bar{\mathbf{D}}_f(s)$ and $\mathbf{D}(s)$ have the same column degrees and the same column-degree-coefficient matrix. Note that the latter condition is not required in Corollary 9-24. If the state is not available for feedback, we may design a state estimator with arbitrary eigenvalues, which are equivalent to the roots of det $\mathbf{D}_c(s)$, to generate an estimate of the state. Now the application of the state feedback from the output of the state estimator yields $\mathbf{N}(s)\bar{\mathbf{D}}_f^{-1}(s)$. Note that the eigenvalues of the estimator are not controllable as can be seen from (7-86) and will not appear in the transfer matrix from $\mathbf{r}$ to $\mathbf{y}$.

A remark is in order regarding the degrees of compensators. In the state-variable approach, the dimension of the state estimator is $n-q$ (Theorem 7-13). In the transfer-function approach, the degree of the compensators is $p(v-1)$. Note that the row index v of $\hat{\mathbf{G}}(s)$ is equal to the observability index of any irreducible realization of $\hat{\mathbf{G}}(s)$, and has the property $v \geq n/q$. If $p=q=1$, then $n=v$ and the dimension of the state estimator is equal to the degree of compensators. If $p \geq q$, then $p(v-1) \geq n-q$; if $p<q$, $p(v-1)$ can be greater than, equal to, or less than $n-q$. Hence the results in the state-variable and transfer-function approaches are not exactly identical. Similar to the single-variable case, the design procedure in the transfer-function approach appears to be simpler, conceptually and computationally, than the one in the state-variable approach.

In the following, the input-output feedback system in Figure 9-30 will be extended to the one shown in Figure 9-31, in which $\mathbf{Q}(s)$ is a polynomial matrix and $\mathbf{Q}^{-1}(s)$ is required to be proper. The transfer matrix from $\mathbf{r}$ to $\mathbf{y}$ in Figure 9-31 can be readily computed as

$$\hat{\mathbf{G}}_f(s) = \hat{\mathbf{G}}(s)\mathbf{Q}^{-1}(s)[\mathbf{I} + \mathbf{C}_0(s)\mathbf{Q}^{-1}(s) + \mathbf{C}_1(s)\hat{\mathbf{G}}(s)\mathbf{Q}^{-1}(s)]^{-1} \qquad \textbf{(9-148)}$$

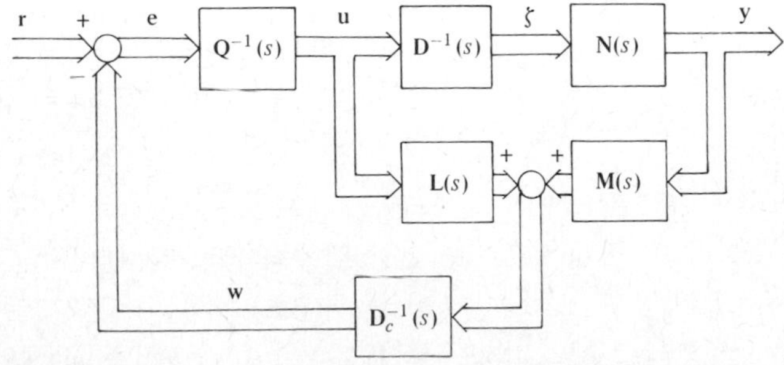

Figure 9-31 An input-output feedback system.

Using $\hat{\mathbf{G}}(s)=\mathbf{N}(s)\mathbf{D}^{-1}(s)$, $\mathbf{C}_0(s)=\mathbf{D}_c^{-1}(s)\mathbf{L}(s)$ and $\mathbf{C}_1(s)=\mathbf{D}_c^{-1}(s)\mathbf{M}(s)$, $\hat{\mathbf{G}}_f(s)$ can be written as

$$\hat{\mathbf{G}}_f(s)=\mathbf{N}(s)[\mathbf{D}_c(s)\mathbf{Q}(s)\mathbf{D}(s)+\mathbf{L}(s)\mathbf{D}(s)+\mathbf{M}(s)\mathbf{N}(s)]^{-1}\mathbf{D}_c(s) \tag{9-149}$$

If we define

$$\mathbf{D}_f(s)\triangleq\mathbf{D}_c(s)\mathbf{Q}(s)\mathbf{D}(s)+\mathbf{L}(s)\mathbf{D}(s)+\mathbf{M}(s)\mathbf{N}(s) \tag{9-150}$$

and

$$\mathbf{E}(s)\triangleq\mathbf{D}_f(s)-\mathbf{D}_c(s)\mathbf{Q}(s)\mathbf{D}(s)=\mathbf{L}(s)\mathbf{D}(s)+\mathbf{M}(s)\mathbf{N}(s) \tag{9-151}$$

then $\hat{\mathbf{G}}_f(s)=\mathbf{N}(s)\mathbf{D}_f^{-1}(s)\mathbf{D}_c(s)$ and has the same form as (9-132). The design in Theorem 9-24 to achieve (9-132) is accomplished without using $\mathbf{Q}^{-1}(s)$; however, the row degrees of $\mathbf{D}_c(s)$ are generally different. In the following, all row degrees of $\mathbf{D}_c(s)$ will be required to be the same. This is possible because of the introduction of $\mathbf{Q}^{-1}(s)$. In this case, $\mathbf{H}_c(s)$ in (9-134) becomes diag $\{s^m, s^m, \dots, s^m\}$.

Theorem 9-25

Consider a plant with $q\times p$ proper rational matrix $\hat{\mathbf{G}}(s)=\mathbf{N}(s)\mathbf{D}^{-1}(s)$. It is assumed that $\mathbf{D}(s)$ and $\mathbf{N}(s)$ are right coprime and $\mathbf{D}(s)$ is column reduced and of column degrees μ_i, $i=1,2,\dots,p$. Let v be the row index of $\hat{\mathbf{G}}(s)$. Let $\mathbf{D}_c(s)$ be a $p\times p$ arbitrary polynomial matrix of row degrees all equal to m with $m\geq v-1$ and be row reduced. Then for any $\mathbf{D}_f(s)$ with the property that

$$\mathbf{H}(s)\mathbf{D}_f^{-1}(s)\mathbf{H}_c(s) \tag{9-152}$$

is a proper rational matrix, there exist proper compensators $\mathbf{D}_c^{-1}(s)\mathbf{L}(s)$, $\mathbf{D}_c^{-1}(s)\mathbf{M}(s)$, and $\mathbf{Q}^{-1}(s)$ so that the feedback system in Figure 9-31 is well posed and has transfer matrix $\mathbf{N}(s)\mathbf{D}_f^{-1}(s)\mathbf{D}_c(s)$. Furthermore, $\mathbf{Q}(s)$ can be computed from

$$\mathbf{D}_f(s)=\mathbf{Q}_1(s)\mathbf{D}(s)+\mathbf{R}_1(s) \tag{9-153}$$

with $\delta_{ci}\mathbf{R}_1(s)<\delta_{ci}\mathbf{D}(s)=\mu_i$, and

$$\mathbf{Q}_1(s)=\mathbf{D}_c(s)\mathbf{Q}(s)+\mathbf{R}_2(s) \tag{9-154}$$

with $\delta_{ri}\mathbf{R}_2(s)<\delta_{ri}\mathbf{D}_c(s)=m$, and $\mathbf{L}(s)$ and $\mathbf{M}(s)$ are solutions of

$$\mathbf{R}_2(s)\mathbf{D}(s)+\mathbf{R}_1(s)=\mathbf{L}(s)\mathbf{D}(s)+\mathbf{M}(s)\mathbf{N}(s) \tag{9-155}$$

Proof

The column degrees of $\mathbf{R}_2(s)\mathbf{D}(s)+\mathbf{R}_1(s)$ are clearly at most $m+\mu_i$. Hence $\mathbf{L}(s)$ and $\mathbf{M}(s)$ of row degrees at most m exist in (9-155) (Theorem 9-17). Thus $\mathbf{D}_c^{-1}(s)\mathbf{L}(s)$ and $\mathbf{D}_c^{-1}(s)\mathbf{M}(s)$ are proper.

Next we show that $\mathbf{D}(s)\mathbf{D}_f^{-1}(s)\mathbf{D}_c(s)$ is proper under the assumption of (9-152). We use (9-138) and (9-139) with $m=m_i$, for all i, to write

$$\mathbf{D}(s)\mathbf{D}_f^{-1}(s)\mathbf{D}_c(s)=[\mathbf{D}_h+\mathbf{D}_l(s)]\mathbf{H}(s)\mathbf{D}_f^{-1}(s)\mathbf{H}_c(s)[\mathbf{D}_{ch}+\mathbf{D}_{cl}(s)] \tag{9-156}$$

Since $\mathbf{D}_h+\mathbf{D}_l(s)$ and $\mathbf{D}_{ch}+\mathbf{D}_{cl}(s)$ are proper, if (9-152) holds, then $\mathbf{D}(s)\mathbf{D}_f^{-1}(s)\mathbf{D}_c(s)$ is proper.

From (9-153) and (9-154), we have

$$\mathbf{D}_f(s) = \mathbf{D}_c(s)\mathbf{Q}(s)\mathbf{D}(s) + \mathbf{R}_2(s)\mathbf{D}(s) + \mathbf{R}_1(s) \tag{9-157}$$

which implies

$$\mathbf{D}_c^{-1}(s)\mathbf{D}_f(s)\mathbf{D}^{-1}(s) = \mathbf{Q}(s) + \mathbf{D}_c^{-1}(s)\mathbf{R}_2(s) + \mathbf{D}_c^{-1}(s)\mathbf{R}_1(s)\mathbf{D}^{-1}(s) \tag{9-158}$$

Because $\mathbf{D}(s)$ is column reduced and $\delta_{ci}\mathbf{D}(s) > \delta_{ci}\mathbf{R}_1(s)$, $\mathbf{R}_1(s)\mathbf{D}^{-1}(s)$ is strictly proper. Because $\mathbf{D}_c(s)$ is row reduced and $\delta_{ri}\mathbf{D}_c(s) > \delta_{ri}\mathbf{R}_2(s)$, $\mathbf{D}_c^{-1}(s)\mathbf{I}$ is proper and $\mathbf{D}_c^{-1}(s)\mathbf{R}_2(s)$ is strictly proper. Hence the polynomial part of $\mathbf{D}_c^{-1}(s)\mathbf{D}_f(s)\mathbf{D}^{-1}(s)$ is $\mathbf{Q}(s)$. Since $[\mathbf{D}_c^{-1}(s)\mathbf{D}_f(s)\mathbf{D}^{-1}(s)]^{-1}$ is proper, it follows from Theorem 3-4 that $\mathbf{Q}^{-1}(s)$ is proper.

The feedback system is well posed if and only if the matrix

$$\begin{aligned}\mathbf{P}^{-1}(s) &\triangleq [\mathbf{I} + \mathbf{D}_c^{-1}(s)\mathbf{L}(s)\mathbf{Q}^{-1}(s) + \mathbf{D}_c^{-1}(s)\mathbf{M}(s)\mathbf{N}(s)\mathbf{D}^{-1}(s)\mathbf{Q}^{-1}(s)]^{-1} \\ &= \mathbf{Q}(s)\mathbf{D}(s)[\mathbf{D}_c(s)\mathbf{Q}(s)\mathbf{D}(s) + \mathbf{L}(s)\mathbf{D}(s) + \mathbf{M}(s)\mathbf{N}(s)]^{-1}\mathbf{D}_c(s)\end{aligned} \tag{9-159}$$

is proper (Theorem 3-6). The substitution of (9-150) into (9-159) yields

$$\mathbf{P}^{-1}(s) = \mathbf{Q}(s)\mathbf{D}(s)\mathbf{D}_f^{-1}(s)\mathbf{D}_c(s) \tag{9-160}$$

which implies

$$\mathbf{P}(s) = \mathbf{D}_c^{-1}(s)\mathbf{D}_f(s)\mathbf{D}^{-1}(s)\mathbf{Q}^{-1}(s) \tag{9-161}$$

Using (9-153) and (9-154), we have

$$\begin{aligned}\mathbf{P}(s) &= \mathbf{D}_c^{-1}(s)[\mathbf{D}_c(s)\mathbf{Q}(s)\mathbf{D}(s) + \mathbf{R}_2(s)\mathbf{D}(s) + \mathbf{R}_1(s)]\mathbf{D}^{-1}(s)\mathbf{Q}^{-1}(s) \\ &= \mathbf{I} + \mathbf{D}_c^{-1}(s)\mathbf{R}_2(s)\mathbf{Q}^{-1}(s) + \mathbf{D}_c^{-1}(s)\mathbf{R}_1(s)\mathbf{D}^{-1}(s)\mathbf{Q}^{-1}(s)\end{aligned} \tag{9-162}$$

Because $\mathbf{D}_c^{-1}(s)\mathbf{R}_2(s)$ and $\mathbf{R}_1(s)\mathbf{D}^{-1}(s)$ are strictly proper, $\mathbf{D}_c^{-1}(s)$ and $\mathbf{Q}^{-1}(s)$ are proper, the polynomial part of $\mathbf{P}(s)$ is $\mathbf{I}$. Hence $\mathbf{P}^{-1}(s)$ is, following Theorem 3-4, proper. This establishes the well-posedness of the feedback system. Q.E.D.

We remark the condition in (9-152). One way to check the properness of $\mathbf{H}(s)\mathbf{D}_f^{-1}(s)\mathbf{H}_c(s)$ is by direct computation. The computation of the inverse of $\mathbf{D}_f(s)$ is, however, complicated. Instead, we may compute

$$\mathbf{V}(s) \triangleq \text{the polynomial part of } \mathbf{H}_c^{-1}(s)\mathbf{D}_f(s)\mathbf{H}^{-1}(s)$$

which can be obtained by inspection because of the forms of $\mathbf{H}_c(s)$ and $\mathbf{H}(s)$. Clearly $\mathbf{V}(s)$ is much simpler than $\mathbf{D}_f(s)$. Now $\mathbf{H}(s)\mathbf{D}_f^{-1}(s)\mathbf{H}_c(s)$ is proper if and only if $\mathbf{V}^{-1}(s)$ is proper (Theorem 3-4). This is a simpler way of checking the condition in (9-152). Because of $\mathbf{H}_c(s) = \text{diag}\{s^m, s^m, \ldots, s^m\}$, a sufficient condition for $\mathbf{V}^{-1}(s)$ to be proper is that $\mathbf{D}_f(s)$ has column degrees $m + \mu_i$ and is column reduced. Even if $\mathbf{D}_f(s)$ is not column reduced, it is still possible for $\mathbf{V}(s)$ to have a proper inverse. See the footnote on page 115.

The designs in Theorems 9-24 and 9-25 yield

$$\begin{aligned}\hat{\mathbf{G}}_f(s) &= \mathbf{N}(s)\mathbf{D}_f^{-1}(s)\mathbf{D}_c(s) = \mathbf{N}(s)\mathbf{D}^{-1}(s)\mathbf{D}(s)\mathbf{D}_f^{-1}(s)\mathbf{D}_c(s) \\ &= \hat{\mathbf{G}}(s)\mathbf{T}(s)\end{aligned} \tag{9-163}$$

where $\mathbf{T}(s) \triangleq \mathbf{D}(s)\mathbf{D}_f^{-1}(s)\mathbf{D}_c(s)$ is, as proved in (9-156), a proper rational matrix. It turns out that every system expressible as $\hat{\mathbf{G}}(s)\mathbf{T}(s)$, where $\mathbf{T}(s)$ is an arbitrary proper rational matrix, can be implemented as shown in Figure 9-31. This problem will be studied in the next section.

Implementations of open-loop compensators.[19] Consider the open-loop system shown in Figure 9-32, where the plant is denoted by the $q \times p$ proper rational matrix $\hat{\mathbf{G}}(s)$ and $\mathbf{T}(s)$ is an arbitrary proper compensator. In this subsection, we discuss the implementations of this open loop system by feedback configurations. Before proceeding, we mention that if $\hat{\mathbf{G}}(s) = \mathbf{N}(s)\mathbf{D}^{-1}(s)$ and if $\mathbf{T}(s)$ is chosen as $\mathbf{D}(s)\mathbf{D}_f^{-1}(s)$ and is proper, then the open-loop transfer matrix is $\hat{\mathbf{G}}_0(s) = \hat{\mathbf{G}}(s)\mathbf{T}(s) = \mathbf{N}(s)\mathbf{D}_f^{-1}(s)$. This is the problem of arbitrary assignment of denominator matrix discussed in Corollary 9-24. If $\hat{\mathbf{G}}(s)$ is square and nonsingular and if we choose $\mathbf{T}(s)$ as

$$\mathbf{T}(s) = \hat{\mathbf{G}}^{-1}(s) \operatorname{diag} \{d_1(s), d_2(s), \ldots, d_p(s)\} \tag{9-164}$$

where $d_i(s)$ are Hurwitz polynomials of smallest possible degrees to make $\mathbf{T}(s)$ proper, then the resulting system is decoupled. Thus the problem to be discussed may find several applications in the design of multivariable systems.

Let $\hat{\mathbf{G}}(s) = \mathbf{N}(s)\mathbf{D}^{-1}(s)$ be a coprime fraction with $\mathbf{D}(s)$ column reduced. Then the open-loop transfer matrix is

$$\hat{\mathbf{G}}_o(s) = \hat{\mathbf{G}}(s)\mathbf{T}(s) = \mathbf{N}(s)\mathbf{D}^{-1}(s)\mathbf{T}(s) \triangleq \mathbf{N}(s)\mathbf{D}_f^{-1}(s)\mathbf{N}_f(s) \tag{9-165}$$

where

$$\mathbf{D}^{-1}(s)\mathbf{T}(s) = \mathbf{D}_f^{-1}(s)\mathbf{N}_f(s) \tag{9-166}$$

and $\mathbf{D}_f(s)$ and $\mathbf{N}_f(s)$ are left coprime. If we use the unity feedback system shown in Figure 9-16, the overall transfer matrix is $\mathbf{N}(s)\mathbf{D}_f^{-1}(s)\mathbf{N}_c(s)$ and $\mathbf{N}_c(s)$ is to be solved from a Diophantine equation and cannot be arbitrarily chosen; hence the unity feedback system in Figure 9-16 cannot be used to implement the open-loop system. If we use the input-output feedback system in Figure 9-31, the overall transfer matrix is $\mathbf{N}(s)\mathbf{D}_f^{-1}(s)\mathbf{D}_c(s)$, where $\mathbf{D}_c(s)$ is arbitrarily assignable. A comparison of this with (9-165) reveals immediately the possibility of implementing $\hat{\mathbf{G}}_0(s)$ in Figure 9-31 as long as the conditions on $\mathbf{D}_c(s)$ and $\mathbf{D}_f(s)$ are met. This will be done in Implementation I. We shall also introduce a different implementation in Implementation II.

Implementation I. Consider the input-output feedback system shown in Figure 9-31. Its overall transfer matrix is, from (9-148) and (9-149),

$$\hat{\mathbf{G}}_f(s) = \hat{\mathbf{G}}(s)\mathbf{Q}^{-1}(s)[\mathbf{I} + \mathbf{D}_c^{-1}(s)\mathbf{L}(s)\mathbf{Q}^{-1}(s) + \mathbf{D}_c^{-1}(s)\mathbf{M}(s)\mathbf{N}(s)\mathbf{D}^{-1}(s)\mathbf{Q}^{-1}(s)]^{-1} \tag{9-167}$$

[19] This problem was first formulated and solved in Reference S218. This presentation follows Reference S50.

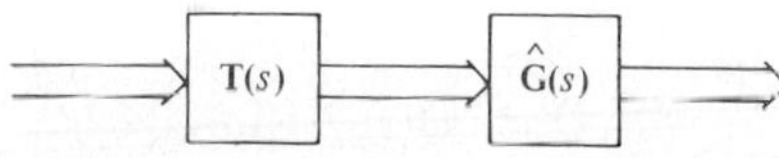

Figure 9-32 An open-loop system.

or $$\hat{\mathbf{G}}_f(s)=\mathbf{N}(s)[\mathbf{D}_c(s)\mathbf{Q}(s)\mathbf{D}(s)+\mathbf{L}(s)\mathbf{D}(s)+\mathbf{M}(s)\mathbf{N}(s)]^{-1}\mathbf{D}_c(s) \tag{9-168}$$

where $\mathbf{D}_c(s)$ can be arbitrarily chosen. Hence this configuration may be used to implement $\hat{\mathbf{G}}(s)\mathbf{T}(s)=\mathbf{N}(s)\mathbf{D}_f^{-1}(s)\mathbf{N}_f(s)$. In this implementation, we require $\mathbf{T}(s)$ to be nonsingular. The nonsingularity of $\mathbf{T}(s)$ implies the nonsingularity of $\mathbf{N}_f(s)$. Consequently we can always find a unimodular $\mathbf{U}(s)$ such that $\mathbf{U}(s)\mathbf{N}_f(s)$ is row reduced and $\mathbf{U}(s)\mathbf{D}_f(s)$ and $\mathbf{U}(s)\mathbf{N}_f(s)$ remain to be let coprime (Theorem G-11). Thus we assume without loss of generality that $\mathbf{N}_f(s)$ is row reduced.

In the following, we present a procedure so that $\hat{\mathbf{G}}_f(s)$ will implement $\hat{\mathbf{G}}(s)\mathbf{T}(s)$.

Step 1. Compute a fraction $\hat{\mathbf{G}}(s)=\mathbf{N}(s)\mathbf{D}^{-1}(s)$, where $\mathbf{D}(s)$ and $\mathbf{N}(s)$ are right coprime and $\mathbf{D}(s)$ is column reduced. Compute

$$\mathbf{D}^{-1}(s)\mathbf{T}(s)=\mathbf{D}_f^{-1}(s)\mathbf{N}_f(s) \tag{9-169}$$

where $\mathbf{D}_f(s)$ and $\mathbf{N}_f(s)$ are left coprime and $\mathbf{N}_f(s)$ is row reduced. Let

$$\delta_{ri}\mathbf{N}_f(s)=m_i, \quad i=1,2,\ldots,p$$

where δ_{ri} denotes the ith row degree.

Step 2. Compute the row index, v, of $\hat{\mathbf{G}}(s)$. Define

$$m=\max\{v-1, m_1, m_2, \ldots, m_p\}$$

Let

$$\bar{\mathbf{D}}_c(s)=\text{diag}\{\alpha_1(s), \alpha_2(s), \ldots, \alpha_p(s)\}$$

where $\alpha_i(s)$ is an arbitrary polynomial of degree $m-m_i$. Then the matrix

$$\mathbf{D}_c(s)=\bar{\mathbf{D}}_c(s)\mathbf{N}_f(s) \tag{9-170}$$

has row degrees all equal to $m \geq v-1$ and is row reduced.

Step 3. If

$$\delta_{ci}[\bar{\mathbf{D}}_c(s)\mathbf{D}_f(s)] \leq \mu_i+m \qquad i=1,2,\ldots,p \tag{9-171}$$

set $\mathbf{Q}(s)=\mathbf{I}$ and go to step 4. If not, compute

$$\mathbf{D}_f(s)=\mathbf{Q}_1(s)\mathbf{D}(s)+\mathbf{R}_1(s) \tag{9-172}$$

with $\delta_{ci}\mathbf{R}_1(s)<\delta_{ci}\mathbf{D}(s)=\mu_i$, for all i, and compute

$$\mathbf{Q}_1(s)=\mathbf{N}_f(s)\mathbf{Q}(s)+\mathbf{R}_2(s) \tag{9-173}$$

with $\delta_{ri}\mathbf{R}_2(s)<\delta_{ri}\mathbf{N}_f(s)=m_i$, $i=1,2,\ldots,p$. These decompositions are unique.

Step 4. Solve $\mathbf{L}(s)$ and $\mathbf{M}(s)$ from

$$\bar{\mathbf{D}}_c(s)[\mathbf{D}_f(s)-\mathbf{N}_f(s)\mathbf{D}(s)]=\mathbf{L}(s)\mathbf{D}(s)+\mathbf{M}(s)\mathbf{N}(s) \tag{9-174a}$$

or $$\bar{\mathbf{D}}_c(s)[\mathbf{R}_2(s)\mathbf{D}(s)+\mathbf{R}_1(s)]=\mathbf{L}(s)\mathbf{D}(s)+\mathbf{M}(s)\mathbf{N}(s) \tag{9-174b}$$

Theorem 9-26

The input-output feedback system in Figure 9-31 with $\mathbf{D}_c(s)$, $\mathbf{Q}(s)$, $\mathbf{L}(s)$, and $\mathbf{M}(s)$, computed from (9-170) and (9-174) implements the open-loop system $\hat{\mathbf{G}}_0(s)=\hat{\mathbf{G}}(s)\mathbf{T}(s)=\mathbf{N}(s)\mathbf{D}_f^{-1}(s)\mathbf{N}_f(s)$ and is well posed.

Proof

The substitution of (9-170) to (9-174) into (9-168) yields

$$\hat{\mathbf{G}}_f(s)=\mathbf{N}(s)[\bar{\mathbf{D}}_c(s)\mathbf{D}_f(s)]^{-1}\bar{\mathbf{D}}_c(s)\mathbf{N}_f(s)=\mathbf{N}(s)\mathbf{D}_f^{-1}(s)\mathbf{N}_f(s) \tag{9-175}$$

Hence $\hat{\mathbf{G}}_f(s)$ implements the open-loop system.

Equations (9-169) and (9-170) imply

$$\mathbf{T}(s)=\mathbf{D}(s)\mathbf{D}_f^{-1}(s)\mathbf{N}_f(s)=\mathbf{D}(s)[\bar{\mathbf{D}}_c(s)\mathbf{D}_f(s)]^{-1}\mathbf{D}_c(s) \tag{9-176}$$

Note that $\bar{\mathbf{D}}_c(s)\mathbf{D}_f(s)$ in (9-175) corresponds to the $\mathbf{D}_f(s)$ in Theorem 9-25. Using (9-138) and (9-139) with $m_i=m$, for all i, we can write (9-176) as

$$[\mathbf{D}_h+\mathbf{D}_l(s)]^{-1}\mathbf{T}(s)[\mathbf{D}_{ch}+\mathbf{D}_{cl}(s)]^{-1}=\mathbf{H}(s)[\bar{\mathbf{D}}_c(s)\mathbf{D}_f(s)]^{-1}\mathbf{H}_c(s) \tag{9-177}$$

Since $\mathbf{D}_h$ and $\mathbf{D}_{ch}$ are nonsingular by assumption, $[\mathbf{D}_h+\mathbf{D}_l(s)]^{-1}$ and $[\mathbf{D}_{ch}+\mathbf{D}_{cl}(s)]^{-1}$ are, following Corollary 3-4, proper. Hence, if $\mathbf{T}(s)$ is proper, the condition in (9-152) is satisfied and the theorem follows directly from Theorem 9-25. Q.E.D.

We discuss now the stability of the implementation. Clearly, the stability depends on $\mathbf{T}(s)$. If $\mathbf{T}(s)$ is chosen so that $\mathbf{T}(s)$ and $\hat{\mathbf{G}}(s)\mathbf{T}(s)$ are BIBO stable, then $\det \mathbf{D}_f(s)$ in (9-169) is a Hurwitz polynomial, and the feedback implementation will also be BIBO stable. The design involves the cancellation of the roots of $\det \bar{\mathbf{D}}_c(s)$ which are however arbitrarily assignable. If $\det \bar{\mathbf{D}}_c(s)$ is chosen as Hurwitz, then the feedback implementation is asymptotically stable as well.

The decompositions in (9-172) and (9-173) can be carried out by using the algorithms in References S34, S137 and S236. See also Problem G-15. In the design, the conditions in (9-172) and (9-173) can be relaxed. We may replace (9-172) by

$$\mathbf{D}_f(s)=\bar{\mathbf{Q}}_1(s)\mathbf{D}(s)+\bar{\mathbf{R}}_1(s) \tag{9-178}$$

with $\delta_{ci}\bar{\mathbf{R}}_1(s)\le\delta_{ci}\mathbf{D}(s)$. In this case, the decomposition is not unique. For example, from the unique $\mathbf{Q}_1(s)$ and $\mathbf{R}_1(s)$ in (9-172), we can obtain

$$\bar{\mathbf{Q}}_1(s)=\mathbf{Q}_1(s)+\mathbf{B} \qquad \bar{\mathbf{R}}_1(s)=\mathbf{R}(s)-\mathbf{B}\mathbf{D}(s) \tag{9-179}$$

for any constant matrix $\mathbf{B}$, to meet (9-178). Similarly, (9-173) can be replaced by

$$\bar{\mathbf{Q}}_1(s)=\mathbf{N}_f(s)\bar{\mathbf{Q}}(s)+\bar{\mathbf{R}}_2(s) \tag{9-180}$$

with $\delta_{ri}\bar{\mathbf{R}}_2(s)\le\delta_{ri}\mathbf{N}_f(s)$. In this case, we cannot expect every $\bar{\mathbf{Q}}_1^{-1}(s)$ to be proper. However, for almost all $\bar{\mathbf{Q}}_1(s)$ so computed, $\bar{\mathbf{Q}}_1^{-1}(s)$ is still proper. Furthermore, the resulting $\mathbf{D}_c(s)[\mathbf{D}_f(s)-\mathbf{N}_f(s)\bar{\mathbf{Q}}(s)\mathbf{D}(s)]$ still meets the column degree conditions in Theorem 9-17; hence solutions $\mathbf{L}(s)$ and $\mathbf{M}(s)$ of row degrees at most m still exist in (9-174).

Implementation II. We introduce a different implementation in this subsection. In this implementation, the open-loop compensator $\mathbf{T}(s)$ may be nonsquare and may not have a full rank. The configuration is identical to the one in Figure 9-31 except that a compensator $\mathbf{D}_c^{-1}(s)\mathbf{K}(s)$ is placed at the input terminal as shown in Figure 9-33. The transfer matrix of the system is clearly

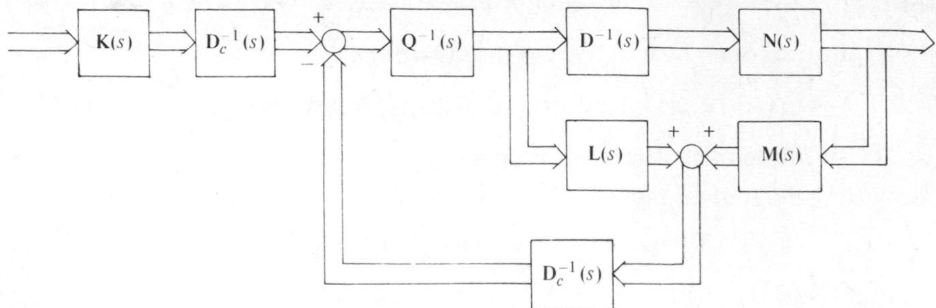

Figure 9-33 A plant input-output feedback system.

equal to, following (9-167) and (9-168),

$$\mathbf{G}_f(s)=\mathbf{G}(s)\mathbf{Q}^{-1}(s)[\mathbf{I}+\mathbf{D}_c^{-1}(s)\mathbf{L}(s)\mathbf{Q}^{-1}(s)+\mathbf{D}_c^{-1}(s)\mathbf{M}(s)\mathbf{G}(s)\mathbf{Q}^{-1}(s)]^{-1}\mathbf{D}_c^{-1}(s)\mathbf{K}(s) \tag{9-181}$$

or
$$\mathbf{G}_f(s)=\mathbf{N}(s)[\mathbf{D}_c(s)\mathbf{Q}(s)\mathbf{D}(s)+\mathbf{L}(s)\mathbf{D}(s)+\mathbf{M}(s)\mathbf{N}(s)]^{-1}\mathbf{K}(s) \tag{9-182}$$

The implementation consists of the following steps:

Step 1. Compute a fraction $\hat{\mathbf{G}}(s)=\mathbf{N}(s)\mathbf{D}^{-1}(s)$, where $\mathbf{D}(s)$ and $\mathbf{N}(s)$ are right coprime and $\mathbf{D}(s)$ is column reduced. Compute

$$\mathbf{D}^{-1}(s)\mathbf{T}(s)=\mathbf{D}_f^{-1}(s)\mathbf{N}_f(s) \tag{9-183}$$

where $\mathbf{D}_f(s)$ and $\mathbf{N}_f(s)$ are left coprime. Furthermore, we require $\mathbf{D}_f(s)$ to be column reduced and to have $\mathbf{D}_{fh}=\mathbf{I}$ (the column-degree-coefficient matrix is the unit matrix).

Step 2. Let $\delta_{ci}\mathbf{D}_f(s)=f_i$, and $\delta_{ci}\mathbf{D}(s)=\mu_i$, $i=1, 2, \ldots, p$. Define

$$\mathbf{V}(s)=\text{diag}\,\{v_1(s), v_2(s), \ldots, v_p(s)\} \tag{9-184}$$

If $f_i \geq \mu_i$, set $v_i(s)=1$. If $f_i<\mu_i$, set $v_i(s)$ as a monic arbitrary Hurwitz polynomial of degree μ_i-f_i. We then write

$$\begin{aligned}\mathbf{D}^{-1}(s)\mathbf{T}(s)&=\mathbf{D}_f^{-1}(s)\mathbf{V}^{-1}(s)\mathbf{V}(s)\mathbf{N}_f(s)=(\mathbf{V}(s)\mathbf{D}_f(s))^{-1}(\mathbf{V}(s)\mathbf{N}_f(s))\\&\triangleq\bar{\mathbf{D}}_f^{-1}(s)\bar{\mathbf{N}}_f(s)\end{aligned} \tag{9-185}$$

where $\bar{\mathbf{D}}_f(s)\triangleq\mathbf{V}(s)\mathbf{D}_f(s)$ and $\bar{\mathbf{N}}_f(s)=\mathbf{V}(s)\mathbf{N}_f(s)$. Let $\delta_{ci}\bar{\mathbf{D}}_f(s)=\bar{f}_i$. Because of the assumption $\mathbf{D}_{fh}=\mathbf{I}$, we may conclude $\bar{f}_i\geq\mu_i$.[20]

Step 3. Compute the row index, v, of $\hat{\mathbf{G}}(s)$. Let $\mathbf{D}_c(s)$ be an arbitrary polynomial matrix of row degrees all equal to m and row reduced such that $m\geq v-1$ and $\mathbf{D}_c^{-1}\bar{\mathbf{N}}_f(s)$ is proper.

[20] Because the properness of $\mathbf{T}(s)=\mathbf{D}(s)\mathbf{D}_f^{-1}(s)\mathbf{N}_f(s)$ does not imply the properness of $\mathbf{D}(s)\mathbf{D}_f^{-1}(s)$ (see Problem 3-42), we cannot conclude that $f_i\geq\mu_i$. Consequently, we must introduce $v_i(s)$ to conclude $\bar{f}_i\geq\mu_i$. However, in most cases, for example, when $\mathbf{T}(s)$ has a full rank, we have $v_i(s)=1$, for all i.

Step 4. Define

$$\mathbf{W}(s) = \text{diag}\,\{w_1(s), w_2(s), \ldots, w_p(s)\} \tag{9-186}$$

If $\bar{f}_i - \mu_i - m \geq 0$, set $w_i(s) = 1$. If $\bar{f}_i - \mu_i - m < 0$, set $w_i(s)$ as a monic arbitrary Hurwitz polynomial of degree $(\mu_i + m - \bar{f}_i)$. Because of $\bar{f}_i \geq \mu_i$, we have $\delta_{ri}\mathbf{W}(s) \leq m$.

Step 5. If

$$\delta_{ci}[\mathbf{W}(s)\bar{\mathbf{D}}_f(s)] \leq \mu_i + m \qquad i = 1, 2, \ldots, p \tag{9-187}$$

set $\mathbf{Q}(s) = \mathbf{I}$ and go to step 6. If not, compute

$$\bar{\mathbf{D}}_f(s) = \mathbf{Q}_1(s)\mathbf{D}(s) + \mathbf{R}_1(s) \tag{9-188}$$

with $\delta_{ci}\mathbf{R}_1(s) < \delta_{ci}\mathbf{D}(s) = \mu_i$, and compute

$$\mathbf{W}(s)\mathbf{Q}_1(s) = \mathbf{D}_c(s)\mathbf{Q}(s) + \mathbf{R}_2(s) \tag{9-189}$$

with $\delta_{ri}\mathbf{R}_2(s) < \delta_{ri}\mathbf{D}_c(s) = m$.

Step 6. Let

$$\mathbf{K}(s) = \mathbf{W}(s)\bar{\mathbf{N}}_f(s) = \mathbf{W}(s)\mathbf{V}(s)\mathbf{N}_f(s) \tag{9-190}$$

and solve $\mathbf{L}(s)$ and $\mathbf{M}(s)$ from

$$\mathbf{W}(s)\bar{\mathbf{D}}_f(s) - \mathbf{D}_c(s)\mathbf{Q}(s)\mathbf{D}(s) = \mathbf{L}(s)\mathbf{D}(s) + \mathbf{M}(s)\mathbf{N}(s) \tag{9-191}$$

Theorem 9-27

The feedback system in Figure 9-33 obtained from (9-184) to (9-191) implements the open-loop system $\hat{\mathbf{G}}(s)\mathbf{T}(s) = \mathbf{N}(s)\mathbf{D}_f^{-1}(s)\mathbf{N}_f(s)$ and is well posed.

Proof

The substitution of $\mathbf{K}(s) = \mathbf{W}(s)\bar{\mathbf{N}}_f(s)$, (9-186), and (9-191) into (9-182) yields

$$\begin{aligned}\hat{\mathbf{G}}_f(s) &= \mathbf{N}(s)[\mathbf{W}(s)\bar{\mathbf{D}}_f(s)]^{-1}\mathbf{W}(s)\bar{\mathbf{N}}_f(s) = \mathbf{N}(s)\bar{\mathbf{D}}_f^{-1}(s)\bar{\mathbf{N}}_f(s)\\ &= \mathbf{N}(s)\mathbf{D}^{-1}(s)\mathbf{T}(s) = \mathbf{G}(s)\mathbf{T}(s)\end{aligned}$$

This shows that the feedback system in Figure 9-33 does implement the open-loop compensator $\mathbf{T}(s)$.

From (9-188) and (9-189), we have

$$\mathbf{W}(s)\bar{\mathbf{D}}_f(s) - \mathbf{D}_c(s)\mathbf{Q}(s)\mathbf{D}(s) = \mathbf{R}_2(s)\mathbf{D}(s) + \mathbf{W}(s)\mathbf{R}_1(s) \tag{9-192}$$

Because $\delta_{ri}\mathbf{R}_2(s) < m$, $\delta_{ci}\,\mathbf{D}(s) = \mu_i$, $\delta_{ri}\mathbf{W}(s) \leq m$, and $\delta_{ci}\mathbf{R}_1(s) < \mu_i$, we have

$$\delta_{ci}[\mathbf{W}(s)\bar{\mathbf{D}}_f(s) - \mathbf{D}_c(s)\mathbf{Q}(s)\mathbf{D}(s)] \leq m + \mu_i \qquad \text{for all } i \tag{9-193}$$

Hence under the assumption that $\mathbf{D}(s)$ and $\mathbf{N}(s)$ right coprime, $\mathbf{D}(s)$ column reduced and $m \geq \nu - 1$, solutions $\mathbf{L}(s)$ and $\mathbf{M}(s)$ of degrees at most m exist in (9-191). Hence the compensators $\mathbf{D}_c^{-1}(s)\mathbf{L}(s)$ and $\mathbf{D}_c^{-1}(s)\mathbf{M}(s)$ are proper.

The polynomial equation in (9-192) can be written as

$$\mathbf{D}_c^{-1}(s)\mathbf{W}(s)\bar{\mathbf{D}}_f(s)\mathbf{D}^{-1}(s)=\mathbf{Q}(s)+\mathbf{D}_c^{-1}(s)\mathbf{R}_2(s)+\mathbf{D}_c^{-1}(s)\mathbf{W}(s)\mathbf{R}_1(s)\mathbf{D}^{-1}(s) \tag{9-194}$$

Because $\mathbf{D}_c^{-1}(s)\mathbf{R}_2(s)$ and $\mathbf{R}_1(s)\mathbf{D}^{-1}(s)$ are strictly proper and $\mathbf{D}_c^{-1}(s)\mathbf{W}(s)$ is proper, the polynomial part of $\mathbf{D}_c^{-1}(s)\mathbf{W}(s)\bar{\mathbf{D}}_f(s)\mathbf{D}^{-1}(s)$ is $\mathbf{Q}(s)$. To show $\mathbf{Q}^{-1}(s)$ is proper, it is sufficient to show, by using Theorem 3-4, that $(\mathbf{D}_c^{-1}(s)\mathbf{W}(s)\bar{\mathbf{D}}_f(s)\mathbf{D}^{-1}(s))^{-1}$ is proper. Using (9-133), (9-134), (9-138), (9-139), and similar formulas for $\mathbf{D}_f(s)$ and $\mathbf{W}(s)$, we have

$$\begin{aligned}\mathbf{D}(s)\bar{\mathbf{D}}_f^{-1}(s)\mathbf{W}^{-1}(s)\mathbf{D}_c(s)=[\mathbf{D}_h+\mathbf{D}_l(s)]\mathbf{H}(s)\mathbf{H}_{\bar{D}_f}^{-1}(s)[\mathbf{I}+\bar{\mathbf{D}}_{fl}(s)]^{-1}\\ [\mathbf{I}+\mathbf{W}_l(s)]^{-1}\mathbf{H}_{\bar{W}}^{-1}(s)\mathbf{H}_c(s)[\mathbf{D}_{ch}+\mathbf{D}_{cl}(s)]\end{aligned} \tag{9-195}$$

Because $\mathbf{D}_l(s)$, $\bar{\mathbf{D}}_{fl}(s)$, $\mathbf{W}_l(s)$, and $\mathbf{D}_{cl}(s)$ are all strictly proper, we have, as $s\to\infty$,

$$\mathbf{D}(s)\bar{\mathbf{D}}_f^{-1}(s)\mathbf{W}^{-1}(s)\mathbf{D}_c(s)\to\mathbf{D}_h\mathbf{H}(s)\mathbf{H}_{\bar{D}_f}^{-1}(s)\mathbf{H}_{\bar{W}}^{-1}(s)\mathbf{H}_c(s)\mathbf{D}_{ch} \tag{9-196}$$

which approaches a zero or nonzero finite constant matrix. Hence $\mathbf{D}(s)\bar{\mathbf{D}}_f^{-1}(s)\bar{\mathbf{W}}^{-1}(s)\mathbf{D}_c(s)$ is proper, and consequently, $\mathbf{Q}^{-1}(s)$ is proper.

Now we show that the compensator $\mathbf{D}_c^{-1}(s)\mathbf{K}(s)=\mathbf{D}_c^{-1}(s)\bar{\mathbf{W}}(s)\bar{\mathbf{N}}_f(s)$ is proper. Consider

$$\begin{aligned}\mathbf{T}(s)&=\mathbf{D}(s)\bar{\mathbf{D}}_f^{-1}(s)\bar{\mathbf{N}}_f(s)\\ &=[\mathbf{D}_h+\mathbf{D}_l(s)]\mathbf{H}(s)\mathbf{H}_{\bar{D}_f}^{-1}(s)[\mathbf{I}+\bar{\mathbf{D}}_{fl}(s)]^{-1}\mathbf{H}_{\bar{N}_f}(s)[\bar{\mathbf{N}}_{fh}+\bar{\mathbf{N}}_{fl}(s)]\end{aligned}$$

As $s\to\infty$, we have $\mathbf{T}(s)\to\mathbf{D}_h\mathbf{H}(s)\mathbf{H}_{\bar{D}_f}^{-1}(s)\mathbf{H}_{\bar{N}_f}(s)\bar{\mathbf{N}}_{fh}$. Because $\mathbf{T}(s)$ is proper, we conclude

$$\delta_{ri}\bar{\mathbf{N}}_f(s)\le\bar{f}_i-\mu_i$$

The row degree m of $\mathbf{D}_c(s)$ is chosen so that $\mathbf{D}_c^{-1}(s)\bar{\mathbf{N}}_f(s)$ is proper. Now we claim that $\mathbf{D}_c^{-1}(s)\mathbf{W}(s)\bar{\mathbf{N}}_f(s)$ remains to be proper. If $\bar{f}_i-\mu_i-m\ge0$, $\delta_{ri}\mathbf{W}(s)=0$, and $\delta_{ri}(\mathbf{W}(s)\bar{\mathbf{N}}_f(s))=\delta_{ri}\bar{\mathbf{N}}_f(s)\le\delta_{ri}\mathbf{D}_c(s)$. If $\bar{f}_i-\mu_i-m<0$, $\delta_{ri}\mathbf{W}(s)=\mu_i+m-\bar{f}_i$, and we have $\delta_{ri}(\mathbf{W}(s)\bar{\mathbf{N}}_f(s))\le\mu_i+m-\bar{f}_i+\bar{f}_i-\mu_i=m=\delta_{ri}\mathbf{D}_c(s)$. Hence $\mathbf{D}_c^{-1}(s)\mathbf{K}(s)$ is proper.

We have shown that all compensators are proper. What remains to be shown is that the overall system is well posed. The overall system is well posed if and only if the matrix

$$\begin{aligned}\mathbf{P}^{-1}(s)&\triangleq[\mathbf{I}+\mathbf{D}_c^{-1}(s)\mathbf{L}(s)\mathbf{Q}^{-1}(s)+\mathbf{D}_c^{-1}(s)\mathbf{M}(s)\mathbf{N}(s)\mathbf{D}^{-1}(s)\mathbf{Q}^{-1}(s)]^{-1}\\ &=\mathbf{Q}(s)\mathbf{D}(s)[\mathbf{D}_c(s)\mathbf{Q}(s)\mathbf{D}(s)+\mathbf{L}(s)\mathbf{D}(s)+\mathbf{M}(s)\mathbf{N}(s)]^{-1}\mathbf{D}_c(s)\end{aligned} \tag{9-197}$$

is proper. The substitution of (9-191) into (9-197) yields

$$\mathbf{P}^{-1}(s)=\mathbf{Q}(s)\mathbf{D}(s)[\mathbf{W}(s)\bar{\mathbf{D}}_f(s)]^{-1}\mathbf{D}_c(s)$$

which implies, by using (9-192),

$$\begin{aligned}\mathbf{P}(s)&=\mathbf{D}_c^{-1}(s)[\mathbf{D}_c(s)\mathbf{Q}(s)\mathbf{D}(s)+\mathbf{R}_2(s)\mathbf{D}(s)+\mathbf{W}(s)\mathbf{R}_1(s)]\mathbf{D}^{-1}(s)\mathbf{Q}^{-1}(s)\\ &=\mathbf{I}+\mathbf{D}_c^{-1}(s)\mathbf{R}_2(s)\mathbf{Q}^{-1}(s)+\mathbf{D}_c^{-1}(s)\mathbf{W}(s)\mathbf{R}_1(s)\mathbf{D}^{-1}(s)\mathbf{Q}^{-1}(s)\end{aligned}$$

Because $\mathbf{D}_c^{-1}(s)\mathbf{R}_2(s)$, $\mathbf{R}_1(s)\mathbf{D}^{-1}(s)$ are strictly proper and because $\mathbf{Q}^{-1}(s)$ and $\mathbf{D}_c^{-1}(s)\mathbf{W}(s)$ are proper, the polynomial part of $\mathbf{P}(s)$ is $\mathbf{I}$, which is nonsingular. Hence $\mathbf{P}^{-1}(s)$ is proper. Q.E.D.

A remark is in order regarding the degrees of compensators in this design. In general, the degree of det $\mathbf{Q}(s)$ in this design is smaller than the one in the previous subsection. Because $\mathbf{D}_c^{-1}(s)\mathbf{K}(s)$, $\mathbf{D}_c^{-1}(s)\mathbf{L}(s)$ and $\mathbf{D}_c^{-1}(s)\mathbf{M}(s)$ all have the same denominator matrix, they can be implemented with a degree equal to deg det $\mathbf{D}_c(s)$. Hence the total degree of compensators in this design is deg det $\mathbf{Q}(s)+$ deg det $\mathbf{D}_c(s)$, which is generally smaller than the ones in the previous subsection. If det $\mathbf{D}_f(s)$, det $\mathbf{V}(s)$, det $\mathbf{D}_c(s)$ and det $\mathbf{W}(s)$ are Hurwitz, the implementation is, similar to Implementation I, asymptotically stable.

The remarks in (9-178) and (9-180) regarding (9-172) and (9-173) in the previous subsection are equally applicable to (9-188) and (9-189) in this subsection and will not be repeated.

In addition to the two implementations, it is possible to obtain other feedback implementations of the open-loop compensator $\mathbf{T}(s)$. For example, we may use the single-loop feedback system in Figure P9-32 or the input-output feedback system in Figure 9-30 [without using $\mathbf{Q}^{-1}(s)$] to implement $\mathbf{T}(s)$. The basic idea and design procedures will be similar to the ones in Implementation I; however, they may require less computation because the division steps in (9-172) and (9-173) are not needed. For yet another implementation, see Reference S218.

Applications.[21] In this subsection, the results of the previous subsection will be employed to design feedback systems. We consider first the decoupling problem and then decoupling together with asymptotic tracking and disturbance rejection. Finally, we study the model matching problem.

Decoupling. Consider a $p \times p$ nonsingular proper rational matrix $\hat{\mathbf{G}}(s)=\mathbf{N}(s)\mathbf{D}^{-1}(s)$. If an open-loop compensator $\mathbf{T}(s)$ is chosen as $\mathbf{D}(s)\mathbf{N}^{-1}(s)\tilde{\mathbf{D}}_f^{-1}(s)$, where $\tilde{\mathbf{D}}_f(s)=\text{diag}\,\{d_1(s), d_2(s), \ldots, d_p(s)\}$ and $d_i(s)$ are Hurwitz polynomials of minimum degrees to make $\mathbf{T}(s)$ proper, then the implementation of $\hat{\mathbf{G}}(s)\mathbf{T}(s)$ in the input-output feedback system in Figure 9-31 will yield a decoupled system $\tilde{\mathbf{D}}_f^{-1}(s)$. We note that in this design, the roots of det $\mathbf{D}(s)$ are *shifted* to the roots of det $(\tilde{\mathbf{D}}_f(s)\mathbf{N}(s))$; hence there are no cancellations involving the roots of det $\mathbf{D}(s)$. Since the input-output feedback has no effect on the numerator matrix $\mathbf{N}(s)$, the disappearance of $\mathbf{N}(s)$ from the decoupled system is accomplished by exact cancellations. Thus if det $\mathbf{N}(s)$ is not a Hurwitz polynomial, the design will involve unstable pole-zero cancellations. Thus the choice of $\mathbf{T}(s)=\mathbf{D}(s)\mathbf{N}^{-1}(s)\tilde{\mathbf{D}}_f^{-1}(s)$ cannot always be employed to design decoupled systems.

We discuss in the following a method of designing decoupled systems without involving any unstable pole-zero cancellations. Consider a $p \times p$ nonsingular proper rational matrix $\hat{\mathbf{G}}(s)=\mathbf{N}(s)\mathbf{D}^{-1}(s)$, where $\mathbf{D}(s)$ and $\mathbf{N}(s)$ are right coprime and $\mathbf{D}(s)$ is column reduced. We factor $\mathbf{N}(s)$ as

$$\mathbf{N}(s)=\mathbf{N}_1(s)\mathbf{N}_2(s)$$

with

$$\mathbf{N}_1(s)=\text{diag}\,\{\beta_{11}(s), \beta_{12}(s), \ldots, \beta_{1p}(s)\}$$

[21] Follows Reference S54.

where $\beta_{1i}(s)$ is the greatest common divisor of the ith row of $\mathbf{N}(s)$. Let $\beta_{2i}(s)$ be the least common denominator of the *unstable* poles of the ith column of $\mathbf{N}_2^{-1}(s)$. Define $\mathbf{N}_{2d}(s) \triangleq \operatorname{diag}\{\beta_{21}(s), \beta_{22}(s), \ldots, \beta_{2p}(s)\}$ and define

$$\bar{\mathbf{N}}_2(s) \triangleq \mathbf{N}_2^{-1}(s)\mathbf{N}_{2d}(s)$$

It is a rational matrix with only stable poles, that is, poles with negative real parts. Then we have

$$\mathbf{N}_2(s)\bar{\mathbf{N}}_2(s) = \mathbf{N}_{2d}(s) = \operatorname{diag}\{\beta_{21}(s), \beta_{22}(s), \ldots, \beta_{2p}(s)\}$$

Now we choose an open-loop compensator $\mathbf{T}(s)$ as

$$\mathbf{T}(s) = \mathbf{D}(s)\bar{\mathbf{N}}_2(s)\mathbf{D}_t^{-1}(s) \tag{9-198}$$

with

$$\mathbf{D}_t(s) = \operatorname{diag}\{\alpha_1(s), \alpha_2(s), \ldots, \alpha_p(s)\}$$

where $\alpha_i(s)$ are Hurwitz polynomials of minimum degrees to make $\mathbf{T}(s)$ proper. Then we have

$$\begin{aligned}\hat{\mathbf{G}}(s)\mathbf{T}(s) &= \mathbf{N}_1(s)\mathbf{N}_2(s)\mathbf{D}^{-1}(s)\mathbf{D}(s)\bar{\mathbf{N}}_2(s)\mathbf{D}_t^{-1}(s) = \mathbf{N}_1(s)\mathbf{N}_{2d}(s)\mathbf{D}_t^{-1}(s) \\ &= \operatorname{diag}\left\{\frac{\beta_1(s)}{\alpha_1(s)}, \frac{\beta_2(s)}{\alpha_2(s)}, \ldots, \frac{\beta_p(s)}{\alpha_p(s)}\right\}\end{aligned} \tag{9-199}$$

where $\beta_i(s) = \beta_{1i}(s)\beta_{2i}(s)$. Hence the implementation of $\hat{\mathbf{G}}(s)\mathbf{T}(s)$ will yield a decoupled system. We note that this design involves only the cancellations of the stable roots of det $\mathbf{N}_2(s)$, and the resulting feedback system will be asymptotically stable.

Example 2

Consider the plant

$$\hat{\mathbf{G}}(s) = \mathbf{N}(s)\mathbf{D}^{-1}(s) = \begin{bmatrix} s^2 & 1 \\ s+1 & s+1 \end{bmatrix}\begin{bmatrix} s^2+1 & 1 \\ 0 & s \end{bmatrix}^{-1}$$

We compute $\beta_{11}(s) = 1$, and $\beta_{12}(s) = s+1$. Hence we may factor $\mathbf{N}(s)$ as

$$\mathbf{N}(s) = \mathbf{N}_1(s)\mathbf{N}_2(s) = \begin{bmatrix} 1 & 0 \\ 0 & s+1 \end{bmatrix}\begin{bmatrix} s^2 & 1 \\ 1 & 1 \end{bmatrix}$$

We compute

$$\mathbf{N}_2^{-1}(s) = \frac{1}{(s+1)(s-1)}\begin{bmatrix} 1 & -1 \\ -1 & s^2 \end{bmatrix}$$

Hence we have

$$\mathbf{N}_{2d}(s) = \operatorname{diag}\{(s-1), (s-1)\} \tag{9-200}$$

and

$$\bar{\mathbf{N}}_2(s) = \mathbf{N}_2^{-1}(s)\mathbf{N}_{2d}(s) = \begin{bmatrix} 1 & -1 \\ -1 & s^2 \end{bmatrix}\frac{1}{s+1}$$

Now we choose $\mathbf{T}(s) = \mathbf{D}(s)\bar{\mathbf{N}}_2(s)\mathbf{D}_t^{-1}(s)$, with $\mathbf{D}_t(s) = \operatorname{diag}\{\alpha_1(s), \alpha_2(s)\}$, where $\alpha_i(s)$ are Hurwitz polynomials of minimum degrees to make $\mathbf{T}(s)$ proper. We compute

$$\mathbf{D}(s)\bar{\mathbf{N}}_2(s)=\begin{bmatrix} s^2+1 & 1 \\ 0 & s \end{bmatrix}\begin{bmatrix} 1 & -1 \\ -1 & s^2 \end{bmatrix}\frac{1}{s+1}=\begin{bmatrix} s^2 & -1 \\ -s & s^3 \end{bmatrix}\frac{1}{s+1}$$

Hence we may choose

$$\mathbf{D}_t(s)=\begin{bmatrix} s+2 & 0 \\ 0 & (s+1)^2 \end{bmatrix} \tag{9-201}$$

The implementation of $\hat{\mathbf{G}}(s)\mathbf{T}(s)$ will yield a decoupled system with transfer matrix

$$\begin{aligned}\hat{\mathbf{G}}(s)\mathbf{T}(s)&=\text{diag}\left\{\frac{s-1}{s+2}, \frac{(s+1)(s-1)}{(s+1)^2}\right\}\\ &=\text{diag}\left\{\frac{s-1}{s+2}, \frac{s-1}{s+1}\right\}\end{aligned} \tag{9-202}$$

Implementation I. We implement $\hat{\mathbf{G}}(s)\mathbf{T}(s)$ in Figure 9-31. We compute

$$\begin{aligned}\mathbf{D}^{-1}(s)\mathbf{T}(s)=\bar{\mathbf{N}}_2(s)\mathbf{D}_t^{-1}(s)&=\begin{bmatrix} 1 & -1 \\ -1 & s^2 \end{bmatrix}\begin{bmatrix} (s+2)(s+1) & 0 \\ 0 & (s+1)^3 \end{bmatrix}^{-1}\\ &=\begin{bmatrix} s^2+\frac{9}{4}s+\frac{3}{4} & -\frac{3}{4}s-\frac{5}{4} \\ -\frac{1}{4}s+\frac{1}{4} & s^2+\frac{11}{4}s+\frac{9}{4} \end{bmatrix}^{-1}\begin{bmatrix} 1 & -\frac{3}{4} \\ -1 & s-\frac{1}{4} \end{bmatrix} \triangleq \mathbf{D}_f^{-1}(s)\mathbf{N}_f(s)\end{aligned}$$

where $\mathbf{D}_f(s)$ and $\mathbf{N}_f(s)$ are left coprime. This is computed by using the procedure discussed in Appendix G. The row index of $\hat{\mathbf{G}}(s)$ can be computed as $v=2$. Clearly we have $m=\max\{v-1, m_1, m_2\}=\max\{2-1, 0, 1\}=1$ and, by choosing $\bar{\mathbf{D}}_c(s)$ arbitrarily as $\bar{\mathbf{D}}_c(s)=\text{diag}\{s+3, 1\}$,

$$\mathbf{D}_c(s)=\bar{\mathbf{D}}_c(s)\mathbf{N}_f(s)=\begin{bmatrix} s+3 & 0 \\ 0 & 1 \end{bmatrix}\begin{bmatrix} 1 & -\frac{3}{4} \\ -1 & s-\frac{1}{4} \end{bmatrix}=\begin{bmatrix} s+3 & -\dfrac{3(s+3)}{4} \\ -1 & s-\frac{1}{4} \end{bmatrix}$$

which has row degrees all equal to $m=1$ and is row reduced. Clearly we have $\delta_{ci}(\bar{\mathbf{D}}_c(s)\mathbf{D}_f(s))\le\mu_i+m$, $i=1, 2$, hence we may set $\mathbf{Q}(s)=\mathbf{I}$. We compute

$$\bar{\mathbf{D}}_c(s)[\mathbf{D}_f(s)-\mathbf{N}_f(s)\mathbf{D}(s)]=\begin{bmatrix} \frac{-9}{4}s^2+\frac{13}{2}s-\frac{3}{4} & \frac{-9}{4}s-\frac{27}{4} \\ s^2-\frac{1}{4}s+\frac{5}{4} & 3s+\frac{13}{4} \end{bmatrix}$$

Thus, the $\mathbf{L}(s)$ and $\mathbf{M}(s)$ in (9-174) can be solved from

$$[\mathbf{L}_0 \quad \mathbf{M}_0 \quad \mathbf{L}_1 \quad \mathbf{M}_1]\left[\begin{array}{cccccccc} 1 & 1 & 0 & 0 & 1 & 0 & & \\ 0 & 0 & 0 & 1 & 0 & 0 & & \\ \hdashline 0 & 1 & 0 & 0 & 1 & 0 & & \\ 1 & 1 & 1 & 1 & 0 & 0 & & \\ \hdashline & & 1 & 1 & 0 & 0 & 1 & 0 \\ & & 0 & 0 & 0 & 1 & 0 & 0 \\ \hdashline & & 0 & 1 & 0 & 0 & 1 & 0 \\ & & 1 & 1 & 1 & 1 & 0 & 0 \end{array}\right]$$

$$=\left[\begin{array}{cc:cc:cc:cc} -3/4 & -27/4 & 13/2 & -9/4 & 9/4 & 0 & 0 & 0 \\ 5/4 & 13/4 & -1/4 & 3 & 1 & 0 & 0 & 0 \end{array}\right]$$

as

$$\mathbf{L}(s)=\begin{bmatrix}(31/2)s+33/4 & 27/4\\ (-7/2)s-2 & -1/4\end{bmatrix} \qquad \mathbf{M}(s)=\begin{bmatrix}(-31/2)s-6 & -9\\ (7/2)s+2 & 13/4\end{bmatrix}$$

This completes the first implementation.

Implementation II. We implement $\hat{\mathbf{G}}(s)\mathbf{T}(s)$ in Figure 9-33. We have $\mathbf{D}_{fh}=\mathbf{I}$ and $f_i \geq \mu_i$; hence $\mathbf{V}(s)=\mathbf{I}$, $\bar{\mathbf{D}}_f(s)=\mathbf{D}_f(s)$, and $\bar{\mathbf{N}}_f(s)=\mathbf{N}_f(s)$. We choose $m=v-1=1$ and

$$\mathbf{D}_c(s)=\begin{bmatrix}s+3 & 0\\ 0 & s+3\end{bmatrix}$$

Because $f_1-\mu_1-m=2-2-1=-1$, we choose $w_1(s)$ arbitrarily as $s+2$, of degree 1. Because $f_2-\mu_2-m=2-1-1=0$, we set $w_2(s)=1$. Hence we have

$$\mathbf{W}(s)=\begin{bmatrix}s+2 & 0\\ 0 & 1\end{bmatrix}$$

Clearly, we have $\delta_{ci}(\mathbf{W}(s)\mathbf{D}_f(s)) \leq \mu_i+m$, $i=1,2$; hence we set $\mathbf{Q}(s)=\mathbf{I}$ and

$$\mathbf{K}(s)=\mathbf{W}(s)\mathbf{N}_f(s)=\begin{bmatrix}s+2 & -\dfrac{3(s+2)}{4}\\ -1 & s-\frac{1}{4}\end{bmatrix}$$

The $\mathbf{L}(s)$ and $\mathbf{M}(s)$ in (9-191) can be solved as

$$\mathbf{L}(s)=\begin{bmatrix}11s+21/4 & (-3/4)s+3\\ (-5/2)s-2 & -5/2\end{bmatrix} \qquad \mathbf{M}(s)=\begin{bmatrix}-11s-4 & -27/4\\ (5/2)s+2 & 9/4\end{bmatrix}$$

This completes the second implementation. ∎

For this example, the total degrees of compensators of these two implementations are the same. In general, the total degree of compensators in the second implementation is less than or equal to the one in the first implementation.

The design procedure discussed in this subsection can be modified in several ways. For example, if a stable root of det $\mathbf{N}_2(s)$ is very close to the imaginary axis, it may be retained in $\beta_{2i}(s)$, instead of being canceled. Instead of decoupling the plant for each pair of input and output, we may decouple it for a group of inputs and a group of outputs. In this case, the plant is to be decoupled into a *block* diagonal matrix. These modifications are straightforward and will not be discussed.

Asymptotic Tracking, Disturbance Rejection, and Decoupling. In this subsection, we shall design a robust system to achieve decoupling, asymptotic tracking and disturbance rejection. Let $\hat{\mathbf{G}}(s)=\mathbf{N}(s)\mathbf{D}^{-1}(s)$ be a $p \times p$ nonsingular proper transfer matrix. It can be decoupled, without involving any unstable pole-zero cancellations, as

$$\text{diag}\left\{\frac{\beta_1(s)}{\alpha_1(s)}, \frac{\beta_2(s)}{\alpha_2(s)}, \ldots, \frac{\beta_p(s)}{\alpha_p(s)}\right\}$$

where $\beta_i(s)$ are uniquely determinable from $\mathbf{N}(s)$ and the degrees of $\alpha_i(s)$ are to be chosen to make the open-loop compensator $\mathbf{T}(s)$ proper. If the plant is to be decoupled only, $\alpha_i(s)$ can be arbitrarily assigned. If the plant is to be designed, in addition to decoupling, to achieve tracking and disturbance rejection, $\alpha_i(s)$ cannot be arbitrarily assigned. They must be used to stabilize the additional feedback connection to be introduced for tracking and disturbance rejection.

In order to achieve asymptotic tracking and disturbance rejection, and to be robust with respect to parameter perturbations, we must introduce, as discussed in Section 9-6, an internal model $\phi^{-1}(s)\mathbf{I}_p$ as shown in Figure 9-34(a). If we introduce a diagonal polynomial matrix $\mathbf{H}(s) = \text{diag}\,\{h_1(s), h_2(s), \ldots, h_p(s)\}$ with $h_i(s)/\phi(s)$ proper or strictly proper, then the system in Figure 9-34(a) reduces to p number of single-variable systems shown in Figure 9-34(b). The transfer function of the system in Figure 9-34(b) is clearly equal to

$$\frac{h_i(s)\beta_i(s)}{\phi(s)\alpha_i(s) + h_i(s)\beta_i(s)} \triangleq \frac{h_i(s)\beta_i(s)}{f_i(s)} \tag{9-203}$$

Its denominator is a polynomial of degree $\deg \phi(s) + \deg \alpha_i(s)$. If we require $h_i(s)/\phi(s)$ to be strictly proper, there are $\deg \phi(s)$ number of free parameters in $h_i(s)$; there are $\deg \alpha_i(s) + 1$ number of free parameters in $\alpha_i(s)$. Hence if $\phi(s)$ and $\beta_i(s)$ are coprime, the roots of $f_i(s)$ in (9-203) can be arbitrarily assigned by proper choices of $h_i(s)$ and $\alpha_i(s)$. The condition for asymptotic stability, tracking and disturbance rejection is that no root of $\phi(s)$ is a transmission zero of $\hat{\mathbf{G}}(s)$ or, because $\mathbf{N}(s)$ in $\hat{\mathbf{G}}(s) = \mathbf{N}(s)\bar{\mathbf{D}}^{-1}(s)$ is square, a root of $\det \mathbf{N}(s)$. Because $\beta_i(s)$ is a factor of $\det \mathbf{N}(s)$, we conclude that if no root of $\phi(s)$ is a transmission

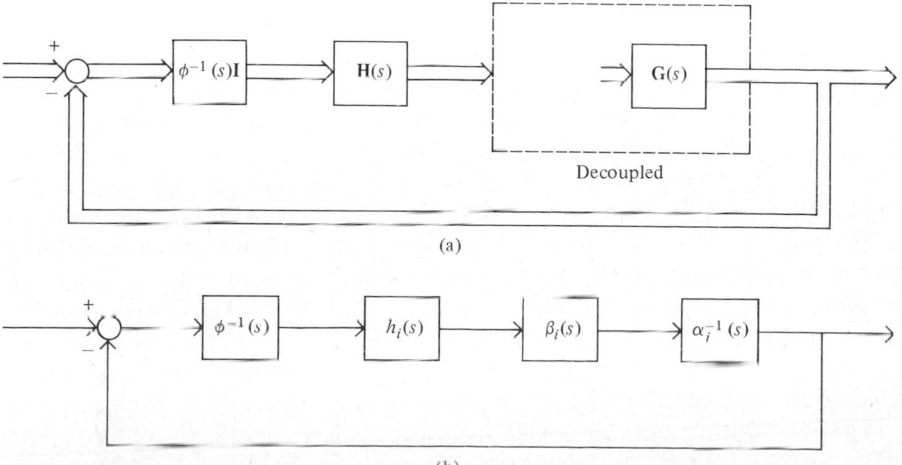

Figure 9-34 Design of robust system.

zero of $\hat{\mathbf{G}}(s)$, then $\phi(s)$ and $\beta_i(s)$ are coprime, and the roots of $f_i(s)$ can be arbitrarily assigned. From the assignment of $f_i(s)$, we can compute $h_i(s)$ and $\alpha_i(s)$. We then design an input-output feedback system to decouple the plant as diag $\{\beta_1(s)/\alpha_1(s), \beta_2(s)/\alpha_2(s), \ldots, \beta_p(s)/\alpha_p(s)\}$. The resulting system is asymptotically stable and is decoupled. It will also track asymptotically the reference input and reject plant-entered disturbances. In this design, if there are parameter perturbations in the plant and compensators (excluding the internal model), the decoupling property will be destroyed. However, if the overall system remains to be asymptotically stable, the property of asymptotic tracking and disturbance rejection will be preserved.

Example 3

Consider the plant in Example 2 on page 524. We have $\beta_1(s)=s-1, \beta_2(s)=s^2-1$, $\deg \alpha_1(s)=1$ and $\deg \alpha_2(s)=2$. Suppose the plant is to be designed to track step inputs $1/s$ and reject plant-entered disturbance of the form e^{2t}, then we have

$$\phi(s)=s(s-2)$$

Since $\phi(s)$ and $\beta_i(s)$ are coprime, $h_i(s)$ and $\alpha_i(s)$ can be found to stabilize the system in Figure 9-34(b). Rather arbitrarily, we choose $f_1(s)=(s+2)^3$. Then the solutions of

$$\phi(s)\alpha_1(s)+\beta_1(s)h_1(s)=f_1(s)=(s+2)^3$$

are $h_1(s)=36s-8$ and $\alpha_1(s)=s-28$. If $f_2(s)=(s+2)^4$, the solutions of

$$\phi(s)\alpha_2(s)+\beta_2(s)h_2(s)=f_2(s)=(s+2)^4$$

are $h_2(s)=\frac{152}{3}s-16$ and $\alpha_2(s)=s^2-\frac{122}{3}s-\frac{124}{3}$. Next we replace (9-201) and (9-202) by

$$\mathbf{D}_t(s)=\begin{bmatrix}\alpha_1(s) & 0\\ 0 & \alpha_2(s)\end{bmatrix}=\begin{bmatrix}s-28 & 0\\ 0 & s^2-\frac{122}{3}s-\frac{124}{3}\end{bmatrix}$$

and
$$\hat{\mathbf{G}}(s)\mathbf{T}(s)=\operatorname{diag}\left\{\frac{s-1}{s-28}, \frac{s^2-1}{s^2-\frac{122}{3}s-\frac{124}{3}}\right\}$$

Once $\hat{\mathbf{G}}(s)\mathbf{T}(s)$ is implemented as an input-output feedback system inside the box in Figure 9-34(a), the design is completed. This part of design is similar to the one in Example 2 and will not be repeated. ∎

Model matching. The design problems discussed so far concern only with the assignment of poles and denominator matrix; the numerator matrix is left unspecified. In this subsection, we discuss the assignment of denominator matrix as well as the numerator matrix or, equivalently, the entire overall transfer matrix. This problem is often referred to as the *exact model matching problem.*

Consider a plant with $q \times p$ proper rational matrix $\hat{\mathbf{G}}(s)$. The desired model is assumed to have the $q \times r$ proper rational matrix $\hat{\mathbf{G}}_m(s)$. The problem is to find a configuration and compensators for the plant so that the resulting overall system has $\hat{\mathbf{G}}_m(s)$ as its transfer matrix. We study the problem by using an open-loop compensator. Let $\mathbf{T}(s)$ be a $p \times r$ open-loop compensator so that

$$\hat{\mathbf{G}}(s)\mathbf{T}(s) = \hat{\mathbf{G}}_m(s) \tag{9-204}$$

Now if a solution $\mathbf{T}(s)$ exists in (9-204) and is a proper rational matrix, then the design can be accomplished by implementing $\mathbf{T}(s)$ in the input-output feedback configuration shown in Figure 9-31 or 9-33. Thus the design problem hinges on the solution of (9-204).

Let $\mathbf{T}_i(s)$ and $\hat{\mathbf{G}}_{mi}(s)$ be the ith column of $\mathbf{T}(s)$ and $\hat{\mathbf{G}}_m(s)$. Then (9-204) can be written, following (2-2), as $\hat{\mathbf{G}}(s)\mathbf{T}_i(s) = \hat{\mathbf{G}}_{mi}(s)$, $i = 1, 2, \ldots, r$. Each equation is a linear algebraic equation studied in (2-36) with entries in $\mathbb{R}(s)$, the field of real rational functions. Hence Theorem 2-4 is directly applicable. Thus the necessary and sufficient condition for the existence of a solution $\mathbf{T}(s)$ in (9-204) is

$$\text{rank } \hat{\mathbf{G}}(s) = \text{rank } [\hat{\mathbf{G}}(s) \quad \hat{\mathbf{G}}_m(s)]$$

over $\mathbb{R}(s)$. The solution $\mathbf{T}(s)$ is generally a rational matrix, proper or improper. In the model matching problem, we are interested in only solutions which are proper rational matrices. Furthermore, we require the degree of $\mathbf{T}(s)$ to be as small as possible. This problem of finding a proper $\mathbf{T}(s)$ with a minimal degree to meet (9-204) is called the *minimal design problem*[22] in References S95, S125, S172 and S210.

Before proceeding, we digress to introduce some concepts. Recall that $\mathbf{A}(s)$ and $\mathbf{B}(s)$ are left coprime if and only if their greatest common left divisor is unimodular, that is, the $\mathbf{R}(s)$ in any factorization $[\mathbf{A}(s) \quad \mathbf{B}(s)] = \mathbf{R}(s)[\bar{\mathbf{A}}(s) \quad \bar{\mathbf{B}}(s)]$ is unimodular. Following this, we may define a $q \times n$, with $n \geq q$, polynomial matrix $\mathbf{M}(s)$ to be *row irreducible*, or its column submatrices to be left coprime, if the $\mathbf{R}(s)$ in any factorization $\mathbf{M}(s) = \mathbf{R}(s)\bar{\mathbf{M}}(s)$ is unimodular. Following Theorem G-8′, $\mathbf{M}(s)$ is row irreducible if and only if rank $\mathbf{M}(s) = q$ for every s in $\mathbb{C}$. We may also extend the concept of row reducedness to nonsquare matrix. A square polynomial matrix $\mathbf{A}(s)$ is row reduced if deg det $\mathbf{A}(s)$ is equal to the sum of all row degrees of $\mathbf{A}(s)$. This definition implies that $\mathbf{A}(s)$ is row reduced if and only if its row-degree-coefficient matrix is nonsingular. Following this, we define a $q \times n$ polynomial matrix $\mathbf{M}(s)$ to be row reduced if its row-degree-coefficient matrix is of rank q or of full row rank. Similar to Theorem G-11, if $\mathbf{M}(s)$ is of full row rank in $\mathbb{R}(s)$, there exists a unimodular polynomial matrix $\mathbf{U}(s)$ such that $\mathbf{U}(s)\mathbf{M}(s)$ is row reduced. We may similarly define column irreducibility and column reducedness for nonsquare polynomial matrices. With this preliminary, we are ready to study the minimal design problem.

We discuss first the condition for the solution $\mathbf{T}(s)$ in (9-204) to be proper. If $\hat{\mathbf{G}}(s)$ is square and nonsingular, the answer is very simple: $\mathbf{T}(s)$ is proper if and only if $\hat{\mathbf{G}}^{-1}(s)\hat{\mathbf{G}}_m(s)$ is proper. If $\hat{\mathbf{G}}(s)$ is not square, the situation is slightly more complicated. Let $\phi(s)$ be the least common denominator of all entries of $\hat{\mathbf{G}}(s)$ and $\hat{\mathbf{G}}_m(s)$. The multiplication of $\phi(s)$ to (9-204) yields

$$\mathbf{A}(s)\mathbf{T}(s) = \mathbf{B}(s) \tag{9-205}$$

[22] If $\hat{\mathbf{G}}_m(s) = \mathbf{I}$, then the solution $\mathbf{T}(s)$ in (9-204) is a right inverse of $\mathbf{G}(s)$. Hence the inverse problem reduces also to the minimum design problem.

where $\mathbf{A}(s)=\phi(s)\hat{\mathbf{G}}(s)$ and $\mathbf{B}(s)=\phi(s)\hat{\mathbf{G}}_m(s)$ are $q\times p$ and $q\times r$ polynomial matrices. If $\hat{\mathbf{G}}(s)$ and, consequently, $\mathbf{A}(s)$ are of full row rank, there exists a unimodular polynomial matrix $\mathbf{U}(s)$ such that $\mathbf{U}(s)\mathbf{A}(s)$ is row reduced. Consider

$$\mathbf{U}(s)\mathbf{A}(s)\mathbf{T}(s)=\mathbf{U}(s)\mathbf{B}(s) \tag{9-206}$$

We assert that if $\hat{\mathbf{G}}(s)$ is of full row rank in $\mathbb{R}(s)$, then $\mathbf{T}(s)$ is proper if and only if

$$\delta_{ri}(\mathbf{U}(s)\mathbf{A}(s))\geq\delta_{ri}(\mathbf{U}(s)\mathbf{B}(s)) \qquad i=1,2,\ldots,q \tag{9-207}$$

where δ_{ri} denotes the ith row degree. If $\mathbf{A}(s)$ is square, this assertion is essentially Theorem G-10. The proof for the nonsquare case is similar to the one of Theorem G-10 and will be omitted.

The properness condition in (9-207) does not tell us how to find a $\mathbf{T}(s)$ with a minimum degree. In the following, we shall introduce such a method. The method also gives the properness condition of $\mathbf{T}(s)$; hence the condition in (9-207) is not really needed. Let $\mathbf{T}(s)=\mathbf{N}_T(s)\mathbf{D}_T^{-1}(s)$. Then (9-205) becomes $\mathbf{A}(s)\mathbf{N}_T(s)=\mathbf{B}(s)\mathbf{D}_T(s)$ or

$$[\mathbf{A}(s) \quad \mathbf{B}(s)]\begin{bmatrix}-\mathbf{N}_T(s)\\ \mathbf{D}_T(s)\end{bmatrix}=\mathbf{0} \tag{9-208}$$

This is the polynomial equation studied in Equation (G-90); hence all discussion in Section G-6 is directly applicable. Let $(\mathbb{V}, \mathbb{R}(s))$ denote the right null space of (9-208). Its dimension is, following Theorem 2-5, equal to

$$\bar{r}\triangleq p+r-\text{rank}\,[\mathbf{A}(s) \quad \mathbf{B}(s)]=p+r-\text{rank}\,[\hat{\mathbf{G}}(s) \quad \hat{\mathbf{G}}_m(s)]$$

Now we may apply Theorem G-14′ to solve (9-208). We form the generalized resultant $\mathbf{T}_k$ from the coefficient matrices of $\mathbf{A}(s)$ and $\mathbf{B}(s)$ as in Theorem G-14′. We then search its linearly dependent columns in order from left to right by using the column searching algorithm. There will be exactly $\bar{r}$ primary dependent columns in $\mathbf{T}_k$. Let the $(p+r)\times\bar{r}$ polynomial matrix $\mathbf{Y}(s)$ be the solutions corresponding to these $\bar{r}$ primary dependent columns. Then $\mathbf{Y}(s)$ is, as in Theorem G-14′, column irreducible and column reduced, and is a minimal polynomial basis of the right null space of (9-208).

Let $\mathbf{Y}_{hc}$ be the column-degree-coefficient matrix of $\mathbf{Y}(s)$. For convenience of discussion, we assume $\bar{r}=r$ and partition $\mathbf{Y}_{hc}$ and $\mathbf{Y}(s)$ as

$$\mathbf{Y}_{hc}=\begin{bmatrix}\bar{\mathbf{Y}}_{hc}\\ \tilde{\mathbf{Y}}_{hc}\end{bmatrix} \qquad \mathbf{Y}(s)=\begin{bmatrix}-\mathbf{N}_T(s)\\ \mathbf{D}_T(s)\end{bmatrix} \tag{9-209}$$

where $\bar{\mathbf{Y}}_{hc}$ and $\mathbf{N}_T(s)$ are $p\times r$ and $\tilde{\mathbf{Y}}_{hc}$ and $\mathbf{D}_T(s)$ are $r\times r$ matrices. We note that, because of the properness assumption of $\hat{\mathbf{G}}(s)=\mathbf{A}^{-1}(s)\mathbf{B}(s)$, the $[-\mathbf{N}'(s) \quad \mathbf{D}'(s)]'$ in Theorem G-14′ always has the properties that $\delta_{ci}\mathbf{N}(s)\leq\delta_{ci}\mathbf{D}(s)$ and $\mathbf{D}(s)$ is column reduced or $\mathbf{D}_{hc}$ has a full column rank. In the minimal design problem, $\mathbf{A}(s)$ in (9-208) is not necessarily square and $\mathbf{A}^{-1}(s)\mathbf{B}(s)$ may not be defined; therefore there is no guarantee that $\mathbf{D}_{Thc}$, the column-degree-coefficient matrix of $\mathbf{D}_T(s)$ in (9-209), is of full column rank. We note that $\mathbf{Y}_{hc}$

does have a full column rank as a consequence of the column searching algorithm.

Theorem 9-28

If $\mathbf{Y}(s)$ in (9-209) is a minimal polynomial basis of (9-208) obtained as in Theorem G-14′, then $\mathbf{N}_T(s)\mathbf{D}_T^{-1}(s)$ is proper if and only if $\tilde{\mathbf{Y}}_{hc}$ has rank r.

Proof

If $\tilde{\mathbf{Y}}_{hc}$ has rank r, then $\mathbf{D}_T(s)$ is column reduced and $\delta_{ci}\mathbf{N}_T(s) \leq \delta_{ci}\mathbf{D}_T(s)$, for all i; hence $\mathbf{N}_T(s)\mathbf{D}_T^{-1}(s)$ is proper. To show the converse, we consider

$$\mathbf{T}(s) = \mathbf{N}_T(s)\mathbf{D}_T^{-1}(s) \qquad \text{or} \qquad \mathbf{T}(s)\mathbf{D}_T(s) = \mathbf{N}_T(s) \tag{9-210}$$

If $\tilde{\mathbf{Y}}_{hc}$ has no full column rank, there is, as a consequence of the column searching algorithm, at least one column, say the jth column, such that $\delta_{cj}\mathbf{N}_T(s) > \delta_{cj}\mathbf{D}_T(s)$ and $\tilde{\mathbf{Y}}_{hcj} = \mathbf{0}$, where $\tilde{\mathbf{Y}}_{hcj}$ denotes the jth column of $\tilde{\mathbf{Y}}_{hc}$.[23] Since $\mathbf{Y}(s)$ is a minimal basis, $\mathbf{Y}_{hc}$ has a full column rank. Hence if $\tilde{\mathbf{Y}}_{hcj} = \mathbf{0}$, then $\bar{\mathbf{Y}}_{hcj} \neq \mathbf{0}$. Let s^{kj} be the jth column degree of $\mathbf{Y}(s)$. Consider the jth column equation of (9-210):

$$\mathbf{T}(s)\mathbf{D}_{Tj}(s)s^{-kj} = \mathbf{N}_{Tj}(s)s^{-kj}$$

If $\mathbf{T}(s)$ is proper, as $s \to \infty$, we have $\mathbf{T}(\infty)\tilde{\mathbf{Y}}_{hcj} = \mathbf{0} = \bar{\mathbf{Y}}_{hcj} \neq \mathbf{0}$. This is not possible. Hence if $\mathbf{Y}_{hc}$ has no full column rank, then $\mathbf{T}(s)$ is improper. This establishes the theorem. Q.E.D.

In this theorem, the condition that $\mathbf{Y}_{hc}$ is of full column rank or $\mathbf{Y}(s)$ is a minimal basis is essential; otherwise the theorem does not hold (see Problem 9-33). The theorem is developed for the case $\bar{r} = r$. We discuss now the general case. If $\bar{r} < r$, clearly no solution $\mathbf{T}(s)$, proper or improper, exists in (9-204). If $\bar{r} \geq r$, $\mathbf{Y}(s)$ and $\mathbf{Y}_{hc}$ are $(p+r) \times \bar{r}$ matrices of rank $\bar{r}$, and Theorem 9-28 still holds. In this case, in order to obtain a minimal solution, we arrange $\mathbf{Y}(s)$ to have increasing column degrees, that is, $\delta_{c1}\mathbf{Y}(s) \leq \delta_{c2}\mathbf{Y}(s) \leq \cdots \leq \delta_{c\bar{r}}\mathbf{Y}(s)$. Then the first r columns with a nonsingular $\tilde{\mathbf{Y}}_{hc}$ will give the minimal proper solution of (9-204). This completes the discussion of the minimal design problem.

Example 4

Find a 3×2 minimal proper solution of

$$\begin{bmatrix} \dfrac{1}{s+1} & 0 & \dfrac{-s}{s+1} \\ \dfrac{1}{s} & -2 & -1 \end{bmatrix} \mathbf{T}(s) = \begin{bmatrix} \dfrac{s}{s+3} & \dfrac{-s}{s+3} \\ \dfrac{s+1}{s+3} & \dfrac{-3s-7}{s+3} \end{bmatrix}$$

---

[23] If $\mathbf{Y}_{hc}$ is not computed by using the column searching algorithm, we may not have $\tilde{\mathbf{Y}}_{hcj} = 0$, for some j. However, if rank $\tilde{\mathbf{Y}}_{hc} < r$, one of its column can be transformed into a zero column by elementary column operations.

The multiplication of $s(s+1)(s+3)$ and the substitution of $\mathbf{T}(s)=\mathbf{N}_T(s)\mathbf{D}_T^{-1}(s)$ yields

$$\begin{bmatrix} s^2+3s & 0 & -s^3-3s^2 \\ s^2+4s+3 & -2s^3-8s^2-6s & -s^3-4s^2-3s \end{bmatrix}\mathbf{N}_T(s)$$
$$=\begin{bmatrix} s^3+s^2 & -s^3-s^2 \\ s^3+2s^2+s & -3s^3-10s^2-7s \end{bmatrix}\mathbf{D}_T(s)$$

Clearly we have $\bar{r}=p+r-\text{rank}\,[\hat{\mathbf{G}}(s)\quad \hat{\mathbf{G}}_m(s)]=3+2-2=3$. We form $\mathbf{T}_1$ as in Theorem G-14′ and then apply the column searching algorithm to compute

$$\left[\begin{array}{ccc:cc:ccc:cc}
0 & 0 & 0 & 0 & 0 & & & & & \\
3 & 0 & 0 & 0 & 0 & & & \bigcirc & & \\
\hdashline
3 & 0 & 0 & 0 & 0 & 0 & 0 & 0 & 0 & 0 \\
4 & -6 & -3 & 1 & -7 & 3 & 0 & 0 & 0 & 0 \\
\hdashline
1 & 0 & -3 & 1 & -1 & 3 & 0 & 0 & 0 & 0 \\
1 & -8 & -4 & 2 & -10 & 4 & -6 & -3 & 1 & -7 \\
\hdashline
0 & 0 & -1 & 1 & -1 & 1 & 0 & -3 & 1 & -1 \\
0 & -2 & -1 & 1 & -3 & 1 & -8 & -4 & 2 & -10 \\
\hdashline
 & & & & & 0 & 0 & -1 & 1 & -1 \\
 & & \bigcirc & & & 0 & -2 & -1 & 1 & -3
\end{array}\right]$$

$$* \left[\begin{array}{ccccc:ccccc}
1 & 0 & 0 & 0 & 0 & 0 & 0 & 0 & 0 & 0 \\
0 & 1 & -1/2 & 1/6 & -7/6 & 1/2 & 0 & 0 & 0 & 0 \\
0 & 0 & 1 & 1 & -1 & 1 & 0 & -3 & 1 & -1 \\
0 & 0 & 0 & 1 & 1 & 0 & 0 & 9/2 & -3/2 & 3/2 \\
0 & 0 & 0 & 0 & 1 & 0 & 0 & 0 & 0 & 0 \\
\hdashline
0 & 0 & 0 & 0 & 0 & 1 & 0 & 0 & 0 & 0 \\
0 & 0 & 0 & 0 & 0 & 0 & 1 & 0 & 0 & -1 \\
0 & 0 & 0 & 0 & 0 & 0 & 0 & 1 & 0 & -1 \\
0 & 0 & 0 & 0 & 0 & 0 & 0 & 0 & 1 & 0 \\
0 & 0 & 0 & 0 & 0 & 0 & 0 & 0 & 0 & -1
\end{array}\right]$$

$$=\left[\begin{array}{ccc:cc:ccc:cc} \mathbf{x} & \mathbf{x} & \mathbf{x} & \mathbf{x} & \mathbf{0} & \mathbf{0} & \mathbf{x} & \mathbf{x} & \mathbf{0} & \mathbf{0} \\ & & & & \uparrow & \uparrow & & & \uparrow & \end{array}\right]$$

Where $\mathbf{x}$ denotes the nonzero column, and $\mathbf{0}$ the zero column. This computation is very easy to carry out by hand. The notation $*$ is explained in Appendix A. There are four linearly dependent columns in $\mathbf{T}_1$; three of them are primary dependent columns as indicated by the arrows. Note that in Theorem G-14′, linearly dependent columns will appear only in B-rows (see dual of Theorem G-13). In the minimal design problem, dependent columns may however appear in B- as well as A-columns as shown in this example. Corresponding to the three primary dependent columns, we use the dual formula of (A-11) to compute

$$\left[\begin{array}{ccc} 0 & 0 & 0 \\ -1 & 0 & 0 \\ 0 & 1 & 1 \\ 1 & 0 & 3 \\ 1 & 0 & 0 \\ \hdashline 0 & 1 & 0 \\ 0 & 0 & 0 \\ 0 & 0 & 1 \\ 0 & 0 & 1 \\ 0 & 0 & 0 \end{array}\right]\begin{array}{l} \left.\vphantom{\begin{array}{c}0\\0\\0\\0\\0\end{array}}\right\} s^0 \\ \left.\vphantom{\begin{array}{c}0\\0\\0\\0\\0\end{array}}\right\} s^1 \end{array}$$

which yields immediately

$$\mathbf{Y}(s)=\left[\begin{array}{ccc} 0 & s & 0 \\ -1 & 0 & 0 \\ 0 & 1 & s+1 \\ \hdashline 1 & 0 & s+3 \\ 1 & 0 & 0 \end{array}\right] \qquad \text{with } \mathbf{Y}_{hc}=\left[\begin{array}{ccc} 0 & 1 & 0 \\ -1 & 0 & 0 \\ 0 & 0 & 1 \\ \hdashline 1 & 0 & 1 \\ 1 & 0 & 0 \end{array}\right]$$

$\mathbf{Y}(s)$ has rank 3 for every s in $\mathbb{C}$, hence it is column irreducible. It is also column reduced, for $\mathbf{Y}_{hc}$ is of full column rank. Hence $\mathbf{Y}(s)$ is a minimal polynomial basis of the right null space. The first and third columns of $\mathbf{Y}_{hc}$ yield a $\tilde{\mathbf{Y}}_{hc}$ of rank 2; hence we have

$$\begin{bmatrix} -\mathbf{N}_T(s) \\ \mathbf{D}_T(s) \end{bmatrix}=\left[\begin{array}{cc} 0 & 0 \\ -1 & 0 \\ 0 & s+1 \\ \hdashline 1 & s+3 \\ 1 & 0 \end{array}\right]$$

and $$\mathbf{T}(s)=\mathbf{N}_T(s)\mathbf{D}_T^{-1}(s)=\begin{bmatrix} 0 & 0 \\ 1 & 0 \\ 0 & -s-1 \end{bmatrix}\begin{bmatrix} 1 & s+3 \\ 1 & 0 \end{bmatrix}^{-1}=\begin{bmatrix} 0 & 0 \\ 0 & 1 \\ \dfrac{-s-1}{s+3} & \dfrac{s+1}{s+3} \end{bmatrix}$$

This is a minimal proper solution. ∎

The numerator matrix of a plant cannot be affected by state feedback, as discussed in Section 7-3, nor by output feedback; hence the only way to affect the numerator matrix is by direct cancellation. In the design of compensators to achieve pole or denominator matrix assignment, we have been careful not to introduce any undesirable pole-zero cancellation. This is possible because the numerator matrices in these designs are not specified. In the model matching, because the numerator matrix is also specified, undesirable pole-zero cancellation may be unavoidable. Therefore in the model matching, the choice of $\hat{\mathbf{G}}_m(s)$ must be very careful. Otherwise, even if we can find a proper $\mathbf{T}(s)$ and implement it in a well-posed input-output configuration, the design is still unacceptable.

9-8 Concluding Remarks

In this chapter, we studied a number of topics: characterization of composite systems by their transfer matrices, stability of feedback systems, and the design of compensators in the unity feedback systems and the plant input-output feedback systems to achieve arbitrary pole placements, asymptotic tracking, disturbance rejection, and decoupling. We also studied the implementation of open-loop compensators by using plant input-output feedback systems.

All the design problems in this chapter hinge essentially on the solution of the Diophantine equation:

$$\mathbf{L}(s)\mathbf{D}(s)+\mathbf{M}(s)\mathbf{N}(s)=\mathbf{F}(s) \tag{9-211a}$$

or

$$\mathbf{D}(s)\mathbf{L}(s)+\mathbf{N}(s)\mathbf{M}(s)=\mathbf{F}(s) \tag{9-211b}$$

Several methods are available to solve this equation. First, we may apply a sequence of polynomial elementary matrices to transform $[\mathbf{D}'(s) \quad \mathbf{N}'(s)]'$ into a Hermite row form or a upper triangular form as in (G-33):

$$\begin{bmatrix} \mathbf{U}_{11}(s) & \mathbf{U}_{12}(s) \\ \mathbf{U}_{21}(s) & \mathbf{U}_{22}(s) \end{bmatrix}\begin{bmatrix} \mathbf{D}(s) \\ \mathbf{N}(s) \end{bmatrix}=\begin{bmatrix} \mathbf{R}(s) \\ \mathbf{0} \end{bmatrix}$$

Then a solution exists in (9-211a) if and only if $\mathbf{R}(s)$ is a right divisor of $\mathbf{F}(s)$. A general solution $\mathbf{L}(s)$ and $\mathbf{M}(s)$ can then be obtained as in Problem G-14. See References S34, S139, and S189. Another method is to find an observable realization $\{\mathbf{A}, \mathbf{B}, \mathbf{C}\}$ of $\mathbf{D}^{-1}(s)\mathbf{N}(s)$ as in (6-131). We then use the operator Π_D defined in (6-225) to compute $\Pi_D(\mathbf{F}(s))$. If we use the method in Problem G-15 or any other method to compute

$$\mathbf{F}(s)=\mathbf{D}(s)\mathbf{Q}_1(s)+\mathbf{F}_1(s)$$

with $\delta_{ri}\mathbf{F}_1(s)<\delta_{ri}\mathbf{D}(s)$, then we have $\Pi_D(\mathbf{F}(s))=\mathbf{F}_1(s)$. We then express $\mathbf{F}_1(s)$ in terms of the basis in (6-223) as

$$\mathbf{F}_1(s)=\bar{\mathbf{L}}(s)\mathbf{E}_1$$

Then a solution $\mathbf{L}(s)$ and $\mathbf{M}(s)$ exists in (9-211b) if and only if

$$\text{rank}\,[\mathbf{B} \quad \mathbf{AB} \quad \cdots \quad \mathbf{A}^{n-1}\mathbf{B}]=\text{rank}\,[\mathbf{B} \quad \mathbf{AB} \quad \cdots \quad \mathbf{A}^{n-1}\mathbf{B} \quad \mathbf{E}_1]$$

or, equivalently, $\mathbf{E}_1$ lies in the space spanned by the column of $[\mathbf{B} \quad \mathbf{AB} \quad \cdots \quad \mathbf{A}^{n-1}\mathbf{B}]$. The solution of $\mathbf{E}_1=[\mathbf{B} \quad \mathbf{AB} \quad \cdots \quad \mathbf{A}^{n-1}\mathbf{B}]\mathbf{P}$ is the representation of $\mathbf{M}(s)$ in (9-211b) with respect to the basis $\bar{\mathbf{L}}(s)$, that is, $\mathbf{M}(s)=\bar{\mathbf{L}}(s)\mathbf{P}$. Next we compute

$$\mathbf{N}(s)\mathbf{M}(s)=\mathbf{D}(s)\mathbf{Q}_2(s)+\mathbf{F}_2(s)$$

Then it can be shown that $\mathbf{L}(s)$ in (9-211b) is given by $\mathbf{L}(s)=\mathbf{Q}_1(s)-\mathbf{Q}_2(s)$. See Reference S85, where the method is developed only for strictly proper $\mathbf{D}^{-1}(s)\mathbf{N}(s)$; the method, however, has been extended to linear systems over rings. The method presented in this chapter follows the line of References 18, S51, and S218. We first translate (9-211) into a set of linear algebraic equations as

$$[\mathbf{L}_0 \quad \mathbf{M}_0 \vdots \mathbf{L}_1 \quad \mathbf{M}_1 \vdots \cdots \vdots \mathbf{L}_m \quad \mathbf{M}_m]\mathbf{S}_m=[\mathbf{F}_0 \quad \mathbf{F}_1 \quad \cdots \quad \mathbf{F}_{\mu+m}] \triangleq \mathbf{F}$$

Then a solution exists in (9-211) if and only if $\mathbf{F}$ is the row space of $\mathbf{S}_m$. The solution of the equation yields immediately the required $\mathbf{L}(s)$ and $\mathbf{M}(s)$. Numerical comparisons of these three approaches are not available at present.

In addition to Equation (9-204), the equation

$$\mathbf{L}(s)\mathbf{D}(s)+\mathbf{N}(s)\mathbf{M}(s)=\mathbf{F}(s) \tag{9-212}$$

also arises in the design of multivariable systems. This equation is quite different from the Diophantine equation, and its solutions are much more complicated. See References S88, S139 and S220. The equation is essential in the study of asymptotic tracking and disturbance rejection (without robust). In our design, the internal model is chosen as $\phi^{-1}(s)\mathbf{I}_p$. This results in a robust design and also bypasses the equation. Consequently, the equation in (9-212) does not arise in this text.

In this chapter we studied two feedback configurations: unity feedback and plant input-output feedback. In the former, the denominator matrix of the compensator is not assignable, and the resulting overall transfer matrix is $\mathbf{N}(s)\mathbf{D}_f^{-1}(s)\mathbf{N}_c(s)$. In the latter, the denominator matrix of the compensators is assignable, and the resulting overall transfer matrix is $\mathbf{N}(s)\mathbf{D}_f^{-1}(s)\mathbf{D}_c(s)$. The $\mathbf{D}_c(s)$ in $\mathbf{N}(s)\mathbf{D}_f^{-1}(s)\mathbf{D}_c(s)$ can be used to generate the desired numerator or be completely concelled by $\mathbf{D}_f(s)$. The reason that we have freedom in assigning $\mathbf{D}_c(s)$ in the plant input-output feedback configuration but not in the unity feedback configuration is very simple. There are three sets of parameters $\mathbf{L}(s)$, $\mathbf{M}(s)$, and $\mathbf{D}_c(s)$ in the former, but only two sets of parameters $\mathbf{N}_c(s)$ and $\mathbf{D}_c(s)$ in the latter, and two sets of parameters are needed to achieve the design.

In the design, the transfer matrix $\hat{\mathbf{G}}(s)$ must be expressed in a coprime fractional form. Two such procedures are developed in Appendix G. In the first method, we first find a noncoprime *right* fraction and then use a sequence of polynomial elementary operations to reduce it to a right coprime one as in (G-33). The $\mathbf{D}(s)$ obtained by this process may not be column reduced. In this case, we must find a unimodular matrix to transform $\mathbf{D}(s)$ to a column reduced one. In the second method, we first find a noncoprime *left* fraction $\mathbf{A}^{-1}(s)\mathbf{B}(s)$. We form a generalized resultant from the coefficient matrices of $\mathbf{A}(s)$ and $\mathbf{B}(s)$, and then search its linearly independent columns in order from left to right. From the primary linearly dependent columns, we can readily obtain a right-coprime fraction. Furthermore, the resulting $\mathbf{D}(s)$ is always column reduced; in fact, it is also row reduced and in the polynomial echelon form.

Once a right-coprime fraction $\hat{\mathbf{G}}(s)=\mathbf{N}(s)\mathbf{D}^{-1}(s)$ is obtained, the next step is to compute the row index of $\hat{\mathbf{G}}(s)$. In the scalar case, this step is unnecessary because we have $v=n=\deg \mathbf{D}(s)$. In the multivariable case, we search linearly independent rows of $\mathbf{S}_m$ in order from top to bottom. Once all

N rows of the last block row of $\mathbf{S}_m$ become linearly dependent, the m is the row index. The total number of linearly independent N rows in $\mathbf{S}_v$ should be equal to deg det $\mathbf{D}(s)$. This can be used to check the correctness of the computation. With the computed v, the chosen $\mathbf{F}$ will be in the row space of $\mathbf{S}_{v-1}$, and the solutions of (9-211) yield immediately the required compensators.

The search of linearly dependent rows or columns of a matrix is basically an ill-conditioned problem, because a zero row can be made to be nonzero by introducing a very small number. Consequently, no matter what method, singular value decomposition or Householder transformations with pivoting, is used, difficulties remain. In our design, the order of rows cannot be changed. Hence the singular value decomposition or Householder transformations with pivoting cannot be used. However we may use Householder transformations without pivoting and back substitution. These two methods are numerically stable and can be used in the design. We emphasize that the numerical stability of an algorithm and the ill- or well-condition of a problem are two distinct properties. Since the search of dependent rows is an ill-conditioned problem, even though we use only numerically stable methods, difficulties may still arise in the computation.

Problems

9-1 Consider

$$\hat{g}_1(s) = \frac{1}{(s+1)(s+2)} \qquad \text{and} \qquad \hat{g}_2(s) = \frac{s+2}{s+3}$$

Is the tandem connection of $\hat{g}_1(s)$ followed by $\hat{g}_2(s)$ completely characterized by its overall transfer function? Verify your result by showing the controllability and observability of its composite dynamical equation.

9-2 Is the feedback system with $\hat{g}_1(s)$, given in Problem 9-1, in the forward path and $g_2(s)$ in the feedback path completely characterized by its overall transfer function?

9-3 Consider

$$\hat{\mathbf{G}}_1(s) = \begin{bmatrix} \dfrac{s+2}{s+1} & 0 \\ 0 & \dfrac{s+1}{s+2} \end{bmatrix} \qquad \hat{\mathbf{G}}_2(s) = \begin{bmatrix} \dfrac{1}{s-1} & \dfrac{s+1}{s+2} \\ 0 & \dfrac{1}{s+2} \end{bmatrix}$$

Are their parallel and tandem [$\hat{\mathbf{G}}_1(s)$ followed by $\hat{\mathbf{G}}_2(s)$] connections controllable? observable? completely characterized by their transfer matrices?

9-4 Is the parallel connection of

$$\hat{\mathbf{G}}_1(s) = \begin{bmatrix} \dfrac{1}{s+1} & 0 \\ 0 & \dfrac{1}{s+2} \end{bmatrix} \qquad \text{and} \qquad \hat{\mathbf{G}}_2(s) = \begin{bmatrix} \dfrac{1}{s+2} & 0 \\ 0 & \dfrac{1}{s+1} \end{bmatrix}$$

controllable and observable? Note that $\hat{\mathbf{G}}_1(s)$ and $\hat{\mathbf{G}}_2(s)$ have the same set of poles.

9-5 Are the feedback systems shown in Figure P9-5 BIBO stable? asymptotically stable?

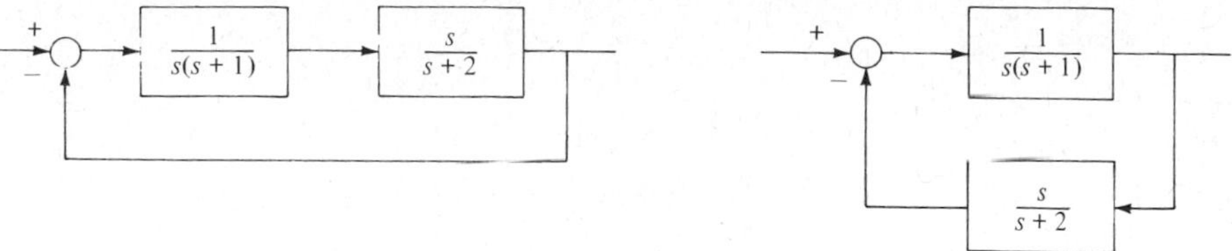

Figure P9-5

9-6 Is the feedback system shown in Figure 9-11 with

$$G_1(s)=\begin{bmatrix} \dfrac{1}{s^2-1} & \dfrac{1}{s+1} \\ \dfrac{1}{s-1} & 1 \end{bmatrix} \qquad G_2(s)=\begin{bmatrix} \dfrac{1}{s+2} & \dfrac{1}{s+3} \\ \dfrac{1}{s+1} & \dfrac{1}{s+3} \end{bmatrix}$$

BIBO stable? asymptotically stable? Use both the coprime fraction formula and $\Delta_1(s)\Delta_2(s)\det(\mathbf{I}+\hat{\mathbf{G}}_1(s)\hat{\mathbf{G}}_2(s))$. Which one is simpler to use for this problem?

9-7 Given the plant $\hat{g}(s)=(s-1)/s(s-2)$. Find a compensator in the unity feedback system so that the poles of the resulting system are -1, -2, and -3.

9-8 Given the plant $\hat{g}(s)=(s^2-1)/(s^2-3s+1)$. Find a proper compensator $C(s)$ of degree 1 so that the poles of the unity feedback system are -1, -2, and -3. Find a strictly proper compensator $C(s)$ of degree 2 so that the poles of the unity feedback system are -1, -2, -3, and -4.

9-9 Consider the plant given in Problem 9-8. Find a set of two poles so that the compensator in the unity feedback system is a constant. Find a set of three poles so that the compensator of degree 1 in the unity feedback system is improper.

9-10 Consider the feedback system shown in Figure P9-10. What is its overall transfer function? Let $N(s)=N_1(s)N_2(s)$, where $N_2(s)$ can be cancelled. Show that, for any $g_f(s)=N_1(s)N_f(s)/D_f(s)$, with $\deg D_f(s)-\deg(N_1(s)N_f(s))\geq \deg D(s)-\deg N(s)$, there exist proper compensator $P(s)/D_c(s)$ and $N_c(s)/D_c(s)$ to achieve the design. If $D_f(s)$ is Hurwitz, will $D_c(s)$ be Hurwitz? If $D_c(s)$ is not Hurwitz and if the system is implemented as shown, will the system be acceptable? Can you find a different but equivalent implementation which will be acceptable? See Problem 9-32 and Reference S34.

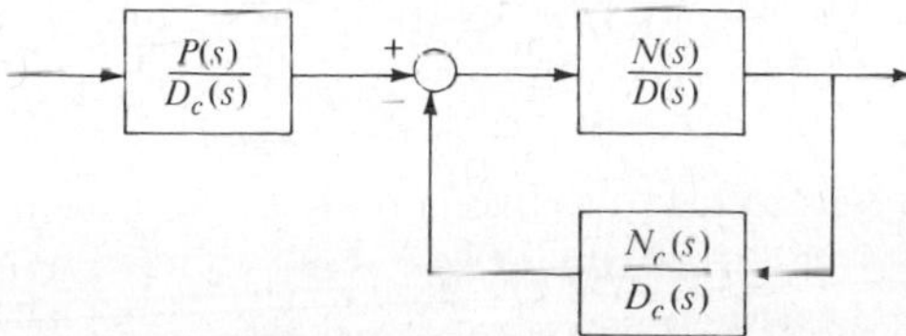

Figure P9-10

9-11 Consider

$$\hat{\mathbf{G}}(s)=\begin{bmatrix}\dfrac{s+1}{s(s-1)}\\[2mm] \dfrac{1}{s^2-1}\end{bmatrix}$$

Find a compensator so that the poles of the overall unity feedback system are -1, $-2\pm j$, and the rest chosen as -2.

9-12 Repeat Problem 9-11 for

$$\hat{\mathbf{G}}(s)=\begin{bmatrix}\dfrac{s+1}{s(s-1)} & \dfrac{1}{s^2-1}\end{bmatrix}$$

9-13 Which of the following matrices are cyclic?

$$\hat{\mathbf{G}}_1(s)=\begin{bmatrix}\dfrac{1}{s} & \dfrac{s+1}{s-2}\\[2mm] \dfrac{1}{s+3} & \dfrac{1}{(s-1)^2}\end{bmatrix} \qquad \hat{\mathbf{G}}_2(s)=\begin{bmatrix}\dfrac{s+2}{(s-1)(s+1)} & \dfrac{1}{s+1}\\[2mm] \dfrac{1}{s-1} & \dfrac{1}{s-1}\end{bmatrix}$$

$$\hat{\mathbf{G}}_3(s)=\begin{bmatrix}\dfrac{2s}{(s-1)(s+1)} & \dfrac{1}{(s-1)(s+1)}\\[2mm] \dfrac{2}{s-1} & \dfrac{1}{s-1}\end{bmatrix} \qquad \hat{\mathbf{G}}_4(s)=\begin{bmatrix}\dfrac{1}{s} & 0\\[2mm] 0 & \dfrac{1}{s}\end{bmatrix}$$

9-14 Show that if all elements of $\hat{\mathbf{G}}(s)$ have no poles in common, then $\hat{\mathbf{G}}(s)$ is cyclic. Note that repeated poles in individual $\hat{g}_{ij}(s)$ are, however, permitted.

9-15 Find the set of 2×1 vectors $\mathbf{t}$ such that $\Delta(\hat{\mathbf{G}}(s))\neq\Delta(\hat{\mathbf{G}}(s)\mathbf{t})$, where

$$\hat{\mathbf{G}}(s)=\begin{bmatrix}\dfrac{s+1}{s-1} & \dfrac{-1}{s}\\[2mm] \dfrac{-2}{s-1} & \dfrac{2}{s}\end{bmatrix}$$

9-16 Show that if the rational matrix $\mathbf{C}(s)$ is cyclic, so is $\bar{\mathbf{C}}(s)=\mathbf{C}(s)+\mathbf{K}$ for any constant $\mathbf{K}$.

9-17 Consider the system in Example 3 on page 484; that is,

$$\hat{\mathbf{G}}(s)=\begin{bmatrix}\dfrac{s^2+1}{s^2-1} & \dfrac{s}{s^2-1}\\[2mm] 0 & \dfrac{s^2+s+1}{s^2-1}\end{bmatrix}$$

Find a compensator in the unity feedback system so that the poles of the resulting system consists of the roots of s^2+s+1 and the rest from -1. Compare your result with the one in the text.

9-18 Consider the system in Example 2 on page 476, that is,

$$\hat{\mathbf{G}}(s)=\begin{bmatrix}\frac{1}{s^2} & \frac{1}{s} & 0\\ 0 & 0 & \frac{1}{s}\end{bmatrix}=\begin{bmatrix}s^2 & 0\\ 0 & s\end{bmatrix}^{-1}\begin{bmatrix}1 & s & 0\\ 0 & 0 & 1\end{bmatrix}$$

Find a compensator in the unity feedback system so that the resulting denominator is

$$\mathbf{D}_f(s)=\begin{bmatrix}(s+1)^2 & 0\\ 0 & s+1\end{bmatrix}$$

Is the resulting $\hat{\mathbf{G}}_f(s)$ cyclic? Compare your result with the one in the text.

9-19 Consider $\hat{\mathbf{G}}(s)=\mathbf{N}(s)\mathbf{D}^{-1}(s)\mathbf{M}(s)$, where $\mathbf{N}(s)$, $\mathbf{D}(s)$, and $\mathbf{M}(s)$ are polynomial matrices. Let $\mathbf{N}(s)\mathbf{D}^{-1}(s)=\bar{\mathbf{D}}^{-1}(s)\bar{\mathbf{N}}(s)$, where $\bar{\mathbf{D}}(s)$ and $\bar{\mathbf{N}}(s)$ are left coprime. Show that if $\mathbf{D}(s)$ and $\mathbf{M}(s)$ are left coprime, so are $\bar{\mathbf{D}}(s)$ and $\bar{\mathbf{N}}(s)\mathbf{M}(s)$.

9-20 Let $\mathbf{D}^{-1}(s)\mathbf{N}(s)$ be a left coprime fraction. Show that $\mathbf{M}(s)\mathbf{D}^{-1}(s)\mathbf{N}(s)$ is a polynomial matrix if and only if $\mathbf{D}(s)$ is a right divisor of $\mathbf{M}(s)$, that is; there exists a polynomial matrix $\bar{\mathbf{M}}(s)$ such that $\mathbf{M}(s)=\bar{\mathbf{M}}(s)\mathbf{D}(s)$.

9-21 Consider the feedback system shown in Figure P9-21. It is assumed that the feedback system is asymptotically stable. Let $\hat{\mathbf{G}}(s)=\mathbf{D}^{-1}(s)\mathbf{N}(s)$ and $\hat{\mathbf{r}}(s)=\mathbf{D}_r^{-1}(s)\mathbf{N}_r(s)\mathbf{r}_0$, where all poles of $\hat{\mathbf{r}}(s)$ lie inside the closed right-half s plane and $\mathbf{D}_r(s)$ and $\mathbf{N}_r(s)$ are left coprime. Show that

$$\hat{\mathbf{e}}(s)=(\mathbf{I}+\hat{\mathbf{G}}(s))^{-1}\hat{\mathbf{r}}(s)=(\mathbf{D}(s)+\mathbf{N}(s))^{-1}\mathbf{D}(s)\mathbf{D}_r^{-1}(s)\mathbf{N}_r(s)\mathbf{r}_0$$

has no closed right-half-plane poles if and only if $\mathbf{D}_r(s)$ is a right divisor of $\mathbf{D}(s)$.

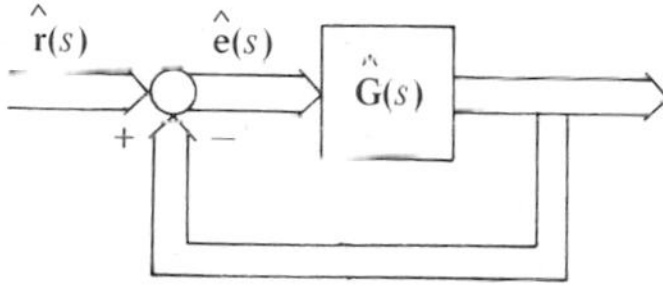

Figure P9-21

9-22 Consider the unity feedback system shown in Figure 9-16. Let $\hat{\mathbf{G}}(s)=\mathbf{N}(s)\mathbf{D}^{-1}(s)=\bar{\mathbf{D}}^{-1}(s)\bar{\mathbf{N}}(s)$ be coprime fractions of the plant transfer matrix. Let $\mathbf{X}(s)$ and $\mathbf{Y}(s)$ be polynomial matrices such that $\mathbf{X}(s)\mathbf{D}(s)+\mathbf{Y}(s)\mathbf{N}(s)=\mathbf{I}$. Show that for any rational matrix $\mathbf{H}(s)$ with all poles inside the open left-half s plane, the compensator

$$\mathbf{C}(s)=[\mathbf{X}(s)-\mathbf{H}(s)\bar{\mathbf{N}}(s)]^{-1}[\mathbf{Y}(s)+\mathbf{H}(s)\bar{\mathbf{D}}(s)]$$

stabilizes the unity feedback system. If $\mathbf{H}(s)$ is proper, will $\mathbf{C}(s)$ be proper? [*Answer:* No. Try $\mathbf{G}(s)=(s+1)/s(s+2)$ and $\mathbf{H}(s)=0$.]

9-23 Given $\hat{g}(s)=(s-1)/s(s-2)$ and $\hat{g}_f(s)=(s-1)/(2s^2+4s+3)$, design an input-output feedback system to have $\hat{g}_f(s)$ as its overall transfer function.

9-24 Consider the plant transfer matrix

$$\hat{\mathbf{G}}(s) = \begin{bmatrix} s & 1 \\ 1 & 1 \end{bmatrix} \begin{bmatrix} s-1 & s \\ 0 & s^2 \end{bmatrix}^{-1}$$

Verify that the row index of $\hat{\mathbf{G}}(s)$ is 2. Design an input-output feedback system to have

$$\mathbf{D}_f(s) = \begin{bmatrix} (s+1)^2 & 0 \\ 0 & (s+1)^3 \end{bmatrix}$$

as its denominator matrix. Choose $\mathbf{D}_c(s)$ as diag $\{s+2, s+2\}$.

9-25 Repeat Problem 9-24 to have

$$\bar{\mathbf{D}}_f(s) = \begin{bmatrix} s+1 & 0 \\ 0 & (s+1)^2 \end{bmatrix}$$

as its denominator matrix.

9-26 Repeat Problem 9-24 to have

$$\mathbf{D}_f(s) = \begin{bmatrix} s(s+2) & s+2 \\ s+2 & s+2 \end{bmatrix}$$

as its denominator matrix. Is the system decoupled? Are all compensators proper? Is the system well posed? Is there any unstable pole-zero cancellation?

9-27 Consider the plant given in Problem 9-24. Design a well-posed input-output feedback system to decouple it without involving any unstable pole-zero cancellations.

9-28 Consider Corollary 9-24 with $\hat{\mathbf{G}}(s)$ strictly proper. Show that if $\delta_{ci}\bar{\mathbf{D}}_f(s) = \delta_{ci}\mathbf{D}(s)$ and $\bar{\mathbf{D}}_{fhc} = \mathbf{D}_{hc}$, that is, column degrees and column-degree-coefficient matrices of $\bar{\mathbf{D}}_f(s)$ and $\mathbf{D}(s)$ are all equal, then the compensator $\mathbf{C}_0(s) = \mathbf{D}_c^{-1}(s)\mathbf{L}(s)$ is strictly proper. The compensator $\mathbf{C}_1(s) = \mathbf{D}_c^{-1}(s)\mathbf{M}(s)$ is generally proper.

9-29 Consider Equations (9-70) and (9-74). Show that if $m \geq v - 1$, for any real k_i, $i = 1, 2, \ldots, \alpha$, the following

$$[\mathbf{D}_{c0} \quad \mathbf{N}_{c0} \,\vdots \cdots \vdots\, \mathbf{D}_{cm} \quad \mathbf{N}_{cm}] + [k_1 \quad k_2 \quad \cdots \quad k_\alpha][-\mathbf{B}_0 \quad \mathbf{A}_0 \,\vdots \cdots \vdots\, -\mathbf{B}_m \quad \mathbf{A}_m]$$

meets (9-74), where $\alpha = (m+1)q - n$ and $[-\mathbf{B}_0 \quad \mathbf{A}_0 \,\vdots \cdots \vdots\, -\mathbf{B}_m \quad \mathbf{A}_m]$ is a basis of the α-dimensional left null space of $\mathbf{S}_m$ computed as in Appendix G. This is a parameterization of all solutions of (9-74).

9-30 Consider $\hat{\mathbf{G}}(s) = \mathbf{N}(s)\mathbf{D}^{-1}(s)$, $\mathbf{T}(s) = \mathbf{D}_t^{-1}\mathbf{N}_t(s)$, and

$$\hat{\mathbf{G}}_0(s) = \hat{\mathbf{G}}(s)\mathbf{T}(s) = \mathbf{N}(s)[\mathbf{D}_t(s)\mathbf{D}(s)]^{-1}\mathbf{N}_t(s)$$

Show that if $\mathbf{D}_t(s)\mathbf{D}(s)$ and $\mathbf{N}_t(s)$ are left coprime, any feedback implementation of $\hat{\mathbf{G}}_0(s)$ reduces essentially to an open-loop system.

9-31 Consider

$$\hat{\mathbf{G}}(s) = \begin{bmatrix} 0 & 1 \\ 1 & s+1 \end{bmatrix} \begin{bmatrix} s^2+1 & 1 \\ 1 & s \end{bmatrix}^{-1}$$

Design a decoupled system to track step reference inputs and reject plant entered disturbance of the type e^t.

9-32 Show that the transfer matrix of the feedback system in Figure P9-32 is $\hat{\mathbf{G}}_f(s) = \mathbf{N}(s)[\mathbf{D}_c(s)\mathbf{D}(s)+\mathbf{N}_c(s)\mathbf{N}(s)]^{-1}\mathbf{K}(s)$. Let $\hat{\mathbf{G}}(s)=\mathbf{N}(s)\mathbf{D}^{-1}(s)=\mathbf{N}_1(s)\mathbf{N}_2(s)\mathbf{D}^{-1}(s)$, where $\mathbf{N}_2(s)$ is nonsingular and can be cancelled. Given $\hat{\mathbf{G}}_f(s)=\mathbf{N}_1(s)\mathbf{D}_f^{-1}(s)\mathbf{N}_f(s)$, under what conditions, will there exist proper compensators $\mathbf{D}_c^{-1}(s)\mathbf{K}(s)$ and $\mathbf{D}_c^{-1}(s)\mathbf{N}_c(s)$ to achieve the design?

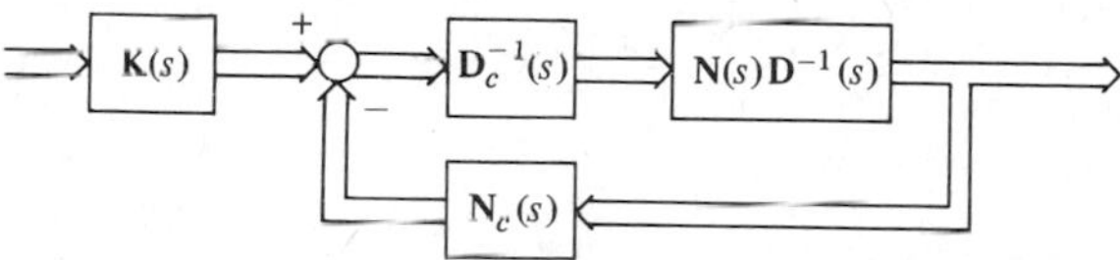

Figure P9-32

9-33 Consider $\hat{\mathbf{G}}(s)=\mathbf{N}(s)\mathbf{D}^{-1}(s)$ with $\delta_{ci}\mathbf{N}(s)\le\delta_{ci}\mathbf{D}(s)$, for all i. Is it true that $\hat{\mathbf{G}}(s)$ is proper if and only if $\mathbf{D}(s)$ is column reduced? If not, find a counterexample.

9-34 Find a minimal proper solution of

$$\begin{bmatrix} \dfrac{1}{s-1} & \dfrac{s-1}{s+3} \\ \dfrac{s}{s+2} & 0 \end{bmatrix}\mathbf{T}(s)=\begin{bmatrix} \dfrac{s+1}{(s+2)(s+3)} & \dfrac{2(s+1)}{(s-1)(s+3)} \\ \dfrac{s(s-1)}{(s+2)^2} & \dfrac{s}{s+2} \end{bmatrix}$$

9-35 Let $\mathbf{Y}(s)$ be a $n\times p$ polynomial matrix of full column rank with column degrees μ_i, $i=1,2,\ldots,p$, and let $\mathbf{x}(s)=[x_1(s) \quad x_2(s) \quad \cdots \quad x_p(s)]'$ be any $p\times 1$ polynomial vector. Consider

$$\mathbf{y}(s)=\mathbf{Y}(s)\mathbf{x}(s)$$

Show that

$$\deg \mathbf{y}(s)=\max_{i:x_i(s)\neq 0}\{\mu_i+\deg x_i(s)\}$$

if and only if $\mathbf{Y}(s)$ is column reduced. This is called the *predictable degree property* in Reference S95 and can be used to show the degrees of a minimal polynomial basis to be minimal. *Hint*: Let $d=\max\{\mu_i+\deg x_i(s)\}$. Then clearly we have $\deg \mathbf{y}(s)\le d$. To show the equality, we show the $\mathbf{y}_d$ in $\mathbf{y}(s)=\mathbf{y}_0+\mathbf{y}_1 s+\cdots+\mathbf{y}_d s^d$ is nonzero by using the relation $\mathbf{y}_d=\mathbf{Y}_{hc}\alpha$, where $x_i(s)=\alpha_i s^{d-\mu_i}+\cdots$ and $\alpha=[\alpha_1 \quad \alpha_2 \quad \cdots \quad \alpha_p]'$.

A

Elementary Transformations

Consider a matrix $\mathbf{A}$ with elements in $\mathbb{R}$ or $\mathbb{C}$. The following operations on $\mathbf{A}$ are called the *elementary operations* or *transformations*: (1) multiplying a row or a column by a nonzero real or complex number, (2) interchanging two rows or two columns, and (3) adding the product of one row or column and a number to another row or column. These transformations can be achieved by using the *elementary matrices*, for $n=5$, of the forms

$$\mathbf{E}_1=\begin{bmatrix}1&0&0&0&0\\0&1&0&0&0\\0&0&1&0&0\\0&0&0&c&0\\0&0&0&0&1\end{bmatrix}\qquad \mathbf{E}_2=\begin{bmatrix}1&0&0&0&0\\0&0&0&0&1\\0&0&1&0&0\\0&0&0&1&0\\0&1&0&0&0\end{bmatrix}\qquad \mathbf{E}_3=\begin{bmatrix}1&0&0&0&0\\0&1&0&0&0\\0&0&1&0&0\\0&d&0&1&0\\0&0&0&0&1\end{bmatrix} \tag{A-1}$$

where c and d are real or complex numbers and $c\neq 0$. Elementary matrices are square and nonsingular. Their inverses are

$$\mathbf{E}_1^{-1}=\begin{bmatrix}1&0&0&0&0\\0&1&0&0&0\\0&0&1&0&0\\0&0&0&1/c&0\\0&0&0&0&1\end{bmatrix}\qquad \mathbf{E}_2^{-1}=\mathbf{E}_2\qquad \mathbf{E}_3^{-1}=\begin{bmatrix}1&0&0&0&0\\0&1&0&0&0\\0&0&1&0&0\\0&-d&0&1&0\\0&0&0&0&1\end{bmatrix} \tag{A-2}$$

They are again elementary matrices. The premultiplication of $\mathbf{E}_i$ on $\mathbf{A}$ operates on the rows of $\mathbf{A}$, whereas the postmultiplication of $\mathbf{E}_i$ on $\mathbf{A}$ operates on the columns of $\mathbf{A}$. For example, $\mathbf{E}_2\mathbf{A}$ interchanges the second and fifth rows; $\mathbf{E}_3\mathbf{A}$ adds the product of the second row and d to the fourth row of $\mathbf{A}$; whereas $\mathbf{A}\mathbf{E}_3$ adds the product of the fourth column and d to the second column of $\mathbf{A}$.

A-1 Gaussian Elimination

Consider the $n \times m$ matrix

$$\mathbf{A} = \begin{bmatrix} a_{11} & a_{12} & a_{13} & \cdots & a_{1m} \\ a_{21} & a_{22} & a_{23} & \cdots & a_{2m} \\ \vdots & \vdots & \vdots & & \vdots \\ a_{n1} & a_{n2} & a_{n3} & \cdots & a_{nm} \end{bmatrix} \tag{A-3}$$

First we assume $a_{11} \neq 0$. We add the product of the first row and $(-a_{i1}/a_{11})$ to the ith row, for $i = 2, 3, \ldots, n$; then we have

$$\mathbf{K}_1\mathbf{A} = \begin{bmatrix} 1 & 0 & 0 & \cdots & 0 \\ e_{21} & 1 & 0 & \cdots & 0 \\ e_{31} & 0 & 1 & \cdots & 0 \\ \vdots & \vdots & \vdots & & \vdots \\ e_{n1} & 0 & 0 & \cdots & 1 \end{bmatrix} \mathbf{A} = \begin{bmatrix} a_{11} & a_{12} & a_{13} & \cdots & a_{1m} \\ 0 & a_{22}^1 & a_{23}^1 & \cdots & a_{2m}^1 \\ 0 & a_{32}^1 & a_{33}^1 & \cdots & a_{3m}^1 \\ \vdots & \vdots & \vdots & & \vdots \\ 0 & a_{n2}^1 & a_{n3}^1 & \cdots & a_{nm}^1 \end{bmatrix} \tag{A-4}$$

where $e_{i1} = -a_{i1}/a_{11}$, $i = 2, 3, \ldots, n$ and $a_{ij}^1 = a_{ij} + e_{i1}a_{1j}$. Note that the $n \times n$ matrix $\mathbf{K}_1$ is the product of $n-1$ number of elementary matrices of the form of $\mathbf{E}_3$. If a_{22}^1 is different from zero, then the addition of the product of the second row of $\mathbf{K}_1\mathbf{A}$ and $-a_{i2}^1/a_{22}^1$ to the ith row, $i = 3, 4, \ldots, n$, yields

$$\mathbf{K}_2\mathbf{K}_1\mathbf{A} = \begin{bmatrix} 1 & 0 & 0 & \cdots & 0 \\ 0 & 1 & 0 & \cdots & 0 \\ 0 & e_{32} & 1 & \cdots & 0 \\ \vdots & \vdots & \vdots & & \vdots \\ 0 & e_{n2} & 0 & \cdots & 1 \end{bmatrix} \mathbf{K}_1\mathbf{A} = \begin{bmatrix} a_{11} & a_{12} & a_{13} & \cdots & a_{1m} \\ 0 & a_{22}^1 & a_{23}^1 & \cdots & a_{2m}^1 \\ 0 & 0 & a_{33}^2 & \cdots & a_{3m}^2 \\ \vdots & \vdots & \vdots & & \vdots \\ 0 & 0 & a_{n3}^2 & \cdots & a_{nm}^2 \end{bmatrix} \tag{A-5}$$

where $e_{i2} = -a_{i2}^1/a_{22}^1$ and $a_{ij}^2 = a_{ij}^1 + e_{i2}a_{2j}^1$. Proceeding in this manner, matrix $\mathbf{A}$ can be transformed into a matrix whose elements below the diagonal are zero as

$$\begin{bmatrix} x & x & x & \cdots & x & x & \cdots & x \\ 0 & x & x & \cdots & x & x & \cdots & x \\ 0 & 0 & x & \cdots & x & x & \cdots & x \\ \vdots & \vdots & \vdots & & \vdots & \vdots & & \vdots \\ 0 & 0 & 0 & \cdots & x & x & \cdots & x \\ 0 & 0 & 0 & \cdots & 0 & 0 & \cdots & 0 \\ \vdots & \vdots & \vdots & & \vdots & \vdots & & \vdots \\ 0 & 0 & 0 & \cdots & 0 & 0 & \cdots & 0 \end{bmatrix} \quad \text{or} \quad \begin{bmatrix} x & x & x & \cdots & x & x & \cdots & x \\ 0 & x & x & \cdots & x & x & \cdots & x \\ 0 & 0 & x & \cdots & x & x & \cdots & x \\ \vdots & \vdots & \vdots & & \vdots & \vdots & & \vdots \\ 0 & 0 & 0 & \cdots & x & x & \cdots & x \\ 0 & 0 & 0 & \cdots & 0 & x & \cdots & x \end{bmatrix} \tag{A-6}$$

where x denotes possible nonzero elements. This process of transformations is called the *gaussian elimination*. The matrices in (A-6) are upper right triangular matrices.[1]

---

[1] All diagonal elements in (A-6) are not necessarily nonzero. It is possible to make them nonzero by using additional column operations. See also Equation (G-28).

This process will fail if any of a_{11}, a_{22}^1, a_{33}^2, or other diagonal element is zero. Even if all of them are nonzero, the process may still encounter some computational difficulty. For example, if a_{11} is very small in magnitude, then $(-a_{i1}/a_{11}) = e_{i1}$ will be a very large number. If there are errors in representing a_{1j}, these errors will be greatly amplified by e_{i1} in the operations. Furthermore, these errors may propagate and be reamplified at later stages. Consequently, the final result may be overwhelmed by errors. Because of this phenomenon, the gaussian elimination is said to be numerically unstable.

To overcome this difficulty, before carrying out the elimination, we first search the element with the largest absolute value[2] in the first column of **A**, and then interchange this row with the first row by using a matrix of the form of $\mathbf{E}_2$. We then carry out the elimination for the first column. Before carrying out the elimination for the second column, we search the largest element in magnitude in the second column of $\mathbf{K}_1\mathbf{A}$ excluding the first element, and bring it to the a_{22}^1 position. This process of elimination is called the *gaussian elimination with partial pivoting*. In this process, we have $|e_{ij}| \leq 1$, and the errors will not be amplified. Hence this process is more stable numerically than the gaussian elimination. By using partial pivoting, we conclude that every matrix **A** can be transformed into the form in (A-6) by using a sequence of elementary row operations.

The stability of the gaussian elimination can be further improved by using *complete pivoting*. In this case, we must use elementary row operations as well as elementary column operations. We first search the largest element in magnitude among all elements of **A**. We then move the element to the first column by an elementary column operation and then to the first row by an elementary row operation. We then carry out the elimination of the first column of the transformed **A**. After this elimination, we repeat the process for the remaining matrix. This process is called the gaussian elimination with *complete pivoting*. This process is more stable numerically than the one with partial pivoting; however, it is more costly because the search of the largest element in magnitude is a time-consuming process. Hence complete pivoting is not used as often as partial pivoting. According to Reference S138, from the point of view of overall performance which includes efficiency, accuracy, reliability, generality, and ease of use, the gaussian elimination with partial pivoting is satisfactory for most general matrices.

*A-2 Householder Transformation

The numerical stability of the elimination can be further improved by using the Householder transformation. Let **I** be a unit matrix. A Householder transformation is a square matrix of the form[3]

$$\mathbf{H} = \mathbf{I} - 2\mathbf{x}\mathbf{x}'$$

---

[2] For conciseness, we call such element the largest element in magnitude.

[3] All arguments still hold if **H** is defined as $\mathbf{H} = \mathbf{I} - 2\mathbf{x}\mathbf{x}^*$, where $\mathbf{x}^*$ is the complex conjugate transpose of **x**. In this case, we have $\mathbf{H}^{-1} = \mathbf{H}^*$ and **H** is unitary.

with $\mathbf{x}'\mathbf{x} = 1$, where $\mathbf{x}$ is a column vector and $\mathbf{x}'$ is its transpose. Clearly $\mathbf{H}' = (\mathbf{I} - 2\mathbf{x}\mathbf{x}')' = \mathbf{I} - 2\mathbf{x}\mathbf{x}' = \mathbf{H}$; hence $\mathbf{H}$ is symmetric. Because, by using $\mathbf{x}'\mathbf{x} = 1$,

$$\mathbf{H}'\mathbf{H} = (\mathbf{I} - 2\mathbf{x}\mathbf{x}')(\mathbf{I} - 2\mathbf{x}\mathbf{x}') = \mathbf{I} - 4\mathbf{x}\mathbf{x}' + 4\mathbf{x}\mathbf{x}'\mathbf{x}\mathbf{x}' = \mathbf{I}$$

we have $\mathbf{H}^{-1} = \mathbf{H}'$; hence $\mathbf{H}$ is orthogonal (see Appendix E). Because all the singular values of $\mathbf{H}$ are 1 (Problem E-9), its condition number, defined as the ratio of the largest singular value and the smallest singular value, is 1. Hence the use of Householder transformations will not impair the numerical property of a problem (see Problem E-10).

An important property of Householder transformations is that given any two vectors $\mathbf{a}$ and $\mathbf{b}$ of equal euclidean norms, that is, $\|\mathbf{a}\|_2^2 = \mathbf{a}'\mathbf{a} = \mathbf{b}'\mathbf{b}$, there exists a Householder transformation $\mathbf{H}$ such that $\mathbf{Ha} = \mathbf{b}$. Indeed, if we choose $\mathbf{x} = (\mathbf{a} - \mathbf{b})/\|\mathbf{a} - \mathbf{b}\|_2$, then we have

$$\mathbf{Ha} = \left[\mathbf{I} - \frac{2(\mathbf{a} - \mathbf{b})(\mathbf{a} - \mathbf{b})'}{\|\mathbf{a} - \mathbf{b}\|_2^2}\right]\mathbf{a} = \mathbf{a} - \frac{2(\mathbf{a} - \mathbf{b})[(\mathbf{a} - \mathbf{b})'\mathbf{a}]}{\mathbf{a}'\mathbf{a} - \mathbf{a}'\mathbf{b} - \mathbf{b}'\mathbf{a} + \mathbf{b}'\mathbf{b}}$$

Since $\mathbf{a}'\mathbf{a} = \mathbf{b}'\mathbf{b}$ by assumption and $\mathbf{a}'\mathbf{b} = \mathbf{b}'\mathbf{a}$ for this is a scalar quantity which is invariant under transposition, we have

$$\mathbf{Ha} = \mathbf{a} - \frac{2(\mathbf{a}'\mathbf{a} - \mathbf{b}'\mathbf{a})}{2(\mathbf{a}'\mathbf{a} - \mathbf{b}'\mathbf{a})}(\mathbf{a} - \mathbf{b}) = \mathbf{b}$$

This establishes the assertion. Now we use this property to show that the $\mathbf{K}_1$ and $\mathbf{K}_2$ in (A-4) and (A-5) can be chosen as Householder transformations.

Consider the $n \times m$ matrix $\mathbf{A}$ given in (A-3). First we compute the norm σ of the first column $\mathbf{a}_1$ of $\mathbf{A}$. We choose $\mathbf{b}_1$ as $[\pm\sigma \quad 0 \quad 0 \quad \cdots \quad 0]'$.[4] Then there exists a Householder transformation $\mathbf{H}_1$ such that $\mathbf{H}_1\mathbf{A}$ is in the form of the right-hand-side matrix in (A-4). Next we delete the first column and first row of $\mathbf{H}_1\mathbf{A}$ and repeat the process for the first column of the remaining submatrix. Proceeding in this manner, the matrix $\mathbf{A}$ can be transformed by a sequence of Householder transformations into the form shown in (A-6).

The numerical stability of this process can be further improved by the permutation of columns. First we compute the norms of all columns of $\mathbf{A}$. Next we permute the column with the largest norm with the first column. We then apply the Householder transformation to carry out the elimination. By this process, the diagonal elements of the resulting matrix will be in the decreasing order of magnitude. We call this process *Householder transformations with pivoting*.

We note that Householder transformations are not triangular matrices as $\mathbf{K}_1$ and $\mathbf{K}_2$ in (A-4) and (A-5). Hence the rows of $\mathbf{A}$ are completely scrambled after the application of Householder transformations.

[4]The sign may be chosen to be equal to $-\text{sign}\, a_{11}$ to reduce the roundoff errors. In the QR factorization (see Problem A-5), we must choose however the positive sign.

If a matrix is transformed into the form in (A-6), the rank of the matrix is equal to the number of nonzero rows in (A-6). This is a simple way of computing the rank of a matrix.[5]

A-3 Row-Searching Algorithm[6]

In our application, it is often required to search the linearly independent rows of matrix **A** in order from top to bottom. By this, we mean that we first check whether or not the first row of **A** is nonzero, or equivalently, linearly independent by itself. We then check whether or not the second row is linearly independent of the first row. In the kth step, we check whether or not the kth row of **A** is linearly independent of its previous $k-1$ rows. If a row is linearly dependent on its previous rows, the row is to be eliminated from subsequent consideration. Furthermore, we may want to find the coefficients of the linear combination of the dependent row.

Let the (i, j)th element of the $n \times m$ matrix **A** be denoted by a_{ij}. Let a_{1k} be any nonzero element in the first row of **A**. This element will be called *the pivot element* or simply *the pivot*. Let $\mathbf{K}_1$ be of the form shown in (A-4) with $e_{i1} = -a_{ik}/a_{1k}$, $i = 2, 3, \ldots, n$. Then the kth column, except the first element, of $\mathbf{K}_1\mathbf{A} = (a_{ij}^1)$ is a zero column, where $a_{ij}^1 = a_{ij} + e_{i1}a_{1j}$. Let a_{2j}^1 be any nonzero element in the second row of $\mathbf{K}_1\mathbf{A}$. Let $\mathbf{K}_2$ be of the form shown in (A-5) with $e_{i2} = -a_{ij}^1/a_{2j}^1$. Then the jth column, except the first two elements, of $\mathbf{K}_2\mathbf{K}_1\mathbf{A} = (a_{ij}^2)$ is a zero column, where $a_{ij}^2 = a_{ij}^1 + e_{i2}a_{2j}^1$. In this process, if there

[5]The computation of the rank of a matrix is a difficult problem. For example, the matrix

$$\mathbf{A}_5 = \begin{bmatrix} 1 & -1 & -1 & -1 & -1 \\ 0 & 1 & -1 & -1 & -1 \\ 0 & 0 & 1 & -1 & -1 \\ 0 & 0 & 0 & 1 & -1 \\ 0 & 0 & 0 & 0 & 1 \end{bmatrix}$$

is clearly of rank 5 and nonsingular. However, we have

$$\mathbf{A}_5\mathbf{x} \triangleq \mathbf{A}_5 \begin{bmatrix} 1 \\ 2^{-1} \\ 2^{-2} \\ 2^{-3} \\ 2^{-4} \end{bmatrix} = \begin{bmatrix} 2^{-4} \\ 2^{-4} \\ 2^{-4} \\ 2^{-4} \\ 2^{-4} \end{bmatrix} = 2^{-4} \begin{bmatrix} 1 \\ 1 \\ 1 \\ 1 \\ 1 \end{bmatrix} \triangleq 2^{-4}\mathbf{e}$$

Thus, if n in $\mathbf{A}_n$ is very large, there exists a nonzero $\mathbf{x}$ such that $\mathbf{A}_n\mathbf{x} = 2^{1-n}\mathbf{e} \to \mathbf{0}$ and $\mathbf{A}_n$ is nearly singular. This information cannot be detected from the determinant of $\mathbf{A}_n$ (which is equal to 1 for all n), nor from its eigenvalues [which are all equal to 1 (see Problem 2-23)], nor from the form of $\mathbf{A}_n$. However, if we compute the singular values of $\mathbf{A}_n$ (see the singular value decomposition in Appendix E), then the smallest singular value can be shown to behave as 2^{-n}, for large n (see Problem E-13), and the rank degeneracy of $\mathbf{A}_n$ can therefore be detected. Thus the singular value decomposition is considered the most reliable method of computing the rank of a matrix.

[6]This algorithm is a simplified version of the one in Reference S22.

is no nonzero element in a row, we assign $\mathbf{K}_i$ as a unit matrix and then proceed to the next row. If we carry this process to the last row of $\mathbf{A}$, then we have[7]

$$\mathbf{K}_{n-1}\mathbf{K}_{n-2}\cdots\mathbf{K}_2\mathbf{K}_1\mathbf{A} \triangleq \mathbf{K}\mathbf{A} = \bar{\mathbf{A}} \tag{A-7}$$

where $\mathbf{K} \triangleq \mathbf{K}_{n-1}\mathbf{K}_{n-2}\cdots\mathbf{K}_1$. Since every $\mathbf{K}_i$ is a lower triangular matrix with 1 on its diagonal, so is $\mathbf{K}$. The number of nonzero rows in $\bar{\mathbf{A}}$ gives the rank of $\mathbf{A}$. If the jth row of $\bar{\mathbf{A}}$ is a zero row vector, then the jth row of $\mathbf{A}$ is linearly dependent on its previous rows. Furthermore, the coefficients of combination

$$[b_{j1} \quad b_{j2} \quad \cdots \quad b_{j(j-1)} \quad b_{jj} \quad 0 \quad \cdots \quad 0]\mathbf{A} = \mathbf{0} \tag{A-8}$$

with $b_{jj} = 1$, is just the jth row of $\mathbf{K}$.

The matrix $\mathbf{K}$ can be computed by direct multiplications of $\mathbf{K}_i$, $i = 1, 2, \ldots, n-1$. This is complicated. We introduce in the following a recursive way of computing any row of $\mathbf{K}$. First we store the ith column of $\mathbf{K}_i$ in the ith column of

$$\mathbf{F} = \begin{bmatrix} 1 & 0 & 0 & \cdots & 0 \\ e_{21} & 1 & 0 & \cdots & 0 \\ e_{31} & e_{32} & 1 & \cdots & 0 \\ \vdots & \vdots & \vdots & & \vdots \\ e_{n1} & e_{n2} & e_{n3} & \cdots & 1 \end{bmatrix} \tag{A-9}$$

To compute the jth row of $\mathbf{K}$, we take the first j rows of $\mathbf{F}$ and then arrange the jth row of $\mathbf{K}$ under it as

$$\begin{array}{l} \left.\begin{bmatrix} 1 & 0 & 0 & \cdots & & & & & 0 \\ e_{21} & 1 & 0 & \cdots & & & & & 0 \\ e_{31} & e_{32} & 1 & \cdots & & & & & 0 \\ \vdots & \vdots & \vdots & & & & & & \vdots \\ e_{j1} & e_{j2} & e_{j3} & \cdots & e_{j(j-1)} & 1 & 0 & \cdots & 0 \end{bmatrix}\right\} \text{First } j \text{ rows of } \mathbf{F} \\ \quad\; b_{j2} \quad b_{j3} \quad \cdots \quad b_{j(j-1)} \quad b_{jj} \quad 0 \quad \cdots \quad 0 \quad \leftarrow j\text{th row of } \mathbf{K} \end{array} \tag{A-10}$$

Then it is straightforward to verify that

$$\begin{aligned} b_{jj} &= 1. \\ b_{jk} &= [b_{j(k+1)} \quad b_{j(k+2)} \quad \cdots \quad b_{jj}] \begin{bmatrix} e_{(k+1)k} \\ e_{(k+2)k} \\ \vdots \\ e_{jk} \end{bmatrix} \\ &= \sum_{p=k+1}^{j} b_{jp} e_{pk} \qquad k = j-1,\ j-2, \ldots, 1 \end{aligned} \tag{A-11}$$

[7] $\mathbf{K}_i$ cannot be chosen as Householder transformations because they will scramble the order of the rows of $\mathbf{A}$.

We see that b_{jk} is just the inner product of the vector on its right-hand side and the vector above it as shown in (A-10).[8] Hence the coefficients of combination in (A-8) or, equivalently, the jth row of $\mathbf{K}$ can be readily computed by this simple procedure.

In the application of this algorithm in this text, the information of the entire $\mathbf{K}=\mathbf{K}_{n-1}\cdots\mathbf{K}_1$ is never needed. We need only a few rows of $\mathbf{K}$. Therefore it is better to store $\mathbf{K}_i$ in $\mathbf{F}$ as in (A-9) and write (A-7) as

$$\mathbf{F} \overset{*}{*} \mathbf{A}=\mathbf{K}_{n-1}\mathbf{K}_{n-2}\cdots\mathbf{K}_1\mathbf{A}=\mathbf{K}\mathbf{A}=\bar{\mathbf{A}} \qquad \textbf{(A-12)}$$

Whenever a row of $\mathbf{K}$ is needed we then use the procedure in (A-10) and (A-11) to compute that row of $\mathbf{K}$ from $\mathbf{F}$.

Example 1

Find the linearly independent rows of

$$\mathbf{A}=\begin{bmatrix} -1 & -1 & \circled{2} & -2 & 1 \\ 1 & -1 & 4 & -2 & 4 \\ -1 & -3 & 8 & -6 & 6 \\ 5 & 1 & -4 & 10 & 1 \\ 7 & 1 & -2 & 10 & 4 \end{bmatrix}$$

We choose the (1, 3) element as the pivot and compute

$$\mathbf{K}_1\mathbf{A}=\begin{bmatrix} 1 & & & & \\ -2 & 1 & & & \\ -4 & 0 & 1 & & \\ 2 & 0 & 0 & 1 & \\ 1 & 0 & 0 & 0 & 1 \end{bmatrix}\mathbf{A}=\begin{bmatrix} -1 & -1 & \circled{2} & -2 & 1 \\ \circled{3} & 1 & 0 & 2 & 2 \\ 3 & 1 & 0 & 2 & 2 \\ 3 & -1 & 0 & 6 & 3 \\ 6 & 0 & 0 & 8 & 5 \end{bmatrix} \triangleq \mathbf{A}_1$$

Except the first element, the first column of $\mathbf{K}_1$ is the third column of $\mathbf{A}$ divided by -2. We next choose the (2, 1) element of $\mathbf{A}_1$ as the pivot and compute

$$\mathbf{K}_2\mathbf{A}_1=\begin{bmatrix} 1 & & & & \\ 0 & 1 & & & \\ 0 & -1 & 1 & & \\ 0 & -1 & 0 & 1 & \\ 0 & -2 & 0 & 0 & 1 \end{bmatrix}\mathbf{A}_1=\begin{bmatrix} -1 & -1 & \circled{2} & -2 & 1 \\ \circled{3} & 1 & 0 & 2 & 2 \\ 0 & 0 & 0 & 0 & 0 \\ 0 & -2 & 0 & 4 & 1 \\ 0 & -2 & 0 & 4 & 1 \end{bmatrix} \triangleq \mathbf{A}_2$$

Since the third row of $\mathbf{A}_2$ is a zero row, we set $\mathbf{K}_3=\mathbf{I}$ and proceed to the next row. The pivot element of the fourth row of $\mathbf{A}_2$ is chosen as shown. We compute

[8] This presentation was suggested by Professor J. W. Wang.

$$\mathbf{K}_4\mathbf{K}_3\mathbf{A}_2 = \mathbf{K}_4\mathbf{A}_2 = \begin{bmatrix} 1 & & & & \\ 0 & 1 & & & \\ 0 & 0 & 1 & & \\ 0 & 0 & 0 & 1 & \\ 0 & 0 & 0 & -1 & 1 \end{bmatrix} \mathbf{A}_2 = \begin{bmatrix} -1 & -1 & \textcircled{2} & -2 & 1 \\ \textcircled{3} & 1 & 0 & 2 & 2 \\ 0 & 0 & 0 & 0 & 0 \\ 0 & -2 & 0 & \textcircled{4} & 1 \\ 0 & 0 & 0 & 0 & 0 \end{bmatrix} \triangleq \bar{\mathbf{A}}$$

The last row of $\bar{\mathbf{A}}$ is a zero row, and the search is completed. We use (A-12) to write

$$\mathbf{F}{}^*_*\mathbf{A} = \begin{bmatrix} 1 & & & & \\ -2 & 1 & & & \\ -4 & -1 & 1 & & \\ 2 & -1 & 0 & 1 & \\ 1 & -2 & 0 & -1 & 1 \end{bmatrix} {}^*_*\mathbf{A} = \begin{bmatrix} -1 & -1 & 2 & -2 & 1 \\ 3 & 1 & 0 & 2 & 2 \\ 0 & 0 & 0 & 0 & 0 \\ 0 & -2 & 0 & 4 & 1 \\ 0 & 0 & 0 & 0 & 0 \end{bmatrix} \triangleq \bar{\mathbf{A}}$$

Note that the ith column of $\mathbf{F}$ is the ith column of $\mathbf{K}_i$, $i = 1, 2, 3, 4$. Since the third row of $\bar{\mathbf{A}}$ is a zero row, the third row of $\mathbf{A}$ is linearly dependent on its previous two rows. Similarly, the fifth row of $\mathbf{A}$ is linearly dependent on its previous four rows.

To find the coefficients of combination for the third row of $\mathbf{A}$, we compute, by using (A-11), from the first three rows of $\mathbf{F}$ as

$$\begin{bmatrix} 1 & 0 & 0 & 0 & 0 \\ \boxed{-2} & 1 & 0 & 0 & 0 \\ \boxed{-4} & \textcircled{-1} & 1 & 0 & 0 \end{bmatrix}$$
$$\boxed{-2} \quad \textcircled{-1} \quad \textcircled{1}$$

Hence we have

$$[-2 \quad -1 \quad 1 \quad 0 \quad 0]\mathbf{A} = \mathbf{0}$$

or

$$-2\mathbf{a}_1 - \mathbf{a}_2 + \mathbf{a}_3 = \mathbf{0} \tag{A-13}$$

where $\mathbf{a}_i$ is the ith row of $\mathbf{A}$.

To find the coefficients of combination for the fifth row of $\mathbf{A}$, we compute

$$\begin{bmatrix} 1 & & & & \\ -2 & 1 & & & \\ -4 & -1 & 1 & & \\ 2 & -1 & \boxed{0} & 1 & \\ 1 & -2 & \boxed{0} & -1 & 1 \end{bmatrix}$$
$$1 \quad -1 \quad 0 \quad \boxed{-1 \quad 1}$$

Hence we have

$$[1 \quad -1 \quad 0 \quad -1 \quad 1]\mathbf{A} = \mathbf{0}$$

or

$$\mathbf{a}_1 - \mathbf{a}_2 + 0\mathbf{a}_3 - \mathbf{a}_4 + \mathbf{a}_5 = \mathbf{0} \tag{A-14}$$

∎

This algorithm of searching linearly independent rows of **A** in the order of the first row, the second row, and so forth will be referred to as the *row-searching algorithm*. In this algorithm, if the ith row is linearly dependent, then the ith row will not contribute to the linear combination for the jth linear dependent row for $j > i$. For example, since $\mathbf{a}_3$ is linearly dependent, its coefficient is zero in (A-14); hence $\mathbf{a}_5$ is expressed as a linear combination of its previous linearly independent rows. In general, in (A-8), if the ith row, with $i < j$, of **A** is linearly dependent, then $b_{ji} = 0$. Indeed, if the ith row is linearly dependent, then the ith row of $\bar{\mathbf{A}}$ is a zero row, and the $\mathbf{K}_i$ in (A-12) is a unit matrix. Consequently, all elements below e_{ii} in the ith column of **F** are zero. That is, $e_{pi} = 0$, for $p = i+1, i+2, \ldots, j$. Hence from (A-11), we have $b_{ji} = 0$. Because of this property, the b_{jk} computed in this algorithm are unique. Without $b_{ji} = 0$, the coefficients of combination may not be unique. For example, the addition of (A-13) and (A-14) yields

$$-\mathbf{a}_1 - 2\mathbf{a}_2 + \mathbf{a}_3 - \mathbf{a}_4 + \mathbf{a}_5 = [-1 \quad -2 \quad 1 \quad -1 \quad 1]\mathbf{A} = \mathbf{0}$$

which does not have $b_{53} = 0$, and is a different combination. The property $b_{ji} = 0$ if the ith row is dependent is essential in establishing Theorem G-14 in Appendix G.

A remark is in order regarding the numerical stability of the row-searching algorithm. If the pivot element of each row is chosen as the leftmost nonzero element, then this algorithm reduces essentially to the gaussian elimination (without any pivoting). Though the pivot is chosen as the largest element in magnitude of each row, the pivot is not necessarily the largest element in magnitude of its column. Hence the row-searching algorithm by choosing the largest element in magnitude as the pivot is not exactly equivalent to the gaussian elimination with partial pivoting. Consequently, there is no guarantee that the row-searching algorithm will be numerically stable.

In the search of linearly independent rows of **A**, it is essential not to alter the order of rows. The column positions, however, can be arbitrarily altered. Hence we may apply to the columns of **A** the gaussian elimination with partial pivoting or the Householder transformations to transform **A** into the form

$$\mathbf{A}\mathbf{L}_1\mathbf{L}_2\cdots \overset{\text{say}}{=} \begin{bmatrix} \bar{a}_{11} & 0 & 0 & 0 & 0 \\ \bar{a}_{21} & \bar{a}_{22} & 0 & 0 & 0 \\ \bar{a}_{31} & \bar{a}_{32} & 0 & 0 & 0 \\ \bar{a}_{41} & \bar{a}_{42} & \bar{a}_{43} & 0 & 0 \\ \bar{a}_{51} & \bar{a}_{52} & \bar{a}_{53} & 0 & 0 \end{bmatrix} \triangleq \bar{\mathbf{A}} \tag{A-15}$$

where $\mathbf{L}_i$ are elementary matrices or the Householder transformations.[9] It is assumed that $\bar{a}_{11} \neq 0$, $\bar{a}_{22} \neq 0$, and $\bar{a}_{43} \neq 0$. Clearly (A-15) implies that the first, second, and fourth rows of **A** are linearly independent of their previous rows and the third and fifth rows of **A** are linearly dependent on their previous rows.

[9] Householder transformations with pivoting (permutation of rows), however, are not permitted.

To find the coefficients of linear combination for the third row, we solve

$$[b_{31} \quad b_{32} \quad 1 \quad 0 \quad 0]\bar{\mathbf{A}} = \mathbf{0}$$

which implies

$$b_{32} = \frac{-\bar{a}_{32}}{\bar{a}_{22}}$$

and

$$b_{31} = \frac{-(\bar{a}_{31} + b_{32}\bar{a}_{21})}{\bar{a}_{11}}$$

To find the coefficients of linear combination for the fifth row, we solve

$$[b_{51} \quad b_{52} \quad 0 \quad b_{54} \quad 1]\bar{\mathbf{A}} = \mathbf{0} \tag{A-16}$$

which implies

$$b_{54} = \frac{-\bar{a}_{53}}{\bar{a}_{43}} \qquad b_{52} = \frac{-(\bar{a}_{52} + b_{54}\bar{a}_{42})}{\bar{a}_{22}}$$

and

$$b_{51} = \frac{-(\bar{a}_{51} + b_{54}\bar{a}_{41} + b_{52}\bar{a}_{21})}{\bar{a}_{11}}$$

We see that the b_{ij} are obtained by back substitution. This is a numerically stable method (see Reference S212). Hence we conclude that once **A** is transformed into the form in (A-15), the coefficients of linear combination can be easily obtained. In solving these coefficients, the elements corresponding to linearly dependent rows are set to zero as in (A-16). In so doing, the result will be identical to the one obtained by the row-searching algorithm. Although this method of searching linearly dependent rows is less direct than the row searching algorithm, the method is stable numerically.

The row searching algorithm, however, will be used exclusively in this text for the following two reasons. First, the concept is very simple. Its use will not complicate or obscure the basic issue of the problem. Second, the method can be easily carried out by hand for simple problems. After solving one or two problems by hand, one would understand the material better and gain a confidence in using a digital computer. For hand calculation, the pivot should be chosen for the convenience of computation. For example, we may choose the element with value +1 or the element whose column has the largest number of zero elements as the pivot. In summary, the use of the row searching algorithm is for pedagogical reasons. In actual digital computer computation, one should use the method discussed in (A-15) or other method which is numerically stable.

*A-4 Hessenberg Form

Every square matrix can be transformed into a Jordan canonical form by a similarity transformation. However, this requires the computation of eigenvalues and eigenvectors and is a numerically unstable problem. In the following, we shall use elementary matrices in the similarity transformation to transform a matrix into a special form, called the *Hessenberg form*. This form is very important in computer computation and its generalized form will be used in Section 5-8.

Let $\mathbf{A}$ be an $n \times n$ matrix. Consider

$$\mathbf{P}_1\mathbf{A} \triangleq \left[\begin{array}{c|ccccc} 1 & 0 & 0 & 0 & \cdots & 0 \\ \hline 0 & & & & & \\ 0 & & & & & \\ 0 & & & \mathbf{P}_{11} & & \\ \vdots & & & & & \\ 0 & & & & & \end{array}\right] \mathbf{A} = \begin{bmatrix} \mathrm{x} & \mathrm{x} & \mathrm{x} & \mathrm{x} & \cdots & \mathrm{x} \\ \mathrm{x} & \mathrm{x} & \mathrm{x} & \mathrm{x} & \cdots & \mathrm{x} \\ 0 & \mathrm{x} & \mathrm{x} & \mathrm{x} & \cdots & \mathrm{x} \\ 0 & \mathrm{x} & \mathrm{x} & \mathrm{x} & \cdots & \mathrm{x} \\ \vdots & \vdots & & \vdots & & \vdots \\ 0 & \mathrm{x} & \mathrm{x} & \mathrm{x} & \cdots & \mathrm{x} \end{bmatrix}$$

The matrix $\mathbf{P}_{11}$ is chosen to make the first column of $\mathbf{A}$, except the first two elements, zeros as shown. This can be achieved by using gaussian elimination with partial pivoting or a Householder transformation. The inverse of $\mathbf{P}_1$ has the same form as $\mathbf{P}_1$. The postmultiplication of $\mathbf{P}_1\mathbf{A}$ by $\mathbf{P}_1^{-1}$ will not operate on the first column of $\mathbf{P}_1\mathbf{A}$; hence the pattern of zeros in the first column of $\mathbf{P}_1\mathbf{A}$ is preserved in $\mathbf{P}_1\mathbf{A}\mathbf{P}_1^{-1}$. Next we find a $\mathbf{P}_2$ so that

$$\mathbf{P}_2\mathbf{P}_1\mathbf{A}\mathbf{P}_1^{-1} \triangleq \left[\begin{array}{cc|cccc} 1 & 0 & 0 & 0 & \cdots & 0 \\ 0 & 1 & 0 & 0 & \cdots & 0 \\ \hline 0 & 0 & & & & \\ 0 & 0 & & & & \\ \vdots & & & \mathbf{P}_{22} & & \\ 0 & 0 & & & & \end{array}\right] \mathbf{P}_1\mathbf{A}\mathbf{P}_1^{-1} = \begin{bmatrix} \mathrm{x} & \mathrm{x} & \mathrm{x} & \mathrm{x} & \cdots & \mathrm{x} \\ \mathrm{x} & \mathrm{x} & \mathrm{x} & \mathrm{x} & \cdots & \mathrm{x} \\ 0 & \mathrm{x} & \mathrm{x} & \mathrm{x} & \cdots & \mathrm{x} \\ 0 & 0 & \mathrm{x} & \mathrm{x} & \cdots & \mathrm{x} \\ \vdots & \vdots & \vdots & \vdots & & \vdots \\ 0 & 0 & \mathrm{x} & \mathrm{x} & \cdots & \mathrm{x} \end{bmatrix}$$

Again the inverse of $\mathbf{P}_2$ has the same form as $\mathbf{P}_2$ and the postmultiplication of $\mathbf{P}_2\mathbf{P}_1\mathbf{A}\mathbf{P}_1^{-1}$ by $\mathbf{P}_2^{-1}$ will not operate on the first two columns of $\mathbf{P}_2\mathbf{P}_1\mathbf{A}\mathbf{P}_1^{-1}$. Hence the pattern of zeros in the first two columns of $\mathbf{P}_2\mathbf{P}_1\mathbf{A}\mathbf{P}_1^{-1}$ is preserved in $\mathbf{P}_2\mathbf{P}_1\mathbf{A}\mathbf{P}_1^{-1}\mathbf{P}_2^{-1}$. Proceeding similarly, we can transform a matrix, by a sequence of similarity transformations, into the form

$$\begin{bmatrix} \mathrm{x} & \mathrm{x} & \mathrm{x} & \mathrm{x} & \cdots & \mathrm{x} & \mathrm{x} & \mathrm{x} \\ \mathrm{x} & \mathrm{x} & \mathrm{x} & \mathrm{x} & \cdots & \mathrm{x} & \mathrm{x} & \mathrm{x} \\ 0 & \mathrm{x} & \mathrm{x} & \mathrm{x} & \cdots & \mathrm{x} & \mathrm{x} & \mathrm{x} \\ 0 & 0 & \mathrm{x} & \mathrm{x} & \cdots & \mathrm{x} & \mathrm{x} & \mathrm{x} \\ 0 & 0 & 0 & \mathrm{x} & \cdots & \mathrm{x} & \mathrm{x} & \mathrm{x} \\ \vdots & \vdots & \vdots & \vdots & & \vdots & \vdots & \vdots \\ 0 & 0 & 0 & 0 & \cdots & \mathrm{x} & \mathrm{x} & \mathrm{x} \\ 0 & 0 & 0 & 0 & \cdots & 0 & \mathrm{x} & \mathrm{x} \end{bmatrix} \qquad \textbf{(A-17)}$$

This is called an *upper Hessenberg form* and can be obtained by using numerically stable methods. This form is very useful in many computer computations.

Problems

A-1 Use the row searching algorithm to find linearly dependent rows of

$$\begin{bmatrix} 3 & \textcircled{1} & 3 \\ 5 & 2 & \textcircled{7} \\ 4 & 1.5 & 5 \end{bmatrix}$$

Use the circled elements as pivot elements. Use the procedure in (A-9) to (A-12) to find the coefficients of its linear combination.

A-2 Repeat Problem A-1 for the matrix

$$\begin{bmatrix} 3 & \textcircled{1} & 0 & -1 & -2 \\ 4 & 1 & \textcircled{1} & 2 & 1 \\ 10 & 3 & 1 & 0 & -3 \\ \textcircled{2} & 1 & 0 & 3 & 0 \\ 20 & 6 & 3 & 7 & -1 \end{bmatrix}$$

Verify that $b_{ji}=0$ if the ith row, with $i<j$, is linearly dependent.

A-3 Is it possible to obtain a triangular form by using the procedure of obtaining a Hessenberg form?

A-4 Show that if $\mathbf{Q}_i$, $i=1, 2, \ldots, m$, are unitary matrices, that is, $\mathbf{Q}_i^*\mathbf{Q}_i=\mathbf{Q}_i\mathbf{Q}_i^*=\mathbf{I}$, then so is $\mathbf{Q} \triangleq \mathbf{Q}_m\mathbf{Q}_{m-1}\cdots\mathbf{Q}_2\mathbf{Q}_1$.

A-5 Let $\mathbf{A}$ be an $n\times m$ matrix of rank m. Show that there exists a unitary matrix $\mathbf{Q}$ such that

$$\mathbf{QA}=\begin{bmatrix}\mathbf{R}\\ \mathbf{0}\end{bmatrix} \qquad \text{or} \qquad \mathbf{A}=\mathbf{Q}^*\begin{bmatrix}\mathbf{R}\\ \mathbf{0}\end{bmatrix}=[\mathbf{Q}_1 \quad \mathbf{Q}_2]\begin{bmatrix}\mathbf{R}\\ \mathbf{0}\end{bmatrix}=\mathbf{Q}_1\mathbf{R}$$

where $\mathbf{R}$ is an upper triangular matrix with nonnegative elements on the diagonal and $\mathbf{Q}_1$ is the first m columns of $\mathbf{Q}^*$. This is called the *QR factorization* of $\mathbf{A}$. (*Hint*: Use Householder transformations and Problem A-4. Note that the only difference between the *QR* factorization and the Householder transformation is that the signs of the diagonal elements of $\mathbf{R}$ may be different. In the former, the signs must be chosen as positive; in the latter, they can be chosen as positive or negative to reduce the roundoff errors.)

A-6 Let $\mathbf{A}$ be a square matrix and let $\mathbf{A}=\mathbf{QR}$ be its *QR* factorization. Show that the matrix $\bar{\mathbf{A}} \triangleq \mathbf{RQ}$ has the same set of eigenvalues of $\mathbf{A}=\mathbf{QR}$. (*Hint*: Show that $\mathbf{A}$ and $\bar{\mathbf{A}}$ are similar. *The QR algorithm*, which is the most reliable method of computing the eigenvalues of $\mathbf{A}$, is based on this property. See References S181 and S200.)

B

Analytic Functions of a Real Variable

Let D be an open interval in the real line $\mathbb{R}$ and let $f(\cdot)$ be a function defined on D; that is, to each point in D, a unique number is assigned to f. The function $f(\cdot)$ may be real-valued or complex-valued.

A function $f(\cdot)$ of a real variable is said to be an element of class C^n on D if its nth derivative, $f^{(n)}(\cdot)$, exists and is continuous for all t in D. C^∞ is the class of functions having derivatives of all orders.

A function of a real variable, $f(\cdot)$, is said to be *analytic* on D if f is an element of C^∞ *and* if for each t_0 in D there exists a positive real number ε_0 such that, for all t in $(t_0 - \varepsilon_0, t_0 + \varepsilon_0)$, $f(t)$ is representable by a Taylor series about the point t_0:

$$f(t) = \sum_{n=0}^{\infty} \frac{(t - t_0)^n}{n!} f^{(n)}(t_0) \tag{B-1}$$

A remark is in order at this point. If f is a function of a complex variable and if f has continuous derivative, then it can be shown that f has continuous second derivative, third derivative, . . . , and in fact a Taylor-series expansion. Therefore, a function of a *complex* variable may be defined as analytic if it has a continuous derivative. However, for functions of a *real* variable, even if a function possesses derivatives of all order, it may still not be analytic. For example, the function

$$f(t) = \begin{cases} e^{1/t^2} & \text{for } t \neq 0 \\ 0 & \text{for } t = 0 \end{cases}$$

is not analytic at $t = 0$, even though it is infinitely differentiable at $t = 0$; see Reference 9.

The sum, product, or quotient (provided the denominator is not equal to 0 at any point) of analytic functions of a real variable is analytic. All polynomials, exponential functions, and sinusoidal functions are analytic in the entire real line.

If a function is known to be analytic in D, then the function is completely determinable from an arbitrary point in D if all the derivatives at that point are known. This can be argued as follows: Suppose that the value of f and its derivatives at t_0 are known, then by using (B-1), we can compute f over $(t_0 - \varepsilon_0, t_0 + \varepsilon_0)$. Next we choose a point t_1 that is almost equal to $t_0 + \varepsilon_0$; again, from (B-1), we can compute f over $(t_0 + \varepsilon_0, t_0 + \varepsilon_0 + \varepsilon_1)$. Proceeding in both directions, we can compute the function f over D. This process is called *analytic continuation.*

Theorem B-1

If a function f is analytic on D and if f is known to be identically zero on an arbitrarily small nonzero interval in D, then the function f is identically zero on D.

Proof

If the function is identically zero on an arbitrarily small nonzero interval, say (t_0, t_1), then the function and its derivatives are all equal to zero on (t_0, t_1). By analytic continuation, the function can be shown to be identically zero. Q.E.D.

C

Minimum-Energy Control[1]

Consider the n-dimensional linear time-varying state equation

$$E: \qquad \dot{\mathbf{x}} = \mathbf{A}(t)\mathbf{x} + \mathbf{B}(t)\mathbf{u}$$

where $\mathbf{x}$ is the $n \times 1$ state vector, $\mathbf{u}$ is the $p \times 1$ input vector, and $\mathbf{A}$ and $\mathbf{B}$ are, respectively, $n \times n$ and $n \times p$ matrices whose entries are continuous functions of t defined over $(-\infty, \infty)$. If the state equation is controllable at time t_0, then given any initial state $\mathbf{x}_0$ at time t_0 and any desired final state $\mathbf{x}_1$, there exists a finite $t_1 > t_0$ and an input $\mathbf{u}_{[t_0, t_1]}$ that will transfer the state $\mathbf{x}_0$ at time t_0 to $\mathbf{x}_1$ at time t_1, denoted as $(\mathbf{x}_0, t_0)$ to $(\mathbf{x}_1, t_1)$. Define

$$\mathbf{W}(t_0, t_1) \triangleq \int_{t_0}^{t_1} \mathbf{\Phi}(t_0, \tau)\mathbf{B}(\tau)\mathbf{B}^*(\tau)\mathbf{\Phi}^*(t_0, \tau)\, d\tau$$

Then the input $\mathbf{u}^0_{[t_0, t_1]}$ defined by

$$\mathbf{u}^0(t) = (\mathbf{\Phi}(t_0, t)\mathbf{B}(t))^* \mathbf{W}^{-1}(t_0, t_1)[\mathbf{\Phi}(t_0. t_1)\mathbf{x}_1 - \mathbf{x}_0] \qquad \text{for all } t \text{ in } [t_0, t_1] \qquad \textbf{(C-1)}$$

will transfer $(\mathbf{x}_0, t_0)$ to $(\mathbf{x}_1, t_1)$. This is proved in Theorem 5-4. Now we show that the input $\mathbf{u}^0_{[t_0, t_1]}$ consumes the minimal amount of energy, among all the $\mathbf{u}$'s that can transfer $(\mathbf{x}_0, t_0)$ to $(\mathbf{x}_1, t_1)$.

Theorem C-1

Let $\mathbf{u}^1_{[t_0, t_1]}$ be any control that transfers $(\mathbf{x}_0, t_0)$ to $(\mathbf{x}_1, t_1)$, and let $\mathbf{u}^0$ be the control defined in (C-1) that accomplishes the same transfer; then

$$\int_{t_0}^{t_1} \|\mathbf{u}^1(t)\|^2\, dt \geq \int_{t_0}^{t_1} \|\mathbf{u}^0(t)\|^2\, dt$$

[1] This appendix closely follows Reference 69.

where $\|\mathbf{u}(t)\| \triangleq (\langle \mathbf{u}(t), \mathbf{u}(t)\rangle)^{1/2} \triangleq (\mathbf{u}^*(t)\mathbf{u}(t))^{1/2}$, the euclidean norm of $\mathbf{u}(t)$. (See Section 2-8.)

Proof

The solution of the state equation E is

$$\mathbf{x}(t_1) = \mathbf{\Phi}(t_1, t_0)\left[\mathbf{x}(t_0) + \int_{t_0}^{t_1} \mathbf{\Phi}(t_0, \tau)\mathbf{B}(\tau)\mathbf{u}(\tau)\, d\tau\right] \tag{C-2}$$

Define

$$\bar{\mathbf{x}} \triangleq \mathbf{\Phi}^{-1}(t_1, t_0)\mathbf{x}(t_1) - \mathbf{x}(t_0) = \mathbf{\Phi}(t_0, t_1)\mathbf{x}_1 - \mathbf{x}_0$$

Then the assumptions that $\mathbf{u}^1$ and $\mathbf{u}^0$ transfer $(\mathbf{x}_0, t_0)$ to $(\mathbf{x}_1, t_1)$ imply that

$$\bar{\mathbf{x}} = \int_{t_0}^{t_1} \mathbf{\Phi}(t_0, \tau)\mathbf{B}(\tau)\mathbf{u}^1(\tau)\, d\tau = \int_{t_0}^{t_1} \mathbf{\Phi}(t_0, \tau)\mathbf{B}(\tau)\mathbf{u}^0(\tau)\, d\tau$$

Subtracting both sides, we obtain

$$\int_{t_0}^{t_1} \mathbf{\Phi}(t_0, \tau)\mathbf{B}(\tau)(\mathbf{u}^1(\tau) - \mathbf{u}^0(\tau))\, d\tau = \mathbf{0}$$

which implies that

$$\left\langle \int_{t_0}^{t_1} \mathbf{\Phi}(t_0, \tau)\mathbf{B}(\tau)(\mathbf{u}^1(\tau) - \mathbf{u}^0(\tau))\, d\tau, \mathbf{W}^{-1}(t_0, t_1)\bar{\mathbf{x}} \right\rangle = 0$$

By using (2-94), we can write this equation as

$$\int_{t_0}^{t_1} \langle \mathbf{u}^1(\tau) - \mathbf{u}^0(\tau), (\mathbf{\Phi}(t_0, \tau)\mathbf{B}(\tau))^*\mathbf{W}^{-1}(t_0, t_1)\bar{\mathbf{x}} \rangle\, d\tau = 0 \tag{C-3}$$

With the use of (C-1), Equation (C-3) becomes

$$\int_{t_0}^{t_1} \langle \mathbf{u}^1(\tau) - \mathbf{u}^0(\tau), \mathbf{u}^0(\tau) \rangle\, d\tau = 0 \tag{C-4}$$

Consider now

$$\int_{t_0}^{t_1} \|\mathbf{u}^1\|^2\, d\tau$$

By some manipulation and using (C-4), we obtain

$$\begin{aligned}
\int_{t_0}^{t_1} \|\mathbf{u}^1(\tau)\|^2\, d\tau &= \int_{t_0}^{t_1} \|\mathbf{u}^1(\tau) - \mathbf{u}^0(\tau) + \mathbf{u}^0(\tau)\|^2\, d\tau \\
&= \int_{t_0}^{t_1} \|\mathbf{u}^1(\tau) - \mathbf{u}^0(\tau)\|^2\, d\tau + \int_{t_0}^{t_1} \|\mathbf{u}^0(\tau)\|^2\, d\tau \\
&\quad + 2\int_{t_0}^{t_1} \langle \mathbf{u}^1(\tau) - \mathbf{u}^0(\tau), \mathbf{u}^0(\tau) \rangle\, d\tau \\
&= \int_{t_0}^{t_1} \|\mathbf{u}^1(\tau) - \mathbf{u}^0(\tau)\|^2\, d\tau + \int_{t_0}^{t_1} \|\mathbf{u}^0(\tau)\|^2\, d\tau
\end{aligned}$$

Since

$$\int_{t_0}^{t_1} \|\mathbf{u}^1(\tau) - \mathbf{u}^0(\tau)\|^2 \, d\tau$$

is always nonnegative, we conclude that

$$\int_{t_0}^{t_1} \|\mathbf{u}^1(\tau)\|^2 \, d\tau \geq \int_{t_0}^{t_1} \|\mathbf{u}^0(\tau)\|^2 \, d\tau \qquad \text{Q.E.D.}$$

We see that in transferring $(\mathbf{x}_0, t_0)$ to $(\mathbf{x}_1, t_1)$, if minimizing

$$\int_{t_0}^{t_1} \|\mathbf{u}(\tau)\|^2 \, d\tau$$

is used as a criterion, the control defined in (C-1) is optimal. Since, in many instances,

$$\int_{t_0}^{t_1} \|\mathbf{u}(\tau)\|^2 \, d\tau$$

is related to energy, the control $\mathbf{u}^0$ is called the *minimum-energy control*.

D

Controllability after the Introduction of Sampling

Consider the linear time-invariant dynamical equation

$$FE: \qquad \dot{\mathbf{x}} = \mathbf{A}\mathbf{x} + \mathbf{B}\mathbf{u} \tag{D-1a}$$

$$\mathbf{y} = \mathbf{C}\mathbf{x} \tag{D-1b}$$

where $\mathbf{x}$ is the $n \times 1$ state vector, $\mathbf{u}$ is the $p \times 1$ input vector, $\mathbf{y}$ is the $q \times 1$ output vector; $\mathbf{A}$, $\mathbf{B}$, and $\mathbf{C}$ are $n \times n$, $n \times p$, and $q \times n$ constant matrices, respectively. The response of FE is given by

$$\mathbf{x}(t) = e^{\mathbf{A}t}\mathbf{x}_0 + \int_0^t e^{\mathbf{A}(t-\tau)}\mathbf{B}\mathbf{u}(\tau)\, d\tau \tag{D-2a}$$

$$\mathbf{y}(t) = \mathbf{C}\mathbf{x}(t) \tag{D-2b}$$

where $\mathbf{x}_0$ is the initial state at $t = 0$.

We consider now the case in which the input $\mathbf{u}$ is piecewise constant; that is, the input $\mathbf{u}$ changes values only at discrete instants of time. Inputs of this type occur in sampled-data systems or in systems in which digital computers are used to generate $\mathbf{u}$. A piecewise-constant function is often generated by a sampler and a filter, called zero-order hold, as shown in Figure D-1. Let

$$\mathbf{u}(t) = \mathbf{u}(k) \qquad \text{for } kT \leq t < (k+1)T;\ k = 0, 1, 2, \ldots \tag{D-3}$$

where T is a positive constant, called the *sampling period*. The discrete times $0, T, 2T, \ldots$, are called *sampling instants*. The behavior of FE with the piecewise-constant inputs given in (D-3) can be computed from (D-2). However, if only the behavior at sampling instants $0, T, 2T, \ldots$, is of interest, a discrete-time dynamical equation can be written to give the response of $\mathbf{x}(k) \triangleq \mathbf{x}(kT)$ at

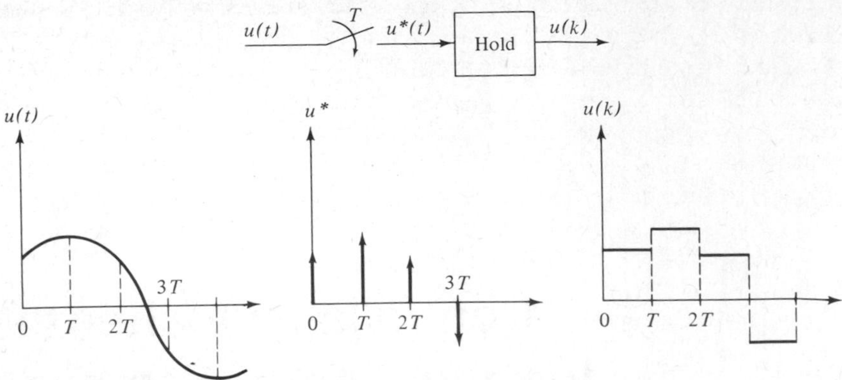

Figure D-1

$k = 0, 1, 2, \ldots$. From (D-2a), we have

$$\begin{aligned}\mathbf{x}(k+1) &= e^{\mathbf{A}(k+1)T}\mathbf{x}_0 + \int_0^{(k+1)T} e^{\mathbf{A}[kT+T-\tau]}\mathbf{B}\mathbf{u}(\tau)\,d\tau \\ &= e^{\mathbf{A}T}\left[e^{\mathbf{A}kT}\mathbf{x}_0 + \int_0^{kT} e^{\mathbf{A}(kT-\tau)}\mathbf{B}\mathbf{u}(\tau)\,d\tau\right] \\ &\quad + \int_{kT}^{(k+1)T} e^{\mathbf{A}(kT+T-\tau)}\mathbf{B}\mathbf{u}(\tau)\,d\tau \end{aligned} \tag{D-4}$$

The term in brackets in (D-4) is equal to $\mathbf{x}(k)$; the input $\mathbf{u}(\tau)$ is constant in the interval $(kT, kT+T)$ and is equal to $\mathbf{u}(k)$; hence (D-4) becomes, after the change of variable $\alpha = kT + T - \tau$,

$$\mathbf{x}(k+1) = e^{\mathbf{A}T}\mathbf{x}(k) + \left(\int_0^T e^{\mathbf{A}\alpha}\,d\alpha\right)\mathbf{B}\mathbf{u}(k)$$

which is a discrete-time state equation. Therefore, if the input is piecewise constant over the same interval T, and if only the response at the sampling instants is of interest, the dynamical equation FE in (D-1) can be replaced by the following discrete-time linear time-invariant dynamical equation

$$DFE: \quad \mathbf{x}(k+1) = \tilde{\mathbf{A}}\mathbf{x}(k) + \tilde{\mathbf{B}}\mathbf{u}(k) \tag{D-5a}$$

$$\mathbf{y}(k) = \tilde{\mathbf{C}}\mathbf{x}(k) \tag{D-5b}$$

where

$$\tilde{\mathbf{A}} = e^{\mathbf{A}T} \tag{D-6}$$

$$\tilde{\mathbf{B}} = \left(\int_0^T e^{\mathbf{A}\tau}\,d\tau\right)\mathbf{B} \triangleq \mathbf{M}\mathbf{B} \tag{D-7}$$

$$\tilde{\mathbf{C}} = \mathbf{C} \tag{D-8}$$

If the dynamical equation FE is controllable, it is of interest to study whether the system remains controllable after the introduction of sampling or, correspondingly, whether the discrete-time dynamical equation DFE is controllable.

This problem is important in the design of dead-beat sampled-data systems. See Problem 7-30.

A discrete-time dynamical equation *DFE* is defined to be controllable if for any initial state $\mathbf{x}_0$ and any $\mathbf{x}_1$ in the state space, there exists an input sequence $\{\mathbf{u}(n)\}$ of finite length that transfers $\mathbf{x}_0$ to $\mathbf{x}_1$. The controllability conditions in statements 3 and 4 of Theorem 5-7 and in Theorem 5-21 and its corollary are directly applicable to the discrete-time case without any modification.

Let $\lambda_i(\mathbf{A})$ denote an eigenvalue of $\mathbf{A}$, and let "Im" and "Re" stand for the imaginary part and the real part, respectively.

Theorem D-1

Assume that the dynamical equation *FE* given in (D-1) is controllable. A sufficient condition for the discrete-time dynamical equation *DFE* given in (D-5) to be controllable is that $\text{Im}\,[\lambda_i(\mathbf{A})-\lambda_j(\mathbf{A})] \neq 2\pi\alpha/T$ for $\alpha = \pm 1, \pm 2, \ldots,$ whenever $\text{Re}\,[\lambda_i(\mathbf{A})-\lambda_j(\mathbf{A})]=0$.

For the single-input case ($p=1$), the condition is necessary as well. ∎

First we remark on the conditions of the theorem. Because of $\tilde{\mathbf{A}} = e^{\mathbf{A}T}$, if λ_i is an eigenvalue of $\mathbf{A}$, $\tilde{\lambda}_i \triangleq e^{\lambda_i T}$ is an eigenvalue of $\tilde{\mathbf{A}}$ (Problem 2-32). Let $\lambda_1 = \tau + j\beta$ and $\lambda_2 = \tau + j(\beta + 2\pi\alpha/T)$, for $\alpha = \pm 1, \pm 2, \ldots.$ Then we have $\tilde{\lambda}_2 = e^{(\tau + j(\beta + 2\pi\alpha/T))T} = e^{(\tau + j\beta)T + j2\pi\alpha} = e^{(\tau + j\beta)T} = \tilde{\lambda}_1$. In other words, if the conditions of the theorem are violated, different λ_i may yield the same $\tilde{\lambda}_i$. Hence the conditions of the theorem ensure that if $\{\lambda_i, i=1, 2, \ldots, m\}$ are distinct eigenvalues of $\mathbf{A}$, then $\{\tilde{\lambda}_i, i=1, 2, \ldots, m\}$ are distinct eigenvalues of $\tilde{\mathbf{A}}$.

Proof of Theorem D-1

We assume, without loss of generality, that $\mathbf{A}$ is in the Jordan canonical form shown in Table 5-1. Then, from (D-6), (2-70), and (2-71), we have

$$\tilde{\mathbf{A}} = e^{\mathbf{A}T} = \begin{bmatrix} e^{\mathbf{A}_{11}T} & \mathbf{0} & \cdots & \mathbf{0} \\ \mathbf{0} & e^{\mathbf{A}_{12}T} & \cdots & \mathbf{0} \\ \vdots & \vdots & & \vdots \\ \mathbf{0} & \mathbf{0} & \cdots & e^{\mathbf{A}_{mr(m)}T} \end{bmatrix} \tag{D-9}$$

where

$$\tilde{\mathbf{A}}_{ij} \triangleq e^{\mathbf{A}_{ij}T} = \tilde{\lambda}_i \begin{bmatrix} 1 & T & \cdots & \dfrac{T^{(n_{ij}-1)}}{(n_{ij}-1)!} \\ 0 & 1 & \cdots & \dfrac{T^{(n_{ij}-2)}}{(n_{ij}-2)!} \\ 0 & 0 & \cdots & \dfrac{T^{(n_{ij}-3)}}{(n_{ij}-3)!} \\ \vdots & \vdots & & \vdots \\ 0 & 0 & \cdots & 1 \end{bmatrix} \tag{D-10}$$

and $\tilde{\lambda}_i = e^{\lambda_i T}$. Although $\mathbf{A}_{ij}$ is a Jordan block, $\tilde{\mathbf{A}}_{ij}$ is not in a Jordan form. Now we shall transform it into a Jordan form. We compute

$$\tilde{\mathbf{A}}_{ij}-\tilde{\lambda}_i\mathbf{I}=\begin{bmatrix} 0 & x & x & \cdots & x & x \\ 0 & 0 & x & \cdots & x & x \\ 0 & 0 & 0 & \cdots & x & x \\ \vdots & \vdots & \vdots & & \vdots & \vdots \\ 0 & 0 & 0 & \cdots & 0 & x \\ 0 & 0 & 0 & \cdots & 0 & 0 \end{bmatrix}$$

where x denotes possible nonzero element. Because of its special form, it is easy to verify that $(\tilde{\mathbf{A}}_{ij}-\tilde{\lambda}_i\mathbf{I})^2$ has one more zero superdiagonal row than $(\tilde{\mathbf{A}}_{ij}-\tilde{\lambda}_i\mathbf{I})$; and finally, similar to (2-64), $(\tilde{\mathbf{A}}_{ij}-\tilde{\lambda}_i\mathbf{I})^{n_{ij}-1}$ has only one nonzero element at the upper right corner and $(\tilde{\mathbf{A}}_{ij}-\tilde{\lambda}_i\mathbf{I})^{n_{ij}}$ is a zero matrix. Consequently, $\mathbf{e}=[0 \;\; 0 \;\; \cdots \;\; 0 \;\; 1]'$ is a generalized eigenvector of grade n_{ij} (see Section 2-6). From this vector we can obtain a similarity transformation of the form

$$\mathbf{Q}_{ij}=\begin{bmatrix} x & x & x & \cdots & x & 0 \\ 0 & x & x & \cdots & x & 0 \\ 0 & 0 & x & \cdots & x & 0 \\ \vdots & \vdots & \vdots & & \vdots & \vdots \\ 0 & 0 & 0 & \cdots & x & 0 \\ 0 & 0 & 0 & \cdots & 0 & 1 \end{bmatrix} \tag{D-11}$$

such that

$$\mathbf{P}_{ij}\tilde{\mathbf{A}}_{ij}\mathbf{Q}_{ij}=\begin{bmatrix} \tilde{\lambda}_i & 1 & 0 & \cdots & 0 & 0 \\ 0 & \tilde{\lambda}_i & 1 & \cdots & 0 & 0 \\ 0 & 0 & \tilde{\lambda}_i & \cdots & 0 & 0 \\ \vdots & \vdots & \vdots & & \vdots & \vdots \\ 0 & 0 & 0 & \cdots & \tilde{\lambda}_i & 1 \\ 0 & 0 & 0 & \cdots & 0 & \tilde{\lambda}_i \end{bmatrix} \tag{D-12}$$

where $\mathbf{P}_{ij}\triangleq\mathbf{Q}_{ij}^{-1}$ and has exactly the same form as $\mathbf{Q}_{ij}$. We see that the Jordan form of $\tilde{\mathbf{A}}_{ij}$ is the same as $\mathbf{A}_{ij}$ with λ_i replaced by $\tilde{\lambda}_i$. Define

$$\mathbf{P}=\text{diag}\,\{\mathbf{P}_{11}, \mathbf{P}_{12}, \ldots, \mathbf{P}_{mr(m)}\} \tag{D-13}$$

then (D-5a) can be transformed into

$$\mathbf{x}(k+1)=\mathbf{P}\tilde{\mathbf{A}}\mathbf{P}^{-1}\mathbf{x}(k)+\mathbf{PMBu}(k) \tag{D-14}$$

with $\mathbf{P}\tilde{\mathbf{A}}\mathbf{P}^{-1}$ exactly in the form in Table 5-1 with λ_i replaced by $\tilde{\lambda}_i$. Now we shall use this equation to establish the theorem.

First we note that, because of (D-9) and (D-10), the matrix $\mathbf{M}$ defined in (D-7) is of block diagonal form, each block is a triangular form of order n_{ij} with all diagonal elements equal to $(1-e^{-\lambda_i T})/\lambda_i$ if $\lambda_i \neq 0$ or T if $\lambda_i=0$. The condition of the theorem implies $\lambda_i T \neq 2\pi\alpha$, if Re $\lambda_i = 0$ or $e^{-\lambda_i T} \neq 1$. Hence if $T > 0$, **M** is nonsingular.

If $\tilde{\mathbf{A}}$ is in the form in Table 5-1 and if $\{\mathbf{A}, \mathbf{B}\}$ is controllable, then the rows $\mathbf{b}_{lij}$, $j = 1, 2, \ldots, r(i)$, in (5-71a) are linearly independent. Now under the conditions on the eigenvalues stated in Theorem D-1, $\bar{\lambda}_i$, $i = 1, 2, \ldots, m$, are distinct. Hence the necessary and sufficient conditions for $\{\mathbf{P}\tilde{\mathbf{A}}\mathbf{P}^{-1}, \mathbf{PMB}\}$ to be controllable is that the rows of **PMB** corresponding to the rows of $\mathbf{b}_{lij}$, $l = 1, 2, \ldots, r(i)$, are linearly independent. Because of the special forms of **P** and **M**, this is the case if and only if $\mathbf{b}_{lij}$, $j = 1, 2, \ldots, r(i)$ are linearly independent. Hence we conclude that under the conditions on the eigenvalues stated in Theorem D-1, the discrete-time equation in (D-5a) is controllable if and only if the continuous-time equation in (D-1a) is controllable.

In the single-input case, if the conditions on the eigenvalues are violated, there will be two or more Jordan blocks associated with the same eigenvalue in Equation (D-14). Hence the single-input equation (D-14) is, following Corollary 5-21, not controllable. This establishes the theorem. Q.E.D.

In the proof of Theorem D-1, we have essentially established the following theorem.

Theorem D-2

If a continuous-time linear time-invariant state equation is not controllable, its discretized state equation for any sampling period is not controllable. ∎

This theorem is intuitively obvious. If a state equation is not controllable by using an input defined for all time, its discretized equation certainly cannot be controlled by an input defined only at discrete instants of time.

Example

Consider the sampled-data system shown in Figure D-2. By partial fraction expansion, we have

$$\frac{8}{(s+1)(s+1+2i)(s+1-2i)} = \frac{2}{s+1} - \frac{1}{(s+1+2i)} - \frac{1}{(s+1-2i)}$$

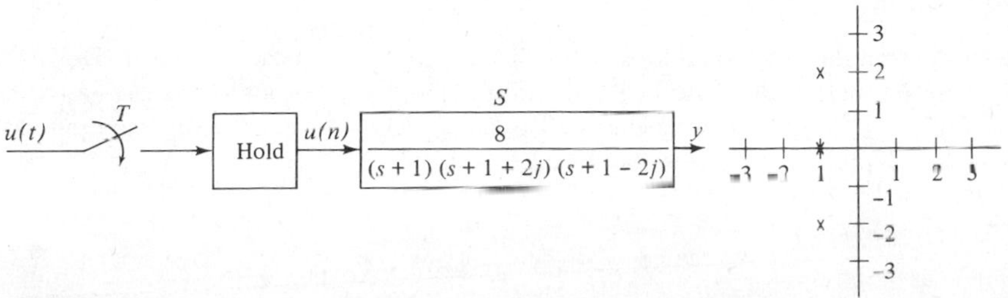

Figure D-2

Consequently, an irreducible Jordan-form realization of the system S can be found as

$$\begin{bmatrix}\dot{x}_1\\ \dot{x}_2\\ \dot{x}_3\end{bmatrix}=\begin{bmatrix}-1 & 0 & 0\\ 0 & -1-2i & 0\\ 0 & 0 & -1+2i\end{bmatrix}\begin{bmatrix}x_1\\ x_2\\ x_3\end{bmatrix}+\begin{bmatrix}1\\ 1\\ 1\end{bmatrix}u$$

$$y=[2 \quad -1 \quad -1]\mathbf{x}$$

By using (D-5) through (D-8), a discrete-time state equation can be computed as

$$\mathbf{x}(k+1)=\begin{bmatrix}e^{-T} & 0 & 0\\ 0 & e^{-T-2iT} & 0\\ 0 & 0 & e^{-T+2iT}\end{bmatrix}\mathbf{x}(k)+\begin{bmatrix}1-e^{-T}\\ \dfrac{1}{1+2i}(1-e^{-T-2iT})\\ \dfrac{1}{1-2i}(1-e^{-T+2iT})\end{bmatrix}u(k) \qquad \textbf{(D-15)}$$

We conclude from Theorem D-1 that the discrete-time state equation (D-15) is controllable if and only if

$$T\neq\frac{2\pi\alpha}{2}=\pi\alpha$$

and

$$T\neq\frac{2\pi\alpha}{4}=\frac{\pi}{2}\alpha \qquad \alpha=\pm 1, \pm 2, \ldots$$

This fact can also be verified directly from (D-15) by using either the criterion $\rho[\tilde{\mathbf{B}} \quad \tilde{\mathbf{A}}\tilde{\mathbf{B}} \quad \tilde{\mathbf{A}}^2\tilde{\mathbf{B}}]=3$ or Corollary 5-21.

Problems

D-1 Consider the continuous-time state equation

$$\dot{\mathbf{x}}(t)=\begin{bmatrix}-1 & 0 & 0\\ 0 & -1-2i & 0\\ 0 & 0 & -1+2i\end{bmatrix}\mathbf{x}(t)+\begin{bmatrix}1 & 0 & 0\\ 0 & 1 & 0\\ 0 & 0 & 1\end{bmatrix}\mathbf{u}(t)$$

Show that its discretized state equation is always controllable for any T including $T=2\pi\alpha/4$ for $\alpha=\pm 1, \pm 2, \ldots$. This shows that the conditions on the eigenvalues in Theorem D-1 are not necessary for a multiple-input discretized equation to be controllable.

D-2 Show that sufficient conditions for the discrete-time dynamical equation in (D-5) to be observable are that the dynamical equation in (D-1) is observable and Im $[\lambda_i(\mathbf{A})-\lambda_j(\mathbf{A})]\neq 2\pi\alpha/T$ for $\alpha=\pm 1, \pm 2, \ldots$, whenever Re $[\lambda_i(\mathbf{A})-\lambda_j(\mathbf{A})]=0$. For the single-output case $(q=1)$, the conditions are necessary as well.

E

Hermitian Forms and Singular Value Decomposition

A *hermitian form* of n complex variables $x_1, x_2, \ldots, x_n$ is a *real-valued* homogeneous polynomial of the form

$$\sum_{i,j=1}^{n} m_{ij}\bar{x}_i x_j$$

or, in matrix form,

$$[\bar{x}_1 \quad \bar{x}_2 \quad \cdots \quad \bar{x}_n]\begin{bmatrix} m_{11} & m_{12} & \cdots & m_{1n} \\ m_{21} & m_{22} & \cdots & m_{2n} \\ \vdots & \vdots & & \vdots \\ m_{n1} & m_{n2} & \cdots & m_{nn} \end{bmatrix}\begin{bmatrix} x_1 \\ x_2 \\ \vdots \\ x_n \end{bmatrix} \triangleq \mathbf{x}^*\mathbf{M}_1\mathbf{x} \tag{E-1}$$

where the m_{ij}'s are any complex numbers and $\bar{x}_i$ is the complex conjugate of x_i. Since every hermitian form is assumed to be real-valued, we have

$$\mathbf{x}^*\mathbf{M}_1\mathbf{x} = (\mathbf{x}^*\mathbf{M}_1\mathbf{x})^* = \mathbf{x}^*\mathbf{M}_1^*\mathbf{x}$$

where $\mathbf{M}_1^*$ is the complex conjugate transpose of $\mathbf{M}_1$; hence

$$\mathbf{x}^*\mathbf{M}_1\mathbf{x} = \mathbf{x}^*(\tfrac{1}{2}(\mathbf{M}_1 + \mathbf{M}_1^*))\mathbf{x} \triangleq \mathbf{x}^*\mathbf{M}\mathbf{x} \tag{E-2}$$

where $\mathbf{M} = \frac{1}{2}(\mathbf{M}_1 + \mathbf{M}_1^*)$. It is clear that $\mathbf{M} = \mathbf{M}^*$. Thus *every hermitian form can be written as* $\mathbf{x}^*\mathbf{M}\mathbf{x}$ *with* $\mathbf{M} = \mathbf{M}^*$. A matrix $\mathbf{M}$ with the property $\mathbf{M} = \mathbf{M}^*$ is called a *hermitian matrix*.

In the study of hermitian forms, it is convenient to use the notation of inner product. Observe that the hermitian matrix $\mathbf{M}$ can be considered as a linear operator that maps the n-dimensional complex vector space $(\mathbb{C}^n, \mathbb{C})$ into itself. If the inner product of $(\mathbb{C}^n, \mathbb{C})$ is chosen as

$$\langle \mathbf{x}, \mathbf{y} \rangle \triangleq \mathbf{x}^*\mathbf{y} \tag{E-3}$$

where **x** and **y** are any vectors in $(\mathbb{C}^n, \mathbb{C})$, then the hermitian form can be written as

$$\mathbf{x}^*\mathbf{M}\mathbf{x} = \langle \mathbf{x}, \mathbf{M}\mathbf{x} \rangle = \langle \mathbf{M}^*\mathbf{x}, \mathbf{x} \rangle = \langle \mathbf{M}\mathbf{x}, \mathbf{x} \rangle \tag{E-4}$$

where, in the last step, we have used the fact that $\mathbf{M}^* = \mathbf{M}$.

Theorem E-1

All the eigenvalues of a hermitian matrix **M** are real.

Proof

Let λ be any eigenvalue of **M** and let **e** be an eigenvector of **M** associated with λ; that is, $\mathbf{M}\mathbf{e} = \lambda\mathbf{e}$. Consider

$$\langle \mathbf{e}, \mathbf{M}\mathbf{e} \rangle = \langle \mathbf{e}, \lambda\mathbf{e} \rangle = \lambda \langle \mathbf{e}, \mathbf{e} \rangle \tag{E-5}$$

Since $\langle \mathbf{e}, \mathbf{M}\mathbf{e} \rangle$ is a real number and $\langle \mathbf{e}, \mathbf{e} \rangle$ is a positive real number, from (E-5) we conclude that λ is a real number. Q.E.D.

Theorem E-2

The Jordan-form representation of a hermitian matrix **M** is a diagonal matrix.

Proof

Recall from Section 2-6 that every square matrix which maps $(\mathbb{C}^n, \mathbb{C})$ into itself has a Jordan-form representation. The basis vectors that give a Jordan-form representation consist of eigenvectors and generalized eigenvectors of the matrix. We show that if a matrix is hermitian, then there is no generalized eigenvector of grade $k \geq 2$; we show this by contradiction. Suppose there exists a vector **e** such that $(\mathbf{M} - \lambda_i \mathbf{I})^k \mathbf{e} = \mathbf{0}$ and $(\mathbf{M} - \lambda_i \mathbf{I})^{k-1} \mathbf{e} \neq \mathbf{0}$ for some eigenvalue λ_i of **M**. Consider now, for $k \geq 2$,

$$\begin{aligned}\mathbf{0} &= \langle (\mathbf{M} - \lambda_i \mathbf{I})^{k-2}\mathbf{e}, (\mathbf{M} - \lambda_i \mathbf{I})^k \mathbf{e} \rangle = \langle (\mathbf{M} - \lambda_i \mathbf{I})^{k-1}\mathbf{e}, (\mathbf{M} - \lambda_i \mathbf{I})^{k-1}\mathbf{e} \rangle \\ &= \|(\mathbf{M} - \lambda_i \mathbf{I})^{k-1}\mathbf{e}\|^2\end{aligned}$$

which implies $(\mathbf{M} - \lambda_i \mathbf{I})^{k-1}\mathbf{e} = \mathbf{0}$. This is a contradiction. Hence there is no generalized eigenvector of grade $k \geq 2$. Consequently, there is no Jordan block whose order is greater than one. Hence the Jordan-form representation of a hermitian matrix is a diagonal matrix. In other words, there exists a nonsingular matrix **P** such that $\mathbf{P}\mathbf{M}\mathbf{P}^{-1} = \hat{\mathbf{M}}$ and $\hat{\mathbf{M}}$ is a diagonal matrix with eigenvalues on the diagonal. Q.E.D.

Two vectors **x**, **y** are said to be *orthogonal* if and only if $\langle \mathbf{x}, \mathbf{y} \rangle = \mathbf{0}$. A vector **x** is said to be *normalized* if and only if $\langle \mathbf{x}, \mathbf{x} \rangle \triangleq \|\mathbf{x}\|^2 = 1$. It is clear that every vector **x** can be normalized by choosing $\hat{\mathbf{x}} = (1/\|\mathbf{x}\|)\mathbf{x}$. A set of basis vectors $\{\mathbf{q}_1, \mathbf{q}_2, \ldots, \mathbf{q}_n\}$ is said to be an *orthonormal basis* if and only if

$$\langle \mathbf{q}_i, \mathbf{q}_j \rangle = \begin{cases} 0 & i \neq j \\ 1 & i = j \end{cases} \tag{E-6}$$

Now we show that the basis of a Jordan-form representation of a hermitian matrix can be chosen as an orthonormal basis. This is derived in part from the following theorem.

Theorem E-3

The eigenvectors of a hermitian matrix $\mathbf{M}$ corresponding to different eigenvalues are orthogonal.

Proof

Let $\mathbf{e}_i$ and $\mathbf{e}_j$ be the eigenvectors of $\mathbf{M}$ corresponding to the distinct eigenvalues λ_i and λ_j, respectively; that is, $\mathbf{Me}_i = \lambda_i \mathbf{e}_i$ and $\mathbf{Me}_j = \lambda_j \mathbf{e}_j$. Consider

$$\langle \mathbf{e}_j, \mathbf{Me}_i \rangle = \langle \mathbf{e}_j, \lambda_i \mathbf{e}_i \rangle = \lambda_i \langle \mathbf{e}_j, \mathbf{e}_i \rangle \tag{E-7}$$

and

$$\langle \mathbf{e}_j, \mathbf{Me}_i \rangle = \langle \mathbf{Me}_j, \mathbf{e}_i \rangle = \langle \lambda_j \mathbf{e}_j, \mathbf{e}_i \rangle = \lambda_j \langle \mathbf{e}_j, \mathbf{e}_i \rangle \tag{E-8}$$

where we have used the fact that the eigenvalues are real. Subtracting (E-8) from (E-7), we obtain $(\lambda_i - \lambda_j)\langle \mathbf{e}_j, \mathbf{e}_i \rangle = 0$. Since $\lambda_i \neq \lambda_j$, we conclude that $\langle \mathbf{e}_i, \mathbf{e}_j \rangle = 0$. Q.E.D.

Since every eigenvector can be normalized and since eigenvectors of a hermitian matrix $\mathbf{M}$ associated with distinct eigenvalues are orthogonal, then the eigenvectors associated with different eigenvalues can be made to be orthonormal. We consider now the linearly independent eigenvectors associated with the same eigenvalue. Let $\{\mathbf{e}_1, \mathbf{e}_2, \ldots, \mathbf{e}_m\}$ be a set of linearly independent eigenvectors associated with the same eigenvalue. Now we shall obtain a set of orthonormal vectors from the set $\{\mathbf{e}_1, \mathbf{e}_2, \ldots, \mathbf{e}_m\}$. Let

$$\mathbf{u}_1 = \mathbf{e}_1 \qquad \mathbf{q}_1 = \frac{\mathbf{u}_1}{\|\mathbf{u}_1\|}$$

$$\mathbf{u}_2 = \mathbf{e}_2 - \langle \mathbf{q}_1, \mathbf{e}_2 \rangle \mathbf{q}_1 \qquad \mathbf{q}_2 = \frac{\mathbf{u}^2}{\|\mathbf{u}_2\|}$$

$$\mathbf{u}_3 = \mathbf{e}_3 - \langle \mathbf{q}_1, \mathbf{e}_3 \rangle \mathbf{q}_1 - \langle \mathbf{q}_2, \mathbf{e}_3 \rangle \mathbf{q}_2 \qquad \mathbf{q}_3 = \frac{\mathbf{u}_3}{\|\mathbf{u}_3\|}$$

$$\cdots\cdots\cdots\cdots \qquad \cdots\cdots$$

$$\mathbf{u}_m = \mathbf{e}_m - \sum_{k=1}^{m-1} \langle \mathbf{q}_k, \mathbf{e}_m \rangle \mathbf{q}_k \qquad \mathbf{q}_m = \frac{\mathbf{u}_m}{\|\mathbf{u}_m\|}$$

The procedure for defining $\mathbf{q}_i$ is illustrated in Figure E-1. It is called the *Schmidt orthonormalization procedure*. By direct verification, it can be shown that $\langle \mathbf{q}_i, \mathbf{q}_j \rangle = 0$ for $i \neq j$.

From Theorem E-3 and the Schmidt orthonormalization procedure we conclude that for any hermitian matrix there exists a set of orthonormal vectors with respect to which the hermitian matrix has a diagonal-form representation; or equivalently, for any hermitian matrix $\mathbf{M}$, there exists a nonsingular matrix $\mathbf{Q}$

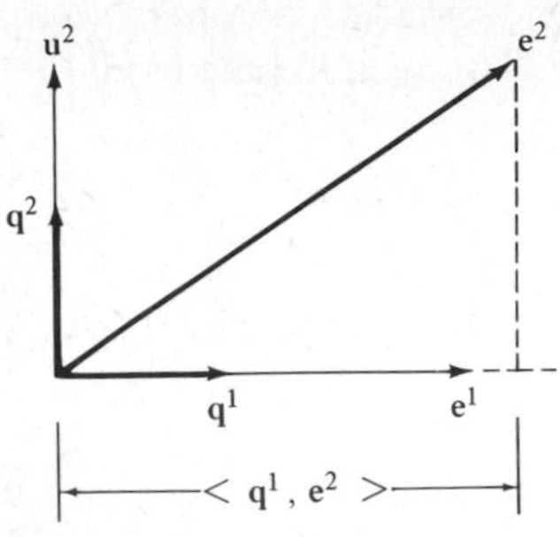

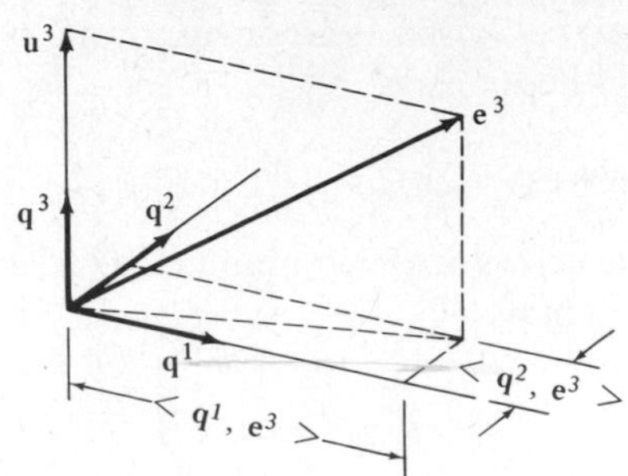

Figure E-1

whose columns are orthonormal, such that

$$\hat{\mathbf{M}} = \mathbf{Q}^{-1}\mathbf{M}\mathbf{Q} \triangleq \mathbf{P}\mathbf{M}\mathbf{P}^{-1}$$

where $\hat{\mathbf{M}}$ is a diagonal matrix and $\mathbf{P} \triangleq \mathbf{Q}^{-1}$. Let $\mathbf{Q} = [\mathbf{q}_1 \quad \mathbf{q}_2 \quad \cdots \quad \mathbf{q}_n]$. Because of the orthonormal assumption, we have

$$\mathbf{Q}^*\mathbf{Q} = \begin{bmatrix} (\mathbf{q}_1)^* \\ (\mathbf{q}_2)^* \\ \vdots \\ (\mathbf{q}_n)^* \end{bmatrix} [\mathbf{q}_1 \quad \mathbf{q}_2 \quad \cdots \quad \mathbf{q}_n] = \mathbf{I}$$

Hence $\mathbf{Q}^{-1} = \mathbf{Q}^*$, and $\mathbf{P}^{-1} = \mathbf{P}^*$. A matrix $\mathbf{Q}$ with the property $\mathbf{Q}^* = \mathbf{Q}^{-1}$ is called a *unitary matrix*. We summarize what we have achieved in the following.

Theorem E-4

A hermitian matrix $\mathbf{M}$ can be transformed by a unitary matrix into a diagonal matrix with real elements; or, equivalently, for any hermitian matrix $\mathbf{M}$ there exists a nonsingular matrix $\mathbf{P}$ with the property $\mathbf{P}^{-1} = \mathbf{P}^*$ such that

$$\hat{\mathbf{M}} = \mathbf{P}\mathbf{M}\mathbf{P}^*$$

where $\hat{\mathbf{M}}$ is a diagonal matrix with real eigenvalues of $\mathbf{M}$ on the diagonal. ∎

In the following, we shall utilize Theorem E-4 to develop the singular value decompositions for matrices. Let $\mathbf{H}$ be an $m \times n$ matrix. Then the matrix $\mathbf{H}^*\mathbf{H}$ is a square matrix of order n. Clearly $\mathbf{H}^*\mathbf{H}$ is hermitian; hence its eigenvalues are all real. Because $\mathbf{H}^*\mathbf{H}$ is positive semidefinite, its eigenvalues are all nonnegative (Theorem 8-19). Let λ_i^2, $i = 1, 2, \ldots, n$ be the eigenvalues of $\mathbf{H}^*\mathbf{H}$. The set $\{\lambda_i \geq 0, i = 1, 2, \ldots, n\}$ is called the *singular values of* $\mathbf{H}$. If $\mathbf{H}$ is hermitian, its singular values are equal to the absolute values of the eigenvalues of $\mathbf{H}$. For convenience, we arrange $\{\lambda_i\}$ such that

$$\lambda_1^2 \geq \lambda_2^2 \cdots \geq \lambda_n^2 \geq 0 \qquad \textbf{(E-9)}$$

If the rank of $\mathbf{H}$ is r, so is the rank of $\mathbf{H}^*\mathbf{H}$ (Problem E-3). Hence we have $\lambda_1^2 \geq \lambda_2^2 \geq \cdots \lambda_r^2 > 0$ and $\lambda_{r+1}^2 = \lambda_{r+2}^2 = \cdots = \lambda_n^2 = 0$. Let $\mathbf{q}_i$, $i = 1, 2, \ldots, n$, be the orthonormal eigenvectors of $\mathbf{H}^*\mathbf{H}$ associated with λ_i^2. Define

$$\mathbf{Q} = [\mathbf{q}_1 \quad \mathbf{q}_2 \quad \cdots \quad \mathbf{q}_r \ \vdots\ \mathbf{q}_{r+1} \quad \cdots \quad \mathbf{q}_n] = [\mathbf{Q}_1 \ \vdots\ \mathbf{Q}_2] \tag{E-10}$$

Then Theorem E-4 implies

$$\mathbf{Q}^*\mathbf{H}^*\mathbf{H}\mathbf{Q} = \begin{bmatrix} \boldsymbol{\Sigma}^2 & \mathbf{0} \\ \mathbf{0} & \mathbf{0} \end{bmatrix} \tag{E-11}$$

where $\boldsymbol{\Sigma}^2 = \text{diag}\{\lambda_1^2, \lambda_2^2, \ldots, \lambda_r^2\}$. Using $\mathbf{Q} = [\mathbf{Q}_1 \quad \mathbf{Q}_2]$, (E-11) can be written as

$$\mathbf{Q}_2^*\mathbf{H}^*\mathbf{H}\mathbf{Q}_2 = \mathbf{0} \tag{E-12}$$

and

$$\mathbf{Q}_1^*\mathbf{H}^*\mathbf{H}\mathbf{Q}_1 = \boldsymbol{\Sigma}^2$$

which implies

$$\boldsymbol{\Sigma}^{-1}\mathbf{Q}_1^*\mathbf{H}^*\mathbf{H}\mathbf{Q}_1\boldsymbol{\Sigma}^{-1} = \mathbf{I} \tag{E-13}$$

where $\boldsymbol{\Sigma} = \text{diag}\{\lambda_1, \lambda_2, \ldots, \lambda_r\}$. Define the $m \times r$ matrix $\mathbf{R}_1$ by

$$\mathbf{R}_1 \triangleq \mathbf{H}\mathbf{Q}_1\boldsymbol{\Sigma}^{-1} \tag{E-14}$$

Then (E-13) becomes $\mathbf{R}_1^*\mathbf{R}_1 = \mathbf{I}$ which implies that the columns of $\mathbf{R}_1$ are orthonormal. Let $\mathbf{R}_2$ be chosen so that $\mathbf{R} = [\mathbf{R}_1 \quad \mathbf{R}_2]$ is unitary. Consider

$$\mathbf{R}^*\mathbf{H}\mathbf{Q} = \begin{bmatrix} \mathbf{R}_1^* \\ \mathbf{R}_2^* \end{bmatrix} \mathbf{H}[\mathbf{Q}_1 \quad \mathbf{Q}_2] = \begin{bmatrix} \mathbf{R}_1^*\mathbf{H}\mathbf{Q}_1 & \mathbf{R}_1^*\mathbf{H}\mathbf{Q}_2 \\ \mathbf{R}_2^*\mathbf{H}\mathbf{Q}_1 & \mathbf{R}_2^*\mathbf{H}\mathbf{Q}_2 \end{bmatrix} \tag{E-15}$$

Clearly, (E-14) implies $\mathbf{H}\mathbf{Q}_1 = \mathbf{R}_1\boldsymbol{\Sigma}$ and (E-12) implies $\mathbf{H}\mathbf{Q}_2 = \mathbf{0}$. Because $\mathbf{R}$ is orthonormal, we have $\mathbf{R}_1^*\mathbf{R}_1 = \mathbf{I}$ and $\mathbf{R}_2^*\mathbf{R}_1 = \mathbf{0}$. Consequently, (E-15) becomes

$$\mathbf{R}^*\mathbf{H}\mathbf{Q} = \begin{bmatrix} \boldsymbol{\Sigma} & \mathbf{0} \\ \mathbf{0} & \mathbf{0} \end{bmatrix}$$

This is stated as a theorem.

Theorem E-5 (Singular value decomposition)

Every $m \times n$ matrix $\mathbf{H}$ of rank r can be transformed into the form

$$\mathbf{R}^*\mathbf{H}\mathbf{Q} = \begin{bmatrix} \boldsymbol{\Sigma} & \mathbf{0} \\ \mathbf{0} & \mathbf{0} \end{bmatrix} \quad \text{or} \quad \mathbf{H} = \mathbf{R}\begin{bmatrix} \boldsymbol{\Sigma} & \mathbf{0} \\ \mathbf{0} & \mathbf{0} \end{bmatrix}\mathbf{Q}^*$$

where $\mathbf{R}^*\mathbf{R} = \mathbf{R}\mathbf{R}^* = \mathbf{I}_m$, $\mathbf{Q}^*\mathbf{Q} = \mathbf{Q}\mathbf{Q}^* = \mathbf{I}_n$, and $\boldsymbol{\Sigma} = \text{diag}\{\lambda_1, \lambda_2, \ldots, \lambda_r\}$ with $\lambda_1 \geq \lambda_2 \geq \cdots \geq \lambda_r > 0$. ∎

Although $\boldsymbol{\Sigma}$ is uniquely determined by $\mathbf{H}$, the unitary matrices $\mathbf{R}$ and $\mathbf{Q}$ are not necessarily unique. Indeed let λ_i^2 be a multiple eigenvalue of $\mathbf{H}^*\mathbf{H}$; then the corresponding columns of $\mathbf{Q}$ may be chosen as any orthonormal basis for the space spanned by the eigenvectors of $\mathbf{H}^*\mathbf{H}$ corresponding to λ_i^2. Hence $\mathbf{Q}$

is not unique. Once $\mathbf{Q}$ is chosen, $\mathbf{R}_1$ can be computed from (E-14). The choice of $\mathbf{R}_2$ again may not be unique so long as $[\mathbf{R}_1 \quad \mathbf{R}_2]$ is unitary.

The singular value decomposition has found many applications in linear systems. In Section 6-5, we use it to find an irreducible realization from a Hankel matrix. It can also be used to find simplified or approximated models of systems. See References S141, S161, and S171. The singular value decomposition is also essential in the study of sensitivity and stability margin of multivariable systems. See References S34 and S194. For computer programs, see Reference S82.

The elements of matrices in this appendix are permitted to assume real or complex numbers. Certainly all results still apply if they are limited to real numbers. For a real matrix $\mathbf{M}$ we have $\mathbf{M}^* = \mathbf{M}'$, where the prime denotes the transpose. A real matrix with $\mathbf{M} = \mathbf{M}'$ is called a *symmetric matrix*; a real matrix with $\mathbf{M}^{-1} = \mathbf{M}'$ is called an *orthogonal matrix*. With these modifications in nomenclature, all theorems in this appendix apply directly to real matrices.

Problems

E-1 If $\mathbf{M}$ is an $n \times n$ matrix with complex coefficients, verify that $\mathbf{x}^*\mathbf{M}\mathbf{x}$ is a real number for any $\mathbf{x}$ in $\mathbb{C}^n$ if and only if $\mathbf{M}^* = \mathbf{M}$. If $\mathbf{M}$ is an $n \times n$ matrix with real coefficients, is it true that $\mathbf{x}'\mathbf{M}\mathbf{x}$ is a real number for any $\mathbf{x}$ in $\mathbb{R}^n$ if and only if $\mathbf{M}' = \mathbf{M}$?

E-2 Find an orthonormal set from the vectors

$$\begin{bmatrix} 1 \\ 1 \\ 1 \end{bmatrix} \quad \begin{bmatrix} 1 \\ 2 \\ 1 \end{bmatrix} \quad \begin{bmatrix} 3 \\ 1 \\ -1 \end{bmatrix}$$

E-3 Show that if the rank of $\mathbf{H}$ is r, so are the ranks of $\mathbf{H}^*\mathbf{H}$ and $\mathbf{H}\mathbf{H}^*$. (*Hint*: Find a $\mathbf{P}$ such that $\mathbf{A}\mathbf{P} = [\bar{\mathbf{A}} \quad \mathbf{0}]$ and rank $\mathbf{A} =$ rank $\bar{\mathbf{A}}$.)

E-4 Find the eigenvalues and singular values of the matrices

$$\mathbf{H}_1 = \begin{bmatrix} 0 & 1 & 0 \\ 0 & 0 & 1 \\ 2 & 0 & -1 \end{bmatrix} \qquad \mathbf{H}_2 = \begin{bmatrix} 1 & 0 & 1 \\ 0 & 0 & 0 \\ 1 & 0 & 1 \end{bmatrix}$$

E-5 Find the singular value decompositions of the matrices in Problem E-4.

E-6 Show that

$$\begin{aligned} \|\mathbf{A}\|_2 &= \text{largest singular value of } \mathbf{A} \\ &= [\lambda_{\max}(\mathbf{A}^*\mathbf{A})]^{1/2} \end{aligned}$$

E-7 In Theorem E-5, we have $\mathbf{Q}^*\mathbf{Q} = \mathbf{Q}\mathbf{Q}^* = \mathbf{I}_n$. If we write $\mathbf{Q} = [\mathbf{Q}_1 \quad \mathbf{Q}_2]$, do we have $\mathbf{Q}_1{}^*\mathbf{Q}_1 = \mathbf{I}_r$ and $\mathbf{Q}_1\mathbf{Q}_1^* = \mathbf{I}_n$?

E-8 Are all the eigenvalues of a unitary matrix (including orthogonal matrix) real? Show that all the eigenvalues of a unitary matrix have magnitudes equal to 1.

E-9 Show that all the singular values of a unitary matrix (including orthogonal matrix) are equal to 1.

E-10 What are the eigenvalues and singular values of the elementary matrices

$$\mathbf{E}_1 = \begin{bmatrix} 1 & 0 & 0 \\ 5 & 1 & 0 \\ -3 & 0 & 1 \end{bmatrix} \qquad \mathbf{E}_2 = \begin{bmatrix} 1 & 0 & 0 \\ 0.5 & 1 & 0 \\ 0.8 & 0 & 1 \end{bmatrix}$$

$\mathbf{E}_1$ may arise in gaussian elimination without any pivoting; $\mathbf{E}_2$ may arise in gaussian elimination with partial pivoting. If the condition number of a matrix is defined as the ratio of the largest and smallest singular values, which matrix has a larger condition number? Roughly speaking, a condition number gives the amplification factor of the relative errors in the computation.

E-11 Show that the controllability grammian

$$\mathbf{W}_{ct} = \int_0^t e^{\mathbf{A}\tau}\mathbf{B}\mathbf{B}^* e^{\mathbf{A}^*\tau}\, d\tau$$

is positive definite if and only if $\{\mathbf{A}, \mathbf{B}\}$ is controllable.

E-12 Show that Theorem E-5 reduces to Theorem E-4 if $\mathbf{H}$ is square, hermitian, and positive semidefinite. If $\mathbf{H}$ is square and hermitian (without being positive semidefinite), what are the differences between Theorem E-5 and Theorem E-4?

E-13 What are the singular values of the following matrices?

$$\mathbf{A}_1 = 1 \qquad \mathbf{A}_2 = \begin{bmatrix} 1 & -1 \\ 0 & 1 \end{bmatrix} \qquad \mathbf{A}_3 = \begin{bmatrix} 1 & -1 & -1 \\ 0 & 1 & -1 \\ 0 & 0 & 1 \end{bmatrix}$$

F

On the Matrix Equation AM+MB=N

In this appendix we shall study the matrix equation $\mathbf{AM}+\mathbf{MB}=\mathbf{N}$, where $\mathbf{A}$, $\mathbf{B}$, $\mathbf{M}$, and $\mathbf{N}$ are $n \times n$ complex-valued matrices. We observe that all the $n \times n$ complex-valued matrices, with the usual rules of multiplication and addition, form a linear space (Problem 2-10). Let us denote this space by $(\chi, \mathbb{C})$. The dimension of $(\chi, \mathbb{C})$ is n^2. Consider the operator $\mathscr{A}$ defined by

$$\mathscr{A}(\mathbf{M}) \triangleq \mathbf{AM}+\mathbf{MB} \qquad \text{for all } \mathbf{M} \text{ in } \chi$$

It is clear that the operator $\mathscr{A}$ maps $(\chi, \mathbb{C})$ into itself and is a linear operator. The equation $\mathbf{AM}+\mathbf{MB}=\mathbf{N}$ is often called a *Lyapunov matrix equation.*

Theorem F-1

Let $\mathscr{A}:(\chi, \mathbb{C}) \rightarrow (\chi, \mathbb{C})$ be the operator defined by $\mathscr{A}(\mathbf{M})=\mathbf{AM}+\mathbf{MB}$ for all $\mathbf{M}$ in χ. Let λ_i, for $i=1, 2, \ldots, l \leq n$, be the distinct eigenvalues of $\mathbf{A}$ and let μ_j, for $j=1, 2, \ldots, m \leq n$, be the distinct eigenvalues of $\mathbf{B}$. Then $(\lambda_i+\mu_j)$ is an eigenvalue of $\mathscr{A}$. Conversely, let η_k, $k=1, 2, \ldots, p \leq n^2$, be the distinct eigenvalues of $\mathscr{A}$, then for each k,

$$\eta_k=\lambda_i+\mu_j$$

for some i and some j.

Proof

We prove, first, that $\lambda_i+\mu_j$ is an eigenvalue of $\mathscr{A}$. Let $\mathbf{x}$ and $\mathbf{y}$ be nonzero $n \times 1$ and $1 \times n$ vectors such that

$$\mathbf{Ax}=\lambda_i\mathbf{x} \qquad \mathbf{yB}=\mu_j\mathbf{y}$$

clearly $\mathbf{x}$ is a right eigenvector of $\mathbf{A}$ and $\mathbf{y}$ is a left eigenvector of $\mathbf{B}$. Applying the operator $\mathscr{A}$ on the $n \times n$ matrix $\mathbf{xy}$ that is clearly a nonzero matrix, we obtain

$$\mathscr{A}(\mathbf{xy}) = \mathbf{Axy} + \mathbf{xyB} = \lambda_i \mathbf{xy} + \mu_j \mathbf{xy} = (\lambda_i + \mu_j)\mathbf{xy}$$

Hence $\lambda_i + \mu_j$ is an eigenvalue of $\mathscr{A}$ for $i = 1, 2, \ldots, l$, and $j = 1, 2, \ldots, m$. See Definition 2-12.

Next we prove that all eigenvalues of $\mathscr{A}$ are of the form $\lambda_i + \mu_j$. Let us assume that η_k is an eigenvalue of $\mathscr{A}$. Then, by definition, there exists a $\mathbf{M} \neq \mathbf{0}$ such that

$$\mathscr{A}(\mathbf{M}) = \mathbf{AM} + \mathbf{MB} = \eta_k \mathbf{M}$$

or

$$(\eta_k \mathbf{I} - \mathbf{A})\mathbf{M} = \mathbf{MB} \tag{F-1}$$

We now show that the matrices $\eta_k \mathbf{I} - \mathbf{A}$ and $\mathbf{B}$ have at least one eigenvalue in common. We prove this by contradiction. Let

$$\Delta(s) = s^n + \alpha_1 s^{n-1} + \cdots + \alpha_{n-1} s + \alpha_n$$

be the characteristic polynomial of $\eta_k \mathbf{I} - \mathbf{A}$, that is $\Delta(s) = \det(s\mathbf{I} - \eta_k \mathbf{I} + \mathbf{A})$. Then we have

$$\Delta(\eta_k \mathbf{I} - \mathbf{A}) = \mathbf{0} \tag{F-2}$$

If $\eta_k \mathbf{I} - \mathbf{A}$ and $\mathbf{B}$ have no common eigenvalue, then the matrix

$$\Delta(\mathbf{B})$$

is nonsingular. Indeed, if μ_i, $i = 1, 2, \ldots, n$, are eigenvalues of $\mathbf{B}$, then $\Delta(\mu_i)$ are eigenvalues of $\Delta(\mathbf{B})$ (Problem 2-32) and $\det \Delta(\mathbf{B}) = \Pi \Delta(\mu_i)$ (Problem 2-22). If $\eta_k \mathbf{I} - \mathbf{A}$ and $\mathbf{B}$ have no common eigenvalue, then $\Delta(\mu_i) \neq 0$, for all i, and $\det \Delta(\mathbf{B}) \neq 0$. Hence $\Delta(\mathbf{B})$ is nonsingular.

From (F-1), we can develop the following equalities

$$(\eta_k \mathbf{I} - \mathbf{A})^2 \mathbf{M} = (\eta_k \mathbf{I} - \mathbf{A})\mathbf{MB} = \mathbf{MB}^2$$
$$\cdots\cdots\cdots\cdots\cdots\cdots\cdots$$
$$(\eta_k \mathbf{I} - \mathbf{A})^n \mathbf{M} = \mathbf{MB}^n$$

The summation of the products of $(\eta_k \mathbf{I} - \mathbf{A})^i \mathbf{M} = \mathbf{MB}^i$ and α_i, for $i = 0, 1, 2, \ldots, n$, with $\alpha_0 = 1$, yields

$$\Delta(\eta_k \mathbf{I} - \mathbf{A})\mathbf{M} = \mathbf{M}\Delta(\mathbf{B})$$

which, together with (F-2), implies

$$\mathbf{0} = \mathbf{M}\Delta(\mathbf{B}) \tag{F-3}$$

Since $\Delta(\mathbf{B})$ is nonsingular, (F-3) implies $\mathbf{M} = \mathbf{0}$. This contradicts the assumption that $\mathbf{M} \neq \mathbf{0}$. Hence the matrices $\eta_k \mathbf{I} - \mathbf{A}$ and $\mathbf{B}$ have at least one common eigenvalue. Now the eigenvalue of $\eta_k \mathbf{I} - \mathbf{A}$ is of the form $\eta_k - \lambda_i$. Consequently, for some i and for some j,

$$\eta_k - \lambda_i = \mu_j \qquad \text{or} \qquad \eta_k = \lambda_i + \mu_j$$

Q.E.D.

Corollary F-1a

Any matrix representation of the operator $\mathscr{A}$ is nonsingular if and only if $\lambda_i + \mu_j \neq 0$ for all i, j.

Proof

Since the linear operator $\mathscr{A}$ maps an n^2-dimensional linear space into itself, it has a matrix representation (Theorem 2-3). A matrix representation can be easily obtained by writing the n^2 equations $\mathbf{AM} + \mathbf{MB} = \mathbf{C}$ in the form of $\bar{\mathbf{A}}\bar{\mathbf{m}} = \bar{\mathbf{c}}$, where $\bar{\mathbf{m}}$ is an $n^2 \times 1$ column vector consisting of all the n^2 elements of $\mathbf{M}$. The corollary follows directly from the fact that the determinant of $\bar{\mathbf{A}}$ is the product of its eigenvalues (Problem 2-22). Q.E.D.

Corollary F-1b

If all the eigenvalues of $\mathbf{A}$ have negative real parts, then for any $\mathbf{N}$ there exists a unique $\mathbf{M}$ that satisfies the matrix equation

$$\mathbf{A}^*\mathbf{M} + \mathbf{MA} = -\mathbf{N} \tag{F-4}$$

Furthermore, the solution $\mathbf{M}$ can be expressed as

$$\mathbf{M} = \int_0^\infty \mathbf{e}^{\mathbf{A}^*t}\mathbf{N}e^{\mathbf{A}t}\, dt \tag{F-5}$$

Proof

Since the eigenvalues of $\mathbf{A}^*$ are the complex conjugates of the eigenvalues of $\mathbf{A}$, all the eigenvalues of $\mathbf{A}$ and $\mathbf{A}^*$ have negative real parts. Consequently, $\lambda_i + \mu_j \neq 0$ for all i, j, which implies that the matrix representation of $\mathscr{A}(\mathbf{M}) = \mathbf{A}^*\mathbf{M} + \mathbf{MA}$ is nonsingular. Hence, for any $\mathbf{N}$ there exists a unique $\mathbf{M}$ satisfying $\mathscr{A}(\mathbf{M}) = -\mathbf{N}$.

Now we show that the $\mathbf{M}$ in (F-5) is the solution of (F-4). Because all the eigenvalues of $\mathbf{A}$ have negative real parts, the integral in (F-5) converges. Consider

$$\mathbf{A}^*\mathbf{M} + \mathbf{MA} = \int_0^\infty (\mathbf{A}^*e^{\mathbf{A}^*t}\mathbf{N}e^{\mathbf{A}t} + e^{\mathbf{A}^*t}\mathbf{N}e^{\mathbf{A}t}\mathbf{A})\, dt = \int_0^\infty \frac{d}{dt}(e^{\mathbf{A}^*t}\mathbf{N}e^{\mathbf{A}t})\, dt$$

$$= e^{\mathbf{A}^*t}\mathbf{N}e^{\mathbf{A}t}\Big|_{t=0}^{t=\infty} = -\mathbf{N}$$

This establishes the corollary. Q.E.D.

Although the matrix $\mathbf{M}$ can be solved from (F-5), the formula is not suitable for computer computation. The solution of $\mathbf{AM} + \mathbf{MB} = \mathbf{N}$ has been extensively studied in the literature. The reader is referred to References S9, S18, and S107.

We list in the following the discrete-time version of Theorem F-1.

Theorem F-2

Let $\mathscr{A}$: $(\chi, \mathbb{C}) \rightarrow (\chi, \mathbb{C})$ be the operator defined by $\mathscr{A}(\mathbf{M}) = \mathbf{AMB} - \mathbf{M}$ for all $\mathbf{M}$ in χ. Let λ_i, for $i = 1, 2, \ldots, l \leq n$, be the distinct eigenvalues of $\mathbf{A}$ and let μ_j, for $j = 1, 2, \ldots, m \leq n$, be the distinct eigenvalues of $\mathbf{B}$. Then $(\lambda_i \mu_j - 1)$ is an eigenvalue of $\mathscr{A}$. Conversely, let η_k, $k = 1, 2, \ldots, p \leq n^2$, be the distinct eigenvalues of $\mathscr{A}$, then for each k,

$$\eta_k = \lambda_i \mu_j - 1 \tag{F-6}$$

for some i and some j.

Corollary F-2a

Any matrix representation of the operator $\mathscr{A}$ is nonsingular if and only if $\lambda_i \mu_j \neq 1$ for all i, j.

Corollary F-2b

If all the eigenvalues of $\mathbf{A}$ have magnitudes less than 1, then for any $\mathbf{N}$ there exists a unique $\mathbf{M}$ that satisfies the matrix equation

$$\mathbf{A}^*\mathbf{M}\mathbf{A} - \mathbf{M} = -\mathbf{N} \tag{F-7}$$

Furthermore, the solution $\mathbf{M}$ can be expressed as

$$\mathbf{M} = \sum_{k=0}^{\infty} (\mathbf{A}^*)^k \mathbf{N} \mathbf{A}^k \tag{F-8}$$

∎

The procedure used to prove Theorem F-1 can be used to establish Theorem F-2. Corollary F-2b can be readily verified by direct substitution. We discuss three methods of solving (F-7) to conclude this appendix:

1. By equating the corresponding elements of $\mathbf{A}^*\mathbf{MA} - \mathbf{M}$ and $-\mathbf{N}$, we can obtain a set of linear algebraic equations. If $\mathbf{A}$ is an $n \times n$ matrix, the set consists of n^2 equations. The number of equations can be reduced by using the hermitian or symmetry property of $\mathbf{M}$ (see Reference S16). The set of linear algebraic equations can then be solved by using existing subroutines in a computing center.
2. Compute the infinite power series in (F-8) directly. If all the eigenvalues of $\mathbf{A}$ have magnitudes much less than 1, the series will converge rapidly. The infinite series can be computed recursively as follows:

$$\mathbf{M}_{k+1} = \mathbf{A}^*\mathbf{M}_k\mathbf{A} + \mathbf{N} \qquad \text{with } \mathbf{M}_0 = \mathbf{0}$$

The computation may be stopped when $\|\mathbf{M}_{k+1} - \mathbf{M}_k\|$ first becomes smaller than a predetermined number.

3. Define

$$\mathbf{P}_{k+1} = \mathbf{P}_k^2 \qquad \mathbf{P}_1 = \mathbf{A}$$
$$\mathbf{M}_{k+1} = \mathbf{P}_k^* \mathbf{M}_k \mathbf{P}_k + \mathbf{M}_k \qquad \mathbf{M}_1 = \mathbf{N}$$

By this process, the convergence in method 2 can be speeded up (see Reference S4).

Problems

F-1 Find a matrix representation of the operator $\mathscr{A}(\mathbf{M}) = \mathbf{AM} + \mathbf{MB}$ with

$$\mathbf{A} = \begin{bmatrix} 0 & 1 \\ -2 & -1 \end{bmatrix} \qquad \mathbf{B} = \begin{bmatrix} 0 & 1 \\ 0 & 0 \end{bmatrix}$$

and verify Theorem F-1.

F-2 Find the $\mathbf{M}$ to meet $\mathbf{A}^*\mathbf{M} + \mathbf{MA} = -\mathbf{N}$ with $\mathbf{A}$ given in Problem F-1 and $\mathbf{N} = \mathbf{I}$, by setting up a set of linear algebraic equations.

F-3 Solve Problem F-2 by using (F-5). Which method, by solving algebraic equations or by direct integration, is simpler?

F-4 Establish, without using Corollary F-1a, the uniqueness part of Corollary F-1b. (*Hint*: Use $d(e^{\mathbf{A}^*t}(\mathbf{M}_1 - \mathbf{M}_2)e^{\mathbf{A}t})/dt = \mathbf{0}$.)

F-5 Under what conditions can the solution of $\mathbf{AM} + \mathbf{MB} = -\mathbf{N}$ be expressed as

$$\mathbf{M} = \int_0^\infty e^{\mathbf{A}t}\mathbf{N}e^{\mathbf{B}t}\,dt$$

F-6 Prove Theorem F-2 and its corollaries.

F-7 Find a matrix representation of $\mathscr{A}(\mathbf{M}) = \mathbf{AMB} - \mathbf{M}$, with $\mathbf{A}$ and $\mathbf{B}$ given in Problem F-1 and verify Theorem F-2.

F-8 Solve (F-7) with $\mathbf{A}$ and $\mathbf{N}$ given in Problems F-1 and F-2.

F-9 Transform $\mathbf{A}^*\mathbf{M} + \mathbf{MA} = -\mathbf{N}$ into a discrete-time Lyapunov equation by using the transformation

$$\mathbf{A} \rightarrow (\mathbf{A}_d - \mathbf{I})(\mathbf{A}_d + \mathbf{I})^{-1}$$

G
Polynomials and Polynomial Matrices

In this appendix, we shall study the coprimeness of two polynomials, its extension to the polynomial matrix case and the factorization of a rational matrix into two coprime polynomial matrices. The coprimeness of polynomials was studied by the French mathematician E. Bezout in 1764; its application to linear system problems, however, was quite recent. The concept of coprimeness in the fractions of transfer-function matrices is, roughly speaking, equivalent to the concepts of controllability and observability in dynamical equations; hence its importance cannot be overstated. In this appendix, the concepts and results in the scalar case will be developed in such a way that they can be readily extended to the matrix case. The material will be presented in a manner suitable for numerical computation. The row-searching algorithm discussed in Appendix A will be constantly used.

G-1 Coprimeness of Polynomials

Let $D(s)$ be a polynomial with real coefficients and indeterminate s expressed as[1]

$$D(s) = D_n s^n + D_{n-1} s^{n-1} + \cdots + D_1 s + D_0$$

The polynomial $D(s)$ is said to be of *degree* n if $D_n \neq 0$. The coefficient D_n associated with the highest power of s is called the *leading coefficient*. A polynomial is called *monic* if its leading coefficient is equal to 1.

[1] In this appendix, capital letters without boldface are also used to denote scalars.

Theorem G-1

Let $D(s)$ and $N(s)$ be two polynomials and let $D(s) \neq 0$. Then there exist unique polynomials $Q(s)$ and $R(s)$ such that

$$N(s) = Q(s)D(s) + R(s)$$

and

$$\deg R(s) < \deg D(s) \tag{G-1}$$

Proof

If $\deg N(s) < \deg D(s)$, then $Q(s) = 0$ and $R(s) = N(s)$. Let $\deg N(s) \geq \deg D(s)$. Then by direct division, we have the relation in (G-1) with $Q(s)$ as the quotient and $R(s)$ as the remainder with $\deg R(s) < \deg D(s)$. For example, if $N(s) = 2s^3 + 3s - 1$ and $D(s) = s^2 + s - 2$, then by long division, we have

$$\begin{array}{r@{}l} & 2s \quad -2 \\ \cline{2-2} s^2 + s - 2 \,)\! & 2s^3 + 0 \quad + 3s - 1 \\ & 2s^3 + 2s^2 - 4s \\ \cline{2-2} & -2s^2 + 7s - 1 \\ & -2s^2 - 2s + 4 \\ \cline{2-2} & 9s - 5 \end{array}$$

and

$$N(s) = (2s - 2)D(s) + (9s - 5)$$

Now we show that $Q(s)$ and $R(s)$ are unique. Suppose there are other $Q_1(s)$ and $R_1(s)$ such that

$$N(s) = Q(s)D(s) + R(s) = Q_1(s)D(s) + R_1(s)$$

which implies

$$(Q(s) - Q_1(s))D(s) = R_1(s) - R(s) \tag{G-2}$$

If $Q(s) - Q_1(s) \neq 0$, the degree of the left-hand-side polynomial of (G-2) is equal to or larger than that of $D(s)$, whereas the degree of the right-hand-side polynomial of (G-2) is smaller than that of $D(s)$. This is not possible. Hence we have

$$Q(s) = Q_1(s) \qquad \text{and} \qquad R_1(s) = R(s)$$

This completes the proof of the theorem. Q.E.D.

The division procedure in this theorem can be put into a programmatic format as follows:

1. $Q(s) = 0$.
2. If $\deg N(s) < \deg D(s)$, stop.
3. Let N_m and D_n be the leading coefficients of $N(s)$ and $D(s)$ with $m > n$:

$$N(s) \leftarrow N(s) - \frac{N_m}{D_n} s^{m-n} D(s)$$

4. $Q(s) \leftarrow Q(s) + \dfrac{N_m}{D_n} s^{m-n}$

5. Go to step 2.

At the end, the resulting $N(s)$ is the remainder $R(s)$.[2]

In Equation (G-1) if $R(s)=0$, $N(s)$ is said to be divisible (without remainder) by $D(s)$, and $D(s)$ is a *factor*, or a *divisor*, of $N(s)$. If $R(s)$ is a divisor of $D(s)$ and a divisor of $N(s)$, then $R(s)$ is called a *common divisor* or common factor of $D(s)$ and $N(s)$. Note that a nonzero constant is a common factor of every nonzero $D(s)$ and $N(s)$, and is called a *trivial* common factor of $D(s)$ and $N(s)$. A nontrivial common factor will be a polynomial of degree 1 or higher.

Definition G-1

A polynomial $R(s)$ is a *greatest common divisor* (gcd) of $D(s)$ and $N(s)$ if $R(s)$ is a common divisor of $D(s)$ and $N(s)$, and is divisible by every common divisor of $D(s)$ and $N(s)$. If a gcd of $D(s)$ and $N(s)$ is a nonzero constant (independent of s), then $D(s)$ and $N(s)$ are said to be *relatively prime* or *coprime*. ∎

In other words, two polynomials are coprime if they have only trivial common factors. If they have nontrivial common factors, they are not coprime.

The gcd is unique only up to a constant; that is, if $R(s)$ is a gcd, then $cR(s)$, for any nonzero number c, is also a gcd. If we require a gcd to be monic, then the gcd is unique.

Given two polynomials $D(s)$ and $N(s)$, by a sequence of long divisions, often called the euclidean algorithm, we can write

$$
\begin{array}{rll}
N(s) = Q_1(s)D(s) + R_1(s) & \deg R_1 < \deg D & \\
D(s) = Q_2(s)R_1(s) + R_2(s) & \deg R_2 < \deg R_1 & \\
R_1(s) = Q_3(s)R_2(s) + R_3(s) & \deg R_3 < \deg R_2 & \\
\cdots\cdots\cdots\cdots\cdots & \cdots\cdots\cdots & \text{(G-3)} \\
R_{p-2}(s) = Q_p(s)R_{p-1}(s) + R_p(s) & \deg R_p < \deg R_{p-1} & \\
R_{p-1}(s) = Q_{p+1}(s)R_p(s) + 0 & &
\end{array}
$$

This process will eventually stop because the degree of $R_i(s)$ decreases at each step. We claim that $R_p(s)$ is a gcd of $D(s)$ and $N(s)$.

From the last equation of (G-3) we see that $R_{p-1}(s)$ is divisible, without remainder, by $R_p(s)$. The next to the last equation can be written as $R_{p-2}(s) = (1+Q_pQ_{p+1})R_p(s)$; hence $R_{p-2}(s)$ is also divisible by $R_p(s)$. Proceeding upward, it can be shown that $R_{p-3}, \ldots, R_1$, $D(s)$, and $N(s)$ are all divisible by $R_p(s)$. Hence $R_p(s)$ is a common divisor of $D(s)$ and $N(s)$.

Now we claim that for each $R_i(s)$ in (G-3), there exist polynomials $X_i(s)$ and $Y_i(s)$ such that $R_i(s) = X_i(s)D(s) + Y_i(s)N(s)$. This is clearly true for $R_1(s)$ with $X_1(s) = -Q_1(s)$ and $Y_1(s)=1$. The substitution of $R_1(s)$ into the second equation shows that the claim holds for $R_2(s)$. Proceeding downward, the

[2] If D_n has a very small absolute value, large errors may arise on a digital computer implementation of this algorithm. Hence this method may not be numerically stable. This situation is similar to the gaussian elimination without any pivoting.

claim can be verified for every R_i, $i = 1, 2, \ldots, p$. Hence there exist polynomials $X(s)$ and $Y(s)$ such that

$$R_p(s) = X(s)D(s) + Y(s)N(s) \tag{G-4}$$

This equation implies that every common divisor of $D(s)$ and $N(s)$ divides $R_p(s)$. Indeed if $C(s)$ is a common factor, that is, $D(s) = \bar{D}(s)C(s)$ and $N(s) = \bar{N}(s)C(s)$, then we have $R_p(s) = [X(s)\bar{D}(s) + Y(s)\bar{N}(s)]C(s)$. Hence $R_p(s)$ is divisible by every common divisor of $D(s)$ and $N(s)$. Consequently, $R_p(s)$ is a gcd. All gcd differ at most by a constant; hence every gcd can be expressed in the form of (G-4). This is stated as a theorem.

Theorem G-2

Every gcd of the polynomials $D(s)$ and $N(s)$ is expressible in the form

$$R(s) = X(s)D(s) + Y(s)N(s)$$

where $X(s)$ and $Y(s)$ are polynomials. ∎

Although every gcd is expressible in the form of (G-4), the converse is not true. That is, a polynomial expressible as in (G-4) is not necessarily a gcd of $D(s)$ and $N(s)$ (why?).

Theorem G-3

Consider two polynomials $D(s)$ and $N(s)$ with $D(s) \neq 0$. Then $D(s)$ and $N(s)$ are coprime if and only if any one of the following conditions holds:

1. For every s in $\mathbb{C}$, the field of complex numbers, or for every root of $D(s)$, the 2×1 matrix $\begin{bmatrix} D(s) \\ N(s) \end{bmatrix}$ has rank 1.
2. There exist two polynomials $X(s)$ and $Y(s)$ such that[3]

$$X(s)D(s) + Y(s)N(s) = 1 \tag{G-5}$$

3. There exist no polynomials $A(s)$ and $B(s)$ such that

$$\frac{N(s)}{D(s)} = \frac{B(s)}{A(s)} \tag{G-6}$$

or, equivalently,

$$-B(s)D(s) + A(s)N(s) = [-B(s) \quad A(s)]\begin{bmatrix} D(s) \\ N(s) \end{bmatrix} = 0 \tag{G-7}$$

and

$$\deg A(s) < \deg D(s)$$

[3]It can be shown that $\deg X(s) < \deg N(s)$ and $\deg Y(s) < \deg D(s)$ (see Reference S125). This property is not needed in this text.

Proof

1. If $D(s)$ and $N(s)$ are coprime, there is no s in $\mathbb{C}$ such that $D(s)=0$ and $N(s)=0$. Hence the matrix $[D(s) \quad N(s)]'$ has rank 1 for every s in $\mathbb{C}$. If $D(s)$ and $N(s)$ are not coprime, there exists at least one s in $\mathbb{C}$ such that $D(s)=0$ and $N(s)=0$ and the rank of $[D(s) \quad N(s)]'$ is zero at that s. The matrix $[D(s) \quad N(s)]'$ has rank 1 at every s except at the roots of $D(s)$; hence it is necessary to check its rank only at the roots of $D(s)$.
2. This follows directly from Equation (G-4).
3. If $D(s)$ and $N(s)$ are not coprime, by canceling their nontrivial common divisor, we obtain $A(s)$ and $B(s)$ with deg $A(s) <$ deg $D(s)$. If $D(s)$ and $N(s)$ are coprime, there exist polynomials $X(s)$ and $Y(s)$ such that $X(s)D(s)+Y(s)N(s) = 1$. The substitution of $N(s)=B(s)D(s)/A(s)$ into it yields

$$[X(s)A(s)+Y(s)B(s)]D(s)=A(s)$$

which implies deg $A(s) \geq$ deg $D(s)$. This completes the proof. Q.E.D.

We give a remark concerning the rank of $[D(s) \quad N(s)]'$. Let $R(s)$ be the gcd of $D(s)$ and $N(s)$. Then the matrix $[D(s) \quad N(s)]'$ has a rank of 0 at those s which are roots of $R(s)$. The matrix however still has a rank of 1 at all other s in $\mathbb{C}$. Hence if $D(s)$ and $N(s)$ are not coprime, the matrix $[D(s) \quad N(s)]'$ has rank 1 for *almost all* s in $\mathbb{C}$. If they are coprime, the matrix has rank 1 for *all* s in $\mathbb{C}$.

In the following, we discuss a method of solving Equation (G-7). Let

$$D(s)=D_0+D_1s+D_2s^2+\cdots+D_ns^n \qquad D_n \neq 0 \tag{G-8}$$

$$N(s)=N_0+N_1s+N_2s^2+\cdots+N_ms^m \qquad N_m \neq 0 \tag{G-9}$$

and let

$$A(s)=A_0+A_1s+\cdots+A_{n-1}s^{n-1} \tag{G-10}$$

$$B(s)=B_0+B_1s+\cdots+B_{m-1}s^{m-1} \tag{G-11}$$

with no assumption of $A_{n-1}\neq 0$ and $B_{m-1}\neq 0$. Define

$$\mathbf{S} \triangleq \left[\begin{array}{ccccccccccc}
D_0 & D_1 & D_2 & \cdots & D_{n-1} & D_n & 0 & 0 & \cdots & 0 & 0 \\
0 & D_0 & D_1 & \cdots & D_{n-2} & D_{n-1} & D_n & 0 & \cdots & 0 & 0 \\
\vdots & \vdots & \vdots & & \vdots & \vdots & \vdots & \vdots & & \vdots & \vdots \\
0 & 0 & 0 & \cdots & 0 & D_0 & D_1 & D_2 & \cdots & D_{n-1} & D_n \\
\hdashline
N_0 & N_1 & N_2 & \cdots & N_m & 0 & 0 & 0 & \cdots & 0 & 0 \\
0 & N_0 & N_1 & \cdots & N_{m-1} & N_m & 0 & 0 & \cdots & 0 & 0 \\
\vdots & \vdots & \vdots & & \vdots & \vdots & \vdots & \vdots & & \vdots & \vdots \\
0 & 0 & 0 & \cdots & 0 & 0 & N_0 & N_1 & \cdots & N_{m-1} & N_m
\end{array}\right]
\begin{array}{l} \left.\vphantom{\begin{array}{c}0\\0\\0\\0\end{array}}\right\} m \text{ rows} \\ \left.\vphantom{\begin{array}{c}0\\0\\0\\0\end{array}}\right\} n \text{ rows} \end{array} \tag{G-12}$$

It is a square matrix of order $n+m$ and is called the *Sylvester matrix* or the *resultant* of $D(s)$ and $N(s)$. The substitution of (G-8) to (G-11) into (G-7) and equating the coefficients of s^i, $i=0, 1, \ldots, n+m-1$, yields

$$[-B_0 \ -B_1 \ -B_2 \ \cdots \ -B_{m-1} \vdots A_0 \ A_1 \ A_2 \ \cdots \ A_{n-1}]\mathbf{S}=[-\mathbf{B} \vdots \mathbf{A}]\mathbf{S}=\mathbf{0} \tag{G-13}$$

This is a set of $n+m$ linear algebraic homogeneous equations. We see that the polynomial equation $-B(s)D(s)+A(s)N(s)=0$ has been transformed into the equation in (G-13). If the resultant $\mathbf{S}$ is nonsingular, then the only solution of (G-13) is the trivial solution $A(s)=B(s)=0$. In other words, there exists no $A(s)$ of degree $n-1$ or less to meet $N(s)/D(s)=B(s)/A(s)$; hence $D(s)$ and $N(s)$ are, following Theorem G-3, coprime. If the resultant $\mathbf{S}$ is singular, a nontrivial solution $[-\mathbf{B} \ \ \mathbf{A}]$ exists in (G-13). In other words, polynomials $A(s)$ of degree $n-1$ or less and $B(s)$ exist such that $N(s)/D(s)=B(s)/A(s)$, and $D(s)$ and $N(s)$ are not coprime. Hence we have established the following corollary.

Corollary G-3

The polynomials $D(s)$ and $N(s)$ are coprime if and only if their Sylvester matrix $\mathbf{S}$ defined in (G-12) is nonsingular. ∎

Whether or not $D(s)$ and $N(s)$ are coprime hinges on the existence of a nontrivial solution in (G-13). Many numerically stable methods and "canned" subroutines are available in the literature and computing centers to solve this problem. However, we are interested in only the special solution which yields the smallest degree in $A(s)$. The row searching algorithm discussed in Appendix A turns out to yield such a solution. This is illustrated by an example

Example 1

Consider the polynomials $D(s)=-2s^4+2s^3-s^2-s+1$ and $N(s)=s^3+2s^2-2s+3$. We form their resultant:

$$\mathbf{S}=\left[\begin{array}{ccccccc} 1 & -1 & -1 & 2 & -2 & 0 & 0 \\ 0 & 1 & -1 & -1 & 2 & -2 & 0 \\ 0 & 0 & 1 & -1 & -1 & 2 & -2 \\ \hdashline 3 & -2 & 2 & 1 & 0 & 0 & 0 \\ 0 & 3 & -2 & 2 & 1 & 0 & 0 \\ 0 & 0 & 3 & -2 & 2 & 1 & 0 \\ 0 & 0 & 0 & 3 & -2 & 2 & 1 \end{array}\right] \tag{G-14}$$

We use the row searching algorithm discussed in Appendix A to search the linearly dependent rows of $\mathbf{S}$. The circled pivots are chosen as shown. The result is as follows:

$$\mathbf{F}_*^*\mathbf{S}=\begin{bmatrix}1 & & & & & & \\ 0 & 1 & & & & & \\ 0 & 0 & 1 & & & & \\ -3 & -1 & 0 & 1 & & & \\ 0 & -3 & 0 & -3 & 1 & & \\ 0 & 0 & 0 & -0.5 & 0 & 1 & \\ \hdashline 0 & 0 & 0.5 & -1.5 & -0.5 & 0 & 1\end{bmatrix}{}_*^*\mathbf{S}$$

$$=\begin{bmatrix}\boxed{1} & -1 & -1 & 2 & -2 & 0 & 0\\ 0 & \boxed{1} & -1 & -1 & 2 & -2 & 0\\ 0 & 0 & 1 & -1 & -1 & 2 & \boxed{-2}\\ 0 & 0 & 6 & -4 & 4 & \boxed{2} & 0\\ 0 & 0 & \boxed{-17} & 17 & -17 & 0 & 0\\ 0 & 0 & 0 & 0 & 0 & 0 & 0\\ 0 & 0 & 0 & 0 & 0 & 0 & 0\end{bmatrix}\triangleq\bar{\mathbf{S}} \qquad \textbf{(G-15)}$$

Note that the matrix **F** is the one defined in Equation (A-9) of Appendix A. There are five nonzero rows in $\bar{\mathbf{S}}$; hence the resultant **S** has a rank of 5 and is singular. Corresponding to the first zero row of $\bar{\mathbf{S}}$, we can obtain from the first six rows of **F**, by using the recursive formula in (A-11), the equation

$$[1.5 \quad 0.5 \quad 0 \,\vdots\, -0.5 \quad 0 \quad 1 \quad 0]\mathbf{S}=\mathbf{0}$$

Using (G-10), (G-11), and (G-13), we have

$$A(s)=s^2-0.5 \qquad B(s)=-0.5s-1.5$$

and

$$\frac{N(s)}{D(s)}=\frac{B(s)}{A(s)}=\frac{-0.5s-1.5}{s^2-0.5}=\frac{s+3}{-2s^2+1} \qquad \textbf{(G-16)}$$

Hence $D(s)$ and $N(s)$ are not coprime. ∎

We note that the $B(s)$ and $A(s)$ computed from $[-\mathbf{B} \quad \mathbf{A}]\mathbf{S}=\mathbf{0}$ by using the *first* linearly dependent row of **S**, which corresponds to the *first* zero row of the right-hand-side matrix of (G-15), are coprime. Suppose not, then there exist $\bar{B}(s)$ and $\bar{A}(s)$ of smaller degrees to meet $[-\bar{\mathbf{B}} \quad \bar{\mathbf{A}}]\mathbf{S}=0$. This implies that a linearly dependent row of **S** will appear before the appearance of the zero rows in (G-15). This is not possible. Hence we conclude that the $B(s)$ and $A(s)$ in (G-16) are coprime.

In the Sylvester matrix in (G-12), the coefficients of $D(s)$ and $N(s)$ are arranged in the ascending power of s. Clearly, we can also arrange the coefficients in the descending power of s as

$$\bar{\mathbf{S}} = \left[\begin{array}{cccccccccc} D_n & D_{n-1} & \cdots & D_2 & D_1 & D_0 & 0 & 0 & \cdots 0 & 0 \\ \vdots & \vdots & & \vdots & \vdots & \vdots & \vdots & \vdots & \vdots & \vdots \\ 0 & 0 & \cdots & 0 & 0 & D_n & D_{n-1} & D_{n-2} & \cdots D_1 & D_0 \\ \hline N_m & N_{m-1} & \cdots & N_0 & 0 & 0 & 0 & 0 & \cdots 0 & 0 \\ \vdots & \vdots & & \vdots & \vdots & \vdots & \vdots & \vdots & \vdots & \vdots \\ 0 & 0 & \cdots & 0 & 0 & 0 & N_m & N_{m-1} & \cdots N_1 & N_0 \end{array}\right] \tag{G-17}$$

Using this Sylvester matrix, Corollary G-3 still holds. However, if we use (G-17) to compute $\bar{A}(s)$ and $\bar{B}(s)$, care must be taken in determining their degrees. To see this, we write

$$[-\bar{\mathbf{B}} \quad \bar{\mathbf{A}}]\bar{\mathbf{S}} = [-\bar{\mathbf{B}}_{m-1} \quad \cdots \quad -\bar{\mathbf{B}}_0 \,\vdots\, \bar{\mathbf{A}}_{n-1} \quad \cdots \quad \bar{\mathbf{A}}_0]\bar{\mathbf{S}} = \mathbf{0}$$

In general, $\bar{B}_{m-1} \neq 0$. Hence the degree of $\bar{B}(s)$ is $m-1$, and consequently, the degree of $\bar{A}(s)$ is $n-1$. However, if the degree of the gcd of $D(s)$ and $N(s)$ is 2 or higher, then $\bar{B}(s)$ and $\bar{A}(s)$ have a common factor of form s^k. No such problem will arise in using (G-12). Hence we shall arrange the coefficients of polynomials in ascending order throughout this appendix.

G-2 Reduction of Reducible Rational Functions

Consider a rational function $N(s)/D(s) = N(s)D^{-1}(s)$, where $D(s)$ and $N(s)$ are polynomials. If $D(s)$ and $N(s)$ are coprime, the rational function $N(s)/D(s)$ is said to be *irreducible*. Otherwise, it is said to be *reducible*. In application, it is often required to reduce a reducible rational function to an irreducible one. Clearly the Sylvester matrix in (G-12) can be used for this reduction, as shown in the example of the previous section. In this section, we shall modify the procedure to improve its computability and to lay ground for the extension to the matrix case.

Consider the two polynomials

$$D(s) = D_n s^n + D_{n-1}s^{n-1} + \cdots + D_1 s + D_0 \qquad D_n \neq 0 \tag{G-18}$$

and

$$N(s) = N_n s^n + N_{n-1}s^{n-1} + \cdots + N_1 s + N_0 \tag{G-19}$$

No assumption of $N_n \neq 0$ is imposed; hence the degree of $N(s)$ can be smaller than n. We define, for $k = 0, 1, 2, \ldots,$

$$\mathbf{S}_k = \left[\begin{array}{ccccccccc} D_0 & D_1 & \cdots & D_{n-1} & D_n & 0 & 0 & \cdots & 0 \\ N_0 & N_1 & \cdots & N_{n-1} & N_n & 0 & 0 & \cdots & 0 \\ \hline 0 & D_0 & D_1 & \cdots & D_{n-1} & D_n & 0 & \cdots & 0 \\ 0 & N_0 & N_1 & \cdots & N_{n-1} & N_n & 0 & \cdots & 0 \\ \hline & & & & \vdots & & & & \\ \hline 0 & 0 & \cdots & 0 & D_0 & D_1 & \cdots & D_{n-1} & D_n \\ 0 & 0 & \cdots & 0 & N_0 & N_1 & \cdots & N_{n-1} & N_n \end{array}\right] \begin{array}{l} \}\ 1 \text{ block row} \\ \\ k+1 \text{ block rows} \end{array} \tag{G-20}$$

It is a $2(k+1)\times(n+k+1)$ matrix and consists of $k+1$ block rows. Each block has two rows formed from the coefficients, in the ascending power of s, of $D(s)$ and $N(s)$. We note that each block is the shifting of its previous block to the right by one column. If $k<n-1$, $\mathbf{S}_k$ has more columns than rows; if $k=n-1$, $\mathbf{S}_k$ is a square matrix; if $k>n-1$, then $\mathbf{S}_k$ has more rows than columns.

We search now linearly independent rows of $\mathbf{S}_k$ in order from top to bottom. For convenience of discussion, the rows formed from D_i are called *D rows*; the rows formed from N_i are called *N rows*. First we note that all D rows in $\mathbf{S}_k$ are linearly independent of their previous rows. This follows from the assumption $D_n\neq 0$ and the structure of $\mathbf{S}_k$. For example, if a new block is added to $\mathbf{S}_k$, all elements above D_n in the last column of $\mathbf{S}_{k+1}$ are zeros; hence the new D row is linearly independent of its previous rows. The new N row, however, may or may not be linearly independent of its previous rows. From the structure of $\mathbf{S}_k$, we can readily see that once an N row becomes linearly dependent on its previous rows, then all N rows in subsequent blocks will be linearly dependent. Hence the total number of linearly independent N rows in $\mathbf{S}_k$ will increase monotonically as k increases. However, once the number ceases to increase, the total number of linearly independent N rows will remain to be the same no matter how many more block rows are added to $\mathbf{S}_k$.

Let v be the total number of linear independent N rows in $\mathbf{S}_\infty$. In other words, the v N rows in $\mathbf{S}_{v-1}$ are all linearly independent of their previous rows, and all N rows, not in the first v block rows of $\mathbf{S}_k$, $k\geq v$, are linearly dependent on their previous rows. Note that all $(k+1)$ D rows in $\mathbf{S}_k$ are linearly independent; hence we have

$$\text{rank } \mathbf{S}_k=\begin{cases}2(k+1) & \text{for } k\leq v-1\\ (k+1)+v & \text{for } k\geq v\end{cases} \tag{G-21}$$

Since $\mathbf{S}_k$ is a $2(k+1)\times(n+k+1)$ matrix, if $k\leq v-1$, then $\mathbf{S}_k$ has a full row rank. A necessary condition for $\mathbf{S}_k$ to have a full row rank is that $2(k+1)\leq n+k+1$ or $k\leq n-1$. Hence we conclude that $v\leq n$ or, equivalently, the total number of linearly independent N rows in $\mathbf{S}_k$ is at most equal to n.

Let $A(s)$ and $B(s)$ be two polynomials defined as

$$\begin{aligned}&A(s)=A_0+A_1s+\cdots+A_ks^k\\ \text{and}\quad &B(s)=B_0+B_1s+\cdots+B_ks^k\end{aligned} \tag{G-22}$$

Then from the equation

$$-B(s)D(s)+A(s)N(s)=0 \tag{G-23}$$

we can obtain, similar to (G-13),

$$[-B_0 \quad A_0 \,\vdots\, -B_1 \quad A_1 \,\vdots\, \cdots \,\vdots\, -B_k \quad A_k]\mathbf{S}_k=\mathbf{0} \tag{G-24}$$

If $\mathbf{S}_k$ has a full row rank, the only solution in (G-24) or, equivalently, (G-23) is the trivial solution $A(s)=0$, $B(s)=0$. As k increases from 0, 1, 2, . . . , the first nontrivial solution will appear at $k=v$. Hence the v in (G-21) yields the smallest degree among all $A(s)$ and $B(s)$ which satisfy (G-23). In other words, the smallest degree of $A(s)$ to meet (G-23) is equal to the total number of linear independent N rows in $\mathbf{S}_v$.

Theorem G-4

Consider two polynomials $D(s)$ and $N(s)$ with deg $N(s) \leq$ deg $D(s) = n$. The $A(s)$ and $B(s)$ solved from (G-24) by using the first linearly dependent row of $\mathbf{S}_k$ are coprime, where $\mathbf{S}_k$ is defined as in (G-20).

Proof

Let v be the least integer such that the last N row of $\mathbf{S}_v$ is linearly dependent of its previous rows. Since all D rows are linearly independent, this N row is the first linearly dependent row in $\mathbf{S}_k$, for $k \geq v$. Corresponding to this dependent row, the solution of (G-24) yields an $A(s)$ of degree v and $B(s)$ such that $N(s)/D(s) = B(s)/A(s)$. Now if $A(s)$ and $B(s)$ are not coprime, there exist $\bar{A}(s)$ of a degree smaller than v and $\bar{B}(s)$ such that

$$\frac{N(s)}{D(s)} = \frac{B(s)}{A(s)} = \frac{\bar{B}(s)}{\bar{A}(s)}$$

This implies that a linearly dependent row appears before the last N row of $\mathbf{S}_v$. This is not possible. Hence $A(s)$ and $B(s)$ are coprime. Q.E.D.

We note that if $k \geq n$, the resultant $\mathbf{S}_k$ has more rows than columns and solutions always exist in (G-24). For example, if $k = n$, then $A(s) = D(s)$ and $B(s) = N(s)$ are solutions of (G-24) and (G-23); if $k = n + 1$, then $A(s) = D(s)(s + c)$ and $B(s) = N(s)(s + c)$, for any real c, are solutions of (G-24). Clearly these solutions are of no interest to us.

Corollary G-4

The two polynomials $D(s)$ and $N(s)$ with deg $N(s) \leq$ deg $D(s) = n$ are coprime if and only if the square matrix $\mathbf{S}_{n-1}$ of order $2n$ defined in (G-20) is nonsingular or if and only if the total number of linear independent N rows in $\mathbf{S}_{n-1}$ is equal to n. ∎

This corollary follows directly from Theorem G-4, and its proof is left as an exercise. From Theorem G-4, we see that the reduction of $N(s)/D(s)$ hinges on the search of the first linearly dependent row in $\mathbf{S}_k$. The row searching algorithm in Appendix A is developed exactly for this purpose. It is illustrated in Example 1 of Section G-1 and will not be repeated.

The matrix $\mathbf{S}_k$ has a special structure: Every block row is a shift of its previous block row. By using this shifting property, a very efficient method is developed in Reference S140 to search the first linearly dependent row of $\mathbf{S}_k$. To fully utilize the shifting property, the elimination must be carried out from left to right and the method is generally not numerically stable. For hand calculation, the method can definitely be used in place of the row searching algorithm. The result, however, may not be in an echelon form in the matrix case.

G-3 Polynomial Matrices

A matrix $\mathbf{A}(s)$ with polynomials as elements is called a *polynomial matrix*. Similar to matrices with elements in $\mathbb{R}$ or $\mathbb{C}$, we may introduce the following elementary operations on $\mathbf{A}(s)$:

1. Multiplication of a row or column by a nonzero real or complex number.
2. Interchange any two rows or two columns.
3. Addition of the product of one row or column and a polynomial to another row or column.

These operations can be carried out by using the elementary matrices, for $n=5$, of the form

$$\mathbf{E}_1=\begin{bmatrix}1&0&0&0&0\\0&1&0&0&0\\0&0&1&0&0\\0&0&0&c&0\\0&0&0&0&1\end{bmatrix}\qquad \mathbf{E}_2=\begin{bmatrix}1&0&0&0&0\\0&0&0&0&1\\0&0&1&0&0\\0&0&0&1&0\\0&1&0&0&0\end{bmatrix}\qquad \mathbf{E}_3=\begin{bmatrix}1&0&0&0&0\\0&1&0&0&0\\0&0&1&0&0\\0&d(s)&0&1&0\\0&0&0&0&1\end{bmatrix} \tag{G-25}$$

with $c\neq 0$ and $d(s)$ is a polynomial. We note that the determinants of these elementary matrices are nonzero constants and are independent of s. Their inverses are

$$\mathbf{E}_1^{-1}=\begin{bmatrix}1&0&0&0&0\\0&1&0&0&0\\0&0&1&0&0\\0&0&0&c^{-1}&0\\0&0&0&0&1\end{bmatrix}\qquad \mathbf{E}_2^{-1}=\mathbf{E}_2\qquad \mathbf{E}_3^{-1}=\begin{bmatrix}1&0&0&0&0\\0&1&0&0&0\\0&0&1&0&0\\0&-d(s)&0&1&0\\0&0&0&0&1\end{bmatrix} \tag{G-26}$$

They are again elementary matrices. The premultiplication of $\mathbf{E}_i$ on $\mathbf{A}(s)$ operates on the rows of $\mathbf{A}(s)$, whereas the postmultiplication of $\mathbf{E}_i$ on $\mathbf{A}(s)$ operates on the columns of $\mathbf{A}(s)$. For example, $\mathbf{E}_2\mathbf{A}(s)$ interchanges the second and fifth row. $\mathbf{E}_3\mathbf{A}(s)$ adds the product of the second row of $\mathbf{A}(s)$ and $d(s)$ to the fourth row of $\mathbf{A}(s)$; whereas $\mathbf{A}(s)\mathbf{E}_3$ adds the product of the fourth column of $\mathbf{A}(s)$ and $d(s)$ to the second column of $\mathbf{A}(s)$. We call $\mathbf{E}_i\mathbf{A}(s)$ elementary row operations and $\mathbf{A}(s)\mathbf{E}_i$ elementary column operations.

The set of polynomials does not form a field because its inverse in multiplication is not a polynomial. If we extend the set to include all rational functions, then the set becomes a field. For matrices with elements in the field of rational functions, the concepts of linear independence, rank, and singularity developed for matrices with elements in the field of real or complex numbers are equally applicable. Hence, if we consider polynomials as elements of the field of rational functions, then we may apply the concept of linear dependence and rank to polynomial matrices. For example, the determinant of the polynomial matrix

$$\begin{bmatrix} s+2 & s-1 \\ s-1 & s+2 \end{bmatrix}$$

is $(s+2)^2-(s-1)^2=6s+3$ which is not the zero element in the field of rational functions. Hence, the matrix is nonsingular and has a full rank. The nonsingularity of the matrix in the field of rational functions does not imply that the matrix is nonsingular for all s in $\mathbb{C}$. For example, the matrix has rank 1, rather than 2 at $s=-0.5$.

Conversely, if the determinant of a polynomial matrix, which is a special case of rational matrices, is equal to the zero element of $\mathbb{R}(s)$, the field of rational functions, then the polynomial matrix is singular. For example, the polynomial matrix

$$\begin{bmatrix} s+2 & s-1 \\ s^2+3s+2 & s^2-1 \end{bmatrix}=\begin{bmatrix} \mathbf{a}_1(s) \\ \mathbf{a}_2(s) \end{bmatrix}$$

has a determinant of $(s+2)(s^2-1)-(s-1)(s^2+3s+2)=0$. Hence the matrix is singular. Consequently, there exist rational functions $\alpha_1(s)$ and $\alpha_2(s)$ such that

$$\alpha_1(s)\mathbf{a}_1(s)+\alpha_2(s)\mathbf{a}_2(s)=\mathbf{0} \tag{G-27}$$

For this example, we may choose $\alpha_1(s)=1$ and $\alpha_2(s)=-1/(s+1)$.

Let $\alpha(s)$ be the least common denominator of $\alpha_1(s)$ and $\alpha_2(s)$, and let $\bar{\alpha}_1(s)=\alpha(s)\alpha_1(s)$ and $\bar{\alpha}_2(s)=\alpha(s)\alpha_2(s)$. Then (G-27) implies

$$\bar{\alpha}_1(s)\mathbf{a}_1(s)+\bar{\alpha}_2(s)\mathbf{a}_2(s)=\mathbf{0}.$$

where $\bar{\alpha}_1(s)$ and $\bar{\alpha}_2(s)$ are polynomials. Hence we conclude that polynomial vectors are linearly dependent in the field of rational functions if and only if they can be made dependent by using only polynomials as coefficients[4] (see Problem 2-9).

Let $\mathbf{A}(s)$ be a polynomial matrix with rank r in the field of rational functions $\mathbb{R}(s)$. We show that, by using exclusively elementary row operations, $\mathbf{A}(s)$ can be transformed into the form

[4]In Definition 2-2 if a field is replaced by a ring $\mathbb{R}_i$ (see Footnote 3 of chapter 2), then $(\mathscr{X}, \mathbb{R}_i)$ is called a *module* over the ring. A $n\times 1$ or $1\times n$ polynomial vector can be considered as an element of the rational vector space $(\mathbb{R}^n(s), \mathbb{R}(s))$, or an element of the module $(\mathbb{R}^n[s], \mathbb{R}[s])$. A set of polynomial vectors is linearly independent over the field $\mathbb{R}(s)$ if and only if the set is linearly independent over the ring $\mathbb{R}[s]$. See Reference S34.

$$\left[\begin{array}{ccccccccccccc}
0 & \cdots & 0 & \overset{\substack{k_1\text{th}\\ \text{column}\\ \downarrow}}{a_{1,k_1}} & a_{1,k_1+1} & \cdots & a_{1,k_2-1} & \overset{\substack{k_2\text{th}\\ \text{column}\\ \downarrow}}{a_{1,k_2}} & \overset{\substack{k_3\text{th}\\ \text{column}\\ \downarrow}}{a_{1,k_2+1}} & \cdots & \overset{\substack{k_r\text{th}\\ \text{column}\\ \downarrow}}{a_{1,k_r}} & a_{1,k_r+1} & \cdots \\
0 & \cdots & 0 & 0 & 0 & \cdots & 0 & a_{2,k_2} & a_{2,k_2+1} & \cdots & a_{2,k_r} & a_{2,k_r+1} & \cdots \\
0 & \cdots & 0 & 0 & 0 & \cdots & 0 & 0 & a_{3,k_3} & \cdots & a_{3,k_r} & a_{3,k_r+1} & \cdots \\
\vdots & & \vdots & \vdots & \vdots & & \vdots & \vdots & \vdots & & \vdots & \vdots & \\
0 & \cdots & 0 & 0 & 0 & \cdots & 0 & 0 & 0 & \cdots & a_{r,k_r} & a_{r,k_r+1} & \cdots \\
0 & \cdots & 0 & 0 & 0 & \cdots & 0 & 0 & 0 & \cdots & 0 & 0 & \cdots \\
\vdots & & \vdots & \vdots & \vdots & & \vdots & \vdots & \vdots & & \vdots & \vdots & \\
0 & \cdots & 0 & 0 & 0 & \cdots & 0 & 0 & 0 & \cdots & 0 & 0 & \cdots
\end{array}\right] \tag{G-28}$$

The first r rows are nonzero rows with polynomial elements. The left-most nonzero element, a_{i,k_i}, of each row is a monic polynomial. The column position of a_{i,k_i} must be on the right hand side of the column position of $a_{i-1,k_{i-1}}$, that is, $k_1 < k_2 < \cdots < k_r$.[5] The degrees of all elements above a_{i,k_i} are smaller than the degree of a_{i,k_i}, that is, $\deg a_{j,k_i} < \deg a_{i,k_i}$, for $i = 2, 3, \ldots, r$; $j = 1, 2, \ldots, i-1$. If $\deg a_{i,k_i} = 0$, then $a_{j,k_i} = 0$, for $j = 1, 2, \ldots, i-1$. The matrix in (G-28) with these properties is said to be in the *Hermite (row) form*. A matrix in the form of (G-28) without the property $\deg a_{j,k_i} < \deg a_{i,k_i}$ has no special name; it is just a upper right triangular matrix.

If **A** is a matrix with real elements, the Hermite form is said to be in the *echelon (row) form*. In this case, the leftmost nonzero element of every nonzero row is 1 and all elements above it are zero (because $\deg a_{i,k_i} = 0$). In other words, the leftmost nonzero element 1 is the only nonzero element in that column. By reversing the order of rows and the order of columns, we will obtain a different, but equivalent echelon (row) form. This form will often appear in the remainder of this appendix.

Every polynomial matrix can be triangularized by using exclusively elementary row operations. This is presented as an algorithm.

Triangularization Procedure

Step 1. $\mathbf{M}(s) = \mathbf{A}(s)$ and delete its columns from left until the first column is nonzero.

Step 2. If all elements, except the first element, of the first column of $\mathbf{M}(s)$ are zero, go to step 6; otherwise, go to step 3.

Step 3. Search the element with the smallest degree in the first column of $\mathbf{M}(s)$ and bring it to the (1, 1) position by the interchange of two rows. Make the element monic. Call the resulting matrix $\mathbf{M}_1(s) = (m_{ij}^1(s))$

Step 4. Compute $m_{i1}^1(s) = q_{i1}(s) m_{11}^1(s) + m_{i1}^2(s)$ with $\deg m_{i1}^2(s) < \deg m_{11}^1(s)$. Add the product of the first row of $\mathbf{M}_1(s)$ and $-q_{i1}(s)$ to the ith row of $\mathbf{M}_1(s)$, $i = 2, 3, \ldots, n$. Call the resulting matrix $\mathbf{M}_2(s) = (m_{ij}^2(s))$.

---

[5] If additional column operations are employed, we can always have $k_i = i$ and transform the matrix into a diagonal matrix called the Smith form. The interested reader is referred to References S34. In (G-28), we set, for reducing the size of the matrix, $k_3 = k_2 + 1$.

Step 5. $\mathbf{M}(s) \leftarrow \mathbf{M}_2(s)$ and go to step 2.
Step 6. Delete the first row and the first column of $\mathbf{M}(s)$ and rename it $\mathbf{M}(s)$; go to step 1. ∎

Steps 2 to 5 will reduce the degree of the (1, 1) element of $\mathbf{M}(s)$ by at least 1; hence, after a finite number of iterations, we will go to step 6. We repeat the process for the submatrices of $\mathbf{M}$, and eventually we will transform $\mathbf{A}(s)$ into the form in (G-28). This completes the triangularization of the matrix.

In order to have the property $\deg a_{j,k_i} < \deg a_{i,k_i}$, for $j=1, 2, \ldots, i-1$, we need some additional row operations. Let $a_{j,k_i} = q_j(s) a_{i,k_i} + a^1_{j,k_i}$ with $\deg a^1_{j,k_i} < \deg a_{i,k_i}$. We add the product of the ith row and $-q_j(s)$ to the jth row, the resulting matrix will retain the form of (G-28) and has the property $\deg a_{j,k_i} < \deg a_{i,k_i}$, $j=1, 2, \ldots, i-1$. Thus we have established the following theorem.

Theorem G-5

Every polynomial matrix can be transformed into the Hermite row form in (G-28) by a sequence of elementary row operations. ∎

Example 1

We give an example to illustrate this theorem. Consider

$$\mathbf{A}(s) = \begin{bmatrix} s & 3s+1 \\ -1 & s^2+s-2 \\ -1 & s^2+2s-1 \end{bmatrix} \xrightarrow{1} \begin{bmatrix} -1 & s^2+s-2 \\ s & 3s+1 \\ -1 & s^2+2s-1 \end{bmatrix} \xrightarrow{2} \begin{bmatrix} 1 & -s^2-s+2 \\ 0 & s^3+s^2+s+1 \\ 0 & s+1 \end{bmatrix}$$

$$\xrightarrow{3} \begin{bmatrix} 1 & -s^2-s+2 \\ 0 & s+1 \\ 0 & s^3+s^2+s+1 \end{bmatrix} \xrightarrow{4} \begin{bmatrix} 1 & -s^2-s+2 \\ 0 & s+1 \\ 0 & 0 \end{bmatrix} \xrightarrow{5} \begin{bmatrix} 1 & 2 \\ 0 & s+1 \\ 0 & 0 \end{bmatrix} \quad \text{(G-29)}$$

In the first step, we interchange the first and second row. This is achieved by the multiplication of $\mathbf{A}(s)$ by the elementary matrix denoted by 1 listed at the end of this paragraph. In the second step, we multiply the first row by -1, add the product of the first row and s to the second row and add the product of the first row and -1 to the third row. This is achieved by the matrix denoted by 2. In the third step, we interchange the second and third row. We then add the product of the second row and $-(s^2+1)$ to the third row in step 4. In the last step, we add the product of the second row and s to the first row so that the degree of $a_{12}(s)$ is smaller than that of $a_{11}(s)$. The corresponding elementary matrices are

$$\overset{5}{\begin{bmatrix} 1 & s & 0 \\ 0 & 1 & 0 \\ 0 & 0 & 1 \end{bmatrix}} \overset{4}{\begin{bmatrix} 1 & 0 & 0 \\ 0 & 1 & 0 \\ 0 & -(s^2+1) & 1 \end{bmatrix}} \overset{3}{\begin{bmatrix} 1 & 0 & 0 \\ 0 & 0 & 1 \\ 0 & 1 & 0 \end{bmatrix}} \overset{2}{\begin{bmatrix} -1 & 0 & 0 \\ s & 1 & 0 \\ -1 & 0 & 1 \end{bmatrix}} \overset{1}{\begin{bmatrix} 0 & 1 & 0 \\ 1 & 0 & 0 \\ 0 & 0 & 1 \end{bmatrix}}$$

The product of these five matrices is

$$\begin{bmatrix} 0 & -s-1 & s \\ 0 & -1 & 1 \\ 1 & s^2+s+1 & -(s^2+1) \end{bmatrix} \tag{G-30}$$

Hence, we have

$$\left[\begin{array}{cc:c} 0 & -s-1 & s \\ 0 & -1 & 1 \\ \hdashline 1 & s^2+s+1 & -(s^2+1) \end{array}\right] \begin{bmatrix} s & 3s+1 \\ -1 & s^2+s-2 \\ -1 & s^2+2s-1 \end{bmatrix} = \left[\begin{array}{cc} 1 & 2 \\ 0 & s+1 \\ \hdashline 0 & 0 \end{array}\right] \tag{G-31}$$

∎

In this example, if we are interested in only the triangularization, we may stop at step 4. Dual to Theorem G-5, every polynomial matrix can be transformed, by a sequence of elementary column operations, into a lower left triangular matrix or the Hermite column form which is the transpose of (G-28).

Since the determinants of elementary matrices are nonzero constants and independent of s, so is the polynomial matrix in (G-30). Such polynomial matrices are called *unimodular*.

Definition G-2

A square polynomial matrix $\mathbf{M}(s)$ is called a *unimodular matrix* if its determinant is nonzero and independent of s. ∎

Theorem G-6

A square polynomial matrix is unimodular if and only if its inverse is a polynomial matrix.

Proof

Let $\mathbf{M}(s)$ be unimodular. Then det $\mathbf{M}(s)$ is a constant. Hence $\mathbf{M}^{-1}(s) = [\text{Adj } \mathbf{M}(s)]/\det \mathbf{M}(s)$ is clearly a polynomial matrix. Let $\mathbf{M}(s)$ and $\mathbf{M}^{-1}(s)$ be polynomial matrices. Then, det $\mathbf{M}(s) = a(s)$ and det $\mathbf{M}^{-1}(s) = b(s)$ are polynomials. Since $\mathbf{M}(s)\mathbf{M}^{-1}(s) = \mathbf{I}$, we have $\det \mathbf{M}(s) \cdot \det \mathbf{M}^{-1}(s) = a(s)b(s) = 1$. The only way for polynomials $a(s)$ and $b(s)$ to meet $a(s)b(s) = 1$ is that $a(s)$ and $b(s)$ are both nonzero constants. Hence, $\mathbf{M}(s)$ and $\mathbf{M}^{-1}(s)$ are unimodular. Q.E.D.

In the proof, we have shown that the inverse of a unimodular matrix is also a unimodular matrix. It can be shown that every unimodular matrix can be written as a product of elementary matrices in (G-25).

Unimodular matrices are clearly nonsingular in the field of rational functions. For every s in $\mathbb{C}$, they are also nonsingular in the field of complex numbers. In general, a nonsingular polynomial matrix is nonsingular in the field of complex numbers for *almost all* s in $\mathbb{C}$. It becomes singular only at those s which are the roots of its determinant.

The rank of a polynomial matrix in $\mathbb{R}(s)$ will not change if it is premultiplied or postmultiplied by nonsingular polynomial matrices, in particular, unimodular matrices.

G-4 Coprimeness of Polynomial Matrices

In this section, the concept of coprimeness for scalar polynomials will be extended to polynomial matrices. Since the multiplication of matrices does not commute in general, the situation here is more complicated.

Consider $\mathbf{A}(s)=\mathbf{B}(s)\mathbf{C}(s)$, where $\mathbf{A}(s)$, $\mathbf{B}(s)$, and $\mathbf{C}(s)$ are polynomial matrices of appropriate orders. We call $\mathbf{C}(s)$ a *right divisor* of $\mathbf{A}(s)$ and $\mathbf{A}(s)$ a *left multiple* of $\mathbf{C}(s)$. Similarly we call $\mathbf{B}(s)$ a *left divisor* of $\mathbf{A}(s)$ and $\mathbf{A}(s)$ a *right multiple* of $\mathbf{B}(s)$.

Consider two polynomial matrices $\mathbf{N}(s)$ and $\mathbf{D}(s)$. The square polynomial matrix $\mathbf{R}(s)$ is called a common *right* divisor of $\mathbf{N}(s)$ and $\mathbf{D}(s)$ if there exist polynominal matrices $\bar{\mathbf{N}}(s)$ and $\bar{\mathbf{D}}(s)$ such that

$$\mathbf{N}(s)=\bar{\mathbf{N}}(s)\mathbf{R}(s) \qquad \mathbf{D}(s)=\bar{\mathbf{D}}(s)\mathbf{R}(s) \tag{G-32}$$

In this definition, $\mathbf{N}(s)$ and $\mathbf{D}(s)$ are required to have the same number of columns. Their numbers of rows, however, can be different.

Definition G-3

A square polynomial matrix $\mathbf{R}(s)$ is a *greatest common right divisor* (gcrd) of $\mathbf{N}(s)$ and $\mathbf{D}(s)$ if $\mathbf{R}(s)$ is a common right divisor of $\mathbf{N}(s)$ and $\mathbf{D}(s)$ and is a left multiple of every common right divisor of $\mathbf{N}(s)$ and $\mathbf{D}(s)$. If a gcrd is a unimodular matrix, then $\mathbf{N}(s)$ and $\mathbf{D}(s)$ are said to be *right coprime*. ∎

Dual to this definition, a square polynomial matrix $\mathbf{Q}(s)$ is called a *greatest common left divisor* (gcld) of $\mathbf{A}(s)$ and $\mathbf{B}(s)$ if $\mathbf{Q}(s)$ is a common left divisor of $\mathbf{A}(s)$ and $\mathbf{B}(s)$ [that is, there exist polynomial matrices $\bar{\mathbf{A}}(s)$ and $\bar{\mathbf{B}}(s)$ such that $\mathbf{A}(s)=\mathbf{Q}(s)\bar{\mathbf{A}}(s)$, $\mathbf{B}(s)=\mathbf{Q}(s)\bar{\mathbf{B}}(s)$] and $\mathbf{Q}(s)$ is a right multiple of every common left divisor $\mathbf{Q}_1(s)$ of $\mathbf{A}(s)$ and $\mathbf{B}(s)$ [that is, there exists a polynomial matrix $\mathbf{W}(s)$ such that $\mathbf{Q}(s)=\mathbf{Q}_1(s)\mathbf{W}(s)$].

Now we shall extend Theorem G-2 to the matrix case.

Theorem G-7

Consider the $p \times p$ and $q \times p$ polynomial matrices $\mathbf{D}(s)$ and $\mathbf{N}(s)$. Then they have a gcrd $\mathbf{R}(s)$ expressible in the form

$$\mathbf{R}(s)=\mathbf{X}(s)\mathbf{D}(s)+\mathbf{Y}(s)\mathbf{N}(s)$$

where $\mathbf{X}(s)$ and $\mathbf{Y}(s)$ are $p \times p$ and $p \times q$ polynomial matrices, respectively.

Proof

We form the composite polynomial matrix $[\mathbf{D}'(s) \quad \mathbf{N}'(s)]'$, where the prime denotes the transpose. Then Theorem G-5 implies that there exists a unimodular matrix $\mathbf{U}(s)$ such that

$$\underbrace{\begin{matrix} p \\ q \end{matrix}\begin{bmatrix} \mathbf{U}_{11}(s) & \mathbf{U}_{12}(s) \\ \mathbf{U}_{21}(s) & \mathbf{U}_{22}(s) \end{bmatrix}}_{\mathbf{U}(s)}\begin{bmatrix} \mathbf{D}(s) \\ \mathbf{N}(s) \end{bmatrix}=\begin{bmatrix} \mathbf{R}(s) \\ \mathbf{0} \end{bmatrix}\begin{matrix} \}p \\ \}q \end{matrix} \tag{G-33}$$

where $\mathbf{R}(s)$ is an upper triangular polynomial matrix of the form shown in (G-28). The form of $\mathbf{R}(s)$ is immaterial here. What is important is that the last q rows of the right-hand-side matrix are zero rows. This follows from the fact that the composite matrix $[\mathbf{D}'(s) \quad \mathbf{N}'(s)]'$ is of dimension $(q+p) \times p$. Hence it has a rank of at most p (in the field of rational functions). Consequently, it has at least q linearly dependent rows. Therefore, by a sequence of elementary row operations, it is always possible to achieve (G-33).

Since $\mathbf{U}(s)$ is unimodular, its inverse is a polynomial matrix. Let

$$\begin{bmatrix} \mathbf{U}_{11}(s) & \mathbf{U}_{12}(s) \\ \mathbf{U}_{21}(s) & \mathbf{U}_{22}(s) \end{bmatrix}^{-1} \triangleq \begin{bmatrix} \mathbf{V}_{11}(s) & \mathbf{V}_{12}(s) \\ \mathbf{V}_{21}(s) & \mathbf{V}_{22}(s) \end{bmatrix} \triangleq \mathbf{V}(s) \tag{G-34}$$

Then we have

$$\begin{bmatrix} \mathbf{D}(s) \\ \mathbf{N}(s) \end{bmatrix} = \begin{bmatrix} \mathbf{V}_{11}(s) & \mathbf{V}_{12}(s) \\ \mathbf{V}_{21}(s) & \mathbf{V}_{22}(s) \end{bmatrix} \begin{bmatrix} \mathbf{R}(s) \\ \mathbf{0} \end{bmatrix} \tag{G-35}$$

and

$$\mathbf{D}(s) = \mathbf{V}_{11}(s)\mathbf{R}(s) \qquad \mathbf{N}(s) = \mathbf{V}_{21}(s)\mathbf{R}(s)$$

Hence, $\mathbf{R}(s)$ is a common right divisor of $\mathbf{D}(s)$ and $\mathbf{N}(s)$. From (G-33) we have

$$\mathbf{R}(s) = \mathbf{U}_{11}(s)\mathbf{D}(s) + \mathbf{U}_{12}(s)\mathbf{N}(s) \tag{G-36}$$

Let $\mathbf{R}_1(s)$ be any common right divisor of $\mathbf{D}(s)$ and $\mathbf{N}(s)$; that is, $\mathbf{D}(s) = \bar{\mathbf{D}}(s)\mathbf{R}_1(s)$ and $\mathbf{N}(s) = \bar{\mathbf{N}}(s)\mathbf{R}_1(s)$. The substitution of these into (G-36) yields $\mathbf{R}(s) = [\mathbf{U}_{11}(s)\bar{\mathbf{D}}(s) + \mathbf{U}_{12}(s)\bar{\mathbf{N}}(s)]\mathbf{R}_1(s)$; that is, $\mathbf{R}(s)$ is a left multiple of $\mathbf{R}_1(s)$. Hence, we conclude that $\mathbf{R}(s)$ is a gcrd. This establishes the theorem. Q.E.D.

Example 2

Find a gcrd of the polynomial matrices

$$\mathbf{D}(s) = \begin{bmatrix} s & 3s+1 \\ -1 & s^2+s-2 \end{bmatrix} \qquad \mathbf{N}(s) = [-1 \quad s^2+2s-1]$$

From (G-30), we have

$$\left[\begin{array}{cc:c} 0 & -s-1 & s \\ 0 & 1 & 1 \\ \hdashline 1 & s^2+s+1 & -(s^2+1) \end{array}\right] \begin{bmatrix} \mathbf{D}(s) \\ \mathbf{N}(s) \end{bmatrix} = \begin{bmatrix} \mathbf{R}(s) \\ \hdashline \mathbf{0} \end{bmatrix} = \left[\begin{array}{cc} 1 & 2 \\ 0 & s+1 \\ \hdashline 0 & 0 \end{array}\right]$$

Hence a gcrd of $\mathbf{D}(s)$ and $\mathbf{N}(s)$ is

$$\mathbf{R}(s) = \begin{bmatrix} 1 & 2 \\ 0 & s+1 \end{bmatrix}$$

which is not a unimodular matrix. Hence $\mathbf{D}(s)$ and $\mathbf{N}(s)$ are not right coprime. ■

Let $\mathbf{W}(s)$ be any $p \times p$ unimodular matrix, and let $\mathbf{R}(s)$ be a gcrd of $\mathbf{D}(s)$ and $\mathbf{N}(s)$. Then $\mathbf{W}(s)\mathbf{R}(s)$ is also a gcrd of $\mathbf{N}(s)$ and $\mathbf{D}(s)$. This can be proved by premultiplication of the unimodular matrix diag$\{\mathbf{W}(s), \mathbf{I}_q\}$ to (G-33). Hence the gcrd of $\mathbf{D}(s)$ and $\mathbf{N}(s)$ is not unique. For example, the polynomial matrix

$$\mathbf{W}(s)=\begin{bmatrix} s^k+1 & 1 \\ s^k & 1 \end{bmatrix}$$

is unimodular for any positive integer k; hence

$$\mathbf{R}_1(s)=\mathbf{W}(s)\mathbf{R}(s)=\begin{bmatrix} s^k+1 & 1 \\ s^k & 1 \end{bmatrix}\begin{bmatrix} 1 & 2 \\ 0 & s+1 \end{bmatrix}=\begin{bmatrix} s^k+1 & 2s^k+s+3 \\ s^k & 2s^k+s+1 \end{bmatrix}$$

is also a gcrd of the $\mathbf{D}(s)$ and $\mathbf{N}(s)$ in the example. We see that the degrees of the elements of $\mathbf{R}_1(s)$ may be larger than those of $\mathbf{D}(s)$ and $\mathbf{N}(s)$. This phenomenon can never arise in the scalar case.

Let $\mathbf{R}_1(s)$ and $\mathbf{R}_2(s)$ be two different gcrds of $\mathbf{D}(s)$ and $\mathbf{N}(s)$. Can they always be related by a unimodular matrix? The answer is affirmative if the matrix $[\mathbf{D}'(s) \quad \mathbf{N}'(s)]'$ is of full column rank.

Corollary G-7

If $[\mathbf{D}'(s) \quad \mathbf{N}'(s)]'$ is of full column rank, in particular, if $\mathbf{D}(s)$ is nonsingular, then all gcrds of $\mathbf{D}(s)$ and $\mathbf{N}(s)$ are nonsingular and are related by unimodular matrices.

Proof

From (G-33), if $[\mathbf{D}'(s) \quad \mathbf{N}'(s)]'$ is of full column rank, so is $[\mathbf{R}'(s) \quad \mathbf{0}']'$. Hence, $\mathbf{R}(s)$ is nonsingular. Let $\mathbf{R}_1(s)$ be any gcrd of $\mathbf{D}(s)$ and $\mathbf{N}(s)$. Then by definition, we have two polynomial matrices $\mathbf{W}_1(s)$ and $\mathbf{W}_2(s)$ such that

$$\mathbf{R}(s)=\mathbf{W}_1(s)\mathbf{R}_1(s) \qquad \mathbf{R}_1(s)=\mathbf{W}_2(s)\mathbf{R}(s).$$

which imply

$$\mathbf{R}(s)=\mathbf{W}_1(s)\mathbf{W}_2(s)\mathbf{R}(s) \tag{G-37}$$

Since $\mathbf{R}(s)$ is nonsingular, we have $\mathbf{W}_1(s)\mathbf{W}_2(s)=\mathbf{I}$. Hence, both $\mathbf{W}_1(s)$ and $\mathbf{W}_2(s)$ are unimodular matrices. Consequently, $\mathbf{R}_1(s)=\mathbf{W}_2(s)\mathbf{R}(s)$ is also nonsingular. Q.E.D.

In application, we often have the condition that $\mathbf{D}(s)$ is nonsingular. With this condition, the condition in Corollary G-7 is always met. In this case, the gcrd of $\mathbf{D}(s)$ and $\mathbf{N}(s)$ is unique in the sense that all gcrds can be obtained from a single gcrd by premultiplying unimodular matrices.

Theorem G-8

Let $\mathbf{D}(s)$ and $\mathbf{N}(s)$ be $p \times p$ and $q \times p$ polynomial matrices, and let $\mathbf{D}(s)$ be nonsingular. Then $\mathbf{D}(s)$ and $\mathbf{N}(s)$ are right coprime if and only if any one of the following conditions holds:

1. For every s in $\mathbb{C}$, or for every root of the determinant of $\mathbf{D}(s)$, the $(p+q)\times p$ matrix

$$\begin{bmatrix} \mathbf{D}(s) \\ \mathbf{N}(s) \end{bmatrix}$$

has rank p (in the field of complex numbers).

2. There exist polynomial matrices $\mathbf{X}(s)$ and $\mathbf{Y}(s)$ of order $p \times p$ and $p \times q$ such that

$$\mathbf{X}(s)\mathbf{D}(s) + \mathbf{Y}(s)\mathbf{N}(s) = \mathbf{I} \tag{G-38}$$

This is called the *Bezout identity* in References S34 and S125.

3. There exist no polynomial matrices $\mathbf{B}(s)$ and $\mathbf{A}(s)$ of order $q \times p$ and $q \times q$ such that $\mathbf{B}(s)\mathbf{D}(s) = \mathbf{A}(s)\mathbf{N}(s)$ or, equivalently,

$$-\mathbf{B}(s)\mathbf{D}(s) + \mathbf{A}(s)\mathbf{N}(s) = [-\mathbf{B}(s) \quad \mathbf{A}(s)] \begin{bmatrix} \mathbf{D}(s) \\ \mathbf{N}(s) \end{bmatrix} = \mathbf{0} \tag{G-39}$$

and $$\deg \det \mathbf{A}(s) < \deg \det \mathbf{D}(s).$$

Proof

1. For convenience we rewrite (G-33) in the following

$$\begin{bmatrix} \mathbf{U}_{11}(s) & \mathbf{U}_{12}(s) \\ \mathbf{U}_{21}(s) & \mathbf{U}_{22}(s) \end{bmatrix} \begin{bmatrix} \mathbf{D}(s) \\ \mathbf{N}(s) \end{bmatrix} = \begin{bmatrix} \mathbf{R}(s) \\ \mathbf{0} \end{bmatrix} \tag{G-40}$$

For every s in $\mathbb{C}$, the unimodular matrix $\mathbf{U}(s)$ is nonsingular in the field of complex numbers. Hence, for every s in $\mathbb{C}$, the rank of $[\mathbf{D}'(s) \quad \mathbf{N}'(s)]'$ is the same as the rank of $\mathbf{R}(s)$. If $\mathbf{D}(s)$ and $\mathbf{N}(s)$ are right coprime, then $\mathbf{R}(s)$ is unimodular and has rank p for every s in $\mathbb{C}$; hence $[\mathbf{D}'(s) \quad \mathbf{N}'(s)]'$ also has rank p for every s in $\mathbb{C}$. If $\mathbf{D}(s)$ and $\mathbf{N}(s)$ are not right coprime, then det $\mathbf{R}(s)$ is a polynomial of degree one or higher [Note that $\mathbf{R}(s)$ is nonsingular following the nonsingularity of $\mathbf{D}(s)$]. Therefore, there is at least one s in $\mathbb{C}$ such that det $\mathbf{R}(s) = 0$ or, equivalently, the rank of $\mathbf{R}(s)$ is smaller than p. Hence, we conclude that if $\mathbf{D}(s)$ and $\mathbf{N}(s)$ are not right coprime, then the rank of $[\mathbf{D}'(s) \quad \mathbf{N}'(s)]'$ is smaller than p for some s in $\mathbb{C}$. This completes the proof of the first statement. Since $\mathbf{D}(s)$ is nonsingular by assumption, the rank condition is met at every s in $\mathbb{C}$ except the roots of det $\mathbf{D}(s) = 0$. Hence we check the rank only at the roots of det $\mathbf{D}(s) = 0$.

2. To show the second statement, we write the first p equations of (G-40) as

$$\mathbf{U}_{11}(s)\mathbf{D}(s) + \mathbf{U}_{12}(s)\mathbf{N}(s) = \mathbf{R}(s) \tag{G-41}$$

If $\mathbf{D}(s)$ and $\mathbf{N}(s)$ are right coprime, $\mathbf{R}(s)$ is unimodular, and $\mathbf{R}^{-1}(s)$ is a polynomial matrix. The premultiplication of $\mathbf{R}^{-1}(s)$ on both sides of (G-41) yields (G-38) with $\mathbf{X}(s) = \mathbf{R}^{-1}(s)\mathbf{U}_{11}(s)$ and $\mathbf{Y}(s) = \mathbf{R}^{-1}(s)\mathbf{U}_{12}(s)$.

Conversely, we show that (G-38) implies the right coprimeness of $\mathbf{D}(s)$ and $\mathbf{N}(s)$. Let $\mathbf{R}(s)$ be a gcrd of $\mathbf{D}(s)$ and $\mathbf{N}(s)$; that is, $\mathbf{D}(s) = \bar{\mathbf{D}}(s)\mathbf{R}(s)$ and $\mathbf{N}(s) = \bar{\mathbf{N}}(s)\mathbf{R}(s)$. The substitution of these into (G-38) yields

$$[\mathbf{X}(s)\bar{\mathbf{D}}(s) + \mathbf{Y}(s)\bar{\mathbf{N}}(s)]\mathbf{R}(s) = \mathbf{I}$$

which implies

$$\mathbf{R}^{-1}(s) = \mathbf{X}(s)\bar{\mathbf{D}}(s) + \mathbf{Y}(s)\bar{\mathbf{N}}(s) = \text{polynomial matrix}$$

It follows from Theorem G-6 that $\mathbf{R}(s)$ is unimodular. Hence, $\mathbf{D}(s)$ and $\mathbf{N}(s)$ are right coprime.

3. We write the bottom q equations of (G-40) as

$$\mathbf{U}_{21}(s)\mathbf{D}(s) + \mathbf{U}_{22}(s)\mathbf{N}(s) = \mathbf{0}$$

This is already in the form of (G-39) if we identify $\mathbf{B}(s) = -\mathbf{U}_{21}(s)$ and $\mathbf{A}(s) = \mathbf{U}_{22}(s)$. Hence what remains to be proved is the following inequality:

$$\deg \det \mathbf{A}(s) = \deg \det \mathbf{U}_{22}(s) < \deg \det \mathbf{D}(s)$$

From (G-34) and (G-35), we have

$$\mathbf{D}(s) = \mathbf{V}_{11}(s)\mathbf{R}(s) \tag{G-42}$$

Hence, if $\mathbf{D}(s)$ is nonsingular, so are $\mathbf{V}_{11}(s)$ and $\mathbf{R}(s)$. Using the identity

$$\begin{bmatrix} \mathbf{I} & \mathbf{0} \\ -\mathbf{V}_{21}(s)\mathbf{V}_{11}^{-1}(s) & \mathbf{I} \end{bmatrix} \begin{bmatrix} \mathbf{V}_{11}(s) & \mathbf{V}_{12}(s) \\ \mathbf{V}_{21}(s) & \mathbf{V}_{22}(s) \end{bmatrix}$$

$$= \begin{bmatrix} \mathbf{V}_{11}(s) & \mathbf{V}_{12}(s) \\ \mathbf{0} & \mathbf{V}_{22}(s) - \mathbf{V}_{21}(s)\mathbf{V}_{11}^{-1}(s)\mathbf{V}_{12}(s) \end{bmatrix} \tag{G-43}$$

we can write the determinant of $\mathbf{V}(s)$ defined in (G-34) as

$$\det \mathbf{V}(s) = \det \mathbf{V}_{11}(s) \det [\mathbf{V}_{22}(s) - \mathbf{V}_{21}(s)\mathbf{V}_{11}^{-1}(s)\mathbf{V}_{12}(s)] \neq 0 \tag{G-44}$$

for all s in $\mathbb{C}$. Taking the inverse of (G-43) yields

$$\begin{bmatrix} \mathbf{V}_{11}(s) & \mathbf{V}_{12}(s) \\ \mathbf{V}_{21}(s) & \mathbf{V}_{22}(s) \end{bmatrix}^{-1} \begin{bmatrix} \mathbf{I} & \mathbf{0} \\ -\mathbf{V}_{21}(s)\mathbf{V}_{11}^{-1}(s) & \mathbf{I} \end{bmatrix}^{-1} = \begin{bmatrix} \mathbf{V}_{11}(s) & \mathbf{V}_{12}(s) \\ \mathbf{0} & \Delta \end{bmatrix}^{-1} \tag{G-45}$$

where $\Delta = \mathbf{V}_{22}(s) - \mathbf{V}_{21}(s)\mathbf{V}_{11}^{-1}(s)\mathbf{V}_{12}(s)$ is nonsingular following (G-44). The inverse of a triangular matrix is again triangular and can be readily computed. After computing the inverse of the right-hand-side matrix of (G-45), we then move the second inverse on the left-hand side of the equality in (G-45) to the right-hand side:

$$\begin{bmatrix} \mathbf{V}_{11}(s) & \mathbf{V}_{12}(s) \\ \mathbf{V}_{21}(s) & \mathbf{V}_{22}(s) \end{bmatrix}^{-1} = \begin{bmatrix} \mathbf{V}_{11}^{-1}(s) & -\mathbf{V}_{11}^{-1}(s)\mathbf{V}_{12}(s)\Delta^{-1} \\ \mathbf{0} & \Delta^{-1} \end{bmatrix} \begin{bmatrix} \mathbf{I} & \mathbf{0} \\ -\mathbf{V}_{21}(s)\mathbf{V}_{11}^{-1}(s) & \mathbf{I} \end{bmatrix}$$

$$= \begin{bmatrix} \mathrm{X} & \mathrm{X} \\ \mathrm{X} & \Delta^{-1} \end{bmatrix}$$

where X denotes elements which are not needed in the following. The comparison of this equation with (G-34) yields

$$\mathbf{U}_{22}(s) = \Delta^{-1} = [\mathbf{V}_{22}(s) - \mathbf{V}_{21}(s)\mathbf{V}_{11}^{-1}(s)\mathbf{V}_{12}(s)]^{-1}$$

The substitution of this equation to (G-44) yields

$$\det \mathbf{V}(s) = \frac{\det \mathbf{V}_{11}(s)}{\det \mathbf{U}_{22}(s)} \tag{G-46}$$

Since $\mathbf{V}(s)$ is unimodular, we have $\deg \det \mathbf{V}(s) = 0$. Hence, (G-46) implies

$$\deg \det \mathbf{V}_{11}(s) = \deg \det \mathbf{U}_{22}(s) \tag{G-47}$$

This relation holds for any unimodular matrix $\mathbf{U}(s)$ and its inverse $\mathbf{V}(s) = \mathbf{U}^{-1}(s)$. We shall now use (G-47) and (G-42) to establish statement 3 of Theorem G-8. If $\mathbf{D}(s)$ and $\mathbf{N}(s)$ are not right coprime, there exists a gcrd $\mathbf{R}(s)$ with $\deg \det \mathbf{R}(s) > 0$. From (G-42), we have $\det \mathbf{D}(s) = \det \mathbf{V}_{11}(s) \times \det \mathbf{R}(s)$, which implies

$$\deg \det \mathbf{D}(s) > \deg \det \mathbf{V}_{11}(s)$$

Because of (G-47) and $\mathbf{A}(s) = -\mathbf{U}_{22}(s)$, we conclude $\deg \det \mathbf{A}(s) < \deg \det \mathbf{D}(s)$.

Conversely, if $\deg \det \mathbf{A}(s) < \deg \det \mathbf{D}(s)$, we may reverse the above argument and conclude that $\mathbf{D}(s)$ and $\mathbf{N}(s)$ are not right coprime. Q.E.D.

In the following, we develop a dual of Theorem G-8 for matrices which are left coprime. Given two polynomial matrices $\mathbf{A}(s)$ and $\mathbf{B}(s)$ of the same number of rows, then, similar to (G-33), there exists a sequence of elementary transformations so that

$$q\{[\underbrace{\mathbf{Q}(s)}_{q} \quad \underbrace{\mathbf{0}}_{p}] = q\{[\underbrace{\mathbf{A}(s)}_{q} \quad \underbrace{\mathbf{B}(s)}_{p}] \begin{bmatrix} \overbrace{\mathbf{V}_{11}(s)}^{q} & \overbrace{\mathbf{V}_{12}(s)}^{p} \\ \mathbf{V}_{21}(s) & \mathbf{V}_{22}(s) \end{bmatrix} \begin{matrix} \}q \\ \}p \end{matrix}$$

Based on this, we have the following.

Theorem G-8′

Let $\mathbf{A}(s)$ and $\mathbf{B}(s)$ be $q \times q$ and $q \times p$ polynomial matrices and let $\mathbf{A}(s)$ be nonsingular. Then $\mathbf{A}(s)$ and $\mathbf{B}(s)$ are left coprime if and only if any one of the following conditions holds:

1. For every s in $\mathbb{C}$, or for every root of the determinant of $\mathbf{A}(s)$, the $q \times (q+p)$ matrix

$$[\mathbf{A}(s) \quad \mathbf{B}(s)] \tag{G-48}$$

has rank q in the field of complex numbers.

2. There exist polynomial matrices $\bar{\mathbf{X}}(s)$ and $\bar{\mathbf{Y}}(s)$ of order $q \times q$ and $p \times q$ such that

$$\mathbf{A}(s)\bar{\mathbf{X}}(s) + \mathbf{B}(s)\bar{\mathbf{Y}}(s) = \mathbf{I} \tag{G-49}$$

3. There exists no polynomial matrices $\mathbf{N}(s)$ and $\mathbf{D}(s)$ of order $q \times p$ and $p \times p$ such that

$$\mathbf{A}(s)\mathbf{N}(s) = \mathbf{B}(s)\mathbf{D}(s)$$

or, equivalently,

$$-\mathbf{A}(s)\mathbf{N}(s)+\mathbf{B}(s)\mathbf{D}(s)=[\mathbf{A}(s) \quad \mathbf{B}(s)]\begin{bmatrix}-\mathbf{N}(s)\\ \mathbf{D}(s)\end{bmatrix}=\mathbf{0} \tag{G-50}$$

and deg det $\mathbf{D}(s) <$ deg det $\mathbf{A}(s)$. ∎

We give the following corollary to conclude this section.

Corollary G-8

Let $\mathbf{D}(s)$ and $\mathbf{N}(s)$ be $p \times p$ and $q \times p$ polynomial matrices and let $\mathbf{D}(s)$ be nonsingular. Let $\mathbf{U}(s)$ be a unimodular matrix such that

$$\mathbf{U}(s)\begin{bmatrix}\mathbf{D}(s)\\ \mathbf{N}(s)\end{bmatrix} \triangleq \begin{bmatrix}\mathbf{U}_{11}(s) & \mathbf{U}_{12}(s)\\ \mathbf{U}_{21}(s) & \mathbf{U}_{22}(s)\end{bmatrix}\begin{bmatrix}\mathbf{D}(s)\\ \mathbf{N}(s)\end{bmatrix}=\begin{bmatrix}\mathbf{R}(s)\\ \mathbf{0}\end{bmatrix}$$

Then we have

1. $\mathbf{U}_{22}(s)$ and $\mathbf{U}_{21}(s)$ are left coprime.
2. $\mathbf{U}_{22}(s)$ is nonsingular and $\mathbf{N}(s)\mathbf{D}^{-1}(s) = -\mathbf{U}_{22}^{-1}(s)\mathbf{U}_{21}(s)$.
3. $\mathbf{D}(s)$ and $\mathbf{N}(s)$ are right coprime if and only if deg det $\mathbf{D}(s)=$deg det $\mathbf{U}_{22}(s)$.

Proof

Since $\mathbf{U}(s)$ is unimodular, it has rank $p+q$ for every s in $\mathbb{C}$. This implies that for every s in $\mathbb{C}$, its submatrix $[\mathbf{U}_{21}(s) \quad \mathbf{U}_{22}(s)]$, a $q \times (p+q)$ polynomial matrix, has rank q. Hence $\mathbf{U}_{22}(s)$ and $\mathbf{U}_{21}(s)$ are, following Theorem G-8′, left coprime.

We show the nonsingularity of $\mathbf{U}_{22}(s)$ by contradiction. Suppose $\mathbf{U}_{22}(s)$ is not nonsingular, then there exists a $1 \times q$ polynomial vector $\mathbf{a}(s)$, not identically zero, such that

$$\mathbf{a}(s)\mathbf{U}_{22}(s)=0$$

which, together with $\mathbf{U}_{21}(s)\mathbf{D}(s)+\mathbf{U}_{22}(s)\mathbf{N}(s)=\mathbf{0}$, implies

$$\mathbf{a}(s)\mathbf{U}_{21}(s)\mathbf{D}(s)=\mathbf{0}$$

Since $\mathbf{D}(s)$ is nonsingular by assumption, $\mathbf{a}(s)\mathbf{U}_{21}(s)\mathbf{D}(s)=0$ implies

$$\mathbf{a}(s)\mathbf{U}_{21}(s)=\mathbf{0}$$

Hence we have $\mathbf{a}(s)[\mathbf{U}_{21}(s) \quad \mathbf{U}_{22}(s)]=\mathbf{0}$. This contradicts with the fact that $[\mathbf{U}_{21} \quad \mathbf{U}_{22}]$ has rank q in the field of rational functions. Hence we conclude that $\mathbf{U}_{22}(s)$ is nonsingular. Consequently, from $\mathbf{U}_{21}\mathbf{D}+\mathbf{U}_{22}\mathbf{N}=\mathbf{0}$, we have

$$\mathbf{N}(s)\mathbf{D}^{-1}(s) = -\mathbf{U}_{22}^{-1}(s)\mathbf{U}_{21}(s)$$

Part 3 of this theorem has been essentially established in the proof of Theorem G-8. Indeed, if $\mathbf{D}(s)$ and $\mathbf{N}(s)$ are right coprime, then $\mathbf{R}(s)$ in (G-42) is unimodular and deg det $\mathbf{D}(s)=$deg det $\mathbf{V}_{11}(s)$. This, together with (G-47), implies deg det $\mathbf{D}(s)=$deg det $\mathbf{U}_{22}(s)$. Conversely, if deg det $\mathbf{D}(s)=$deg det $\mathbf{U}_{22}(s)$, we may reverse the above argument to conclude deg det $\mathbf{R}(s)=0$. Since $\mathbf{R}(s)$ is nonsinguar and deg det $\mathbf{R}(s)=0$, $\mathbf{R}(s)$ must be unimodular. Hence $\mathbf{D}(s)$ and $\mathbf{N}(s)$ are right coprime. Q.E.D.

G-5 Column- and Row-Reduced Polynomial Matrices

Consider two polynomial matrices $\mathbf{N}(s)$ and $\mathbf{D}(s)$. If $\mathbf{D}(s)$ is square and nonsingular, then the matrix $\mathbf{N}(s)\mathbf{D}^{-1}(s)$ is generally a rational matrix. Conversely, given a $q \times p$ rational matrix $\hat{\mathbf{G}}(s)$, we can always factor $\hat{\mathbf{G}}(s)$ as

$$\hat{\mathbf{G}}(s)=\mathbf{N}(s)\mathbf{D}^{-1}(s) \qquad \textbf{(G-51)}$$

or

$$\hat{\mathbf{G}}(s)=\mathbf{A}^{-1}(s)\mathbf{B}(s) \qquad \textbf{(G-52)}$$

where $\mathbf{N}(s)$, $\mathbf{D}(s)$, $\mathbf{A}(s)$, and $\mathbf{B}(s)$ are, respectively, $q \times p$, $p \times p$, $q \times q$, and $q \times p$ polynomial matrices. For example, we have

$$\begin{bmatrix} \frac{n_{11}}{d_{11}} & \frac{n_{12}}{d_{12}} & \frac{n_{13}}{d_{13}} \\ \frac{n_{21}}{d_{21}} & \frac{n_{22}}{d_{22}} & \frac{n_{23}}{d_{23}} \end{bmatrix} = \begin{bmatrix} \frac{\bar{n}_{11}}{d_{c1}} & \frac{\bar{n}_{12}}{d_{c2}} & \frac{\bar{n}_{13}}{d_{c3}} \\ \frac{\bar{n}_{21}}{d_{c1}} & \frac{\bar{n}_{22}}{d_{c2}} & \frac{\bar{n}_{23}}{d_{c3}} \end{bmatrix} = \begin{bmatrix} \bar{n}_{11} & \bar{n}_{12} & \bar{n}_{13} \\ \bar{n}_{21} & \bar{n}_{22} & \bar{n}_{23} \end{bmatrix} \times \begin{bmatrix} d_{c1} & 0 & 0 \\ 0 & d_{c2} & 0 \\ 0 & 0 & d_{c3} \end{bmatrix}^{-1} \qquad \textbf{(G-53)}$$

$$= \begin{bmatrix} \frac{\tilde{n}_{11}}{d_{r1}} & \frac{\tilde{n}_{12}}{d_{r1}} & \frac{\tilde{n}_{13}}{d_{r1}} \\ \frac{\tilde{n}_{21}}{d_{r2}} & \frac{\tilde{n}_{22}}{d_{r2}} & \frac{\tilde{n}_{23}}{d_{r2}} \end{bmatrix} = \begin{bmatrix} d_{r1} & 0 \\ 0 & d_{r2} \end{bmatrix}^{-1} \begin{bmatrix} \tilde{n}_{11} & \tilde{n}_{12} & \tilde{n}_{13} \\ \tilde{n}_{21} & \tilde{n}_{22} & \tilde{n}_{23} \end{bmatrix}$$

where n_{ij} and d_{ij} are polynomials, d_{ci} is the least common denominator of the ith column of $\hat{\mathbf{G}}(s)$, and d_{ri} is the least common denominator of the ith row of $\hat{\mathbf{G}}(s)$. These fractions are easy to carry out; however, $\mathbf{N}(s)$ and $\mathbf{D}(s)$ are generally not right coprime and $\mathbf{A}(s)$ and $\mathbf{B}(s)$ are generally not left coprime.

A rational function $\hat{\mathbf{G}}(s)$ is called *strictly proper* if $\hat{\mathbf{G}}(\infty)<\infty$ (see Definition 3-5). In terms of the elements of $\hat{\mathbf{G}}(s)$, the properness of $\hat{\mathbf{G}}(s)$ can be easily determined. For example, the rational matrix $\hat{\mathbf{G}}(s)$ is proper if and only if the degree of the numerator of *every element* of $\hat{\mathbf{G}}(s)$ is smaller than or equal to that of its denominator. In terms of the fractions in (G-51) or (G-52) the situation is more complicated. We shall study this problem in this section.

Given a polynomial column or row vector, its degree is defined as the highest power of s in all entries of the vector. We define

$$\delta_{ci}\mathbf{M}(s)=\text{the degree of the } i\text{th column of } \mathbf{M}(s)$$

$$\delta_{ri}\mathbf{M}(s)=\text{the degree of the } i\text{th row of } \mathbf{M}(s)$$

and call δ_{ci} *column degree* and δ_{ri} *row degree*. For example, for

$$\mathbf{M}(s)=\begin{bmatrix} s+1 & s^3+2s+1 & s \\ s-1 & s^3 & 0 \end{bmatrix} \tag{G-54}$$

we have $\delta_{c1}=1$, $\delta_{c2}=3$, $\delta_{c3}=1$, and $\delta_{r1}=3$, $\delta_{r2}=3$.

Theorem G-9

If $\hat{\mathbf{G}}(s)$ is a $q \times p$ proper (strictly proper) rational matrix and if $\hat{\mathbf{G}}(s)=\mathbf{N}(s)\mathbf{D}^{-1}(s)$ $=\mathbf{A}^{-1}(s)\mathbf{B}(s)$, then

$$\delta_{ci}\mathbf{N}(s) \leq \delta_{ci}\mathbf{D}(s) \qquad [\delta_{ci}\mathbf{N}(s) < \delta_{ci}\mathbf{D}(s)]$$

for $i=1, 2, \ldots, p$, and

$$\delta_{rj}\mathbf{B}(s) \leq \delta_{rj}\mathbf{A}(s) \qquad [\delta_{rj}\mathbf{B}(s) < \delta_{rj}\mathbf{A}(s)]$$

for $j=1, 2, \ldots, q$.

Proof

We write $\mathbf{N}(s)=\hat{\mathbf{G}}(s)\mathbf{D}(s)$. Let $n_{ij}(s)$ be the ijth element of $\mathbf{N}(s)$. Then we have

$$n_{ij}(s)=\sum_{k=1}^{p} g_{ik}(s)d_{kj}(s) \qquad i=1, 2, \ldots, q$$

Note that, for every element in the jth column of $\mathbf{N}(s)$, the summation is carried over the jth column of $\mathbf{D}(s)$. If $\hat{\mathbf{G}}(s)$ is proper, the degree of $n_{ij}(s)$, $i=1, 2, \ldots, q$, is smaller than or equal to the highest degree in $d_{kj}(s)$, $k=1, 2, \ldots, p$. Hence we have

$$\delta_{cj}\mathbf{N}(s) \leq \delta_{cj}\mathbf{D}(s) \qquad j=1, 2, \ldots, p$$

The rest of the theorem can be similarly proved. Q.E.D.

We showed in Theorem G-9 that if $\hat{\mathbf{G}}(s)=\mathbf{N}(s)\mathbf{D}^{-1}(s)$ is strictly proper, then column degrees of $\mathbf{N}(s)$ are smaller than the corresponding column degrees of $\mathbf{D}(s)$. It is natural to ask whether the converse is also true. In general, the answer is negative, as can be seen from

$$\mathbf{N}(s)=[1 \quad 2] \qquad \mathbf{D}(s)=\begin{bmatrix} s^2 & s-1 \\ s+1 & 1 \end{bmatrix} \tag{G-55}$$

where $\delta_{ci}\mathbf{N}(s) < \delta_{ci}\mathbf{D}(s)$, $i=1, 2$. However, we have

$$\mathbf{N}(s)\mathbf{D}^{-1}(s)=\begin{bmatrix} \dfrac{-2s-1}{1} & \dfrac{2s^2-s+1}{1} \end{bmatrix}$$

which is neither strictly proper nor proper.

In order to resolve this difficulty, we need a new concept.

Definition G-4

A nonsingular $p \times p$ polynomial matrix $\mathbf{M}(s)$ is called *column reduced* if

$$\deg \det \mathbf{M}(s) = \sum_{i=1}^{p} \delta_{ci}\mathbf{M}(s)$$

It is called *row reduced* if

$$\deg \det \mathbf{M}(s) = \sum_{i=1}^{p} \delta_{ri}\mathbf{M}(s) \qquad \blacksquare$$

The matrix $\mathbf{D}(s)$ in (G-55) is not column reduced because $\deg \det \mathbf{D}(s) = 0 < \delta_{c1}\mathbf{D}(s) + \delta_{c2}\mathbf{D}(s) = 2 + 1 = 3$; neither is it row reduced. A matrix may be column reduced but not row reduced or vice versa. For example, the matrix

$$\mathbf{M}(s) = \begin{bmatrix} 3s^2 + 2s & 2s + 1 \\ s^2 + s - 3 & s \end{bmatrix} \tag{G-56}$$

is column reduced but not row reduced (verify!). A diagonal polynomial matrix is always both column and row reduced.

Let $\delta_{ci}\mathbf{M}(s) = k_{ci}$. Then the polynomial matrix $\mathbf{M}(s)$ can be written as

$$\mathbf{M}(s) = \mathbf{M}_{hc}\mathbf{H}_c(s) + \mathbf{M}_{lc}(s) \tag{G-57}$$

where $\mathbf{H}_c(s) = \text{diag}\{s^{k_{ci}}, i = 1, 2, \ldots, p\}$. The constant matrix $\mathbf{M}_{hc}$ will be called the *column-degree coefficient matrix*; its ith column is the coefficients of the ith column of $\mathbf{M}(s)$ associated with $s^{k_{ci}}$. The polynomial matrix $\mathbf{M}_{lc}(s)$ contains the remaining terms and its ith column has a degree smaller than k_{ci}. For example, the $\mathbf{M}(s)$ in (G-56) can be written as

$$\mathbf{M}(s) = \begin{bmatrix} 3 & 2 \\ 1 & 1 \end{bmatrix}\begin{bmatrix} s^2 & 0 \\ 0 & s \end{bmatrix} + \begin{bmatrix} 2s & 1 \\ s-3 & 0 \end{bmatrix}$$

In terms of (G-57), we have

$$\det \mathbf{M}(s) = (\det \mathbf{M}_{hc})s^{\Sigma k_{ci}} + \text{terms with degrees smaller than } \Sigma k_{ci}$$

Hence, we conclude that $\mathbf{M}(s)$ *is column-reduced if and only if its column-degree coefficient matrix* $\mathbf{M}_{hc}$ *is nonsingular.*

Similar to (G-57), we can also write $\mathbf{M}(s)$ as

$$\mathbf{M}(s) = \mathbf{H}_r(s)\mathbf{M}_{hr} + \mathbf{M}_{lr}(s) \tag{G-58}$$

where $\mathbf{H}_r(s) = \text{diag}\{s^{k_{ri}}, i = 1, 2, \ldots, p\}$ and $k_{ri} = \delta_{ri}\mathbf{M}(s)$ is the degree of the ith row. $\mathbf{M}_{hr}$ will be called the *row-degree coefficient matrix*; its ith row is the coefficients of the ith row of $\mathbf{M}(s)$ associated with $s^{k_{ri}}$. The polynomial matrix $\mathbf{M}_{lr}(s)$ contains the remaining terms and its ith row has a degree smaller than k_{ri}. For example, the matrix in (G-56) can be written as

$$\mathbf{M}(s) = \begin{bmatrix} s^2 & 0 \\ 0 & s^2 \end{bmatrix}\begin{bmatrix} 3 & 0 \\ 1 & 0 \end{bmatrix} + \begin{bmatrix} 2s & 2s+1 \\ s-3 & s \end{bmatrix}$$

In terms of (G-58), we have that $\mathbf{M}(s)$ *is row reduced if and only if its row-degree coefficient matrix* $\mathbf{M}_{hr}$ *is nonsingular.*

With the concept, we can now generalize Theorem G-9 to the following.

Theorem G-10

Let $\mathbf{N}(s)$ and $\mathbf{D}(s)$ be $q \times p$ and $p \times p$ polynomial matrices, and let $\mathbf{D}(s)$ be column reduced. Then the rational function $\mathbf{N}(s)\mathbf{D}^{-1}(s)$ is proper (strictly proper) if and only if

$$\delta_{ci}\mathbf{N}(s) \le \delta_{ci}\mathbf{D}(s) \qquad [\delta_{ci}\mathbf{N}(s) < \delta_{ci}\mathbf{D}(s)]$$

for $i = 1, 2, \ldots, p$.

Proof

The necessity part has been established in Theorem G-9. We now show the sufficient part. Following (G-57), we write

$$\mathbf{D}(s) = \mathbf{D}_{hc}\mathbf{H}_c(s) + \mathbf{D}_{lc}(s) = [\mathbf{D}_{hc} + \mathbf{D}_{lc}(s)\mathbf{H}_c^{-1}(s)]\mathbf{H}_c(s)$$
$$\mathbf{N}(s) = \mathbf{N}_{hc}\mathbf{H}_c(s) + \mathbf{N}_{lc}(s) = [\mathbf{N}_{hc} + \mathbf{N}_{lc}(s)\mathbf{H}_c^{-1}(s)]\mathbf{H}_c(s)$$

where $\delta_{ci}\mathbf{D}_{lc}(s) < \delta_{ci}\mathbf{D}(s) \triangleq \mu_i$, $\mathbf{H}_c(s) \triangleq \text{diag}\{s^{\mu_1}, s^{\mu_2}, \ldots, s^{\mu_p}\}$, and $\delta_{ci}\mathbf{N}_{lc}(s) < \mu_i$. Then we have

$$\hat{\mathbf{G}}(s) \triangleq \mathbf{N}(s)\mathbf{D}^{-1}(s) = [\mathbf{N}_{hc} + \mathbf{N}_{lc}(s)\mathbf{H}_c^{-1}(s)][\mathbf{D}_{hc} + \mathbf{D}_{lc}(s)\mathbf{H}_c^{-1}(s)]^{-1}$$

Clearly $\mathbf{N}_{lc}(s)\mathbf{H}_c^{-1}(s)$ and $\mathbf{D}_{lc}(s)\mathbf{H}_c^{-1}(s)$ both approach zero as $s \to \infty$. Hence we have

$$\lim_{s \to \infty} \hat{\mathbf{G}}(s) = \mathbf{N}_{hc}\mathbf{D}_{hc}^{-1}$$

where $\mathbf{D}_{hc}$ is nonsingular by the column reducedness assumption of $\mathbf{D}(s)$. Now if $\delta_{ci}\mathbf{N}(s) \le \delta_{ci}\mathbf{D}(s)$, $\mathbf{N}_{hc}$ is a nonzero matrix and $\hat{\mathbf{G}}(s)$ is proper. If $\delta_{ci}\mathbf{N}(s) < \delta_{ci}\mathbf{D}(s)$, $\mathbf{N}_{hc}$ is a zero matrix and $\hat{\mathbf{G}}(s)$ is strictly proper. Q.E.D.

The Hermite-form polynomial matrix shown in (G-28) is column reduced. Since every polynomial matrix can be transformed into the Hermite form by a sequence of elementary row transformations (Theorem G-5), we conclude that every nonsingular polynomial matrix can be transformed to be column reduced by a sequence of elementary *row* operations. It turns out that the same can be achieved by a sequence of *columns* operations. This will be illustrated by an example.

Example

Consider

$$\mathbf{M}(s) = \begin{bmatrix} s+1 & s^2+2s+1 & 2 \\ 2s-2 & -2s^2+1 & 2 \\ -s & 5s^2-2s & 1 \end{bmatrix}$$

The column degrees are 1, 2, and 0. The column-degree coefficient matrix is

$$\mathbf{M}_{hc} = \begin{bmatrix} 1 & 1 & 2 \\ 2 & -2 & 2 \\ -1 & 5 & 1 \end{bmatrix}$$

and is singular. Hence, $\mathbf{M}(s)$ is not column reduced. Since $\mathbf{M}_{hc}$ is singular there exist α_1, α_2, and α_3 such that

$$\mathbf{M}_{hc} \begin{bmatrix} \alpha_1 \\ \alpha_2 \\ \alpha_3 \end{bmatrix} = 0$$

We normalize the α_i associated with the column with the highest column degree to be 1. In this example, the second column has the highest degree; hence, we choose $\alpha_2 = 1$, and α_1 and α_3 can be computed as $\alpha_1 = 3$, $\alpha_3 = -2$. Now if we postmultiply the unimodular matrix

$$\mathbf{U}_1(s) = \begin{bmatrix} 1 & \alpha_1 s & 0 \\ 0 & 1 & 0 \\ 0 & \alpha_3 s^2 & 1 \end{bmatrix} = \begin{bmatrix} 1 & 3s & 0 \\ 0 & 1 & 0 \\ 0 & -2s^2 & 1 \end{bmatrix}$$

to $\mathbf{M}(s)$, we obtain

$$\mathbf{M}(s)\mathbf{U}_1(s) = \begin{bmatrix} s+1 & 5s+1 & 2 \\ 2s-2 & -6s+1 & 2 \\ -s & -2s & 1 \end{bmatrix} \triangleq \mathbf{M}_1(s)$$

where the degree of the second column is reduced by one. It can be readily verified that $\mathbf{M}_1(s)$ is column reduced. ∎

From the example, we see that by a proper elementary column operation, the column degree can be reduced, whereas the determinantal degree remains unchanged. Hence, by a sequence of elementary column operations, a polynomial matrix can be reduced to be column reduced. We summarize this with the earlier statement as a theorm.

Theorem G-11

For every nonsingular polynomial matrix $\mathbf{M}(s)$, there exist unimodular matrices $\mathbf{U}(s)$ and $\mathbf{V}(s)$ such that $\mathbf{M}(s)\mathbf{U}(s)$ and $\mathbf{V}(s)\mathbf{M}(s)$ are column reduced or row reduced. ∎

An algorithm is available in Reference S137 to transform a matrix into a column- or row-reduced one. In the following, we shall extend Theorem G-1 to the matrix case.

Theorem G-12

Let $\mathbf{D}(s)$ and $\mathbf{N}(s)$ be $p \times p$ and $q \times p$ polynomial matrices and let $\mathbf{D}(s)$ be nonsingular. Then there exist unique $q \times p$ polynomial matrices $\mathbf{Q}(s)$ and $\mathbf{R}(s)$ such that

$$\mathbf{N}(s) = \mathbf{Q}(s)\mathbf{D}(s) + \mathbf{R}(s)$$

and $\quad \mathbf{R}(s)\mathbf{D}^{-1}(s)$ is strictly proper

which can be replaced by, if $\mathbf{D}(s)$ is column reduced,

$$\delta_{ci}\mathbf{D}(s) > \delta_{ci}\mathbf{R}(s) \qquad i = 1, 2, \ldots, p$$

Proof

Consider the rational matrix $\hat{\mathbf{G}}(s) = \mathbf{N}(s)\mathbf{D}^{-1}(s)$. This rational matrix is not necessarily proper. If every element $\hat{g}_{ij}(s)$ of $\hat{\mathbf{G}}(s)$ is decomposed as $\hat{g}_{ij}(s) = \hat{g}_{ijsp}(s) + q_{ij}(s)$, where $\hat{g}_{ijsp}(s)$ is a strictly proper rational function and $q_{ij}(s)$ is a polynomial, then we can write $\hat{\mathbf{G}}(s)$ as

$$\hat{\mathbf{G}}(s) = \mathbf{N}(s)\mathbf{D}^{-1}(s) = \hat{\mathbf{G}}_{sp}(s) + \mathbf{Q}(s) \tag{G-59}$$

where $\hat{\mathbf{G}}_{sp}(s)$ is a strictly proper rational matrix and $\mathbf{Q}(s)$ is a polynomial matrix. The postmultiplication of $\mathbf{D}(s)$ to (G-59) yields

$$\mathbf{N}(s) = \mathbf{Q}(s)\mathbf{D}(s) + \mathbf{R}(s)$$

with $\quad \mathbf{R}(s) = \hat{\mathbf{G}}_{sp}(s)\mathbf{D}(s) \quad$ or $\quad \hat{\mathbf{G}}_{sp}(s) = \mathbf{R}(s)\mathbf{D}^{-1}(s)$

Since $\mathbf{R}(s)$ is equal to the difference of two polynomial matrices $[\mathbf{R}(s) = \mathbf{N}(s) - \mathbf{Q}(s)\mathbf{D}(s)]$, it must be a polynomial matrix.

To show uniqueness, suppose there are other $\bar{\mathbf{Q}}(s)$ and $\bar{\mathbf{R}}(s)$ such that

$$\mathbf{N}(s) = \mathbf{Q}(s)\mathbf{D}(s) + \mathbf{R}(s) = \bar{\mathbf{Q}}(s)\mathbf{D}(s) + \bar{\mathbf{R}}(s) \tag{G-60}$$

and $\bar{\mathbf{R}}(s)\mathbf{D}^{-1}(s)$ is strictly proper. Then Equation (G-60) implies

$$[\mathbf{R}(s) - \bar{\mathbf{R}}(s)]\mathbf{D}^{-1}(s) = \bar{\mathbf{Q}}(s) - \mathbf{Q}(s).$$

Its right-hand side is a polynomial matrix; whereas its left-hand side is a strictly proper rational matrix. This is possible only if $\bar{\mathbf{Q}}(s) = \mathbf{Q}(s)$ and $\mathbf{R}(s) = \bar{\mathbf{R}}(s)$. The column degree inequality follows directly from Theorem G-10. Q.E.D.

Theorem G-12′

Let $\mathbf{A}(s)$ and $\mathbf{B}(s)$ be $q \times q$ and $q \times p$ polynomial matrices. If $\mathbf{A}(s)$ is nonsingular, there exist unique $q \times p$ polynomial matrices $\mathbf{Q}(s)$ and $\mathbf{R}(s)$ such that

$$\mathbf{B}(s) = \mathbf{A}(s)\mathbf{Q}(s) + \mathbf{R}(s)$$

and $\quad \mathbf{A}^{-1}(s)\mathbf{R}(s)$ is strictly proper

which can be replaced by, if $\mathbf{A}(s)$ is row reduced,

$$\delta_{ri}\mathbf{A}(s) > \delta_{ri}\mathbf{R}(s) \qquad i = 1, 2, \ldots, q$$

■

This theorem is dual to Theorem G-12, and its proof is omitted. The proof of Theorem G-12 is constructive in nature, and its procedure can be used to compute $\mathbf{Q}(s)$ and $\mathbf{R}(s)$. If $\mathbf{D}(s)$ is column reduced, different procedures are available in References S34, S137, and S236 (see also Problem G-15).

In the following, we discuss two special cases of Theorems G-12 and G-12'. Consider $p \times p$ polynomial matrices $\mathbf{D}(s)$ and $\mathbf{N}(s)$. Let $\mathbf{D}(s) = s\mathbf{I} - \mathbf{A}$, where $\mathbf{A}$ is a $p \times p$ constant matrix. Clearly $\mathbf{D}(s)$ is nonsingular and $\delta_{ci}\mathbf{D}(s) = \delta_{ri}\mathbf{D}(s) = 1$, $i = 1, 2, \ldots, p$. We write $\mathbf{N}(s)$ as

$$\mathbf{N}(s) = \mathbf{N}_n s^n + \mathbf{N}_{n-1}s^{n-1} + \cdots + \mathbf{N}_0$$

Define

$$\mathbf{N}_r(\mathbf{A}) = \mathbf{N}_n\mathbf{A}^n + \mathbf{N}_{n-1}\mathbf{A}^{n-1} + \cdots + \mathbf{N}_0\mathbf{I}$$

and

$$\mathbf{N}_l(\mathbf{A}) = \mathbf{A}^n\mathbf{N}_n + \mathbf{A}^{n-1}\mathbf{N}_{n-1} + \cdots + \mathbf{I}\mathbf{N}_0$$

Corollary G-12

Let $\mathbf{D}(s) = s\mathbf{I} - \mathbf{A}$, and let $\mathbf{N}(s)$ be an arbitrary polynomial matrix. Then there exist unique polynomial matrices $\mathbf{Q}_r(s)$ and $\mathbf{Q}_l(s)$ such that

$$\mathbf{N}(s) = \mathbf{Q}_r(s)(s\mathbf{I} - \mathbf{A}) + \mathbf{N}_r(\mathbf{A})$$

and

$$\mathbf{N}(s) = (s\mathbf{I} - \mathbf{A})\mathbf{Q}_l(s) + \mathbf{N}_l(\mathbf{A})$$ ∎

This corollary can be readily verified by using

$$\mathbf{Q}_r(s) = \mathbf{N}_n s^{n-1} + (\mathbf{N}_n\mathbf{A} + \mathbf{N}_{n-1})s^{n-2} + \cdots + (\mathbf{N}_n\mathbf{A}^{n-1} + \mathbf{N}_{n-1}\mathbf{A}^{n-2} + \cdots + \mathbf{N}_1)$$

and is left as an exercise. Note that $\mathbf{N}_r(\mathbf{A})$ and $\mathbf{N}_l(\mathbf{A})$ are constant matrices and their column and row degrees are all equal to zero.

G-6 Coprime Fractions of Proper Rational Matrices

Consider a $q \times p$ proper rational matrix $\hat{\mathbf{G}}(s)$. The fraction $\hat{\mathbf{G}}(s) = \mathbf{N}(s)\mathbf{D}^{-1}(s)$ is called a *right-coprime fraction* if $\mathbf{N}(s)$ and $\mathbf{D}(s)$ are right coprime; $\hat{\mathbf{G}}(s) = \mathbf{A}^{-1}(s)\mathbf{B}(s)$ a *left-coprime fraction* if $\mathbf{A}(s)$ and $\mathbf{B}(s)$ are left coprime. Either one will also be called an *irreducible fraction.* Given a $\hat{\mathbf{G}}(s)$, it is possible to obtain many fractions, some are irreducible and some are not. However, they are all related by the following theorem.

Theorem G-13

Consider a $q \times p$ proper rational matrix $\hat{\mathbf{G}}(s)$ with the right-coprime fraction $\hat{\mathbf{G}}(s) = \mathbf{N}(s)\mathbf{D}^{-1}(s)$. Then for any other fraction $\hat{\mathbf{G}}(s) = \bar{\mathbf{N}}(s)\bar{\mathbf{D}}^{-1}(s)$ there exists a $p \times p$ nonsingular polynomial matrix $\mathbf{T}(s)$ such that

$$\bar{\mathbf{N}}(s) = \mathbf{N}(s)\mathbf{T}(s) \qquad \text{and} \qquad \bar{\mathbf{D}}(s) = \mathbf{D}(s)\mathbf{T}(s)$$

If the fraction $\bar{\mathbf{N}}(s)\bar{\mathbf{D}}^{-1}(s)$ is also right coprime, then $\mathbf{T}(s)$ is unimodular. ∎

Proof[6]

Let $\mathbf{D}^{-1}(s) = \text{Adj }\mathbf{D}(s)/\det \mathbf{D}(s)$, where Adj stands for the adjoint and det stands for the determinant of a matrix. Then, $\mathbf{N}(s)\mathbf{D}^{-1}(s) = \bar{\mathbf{N}}(s)\bar{\mathbf{D}}^{-1}(s)$ and $\mathbf{D}(s)\mathbf{D}^{-1}(s) = \bar{\mathbf{D}}(s)\bar{\mathbf{D}}^{-1}(s) = \mathbf{I}$ imply

$$\mathbf{N}(s) \text{ Adj } \mathbf{D}(s) \det \bar{\mathbf{D}}(s) = \bar{\mathbf{N}}(s) \text{ Adj } \bar{\mathbf{D}}(s) \det \mathbf{D}(s)$$

and

$$\mathbf{D}(s) \text{ Adj } \mathbf{D}(s) \det \bar{\mathbf{D}}(s) = \bar{\mathbf{D}}(s) \text{ Adj } \bar{\mathbf{D}}(s) \det \mathbf{D}(s) \tag{G-61}$$

Let $\mathbf{R}(s)$ be a gcrd of $\mathbf{N}(s)$ and $\mathbf{D}(s)$, and let $\bar{\mathbf{R}}(s)$ be a gcrd of $\bar{\mathbf{N}}(s)$ and $\bar{\mathbf{D}}(s)$. Then, it is clear that $\mathbf{R}(s)$ Adj $\mathbf{D}(s)$ det $\bar{\mathbf{D}}(s)$ is a gcrd of the two left-hand-side polynomial matrices in (G-61) and $\bar{\mathbf{R}}(s)$ Adj $\bar{\mathbf{D}}(s)$ det $\mathbf{D}(s)$ is a gcrd of the right-hand-side polynomial matrices in (G-61). Because of the equalities in (G-61), $\mathbf{R}(s)$ Adj $\mathbf{D}(s)$ det $\bar{\mathbf{D}}(s)$ and $\bar{\mathbf{R}}(s)$ Adj $\bar{\mathbf{D}}(s)$ det $\mathbf{D}(s)$ are two different gcrds of $\mathbf{N}(s)$ Adj $\mathbf{D}(s)$ det $\bar{\mathbf{D}}(s)$ and $\mathbf{D}(s)$ Adj $\mathbf{D}(s)$ det $\bar{\mathbf{D}}(s)$. We claim that the polynomial matrix $\mathbf{D}(s)$ Adj $\mathbf{D}(s)$ det $\bar{\mathbf{D}}(s)$ is nonsingular. Indeed, because of $\det \mathbf{D}(s)\mathbf{D}^{-1}(s) = \det [\mathbf{D}(s) \text{ Adj } \mathbf{D}(s)/\det \mathbf{D}(s)] = \det [\mathbf{D}(s) \text{ Adj } \mathbf{D}(s)]/(\det \mathbf{D}(s))^p = 1$, we have $\det [\mathbf{D}(s) \text{ Adj } \mathbf{D}(s) \det \bar{\mathbf{D}}(s)] = [\det \mathbf{D}(s) \det \bar{\mathbf{D}}(s)]^p \neq 0$. Hence Corollary G-7 implies the existence of a unimodular matrix $\mathbf{W}(s)$ such that

$$\mathbf{R}(s) \text{ Adj } \mathbf{D}(s) \det \bar{\mathbf{D}}(s) = \mathbf{W}(s)\bar{\mathbf{R}}(s) \text{ Adj } \bar{\mathbf{D}}(s) \det \mathbf{D}(s)$$

which implies

$$\mathbf{R}(s)\mathbf{D}^{-1}(s) = \mathbf{W}(s)\bar{\mathbf{R}}(s)\bar{\mathbf{D}}^{-1}(s)$$

or

$$\mathbf{D}^{-1}(s) = \mathbf{R}^{-1}(s)\mathbf{W}(s)\bar{\mathbf{R}}(s)\bar{\mathbf{D}}^{-1}(s) \tag{G-62}$$

Since $\mathbf{R}(s)$ is unimodular following the irreducibility assumption of $\mathbf{N}(s)\mathbf{D}^{-1}(s)$, the matrix

$$\mathbf{T}(s) = \mathbf{R}^{-1}(s)\mathbf{W}(s)\bar{\mathbf{R}}(s) \tag{G-63}$$

is a polynomial matrix. The substitution of (G-63) into (G-62) yields $\mathbf{D}^{-1}(s) = \mathbf{T}(s)\bar{\mathbf{D}}^{-1}(s)$ or $\bar{\mathbf{D}}(s) = \mathbf{D}(s)\mathbf{T}(s)$. The substitution of $\mathbf{D}^{-1}(s) = \mathbf{T}(s)\bar{\mathbf{D}}^{-1}(s)$ into $\bar{\mathbf{N}}(s)\bar{\mathbf{D}}^{-1}(s) = \mathbf{N}(s)\mathbf{D}^{-1}(s)$ yields immediately $\bar{\mathbf{N}}(s) = \mathbf{N}(s)\mathbf{T}(s)$. The nonsingularity of $\mathbf{T}(s)$ follows from the nonsingularities of $\mathbf{R}(s)$, $\mathbf{W}(s)$, and $\bar{\mathbf{R}}(s)$.

If $\bar{\mathbf{N}}(s)$ and $\bar{\mathbf{D}}(s)$ are right coprime, then $\bar{\mathbf{R}}(s)$ is unimodular. Consequently, the $\mathbf{T}(s)$ in (G-63) is also unimodular. This completes the proof of the theorem. Q.E.D.

From this theorem, we see that all irreducible fractions of a proper rational matrix are related by unimodular matrices. Hence, the irreducible fraction is unique in the sense that all irreducible fractions can be generated from a single irreducible fraction.

Consider a fraction $\hat{\mathbf{G}}(s) = \mathbf{N}(s)\mathbf{D}^{-1}(s)$. If it is not irreducible, we may use the procedure in (G-33) to compute the gcrd $\mathbf{R}(s)$ of $\mathbf{N}(s)$ and $\mathbf{D}(s)$. We then

[6] For a different proof, see Problems G-12 and G-13.

compute $\mathbf{R}^{-1}(s)$ and

$$\bar{\mathbf{D}}(s)=\mathbf{D}(s)\mathbf{R}^{-1}(s) \qquad \bar{\mathbf{N}}(s)=\mathbf{N}(s)\mathbf{R}^{-1}(s)$$

Then we have

$$\begin{aligned}\hat{\mathbf{G}}(s)&=\mathbf{N}(s)\mathbf{D}^{-1}(s)=\bar{\mathbf{N}}(s)\mathbf{R}(s)[\bar{\mathbf{D}}(s)\mathbf{R}(s)]^{-1}=\bar{\mathbf{N}}(s)\mathbf{R}(s)\mathbf{R}^{-1}(s)\bar{\mathbf{D}}^{-1}(s)\\&=\bar{\mathbf{N}}(s)\bar{\mathbf{D}}^{-1}(s)\end{aligned}$$

and $\bar{\mathbf{N}}(s)\bar{\mathbf{D}}^{-1}(s)$ is irreducible. This is one way to obtain an irreducible fraction. This procedure, however, requires the computation of the inverse of a polynomial matrix and is rather complicated. If in the process of generating (G-33), we also compute $\mathbf{V}(s)$ in (G-35), then we have

$$\begin{bmatrix}\mathbf{D}(s)\\\mathbf{N}(s)\end{bmatrix}=\mathbf{V}(s)\begin{bmatrix}\mathbf{R}(s)\\\mathbf{0}\end{bmatrix}=\begin{bmatrix}\mathbf{V}_{11}(s)\\\mathbf{V}_{21}(s)\end{bmatrix}\mathbf{R}(s)$$

and $\mathbf{N}(s)\mathbf{D}^{-1}(s)=\mathbf{V}_{21}(s)\mathbf{V}_{11}^{-1}(s)$. Since $\mathbf{V}(s)$ is unimodular, $\mathbf{V}_{21}(s)\mathbf{V}_{11}^{-1}(s)$ is a right coprime fraction (why?). By this method, the computation of the inverse of $\mathbf{R}(s)$ can be avoided. Note that $\mathbf{V}_{11}(s)$ obtained in this process is not necessarily column reduced.

In this section, we shall introduce a method of obtaining a left-coprime fraction from a right fraction, not necessarily coprime, and vice versa. The procedure is similar to the scalar case discussed in Section G-2.

Consider the $q \times p$ proper rational matrix $\hat{\mathbf{G}}(s)$. Let $\hat{\mathbf{G}}(s)=\mathbf{A}^{-1}(s)\mathbf{B}(s)=\mathbf{N}(s)\mathbf{D}^{-1}(s)$, where $\mathbf{A}(s)$, $\mathbf{B}(s)$, $\mathbf{N}(s)$, and $\mathbf{D}(s)$ are, respectively, $q\times q$, $q\times p$, $q\times p$, and $p\times p$ polynomial matrices. The equality $\mathbf{A}^{-1}(s)\mathbf{B}(s)=\mathbf{N}(s)\mathbf{D}^{-1}(s)$ can be written as $\mathbf{B}(s)\mathbf{D}(s)=\mathbf{A}(s)\mathbf{N}(s)$ or

$$[-\mathbf{B}(s) \quad \mathbf{A}(s)]\begin{bmatrix}\mathbf{D}(s)\\\mathbf{N}(s)\end{bmatrix}=0 \tag{G-64}$$

If we consider polynomials as elements of the field of real rational functions $\mathbb{R}(s)$, then Equation (G-64) is a homogeneous linear algebraic equation. Consequently, all $1\times(p+q)$ vectors $\mathbf{x}(s)$ with elements in $\mathbb{R}(s)$ (including polynomials) satisfying

$$\mathbf{x}(s)\begin{bmatrix}\mathbf{D}(s)\\\mathbf{N}(s)\end{bmatrix}=\mathbf{0}$$

is a linear space over $\mathbb{R}(s)$, denoted as $(\mathbb{V}, \mathbb{R}(s))$. It is a subspace of $(\mathbb{R}^{p+q}(s), \mathbb{R}(s))$. Following Definition 2-11, we call it the *left null space*. Its dimension is equal to $(p+q)-\text{rank}\ [\mathbf{D}'(s) \quad \mathbf{N}'(s)]'=p+q-p=q$ (Problem 2-51). In this q-dimensional null space $(\mathbb{V}, \mathbb{R}(s))$, any set of q linearly independent vectors in $\mathbb{V}$ qualifies as a basis (Theorem 2-1). In our study we are however interested in only the polynomial solutions of (G-64). We use $\mathbb{V}_p$ to denote the polynomial part of $\mathbb{V}$. A set of q vectors in $\mathbb{V}_p$ will be called a *polynomial basis*, or a basis in $(\mathbb{V}_p, \mathbb{R}[s])$,[7] if every vector in $\mathbb{V}_p$ can be expressed as a unique combination

[7] It is in fact a free module over the polynomial ring $\mathbb{R}[s]$ with dimension q. See footnote 4 of this chapter and Reference S34.

of the q vectors by using only polynomials as coefficients. We note that every basis of $(\mathbb{V}_p, \mathbb{R}[s])$ is a polynomial basis of $(\mathbb{V}, \mathbb{R}(s))$; the converse however is not true in general.[8] It turns out that $\mathbf{A}(s)$ and $\mathbf{B}(s)$ are left coprime if and only if the set of the q rows of $[-\mathbf{B}(s) \quad \mathbf{A}(s)]$ is a basis of $(\mathbb{V}_p, \mathbb{R}[s])$ (Problem G-16). Among all bases in $(\mathbb{V}_p, \mathbb{R}[s])$, some have the additional property that the row degrees of the basis vectors are smallest possible. This type of basis is called a *minimal polynomial basis*. It will be shown that the set of q rows of $[-\mathbf{B}(s) \quad \mathbf{A}(s)]$ is a minimal polynomial basis if and only if $\mathbf{A}(s)$ and $\mathbf{B}(s)$ are left coprime and $\mathbf{A}(s)$ is row reduced. In the following, we discuss a method to find a minimal polynomial basis.

Instead of solving (G-64) directly, we shall translate it into a homogeneous linear algebraic equation with real numbers as entries. Let

$$
\begin{aligned}
\mathbf{D}(s) &= \mathbf{D}_0 + \mathbf{D}_1 s + \cdots + \mathbf{D}_d s^d \\
\mathbf{N}(s) &= \mathbf{N}_0 + \mathbf{N}_1 s + \cdots + \mathbf{N}_d s^d \\
\mathbf{A}(s) &= \mathbf{A}_0 + \mathbf{A}_1 s + \cdots + \mathbf{A}_m s^m \\
\text{and} \qquad \mathbf{B}(s) &= \mathbf{B}_0 + \mathbf{B}_1 s + \cdots + \mathbf{B}_m s^m
\end{aligned}
\tag{G-65}
$$

where $\mathbf{D}_i$, $\mathbf{N}_i$, $\mathbf{A}_i$, and $\mathbf{B}_i$ are $p \times p$, $q \times p$, $q \times q$, $q \times p$ constant matrices. By substituting (G-65) into (G-64) and equating the coefficient of s^i to zero yield, similar to (G-22) and (G-24), we obtain

$$
[-\mathbf{B}_0 \;\; \mathbf{A}_0 \,\vdots\, -\mathbf{B}_1 \;\; \mathbf{A}_1 \,\vdots \cdots \vdots\, -\mathbf{B}_m \;\; \mathbf{A}_m]
\left[\begin{array}{cccccccc}
\mathbf{D}_0 & \mathbf{D}_1 & \cdots & \mathbf{D}_d & \mathbf{0} & \mathbf{0} & \cdots & \mathbf{0} \\
\mathbf{N}_0 & \mathbf{N}_1 & \cdots & \mathbf{N}_d & \mathbf{0} & \mathbf{0} & \cdots & \mathbf{0} \\
\hdashline
\mathbf{0} & \mathbf{D}_0 & \cdots & \mathbf{D}_{d-1} & \mathbf{D}_d & \mathbf{0} & \cdots & \mathbf{0} \\
\mathbf{0} & \mathbf{N}_0 & \cdots & \mathbf{N}_{d-1} & \mathbf{N}_d & \mathbf{0} & \cdots & \mathbf{0} \\
\hdashline
\vdots & \vdots & & & & & & \vdots \\
\hdashline
\mathbf{0} & \mathbf{0} & \cdots & \mathbf{D}_0 & \mathbf{D}_1 & \cdots & \cdots & \mathbf{D}_d \\
\mathbf{0} & \mathbf{0} & \cdots & \mathbf{N}_0 & \mathbf{N}_1 & \cdots & \cdots & \mathbf{N}_d
\end{array}\right] = \mathbf{0}
\tag{G-66}
$$

We shall call the matrix formed from $\mathbf{D}_i$ and $\mathbf{N}_i$ the *generalized resultant* of $\mathbf{D}(s)$ and $\mathbf{N}(s)$. If $\mathbf{D}(s)$ and $\mathbf{N}(s)$ are known, this equation can be used to solve $\mathbf{B}_i$ and $\mathbf{A}_i$ and, consequently, $\mathbf{B}(s)$ and $\mathbf{A}(s)$. Conversely, if $\mathbf{A}(s)$ and $\mathbf{B}(s)$ are given, a similar equation can be set up to solve for $\mathbf{D}(s)$ and $\mathbf{N}(s)$. In Equation (G-66), there are q rows of unknown $[-\mathbf{B}_0 \quad \mathbf{A}_0 \quad \cdots \quad -\mathbf{B}_m \quad \mathbf{A}_m]$. In order to have q rows of nontrivial solutions, there must be, roughly speaking, q linearly dependent rows in the resultant in (G-66). Since it is desirable to have m, the degree of $\mathbf{A}(s)$ and $\mathbf{B}(s)$, as small as possible, we shall try to use, roughly speaking,

[8] The discussion is brief and the reader needs not be concerned because the subsequent development is independent of the discussion. For a complete discussion, see Reference S95, where $[\mathbf{D}'(s) \quad \mathbf{N}'(s)]'$ is assumed to have a full rank. Our problem assumes $\mathbf{D}(s)$ to be nonsingular and the development can be simplified slightly.

the first q linearly dependent rows in the resultant in (G-66). In order to do so, we define

$$\mathbf{S}_k = \begin{array}{r} \text{1st block} \left\{ \begin{array}{l} \\ \\ \end{array} \right. \\ \text{2nd block} \left\{ \begin{array}{l} \\ \\ \end{array} \right. \\ \\ (k+1)\text{th block} \left\{ \begin{array}{l} \\ \\ \end{array} \right. \end{array} \begin{bmatrix} \mathbf{D}_0 & \mathbf{D}_1 & \cdots & \mathbf{D}_d & \mathbf{0} & \mathbf{0} & \cdots & \mathbf{0} \\ \mathbf{N}_0 & \mathbf{N}_1 & \cdots & \mathbf{N}_d & \mathbf{0} & \mathbf{0} & \cdots & \mathbf{0} \\ \hdashline \mathbf{0} & \mathbf{D}_0 & \cdots & \mathbf{D}_{d-1} & \mathbf{D}_d & \mathbf{0} & \cdots & \mathbf{0} \\ \mathbf{0} & \mathbf{N}_0 & \cdots & \mathbf{N}_{d-1} & \mathbf{N}_d & \mathbf{0} & \cdots & \mathbf{0} \\ \hdashline \vdots & & & & & & & \\ \hdashline \mathbf{0} & \mathbf{0} & \cdots & \mathbf{D}_0 & \mathbf{D}_1 & \cdots & \cdots & \mathbf{D}_d \\ \mathbf{0} & \mathbf{0} & \cdots & \mathbf{N}_0 & \mathbf{N}_1 & \cdots & \cdots & \mathbf{N}_d \end{bmatrix} \begin{array}{l} \\ \} \; r_0 \text{ (number of dependent rows)} \\ \\ \} \; r_1 \\ \\ \\ \} \; r_k \end{array} \tag{G-67}$$

The rows formed from $\mathbf{D}_i$ will be called D rows; those from $\mathbf{N}_i$, N rows. The matrix $\mathbf{S}_k$ has $k+1$ block rows; each block row has p D rows and q N rows. Now it is assumed that the row-searching algorithm has been applied to $\mathbf{S}_k$ and its linearly dependent rows in order from top to bottom have been identified.

Lemma G-1

If $\hat{\mathbf{G}}(s) = \mathbf{N}(s)\mathbf{D}^{-1}(s)$ is proper, all D rows in $\mathbf{S}_k$, $k = 0, 1, \ldots$, are linearly independent of their previous rows. ∎

This lemma will be proved later. This lemma does not require that $\mathbf{N}(s)\mathbf{D}^{-1}(s)$ be a right-coprime fraction nor $\mathbf{D}(s)$ be column reduced. If $\mathbf{N}(s)\mathbf{D}^{-1}(s)$ is not proper, then the statement is not true in general. For example, a linearly dependent row will appear in the D rows of $\mathbf{S}_k$ formed for

$$\begin{bmatrix} s^2 & 0 \\ s & 1 \end{bmatrix} \begin{bmatrix} s^2+2s & 1 \\ s^2+s & 1 \end{bmatrix}^{-1} = \begin{bmatrix} s & -s \\ -s & s+1 \end{bmatrix}$$

even though we have $\delta_{ci}\mathbf{N}(s) \leq \delta_{ci}\mathbf{D}(s)$, where δ_{ci} denotes the column degree.

If, instead of $\mathbf{S}_k$, we arrange $\mathbf{D}_i$ and $\mathbf{N}_i$ in the descending power of s as

$$\hat{\mathbf{S}}_k = \begin{bmatrix} \mathbf{D}_d & \mathbf{D}_{d-1} & \cdots & \mathbf{D}_0 & \mathbf{0} & \cdots & \mathbf{0} \\ \mathbf{N}_d & \mathbf{N}_{d-1} & \cdots & \mathbf{N}_0 & \mathbf{0} & \cdots & \mathbf{0} \\ \hdashline \mathbf{0} & \mathbf{D}_d & \cdots & \mathbf{D}_1 & \mathbf{D}_0 & \cdots & \mathbf{0} \\ \mathbf{0} & \mathbf{N}_d & \cdots & \mathbf{N}_1 & \mathbf{N}_0 & \cdots & \mathbf{0} \\ \hdashline \vdots & \vdots & & \vdots & \vdots & & \vdots \end{bmatrix} \tag{G-68}$$

then the statement is again not valid even if $\mathbf{N}(s)\mathbf{D}^{-1}(s)$ is proper. For example, consider

$$[1 \quad 1] \begin{bmatrix} s & s \\ 1 & s \end{bmatrix}^{-1}$$

If we use $\hat{\mathbf{S}}_k$, then a linearly dependent row will appear in a D row.

Because of Lemma G-1, the linearly dependent rows of $\mathbf{S}_k$ will appear only in the N rows. Let r_i be the number of linearly dependent N rows in the $(i+1)$th block row. Because of the structure of $\mathbf{S}_k$, we have

$$0 \leq r_0 \leq r_1 \leq \cdots \leq q \tag{G-69}$$

Let v be the least integer such that $r_v = q$ or, equivalently,

$$\begin{aligned} 0 \leq r_0 \leq r_1 \leq \cdots \leq r_{v-1} < q \\ r_v = r_{v+1} = \cdots = q \end{aligned} \tag{G-70}$$

This implies that as k increases, the total number of linearly independent N rows in $\mathbf{S}_k$ will increase monotonically. However, once the number ceases to increase, no matter how many more block rows are added, the number of linearly independent N rows will remain the same. Define

$$n = (q - r_0) + (q - r_1) + \cdots + (q - r_{v-1}) \tag{G-71}$$

It is the total number of linearly independent N rows in $\mathbf{S}_k$ for $k \geq v - 1$. It turns out that n is the degree of $\hat{\mathbf{G}}(s)$ or the dimension of any irreducible realization of $\hat{\mathbf{G}}(s)$.

The number of linearly dependent N rows in $\mathbf{S}_v$ is equal to $r_0 + r_1 + \cdots + r_v$, which is clearly larger than q. However, there are only q primary linearly dependent rows in $\mathbf{S}_v$. A dependent row is called *primary* if the corresponding row in the previous block is independent of its previous rows. For example, all the r_0 dependent rows in the first block of $\mathbf{S}_v$ are primary linearly dependent rows. However, the r_1 dependent rows in the second block of $\mathbf{S}_v$ are not all primary because r_0 of the corresponding rows in the first block have already appeared as linearly dependent rows. Hence, in the second block of $\mathbf{S}_v$, there are only $r_1 - r_0$ primary linearly dependent rows. Similarly, there are $r_2 - r_1$ primary linearly dependent rows in the third block of $\mathbf{S}_v$. Proceeding in this manner, we conclude that the number of primary linearly dependent rows in $\mathbf{S}_v$ is equal to

$$r_0 + (r_1 - r_0) + (r_2 - r_1) + \cdots + (r_v - r_{v-1}) = r_v = q$$

Consider Equation (G-66) with m replaced by v:

$$[-\mathbf{B}_0 \quad \mathbf{A}_0 \vdots -\mathbf{B}_1 \quad \mathbf{A}_1 \vdots \cdots \vdots -\mathbf{B}_v \quad \mathbf{A}_v]\mathbf{S}_v = \mathbf{0} \tag{G-72}$$

These $\mathbf{A}_i$ and $\mathbf{B}_i$ are to be obtained by using the row-searching algorithm. In other words, they are the q rows of $\mathbf{K}$ in $\mathbf{KS}_v$, computed as in (A-7), corresponding to the q primary dependent rows of $\mathbf{S}_v$.

Theorem G-14

Consider a $q \times p$ proper rational matrix $\hat{\mathbf{G}}(s)$ factored as $\hat{\mathbf{G}}(s) = \mathbf{N}(s)\mathbf{D}^{-1}(s)$. We form $\mathbf{S}_k$ and search its linearly dependent rows by using the row-searching algorithm.[9] Let $[-\mathbf{B}_0 \quad \mathbf{A}_0 \quad \cdots \quad -\mathbf{B}_v \quad \mathbf{A}_v]$ be the q rows of $\mathbf{K}$ in $\mathbf{KS}_v = \bar{\mathbf{S}}_v$ corresponding to the q primary dependent rows of $\mathbf{S}_v$. Then the polynomial

[9]On a digital computer computation, this algorithm should be replaced by a numerically stable method. See Appendix A. If the row searching algorithm is not employed, the result is generally not in the polynomial echelon form.

matrices

$$\mathbf{A}(s) = \sum_{i=0}^{v} \mathbf{A}_i s^i \qquad \text{and} \qquad \mathbf{B}(s) = \sum_{i=0}^{v} \mathbf{B}_i s^i \tag{G-73}$$

are left coprime, and $\mathbf{A}(s)$ is in a canonical form called the *polynomial echelon form.*

Proof

In order not to be overwhelmed by notations, we assume $p=q=3$ and

$$\mathbf{KS}_3 = \begin{matrix} \\ \\ \\ \\ \\ \\ \\ \\ \\ \\ \\ \\ \\ \rightarrow \\ \rightarrow \\ \\ \end{matrix} \begin{bmatrix} D \\ \text{x} \\ \text{x} \\ \text{x} \\ \hdashline D \\ \text{x} \\ \text{x} \\ 0 \\ \hdashline D \\ \text{x} \\ \text{x} \\ 0 \\ \hdashline D \\ 0 \\ 0 \\ 0 \end{bmatrix} \begin{matrix} \\ \left.\begin{matrix} \\ \\ \\ \end{matrix}\right\} r_0 = 0 \\ \\ \left.\begin{matrix} \\ \\ \\ \end{matrix}\right\} r_1 = 1 \\ \\ \left.\begin{matrix} \\ \\ \\ \end{matrix}\right\} r_2 = 1 \\ \\ \left.\begin{matrix} \\ \\ \\ \end{matrix}\right\} r_3 = 3 \end{matrix} \tag{G-74}$$

where x denotes nonzero row and 0 zero rows. Since all D rows are linearly independent of their previous rows, they are not written out explicitly in (G-74). From (G-74), we have $r_0=0$, $r_1=1$, $r_2=1$, $r_3=3=q$, and $v=3$. In order to study the structure of $\mathbf{A}(s)$, we write (G-72) as

$$\left[\begin{array}{cccc:cccc:cccc:cccc} & a_{11}^0 & a_{12}^0 & a_{13}^0 & & a_{11}^1 & a_{12}^1 & a_{13}^1 & & a_{11}^2 & a_{12}^2 & a_{13}^2 & & a_{11}^3 & a_{12}^3 & a_{13}^3 \\ -\mathbf{B}_0 & a_{21}^0 & a_{22}^0 & a_{23}^0 & -\mathbf{B}_1 & a_{21}^1 & a_{22}^1 & a_{23}^1 & -\mathbf{B}_2 & a_{21}^2 & a_{22}^2 & a_{23}^2 & -\mathbf{B}_3 & a_{21}^3 & a_{22}^3 & a_{23}^3 \\ & a_{31}^0 & a_{32}^0 & a_{33}^0 & & a_{31}^1 & a_{32}^1 & a_{33}^1 & & a_{31}^2 & a_{32}^2 & a_{33}^2 & & a_{31}^3 & a_{32}^3 & a_{33}^3 \end{array}\right] \mathbf{S}_3 = \mathbf{0} \tag{G-75}$$

The primary dependent rows of $\mathbf{S}_3$ in (G-74) are indicated by the arrows as shown. Corresponding to these primary dependent rows, the $\mathbf{B}_i$ and $\mathbf{A}_i$ in (G-75) assume the form

$$\left[\begin{array}{cccc:cccc:cccc:cccc} \multicolumn{4}{c}{\overbrace{\hphantom{xxxxxxxxxx}}^{s^0}} & \multicolumn{4}{c}{\overbrace{\hphantom{xxxxxxxxxx}}^{s^1}} & \multicolumn{4}{c}{\overbrace{\hphantom{xxxxxxxxxx}}^{s^2}} & \multicolumn{4}{c}{\overbrace{\hphantom{xxxxxxxxxx}}^{s^3}} \\ & a_{11}^0 & a_{12}^0 & a_{13}^0 & & a_{11}^1 & a_{12}^1 & \textcircled{1} & & 0 & 0 & 0 & & 0 & 0 & 0 \\ -\mathbf{B}_0 & a_{21}^0 & a_{22}^0 & a_{23}^0 & -\mathbf{B}_1 & a_{21}^1 & a_{22}^1 & 0 & -\mathbf{B}_2 & a_{21}^2 & a_{22}^2 & 0 & -\mathbf{B}_3 & \textcircled{1} & 0 & 0 \\ & a_{31}^0 & a_{32}^0 & a_{33}^0 & & a_{31}^1 & a_{32}^1 & 0 & & a_{31}^2 & a_{32}^2 & 0 & & 0 & \textcircled{1} & 0 \end{array}\right] \tag{G-76}$$

This is in an echelon form. The column positions of the three $\boxed{1}$ are determined by the primary linearly dependent rows. For example, the last row of the second block of $\mathbf{S}_3$ is a primary dependent row; hence the last column of the second block in (G-76) has the element $\boxed{1}$. If (G-76) is obtained by the row searching algorithm, then (G-76) has the following properties:

1. All elements (including $-\mathbf{B}_i$) on the right-hand side of $\boxed{1}$ are zeros.
2. All elements, except element $\boxed{1}$, of the columns corresponding to primary linearly dependent rows of $\mathbf{S}_v$ are zeros.
3. All elements of the column corresponding to nonprimary linearly dependent rows of $\mathbf{S}_v$ are zeros.

Property 3 is the same as saying that all columns which are on the right-hand side and occupy the same positions in each block as those columns with the $\boxed{1}$ elements are zero columns. For example, in (G-76), the sixth column of the third and fourth blocks are zero columns because they are on the right-hand side of the sixth column, with element $\boxed{1}$, of the second block. We note that the above three properties may overlap. For example, the rightmost column of (G-76) is a zero column following 1 alone or following 2 alone. Because of these properties, $\mathbf{A}(s)$ becomes

$$\mathbf{A}(s)=\begin{bmatrix} a_{11}^0+a_{11}^1 s & a_{12}^0+a_{12}^1 s & \boxed{a_{13}^0+s} \\ \boxed{a_{21}^0+a_{21}^1 s+a_{21}^2 s^2+s^3} & a_{22}^0+a_{22}^1 s+a_{22}^2 s^2 & a_{23}^0 \\ a_{31}^0+a_{31}^1 s+a_{31}^2 s^2 & \boxed{a_{32}^0+a_{32}^1 s+a_{32}^2 s^2+s^3} & a_{33}^0 \end{bmatrix} \quad \text{(G-77)}$$

The elements encircled by dotted lines will be called pivot elements. Their positions are determined by the elements $\boxed{1}$ in (G-76). We note that every row and every column has only one pivot element. Because of Property 1, the degree of a pivot element is larger than the degree of every right-hand-side element in the same row and is larger than or equal to the degree of every left-hand-side element in the same row. Because of properties 2 and 3, the degree of a pivot element is larger than the degree of every other element in the same column. A polynomial matrix with these properties is said to be in the *row polynomial echelon form* or *Popov form.*[10] It is clear that a polynomial matrix in the echeleon form is column reduced and row reduced.

We discuss now the row degrees of $\mathbf{A}(s)$. We note that in each block row of $\mathbf{S}_v$ there are q N rows, and the ith N row, $i=1, 2, \ldots, q$, appears $v+1$ times in $\mathbf{S}_v$. Define, for $i=1, 2, \ldots, q$,

$$v_i \triangleq \text{number of linearly independent } i\text{th } N \text{ row in } \mathbf{S}_v \quad \text{(G-78)}$$

They will be called the row indices of $\hat{\mathbf{G}}(s)=\mathbf{N}(s)\mathbf{D}^{-1}(s)$. For the example in (G-74), we have $v_1=3$, $v_2=3$, and $v_3=1$. Clearly the n defined in (G-71) is also given by

$$n=v_1+v_2+\cdots+v_q \quad \text{(G-79)}$$

[10] The numerical matrices $[-\mathbf{B}_0 \ \ \mathbf{A}_0 \ \ \cdots \ \ -\mathbf{B}_v \ \ \mathbf{A}_v]$ and $[\mathbf{A}_0 \ \ \mathbf{A}_1 \ \ \cdots \ \ \mathbf{A}_v]$ obtained by the row searching algorithm will always be in the echelon form. Hence the corresponding polynomial matrix $[-\mathbf{B}(s) \ \ \mathbf{A}(s)]$ and $\mathbf{A}(s)$ are said to be in the polynomial echelon form.

and the v defined in (G-70) is

$$v = \max \{v_i, i = 1, 2, \ldots, q\} \tag{G-80}$$

We shall call v the *row index* of $\hat{\mathbf{G}}(s)$. If we rename v_i as $\bar{v}_i$ such that $\bar{v}_1 \leq \bar{v}_2 \leq \bar{v}_3 \leq \cdots$, then (G-76) and (G-77) imply that

$$\text{The } i\text{th row degree of } \mathbf{A}(s) = \bar{v}_i \qquad i = 1, 2, \ldots, q.$$

Hence we have

$$\{\text{set of row degrees of } \mathbf{A}(s)\} = \{\bar{v}_i, i = 1, 2, \ldots, q\} = \{v_i, i = 1, 2, \ldots, q\} \tag{G-81}$$

Since $\mathbf{A}(s)$ is row reduced, we have

$$\deg \det \mathbf{A}(s) = \sum_{i=1}^{q} v_i = n = \text{total number of linear independent } N \text{ rows in } \mathbf{S}_v \tag{G-82}$$

where deg det stands for the degree of the determinant. With this background, we are ready to show that $\mathbf{A}(s)$ and $\mathbf{B}(s)$ are left coprime. Suppose they are not, then there exists a polynomial matrix $\mathbf{Q}(s)$ such that

$$\deg \det \mathbf{Q}(s) > 0$$

$$\mathbf{A}(s) = \mathbf{Q}(s)\bar{\mathbf{A}}(s) \qquad \mathbf{B}(s) = \mathbf{Q}(s)\bar{\mathbf{B}}(s) \tag{G-83}$$

and

$$\mathbf{A}^{-1}(s)\mathbf{B}(s) = \bar{\mathbf{A}}^{-1}(s)\bar{\mathbf{B}}(s) \tag{G-84}$$

Because of $\deg \det \mathbf{Q}(s) > 0$, we have

$$\deg \det \bar{\mathbf{A}}(s) < \deg \det \mathbf{A}(s) = n$$

This implies that the number of linearly independent N rows in $\mathbf{S}_v$ is smaller than n. This is not possible. Hence $\mathbf{A}(s)$ and $\mathbf{B}(s)$ are left coprime. This completes the proof of this theorem. Q.E.D.

Proof of Lemma G-1

If the linearly dependent rows of $\mathbf{S}_v$ appear only in the N rows, then from (G-76) we can see that the row degrees of $\mathbf{B}(s)$ are smaller than or equal to those of $\mathbf{A}(s)$. Since $\mathbf{A}(s)$ is row reduced, the computed $\mathbf{A}^{-1}(s)\mathbf{B}(s)$ is proper. Now if a linearly dependent row of $\mathbf{S}_v$ shows up in a D row, then $\fbox{1}$ element in (G-76) will appear in the column of $-\mathbf{B}_i$. Now because of property 1, the row degree of $\mathbf{B}(s)$ will be larger than the corresponding row degree of $\mathbf{A}(s)$. This violates the assumption that $\hat{\mathbf{G}}(s)$ is proper. Hence all linearly dependent rows of $\mathbf{S}_v$ must appear in the N rows. Q.E.D.

Example 1

Consider

$$\hat{\mathbf{G}}(s) = \begin{bmatrix} \dfrac{1}{s+1} & \dfrac{1}{s-1} \\ \dfrac{s^2}{s^2-1} & \dfrac{2}{s-1} \end{bmatrix} = \begin{bmatrix} \dfrac{s-1}{s^2-1} & \dfrac{1}{s-1} \\ \dfrac{s^2}{s^2-1} & \dfrac{2}{s-1} \end{bmatrix} = \begin{bmatrix} s-1 & 1 \\ s^2 & 2 \end{bmatrix} \begin{bmatrix} s^2-1 & 0 \\ 0 & s-1 \end{bmatrix}^{-1}$$

The fraction $\mathbf{N}(s)\mathbf{D}^{-1}(s)$ is carried out by taking the least common denominator of each column of $\hat{\mathbf{G}}(s)$ as the corresponding diagonal element of $\mathbf{D}(s)$. This fraction happens to be right coprime. In general, a fraction obtained by this process will not be coprime.

We form $\mathbf{S}_2$ and apply the row-searching algorithm:

$$
\mathbf{F}_*^*\mathbf{S}_2 = \left[\begin{array}{cccc|cccc|cccc}
1 &&&&&&&&&&& \\
0 & 1 &&&&&&&&&& \\
-1 & 1 & 1 &&&&&&&&& \\
0 & 2 & 0 & 1 &&&&&&&& \\
\hline
0 & 0 & 1 & 1 & 1 &&&&&&& \\
0 & 0 & 0 & 0 & 0 & 1 &&&&&& \\
0 & 0 & 1 & 0 & 0 & 0 & 1 &&&&& \\
0 & 0 & 0 & 0 & -1 & 0 & 0.5 & 1 &&&& \\
\hline
0 & 0 & 0 & 1 & 0 & 0 & -1 & 0 & 1 &&& \\
0 & 0 & 0 & 0 & 0 & 1 & 0.5 & 0 & 0 & 1 && \\
0 & 0 & 0 & 1 & -1 & -1 & 0 & 0 & 0 & 0 & 1 & \\
0 & 0 & 0 & 0 & 0 & -2 & -1 & 0 & -1 & 0 & 0 & 1
\end{array}\right]
$$

$$
** \left[\begin{array}{cccccccccc}
\textcircled{-1} & 0 & 0 & 0 & 1 & 0 &&&& \\
0 & \textcircled{-1} & 0 & 1 & 0 & 0 &&&& \\
-1 & 1 & \textcircled{1} & 0 & 0 & 0 &&&& \\
0 & 2 & 0 & 0 & \textcircled{1} & 0 &&&& \\
\hline
&& -1 & 0 & 0 & 0 & \textcircled{1} & 0 && \\
&& 0 & -1 & 0 & \textcircled{1} & 0 & 0 && \\
&& -1 & \textcircled{1} & 1 & 0 & 0 & 0 && \\
&& 0 & 2 & 0 & 0 & 1 & 0 && \\
\hline
&&&& -1 & 0 & 0 & 0 & \textcircled{1} & 0 \\
&&&& 0 & -1 & 0 & \textcircled{1} & 0 & 0 \\
&&&& -1 & 1 & 1 & 0 & 0 & 0 \\
&&&& 0 & 2 & 0 & 0 & 1 & 0
\end{array}\right]
$$

$$
= \left[\begin{array}{cccccccccc}
\textcircled{-1} & 0 & 0 & 0 & 1 & 0 &&&& \\
0 & \textcircled{-1} & 0 & 1 & 0 & 0 &&&& \\
0 & 0 & \textcircled{1} & 1 & -1 & 0 &&&& \\
0 & 0 & 0 & 2 & \textcircled{1} & 0 &&&& \\
\hline
&& 0 & 3 & 0 & 0 & \textcircled{1} & 0 && \\
&& 0 & -1 & 0 & 1 & 0 & 0 && \\
&& 0 & \textcircled{2} & 0 & 0 & 0 & 0 && \\
&& 0 & 0 & 0 & 0 & 0 & 0 && \\
\hline
&& 0 & 0 & 0 & 0 & 0 & 0 & \textcircled{1} & 0 \\
&& 0 & 0 & 0 & 0 & 0 & \textcircled{1} & 0 & 0 \\
&& 0 & 0 & 0 & 0 & 0 & 0 & 0 & 0 \\
&& 0 & 0 & 0 & 0 & 0 & 0 & 0 & 0
\end{array}\right]
\begin{array}{l}
\\ \\ \\ \\ \\ \\ \triangleq \bar{\mathbf{S}}_2 \\ \leftarrow \\ \\ \\ \leftarrow \\ \\
\end{array}
$$

Note that the pivots are chosen for the convenience of hand calculation. Since $r_2 - 2 = q$, we have $v = 2$. The primary dependent rows of $\mathbf{S}_2$ are indicated by the arrows shown. Corresponding to these primary dependent rows, we use the formula in (A-11) to compute

$$\left[\begin{array}{cc:cc:cc:cc:cc:cc} 0.5 & -2.5 & -0.5 & -1 & -1 & 0 & 0.5 & \boxed{1} & 0 & 0 & 0 & 0 \\ 1 & -1 & -1 & 0 & -1 & -1 & 0 & 0 & 0 & 0 & \boxed{1} & 0 \end{array}\right]\mathbf{S}_2 = \mathbf{0}$$

Note that the solution is in the echelon form. Hence, we have

$$\mathbf{B}(s) = -\begin{bmatrix} 0.5 & -2.5 \\ 1 & -1 \end{bmatrix} - \begin{bmatrix} -1 & 0 \\ -1 & -1 \end{bmatrix} s - \begin{bmatrix} 0 & 0 \\ 0 & 0 \end{bmatrix} s^2 = \begin{bmatrix} s - 0.5 & 2.5 \\ s - 1 & s + 1 \end{bmatrix}$$

and

$$\mathbf{A}(s) = \begin{bmatrix} -0.5 & -1 \\ -1 & 0 \end{bmatrix} + \begin{bmatrix} 0.5 & 1 \\ 0 & 0 \end{bmatrix} s + \begin{bmatrix} 0 & 0 \\ 1 & 0 \end{bmatrix} s^2 = \begin{bmatrix} 0.5s - 0.5 & \boxed{s - 1} \\ \boxed{s^2 - 1} & 0 \end{bmatrix}$$

and $\hat{\mathbf{G}}(s) = \mathbf{A}^{-1}(s)\mathbf{B}(s)$. Note that deg det $\mathbf{A}(s)$ is equal to the total number of linearly independent N rows in $\mathbf{S}_2$, $\mathbf{A}(s)$ and $\mathbf{B}(s)$ are left coprime, and $\mathbf{A}(s)$ is in the polynomial echelon form. ∎

Combining Theorems G-8 and G-14, we have the following corollary. The corollary reduces to Corollary G-4 for the scalar case.

Corollary G-14

Consider a proper rational matrix $\hat{\mathbf{G}}(s)$. The fraction $\hat{\mathbf{G}}(s) = \mathbf{N}(s)\mathbf{D}^{-1}(s)$ is right coprime if and only if deg det $\mathbf{D}(s) = n$, where n is the total number of linearly independent N rows in $\mathbf{S}_{v-1}$ or $\mathbf{S}_k$ for $k \geq v - 1$. ∎

In the following we show that the row indices defined in (G-78) are invariant properties of $\hat{\mathbf{G}}(s)$ and are independent of the $\mathbf{N}(s)$ and $\mathbf{D}(s)$ used in the computation. First, we note that the order of the rows of the matrix in (G-76) can be altered without affecting Equation (G-75). This can also be seen from (G-83) and (G-84) by choosing $\mathbf{Q}(s)$ as an elementary matrix which interchanges the row positions of $\mathbf{A}(s)$ and $\mathbf{B}(s)$. Hence what is important is the set $\{v_i, i = 1, 2, \ldots, q\}$ rather than the individual v_i. We need the following lemma to establish the main result.

Lemma G-2

The $m \times m$ polynomial matrix $\mathbf{T}(s) = \mathbf{T}_0 + \mathbf{T}_1 s + \cdots + \mathbf{T}_j s^j$ is nonsingular (in the field of rational functions) if and only if the numerical matrix

$$\mathbf{V}_k = \begin{bmatrix} \mathbf{T}_0 & \mathbf{T}_1 & \cdots & \mathbf{T}_j & \mathbf{0} & \cdots & \mathbf{0} \\ \mathbf{0} & \mathbf{T}_0 & \cdots & \mathbf{T}_{j-1} & \mathbf{T}_j & \cdots & \mathbf{0} \\ \vdots & & & \vdots & & & \vdots \\ \mathbf{0} & \mathbf{0} & \cdots & \mathbf{T}_0 & \mathbf{T}_1 & \cdots & \mathbf{T}_j \end{bmatrix} \quad k + 1 \text{ block rows} \qquad \textbf{(G-85)}$$

is of full row rank (in the field of complex numbers) for $k = 0, 1, 2, \ldots$.

Proof

If $\mathbf{T}(s)$ is singular, there exists a $1 \times m$ nonzero polynomial vector

$$\boldsymbol{\alpha}(s) = \boldsymbol{\alpha}_0 + \boldsymbol{\alpha} s + \cdots + \boldsymbol{\alpha}_k s^k$$

such that

$$\boldsymbol{\alpha}(s)\mathbf{T}(s) = \mathbf{0}$$

This equation implies

$$[\boldsymbol{\alpha}_0 \quad \boldsymbol{\alpha}_1 \quad \cdots \quad \boldsymbol{\alpha}_k]\mathbf{V}_k = \mathbf{0}$$

Hence, if $\mathbf{T}(s)$ is singular, $\mathbf{V}_k$ does not have a full row rank for every k.

By reversing the above argument, we can show that if $\mathbf{V}_k$ does not have a full row rank for every k, then $\mathbf{T}(s)$ is singular. This completes the proof of the lemma. Q.E.D.

With this lemma, we are ready to establish the following theorem.

Theorem G-15

Let $\hat{\mathbf{G}}(s) = \mathbf{A}^{-1}(s)\mathbf{B}(s)$ be a left-coprime fraction and $\mathbf{A}(s)$ be row reduced. Then the row degrees of $\mathbf{A}(s)$ are intrinsic properties of $\hat{\mathbf{G}}(s)$ and are independent of the $\mathbf{N}(s)$ and $\mathbf{D}(s)$ used in Theorem G-14 in the computation.

Proof

Because of (G-81), it is sufficient to show that the row indices defined in (G-78) are independent of the $\mathbf{N}(s)$ and $\mathbf{D}(s)$ used in the computation. We recall that every right fraction $\mathbf{N}(s)\mathbf{D}^{-1}(s)$ of $\hat{\mathbf{G}}(s)$ can be obtained from a single right-coprime fraction $\bar{\mathbf{N}}(s)\bar{\mathbf{D}}^{-1}(s)$ by the relationship

$$\begin{bmatrix} \mathbf{D}(s) \\ \mathbf{N}(s) \end{bmatrix} = \begin{bmatrix} \bar{\mathbf{D}}(s) \\ \bar{\mathbf{N}}(s) \end{bmatrix} \mathbf{T}(s) \tag{G-86}$$

where $\mathbf{T}(s)$ is a $p \times p$ nonsingular polynomial matrix (Theorem G-13). If we write $\mathbf{T}(s)$ as $\mathbf{T}(s) = \mathbf{T}_0 + \mathbf{T}_1 s + \cdots + \mathbf{T}_j s^j$, then (G-86) implies, for $k = 0, 1, 2, \ldots$,

$$\mathbf{S}_k = \bar{\mathbf{S}}_k \mathbf{V}_{\bar{d}+k} \tag{G-87}$$

where $\mathbf{S}_k$ and $\bar{\mathbf{S}}_k$ are defined similarly as in (G-67), $\bar{d}$ is the degree of $\bar{\mathbf{D}}(s)$ as in (G-65) and $\mathbf{V}_{\bar{d}+k}$ is defined as in (G-85). Since $\mathbf{V}_{\bar{d}+k}$ has a full row rank for every k, (G-87) implies

$$\text{rank } \mathbf{S}_k = \text{rank } \bar{\mathbf{S}}_k \qquad k = 0, 1, 2, \ldots \tag{G-88}$$

Consequently, $r_i = \bar{r}_i$, $i = 0, 1, 2, \ldots$, where $\bar{r}_i$ are defined as in (G-67). Hence r_i, $i = 0, 1, 2, \ldots, v$ are independent of $\mathbf{N}(s)$ and $\mathbf{D}(s)$ used in the computation.

Now we shall show that the set $\{v_i, i = 1, 2, \ldots, q\}$ is uniquely determinable from $\{r_i, i = 0, 1, 2, \ldots, v\}$. We observe that r_i gives the number of N rows with row index i or smaller. Define $r_{-1} = 0$. Then

$$\beta_i \triangleq r_i - r_{i-1} \qquad i = 0, 1, \ldots, v \tag{G-89}$$

yields the number of N rows with row index exactly equal to i. For example, consider (G-74). Since $r_0 - r_{-1} = 0$, $r_1 - r_0 = 1$, $r_2 - r_1 = 0$, and $r_3 - r_2 = 2$, we have one N row with row index 1, no N row with row index 2, and two N rows with row index 3. In other words, the row indices are $\{1, 3, 3\}$. This shows that the row indices are uniquely determined by r_i, $i = 0, 1, \ldots \nu$. Hence the row indices are also intrinsic properties of $\hat{\mathbf{G}}(s)$ and are independent of the $\mathbf{N}(s)$ and $\mathbf{D}(s)$ used in the computation. Q.E.D.

As implied by (G-81), the set of row indices is equal to the set of row degrees. Hence we conclude that the row degrees are also intrinsic properties of $\hat{\mathbf{G}}(s)$. In other words, if $\hat{\mathbf{G}}(s)$ is factored as $\mathbf{A}^{-1}(s)\mathbf{B}(s)$, where $\mathbf{A}(s)$ and $\mathbf{B}(s)$ are left coprime and $\mathbf{A}(s)$ is row reduced, then the set of row degrees of $\mathbf{A}(s)$ is unique.

As discussed following Equation (G-64), all solutions $\mathbf{x}(s)$ of

$$\mathbf{x}(s)\begin{bmatrix} \mathbf{D}(s) \\ \mathbf{N}(s) \end{bmatrix} = \mathbf{0}$$

form a q-dimensional left null space. The $\mathbf{A}(s)$ and $\mathbf{B}(s)$ computed in Theorem G-14 are left coprime; hence the set of the q rows of $[-\mathbf{B}(s) \quad \mathbf{A}(s)]$ is a polynomial basis of the null space (Problem G-16). Because the linearly dependent rows of $\mathbf{S}_k$ are searched in order from top to bottom, the row degrees of $[-\mathbf{B}(s) \quad \mathbf{A}(s)]$ are the smallest possible. The properness of $\mathbf{N}(s)\mathbf{D}^{-1}(s)$ ensures that $\delta_{ri}\mathbf{B}(s) \leq \delta_{ri}\mathbf{A}(s)$ and that all pivot elements appear in $\mathbf{A}(s)$; thus the row degrees of $[-\mathbf{B}(s) \quad \mathbf{A}(s)]$ are equal to those of $\mathbf{A}(s)$ and $\mathbf{A}(s)$ is row reduced. Since the row degrees of $\mathbf{A}(s)$ are smallest possible and since the set of row degrees is unique (Theorem G-15), the set of q rows of $[-\mathbf{B}(s) \quad \mathbf{A}(s)]$ is indeed a *minimal polynomial basis*. We show that if $\bar{\mathbf{A}}(s)$ and $\bar{\mathbf{B}}(s)$ are left coprime but $\bar{\mathbf{A}}(s)$ is not row reduced, then $[-\bar{\mathbf{B}}(s) \quad \bar{\mathbf{A}}(s)]$ is not a minimal basis. Indeed, we have deg det $\bar{\mathbf{A}}(s) =$ deg det $\mathbf{A}(s)$ and

$$\sum \delta_{ri}\bar{A}(s) > \deg\det \bar{\mathbf{A}}(s) = \deg\det \mathbf{A}(s) = \sum \delta_{ri}A(s)$$

Thus the row degrees of $[-\bar{\mathbf{B}}(s) \quad \bar{\mathbf{A}}(s)]$ are not minimum. This establishes that $[-\mathbf{B}(s) \quad \mathbf{A}(s)]$ is a minimal polynomial basis if and only if $\mathbf{A}(s)$ and $\mathbf{B}(s)$ are left coprime and $\mathbf{A}(s)$ is row reduced. The basis obtained in Theorem G-14 has one additional nice property; it is in the polynomial echelon form. This is a consequence of the row searching algorithm. In conclusion, the solution of (G-66) obtained by using the row searching algorithm is a polynomial basis, a minimal basis and in a canonical form.

We discuss briefly the dual case of Equation (G-64). Consider

$$[\mathbf{A}(s) \quad \mathbf{B}(s)]\begin{bmatrix} -\mathbf{N}(s) \\ \mathbf{D}(s) \end{bmatrix} = \mathbf{0}$$

All solutions $\mathbf{y}(s)$ of $[\mathbf{A}(s) \quad \mathbf{B}(s)]\mathbf{y}(s) = \mathbf{0}$ form a p-dimensional right null space. If $\mathbf{D}(s)$ and $\mathbf{N}(s)$ are right coprime, the set of the p columns of $[-\mathbf{N}'(s) \quad \mathbf{D}'(s)]'$ is a polynomial basis of the null space. If, in addition, $\mathbf{D}(s)$ is column reduced, the set is a minimal polynomial basis. If the solution is obtained by using the column searching algorithm (dual to the row searching algorithm), then the minimal polynomial basis is in an echelon form. These results are dual to Theorems G-14 and G-15 and are stated as theorems.

Theorem G-14′

Consider a $q \times p$ proper rational matrix $\hat{\mathbf{G}}(s)$ factored as $\hat{\mathbf{G}}(s) = \mathbf{A}^{-1}(s)\mathbf{B}(s)$ with $\mathbf{A}(s) = \sum_{i=0}^{m} \mathbf{A}_i s^i$ and $\mathbf{B}(s) = \sum_{i=0}^{m} \mathbf{B}_i s^i$. We form

$$\mathbf{T}_k = \left[\begin{array}{cc|cc|c|cc} \mathbf{A}_0 & \mathbf{B}_0 & \mathbf{0} & 0 & & 0 & 0 \\ \mathbf{A}_1 & \mathbf{B}_1 & \mathbf{A}_0 & \mathbf{B}_0 & \cdots & 0 & 0 \\ \vdots & \vdots & \vdots & \vdots & & \vdots & \vdots \\ \mathbf{A}_{m-1} & \mathbf{B}_{m-1} & \mathbf{A}_{m-2} & \mathbf{B}_{m-2} & & \mathbf{A}_0 & \mathbf{B}_0 \\ \mathbf{A}_m & \mathbf{B}_m & \mathbf{A}_{m-1} & \mathbf{B}_{m-1} & & \mathbf{A}_1 & \mathbf{B}_1 \\ \mathbf{0} & \mathbf{0} & \mathbf{A}_m & \mathbf{B}_m & & \mathbf{A}_2 & \mathbf{B}_2 \\ & & 0 & 0 & & & \\ \vdots & \vdots & \vdots & \vdots & & \vdots & \vdots \\ \mathbf{0} & \mathbf{0} & \mathbf{0} & 0 & & \mathbf{A}_m & \mathbf{B}_m \end{array}\right] \quad \begin{array}{l} k+1 \text{ block columns (each} \\ \text{block has } q+p \text{ columns)} \end{array}$$

and search linearly dependent columns in order from left to right. Let r_i be the number of linearly dependent B columns in the $(i+1)$th block, and let μ be the least integer such that $r_\mu = p$. Then $\mathbf{D}(s) = \sum_{i=0}^{\mu} \mathbf{D}_i s^i$ and $\mathbf{N}(s) = \sum_{i=0}^{\mu} \mathbf{N}_i s^i$, solved from

$$\mathbf{T}_\mu \begin{bmatrix} -\mathbf{N}_0 \\ \mathbf{D}_0 \\ \vdots \\ -\mathbf{N}_\mu \\ \mathbf{D}_\mu \end{bmatrix} = \mathbf{0}$$

by using the column-searching algorithm corresponding to the p primary linearly dependent columns of $\mathbf{T}_\mu$, are right coprime. Furthermore, $\mathbf{D}(s)$ is in the column polynomial echelon form, that is, the degree of every pivot element of $\mathbf{D}(s)$ is larger than the degree of every other element in the same row, larger than the degree of every lower element in the same column, larger than or equal to the degree of every upper element in the same column. We call μ the *column index* of $\hat{\mathbf{G}}(s)$.

Theorem G-15′

Let $\hat{\mathbf{G}}(s) = \mathbf{N}(s)\mathbf{D}^{-1}(s)$ be a right-coprime fraction and $\mathbf{D}(s)$ be column reduced. Then the column degrees of $\mathbf{D}(s)$ are intrinsic properties of $\hat{\mathbf{G}}(s)$ and the set of the column degrees is unique. ∎

Problems

G-1 Apply the row searching algorithm to $\mathbf{S}_k$ in (G-20) to reduce the following rational functions to irreducible ones:

$$\frac{N(s)}{D(s)} = \frac{s^3 + s^2 - s + 2}{2s^3 - s^2 + s + 1}$$

$$\frac{N(s)}{D(s)} = \frac{2s^3 + 2s^2 - s - 1}{2s^4 + s^2 \quad 1}$$

G-2 The polynomials $D(s)$ of degree n and $N(s)$ of degree $m < n$ are coprime if and only if the square matrix $\mathbf{S}$ of order $n+m$ in (G-12) is nonsingular or if and only if the square matrix $\mathbf{S}_{n-1}$ of order $2n$ in (G-20) is nonsingular. Show the equivalence of these two statements.

G-3 Transform the matrix

$$\begin{bmatrix} 0 & 0 & (s+1)^2 & -s^2+s+1 \\ 0 & 0 & -s-1 & s-1 \\ s+1 & s^2 & s^2+s+1 & s \end{bmatrix}$$

into the Hermite row form.

$$\left(\text{Solution: } \begin{bmatrix} s+1 & s^2 & 1 & 0 \\ 0 & 0 & s+1 & 1 \\ 0 & 0 & 0 & s \end{bmatrix}\right)$$

G-4 Find a gcrd of the following two matrices by using (G-33):

$$\mathbf{D}(s) = \begin{bmatrix} s^2 + 2s & s + 3 \\ 2s^2 - s & 3s - 2 \end{bmatrix} \qquad \mathbf{N}(s) = [s \quad 1]$$

G-5 Are the following pairs of polynomial matrices right coprime?

a. $\mathbf{D}_1(s) = \begin{bmatrix} s+1 & 0 \\ (s-1)(s+2) & s-1 \end{bmatrix} \qquad \mathbf{N}_1(s) = [s+2 \quad s+1]$

b. $\mathbf{D}_2(s) = \mathbf{D}_1(s) \qquad \mathbf{N}_2(s) = [s-1 \quad s+1]$

c. $\mathbf{D}_3(s) = \mathbf{D}_1(s) \qquad \mathbf{N}_3(s) = [s+1 \quad s-1]$

d. $\mathbf{D}_4(s) = \mathbf{D}_1(s) \qquad \mathbf{N}_4(s) = [s \quad s]$

G-6 Are the pairs $\{\mathbf{A}_i(s), \mathbf{B}_i(s)\}$, $i = 1, 2, 3, 4$, left coprime if $\mathbf{A}_i(s) = \mathbf{D}_i'(s)$ and $\mathbf{B}_i(s) = \mathbf{N}_i'(s)$ in Problem G-5? Are they left coprime if $\mathbf{A}_i(s) = \mathbf{D}_i(s)$ and $\mathbf{B}_i(s) = \mathbf{N}_i'(s)$? (The prime denotes the transpose.)

G-7. Is the matrix

$$\mathbf{M}(s) = \begin{bmatrix} s^3 + s^2 + 1 & 2s + 1 & 3s^2 + s + 1 \\ 2s^3 + s - 1 & 0 & 2s^2 + s \\ 1 & s - 1 & s^2 - s \end{bmatrix}$$

column reduced? If not, find unimodular matrices $\mathbf{U}(s)$ and $\mathbf{V}(s)$ such that $\mathbf{M}(s)\mathbf{U}(s)$ and $\mathbf{V}(s)\mathbf{M}(s)$ are column reduced.

G-8 Is the $\mathbf{M}(s)$ in Problem G-7 row reduced? If not, transform it to a row-reduced one.

G-9 Find, by applying the row searching algorithm to $\mathbf{S}_k$ in (G-67), left coprime fractions from the following three different right fractions:

$$\hat{\mathbf{G}}(s)=\begin{bmatrix} s^3+s^2+s+1 & s^2+s \\ s^2+1 & 2s \end{bmatrix}\begin{bmatrix} s^4+s^2 & s^3 \\ s^2+1 & -s^2+2s \end{bmatrix}^{-1}$$
$$=\begin{bmatrix} s+1 & 0 \\ 1 & 1 \end{bmatrix}\begin{bmatrix} s^2 & 0 \\ 1 & -s+1 \end{bmatrix}^{-1}$$
$$=\begin{bmatrix} s^2+s & 0 \\ 2s+1 & 1 \end{bmatrix}\begin{bmatrix} s^3 & 0 \\ -s^2+s+1 & -s+1 \end{bmatrix}^{-1}$$

Are the results the same? Which right fraction, if there is any, is right coprime?

G-10 Find a right fraction from the left fraction

$$\hat{\mathbf{G}}(s)=\begin{bmatrix} s^2-1 & 0 \\ 0 & s^2-1 \end{bmatrix}^{-1}\begin{bmatrix} s-1 & s+1 \\ s^2 & 2s+2 \end{bmatrix}$$

$$\left(\text{Solution: } \begin{bmatrix} 1 & s-1 \\ 2 & s^2 \end{bmatrix}\begin{bmatrix} 0 & s^2-1 \\ s-1 & 0 \end{bmatrix}^{-1}\right)$$

G-11 Let $\hat{\mathbf{G}}(s)=\mathbf{A}^{-1}(s)\mathbf{B}(s)=\mathbf{N}(s)\mathbf{D}^{-1}(s)$ be two coprime fractions. Show that there exists a unimodular matrix of the form

$$\begin{bmatrix} \mathbf{U}_{11}(s) & \mathbf{U}_{12}(s) \\ \mathbf{B}(s) & \mathbf{A}(s) \end{bmatrix}$$

such that

$$\begin{bmatrix} \mathbf{U}_{11}(s) & \mathbf{U}_{12}(s) \\ \mathbf{B}(s) & \mathbf{A}(s) \end{bmatrix}\begin{bmatrix} \mathbf{D}(s) \\ -\mathbf{N}(s) \end{bmatrix}=\begin{bmatrix} \mathbf{I} \\ \mathbf{0} \end{bmatrix} \quad \text{or} \quad \begin{bmatrix} \mathbf{U}_{12}(s) & \mathbf{U}_{11}(s) \\ \mathbf{A}(s) & \mathbf{B}(s) \end{bmatrix}\begin{bmatrix} -\mathbf{N}(s) \\ \mathbf{D}(s) \end{bmatrix}=\begin{bmatrix} \mathbf{I} \\ \mathbf{0} \end{bmatrix}$$

(*Hint*: If **D** and **N** are right coprime, we have

$$\begin{bmatrix} \mathbf{U}_{11} & \mathbf{U}_{12} \\ \mathbf{U}_{21} & \mathbf{U}_{22} \end{bmatrix}\begin{bmatrix} \mathbf{D} \\ -\mathbf{N} \end{bmatrix}=\begin{bmatrix} \mathbf{I} \\ \mathbf{0} \end{bmatrix}$$

and $\mathbf{U}_{21}$ and $\mathbf{U}_{22}$ are left coprime and $\mathbf{U}_{22}^{-1}\mathbf{U}_{21}=\mathbf{N}\mathbf{D}^{-1}$. Using the dual of Theorem G-13, there exists a unimodular matrix $\mathbf{M}$ such that $\mathbf{M}\mathbf{U}_{21}=\mathbf{B}$ and $\mathbf{M}\mathbf{U}_{22}=\mathbf{A}$.)

G-12 Show that if $\mathbf{N}(s)\mathbf{D}^{-1}(s)=\bar{\mathbf{N}}(s)\bar{\mathbf{D}}^{-1}(s)$ are two right-coprime fractions, then the matrix $\mathbf{U}(s)=\bar{\mathbf{D}}^{-1}(s)\mathbf{D}(s)$ is unimodular. (*Hint*: The equation $\mathbf{XN}+\mathbf{YD}=\mathbf{I}$ implies $\mathbf{X}\bar{\mathbf{N}}\bar{\mathbf{D}}^{-1}\mathbf{D}+\mathbf{Y}\bar{\mathbf{D}}\bar{\mathbf{D}}^{-1}\mathbf{D}=(\mathbf{X}\bar{\mathbf{N}}+\mathbf{Y}\bar{\mathbf{D}})\mathbf{U}=\mathbf{I}$ which implies $\mathbf{U}^{-1}$ to be a polynomial matrix.)

G-13 Prove Theorem G-13 by using Problem G-12.

G-14 Consider the Diophantine polynomial equation

$$\mathbf{D}_c(s)\mathbf{D}(s)+\mathbf{N}_c(s)\mathbf{N}(s)=\mathbf{F}(s)$$

where $\mathbf{D}_c$, $\mathbf{D}$, $\mathbf{N}_c$, $\mathbf{N}$, and $\mathbf{F}$ are, respectively, $p\times p$, $p\times p$, $p\times q$, $q\times p$, and $p\times p$ polynomial matrices. Let

$$\begin{bmatrix} \mathbf{U}_{11}(s) & \mathbf{U}_{12}(s) \\ \mathbf{U}_{21}(s) & \mathbf{U}_{22}(s) \end{bmatrix}\begin{bmatrix} \mathbf{D}(s) \\ \mathbf{N}(s) \end{bmatrix}=\begin{bmatrix} \mathbf{R}(s) \\ \mathbf{0} \end{bmatrix}$$

where the leftmost matrix is unimodular and $\mathbf{R}(s)$ is a gcrd of $\mathbf{D}(s)$ and $\mathbf{N}(s)$. Given $\mathbf{D}(s)$

and $\mathbf{N}(s)$, show that for any $\mathbf{F}(s)$, there exist solutions $\mathbf{D}_c(s)$ and $\mathbf{N}_c(s)$ to meet the Diophantine equation if and only if $\mathbf{R}(s)$ is a right divisor of $\mathbf{F}(s)$, that is, there exists a polynomial matrix $\bar{\mathbf{F}}(s)$ such that $\mathbf{F}(s)=\bar{\mathbf{F}}(s)\mathbf{R}(s)$. Show also that

$$\mathbf{D}_c^0(s)=\bar{\mathbf{F}}(s)\mathbf{U}_{11}(s) \qquad \mathbf{N}_c^0(s)=\bar{\mathbf{F}}(s)\mathbf{U}_{12}(s)$$

are solutions of the Diophantine equation. Finally, show that, for any $p \times p$ polynomial matrix $\mathbf{T}(s)$, the matrices

$$\mathbf{D}_c(s)=\mathbf{D}_c^0(s)+\mathbf{T}(s)\mathbf{U}_{21}(s)$$

and

$$\mathbf{N}_c(s)=\mathbf{N}_c^0(s)+\mathbf{T}(s)\mathbf{U}_{22}(s)$$

are solutions of the Diophantine equation (see Reference S139).

G-15 Consider

$$\mathbf{N}(s)=\mathbf{Q}(s)\mathbf{D}(s)+\mathbf{R}(s)$$

with $\delta_{ci}\mathbf{R}(s)<\delta_{ci}\mathbf{D}(s)=\mu_i$. Let $\mathbf{H}(s)=\text{diag}\{s^{\mu_1}, s^{\mu_2}, \ldots, s^{\mu_p}\}$ and let $\mathbf{D}(s)=\mathbf{D}_{hc}\mathbf{H}(s)+\mathbf{D}_{lc}(s)$. If $\mathbf{D}(s)$ is column reduced, then $\mathbf{D}_{hc}$ is nonsingular. We rewrite the equation as

$$\begin{aligned}\mathbf{N}(s)\mathbf{H}^{-1}(s) \triangleq \mathbf{N}_p(s)+\mathbf{N}_r(s) &= \mathbf{Q}(s)(\mathbf{D}_{hc}\mathbf{H}(s)+\mathbf{D}_{lc}(s))\mathbf{H}^{-1}(s)+\mathbf{R}(s)\mathbf{H}^{-1}(s)\\ &= \mathbf{Q}(s)\mathbf{D}_{hc}(\mathbf{I}+\mathbf{D}_{hc}^{-1}\mathbf{D}_{lc}(s)\mathbf{H}^{-1}(s))+\mathbf{R}(s)\mathbf{H}^{-1}(s)\end{aligned}$$

where $\mathbf{N}_p(s)$ is a polynomial matrix and $\mathbf{N}_r(s)$ is strictly proper. Let v_i be the row degrees of $\mathbf{N}_p(s)$. Define $\bar{\mathbf{H}}(s)=\text{diag}\{s^{v_1}, s^{v_2}, \ldots, s^{v_p}\}$ and $\mu=\max\{\mu_i\}$, $v=\max\{v_i\}$. We write

$$\begin{aligned}\mathbf{N}_p(s) &= \bar{\mathbf{N}}(s)[\mathbf{N}_{p0}+\mathbf{N}_{p1}s^{-1}+\cdots+\mathbf{N}_{pv}s^{-v}]\\ \bar{\mathbf{Q}}(s) &\triangleq \mathbf{Q}(s)\mathbf{D}_{hc}\triangleq\bar{\mathbf{H}}(s)[\bar{\mathbf{Q}}_0+\bar{\mathbf{Q}}_1s^{-1}+\cdots+\bar{\mathbf{Q}}_v s^{-v}]\\ \bar{\mathbf{D}}(s) &\triangleq \mathbf{D}_{hc}^{-1}\mathbf{D}_{lc}(s)\mathbf{H}^{-1}(s)\triangleq\bar{\mathbf{D}}_1s^{-1}+\bar{\mathbf{D}}_2s^{-2}+\cdots+\bar{\mathbf{D}}_\mu s^{-\mu}\end{aligned}$$

where the ith row of $\bar{\mathbf{Q}}_k$ is a zero row if $k>v_i$. Show that the nonzero rows of $\bar{\mathbf{Q}}_k$ can be computed recursively as

$$\bar{\mathbf{Q}}_0=\mathbf{N}_{p0}$$

$$\bar{\mathbf{Q}}_k=\mathbf{N}_{pk}-\sum_{l=0}^{k-1}\bar{\mathbf{Q}}_l\bar{\mathbf{D}}_{k-l} \qquad k=1,2,\ldots v$$

where $\mathbf{D}_l\triangleq\mathbf{0}$ if $l>\mu$. From $\bar{\mathbf{Q}}(s)$, we can compute $\mathbf{Q}(s)=\bar{\mathbf{Q}}(s)\mathbf{D}_{hc}^{-1}$ and then compute $\mathbf{R}(s)=\mathbf{N}(s)-\mathbf{Q}(s)\mathbf{D}(s)$. See Reference S137.

G-16 Consider $\hat{\mathbf{G}}(s)=\mathbf{A}^{-1}(s)\mathbf{B}(s)=\mathbf{N}(s)\mathbf{D}^{-1}(s)$. Let $\mathbb{V}$ denote the left null space of all rational function solutions of

$$\mathbf{x}(s)\begin{bmatrix}\mathbf{D}(s)\\ \mathbf{N}(s)\end{bmatrix}=\mathbf{0}$$

and let $\mathbb{V}_p$ denote the polynomial part of $\mathbb{V}$. A set of polynomial vectors is called a polynomial basis of $\mathbb{V}$ if every vector in $\mathbb{V}_p$ can be expressed as a unique combination of the basis by using only polynomials as coefficients. Show that the set of the rows of $[-\mathbf{B}(s) \quad \mathbf{A}(s)]$ is a polynomial basis of $\mathbb{V}$ if and only if $\mathbf{A}(s)$ and $\mathbf{B}(s)$ are left coprime. [*Hint*: ($\Rightarrow$) Let $[-\bar{\mathbf{B}}(s) \quad \bar{\mathbf{A}}(s)]$ be left coprime. If $[-\mathbf{B}(s) \quad \mathbf{A}(s)]$ is a basis, there exists a polynomial matrix $\mathbf{T}(s)$ such that

$$[-\bar{\mathbf{B}}(s) \quad \bar{\mathbf{A}}(s)]=\mathbf{T}(s)[-\mathbf{B}(s) \quad \mathbf{A}(s)]$$

Show that $\mathbf{T}(s)$ is unimodular and hence $\mathbf{A}(s)$ and $\mathbf{B}(s)$ are left coprime. ($\Leftarrow$) Let $[-\bar{\mathbf{b}}(s) \quad \bar{\mathbf{a}}(s)]$ be any vector in $\mathbb{V}_p$. Show that $\bar{\mathbf{a}}(s)\neq\mathbf{0}$, and then append $(q-1)$ vectors in $\mathbb{V}_p$ to it to form

$[-\bar{\mathbf{B}}(s) \quad \bar{\mathbf{A}}(s)]$ with $\bar{\mathbf{A}}(s)$ nonsingular and

$$[-\bar{\mathbf{B}}(s) \quad \bar{\mathbf{A}}(s)]\begin{bmatrix}\mathbf{D}(s)\\ \mathbf{N}(s)\end{bmatrix}=\mathbf{0}$$

Then use Theorem G-13 to show the existence of a unique polynomial matrix $\mathbf{T}(s)$ such that $[-\bar{\mathbf{B}}(s) \quad \bar{\mathbf{A}}(s)]=\mathbf{T}(s)[-\mathbf{B}(s) \quad \mathbf{A}(s)]$ and $[-\bar{\mathbf{b}}(s) \quad \bar{\mathbf{a}}(s)]=\mathbf{t}(s)[-\mathbf{B}(s) \quad \mathbf{A}(s)]$.]

G-17 Let $\mathbf{M}_p(s)$ be a polynomial matrix and let $\mathbf{U}(s)$ be a unimodular matrix such that $\mathbf{M}_p(s)\mathbf{U}(s)$ is column reduced. Show that $\mathbf{M}_p^{-1}(s)$ is proper if and only if $\delta_{ci}\mathbf{U}(s)\leq \delta_{ci}(\mathbf{M}_p(s)\mathbf{U}(s))$, for all i.

H
Poles and Zeros

In this appendix we shall introduce the concepts of pole and zero for transfer functions. We discuss first the single-variable case and then the multivariable case.

Definition H-1

A number λ (real or complex) is said to be a *pole* of a proper rational transfer function $\hat{g}(s)$ if $|\hat{g}(\lambda)| = \infty$. It is said to be a *zero* of $\hat{g}(s)$ if $\hat{g}(\lambda) = 0$. ∎

If a proper rational transfer function is irreducible (that is, there is no nontrivial common factor between its numerator and denominator), then it is clear that every root of the denominator of $\hat{g}(s)$ is a pole of $\hat{g}(s)$ and every root of the numerator of $\hat{g}(s)$ is a zero of $\hat{g}(s)$. Without this irreducibility assumption, a root of the denominator of $\hat{g}(s)$ may not be a pole of $\hat{g}(s)$. For example, -1 is not a pole of

$$\hat{g}(s) = \frac{s+1}{s^2+3s+2}$$

although it is a root of s^2+3s+2.

Consider a proper rational function $\hat{g}(s)$ with the following *irreducible* realization

$$\dot{\mathbf{x}} = \mathbf{A}\mathbf{x} + \mathbf{b}u \qquad y = \mathbf{c}\mathbf{x} + eu$$

If we write $\hat{g}(s) = N(s)/D(s)$ and $N(s)$ and $D(s)$ are coprime (have no nontrivial common factor), then we have

$$D(s) = k \det (s\mathbf{I} - \mathbf{A})$$

with some constant k. If $D(s)$ is monic (the coefficient associated with the highest power of s is equal to 1), then $k=1$. We discuss in the following the implications of poles and zeros.

Theorem H-1

Consider a single-variable system with proper transfer function $\hat{g}(s)$ and an irreducible realization $\{\mathbf{A}, \mathbf{b}, \mathbf{c}, e\}$. Then a number λ is a pole of $\hat{g}(s)$ if and only if there exists an initial state $\mathbf{x}_0$ such that the zero-input response at the output of the system is equal to

$$y(t) = re^{\lambda t} \qquad \text{for all } t \geq 0$$

for some nonzero constant r.

Proof

The zero-input response of the system is given by

$$\hat{y}(s) = \mathbf{c}(s\mathbf{I} - \mathbf{A})^{-1}\mathbf{x}(0)$$

If λ is a pole of $\hat{g}(s)$, then it is an eigenvalue of $\mathbf{A}$. Let $\mathbf{v}$ be an eigenvector of $\mathbf{A}$ associated with λ; that is, $\mathbf{A}\mathbf{v} = \lambda\mathbf{v}$. Then it can be verified (Problem 2-32) that $\mathbf{v}$ is an eigenvector of $(s\mathbf{I} - \mathbf{A})^{-1}$ associated with eigenvalue $(s-\lambda)^{-1}$. Hence we have

$$\hat{y}(s) = \mathbf{c}(s\mathbf{I} - \mathbf{A})^{-1}\mathbf{v} = \mathbf{c}\mathbf{v}(s-\lambda)^{-1}$$

or

$$y(t) = \mathbf{c}\mathbf{v}e^{\lambda t} \qquad \text{for all } t \geq 0$$

What remains to be shown is that the constant $r = \mathbf{c}\mathbf{v}$ is different from zero. The realization $\{\mathbf{A}, \mathbf{c}\}$ is observable by assumption; hence the matrix $[s\mathbf{I}' - \mathbf{A}' \,\vdots\, \mathbf{c}']'$ has a full rank at every s in $\mathbb{C}$. Consequently, for every nonzero vector, in particular, the vector $\mathbf{v}$, we have

$$\begin{bmatrix} \lambda\mathbf{I} - \mathbf{A} \\ \mathbf{c} \end{bmatrix} \mathbf{v} \neq \mathbf{0}$$

Since $(\lambda\mathbf{I} - \mathbf{A})\mathbf{v} = \mathbf{0}$, we must have $r = \mathbf{c}\mathbf{v} \neq 0$; otherwise, it would have violated the above condition. This completes the proof of the necessity of the theorem.

To show the converse, we show that if $y = re^{\lambda t}$, $t \geq 0$, then λ is a pole of $\hat{g}(s)$. If $y(t) = re^{\lambda t}$, then we have

$$\hat{y}(s) = \mathbf{c}(s\mathbf{I} - \mathbf{A})^{-1}\mathbf{x}(0) = r(s-\lambda)^{-1}$$

or

$$\frac{1}{\det(s\mathbf{I} - \mathbf{A})}\mathbf{c}\,[\operatorname{Adj}(s\mathbf{I} - \mathbf{A})]\mathbf{x}(0) = \frac{r}{s-\lambda}$$

or

$$(s-\lambda)\mathbf{c}\,[\operatorname{Adj}(s\mathbf{I} - \mathbf{A})]\mathbf{x}(0) = r\det(s\mathbf{I} - \mathbf{A})$$

which implies $\det(\lambda\mathbf{I} - \mathbf{A}) = 0$. Hence λ is an eigenvalue of $\mathbf{A}$ and, consequently, a pole of $\hat{g}(s)$. This completes the proof of the theorem. Q.E.D.

This theorem states that if λ is a pole of $\hat{g}(s)$, the mode $e^{\lambda t}$ can be generated at the output by an initial state without the application of any input. If λ is not a pole of $\hat{g}(s)$, then this is not possible; the only way to generate $e^{\lambda t}$ at the output is to apply $e^{\lambda t}$ at the input.

Theorem H-2

Consider a system with proper transfer function $\hat{g}(s)$ and an irreducible realization $\{\mathbf{A}, \mathbf{b}, \mathbf{c}, e\}$. If the input $u(t)$ is of the form $e^{\lambda t}$, where λ, real or complex, is not a pole of $\hat{g}(s)$, then the output due to the initial state $\mathbf{x}(0) = -(\mathbf{A}-\lambda\mathbf{I})^{-1}\mathbf{b}$ and the input $u = e^{\lambda t}$ is equal to $y(t) = \hat{g}(\lambda)e^{\lambda t}$ for $t \geq 0$. ■

To prove this theorem, we need the following identity:

$$(s\mathbf{I}-\mathbf{A})^{-1}(s-\lambda)^{-1} = (\lambda\mathbf{I}-\mathbf{A})^{-1}(s-\lambda)^{-1} + (s\mathbf{I}-\mathbf{A})^{-1}(\mathbf{A}-\lambda\mathbf{I})^{-1} \quad \textbf{(H-1)}$$

for any λ that is not an eigenvalue of $\mathbf{A}$. Note the similarity of this identity to the partial fraction expansion

$$\frac{1}{(s-a)(s-\lambda)} = \frac{1}{(\lambda-a)}\cdot\frac{1}{(s-\lambda)} + \frac{1}{(s-a)}\cdot\frac{1}{(a-\lambda)}$$

The identity in (H-1) can be readily verified by post- and premultiplication of $(\mathbf{A}-\lambda\mathbf{I})$ and $(s\mathbf{I}-\mathbf{A})$. (Problem H-1.)

Proof of Theorem H-2

The response of the system due to the initial state $\mathbf{x}(0)$ and the input $u(t)$ is, as derived in (4-25),

$$\begin{aligned}\hat{y}(s) &= \mathbf{c}(s\mathbf{I}-\mathbf{A})^{-1}\mathbf{x}(0) + \mathbf{c}(s\mathbf{I}-\mathbf{A})^{-1}\mathbf{b}\hat{u}(s) + e\hat{u}(s)\\ &= \mathbf{c}(s\mathbf{I}-\mathbf{A})^{-1}\mathbf{x}(0) + \mathbf{c}(s\mathbf{I}-\mathbf{A})^{-1}\mathbf{b}(s-\lambda)^{-1} + e(s-\lambda)^{-1} \end{aligned} \quad \textbf{(H-2)}$$

The substitution of (H-1) into (H-2) yields

$$\begin{aligned}\hat{y}(s) &= \mathbf{c}(s\mathbf{I}-\mathbf{A})^{-1}\mathbf{x}(0) + \mathbf{c}(\lambda\mathbf{I}-\mathbf{A})^{-1}\mathbf{b}(s-\lambda)^{-1}\\ &\qquad + \mathbf{c}(s\mathbf{I}-\mathbf{A})^{-1}(\mathbf{A}-\lambda\mathbf{I})^{-1}\mathbf{b} + e(s-\lambda)^{-1}\\ &= \mathbf{c}(s\mathbf{I}-\mathbf{A})^{-1}[\mathbf{x}(0) + (\mathbf{A}-\lambda\mathbf{I})^{-1}\mathbf{b}] + [\mathbf{c}(\lambda\mathbf{I}-\mathbf{A})^{-1}\mathbf{b} + e](s-\lambda)^{-1}\end{aligned}$$

Since $\hat{g}(s) = \mathbf{c}(s\mathbf{I}-\mathbf{A})^{-1}\mathbf{b} + e$, if $\mathbf{x}(0) = -(\mathbf{A}-\lambda\mathbf{I})^{-1}\mathbf{b}$, then

$$\hat{y}(s) = \hat{g}(\lambda)(s-\lambda)^{-1}$$

or $$y(t) = \hat{g}(\lambda)e^{\lambda t} \qquad \text{for } t \geq 0$$

This completes the proof. Q.E.D.

In this theorem, the assumption that λ is not a pole of $\hat{g}(s)$ is essential. Otherwise, the theorem does not hold. This theorem is very useful in establishing that the impedance of a linear time-invariant, lumped, passive network is a positive real function (see Reference 31). This theorem can also be used to give a physical interpretation of the zeros of a transfer function. If λ is a zero of $\hat{g}(s)$, that is, $\hat{g}(\lambda) = 0$, then for a certain initial state, the output of the system is identically zero even if the input $e^{\lambda t}$ is applied. In other words, the input $e^{\lambda t}$ is blocked by the system.

Corollary H-2 (Transmission-blocking property)

Consider a system with proper transfer function $\hat{g}(s)$ and an irreducible realization $\{\mathbf{A}, \mathbf{b}, \mathbf{c}, e\}$. If λ is a zero of $\hat{g}(s)$, then the response at the output due to the initial state $\mathbf{x}(0) = -(\mathbf{A} - \lambda\mathbf{I})^{-1}\mathbf{b}$ and the input $u(t) = e^{\lambda t}$ is identically zero. ∎

We shall now extend the concepts of poles and zeros to the multivariable case. Consider a $q \times p$ proper rational matrix $\hat{\mathbf{G}}(s)$ with the following coprime fraction

$$\hat{\mathbf{G}}(s) = \mathbf{D}_l^{-1}(s)\mathbf{N}_l(s) = \mathbf{N}_r(s)\mathbf{D}_r^{-1}(s)$$

where $\mathbf{D}_l$, $\mathbf{N}_l$, $\mathbf{N}_r$, and $\mathbf{D}_r$ are, respectively, $q \times q$, $q \times p$, $q \times p$, and $p \times p$ polynomial matrices. Furthermore, $\mathbf{N}_l(s)$ and $\mathbf{D}_l(s)$ are left coprime; $\mathbf{N}_r(s)$ and $\mathbf{D}_r(s)$ are right coprime. Let

$$\dot{\mathbf{x}} = \mathbf{A}\mathbf{x} + \mathbf{B}\mathbf{u} \qquad \mathbf{y} = \mathbf{C}\mathbf{x} + \mathbf{E}\mathbf{u}$$

be an irreducible realization of $\hat{\mathbf{G}}(s)$. Then we have

$$\det(s\mathbf{I} - \mathbf{A}) = k_1 \det \mathbf{D}_l(s) = k_2 \det \mathbf{D}_r(s)$$

where k_1 and k_2 are constants. In view of this relationship, we may give the following definition.

Definition H-2

A number λ, real or complex, is said to be a *pole* of a proper rational matrix $\hat{\mathbf{G}}(s)$ if and only if it is a root of $\det \mathbf{D}(s) = 0$, where $\mathbf{D}(s)$ is the denominator matrix of any right- or left-coprime fraction of $\hat{\mathbf{G}}(s)$. ∎

Similar to Theorem H-1, we have the following theorem to characterize the poles of $\hat{\mathbf{G}}(s)$.

Theorem H-3

Consider a multivariable system with proper transfer matrix $\hat{\mathbf{G}}(s)$ and an irreducible realization $\{\mathbf{A}, \mathbf{B}, \mathbf{C}, \mathbf{E}\}$. Then a number λ is a pole of $\hat{\mathbf{G}}(s)$ if and only if there exists an initial state $\mathbf{x}_0$ such that the zero-input response at the output of the system is equal to

$$\mathbf{y}(t) = \mathbf{r}e^{\lambda t}$$

for some nonzero vector $\mathbf{r}$. ∎

The proof of this theorem is identical to the one of Theorem H-1 and will not be repeated. We note that every pole of $\hat{\mathbf{G}}(s)$ must be a pole of some element of $\hat{\mathbf{G}}(s)$, and every pole of every element of $\hat{\mathbf{G}}(s)$ must be a pole of $\hat{\mathbf{G}}(s)$. This fact follows from the fact that $\det \mathbf{D}(s)$ is equal to the least common denominator of all minors of $\hat{\mathbf{G}}(s)$ (see Definition 6-1 and Theorem 6-2).

We shall now extend the concept of zeros to the multivariable case. First, we assume that the $q \times p$ proper rational matrix $\hat{\mathbf{G}}(s)$ has a full rank in the field of rational functions. By this, we mean that if $q \le p$, then $\rho\hat{\mathbf{G}}(s) = q$; if $q > p$, then $\rho\hat{\mathbf{G}}(s) = p$, where ρ denotes the rank. If $\hat{\mathbf{G}}(s)$ does not have a full rank in the field of rational functions, then there exists a $1 \times q$ rational vector $\mathbf{M}(s)$ or a $p \times 1$ rational vector $\mathbf{P}(s)$ such that

$$\mathbf{M}(s)\hat{\mathbf{G}}(s) = \mathbf{0} \qquad \text{or} \qquad \hat{\mathbf{G}}(s)\mathbf{P}(s) = \mathbf{0} \tag{H-3}$$

If we multiply the least common denominator of $\mathbf{M}(s)$ or $\mathbf{P}(s)$ to (H-3), they become

$$\bar{\mathbf{M}}(s)\hat{\mathbf{G}}(s) = \mathbf{0} \qquad \text{or} \qquad \hat{\mathbf{G}}(s)\bar{\mathbf{P}}(s) = \mathbf{0}$$

where $\bar{\mathbf{M}}(s)$ and $\bar{\mathbf{P}}(s)$ are polynomial vectors. Since $\hat{\mathbf{y}}(s) = \hat{\mathbf{G}}(s)\hat{\mathbf{u}}(s)$, we have

$$\bar{\mathbf{M}}(s)\hat{\mathbf{y}}(s) = \bar{\mathbf{M}}(s)\hat{\mathbf{G}}(s)\hat{\mathbf{u}}(s) = \mathbf{0}$$

for all possible inputs. This implies that the number of effective outputs is smaller than q. Similarly, if $\hat{\mathbf{G}}(s)\bar{\mathbf{P}}(s) = 0$, and if we connect a precompensator $\bar{\mathbf{P}}(s)$ to the system, then for all possible inputs $\hat{\mathbf{u}}(s)$ we have

$$\hat{\mathbf{G}}(s)\bar{\mathbf{P}}(s)\hat{\mathbf{u}}(s) = \mathbf{0}$$

Hence the number of effective inputs is smaller than p. Hence, if $\hat{\mathbf{G}}(s)$ has a full rank, there are no redundant or noneffective input and output terminals.

Consider $\hat{\mathbf{G}}(s)$ with the following coprime fractions:

$$\hat{\mathbf{G}}(s) = \mathbf{D}_l^{-1}(s)\mathbf{N}_l(s) = \mathbf{N}_r(s)\mathbf{D}_r^{-1}(s) \tag{H-4}$$

If $\hat{\mathbf{G}}(s)$ is of full rank in the field of rational functions, so are $\mathbf{N}_l(s)$ and $\mathbf{N}_r(s)$. This implies that for almost every λ in $\mathbb{C}$, the $q \times p$ complex matrix $\mathbf{N}_l(\lambda)$ has a rank equal to min (p, q) in the field of complex numbers.

Definition H-3

Consider a $q \times p$ proper rational function $\hat{\mathbf{G}}(s)$ with the coprime fraction $\hat{\mathbf{G}}(s) = \mathbf{D}_l^{-1}(s)\mathbf{N}_l(s)$. It is assumed[1] that $\hat{\mathbf{G}}(s)$ and, consequently, $\mathbf{N}_l(s)$ have a full rank (in the field of rational function). Then a number λ, real or complex, is said to be a *transmission zero* of $\hat{\mathbf{G}}(s)$ if and only if rank $\mathbf{N}_l(\lambda) < \min(p, q)$, in $\mathbb{C}$, the field of complex numbers. ∎

Example 1

Consider the left coprime fraction

$$\hat{\mathbf{G}}_1(s) = \begin{bmatrix} \dfrac{s}{s+2} & 0 & \dfrac{s+1}{s+2} \\ 0 & \dfrac{s+1}{s^2} & \dfrac{1}{s} \end{bmatrix} = \begin{bmatrix} s+2 & 0 \\ 0 & s^2 \end{bmatrix}^{-1} \begin{bmatrix} s & 0 & s+1 \\ 0 & s+1 & s \end{bmatrix}$$

This $\mathbf{N}_l(s)$ has rank 2 for every s in $\mathbb{C}$; hence $\hat{\mathbf{G}}_1(s)$ has no transmission zero. ∎

---

[1] If $\hat{\mathbf{G}}(s)$ does not have a full rank, then $\rho\mathbf{N}(\lambda) < \min(p, q)$ for every λ in $\mathbb{C}$. In other words, every λ in $\mathbb{C}$ is a transmission zero of $\hat{\mathbf{G}}(s)$. This is a degenerate case and will not be considered.

Example 2

Consider the left coprime factorization

$$\hat{\mathbf{G}}_2(s)=\begin{bmatrix}\dfrac{s}{s+2} & 0\\ 0 & \dfrac{s+2}{s}\end{bmatrix}=\begin{bmatrix}s+2 & 0\\ 0 & s\end{bmatrix}^{-1}\begin{bmatrix}s & 0\\ 0 & s+2\end{bmatrix}$$

This $\mathbf{N}_l(s)$ has rank 1 at $s=0$ and $s=-2$. Hence 0 and -2 are two transmission zeros of $\hat{\mathbf{G}}(s)$. Note that 0 and -2 are also poles of $\hat{\mathbf{G}}_2(s)$. ∎

From Example 2, we see that $\hat{\mathbf{G}}(s)$ may not be well defined at its transmission zeros. Hence we cannot use $\rho\hat{\mathbf{G}}(\lambda)<\min(p, q)$ to define its transmission zero. We note that $\mathbf{N}_l(s)$ and $\mathbf{N}_r(s)$ in (H-4) are both $q\times p$ polynomial matrices. If λ is not a pole of $\hat{\mathbf{G}}(s)$, then it is clear that $\rho\mathbf{N}_l(\lambda)<\min(p, q)$ if and only if $\rho\mathbf{N}_r(\lambda)<\min(p, q)$. In fact, it is shown in Reference S34 by using the Smith-McMillan form that there exist unimodular matrices $\mathbf{V}(s)$ and $\mathbf{U}(s)$ such that

$$\mathbf{N}_l(s)=\mathbf{U}(s)\mathbf{N}_r(s)\mathbf{V}(s) \tag{H-5}$$

Hence the transmission zeros of $\hat{\mathbf{G}}(s)$ can be defined by using $\mathbf{N}_l(s)$ or $\mathbf{N}_r(s)$. We note that if $\mathbf{N}_l(s)$ and $\mathbf{N}_r(s)$ are square, the transmission zeros of $\hat{\mathbf{G}}(s)$ are the roots of det $\mathbf{N}_l(s)$ or det $\mathbf{N}_r(s)$, where det stands for the determinant.

From the above two examples, we see that the transmission zeros of $\hat{\mathbf{G}}(s)$ exhibits some phenomena which do not exist in the scalar case. A transmission zero may appear as a pole of the same $\hat{\mathbf{G}}(s)$. Even though elements of $\hat{\mathbf{G}}(s)$ have zeros, $\hat{\mathbf{G}}(s)$ may not have any transmission zero. The transmission zeros of a square $\hat{\mathbf{G}}(s)$ may be different from the zeros of det $\hat{\mathbf{G}}(s)$. In spite of these differences, the transmission properties of the transmission zeros of $\hat{\mathbf{G}}(s)$ are quite analogous to the scalar case. We shall establish these in the following.

Theorem H-4

Consider a multivariable system with $q\times p$ proper rational matrix $\hat{\mathbf{G}}(s)$ and an irreducible realization $\{\mathbf{A}, \mathbf{B}, \mathbf{C}, \mathbf{E}\}$. If the input $\mathbf{u}(t)$ is of the form $\mathbf{k}e^{\lambda t}$, where λ is real or complex and is not a pole of $\hat{\mathbf{G}}(s)$ and $\mathbf{k}$ is an arbitrary $p\times 1$ constant vector, then the output due to this input and the initial state $\mathbf{x}(0)=-(\mathbf{A}-\lambda\mathbf{I})^{-1}\mathbf{Bk}$ is equal to

$$\mathbf{y}(t)=\hat{\mathbf{G}}(\lambda)\mathbf{k}e^{\lambda t} \qquad \text{for } t\geq 0 \tag{H-6}$$

∎

The proof of this theorem is identical to the one of Theorem H-2 and will not be repeated.

The substitution of $\hat{\mathbf{G}}(\lambda)=\mathbf{D}_l^{-1}(\lambda)\mathbf{N}_l(\lambda)$ into (H-6) yields

$$\mathbf{y}(t)=\mathbf{D}_l^{-1}(\lambda)\mathbf{N}_l(\lambda)\mathbf{k}e^{\lambda t} \tag{H-7}$$

where $\mathbf{N}_l(s)$ is a $q\times p$ polynomial matrix of full rank. In the following, we discuss separately the case $q\geq p$ and the case $q<p$. We use ρ to denote the rank.

Case I: $q \geq p$

In this case, we have $\rho \mathbf{N}_l(s) = p$ in $\mathbb{R}(s)$ by assumption. Hence we have $\rho \mathbf{N}_l(\lambda) = p$ in $\mathbb{C}$ for all λ in $\mathbb{C}$ except the transmission zeros of $\hat{\mathbf{G}}(s)$. If λ is a transmission zero, then $\rho \mathbf{N}_l(\lambda) < p$. Consequently, there exists a $p \times 1$ nonzero constant vector $\mathbf{k}$ such that

$$\mathbf{N}_l(\lambda)\mathbf{k} = 0$$

and consequently,

$$\mathbf{y}(t) = \mathbf{D}_l^{-1}(\lambda)\mathbf{N}_l(\lambda)\mathbf{k}e^{\lambda t} = \mathbf{0} \qquad \text{for all } t \geq 0 \tag{H-8}$$

This property is similar to the transmission-blocking property of Corollary H-2. Note that if λ is neither a transmission zero nor a pole of $\hat{\mathbf{G}}(s)$, then $\rho \mathbf{N}(\lambda) = p$ and $\mathbf{N}(\lambda)\mathbf{k} \neq \mathbf{0}$, for all constant $\mathbf{k}$, which implies

$$\mathbf{y}(t) = \mathbf{D}_l^{-1}(\lambda)\mathbf{N}_l(\lambda)\mathbf{k}e^{\lambda t} \neq 0$$

Case II: $q < p$

In this case, we have $\rho \mathbf{N}_l(s) = q$ in $\mathbb{R}(s)$ by assumption. In this case, for *any* λ in $\mathbb{C}$, we have $\rho \mathbf{N}_l(\lambda) \leq q < p$ in $\mathbb{C}$. Consequently, there exists a nonzero $\mathbf{k}$ such that $\mathbf{N}_l(\lambda)\mathbf{k} = \mathbf{0}$ and

$$\mathbf{y}(t) = \mathbf{D}_l^{-1}(\lambda)\mathbf{N}_l(\lambda)\mathbf{k}e^{\lambda t} = \mathbf{0} \qquad \text{for all } t \geq 0 \tag{H-9}$$

Since this equation holds for every λ in $\mathbb{C}$, it cannot be used to characterize the transmission zero of $\hat{\mathbf{G}}(s)$ as in the case of $q \geq p$. For this reason, we must consider this case separately from the case $q \geq p$.

If $q < p$, then λ is a transmission zero of $\hat{\mathbf{G}}(s)$ if and only if there exists a nonzero $1 \times q$ constant vector $\mathbf{h}$ such that

$$\mathbf{h}\mathbf{N}_l(\lambda) = \mathbf{0} \tag{H-10}$$

If, in addition, λ is not a pole of $\hat{\mathbf{G}}(s)$, the $1 \times q$ vector $\mathbf{f}$ defined by $\mathbf{f} = \mathbf{h}\mathbf{D}_l(\lambda)$ is also a nonzero vector. Consider

$$\begin{aligned}\mathbf{f}\mathbf{y}(t) &= \mathbf{f}\mathbf{G}(\lambda)\mathbf{k}e^{\lambda t} = \mathbf{h}\mathbf{D}_l(\lambda)\mathbf{D}_l^{-1}(\lambda)\mathbf{N}_l(\lambda)\mathbf{k}e^{\lambda t} \\ &= \mathbf{h}\mathbf{N}_l(\lambda)\mathbf{k}e^{\lambda t}\end{aligned} \tag{H-11}$$

which is, because of (H-10), identically zero for any $\mathbf{k}$. This property, namely $\mathbf{f}\mathbf{y}(t) = \mathbf{0}$ for $t \geq 0$, exists only if λ is a transmission zero of $\hat{\mathbf{G}}(s)$. Hence this can be used to characterize the transmission zero of $\hat{\mathbf{G}}(s)$ for the case $q < p$.

Corollary H-4 (Transmission-blocking property)

Consider a multivariable system with $q \times p$ proper rational matrix $\hat{\mathbf{G}}(s)$. It is assumed that $\hat{\mathbf{G}}(s)$ has a full rank, the coprime fraction $\hat{\mathbf{G}}(s) = \mathbf{D}_l^{-1}(s)\mathbf{N}_l(s)$ and the irreducible realization $\{\mathbf{A}, \mathbf{B}, \mathbf{C}, \mathbf{E}\}$. (1) If $q \geq p$, and if λ is a transmission zero of $\hat{\mathbf{G}}(s)$, then there exists a nonzero $p \times 1$ constant vector $\mathbf{k}$ such that the output of the system due to the initial state $\mathbf{x}(0) = -(\mathbf{A} - \lambda\mathbf{I})^{-1}\mathbf{B}\mathbf{k}$ and the input $\mathbf{u}(t)=\mathbf{k}e^{\lambda t}$ is identically equal to zero. (2) If $q<p$ and if λ is a transmission zero but not a pole of $\hat{\mathbf{G}}(s)$, then for the input $\mathbf{u}(t)=\mathbf{k}e^{\lambda t}$, where $\mathbf{k}$ is an arbitrary $p\times 1$ constant vector, there exists a nonzero $1\times q$ vector $\mathbf{f}$ such that the output $\mathbf{y}(t)$ due to $\mathbf{u}(t)=\mathbf{k}e^{\lambda t}$ and the initial state $\mathbf{x}(0)=-(\mathbf{A}-\lambda\mathbf{I})^{-1}\mathbf{Bk}$ has the property

$$\mathbf{fy}(t)=\mathbf{f}\hat{\mathbf{G}}(\lambda)\mathbf{k}e^{\lambda t}=\mathbf{0} \qquad \text{for all } t\geq 0$$ ∎

This corollary is just a restatement of (H-8) and (H-10). We see that this corollary is quite similar to the one for the scalar case. Because of this transmission-blocking property, the zeros defined in Definition H-3 are called the *transmission zeros*.

The transmission zero can also be defined by using dynamical equations. Consider a system with a coprime fraction $\hat{\mathbf{G}}(s)=\mathbf{D}_l^{-1}(s)\mathbf{N}_l(s)$ and an irreducible dynamical equation $\{\mathbf{A}, \mathbf{B}, \mathbf{C}, \mathbf{E}\}$. Consider the system matrix

$$\mathbf{S}(s)\triangleq\begin{bmatrix} s\mathbf{I}-\mathbf{A} & \mathbf{B} \\ -\mathbf{C} & \mathbf{E}\end{bmatrix} \tag{H-12}$$

If $\mathbf{A}$, $\mathbf{B}$, $\mathbf{C}$, and $\mathbf{E}$ are, respectively, $n\times n$, $n\times p$, $q\times n$, and $q\times p$ matrices, then $\mathbf{S}(s)$ is a $(n+q)\times(n+p)$ matrix. Though elements of $\mathbf{S}(s)$ are polynomials, we shall consider them as elements of the field of rational functions. Because of

$$\begin{bmatrix} \mathbf{I}_n & \mathbf{0} \\ \mathbf{C}(s\mathbf{I}-\mathbf{A})^{-1} & \mathbf{I}_q\end{bmatrix}\begin{bmatrix} s\mathbf{I}-\mathbf{A} & \mathbf{B} \\ -\mathbf{C} & \mathbf{E}\end{bmatrix}=\begin{bmatrix} s\mathbf{I}-\mathbf{A} & \mathbf{B} \\ \mathbf{0} & \mathbf{C}(s\mathbf{I}-\mathbf{A})^{-1}\mathbf{B}+\mathbf{E}\end{bmatrix}$$

$$=\begin{bmatrix} s\mathbf{I}-\mathbf{A} & \mathbf{B} \\ \mathbf{0} & \hat{\mathbf{G}}(s)\end{bmatrix}=\begin{bmatrix} s\mathbf{I}-\mathbf{A} & \mathbf{B} \\ 0 & \mathbf{D}_l^{-1}(s)\mathbf{N}_l(s)\end{bmatrix}$$

we have

$$\rho\begin{bmatrix} s\mathbf{I}-\mathbf{A} & \mathbf{B} \\ -\mathbf{C} & \mathbf{E}\end{bmatrix}=\rho\begin{bmatrix} s\mathbf{I}-\mathbf{A} & \mathbf{B} \\ \mathbf{0} & \hat{\mathbf{G}}(s)\end{bmatrix}=\rho(s\mathbf{I}-\mathbf{A})+\rho\hat{\mathbf{G}}(s)$$
$$=n+\rho[\mathbf{D}_l^{-1}(s)\mathbf{N}_l(s)]=n+\rho\mathbf{N}_l(s) \tag{H-13}$$

where ρ denotes the rank in the field of rational functions. If $\hat{\mathbf{G}}(s)$ is of full rank, that is, $\rho\hat{\mathbf{G}}(s)=\min(p, q)$, then

$$\rho\begin{bmatrix} s\mathbf{I}-\mathbf{A} & \mathbf{B} \\ -\mathbf{C} & \mathbf{E}\end{bmatrix}=n+\min(p, q) \tag{H-14}$$

Now if s is replaced by an element in $\mathbb{C}$, the field of complex numbers, say $s=\lambda$, then $\mathbf{S}(\lambda)$ is a matrix with elements in $\mathbb{C}$ and its rank must be computed in $\mathbb{C}$.

Theorem H-5

Consider a $q\times p$ proper rational matrix $\hat{\mathbf{G}}(s)$ with full rank and with an irreducible realization $\{\mathbf{A}, \mathbf{B}, \mathbf{C}, \mathbf{E}\}$. If λ is not a pole of $\hat{\mathbf{G}}(s)$, then λ is a transmission zero of $\hat{\mathbf{G}}(s)$ if and only if

$$\rho\mathbf{S}(\lambda)=\rho\begin{bmatrix}\lambda\mathbf{I}-\mathbf{A} & \mathbf{B}\\ -\mathbf{C} & \mathbf{E}\end{bmatrix}<n+\min(p,q) \tag{H-15}$$

■

This theorem follows directly from Definition H-3 and Equation (H-13). In establishing (H-13), the nonsingularity of $s\mathbf{I}-\mathbf{A}$ is implicitly assumed. Hence the proof is applicable only if λ is not a pole of $\hat{\mathbf{G}}(s)$. Before removing this restriction, we shall reestablish Theorem H-5 from Corollary H-4. If $q \geq p$, and if λ is a transmission zero, Corollary H-4 states that there exists a nonzero $\mathbf{k}$ such that the output due to $\mathbf{u}(t)=\mathbf{k}e^{\lambda t}$ and the initial condition

$$\mathbf{x}(0)=-(\mathbf{A}-\lambda\mathbf{I})^{-1}\mathbf{Bk} \qquad \text{or} \qquad (\mathbf{A}-\lambda\mathbf{I})\mathbf{x}(0)+\mathbf{Bk}=0 \tag{H-16}$$

are identically zero. Since $\mathbf{y}(t)=\mathbf{Cx}(t)+\mathbf{Eu}(t)=\mathbf{Cx}(t)+\mathbf{Ek}e^{\lambda t}$, we have

$$\mathbf{y}(0)=\mathbf{Cx}(0)+\mathbf{Ek}=\mathbf{0} \tag{H-17}$$

The combination of (H-16) and (H-17) yields

$$\begin{bmatrix}\lambda\mathbf{I}-\mathbf{A} & \mathbf{B}\\ -\mathbf{C} & \mathbf{E}\end{bmatrix}\begin{bmatrix}-\mathbf{x}(0)\\ \mathbf{k}\end{bmatrix}=\mathbf{0}$$

Since $[-\mathbf{x}'(0)\quad \mathbf{k}']'$ is a nonzero vector, we have $\rho\mathbf{S}(\lambda)<n+p$.

Now consider the case $q<p$. Corollary H-4 states that if λ is a transmission zero, for any $\mathbf{k}$ there exists a nonzero $1\times q$ vector $\mathbf{f}$ such that $\mathbf{fy}(t)=\mathbf{f}\hat{\mathbf{G}}(\lambda)\mathbf{k}e^{\lambda t}=\mathbf{0}$ for $t\geq 0$, or $\mathbf{f}\hat{\mathbf{G}}(\lambda)\mathbf{k}=\mathbf{0}$. From (H-11), we can see that, for the same $\mathbf{f}$, $\mathbf{f}\hat{\mathbf{G}}(\lambda)\mathbf{k}=\mathbf{0}$ holds for every $\mathbf{k}$. Hence we have

$$\mathbf{fG}(\lambda)=\mathbf{0}$$

or

$$\mathbf{f}[\mathbf{C}(\lambda\mathbf{I}-\mathbf{A})^{-1}\mathbf{B}+\mathbf{E}]=\mathbf{fC}(\lambda\mathbf{I}-\mathbf{A})^{-1}\mathbf{B}+\mathbf{fE}=\mathbf{0}$$

Define $\mathbf{f}_1-\mathbf{fC}(\lambda\mathbf{I}-\mathbf{A})^{-1}$ Then we have

$$\mathbf{f}_1(\lambda\mathbf{I}-\mathbf{A})=\mathbf{fC}$$

and

$$\mathbf{f}_1\mathbf{B}+\mathbf{fE}=\mathbf{0}$$

which can be combined as

$$[\mathbf{f}_1\quad \mathbf{f}]\begin{bmatrix}\lambda\mathbf{I}-\mathbf{A} & \mathbf{B}\\ -\mathbf{C} & \mathbf{E}\end{bmatrix}=\mathbf{0} \tag{H-18}$$

Since $[\mathbf{f}_1\quad \mathbf{f}]$ is a nonzero vector, (H-18) implies $\rho\mathbf{S}(\lambda)<n+q$. This completes the link between Corollary H-4 and Theorem H-5.

In the following, the restriction that λ is not a pole of $\hat{\mathbf{G}}(s)$ in Theorem H-5 will be removed. The proof relies heavily on the result of Section 6-6 and may be skipped in the first reading

Theorem H-6

Consider a $q\times p$ proper rational matrix $\hat{\mathbf{G}}(s)$ with a full rank and with an n-dimensional irreducible realization $\{\mathbf{A},\mathbf{B},\mathbf{C},\mathbf{E}\}$. Then λ is a transmission zero of $\hat{\mathbf{G}}(s)$ if and only if

$$\rho\mathbf{S}(\lambda) = \rho\begin{bmatrix} \lambda\mathbf{I} - \mathbf{A} & \mathbf{B} \\ -\mathbf{C} & \mathbf{E} \end{bmatrix} < n + \min(p, q)$$

Proof[2]

Let $\mathbf{E} = \hat{\mathbf{G}}(\infty)$. We factor $\hat{\mathbf{G}}(s)$ as

$$\hat{\mathbf{G}}(s) = \mathbf{E} + \mathbf{D}^{-1}(s)\bar{\mathbf{N}}(s) = \mathbf{D}^{-1}(s)[\mathbf{D}(s)\mathbf{E} + \bar{\mathbf{N}}(s)] \triangleq \mathbf{D}^{-1}(s)\mathbf{N}(s) \tag{H-19}$$

where $\mathbf{D}^{-1}(s)\bar{\mathbf{N}}(s)$ is strictly proper and left coprime and $\mathbf{D}(s)$ is row reduced. We then apply the procedure in Section 6-6 to find an irreducible $\{\mathbf{A}_0, \mathbf{B}_0, \mathbf{C}_0\}$ such that, similar to (6-133) and (6-130),

$$\mathbf{C}_0(s\mathbf{I} - \mathbf{A}_0)^{-1} = \mathbf{D}^{-1}(s)\bar{\mathbf{L}}(s) \tag{H-20}$$

and

$$\bar{\mathbf{N}}(s) = \bar{\mathbf{L}}(s)\mathbf{B}_0 \tag{H-21}$$

where $\bar{\mathbf{L}}(s)$ is defined as in (6-129). Furthermore, we have $\{\mathbf{D}(s), \bar{\mathbf{L}}(s)\}$ left coprime and $\{s\mathbf{I} - \mathbf{A}_0, \mathbf{C}_0\}$ right coprime. Hence, Theorems G-8 and G-8′ imply that there exist polynomial matrices $\mathbf{X}(s)$, $-\mathbf{Y}(s)$, $\bar{\mathbf{X}}(s)$, and $\bar{\mathbf{Y}}(s)$ such that

$$\mathbf{X}(s)(s\mathbf{I} - \mathbf{A}_0) - \mathbf{Y}(s)\mathbf{C}_0 = \mathbf{I}_n$$
$$\bar{\mathbf{L}}(s)\bar{\mathbf{X}}(s) + \mathbf{D}(s)\bar{\mathbf{Y}}(s) = \mathbf{I}_q$$

which, together with $\mathbf{D}(s)\mathbf{C}_0 = \bar{\mathbf{L}}(s)(s\mathbf{I} - \mathbf{A}_0)$, can be written in matrix form as

$$\begin{bmatrix} \mathbf{X}(s) & \mathbf{Y}(s) \\ \bar{\mathbf{L}}(s) & \mathbf{D}(s) \end{bmatrix}\begin{bmatrix} s\mathbf{I} - \mathbf{A}_0 & \bar{\mathbf{X}}(s) \\ -\mathbf{C}_0 & \bar{\mathbf{Y}}(s) \end{bmatrix} = \begin{bmatrix} \mathbf{I}_n & \mathbf{X}(s)\bar{\mathbf{X}}(s) + \mathbf{Y}(s)\bar{\mathbf{Y}}(s) \\ \mathbf{0} & \mathbf{I}_q \end{bmatrix} \tag{H-22}$$

The right-hand-side matrix is clearly unimodular; hence the left-hand-side matrices are also unimodular (Why? Note that the left-hand-side matrices are polynomial matrices). We compute

$$\begin{bmatrix} \mathbf{X}(s) & \mathbf{Y}(s) \\ \bar{\mathbf{L}}(s) & \mathbf{D}(s) \end{bmatrix}\begin{bmatrix} s\mathbf{I} - \mathbf{A}_0 & \mathbf{B}_0 \\ -\mathbf{C}_0 & \mathbf{E} \end{bmatrix} = \begin{bmatrix} \mathbf{I}_n & \mathbf{X}(s)\mathbf{B}_0 + \mathbf{Y}(s)\mathbf{E} \\ \mathbf{0} & \bar{\mathbf{L}}(s)\mathbf{B}_0 + \mathbf{D}(s)\mathbf{E} \end{bmatrix}$$
$$= \begin{bmatrix} \mathbf{I}_n & \mathbf{X}(s)\mathbf{B}_0 + \mathbf{Y}(s)\mathbf{E} \\ \mathbf{0} & \mathbf{N}(s) \end{bmatrix} \tag{H-23}$$

In the last step, we have used $\mathbf{N}(s) = \mathbf{D}(s)\mathbf{E} + \bar{\mathbf{N}}(s) = \mathbf{D}(s)\mathbf{E} + \bar{\mathbf{L}}(s)\mathbf{B}_0$ in (H-19) and (H-21). Since the leftmost matrix in (H-23) is unimodular, we have, for every s in $\mathbb{C}$,

$$\rho\begin{bmatrix} s\mathbf{I} - \mathbf{A}_0 & \mathbf{B}_0 \\ -\mathbf{C}_0 & \mathbf{E} \end{bmatrix} = \rho\begin{bmatrix} \mathbf{I}_n & \mathbf{X}(s)\mathbf{B}_0 + \mathbf{Y}(s)\mathbf{E} \\ \mathbf{0} & \mathbf{N}(s) \end{bmatrix} = n + \rho\mathbf{N}(s)$$

Since $\{\mathbf{A}, \mathbf{B}, \mathbf{C}, \mathbf{E}\}$ and $\{\mathbf{A}_0, \mathbf{B}_0, \mathbf{C}_0, \mathbf{E}\}$ are both irreducible realizations of $\hat{\mathbf{G}}(s)$, they are strictly system equivalent. Hence we have, for every s in $\mathbb{C}$,

[2] This proof was provided by Professor L. S. Chang of China University of Sciences and Technology, Ho Fei.

$$\rho\begin{bmatrix} s\mathbf{I}-\mathbf{A} & \mathbf{B} \\ -\mathbf{C} & \mathbf{E} \end{bmatrix} = \rho\begin{bmatrix} s\mathbf{I}-\mathbf{A}_0 & \mathbf{B}_0 \\ -\mathbf{C}_0 & \mathbf{E} \end{bmatrix} = n + \rho\mathbf{N}(s) \tag{H-24}$$

and the theorem follows from the definition of transmission zero. Q.E.D.

The concept of transmission zeros introduced above is only one of many possible ways of defining zeros for rational matrices. In the following, we introduce one more definition.

Definition H-4

Consider a $q \times p$ proper rational matrix $\hat{\mathbf{G}}(s)$. Let $\beta(s)$ be the greatest common divisor (gcd) of the numerators of all elements of $\hat{\mathbf{G}}(s)$. Then the roots of $\beta(s)$ are called the *blocking* zeros of $\hat{\mathbf{G}}(s)$.

Example 3

Consider

$$\hat{\mathbf{G}}_3(s) = \begin{bmatrix} \dfrac{s(s+1)}{s^2+1} & \dfrac{s+1}{s+2} \\ 0 & \dfrac{(s+2)(s+1)}{s^2+2s+2} \end{bmatrix}$$

The gcd of the three numerators of $\hat{\mathbf{G}}_3(s)$ is $s+1$. Hence -1 is the only blocking zero of $\hat{\mathbf{G}}(s)$. It can be shown that the transmission zeros of $\hat{\mathbf{G}}(s)$ are 0, -1, and -2 (Problem H-5). ∎

From this example, we see that the definition of transmission zero and that of blocking zero are not equivalent. If every element of $\hat{\mathbf{G}}(s)$ is irreducible, a blocking zero can never be a pole of $\hat{\mathbf{G}}(s)$. On the other hand, a transmission zero can also be a pole. If we factor $\hat{\mathbf{G}}(s) = \mathbf{N}_r(s)\mathbf{D}_r^{-1}(s)$, then every blocking zero of $\hat{\mathbf{G}}(s)$ will appear in every element of $\mathbf{N}_r(s)$; hence every blocking zero is a transmission zero. The converse is, of course, not necessarily true.

If $\hat{\mathbf{G}}(s)$ is a scalar transfer function, then there is no difference between transmission zeros and blocking zeros.

Similar to transmission zeros, blocking zeros also have the property of blocking the transmission of certain mode. Let $\hat{u}_i(s)$, $i = 1, 2, \ldots, p$, be the ith component of the input vector $\hat{\mathbf{u}}(s)$. Let

$$\hat{u}_i(s) = \frac{f_i(s)}{\phi(s)} \qquad i = 1, 2, \ldots, p \tag{H-25}$$

where $\phi(s)$ and $f_i(s)$ are polynomials. Let $s - \alpha$ be a root of $\phi(s)$. If $e^{\alpha t}$ appears in the output $y_i(t)$, then $s - \alpha$ is said to appear as a mode of $y_i(t)$. Otherwise, it is blocked from $y_i(t)$.

Theorem H-7

Consider a system with proper transfer matrix $\hat{\mathbf{G}}(s)$. Let the input $\hat{\mathbf{u}}(s)$ be of the form shown in (H-25). Then, for any initial state, no root of $\phi(s)$ will appear as a mode at any output terminal if and only if every root of $\phi(s)$ is a blocking zero of $\hat{\mathbf{G}}(s)$.

Proof

If every root of $\phi(s)$ is a blocking zero of $\hat{\mathbf{G}}(s)$, then $\hat{\mathbf{G}}(s)$ contains no root of $\phi(s)$ as a pole. Consequently, for any initial state, the zero-input response of the system will not contain any mode due to the roots of $\phi(s)$. Let $\hat{y}_i(s)$ be the ith component of the output vector $\hat{\mathbf{y}}(s)$ and let $\hat{\mathbf{G}}(s) = (\hat{g}_{ij}(s))$. Then the zero-state response of the system is given by

$$\hat{y}_i(s) = \sum_{j=1}^{p} \hat{g}_{ij}(s)\hat{u}_j(s) \tag{H-26}$$

Let $\beta(s)$ be the gcd of the numerators of all $\hat{g}_{ij}(s)$. Then we can write $\hat{g}_{ij}(s)$ as

$$\hat{g}_{ij}(s) = \beta(s)\frac{n_{ij}(s)}{d_{ij}(s)}$$

and (H-26) becomes, by using (H-25),

$$\hat{y}_i(s) = \sum_{j=1}^{p} \frac{\beta(s)}{\phi(s)} \frac{n_{ij}(s)f_j(s)}{d_{ij}(s)}$$

Now if every root of $\phi(s)$ is a blocking zero, then $\phi(s)$ will divide $\beta(s)$. Hence no root of $\phi(s)$ will appear as a mode of $y_i(t)$.

Now suppose the root, $s-\alpha$, of $\phi(s)$ is not a blocking zero. Then there exists at least one $n_{kl}(s)$ which does not contain $s-\alpha$ as a factor; otherwise, $s-\alpha$ would be a blocking zero. We choose $f_i(s)=0$ for $j=1, 2, \ldots, l-1, l+1, \ldots, p$ and $f_l(s)=1$, then

$$\hat{y}_k(s) = \frac{\beta(s)}{\phi(s)} \frac{n_{kl}(s)}{d_{kl}(s)}$$

and $s-\alpha$ will appear as a mode of $y_k(t)$. This completes the proof of this theorem Q.E.D.

Let $\hat{\mathbf{u}}(s) = \mathbf{k}/(s-\lambda)$. If λ is a transmission zero of $\hat{\mathbf{G}}(s)$, then $e^{\lambda t}$ will not appear as a mode at the output vector only for a certain $\mathbf{k}$. If λ is a blocking zero of $\hat{\mathbf{G}}(s)$, then $e^{\lambda t}$ will not appear at the output vector for any $\mathbf{k}$. These concepts of zeros will be used in the study of tracking control problems in Chapter 9.

The transmission zeros of a proper transfer matrix $\hat{\mathbf{G}}(s)$ are defined from a coprime fraction of $\hat{\mathbf{G}}(s)$ in Definition H-3. We see from Theorem H-6 that they can also be defined from an irreducible realization of $\hat{\mathbf{G}}(s)$ by using the system matrix in (H-15). They can again be defined from the Smith-McMillan form of $\hat{\mathbf{G}}(s)$, see References S34 and S185. For discussions of various definitions of zeros, see References S152 and S186; for their computation, see References S72 and S148. If all transmission zeros lie inside the open left half s-plane, $\hat{\mathbf{G}}(s)$ is said to be minimum phase. See Reference S67. The presentation of this appendix follows closely References S78 and S94.

Problems

H-1 Prove the identity in Equation (H-1).

H-2 Consider the dynamical equation

$$\dot{\mathbf{x}} = \begin{bmatrix} 0 & 1 \\ -1 & -2 \end{bmatrix}\mathbf{x} + \begin{bmatrix} 0 \\ 1 \end{bmatrix}u \qquad y = [2 \quad 1]\mathbf{x}$$

Find an initial state so that its zero-input response is $y(t) = 5e^{-t}$ for all $t \geq 0$.

H-3 Consider the equation in Problem H-2. Find the initial state so that the response $y(t)$ due to $u(t) = e^{3t}$ and the initial state is of the form e^{3t} for $t \geq 0$.

H-4 Prove Theorems H-3 and H-4.

H-5 What are the poles and transmission zeros of the transfer matrices:

$$\hat{\mathbf{G}}_1(s) = \begin{bmatrix} \dfrac{1}{s-1} & \dfrac{s+10}{2(s-1)^3} \\ 0 & \dfrac{s+1}{(s-1)^2} \end{bmatrix} \qquad \hat{\mathbf{G}}_2(s) = \begin{bmatrix} \dfrac{s(s+1)}{s^2+1} & \dfrac{s+1}{s+2} \\ 0 & \dfrac{(s+2)(s+1)}{s^2+2s+2} \end{bmatrix}$$

$$\hat{\mathbf{G}}_3(s) = \begin{bmatrix} 0 & s^2-1 \\ s-1 & 0 \end{bmatrix}^{-1} \begin{bmatrix} s-1 & 1 & s-1 \\ s+1 & 0 & s+1 \end{bmatrix}$$

$$\hat{\mathbf{G}}_4(s) = \begin{bmatrix} s^2-1 & 0 \\ 0 & s-1 \end{bmatrix}^{-1} \begin{bmatrix} 1 & s-1 \\ 2 & s^2 \end{bmatrix}$$

H-6 Let $\mathbf{N}(s)\mathbf{D}^{-1}(s) = \bar{\mathbf{N}}(s)\bar{\mathbf{D}}^{-1}(s)$ be two right-coprime fractions of $\hat{\mathbf{G}}(s)$. Show that the set of transmission zeros defined from $\mathbf{N}(s)$ and the one from $\bar{\mathbf{N}}(s)$ are the same.

H-7 Show that

$$\text{rank}\begin{bmatrix} \mathbf{A}-\lambda\mathbf{I} & \mathbf{B} \\ -\mathbf{C} & \mathbf{E} \end{bmatrix} = \text{rank}\begin{bmatrix} \mathbf{T}(\mathbf{A}+\mathbf{B}\mathbf{K}-\lambda\mathbf{I})\mathbf{T}^{-1} & \mathbf{T}\mathbf{B} \\ -(\mathbf{C}+\mathbf{E}\mathbf{K})\mathbf{T}^{-1} & \mathbf{E} \end{bmatrix}$$

for any nonsingular $\mathbf{T}$ and $\mathbf{K}$. Can you conclude that the transmission zeros are invariant under any state feedback and any equivalence transformation? Can you arrive at the same conclusion from the facts that a state feedback does not affect the numerator matrix of a transfer-function matrix and that an equivalence transformation does not affect a transfer-function matrix?

H-8 What are the blocking zeros of the transfer matrices in Problem H-5?

References

There are two lists of references. The first one was used in the preparation of the original edition and the second one, the present edition. The lists are not necessarily exhaustive nor do they indicate the original sources.

1. Anderson, B. D. O., and D. G. Luenberger, "Design of multivariable feedback systems," *Proc. IEE* (*London*), vol. 114, 1967, pp. 295–399, 1967.

2. ——, R. W. Newcomb, R. E. Kalman, and D. C. Youla, "Equivalence of linear time-invariant dynamical systems," *J. Franklin Inst.*, vol. 281, pp. 371–378, 1966.

3. Athans, M., and P. L. Falb, *Optimal Control.* New York: McGraw-Hill, 1966.

4. Bass, R. W., and I. Gura, "High order system design via state-space considerations," *Preprints 1965 JACC*, pp. 311–318.

5. Bellman, R., *Introduction to Matrix Analysis.* New York: McGraw-Hill, 1960.

6. Brand, L., "The companion matrix and its properties," *Am. Math Monthly*, vol. 71, pp. 629–634, 1964; vol. 75, pp. 146–152, 1968.

7 Brockett, R. W., "Poles, zeros, and feedback: State space interpretation," *IEEE Trans. Automatic Control*, vol. AC-10, pp. 129–135, 1965.

8. ——, and M. Mesarovic, "The reproducibility of multivariable systems," *J. Math. Anal. Appl.*, vol. 11, pp. 548–563, 1965.

9. Buck, R. C., *Advanced Calculus.* New York: McGraw-Hill, 1956.

10. Bucy, R. S., "Canonical forms for multivariable systems," *IEEE Trans. Automatic Control*, vol. AC-13, pp. 567–569, 1968.

11. Chang, A., "An algebraic characterization of controllability," *IEEE Trans. Automatic Control*, vol. AC-10, pp. 112–113, 1965.

12. Chang, S. S. L., *Synthesis of Optimal Control Systems.* New York: McGraw-Hill, 1961.

13. Chen, C. T., "Output controllability of composite systems," *IEEE Trans. Automatic Control*, vol. AC-12, p. 201, 1967.

14. ——, "Linear independence of analytic functions," *SIAM J. Appl. Math.*, vol. 15, pp. 1272–1274, 1967.

15. ——, "Representation of linear time-invariant composite systems," *IEEE Trans. Automatic Control*, vol. AC-13, pp. 277–283, 1968.

16. ——, "Stability of linear multivariable feedback systems," *Proc. IEEE*, vol. 56, pp. 821–828, 1968.

17. ——, "A note on pole assignment," *IEEE Trans. Automatic Control*, vol. AC-13, pp. 597–598, 1968.

18. ——, "Design of feedback control systems," *Proc. Natl. Electron. Conf.* vol. 57, pp. 46–51, 1969.

19. ——, "Design of pole-placement compensators for multivariable systems," *Preprints 1970 JACC.*

20. ——, and Desoer, C. A., "Controllability and observability of composite systems," *IEEE Trans. Automatic Control*, vol. AC-12, pp. 402–409, 1967.

21. ——, and ——, "A proof of controllability of Jordan form state equations," *IEEE Trans. Automatic Control*, vol. AC-13, pp. 195–196, 1968.

22. ——, ——, and A. Niederlinski, "Simplified condition for controllability and observability of linear time-invariant systems," *IEEE Trans. Automatic Control*, vol. AC-11, pp. 613–614, 1966.

23. Chi, H. H., and C. T. Chen, "A sensitivity study of analog computer simulation," *Proc. Allerton Conf.*, pp. 845–854, 1969.

24. Coddington, E. A., and N. Levinson, *Theory of Ordinary Differential Equations.* New York: McGraw-Hill, 1955.

25. Dellon, F., and P. E. Sarachik, "Optimal control of unstable linear plants with inaccessible states," *IEEE Trans. Automatic Control*, vol. AC-13, pp. 491–495, 1968.

26. Dertouzos, M. L., M. E. Kaliski, and K. P. Polzen, "On-line simulation of block-diagram systems," *IEEE Trans. Computers*, vol. C-18, pp. 333–342, 1969.

27. DeRusso, P. M., R. J. Roy, and C. M. Close, *State Variables for Engineers.* New York: Wiley, 1965.

28. Desoer, C. A., "A general formulation of the Nyquist criterion," *IEEE Trans. Circuit Theory*, vol. CT-12, pp. 230–234, 1965.

29. ——, EECS 222 Lecture Notes, University of California, Berkeley, Fall 1968.

30. ——, and C. T. Chen, "Controllability and observability of feedback systems," *IEEE Trans. Automatic Control*, vol. AC-12, pp. 474–475, 1967.

31. ——, and E. S. Kuh, *Basic Circuit Theory.* New York: McGraw-Hill, 1969.

32. ——, and P. Varaiya, "The minimal realization of a nonanticipative impulse response matrix," *SIAM J. Appl. Math.*, vol. 15, pp. 754–763, 1967.

33. ——, and M. Y. Wu, "Stability of linear time-invariant systems," *IEEE Trans. Circuit Theory*, vol. CT-15, pp. 245–250, 1968.

34. ——, and ——, "Stability of multiple-loop feedback linear time-invariant systems," *J. Math. Anal. Appl.*, vol. 23, pp. 121–129, 1968.

35. Duffin, R. J., and D. Hazony, "The degree of a rational matrix function," *J. SIAM*, vol. 11, pp. 645–658, 1963.

36. Falb, P. L., and W. A. Wolovich, "Decoupling in the design of multivariable control systems," *IEEE Trans. Automatic Control*, vol. AC-12, pp. 651–659, 1967.

37. Ferguson, J. D., and Z. V. Rekasius, "Optimal linear control systems with incomplete state measurements," *IEEE Trans. Automatic Control*, vol. AC-14, pp. 135–140, 1969.

38. Friedman, B., *Principles and Techniques of Applied Mathematics.* New York: Wiley, 1956.

39. Gantmacher, F. R., *The Theory of Matrices*, vols. 1 and 2. New York: Chelsea, 1959.

40. Gilbert, E. G., "Controllability and observability in multivariable control systems," *SIAM J. Control*, vol. 1, pp. 128–151, 1963.

41. ——, "The decoupling of multivariable systems by state feedback," *SIAM J. Control*, vol. 7, pp. 50–63, 1969.

42. Gueguen, C. J., and E. Toumire, "Comments on 'Irreducible Jordan form realization of a rational matrix,'" *IEEE Trans. Automatic Control*, vol. AC-15, 1970.

43. Hadley, G., *Linear Algebra.* Reading, Mass.: Addison-Wesley, 1961.

44. Halmos, P. R., *Finite Dimensional Vector Spaces*, 2d ed. Princeton, N.J.: Van Nostrand, 1950.

45. Herstein, I. N., *Topics in Algebra.* Waltham, Mass.: Blaisdell, 1964.

46. Heymann, M., "Comments 'On pole assignment in multi-input controllable linear systems,'" *IEEE Trans. Automatic Control*, vol. AC-13, pp. 748–749, 1968.

47. Ho, B. L., and R. E. Kalman, "Effective construction of linear state variable models from input/output data," *Proc. Third Allerton Conf.*, pp. 449–459, 1965.

48. Ho, Y. C., "What constitutes a controllable system," *IRE Trans. Automatic Control*, vol. AC-7, p. 76, 1962.

49. Hsu, C. H., and C. T. Chen, "A proof of stability of multivariable feedback systems," *Proc. IEEE*, vol. 56, pp. 2061–2062, 1968.

50. Jacob, J. P., and E. Polak, "On the inverse of the operator $(\cdot) = A(\cdot) + B(\cdot)$," *Am. Math. Monthly*, vol. 73, pp. 388–390, 1966.

51. Johnson, C. D., and W. M. Wonham, "A note on the transformation to canonical (phase-variable) form," *IEEE Trans. Automatic Control*, vol. AC-9, pp. 312–313, 1964.

52. Joseph, P. D., and J. T. Tou, "On linear control theory," *AIEE Trans. Applications and Industry*, vol. 80, pt. II, pp. 193–196, 1961.

53. Jury, E. I., *Sampled-Data Control Systems.* New York: Wiley, 1958.

54. Kalman, R. E., "A new approach to linear filtering and prediction problems," *Trans. ASME*, ser. D, vol. 82, pp. 35–45, 1960.

55. ——, "On the general theory of control systems," *Proc. First Intern. Congr. Autom. Control*, Butterworth, London, pp. 481–493, 1960.

56. ——, "Contribution to the theory of optimal control," *Bol. Soc. Mat. Mex.*, vol. 5, pp. 102–119, 1960.

57. ——, "Canonical structure of linear dynamical systems," *Proc. Natl. Acad. Sci. U.S.*, vol. 48, no. 4, pp. 596–600, 1962.

58. ——, "On the stability of linear time-varying systems," *IRE Trans. Circuit Theory*, vol. CT-9, pp. 420–423, 1962.

59. ——, "Lyapunov functions for the problems of Lur'e in automatic control," *Proc. Natl. Acad. Sci. U.S.*, vol. 49, pp. 201–205, 1962.

60. ——, "Mathematical description of linear dynamical system," *SIAM J. Control*, vol. 1, pp. 152–192, 1963.

61. ——, "When is a linear control system optimal?" *Trans. ASME*, ser. D, vol. 86, pp. 51–60, 1964.

62. ——, "Irreducible realizations and the degree of a rational matrix," *SIAM J.*, vol. 13, pp. 520–544, 1965.

63. ——, "Toward a theory of difficulty of computation in optimal control," *Proc. 4th IBM Sci. Comput. Symp.*, pp. 25–43, 1964.

64. ——, "On structural properties of linear constant multivariable systems," reprint of paper 6A, Third Congress of the International Federation of Automatic Control, 1966.

65. ——, and J. E. Bertram, "Control system analysis and design via the 'second method' of Lyapunov," *Trans. ASME*, ser. D, vol. 82, pp. 371–393, 1960.

66. ——, and R. S. Bucy, "New results in linear filtering and prediction theory," *Trans. ASME*, ser. D, vol. 83, pp. 95–108, 1961.

67. ——, and T. S. Engler, *A User's Manual for Automatic Synthesis Programs*, NASA, Washington, D.C., 1966.

68. ——, P. L. Falb, and M. A. Arbib, *Topics in Mathematical System Theory*. New York: McGraw-Hill, 1969.

69. ——, Y. C. Ho, and K. S. Narendra, "Controllability of linear dynamical systems," *Contrib. Differential Equations*, vol. 1, pp. 189–213, 1961.

70. Kaplan, W., *Operational Methods for Linear Systems.* Reading, Mass.: Addison-Wesley, 1962.

71. Kreindler, A., and P. E. Sarachik, "On the concepts of controllability and observability of linear systems," *IEEE Trans. Automatic Control*, vol. AC-9, pp. 129–136, 1964.

72. Kuh, E. S., and R. A. Rohrer, "The state-variable approach to network analysis," *Proc. IEEE*, vol. 53, pp. 672–686, 1965.

73. ——, and ——, *Theory of Linear Active Networks*. San Francisco: Holden-Day, 1967.

74. Kumar, S., "Computer-aided design of multivariable systems," M.S. thesis, State University of New York, Stony Brook, 1969.

75. Kuo, F. F., and J. F. Kaiser, *System Analysis by Digital Computer*. New York: Wiley, 1966.

76. LaSalle, J., and S. Lefschetz, *Stability of Liapunov's Direct Method*. New York: Academic, 1961.

77. Lefschetz, S., *Differential Equations: Geometric Theory*. New York: Interscience, 1957.

78. Luenberger, D. G., "Observing the state of a linear system," *IEEE Trans. Military Electronics*, vol. MIL-8, pp. 74–80, 1964.

79. ——, "Observers for multivariable systems," *IEEE Trans. Automatic Control*, vol. AC-11, pp. 190–197, 1966.

80. ——, "Canonical forms for linear multivariable systems," *IEEE Trans. Automatic Control*, vol. AC-12, pp. 290–293, 1967.

81. Mantey, P. E., "Eigenvalue sensitivity and state-variable selection," *IEEE Trans. Automatic Control*, vol. AC-13, pp. 263–269, 1968.

82. Mayne, D. Q., "Computational procedure for the minimal realization of transfer-function matrices," *Proc. IEE (London)*, vol. 115, pp. 1363–1368, 1968.

83. McMillan, B., "Introduction to formal realizability theory," *Bell System Tech. J.*, vol. 31, pp. 217–279, 541–600, 1952.

84. Morgan, B. S., Jr., "The synthesis of linear multivariable systems by state variable feedback," *Proc. 1964 JACC*, pp. 468–472.

85. Narendra, K. S., and C. P. Neuman, "Stability of a class of differential equations with a single monotone linearity," *SIAM J. Control*, vol. 4, pp. 295–308, 1966.

86. Nering, E. D., *Linear Algebra and Matrix Theory*. New York: Wiley, 1963.

87. Newcomb, R. W., *Active Integrated Circuit Synthesis*. Englewood Cliffs, N.J.: Prentice-Hall, 1968.

88. Ogata, K., *State Space Analysis of Control Systems*. Englewood Cliffs, N.J.: Prentice-Hall, 1967.

89. Panda, S. P., and C. T. Chen, "Irreducible Jordan form realization of a rational matrix," *IEEE Trans. Automatic Control*, vol. AC-14, pp. 66–69, 1969.

90. Parks, P. C., "A new proof of the Routh–Hurwitz stability criterion using the 'second method' of Lyapunov," *Proc. Cambridge Phil. Soc.*, vol. 58, pt. 4, pp. 694–720, 1962.

91. Polak, E., "An algorithm for reducing a linear, time-invariant differential system to state form," *IEEE Trans. Automatic Control*, vol. AC-11, pp. 577–579, 1966.

92. Pontryagin, L. S., *Ordinary Differential Equations.* Reading, Mass.: Addison-Wesley, 1962.

93. Rekasisu, Z. V., "Decoupling of multivariable systems by means of state variable feedback," *Proc. Third Allerton Conf.*, pp. 439–447, 1965.

94. Sandberg, I. W., "Linear multiloop feedback systems," *Bell System Tech. J.*, vol. 42, pp. 355–382, 1963.

95. ——, "On the L_2-boundness of solutions of nonlinear functional equations," *Bell System Tech. J.*, vol. 43, pp. 1581–1599, 1964.

96. Schwartz, L., *Théorie des distributions.* Paris: Hermann & Cie, 1951, 1957.

97. Schwarz, R. J., and B. Friedland, *Linear Systems.* New York: McGraw-Hill, 1965.

98. Silverman, L. M., "Structural properties of time-variable linear systems," Ph.D. dissertation, Dept. of Elec. Eng., Columbia University, 1966.

99. ——, "Transformation of time-variable systems to canonical (phase-variable) form," *IEEE Trans. Automatic Control*, vol. AC-11, pp. 300–303, 1966.

100. ——, "Stable realization of impulse response matrices," *1967 IEEE Intern. Conv. Record*, vol. 15, pt. 5, pp. 32–37.

101. ——, "Synthesis of impulse response matrices by internally stable and passive realizations," *IEEE Trans. Circuit Theory*, vol. CT-15, pp. 238–245, 1968.

102. ——, and B. D. O. Anderson, "Controllability, observability and stability of linear systems," *SIAM J. Control*, vol. 6, pp. 121–129, 1968.

103. ——, and H. E. Meadows, "Controllability and observability in time-variable linear systems," *SIAM J. Control*, vol. 5, pp. 64–73, 1967.

104. Truxal, J. G., *Control System Synthesis.* New York: McGraw-Hill, 1955.

105. Weiss, L., "The concepts of differential controllability and differential observability," *J. Math. Anal. Appl.*, vol. 10, pp. 442–449, 1965.

106. ——, "On the structure theory of linear differential systems," *Proc. Second Princeton Conf.*, pp. 243–249, 1968; also *SIAM J. Control*, vol. 6, pp. 659–680, 1968.

107. ——, "Lectures on controllability and observability," Tech. Note BN-590, University of Maryland, Jan. 1969.

108. ——, and P. L. Falb, "Dolezal's theorem, linear algebra with continuously parametrized elements, and time-varying systems," *Math. System Theory*, vol. 3, pp. 67–75, 1969.

109. ——, and R. E. Kalman, "Contributions to linear system theory," *Int. J. Eng. Soc.*, vol. 3, pp. 141–171, 1965.

110. Wolovich, N. A., "One state estimation of observable systems," *Preprints 1968 JACC*, pp. 210–220.

111. Wonham, W. M., "On pole assignment in multi-input controllable linear systems," *IEEE Trans. Automatic Control*, vol. AC-12, pp. 660–665, 1967.

112. ——, and A. S. Morse, "Decoupling and pole-assignment in linear multivariable systems—a geometric approach," *SIAM J. Control*, vol. 8, pp. 1–18, 1970.

113. Wylie, C. R., Jr., *Advanced Engineering Mathematics*. New York: McGraw-Hill, 1951.

114. Youla, D. C., "The synthesis of linear dynamical systems from prescribed weighting patterns," *SIAM J. Appl. Math.*, vol. 14, pp. 527–549, 1966.

115. ——, and Plinio Tissi, "n-port synthesis via reactance extraction—part I," *1966 IEEE Intern. Conv. Record*, vol. 14, pt. 7, pp. 183–208.

116. Zadeh, L. A., and C. A. Desoer, *Linear System Theory*. New York: McGraw-Hill, 1963.

117. Zames, G., "On the input-output stability of time-varying nonlinear feedback systems," pts. I and II, *IEEE Trans. Automatic Control*, vol. AC-11, pp. 228–238 and 465–476, 1966.

Supplementary References

S1. Anderson, B. D. O., "Internal and external stability of linear time-varying systems," *SIAM J. Control and Optimization*, vol. 20, pp. 408–413, 1982.

S2. ——, and M. R. Gevers, "On multivariable pole-zero cancellations and the stability of feedback systems," *IEEE Trans. Circuits and Systems*, vol. CAS-28, pp. 830–833, 1981.

S3. ——, and Jury, E. I., "Generalized Bezoutian and Sylvester matrices in multivariable linear control," *IEEE Trans. Automatic Control*, vol. AC-21, pp. 551–556, 1976.

S4. ——, and J. B. Moore, *Optimal Filtering*. Englewood Cliffs, N.J.: Prentice-Hall, 1979.

S5. Antsaklis, P. J., and J. B. Pearson, "Stabilization and regulation in linear multivariable systems," *IEEE Trans. Automatic Control*, vol. AC-23, pp. 928–930, 1978.

S6. Aplevich, J. D., "Direct computation of canonical forms of linear systems by elementary matrix operations," *IEEE Trans. Automatic Control*, vol. AC-19, pp. 124–126, 1974.

S7. ——, "Tableau methods for analysis and design of linear systems," *Automatica*, vol. 15, pp. 419–429, 1979.

S8. Araki, M., and M. Saeki, "A quantitative condition for the well-posedness of interconnected dynamical systems," *IEEE Trans. Automatic Control*, vol. AC-28, pp. 569–577, 1983.

S9. Armstrong, E. S., *ORACLS: A design system for linear multivariable control.* New York: Dekker, 1980.

S10. Aström, K. J., *Introduction to Stochastic Control Theory.* New York: Academic, 1970.

S11. ——, "Algebraic system theory as a tool for regular design," *ACTA Polytechnica Scandinavica*, Ma 31, pp. 52–65, 1979.

S12. Athans, M., M. L. Destouzos, R. N. Spann, and S. J. Mason, *Systems, Networks and Computational Multivariable Methods.* New York: McGraw-Hill, 1974.

S13. Balestrino, A., and G. Celentano, "Pole assignment in linear multivariable systems using observers of reduced order," *IEEE Trans. Automatic Control*, vol. AC-24, pp. 144–146, 1979.

S14. Bar-ness, Y., and G. Langhotz, "Preservation of controllability under sampling," *Int. J. Control*, vol. 22, pp. 39–47, 1975.

S15. Barnett, S., *Matrices in Control Theory.* London: Van Nostrand Reinhold, 1971.

S16. ——, "Regular greatest common divisor of two polynomial matrices," *Proc. Cambridge Philos. Soc.*, 72, pp. 161–165, 1972.

S17. Barry, P. E., "State feedback, pole placement and transient response," unpublished notes.

S18. Bartels, R. H., and G. H. Stewart, "Algorithm 432, solution of the matrix equation AX + XB = C," *Commun. Ass. Comput Mach.*, vol. 15, pp. 820–826, 1972.

S19. Bengtsson, G., "Output regulation and internal models—a frequency domain approach," *Automatica*, vol. 13, pp. 333–345, July 1977.

S20. Bhattacharyya, S. P., "The Structure of Robust Observers," *IEEE Trans. Automatic Control*, vol. AC-21 (4), pp. 581–588, 1976.

S21. Bingulac, S. P., and M. A. Farias, "Identification and minimal realization of multivariable systems," in *Multivariable Technological Systems* (D. P. Atherton, Ed.), New York: Pergamon, 1979.

S22. Birkoff, G. and S. MacLane, *A Survey of Modern Algebra*, 3d ed. New York: Macmillan, 1965.

S23. Bitmead, R. R., S. Y. Kung, B. D. O. Anderson, and T. Kailath, "Greatest common divisors via generalized Sylvester and Bezout matrices," *IEEE Trans. Automatic Control*, vol. AC-23, pp. 1043–1047, 1978.

S24. Bongiorno, J. J., and D. C. Youla, "On observers in multivariable control system," *Int. J. Control*, vol. 8, pp. 221–243, 1968; its discussion, vol. 12, pp. 183–190, 1970.

S25. ——, and ——, "On the design of single-loop single-input-output feedback systems in the complex-frequency domain," *IEEE Trans. Automatic Control*, vol. AC-22, pp. 416–423, 1977.

S26. Brasch, F. M., Jr., and J. B. Pearson, "Pole assignment using dynamic compensator," *IEEE Trans. Automatic Control*, vol. AC-15 (1), pp. 34–43, 1970.

S27. Brockett, R. W., *Finite-dimensional Linear Systems*. New York: Wiley, 1970.

S28. Bruni, C., A. Isidori, and A. Ruberts, "A method of realization based on the moments of the impulse response matrix," *IEEE Trans. Automatic Control*, vol. AC-14, pp. 203–204, 1969.

S29. Brunovsky, P., "A classification of linear controllable systems," *Kybernetika*, vol. 3, pp. 173–188, 1970.

S30. Budin, M. A., "Minimal realization of discrete systems from input-output observations," *IEEE Trans. Automatic Control*, vol. AC-16, pp. 395–401, 1971.

S31. Byrnes, C. I., and P. K. Stevens, "Pole placement by static and dynamic output feedback," *Proc. 21st IEEE Conf. Decision and Control*, pp. 130–133, 1982.

S32. Callier, F. M. and C. A. Desoer, "An algebra of transfer functions for distributed linear time-invariant systems" *IEEE Trans. Circuits and Systems*, vol. CAS-25, pp. 651–662, 1978. Its simplification and clarification, vol. CAS-27, pp. 320–323, 1980.

S33. ——, and ——, "Stabilization, tracking and disturbance rejection in multivariable convolution systems," *1978 IEEE Conf. on Decision and Control*, San Diego, Calif., 1979.

S34. ——, and ——, *Multivariable Feedback Systems*. New York: Springer-Verlag, 1982.

S35. ——, and C. D. Nahum, "Necessary and sufficient conditions for the complete controllability and observability of systems in series using the coprime factorization of a rational matrix," *IEEE Trans. Circuits and Systems*, CAS-22, pp. 90–95, 1975.

S36. Chang, F. C., "The inverse of generalized Vandermonde matrix through the partial fraction expansion," *IEEE Trans. Automatic Control*, vol. 19, pp. 151–152, 1974.

S37. Chang, T. S., and C. T. Chen, "On the Routh-Hurwitz criterion," *IEEE Trans. Automatic Control*, vol. AC-19, pp. 250–251, 1974.

S38. Chen, C. T., *Introduction to Linear System Theory*. New York: Holt, Rinehart and Winston, 1970.

S39. ——, "Irreducibility of dynamical equation realizations of sets of differential equations," *IEEE Trans. Automatic Control*, vol. AC-15, p. 131, 1970.

S40. ——, "A new look at transfer function design," *Proc. IEEE*, vol. 59, pp. 1580–1585, Nov. 1971.

S41. ——, "A generalization of the inertia theorem," *SIAM J. Appl. Math.*, vol. 25, no. 2, pp. 158–161, 1973.

S42. ——, "Minimization of linear sequential machines," *IEEE Trans. Computers*, vol. C-23, pp. 93–95, 1974.

S43. ——, "An algorithm for Jordan form transformation," *J. Franklin Inst.*, vol. 297, pp. 449–455, 1974.

S44. ——, "Inertia Theorem for general matrix equations," *J. Math. Anal. Appl.*, vol. 49, pp. 207–210, 1975.

S45. ——, "Synthesis of linear sequential machines," *Information Control*, vol. 32, pp. 112–127, 1976.

S46. ——, *Analysis and Synthesis of linear control systems.* Stony Brook, N.Y.: Pond Woods, 1978.

S47. ——, *One-dimensional Digital Signal Processing.* New York: Dekker, 1979.

S48. ——, "Identification of linear time-invariant multivariable systems," *Proc. 1982 IFAC Symposium on Identification*, June 1982.

S49. ——, "A contribution to the design of linear time-invariant multivariable systems," *Proc. Am. Automatic Control Conf.*, June, 1982.

S50. ——, "Feedback implementations of open-loop compensators." Tech. Report, SUNY, Stony Brook, N.Y., 1983.

S51. ——, and C. H. Hsu, "Design of dynamic compensator for multivariable systems," *Preprint JACC*, pp. 893–900, 1971.

S52. ——, and Y. S. Kao, "Identification of two-dimensional transfer functions from finite input-output data," *IEEE Trans. Automatic Control*, vol. AC-24, pp. 748–752, 1979.

S53. ——, and D. P. Mital, "A simplified irreducible realization algorithm, *IEEE Trans. Automatic Control*, vol. AC-17, pp. 535–537, 1972.

S54. ——, and S. Y. Zhang, "Multivariable multipurpose controllers." *System and Control letters*, vol. 3, pp. 73–76, 1983.

S55. Cheng, L., and J. B. Pearson, Jr., "Frequency domain synthesis of multivariable linear regulators," *IEEE Trans. Automatic Control*, vol. AC-23, pp. 3–15, 1978.

S56. ——, and ——, "Synthesis of linear multivariable regulators," *IEEE Trans. Automatic Control*, vol. AC-26, pp. 194–202, 1981.

S57. Cheng, V. H. L., and C. A. Desoer, "Limitations on the closed-loop transfer function due to right-half plane transmission zeros of the plant," *IEEE Trans. Automatic Control*, vol. AC-25, pp. 1218–1220, 1980.

S58. Csaki, F. G., "Some notes on the inversion of confluent Vandermonde matrices," *IEEE Trans. Automatic Control*, vol. AC-20, pp. 154–157, 1973.

S59. D'Alessandro, O. P., and S. Guilianelli, "A direct procedure for irreducible Jordan form realization," *Richerche Di Automatica*, vol. 2, pp. 193–198, 1971.

S60. Daly, K. C., "The computation of Luenberger canonical forms using elementary similarity transformations," *Int. J. Systems Sci.*, vol. 7, pp. 1–15, 1976.

S61. Damen, A. A. H., and A. K. Hajdasinski, "Practical tests with different approximate realizations based on the singular value decomposition of the Hankel matrix," *Proc. 6th IFAC Symposium on Identification*, Washington, D.C., 1982.

S62. Datta, K. B., "An algorithm to compute canonical forms in multivariable control systems," *IEEE Trans. Automatic Control*, vol. AC-22, pp. 129–132, 1977.

S63. ——, "Minimal realization in companion forms," *J. Franklin Inst.*, vol. 309, pp. 103–123, 1980.

S64. Davison, E. J., "The output control of linear time-invariant multivariable systems with unmeasurable arbitrary disturbances," *IEEE Trans. Automatic Control*, vol. AC-17 (5), pp. 621–630, 1972; its correction, vol. AC-20, p. 824, 1975.

S65. ——, "A generalization of the output control of linear multivariable systems with unmeasurable arbitrary disturbances," *IEEE Trans. Automatic Control*, vol. AC-20 (6), pp. 788–792, 1975.

S66. ——, "The robust control of a servomechanism problem for linear time invariant multivariable systems," *IEEE Trans. Automatic Control*, vol. AC-21 (1), pp. 25–34, 1976; addendum, vol. AC-22, p. 283, 1977.

S67. ——, "Some properties of minimum phase systems and 'squared-down' systems," *IEEE Trans. Automatic Control*, vol. AC-28, pp. 221–222, 1983.

S68. ——, and A. Goldenberg, "Robust control of a general servomechanism problem: The servo compensator," *Automatica*, vol. 11 (5), pp. 461–471, 1975.

S69. ——, and S. H. Wang, "Properties of linear time-invariant multivariable systems subject to arbitrary output and state feedback," *IEEE Trans. Automatic Control*, vol. AC-18, pp. 24–32, 1973; its correction, vol. AC-18, p. 563, 1973.

S70. ——, and ——, "Properties and calculation of transmission zeros of linear multivariable systems," *Automatica*, vol. 10, pp. 643–658, 1974.

S71. ——, and ——, "On pole assignment in linear multivariable systems using output feedback," *IEEE Trans. Automatic Control*, vol. AC-20, pp. 516–518, 1975.

S72. ——, and ——, "An algorithm for the calculation of transmission zeros of (C, A, B, D) using high-gain output feedback," *IEEE Trans. Automatic Control*, vol. AC-23, pp. 738–741, 1978.

S73. Desoer, C. A., *Notes for a Second Course on Linear Systems.* New York: Van Nostrand, 1970.

S74. ——, F. M. Callier, and W. S. Chan, "Robustness of stability conditions for linear time-invariant feedback systems," *IEEE Trans. Automatic Control*, vol. AC-22, pp. 586–590, 1977.

S75. ——, and W. S. Chan, "The feedback interconnection of lumped linear time-invariant systems," *J. Franklin Inst.*, vol. 300, pp. 335–351, 1975.

S76. ——, and M. J. Chen, "Design of multivariable feedback systems with stable plant, and its extension," *IEEE Trans. Automatic Control*, vol. AC-26, pp. 408–415, pp. 526–527, 1981.

S77. ——, R. W. Liu, John Murray, and Richard Sacks, "Feedback system design: The fractional representation approach to analysis and synthesis," *IEEE Trans. Automatic Control*, vol. AC-25, pp. 399–412, 1980.

S78. ——, and J. D. Schulman, "Zeros and poles of matrix transfer functions and their dynamic interpretation," *IEEE Trans. Circuits and Systems*, vol. CAS-21 (1), pp. 1–8, 1974.

S79. ——, and M. Vidyasagar, *Feedback Systems:* Input-Output Properties. New York: Academic, 1975.

S80. ——, and Y. W. Wang, "On the minimum order of a robust servo-compensator," *IEEE Trans. Automatic Control*, vol. AC-23 (1), pp. 70–73, 1976.

S81. ——, and Y. T. Wang, "Linear time-invariant robust servomechanism problem: A self-contained exposition," in *Control and Dynamic Systems*, vol 16 (C. T. Leondes, Ed.). New York: Academic, pp. 81–129, 1980.

S82. Dongarra, J. J., C. B. Moler, J. R. Bunch and G. W. Stewart, *LINPACK User's Guide*. Philadelphia: SIAM, 1979.

S83. Emami-Naeini, A., and G. F. Franklin, "Zero assignment in the multivariable robust servomechanism," *Proc. 21st IEEE Conf. on Decision and Control, pp.* 891–893, 1982.

S84. ——, and P. VanDooren, "On computation of transmission zeros and transfer functions," *Proc. 21st IEEE Conf. Decision and Control*, pp. 51–55, 1982.

S85. Emre, E., "The polynomial equation $QQ_c + RP_c = \Phi$ with application to dynamic feedback," *SIAM J. Control & Optimization*, vol. 18, pp. 611–620, 1980.

S86. ——, "Pole-zero cancellations in dynamic feedback systems," *Proc. 21st IEEE Conf. Decision and Control*, pp. 422–423, 1982.

S87. ——, and L. H. Silverman, "New criteria and system theoretic interpretation for relatively prime polynomial matrices," *IEEE Trans. Automatic Control*, vol. AC-22, pp. 239–242, 1977.

S88. ——, and ——, "The equation $XR + QY = \Phi$: A characterization of solutions, "*SIAM J. Control & Optimization*, vol. 19, pp. 32–38, 1981.

S89. Eykhoff, P., *System Identification.* New York: Wiley, 1974.

S90. Fadeev, D. K., and V. N. Faddeeva, *Computational Methods in Linear Algebra.* San Francisco: Freeman, 1963.

S91. Fahmy, M. M., and J. O'Reilly, "On eigenstructure assignment in linear multivariable systems," *IEEE Trans. Automatic Control*, vol. AC-27, pp. 690–693, 1982.

S92. Fairman, F. W., and R. D. Gupta, "Design of multifunctional reduced order observers," *Int. J. Systems Sci.*, vol. 11, pp. 1083–1094, 1980.

S93. Fernando, K. V., and H. Nicholson, "Singular perturbational model reduction of balanced systems," *IEEE Trans. Automatic Control*, vol. AC-29, pp. 466–468, 1982.

S94. Ferreira, P. G., and S. P. Bhattacharyya, "On blocking zeros," *IEEE Trans. Automatic Control*, vol. AC-22 (2), pp. 258–259, 1977.

S95. Forney, G. D., Jr., "Minimal bases of rational vector spaces with applications to multivariable linear systems," *SIAM J. Control*, vol. 13, pp. 493–520, 1975.

S96. Fortmann, T. E., and K. L. Hitz, *An Introduction to Linear Control Systems.* New York: Dekker, 1977.

S97. ——, and D. Williamson, "Design of low-order observers for linear feedback control laws," *IEEE Trans. Automatic Control*, vol. AC-17, pp. 301–308, 1972.

S98. Francis, B. A., "The multivariable servomechanism problem from the input-output viewpoint," *IEEE Trans. Automatic Control*, vol. AC-22 (3), pp. 322–328, 1977.

S99. ——, and W. M. Wonham, "The role of transmission zeros in linear multivariable regulators," *Int. J. Control*, vol. 22 (5), pp. 657–681, 1975.

S100. ——, and ——, "The internal model principle of control theory," *Automatica*, vol. 12, pp. 457–465, 1976.

S101. Franklin, G. F., and C. R. Johnson, "A condition for full zero assignment in linear control systems," *IEEE Trans. Automatic Control*, vol. AC-26, pp. 521–523, 1981.

S102. Fuhrmann, P. A., "Algebraic system theory: An analyst's point of view," *J. Franklin Inst.*, vol. 301, pp. 521–540, 1976.

S103. Garbow, B. S., J. M. Boyle, J. J. Dongarra, and C. B. Moler, "Matrix eigensystem routines—Eispack guide extensions," *Lecture Notes in Computer Sciences*, vol. 51. New York: Springer-Verlag, 1977.

S104. Gentin, Y. V., and S. Y. Kung, "A two-variable approach to the model reduction problem with Hankel norm-criterion," *IEEE Trans. Circuits Systems*, vol. CAS-28, pp. 912–924, 1981.

S105. Gibson, J. A., and T. T. Ha, "Further to the preservation of controllability under sampling," *Int. J. Control*, vol. 31, pp. 1013–1026, 1980.

S106. Gohberg, I. C., and L. E. Lerer, "Resultants of matrix polynomials," *Bull. Amer. Math. Soc.*, vol. 82, pp. 565–567, 1976.

S107. Golub, G. H., S. Nash and C. Van Loan, "A Hessenberg-Scheer method for the problem AX + XB = C," *IEEE Trans. Automatic Control*, vol. AC-24, pp. 909–913, 1979.

S108. ——, and J. H. Wilkinson, "Ill-conditioned eigensystems and the computation of the Jordan canonical form," *SIAM Rev.*, vol. 18, pp. 578–619, 1976.

S109. Gopinath, B., "On the control of linear multiple input-output systems," *Bell Syst. Tech. J.*, vol. 50, pp. 1063–1081, 1971.

S110. Gupta, R. D., F. W. Fairman, and T. Hinamoto, "A direct procedure for the design of single functional observers," *IEEE Trans. Circuits Systems*, vol. CAS-28, pp. 294–300, 1981.

S111. Harn, Y. P., and C. T. Chen, "A proof of a discrete stability test via the Liapunov theorem," *IEEE Trans. Automatic Control*, vol. AC-26, pp. 733–734, 1981.

S112. Hayton, G. E., "The generalized resultant matrix," *Int. J. Control*, vol. 32, pp. 567–579, 1980.

S113. Heymann, M., *Structure and Realization Problems in the Theory of Dynamical Systems.* New York: Springer-Verlag, 1975.

S114. ——, "The pole shifting theorem revisited," *IEEE Trans. Automatic Control*, vol. AC-24, pp. 479–480, 1979.

S115. Hikita, H., "Design of exact model matching systems and its applications to output tracking problems," *Int. J. Control*, vol. 34, pp. 1095–1111, 1981.

S116. Householder, A. S., *The Theory of Matrices in Numerical Analysis.* Waltham Mass.: Blaisdell, 1964.

S117. Hsia, T. C., *System Identification.* Boston: Lexington, 1977.

S118. Hsu, C. H., and C. T. Chen, "Conversion of cyclicity for transfer function matrices," *Int. J. Control*, vol. 16, no. 3, pp. 451–463, 1972.

S119. Ikeda, M., H. Malda, and S. Kodama, "Estimation and Feedback in linear time-varying systems: A deterministic theory," *SIAM J. Control*, Vol. 13, pp. 304–326, 1975.

S120. Inouye, Y., "Notes on controllability and constructability of linear discrete-time systems," *Int. J. Control*, vol. 35, pp. 1081–1084, 1982.

S121. Johnson, C. D., "Accommodation of External Disturbances in Linear Regulator and Servomechanism Problems," *IEEE Trans. Automatic Control*, vol. AC-16 (6), pp. 635–644 (1971).

S122. ——, "Theory of disturbance-accommodating controllers," in *Control and Dynamic Systems*, vol. 12 (C. T. Leondes, ed.). New York: Academic, pp. 287–489, 1976.

S123. Jury, E. I., *Inners and Stability of Dynamic Systems.* New York: Wiley, 1974.

S124. Kagstrom, B., and A. Rule, "An algorithm for numerical computation of the Jordan normal form of a complex matrix [F2]," *ACM Trans. on Math. Software*, vol. 6, pp. 437–443, 1980.

S125. Kailath, T., *Linear Systems.* Englewood Cliffs, N.J.: Prentice-Hall, 1980.

S126. Kalman, R. E., "On partial realizations, transfer functions, and canonical forms," *ACTA, Polytechnica Scandinavica*, Ma 31, pp. 9–32, 1979.

S127. ——, "On the computation of the reachable/observable canonical form," *SIAM J. Control & Optimization*, vol. 20, pp. 408–413, 1982.

S128. Kamen, E. W., "New results in realization theory for linear time-varying analytic systems," *IEEE Trans. Automatic Control*, vol. AC-24, pp. 866–878, 1979.

S129. Kimura, H., "Pole assignment by gain output feedback," *IEEE Trans. Automatic Control*, vol. AC-20, pp. 509–516, 1975.

S130. ——, "Pole assignment by gain output feedback," *IEEE Trans. Automatic Control*, vol. AC-22, pp. 458–463, 1977.

S131. ——, "Geometric structure of observers for linear feedback control laws," *IEEE Trans. Automatic Control*, vol. AC-22, pp. 846–854, 1977.

S132. ——, "On pole assignment by output feedback," *Int. J. Control*, vol. 28, pp. 11–22, 1978.

S133. Klein, G., and B. C. Moore, "Eigenvalue-generalized eigenvector assignment with state feedback," *IEEE Trans. Automatic Control*, vol. AC-22, pp. 140–141, 1977.

S134. Klema, V. C., and A. J. Laub, "The singular value decomposition: Its computation and some application," *IEEE Trans. Automatic Control*, vol. AC-25, pp. 164–176, 1980.

S135. Krishnamurthi, V., "Implications of Routh stability criteria," *IEEE Trans. Automatica Control*, vol. AC-25, pp. 554–555, 1980.

S136. Krishnarao, I. S., and C. T. Chen, "Properness of feedback transfer function matrices," *Int. J. Control*, vol. 39, pp. 57–61, 1984.

S137. ——, and ——, "Two polynomial matrix operations," *IEEE Trans. Automatic Control*, vol. AC-29, No. 3, March, 1984.

S138. Kronsjo, L., *Algorithms—their complexity and efficiency*. New York: Wiley, 1979.

S139. Kucera, V., *Discrete Linear Control—The Polynomial Equation Approach*. Chichester: Wiley, 1979.

S140. Kung, S., "Multivariable and multidimensional systems: Analysis and Design," Ph.D. dissertation, Stanford University, 1977.

S141. ——, "A new low-order approximation algorithm via singular value decomposition," *Proc. 12th Amer. Asilomar Conf. Circuits, Systems and Computers*, 1978.

S142. ——, and T. Kailath, "Fast projection methods for minimal design problems in linear system theory," *Automatica*, vol. 16, pp. 399–403, 1980.

S143. ——, ——, and M. Morf, "A generalized resultant matrix for polynomial matrices," *Proc. IEEE Conf. on Decision and Control*, pp. 892–895, 1976.

S144. ——, ——, and ——, "Fast and stable algorithms for minimal design problems," *Proc. Fourth IFAC Int. Symposium on Multivariable Tech. Systems* (D. P. Altherton, ed.). London: Pergamon, pp. 97–104, 1977.

S145. Kuo, Y. L., "On the irreducible Jordan form realization and the degree of a rational matrix," *IEEE Trans. Circuit Theory*, vol. CT-17, pp. 322–332, 1970.

S146. Lal, M., H. Singh, and K. A. Khan, "A simplified minimal realization algorithm for symmetric impulse response matrix using moments," *IEEE Trans.*, vol. AC-18, pp. 683–684, 1973.

S147. Laub, A. J., "Linear multivariable control: numerical considerations," MIT Rep. ESL-P-833, July 1978.

S148. ——, and B. C. Moore, "Calculation of transmission zeros using QZ techniques, Automatica, vol. 14, pp. 557–566, 1978.

S149. Liu, R., and L. C. Suen, "Minimal dimensional realization and identifiability of input-output sequences," *IEEE Trans. Automatic Control*, vol. AC-22, pp. 227–232, 1977.

S150. Luenberger, D. G., "An introduction to observers," *IEEE Trans. Automatic Control*, vol. AC-16, pp. 596–603, 1971.

S151. MacFarlane, A. G. J., "Gains, phase and angles in multivariable systems," *Proc. 21st IEEE Conf. on Decision and Control*, pp. 944–947, 1982.

S152. MacFarlane, A. G., and N. Karcanias, "Poles and zeros of linear multivariable systems: A survey of the algebraic geometric and complex-variable theory," *Int. J. Control*, vol. 24, pp. 33–74, 1976.

S153. Mansour, M., "A note on the stability of linear discrete systems and Lyapunov method," *IEEE Trans. Automatic Control*, vol. AC-27, pp. 707–708, 1982.

S154. Maronlas, J., and S. Barnett, "Canonical forms of time-invariant linear control systems: a survey with extension. Part I, single-input cases," *Int. J. Systems Sci.*, vol. 9, pp. 497–514, 1978; Part II, multivariable case, vol. 10, pp. 33–50, 1979.

S155. Mayne, D. Q., "An elementary derivation of Rosenbrock's minimal realization algorithm," *IEEE Trans. Automatic Control*, vol. AC-18, pp. 306–307, 1973.

S156. Miminis, G. S., and C. C. Paige, "An algorithm for pole assignment of time-invariant linear systems," *Int. J. Control*, vol. 35, pp. 341–354, 1982.

S157. ——, and ——, "An algorithm for pole-assignment of time-invariant multi-input linear systems," *Proc. 21st IEEE Conf. Decision and Control*, pp. 62–67, 1982.

S158. Mital, D. T., and C. T. Chen, "Canonical form realization of linear time-invariant systems," *Int. J. Control*, vol. 18, pp. 881–887, 1973.

S159. Moler, C. B., and C. F. Van Loan, "Nineteen dubious ways to compute the exponential of a matrix," *SIAM Rev.*, vol. 20, pp. 801–836, 1978.

S160. Moore, B. C., "On the flexibility offered by state feedback in multivariable systems beyond closed loop eigenvalue assignment," *IEEE Trans. Automatic Control*, vol. AC-21, pp. 689–692, 1976; its comments, vol. AC-22, pp. 888–889 by Porter and D'Azzo, p. 889 by Gonrishankar, 1977.

S161. ——, "Singular value analysis of linear systems," in *Proc. 17th IEEE Conf. Decision Control*, pp. 66–73, Jan. 1979.

S162. ——, "Principal component analysis in linear systems: Controllability, observability and model reduction," *IEEE Trans. Automatic Control*, vol. AC-26, pp. 17–32, 1981.

S163. Morse, A. S., "Minimal solutions to transfer function equations," *IEEE Trans. Automatic Control*, vol. AC-21, pp. 131–133, 1976.

S164. ——, W. A. Wolovich and B. D. O. Anderson, "Generic pole assignment: Preliminary results," *IEEE Trans. Automatic Control*, vol. AC-28, pp. 503–506, 1983.

S165. ——, and W. M. Wonham, "Decoupling and pole assignment compensation," *SIAM J. Control*, vol. 8, pp. 317–337, 1970.

S166. Owens, D. H., *Multivariable and Optimal Systems.* New York: Academic, 1981.

S167. Pace, I. S., and S. Barnett, "Comparison of algorithms of calculations of g.c.d. of polynomials," *Int. J. Systems Sci.*, vol. 4, pp. 211–226, 1973.

S168. Pace, I. S., and S. Barnett, "Efficient algorithms for linear system calculations Part I—Smith form and common divisor of polynomial matrices," *Int. J. Systems Sci.*, vol. 5, pp. 68–72, 1974.

S169. Paige, C. C., "Properties of numerical algorithms related to computing controllability," *IEEE Trans. Automatic Control*, vol. AC-26, pp. 130–138, 1981.

S170. Patel, R. V., "Computation of matrix fraction descriptions of linear time-invariant systems," *IEEE Trans. Automatic Control*, vol. AC-26, pp. 148–161, 1981.

S171. ——, "Computation of numerical-order state-space realizations and observability indices using orthogonal transformation," *Int. J. Control*, vol. 33, pp. 227–246, 1981.

S172. ——, and N. Munro, *Multivariable System Theory and Design.* New York: Pergamon, 1982.

S173. Pearson, J. B., R. W. Shields, and P. W. States, Jr., "Robust solutions to linear multivariable control problems," *IEEE Trans. Automatic Control*, vol. AC-19, pp. 508–517, 1974.

S174. ——, and P. W. Staats, Jr., "Robust controllers for linear regulators," *IEEE Trans. Automatic Control*, vol. AC-19 (3), pp. 231–234, 1974.

S175. Pernebo, L., "An algebraic theory for the design of controllers for linear multivariable system. Part I: Structure matrices and feedforward design. Part II: Feedback realizations and feedback design, *IEEE Trans. Automatic Control*, vol. AC-26, pp. 171–172, pp. 183–194, 1981.

S176. ——, and L. M. Silverman, "Model reduction via balanced state space representation," *IEEE Trans. Automatic Control*, vol. AC-21, pp. 382–387, 1982.

S177. Popov, V. M., "Invariant description of linear time-invariant controllable systems," *SIAM J. Control*, vol. 10, pp. 252–264, 1972.

S178. ——, *Hyperstability of Control Systems.* Berlin: Springer-Verlag, 1973.

S179. Porter, B., and A. Bradshaw, "Design of linear multivariable continuous-time tracking systems," *Int. J. Systems Sci.*, vol. 5 (12), pp. 1155–1164, 1974.

S180. Pugh, A. C., "Transmission and system zeros," *Int. J. Control*, vol. 26, pp. 315–324, 1979.

S181. Ralson, A., and P. Rabinowitz, *A First Course in Numerical Analysis.* New York: McGraw-Hill, 1978.

S182. Rice, J., *Matrix Computation and Mathematical Software.* New York: McGraw-Hill, 1981.

S183. Rissanen, J., "Recursive identification of linear systems," *SIAM J. Control*, vol. 9, pp. 420–430, 1971.

S184. Roman, J. R., and T. E. Bullock, "Design of minimum order stable observers for linear functions of the state via realization theory," *IEEE Trans. Automatic Control*, vol. AC-20, pp. 613–622, 1975.

S185. Rosenbrock, H. H., *State-Space and Multivariable Theory.* New York: Wiley-Interscience, 1970.

S186. ——, "The zeros of a system," *Int. J. Control*, vol. 18, pp. 297–299, 1973.

S187. ——, "The transformation of strict system equivalues," *Int. J. Control*, vol. 25, pp. 11–19, 1977.

S188. ——, "Comments on 'Poles and zeros of linear multivariable systems: A survey of the algebraic, geometric and complex variable theory,'" *Int. J. Control*, vol. 26, pp. 157–161, 1977.

S189. ——, and G. E. Hayton, "The general problem of pole placement," *Int. J. Control*, vol. 27, pp. 837–852, 1978.

S190. ——, and A. J. J. van der Weiden, "Inverse System," *Int. J. Control*, vol. 25, pp. 389–392, 1977.

S191. Rozsa, P., and N. K. Sinha, "Efficient algorithm for irreducible realization of a rational matrix," *Int. J. Control*, vol. 20, pp. 739–748, 1974. Its comments by Y. S. Kao and C. T. Chen, *Int. J. Control*, vol. 28, pp. 325–326, 1978.

S192. Saeks, R., and J. Murray, "Feedback system design: The tracking and disturbance rejection problems," *IEEE Trans. Automatic Control*, vol. AC-26, pp. 203–217, 1981.

S193. Sain, M. K., *Introduction to Algebraic System Theory.* New York: Academic, 1981.

S194. Safonov, M. G., A. J. Laub, and G. L. Hartmann, "Feedback properties of multivariable systems. The role and use of the return difference matrix," *IEEE Trans.*, vol. AC-26, pp. 47–65, 1981.

S195. Schumacher, J. M., "Compensator synthesis using (**C**, **A**, **B**)-pairs," *IEEE Trans. Automatic Control*, vol. AC-25, pp. 1133–1137, 1980.

S196. Scott, R. W., and B. D. O. Anderson, "Comments on 'Conditions for a feedback transfer matrix to be proper,'" *IEEE Trans. Automatic Control*, vol. AC-21, pp. 632–634, 1976.

S197. Shieh, L. S., and Y. T. Tsay, "Transformations of a class of multivariable control systems to block companion forms," *IEEE Trans. Automatic Control*, vol. AC-27, pp. 199–203, 1982.

S198. Silverman, L., "Realization of linear dynamical systems," *IEEE Trans. Automatic Control*, vol. AC-16, pp. 554–567, 1971.

S199. Smith, M. C., "Matrix fractions and strict system equivalence," *Int. J. Control*, vol. 34, pp. 869–884, 1981.

S200. Stewart, G. W., *Introduction to Matrix Computations.* New York: Academic, 1973.

S201. Tsui, C. C., and C. T. Chen, "An algorithm for companion form realization," *Int. J. Control*, vol. 38, pp. 769–777, 1983.

S202. ——, "Computational aspects in linear system theory," Ph.D. Dissertation, SUNY, Stony Brook, 1983.

S203. Van Dooren, P. M., "The generalized eigenstructure problem in linear system theory," *IEEE Trans. Automatic Control*, vol. AC-26, pp. 111–129, 1981.

S204. Varga, A., and V. Sima, "Numerically stable algorithm for transfer function matrix evaluation," *Int. J. Control*, vol. 33, pp. 1123–1133, 1981.

S205. Vidyasagar, M., "Conditions for a feedback transfer function matrix to be proper," *IEEE Trans. Automatic Control*, vol. AC-20, pp. 570–571, 1975.

S206. ——, *Nonlinear Systems Analysis.* Englewood Cliffs, N.J.: Prentice-Hall, 1978.

S207. ——, "On the well-posedness of large-scale interconnected systems," *IEEE Trans. Automatic Control*, vol. AC-25, pp. 413–421, 1980.

S208. Wang, J. W., and C. T. Chen, "On the computation of the characteristic polynomial of a matrix," *IEEE Trans. Automatic Control*, vol. AC-27, pp. 449–551, 1982.

S209. Wang, S. H., *Design of Linear Multivariable Systems*, Memo. No. ERL M309, University of California, Berkeley, 1971.

S210. ——, and E. J. Davison, "A minimization algorithm for the design of linear multivariable systems," *IEEE Trans. Automatic Control*, vol. AC-18, pp. 220–225, 1973.

S211. ——, and ——, "A new invertability criterion for linear multivariable systems," *IEEE Trans. Automatic Control*, vol. AC-18, pp. 538–539, 1973.

S212. Wilkinson, J. H., *The Algebraic Eigenvalue Problem.* London: Oxford University Press, 1965.

S213. ——, and C. Reinsch (Eds.), *Handbook for Automatic Computation*, Linear Algebra, Vol. II. New York: Springer-Verlag, 1971.

S214. Willems, J. C., *The Analysis of Feedback Systems.* Cambridge: MIT Press, 1971.

S215. Willems, J. L., "Design of state observers for linear discrete-time systems," *Int. J. Systems Sci.*, vol. 11, pp. 139–147, 1980.

S216. Wolovich, W. A., "On determining the zeros of state-space systems," *IEEE Trans. Automatic Control*, vol. AC-18 (5), pp. 542–544, 1973.

S217. ——, "Multivariable system synthesis with step disturbances rejection," *IEEE Trans. Automatic Control*, vol. AC-19 (1), pp. 127–130, 1974.

S218. ——, *Linear Multivariable Systems.* New York: Springer-Verlag, 1974.

S219. ——, "Output feedback decoupling," *IEEE Trans. Automatic Control*, vol. AC-20, pp. 148–149, 1975.

S220. ——, "Sket prime polynomial matrices," *IEEE Trans. Automatic Control*, vol. AC-23, pp. 880–887, 1978.

S221. ——, "On the stabilization of closed-loop stabilizable systems," *IEEE Trans. Automatic Control*, vol. AC-23, pp. 1103–1104, 1978.

S222. ——, "Multipurpose controllers for multivariable systems," *IEEE Trans. Automatic Control*, vol. AC-26, pp. 162–170, 1981.

S223. ——, P. Antsaklis, and H. Elliott, "On the stability of solutions to minimal and nonminimal design problems," *IEEE Trans. Automatic Control*, vol. AC-22, pp. 88–94, 1977.

S224. ——, and P. Ferreira, "Output regulation and tracking in linear multivariable systems," *IEEE Trans. Automatic Control*, vol. AC-24, pp. 460–465, 1979.

S225. Wonham, W. M., "Tracking and regulation in linear multivariable systems," *SIAM J. Control*, vol. 11, pp. 424–437, 1973.

S226. ——, "Towards an abstract internal model principle," *IEEE Trans. System, Man and Cybernetics*, vol. SMC-6 (11), pp. 735–740, 1976.

S227. ——, *Linear Multivariable Control: A Geometric Approach*, 2d ed. New York: Springer-Verlag, 1979.

S228. ——, and J. B. Pearson, "Regulation and internal stabilization in linear multivariable systems," *SIAM J. Control*, vol. 12, pp. 5–8, 1974.

S229. Wu, M. Y., "A note on stability of linear time-varying systems," *IEEE Trans. Automatic Control*, vol. AC-19, p. 162, 1974.

S230. ——, "Some new results in linear time-varying systems" (tech. corresp.), *IEEE Trans. Automatic Control*, vol. AC-20, pp. 159–161, 1975.

S231. ——, "Solutions of certain classes of linear time varying systems," *Int. J. Control*, vol. 31, pp. 11–20, 1980.

S232. ——, "A successive decomposition method for the solution of linear time varying systems," *Int. J. Control*, vol. 33, pp. 181–186, 1981.

S233. Wu, Y. C., and Z. V. Rekasius, "Deterministic identification of linear dynamic systems," *IEEE Trans. Automatic Control*, vol. AC-25, pp. 501–504, 1980.

S234. Yeung, K. S., "A necessary condition for Hurwitz polynomials," *IEEE Trans. Automatic Control*, vol. AC-27, pp. 251–252, 1982.

S235. Youla, D. C., J. J. Bongiorno, and H. A. Jabr, "Modern Wiener-Hopf design of optimal controllers—Part I and Part II," *IEEE Trans. Auto-*

matic Control, vol. AC-21, pp. 3–15, pp. 319–338, 1976.

S236. Zhang, S. Y., and C. T. Chen, "An algorithm for the division of two polynomial matrices," *IEEE Trans. Automatic Control*, vol. AC-28, pp. 238–240, 1983.

S237. ——, and ——, "Design of unity feedback systems for arbitrary denominator matrix," *IEEE Trans. Automatic Control*, vol. AC-28, pp. 518–521, 1983.

S238. ——, and ——, "Design of robust asymptotic tracking and disturbance rejection," *Proc. 22nd IEEE Conf. Decision and Control*, Dec. 1983.

S239. *IEEE Trans. Automatic Control 1956–1980 Cumulative Index*, vol. AC-26, no. 4, August 1981.

Index